Arithmetic Operations:

$$ab + ac = a(b+c)$$

$$\frac{a}{b} + \frac{c}{d} = \frac{ad+bc}{bd}$$

$$\frac{a+b}{c} = \frac{a}{c} + \frac{b}{c}$$

$$\frac{\left(\frac{a}{b}\right)}{\left(\frac{c}{d}\right)} = \frac{ad}{bc}$$

$$a\left(\frac{b}{c}\right) = \frac{ab}{c}$$

$$\frac{a-b}{c-d} = \frac{b-a}{d-c}$$

$$\frac{ab+ac}{a} = b+c,\ a \neq 0$$

$$\frac{\left(\frac{a}{b}\right)}{c} = \frac{a}{bc}$$

$$\frac{a}{\left(\frac{b}{c}\right)} = \frac{ac}{b}$$

Exponents and Radicals:

$$a^0 = 1,\ a \neq 0$$

$$\frac{a^x}{a^y} = a^{x-y}$$

$$\left(\frac{a}{b}\right)^x = \frac{a^x}{b^x}$$

$$\sqrt[n]{a^m} = a^{m/n} = \left(\sqrt[n]{a}\right)^m$$

$$a^{-x} = \frac{1}{a^x}$$

$$(a^x)^y = a^{xy}$$

$$\sqrt{a} = a^{1/2}$$

$$\sqrt[n]{ab} = \sqrt[n]{a}\,\sqrt[n]{b}$$

$$a^x a^y = a^{x+y}$$

$$(ab)^x = a^x b^x$$

$$\sqrt[n]{a} = a^{1/n}$$

$$\sqrt[n]{\left(\frac{a}{b}\right)} = \frac{\sqrt[n]{a}}{\sqrt[n]{b}}$$

Algebraic Errors to Avoid:

$\dfrac{a}{x+b} \neq \dfrac{a}{x} + \dfrac{a}{b}$ (To see this error, let $a = b = x = 1$.)

$\sqrt{x^2+a^2} \neq x + a$ (To see this error, let $x = 3$ and $a = 4$.)

$a - b(x-1) \neq a - bx - b$ (Remember to distribute negative signs. The equation should be $a - b(x-1) = a - bx + b$.)

$\dfrac{\left(\frac{x}{a}\right)}{b} \neq \dfrac{bx}{a}$ (To divide fractions, invert and multiply. The equation should be

$$\frac{\frac{x}{a}}{b} = \frac{\frac{x}{a}}{\frac{b}{1}} = \left(\frac{x}{a}\right)\left(\frac{1}{b}\right) = \frac{x}{ab}.)$$

$\sqrt{-x^2+a^2} \neq -\sqrt{x^2-a^2}$ (We can't factor a negative sign outside of the square root.)

$\dfrac{\cancel{a}+bx}{\cancel{a}} \neq 1+bx$ (This is one of many examples of incorrect cancellation. The equation should be $\dfrac{a+bx}{a} = \dfrac{a}{a} + \dfrac{bx}{a} = 1 + \dfrac{bx}{a}$.)

$\dfrac{1}{x^{1/2}-x^{1/3}} \neq x^{-1/2} - x^{-1/3}$ (This error is a sophisticated version of the first error.)

$(x^2)^3 \neq x^5$ (The equation should be $(x^2)^3 = x^2 x^2 x^2 = x^6$.)

Conversion Table:

1 centimeter = 0.394 inches	1 joule = 0.738 foot-pounds	1 mile = 1.609 kilometers
1 meter = 39.370 inches	1 gram = 0.035 ounces	1 gallon = 3.785 liters
= 3.281 feet	1 kilogram = 2.205 pounds	1 pound = 4.448 newtons
1 kilometer = 0.621 miles	1 inch = 2.540 centimeters	1 foot-lb = 1.356 joules
1 liter = 0.264 gallons	1 foot = 30.480 centimeters	1 ounce = 28.350 grams
1 newton = 0.225 pounds	= 0.305 meters	1 pound = 0.454 kilograms

GRAPHS OF COMMON FUNCTIONS

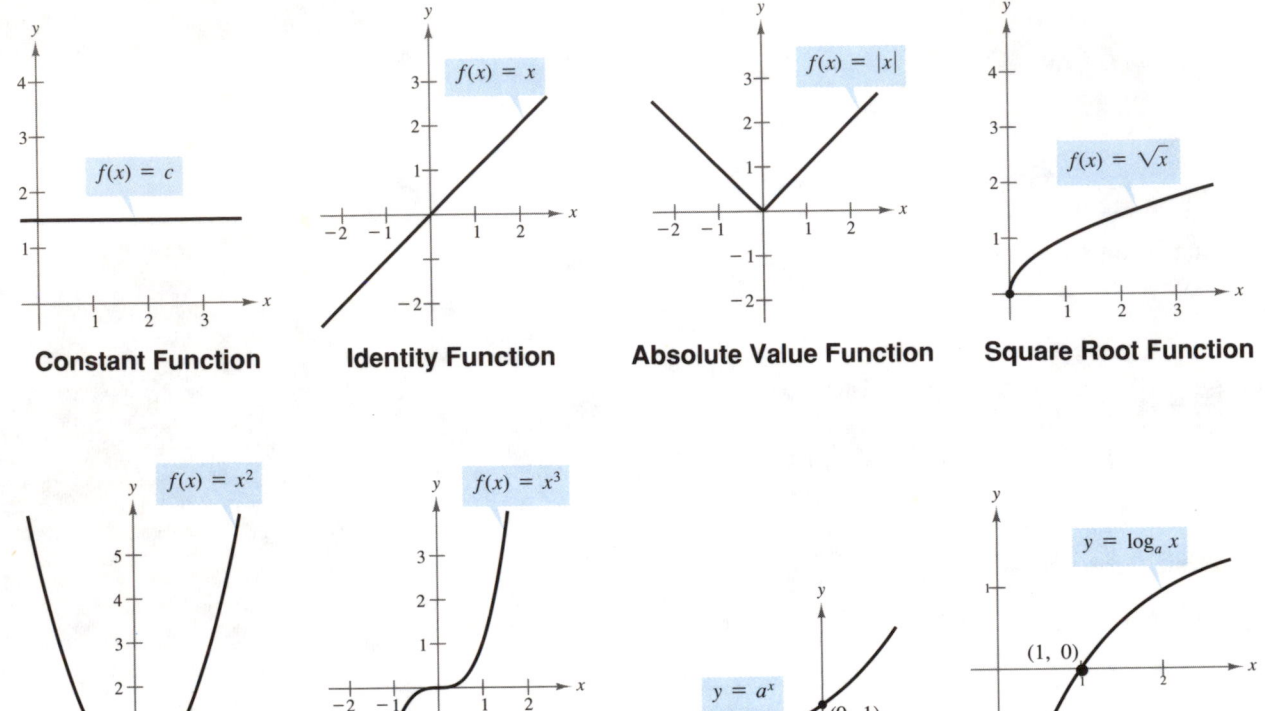

Constant Function   **Identity Function**   **Absolute Value Function**   **Square Root Function**

Squaring Function   **Cubing Function**   **Exponential Function**   **Logarithmic Function**

SYMMETRY

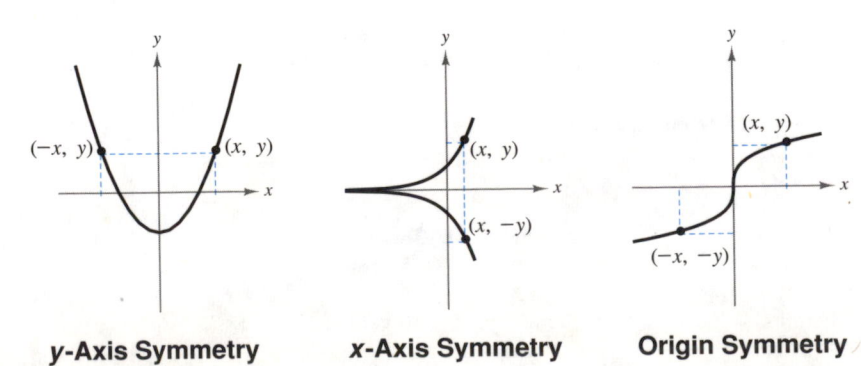

y-Axis Symmetry   **x-Axis Symmetry**   **Origin Symmetry**

Algebra and Trigonometry

FOURTH EDITION

Roland E. Larson   **Robert P. Hostetler**

The Pennsylvania State University
The Behrend College

With the assistance of
David E. Heyd

The Pennsylvania State University
The Behrend College

HOUGHTON MIFFLIN COMPANY   Boston   New York

Sponsoring Editor: Christine B. Hoag
Associate Editor: Maureen Brooks
Managing Editor: Catherine B. Cantin
Senior Project Editor: Karen Carter
Assistant Project Editor: Rachel D'Angelo Wimberly
Production Supervisor: Lisa Merrill
Art Supervisor: Gary Crespo
Marketing Manager: Charles Cavaliere

Cover design by Harold Burch Design, NYC

Composition: Meridian Creative Group

Preface

A firm foundation in algebra is necessary for success in college-level mathematics courses. *Algebra and Trigonometry,* Fourth Edition, is designed to help students develop their proficiency in algebra, and so strengthen their understanding of the underlying concepts. Although the basic concepts of algebra are reviewed in the text, it is assumed that most students taking this course have completed two years of high school algebra.

The text takes every opportunity to show how algebra is a modern modeling language for real-life problems. Examples, exercises, and group activities —many using real data—provide a real-life context to help students grasp mathematical concepts. As appropriate, graphing technology is utilized throughout the text to enhance student understanding of mathematical concepts.

New to the Fourth Edition

All text elements in the previous edition were considered for revision, and many new examples, exercises, and applications were added to the text. Following are the major changes in the Fourth Edition.

Improved Coverage The Fourth Edition begins with Prerequisites, a review chapter. All or part of this material may be covered or it can be omitted, offering greater flexibility in designing the course syllabus. Graphing is introduced earlier than in the previous edition, with the discussion of the Cartesian plane in Chapter P and graphs of equations in Chapter 1. The ability to visualize a mathematical concept is an important skill in algebra that is made available to the student as early in the course as is possible and then reinforced throughout. In response to user feedback, the discussions of quadratic equations and the quadratic formula are now combined in Chapter 1.

In the Fourth Edition, several topics receive increased attention. A new discussion of least squares regression lines is included in Chapter 3. The discussion of rational functions is expanded and appears earlier in the text in Chapter 4. In Chapter 5, "Exponential and Logarithmic Functions," there is increased emphasis on growth and decay and on exponential and logarithmic modeling. Mathematical modeling of the sine and cosine functions was added to Section 6.4, and damped trigonometric graphs were combined with the graphs of other trigonometric functions in Section 6.5. In Chapter 8, a new section (8.4) on vectors and dot products was added, and Section 8.5 now combines the discussion of the powers and roots of complex functions with that of the trigonometric form of complex numbers. In this new edition of the text, the focus in Chapter 10, "Matrices and Determinants," is shifted more toward the application and interpretation of matrices. A new discussion of pattern recognition and finite differences is included in Chapter 11. In general, throughout the text, greater emphasis is given to geometry, numeracy, collecting and interpreting data and statistics, updated data analysis, and creating models.

CD-ROM To accommodate a variety of teaching and learning styles, *Algebra and Trigonometry,* Fourth Edition, is also available in a multimedia, CD-ROM format. *Interactive Algebra and Trigonometry* offers students a variety of additional tutorial assistance, including examples and exercises with detailed solutions; pre-, post-, and self-tests with answers; and *TI-82* and *TI-83* graphing calculator emulators. (See pages xvi–xviii for more detailed information.)

Technology   The new Fourth Edition acknowledges the increasing availability of graphing technology by offering the opportunity to use graphing utilities throughout, without requiring their use. This is achieved through a combination of features, including—at point of use—many opportunities for exploration using technology (see pages 182 and 412); graphing utility instructions in the text margin (see pages 84 and 276); and clearly labeled exercises that require the use of a graphing utility (see pages 234 and 267). In addition, *Interactive Algebra and Trigonometry* offers the text in a CD-ROM format, as well as additional tutorial and technology enhanced features.

Data Analysis and Modeling   Throughout the Fourth Edition, students are offered many more opportunities to collect and interpret data, to make conjectures, and to construct mathematical models. Students are exposed to modeling problems with experimental and theoretical probabilities (see pages 866 and 871); encouraged to use mathematical models to make predictions or draw conclusions from real data (see pages 95 and 222); invited to compare models (see pages 447 and 543); and asked to use curve-fitting techniques to write models from data (see pages 317 and 694). This edition encourages greater use of charts, tables, scatter plots, and graphs to summarize, analyze, and interpret data.

Applications   To emphasize for students the connection between mathematical concepts and real-world situations, up-to-date, real-life applications are integrated throughout the text. Appearing as chapter introductions with related exercises (see pages 181 and 210), examples (see pages 231 and 441), exercises (see pages 192 and 446), Group Activities (see page 533), and Chapter Projects (see pages 178–179), these applications offer students frequent opportunities to use and review their problem-solving skills. The applications cover a wide range of disciplines including areas such as physics, chemistry, the social sciences, biology, and business.

Group Activities   Each section ends with a Group Activity. These exercises reinforce students' understanding by exploring mathematical concepts in a variety of ways, including interpretation of mathematical concepts and results (see pages 204 and 297); problem posing and error analysis (see pages 138 and 276); and constructing mathematical models, tables, and graphs (see pages 51, 317, 347, and 430). Designed to be completed in class or as homework assignments, the Group Activities give the students the opportunity to work cooperatively as they think, talk, and write about mathematics.

Connections   In addition to highlighting the connections between algebra and areas outside mathematics through real-world applications, this text emphasizes the connections between algebra and other branches of mathematics, such as probability (see page 865), geometry (see page 31), and statistics (see Appendix A). Many examples and exercises throughout the text also reinforce the connections through graphical, numerical, and analytical representations of important algebraic concepts (see pages 90 and 152).

There are many other new features of the Fourth Edition as well, including Exploration, Study Tips, Historical Notes, Focus on Concepts, and Chapter Projects. These and other features of the Fourth Edition are described in greater detail on the following pages.

Features of the Fourth Edition

Chapter Opener   Each chapter opens with a look at a real-life application. Real data is presented using graphical, numerical, and algebraic techniques. In addition, a list of the section titles shows students how the topics fit into the overall development of algebra and trigonometry.

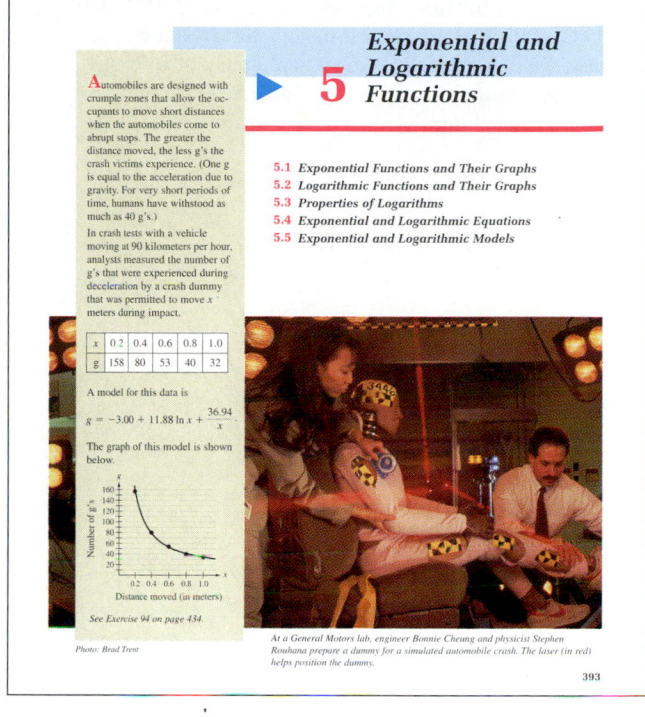

Section Outline   Each section begins with a list of the major topics covered in the section. These topics are also the subsection titles and can be used for easy reference and review by students. In addition, an exercise application that uses a skill or illustrates a concept covered in the section is highlighted to emphasize the connection between mathematical concepts and real-life situations.

Graphics   Visualization is a critical problem-solving skill. To encourage the development of this ability, the text has over 2800 figures in examples, exercises, and answers to exercises. Included are graphs of equations and functions, geometric figures, displays of statistical information, scatter plots, and numerous screen outputs from graphing technology. All graphs of equations and functions are computer- or calculator-generated for accuracy, and they are designed to resemble students' actual screen outputs as closely as possible. Graphics are also used to emphasize graphical interpretation, comparison, and estimation.

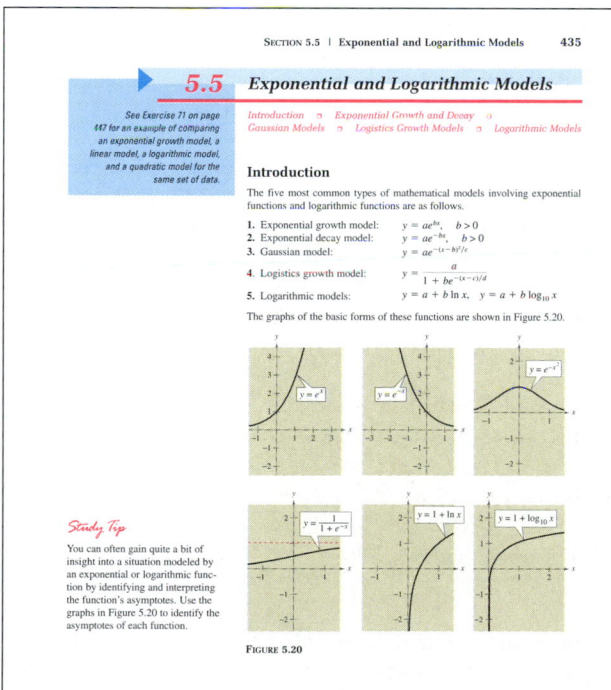

Algebra of Calculus   Special emphasis is given to the algebraic skills that are needed in calculus. In addition to the material in Section P.6 shown here, many other examples in the Fourth Edition discuss algebraic techniques that are used in calculus.

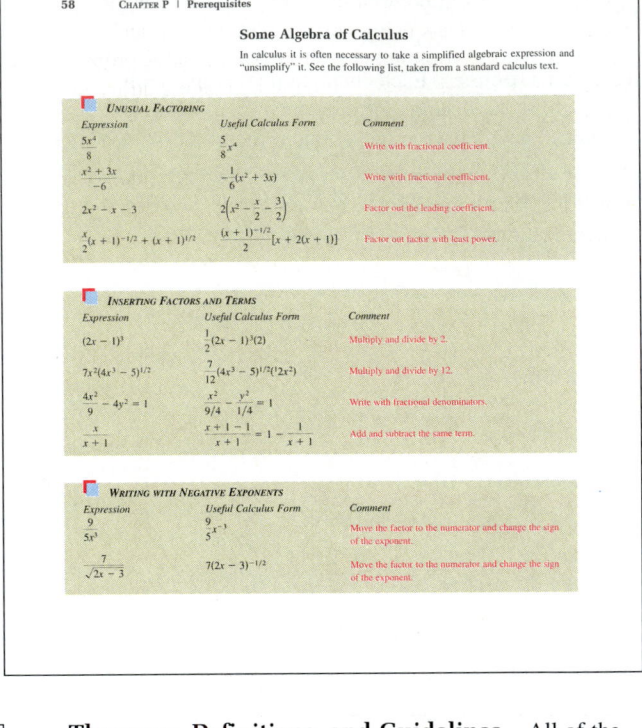

Theorems, Definitions, and Guidelines   All of the important rules, formulas, theorems, guidelines, properties, definitions, and summaries are highlighted for emphasis. Each is also titled for easy reference.

Think About the Proof   Located in the margin adjacent to the corresponding theorem, each Think About the Proof feature offers strategies for proving the theorem. Detailed proofs for all theorems are given in Appendix B.

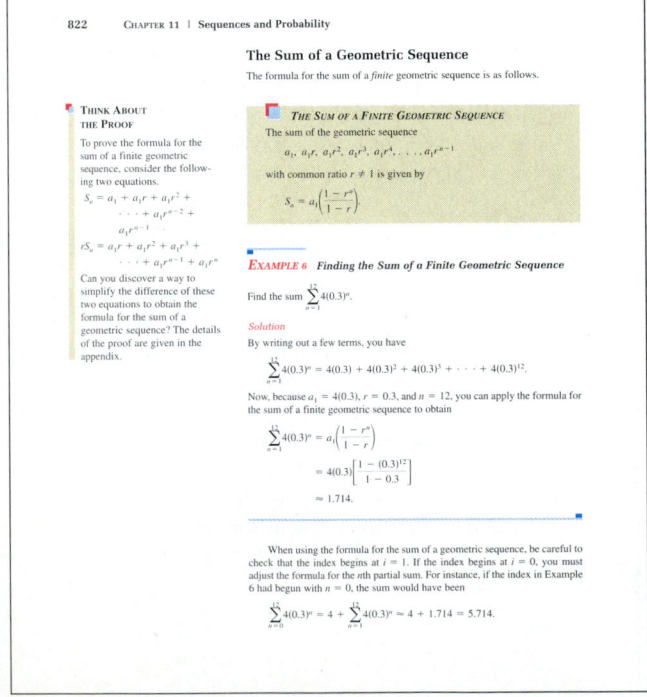

Technology   Instructions for using graphing utilities appear in the text at point of use. They offer convenient reference for students using graphing technology, and they can easily be omitted if desired. Additionally, problems in the Exercise Sets that require a graphing utility have been identified with the icon ⊞ .

Notes   Notes anticipate students' needs by offering additional insights, pointing out common errors, and describing generalizations.

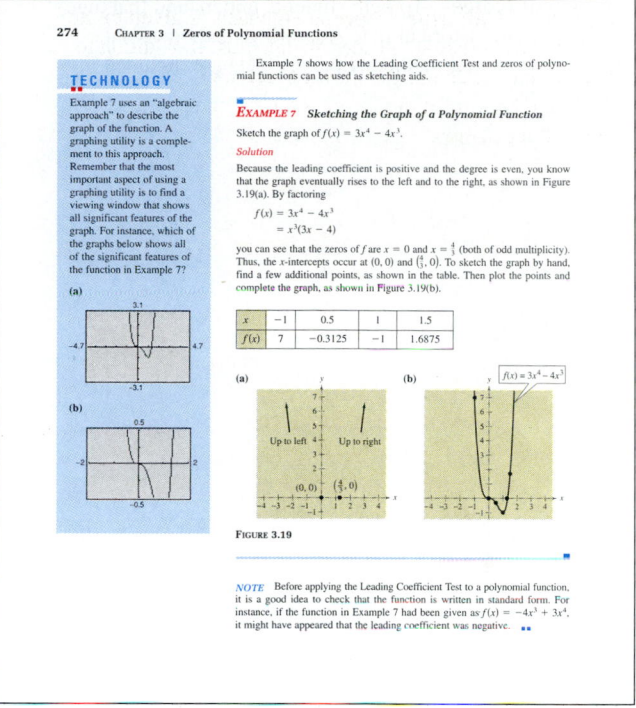

Example 7 shows how the Leading Coefficient Test and zeros of polynomial functions can be used as sketching aids.

TECHNOLOGY

Example 7 uses an "algebraic approach" to describe the graph of the function. A graphing utility is a complement to this approach. Remember that the most important aspect of using a graphing utility is to find a viewing window that shows all significant features of the graph. For instance, which of the graphs below shows all of the significant features of the function in Example 7?

(a)

(b)

EXAMPLE 7   *Sketching the Graph of a Polynomial Function*

Sketch the graph of $f(x) = 3x^4 - 4x^3$.

Solution

Because the leading coefficient is positive and the degree is even, you know that the graph eventually rises to the left and to the right, as shown in Figure 3.19(a). By factoring

$$f(x) = 3x^4 - 4x^3$$
$$= x^3(3x - 4)$$

you can see that the zeros of f are $x = 0$ and $x = \frac{4}{3}$ (both of odd multiplicity). Thus, the x-intercepts occur at $(0, 0)$ and $(\frac{4}{3}, 0)$. To sketch the graph by hand, find a few additional points, as shown in the table. Then plot the points and complete the graph, as shown in Figure 3.19(b).

x	-1	0.5	1	1.5
$f(x)$	7	-0.3125	-1	1.6875

(a)   (b)

FIGURE 3.19

NOTE   Before applying the Leading Coefficient Test to a polynomial function, it is a good idea to check that the function is written in standard form. For instance, if the function in Example 7 had been given as $f(x) = -4x^3 + 3x^4$, it might have appeared that the leading coefficient was negative.

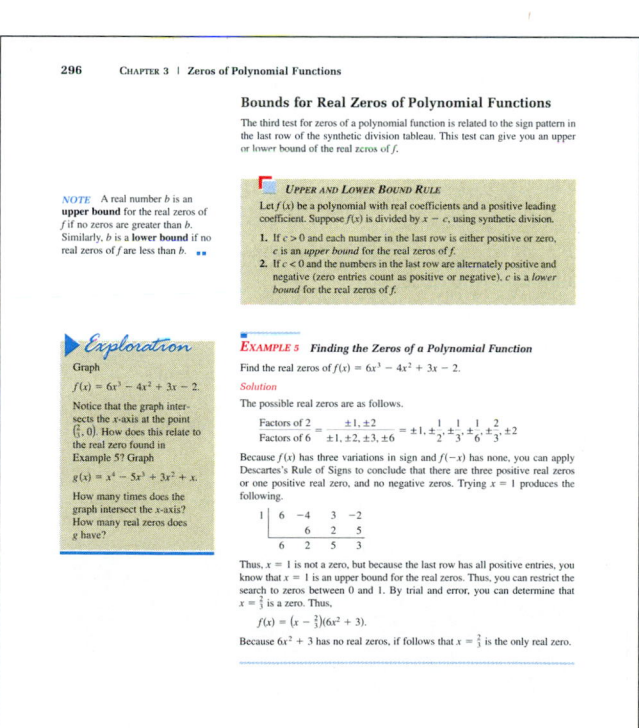

Bounds for Real Zeros of Polynomial Functions

The third test for zeros of a polynomial function is related to the sign pattern in the last row of the synthetic division tableau. This test can give you an upper or lower bound of the real zeros of f.

NOTE   A real number b is an **upper bound** for the real zeros of f if no zeros are greater than b. Similarly, b is a **lower bound** if no real zeros of f are less than b.

UPPER AND LOWER BOUND RULE

Let $f(x)$ be a polynomial with real coefficients and a positive leading coefficient. Suppose $f(x)$ is divided by $x - c$, using synthetic division.

1. If $c > 0$ and each number in the last row is either positive or zero, c is an *upper bound* for the real zeros of f.
2. If $c < 0$ and the numbers in the last row are alternately positive and negative (zero entries count as positive or negative), c is a *lower bound* for the real zeros of f.

▶ *Exploration*

Graph

$$f(x) = 6x^3 - 4x^2 + 3x - 2.$$

Notice that the graph intersects the x-axis at the point $(\frac{2}{3}, 0)$. How does this relate to the real zero found in Example 5? Graph

$$g(x) = x^4 - 5x^3 + 3x^2 + x.$$

How many times does the graph intersect the x-axis? How many real zeros does g have?

EXAMPLE 5   *Finding the Zeros of a Polynomial Function*

Find the real zeros of $f(x) = 6x^3 - 4x^2 + 3x - 2$.

Solution

The possible real zeros are as follows.

$$\frac{\text{Factors of 2}}{\text{Factors of 6}} = \frac{\pm 1, \pm 2}{\pm 1, \pm 2, \pm 3, \pm 6} = \pm 1, \pm \frac{1}{2}, \pm \frac{1}{3}, \pm \frac{1}{6}, \pm \frac{2}{3}, \pm 2$$

Because $f(x)$ has three variations in sign and $f(-x)$ has none, you can apply Descartes's Rule of Signs to conclude that there are three positive real zeros or one positive real zero, and no negative zeros. Trying $x = 1$ produces the following.

```
1 | 6   -4    3   -2
  |       6    2    5
  ---------------------
    6     2    5    3
```

Thus, $x = 1$ is not a zero, but because the last row has all positive entries, you know that $x = 1$ is an upper bound for the real zeros. Thus, you can restrict the search to zeros between 0 and 1. By trial and error, you can determine that $x = \frac{2}{3}$ is a zero. Thus,

$$f(x) = (x - \frac{2}{3})(6x^2 + 3).$$

Because $6x^2 + 3$ has no real zeros, if follows that $x = \frac{2}{3}$ is the only real zero.

Exploration   Throughout the text, the Exploration features encourage active participation by students, strengthening their intuition and critical thinking skills by exploring mathematical concepts and discovering mathematical relationships. Using a variety of approaches, including visualization, verification, use of graphing utilities, pattern recognition, and modeling, students develop conceptual understanding of theoretical topics.

Historical Notes   To help students understand that algebra has a past, historical notes featuring mathematicians and their work and mathematical artifacts are included in each chapter.

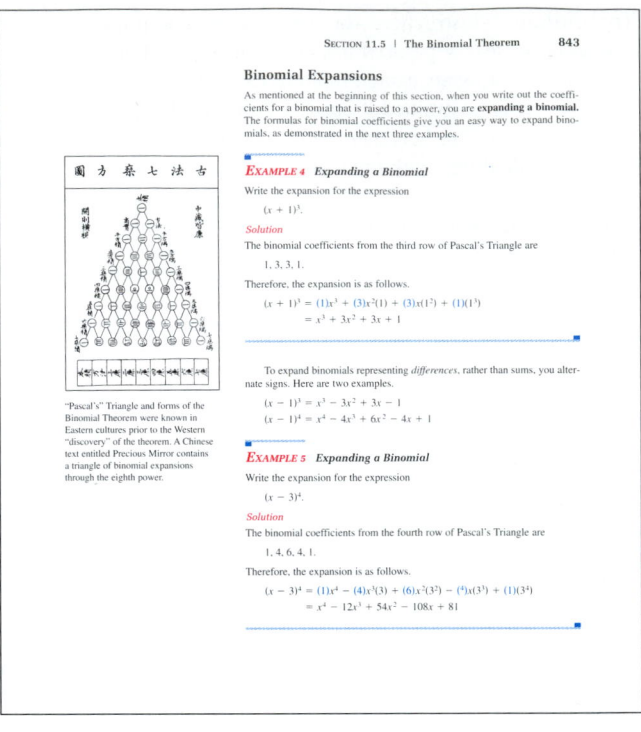

Binomial Expansions

As mentioned at the beginning of this section, when you write out the coefficients for a binomial that is raised to a power, you are **expanding a binomial.** The formulas for binomial coefficients give you an easy way to expand binomials, as demonstrated in the next three examples.

EXAMPLE 4   *Expanding a Binomial*

Write the expansion for the expression

$(x + 1)^3$.

Solution

The binomial coefficients from the third row of Pascal's Triangle are

1, 3, 3, 1.

Therefore, the expansion is as follows.

$(x + 1)^3 = (1)x^3 + (3)x^2(1) + (3)x(1^2) + (1)(1^3)$
$= x^3 + 3x^2 + 3x + 1$

To expand binomials representing *differences*, rather than sums, you alternate signs. Here are two examples.

$(x - 1)^3 = x^3 - 3x^2 + 3x - 1$
$(x - 1)^4 = x^4 - 4x^3 + 6x^2 - 4x + 1$

EXAMPLE 5   *Expanding a Binomial*

Write the expansion for the expression

$(x - 3)^4$.

Solution

The binomial coefficients from the fourth row of Pascal's Triangle are

1, 4, 6, 4, 1.

Therefore, the expansion is as follows.

$(x - 3)^4 = (1)x^4 - (4)x^3(3) + (6)x^2(3^2) - (4)x(3^3) + (1)(3^4)$
$= x^4 - 12x^3 + 54x^2 - 108x + 81$

"Pascal's" Triangle and forms of the Binomial Theorem were known in Eastern cultures prior to the Western "discovery" of the theorem. A Chinese text entitled Precious Mirror contains a triangle of binomial expansions through the eighth power.

Applications   Real-life applications are integrated throughout the text in examples and exercises. These applications offer students constant review of problem-solving skills, and they emphasize the relevance of the mathematics. Many of the applications use recent, real data, and all are titled for easy reference. Photographs with captions in the introduction to the chapter and throughout the text also encourage students to see the link between mathematics and real life.

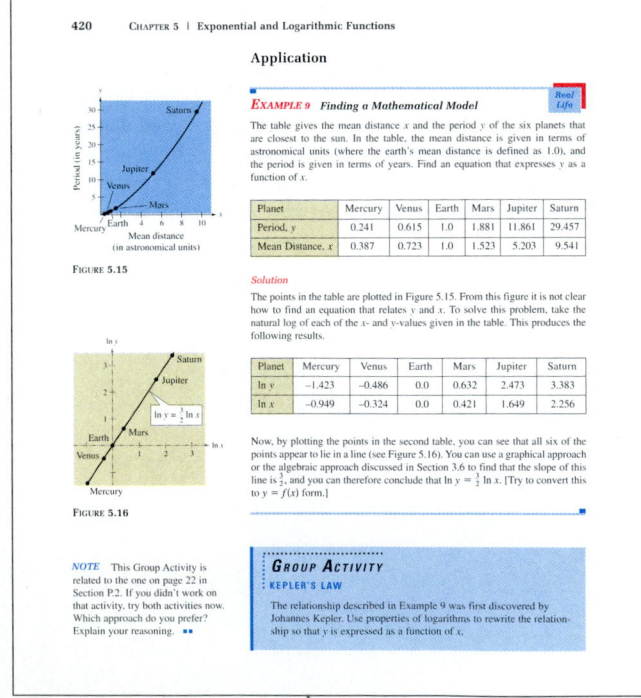

Application

EXAMPLE 9   *Finding a Mathematical Model*

Real Life

The table gives the mean distance x and the period y of the six planets that are closest to the sun. In the table, the mean distance is given in terms of astronomical units (where the earth's mean distance is defined as 1.0), and the period is given in terms of years. Find an equation that expresses y as a function of x.

Planet	Mercury	Venus	Earth	Mars	Jupiter	Saturn
Period, y	0.241	0.615	1.0	1.881	11.861	29.457
Mean Distance, x	0.387	0.723	1.0	1.523	5.203	9.541

Solution

The points in the table are plotted in Figure 5.15. From this figure it is not clear how to find an equation that relates y and x. To solve this problem, take the natural log of each of the x- and y-values given in the table. This produces the following results.

Planet	Mercury	Venus	Earth	Mars	Jupiter	Saturn
$\ln y$	−1.423	−0.486	0.0	0.632	2.473	3.383
$\ln x$	−0.949	−0.324	0.0	0.421	1.649	2.256

Now, by plotting the points in the second table, you can see that all six of the points appear to lie in a line (see Figure 5.16). You can use a graphical approach or the algebraic approach discussed in Section 3.6 to find that the slope of this line is $\frac{3}{2}$, and you can therefore conclude that $\ln y = \frac{3}{2} \ln x$. [Try to convert this to $y = f(x)$ form.]

FIGURE 5.15

FIGURE 5.16

NOTE   This Group Activity is related to the one on page 22 in Section P.2. If you didn't work on that activity, try both activities now. Which approach do you prefer? Explain your reasoning.

GROUP ACTIVITY
KEPLER'S LAW

The relationship described in Example 9 was first discovered by Johannes Kepler. Use properties of logarithms to rewrite the relationship so that y is expressed as a function of x.

Study Tips   Study Tips appear in the margin at point of use and offer students specific suggestions for studying algebra.

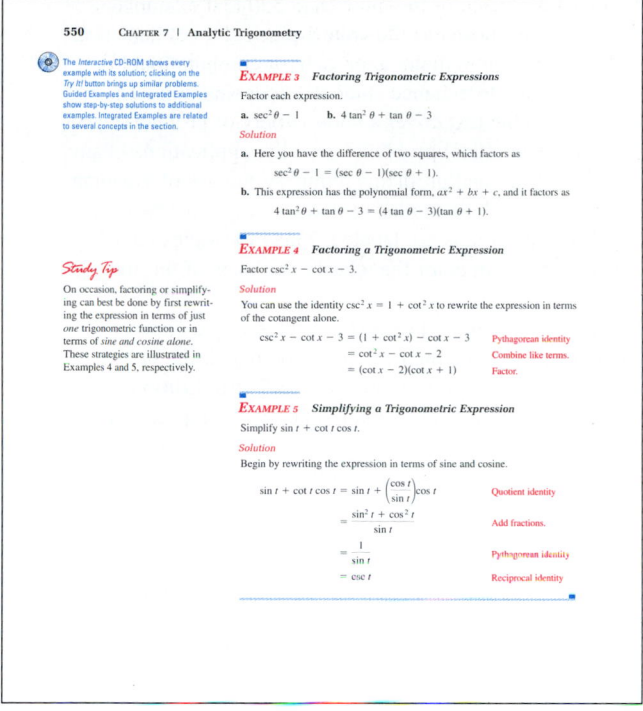

CD-ROM   The icon 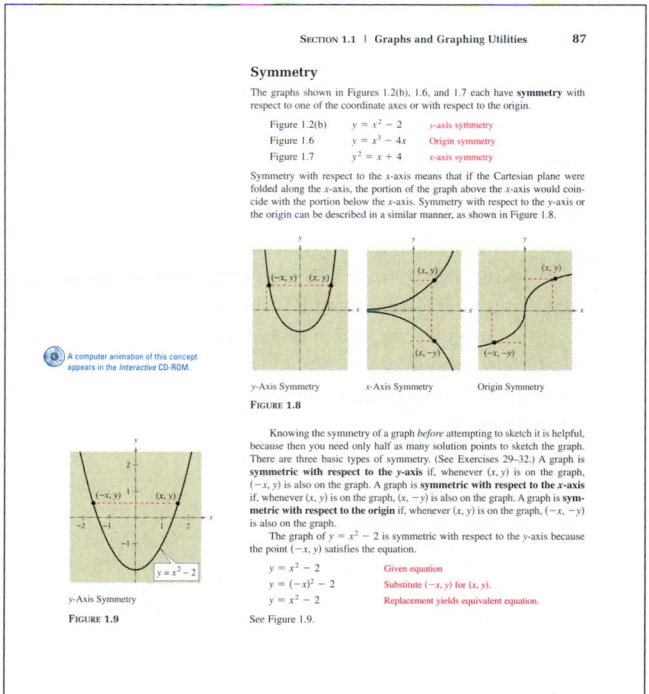 refers to additional features of *Interactive Algebra and Trigonometry* that enhance the text presentation, such as exercises, computer animations, examples, tests, and *TI-82* and *TI-83* graphing calculator emulators.

Examples   Each of the more than 550 text examples was carefully chosen to illustrate a particular mathematical concept, problem-solving approach, or computational technique, and to enhance students' understanding. The examples in the text cover a wide variety of problem types, including theoretical problems, real-life applications (many with real data), and problems requiring the use of graphing technology. Each example is titled for easy reference, and real-life applications are labeled. Many examples include side comments in color that clarify the steps of the solution.

Problem Solving   The text provides ample opportunity for students to hone their problem-solving skills. In both the exercises and the examples in the Fourth Edition, students are asked to apply verbal, analytical, graphical, and numerical approaches to problem solving. Students are also encouraged to use a graphing utility as a tool for solving problems. Students are taught the following approach to solving applied problems: (1) construct a verbal model; (2) label variable and constant terms; (3) construct an algebraic model; (4) using the model, solve the problem; and (5) check the answer in the original statement of the problem.

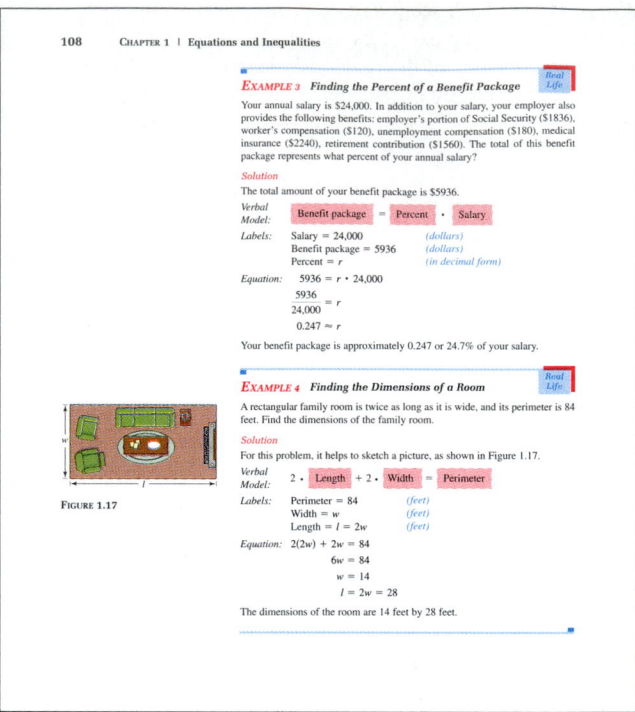

108   CHAPTER 1 | Equations and Inequalities

EXAMPLE 3   *Finding the Percent of a Benefit Package*   Real Life

Your annual salary is $24,000. In addition to your salary, your employer also provides the following benefits: employer's portion of Social Security ($1836), worker's compensation ($120), unemployment compensation ($180), medical insurance ($2240), retirement contribution ($1560). The total of this benefit package represents what percent of your annual salary?

Solution

The total amount of your benefit package is $5936.

Verbal Model: | Benefit package | = | Percent | · | Salary |

Labels: Salary = 24,000   *(dollars)*
Benefit package = 5936   *(dollars)*
Percent = r *(in decimal form)*

Equation: $5936 = r \cdot 24{,}000$

$$\frac{5936}{24{,}000} = r$$

$$0.247 \approx r$$

Your benefit package is approximately 0.247 or 24.7% of your salary.

EXAMPLE 4   *Finding the Dimensions of a Room*   Real Life

A rectangular family room is twice as long as it is wide, and its perimeter is 84 feet. Find the dimensions of the family room.

Solution

For this problem, it helps to sketch a picture, as shown in Figure 1.17.

Verbal Model: $2 \cdot$ | Length | $+ 2 \cdot$ | Width | = | Perimeter |

Labels: Perimeter = 84   *(feet)*
Width = w *(feet)*
Length = $l = 2w$ *(feet)*

Equation: $2(2w) + 2w = 84$

$$6w = 84$$
$$w = 14$$
$$l = 2w = 28$$

The dimensions of the room are 14 feet by 28 feet.

FIGURE 1.17

Geometry   Geometric formulas and concepts are reviewed throughout the text in examples, group activities, and exercises. For reference, common formulas are listed inside the back cover of this text.

Group Activities   The Group Activities that appear at the ends of sections reinforce students' understanding by studying mathematical concepts in a variety of ways, including talking and writing about mathematics, creating and solving problems, analyzing errors, and developing and using mathematical models. Designed to be completed as group projects in class or as homework assignments, the Group Activities give students opportunities to do interactive learning and to think, talk, and write about mathematics.

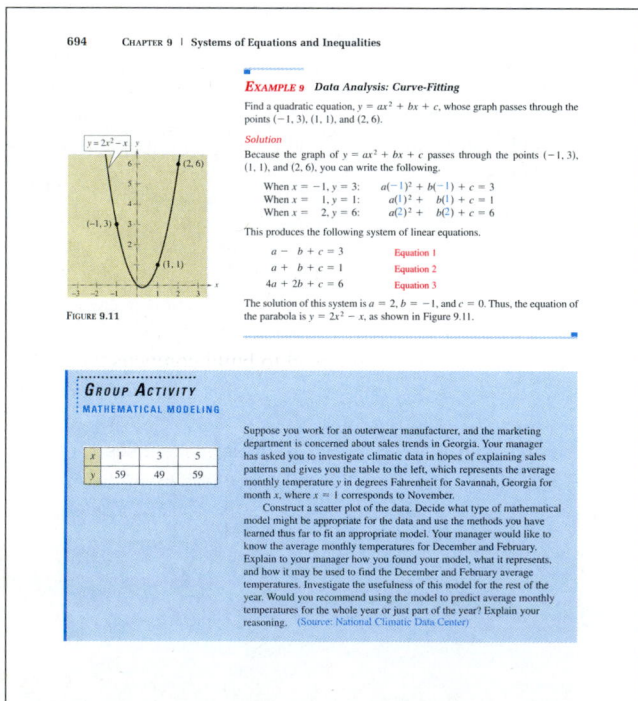

694   CHAPTER 9 | Systems of Equations and Inequalities

EXAMPLE 9   *Data Analysis: Curve-Fitting*

Find a quadratic equation, $y = ax^2 + bx + c$, whose graph passes through the points $(-1, 3)$, $(1, 1)$, and $(2, 6)$.

Solution

Because the graph of $y = ax^2 + bx + c$ passes through the points $(-1, 3)$, $(1, 1)$, and $(2, 6)$, you can write the following.

When $x = -1$, $y = 3$: $a(-1)^2 + b(-1) + c = 3$
When $x = 1$, $y = 1$: $a(1)^2 + b(1) + c = 1$
When $x = 2$, $y = 6$: $a(2)^2 + b(2) + c = 6$

This produces the following system of linear equations.

$a - b + c = 3$ Equation 1
$a + b + c = 1$ Equation 2
$4a + 2b + c = 6$ Equation 3

The solution of this system is $a = 2$, $b = -1$, and $c = 0$. Thus, the equation of the parabola is $y = 2x^2 - x$, as shown in Figure 9.11.

FIGURE 9.11

GROUP ACTIVITY
MATHEMATICAL MODELING

| x | 1 | 3 | 5 |
| y | 59 | 49 | 59 |

Suppose you work for an outerwear manufacturer, and the marketing department is concerned about sales trends in Georgia. Your manager has asked you to investigate climatic data in hopes of explaining sales patterns and gives you the table to the left, which represents the average monthly temperature y in degrees Fahrenheit for Savannah, Georgia for month x, where $x = 1$ corresponds to November.

Construct a scatter plot of the data. Decide what type of mathematical model might be appropriate for the data and use the methods you have learned thus far to fit an appropriate model. Your manager would like to know the average monthly temperatures for December and February. Explain to your manager how you found your model, what it represents, and how it may be used to find the December and February average temperatures. Investigate the usefulness of this model for the rest of the year. Would you recommend using the model to predict average monthly temperatures for the whole year or just part of the year? Explain your reasoning. (Source: National Climatic Data Center)

Warm Ups   Each section (except Section P.1) contains a set of 10 warm-up exercises that students can use for review and practice of the previously learned skills that are necessary for mastery of the new skills and concepts presented in the section. All warm-up exercises are answered in the back of the text.

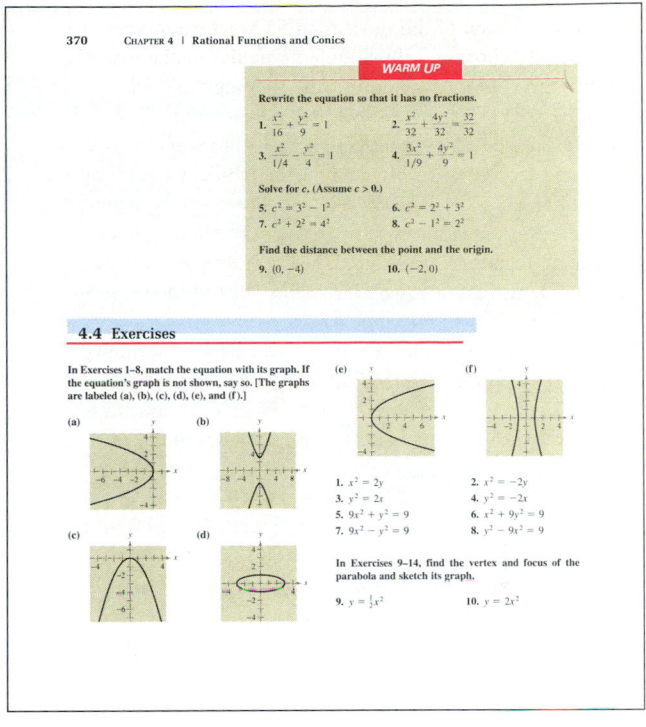

Exercises   The exercise sets were completely revised—and expanded by over 20%—for the Fourth Edition. The text now offers over 7500 exercises with a broad range of conceptual, computational, and applied problems to accommodate a variety of teaching and learning styles. Included in the section and review exercise sets are multi-part, writing, and more challenging problems with extensive graphics that encourage exploration and discovery, enhance students' skills in mathematical modeling, estimation, and data interpretation and analysis, and encourage the use of graphing technology for conceptual understanding. Applications are labeled for easy reference. The exercise sets are designed to build competence, skill, and understanding; each exercise set is graded in difficulty to allow students to gain confidence as they progress. Detailed solutions to all odd-numbered exercises are given in the *Study and Solutions Guide*; answers to all odd-numbered exercises appear in the back of the text.

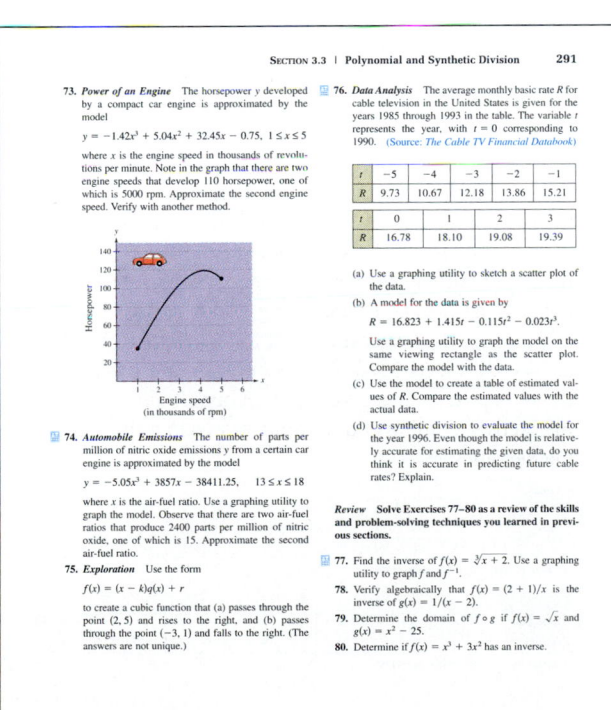

Focus on Concepts   Each Focus on Concepts feature is a set of exercises that test students' understanding of the basic concepts covered in the chapter. Answers to all questions are given in the back of the text.

Chapter Projects   Chapter Projects are extended applications that use real data, graphs, and modeling to enhance students' understanding of mathematical concepts. Designed as individual or group projects, they offer additional opportunities to think, discuss, and write about mathematics. Many projects give students the opportunity to collect, analyze, and interpret data.

---

448   CHAPTER 5 | Exponential and Logarithmic Functions

FOCUS ON CONCEPTS

In this chapter, you studied several concepts related to exponential and logarithmic functions. Answer the following questions to check your understanding of several of the basic concepts discussed in this chapter. The answers to these questions are given in the back of the book.

1. *Comparing Graphs*   The graphs of $y = e^{kx}$ are shown for $k = a, b, c,$ and d. Use the graphs to order $a, b, c,$ and d. Which of the four values are negative? Which are positive?

(a)   (b)   (c)   (d)

2. *True or False?*   Rewrite each verbal statement as an equation. Then decide whether the statement is true or false. If it is false, give an example that shows it is false.

(a) The logarithm of the product of two numbers is equal to the sum of the logarithms of the numbers.

(b) The logarithm of the sum of two numbers is equal to the product of the logarithms of the numbers.

(c) The logarithm of the difference of two numbers is equal to the difference of the logarithms of the numbers.

(d) The logarithm of the quotient of two numbers is equal to the difference of the logarithms of the numbers.

3. *Investing Money*   You are investing P dollars at an annual rate of r, compounded continuously, for t years. Which of the following would be most advantageous? Explain your reasoning.

(a) Double the amount you invest.

(b) Double your interest rate.

(c) Double the number of years.

4. Identify the model as linear, logarithmic, exponential, logistic, or none of the above. Explain your reasoning.

(a)   (b)   (c)   (d)   (e)   (f)

---

726   CHAPTER 9 | Systems of Equations and Inequalities

CHAPTER PROJECT: *Fitting Models to Data*

Many of the models in this book were created with a statistical method called *least squares regression analysis*. This procedure is tedious to perform by hand, but can be performed easily with a computer or graphing calculator.

EXAMPLE 1   Fitting a Line to Data   Real Life

The numbers (in millions) of morning and evening newspapers sold each day in the United States from 1978 through 1992 are shown in the table. Use the data to project the numbers of morning and evening newspapers that will be sold each day in 1998. In the table, $t = 0$ represents 1980. (Source: Editor and Publisher Company)

Year, t	−2	−1	0	1	2	3	4	5
Morning	27.7	28.6	29.4	30.6	33.2	33.8	35.4	36.4
Evening	34.3	33.6	32.8	30.9	29.3	28.8	27.7	26.4

Year, t	6	7	8	9	10	11	12
Morning	37.4	39.1	40.4	40.7	41.3	41.5	42.4
Evening	25.1	23.7	22.2	21.8	21.0	19.2	17.8

Solution

Begin by finding a computer or graphing calculator that will perform linear regression analysis. After entering the data and running the program, you should obtain the following models. (Both models have correlation coefficients whose absolute values are greater than 0.98, which means that the models are very good fits for the data.)

$y = 30.246 + 1.123t$ Morning newspaper circulation

$y = 32.225 − 1.184t$ Evening newspaper circulation

With these models, you can project the 1998 newspaper sales. If the sales through 1998 continue to follow the pattern from 1978 through 1992, the 1998 sales of newspapers should be about

$y = 30.246 + 1.123(18) \approx 50.5$ million   Morning newspaper prediction

$y = 32.225 + 1.184(18) \approx 10.9$ million.   Evening newspaper prediction

The graphs of the data and models are shown at the left.

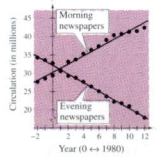

Circulation (in millions)
Morning newspapers
Evening newspapers
Year (0 ↔ 1980)

Chapter Project   727

CHAPTER PROJECT INVESTIGATIONS

1. *Sunday Paper Circulation*   The numbers (in millions) of Sunday newspapers sold each week in the United States from 1981 through 1992 are shown in the table below. Find a linear model that represents the data. Use your model to project the number of Sunday newspapers to be sold each week in 1998.

2. *Newspaper Companies*   The numbers of newspaper companies in the United States from 1981 through 1992 are shown in the table below. Find a linear model for each of these sets of data.

3. *Average Sales*   From 1981 through 1992, the circulation of morning newspapers increased. However, because the number of morning newspaper companies also increased, the competition for morning newspaper readers became keener. Did the average circulation per morning newspaper company increase or decrease? Explain.

4. *Average Sales*   From 1981 through 1992, the circulation of evening newspapers decreased. However, because the number of evening newspaper companies also decreased, the competition for evening newspaper readers became less keen. Did the average circulation per evening newspaper company increase or decrease? Explain.

5. *Average Sales*   From 1981 through 1992, the circulation of Sunday newspapers increased. However, because the number of Sunday newspaper companies also increased, the competition for Sunday newspaper readers became keener. Did the average circulation per Sunday newspaper company increase or decrease? Explain.

6. *Which Would You Choose?*   If you had the opportunity to invest in a company that published only one type of newspaper (morning, evening, or Sunday), which would you choose? Explain your reasoning.

7. *Households and Population*   Figures for population and number of households in the United States (both in millions) from 1981 through 1992 are given in the table below. From this information would you say that the percent of Americans who read newspapers was increasing or decreasing from 1981 through 1992? Explain your reasoning.

Year, t	1	2	3	4	5	6	7	8	9	10	11	12
Sunday	55.2	56.3	56.7	57.5	58.8	58.9	60.1	61.5	62.0	62.6	62.1	62.2

Table for Exercise 1 ($t = 0$ represents 1980.)

Year, t	1	2	3	4	5	6	7	8	9	10	11	12
Morning	408	434	446	458	482	499	511	529	530	559	571	596
Evening	1352	1310	1284	1257	1220	1188	1166	1141	1125	1084	1042	996
Sunday	755	768	772	783	798	802	820	840	847	863	875	891

Table for Exercise 2 ($t = 0$ represents 1980.)

Year, t	1	2	3	4	5	6	7	8	9	10	11	12
Households	81.6	83.0	83.9	85.4	86.8	88.5	89.5	91.1	92.8	93.3	95.7	96.4
Population	230.0	232.2	234.3	236.3	238.5	240.7	242.8	245.0	247.3	249.9	252.6	255.5

Table for Exercise 7 ($t = 0$ represents 1980.)

---

326 CHAPTER 3 | Zeros of Polynomial Functions

Review Exercises

In Exercises 1–4, sketch the graph of the quadratic function. Identify the vertex and the intercepts.

1. $f(x) = (x + \frac{3}{2})^2 + 1$
2. $f(x) = (x - 4)^2 - 4$
3. $f(x) = \frac{1}{3}(x^2 + 5x - 4)$
4. $f(x) = 3x^2 - 12x + 11$

In Exercises 5–8, find the quadratic function that has the indicated vertex and whose graph passes through the given point.

5.
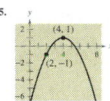

6.
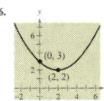

7. Vertex: $(1, -4)$; Point: $(2, -3)$
8. Vertex: $(2, 3)$; Point: $(-1, 6)$

Graphical Reasoning In Exercises 9 and 10, use a graphing utility to graph each equation on the same viewing rectangle. Describe how each graph differs from the graph of $y = x^2$.

9. (a) $y = 2x^2$ (b) $y = -2x^2$
   (c) $y = x^2 + 2$ (d) $y = (x + 2)^2$
10. (a) $y = x^2 - 4$ (b) $y = 4 - x^2$
    (c) $y = (x - 3)^2$ (d) $y = \frac{1}{2}x^2 - 1$

In Exercises 11–18, find the maximum or minimum value of the quadratic function.

11. $g(x) = x^2 - 2x$ 12. $f(x) = x^2 + 8x + 10$
13. $f(x) = 6x - x^2$ 14. $h(x) = 3 + 4x - x^2$

15. $f(t) = -2t^2 + 4t + 1$
16. $h(x) = 4x^2 + 4x + 13$
17. $h(x) = x^2 + 5x - 4$
18. $f(x) = 4x^2 + 4x + 5$

19. *Numerical, Graphical, and Analytical Analysis* A rectangle is inscribed in the region bounded by the x-axis, the y-axis, and the graph of $x + 2y - 8 = 0$ (see figure).

(a) Complete six rows of a table such as the one below.

x	y	Area
1	$4 - \frac{1}{2}(1)$	$(1)[4 - \frac{1}{2}(1)] = \frac{7}{2}$
2	$4 - \frac{1}{2}(2)$	$(2)[4 - \frac{1}{2}(2)] = 6$

(b) Use a graphing utility to generate additional rows of the table. Use the table to estimate the dimensions that will produce the maximum area.

(c) Write the area A as a function of x. Determine the domain of the function in the context of the problem.

(d) Use a graphing utility to graph the area function. Use the graph to approximate the dimensions that will produce the maximum area.

(e) Write the area function in standard form to find analytically the dimensions that will produce the maximum area.

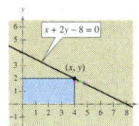

---

454 CUMULATIVE TEST | Chapters 3–5

Cumulative Test for Chapters 3–5

Take this test as you would take a test in class. After you are done, check your work against the answers given in the back of the book.

In Exercises 1–6, sketch the graph of the function without the aid of a graphing utility.

1. $h(x) = -(x^2 + 4x)$ 2. $f(t) = \frac{1}{4}t(t - 2)^2$ 3. $g(s) = \frac{2s}{s - 3}$
4. $g(s) = \frac{2s^2}{s - 3}$ 5. $f(x) = 6(2^{-x})$ 6. $g(x) = \log_3 x$

7. Divide: $\dfrac{6x^3 - 4x^2}{2x^2 + 1}$.

8. Find all the zeros of $f(x) = x^3 + 2x^2 + 4x + 8$.

9. Use a graphing utility to approximate the real zero of the function $g(x) = x^3 + 3x^2 - 6$ to the nearest hundredth.

In Exercises 10 and 11, sketch a graph of the conic.

10. $6x - y^2 = 0$ 11. $\dfrac{(x - 2)^2}{4} + \dfrac{(y + 1)^2}{9} = 1$

12. Find an equation of the parabola in the figure.

13. Find an equation of the hyperbola with foci $(0, 0)$ and $(0, 4)$ and asymptotes $y = \pm \frac{1}{2}x + 2$.

14. Write $2 \ln x - \frac{1}{2} \ln(x + 5)$ as a logarithm of a single quantity.

15. Use a graphing utility to graph $f(x) = \dfrac{1000}{1 + 4e^{-0.2x}}$ and determine the horizontal asymptotes.

In Exercises 16 and 17, solve the equation.

16. $6e^{2x} = 72$ 17. $\log_2 x + \log_2 5 = 6$

18. Let x be the amount (in hundreds of dollars) that a company spends on advertising, and let P be the profit (in thousands of dollars), where
$$P = 230 + 20x - \tfrac{1}{2}x^2.$$
What amount of advertising will yield a maximum profit?

19. On the day a grandchild is born, a grandparent deposits $2500 into a fund earning 7.5%, compounded continuously. Determine the balance in the account at the time of the grandchild's 25th birthday.

FIGURE FOR 12

---

332 CHAPTER 3 | Zeros of Polynomial Functions

Chapter Test

Take this test as you would take a test in class. After you are done, check your work against the answers given in the back of the book.

1. Describe how the graph of g differs from the graph of $f(x) = x^2$.
   (a) $g(x) = 2 - x^2$ (b) $g(x) = \left(x - \frac{3}{2}\right)^2$
2. Identify the vertex and intercepts of the graph of $y = x^2 + 4x + 3$.
3. Find an equation of the parabola shown in the figure.
4. The path of a ball is given by $y = -\frac{1}{20}x^2 + 3x + 5$, where y is the height in feet and x is the horizontal distance (in feet) from where the ball was thrown.
   (a) Find the maximum height of the ball.
   (b) Which constant determines the height at which the ball was thrown? Does changing this constant change the coordinates of the maximum height of the ball? Explain.
5. Determine the right-hand and left-hand behavior of the graph of the function $h(t) = -\frac{3}{4}t^5 + 2t^2$.
6. Divide by long division.   7. Divide by synthetic division.
   $\dfrac{3x^3 + 4x - 1}{x^2 + 1}$ $\dfrac{2x^4 - 5x^2 - 3}{x - 2}$
8. Use synthetic division to show that $x = \sqrt{3}$ is a solution of the equation $4x^3 - x^2 - 12x + 3 = 0$. Use the result to factor the polynomial completely and list all the real solutions of the equation.

In Exercises 9 and 10, list all the possible rational zeros of the function. Use a graphing utility to graph the function and find all the rational zeros.

9. $g(t) = 2t^4 - 3t^3 + 16t - 24$ 10. $h(x) = 3x^5 + 2x^4 - 3x - 2$

In Exercises 11 and 12, use the root-finding capabilities of a graphing utility to approximate the real zeros of the function accurate to three decimal places.

11. $f(x) = x^4 - x^3 - 1$ 12. $f(x) = 3x^5 + 2x^4 - 12x - 8$

In Exercises 13 and 14, find a polynomial function with integer coefficients that has the given zeros.

13. $0, 3, 3 + i, 3 - i$ 14. $1 + \sqrt{3}i, 1 - \sqrt{3}i, 2, 2$

FIGURE FOR 3

The *Interactive* CD-ROM provides answers to the Chapter Tests and Cumulative Tests. It also offers Chapter Pre-Tests (which test key skills and concepts covered in previous chapters) and Chapter Post-Tests, both of which have randomly generated exercises with diagnostic capabilities.

Review Exercises The Review Exercises at the end of each chapter offer students an opportunity for additional practice. Answers to odd-numbered review exercises are given in the back of the text.

Chapter Tests Each chapter that is not followed by a Cumulative Test ends with a Chapter Test, an effective tool for student self-assessment.

Cumulative Tests The Cumulative Tests that follow Chapters 3, 5, 8, and 11 help students judge their mastery of previously covered material, as well as reinforce the knowledge they have been accumulating throughout the text—preparing them for other exams and for future courses.

Supplements

Algebra and Trigonometry, Fourth Edition, by Larson and Hostetler, is accompanied by a comprehensive supplements package. Most items are keyed to the text.

Printed Resources

FOR THE STUDENT

Study and Solutions Guide by Dianna Zook, Indiana University/Purdue University—Fort Wayne

- Section summaries of key concepts
- Detailed, step-by-step solutions to all odd-numbered exercises
- Practice tests with solutions
- Study strategies

Graphing Technology Keystroke Guide: Precalculus

- Keystroke instructions for many graphing calculators from Texas Instruments, Sharp, Casio, and Hewlett-Packard, including *TI-83, TI-92, HP-38G,* and *Casio CFX-9800G.*
- BestGrapher instructions for both IBM and Macintosh
- Examples with step-by-step solutions
- Extensive graphics screen output
- Technology tips

FOR THE INSTRUCTOR

Instructor's Annotated Edition

- Includes the entire student edition of the text, with the student answers section
- Instructor's Answers section: Answers to all even-numbered exercises, and answers to all Explorations, Technology exercises, Chapter Project exercises, and Group Activities
- Annotations at point of use offer specific teaching strategies and suggestions for implementing Group Activities, point out common student errors, and give additional examples, exercises, class activities, and group activities.

Complete Solutions Guide

- Detailed, step-by-step solutions to all section, review, Focus on Concepts, and Chapter Project exercises

Test Item File and Instructor's Resource Guide

- Printed test bank with nearly 2500 test items (multiple-choice, open-ended, and writing) coded by level of difficulty
- Technology-required test items coded for easy reference
- Bank of chapter test forms with answer keys

- Two final exam test forms
- Notes to the instructor, including materials for alternative assessment and managing the multicultural and cooperative-learning classrooms

Problem Solving, Modeling, and Data Analysis Labs by Wendy Metzger, Palomar College

- Multipart, guided discovery activities and applications
- Keystroke instructions for Derive and *TI-82*
- Keyed to the text by topic
- Funded in part by NSF (National Science Foundation, Instrumentation and Laboratory Improvement) and California Community College Fund for Instructional Improvement

Media Resources

FOR THE STUDENT

Interactive Algebra and Trigonometry (See pages xvi–xviii for a description, or visit the Houghton Mifflin home page at http://www.hmco.com for a preview.)

- Interactive, multimedia CD-ROM format
- IBM-PC for Windows

Tutor software

- Interactive tutorial software keyed to the text by section
- Diagnostic feedback
- Chapter self-tests
- Guided exercises with step-by-step solutions
- Glossary

Videotapes by Dana Mosely

- Comprehensive, text-specific coverage keyed to the text by section
- Real-life application vignettes introduced where appropriate
- Computer-generated animation
- For media/resource centers
- Additional explanation of concepts, sample problems, and applications

FOR THE INSTRUCTOR

Computerized Testing (IBM, Macintosh, Windows)

- New on-line testing
- New grade-management capabilities
- Algorithmic test-generating software provides an unlimited number of tests.
- Nearly 2500 test items
- Also available as a printed test bank

Transparency Package

- 50 color transparencies color-coded by topic

Interactive Algebra and Trigonometry

To accommodate a variety of teaching and learning styles, *Algebra and Trigonometry* is also available in a multimedia, CD-ROM format. In this interactive format, the text offers the student additional tutorial assistance with

- Complete solutions to all odd-numbered text exercises.
- Chapter pre-tests, self-tests, and post-tests.
- *TI-82* and *TI-83* emulators.

- Guided examples with step-by-step solutions.
- Editable graphs.
- Animations of mathematical concepts.
- Warm-up, section, and tutorial exercises.
- Glossary of key terms.

These and other pedagogical features of the CD-ROM are illustrated by the screen dumps shown below.

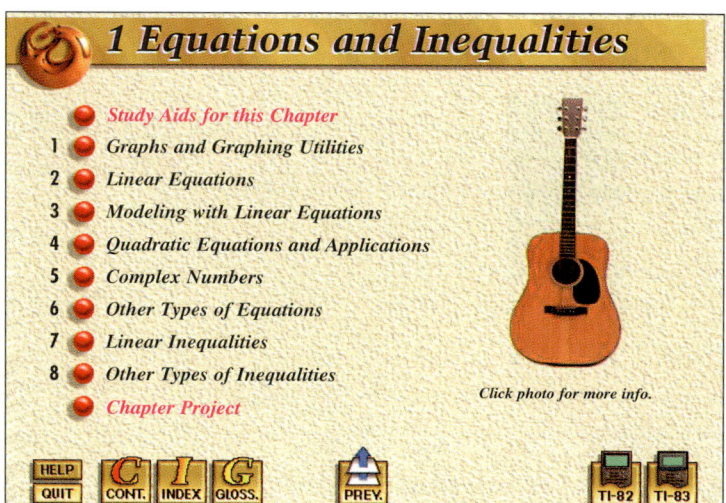

Chapter Topics   Each chapter begins with an outline of the topics to be covered. Using the buttons at the bottom of the screen, the student can quickly move to the appropriate section.

Introductory Chapter Application   Each chapter opens with a real-data application that illustrates the key concepts and techniques to be covered. Clicking on the photo, the student can access additional data and background information that frames the real-world context for a mathematical concept.

Chapter Project   Each chapter is accompanied by a Chapter Project. This offers the student the opportunity to synthesize the algebraic techniques and concepts studied in the chapter. Many projects use real data and emphasize data analysis and mathematical modeling.

Study Aids   Each section offers the student an array of additional study aids, including Chapter Pre-, Post-, and Self-Tests, Review Exercises, and Focus on Concepts. With diagnostics, complete solutions, or answers, these helpful features promote the focused practice needed to master mathematical concepts.

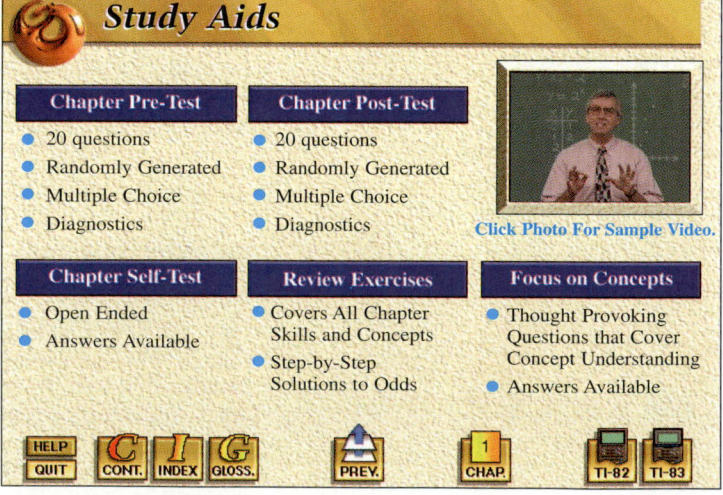

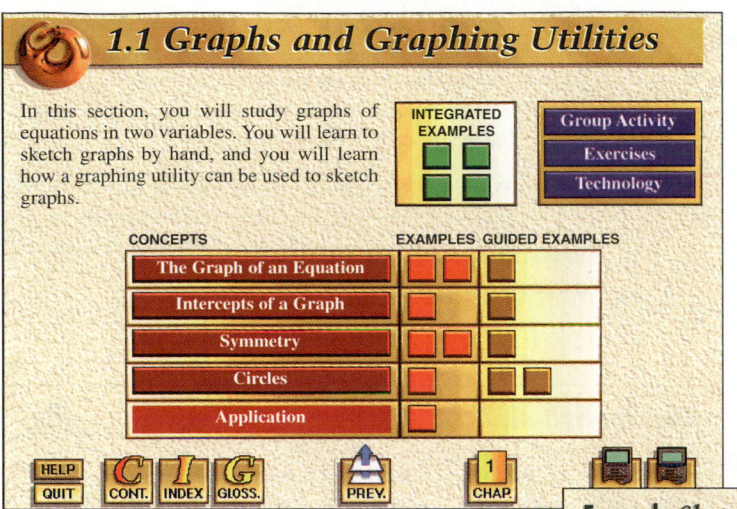

Examples   *Interactive Algebra and Trigonometry* illustrates mathematical concepts by featuring all of the Examples found in the text. Some selected Examples are enhanced by editable graphs. Guided Examples with step-by-step solutions that appear one line at a time offer additional opportunities for practice and skill development. Group Activities, Exercises, and Integrated Examples help synthesize the concepts in the section.

Graphs   Some examples are accompanied by editable graphs for exploration and discovery. Using the keys below the graph, the student can change the function, the graphing window, and the *x*- and *y*-scales, and can trace and zoom.

Example *Sketching the Graph of an Equation*

Sketch the graph of $y = 7 - 3x$.

Solution

The simplest way to sketch the graph of an equation is the *point-plotting method.* With this method, you construct a **table of values** that consists of several solution points of the equation. For instance, when $x = 0$,

$$y = 7 - 3(0) = 7,$$

which implies that $(0, 7)$ is a solution point of the graph.

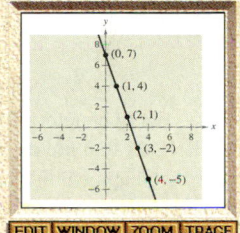

After plotting the points given in the table, you can see that they appear to lie on a line, as shown in the figure. The graph of the equation is the line that passes through the five plotted points.

Try It!

Try It!   After studying the worked-out example, the student can use the Try It! button to access similar examples—with solutions following on separate screens—to test his or her mastery of mathematical concepts and techniques.

TI-82 and TI-83 Emulators   Accessible on every screen, the *TI-82* and *TI-83* emulators give instant access to graphing utilities as tools for computation and exploration. They are also available for working exercises in the text that require the use of a graphing utility, all of which are clearly marked by the icon ▦.

In Exercises 1 and 2, complete a table of values. Use the solution points to sketch the graph of the equation.

1. $y = -\frac{1}{2}x + 2$ **2.** $y = x^2 - 3x$

In Exercises 3–12, sketch the graph *by hand*.

3. $y - 2x - 3 = 0$ **4.** $3x + 2y + 6 = 0$

5. $x - 5 = 0$ **6.** $y = 8 - |x|$

7. $y = \sqrt{5 - x}$ **8.** $y = \sqrt{x + 2}$

9. $y + 2x^2 = 0$ **10.** $y = x^2 - 4x$

11. $y = \sqrt{25 - x^2}$ **12.** $x^2 + y^2 = 10$

Section Exercises   Each section is accompanied by a comprehensive set of exercises promoting skills mastery and conceptual understanding. Solutions to all odd-numbered exercises are available for instant feedback.

Tutorial Exercises   Every section has a set of exercises in a multiple-choice format that offer students additional practice. Examples and diagnostics enhance this guided practice.

Chapter Self-Test

Take this test as you would take a test in class. After you are done, check your work with the answers given by selecting the Answer button.

1   2   3   4   5   6   7   8   9   10

1. Use a graphing utility to graph $y = 4 - \frac{3}{4}x$. Check for symmetry and identify x- and y-intercepts.

Answer

Tests   Every chapter of the interactive text includes tests that are different from those in the textbook: Chapter Pre-Tests (testing key skills and concepts covered in previous chapters), and Chapter Post-Tests test mastery of the material covered in the textbook. The Chapter Self-Tests from the text are also included. Answers to all tests are included.

---

102   CHAPTER 1 | Equations and Inequalities

The *Interactive* CD-ROM provides additional help with Warm-Up exercises by providing a hypertext link to the section in which the concept was introduced.

The *Interactive* CD-ROM contains step-by-step solutions to all odd-numbered Section and Review Exercises. It also provides Tutorial Exercises, which link to Guided Examples for additional help.

WARM UP

Perform the operations and simplify.

1. $(2x - 4) - (5x + 6)$ **2.** $(3x - 5) + (2x - 7)$

3. $2(x + 1) - (x + 2)$ **4.** $-3(2x - 4) + 7(x + 2)$

5. $\frac{x}{3} + \frac{x}{5}$ **6.** $x - \frac{x}{4}$

7. $\frac{1}{x + 1} - \frac{1}{x}$ **8.** $\frac{2}{x} + \frac{3}{x}$

9. $\frac{4}{x} + \frac{3}{x - 2}$ **10.** $\frac{1}{x + 1} - \frac{1}{x - 1}$

1.2 Exercises

Interactive Algebra and Trigonometry supports the mathematical presentation in the text *Algebra and Trigonometry,* Fourth Edition, with a variety of tutorial, diagnostic, and demonstration features. Throughout both the student text and the Instructor's Annotated Edition, CD-ROM icons identify these additional functions of the interactive text, as illustrated by the sample text page (at left).

Acknowledgments

We would like to thank the many people who have helped us at various stages of this project to prepare the text and supplements package. Their encouragement, criticisms, and suggestions have been invaluable to us.

Fourth Edition Reviewers: Joby Milo Anthony, University of Central Florida; Sudhir Kumar Goel, Valdosta State University; Kathy B. Hamrick, Augusta College; Steven Z. Kahn, Anne Arundel Community College; Anne Landry, Dutchess Community College; Sue Little, North Harris Community College; Giles Maloof, Boise State University; Steven E. Martin, Richard Bland College; Carol Paxton, Glendale Community College (CA); Joan N. Powell, Auburn University; Michael P. Scanlon, Fairleigh Dickinson University; Patricia Shelton, North Carolina Agricultural and Technical State University; Laurence Small, Los Angeles Pierce College; Patricia B. Taylor, Thomas Nelson Community College; Gary Thomasson, Palm Beach Community College; Jeffrey X. Watt, Indiana University–Purdue University, Indianapolis; and Jacci Wozniak, Brevard Community College.

Fourth Edition Survey Respondents: Marwan A. Abu-Sawwa, Florida Community College at Jacksonville; Barbara C. Armenta, Pima Community College—East; Gladwin E. Bartel, Otero Junior College; Carole A. Bauer, Triton College; Joyce M. Becker, Luther College; Marybeth Beno, South Suburban College; Charles M. Biles, Humboldt State University; Ruthane Bopp, Lake Forest College; Tim Chappell, North Central Missouri College; Michael Davidson, Cabrillo College; Diane L. Doyle, Adirondack Community College; Donna S. Fatheree, University of Louisiana at Lafayette; John R. Formsma, Los Angeles City College; John S. Frohliger, Saint Norbert College; Gary Glaze, Eastern Washington University; Irwin S. Goldfine, Truman College; Elise M. Grabner, Slippery Rock University; Donnie Hallstone, Green River Community College; Lois E. Higbie, Brookdale Community College; Susan S. Hollar, Kalamazoo Valley Community College; Fran Hopf, Hillsborough Community College; Margaret D. Hovde, Grossmont College; John F. Keating, Massasoit Community College; John G. LaMaster, Indiana University–Purdue University at Fort Wayne; Giles Wilson Maloof, Boise State University, Kenneth Mangels, Concordia State University; Peggy I. Miller, University of Nebraska at Kearney; Gilbert F. Orr, University of Southern Colorado; Jim Paige, Wayne State College; Elise Price, Tarrant County Junior College; Doris Schraeder, McLennan Community College; Linda Schultz, McHenry County College; Fay Sewell, Montgomery County Community College; Patricia G. Shelton, North Carolina A & T State University; Hazal Shows, Hinds Community College; Joseph F. Stokes, Western Kentucky University; Diane Van Nostrand, University of Tulsa; and Raymond D. Wuco, San Joaquin Delta College.

Thanks to all of the people at Houghton Mifflin Company who worked with us in the development and production of the text, especially Chris Hoag, Sponsoring Editor; Cathy Cantin, Managing Editor; Maureen Brooks, Associate Editor; Carolyn Johnson, Assistant Editor; Karen Carter, Senior Project Editor; Rachel Wimberly, Associate Project Editor; Gary Crespo, Art Supervisor; Lisa Merrill, Production Supervisor; Ros Kane, Marketing Associate; and Carrie Lipscomb, Editorial Assistant.

We would also like to thank the staff at Larson Texts, Inc. who assisted with proofreading the manuscript, preparing and proofreading the art package, and checking and typesetting the supplements.

On a personal level, we are grateful to our wives, Deanna Gilbert Larson and Eloise Hostetler, for their love, patience, and support. Also, special thanks go to R. Scott O'Neil.

If you have suggestions for improving the text, please feel free to write to us. Over the past two decades, we have received many useful comments from both instructors and students, and we value these very much.

Roland E. Larson
Robert P. Hostetler

Contents

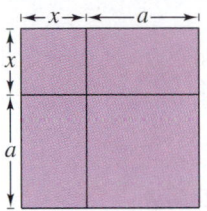

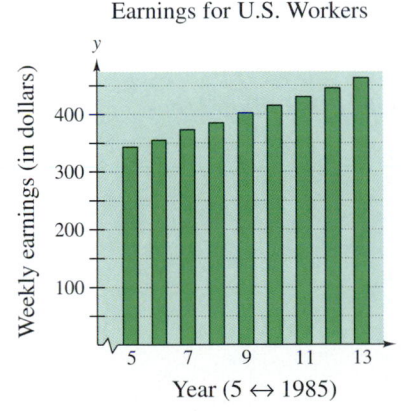

Earnings for U.S. Workers

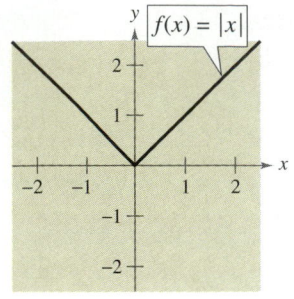

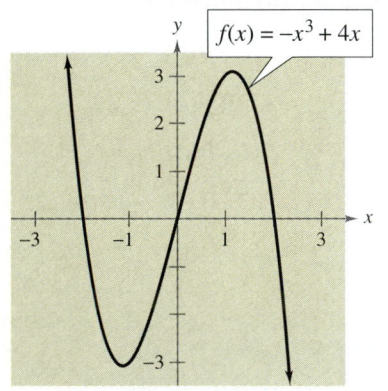

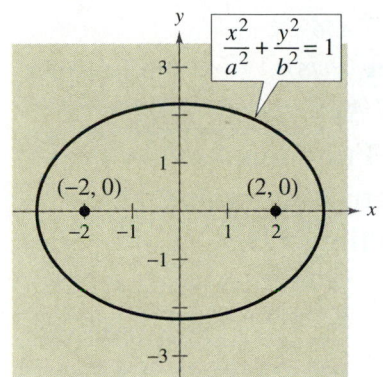

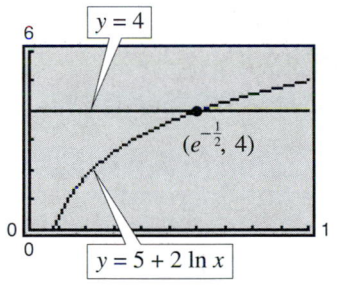

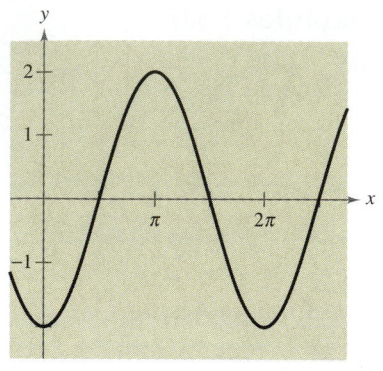

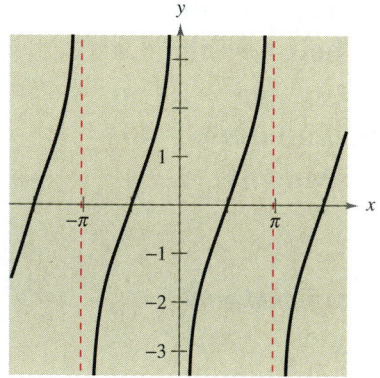

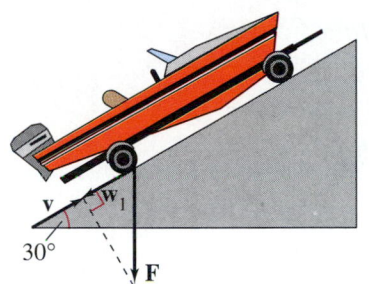

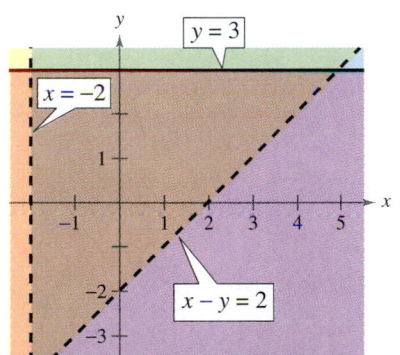

Educational Attainment
of Workers

Less than
high school
12.8%

4 yrs. or
more of
college
26.7%

High
school
39.2%

1–3 yrs.
college
21.3%

The concept for a museum dedicated to the heritage of rock and roll began in 1983 with a group of people from the music industry. This group created the Rock and Roll Hall of Fame Foundation.

The following data shows the number of recording artists who were elected to the Rock and Roll Hall of Fame from 1986 through 1995.

1986 (10)   1987 (15)
1988 (5)   1989 (5)
1990 (8)   1991 (7)
1992 (7)   1993 (8)
1994 (8)   1995 (7)

The artists elected in 1986 were Chuck Berry, James Brown, Ray Charles, Sam Cooke, Fats Domino, The Everly Brothers, Buddy Holly, Jerry Lee Lewis, Elvis Presley, and Little Richard.

This data can be represented by a graph as shown below.

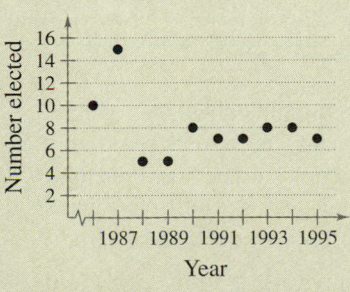

See Exercise 73 on page 73.

▶ P Prerequisites

P.1 *Real Numbers*

P.2 *Exponents and Radicals*

P.3 *Polynomials and Special Products*

P.4 *Factoring*

P.5 *Fractional Expressions*

P.6 *Errors and the Algebra of Calculus*

P.7 *Graphical Representation of Data*

Photos: The Plain Dealer, Cleveland, Ohio; Ingbet Grüttner (inset)

The Rock and Roll Hall of Fame and Museum is located in Cleveland, Ohio. The 150,000 square foot building was designed by I. M. Pei. The museum opened in September, 1995.

1

P.1 — Real Numbers

See Exercises 67–70 on page 11 for an example of how real numbers and absolute value are used to solve a budget variance problem.

Real Numbers ▫ *Ordering Real Numbers* ▫ *Absolute Value and Distance* ▫ *Algebraic Expressions* ▫ *Basic Rules of Algebra*

Real Numbers

Real numbers are used in everyday life to describe quantities such as age, miles per gallon, container size, and population. To represent real numbers, you can use symbols such as

$$9, 0, \frac{4}{3}, 0.666\ \ldots, 28.21, \sqrt{2}, \pi, \text{ and } \sqrt[3]{-32}.$$

Here are some important subsets of the real numbers.

$$\{1, 2, 3, 4, \ldots\}$$ Set of natural numbers
$$\{0, 1, 2, 3, 4, \ldots\}$$ Set of whole numbers
$$\{\ldots -3, -2, -1, 0, 1, 2, 3, \ldots\}$$ Set of integers

A real number is **rational** if it can be written as the ratio p/q of two integers, where $q \neq 0$. For instance, the numbers

$$\frac{1}{3} = 0.3333\ \ldots, \quad \frac{1}{8} = 0.125, \quad \text{and} \quad \frac{125}{111} = 1.126126\ \ldots$$

are rational. The decimal representation of a rational number either *repeats* (as in 3.1454545 . . .) or *terminates* (as in $\frac{1}{2} = 0.5$). A real number that cannot be written as the ratio of two integers is called **irrational.** Irrational numbers have infinite *nonrepeating* decimal representations. For instance, the numbers

$$\sqrt{2} \approx 1.4142136 \quad \text{and} \quad \pi \approx 3.1415927$$

are irrational. (The symbol $\approx$ means "is approximately equal to.")

Real numbers are represented graphically by a **real number line.** The point 0 on the real number line is the **origin.** Numbers to the right of 0 are positive, and numbers to the left of 0 are negative, as shown in Figure P.1. The term **nonnegative** describes a number that is either positive or zero.

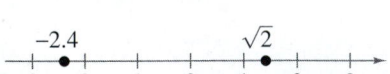

Every real number corresponds to exactly one point on the real number line.

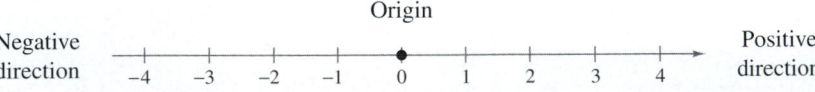

Negative direction · Origin · Positive direction

FIGURE P.1   The Real Number Line

Every point on the real number line corresponds to exactly one real number.

FIGURE P.2   One-to-One Correspondence

As illustrated in Figure P.2, there is a *one-to-one correspondence* between real numbers and points on the real number line.

Ordering Real Numbers

One important property of real numbers is that they are **ordered.**

> **DEFINITION OF ORDER ON THE REAL NUMBER LINE**
>
> If a and b are real numbers, a is **less than** b if $b - a$ is positive. This order is denoted by the **inequality**
>
> $$a < b.$$
>
> This can also be described by saying that b is **greater than** a and writing $b > a$. The inequality $a \le b$ means that a is **less than or equal to** b, and the inequality $b \ge a$ means that b is **greater than or equal to** a. The symbols $<$, $>$, $\le$, and $\ge$ are **inequality symbols.**

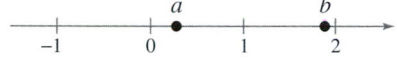

FIGURE P.3 $a < b$ if and only if a lies to the left of b.

Geometrically, this definition implies that $a < b$ if and only if a lies to the *left* of b on the real number line, as shown in Figure P.3.

EXAMPLE 1 *Interpreting Inequalities*

a. The inequality $x \le 2$ denotes all real numbers less than or equal to 2, as shown in Figure P.4(a).

b. The inequality $-2 \le x < 3$ means that $x \ge -2$ *and* $x < 3$. The "double inequality" denotes all real numbers between -2 and 3, including -2 but *not* including 3, as shown in Figure P.4(b).

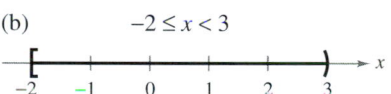

(a) $x \le 2$

(b) $-2 \le x < 3$

FIGURE P.4

Inequalities can be used to describe subsets of real numbers called **intervals.**

NOTE In the bounded intervals at the right, the real numbers a and b are the **endpoints** of each interval. ■■

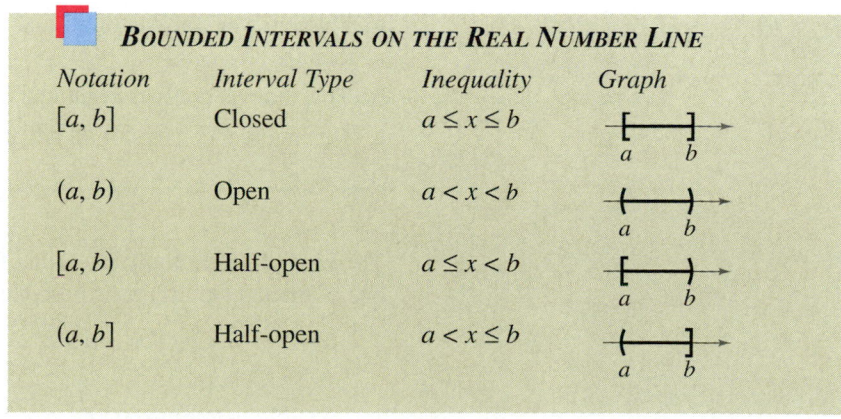

> **BOUNDED INTERVALS ON THE REAL NUMBER LINE**
>
> | Notation | Interval Type | Inequality | Graph |
> |----------|---------------|------------|-------|
> | $[a, b]$ | Closed | $a \le x \le b$ | |
> | (a, b) | Open | $a < x < b$ | |
> | $[a, b)$ | Half-open | $a \le x < b$ | |
> | $(a, b]$ | Half-open | $a < x \le b$ | |

NOTE   The symbols ∞, **positive infinity,** and $-\infty$, **negative infinity,** do not represent real numbers. They are simply convenient symbols used to describe the unboundedness of an interval such as $(1, \infty)$ or $(-\infty, 3]$. ▪▪

UNBOUNDED INTERVALS ON THE REAL NUMBER LINE

Notation	Interval Type	Inequality	Graph
$[a, \infty)$	Half-open	$x \geq a$	
(a, ∞)	Open	$x > a$	
$(-\infty, b]$	Half-open	$x \leq b$	
$(-\infty, b)$	Open	$x < b$	
$(-\infty, \infty)$	Entire real line		

EXAMPLE 2   Using Inequalities to Represent Intervals

Use inequality notation to describe each of the following.

a. c is at most 2.

b. All x in the interval $(-3, 5]$

Solution

a. The statement "c is at most 2" can be represented by $c \leq 2$.

b. "All x in the interval $(-3, 5]$" can be represented by $-3 < x \leq 5$.

EXAMPLE 3   Interpreting Intervals

Give a verbal description of each interval.

a. $(-1, 0)$ **b.** $[2, \infty)$ **c.** $(-\infty, 0)$

Solution

a. This interval consists of all real numbers that are greater than -1 and less than 0.

b. This interval consists of all real numbers that are greater than or equal to 2.

c. This interval consists of all negative real numbers.

The **Law of Trichotomy** states that for any two real numbers a and b, *precisely* one of three relationships is possible:

$$a = b, \quad a < b, \quad \text{or} \quad a > b. \qquad \text{Law of Trichotomy}$$

NOTE The absolute value of a real number is either positive or zero. Moreover, 0 is the only real number whose absolute value is 0. Thus, $|0| = 0$. ■■

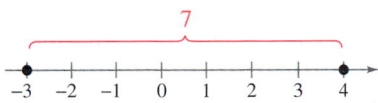

FIGURE P.5 The distance between -3 and 4 is 7.

Absolute Value and Distance

The **absolute value** of a real number is its *magnitude*.

> **DEFINITION OF ABSOLUTE VALUE**
>
> If a is a real number, then the **absolute value** of a is
>
> $$|a| = \begin{cases} a, & \text{if } a \geq 0 \\ -a, & \text{if } a < 0. \end{cases}$$

Notice from this definition that the absolute value of a real number is never negative. For instance if $a = -5$, then $|-5| = -(-5) = 5$.

EXAMPLE 4 *Evaluating the Absolute Value of a Number*

Evaluate $\dfrac{|x|}{x}$ for (a) $x > 0$ and (b) $x < 0$.

Solution

a. If $x > 0$, then $|x| = x$ and $\dfrac{|x|}{x} = \dfrac{x}{x} = 1$.

b. If $x < 0$, then $|x| = -x$ and $\dfrac{|x|}{x} = \dfrac{-x}{x} = -1$.

> **PROPERTIES OF ABSOLUTE VALUES**
>
> **1.** $|a| \geq 0$ **2.** $|-a| = |a|$
>
> **3.** $|ab| = |a||b|$ **4.** $\left|\dfrac{a}{b}\right| = \dfrac{|a|}{|b|}$, $b \neq 0$

Absolute value can be used to define the distance between two numbers on the real number line. For instance, the distance between -3 and 4 is $|-3 - 4| = |-7| = 7$, as shown in Figure P.5.

> **DISTANCE BETWEEN TWO POINTS ON THE REAL LINE**
>
> Let a and b be real numbers. The **distance between a and b** is
>
> $$d(a, b) = |b - a| = |a - b|.$$

Algebraic Expressions

One characteristic of algebra is the use of letters to represent numbers. The letters are **variables,** and combinations of letters and numbers are **algebraic expressions.** Here are a few examples of algebraic expressions.

$$5x, \qquad 2x - 3, \qquad \frac{4}{x^2 + 2}, \qquad 7x + y$$

> **DEFINITION OF AN ALGEBRAIC EXPRESSION**
>
> A collection of letters (**variables**) and real numbers (**constants**) combined using the operations of addition, subtraction, multiplication, division, and exponentiation is an **algebraic expression.**

The **terms** of an algebraic expression are those parts that are separated by *addition.* For example,

$$x^2 - 5x + 8 = x^2 + (-5x) + 8$$

has three terms: x^2 and $-5x$ are the **variable terms** and 8 is the **constant term.** The numerical factor of a variable term is the **coefficient** of the variable term. For instance, the coefficient of $-5x$ is -5, and the coefficient of x^2 is 1.

To **evaluate** an algebraic expression, substitute numerical values for each of the variables in the expression. Here are two examples.

Expression	Value of Variable	Substitute	Value of Expression
$-3x + 5$	$x = 3$	$-3(3) + 5$	$-9 + 5 = -4$
$3x^2 + 2x - 1$	$x = -1$	$3(-1)^2 + 2(-1) - 1$	$3 - 2 - 1 = 0$

Basic Rules of Algebra

There are four arithmetic operations with real numbers: **addition, multiplication, subtraction,** and **division,** denoted by the symbols $+$, $\times$ or $\cdot$, $-$, and $\div$. Of these, addition and multiplication are the two primary operations. Subtraction and division are the inverse operations of addition and multiplication, respectively.

Subtraction	Division
$a - b = a + (-b)$	If $b \neq 0$, then $a \div b = a\left(\dfrac{1}{b}\right) = \dfrac{a}{b}$.

In these definitions, $-b$ is the **additive inverse** (or opposite) of b, and $1/b$ is the **multiplicative inverse** (or reciprocal) of b. In the fractional form a/b, a is the **numerator** of the fraction and b is the **denominator.**

Study Tip

When evaluating an algebraic expression, the Substitution Principle is used. It states, "If $a = b$, then a can be replaced by b in any expression involving a." In the first evaluation shown at the right, for instance, 3 is *substituted* for x in the expression $-3x + 5$.

The French mathematician Nicolas Chuquet (ca. 1500) wrote Triparty en la science des nombres, *in which a form of exponent notation was used. Our expressions* $6x^3$ *and* $10x^2$ *were written as* $.6.^3$ *and* $.10.^2$. *Zero and negative exponents were also represented, so* x^0 *would be written as* $.1.^0$ *and* $3x^{-2}$ *as* $.3.^{2.m}$. *Chuquet wrote that* $.72.^1$ *divided by* $.8.^3$ *is* $.9.^{2.m}$. *That is,* $72x \div 8x^3 = 9x^{-2}$.

Be sure you see that the following **basic rules of algebra** are true for variables and algebraic expressions as well as for real numbers. Try to formulate a verbal description of each property. For instance, the first property states that *the order in which two real numbers are added does not affect their sum.*

BASIC RULES OF ALGEBRA

Let a, b, and c be real numbers, variables, or algebraic expressions.

Property		Example
Commutative Property of Addition:	$a + b = b + a$	$4x + x^2 = x^2 + 4x$
Commutative Property of Multiplication:	$ab = ba$	$(4 - x)x^2 = x^2(4 - x)$
Associative Property of Addition:	$(a + b) + c = a + (b + c)$	$(x + 5) + x^2 = x + (5 + x^2)$
Associative Property of Multiplication:	$(ab)c = a(bc)$	$(2x \cdot 3y)(8) = (2x)(3y \cdot 8)$
Distributive Properties:	$a(b + c) = ab + ac$	$3x(5 + 2x) = 3x \cdot 5 + 3x \cdot 2x$
	$(a + b)c = ac + bc$	$(y + 8)y = y \cdot y + 8 \cdot y$
Additive Identity Property:	$a + 0 = a$	$5y^2 + 0 = 5y^2$
Multiplicative Identity Property:	$a \cdot 1 = a$	$(4x^2)(1) = 4x^2$
Additive Inverse Property:	$a + (-a) = 0$	$5x^3 + (-5x^3) = 0$
Multiplicative Inverse Property:	$a \cdot \dfrac{1}{a} = 1, \quad a \neq 0$	$(x^2 + 4)\left(\dfrac{1}{x^2 + 4}\right) = 1$

NOTE Because subtraction is defined as "adding the opposite," the Distributive Properties are also true for subtraction. For instance, the "subtraction form" of $a(b + c) = ab + ac$ is

$$a(b - c) = ab - ac. \quad \blacksquare\blacksquare$$

NOTE Be sure you see the difference between the *opposite of a number* and a *negative number*. If a is already negative, then its opposite, $-a$, is positive. For instance, if $a = -5$, then $-a = -(-5) = 5$. $\blacksquare\blacksquare$

As well as formulating a verbal description for each of the following basic properties of negation, zero, and fractions, try to gain an *intuitive sense* for the validity of each.

PROPERTIES OF NEGATION

Let a and b be real numbers, variables, or algebraic expressions.

Property	Example
1. $(-1)a = -a$	$(-1)7 = -7$
2. $-(-a) = a$	$-(-6) = 6$
3. $(-a)b = -(ab) = a(-b)$	$(-5)3 = -(5 \cdot 3) = 5(-3)$
4. $(-a)(-b) = ab$	$(-2)(-x) = 2x$
5. $-(a + b) = (-a) + (-b)$	$-(x + 8) = (-x) + (-8)$
	$\quad = -x - 8$

NOTE The "or" in the Zero-Factor Property includes the possibility that either or both factors may be zero. This is an **inclusive or,** and it is the way the word "or" is generally used in mathematics. ■■

PROPERTIES OF ZERO

Let a and b be real numbers, variables, or algebraic expressions.

1. $a + 0 = a$ and $a - 0 = a$
2. $a \cdot 0 = 0$
3. $\dfrac{0}{a} = 0, \qquad a \neq 0$
4. $\dfrac{a}{0}$ is undefined.
5. **Zero-Factor Property:** If $ab = 0$, then $a = 0$ or $b = 0$.

NOTE In Property 1, the phrase "if and only if" implies two statements. One statement is: If $a/b = c/d$, then $ad = bc$. The other statement is: If $ad = bc$, where $b \neq 0$ and $d \neq 0$, then $a/b = c/d$. ■■

PROPERTIES OF FRACTIONS

Let a, b, c, and d be real numbers, variables, or algebraic expressions such that $b \neq 0$ and $d \neq 0$.

1. **Equivalent Fractions:** $\dfrac{a}{b} = \dfrac{c}{d}$ if and only if $ad = bc$.

2. **Rules of Signs:** $-\dfrac{a}{b} = \dfrac{-a}{b} = \dfrac{a}{-b}$ and $\dfrac{-a}{-b} = \dfrac{a}{b}$

3. **Generate Equivalent Fractions:** $\dfrac{a}{b} = \dfrac{ac}{bc}, \qquad c \neq 0$

4. **Add or Subtract with Like Denominators:** $\dfrac{a}{b} \pm \dfrac{c}{b} = \dfrac{a \pm c}{b}$

5. **Add or Subtract with Unlike Denominators:** $\dfrac{a}{b} \pm \dfrac{c}{d} = \dfrac{ad \pm bc}{bd}$

6. **Multiply Fractions:** $\dfrac{a}{b} \cdot \dfrac{c}{d} = \dfrac{ac}{bd}$

7. **Divide Fractions:** $\dfrac{a}{b} \div \dfrac{c}{d} = \dfrac{a}{b} \cdot \dfrac{d}{c} = \dfrac{ad}{bc}, \qquad c \neq 0$

EXAMPLE 5   Properties of Fractions

a. $\dfrac{x}{5} = \dfrac{3 \cdot x}{3 \cdot 5} = \dfrac{3x}{15}$ Generate equivalent fractions.

b. $\dfrac{x}{3} + \dfrac{2x}{5} = \dfrac{5 \cdot x + 3 \cdot 2x}{15}$ Add fractions with unlike denominators.

c. $\dfrac{7}{x} \div \dfrac{3}{2} = \dfrac{7}{x} \cdot \dfrac{2}{3} = \dfrac{14}{3x}$ Divide fractions.

PROPERTIES OF EQUALITY

Let a, b, and c be real numbers, variables, or algebraic expressions.

1. If $a = b$, then $a + c = b + c$.   Add c to both sides.
2. If $a = b$, then $ac = bc$.   Multiply both sides by c.
3. If $a + c = b + c$, then $a = b$.   Subtract c from both sides.
4. If $ac = bc$ and $c \neq 0$, then $a = b$.   Divide both sides by c.

If a, b, and c are integers such that $ab = c$, then a and b are **factors** or **divisors** of c. A **prime number** is an integer that has exactly two positive factors: itself and 1. For example, 2, 3, 5, 7, and 11 are prime numbers. The numbers 4, 6, 8, 9, and 10 are **composite** because they can be written as the product of two or more prime numbers. The number 1 is neither prime nor composite. The **Fundamental Theorem of Arithmetic** states that every positive integer greater than 1 can be written as the product of prime numbers in precisely one way (disregarding order). For instance, the *prime factorization* of 24 is $24 = 2 \cdot 2 \cdot 2 \cdot 3$.

When adding or subtracting fractions with unlike denominators, you have two options. You can use Property 5 of fractions as in Example 5(b), or you can rewrite the fractions with like denominators. Here is an example.

$$\frac{2}{15} - \frac{5}{9} + \frac{4}{5} = \frac{2(3)}{15(3)} - \frac{5(5)}{9(5)} + \frac{4(9)}{5(9)} \qquad \text{The LCD is 45.}$$

$$= \frac{6 - 25 + 36}{45}$$

$$= \frac{17}{45}$$

GROUP ACTIVITY

DECIMAL APPROXIMATIONS OF IRRATIONAL NUMBERS

At the beginning of this section, it was pointed out that $\sqrt{2}$ is not a rational number. There are, however, rational numbers whose squares are very close to 2. For instance, if you square the rational number

$$\frac{140}{99}$$

you obtain 1.9998. Try finding other rational numbers whose squares are even closer to 2. Write a short paragraph explaining how you obtained the numbers.

P.1 Exercises

In Exercises 1–6, determine which numbers are (a) natural numbers, (b) integers, (c) rational numbers, and (d) irrational numbers.

1. $-9, -\frac{7}{2}, 5, \frac{2}{3}, \sqrt{2}, 0, 1$
2. $\sqrt{5}, -7, -\frac{7}{3}, 0, 3.12, \frac{5}{4}$
3. $2.01, 0.666 \ldots, -13, 0.010110111 \ldots$
4. $2.30300030003 \ldots, 0.7575, -4.63, \sqrt{10}$
5. $-\pi, -\frac{1}{3}, \frac{6}{3}, \frac{1}{2}\sqrt{2}, -7.5$
6. $25, -17, -\frac{12}{5}, \sqrt{9}, 3.12, \frac{1}{2}\pi$

In Exercises 7–10, use a calculator to find the decimal form of the rational number. If it is a nonterminating decimal, write the repeating pattern.

7. $\frac{5}{8}$

8. $\frac{1}{3}$

9. $\frac{41}{333}$

10. $\frac{6}{11}$

In Exercises 11 and 12, approximate the numbers and place the correct symbol (< or >) between them.

11.

12.

In Exercises 13–18, plot the two real numbers on the real number line. Then place the appropriate inequality sign (< or >) between them.

13. $\frac{3}{2}, 7$
14. $-3.5, 1$
15. $-4, -8$
16. $1, \frac{16}{3}$
17. $\frac{5}{6}, \frac{2}{3}$
18. $-\frac{8}{7}, -\frac{3}{7}$

In Exercises 19–28, verbally describe the subset of real numbers represented by the inequality. Then sketch the subset on the real number line. State whether the interval is bounded or unbounded.

19. $x \le 5$
20. $x \ge -2$
21. $x < 0$
22. $x > 3$
23. $x \ge 4$
24. $x < 2$
25. $-2 < x < 2$
26. $0 \le x \le 5$
27. $-1 \le x < 0$
28. $0 < x \le 6$

In Exercises 29 and 30, use a calculator to order the numbers from smallest to largest.

29. $\frac{7071}{5000}, \frac{584}{413}, \sqrt{2}, \frac{47}{33}, \frac{127}{90}$

30. $\frac{26}{15}, \sqrt{3}, 1.7320, \frac{381}{220}, \sqrt{10} - \sqrt{2}$

In Exercises 31–36, use inequality notation to describe the set.

31. x is negative.
32. z is at least 10.
33. y is nonnegative.
34. y is no more than 25.
35. The person's age A is at least 30.
36. The annual rate of inflation r is expected to be at least 2.5%, but no more than 5%.

In Exercises 37–46, evaluate the expression.

37. $|-10|$
38. $|0|$
39. $|3 - \pi|$
40. $|4 - \pi|$
41. $\dfrac{-5}{|-5|}$
42. $-3 - |-3|$
43. $-3|-3|$
44. $|-1| - |-2|$
45. $-|16.25| + 20$
46. $2|33|$

In Exercises 47–52, place the correct symbol (<, >, or =) between the pair of real numbers.

47. $|-3|$ ⬜ $-|-3|$ **48.** $|-4|$ ⬜ $|4|$

49. -5 ⬜ $-|5|$ **50.** $-|-6|$ ⬜ $|-6|$

51. $-|-2|$ ⬜ $-|2|$ **52.** $-(-2)$ ⬜ -2

In Exercises 53–60, find the distance between a and b.

53. $a = -1$ $b = 3$

-1   0   1   2   3

54. $a = -4$ $b = -\frac{3}{2}$

-4   -3   -2   -1

55. $a = -\frac{5}{2}$ $b = 0$

-3   -2   -1   0

56.

$a = \frac{1}{4}$ $b = \frac{11}{4}$

0   1   2   3

57. $a = 126, b = 75$ **58.** $a = -126, b = -75$

59. $a = \frac{16}{5}, b = \frac{112}{75}$ **60.** $a = 9.34, b = -5.65$

In Exercises 61–66, use absolute value notation to describe the situation.

61. The distance between x and 5 is no more than 3.

62. The distance between x and -10 is at least 6.

63. While traveling, you pass milepost 7, then milepost 18. How far do you travel during that time period?

64. While traveling, you pass milepost 103, then milepost 86. How far do you travel during that time period?

65. y is at least six units from 0.

66. y is at most two units from a.

Budget Variance   In Exercises 67–70, the accounting department of a company is checking to see whether the actual expenses of a department differ from the budgeted expenses by more than $500 or by more than 5%. Fill in the missing parts of the table, and determine whether the actual expense passes the "budget variance test."

	Budgeted Expense, b	Actual Expense, a	$\lvert a - b \rvert$	$0.05b$
67. Wages	$112,700	$113,356	⬜	⬜
68. Utilities	$9400	$9772	⬜	⬜
69. Taxes	$37,640	$37,335	⬜	⬜
70. Insurance	$2575	$2613	⬜	⬜

Federal Deficit   In Exercises 71–74, use the bar graph, which shows the receipts of the federal government (in billions of dollars) for selected years from 1960 through 1993. In each exercise you are given the outlay of the federal government. Find the magnitude of the surplus or deficit for the year. (Source: U.S. Treasury Department)

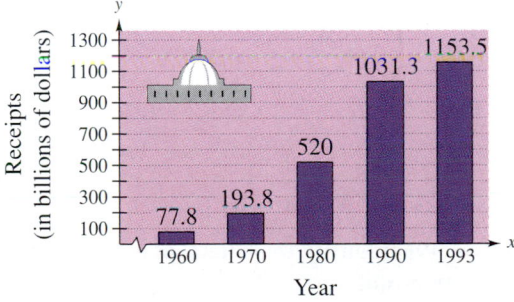

		Income, y	Outlay, x	$\lvert y - x \rvert$
71.	1960	⬜	$92.2 billion	⬜
72.	1980	⬜	$590.9 billion	⬜
73.	1990	⬜	$1252.7 billion	⬜
74.	1993	⬜	$1408.2 billion	⬜

75. *Exploration*   Consider $|u + v|$ and $|u| + |v|$.

    (a) Are the values of the expressions always equal? If not, under what conditions are they unequal?

    (b) If the two expressions are not equal for certain values of u and v, is one of the expressions always greater than the other? Explain.

76. *Think About It*   Is there a difference between saying that a real number is positive and saying that a real number is nonnegative? Explain.

In Exercises 77–80, identify the terms of the expression.

77. $7x + 4$

78. $3x^2 - 8x - 11$

79. $4x^3 + x - 5$

80. $3x^4 + 3x^3$

In Exercises 81–86, evaluate the expression for the values of x. (If not possible, state the reason.)

	Expression	*Values*	
81.	$4x - 6$	(a) $x = -1$	(b) $x = 0$
82.	$9 - 7x$	(a) $x = -3$	(b) $x = 3$
83.	$x^2 - 3x + 4$	(a) $x = -2$	(b) $x = 2$
84.	$-x^2 + 5x - 4$	(a) $x = -1$	(b) $x = 1$
85.	$\dfrac{x + 1}{x - 1}$	(a) $x = 1$	(b) $x = -1$
86.	$\dfrac{x}{x + 2}$	(a) $x = 2$	(b) $x = -2$

In Exercises 87–96, identify the rule(s) of algebra illustrated by the equation.

87. $x + 9 = 9 + x$

88. $2\left(\frac{1}{2}\right) = 1$

89. $\dfrac{1}{h + 6}(h + 6) = 1, \quad h \neq -6$

90. $(x + 3) - (x + 3) = 0$

91. $2(x + 3) = 2x + 6$

92. $(z - 2) + 0 = z - 2$

93. $1 \cdot (1 + x) = 1 + x$

94. $x + (y + 10) = (x + y) + 10$

95. $x(3y) = (x \cdot 3)y = (3x)y$

96. $\frac{1}{7}(7 \cdot 12) = \left(\frac{1}{7} \cdot 7\right)12 = 1 \cdot 12 = 12$

In Exercises 97–100, evaluate the expression. (If not possible, state the reason.)

97. $\dfrac{81 - (90 - 9)}{5}$

98. $10(23 - 30 + 7)$

99. $\dfrac{8 - 8}{-9 + (6 + 3)}$

100. $15 - \dfrac{3 - 3}{5}$

In Exercises 101–110, perform the operations. (Write fractional answers in reduced form.)

101. $(4 - 7)(-2)$

102. $\dfrac{27 - 35}{4}$

103. $\frac{3}{16} + \frac{5}{16}$

104. $\frac{6}{7} - \frac{4}{7}$

105. $\frac{5}{8} - \frac{5}{12} + \frac{1}{6}$

106. $\frac{10}{11} + \frac{6}{33} - \frac{13}{66}$

107. $\frac{4}{5} \cdot \frac{1}{2} \cdot \frac{3}{4}$

108. $\frac{11}{16} \div \frac{3}{4}$

109. $12 \div \frac{1}{4}$

110. $\left(\frac{3}{5} \div 3\right) - \left(6 \cdot \frac{4}{8}\right)$

In Exercises 111–114, use a calculator to evaluate the expression. (Round your answer to two decimal places.)

111. $-3 + \frac{3}{7}$

112. $3\left(-\frac{5}{12} + \frac{3}{8}\right)$

113. $\dfrac{11.46 - 5.37}{3.91}$

114. $\dfrac{\frac{1}{5}(-8 - 9)}{-\frac{1}{3}}$

115. Use a calculator to complete the table.

n	1	0.5	0.01	0.0001	0.000001
$5/n$					

116. *Think About It*   Use the result of Exercise 115 to make a conjecture about the value of $5/n$ as n approaches 0.

117. Use a calculator to complete the table.

n	1	10	100	10,000	100,000
$5/n$					

118. *Think About It*   Use the result of Exercise 117 to make a conjecture about the value of $5/n$ as n increases without bound.

P.2 *Exponents and Radicals*

See Exercise 107 on page 26 for an example of how exponents can be used to find the annual depreciation rate in a declining balances problem.

Exponents ▫ *Scientific Notation* ▫ *Radicals and Their Properties* ▫ *Simplifying Radicals* ▫ *Rationalizing Denominators and Numerators* ▫ *Rational Exponents* ▫ *Radicals and Calculators*

Exponents

Repeated *multiplications* can be written in **exponential form.**

Repeated Multiplication	*Exponential Form*
$a \cdot a \cdot a \cdot a \cdot a$	a^5
$(-4)(-4)(-4)$	$(-4)^3$
$(2x)(2x)(2x)(2x)$	$(2x)^4$

In general, if a is a real number and n is a positive integer, then

$$a^n = \underbrace{a \cdot a \cdot a \cdots a}_{n \text{ factors}}$$

where n is the **exponent** and a is the **base.** The expression a^n is read "a to the nth **power.**"

> ### *Exploration*
>
> Use the definition of exponential form to write each of the expressions as a single power of 2. From the results, can you find a general rule for simplifying expressions of the forms
>
> $$a^m a^n \text{ and } \frac{a^m}{a^n}?$$
>
> **a.** $2^3 \cdot 2^2$ **b.** $2^5 \cdot 2^1$
>
> **c.** $\dfrac{2^3}{2^2}$ **d.** $\dfrac{2^3}{2^1}$

PROPERTIES OF EXPONENTS

Let a and b be real numbers, variables, or algebraic expressions, and let m and n be integers. (All denominators and bases are nonzero.)

Property	*Example*
1. $a^m a^n = a^{m+n}$	$3^2 \cdot 3^4 = 3^{2+4} = 3^6 = 729$
2. $\dfrac{a^m}{a^n} = a^{m-n}$	$\dfrac{x^7}{x^4} = x^{7-4} = x^3$
3. $a^{-n} = \dfrac{1}{a^n} = \left(\dfrac{1}{a}\right)^n$	$y^{-4} = \dfrac{1}{y^4} = \left(\dfrac{1}{y}\right)^4$
4. $a^0 = 1, \quad a \neq 0$	$(x^2 + 1)^0 = 1$
5. $(ab)^m = a^m b^m$	$(5x)^3 = 5^3 x^3 = 125x^3$
6. $(a^m)^n = a^{mn}$	$(y^3)^{-4} = y^{3(-4)} = y^{-12} = \dfrac{1}{y^{12}}$
7. $\left(\dfrac{a}{b}\right)^m = \dfrac{a^m}{b^m}$	$\left(\dfrac{2}{x}\right)^3 = \dfrac{2^3}{x^3} = \dfrac{8}{x^3}$
8. $\|a^2\| = \|a\|^2 = a^2$	$\|(-2)^2\| = \|-2\|^2 = (-2)^2 = 4$

NOTE It is important to recognize the difference between expressions such as $(-2)^4$ and -2^4. In $(-2)^4$, the parentheses indicate that the exponent applies to the negative sign as well as to the 2, but in $-2^4 = -(2^4)$, the exponent applies only to the 2. Hence, $(-2)^4 = 16$, whereas $-2^4 = -16$. ▪▪

Rarely in algebra is there only one way to solve a problem. Don't be concerned if the steps you use to solve a problem are not exactly the same as the steps presented in this text. The important thing is to use steps that you understand *and,* of course, are justified by the rules of algebra. For instance, you might prefer the following steps for Example 2(d).

$$\left(\frac{3x^2}{y}\right)^{-2} = \left(\frac{y}{3x^2}\right)^{2} = \frac{y^2}{9x^4}$$

Note how Property 3 is used in the first step of this solution. The fractional form of this property is

$$\left(\frac{a}{b}\right)^{-m} = \left(\frac{b}{a}\right)^{m}.$$

The properties of exponents listed on the previous page apply to *all* integers m and n, not just positive integers. For instance, by Property 2, you can write

$$\frac{3^4}{3^{-5}} = 3^{4-(-5)} = 3^{4+5} = 3^9.$$

EXAMPLE 1   *Using Properties of Exponents*

a. $(-3ab^4)(4ab^{-3}) = -12(a)(a)(b^4)(b^{-3}) = -12a^2b$

b. $(2xy^2)^3 = 2^3(x)^3(y^2)^3 = 8x^3y^6$

c. $3a(-4a^2)^0 = 3a(1) = 3a, \qquad a \neq 0$

d. $\left(\frac{5x^3}{y}\right)^2 = \frac{5^2(x^3)^2}{y^2} = \frac{25x^6}{y^2}$

EXAMPLE 2   *Rewriting with Positive Exponents*

a. $x^{-1} = \dfrac{1}{x}$ Property 3: $a^{-n} = \dfrac{1}{a^n}$

b. $\dfrac{1}{3x^{-2}} = \dfrac{1(x^2)}{3} = \dfrac{x^2}{3}$ -2 exponent does not apply to 3.

c. $\dfrac{12a^3b^{-4}}{4a^{-2}b} = \dfrac{12a^3 \cdot a^2}{4b \cdot b^4} = \dfrac{3a^5}{b^5}$

d. $\left(\dfrac{3x^2}{y}\right)^{-2} = \dfrac{3^{-2}(x^2)^{-2}}{y^{-2}} = \dfrac{3^{-2}x^{-4}}{y^{-2}} = \dfrac{y^2}{3^2x^4} = \dfrac{y^2}{9x^4}$

e. $\dfrac{x^{-2}}{y^{-2}} = \dfrac{y^2}{x^2}$

TECHNOLOGY

The calculator keystrokes in Example 3 and throughout this text are for the *TI-83* or the *TI-82* graphing calculator. The corresponding scientific calculator keystrokes are given in the appendix.

EXAMPLE 3   *Calculators and Exponents*

Expression	*Graphing Calculator Keystrokes*	*Display*
a. $13^4 + 5$	13 [∧] 4 [+] 5 [ENTER]	28566
b. $3^{-2} + 4^{-1}$	3 [∧] [ ] 2 [+] 4 [∧] [ ] 1 [ENTER]	.3611111111
c. $\dfrac{3^5 + 1}{3^5 - 1}$	(3 [∧] 5 [+] 1) [÷] (3 [∧] 5 [−] 1) [ENTER]	1.008264463

Scientific Notation

Exponents provide an efficient way of writing and computing with very large (or very small) numbers. For instance, a drop of water contains more than 33 billion billion molecules—that is, 33 followed by 18 zeros.

$$33,000,000,000,000,000,000$$

It is convenient to write such numbers in **scientific notation.** This notation has the form $\pm c \times 10^n$, where $1 \leq c < 10$ and n is an integer. Thus, the number of molecules in a drop of water can be written in scientific notation as

$$3.3 \times 10,000,000,000,000,000,000 = 3.3 \times 10^{19}.$$

The *positive* exponent 19 indicates that the number is *large* (10 or more) and that the decimal point has been moved 19 places. A *negative* exponent indicates that the number is *small* (less than 1). For instance, the mass (in grams) of one electron is approximately

$$9.0 \times 10^{-28} = 0.00000000000000000000000000009.$$

28 decimal places

EXAMPLE 4 *Scientific Notation*

a. $1.345 \times 10^2 = 134.5$

b. $0.0000782 = 7.82 \times 10^{-5}$

c. $9.36 \times 10^{-6} = 0.00000936$

d. $836,100,000 = 8.361 \times 10^8$

NOTE Most calculators switch to scientific notation when they are showing large (or small) numbers that exceed the display range. Try evaluating $86,500,000 \times 6000$. If your calculator follows standard conventions, its display should be

$$\boxed{5.19 \quad 11} \quad \text{or} \quad \boxed{5.19 \ \text{E} \ 11}$$

which is 5.19×10^{11}. ■■

EXAMPLE 5 *Using Scientific Notation with a Calculator*

Use a calculator to evaluate $65,000 \times 3,400,000,000$.

Solution

Because $65,000 = 6.5 \times 10^4$ and $3,400,000,000 = 3.4 \times 10^9$, you can multiply the two numbers using the following graphing calculator steps.

$$6.5 \ \boxed{\text{EE}} \ 4 \ \boxed{\times} \ 3.4 \ \boxed{\text{EE}} \ 9 \ \boxed{\text{ENTER}}$$

After entering these keystrokes, the calculator display should read $\boxed{2.21 \ \text{E} \ 14}$. Therefore, the product of the two numbers is

$$(6.5 \times 10^4)(3.4 \times 10^9) = 2.21 \times 10^{14}$$
$$= 221,000,000,000,000.$$

Radicals and Their Properties

A **square root** of a number is one of its two equal factors. For example, 5 is a square root of 25 because 5 is one of the two equal factors of 25. In a similar way, a **cube root** of a number is one of its three equal factors.

DEFINITION OF NTH ROOT OF A NUMBER

Let a and b be real numbers and let $n \geq 2$ be a positive integer. If

$$a = b^n$$

then b is an **nth root of a**. In $n = 2$, the root is a **square root.** If $n = 3$, the root is a **cube root.**

Some numbers have more than one nth root. For example, both 5 and -5 are square roots of 25. The **principal nth root** of a number is defined as follows.

PRINCIPAL NTH ROOT OF A NUMBER

Let a be a real number that has at least one nth root. The **principal nth root of a** is the nth root that has the same sign as a. It is denoted by a **radical symbol**

$$\sqrt[n]{a}. \qquad \color{red}{\text{Principal } n\text{th root}}$$

The positive integer n is the **index** of the radical, and the number a is the **radicand.** If $n = 2$, we omit the index and write $\sqrt{a}$ rather than $\sqrt[2]{a}$. (The plural of index is *indices*.)

EXAMPLE 6 *Evaluating Expressions Involving Radicals*

a. $\sqrt{36} = 6$ because $6^2 = 36$.

b. $-\sqrt{36} = -6$ because $6^2 = 36$.

c. $\sqrt[3]{\dfrac{125}{64}} = \dfrac{5}{4}$ because $\left(\dfrac{5}{4}\right)^3 = \dfrac{5^3}{4^3} = \dfrac{125}{64}$.

d. $\sqrt[5]{-32} = -2$ because $(-2)^5 = -32$.

e. $\sqrt[4]{-81}$ is not a real number because there is no real number that can be raised to the fourth power to produce -81.

Here are some generalizations about the *n*th roots of a real number.

1. If a is a positive real number and n is a positive *even* integer, then a has exactly two real *n*th roots denoted by $\sqrt[n]{a}$ and $-\sqrt[n]{a}$. See Examples 6(a) and 6(b).
2. If a is any real number and n is an *odd* integer, then a has only one real *n*th root denoted by $\sqrt[n]{a}$. See Examples 6(c) and 6(d).
3. If a is a negative real number and n is an *even* integer, then a has no real *n*th root. See Example 6(e).
4. $\sqrt[n]{0} = 0$.

Integers such as 1, 4, 9, 16, 25, and 36 are called **perfect squares** because they have integer square roots. Similarly, integers such as 1, 8, 27, 64, and 125 are called **perfect cubes** because they have integer cube roots.

PROPERTIES OF RADICALS

Let a and b be real numbers, variables, or algebraic expressions such that the indicated roots are real numbers, and let m and n be positive integers.

Property	*Example*
1. $\sqrt[n]{a^m} = \left(\sqrt[n]{a}\right)^m$	$\sqrt[3]{8^2} = \left(\sqrt[3]{8}\right)^2 = (2)^2 = 4$
2. $\sqrt[n]{a} \cdot \sqrt[n]{b} = \sqrt[n]{ab}$	$\sqrt{5} \cdot \sqrt{7} = \sqrt{5 \cdot 7} = \sqrt{35}$
3. $\dfrac{\sqrt[n]{a}}{\sqrt[n]{b}} = \sqrt[n]{\dfrac{a}{b}}, \quad b \neq 0$	$\dfrac{\sqrt[4]{27}}{\sqrt[4]{9}} = \sqrt[4]{\dfrac{27}{9}} = \sqrt[4]{3}$
4. $\sqrt[m]{\sqrt[n]{a}} = \sqrt[mn]{a}$	$\sqrt[3]{\sqrt{10}} = \sqrt[6]{10}$
5. $\left(\sqrt[n]{a}\right)^n = a$	$\left(\sqrt{3}\right)^2 = 3$
6. For n even, $\sqrt[n]{a^n} = \lvert a \rvert$.	$\sqrt{(-12)^2} = \lvert -12 \rvert = 12$
For n odd, $\sqrt[n]{a^n} = a$.	$\sqrt[3]{(-12)^3} = -12$

NOTE A common special case of Property 6 is $\sqrt{a^2} = \lvert a \rvert$. ∎∎

EXAMPLE 7   *Using Properties of Radicals*

a. $\sqrt{8} \cdot \sqrt{2} = \sqrt{8 \cdot 2} = \sqrt{16} = 4$

b. $\left(\sqrt[3]{5}\right)^3 = 5$

c. $\sqrt[3]{x^3} = x$

Simplifying Radicals

An expression involving radicals is in **simplest form** when the following conditions are satisfied.

1. All possible factors have been removed from the radical.
2. All fractions have radical-free denominators (accomplished by a process called *rationalizing the denominator*).
3. The index of the radical is reduced.

To simplify a radical, factor the radicand into factors whose exponents are multiples of the index. The roots of these factors are written outside the radical, and the "leftover" factors make up the new radicand.

EXAMPLE 8   *Simplifying Even Roots*

Perfect   Leftover
4th power   factor

a. $\sqrt[4]{48} = \sqrt[4]{16 \cdot 3} = \sqrt[4]{2^4 \cdot 3} = 2\sqrt[4]{3}$

Perfect   Leftover
square   factor

NOTE In Example 8(b), the expression $\sqrt{75x^3}$ makes sense only for nonnegative values of x. ■■

b. $\sqrt{75x^3} = \sqrt{25x^2 \cdot 3x}$ Find largest square factor.

  $= \sqrt{(5x)^2 \cdot 3x}$

  $= 5x\sqrt{3x}$ Find root of perfect square.

c. $\sqrt[4]{(5x)^4} = |5x| = 5|x|$

EXAMPLE 9   *Simplifying Odd Roots*

Perfect   Leftover
cube   factor

a. $\sqrt[3]{24} = \sqrt[3]{8 \cdot 3} = \sqrt[3]{2^3 \cdot 3} = 2\sqrt[3]{3}$

Perfect   Leftover
cube   factor

b. $\sqrt[3]{24a^4} = \sqrt[3]{8a^3 \cdot 3a}$ Find largest cube factor.

  $= \sqrt[3]{(2a)^3 \cdot 3a}$

  $= 2a\sqrt[3]{3a}$ Find root of perfect cube.

c. $\sqrt[3]{-40x^6} = \sqrt[3]{(-8x^6) \cdot 5} = \sqrt[3]{(-2x^2)^3 \cdot 5} = -2x^2\sqrt[3]{5}$

Rationalizing Denominators and Numerators

To rationalize a denominator or numerator of the form $a - b\sqrt{m}$ or $a + b\sqrt{m}$, multiply both numerator and denominator by a **conjugate**: $a + b\sqrt{m}$ and $a - b\sqrt{m}$ are conjugates of each other. If $a = 0$, then the rationalizing factor for $\sqrt{m}$ is itself, $\sqrt{m}$.

EXAMPLE 10   *Rationalizing Single-Term Denominators*

a. $\dfrac{5}{2\sqrt{3}} = \dfrac{5}{2\sqrt{3}} \cdot \dfrac{\sqrt{3}}{\sqrt{3}} = \dfrac{5\sqrt{3}}{2(3)} = \dfrac{5\sqrt{3}}{6}$

b. $\dfrac{2}{\sqrt[3]{5}} = \dfrac{2}{\sqrt[3]{5}} \cdot \dfrac{\sqrt[3]{5^2}}{\sqrt[3]{5^2}} = \dfrac{2\sqrt[3]{5^2}}{\sqrt[3]{5^3}} = \dfrac{2\sqrt[3]{25}}{5}$

EXAMPLE 11   *Rationalizing a Denominator with Two Terms*

$\dfrac{2}{3 + \sqrt{7}} = \dfrac{2}{3 + \sqrt{7}} \cdot \dfrac{3 - \sqrt{7}}{3 - \sqrt{7}}$ Multiply numerator and denominator by conjugate.

$= \dfrac{2(3 - \sqrt{7})}{(3)^2 - (\sqrt{7})^2}$

$= \dfrac{2(3 - \sqrt{7})}{9 - 7}$

$= \dfrac{2(3 - \sqrt{7})}{2}$

$= 3 - \sqrt{7}$ Cancel like factors.

NOTE Do not confuse the expression $\sqrt{5} + \sqrt{7}$ with the expression $\sqrt{5 + 7}$. In general, $\sqrt{x + y}$ does not equal $\sqrt{x} + \sqrt{y}$. Similarly, $\sqrt{x^2 + y^2}$ does not equal $x + y$. ■■

EXAMPLE 12   *Rationalizing the Numerator*

$\dfrac{\sqrt{5} - \sqrt{7}}{2} = \dfrac{\sqrt{5} - \sqrt{7}}{2} \cdot \dfrac{\sqrt{5} + \sqrt{7}}{\sqrt{5} + \sqrt{7}}$ Multiply numerator and denominator by conjugate.

$= \dfrac{5 - 7}{2(\sqrt{5} + \sqrt{7})}$

$= \dfrac{-2}{2(\sqrt{5} + \sqrt{7})}$

$= \dfrac{-1}{\sqrt{5} + \sqrt{7}}$ Simplify.

Rational Exponents

  DEFINITION OF RATIONAL EXPONENTS

If a is a real number and n is a positive integer such that the principal nth root of a exists, we define $a^{1/n}$ to be

$$a^{1/n} = \sqrt[n]{a}.$$

Moreover, if m is a positive integer that has no common factor with n, then

$$a^{m/n} = (a^{1/n})^m = \left(\sqrt[n]{a}\right)^m \quad \text{and} \quad a^{m/n} = (a^m)^{1/n} = \sqrt[n]{a^m}.$$

The numerator of a rational exponent denotes the *power* to which the base is raised, and the denominator denotes the *index* or the *root* to be taken, as shown below.

$$b^{m/n} = \left(\sqrt[n]{b}\right)^m = \sqrt[n]{b^m}$$

(Power, Index)

When you are working with rational exponents, the properties of integer exponents still apply. For instance,

$$2^{1/2}2^{1/3} = 2^{(1/2)+(1/3)} = 2^{5/6}.$$

NOTE Rational exponents can be tricky, and you must remember that the expression $b^{m/n}$ is not defined unless $\sqrt[n]{b}$ is a real number. This restriction produces some unusual-looking results. For instance, the number $(-8)^{1/3}$ is defined because $\sqrt[3]{-8} = -2$, but the number $(-8)^{2/6}$ is undefined because $\sqrt[6]{-8}$ is not a real number. ■■

EXAMPLE 13   *Changing from Radical to Exponential Form*

a. $\sqrt{3} = 3^{1/2}$

b. $\sqrt{(3xy)^5} = \sqrt[2]{(3xy)^5} = (3xy)^{(5/2)}$

c. $2x\sqrt[4]{x^3} = (2x)(x^{3/4}) = 2x^{1+(3/4)} = 2x^{7/4}$

EXAMPLE 14   *Changing from Exponential to Radical Form*

a. $(x^2 + y^2)^{3/2} = \left(\sqrt{x^2 + y^2}\right)^3 = \sqrt{(x^2 + y^2)^3}$

b. $2y^{3/4}z^{1/4} = 2(y^3z)^{1/4} = 2\sqrt[4]{y^3z}$

c. $a^{-3/2} = \dfrac{1}{a^{3/2}} = \dfrac{1}{\sqrt{a^3}}$

d. $x^{0.2} = x^{1/5} = \sqrt[5]{x}$

Rational exponents are particularly useful for evaluating roots of numbers on a calculator, for reducing the index of a radical, and for simplifying expressions encountered in calculus.

EXAMPLE 15   Simplifying with Rational Exponents

a. $(27)^{2/6} = (27)^{1/3} = \sqrt[3]{27} = 3$

b. $(-32)^{-4/5} = \left(\sqrt[5]{-32}\right)^{-4} = (-2)^{-4} = \dfrac{1}{(-2)^4} = \dfrac{1}{16}$

c. $(-5x^{5/3})(3x^{-3/4}) = -15x^{(5/3)-(3/4)} = -15x^{11/12}, \qquad x \neq 0$

d. $\sqrt[9]{a^3} = a^{3/9} = a^{1/3} = \sqrt[3]{a}$

e. $\sqrt[3]{\sqrt{125}} = \sqrt[6]{125} = \sqrt[6]{(5)^3} = 5^{3/6} = 5^{1/2} = \sqrt{5}$

f. $(2x - 1)^{4/3}(2x - 1)^{-1/3} = (2x - 1)^{(4/3)-(1/3)}$

$$= 2x - 1, \qquad x \neq \dfrac{1}{2}$$

g. $\dfrac{x - 1}{(x - 1)^{-1/2}} = \dfrac{x - 1}{(x - 1)^{-1/2}} \cdot \dfrac{\sqrt{x - 1}}{\sqrt{x - 1}} = \dfrac{(x - 1)^{3/2}}{(x - 1)^0}$

$$= (x - 1)^{3/2}, \qquad x \neq 1$$

Radical expressions can be combined (added or subtracted) if they are **like radicals**—that is, if they have the same index and radicand. For instance, $\sqrt{2}, 3\sqrt{2},$ and $\frac{1}{2}\sqrt{2}$ are like radicals, but $\sqrt{3}$ and $\sqrt{2}$ are unlike radicals. To determine whether two radicals are like radicals, you should first simplify each radical.

EXAMPLE 16   Combining Radicals

a. $2\sqrt{48} - 3\sqrt{27} = 2\sqrt{16 \cdot 3} - 3\sqrt{9 \cdot 3}$ Find square factors.

$= 8\sqrt{3} - 9\sqrt{3}$ Find square roots.

$= (8 - 9)\sqrt{3}$ Combine like terms.

$= -\sqrt{3}$

b. $\sqrt[3]{16x} - \sqrt[3]{54x^4} = \sqrt[3]{8 \cdot 2x} - \sqrt[3]{27 \cdot x^3 \cdot 2x}$

$= 2\sqrt[3]{2x} - 3x\sqrt[3]{2x}$

$= (2 - 3x)\sqrt[3]{2x}$

Radicals and Calculators

There are four methods of evaluating radicals on most graphing calculators. For square roots, you can use the *square root key* $\boxed{\sqrt{\ }}$. For cube roots, you can use the *cube root key* $\boxed{\sqrt[3]{\ }}$. For other roots, you can first convert the radical to exponential form and then use the *exponential key* $\boxed{\wedge}$, or you can use the *nth root key* $\boxed{\sqrt[x]{\ }}$.

EXAMPLE 17   Evaluating Radicals with a Calculator

Use a calculator to evaluate $\sqrt[4]{56} = 56^{1/4}$.

Graphing Calculator Keystrokes

a. 56 $\boxed{\wedge}$ $\boxed{(}$ 1 $\boxed{\div}$ 4 $\boxed{)}$ $\boxed{\text{ENTER}}$

b. 4 $\boxed{\text{MATH}}$ (MATH) $\left(5: \sqrt[x]{\ }\right)$ 56 $\boxed{\text{ENTER}}$

For each of these two keystroke sequences, the display is $\sqrt[4]{56} \approx 2.7355648$.

Saturn, 886 million miles from the sun, completes one rotation every 10.4 earth hours and orbits the sun once in 29.46 earth years. *(Photo: NASA)*

GROUP ACTIVITY

A FAMOUS MATHEMATICAL DISCOVERY

Johannes Kepler (1571–1630), a well-known German astronomer, discovered a relationship between the average distance of a planet from the sun and the time (or period) it takes the planet to orbit the sun. People then knew that planets that are closer to the sun take less time to complete an orbit than planets that are farther from the sun. Kepler discovered that the distance and period are related by an exact mathematical formula. The table shows the average distance x (in astronomical units) and period y (in years) for the six planets that are closest to the sun. By completing the table, can you rediscover Kepler's relationship? Discuss your conclusions.

Planet	Mercury	Venus	Earth	Mars	Jupiter	Saturn
x	0.387	0.723	1.0	1.523	5.203	9.541
$\sqrt{x}$						
y	0.241	0.615	1.0	1.881	11.861	29.457
$\sqrt[3]{y}$						

The *Interactive* CD-ROM provides additional help with Warm-Up exercises by providing a hypertext link to the section in which the concept was introduced.

The *Interactive* CD-ROM contains step-by-step solutions to all odd-numbered Section and Review Exercises. It also provides Tutorial Exercises, which link to Guided Examples for additional help.

P.2 Exercises

In Exercises 1 and 2, write the expression as a repeated multiplication problem.

1. -0.4^6

2. $(-2)^7$

In Exercises 3 and 4, write the expression using exponential notation.

3. $(-10)(-10)(-10)(-10)(-10)$

4. $-\left(\frac{3}{2} \times \frac{3}{2} \times \frac{3}{2} \times \frac{3}{2}\right)$

In Exercises 5–10, evaluate the expression.

5. (a) $4^2 \cdot 3$ (b) $3 \cdot 3^3$

6. (a) $\frac{5^5}{5^2}$ (b) $\frac{3^2}{3^4}$

7. (a) $(3^3)^2$ (b) -3^2

8. (a) $(2^3 \cdot 3^2)^2$ (b) $\left(-\frac{3}{5}\right)^3 \left(\frac{5}{3}\right)^2$

9. (a) $\frac{3}{3^{-4}}$ (b) $24(-2)^{-5}$

10. (a) $\frac{4 \cdot 3^{-2}}{2^{-2} \cdot 3^{-1}}$ (b) $(-2)^0$

In Exercises 11–14, use a calculator to evaluate the expression. (Round to three decimal places.)

11. $(-4)^3(5^2)$ **12.** $(8^{-4})(10^3)$

13. $\frac{3^6}{7^3}$ **14.** $\frac{4^3}{3^{-4}}$

In Exercises 15–18, evaluate the expression for the value of *x*.

Expression	Value
15. $-3x^3$	2
16. $7x^{-2}$	4
17. $6x^0 - (6x)^0$	10
18. $5(-x)^3$	3

In Exercises 19–32, simplify the expression.

19. (a) $(-5z)^3$ (b) $5x^4(x^2)$

20. (a) $(3x)^2$ (b) $(4x^3)^2$

21. (a) $6y^2(2y^4)^2$ (b) $\dfrac{3x^5}{x^3}$

22. (a) $(-z)^3(3z^4)$ (b) $\dfrac{25y^8}{10y^4}$

23. (a) $\dfrac{7x^2}{x^3}$ (b) $\dfrac{12(x+y)^3}{9(x+y)}$

24. (a) $\dfrac{r^4}{r^6}$ (b) $\left(\dfrac{4}{y}\right)^3\left(\dfrac{3}{y}\right)^4$

25. (a) $(x+5)^0,\quad x \neq -5$ (b) $(2x^2)^{-2}$

26. (a) $(2x^5)^0,\quad x \neq 0$ (b) $(z+2)^{-3}(z+2)^{-1}$

27. (a) $(-2x^2)^3(4x^3)^{-1}$ (b) $\left(\dfrac{x}{10}\right)^{-1}$

28. (a) $(4y^{-2})(8y^4)$ (b) $\left(\dfrac{x^{-3}y^4}{5}\right)^{-3}$

29. (a) $(4a^{-2}b^3)^{-3}$ (b) $\left(\dfrac{5x^2}{y^{-2}}\right)^{-4}$

30. (a) $[(x^2y^{-2})^{-1}]^{-1}$ (b) $(5x^2z^6)^3(5x^2z^6)^{-3}$

31. (a) $3^n \cdot 3^{2n}$ (b) $\left(\dfrac{a^{-2}}{b^{-2}}\right)\left(\dfrac{b}{a}\right)^3$

32. (a) $\dfrac{x^2 \cdot x^n}{x^3 \cdot x^n}$ (b) $\left(\dfrac{a^{-3}}{b^{-3}}\right)\left(\dfrac{a}{b}\right)^3$

In Exercises 33–44, fill in the missing description.

Radical Form	Rational Exponent Form
33. $\sqrt{9} = 3$	
34. $\sqrt[3]{64} = 4$	
35.	$32^{1/5} = 2$
36.	$-(144^{1/2}) = -12$
37.	$196^{1/2} = 14$
38. $\sqrt[3]{614.125} = 8.5$	

Radical Form	Rational Exponent Form
39. $\sqrt[3]{-216} = -6$	
40.	$(-243)^{1/5} = -3$
41.	$27^{2/3} = 9$
42. $\left(\sqrt[4]{81}\right)^3 = 27$	
43. $\sqrt[4]{81^3} = 27$	
44.	$16^{5/4} = 32$

In Exercises 45–54, evaluate each expression. (Do not use a calculator.)

45. (a) $\sqrt{9}$ (b) $\sqrt[3]{8}$

46. (a) $\sqrt{49}$ (b) $\sqrt[3]{\dfrac{27}{8}}$

47. (a) $-\sqrt[3]{-27}$ (b) $\dfrac{4}{\sqrt{64}}$

48. (a) $\sqrt[3]{0}$ (b) $\dfrac{\sqrt[4]{81}}{3}$

49. (a) $\left(\sqrt[3]{-125}\right)^3$ (b) $27^{1/3}$

50. (a) $\sqrt[4]{562^4}$ (b) $36^{3/2}$

51. (a) $32^{-3/5}$ (b) $\left(\dfrac{16}{81}\right)^{-3/4}$

52. (a) $100^{-3/2}$ (b) $\left(\dfrac{9}{4}\right)^{-1/2}$

53. (a) $\left(-\dfrac{1}{64}\right)^{-1/3}$ (b) $\left(\dfrac{1}{\sqrt{32}}\right)^{-2/5}$

54. (a) $\left(-\dfrac{125}{27}\right)^{-1/3}$ (b) $-\left(\dfrac{1}{125}\right)^{-4/3}$

In Exercises 55–58, use a calculator to approximate the number. (Round to three decimal places.)

55. (a) $\sqrt{57}$ (b) $\sqrt[5]{-27^3}$

56. (a) $\sqrt[3]{45^2}$ (b) $\sqrt[6]{125}$

57. (a) $(1.2^{-2})\sqrt{75} + 3\sqrt{8}$ (b) $\dfrac{-3+\sqrt{21}}{3}$

58. (a) $(15.25)^{-1.4}$ (b) $(3.4)^{2.5}$

In Exercises 59–64, simplify by removing all possible factors from the radical.

59. (a) $\sqrt{8}$ (b) $\sqrt[3]{24}$

60. (a) $\sqrt[3]{\frac{16}{27}}$ (b) $\sqrt{\frac{75}{4}}$

61. (a) $\sqrt{72x^3}$ (b) $\sqrt{\frac{18^2}{z^3}}$

62. (a) $\sqrt{54xy^4}$ (b) $\sqrt{\frac{32a^4}{b^2}}$

63. (a) $\sqrt[3]{16x^5}$ (b) $\sqrt{75x^2y^{-4}}$

64. (a) $\sqrt[4]{(3x^2)^4}$ (b) $\sqrt[5]{96x^5}$

In Exercises 65–70, perform the operations and simplify.

65. $5^{4/3} \cdot 5^{8/3}$

66. $\dfrac{8^{12/5}}{8^{2/5}}$

67. $\dfrac{(2x^2)^{3/2}}{2^{1/2}x^4}$

68. $\dfrac{x^{4/3}y^{2/3}}{(xy)^{1/3}}$

69. $\dfrac{x^{-3} \cdot x^{1/2}}{x^{3/2} \cdot x^{-1}}$

70. $\dfrac{5^{-1/2} \cdot 5x^{5/2}}{(5x)^{3/2}}$

In Exercises 71–74, rationalize the denominator. Then simplify your answer.

71. (a) $\dfrac{1}{\sqrt{3}}$ (b) $\dfrac{8}{\sqrt[3]{2}}$

72. (a) $\dfrac{5}{\sqrt{10}}$ (b) $\dfrac{5}{\sqrt[3]{(5x)^2}}$

73. (a) $\dfrac{2x}{5 - \sqrt{3}}$ (b) $\dfrac{3}{\sqrt{5} + \sqrt{6}}$

74. (a) $\dfrac{5}{\sqrt{14} - 2}$ (b) $\dfrac{5}{2\sqrt{10} - 5}$

In Exercises 75–78, rationalize the numerator. Then simplify your answer.

75. (a) $\dfrac{\sqrt{8}}{2}$ (b) $\sqrt[3]{\frac{9}{25}}$

76. (a) $\dfrac{\sqrt{2}}{3}$ (b) $\sqrt[4]{\frac{5}{4}}$

77. (a) $\dfrac{\sqrt{5} + \sqrt{3}}{3}$ (b) $\dfrac{\sqrt{7} - 3}{4}$

78. (a) $\dfrac{\sqrt{3} - \sqrt{2}}{2}$ (b) $\dfrac{2\sqrt{3} + \sqrt{3}}{3}$

In Exercises 79 and 80, reduce the index of the radical.

79. (a) $\sqrt[4]{3^2}$ (b) $\sqrt[6]{(x + 1)^4}$

80. (a) $\sqrt[6]{x^3}$ (b) $\sqrt[4]{(3x^2)^4}$

In Exercises 81 and 82, write as a single radical. Then simplify your answer.

81. (a) $\sqrt{\sqrt{32}}$ (b) $\sqrt{\sqrt[4]{2x}}$

82. (a) $\sqrt{\sqrt{243(x + 1)}}$ (b) $\sqrt{\sqrt[3]{10a^7b}}$

In Exercises 83–86, simplify the expression.

83. (a) $2\sqrt{50} + 12\sqrt{8}$ (b) $10\sqrt{32} - 6\sqrt{18}$

84. (a) $4\sqrt{27} - \sqrt{75}$ (b) $\sqrt[3]{16} + 3\sqrt[3]{54}$

85. (a) $5\sqrt{x} - 3\sqrt{x}$ (b) $-2\sqrt{9y} + 10\sqrt{y}$

86. (a) $3\sqrt{x + 1} + 10\sqrt{x + 1}$

     (b) $7\sqrt{80x} - 2\sqrt{125x}$

In Exercises 87–90, fill in the blank with <, =, or >.

87. $\sqrt{5} + \sqrt{3}$ ▢ $\sqrt{5 + 3}$

88. $\sqrt{\dfrac{3}{11}}$ ▢ $\dfrac{\sqrt{3}}{\sqrt{11}}$

89. 5 ▢ $\sqrt{3^2 + 2^2}$

90. 5 ▢ $\sqrt{3^2 + 4^2}$

In Exercises 91–94, write the number in scientific notation.

91. Land Area of Earth: 57,500,000 square miles

92. Light Year: 9,461,000,000,000,000 kilometers

93. Relative Density of Hydrogen: 0.0000899 gram per cm^3

94. One Micron (Millionth of Meter): 0.00003937 inch

In Exercises 95–98, write the number in decimal form.

95. U.S. Daily Coca-Cola Consumption: 5.24×10^8 servings

96. Interior Temperature of Sun: 1.3×10^7 degrees Celsius

97. Charge of Electron: 4.8×10^{-10} electrostatic unit

98. Width of Human Hair: 9.0×10^{-4} meter

In Exercises 99–102, use a calculator to evaluate the expression. (Round to three decimal places.)

99. (a) $750\left(1 + \dfrac{0.11}{365}\right)^{800}$

    (b) $\dfrac{67{,}000{,}000 + 93{,}000{,}000}{0.0052}$

100. (a) $(9.3 \times 10^6)^3(6.1 \times 10^{-4})$

    (b) $\dfrac{(2.414 \times 10^4)^6}{(1.68 \times 10^5)^5}$

101. (a) $\sqrt{4.5 \times 10^9}$ (b) $\sqrt[3]{6.3 \times 10^4}$

102. (a) $(2.65 \times 10^{-4})^{1/3}$ (b) $\sqrt{9 \times 10^{-4}}$

103. **Exploration**   List all possible unit digits of the square of a positive integer. Use that list to determine whether $\sqrt{5233}$ is an integer.

104. **Think About It**   Square the real number $2/\sqrt{5}$ and note that the radical is eliminated from the denominator. Is this equivalent to rationalizing the denominator? Why or why not?

105. **Period of a Pendulum**   The period T in seconds of a pendulum is

$$T = 2\pi\sqrt{\dfrac{L}{32}}$$

where L is the length of the pendulum in feet. Find the period of a pendulum whose length is 2 feet.

106. **Mathematical Modeling**   A funnel is filled with water to a height of h centimeters. The time t (in seconds) for the funnel to empty is

$$t = 0.03[12^{5/2} - (12 - h)^{5/2}], \quad 0 \le h \le 12.$$

Find t for $h = 7$ centimeters.

107. **Declining Balances Depreciation**   Find the annual depreciation rate r for the bar graph below. To find the annual depreciation rate by the **declining balances method,** use the formula

$$r = 1 - \left(\dfrac{S}{C}\right)^{1/n}$$

where n is the useful life of the item (in years), S is the salvage value (in dollars), and C is the original cost (in dollars).

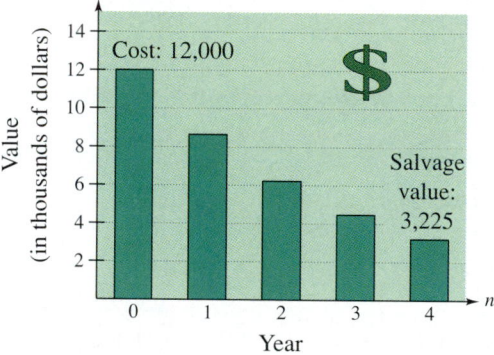

108. **Erosion**   A stream of water moving at the rate of v feet per second can carry particles of size $0.03\sqrt{v}$ inches. Find the size particle that can be carried by a stream flowing at the rate of $\frac{3}{4}$ foot per second.

109. **Speed of Light**   The speed of light is 11,160,000 miles per minute. The distance from the sun to the earth is 93,000,000 miles. Find the time for light to travel from the sun to the earth.

110. **Organizing Data**   There were 1.957×10^8 tons of municipal waste generated in the U.S. in 1990. Find the number of tons for each of the categories in the figure.   (Source: U.S. Environmental Protection Agency)

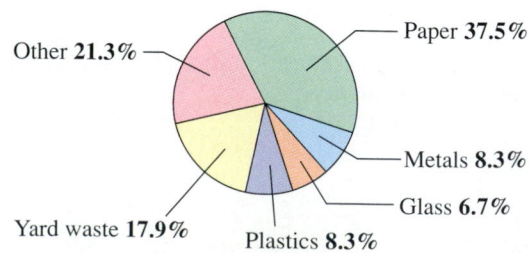

P.3 *Polynomials and Special Products*

See Exercise 87 on page 35 for an example of how a polynomial can be used to model the total stopping distance of an automobile.

Polynomials ❑ *Operations with Polynomials* ❑ *Special Products* ❑ *Application*

Polynomials

The most common type of algebraic expression is the **polynomial.** Some examples are

$$2x + 5, \quad 3x^4 - 7x^2 + 2x + 4, \quad \text{and} \quad 5x^2y^2 - xy + 3.$$

The first two are *polynomials in x* and the third is a *polynomial in x and y.* The terms of a polynomial in x have the form ax^k, where a is the **coefficient** and k is the **degree** of the term. For instance, the third-degree polynomial

$$2x^3 - 5x^2 + 1 = 2x^3 + (-5)x^2 + (0)x + 1$$

has coefficients $2, -5, 0,$ and 1.

DEFINITION OF A POLYNOMIAL IN X

Let $a_0, a_1, a_2, \ldots, a_n$ be *real numbers* and let n be a *nonnegative integer.* A **polynomial in x** is an expression of the form

$$a_n x^n + a_{n-1} x^{n-1} + \cdots + a_1 x + a_0$$

where $a_n \neq 0$. The polynomial is of **degree n**, a_n is the **leading coefficient,** and a_0 is the **constant term.**

NOTE Polynomials with one, two, and three terms are called **monomials, binomials,** and **trinomials,** respectively. ▪▪

In **standard form,** a polynomial is written with descending powers of x.

EXAMPLE 1   *Writing Polynomials in Standard Form*

	Polynomial	Standard Form	Degree
a.	$4x^2 - 5x^7 - 2 + 3x$	$-5x^7 + 4x^2 + 3x - 2$	7
b.	$4 - 9x^2$	$-9x^2 + 4$	2
c.	8	$8 \ (8 = 8x^0)$	0

NOTE For polynomials in more than one variable, the degree of a *term* is the sum of the exponents of the variables in the term. The degree of the *polynomial* is the highest degree of its terms. ▪▪

A polynomial that has all zero coefficients is called the **zero polynomial,** denoted by 0. No degree is assigned to this particular polynomial. Expressions such as $\sqrt{x^2 - 3x}$ and $x^2 + 5x^{-1}$ are not polynomials.

Operations with Polynomials

You can **add** and **subtract** polynomials in much the same way you add and subtract real numbers. Simply add or subtract the *like terms* (terms having the same variables to the same powers) by adding their coefficients. For instance, $-3xy^2$ and $5xy^2$ are like terms and their sum is

$$-3xy^2 + 5xy^2 = (-3 + 5)xy^2 = 2xy^2.$$

EXAMPLE 2   *Sums and Differences of Polynomials*

a. $(5x^3 - 7x^2 - 3) + (x^3 + 2x^2 - x + 8)$

$\qquad = (5x^3 + x^3) + (2x^2 - 7x^2) - x + (8 - 3)$ Group like terms.

$\qquad = 6x^3 - 5x^2 - x + 5$ Combine like terms.

b. $(7x^4 - x^2 - 4x + 2) - (3x^4 - 4x^2 + 3x)$

$\qquad = 7x^4 - x^2 - 4x + 2 - 3x^4 + 4x^2 - 3x$

$\qquad = (7x^4 - 3x^4) + (4x^2 - x^2) + (-3x - 4x) + 2$ Group like terms.

$\qquad = 4x^4 + 3x^2 - 7x + 2$ Combine like terms.

NOTE   A common mistake is to fail to change the sign of *each* term inside parentheses preceded by a negative sign. For instance, note that

$$-(x^2 - x + 3) = -x^2 + x - 3$$

and

$$-(x^2 - x + 3) \neq -x^2 - x + 3.$$

To find the **product** of two polynomials, use the left and right Distributive Properties. For example, if you treat $(5x + 7)$ as a single quantity, you can multiply $(3x - 2)$ by $(5x + 7)$ as follows.

$$(3x - 2)(5x + 7) = 3x(5x + 7) - 2(5x + 7)$$

$$= (3x)(5x) + (3x)(7) - (2)(5x) - (2)(7)$$

$$= 15x^2 + 21x - 10x - 14$$

Product of **First terms**	Product of **Outer terms**	Product of **Inner terms**	Product of **Last terms**

$$= 15x^2 + 11x - 14$$

Note in this **FOIL Method** that for binomials the outer (O) and inner (I) terms are alike and can be combined into one term.

When multiplying two polynomials, be sure to multiply *each* term of one polynomial by *each* term of the other. A vertical arrangement is helpful.

EXAMPLE 3   *A Vertical Arrangement for Multiplication*

Multiply $(x^2 - 2x + 2)$ by $(x^2 + 2x + 2)$.

Solution

$$
\begin{array}{rl}
x^2 - 2x + 2 & \quad\text{Standard form} \\
x^2 + 2x + 2 & \quad\text{Standard form} \\
\hline
x^4 - 2x^3 + 2x^2 & \quad\Longleftarrow\quad x^2(x^2 - 2x + 2) \\
2x^3 - 4x^2 + 4x & \quad\Longleftarrow\quad 2x(x^2 - 2x + 2) \\
2x^2 - 4x + 4 & \quad\Longleftarrow\quad 2(x^2 - 2x + 2) \\
\hline
x^4 + 0x^3 + 0x^2 - 0x + 4 = x^4 + 4 & \quad\text{Combine like terms.}
\end{array}
$$

Thus,

$$(x^2 - 2x + 2)(x^2 + 2x + 2) = x^4 + 4.$$

Special Products

SPECIAL PRODUCTS

Let u and v be real numbers, variables, or algebraic expressions.

Special Product	*Example*
Sum and Difference of Same Terms	
$(u + v)(u - v) = u^2 - v^2$	$(x + 4)(x - 4) = x^2 - 4^2 = x^2 - 16$
Square of a Binomial	
$(u + v)^2 = u^2 + 2uv + v^2$	$(x + 3)^2 = x^2 + 2(x)(3) + 3^2 = x^2 + 6x + 9$
$(u - v)^2 = u^2 - 2uv + v^2$	$(3x - 2)^2 = (3x)^2 - 2(3x)(2) + 2^2 = 9x^2 - 12x + 4$
Cube of a Binomial	
$(u + v)^3 = u^3 + 3u^2v + 3uv^2 + v^3$	$(x + 2)^3 = x^3 + 3x^2(2) + 3x(2^2) + 2^3 = x^3 + 6x^2 + 12x + 8$
$(u - v)^3 = u^3 - 3u^2v + 3uv^2 - v^3$	$(x - 1)^3 = x^3 - 3x^2(1) + 3x(1^2) - 1^3 = x^3 - 3x^2 + 3x - 1$

EXAMPLE 4   Sum and Difference of Same Terms

Find the product of $(5x + 9)$ and $(5x - 9)$.

Solution

The product of a sum and a difference of the *same* two terms has no middle term and it takes the form $(u + v)(u - v) = u^2 - v^2$.

$$(5x + 9)(5x - 9) = (5x)^2 - 9^2 = 25x^2 - 81$$

EXAMPLE 5   Square of a Binomial

Find $(6x - 5)^2$.

Solution

The square of a binomial has the form $(u - v)^2 = u^2 - 2uv + v^2$.

$$(6x - 5)^2 = (6x)^2 - 2(6x)(5) + 5^2$$
$$= 36x^2 - 60x + 25$$

EXAMPLE 6   Cube of a Binomial

Find $(3x + 2)^3$.

Solution

The cube of a binomial has the form $(u + v)^3 = u^3 + 3u^2v + 3uv^2 + v^3$. Note the *decrease* of powers of u and the *increase* of powers of v.

$$(3x + 2)^3 = (3x)^3 + 3(3x)^2(2) + 3(3x)(2)^2 + 2^3$$
$$= 27x^3 + 54x^2 + 36x + 8$$

EXAMPLE 7   The Product of Two Trinomials

Find the product of $(x + y - 2)$ and $(x + y + 2)$.

Solution

By grouping $x + y$ in parentheses, you can write

$$(x + y - 2)(x + y + 2) = [(x + y) - 2][(x + y) + 2]$$
$$= (x + y)^2 - 2^2$$
$$= x^2 + 2xy + y^2 - 4.$$

Application

EXAMPLE 8   *Volume of a Box*

An open box is made by cutting squares out of the corners of a piece of metal that is 16 inches by 20 inches, as shown in Figure P.6. The edge of each cut-out square is x inches. Find the volume when $x = 1$, $x = 2$, and $x = 3$.

Solution

The volume of a rectangular box is equal to the product of its length, width, and height. From the figure, the length is $20 - 2x$, the width is $16 - 2x$, and the height is x. Thus, the volume of the box is

$$\begin{aligned} \text{Volume} &= (20 - 2x)(16 - 2x)(x) \\ &= (320 - 72x + 4x^2)(x) \\ &= 320x - 72x^2 + 4x^3. \end{aligned}$$

When $x = 1$ inch, the volume of the box is

$$\text{Volume} = 320(1) - 72(1^2) + 4(1^3) = 252 \text{ cubic inches.}$$

When $x = 2$ inches, the volume of the box is

$$\text{Volume} = 320(2) - 72(2^2) + 4(2^3) = 384 \text{ cubic inches.}$$

When $x = 3$ inches, the volume of the box is

$$\text{Volume} = 320(3) - 72(3^2) + 4(3^3) = 420 \text{ cubic inches.}$$

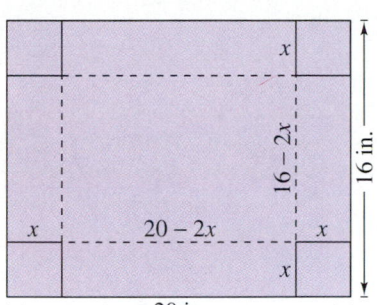

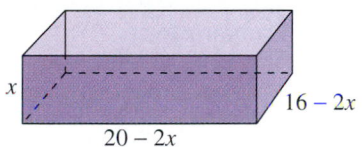

FIGURE P.6

....

GROUP ACTIVITY

A MATHEMATICAL EXPERIMENT

In Example 8, the volume of the open metal box is given by

$$\text{Volume} = 320x - 72x^2 + 4x^3.$$

Suppose you want to create a box that has as much volume as possible. From Example 8, you know that by cutting 1-, 2-, and 3-inch squares from the corners, you can create boxes whose volumes are 252, 384, and 420 cubic inches, respectively. Try several other values of x to see if you can find the size of the square that should be cut from the corners to produce a box that has maximum volume.

WARM UP

Simplify the expression.

1. $(7x^2)(6x)$

2. $(10z^3)(-2z^{-1})$

3. $(-3x^2)^3$

4. $(3x^2y^{-1})^0$

5. $\dfrac{27z^5}{12z^2}$

6. $\sqrt{24} \cdot \sqrt{2}$

7. $\left(\dfrac{2x}{3}\right)^{-2}$

8. $\dfrac{4}{\sqrt{8}}$

9. $16^{3/4}$

10. $\sqrt[3]{-27x^3}$

P.3 Exercises

In Exercises 1–6, match the polynomial with its description. [The polynomials are labeled (a), (b), (c), (d), (e), and (f).]

(a) $3x^2$

(b) $1 - 2x^3$

(c) $x^3 + 3x^2 + 3x + 1$

(d) 12

(e) $-3x^5 + 2x^3 + x$

(f) $\frac{2}{3}x^4 + x^2 + 10$

1. A polynomial of degree zero.

2. A trinomial of degree five.

3. A binomial with leading coefficient -2.

4. A monomial.

5. A trinomial with leading coefficient $\frac{2}{3}$.

6. A third-degree polynomial with leading coefficient 1.

In Exercises 7–12, find the degree and leading coefficient of the polynomial.

7. $2x^2 - x + 1$

8. $-3x^4 + 2x^2 - 5$

9. $x^5 - 1$

10. 3

11. $1 - x + 6x^4 - 4x^5$

12. $3 + 2x$

In Exercises 13–18, is the expression a polynomial? If so, write the polynomial in standard form.

13. $2x - 3x^3 + 8$

14. $2x^3 + x - 3x^{-1}$

15. $\dfrac{3x + 4}{x}$

16. $\dfrac{x^2 + 2x - 3}{2}$

17. $y^2 - y^4 + y^3$

18. $\sqrt{y^2 - y^4}$

In Exercises 19–32, perform the operations and write the result in standard form.

19. $(6x + 5) - (8x + 15)$

20. $(2x^2 + 1) - (x^2 - 2x + 1)$

21. $-(x^3 - 2) + (4x^3 - 2x)$

22. $-(5x^2 - 1) - (-3x^2 + 5)$

23. $(15x^2 - 6) - (-8x^3 - 14x^2 - 17)$

24. $(15x^4 - 18x - 19) - (13x^4 - 5x + 15)$

25. $5z - [3z - (10z + 8)]$

26. $(y^3 + 1) - [(y^2 + 1) + (3y - 7)]$

27. $3x(x^2 - 2x + 1)$

28. $y^2(4y^2 + 2y - 3)$

29. $-5z(3z - 1)$

30. $-4x(3 - x^3)$

31. $(1 - x^3)(4x)$

32. $(-2x)(-3x)(5x + 2)$

In Exercises 33–42, perform the operations using the vertical format.

33. Add:

$$\begin{array}{r} 7x^3 - 2x^2 + 8 \\ -3x^3 \qquad\quad - 4 \\ \hline \end{array}$$

34. Add:

$$\begin{array}{r} 2x^5 - 3x^3 + 2x + 3 \\ 4x^3 + \quad x - 6 \\ \hline \end{array}$$

35. Subtract:

$$\begin{array}{r} 5x^2 - 3x + 8 \\ x - 3 \\ \hline \end{array}$$

36. Subtract:

$$\begin{array}{r} 0.6t^4 - \quad 2t^2 \\ -t^4 + 0.5t^2 - 5.6 \\ \hline \end{array}$$

37. Multiply:

$$\begin{array}{r} -6x^2 + 15x - 4 \\ 5x + 3 \\ \hline \end{array}$$

38. Multiply:

$$\begin{array}{r} 4x^4 + x^3 - 6x^2 + 9 \\ x^2 + 2x \quad + 3 \\ \hline \end{array}$$

39. $(x^2 + 9)(x^2 - x - 4)$

40. $(x - 2)(x^2 + 2x + 4)$

41. $(x^2 - x + 1)(x^2 + x + 1)$

42. $(x^2 + 3x - 2)(x^2 - 3x - 2)$

In Exercises 43–70, find the product.

43. $(x + 3)(x + 4)$

44. $(x - 5)(x + 10)$

45. $(3x - 5)(2x + 1)$

46. $(7x - 2)(4x - 3)$

47. $(2x + 3)^2$

48. $(4x + 5)^2$

49. $(2x - 5y)^2$

50. $(5 - 8x)^2$

51. $[(x - 3) + y]^2$

52. $[(x + 1) - y]^2$

53. $(x + 10)(x - 10)$

54. $(2x + 3)(2x - 3)$

55. $(x + 2y)(x - 2y)$

56. $(2x + 3y)(2x - 3y)$

57. $[(m - 3) + n][(m - 3) - n]$

58. $[(x + y) + 1][(x + y) - 1]$

59. $(2r^2 - 5)(2r^2 + 5)$

60. $(3a^3 - 4b^2)(3a^3 + 4b^2)$

61. $(x + 1)^3$

62. $(x - 2)^3$

63. $(2x - y)^3$

64. $(3x + 2y)^3$

65. $(4x^3 - 3)^2$

66. $(8x + 3)^2$

67. $5x(x + 1) - 3x(x + 1)$

68. $(2x - 1)(x + 3) + 3(x + 3)$

69. $(u + 2)(u - 2)(u^2 + 4)$

70. $(x + y)(x - y)(x^2 + y^2)$

In Exercises 71–74, find the product. The expressions are not polynomials, but the formulas can still be used.

71. $(\sqrt{x} + \sqrt{y})(\sqrt{x} - \sqrt{y})$

72. $(5 + \sqrt{x})(5 - \sqrt{x})$

73. $(x + \sqrt{5})(x - \sqrt{5})(x + 4)$

74. $(x + \sqrt{3})^2$

75. *Think About It* Must the sum of two second-degree polynomials be a second-degree polynomial? If not, give an example.

76. *Think About It* Is the product of two binomials always a binomial? Explain.

77. *Find a Pattern* Perform the multiplications.

(a) $(x - 1)(x + 1)$

(b) $(x - 1)(x^2 + x + 1)$

(c) $(x - 1)(x^3 + x^2 + x + 1)$

From the pattern formed by these products, can you predict the result of $(x - 1)(x^4 + x^3 + x^2 + x + 1)$?

78. *Think About It* When the polynomial $-x^3 + 3x^2 + 2x - 1$ is subtracted from an unknown polynomial, the difference is $5x^2 + 8$. If it is possible, find the unknown polynomial.

79. *Logical Reasoning* Verify that $(x + y)^2$ is not equal to $x^2 + y^2$ by letting $x = 3$ and $y = 4$ and evaluating both expressions.

80. *Profit* A manufacturer can produce and sell x radios per week. The total cost (in dollars) for producing x radios is

$$C = 73x + 25,000$$

and the total revenue is

$$R = 95x.$$

Find the profit obtained by selling 5000 radios per week.

81. *Compound Interest*   After 2 years, an investment of $500 compounded annually at an interest rate r will yield an amount of

$$500(1 + r)^2.$$

(a) Write this polynomial in standard form.

(b) Use a calculator to evaluate the expression for the values of r in the table.

r	$5\frac{1}{2}\%$	7%	8%	$8\frac{1}{2}\%$	9%
$500(1 + r)^2$					

(c) What conclusion can you make from the table?

82. *Compound Interest*   After 3 years, an investment of $1200 compounded annually at an interest rate r will yield an amount of

$$1200(1 + r)^3.$$

(a) Write this polynomial in standard form.

(b) Use a calculator to evaluate the expression for the values of r in the table.

r	6%	7%	$7\frac{1}{2}\%$	8%	$8\frac{1}{2}\%$
$1200(1 + r)^3$					

(c) What conclusion can you make from the table?

83. *Volume of a Box*   An open box is made by cutting squares out of the corners of a piece of metal that is 18 centimeters by 26 centimeters (see figure). If the edge of each cut-out square is x inches, find the volume when $x = 1$, $x = 2$, and $x = 3$.

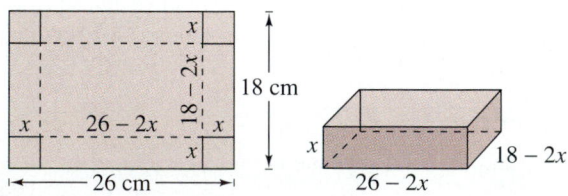

84. *Volume of a Box*   A closed box is constructed by cutting along the solid lines and folding along the broken lines on the rectangular piece of metal shown in the figure. The length and width of the rectangle are 45 centimeters and 15 centimeters, respectively. Find the volume of the box in terms of x. Find the volume when $x = 3$, $x = 5$, and $x = 7$.

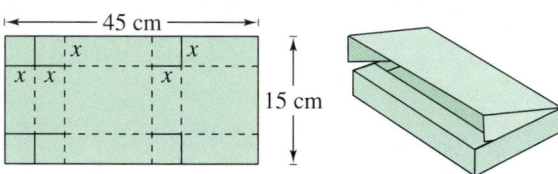

85. *Geometry*   Find the area of the shaded region in the figures. Write your result as a polynomial in standard form.

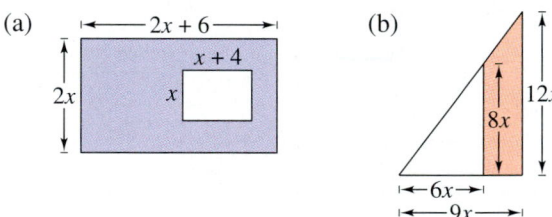

86. *Floor Space*   Find a polynomial that represents the total number of square feet for the floor plan shown in the figure.

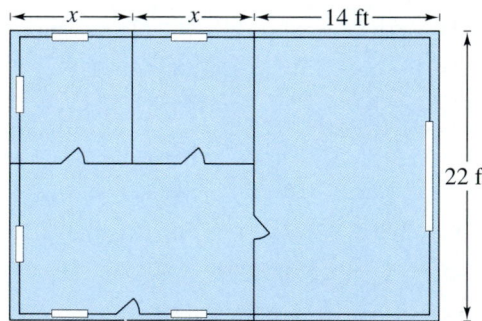

87. *Stopping Distance*   The stopping distance of an automobile is the distance traveled during the driver's reaction time plus the distance traveled after the brakes are applied. In an experiment, these distances were measured (in feet) when the automobile was traveling at a speed of x miles per hour (see figure). The distance traveled during the reaction time was $R = 1.1x$, and the braking distance was $B = 0.14x^2 - 4.43x + 58.40$.

    (a) Determine the polynomial that represents the total stopping distance.

    (b) Use the result of part (a) to estimate the total stopping distance when $x = 30$, $x = 40$, and $x = 55$.

    (c) Use the bar graph to make a statement about the total stopping distance required for increasing speeds.

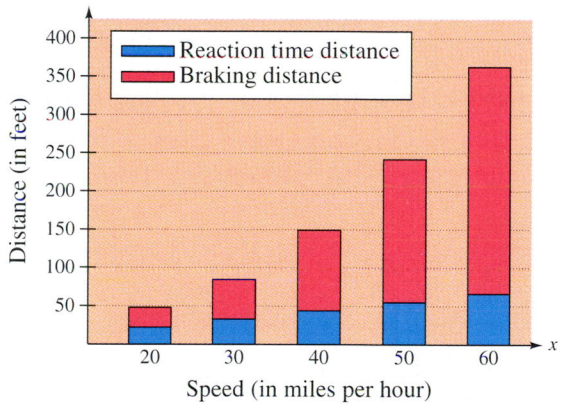

88. *Safe Beam Load*   A uniformly distributed load is placed on a 1-inch-wide steel beam. When the span of the beam is x feet and its depth is 6 inches, the safe load is approximated by

    $$S_6 = (0.06x^2 - 2.42x + 38.71)^2.$$

    When the depth is 8 inches, the safe load is approximated by

    $$S_8 = (0.08x^2 - 3.30x + 51.93)^2.$$

    (a) Estimate the difference in the safe loads of these two beams when the span is 12 feet (see figure).

    (b) How does the difference in safe load change as the span increases?

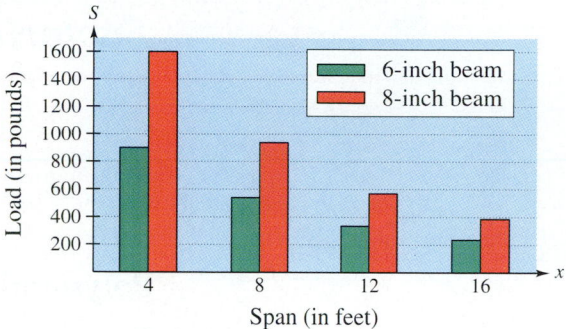

FIGURE FOR 88

Geometrical Modeling   **In Exercises 89 and 90, use the area model to write two different expressions for the area. Then equate the two expressions and name the algebraic property that is illustrated.**

89.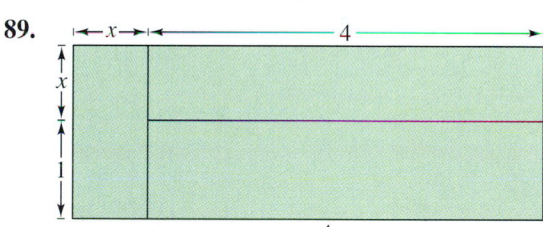

90.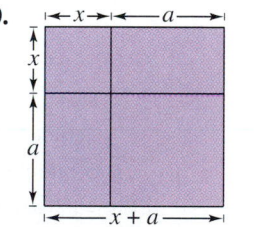

91. Find the degree of the product of two polynomials of degrees m and n.

92. Find the degree of the sum of two polynomials of degrees m and n if $m < n$.

93. *Essay*   A student's homework paper included the following.

    $$(x - 3)^2 = x^2 + 9$$

    Write a paragraph fully explaining the error and give the correct method for squaring a binomial.

> ▶ **P.4**   *Factoring*

See Exercise 109 on page 44 for an example of how factoring can be used to develop an alternative formula for the volume of a cylindrical shell.

Polynomials with Common Factors □ Factoring Special Polynomial Forms □ Trinomials with Binomial Factors □ Factoring by Grouping

Polynomials with Common Factors

The process of writing a polynomial as a product is called **factoring.** It is an important tool for solving equations and for reducing fractional expressions.

Unless noted otherwise, when you are asked to factor a polynomial, you can assume that you are hunting for factors with integer coefficients. If a polynomial cannot be factored using integer coefficients, then it is **prime** or **irreducible over the integers.** For instance, the polynomial $x^2 - 3$ is irreducible over the integers. Over the *real numbers,* this polynomial can be factored as

$$x^2 - 3 = (x + \sqrt{3})(x - \sqrt{3}).$$

A polynomial is **completely factored** when each of its factors is prime. For instance,

$$x^3 - x^2 + 4x - 4 = (x - 1)(x^2 + 4)$$

is completely factored, but

$$x^3 - x^2 - 4x + 4 = (x - 1)(x^2 - 4)$$

is not completely factored. Its complete factorization would be

$$x^3 - x^2 - 4x + 4 = (x - 1)(x + 2)(x - 2).$$

The simplest type of factoring involves a polynomial that can be written as the product of a monomial and another polynomial. The technique used here is the Distributive Property, $a(b + c) = ab + ac$, in the *reverse* direction.

$$ab + ac = a(b + c) \qquad\qquad \textcolor{red}{a \text{ is a common factor.}}$$

Removing (factoring out) a common factor is the first step in completely factoring a polynomial.

■

EXAMPLE 1   Removing Common Factors

a. $6x^3 - 4x = 2x(3x^2) - 2x(2)$ $2x$ is a common factor.

$\qquad\qquad\quad = 2x(3x^2 - 2)$

b. $(x - 2)(2x) + (x - 2)(3) = (x - 2)(2x + 3)$ $x - 2$ is a common factor.

Factoring Special Polynomial Forms

FACTORING SPECIAL POLYNOMIAL FORMS

Factored Form | *Example*

Difference of Two Squares

$u^2 - v^2 = (u + v)(u - v)$ $9x^2 - 4 = (3x)^2 - 2^2 = (3x + 2)(3x - 2)$

Perfect Square Trinomial

$u^2 + 2uv + v^2 = (u + v)^2$ $x^2 + 6x + 9 = x^2 + 2(x)(3) + 3^2 = (x + 3)^2$

$u^2 - 2uv + v^2 = (u - v)^2$ $x^2 - 6x + 9 = x^2 - 2(x)(3) + 3^2 = (x - 3)^2$

Sum or Difference of Two Cubes

$u^3 + v^3 = (u + v)(u^2 - uv + v^2)$ $x^3 + 8 = x^3 + 2^3 = (x + 2)(x^2 - 2x + 4)$

$u^3 - v^3 = (u - v)(u^2 + uv + v^2)$ $27x^3 - 1 = (3x)^3 - 1^3 = (3x - 1)(9x^2 + 3x + 1)$

One of the easiest special polynomial forms to factor is the difference of two squares. Think of the form as follows.

$$u^2 - v^2 = (u + v)(u - v)$$

Difference   Opposite signs

To recognize perfect square terms, look for coefficients that are squares of integers and variables raised to *even powers.*

NOTE In Example 2, note that the first step in factoring a polynomial is to check for common factors. Once the common factor is removed, it is often possible to recognize patterns that were not immediately obvious. ■■

EXAMPLE 2 *Removing a Common Factor First*

$$3 - 12x^2 = 3(1 - 4x^2) = 3[1^2 - (2x)^2] = 3(1 + 2x)(1 - 2x)$$

EXAMPLE 3 *Factoring the Difference of Two Squares*

a. $(x + 2)^2 - y^2 = [(x + 2) + y][(x + 2) - y]$

$= (x + 2 + y)(x + 2 - y)$

b. $16x^4 - 81 = (4x^2)^2 - 9^2$

$= (4x^2 + 9)(4x^2 - 9)$ Difference of two squares

$= (4x^2 + 9)[(2x)^2 - 3^2]$

$= (4x^2 + 9)(2x + 3)(2x - 3)$ Difference of two squares

A perfect square trinomial is the square of a binomial, and it has the following form.

$$u^2 + 2uv + v^2 = (u + v)^2 \qquad \text{or} \qquad u^2 - 2uv + v^2 = (u - v)^2$$

Like signs   Like signs

Note that the first and last terms are squares and the middle term is twice the product of u and v.

EXAMPLE 4   *Factoring Perfect Square Trinomials*

a. $16x^2 + 8x + 1 = (4x)^2 + 2(4x)(1) + 1^2$
$$= (4x + 1)^2$$

b. $x^2 - 10x + 25 = x^2 - 2(x)(5) + 5^2$
$$= (x - 5)^2$$

▶ *Exploration*

Find a formula for completely factoring $u^6 - v^6$ using the formulas from this section. Use your formula to complete factor $x^6 - 1$ and $x^6 - 64$.

The next two formulas show the sums and differences of cubes. Pay special attention to the signs of the terms.

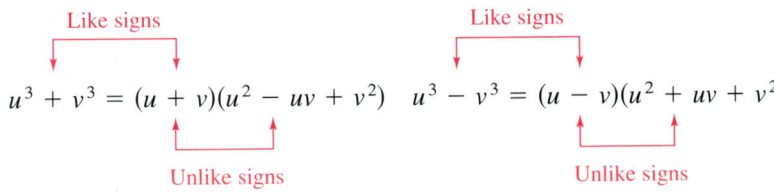

Like signs   Like signs

$$u^3 + v^3 = (u + v)(u^2 - uv + v^2) \quad u^3 - v^3 = (u - v)(u^2 + uv + v^2)$$

Unlike signs   Unlike signs

EXAMPLE 5   *Factoring the Difference of Cubes*

$$x^3 - 27 = x^3 - 3^3 \qquad\qquad \text{Rewrite 27 as } 3^3.$$
$$= (x - 3)(x^2 + 3x + 9) \qquad \text{Factor.}$$

EXAMPLE 6   *Factoring the Sum of Cubes*

a. $y^3 + 8 = y^3 + 2^3 \qquad\qquad\qquad \text{Rewrite 8 as } 2^3.$
$$= (y + 2)(y^2 - 2y + 4) \qquad \text{Factor.}$$

b. $3(x^3 + 64) = 3(x^3 + 4^3) \qquad\qquad \text{Rewrite 64 as } 4^3.$
$$= 3(x + 4)(x^2 - 4x + 16) \qquad \text{Factor.}$$

Trinomials with Binomial Factors

To factor a trinomial of the form $ax^2 + bx + c$, use the following pattern.

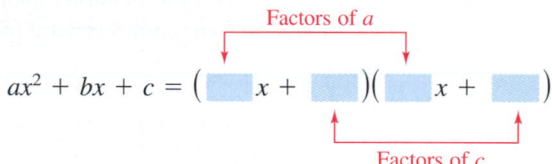

Factors of a

$$ax^2 + bx + c = (\quad x + \quad)(\quad x + \quad)$$

Factors of c

The goal is to find a combination of factors of a and c so that the outer and inner products add up to the middle term bx. For instance, in the trinomial $6x^2 + 17x + 5$, you can write

$$
\begin{array}{cccc}
F & O & I & L \\
\downarrow & \downarrow & \downarrow & \downarrow
\end{array}
$$

$$(2x + 5)(3x + 1) = 6x^2 + 2x + 15x + 5 = 6x^2 + 17x + 5.$$

Note that the outer (O) and inner (I) products add up to $17x$.

EXAMPLE 7   Factoring a Trinomial: Leading Coefficient Is 1

Factor $x^2 - 7x + 12$.

Solution

The possible factorizations are

$$(x - 2)(x - 6), \quad (x - 1)(x - 12), \quad \text{and} \quad (x - 3)(x - 4).$$

Testing the middle term, you will find the correct factorization to be

$$x^2 - 7x + 12 = (x - 3)(x - 4).$$

EXAMPLE 8   Factoring a Trinomial: Leading Coefficient Is Not 1

Factor $2x^2 + x - 15$.

Solution

The eight possible factorizations are as follows.

$$
\begin{array}{ll}
(2x - 1)(x + 15) & (2x + 1)(x - 15) \\
(2x - 3)(x + 5) & (2x + 3)(x - 5) \\
(2x - 5)(x + 3) & (2x + 5)(x - 3) \\
(2x - 15)(x + 1) & (2x + 15)(x - 1)
\end{array}
$$

Testing the middle term, you will find the correct factorization to be

$$2x^2 + x - 15 = (2x - 5)(x + 3).$$

Factoring by Grouping

Sometimes polynomials with more than three terms can be factored by a method called **factoring by grouping.** It is not always obvious which terms to group, and sometimes several different groupings will work.

EXAMPLE 9   *Factoring by Grouping*

$$x^3 - 2x^2 - 3x + 6 = (x^3 - 2x^2) - (3x - 6) \qquad \text{Group terms.}$$
$$= x^2(x - 2) - 3(x - 2) \qquad \text{Factor groups.}$$
$$= (x - 2)(x^2 - 3) \qquad \text{Distributive Property}$$

Factoring a trinomial can involve quite a bit of trial and error. Some of this trial and error can be lessened by using factoring by grouping.

EXAMPLE 10   *Factoring a Trinomial by Grouping*

Use factoring by grouping to factor $2x^2 + 5x - 3$.

Solution

In the trinomial $2x^2 + 5x - 3$, we have $a = 2$ and $c = -3$ which implies that the product ac is -6. Now, because -6 factors as $(6)(-1)$ and $6 - 1 = 5 = b$, we rewrite the middle term as $5x = 6x - x$. This produces the following.

$$2x^2 + 5x - 3 = 2x^2 + 6x - x - 3 \qquad \text{Rewrite middle term.}$$
$$= (2x^2 + 6x) - (x + 3) \qquad \text{Group terms.}$$
$$= 2x(x + 3) - (x + 3) \qquad \text{Factor groups.}$$
$$= (x + 3)(2x - 1) \qquad \text{Distributive Property}$$

Therefore, the trinomial factors as

$$2x^2 + 5x - 3 = (x + 3)(2x - 1).$$

GUIDELINES FOR FACTORING POLYNOMIALS

1. Factor out any common factors by the Distributive Property.
2. Factor according to one of the special polynomial forms.
3. Factor as $ax^2 + bx + c = (mx + r)(nx + s)$.
4. Factor by grouping.

GROUP ACTIVITY

A THREE-DIMENSIONAL VIEW OF A SPECIAL PRODUCT

The figure below shows two cubes.

a. The large cube has a volume of a^3.

b. The small cube has a volume of b^3.

If the smaller cube is removed from the larger, the remaining solid has a volume of $a^3 - b^3$ and is composed of three rectangular boxes, labeled Box 1, Box 2, and Box 3. Find the volume of each box and describe how these results are related to the special product formula.

$$a^3 - b^3 = (a - b)(a^2 + ab + b^2)$$
$$= (a - b)a^2 + (a - b)ab + (a - b)b^2$$

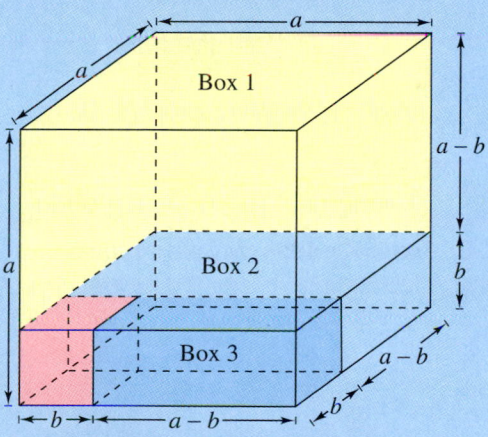

WARM UP

Find the product.

1. $3x(5x - 2)$
2. $-2y(y + 1)$
3. $(2x + 3)^2$
4. $(3x - 8)^2$
5. $(2x - 3)(x + 8)$
6. $(4 - 5z)(1 + z)$
7. $(2y + 1)(2y - 1)$
8. $(x + a)(x - a)$
9. $(x + 4)^3$
10. $(2x - 3)^3$

P.4 Exercises

In Exercises 1–4, find the greatest common factor of the expressions.

1. 90, 300
2. 36, 84, 294
3. $12x^2y^3, 18x^2y, 24x^3y^2$
4. $15(x + 2)^3, 42x(x + 2)^2$

In Exercises 5–10, factor out the common factor.

5. $3x + 6$
6. $5y - 30$
7. $2x^3 - 6x$
8. $4x^3 - 6x^2 + 12x$
9. $(x - 1)^2 + 6(x - 1)$
10. $3x(x + 2) - 4(x + 2)$

In Exercises 11–16, factor the difference of two squares.

11. $x^2 - 36$
12. $x^2 - \frac{1}{4}$
13. $16y^2 - 9$
14. $49 - 9y^2$
15. $(x - 1)^2 - 4$
16. $25 - (z + 5)^2$

In Exercises 17–22, factor the perfect square trinomial.

17. $x^2 - 4x + 4$
18. $x^2 + 10x + 25$
19. $4t^2 + 4t + 1$
20. $9x^2 - 12x + 4$
21. $25y^2 - 10y + 1$
22. $z^2 + z + \frac{1}{4}$

In Exercises 23–26, factor a negative real number from the polynomial and then write the polynomial factor in standard form.

23. $25 - 5x^2$
24. $5 + 3y^2 - y^3$
25. $-2t^3 + 4t + 6$
26. $-3x^5 - 3x^2 + 6x + 9$

In Exercises 27–40, factor the trinomial.

27. $x^2 + x - 2$
28. $x^2 + 5x + 6$
29. $s^2 - 5s + 6$
30. $t^2 - t - 6$
31. $20 - y - y^2$
32. $24 + 5z - z^2$
33. $x^2 - 30x + 200$
34. $x^2 - 13x + 42$
35. $3x^2 - 5x + 2$
36. $2x^2 - x - 1$
37. $-9z^2 + 3z + 2$
38. $12x^2 + 7x + 1$
39. $5x^2 + 26x + 5$
40. $-5u^2 - 13u + 6$

In Exercises 41–46, factor the sum or difference.

41. $x^3 - 8$
42. $x^3 - 27$
43. $y^3 + 64$
44. $z^3 + 125$
45. $8t^3 - 1$
46. $27x^3 + 8$

In Exercises 47–52, factor by grouping.

47. $x^3 - x^2 + 2x - 2$
48. $x^3 + 5x^2 - 5x - 25$
49. $2x^3 - x^2 - 6x + 3$
50. $5x^3 - 10x^2 + 3x - 6$
51. $6 + 2x - 3x^3 - x^4$
52. $x^5 + 2x^3 + x^2 + 2$

In Exercises 53–58, factor the trinomial by grouping.

53. $3x^2 + 10x + 8$
54. $2x^2 + 9x + 9$
55. $6x^2 + x - 2$
56. $6x^2 - x - 15$
57. $15x^2 - 11x + 2$
58. $12x^2 - 13x + 1$

In Exercises 59–90, completely factor the expression.

59. $x^3 - 9x$

60. $12x^2 - 48$

61. $x^3 - 4x^2$

62. $6x^2 - 54$

63. $x^2 - 2x + 1$

64. $16 + 6x - x^2$

65. $1 - 4x + 4x^2$

66. $-9x^2 + 6x - 1$

67. $2x^2 + 4x - 2x^3$

68. $2y^3 - 7y^2 - 15y$

69. $9x^2 + 10x + 1$

70. $13x + 6 + 5x^2$

71. $3x^3 + x^2 + 15x + 5$

72. $5 - x + 5x^2 - x^3$

73. $x^4 - 4x^3 + x^2 - 4x$

74. $3u - 2u^2 + 6 - u^3$

75. $25 - (z + 5)^2$

76. $(t - 1)^2 - 49$

77. $(x^2 + 1)^2 - 4x^2$

78. $(x^2 + 8)^2 - 36x^2$

79. $2t^3 - 16$

80. $5x^3 + 40$

81. $4x(2x - 1) + (2x - 1)^2$

82. $5(3 - 4x)^2 - 8(3 - 4x)(5x - 1)$

83. $2(x + 1)(x - 3)^2 - 3(x + 1)^2(x - 3)$

84. $7(3x + 2)^2(1 - x)^2 + (3x + 2)(1 - x)^3$

85. $7x(2)(x^2 + 1)(2x) - (x^2 + 1)^2(7)$

86. $3(x - 2)^2(x + 1)^4 + (x - 2)^3(4)(x + 1)^3$

87. $2x(x - 5)^4 - x^2(4)(x - 5)^3$

88. $5(x^6 + 1)^4(6x^5)(3x + 2)^3 + 3(3x + 2)^2(3)(x^6 + 1)^5$

89. $\dfrac{x^2}{2}(x^2 + 1)^4 - (x^2 + 1)^5$

90. $5w^3(9w + 1)^4(9) + (2w + 1)^5(3w^2)$

Geometric Modeling **In Exercises 91–94, match the factoring formula with the correct "geometric factoring model." [The models are labeled (a), (b), (c), and (d).]**

91. $a^2 - b^2 = (a + b)(a - b)$

92. $a^2 + 2ab + b^2 = (a + b)^2$

93. $a^2 + 2a + 1 = (a + 1)^2$

94. $ab + a + b + 1 = (a + 1)(b + 1)$

(a)
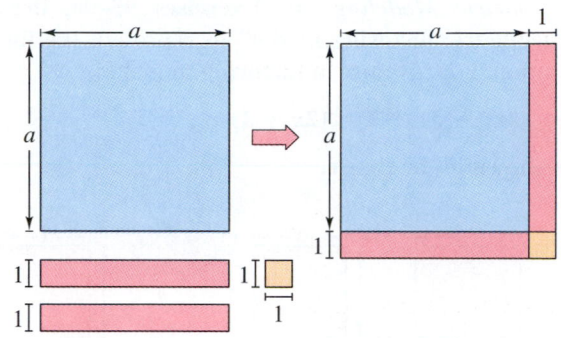

(b)
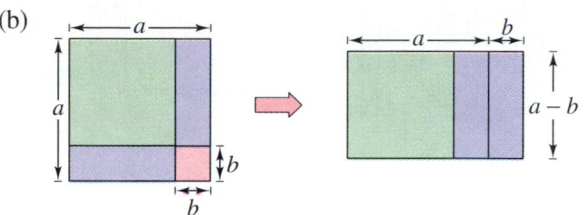

(c)
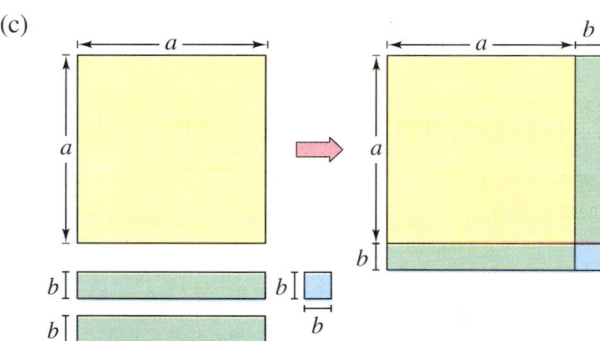

(d)
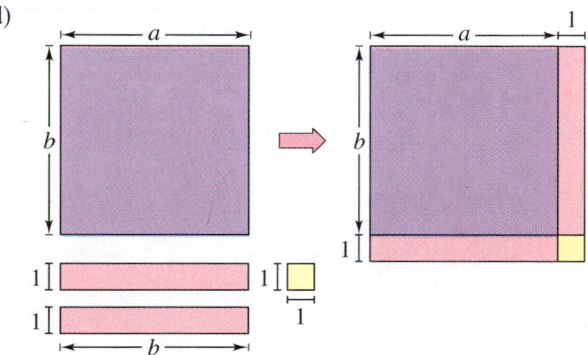

Geometric Modeling In Exercises 95–98, draw a "geometric factoring model" to represent the factorization. For instance, a factoring model for

$$2x^2 + 3x + 1 = (2x + 1)(x + 1)$$

is shown in the figure.

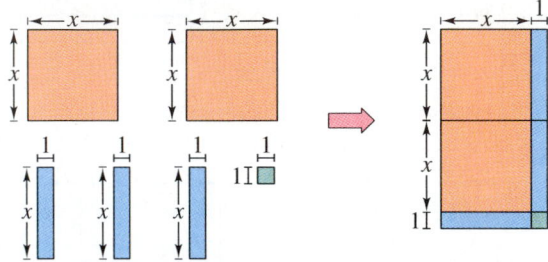

95. $3x^2 + 7x + 2 = (3x + 1)(x + 2)$
96. $x^2 + 4x + 3 = (x + 3)(x + 1)$
97. $2x^2 + 7x + 3 = (2x + 1)(x + 3)$
98. $x^2 + 3x + 2 = (x + 2)(x + 1)$

Geometry In Exercises 99–102, write, in factored form, an expression for the shaded portion of the figure.

99.
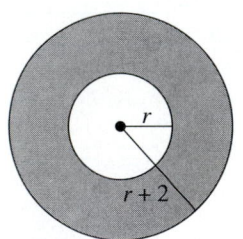

100.
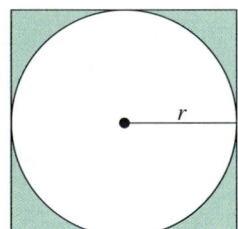

101.
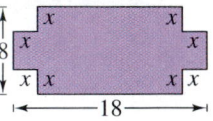

102.
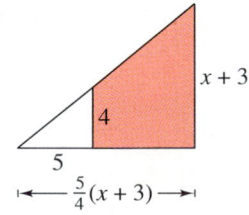

In Exercises 103 and 104, find all values of b for which the trinomial can be factored.

103. $x^2 + bx - 15$
104. $x^2 + bx + 50$

In Exercises 105 and 106, find two integers c such that the trinomial can be factored. (There are many correct answers.)

105. $2x^2 + 5x + c$
106. $3x^2 - 10x + c$

107. *Error Analysis* Describe the error.

$$9x^2 - 9x - 54 = (3x + 6)(3x - 9)$$
$$= 3(x + 2)(x - 3)$$

108. *Think About It* Is $(3x - 6)(x + 1)$ completely factored? Explain.

109. *Geometry* The cylindrical shell shown in the figure has a volume of

$$V = \pi R^2 h - \pi r^2 h.$$

(a) Factor the expression for the volume.

(b) From the result of part (a), show that the volume is 2π (average radius)(thickness of the shell)h.

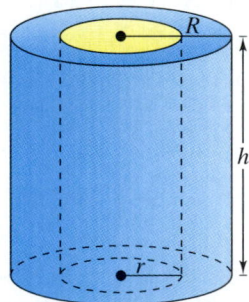

110. *Chemistry* The rate of change of an autocatalytic chemical reaction is $kQx - kx^2$, where Q is the amount of the original substance, x is the amount of substance formed, and k is a constant of proportionality. Factor the expression.

See Exercise 90 on page 55 for an example of how fractional expressions can be used to model the costs per ounce of precious metals from 1988 through 1992.

P.5 *Fractional Expressions*

Domain of an Algebraic Expression ▫ *Simplifying Rational Expressions*
Operations with Rational Expressions ▫ *Compound Fractions*

Domain of an Algebraic Expression

The set of real numbers for which an algebraic expression is defined is the **domain** of the expression. Two algebraic expressions are **equivalent** if they have the same domain and yield the same values for all numbers in their domain. For instance, $(x + 1) + (x + 2)$ and $2x + 3$ are equivalent.

EXAMPLE 1 *Finding the Domain of an Algebraic Expression*

a. The domain of the polynomial

$$2x^3 + 3x + 4$$

is the set of all real numbers. In fact, the domain of any polynomial is the set of all real numbers, *unless* the domain is specifically restricted.

b. The domain of the radical expression

$$\sqrt{x - 2}$$

is the set of real numbers greater than or equal to 2, because the square root of a negative number is not a real number.

c. The domain of the expression

$$\frac{x + 2}{x - 3}$$

is the set of all real numbers except $x = 3$, which would produce an undefined division by zero.

The quotient of two algebraic expressions is a **fractional expression.** Moreover, the quotient of two *polynomials* such as

$$\frac{1}{x}, \quad \frac{2x - 1}{x + 1}, \quad \text{or} \quad \frac{x^2 - 1}{x^2 + 1}$$

is a **rational expression.** Recall that a fraction is in reduced form if its numerator and denominator have no factors in common aside from ±1. To write a fraction in reduced form, apply the following rule.

$$\frac{a \cdot \cancel{c}}{b \cdot \cancel{c}} = \frac{a}{b}, \quad b \neq 0 \quad \text{and} \quad c \neq 0.$$

The key to success in simplifying rational expressions lies in your ability to *factor* polynomials.

EXAMPLE 2   *Reducing a Rational Expression*

Write $\dfrac{x^2 + 4x - 12}{3x - 6}$ in reduced form.

Solution

$$\frac{x^2 + 4x - 12}{3x - 6} = \frac{(x + 6)(x - 2)}{3(x - 2)} \qquad \text{Factor completely.}$$

$$= \frac{x + 6}{3}, \qquad x \neq 2 \qquad \text{Cancel common factors.}$$

Note that the original expression is undefined when $x = 2$ (because division by zero is undefined). To make sure that the reduced expression is *equivalent* to the original expression, you must restrict the domain of the reduced expression by excluding the value $x = 2$.

Study Tip

In Example 2, do not make the mistake of trying to reduce further by dividing *terms*.

$$\frac{x + 6}{3} \neq \frac{\overset{2}{x + 6}}{3} = x + 2$$

Remember that to reduce fractions, divide common *factors*, not terms.

Simplifying Rational Expressions

When simplifying rational expressions, be sure to factor each polynomial completely before concluding that the numerator and denominator have no factors in common. Moreover, changing the sign of a factor may allow further reduction, as shown in part (b) of the next example.

EXAMPLE 3   *Reducing Rational Expressions*

a. $\dfrac{x^3 - 4x}{x^2 + x - 2} = \dfrac{x(x^2 - 4)}{(x + 2)(x - 1)}$

$$= \frac{x(x + 2)(x - 2)}{(x + 2)(x - 1)} \qquad \text{Factor completely.}$$

$$= \frac{x(x - 2)}{(x - 1)}, \qquad x \neq -2 \qquad \text{Cancel common factors.}$$

b. $\dfrac{12 + x - x^2}{2x^2 - 9x + 4} = \dfrac{(4 - x)(3 + x)}{(2x - 1)(x - 4)} \qquad \text{Factor completely.}$

$$= \frac{-(x - 4)(3 + x)}{(2x - 1)(x - 4)} \qquad (4 - x) = -(x - 4)$$

$$= -\frac{3 + x}{2x - 1}, \qquad x \neq 4 \qquad \text{Cancel common factors.}$$

Operations with Rational Expressions

To multiply or divide rational expressions, we use the properties of fractions discussed in Section P.1. Recall that to divide fractions we invert the divisor and multiply.

EXAMPLE 4 Multiplying Rational Expressions

$$\frac{2x^2 + x - 6}{x^2 + 4x - 5} \cdot \frac{x^3 - 3x^2 + 2x}{4x^2 - 6x} = \frac{(2x - 3)(x + 2)}{(x + 5)(x - 1)} \cdot \frac{x(x - 2)(x - 1)}{2x(2x - 3)}$$

$$= \frac{(x + 2)(x - 2)}{2(x + 5)}, \qquad x \neq 0, x \neq 1, x \neq \tfrac{3}{2}$$

EXAMPLE 5 Dividing Rational Expressions

$$\frac{x^3 - 8}{x^2 - 4} \div \frac{x^2 + 2x + 4}{x^3 + 8} = \frac{x^3 - 8}{x^2 - 4} \cdot \frac{x^3 + 8}{x^2 + 2x + 4} \qquad \text{Invert and multiply.}$$

$$= \frac{(x - 2)(x^2 + 2x + 4)}{(x + 2)(x - 2)} \cdot \frac{(x + 2)(x^2 - 2x + 4)}{x^2 + 2x + 4}$$

$$= x^2 - 2x + 4, \qquad x \neq \pm 2$$

To add or subtract rational expressions, you can use the LCD (least common denominator) method or the basic definition

$$\frac{a}{b} \pm \frac{c}{d} = \frac{ad \pm bc}{bd}, \qquad b \neq 0 \text{ and } d \neq 0. \qquad \text{Basic definition}$$

This definition provides an efficient way of adding or subtracting *two* fractions that have no common factors in their denominators.

EXAMPLE 6 Subtracting Rational Expressions

$$\frac{x}{x - 3} - \frac{2}{3x + 4} = \frac{x(3x + 4) - 2(x - 3)}{(x - 3)(3x + 4)} \qquad \text{Basic definition}$$

$$= \frac{3x^2 + 4x - 2x + 6}{(x - 3)(3x + 4)} \qquad \text{Remove parentheses.}$$

$$= \frac{3x^2 + 2x + 6}{(x - 3)(3x + 4)} \qquad \text{Combine like terms.}$$

For three or more fractions, or for fractions with a repeated factor in the denominators, the LCD method works well. Recall that the least common denominator of several fractions consists of the product of all prime factors in the denominators, with each factor given the highest power of its occurrence in any denominator. Here is a numerical example.

$$\frac{1}{6} + \frac{3}{4} - \frac{2}{3} = \frac{1 \cdot 2}{6 \cdot 2} + \frac{3 \cdot 3}{4 \cdot 3} - \frac{2 \cdot 4}{3 \cdot 4} \qquad \textcolor{red}{\text{The LCD is 12.}}$$

$$= \frac{2}{12} + \frac{9}{12} - \frac{8}{12}$$

$$= \frac{3}{12}$$

$$= \frac{1}{4}$$

NOTE Sometimes the numerator of the answer has a factor in common with the denominator. In such cases the answer should be reduced. For instance, in the example above, $\frac{3}{12}$ was reduced to $\frac{1}{4}$. ■■

EXAMPLE 7 **Combining Rational Expressions: The LCD Method**

Perform the operations and simplify.

$$\frac{3}{x - 1} - \frac{2}{x} + \frac{x + 3}{x^2 - 1}$$

Solution

Using the factored denominators $(x - 1)$, x, and $(x + 1)(x - 1)$, you can see that the LCD is $x(x + 1)(x - 1)$.

$$\frac{3}{x - 1} - \frac{2}{x} + \frac{x + 3}{(x + 1)(x - 1)}$$

$$= \frac{3(x)(x + 1)}{x(x + 1)(x - 1)} - \frac{2(x + 1)(x - 1)}{x(x + 1)(x - 1)} + \frac{(x + 3)(x)}{x(x + 1)(x - 1)}$$

$$= \frac{3(x)(x + 1) - 2(x + 1)(x - 1) + (x + 3)(x)}{x(x + 1)(x - 1)}$$

$$= \frac{3x^2 + 3x - 2x^2 + 2 + x^2 + 3x}{x(x + 1)(x - 1)}$$

$$= \frac{2x^2 + 6x + 2}{x(x + 1)(x - 1)}$$

$$= \frac{2(x^2 + 3x + 1)}{x(x + 1)(x - 1)}$$

Compound Fractions

Fractional expressions with separate fractions in the numerator, denominator, or both, are called **compound** or **complex fractions.** Here are two examples.

$$\frac{\left(\dfrac{1}{x}\right)}{x^2+1} \quad \text{and} \quad \frac{\left(\dfrac{1}{x}\right)}{\left(\dfrac{1}{x^2+1}\right)}$$

A compound fraction can be simplified by combining its numerator and denominator into single fractions, then inverting the denominator and multiplying.

EXAMPLE 8   *Simplifying a Compound Fraction*

$$\frac{\left(\dfrac{2}{x}-3\right)}{\left(1-\dfrac{1}{x-1}\right)} = \frac{\left[\dfrac{2-3(x)}{x}\right]}{\left[\dfrac{1(x-1)-1}{x-1}\right]} \qquad \text{Combine fractions.}$$

$$= \frac{\left(\dfrac{2-3x}{x}\right)}{\left(\dfrac{x-2}{x-1}\right)} \qquad \text{Simplify.}$$

$$= \frac{2-3x}{x} \cdot \frac{x-1}{x-2} \qquad \text{Invert and multiply.}$$

$$= \frac{(2-3x)(x-1)}{x(x-2)}, \qquad x \neq 1$$

Another way to simplify a compound fraction is to multiply each term in its numerator and denominator by the LCD of all fractions in its numerator and denominator. This method is applied to the fraction in Example 8 as follows.

$$\frac{\left(\dfrac{2}{x}-3\right)}{\left(1-\dfrac{1}{x-1}\right)} = \frac{\left(\dfrac{2}{x}-3\right)}{\left(1-\dfrac{1}{x-1}\right)} \cdot \frac{x(x-1)}{x(x-1)}$$

$$= \frac{2(x-1)-3x(x-1)}{x(x-1)-x}$$

$$= \frac{-3x^2+5x-2}{x^2-2x}$$

$$= \frac{(2-3x)(x-1)}{x(x-2)}, \qquad x \neq 1$$

The next three examples illustrate some methods for simplifying fractional expressions involving radicals and negative exponents. These types of expressions occur frequently in calculus.

EXAMPLE 9   *Simplifying an Expression with Negative Exponents*

Simplify

$$x(1 - 2x)^{-3/2} + (1 - 2x)^{-1/2}.$$

Solution

By rewriting the expression with positive exponents, you obtain

$$\frac{x}{(1 - 2x)^{3/2}} + \frac{1}{(1 - 2x)^{1/2}}$$

which can then be combined by the LCD method. However, the process can be simplified by first removing the common factor with the *smaller exponent*.

$$x(1 - 2x)^{-3/2} + (1 - 2x)^{-1/2} = (1 - 2x)^{-3/2}[x + (1 - 2x)^{(-1/2)-(-3/2)}]$$
$$= (1 - 2x)^{-3/2}[x + (1 - 2x)^{1}]$$
$$= \frac{1 - x}{(1 - 2x)^{3/2}}$$

NOTE   In Example 9, note that when factoring, you subtract exponents ■■

EXAMPLE 10   *Simplifying a Compound Fraction*

Simplify

$$\frac{(4 - x^2)^{1/2} + x^2(4 - x^2)^{-1/2}}{4 - x^2}.$$

Solution

$$\frac{(4 - x^2)^{1/2} + x^2(4 - x^2)^{-1/2}}{4 - x^2}$$
$$= \frac{(4 - x^2)^{1/2} + x^2(4 - x^2)^{-1/2}}{4 - x^2} \cdot \frac{(4 - x^2)^{1/2}}{(4 - x^2)^{1/2}}$$
$$= \frac{(4 - x^2)^{1} + x^2(4 - x^2)^{0}}{(4 - x^2)^{3/2}}$$
$$= \frac{4 - x^2 + x^2}{(4 - x^2)^{3/2}}$$
$$= \frac{4}{(4 - x^2)^{3/2}}$$

TECHNOLOGY

Some graphing utilities have a table feature that can be used to create tables of values. For instance, to evaluate the expression $x^2 - 4$ for $x = 1, 2, 3, 4, 5, 6,$ and 7 on a *TI-83,* you can use the following keystrokes.

| Y= | | CLEAR |
| X, T, θ, n | x^2 | − | 4 |
| TBLSET |

TblStart=1
ΔTbl=1
Indpnt: Auto
Depend: Auto
| TABLE |

For the *TI-82,* use | X, T, θ | instead of | X, T, θ, n | and set TblMin = 1.

The table produced by these keystrokes is shown below.

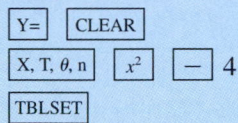

$$X=1$$

EXAMPLE 11   *Simplifying a Compound Fraction*

The expression from calculus

$$\frac{\sqrt{x + h} - \sqrt{x}}{h}$$

is an example of a *difference quotient*. Rewrite this expression by rationalizing its numerator.

Solution

$$\frac{\sqrt{x + h} - \sqrt{x}}{h} = \frac{\sqrt{x + h} - \sqrt{x}}{h} \cdot \frac{\sqrt{x + h} + \sqrt{x}}{\sqrt{x + h} + \sqrt{x}}$$

$$= \frac{\left(\sqrt{x + h}\right)^2 - \left(\sqrt{x}\right)^2}{h\left(\sqrt{x + h} + \sqrt{x}\right)}$$

$$= \frac{h}{h\left(\sqrt{x + h} + \sqrt{x}\right)}$$

$$= \frac{1}{\sqrt{x + h} + \sqrt{x}}, \qquad h \neq 0$$

Notice that the original expression is meaningless when $h = 0$, but the final expression *could* be evaluated when $h = 0$.

GROUP ACTIVITY
COMPARING DOMAINS OF TWO EXPRESSIONS

Complete the following table by evaluating the expressions

$$\frac{x^2 - 3x + 2}{x - 2} \quad \text{and} \quad x - 1$$

for the values of x. If you have a graphing utility with a *table feature*, use it to help create the table. Write a short paragraph describing the equivalence or nonequivalence of the two expressions.

x	−3	−2	−1	0	1	2	3
$\dfrac{x^2 - 3x + 2}{x - 2}$							
$x - 1$							

Completely factor the polynomial.

1. $5x^2 - 15x^3$
2. $16x^2 - 9$
3. $9x^2 - 6x + 1$
4. $9 + 12y + 4y^2$
5. $z^2 + 4z + 3$
6. $x^2 - 15x + 50$
7. $3 + 8x - 3x^2$
8. $3x^2 - 46x + 15$
9. $s^3 + s^2 - 4s - 4$
10. $y^3 + 64$

P.5 Exercises

In Exercises 1–10, find the domain of the expression.

1. $3x^2 - 4x + 7$
2. $2x^2 + 5x - 2$
3. $4x^3 + 3, \quad x \geq 0$
4. $6x^2 - 9, \quad x > 0$
5. $\dfrac{1}{x - 2}$
6. $\dfrac{x + 1}{2x + 1}$
7. $\dfrac{x - 1}{x^2 - 4x}$
8. $\dfrac{2x + 1}{x^2 - 9}$
9. $\sqrt{x + 1}$
10. $\dfrac{1}{\sqrt{x + 1}}$

In Exercises 11–16, find the missing factor in the numerator so that the two fractions will be equivalent.

11. $\dfrac{5}{2x} = \dfrac{5(\quad)}{6x^2}$
12. $\dfrac{3}{4} = \dfrac{3(\quad)}{4(x + 1)}$
13. $\dfrac{x + 1}{x} = \dfrac{(x + 1)(\quad)}{x(x - 2)}$
14. $\dfrac{3y - 4}{y + 1} = \dfrac{(3y - 4)(\quad)}{y^2 - 1}$

15. $\dfrac{3x}{x - 3} = \dfrac{3x(\quad)}{x^2 - 3x}$
16. $\dfrac{1 - z}{z^2} = \dfrac{(1 - z)(\quad)}{z^3 + z^2}$

In Exercises 17–30, write the rational expression in reduced form.

17. $\dfrac{15x^2}{10x}$
18. $\dfrac{18y^2}{60y^5}$
19. $\dfrac{3xy}{xy + x}$
20. $\dfrac{9x^2 + 9x}{2x + 2}$
21. $\dfrac{x - 5}{10 - 2x}$
22. $\dfrac{x^2 - 25}{5 - x}$
23. $\dfrac{x^3 + 5x^2 + 6x}{x^2 - 4}$
24. $\dfrac{x^2 + 8x - 20}{x^2 + 11x + 10}$
25. $\dfrac{y^2 - 7y + 12}{y^2 + 3y - 18}$
26. $\dfrac{3 - x}{x^2 + 11x + 10}$
27. $\dfrac{2 - x + 2x^2 - x^3}{x - 2}$
28. $\dfrac{x^2 - 9}{x^3 + x^2 - 9x - 9}$
29. $\dfrac{z^3 - 8}{z^2 + 2z + 4}$
30. $\dfrac{y^3 - 2y^2 - 3y}{y^3 + 1}$

In Exercises 31 and 32, complete the table. What can you conclude?

31.

x	0	1	2	3	4	5	6
$\dfrac{x^2 - 2x - 3}{x - 3}$							
$x + 1$							

32.

x	0	1	2	3	4	5	6
$\dfrac{x - 3}{x^2 - x - 6}$							
$\dfrac{1}{x + 2}$							

33. *Error Analysis* Describe the error.

$$\frac{5x^3}{2x^3 + 4} = \frac{5x^3}{2x^3 + 4} = \frac{5}{2 + 4} = \frac{5}{6}$$

34. *Think About It* Is the following statement true for all nonzero real numbers a and b? Explain.

$$\frac{ax - b}{b - ax} = -1$$

In Exercises 35 and 36, find the ratio of the area of the shaded portion of the figure to the total area of the figure.

35.

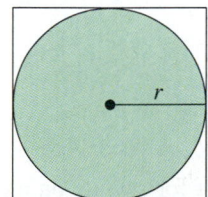

36.

In Exercises 37–50, perform the multiplication or division and simplify.

37. $\dfrac{5}{x - 1} \cdot \dfrac{x - 1}{25(x - 2)}$

38. $\dfrac{x + 13}{x^3(3 - x)} \cdot \dfrac{x(x - 3)}{5}$

39. $\dfrac{(x + 5)(x - 3)}{x + 2} \cdot \dfrac{1}{(x + 5)(x + 2)}$

40. $\dfrac{(x - 9)(x + 7)}{x + 1} \cdot \dfrac{x}{9 - x}$

41. $\dfrac{r}{r - 1} \cdot \dfrac{r^2 - 1}{r^2}$

42. $\dfrac{4y - 16}{5y + 15} \cdot \dfrac{2y + 6}{4 - y}$

43. $\dfrac{t^2 - t - 6}{t^2 + 6t + 9} \cdot \dfrac{t + 3}{t^2 - 4}$

44. $\dfrac{y^3 - 8}{2y^3} \cdot \dfrac{4y}{y^2 - 5y + 6}$

45. $\dfrac{x^2 + xy - 2y^2}{x^3 + x^2y} \cdot \dfrac{x}{x^2 + 3xy + 2y^2}$

46. $\dfrac{x^3 - 1}{x + 1} \cdot \dfrac{x^2 + 1}{x^2 - 1}$

47. $\dfrac{3(x + y)}{4} \div \dfrac{x + y}{2}$

48. $\dfrac{x + 2}{5(x - 3)} \div \dfrac{x - 2}{5(x - 3)}$

49. $\dfrac{\left[\dfrac{x^2}{(x + 1)^2}\right]}{\left[\dfrac{x}{(x + 1)^3}\right]}$

50. $\dfrac{\left(\dfrac{x^2 - 1}{x}\right)}{\left[\dfrac{(x - 1)^2}{x}\right]}$

In Exercises 51–64, perform the addition or subtraction and simplify.

51. $\dfrac{5}{x - 1} + \dfrac{x}{x - 1}$

52. $\dfrac{2x - 1}{x + 3} + \dfrac{1 - x}{x + 3}$

53. $6 - \dfrac{5}{x + 3}$

54. $\dfrac{3}{x - 1} - 5$

55. $\dfrac{3}{x - 2} + \dfrac{5}{2 - x}$

56. $\dfrac{2x}{x - 5} - \dfrac{5}{5 - x}$

57. $\dfrac{2}{x^2 - 4} - \dfrac{1}{x^2 - 3x + 2}$

58. $\dfrac{x}{x^2 + x - 2} - \dfrac{1}{x + 2}$

59. $\dfrac{1}{x^2 - x - 2} - \dfrac{x}{x^2 - 5x + 6}$

60. $\dfrac{2}{x^2 - x - 2} + \dfrac{10}{x^2 + 2x - 8}$

61. $-\dfrac{1}{x} + \dfrac{2}{x^2 + 1} + \dfrac{1}{x^3 + x}$

62. $\dfrac{2}{x + 1} + \dfrac{2}{x - 1} + \dfrac{1}{x^2 - 1}$

63. $x^2(x^2 + 1)^{-5} - (x^2 + 1)^{-4}$

64. $2x(x - 5)^{-3} - 4x^2(x - 5)^{-4}$

Error Analysis **In Exercises 65 and 66, describe the error.**

65. $\dfrac{x + 4}{x + 2} - \dfrac{3x - 8}{x + 2} = \dfrac{x + 4 - 3x - 8}{x + 2}$

$$= \dfrac{-2x - 4}{x + 2}$$

$$= \dfrac{-2(x + 2)}{x + 2} = -2$$

66. $\dfrac{6 - x}{x(x + 2)} + \dfrac{x + 2}{x^2} + \dfrac{8}{x^2(x + 2)}$

$$= \dfrac{x(6 - x) + (x + 2)^2 + 8}{x^2(x + 2)}$$

$$= \dfrac{6x - x^2 + x^2 + 4 + 8}{x^2(x + 2)}$$

$$= \dfrac{6(x + 2)}{x^2(x + 2)} = \dfrac{6}{x^2}$$

In Exercises 67–80, simplify the compound fraction.

67. $\dfrac{\left(\dfrac{x}{2} - 1\right)}{(x - 2)}$

68. $\dfrac{(x - 4)}{\left(\dfrac{x}{4} - \dfrac{4}{x}\right)}$

69. $\dfrac{\left(\dfrac{1}{x} - \dfrac{1}{x + 1}\right)}{\left(\dfrac{1}{x + 1}\right)}$

70. $\dfrac{\left(\dfrac{5}{y} - \dfrac{6}{2y + 1}\right)}{\left(\dfrac{5}{y} + 4\right)}$

71. $\dfrac{\left(\dfrac{x + 3}{x - 3}\right)^2}{\dfrac{1}{x + 3} + \dfrac{1}{x - 3}}$

72. $\dfrac{\left(\dfrac{x + 4}{x + 5} - \dfrac{x}{x + 1}\right)}{4}$

73. $\dfrac{\left[\dfrac{1}{(x + h)^2} - \dfrac{1}{x^2}\right]}{h}$

74. $\dfrac{\left(\dfrac{x + h}{x + h + 1} - \dfrac{x}{x + 1}\right)}{h}$

75. $\dfrac{\left(\sqrt{x} - \dfrac{1}{2\sqrt{x}}\right)}{\sqrt{x}}$

76. $\dfrac{3x^{1/3} - x^{-2/3}}{3x^{-2/3}}$

77. $\dfrac{\left(\dfrac{t^2}{\sqrt{t^2 + 1}} - \sqrt{t^2 + 1}\right)}{t^2}$

78. $\dfrac{-x^3(1 - x^2)^{-1/2} - 2x(1 - x^2)^{1/2}}{x^4}$

79. $\dfrac{x(x + 1)^{-3/4} - (x + 1)^{1/4}}{x^2}$

80. $\dfrac{(2x + 1)^{1/3} - \dfrac{4x}{3(2x + 1)^{2/3}}}{(2x + 1)^{2/3}}$

In Exercises 81 and 82, rationalize the numerator of the expression.

81. $\dfrac{\sqrt{x + 2} - \sqrt{x}}{2}$

82. $\dfrac{\sqrt{z - 3} - \sqrt{z}}{3}$

83. *Rate* A photocopier copies at a rate of 16 pages per minute.

  (a) Find the time required to copy one page.

  (b) Find the time required to copy x pages.

  (c) Find the time required to copy 60 pages.

84. *Rate* After working together for t hours on a common task, two workers have done fractional parts of the job equal to $t/3$ and $t/5$, respectively. What fractional part of the task has been completed?

85. *Average* Determine the average of the two real numbers $x/3$ and $2x/5$.

86. *Partition into Equal Parts* Find three real numbers that divide the real number line between $x/3$ and $3x/4$ into four equal parts.

Monthly Payment In Exercises 87 and 88, use the formula that gives the approximate annual interest rate r of a monthly installment loan:

$$r = \frac{\left[\dfrac{24(NM - P)}{N}\right]}{\left(P + \dfrac{NM}{12}\right)}$$

where N is the total number of payments, M is the monthly payment, and P is the amount financed.

87. (a) Approximate the annual interest rate for a 4-year car loan of $15,000 that has monthly payments of $400.
    (b) Simplify the expression for the annual interest rate r, and then rework part (a).

88. (a) Approximate the annual interest rate for a 5-year car loan of $18,000 that has monthly payments of $400.
    (b) Simplify the expression for the annual interest rate r, and then rework part (a).

89. *Refrigeration* When food (at room temperature) is placed in a refrigerator, the time required for the food to cool depends on the amount of food, the air circulation in the refrigerator, the original temperature of the food, and the temperature of the refrigerator. Consider the model that gives the temperature of the food that is at 75°F and is placed in a 40°F refrigerator

$$T = 10\left(\frac{4t^2 + 16t + 75}{t^2 + 4t + 10}\right)$$

where T is the temperature in degrees Fahrenheit and t is the time in hours.

(a) Complete the table.

t	0	1	2	3	4	5
T						

(b) Create a bar graph showing the temperatures at the times given in the table in part (a).

90. *Precious Metals* The costs per fine ounce of gold and silver for the years 1988 through 1992 are given in the table. (Source: U.S. Bureau of Mines)

Year	1988	1989	1990	1991	1992
Gold	$438	$383	$385	$363	$345
Silver	$6.54	$5.50	$4.82	$4.04	$3.94

Mathematical models for this data are

$$\text{Cost of gold} = \frac{5301t + 37{,}498}{19t + 100}$$

and

$$\text{Cost of silver} = \frac{237t + 4734}{176t + 1000}$$

where $t = 0$ corresponds to the year 1990.

(a) Create a table using the models to estimate the prices of each metal for the given years. Compare the estimates given by the models with the actual prices.

(b) Determine a model for the ratio of the price of gold to the price of silver. Use the model to find this ratio over the given years. Over this period of time, did the price of gold become more expensive or less expensive relative to the price of silver?

Probability In Exercises 91 and 92, consider an experiment in which a marble is tossed into a box whose base is shown in the figure. The probability that the marble will come to rest in the shaded portion of the box is equal to the ratio of the shaded area to the total area of the figure. Find the probability.

91.

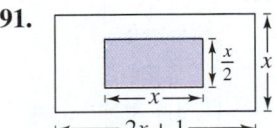

92.

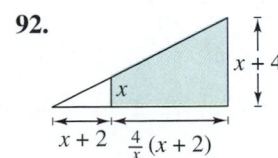

P.6 Errors and the Algebra of Calculus

See the Group Activity on page 61 for an example of how algebra is used in calculus.

Algebraic Errors to Avoid ◻ *Some Algebra of Calculus*

Algebraic Errors to Avoid

This section contains five lists of common algebraic errors: errors involving parentheses, errors involving fractions, errors involving exponents, errors involving radicals, and errors involving cancellation. Many of these errors are made because they seem to be the *easiest* things to do.

ERRORS INVOLVING PARENTHESES

Potential Error	Correct Form	Comment
$a - (x - b) \neq a - x - b$	$a - (x - b) = a - x + b$	Change all signs when distributing minus sign.
$(a + b)^2 \neq a^2 + b^2$	$(a + b)^2 = a^2 + 2ab + b^2$	Remember the middle term when squaring binomials.
$\left(\frac{1}{2}a\right)\left(\frac{1}{2}b\right) \neq \frac{1}{2}(ab)$	$\left(\frac{1}{2}a\right)\left(\frac{1}{2}b\right) = \frac{1}{4}(ab) = \frac{ab}{4}$	$\frac{1}{2}$ occurs twice as a factor.
$(3x + 6)^2 \neq 3(x + 2)^2$	$(3x + 6)^2 = [3(x + 2)]^2$ $= 3^2(x + 2)^2$	When factoring, apply exponents to all factors.

ERRORS INVOLVING FRACTIONS

Potential Error	Correct Form	Comment
$\frac{a}{x + b} \neq \frac{a}{x} + \frac{a}{b}$	Leave as $\frac{a}{x + b}$.	Do not add denominators when adding fractions.
$\frac{\left(\frac{x}{a}\right)}{b} \neq \frac{bx}{a}$	$\frac{\left(\frac{x}{a}\right)}{b} = \left(\frac{x}{a}\right)\left(\frac{1}{b}\right) = \frac{x}{ab}$	Multiply by the reciprocal when dividing fractions.
$\frac{1}{a} + \frac{1}{b} \neq \frac{1}{a + b}$	$\frac{1}{a} + \frac{1}{b} = \frac{b + a}{ab}$	Use the property for adding fractions.
$\frac{1}{3x} \neq \frac{1}{3}x$	$\frac{1}{3x} = \frac{1}{3} \cdot \frac{1}{x}$	Use the property for multiplying fractions.
$(1/3)x \neq \frac{1}{3x}$	$(1/3)x = \frac{1}{3} \cdot x = \frac{x}{3}$	Be careful when using a slash to denote division.
$(1/x) + 2 \neq \frac{1}{x + 2}$	$(1/x) + 2 = \frac{1}{x} + 2 = \frac{1 + 2x}{x}$	Be careful when using a slash to denote division.

ERRORS INVOLVING EXPONENTS

Potential Error	Correct Form	Comment
$(x^2)^3 \neq x^5$	$(x^2)^3 = x^{2 \cdot 3} = x^6$	Multiply exponents when raising a power to a power.
$x^2 \cdot x^3 \neq x^6$	$x^2 \cdot x^3 = x^{2+3} = x^5$	Add exponents when multiplying powers with like bases.
$2x^3 \neq (2x)^3$	$2x^3 = 2(x^3)$	Exponents have priority over coefficients.
$\dfrac{1}{x^2 - x^3} \neq x^{-2} - x^{-3}$	Leave as $\dfrac{1}{x^2 - x^3}$.	Do not move term-by-term from denominator to numerator.

ERRORS INVOLVING RADICALS

Potential Error	Correct Form	Comment
$\sqrt{5x} \neq 5\sqrt{x}$	$\sqrt{5x} = \sqrt{5}\sqrt{x}$	Radicals apply to every factor inside the radical.
$\sqrt{x^2 + a^2} \neq x + a$	Leave as $\sqrt{x^2 + a^2}$.	Do not apply radicals term-by-term.
$\sqrt{-x + a} \neq -\sqrt{x - a}$	Leave as $\sqrt{-x + a}$.	Do not factor minus signs out of square roots.

ERRORS INVOLVING CANCELLATION

Potential Error	Correct Form	Comment
$\dfrac{a + bx}{a} \neq 1 + bx$	$\dfrac{a + bx}{a} = \dfrac{a}{a} + \dfrac{bx}{a} = 1 + \dfrac{b}{a}x$	Cancel common factors, not common terms.
$\dfrac{a + ax}{a} \neq a + x$	$\dfrac{a + ax}{a} = \dfrac{a(1 + x)}{a} = 1 + x$	Factor before canceling.
$1 + \dfrac{x}{2x} \neq 1 + \dfrac{1}{x}$	$1 + \dfrac{x}{2x} = 1 + \dfrac{1}{2} = \dfrac{3}{2}$	Cancel common factors.

For many people, a good way to avoid errors is to *work slowly*, *write neatly*, and *talk to yourself*. Each time you write a step, ask yourself why the step is algebraically legitimate. For instance, when you write

$$\frac{2x}{6} = \frac{2 \cdot x}{2 \cdot 3}$$

$$= \frac{x}{3}$$

you can justify your work because *dividing the numerator and denominator by the same nonzero number produces an equivalent fraction.*

Some Algebra of Calculus

In calculus it is often necessary to take a simplified algebraic expression and "unsimplify" it. See the following list, taken from a standard calculus text.

UNUSUAL FACTORING

Expression	*Useful Calculus Form*	*Comment*
$\dfrac{5x^4}{8}$	$\dfrac{5}{8}x^4$	Write with fractional coefficient.
$\dfrac{x^2 + 3x}{-6}$	$-\dfrac{1}{6}(x^2 + 3x)$	Write with fractional coefficient.
$2x^2 - x - 3$	$2\left(x^2 - \dfrac{x}{2} - \dfrac{3}{2}\right)$	Factor out the leading coefficient.
$\dfrac{x}{2}(x + 1)^{-1/2} + (x + 1)^{1/2}$	$\dfrac{(x + 1)^{-1/2}}{2}[x + 2(x + 1)]$	Factor out factor with least power.

INSERTING FACTORS AND TERMS

Expression	*Useful Calculus Form*	*Comment*
$(2x - 1)^3$	$\dfrac{1}{2}(2x - 1)^3(2)$	Multiply and divide by 2.
$7x^2(4x^3 - 5)^{1/2}$	$\dfrac{7}{12}(4x^3 - 5)^{1/2}(12x^2)$	Multiply and divide by 12.
$\dfrac{4x^2}{9} - 4y^2 = 1$	$\dfrac{x^2}{9/4} - \dfrac{y^2}{1/4} = 1$	Write with fractional denominators.
$\dfrac{x}{x + 1}$	$\dfrac{x + 1 - 1}{x + 1} = 1 - \dfrac{1}{x + 1}$	Add and subtract the same term.

WRITING WITH NEGATIVE EXPONENTS

Expression	*Useful Calculus Form*	*Comment*
$\dfrac{9}{5x^3}$	$\dfrac{9}{5}x^{-3}$	Move the factor to the numerator and change the sign of the exponent.
$\dfrac{7}{\sqrt{2x - 3}}$	$7(2x - 3)^{-1/2}$	Move the factor to the numerator and change the sign of the exponent.

WRITING A FRACTION AS A SUM

Expression	Useful Calculus Form	Comment
$\dfrac{x + 2x^2 + 1}{\sqrt{x}}$	$x^{1/2} + 2x^{3/2} + x^{-1/2}$	Divide each term by $x^{1/2}$.
$\dfrac{1 + x}{x^2 + 1}$	$\dfrac{1}{x^2 + 1} + \dfrac{x}{x^2 + 1}$	Rewrite the fraction as the sum of fractions.
$\dfrac{2x}{x^2 + 2x + 1}$	$\dfrac{2x + 2 - 2}{x^2 + 2x + 1}$	Add and subtract the same term.
	$= \dfrac{2x + 2}{x^2 + 2x + 1} - \dfrac{2}{(x + 1)^2}$	Rewrite the fraction as the difference of fractions.
$\dfrac{x^2 - 2}{x + 1}$	$x - 1 - \dfrac{1}{x + 1}$	Use long division. (See Section 3.3.)
$\dfrac{x + 7}{x^2 - x - 6}$	$\dfrac{2}{x - 3} - \dfrac{1}{x + 2}$	Use the method of partial fractions. (See Section 4.3.)

The next four examples demonstrate many of the steps in the preceding lists.

EXAMPLE 1 *Factors Involving Negative Exponents*

Factor $x(x + 1)^{-1/2} + (x + 1)^{1/2}$.

Solution

When multiplying factors with like bases, you add exponents. When factoring, you are undoing multiplication, and so you *subtract* exponents.

$$x(x + 1)^{-1/2} + (x + 1)^{1/2} = (x + 1)^{-1/2}[x(x + 1)^0 + (x + 1)^1]$$
$$= (x + 1)^{-1/2}[x + (x + 1)]$$
$$= (x + 1)^{-1/2}(2x + 1)$$

Here is another way to factor the expression in Example 1.

$$x(x + 1)^{-1/2} + (x + 1)^{1/2} = x(x + 1)^{-1/2} + (x + 1)^{1/2} \cdot \frac{(x + 1)^{1/2}}{(x + 1)^{1/2}}$$
$$= \frac{x(x + 1)^0 + (x + 1)^1}{(x + 1)^{1/2}}$$
$$= \frac{2x + 1}{\sqrt{x + 1}}$$

EXAMPLE 2   *Rewriting Fractions*

Explain the following.

$$\frac{4x^2}{9} - 4y^2 = \frac{x^2}{9/4} - \frac{y^2}{1/4}$$

Solution

To write the expression on the left side of the equation in the form given on the right, multiply the numerators and denominators of both terms by $\frac{1}{4}$.

$$\frac{4x^2}{9} - 4y^2 = \frac{4x^2}{9}\left(\frac{1/4}{1/4}\right) - 4y^2\left(\frac{1/4}{1/4}\right)$$

$$= \frac{x^2}{9/4} - \frac{y^2}{1/4}$$

EXAMPLE 3   *Rewriting with Negative Exponents*

Rewrite the expression using negative exponents.

$$\frac{2}{5x^3} - \frac{1}{\sqrt{x}} + \frac{3}{5(4x)^2}$$

Solution

Begin by writing the second term in exponential form.

$$\frac{2}{5x^3} - \frac{1}{\sqrt{x}} + \frac{3}{5(4x)^2} = \frac{2}{5x^3} - \frac{1}{x^{1/2}} + \frac{3}{5(4x)^2}$$

$$= \frac{2}{5}x^{-3} - x^{-1/2} + \frac{3}{5}(4x)^{-2}$$

EXAMPLE 4   *Writing a Fraction as a Sum of Terms*

Rewrite the fraction as the sum of three terms.

$$\frac{x + 2x^2 + 1}{\sqrt{x}}$$

Solution

$$\frac{x + 2x^2 + 1}{\sqrt{x}} = \frac{x}{x^{1/2}} + \frac{2x^2}{x^{1/2}} + \frac{1}{x^{1/2}}$$

$$= x^{1/2} + 2x^{3/2} + x^{-1/2}$$

GROUP ACTIVITY

ALGEBRA AND CALCULUS

Suppose you are taking a course in calculus, and for one of the homework problems you obtain the following answer.

$$\frac{1}{10}(2x - 1)^{5/2} + \frac{1}{6}(2x - 1)^{3/2}$$

The answer in the back of the book is given as follows.

$$\frac{1}{15}(2x - 1)^{3/2}(3x + 1)$$

Are these two answers equivalent? If so, show how the second answer can be obtained from the first. Then, use the same technique to simplify the following expressions.

a. $\frac{2}{3}x(2x - 3)^{3/2} - \frac{2}{15}(2x - 3)^{5/2}$

b. $\frac{2}{3}x(4 + x)^{3/2} - \frac{2}{15}(4 + x)^{5/2}$

WARM UP

Factor the expression.

1. $a^3 - 16a$

2. $u^3 + 125v^3$

3. $2 + 5x - 12x^2$

4. $z^3 + 3z^2 - 4z - 12$

Perform the operations and simplify.

5. $\dfrac{8 - z}{4z^3} \cdot \dfrac{8z}{z - 8}$

6. $\dfrac{x^2 - y^2}{2x^2 - 8x} \div \dfrac{(x - y)^2}{2xy}$

7. $\dfrac{1}{x} - \dfrac{3}{y} + \dfrac{3x - y}{xy}$

8. $\dfrac{5}{x - 2} - \dfrac{4}{2 - x}$

9. $\dfrac{\left(16 - \dfrac{1}{x^2}\right)}{\left(\dfrac{1}{4x^2} - 4\right)}$

10. $\dfrac{\left(\dfrac{1}{2 + h} - \dfrac{1}{2}\right)}{h}$

P.6 Exercises

In Exercises 1–24, find and correct any errors.

1. $2x - (3y + 4) = 2x - 3y + 4$

2. $\dfrac{4}{16x - (2x + 1)} = \dfrac{4}{14x + 1}$

3. $5z + 3(x - 2) = 5z + 3x - 2$

4. $\dfrac{x - 1}{(5 - x)(-x)} = \dfrac{1 - x}{x(5 - x)}$

5. $-\dfrac{x - 3}{x - 1} = \dfrac{3 - x}{1 - x}$

6. $x(yz) = (xy)(xz)$

7. $a\left(\dfrac{x}{y}\right) = \dfrac{ax}{ay}$

8. $(5z)(6z) = 30z$

9. $(4x)^2 = 4x^2$

10. $\left(\dfrac{x}{y}\right)^3 = \dfrac{x^3}{y}$

11. $\sqrt{x + 9} = \sqrt{x} + 3$

12. $\sqrt{25 - x^2} = 5 - x$

13. $\dfrac{6x + y}{6x - y} = \dfrac{x + y}{x - y}$

14. $\dfrac{2x^2 + 1}{5x} = \dfrac{2x + 1}{5}$

15. $\dfrac{1}{x + y^{-1}} = \dfrac{y}{x + 1}$

16. $\dfrac{1}{a^{-1} + b^{-1}} = \left(\dfrac{1}{a + b}\right)^{-1}$

17. $x(2x - 1)^2 = (2x^2 - x)^2$

18. $x(x + 5)^{1/2} = (x^2 + 5x)^{1/2}$

19. $\sqrt[3]{x^3 + 7x^2} = x^2\sqrt[3]{x + 7}$

20. $(3x^2 - 6x)^3 = 3x(x - 2)^3$

21. $\dfrac{3}{x} + \dfrac{4}{y} = \dfrac{7}{x + y}$

22. $\dfrac{7 + 5(x + 3)}{x + 3} = 12$

23. $\dfrac{1}{2y} = (1/2)y$

24. $\dfrac{2x + 3x^2}{4x} = \dfrac{2 + 3x^2}{4}$

In Exercises 25–52, insert the required factor in the parentheses.

25. $\dfrac{3x + 2}{5} = \dfrac{1}{5}(\ \blacksquare\ )$

26. $\dfrac{7x^2}{10} = \dfrac{7}{10}(\ \blacksquare\ )$

27. $\dfrac{2}{3}x^2 + \dfrac{1}{3}x + 5 = \dfrac{1}{3}(\ \blacksquare\ )$

28. $\dfrac{3}{4}x + \dfrac{1}{2} = \dfrac{1}{4}(\ \blacksquare\ )$

29. $\dfrac{1}{3}x^3 + 5 = (\ \blacksquare\ )(x^3 + 15)$

30. $\dfrac{5}{2}z^2 - \dfrac{1}{4}z + 2 = (\ \blacksquare\ )(10z^2 - z + 8)$

31. $x(2x^2 + 15) = (\ \blacksquare\ )(2x^2 + 15)(2x)$

32. $x^2(x^3 - 1)^4 = (\ \blacksquare\ )(x^3 - 1)^4(3x^2)$

33. $x(1 - 2x^2)^3 = (\ \blacksquare\ )(1 - 2x^2)^3(-4x)$

34. $5x\sqrt[3]{1 + x^2} = (\ \blacksquare\ )\sqrt[3]{1 + x^2}(2x)$

35. $\dfrac{1}{\sqrt{x}(1 + \sqrt{x})^2} = (\ \blacksquare\ )\dfrac{1}{(1 + \sqrt{x})^2}\left(\dfrac{1}{2\sqrt{x}}\right)$

36. $\dfrac{4x + 6}{(x^2 + 3x + 7)^3} = (\ \blacksquare\ )\dfrac{1}{(x^2 + 3x + 7)^3}(2x + 3)$

37. $\dfrac{x + 1}{(x^2 + 2x - 3)^2} = (\ \blacksquare\ )\dfrac{1}{(x^2 + 2x - 3)^2}(2x + 2)$

38. $\dfrac{1}{(x - 1)\sqrt{(x - 1)^4 - 4}} = \dfrac{(\ \blacksquare\ )}{(x - 1)^2\sqrt{(x - 1)^4 - 4}}$

39. $\dfrac{3}{x} + \dfrac{5}{2x^2} - \dfrac{3}{2}x = (\ \blacksquare\ )(6x + 5 - 3x^3)$

40. $\dfrac{(x - 1)^2}{169} + (y + 5)^2 = \dfrac{(x - 1)^3}{169(\ \blacksquare\ )} + (y + 5)^2$

41. $\dfrac{9x^2}{25} + \dfrac{16y^2}{49} = \dfrac{x^2}{(\ \blacksquare\ )} + \dfrac{y^2}{(\ \blacksquare\ )}$

42. $\dfrac{3x^2}{4} - \dfrac{9y^2}{16} = \dfrac{x^2}{(\ \blacksquare\ )} - \dfrac{y^2}{(\ \blacksquare\ )}$

43. $\dfrac{x^2}{1/12} - \dfrac{y^2}{2/3} = \dfrac{12x^2}{(\ \blacksquare\ )} - \dfrac{3y^2}{(\ \blacksquare\ )}$

44. $\dfrac{x^2}{4/9} + \dfrac{y^2}{7/8} = \dfrac{9x^2}{(\ \blacksquare\ )} + \dfrac{8y^2}{(\ \blacksquare\ )}$

45. $\sqrt{x} + (\sqrt{x})^3 = \sqrt{x}(\ \blacksquare\ )$

46. $x^{1/3} - 5x^{4/3} = x^{1/3}(\ \blacksquare\ )$

47. $3(2x + 1)x^{1/2} + 4x^{3/2} = x^{1/2}(\ \blacksquare\ )$

48. $(1 - 3x)^{4/3} - 4x(1 - 3x)^{1/3} = (1 - 3x)^{1/3}(\ \blacksquare\ )$

49. $\dfrac{x^2}{\sqrt{x^2 + 1}} - \sqrt{x^2 + 1} = \dfrac{1}{\sqrt{x^2 + 1}}(\ \blacksquare\ )$

50. $\dfrac{1}{2\sqrt{x}} + 5x^{3/2} - 10x^{5/2} = \dfrac{1}{2\sqrt{x}}(\ \blacksquare\ )$

51. $\dfrac{1}{10}(2x + 1)^{5/2} - \dfrac{1}{6}(2x + 1)^{3/2} = \dfrac{(2x + 1)^{3/2}}{15}(\ \blacksquare\ )$

52. $\dfrac{3}{7}(t + 1)^{7/3} - \dfrac{3}{4}(t + 1)^{4/3} = \dfrac{3(t + 1)^{4/3}}{28}(\ \blacksquare\ )$

In Exercises 53–58, write the fraction as a sum of two or more terms.

53. $\dfrac{16 - 5x - x^2}{x}$

54. $\dfrac{x^3 - 5x^2 + 4}{x^2}$

55. $\dfrac{4x^3 - 7x^2 + 1}{x^{1/3}}$

56. $\dfrac{2x^5 - 3x^3 + 5x - 1}{x^{3/2}}$

57. $\dfrac{3 - 5x^2 - x^4}{\sqrt{x}}$

58. $\dfrac{x^3 - 5x^4}{3x^2}$

In Exercises 59–66, simplify the expression.

59. $\dfrac{-2(x^2 - 3)^{-3}(2x)(x + 1)^3 - 3(x + 1)^2(x^2 - 3)^{-2}}{[(x + 1)^3]^2}$

60. $\dfrac{x^5(-3)(x^2 + 1)^{-4}(2x) - (x^2 + 1)^{-3}(5)x^4}{(x^5)^2}$

61. $\dfrac{(6x + 1)^3(27x^2 + 2) - (9x^3 + 2x)(3)(6x + 1)^2(6)}{[(6x + 1)^3]^2}$

62. $\dfrac{(4x^2 + 9)^{1/2}(2) - (2x + 3)(\frac{1}{2})(4x^2 + 9)^{-1/2}(8x)}{[(4x^2 + 9)^{1/2}]^2}$

63. $\dfrac{(x + 2)^{3/4}(x + 3)^{-2/3} - (x + 3)^{1/3}(x + 2)^{-1/4}}{[(x + 2)^{3/4}]^2}$

64. $\dfrac{\sqrt{2x - 1} - \dfrac{x + 2}{\sqrt{2x - 1}}}{2x - 1}$

65. $\dfrac{2(3x - 1)^{1/3} - (2x + 1)(\frac{1}{3})(3x - 1)^{-2/3}(3)}{(3x - 1)^{2/3}}$

66. $\dfrac{(x + 1)(\frac{1}{2})(2x - 3x^2)^{-1/2}(2 - 6x) - (2x - 3x^2)^{1/2}}{(x + 1)^2}$

67. (a) Verify that $y_1 = y_2$ analytically.

$$y_1 = x^2(\tfrac{1}{3})(x^2 + 1)^{-2/3}(2x) + (x^2 + 1)^{1/3}(2x)$$

$$y_2 = \frac{2x(4x^2 + 3)}{3(x^2 + 1)^{2/3}}$$

(b) Complete the table and demonstrate the equality of part (a) numerically.

x	-2	-1	$-\frac{1}{2}$	0	1	2	$\frac{5}{2}$
y_1							
y_2							

68. (a) Verify that $y_1 = y_2$ analytically.

$$y_1 = -\frac{\sqrt{9 - x^2}}{x^2} - \frac{1}{\sqrt{9 - x^2}}$$

$$y_2 = \frac{-9}{x^2\sqrt{9 - x^2}}$$

(b) Complete the table and demonstrate the equality of part (a) numerically.

x	-2	-1	$-\frac{1}{2}$	$\frac{1}{4}$	1	2	$\frac{5}{2}$
y_1							
y_2							

69. *Logical Reasoning* Verify that $y_1 \neq y_2$ by letting $x = 0$ and evaluating y_1 and y_2 where

$$y_1 = 2x\sqrt{1 - x^2} - \frac{x^3}{\sqrt{1 - x^2}}$$

and

$$y_2 = \frac{2 - 3x^2}{\sqrt{1 - x^2}}.$$

Change y_2 so that $y_1 = y_2$.

P.7   *Graphical Representation of Data*

See Example 2 on page 65 for an example of how to represent real-life data graphically.

The Cartesian Plane   □   *The Distance Formula*   □   *The Midpoint Formula*   □   *Application*

The Cartesian Plane

Just as you can represent real numbers by points on a real number line, you can represent ordered pairs of real numbers by points in a plane called the **rectangular coordinate system,** or the **Cartesian plane,** after the French mathematician René Descartes (1596–1650).

The Cartesian plane is formed by using two real lines intersecting at right angles, as shown in Figure P.7. The horizontal real line is usually called the *x***-axis,** and the vertical real line is usually called the *y***-axis.** The point of intersection of these two axes is the **origin,** and the two axes divide the plane into four parts called **quadrants.**

Each point in the plane corresponds to an **ordered pair** (x, y) of real numbers x and y, called **coordinates** of the point. The *x***-coordinate** represents the directed distance from the y-axis to the point, and the *y***-coordinate** represents the directed distance from the x-axis to the point, as shown in Figure P.8.

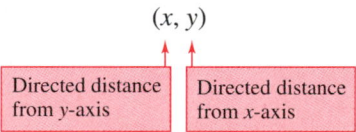

NOTE   The notation (x, y) denotes both a point in the plane and an open interval on the real line. The context will tell you which meaning is intended.
■■

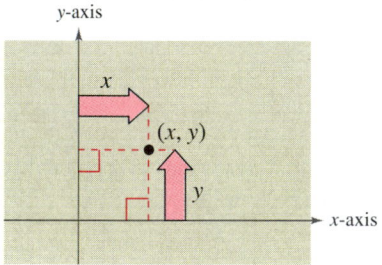

FIGURE P.7

FIGURE P.8

EXAMPLE 1   *Plotting Points in the Cartesian Plane*

Plot the points $(-1, 2)$, $(3, 4)$, $(0, 0)$, $(3, 0)$, and $(-2, -3)$.

Solution

To plot the point

$$(-1, 2),$$

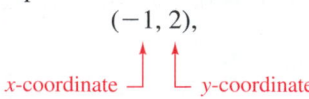

imagine a vertical line through -1 on the x-axis and a horizontal line through 2 on the y-axis. The intersection of these two lines is the point $(-1, 2)$. The other four points can be plotted in a similar way, and are shown in Figure P.9.

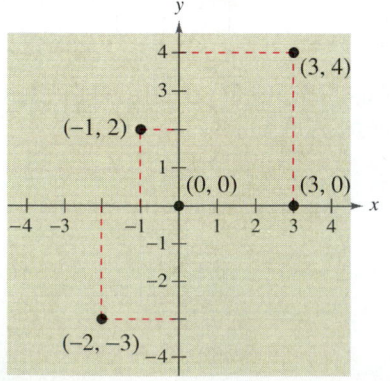

FIGURE P.9

Amount Spent on Fishing Tackle

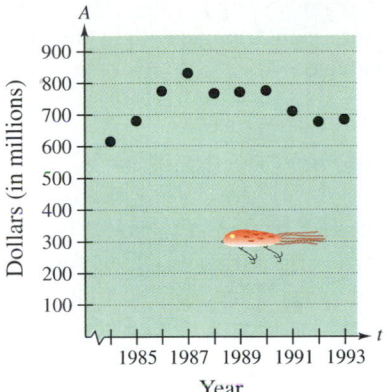

FIGURE P.10

The beauty of a rectangular coordinate system is that it allows you to see relationships between two variables. It would be difficult to overestimate the importance of Descartes's introduction of coordinates to the plane. Today, his ideas are in common use in virtually every scientific and business-related field.

Real Life

EXAMPLE 2 **Sketching a Scatter Plot**

From 1984 through 1993, the amount A (in millions of dollars) spent on fishing tackle in the United States is given in the table below, where t represents the year. Sketch a scatter plot of the data. (Source: National Sporting Goods Association)

t	1984	1985	1986	1987	1988	1989	1990	1991	1992	1993
A	616	681	773	830	766	769	776	711	678	685

Solution

To sketch a *scatter plot* of the data given in the table, you simply represent each pair of values by an ordered pair (t, A) and plot the resulting points, as shown in Figure P.10. For instance, the first pair of values is represented by the ordered pair (1984, 616). Note that the break in the t-axis indicates that the numbers between 0 and 1984 have been omitted.

NOTE In Example 2, you could have let $t = 1$ represent the year 1984. In that case, the horizontal axis would not have been broken, and the tick marks would have been labeled 1 through 10 (instead of 1984 through 1993). ■■

TECHNOLOGY

The scatter plot in Example 2 is only one way to represent the data graphically. Two other techniques are shown at the right. The first is a *bar graph* and the second is a *line graph*. All three graphical representations were created with a computer. If you have access to a graphing utility, try using it to represent graphically the data given in Example 2.

Amount Spent on Fishing Tackle

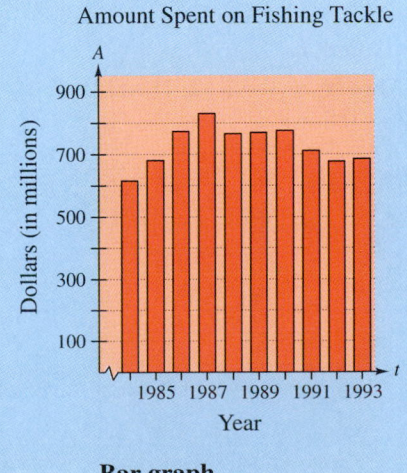

Bar graph

Amount Spent on Fishing Tackle

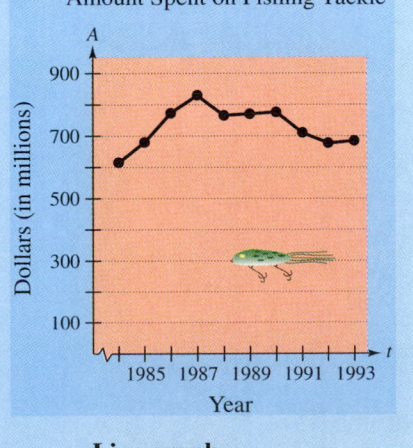

Line graph

The Distance Formula

Recall from the Pythagorean Theorem that, for a right triangle with hypotenuse of length c and sides of lengths a and b, you have

$$a^2 + b^2 = c^2 \qquad \text{Pythagorean Theorem}$$

as shown in Figure P.11. (The converse is also true. That is, if $a^2 + b^2 = c^2$, then the triangle is a right triangle.)

Suppose you want to determine the distance d between two points (x_1, y_1) and (x_2, y_2) in the plane. With these two points, a right triangle can be formed, as shown in Figure P.12. The length of the vertical side of the triangle is $|y_2 - y_1|$, and the length of the horizontal side is $|x_2 - x_1|$. By the Pythagorean Theorem, you can write

$$d^2 = |x_2 - x_1|^2 + |y_2 - y_1|^2$$
$$d = \sqrt{|x_2 - x_1|^2 + |y_2 - y_1|^2}$$
$$d = \sqrt{(x_2 - x_1)^2 + (y_2 - y_1)^2}.$$

This result is the **Distance Formula.**

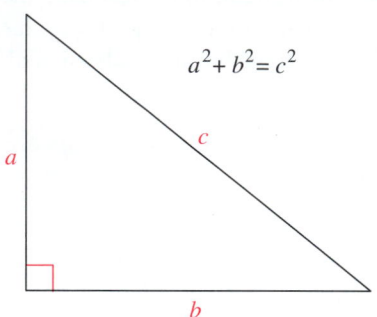

FIGURE P.11

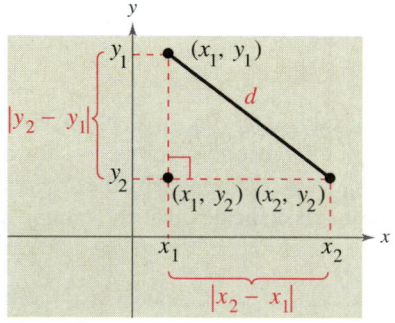

FIGURE P.12

    THE DISTANCE FORMULA

The distance d between the points (x_1, y_1) and (x_2, y_2) in the plane is

$$d = \sqrt{(x_2 - x_1)^2 + (y_2 - y_1)^2}.$$

EXAMPLE 3 *Finding a Distance*

Find the distance between the points $(-2, 1)$ and $(3, 4)$.

Solution

Let $(x_1, y_1) = (-2, 1)$ and $(x_2, y_2) = (3, 4)$. Then apply the Distance Formula as follows.

$$d = \sqrt{(x_2 - x_1)^2 + (y_2 - y_1)^2} \qquad \text{Distance Formula}$$
$$= \sqrt{[3 - (-2)]^2 + (4 - 1)^2} \qquad \text{Substitute for } x_1, y_1, x_2, \text{ and } y_2.$$
$$= \sqrt{(5)^2 + (3)^2} \qquad \text{Simplify.}$$
$$= \sqrt{34}$$
$$\approx 5.83 \qquad \text{Use a calculator.}$$

Note in Figure P.13 that a distance of 5.83 looks about right.

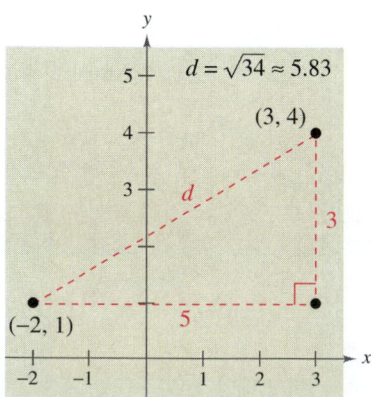

FIGURE P.13

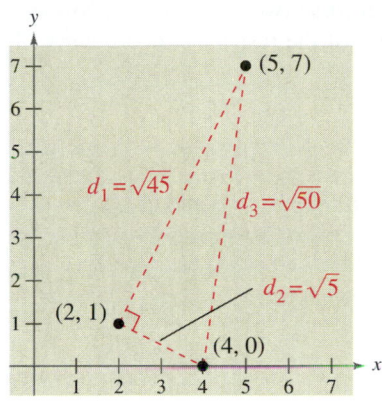

FIGURE P.14

■

EXAMPLE 4 *Verifying a Right Triangle*

Show that the points $(2, 1)$, $(4, 0)$, and $(5, 7)$ are vertices of a right triangle.

Solution

The three points are plotted in Figure P.14. Using the Distance Formula, you can find the lengths of the three sides as follows.

$$d_1 = \sqrt{(5 - 2)^2 + (7 - 1)^2} = \sqrt{9 + 36} = \sqrt{45}$$
$$d_2 = \sqrt{(4 - 2)^2 + (0 - 1)^2} = \sqrt{4 + 1} = \sqrt{5}$$
$$d_3 = \sqrt{(5 - 4)^2 + (7 - 0)^2} = \sqrt{1 + 49} = \sqrt{50}$$

Because

$$d_1{}^2 + d_2{}^2 = 45 + 5 = 50 = d_3{}^2,$$

you can conclude that the triangle must be a right triangle.

■

The figures provided with Examples 3 and 4 were not really essential to the solution. *Nevertheless,* we strongly recommend that you develop the habit of including sketches with your solutions—even if they are not required.

■

EXAMPLE 5 *Finding the Length of a Pass*

A football quarterback throws a pass from the 5-yard line, 20 yards from the sideline. The pass is caught by a wide receiver on the 45-yard line, 50 yards from the same sideline, as shown in Figure P.15. How long is the pass?

Solution

You can find the length of the pass by finding the distance between the points $(20, 5)$ and $(50, 45)$.

$$d = \sqrt{(50 - 20)^2 + (45 - 5)^2} \qquad \text{Distance Formula}$$
$$= \sqrt{900 + 1600}$$
$$= 50 \qquad \text{Simplify.}$$

Thus, the pass is 50 yards long.

■

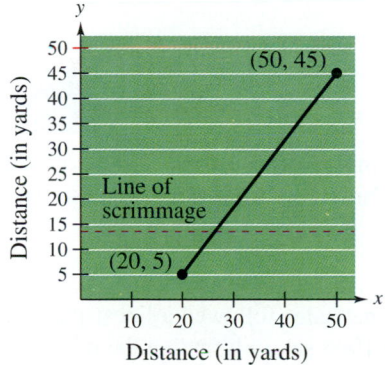

FIGURE P.15

NOTE In Example 5, the scale along the goal line does not normally appear on a football field. However, when you use coordinate geometry to solve real-life problems, you are free to place the coordinate system in any way that is convenient to the solution of the problem. ■ ■

THINK ABOUT
THE PROOF

The Distance Formula can be used to prove the Midpoint Formula. Can you see how to do it? The details of the proof are listed in the appendix.

The Midpoint Formula

To find the **midpoint** of the line segment that joins two points in a coordinate plane, you can simply find the average values of the respective coordinates of the two endpoints.

> **THE MIDPOINT FORMULA**
>
> The midpoint of the segment joining the points (x_1, y_1) and (x_2, y_2) is
>
> $$\text{Midpoint} = \left(\frac{x_1 + x_2}{2}, \frac{y_1 + y_2}{2} \right).$$

EXAMPLE 6   Finding a Segment's Midpoint

Find the midpoint of the line segment joining the points $(-5, -3)$ and $(9, 3)$, as shown in Figure P.16.

Solution

Let $(x_1, y_1) = (-5, -3)$ and $(x_2, y_2) = (9, 3)$.

$$\text{Midpoint} = \left(\frac{x_1 + x_2}{2}, \frac{y_1 + y_2}{2} \right) \qquad \text{Midpoint Formula}$$

$$= \left(\frac{-5 + 9}{2}, \frac{-3 + 3}{2} \right) \qquad \text{Substitute for } x_1, y_1, x_2, \text{ and } y_2.$$

$$= (2, 0) \qquad \text{Simplify.}$$

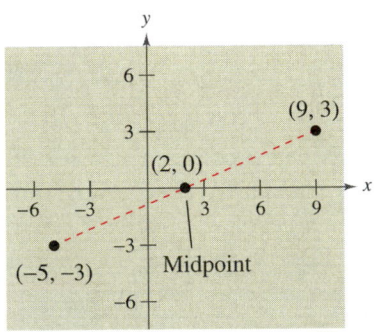

FIGURE P.16

EXAMPLE 7   Estimating Annual Sales

Real
Life

Ben and Jerry's had annual sales of $132.0 million in 1992 and $148.8 million in 1994. Without knowing any additional information, what would you estimate the 1993 sales to have been?   (Source: Ben and Jerry's, Inc.)

Solution

One solution to the problem is to assume that sales followed a linear pattern. With this assumption, you can estimate the 1993 sales by finding the midpoint of the segment connecting the points (1992, 132.0) and (1994, 148.8).

$$\text{Midpoint} = \left(\frac{1992 + 1994}{2}, \frac{132.0 + 148.8}{2} \right) = (1993, 140.4)$$

Hence, you would estimate the 1993 sales to have been about $140.4 million, as shown in Figure P.17. (The actual 1993 sales were $140.3 million.)

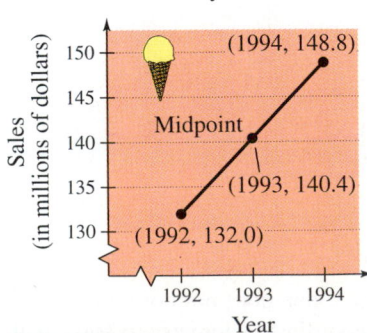

FIGURE P.17

Application

EXAMPLE 8   Translating Points in the Plane

The triangle in Figure P.18(a) has vertices at the points $(-1, 2)$, $(1, -4)$, and $(2, 3)$. Shift the triangle three units to the right and two units up and find the vertices of the shifted triangle, as shown in Figure P.18(b).

(a) 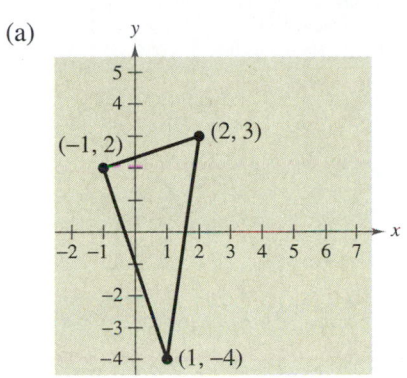   (b)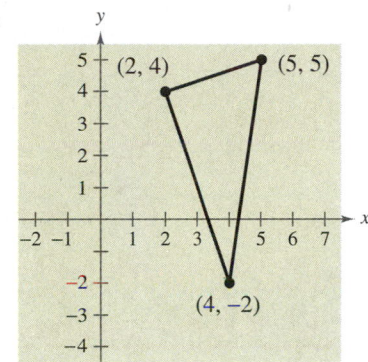

FIGURE P.18

Solution

To shift the vertices three units to the right, add 3 to each x-coordinate. To shift the vertices two units up, add 2 to each of the y-coordinates.

Original Point	Translated Point
$(-1, 2)$	$(-1 + 3, 2 + 2) = (2, 4)$
$(1, -4)$	$(1 + 3, -4 + 2) = (4, -2)$
$(2, 3)$	$(2 + 3, 3 + 2) = (5, 5)$

Much of computer graphics, including this computer-generated goldfish tesselation, consists of transformations of points in a coordinate plane. One type of transformation, a translation, is illustrated in Example 8. Other types include reflections, rotations, and stretches. *(Photo: Paul Morrell)*

GROUP ACTIVITY

EXTENDING THE EXAMPLE

Example 8 shows how to translate points in a coordinate plane. How are the following transformed points related to the original points?

Original Point	Transformed Point
(x, y)	$(-x, y)$
(x, y)	$(x, -y)$
(x, y)	$(-x, -y)$

WARM UP

WARM UP

1. Find the distance between the real numbers -3.5 and 8.
2. Find the distance between the real numbers -20 and -7.

Simplify the expression.

3. $\dfrac{4 + (-2)}{2}$

4. $\dfrac{-1 + (-3)}{2}$

5. $\dfrac{4.2 + 10.5}{2}$

6. $\dfrac{-5.4 - 3.2}{2}$

7. $\sqrt{(2 - 6)^2 + [1 - (-2)]^2}$

8. $\sqrt{(1 - 4)^2 + (-2 - 1)^2}$

9. $\sqrt{18} + \sqrt{45}$

10. $\sqrt{12} + \sqrt{44}$

P.7 Exercises

In Exercises 1–4, sketch the polygon with the indicated vertices.

1. Triangle: $(-1, 1)$, $(2, -1)$, $(3, 4)$
2. Triangle: $(0, 3)$, $(-1, -2)$, $(4, 8)$
3. Square: $(2, 4)$, $(5, 1)$, $(2, -2)$, $(-1, 1)$
4. Parallelogram: $(5, 2)$, $(7, 0)$, $(1, -2)$, $(-1, 0)$

In Exercises 5 and 6, approximate the coordinates of the points.

5.
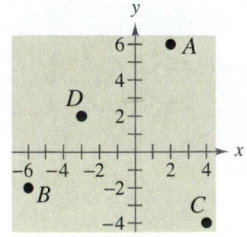

6.
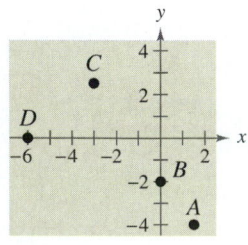

In Exercises 7–10, find the coordinates of the point.

7. The point is located 3 units to the left of the y-axis and 4 units above the x-axis.

8. The point is located 8 units below the x-axis and 4 units to the right of the y-axis.

9. The point is located 5 units below the x-axis and the coordinates of the point are equal.

10. The point is on the x-axis and 12 units to the left of the y-axis.

11. ***Think About It*** What is the y-coordinate of any point on the x-axis? What is the x-coordinate of any point on the y-axis?

12. ***Think About It*** When plotting points on the rectangular coordinate system, is it true that the scales on the x and y axes must be the same? Explain.

In Exercises 13–22, determine the quadrant(s) in which (x, y) is located so that the condition(s) is (are) satisfied.

13. $x > 0$ and $y < 0$

14. $x < 0$ and $y < 0$

15. $x = -4$ and $y > 0$

16. $x > 2$ and $y = 3$

17. $y < -5$

18. $x > 4$

19. $(x, -y)$ is in the second quadrant.

20. $(-x, y)$ is in the fourth quadrant.

21. $xy > 0$

22. $xy < 0$

In Exercises 23 and 24, the polygon is shifted to a new position in the plane. Find the coordinates of the vertices of the polygon in its *new* position.

23.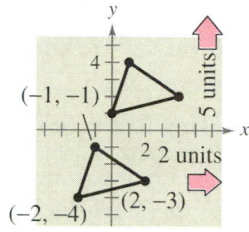

24.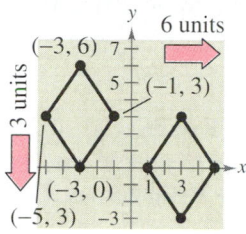

In Exercises 25 and 26, sketch a scatter plot of the data given in the table.

25. *Normal Temperatures*   The normal temperature y (in degrees Fahrenheit) in Duluth, Minnesota, for each month x, where $x = 1$ represents January, is given in the table.   (Source: NCAA)

x	1	2	3	4	5	6
y	6	12	23	38	50	59

x	7	8	9	10	11	12
y	65	63	54	44	28	14

26. *Wal-Mart*   The number y of Wal-Mart stores for each year x from 1985 through 1994 is given in the table.   (Source: Wal-Mart Annual Report for 1994)

x	1985	1986	1987	1988	1989
y	745	859	980	1114	1259

x	1990	1991	1992	1993	1994
y	1399	1568	1714	1850	1953

In Exercises 27 and 28, make a table of values for $x = -2, -1, -\frac{1}{2}, 0, \frac{1}{2}, 1$, and 2. Then plot the points on a rectangular coordinate system.

27. $y = 2 - \frac{1}{2}x$ 28. $y = 2 - \frac{1}{2}x^2$

Milk Prices   In Exercises 29 and 30, refer to the figure, which shows the average price paid to farmers for milk.   (Source: U.S. Department of Agriculture and the National Milk Producers Federation)

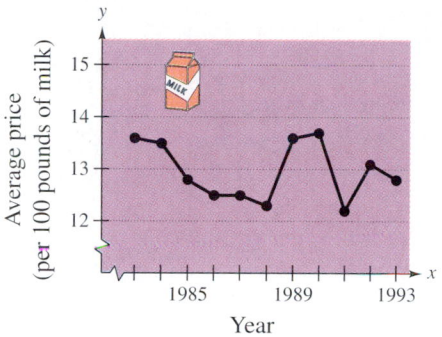

29. Approximate the highest price of milk shown in the graph. When did this occur?

30. Approximate the percent drop in the price of milk from the highest price shown in the graph to the price paid to farmers in January, 1993.

TV Advertising   In Exercises 31 and 32, refer to the figure.   (Source: Nielson Media Research)

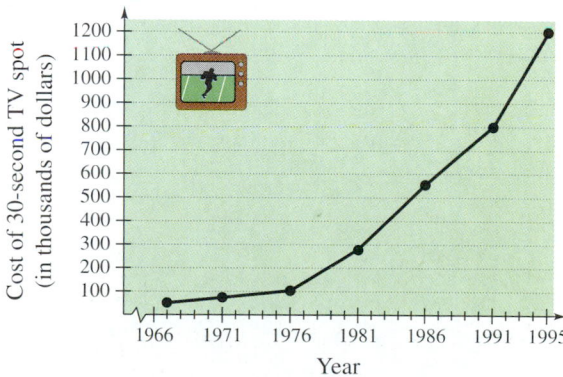

31. Approximate the percent increase in the cost of a 30-second spot from Super Bowl I in 1967 to Super Bowl XXIX in 1995.

32. Estimate the increase in cost of a 30-second spot (a) from Super Bowl V to Super Bowl XV, and (b) from Super Bowl XV to Super Bowl XXV.

Minimum Wage **In Exercises 33 and 34, refer to the figure.** (Source: U.S. Department of Labor)

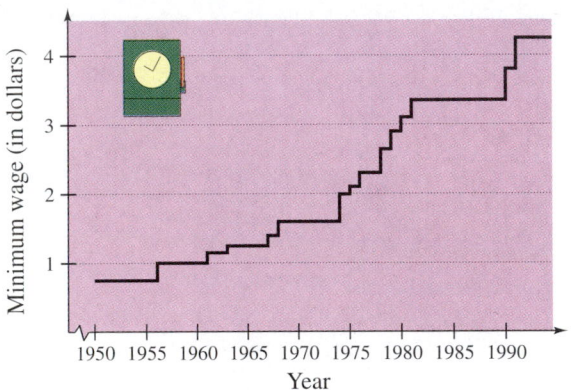

33. During which decade did the minimum wage increase most rapidly?

34. Approximate the percent increase in the minimum wage from 1990 to 1994.

Data Analysis **In Exercises 35 and 36, refer to the figure, which shows the mathematics entrance test scores x, and the final examination scores y, in an algebra course for a sample of 10 students.**

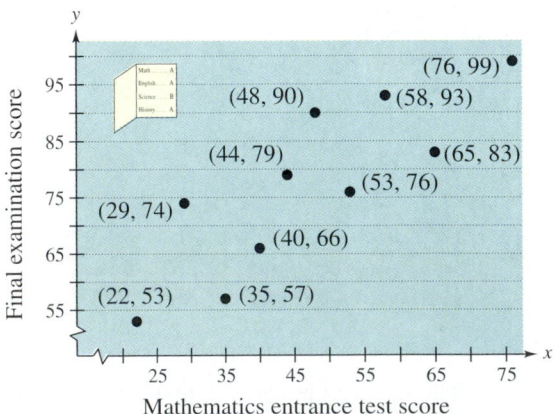

35. Find the entrance exam score of any student with a final exam score in the 80's.

36. Does a higher entrance exam score imply a higher final exam score? Explain.

In Exercises 37–40, find the distance between the points. (*Note:* In each case the two points lie on the same horizontal or vertical line.)

37. $(6, -3), (6, 5)$ **38.** $(1, 4), (8, 4)$

39. $(-3, -1), (2, -1)$ **40.** $(-3, -4), (-3, 6)$

In Exercises 41–44, (a) find the length of each side of a right triangle, and (b) show that these lengths satisfy the Pythagorean Theorem.

41. 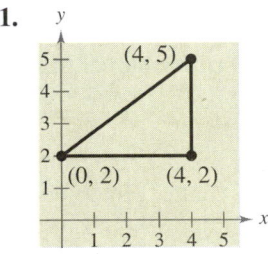   **42.**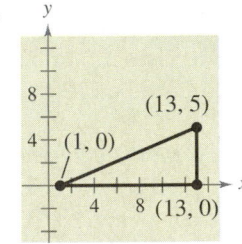

43. 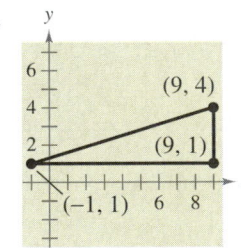   **44.**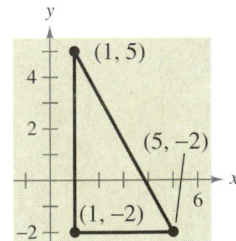

In Exercises 45–56, (a) plot the points, (b) find the distance between the points, and (c) find the midpoint of the line segment joining the points.

45. $(1, 1), (9, 7)$ **46.** $(1, 12), (6, 0)$

47. $(-4, 10), (4, -5)$ **48.** $(-7, -4), (2, 8)$

49. $(-1, 2), (5, 4)$

50. $(2, 10), (10, 2)$

51. $\left(\frac{1}{2}, 1\right), \left(-\frac{5}{2}, \frac{4}{3}\right)$

52. $\left(-\frac{1}{3}, -\frac{1}{3}\right), \left(-\frac{1}{6}, -\frac{1}{2}\right)$

53. $(6.2, 5.4), (-3.7, 1.8)$

54. $(-16.8, 12.3), (5.6, 4.9)$

55. $(-36, -18), (48, -72)$

56. $(1.451, 3.051), (5.906, 11.360)$

In Exercises 57 and 58, use the Midpoint Formula to estimate the sales of a company in 1993, given the sales in 1991 and 1995. Assume the sales followed a linear pattern.

57.

Year	1991	1995
Sales	$520,000	$740,000

58.

Year	1991	1995
Sales	$4,200,000	$5,650,000

In Exercises 59–64, show that the points form the vertices of the polygon.

59. Right triangle: $(4, 0), (2, 1), (-1, -5)$

60. Isosceles triangle: $(1, -3), (3, 2), (-2, 4)$

61. Rhombus: $(0, 0), (1, 2), (2, 1), (3, 3)$
(A rhombus is a parallelogram whose sides are all the same length.)

62. Rhombus: $(4, 0), (0, 6), (-4, 0), (0, -6)$

63. Parallelogram: $(2, 5), (0, 9), (-2, 0), (0, -4)$

64. Parallelogram: $(0, 1), (3, 7), (4, 4), (1, -2)$

65. A line segment has (x_1, y_1) as one endpoint and (x_m, y_m) as its midpoint. Find the other endpoint (x_2, y_2) of the line segment in terms of $x_1, y_1, x_m,$ and y_m.

66. Use the result of Exercise 65 to find the coordinates of the endpoint of a line segment if the coordinates of the other endpoint and midpoint are, respectively,
(a) $(1, -2), (4, -1)$.   (b) $(-5, 11), (2, 4)$.

67. Use the Midpoint Formula three times to find the three points that divide the line segment joining (x_1, y_1) and (x_2, y_2) into four parts.

68. Use the result of Exercise 67 to find the points that divide the line segment joining the given points into four equal parts.
(a) $(1, -2), (4, -1)$ (b) $(-2, -3), (0, 0)$

69. *Football Pass*   In a football game, a quarterback throws a pass from the 15-yard line, 10 yards from the sideline (see figure). The pass is caught on the 40-yard line, 45 yards from the same sideline. How long is the pass?

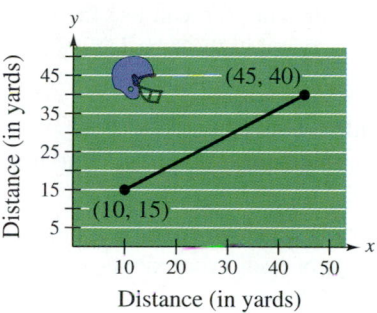

70. *Flying Distance*   A plane flies in a straight line to a city that is 100 kilometers east and 150 kilometers north of the point of departure. How far does it fly?

71. *Make a Conjecture*   Plot the points $(2, 1), (-3, 5),$ and $(7, -3)$ on a rectangular coordinate system. Then change the sign of the x-coordinate of each point and plot the three new points on the same rectangular coordinate system. What conjecture can you make about the location of a point when a sign of the x-coordinate is changed?

72. Prove that the diagonals of the parallelogram in the figure bisect each other.

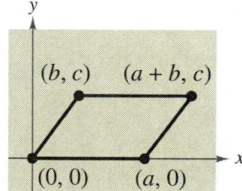

73. *Chapter Opener*   Use the graph on page 1.
(a) Describe any trends in the data. From these trends, predict the number of artists elected in 1996.
(b) Why do you think the numbers elected in 1986 and 1987 were greater than in other years?

FOCUS ON CONCEPTS

In this chapter, you studied several concepts that are required in the study of algebra. You can use the following questions to check your understanding of several of these basic concepts. The answers to these questions are given in the back of the book.

1. Describe the differences among the sets of natural numbers, integers, rational numbers, and irrational numbers.

2. Three real numbers are shown on the real number line. Determine the sign of each expression.

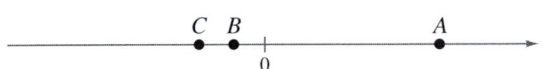

(a) $-A$ (b) $-C$
(c) $B - A$ (d) $A - C$

3. You may hear it said that to take the absolute value of a real number you simply remove any negative sign and make the number positive. Can it ever be true that $|a| = -a$ for a real number a? Explain.

4. Explain why each of the following is *not* equal.

(a) $(3x)^{-1} \neq \dfrac{3}{x}$ (b) $y^3 \cdot y^2 \neq y^6$

(c) $(a^2b^3)^4 \neq a^6b^7$ (d) $(a + b)^2 \neq a^2 + b^2$

(e) $\sqrt{4x^2} \neq 2x$ (f) $\sqrt{2} + \sqrt{3} \neq \sqrt{5}$

5. Is the real number 52.7×10^5 written in scientific notation? Explain.

6. A third-degree polynomial and a fourth-degree polynomial are added.

(a) Can the sum be a fourth-degree polynomial? Explain or give an example.

(b) Can the sum be a second-degree polynomial? Explain or give an example.

(c) Can the sum be a seventh-degree polynomial? Explain or give an example.

7. Explain what is meant when it is said that a polynomial is in factored form.

8. How do you determine whether a rational expression is in reduced form?

In Exercises 9–12, use the plot of the point (x_0, y_0) in the figure. Match the transformation of the point with the correct plot. [The plots are labeled (a), (b), (c), and (d).]

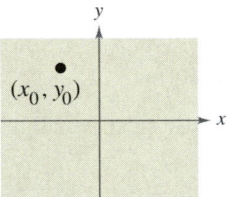

(a) 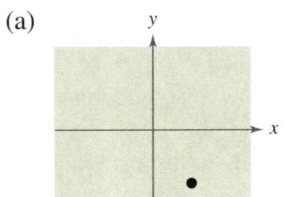   (b)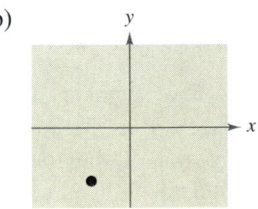

(c) 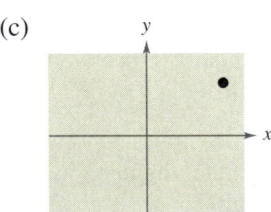   (d)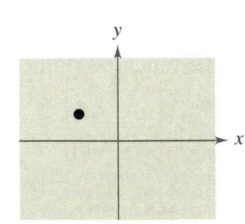

9. $(x_0, -y_0)$ 10. $(-2x_0, y_0)$

11. $\left(x_0, \tfrac{1}{2}y_0\right)$ 12. $(-x_0, -y_0)$

Review Exercises

In Exercises 1 and 2, determine which numbers in the set are (a) natural numbers, (b) integers, (c) rational numbers, and (d) irrational numbers.

1. $\left\{ 11, -14, -\frac{8}{9}, \frac{5}{2}, \sqrt{6}, 0.4 \right\}$
2. $\left\{ \sqrt{15}, -22, -\frac{10}{3}, 0, 5.2, \frac{3}{7} \right\}$

In Exercises 3 and 4, use a calculator to find the decimal form of the rational number. If it is a nonterminating decimal, write the repeating pattern.

3. (a) $\frac{5}{6}$ (b) $\frac{7}{8}$ 4. (a) $\frac{9}{25}$ (b) $\frac{5}{7}$

In Exercises 5 and 6, give a verbal description of the subset of real numbers represented by the inequality, and sketch the subset on the real number line.

5. $x \le 7$ 6. $x > 1$

In Exercises 7–10, use absolute value notation to describe the expression.

7. The distance between x and 7 is at least 4.
8. The distance between x and 25 is no more than 10.
9. The distance between y and -30 is less than 5.
10. The distance between z and -16 is greater than 8.

In Exercise 11–16, perform the operations without the aid of a calculator.

11. $|-3| + 4(-2) - 6$ 12. $\dfrac{|-10|}{-10}$
13. $\frac{5}{18} \div \frac{10}{3}$ 14. $(16 - 8) \div 4$
15. $6[4 - 2(6 + 8)]$ 16. $-4[16 - 3(7 - 10)]$

In Exercises 17–20, identify the rule of algebra illustrated by the equation.

17. $2x + (3x - 10) = (2x + 3x) - 10$

18. $(t + 4)(2t) = (2t)(t + 4)$
19. $\dfrac{2}{y + 4} \cdot \dfrac{y + 4}{2} = 1, \quad y \ne -4$
20. $0 + (a - 5) = a - 5$

In Exercises 21 and 22, simplify the expression.

21. (a) $\dfrac{6^2 u^3 v^{-3}}{12 u^{-2} v}$ (b) $\dfrac{3^{-4} m^{-1} n^{-3}}{9^{-2} m n^{-3}}$

22. (a) $(x + y^{-1})^{-1}$ (b) $\left(\dfrac{x^{-3}}{y}\right)\left(\dfrac{x}{y}\right)^{-1}$

In Exercises 23 and 24, write the number in scientific notation.

23. *Net sales of Procter and Gamble Company in 1994:* $30,296,000,000   (Source: 1994 Annual Report)
24. *Number of Meters in 1 Foot:* 0.3048

In Exercises 25 and 26, write the number in decimal form.

25. *Distance Between Sun and Jupiter:* 4.833×10^8 miles
26. *Ratio of Day to Year:* 2.74×10^{-3}

In Exercises 27 and 28, use a calculator to evaluate the expression. (Round your answer to three decimal places.)

27. (a) $1800(1 + 0.08)^{24}$
    (b) $0.0024(7,658,400)$

28. (a) $50,000\left(1 + \dfrac{0.075}{12}\right)^{48}$
    (b) $\dfrac{28,000,000 + 34,000,000}{87,000,000}$

In Exercises 29 and 30, fill in the missing description.

Radical Form	*Rational Exponent Form*
29. $\sqrt{16} = 4$	$\boxed{} = 4$
30. $\boxed{} = 2$	$16^{1/4} = 2$

In Exercises 31 and 32, simplify by removing all possible factors from the radical.

31. (a) $\sqrt{4x^4}$ (b) $\sqrt{\dfrac{18u^2}{b^3}}$

32. (a) $\sqrt[3]{\dfrac{2x^3}{27}}$ (b) $\sqrt[5]{64x^6}$

In Exercises 33 and 34, rewrite the expression by rationalizing the denominator. Simplify your answer.

33. $\dfrac{1}{2 - \sqrt{3}}$ **34.** $\dfrac{1}{\sqrt{x} - 1}$

In Exercises 35 and 36, simplify the expression.

35. $\sqrt{50} - \sqrt{18}$ **36.** $\sqrt{8x^3} + \sqrt{2x}$

37. *Strength of a Wooden Beam*   The rectangular cross section of a wooden beam cut from a log of diameter 24 inches (see figure) will have a maximum strength if its width w and height h are given by

$$w = 8\sqrt{3} \quad \text{and} \quad h = \sqrt{24^2 - \left(8\sqrt{3}\right)^2}.$$

Find the area of the rectangular cross section and express the answer in simplest form.

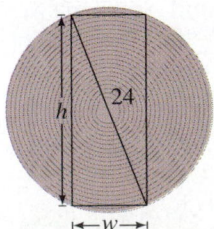

38. *Essay*   Explain why $\sqrt{5u} + \sqrt{3u} \neq 2\sqrt{2u}$.

In Exercises 39–52, describe and correct the error.

39. $\cancel{10(4 \cdot 7) = 40 \cdot 70}$ **40.** $\cancel{\left(\tfrac{1}{3}x\right)\left(\tfrac{1}{3}y\right) = \tfrac{1}{3}xy}$

41. $\cancel{4\left(\tfrac{3}{7}\right) = \tfrac{12}{28}}$ **42.** $\cancel{\tfrac{2}{9} \times \tfrac{4}{9} = \tfrac{8}{9}}$

43. $\cancel{\dfrac{x - 1}{1 - x} = 1}$ **44.** $\cancel{(2x)^4 = 2x^4}$

45. $\cancel{(-x)^6 = -x^6}$ **46.** $\cancel{(3^4)^4 = 3^8}$

47. $\cancel{-x^2(-x^2 + 3) = x^4 + 3x^2}$

48. $\cancel{\sqrt{3^2 + 4^2} = 3 + 4}$ **49.** $\cancel{(5 + 8)^2 = 5^2 + 8^2}$

50. $\cancel{\sqrt{10x} = 10\sqrt{x}}$

51. $\cancel{\sqrt{7x^{3/2}} = \sqrt{14x}}$

52. $\cancel{(9x + 12)^2 = 3(3x + 4)^2}$

In Exercises 53–58, perform the operations and write the result in standard form.

53. $-(3x^2 + 2x) + (1 - 5x)$

54. $8y - [2y^2 - (3y - 8)]$

55. $(2x - 3)^2$ **56.** $\left(3\sqrt{5} + 2\right)\left(3\sqrt{5} - 2\right)$

57. $(x^3 - 3x)(2x^2 + 3x + 5)$

58. $\left(x - \dfrac{1}{x}\right)(x + 2)$

59. *Exploration*   The surface area of a right circular cylinder is $S = 2\pi r^2 + 2\pi rh$.

(a) Draw a right circular cylinder of radius r and height h. Use the figure to explain how the surface area formula was obtained.

(b) Factor the expression for surface area.

60. *Revenue*   The revenue for selling x units of a product at a price of p dollars per unit is $R = xp$. For a particular product the revenue is

$$R = 1600x - 0.50x^2.$$

Factor this expression, and determine an expression that gives the price in terms of x.

In Exercises 61–66, factor completely.

61. $x^3 - x$ **62.** $x(x - 3) + 4(x - 3)$

63. $2x^2 + 21x + 10$ **64.** $3x^2 + 14x + 8$

65. $x^3 - x^2 + 2x - 2$ **66.** $x^3 - 1$

In Exercises 67–70, insert the missing factor.

67. $\frac{2}{3}x^4 - \frac{3}{8}x^3 + \frac{5}{6}x^2 = \frac{1}{24}x^2 \left(\boxed{}\right)$

68. $\dfrac{t}{\sqrt{t+1}} - \sqrt{t+1} = \dfrac{1}{\sqrt{t+1}} \left(\boxed{}\right)$

69. $2x(x^2 - 3)^{1/3} - 5(x^2 - 3)^{4/3} = (x^2 - 3)^{1/3} \left(\boxed{}\right)$

70. $y(y-1)^{5/4} - y^2(y-1)^{1/4} = y(y-1)^{1/4} \left(\boxed{}\right)$

In Exercises 71–76, perform the operations and simplify.

71. $\dfrac{x^2 - 4}{x^4 - 2x^2 - 8} \cdot \dfrac{x^2 + 2}{x^2}$

72. $\dfrac{4x - 6}{(x - 1)^2} \div \dfrac{2x^2 - 3x}{x^2 + 2x - 3}$

73. $2x + \dfrac{3}{2(x - 4)} - \dfrac{1}{2(x + 2)}$

74. $\dfrac{1}{x} - \dfrac{x - 1}{x^2 + 1}$

75. $\dfrac{1}{x - 1} + \dfrac{1 - x}{x^2 + x + 1}$

76. $\dfrac{1}{L}\left(\dfrac{1}{y} - \dfrac{1}{L - y}\right),$ where L is a constant

In Exercises 77 and 78, simplify the compound fraction.

77. $\dfrac{\left[\dfrac{3a}{(a^2/x) - 1}\right]}{\left(\dfrac{a}{x} - 1\right)}$

78. $\dfrac{\left(\dfrac{1}{2x - 3} - \dfrac{1}{2x + 3}\right)}{\left(\dfrac{1}{2x} - \dfrac{1}{2x + 3}\right)}$

Geometry   **In Exercises 79 and 80, plot the points and verify that the points form the polygon.**

79. Right Triangle: $(2, 3), (13, 11), (5, 22)$

80. Parallelogram: $(1, 2), (8, 3), (9, 6), (2, 5)$

In Exercises 81–84, determine the quadrant(s) in which (x, y) is located so that the condition(s) is (are) satisfied.

81. $x > 0$ and $y = -2$

82. $y > 0$

83. $(-x, y)$ is in the third quadrant.

84. $xy = 4$

In Exercises 85 and 86, (a) plot the points, (b) find the distance between the points, and (c) find the midpoint of the line segment joining the points.

85. $(-3, 8), (1, 5)$

86. $(5.6, 0), (0, 8.2)$

87. ***Geometric Modeling***   Use the area model to write two expressions for the total area in the figure. Then equate the two expressions and name the algebraic property illustrated.

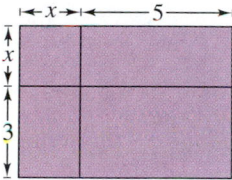

88. ***Geometry***   The four corners are cut from an $8\frac{1}{2}$-by-11-inch piece of paper (see figure). Find the perimeter of the remaining piece of paper.

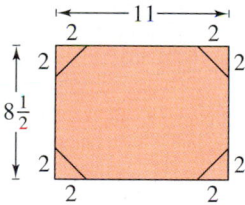

89. Complete the table.

n	1	10	10^2	10^4	10^6	10^{10}
$\dfrac{5}{\sqrt{n}}$						

What number does $5/\sqrt{n}$ approach as n increases without bound?

90. Let m and n be any two integers. Then $2m$ and $2n$ are even integers and $(2m + 1)$ and $(2n + 1)$ are odd integers.

(a) Prove that the sum of two even integers is even.

(b) Prove that the sum of two odd integers is even.

(c) Prove that the product of an even integer and any integer is even.

CHAPTER PROJECT:

A Numerical Approach to Maximizing a Volume

Many mathematical results are discovered by calculating examples and looking for patterns. Prior to the 1950s, this mode of discovery was very time-consuming because the calculations had to be done by hand. The introduction of computer and calculator technology has removed the drudgery of calculation.

Using technology to conduct mathematical experiments usually involves creation of an algebraic **model** to represent the quantity under question.

Real Life

EXAMPLE 1 *Modeling the Volume of a Box*

Consider a rectangular box with a square base and a surface area of 216 square inches. Let x represent the length (in inches) of each side of the base, as shown in the figure at the left. Use the variable x to write a model, or expression, for the volume of the box.

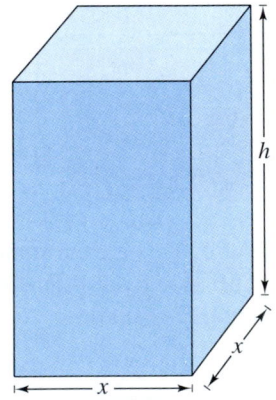

Solution

You can begin by writing a model for the height h (in inches) in terms of x.

Surface area	=	Area of base	+	Area of top	+ 4 ·	Area of side

$$216 = x^2 + x^2 + 4xh$$
$$216 = 2x^2 + 4xh$$
$$216 - 2x^2 = 4xh$$
$$\frac{216 - 2x^2}{4x} = h$$
$$\frac{54}{x} - \frac{x}{2} = h$$

Having written the height in terms of x, you can write the volume in terms of x.

Volume of box	=	Length of box	·	Width of box	·	Height of box

$$V = x \cdot x \cdot \left(\frac{54}{x} - \frac{x}{2}\right)$$
$$V = 54x - \frac{1}{2}x^3, \qquad 0 \le x \le \sqrt{108}$$

You can use the model created in Example 1 to answer questions about the volume of the box. For instance, in Example 2, the model is used to find the maximum volume of a rectangular box with a square base and a surface area of 216 square inches.

EXAMPLE 2   *Finding the Maximum Volume of a Box*

Real
Life

Of all rectangular boxes with square bases and surface areas of 216 square inches, which has the largest volume?

Solution

Base, x	Height	Surface Area	Volume
1.0	53.5	216.0	53.5
1.5	35.3	216.0	79.3
2.0	26.0	216.0	104.0
2.5	20.4	216.0	127.2
3.0	16.5	216.0	148.5
3.5	13.7	216.0	167.6
4.0	11.5	216.0	184.0
4.5	9.8	216.0	197.4
5.0	8.3	216.0	207.5
5.5	7.1	216.0	213.8
6.0	6.0	216.0	216.0
6.5	5.1	216.0	213.7
7.0	4.2	216.0	206.5
7.5	3.5	216.0	194.1
8.0	2.8	216.0	176.0
8.5	2.1	216.0	151.9
9.0	1.5	216.0	121.5
9.5	0.9	216.0	84.3
10.0	0.4	216.0	40.0

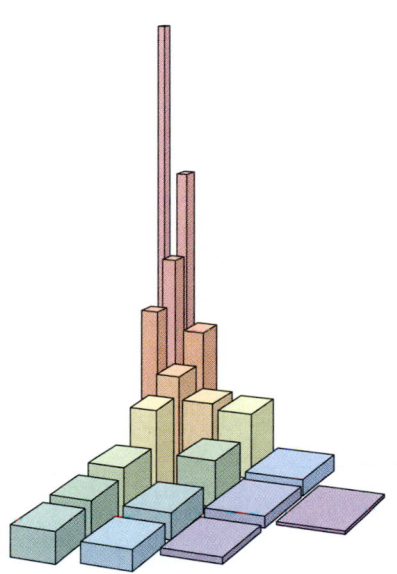

This diagram shows the 19 different boxes whose volumes are calculated in Example 2.

From the results of the experiment, it appears that the cube (the box whose dimensions are 6 by 6 by 6) has the greatest volume.

CHAPTER PROJECT INVESTIGATIONS

1. *Exploration*   In Example 2, what happens to the height of the boxes as x gets closer to 0? Of all boxes with square bases and surface areas of 216 square inches, is there a tallest? Explain your reasoning.

2. *Exploration*   In Example 2, what happens to the height of the boxes as x gets closer and closer to $\sqrt{108}$? Is there a shortest box that has a square base and a surface area of 216 square inches? Explain your reasoning.

3. *Exploration*   Complete the table. Why does it lend further support to Example 2's conclusion?

x	5.9	5.99	5.999	6.001	6.01	6.1
V	?	?	?	?	?	?

4. Of all rectangular boxes with surface areas of 216 square inches and bases of x inches by $2x$ inches, which has the maximum volume? Explain.

Chapter Test

Take this test as you would take a test in class. After you are done, check your work against the answers given in the back of the book.

1. Place < or > between the real numbers $-\frac{10}{3}$ and $-|-4|$.
2. Find the distance between the real numbers -5.4 and $3\frac{3}{4}$.

In Exercises 3–6, evaluate the quantity without the aid of a calculator.

3. (a) $27\left(-\frac{2}{3}\right)$ (b) $\dfrac{5}{18} \div \dfrac{15}{8}$ 4. (a) $\left(-\dfrac{3}{5}\right)^3$ (b) $\left(\dfrac{3^2}{2}\right)^{-3}$

5. (a) $\sqrt{5} \cdot \sqrt{125}$ (b) $\dfrac{\sqrt{72}}{\sqrt{2}}$ 6. (a) $\dfrac{5.4 \times 10^8}{3 \times 10^3}$ (b) $(3 \times 10^4)^3$

In Exercises 7–9, simplify the expression.

7. (a) $3z^2(2z^3)^2$ (b) $(u-2)^{-4}(u-2)^{-3}$

8. (a) $\left(\dfrac{x^{-2}y^2}{3}\right)^{-1}$ (b) $\sqrt[3]{\dfrac{16}{v^5}}$

9. (a) $9z\sqrt{8z} - 3\sqrt{2z^3}$ (b) $-5\sqrt{16y} + 10\sqrt{y}$

In Exercises 10–13, perform the operations and simplify.

10. $(x^2+3) - [3x + (8 - x^2)]$ 11. $\left(x + \sqrt{5}\right)\left(x - \sqrt{5}\right)$

12. $\dfrac{8x}{x-3} + \dfrac{24}{3-x}$ 13. $\dfrac{\left(\dfrac{2}{x} - \dfrac{2}{x+1}\right)}{\left(\dfrac{4}{x^2-1}\right)}$

14. Factor (a) $2x^4 - 3x^3 - 2x^2$ and (b) $x^3 + 2x^2 - 4x - 8$ completely.

15. Rationalize the denominators.

    (a) $\dfrac{16}{\sqrt[3]{16}}$ (b) $\dfrac{6}{1 - \sqrt{3}}$

16. Plot the points $(-2, 5)$ and $(6, 0)$. Find the coordinates of the midpoint of the line segment joining the points and the distance between the points.

17. Write an expression for the area of the shaded region in the figure and simplify the result.

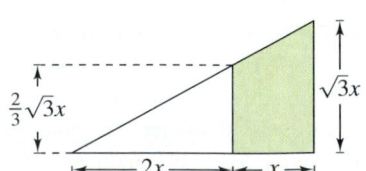

FIGURE FOR 17

Many classical guitarists complain about the sound made by the guitar—the volume is too low, the notes are difficult to sustain, the treble is feeble, and the sound is uneven. Yet, in spite of advances in physics and acoustics, the design of classical guitars has remained unchanged for over 150 years.

To correct these complaints, Michael Kasha of Florida State University used physics and mathematics to design a new classical guitar. As part of his work, he used the equation for the frequency of the vibrations on a circular plate

$$v = \frac{2.6t}{d^2} \sqrt{\frac{E}{\rho}}$$

In this equation, v is the frequency, t is the plate thickness, d is the diameter of the plate, E is the elasticity of the plate material, and ρ is the density of the plate material. For fixed values of d, E, and ρ, the graph of the equation is a line.

See Exercises 97–100 on page 163.

Photos: Richard Schneider; Ruth Goodman (inset)

1.1 *Graphs and Graphing Utilities*
1.2 *Linear Equations*
1.3 *Modeling with Linear Equations*
1.4 *Quadratic Equations and Applications*
1.5 *Complex Numbers*
1.6 *Other Types of Equations*
1.7 *Linear Inequalities*
1.8 *Other Types of Inequalities*

This classical guitar was designed by Michael Kasha of Florida State University. The hole was moved to increase the surface area for bass response, and interior bracing is used for greater resonant range.

1.1   *Graphs and Graphing Utilities*

See Exercise 95 on page 95 for an example of how a graph can be used to estimate the life expectancies of children who are born in the years 1998 and 2000.

The Graph of an Equation   ▫   *Intercepts of a Graph*   ▫   *Symmetry*   ▫
Circles   ▫   *Application*

The Graph of an Equation

In Section P.7, you used a coordinate system to represent graphically the relationship between two quantities. There, the graphical picture consisted of a collection of points in a coordinate plane (see Example 2 in Section P.7).

Frequently, a relationship between two quantities is expressed as an **equation** in two variables. For instance, $y = 7 - 3x$ is an equation in x and y. An ordered pair (a, b) is a **solution** or **solution point** of an equation in x and y if the equation is true when a is substituted for x and b is substituted for y. For instance, $(1, 4)$ is a solution of $y = 7 - 3x$ because $4 = 7 - 3(1)$ is a true statement.

In this section, you will review some basic procedures for sketching the graph of an equation in two variables. The **graph** of an equation is the set of all points that are solutions of the equation.

EXAMPLE 1   *Sketching the Graph of an Equation*

Sketch the graph of $y = 7 - 3x$.

Solution

The simplest way to sketch the graph of an equation is the *point-plotting method.* With this method, you construct a table of values that consists of several solution points of the equation. For instance, when $x = 0$,

$$y = 7 - 3(0) = 7,$$

which implies that $(0, 7)$ is a solution point of the graph.

x	0	1	2	3	4
$y = 7 - 3x$	7	4	1	-2	-5

From the table, it follows that $(0, 7)$, $(1, 4)$, $(2, 1)$, $(3, -2)$, and $(4, -5)$ are solution points of the equation. After plotting these points, you can see that they appear to lie on a line, as shown in Figure 1.1. The graph of the equation is the line that passes through the five plotted points.

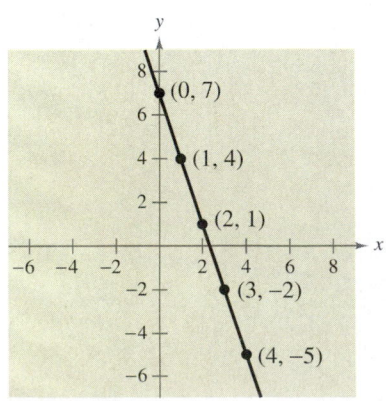

FIGURE 1.1

The *Interactive* CD-ROM offers graphing utility emulators of the *TI-82* and *TI-83*, which can be used with the Examples, Explorations, Technology notes, and Exercises.

EXAMPLE 2   *Sketching the Graph of an Equation*

Sketch the graph of $y = x^2 - 2$.

Solution

Begin by constructing a table of values.

x	-2	-1	0	1	2	3
$y = x^2 - 2$	2	-1	-2	-1	2	7

Next, plot the points given in the table, as shown in Figure 1.2(a). Finally, connect the points with a smooth curve, as shown in Figure 1.2(b).

 A computer animation of this concept appears in the *Interactive* CD-ROM.

NOTE   The graph shown in Example 2 is a **parabola.** The graph of any second-degree equation of the form

$$y = ax^2 + bx + c, \qquad a \neq 0$$

has a similar shape. You will study the graphs of second-degree equations in Section 3.1. ■■

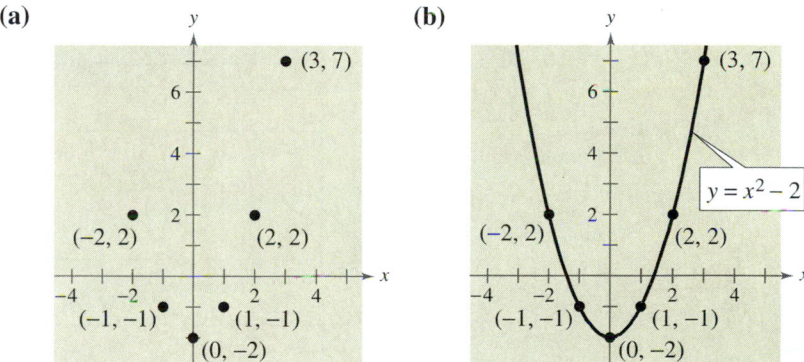

FIGURE 1.2

The point-plotting technique demonstrated in Examples 1 and 2 is easy to use, but it has some shortcomings. With too few solution points, you can badly misrepresent the graph of an equation. For instance, how would you connect the four points in Figure 1.3? Without further information, any one of the three graphs in Figure 1.4 would be reasonable.

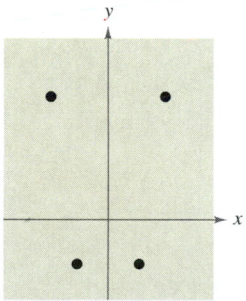

FIGURE 1.3

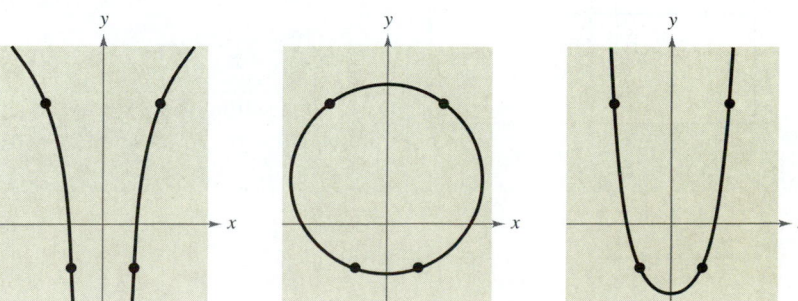

FIGURE 1.4

TECHNOLOGY

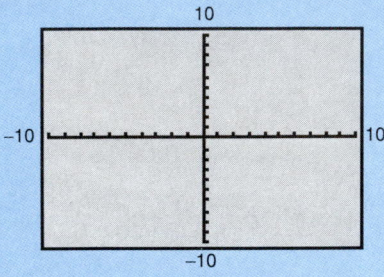

Creating a Viewing Rectangle

A **viewing rectangle** for a graph is a rectangular portion of the coordinate plane. A viewing rectangle is determined by six values: the minimum x-value, the maximum x-value, the x-scale, the minimum y-value, the maximum y-value, and the y-scale. When you enter these six values into a graphing utility, you are setting the **range** or **window.** Some graphing utilities have a standard viewing rectangle, as shown at the left.

By choosing different viewing rectangles for a graph, it is possible to obtain very different impressions of the graph's shape. For instance, below are four different viewing rectangles for the graph of

$$y = 0.1x^4 - x^3 + 2x^2.$$

Of these, the view shown in part (a) is the most complete.

(a)

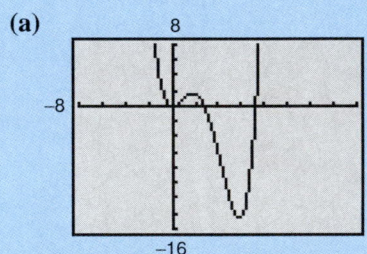

(b)

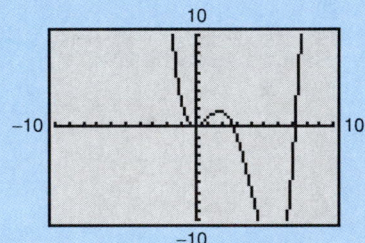

(c)

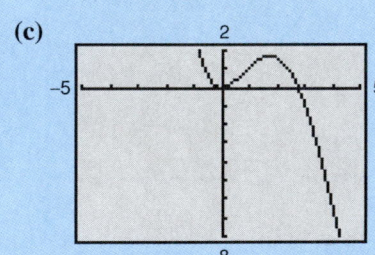

(d)

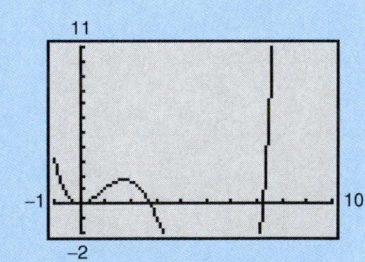

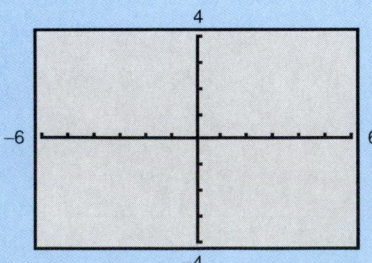

On most graphing utilities, the display screen is two-thirds as high as it is wide. On such screens, you can obtain a graph with a true geometric perspective by using a **square setting**—one in which

$$\frac{Y_{max} - Y_{min}}{X_{max} - X_{min}} = \frac{2}{3}.$$

One such setting is shown at the left. Notice that the x and y tick marks are equally spaced on a square setting, but not on a standard setting.

Intercepts of a Graph

It is often easy to determine the solution points that have zero as either the x-coordinate or the y-coordinate. These points are called **intercepts** because they are the points at which the graph intersects the x- or y-axis.

It is possible for a graph to have no intercepts or several intercepts, as shown in Figure 1.5.

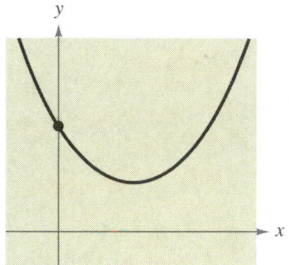

No x-intercept
One y-intercept

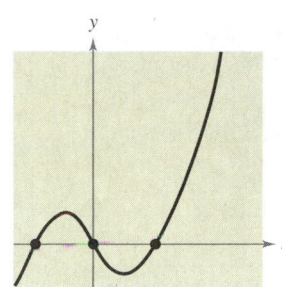

Three x-intercepts
One y-intercept

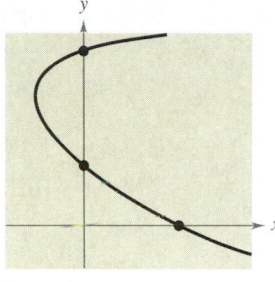

One x-intercept
Two y-intercepts

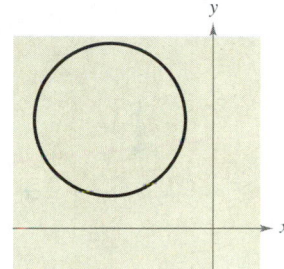

No intercepts

FIGURE 1.5

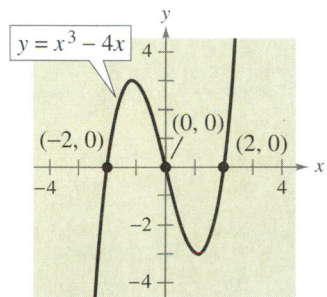

FIGURE 1.6

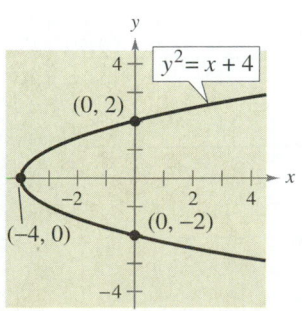

FIGURE 1.7

FINDING INTERCEPTS

1. To find x-intercepts, let y be zero and solve the equation for x.
2. To find y-intercepts, let x be zero and solve the equation for y.

EXAMPLE 3 *Finding x- and y-Intercepts*

Find the x- and y-intercepts for the graph of each equation.

a. $y = x^3 - 4x$ **b.** $y^2 = x + 4$

Solution

a. Let $y = 0$. Then $0 = x(x^2 - 4)$ has solutions $x = 0$ and $x = \pm 2$.

x-intercepts: $(0, 0), (2, 0), (-2, 0)$

Let $x = 0$. Then $y = 0$.

y-intercept: $(0, 0)$ (See Figure 1.6.)

b. Let $y = 0$. Then $-4 = x$.

x-intercept: $(-4, 0)$

Let $x = 0$. Then $y^2 = 4$ has solutions $y = \pm 2$.

y-intercepts: $(0, 2), (0, -2)$ (See Figure 1.7.)

TECHNOLOGY

Zooming in to Find Intercepts

You can use the **zoom** feature of a graphing utility to approximate the x-intercept(s) of a graph. Suppose you want to approximate the x-intercept(s) of the graph of

$$y = 2x^3 - 3x + 2.$$

Begin by graphing the equation, as shown below in part (a). From the viewing rectangle shown, the graph appears to have only one x-intercept. This intercept lies between -2 and -1. By zooming in on the intercept, you can improve the approximation, as shown in part (b). To three decimal places, the solution is $x \approx -1.476$.

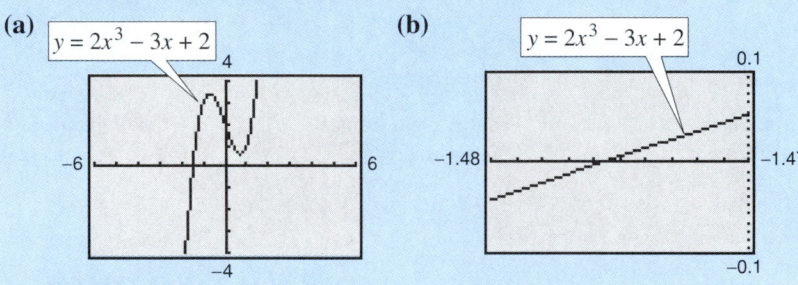

Here are some suggestions for using the zoom feature.

1. With each successive zoom-in, adjust the x-scale so that the viewing rectangle shows at least one tick mark on each side of the x-intercept.
2. The error in your approximation will be less than the distance between two scale marks.
3. The **trace** feature can usually be used to add one more decimal place of accuracy without changing the viewing rectangle.

Part (a) below shows the graph of $y = x^2 - 5x + 3$. Parts (b) and (c) show "zoom-in views" of the two intercepts. From these views, you can approximate the x-intercepts to be $x \approx 0.697$ and $x \approx 4.303$.

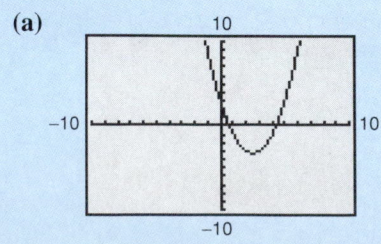

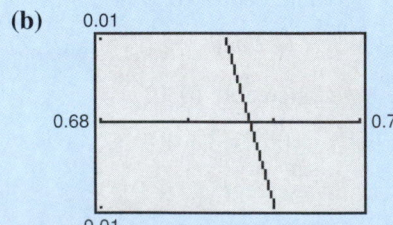

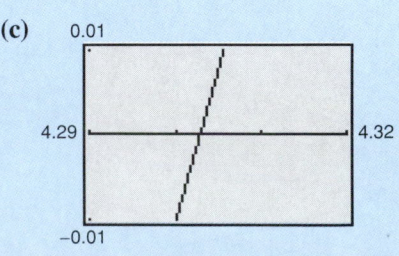

Symmetry

The graphs shown in Figures 1.2(b), 1.6, and 1.7 each have **symmetry** with respect to one of the coordinate axes or with respect to the origin.

Figure 1.2(b)	$y = x^2 - 2$	*y*-axis symmetry
Figure 1.6	$y = x^3 - 4x$	Origin symmetry
Figure 1.7	$y^2 = x + 4$	*x*-axis symmetry

Symmetry with respect to the *x*-axis means that if the Cartesian plane were folded along the *x*-axis, the portion of the graph above the *x*-axis would coincide with the portion below the *x*-axis. Symmetry with respect to the *y*-axis or the origin can be described in a similar manner, as shown in Figure 1.8.

A computer animation of this concept appears in the *Interactive* CD-ROM.

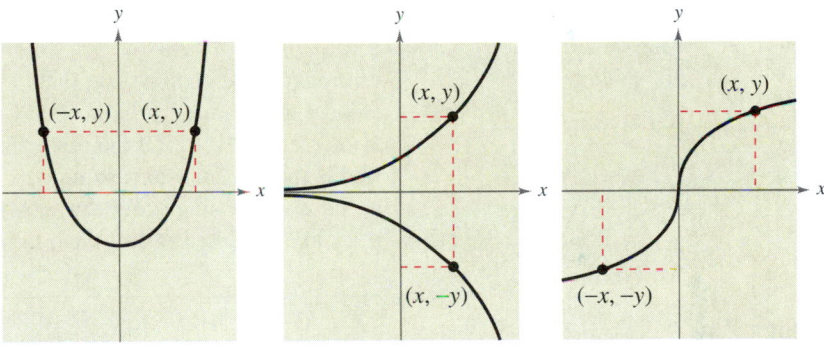

y-Axis Symmetry   x-Axis Symmetry   Origin Symmetry

FIGURE 1.8

Knowing the symmetry of a graph *before* attempting to sketch it is helpful, because then you need only half as many solution points to sketch the graph. There are three basic types of symmetry. (See Exercises 29–32.) A graph is **symmetric with respect to the y-axis** if, whenever (x, y) is on the graph, $(-x, y)$ is also on the graph. A graph is **symmetric with respect to the x-axis** if, whenever (x, y) is on the graph, $(x, -y)$ is also on the graph. A graph is **symmetric with respect to the origin** if, whenever (x, y) is on the graph, $(-x, -y)$ is also on the graph.

The graph of $y = x^2 - 2$ is symmetric with respect to the *y*-axis because the point $(-x, y)$ satisfies the equation.

$y = x^2 - 2$	Given equation
$y = (-x)^2 - 2$	Substitute $(-x, y)$ for (x, y).
$y = x^2 - 2$	Replacement yields equivalent equation.

See Figure 1.9.

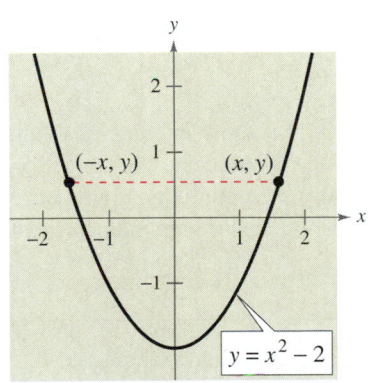

y-Axis Symmetry

FIGURE 1.9

A computer animation of this concept appears in the *Interactive* CD-ROM.

> **TESTS FOR SYMMETRY**
>
> 1. The graph of an equation is symmetric with respect to the *y-axis* if replacing x with $-x$ yields an equivalent equation.
> 2. The graph of an equation is symmetric with respect to the *x-axis* if replacing y with $-y$ yields an equivalent equation.
> 3. The graph of an equation is symmetric with respect to the *origin* if replacing x with $-x$ *and* y with $-y$ yields an equivalent equation.

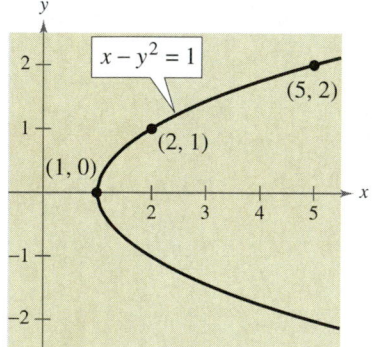

FIGURE 1.10

EXAMPLE 4   *Using Intercepts and Symmetry as Sketching Aids*

Use intercepts and symmetry to sketch the graph of $x - y^2 = 1$.

Solution

Letting $x = 0$, you can see that $-y^2 = 1$ or $y^2 = -1$ has no real solutions. Hence, there are no y-intercepts. Letting $y = 0$, you obtain $x = 1$. Thus, the x-intercept is $(1, 0)$. Of the three tests for symmetry, the only one that is satisfied is the test for x-axis symmetry. Thus, the graph is symmetric with respect to the x-axis. Using symmetry, you need only to find the solution points above the x-axis and then reflect them to obtain the graph, as shown in Figure 1.10.

y	0	1	2
$x = y^2 + 1$	1	2	5

EXAMPLE 5   *Sketching the Graph of an Equation*

Sketch the graph of $y = |x - 1|$.

Solution

Letting $x = 0$ yields $y = 1$, which means that $(0, 1)$ is the y-intercept. Letting $y = 0$ yields $x = 1$, which means that $(1, 0)$ is the x-intercept. This equation fails all three tests for symmetry and consequently its graph is not symmetric with respect to either axis or to the origin. The absolute value sign indicates that y is always nonnegative.

x	-2	-1	0	1	2	3	4
$y = |x - 1|$	3	2	1	0	1	2	3

The graph is shown in Figure 1.11.

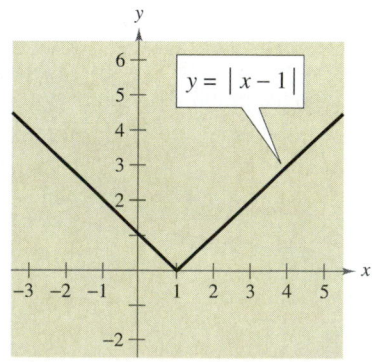

FIGURE 1.11

Circles

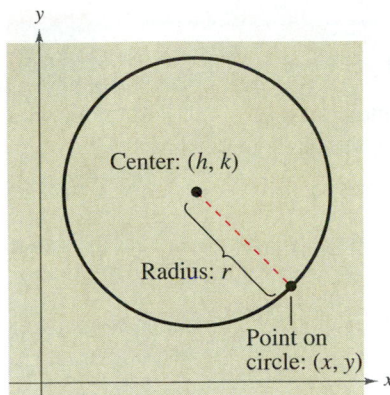

FIGURE 1.12

Throughout this course, you will learn to recognize several types of graphs from their equations. For instance, you will learn to recognize that the graph of a second-degree equation of the form $y = ax^2 + bx + c$ is a parabola (see Example 2). Another easily recognized graph is that of a **circle.**

Consider the circle shown in Figure 1.12. A point (x, y) is on the circle if and only if its distance from the center (h, k) is r. By the Distance Formula,

$$\sqrt{(x - h)^2 + (y - k)^2} = r.$$

By squaring both sides of this equation, you obtain the **standard form of the equation of a circle.**

> ### STANDARD FORM OF THE EQUATION OF A CIRCLE
>
> The point (x, y) lies on the circle of **radius** r and **center** (h, k) if and only if
>
> $$(x - h)^2 + (y - k)^2 = r^2.$$

From this result, you can see that the standard form of the equation of a circle with its center at the origin, $(h, k) = (0, 0)$, is simply

$$x^2 + y^2 = r^2. \qquad \text{Circle with center at origin}$$

EXAMPLE 6 *Finding the Equation of a Circle*

The point $(3, 4)$ lies on a circle whose center is at $(-1, 2)$, as shown in Figure 1.13. Find the standard form of the equation of this circle.

Solution

The radius of the circle is the distance between $(-1, 2)$ and $(3, 4)$.

$$
\begin{aligned}
r &= \sqrt{[3 - (-1)]^2 + (4 - 2)^2} & \text{Distance Formula} \\
&= \sqrt{16 + 4} & \text{Simplify.} \\
&= \sqrt{20} & \text{Radius}
\end{aligned}
$$

Using $(h, k) = (-1, 2)$ and $r = \sqrt{20}$, the equation of the circle is

$$
\begin{aligned}
(x - h)^2 + (y - k)^2 &= r^2 \\
[x - (-1)]^2 + (y - 2)^2 &= \left(\sqrt{20}\right)^2 & \text{Substitute for } h, k, \text{ and } r. \\
(x + 1)^2 + (y - 2)^2 &= 20. & \text{Standard form}
\end{aligned}
$$

FIGURE 1.13

Application

EXAMPLE 7   *Recommended Weight*

The median recommended weight y (in pounds) for men of medium frame who are 25–59 years old can be approximated by the mathematical model

$$y = 0.073x^2 - 6.986x + 288.985, \quad 62 \le x \le 76$$

where x is the person's height in inches.   (Source: Metropolitan Life Insurance Company)

a. Construct a table of values that shows the median recommended weights for heights of 62, 64, 66, 68, 70, 72, 74, and 76 inches.

b. Use the table of values to sketch a graph of the model. Then use the graph to estimate *graphically* the median recommended weight for a man whose height is 71 inches.

c. Use the model to confirm *analytically* the estimate you found in part (b).

Solution

a. You can use a calculator to complete the table, as shown below.

x	62	64	66	68	70	72	74	76
y	136.5	140.9	145.9	151.5	157.7	164.4	171.8	179.7

b. The table of values can be used to sketch the graph of the function, as shown in Figure 1.14. From the graph, you can estimate that a height of 71 inches corresponds to a weight of about 160 pounds.

c. To confirm analytically the estimate found in part (b), you can substitute 71 for x in the model.

$$y = 0.073x^2 - 6.986x + 288.985 \qquad \text{Given model}$$
$$= 0.073(71)^2 - 6.986(71) + 288.985 \qquad \text{Substitute 71 for } x.$$
$$\approx 160.97 \qquad \text{Use a calculator.}$$

Thus, the graphical estimate of 160 pounds is fairly good.

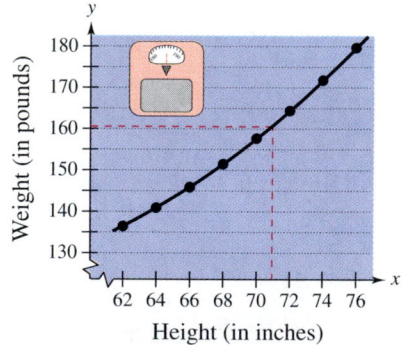

FIGURE 1.14

NOTE   Example 7 demonstrates three approaches to problem solving.

A Numerical Approach: Construct and use a table.

A Graphical Approach: Draw and use a graph.

An Analytical Approach: Use the rules of algebra.

GROUP ACTIVITY

CLASSIFYING GRAPHS OF EQUATIONS

The graphs below represent six common types of models. Match each equation with the corresponding graph.

1. Linear model: $y = x$

2. Quadratic model: $y = x^2$

3. Cubic model: $y = x^3$

4. Square root model: $y = \sqrt{x}$

5. Absolute value model: $y = |x|$

6. Rational model: $y = \dfrac{1}{x}$

(a)

(b)

(c)
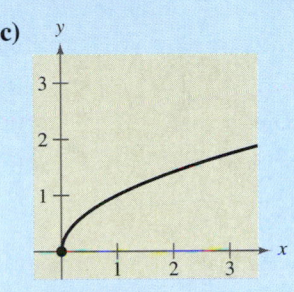

(d)

(e)

(f)
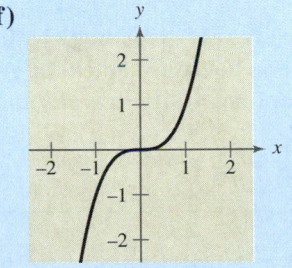

WARM UP

Simplify the expression.

1. $4(x - 3) - 5(6 - 2x)$

2. $-s(5s + 2) + 5(s^2 - 3s)$

3. $3y(-2y^2)^3$

4. $2t(t + 1)^2 - 4(t + 1)$

5. $\sqrt{150x^4}$

6. $\sqrt{4x + 12}$

Simplify the equation.

7. $-y = (-x)^3 + 4(-x)$

8. $(-x)^2 + (-y)^2 = 4$

9. $y = 4(-x)^2 + 8$

10. $(-y)^2 = 3(-x) + 4$

1.1 Exercises

In Exercises 1–8, determine whether the points lie on the graph of the equation.

	Equation	Points	
1.	$y = \sqrt{x + 4}$	(a) $(0, 2)$	(b) $(5, 3)$
2.	$y = x^2 - 3x + 2$	(a) $(2, 0)$	(b) $(-2, 8)$
3.	$y = 4 - |x - 2|$	(a) $(1, 5)$	(b) $(6, 0)$
4.	$y = \frac{1}{3}x^3 - 2x^2$	(a) $\left(2, -\frac{16}{3}\right)$	(b) $(-3, 9)$
5.	$2x - y - 3 = 0$	(a) $(1, 2)$	(b) $(1, -1)$
6.	$x^2 + y^2 = 20$	(a) $(3, -2)$	(b) $(-4, 2)$
7.	$x^2y - x^2 + 4y = 0$	(a) $\left(1, \frac{1}{5}\right)$	(b) $\left(2, \frac{1}{2}\right)$
8.	$y = \dfrac{1}{x^2 + 1}$	(a) $(0, 0)$	(b) $(3, 0.1)$

In Exercises 9–12, complete the table. Use the resulting solution points to sketch the graph of the equation. Use a graphing utility to verify the graph.

9. $y = -2x + 3$

x	-1	0	1	$\frac{3}{2}$	2
y					

10. $y = \frac{3}{2}x - 1$

x	-2	0	$\frac{2}{3}$	1	2
y					

11. $y = x^2 - 2x$

x	-1	0	1	2	3
y					

12. $y = 4 - x^2$

x	-2	-1	0	1	2
y					

In Exercises 13–20, use a graphing utility to graph the equation. Use a standard setting. Approximate any x- or y-intercepts of the graph.

13. $y = x - 5$

14. $y = (x + 1)(x - 3)$

15. $y = x^2 + x - 2$

16. $y = 9 - x^2$

17. $y = x\sqrt{x + 6}$

18. $y = (6 - x)\sqrt{x}$

19. $y = \dfrac{2x}{x - 1}$

20. $y = \dfrac{4}{x}$

In Exercises 21–28, check for symmetry with respect to both axes and the origin.

21. $x^2 - y = 0$

22. $xy^2 + 10 = 0$

23. $x - y^2 = 0$

24. $y = \sqrt{9 - x^2}$

25. $y = x^3$

26. $xy = 4$

27. $y = \dfrac{x}{x^2 + 1}$

28. $y = x^4 - x^2 + 3$

In Exercises 29–32, use symmetry to sketch the complete graph of the equation.

29.

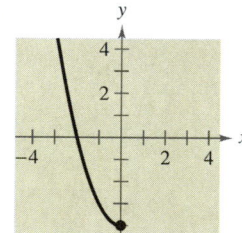

y-Axis Symmetry

30.

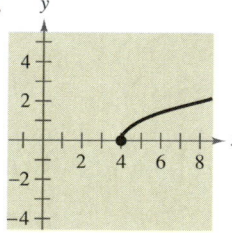

x-Axis Symmetry

31.

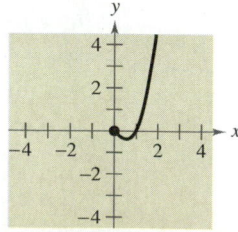

Origin Symmetry

32.

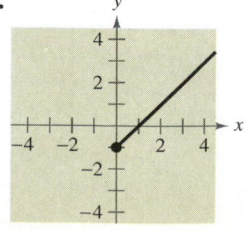

y-Axis Symmetry

In Exercises 33–38, match the equation with its graph. [The graphs are labeled (a), (b), (c), (d), (e), and (f).]

(a)
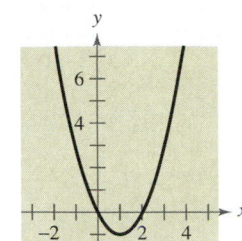

(b)
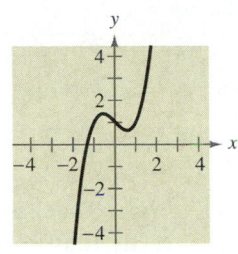

(c)
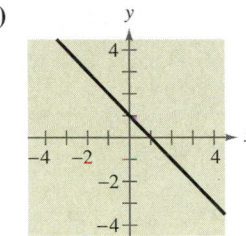

(d)
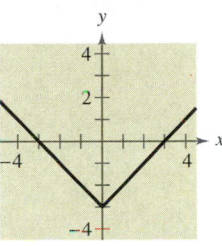

(e)
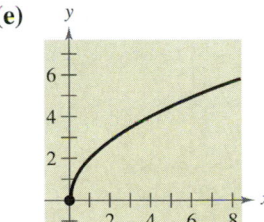

(f)
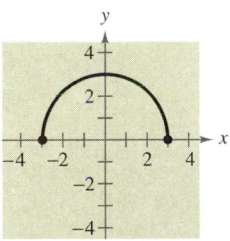

33. $y = 1 - x$

34. $y = x^2 - 2x$

35. $y = \sqrt{9 - x^2}$

36. $y = 2\sqrt{x}$

37. $y = x^3 - x + 1$

38. $y = |x| - 3$

In Exercises 39–52, sketch the graph of the equation. Test for symmetry.

39. $y = -3x + 2$

40. $y = 2x - 3$

41. $y = 1 - x^2$

42. $y = x^2 - 1$

43. $y = x^2 - 3x$

44. $y = -x^2 - 4x$

45. $y = x^3 + 2$

46. $y = x^3 - 1$

47. $y = \sqrt{x - 3}$

48. $y = \sqrt{1 - x}$

49. $y = |x - 2|$

50. $y = 4 - |x|$

51. $x = y^2 - 1$

52. $x = y^2 - 4$

In Exercises 53–60, use a graphing utility to graph the equation. Use a standard setting. Approximate any intercepts.

53. $y = 3 - \frac{1}{2}x$

54. $y = \frac{2}{3}x - 1$

55. $y = x^2 - 4x + 3$

56. $y = \frac{1}{2}(x + 4)(x - 2)$

57. $y = x(x - 2)^2$

58. $y = \dfrac{4}{x^2 + 1}$

59. $y = \sqrt[3]{x}$

60. $y = \sqrt[3]{x + 1}$

In Exercises 61–64, use a graphing utility to sketch the graph of the equation. Begin by using a standard setting. Then graph the equation a second time using the specified setting. Which setting is better? Explain.

61. $y = \frac{5}{2}x + 5$

62. $y = -3x + 50$

Xmin = 0
Xmax = 6
Xscl = 1
Ymin = 0
Ymax = 10
Yscl = 1

Xmin = -1
Xmax = 4
Xscl = 1
Ymin = -5
Ymax = 60
Yscl = 5

63. $y = -x^2 + 10x - 5$

64. $y = 4(x + 5)\sqrt{4 - x}$

Xmin = -1
Xmax = 11
Xscl = 1
Ymin = -5
Ymax = 25
Yscl = 2

Xmin = -6
Xmax = 6
Xscl = 1
Ymin = -5
Ymax = 50
Yscl = 4

In Exercises 65–68, describe the viewing rectangle.

65. $y = 4x^2 - 25$

66. $y = x^3 - 3x^2 + 4$

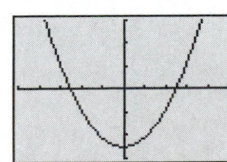

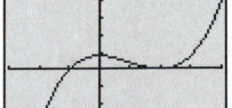

67. $y = |x| + |x - 10|$ **68.** $y = 8\sqrt[3]{x - 6}$

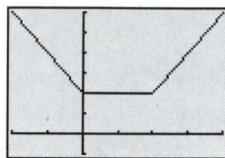   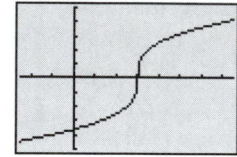

In Exercises 69–78, find the standard form of the equation of the specified circle.

69. Center: $(0, 0)$; radius: 3

70. Center: $(0, 0)$; radius: 5

71. Center: $(2, -1)$; radius: 4

72. Center: $\left(0, \frac{1}{3}\right)$; radius: $\frac{1}{3}$

73. Center: $(-1, 2)$; solution point: $(0, 0)$

74. Center: $(3, -2)$; solution point: $(-1, 1)$

75. Endpoints of a diameter: $(0, 0)$, $(6, 8)$

76. Endpoints of a diameter: $(-4, -1)$, $(4, 1)$

77.   **78.**

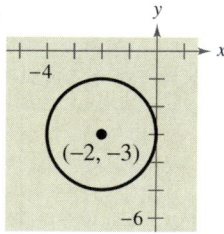   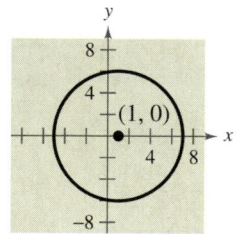

In Exercises 79–84, find the center and radius, and sketch the graph of the equation.

79. $x^2 + y^2 = 4$

80. $x^2 + y^2 = 16$

81. $(x - 1)^2 + (y + 3)^2 = 4$

82. $x^2 + (y - 1)^2 = 1$

83. $\left(x - \frac{1}{2}\right)^2 + \left(y - \frac{1}{2}\right)^2 = \frac{9}{4}$

84. $(x - 2)^2 + (y + 1)^2 = 2$

In Exercises 85 and 86, use a graphing utility to graph y_1 and y_2. Use a square setting. Identify the graph.

85. $y_1 = \sqrt{9 - x^2}$
$y_2 = -\sqrt{9 - x^2}$

86. $y_1 = 2 + \sqrt{16 - (x - 1)^2}$
$y_2 = 2 - \sqrt{16 - (x - 1)^2}$

In Exercises 87–90, explain how to use a graphing utility to verify that $y_1 = y_2$. Identify the rule of algebra that is illustrated.

87. $y_1 = \frac{1}{4}(x^2 - 8)$ **88.** $y_1 = \frac{1}{2}x + (x + 1)$
$y_2 = \frac{1}{4}x^2 - 2$ $y_2 = \frac{3}{2}x + 1$

89. $y_1 = \frac{1}{5}[10(x^2 - 1)]$ **90.** $y_1 = (x - 3) \cdot \dfrac{1}{x - 3}$
$y_2 = 2(x^2 - 1)$ $y_2 = 1$

91. *Depreciation*   A manufacturing plant purchases a new molding machine for \$225,000. The depreciated value y after t years is given by

$$y = 225{,}000 - 20{,}000t, \qquad 0 \le t \le 8.$$

Sketch the graph of the equation.

92. *Dimensions of a Rectangle*   A rectangle of length x and width w has a perimeter of 12 meters.

(a) Draw a rectangle that gives a visual representation of the problem. Use the specified variables to label the sides of the rectangle.

(b) Show that the width of the rectangle is $w = 6 - x$ and its area is $A = x(6 - x)$.

(c) Use a graphing utility to graph the area equation.

(d) From the graph of part (c), estimate the dimensions of the rectangle that yield a maximum area.

93. *Think About It*   Suppose you correctly enter an expression for the variable y on a graphing utility. However, no graph appears on the display when you graph the equation. Give a possible explanation and the steps you could take to remedy the problem. Illustrate your explanation with an example.

Data Analysis In Exercises 94 and 95, (a) sketch a graph to compare the data and the model for the data, (b) use the model to estimate the value of *y* for the year 1998, and (c) repeat part (b) for the year 2000.

94. **Federal Debt** The table gives the per capita federal debt for the United States for several years. (Source: U.S. Treasury Department)

Year	1950	1960	1970
Per Capita Debt	$1688	$1572	$1807

Year	1980	1990	1994
Per Capita Debt	$3981	$12,848	$15,750

A mathematical model for the per capita debt during this period is

$$y = 0.255t^3 - 4.096t^2 + 1570.417$$

where *y* represents the per capita debt and *t* is the time in years, with $t = 0$ corresponding to 1950.

95. **Life Expectancy** The table gives the life expectancy of a child (at birth) in the United States for selected years from 1920 to 1990. (Source: Department of Health and Human Services)

Year	1920	1930	1940	1950
Life Expectancy	54.1	59.7	62.9	68.2

Year	1960	1970	1980	1990
Life Expectancy	69.7	70.8	73.7	75.4

A mathematical model for the life expectancy during this period is

$$y = \frac{t + 66.93}{0.01t + 1}$$

where *y* represents the life expectancy and *t* is the time in years, with $t = 0$ corresponding to 1950.

96. **Think About It** Find *a* and *b* if the graph of $y = ax^2 + bx^3$ is symmetric to (a) the *y*-axis and (b) the origin. (The answer is not unique.)

97. **Dividends Per Share** The dividends per common share of Procter and Gamble Company from 1990 through 1994 can be approximated by the mathematical model

$$y = 0.086t + 0.872, \qquad 0 \le t \le 4$$

where *y* is the dividend and *t* is the calendar year, with $t = 0$ corresponding to 1990. Sketch a graph of this equation. (Source: Procter and Gamble Company 1994 Annual Report)

98. **Copper Wire** The resistance *y* in ohms of 1000 feet of solid copper wire at 77 degrees Fahrenheit can be approximated by the mathematical model

$$y = \frac{10{,}770}{x^2} - 0.37, \qquad 5 \le x \le 100$$

where *x* is the diameter of the wire in mils (0.001 in.). Use the model to estimate the resistance when $x = 50$. (Source: American Wire Gage)

Review Solve Exercises 99–108 as a review of the skills and problem-solving techniques you learned in previous sections.

99. Identify the terms: $9x^5 + 4x^3 - 7$.

100. Write the expression using exponential notation:
$$-(7 \times 7 \times 7 \times 7)$$

101. True or False? $\dfrac{1}{3 \cdot 4^{-1}} = 3 \cdot 4$

102. True or False? $(3 + 4)^2 = 3^2 + 4^2$

103. Simplify: $\sqrt{18x} - \sqrt{2x}$

104. Simplify: $\sqrt[4]{x^5}$

105. Simplify: $\dfrac{70}{\sqrt{7x}}$

106. Simplify: $\dfrac{55}{\sqrt{20} - 3}$

107. Simplify: $\sqrt[6]{t^2}$

108. Simplify: $\sqrt[3]{\sqrt{y}}$

See Exercises 89 and 90 on page 105 for examples of how linear equations can be used to solve problems in anthropology.

1.2 *Linear Equations*

Equations and Solutions of Equations   □   Linear Equations in One Variable   □   Equations Involving Fractional Expressions   □   Applications

Equations and Solutions of Equations

An **equation** is a statement that two algebraic expressions are equal. For example, $3x - 5 = 7$, $x^2 - x - 6 = 0$, and $\sqrt{2x} = 4$ are equations. To **solve** an equation in x means to find all values of x for which the equation is true. Such values are **solutions.** For instance, $x = 4$ is a solution of the equation $3x - 5 = 7$, because $3(4) - 5 = 7$ is a true statement.

The solutions of an equation depend on the kinds of numbers being considered. For instance, in the set of rational numbers $x^2 = 10$ has no solution because there is no rational number whose square is 10. However, in the set of real numbers the equation has the two solutions $\sqrt{10}$ and $-\sqrt{10}$.

An equation that is true for *every* real number in the domain of the variable is called an **identity.** For example, $x^2 - 9 = (x + 3)(x - 3)$ is an identity because it is a true statement for any real value of x, and $x/(3x^2) = 1/(3x)$ where $x \neq 0$, is an identity because it is true for any nonzero real value of x.

An equation that is true for just *some* (or even none) of the real numbers in the domain of the variable is called a **conditional equation.** For example, the equation $x^2 - 9 = 0$ is conditional because $x = 3$ and $x = -3$ are the only values in the domain that satisfy the equation. Learning to solve conditional equations is the primary focus of this chapter.

Linear Equations in One Variable

> **DEFINITION OF LINEAR EQUATION**
>
> A **linear equation** in one variable x is an equation that can be written in the standard form
>
> $$ax + b = 0$$
>
> where a and b are real numbers with $a \neq 0$.

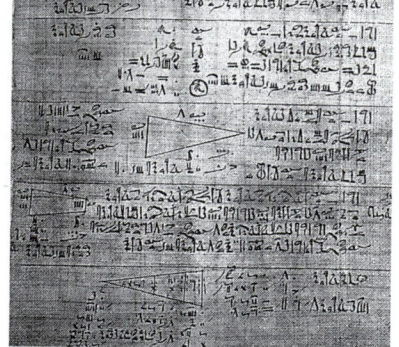

This ancient Egyptian papyrus, discovered in 1858, contains one of the earliest examples of mathematical writing in existence. The papyrus itself dates back to around 1650 B.C., but it is actually a copy of writings from two centuries earlier. The algebraic equations on the papyrus were written in words. Diophantus, a Greek who lived around A.D. 250, is often called the Father of Algebra. He was the first to use abbreviated word forms in equations. *(Photo: © British Museum)*

A linear equation has exactly one solution. To see this, consider the following steps. (Remember that $a \neq 0$.)

$$ax + b = 0 \qquad \text{Original equation}$$

$$ax = -b \qquad \text{Subtract } b \text{ from both sides.}$$

$$x = -\frac{b}{a} \qquad \text{Divide both sides by } a.$$

To solve a conditional equation in x, isolate x on one side of the equation by a sequence of **equivalent** (and usually simpler) equations, each having the same solution(s) as the original equation. The operations that yield equivalent equations come from the Substitution Principle and the simplification techniques studied in Chapter P.

■ **GENERATING EQUIVALENT EQUATIONS**

An equation can be transformed into an *equivalent equation* by one or more of the following steps.

	Given Equation	*Equivalent Equation*
1. Remove symbols of grouping, combine like terms, or reduce fractions on one or both sides of the equation.	$2x - x = 4$	$x = 4$
2. Add (or subtract) the same quantity to (from) *both* sides of the equation.	$x + 1 = 6$	$x = 5$
3. Multiply (or divide) *both* sides of the equation by the same *nonzero* quantity.	$2x = 6$	$x = 3$
4. Interchange the two sides of the equation.	$2 = x$	$x = 2$

■

EXAMPLE 1 *Solving a Linear Equation*

Solve $3x - 6 = 0$.

Solution

$3x - 6 = 0$	Original equation
$3x = 6$	Add 6 to both sides.
$x = 2$	Divide both sides by 3.

Check: After solving an equation, you should **check each solution** in the *original* equation.

$3x - 6 = 0$	Original equation
$3(2) - 6 \overset{?}{=} 0$	Substitute 2 for x.
$0 = 0$	Solution checks. ✔

Exploration

Use a graphing utility to graph the equation $y = 3x - 6$. Use the result to estimate the x-intercept of the graph. Explain how the x-intercept is related to the solution of the equation $3x - 6 = 0$, as shown in Example 1.

Some linear equations have no solutions because all the x-terms sum to zero and a contradictory (false) statement such as $0 = 5$ or $12 = 7$ is obtained. For instance, the linear equation $x = x + 1$ has no solution. Watch for this type of linear equation in the exercises.

Study Tip

Students sometimes tell us that a solution looks easy when we work it out in class, but that they don't see where to begin when trying it alone. Keep in mind that no one— not even great mathematicians— can expect to look at every mathematical problem and immediately know where to begin. Many problems involve some trial and error before a solution is found. To make algebra work for you, you must put in a lot of time, you must expect to try solution methods that end up not working, and you must learn from both your successes and your failures.

EXAMPLE 2 *Solving a Linear Equation*

Solve $6(x - 1) + 4 = 3(7x + 1)$.

Solution

$$6(x - 1) + 4 = 3(7x + 1) \qquad \text{Original equation}$$
$$6x - 6 + 4 = 21x + 3 \qquad \text{Remove parentheses.}$$
$$6x - 2 = 21x + 3 \qquad \text{Simplify.}$$
$$-15x = 5 \qquad \text{Add 2 and subtract } 21x \text{ from both sides.}$$
$$x = -\frac{1}{3} \qquad \text{Divide both sides by } -15.$$

Check: Check this solution by substitution in the original equation.

$$6(x - 1) + 4 = 3(7x + 1) \qquad \text{Original equation}$$
$$6\left(-\tfrac{1}{3} - 1\right) + 4 \stackrel{?}{=} 3\left[7\left(-\tfrac{1}{3}\right) + 1\right] \qquad \text{Replace } x \text{ by } -\tfrac{1}{3}.$$
$$6\left(-\tfrac{4}{3}\right) + 4 \stackrel{?}{=} 3\left[-\tfrac{7}{3} + 1\right] \qquad \text{Add fractions.}$$
$$-\tfrac{24}{3} + 4 \stackrel{?}{=} -\tfrac{21}{3} + 3$$
$$-8 + 4 \stackrel{?}{=} -7 + 3 \qquad \text{Simplify.}$$
$$-4 = -4 \qquad \text{Solution checks.} \checkmark$$

TECHNOLOGY

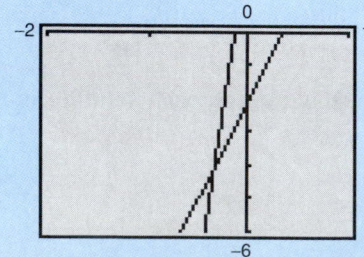

You can use a graphing utility to check that a solution is reasonable. One way to do this is to use a technique called *graph the left side, then graph the right side*. For instance, in Example 2, if you graph the equations

$$y = 6(x - 1) + 4 \qquad \text{The left side}$$
$$y = 3(7x + 1) \qquad \text{The right side}$$

on the same screen, they should intersect when $x = -\tfrac{1}{3}$, as shown in the graph at the left.

Equations Involving Fractional Expressions

To solve an equation involving fractional expressions, find the least common denominator of all terms and multiply every term by this LCD.

EXAMPLE 3 *An Equation Involving Fractional Expressions*

Solve $\dfrac{x}{3} + \dfrac{3x}{4} = 2$.

Solution

$$\frac{x}{3} + \frac{3x}{4} = 2 \qquad \text{Original equation}$$

$$(12)\frac{x}{3} + (12)\frac{3x}{4} = (12)2 \qquad \text{Multiply by the LCD of 12.}$$

$$4x + 9x = 24 \qquad \text{Reduce and multiply.}$$

$$13x = 24 \qquad \text{Combine like terms.}$$

$$x = \frac{24}{13} \qquad \text{Divide both sides by 13.}$$

The solution is $\frac{24}{13}$. Check this in the original equation.

When multiplying or dividing an equation by a *variable* quantity it is possible to introduce an **extraneous** solution.

NOTE An extraneous solution is one that does not satisfy the original equation. ■■

EXAMPLE 4 *An Equation with an Extraneous Solution*

Solve $\dfrac{1}{x - 2} = \dfrac{3}{x + 2} - \dfrac{6x}{x^2 - 4}$.

Solution

The LCD is $x^2 - 4$ or $(x + 2)(x - 2)$. Multiply every term by this LCD.

$$\frac{1}{x - 2}(x + 2)(x - 2) = \frac{3}{x + 2}(x + 2)(x - 2) - \frac{6x}{x^2 - 4}(x + 2)(x - 2)$$

$$x + 2 = 3(x - 2) - 6x, \qquad x \neq \pm 2$$

$$x + 2 = 3x - 6 - 6x$$

$$4x = -8$$

$$x = -2$$

In the original equation, $x = -2$ yields a denominator of zero. Therefore, $x = -2$ is an extraneous solution, and the original equation has *no solution*.

An equation with a *single fraction* on each side can be cleared of denominators by **cross multiplying,** which is equivalent to multiplying by the LCD and then reducing. For instance, in the equation

$$\frac{2}{x-3} = \frac{3}{x+1}$$

the LCD is $(x-3)(x+1)$. Multiply both sides of the equation by this LCD.

$$\frac{2}{x-3}(x-3)(x+1) = \frac{3}{x+1}(x-3)(x+1)$$

$$2(x+1) = 3(x-3), \qquad x \neq -1, x \neq 3$$

By comparing this equation with the original, you can see that the original numerators and denominators have been "cross multiplied." That is, the left numerator was multiplied by the right denominator and the right numerator was multiplied by the left denominator.

Applications

EXAMPLE 5   *Graphical Estimation*

Real Life

The total number A (in thousands) of United States Air Force personnel on active duty from 1986 through 1994 can be approximated by the linear model

$$A = -24t + 768, \qquad 6 \leq t \leq 14$$

where $t = 6$ represents 1986. Use the graph given in Figure 1.15 to estimate graphically (a) the year in which 455,000 people were on active duty and (b) the year in which 600,000 people were on active duty. (Source: U.S. Department of Defense)

Solution

a. From the graph, you can estimate that $A = 455$ corresponds to a t-value of about 13. This implies that there were about 455,000 people on active duty in about 1993.

b. From the graph, you can estimate that $A = 600$ corresponds to a t-value of about 7. This implies that there were about 600,000 people on active duty in about 1987.

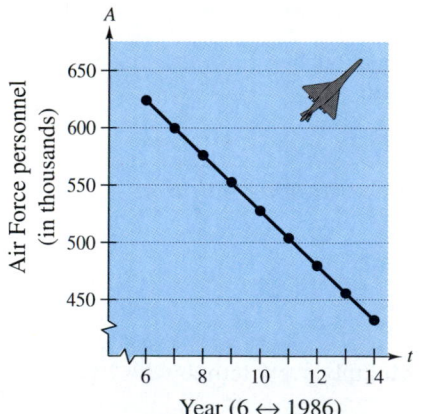

FIGURE 1.15

NOTE   Example 5 uses a graphical approach. You can also answer the question analytically. For instance, to find the year in which there were 455,000 people on active duty, you can solve the equation $455 = -24t + 768$. If you do this, you will find that the solution is $t \approx 13.04$, which agrees with the estimation found in Example 5(a). ▪▪

Earnings for U.S. Workers

Year (5 ↔ 1985)

FIGURE 1.16

EXAMPLE 6   Weekly Earnings

Real
Life

The average weekly earnings y for workers in the United States between 1985 and 1993 can be approximated by the linear model

$$y = 14.9t + 267$$

where $t = 5$ represents 1985. From Figure 1.16, you can see that the average weekly earnings increased in a linear pattern. Assuming that this pattern continues, find the year when the average weekly earnings will reach $500. (Source: U.S. Bureau of Labor Statistics)

Solution

Let $y = 500$ and solve for t.

$y = 14.9t + 267$	Original equation
$500 = 14.9t + 267$	Let $y = 500$.
$233 = 14.9t$	Subtract 267 from both sides.
$t = \dfrac{233}{14.9}$	Divide both sides by 14.9.
$t \approx 16$	Solution

Because $t = 5$ represents 1985, $t = 16$ must represent 1996. Thus, from this model the average weekly earnings will reach $500 by 1996.

GROUP ACTIVITY
SOLVING EQUATIONS

Choose one of the equations below and write a step-by-step explanation of how to solve the equation, without using another equation in the explanation. Exchange your explanation with another student—see if he or she can correctly solve the equation just by following your instructions. Discuss whether any improvements could be made to better communicate the solution process.

a. $x - 2 + \dfrac{3x - 1}{8} = \dfrac{x + 4}{4}$ **b.** $t - \{7 - [t - (7 + t)]\} = 27$

c. $\dfrac{y + 6}{21} = \dfrac{2y}{3} + \dfrac{5y + 1}{7}$ **d.** $-4 + 3(x - 1) = -(x - 1) + 3$

The *Interactive* CD-ROM provides additional help with Warm-Up exercises by providing a hypertext link to the section in which the concept was introduced.

The *Interactive* CD-ROM contains step-by-step solutions to all odd-numbered Section and Review Exercises. It also provides Tutorial Exercises, which link to Guided Examples for additional help.

WARM UP

Perform the operations and simplify.

1. $(2x - 4) - (5x + 6)$ **2.** $(3x - 5) + (2x - 7)$

3. $2(x + 1) - (x + 2)$ **4.** $-3(2x - 4) + 7(x + 2)$

5. $\dfrac{x}{3} + \dfrac{x}{5}$ **6.** $x - \dfrac{x}{4}$

7. $\dfrac{1}{x + 1} - \dfrac{1}{x}$ **8.** $\dfrac{2}{x} + \dfrac{3}{x}$

9. $\dfrac{4}{x} + \dfrac{3}{x - 2}$ **10.** $\dfrac{1}{x + 1} - \dfrac{1}{x - 1}$

1.2 Exercises

In Exercises 1–8, determine whether the values of x are solutions of the equation.

Equation	*Values*	
1. $5x - 3 = 3x + 5$	(a) $x = 0$	(b) $x = -5$
	(c) $x = 4$	(d) $x = 10$
2. $7 - 3x = 5x - 17$	(a) $x = -3$	(b) $x = 0$
	(c) $x = 8$	(d) $x = 3$
3. $3x^2 + 2x - 5$	(a) $x = -3$	(b) $x = 1$
$\quad = 2x^2 - 2$	(c) $x = 4$	(d) $x = -5$
4. $5x^3 + 2x - 3$	(a) $x = 2$	(b) $x = -2$
$\quad = 4x^3 + 2x - 11$	(c) $x = 0$	(d) $x = 10$
5. $\dfrac{5}{2x} - \dfrac{4}{x} = 3$	(a) $x = -\frac{1}{2}$	(b) $x = 4$
	(c) $x = 0$	(d) $x = \frac{1}{4}$
6. $3 + \dfrac{1}{x + 2} = 4$	(a) $x = -1$	(b) $x = -2$
	(c) $x = 0$	(d) $x = 5$
7. $(x + 5)(x - 3) = 20$	(a) $x = 3$	(b) $x = -2$
	(c) $x = 0$	(d) $x = -7$
8. $\sqrt[3]{x - 8} = 3$	(a) $x = 2$	(b) $x = -5$
	(c) $x = 35$	(d) $x = 8$

In Exercises 9–18, determine whether the equation is an identity or a conditional equation.

9. $2(x - 1) = 2x - 2$

10. $3(x + 2) = 5x + 4$

11. $-6(x - 3) + 5 = -2x + 10$

12. $3(x + 2) - 5 = 3x + 1$

13. $4(x + 1) - 2x = 2(x + 2)$

14. $-7(x - 3) + 4x = 3(7 - x)$

15. $x^2 - 8x + 5 = (x - 4)^2 - 11$

16. $x^2 + 2(3x - 2) = x^2 + 6x - 4$

17. $3 + \dfrac{1}{x + 1} = \dfrac{4x}{x + 1}$

18. $\dfrac{5}{x} + \dfrac{3}{x} = 24$

19. *Think About It*   What is meant by equivalent equations? Give an example of two equivalent equations.

20. *Essay*   In your own words, describe the steps used to transform an equation into an equivalent equation.

In Exercises 21 and 22, justify each step of the solution.

21.
$$4x + 32 = 83$$
$$4x + 32 - 32 = 83 - 32$$
$$4x = 51$$
$$\frac{4x}{4} = \frac{51}{4}$$
$$x = \frac{51}{4}$$

22. $3(x - 4) + 10 = 7$
$$3x - 12 + 10 = 7$$
$$3x - 2 = 7$$
$$3x - 2 + 2 = 7 + 2$$
$$3x = 9$$
$$\frac{3x}{3} = \frac{9}{3}$$
$$x = 3$$

In Exercises 23–26, solve the equation mentally.

23. $3x = 15$ **24.** $\frac{1}{2}t = 7$

25. $s + 12 = 18$ **26.** $2u - 3 = 25$

In Exercises 27–30, solve the equation in two ways. Then explain which way is easier.

27. $3(x - 1) = 4$ **28.** $4(x + 3) = 15$

29. $\frac{1}{3}(x + 2) = 5$ **30.** $\frac{3}{4}(z - 4) = 6$

In Exercises 31–46, solve the equation and check your solution.

31. $x + 10 = 15$ **32.** $7 - x = 18$

33. $7 - 2x = 15$ **34.** $7x + 2 = 16$

35. $8x - 5 = 3x + 10$ **36.** $7x + 3 = 3x - 13$

37. $2(x + 5) - 7 = 3(x - 2)$

38. $2(13t - 15) + 3(t - 19) = 0$

39. $6[x - (2x + 3)] = 8 - 5x$

40. $3(x + 3) = 5(1 - x) - 1$

41. $\frac{5x}{4} + \frac{1}{2} = x - \frac{1}{2}$ **42.** $\frac{x}{5} - \frac{x}{2} = 3$

43. $\frac{3}{2}(z + 5) - \frac{1}{4}(z + 24) = 0$

44. $\frac{3x}{2} + \frac{1}{4}(x - 2) = 10$

45. $0.25x + 0.75(10 - x) = 3$

46. $0.60x + 0.40(100 - x) = 50$

Graphical Analysis In Exercises 47–50, use a graphing utility to graph the equation and approximate any x-intercepts. Set $y = 0$ and solve the resulting equation. Compare the results with the graph's x-intercepts.

47. $y = 2(x - 1) - 4$ **48.** $y = \frac{4}{3}x + 2$

49. $y = 20 - (3x - 10)$ **50.** $y = 10 + 2(x - 2)$

In Exercises 51–72, solve the equation (if possible) and check your solution.

51. $x + 8 = 2(x - 2) - x$

52. $8(x + 2) - 3(2x + 1) = 2(x + 5)$

53. $\frac{100 - 4u}{3} = \frac{5u + 6}{4} + 6$

54. $\frac{17 + y}{y} + \frac{32 + y}{y} = 100$

55. $\frac{5x - 4}{5x + 4} = \frac{2}{3}$ **56.** $\frac{10x + 3}{5x + 6} = \frac{1}{2}$

57. $10 - \frac{13}{x} = 4 + \frac{5}{x}$ **58.** $\frac{15}{x} - 4 = \frac{6}{x} + 3$

59. $\frac{1}{x - 3} + \frac{1}{x + 3} = \frac{10}{x^2 - 9}$

60. $\frac{1}{x - 2} + \frac{3}{x + 3} = \frac{4}{x^2 + x - 6}$

61. $\frac{x}{x + 4} + \frac{4}{x + 4} + 2 = 0$

62. $\frac{2}{(x - 4)(x - 2)} = \frac{1}{x - 4} + \frac{2}{x - 2}$

63. $\frac{7}{2x + 1} - \frac{8x}{2x - 1} = -4$

64. $\frac{4}{u - 1} + \frac{6}{3u + 1} = \frac{15}{3u + 1}$

65. $\frac{1}{x} + \frac{2}{x - 5} = 0$ **66.** $\frac{6}{x} - \frac{2}{x + 3} = \frac{3(x + 5)}{x(x + 3)}$

67. $\dfrac{3}{x(x-3)} + \dfrac{4}{x} = \dfrac{1}{x-3}$

68. $3 = 2 + \dfrac{2}{z+2}$

69. $(x+2)^2 + 5 = (x+3)^2$

70. $(x+1)^2 + 2(x-2) = (x+1)(x-2)$

71. $(x+2)^2 - x^2 = 4(x+1)$

72. $(2x+1)^2 = 4(x^2+x+1)$

In Exercises 73–76, solve for x.

73. $4(x+1) - ax = x+5$

74. $4 - 2(x-2b) = ax+3$

75. $6x + ax = 2x+5$

76. $5 + ax = 12 - bx$

In Exercises 77–82, solve the equation for x and round your solution to three decimal places.

77. $0.275x + 0.725(500 - x) = 300$

78. $2.763 - 4.5(2.1x - 5.1432) = 6.32x + 5$

79. $\dfrac{x}{0.6321} + \dfrac{x}{0.0692} = 1000$

80. $\dfrac{x}{2.625} + \dfrac{x}{4.875} = 1$

81. $\dfrac{2}{7.398} - \dfrac{4.405}{x} = \dfrac{1}{x}$

82. $\dfrac{3}{6.350} - \dfrac{6}{x} = 18$

In Exercises 83–86, evaluate each expression in two ways. (a) Calculate entirely on your calculator by storing intermediate results and then rounding the final answer to two decimal places. (b) Round both the numerator and denominator to two decimal places before dividing, and then round the final answer to two decimal places. Does the method from part (b) introduce an additional roundoff error?

83. $\dfrac{1 + 0.73205}{1 - 0.73205}$

84. $\dfrac{1 + 0.86603}{1 - 0.86603}$

85. $\dfrac{3.33 + \frac{1.98}{0.74}}{4 + \frac{6.25}{3.15}}$

86. $\dfrac{1.73205 - 1.19195}{3 - (1.73205)(1.19195)}$

87. *Exploration*

(a) Complete the table.

x	−1	0	1	2	3	4
$3.2x - 5.8$						

(b) Use the table from part (a) to determine the interval in which the solution to the equation $3.2x - 5.8 = 0$ is located. Explain your reasoning.

(c) Complete the table.

x	1.5	1.6	1.7	1.8	1.9	2
$3.2x - 5.8$						

(d) Use the table from part (c) to determine the interval in which the solution to the equation $3.2x - 5.8 = 0$ is located. Explain how this process can be used to approximate the solution to any desired degree of accuracy.

88. *Exploration*   Use the procedure of Exercise 87 to approximate the solution to the equation $0.3(x - 1.5) - 2 = 0$ accurate to two decimal places.

Human Height   **In Exercises 89 and 90, use the following information. The relationship between the length of an adult's thigh bone and the height of the adult can be approximated by the linear equations**

$$y = 0.432x - 10.44 \qquad \text{Female}$$
$$y = 0.449x - 12.15 \qquad \text{Male}$$

where y is the length of the thigh bone in inches and x is the height in inches (see figure below).

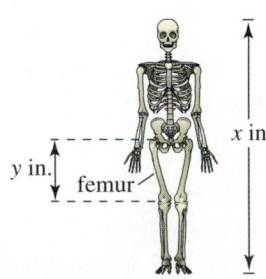

89. An anthropologist discovers a thigh bone belonging to an adult human female. The bone is 16 inches long. Estimate the height of the female.

90. From the foot bones of an adult human male, an anthropologist estimates that the person's height was 69 inches. A few feet away from the site where the foot bone was discovered, the anthropologist discovered a male adult thigh bone that was 19 inches long. Is it likely that both bones came from the same person?

Negative Income Tax In Exercises 91–94, use the following information about a possible negative income tax for a family consisting of two adults and two children (see figure).

Earned income: $I = x$

Subsidy: $S = 10,000 - \frac{1}{2}x,$ $0 \leq x \leq 20,000$

Total income: $T = I + S$

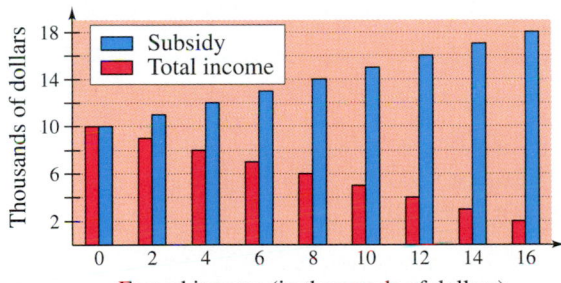

91. Express the total income T in terms of x.

92. Find the earned income x if the subsidy is $6600.

93. Find the earned income x if the total income is $13,800.

94. Find the subsidy S if the total income is $12,500.

95. ***Geometry*** The surface area S of the rectangular solid in the figure is given by

$$S = 2(24) + 2(4x) + 2(6x).$$

Find the length x of the box if the surface area is 248 square centimeters.

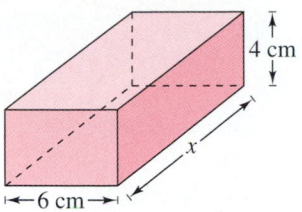

FIGURE FOR 95

96. ***Using a Model*** The number of married women y in the civilian work force (in millions) in the United States from 1988 to 1992 can be approximated by the model

$$y = 0.43t + 30.86$$

where $t = 0$ represents 1990 (see figure). According to this model, during which year did this number reach 30 million? Explain how to answer the question graphically and algebraically. (Source: U.S. Bureau of Labor Statistics)

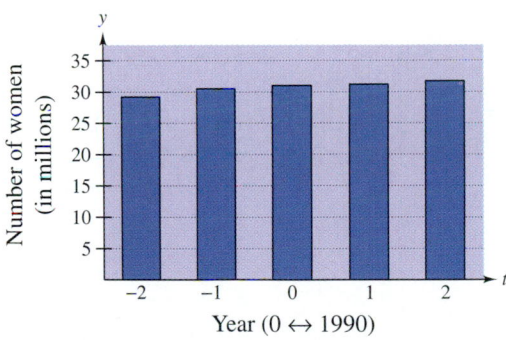

Year (0 ↔ 1990)

97. ***Operating Cost*** A delivery company has a fleet of vans. The annual operating cost per van is

$$C = 0.32m + 2500$$

where m is the number of miles traveled by a van in a year. What number of miles will yield an annual operating cost that is equal to $10,000?

1.3 *Modeling with Linear Equations*

See Exercise 41 on page 114 for an example of how a linear model can be used to analyze the income of the federal government.

Introduction to Problem Solving ❑ *Using Mathematical Models* ❑
Mixture Problems ❑ *Common Formulas*

Introduction to Problem Solving

In this section you will learn how algebra can be used to solve problems that occur in real-life situations. This procedure is **mathematical modeling.** A good approach to mathematical modeling is to use two stages. Begin by using the verbal description of the problem to form a *verbal model*. Then, after assigning labels to the unknown quantities in the verbal model, form a *mathematical model* or *algebraic equation.*

Verbal Description ⟹ Verbal Model ⟹ Algebraic Equation

When you are trying to construct a verbal model, it is helpful to look for a *hidden equality*—a statement that two algebraic expressions are equal.

EXAMPLE 1   *Using a Verbal Model*

Real Life

You have accepted a job for which your annual salary will be $27,236. This salary includes a year-end bonus of $500. If you are paid twice a month, what will your gross pay be for each paycheck?

Solution

Because there are 12 months in a year and you will be paid twice a month, it follows that you will receive 24 paychecks during the year. You can construct an algebraic equation for this problem as follows. Begin with a verbal model, then assign labels, and finally form an algebraic equation.

Verbal Model: 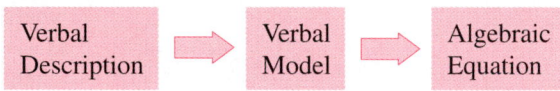 Income for year = 24 paychecks + Bonus

Labels: Income for year = 27,236   *(dollars)*
Amount of each paycheck = x *(dollars)*
Bonus = 500   *(dollars)*

Equation: $27,236 = 24x + 500$

The algebraic equation for this problem is a *linear equation* in the variable x. Using the techniques discussed in Section 1.2, you can solve this equation for x, and you will find that the solution is $x = \$1114$.

TRANSLATING KEY WORDS AND PHRASES

Key Words and Phrases	Verbal Description	Algebraic Statement
Equality:		
Equals, equal to, is, are, was, will be, represents	The sale price S is $10 less than the list price L.	$S = L - 10$
Addition:		
Sum, plus, greater, increased by, more than, exceeds, total of	The sum of 5 and x Seven more than y	$5 + x$ or $x + 5$ $7 + y$ or $y + 7$
Subtraction:		
Difference, minus, less, decreased by, subtracted from, reduced by, the remainder	The difference of 4 and b Three less than z	$4 - b$ $z - 3$
Multiplication:		
Product, multiplied by, twice, times, percent of	Two times x	$2x$
Division:		
Quotient, divided by, ratio, per	The ratio of x and 8	$\dfrac{x}{8}$

Using Mathematical Models

EXAMPLE 2　*Finding the Percent of a Raise*

Real Life

You have accepted a job that pays $8 an hour. You are told that after a 2-month probationary period, your hourly wage will be increased to $9 an hour. What percent raise will you receive after the 2-month period?

Solution

Verbal Model:　　 Raise $=$ Percent $\cdot$ Old wage

Labels:　　Old wage $= 8$　　*(dollars per hour)*
　　　　New wage $= 9$　　*(dollars per hour)*
　　　　Raise $= 9 - 8 = 1$　　*(dollars per hour)*
　　　　Percent $= r$　　*(percent in decimal form)*

Equation:　　$1 = r \cdot 8$
　　　　　　$\frac{1}{8} = r$
　　　　$0.125 = r$

You will receive a raise of 0.125 or 12.5%.

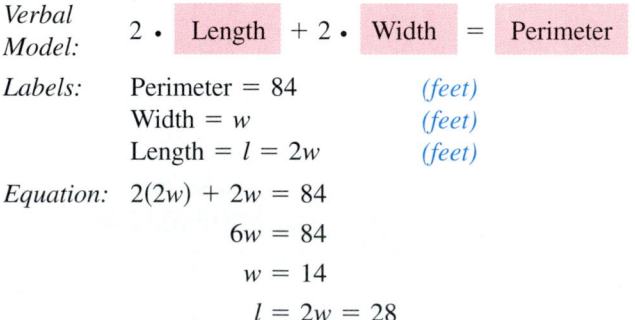

EXAMPLE 3 *Finding the Percent of a Benefit Package*

Your annual salary is $24,000. In addition to your salary, your employer also provides the following benefits: employer's portion of Social Security ($1836), worker's compensation ($120), unemployment compensation ($180), medical insurance ($2240), retirement contribution ($1560). The total of this benefit package represents what percent of your annual salary?

Solution

The total amount of your benefit package is $5936.

Verbal Model: | Benefit package | = | Percent | · | Salary |

Labels: Salary = 24,000   *(dollars)*
           Benefit package = 5936   *(dollars)*
           Percent = r *(in decimal form)*

Equation: $5936 = r \cdot 24{,}000$

$$\frac{5936}{24{,}000} = r$$

$$0.247 \approx r$$

Your benefit package is approximately 0.247 or 24.7% of your salary.

EXAMPLE 4 *Finding the Dimensions of a Room*

A rectangular family room is twice as long as it is wide, and its perimeter is 84 feet. Find the dimensions of the family room.

Solution

For this problem, it helps to sketch a picture, as shown in Figure 1.17.

Verbal Model: | 2 · | Length | + 2 · | Width | = | Perimeter |

Labels: Perimeter = 84   *(feet)*
           Width = w *(feet)*
           Length = $l = 2w$ *(feet)*

Equation: $2(2w) + 2w = 84$

$$6w = 84$$

$$w = 14$$

$$l = 2w = 28$$

The dimensions of the room are 14 feet by 28 feet.

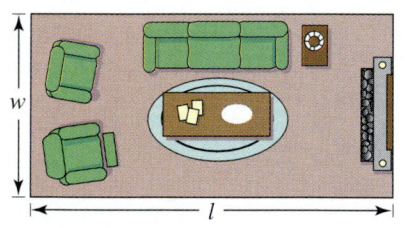

FIGURE 1.17

FIGURE 1.18

Real
Life

EXAMPLE 5   *A Distance Problem*

A plane is flying nonstop from New York to San Francisco, a distance of about 2700 miles, as shown in Figure 1.18. After $1\frac{1}{2}$ hours in the air, the plane flies over Chicago (a distance of 800 miles from New York). Estimate the time it will take the plane to fly from New York to San Francisco.

Solution

Verbal Model: $\boxed{\text{Distance}} = \boxed{\text{Rate}} \cdot \boxed{\text{Time}}$

Labels: Distance = 2700   *(miles)*
            Time = t *(hours)*

            Rate = $\dfrac{\text{Distance to Chicago}}{\text{Time to Chicago}} = \dfrac{800}{1.5}$ *(miles per hour)*

Equation: $2700 = \dfrac{800}{1.5}t$

              $5.06 \approx t$

The trip will take about 5.06 hours or 5 hours and 4 minutes.

Real
Life

EXAMPLE 6   *An Application Involving Similar Triangles*

To measure the height of the twin towers of the World Trade Center, you measure the shadow cast by one of the buildings and find it to be 170.25 feet long, as shown in Figure 1.19. Then you measure the shadow cast by a 4-foot post and find it to be 6 inches long. Estimate the building's height.

Solution

To solve this problem, you use a result from geometry that states that the ratios of corresponding sides of similar triangles are equal.

Verbal Model: $\dfrac{\boxed{\text{Height of building}}}{\boxed{\text{Length of building's shadow}}} = \dfrac{\boxed{\text{Height of post}}}{\boxed{\text{Length of post's shadow}}}$

Labels: Height of building = x *(feet)*
            Length of building's shadow = 170.25   *(feet)*
            Height of post = 4 feet = 48 inches   *(inches)*
            Length of post's shadow = 6   *(inches)*

Equation: $\dfrac{x}{170.25} = \dfrac{48}{6}$

              $x = 1362$

Thus, the World Trade Center is about 1362 feet high.

FIGURE 1.19

x ft

48 in.

6 in.

170.25 ft   (not to scale)

Mixture Problems

EXAMPLE 7   *A Simple Interest Problem*

You invested \$10,000 at $9\frac{1}{2}\%$ and 11% simple interest. During one year, the two accounts earned \$1038.50. How much did you invest in each?

Solution

Verbal Model: | Interest from $9\frac{1}{2}\%$ | $+$ | Interest from 11% | $=$ | Total interest |

Labels:
Amount invested at $9\frac{1}{2}\% = x$ *(dollars)*
Amount invested at $11\% = 10,000 - x$ *(dollars)*
Interest from $9\frac{1}{2}\% = Prt = (x)(0.095)(1)$ *(dollars)*
Interest from $11\% = Prt = (10,000 - x)(0.11)(1)$ *(dollars)*
Total interest $= 1038.50$ *(dollars)*

Equation:
$$0.095x + 0.11(10,000 - x) = 1038.50$$
$$-0.015x = -61.5$$
$$x = \$4100 \text{ at } 9\frac{1}{2}\%$$
$$10,000 - x = \$5900 \text{ at } 11\%$$

NOTE Example 7 uses the simple interest formula $I = Prt$, where I is the interest, P is the principal, r is the annual interest rate (in decimal form), and t is the time in years.

EXAMPLE 8   *An Inventory Problem*

A store has \$30,000 of inventory in 12-inch and 19-inch color televisions. The profit on a 12-inch set is 22% and the profit on a 19-inch set is 40%. The profit for the entire stock is 35%. How much was invested in each type of television?

Solution

Verbal Model: | Profit from 12-inch sets | $+$ | Profit from 19-inch sets | $=$ | Total profit |

Labels:
Inventory of 12-inch sets $= x$ *(dollars)*
Inventory of 19-inch sets $= 30,000 - x$ *(dollars)*
Profit from 12-inch sets $= 0.22x$ *(dollars)*
Profit from 19-inch sets $= 0.40(30,000 - x)$ *(dollars)*
Total profit $= 0.35(30,000) = 10,500$ *(dollars)*

Equation:
$$0.22(x) + 0.40(30,000 - x) = 10,500$$
$$-0.18x = -1500$$
$$x \approx \$8333.33 \text{ in 12-inch sets}$$
$$30,000 - x \approx \$21,666.67 \text{ in 19-inch sets}$$

Common Formulas

Many common types of geometric, scientific, and investment problems use ready-made equations, called **formulas.** Knowing these formulas will help you translate and solve a wide variety of real-life applications.

COMMON FORMULAS FOR AREA A, PERIMETER P, CIRCUMFERENCE C, AND VOLUME V

Square

$A = s^2$

$P = 4s$

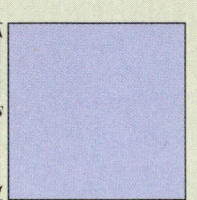

Rectangle

$A = lw$

$P = 2l + 2w$

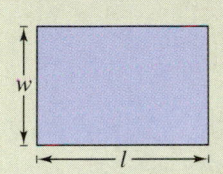

Circle

$A = \pi r^2$

$C = 2\pi r$

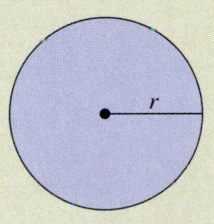

Triangle

$A = \dfrac{1}{2}bh$

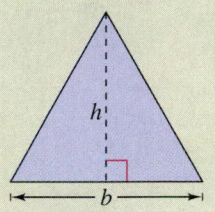

Cube

$V = s^3$

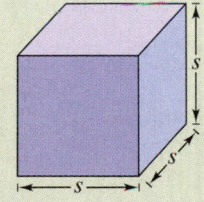

Rectangular Solid

$V = lwh$

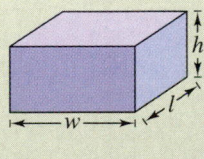

Circular Cylinder

$V = \pi r^2 h$

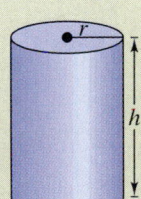

Sphere

$V = \dfrac{4}{3}\pi r^3$

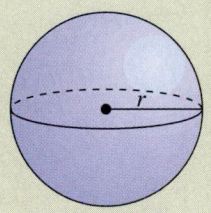

MISCELLANEOUS COMMON FORMULAS

Temperature:	$F = \dfrac{9}{5}C + 32$	F = degrees Fahrenheit, C = degrees Celsius
Simple Interest:	$I = Prt$	I = interest, P = principal, r = annual interest rate, t = time in years
Compound Interest:	$A = P\left(1 + \dfrac{r}{n}\right)^{nt}$	A = balance, P = principal, r = annual interest rate, n = compoundings per year, t = time in years
Distance:	$d = rt$	d = distance traveled, r = rate, t = time

When working with applied problems you often need to rewrite one of the common formulas. For instance, the formula $P = 2l + 2w$, for the perimeter of a rectangle, can be rewritten or solved for w as $w = \frac{1}{2}(P - 2l)$.

Real Life

EXAMPLE 9   *Using a Formula*

A cylindrical can has a volume of 300 cubic centimeters (cm^3) and a radius of 3 centimeters (cm), as shown in Figure 1.20. Find the height of the can.

Solution

The formula for the *volume of a cylinder* is $V = \pi r^2 h$. To find the height of the can, solve for h.

$$h = \frac{V}{\pi r^2}$$

Then, using $V = 300$ cm^3 and $r = 3$ cm, find the height.

$$h = \frac{300 \text{ cm}^3}{\pi(3 \text{ cm})^2} = \frac{300 \text{ cm}^3}{9\pi \text{ cm}^2} \approx 10.61 \text{ cm}$$

FIGURE 1.20

GROUP ACTIVITY

TRANSLATING ALGEBRAIC FORMULAS

Most people use algebraic formulas every day—sometimes without realizing it because they use a verbal form or think of an often-repeated calculation in steps. Translate each of the following verbal descriptions into an algebraic formula, and demonstrate the use of each formula.

a. *The Christmas Tree Rule*   "To find out how many lights your Christmas tree needs, multiply the tree height times the tree width times three." —Michael Spence, lawyer   (Source: *Rules of Thumb* by Tom Parker)

b. *Percent of Calories from Fat*   "To calculate percent of calories from fat, multiply grams of total fat per serving by 9, then divide by the number of calories per serving."   (Source: *Good Housekeeping*)

c. *Building Stairs*   "A set of steps will be comfortable to use if two times the height of one riser plus the width of one tread is equal to 26 inches." —Alice Lukens Bachelder, gardener   (Source: *Rules of Thumb* by Tom Parker)

Solve the equation (if possible) and check your solution.

1. $3x - 42 = 0$

2. $64 - 16x = 0$

3. $2 - 3x = 14 + x$

4. $7 + 5x = 7x - 1$

5. $5[1 + 2(x + 3)] = 6 - 3(x - 1)$

6. $2 - 5(x - 1) = 2[x + 10(x - 1)]$

7. $\dfrac{x}{3} + \dfrac{x}{2} = \dfrac{1}{3}$

8. $\dfrac{2}{x} + \dfrac{2}{5} = 1$

9. $1 - \dfrac{2}{z} = \dfrac{z}{z + 3}$

10. $\dfrac{x}{x + 1} - \dfrac{1}{2} = \dfrac{4}{3}$

1.3 Exercises

In Exercises 1–6, write a verbal description of the algebraic expression without using the variable.

1. $x + 4$

2. $t - 10$

3. $\dfrac{u}{5}$

4. $\dfrac{2}{3}x$

5. $\dfrac{y - 4}{5}$

6. $-3(b + 2)$

In Exercises 7–18, write an algebraic expression for the verbal description.

7. The sum of two consecutive natural numbers

8. The product of two consecutive natural numbers

9. The product of two consecutive odd integers, the first of which is $2n - 1$

10. The sum of the squares of two consecutive even integers, the first of which is $2n$

11. *Distance Traveled*   The distance traveled in t hours by a car traveling at 50 miles per hour

12. *Travel Time*   The travel time for a plane traveling at a rate of r kilometers per hour for 200 kilometers

13. *Acid Solution*   The amount of acid in x liters of a 20% acid solution

14. *Discount*   The sale price for an item that is discounted 20% of its list price L

15. *Perimeter*   The perimeter of a rectangle with a width x and a length that is twice the width

16. *Area of a Triangle*   The area of a triangle with base 20 inches and height h inches

17. *Cost*   The total cost of producing x units for which the fixed costs are \$1200 and the cost per unit is \$25

18. *Total Revenue*   The total revenue obtained by selling x units at \$3.59 per unit

In Exercises 19 and 20, write an expression for the area of the region.

19.

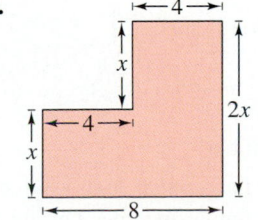

20.

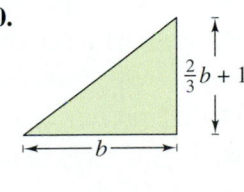

Number Problems   **In Exercises 21–26, write a mathematical model for the number problem and solve.**

21. The sum of two consecutive natural numbers is 525. Find the numbers.

22. The sum of three consecutive natural numbers is 804. Find the numbers.

23. One positive number is 5 times another number. The difference between the two numbers is 148. Find the numbers.

24. One positive number is one-fifth of another number. The difference between the two numbers is 76. Find the numbers.

25. Find two consecutive integers whose product is 5 less than the square of the smaller number.

26. Find two consecutive natural numbers such that the difference of their reciprocals is one-fourth the reciprocal of the smaller number.

In Exercises 27–30, translate the statement into an algebraic expression or equation.

27. Thirty percent of the list price L

28. The amount of water in q quarts of a product that is 35% water

29. The number N is what percent of 500?

30. The percent change in sales from one month to the next if the monthly sales are S_1 and S_2, respectively

31. What is 30% of 45?

32. What is 175% of 360?

33. What is 0.045% of 2,650,000?

34. 432 is what percent of 1600?

35. 459 is what percent of 340?

36. 12 is $\frac{1}{2}$% of what number?

37. 70 is 40% of what number?

38. $825 is 250% of what number?

39. *Private Debt*   A family has annual loan payments equaling 58.6% of their annual income. During the year, their loan payments total $13,077.75. What is their income?

40. *Discount*   The price of a swimming pool has been discounted 16.5%. The sale price is $1210.75. Find the original list price of the pool.

41. *Federal Government Income*   The circle graph shows the sources of income for the federal government in 1993. The total income was $1,147,588,000,000. Find the income for each of the categories. (Round your answers to the nearest billion dollars.) (Source: Office of Management and Budget)

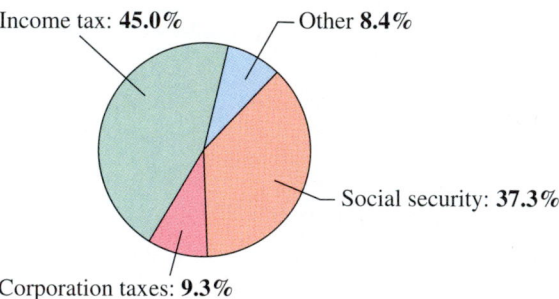

42. *Federal Government Expenses*   The circle graph shows expenses of the federal government in 1993. The total expenses were $1,474,935,000,000. Find the expense for each of the categories. (Round your answers to the nearest billion dollars.) (Source: Office of Management and Budget)

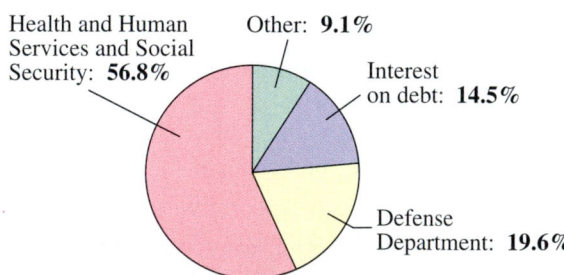

43. *Monthly Profit*   The total profit for a company in February was 20% higher than it was in January. The total profit for the two months was $157,498. Find the profit for each month.

44. *Weekly Paycheck* Your weekly paycheck is 15% less than your coworker's. Your two paychecks total $645. Find the amount of each paycheck.

Then and Now **In Exercises 45–48, the values or prices of different items are given for 1980 and 1992. Find the percent change for each item.** (Source: 1993 Statistical Abstract of the U.S.)

Item	1980	1992
45. Weekly family earnings	$400.00	$688.00
46. Cable TV monthly basic rate	$7.85	$19.08
47. An ounce of gold	$613.00	$350.00
48. One acre of farmland	$737.00	$685.00

49. *Dimensions of a Room* A room is 1.5 times as long as it is wide, and its perimeter is 25 meters.

(a) Create a figure that gives a visual representation of the problem. Identify the length and width as l and w, respectively.

(b) Write l in terms of w and write an equation for the perimeter in terms of w.

(c) Find the dimensions of the room.

50. *Dimensions of a Picture* A picture frame (see figure) has a total perimeter of 2 meters. The height of the frame is 0.62 times its width. Find the dimensions of the frame.

51. *Course Grade* To get an A in a course, you must have an average of at least 90 on four tests of 100 points each. The scores on your first three tests were 87, 92, and 84. What must you score on the fourth test to get an A for the course?

52. *Course Grade* Suppose you are taking a course that has four tests. The first three tests are 100 points each and the fourth test is 200 points. To get an A in the course, you must have an average of at least 90% on the four tests. Your scores on the first three tests were 87, 92, and 84. What must you score on the fourth test to get an A for the course?

53. *Travel Time* Suppose you are driving on a Canadian freeway to a town that is 300 kilometers from your home. After 30 minutes you pass a freeway exit that you know is 50 kilometers from your home. Assuming that you continue at the same constant speed, how long will it take for the entire trip?

54. *Travel Time* On the first part of a 317-mile trip, a salesman averaged 58 miles per hour. He averaged only 52 miles per hour on the last part of the trip because of an increased volume of traffic. Find the amount of time at each of the speeds if the total time was 5 hours and 45 minutes.

55. *Travel Time* Two cars start at a given point and travel in the same direction at average speeds of 40 miles per hour and 55 miles per hour. How much time must elapse before the two cars are 5 miles apart?

56. *Catch-up Time* Students are traveling in two cars to a football game 135 miles away. The first car leaves on time and travels at an average speed of 45 miles per hour. The second car starts $\frac{1}{2}$ hour later and travels at an average speed of 55 miles per hour. How long will it take the second carload of students to catch up to the first car? Will the second car catch up to the first car before the first car arrives at the game?

57. *Travel Time* Two families meet at a park for a picnic. At the end of the day one family travels east at an average speed of 42 miles per hour and the other travels west at an average speed of 50 miles per hour. Both families have approximately 160 miles to travel.

(a) Find the time it takes each family to get home.

(b) Find the time that will have elapsed when they are 100 miles apart.

(c) Find the distance the eastbound family has to travel after the westbound family has arrived home.

58. *Average Speed* A truck driver traveled at an average speed of 55 miles per hour on a 200-mile trip to pick up a load of freight. On the return trip (with the truck fully loaded), the average speed was 40 miles per hour. Find the average speed for the round trip.

59. *Wind Speed* An executive flew in the corporate jet to a meeting in a city 1500 kilometers away. After traveling the same amount of time on the return flight, the pilot mentioned that they still had 300 kilometers to go. If the air speed of the plane was 600 kilometers per hour, how fast was the wind blowing? (Assume that the wind direction was parallel to the flight path and constant all day.)

60. *Speed of Light* Light travels at the speed of 3.0×10^8 meters per second. Find the time in minutes required for light to travel from the sun to the earth (a distance of 1.5×10^{11} meters).

61. *Radio Waves* Radio waves travel at the same speed as light, 3.0×10^8 meters per second. Find the time required for a radio wave to travel from mission control in Houston to NASA astronauts on the surface of the moon 3.86×10^8 meters away.

62. *Height of a Tree* To obtain the height of a tree you measure the tree's shadow and find that it is 8 meters long. You also measure the shadow of a 2-meter lamppost and find that it is 75 centimeters long (see figure). How tall is the tree?

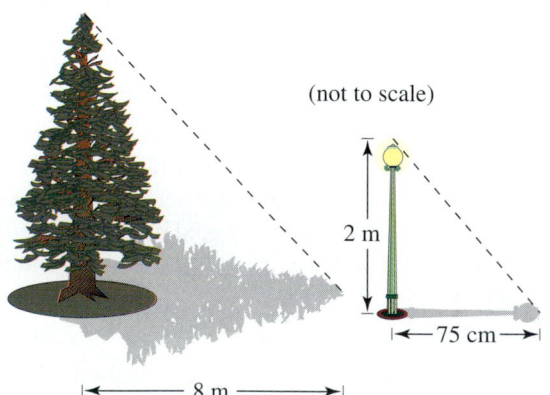

(not to scale)

2 m

|← 75 cm →|

|← 8 m →|

63. *Height of a Building* To obtain the height of a building you measure the building's shadow and find that it is 80 feet long. You also measure the shadow of a 4-foot stake and find that it is $3\frac{1}{2}$ feet long (see figure). How tall is the building?

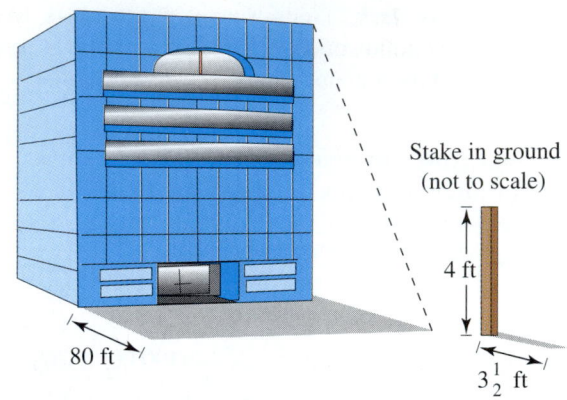

Stake in ground
(not to scale)

4 ft

80 ft

$3\frac{1}{2}$ ft

FIGURE FOR 63

64. *Height of a Flagpole* A person who is 6 feet tall walks away from a flagpole toward the tip of the shadow of the pole. When the person is 30 feet from the pole, the tips of the person's shadow and the shadow cast by the pole coincide at a point 5 feet in front of the person.

(a) Create a figure that gives a visual representation of the problem. Let h represent the height of the pole.

(b) Find the height of the pole.

65. *Walking Distance* A person who is 6 feet tall walks away from a 50-foot silo toward the tip of the silo's shadow. At a distance of 32 feet from the silo, the person's shadow begins to emerge beyond the silo's shadow. How much farther must the person walk to be completely out of the silo's shadow?

66. *Investment Mix* You plan to invest $12,000 in two funds paying $7\frac{1}{2}\%$ and 10% simple interest. (There is more risk in the 10% fund.) Your goal is to obtain a total annual interest income of $1000 from the investments. What is the smallest amount you can invest in the 10% fund in order to meet your objective?

67. *Investment Mix* You plan to invest $25,000 in two funds paying 11% and $12\frac{1}{2}\%$ simple interest. (There is more risk in the $12\frac{1}{2}\%$ fund.) Your goal is to obtain a total annual interest income of $3000 from the investments. What is the smallest amount you can invest in the $12\frac{1}{2}\%$ fund in order to meet your objective?

68. *Comparing Investment Returns* Suppose you invested $12,000 in a fund paying $9\frac{1}{2}$% simple interest and $8000 in a fund where the interest rate is variable. At the end of the year you were notified that the total interest for both funds was $2054.40. Find the equivalent simple interest rate on the variable-rate fund.

69. *Comparing Investment Returns* Suppose you have $10,000 on deposit earning simple interest with the interest rate linked to the *prime rate*. Because of a drop in the prime rate, the rate on your investment dropped by $1\frac{1}{2}$% for the last quarter of the year. Your annual earnings on the fund are $1112.50. Find the interest rate for the first three quarters of the year and the interest rate for the last quarter.

70. *Mixture Problem* Using the values from the table, determine the amounts of solutions 1 and 2, respectively, needed to obtain the desired amount and concentration of the final mixture.

	Concentration			
	Solution 1	Solution 2	Final Solution	Amount of Final Solution
(a)	10%	30%	25%	100 gal
(b)	25%	50%	30%	5 L
(c)	15%	45%	30%	10 qt
(d)	70%	90%	75%	25 gal

71. *Mixture Problem* A 55-gallon barrel contains a mixture with a concentration of 40%. How much of this mixture must be withdrawn and replaced by 100% concentrate to bring the mixture up to 75% concentration?

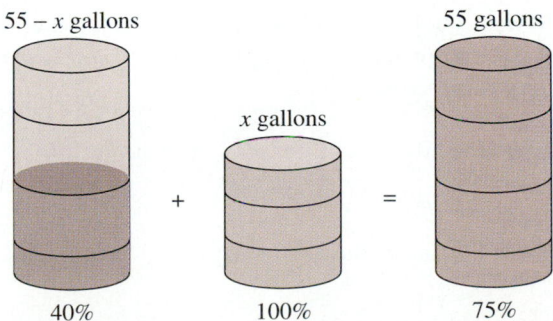

72. *Mixture Problem* A forester mixed gasoline and oil to make 2 gallons of mixture for his two-cycle chainsaw engine. This mixture was 32 parts gasoline and 1 part two-cycle oil. How much gasoline must be added to bring the mixture to 40 parts gasoline and 1 part oil?

73. *Mixture Problem* A grocer mixes two kinds of nuts that cost $2.49 per pound and $3.89 per pound, respectively, to make 100 pounds of a mixture that costs $3.19 per pound. How much of each kind of nut is put into the mixture?

74. *Production Limit* A company has fixed costs of $10,000 per month and variable costs of $8.50 per unit manufactured. The company has $85,000 available to cover the monthly costs. How many units can the company manufacture? (*Fixed costs* are those that occur regardless of the level of production. *Variable costs* depend on the level of production.)

75. *Production Limit* A company has fixed costs of $10,000 per month and variable costs of $9.30 per unit manufactured. The company has $85,000 available to cover the monthly costs. How many units can the company manufacture? (*Fixed costs* are those that occur regardless of the level of production. *Variable costs* depend on the level of production.)

76. *Water Depth* A trough is 12 feet long, 3 feet deep, and 3 feet wide (see figure). Find the depth of the water when the trough contains 70 gallons (1 gallon $\approx$ 0.13368 cubic feet).

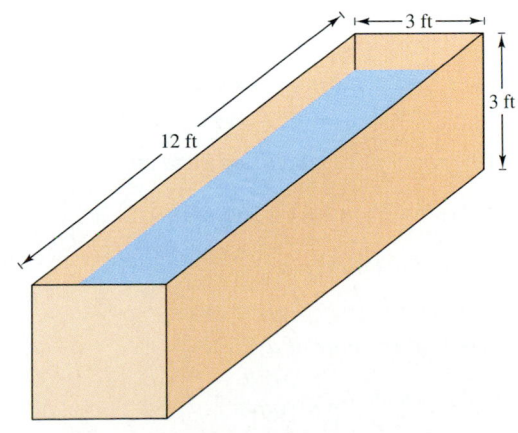

Statics Problems   In Exercises 77 and 78, suppose you have a uniform beam of length L with a fulcrum x feet from one end (see figure). If objects with weights W_1 and W_2 are placed at opposite ends of the beam, then the beam will balance if $W_1 x = W_2(L - x)$. Find x such that the beam will balance.

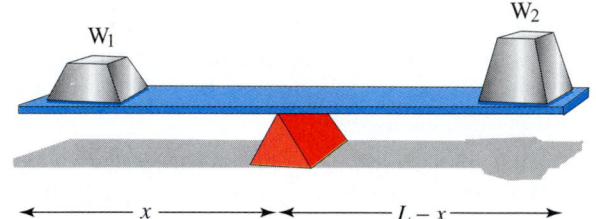

77. Two children weighing 50 pounds and 75 pounds are going to play on a seesaw that is 10 feet long.

78. A person weighing 200 pounds is attempting to move a 550-pound rock with a bar that is 5 feet long.

In Exercises 79–98, solve for the indicated variable.

79. *Area of a Triangle*

Solve for h: $A = \frac{1}{2}bh$

80. *Volume of a Right Circular Cylinder*

Solve for h: $V = \pi r^2 h$

81. *Markup*

Solve for C: $S = C + RC$

82. *Discount*

Solve for L: $S = L - RL$

83. *Investment at Simple Interest*

Solve for r: $A = P + Prt$

84. *Investment at Compound Interest*

Solve for P: $A = P\left(1 + \dfrac{r}{n}\right)^{nt}$

85. *Area of a Trapezoid*

Solve for b: $A = \frac{1}{2}(a + b)h$

86. *Area of a Sector of a Circle*

Solve for θ: $A = \dfrac{\pi r^2 \theta}{360}$

87. *Volume of a Spherical Segment*

Solve for r: $V = \frac{1}{3}\pi h^2(3r - h)$

88. *Volume of an Oblate Spheroid*

Solve for b: $V = \frac{4}{3}\pi a^2 b$

89. *Thermal Expansion*

Solve for α: $L = L_0[1 + \alpha(\Delta t)]$

90. *Free-falling Body*

Solve for a: $h = v_0 t + \frac{1}{2}at^2$

91. *Newton's Law of Universal Gravitation*

Solve for m_2: $F = \alpha\dfrac{m_1 m_2}{r^2}$

92. *Heat Flow*

Solve for t_1: $H = \dfrac{KA(t_2 - t_1)}{L}$

93. *Lensmaker's Equation*

Solve for R_1: $\dfrac{1}{f} = (n - 1)\left(\dfrac{1}{R_1} - \dfrac{1}{R_2}\right)$

94. *Capacitance in Series Circuits*

Solve for C_1: $C = \dfrac{1}{\dfrac{1}{C_1} + \dfrac{1}{C_2}}$

95. *Arithmetic Progression*

Solve for n: $L = a + (n - 1)d$

96. *Arithmetic Progression*

Solve for a: $S = \dfrac{n}{2}[2a + (n - 1)d]$

97. *Geometric Progression*

Solve for r: $S = \dfrac{rL - a}{r - 1}$

98. *Prismoidal Formula*

Solve for S_1: $V = \frac{1}{6}H(S_0 + 4S_1 + S_2)$

Review   Solve Exercises 99–102 as a review of the skills and problem-solving techniques you learned in previous sections.

99. Simplify: $(5x^4)(25x^2)^{-1}$, $x \neq 0$

100. Simplify: $\sqrt{150s^2 t^3}$

101. Rationalize the denominator: $\dfrac{6}{\sqrt{10} - 2}$

102. Add and simplify: $\dfrac{3}{x - 5} + \dfrac{2}{5 - x}$

1.4 *Quadratic Equations and Applications*

See Exercise 99 on page 130 for an example of how a quadratic equation can be used to model the time it takes an object to fall from the top of the CN Tower.

Factoring □ *Extracting Square Roots* □ *Completing the Square* □
The Quadratic Formula □ *Applications*

Factoring

A **quadratic equation** in x is an equation that can be written in the standard form

$$ax^2 + bx + c = 0$$

where a, b, and c are real numbers with $a \neq 0$. A quadratic equation in x is also known as a **second-degree polynomial equation** in x.

In this section, you will study four methods for solving quadratic equations: factoring, extracting square roots, completing the square, and the Quadratic Formula. The first method is based on the Zero-Factor Property given in Section P.1.

If $ab = 0$, then $a = 0$ or $b = 0$.   Zero-Factor Property

To use this property, write the left side of the standard form of a quadratic equation as the product of two linear factors. Then find the solutions of the quadratic equation by setting each linear factor equal to zero.

EXAMPLE 1   *Solving a Quadratic Equation by Factoring*

a. $2x^2 + 9x + 7 = 3$ Original equation

$2x^2 + 9x + 4 = 0$ Standard form

$(2x + 1)(x + 4) = 0$ Factored form

$2x + 1 = 0$ ⟹ $x = -\dfrac{1}{2}$ Set 1st factor equal to 0.

$x + 4 = 0$ ⟹ $x = -4$ Set 2nd factor equal to 0.

The solutions are $-\frac{1}{2}$ and -4. Check these in the original equation.

b. $6x^2 - 3x = 0$ Original equation

$3x(2x - 1) = 0$ Factored form

$3x = 0$ ⟹ $x = 0$ Set 1st factor equal to 0.

$2x - 1 = 0$ ⟹ $x = \dfrac{1}{2}$ Set 2nd factor equal to 0.

The solutions are 0 and $\frac{1}{2}$. Check these in the original equation.

NOTE Be sure you see that the Zero-Factor Property works *only* for equations written in standard form (in which the right side of the equation is zero). Therefore, all terms must be collected on one side *before* factoring. For instance, in the equation

$$(x - 5)(x + 2) = 8$$

it is *incorrect* to set each factor equal to 8. Can you solve this equation correctly? ■■

TECHNOLOGY

You can use a graphing utility to check graphically the real solutions of a quadratic equation. Begin by writing the equation in standard form. Then set y equal to the left side and graph the resulting equation. The x-intercepts of the equation represent the real solutions of the original equation. For example, to check the solutions of $6x^2 - 3x = 0$, graph $y = 6x^2 - 3x$, as shown below. Notice that the x-intercepts occur at $x = 0$ and $x = \frac{1}{2}$, as found in Example 1(b).

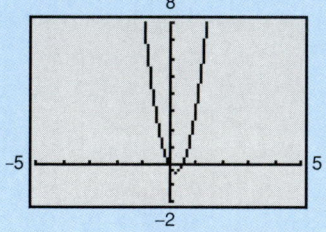

Extracting Square Roots

There is a nice shortcut for solving quadratic equations of the form $u^2 = d$, where $d > 0$ and u is an algebraic expression. By factoring, you can see that this equation has two solutions.

$u^2 = d$	Original equation
$u^2 - d = 0$	Standard form
$\left(u + \sqrt{d}\right)\left(u - \sqrt{d}\right) = 0$	Factored form
$u + \sqrt{d} = 0 \quad \Longrightarrow \quad u = -\sqrt{d}$	Set 1st factor equal to 0.
$u - \sqrt{d} = 0 \quad \Longrightarrow \quad u = \sqrt{d}$	Set 2nd factor equal to 0.

Because the two solutions differ only in sign, you can write the solutions together, using a "plus or minus sign," as

$$u = \pm\sqrt{d}.$$

This form of the solution is read as "u is equal to plus or minus the square root of d." Solving an equation of the form $u^2 = d$ without going through the steps of factoring is **extracting square roots.**

EXTRACTING SQUARE ROOTS

The equation $u^2 = d$, where $d > 0$, has exactly two solutions:

$$u = \sqrt{d} \quad \text{and} \quad u = -\sqrt{d}.$$

These solutions can also be written as

$$u = \pm\sqrt{d}.$$

EXAMPLE 2   *Extracting Square Roots*

a.

$4x^2 = 12$	Original equation
$x^2 = 3$	Divide both sides by 4.
$x = \pm\sqrt{3}$	Extract square roots.

The solutions are $\sqrt{3}$ and $-\sqrt{3}$. Check these in the original equation.

b.

$(x - 3)^2 = 7$	Original equation
$x - 3 = \pm\sqrt{7}$	Extract square roots.
$x = 3 \pm\sqrt{7}$	Add 3 to both sides.

The solutions are $3 \pm\sqrt{7}$. Check these in the original equation.

Completing the Square

The equation in Example 2(b) was given in the form $(x - 3)^2 = 7$ so that you could find the solution by extracting square roots. Suppose, however, that the equation $(x - 3)^2 = 7$ had been given in the standard form

$$x^2 - 6x + 2 = 0. \qquad \text{Standard form}$$

This equation is equivalent to the original and thus has the same two solutions, $x = 3 \pm \sqrt{7}$. However, the left side of the equation is not factorable, and you cannot find its solutions unless you can *reverse* the steps shown above.

> **COMPLETING THE SQUARE**
>
> To **complete the square** for the expression
>
> $$x^2 + bx$$
>
> add $(b/2)^2$, which is the square of half the coefficient of x. Consequently,
>
> $$x^2 + bx + \left(\frac{b}{2}\right)^2 = \left(x + \frac{b}{2}\right)^2.$$

When solving quadratic equations by completing the square, you must add $(b/2)^2$ to *both sides* in order to maintain equality.

EXAMPLE 3 *Completing the Square: Leading Coefficient Is 1*

Solve $x^2 - 6x + 2 = 0$ by completing the square. Compare the solutions to those obtained in Example 2(b).

Solution

NOTE If the leading coefficient is *not* 1, you must divide both sides of the equation by this coefficient *before* completing the square. For instance, to complete the square for $3x^2 - 4x - 5 = 0$, first divide each term by the leading coefficient 3.

$$x^2 - \frac{4}{3}x - \frac{5}{3} = 0$$

Then proceed as in Example 3. ∎∎

$x^2 - 6x + 2 = 0$	Original equation
$x^2 - 6x = -2$	Subtract 2 from both sides.
$x^2 - 6x + 3^2 = -2 + 3^2$	Add 3^2 to both sides.
(half)^2	
$x^2 - 6x + 9 = 7$	Simplify.
$(x - 3)^2 = 7$	Perfect square trinomial
$x - 3 = \pm\sqrt{7}$	Extract square roots.
$x = 3 \pm \sqrt{7}$	Add 3 to both sides.

The solutions are $x = 3 \pm \sqrt{7}$, as found in Example 2(b).

The Quadratic Formula

Often in mathematics you are taught the long way of solving a problem first. Then, the longer method is used to develop shorter techniques. The long way stresses understanding and the short way stresses efficiency.

For instance, you can think of completing the square as a "long way" of solving a quadratic equation. When you use completing the square to solve a quadratic equation, you must complete the square for *each* equation separately. In the following development, you complete the square *once* in a general setting to obtain the **Quadratic Formula**—a shortcut for solving a quadratic equation.

$$ax^2 + bx + c = 0 \qquad\qquad \text{Standard form, } a \neq 0$$

$$ax^2 + bx = -c \qquad\qquad \text{Subtract } c \text{ from both sides.}$$

$$x^2 + \frac{b}{a}x = -\frac{c}{a} \qquad\qquad \text{Divide both sides by } a.$$

$$x^2 + \frac{b}{a}x + \left(\frac{b}{2a}\right)^2 = -\frac{c}{a} + \left(\frac{b}{2a}\right)^2 \qquad\qquad \text{Complete the square.}$$

 (half)^2

$$\left(x + \frac{b}{2a}\right)^2 = \frac{b^2 - 4ac}{4a^2} \qquad\qquad \text{Simplify.}$$

$$x + \frac{b}{2a} = \pm\sqrt{\frac{b^2 - 4ac}{4a^2}} \qquad\qquad \text{Extract square roots.}$$

$$x = -\frac{b}{2a} \pm \frac{\sqrt{b^2 - 4ac}}{2|a|} \qquad\qquad \text{Solutions}$$

Note that since $\pm 2|a|$ represents the same numbers as $\pm 2a$, you can omit the absolute value sign.

THE QUADRATIC FORMULA

The solutions of a quadratic equation in the standard form

$$ax^2 + bx + c = 0, \qquad a \neq 0$$

are given by the **Quadratic Formula**

$$x = \frac{-b \pm \sqrt{b^2 - 4ac}}{2a}.$$

NOTE   The Quadratic Formula is one of the most important formulas in algebra. You should learn the verbal statement of the Quadratic Formula: "Negative *b*, plus or minus the square root of *b* squared minus 4*ac*, all divided by 2*a*." ■■

NOTE If the discriminant of a quadratic equation is negative, as in case 3 at the right, then its square root is imaginary (not a real number) and the quadratic formula yields two complex solutions. You will study complex solutions in Section 1.5. ∎

From each graph, can you tell whether the discriminant is positive, zero, or negative? Explain your reasoning.

a. $x^2 - 2x = 0$

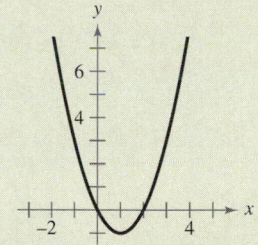

b. $x^2 - 2x + 1 = 0$

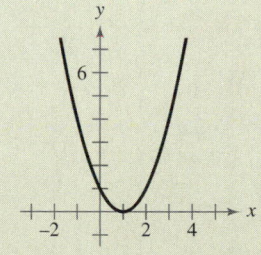

c. $x^2 - 2x + 2 = 0$

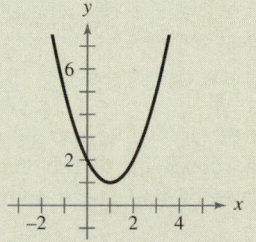

In the Quadratic Formula, the quantity under the radical sign, $b^2 - 4ac$, is the **discriminant** of the quadratic expression $ax^2 + bx + c$. It can be used to determine the nature of the solutions of a quadratic equation.

◼ **SOLUTIONS OF A QUADRATIC EQUATION**

The solutions of a quadratic equation $ax^2 + bx + c = 0$, $a \neq 0$, can be classified as follows.

1. If the discriminant $b^2 - 4ac$ is positive, then the quadratic equation has *two* distinct real solutions.
2. If the discriminant $b^2 - 4ac$ is zero, then the quadratic equation has *one* repeated solution.
3. If the discriminant $b^2 - 4ac$ is negative, then the quadratic equation has *no* real solution.

When using the Quadratic Formula, remember that *before* the formula can be applied, you must first write the quadratic equation in standard form.

EXAMPLE 4   *The Quadratic Formula: Two Distinct Solutions*

Use the Quadratic Formula to solve $x^2 + 3x = 9$.

Solution

$$x^2 + 3x = 9$$ Original equation

$$x^2 + 3x - 9 = 0$$ Standard form with $a = 1, b = 3, c = -9$

$$x = \frac{-b \pm \sqrt{b^2 - 4ac}}{2a}$$ Quadratic Formula

$$x = \frac{-3 \pm \sqrt{(3)^2 - 4(1)(-9)}}{2(1)}$$ Substitute.

$$x = \frac{-3 \pm \sqrt{45}}{2}$$ Simplify.

$$x = \frac{-3 \pm 3\sqrt{5}}{2}$$ Simplify.

The equation has two solutions:

$$x = \frac{-3 + 3\sqrt{5}}{2} \quad \text{and} \quad x = \frac{-3 - 3\sqrt{5}}{2}.$$

Check these in the original equation.

Applications

Quadratic equations often occur in problems dealing with area. Here is a simple example. "A square room has an area of 144 square feet. Find the dimensions of the room." To solve this problem, let x represent the length of each side of the room. Then, by solving the equation $x^2 = 144$, you can conclude that each side of the room is 12 feet long. Note that although the equation $x^2 = 144$ has two solutions, -12 and 12, the negative solution makes no sense, so you choose the positive solution.

Real Life

EXAMPLE 5 **Finding the Dimensions of a Room**

A bedroom is 3 feet longer than it is wide and has an area of 154 square feet, as shown in Figure 1.21. Find the dimensions of the room.

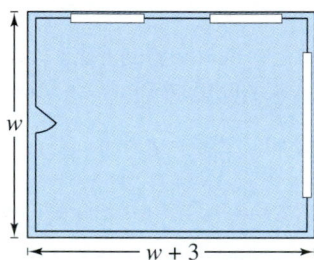

FIGURE 1.21

Solution

Verbal Model: | Width of room | · | Length of room | = | Area of room |

Labels: Area of room $= 154$ *(square feet)*
Width of room $= w$ *(feet)*
Length of room $= w + 3$ *(feet)*

Equation:
$$w(w + 3) = 154$$
$$w^2 + 3w - 154 = 0$$
$$(w - 11)(w + 14) = 0$$
$$w - 11 = 0 \implies w = 11$$
$$w + 14 = 0 \implies w = -14$$

Choosing the positive value, you find that the width is 11 feet and the length is $w + 3$ or 14 feet. You can check this solution by observing that the length is 3 feet longer than the width *and* that the product of the length and width is 154 square feet.

Another common application of quadratic equations involves an object that is falling (or projected into the air). The general equation that gives the height of such an object is called a **position equation,** and on *earth's* surface it has the form

$$s = -16t^2 + v_0 t + s_0.$$

NOTE The position equation at the right ignores air resistance. ▪▪

In this equation, s represents the height of the object (in feet), v_0 represents the original velocity of the object (in feet per second), s_0 represents the original height of the object (in feet), and t represents the time (in seconds).

Real
Life

EXAMPLE 6   *Falling Time*

A construction worker on the 24th floor of a building project (see Figure 1.22) accidentally drops a wrench and yells "Look out below!" Could a person at ground level hear this warning in time to get out of the way?

Solution

Assume that each floor of the building is 10 feet high, so that the wrench is dropped from a height of 240 feet. Because sound travels at about 1100 feet per second, it follows that a person at ground level hears the warning within 1 second of the time the wrench is dropped. To set up a mathematical model for the height of the wrench, use the position equation

$$s = -16t^2 + v_0 t + s_0.$$

Because the object is dropped rather than thrown, the initial velocity is $v_0 = 0$. Moreover, because the initial height is $s_0 = 240$ feet, you have the following model.

$$s = -16t^2 + 240$$

After falling 1 second, the height of the wrench is $-16(1)^2 + 240 = 224$. After falling 2 seconds, the height of the wrench is $-16(2)^2 + 240 = 176$. To find the number of seconds it takes the wrench to hit the ground, let the height s be zero and solve the equation for t.

$s = -16t^2 + 240$	Position equation
$0 = -16t^2 + 240$	Set height equal to 0.
$16t^2 = 240$	Add $16t^2$ to both sides.
$t^2 = 15$	Divide both sides by 16.
$t = \sqrt{15} \approx 3.87$	Extract positive square root.

The wrench will take about 3.87 seconds to hit the ground. If the person hears the warning 1 second after the wrench is dropped, the person still has almost 3 seconds to get out of the way.

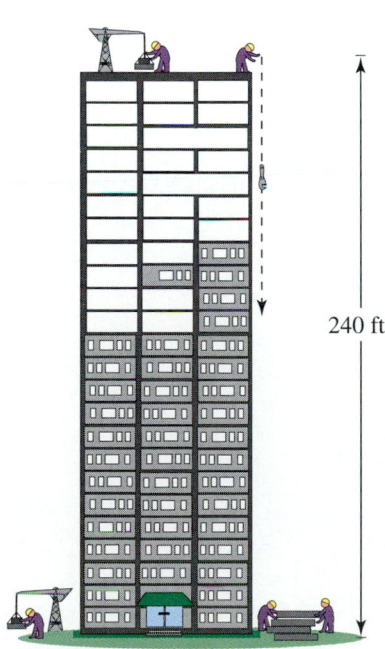

240 ft

FIGURE 1.22

EXAMPLE 7   *Quadratic Modeling: Cable Television*

Real Life

From 1985 to 1992, the number of hours spent annually per person watching basic cable television in the United States closely followed the quadratic model

$$\text{Hours} = 3.8t^2 - 29t + 167$$

where $t = 5$ represents 1985. The number of hours per year is shown graphically in Figure 1.23. According to this model, in which year will the number of hours spent per person reach or surpass 400?   (Source: Veronis, Suhler, & Associates)

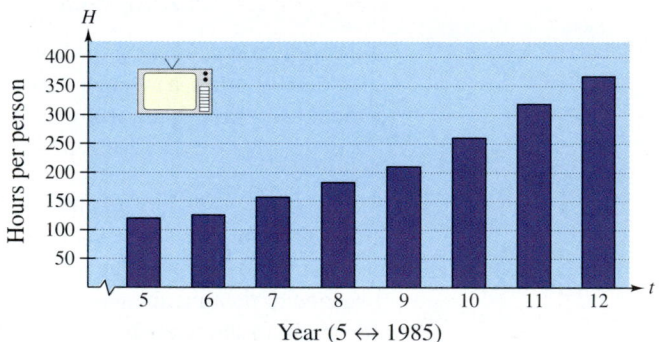

FIGURE 1.23

Solution

To find when the number of hours spent per person will reach 400, you need to solve the equation

$$3.8t^2 - 29t + 167 = 400.$$

To begin, write the equation in standard form.

$$3.8t^2 - 29t - 233 = 0$$

Then apply the Quadratic Formula.

$$t = \frac{-(-29) \pm \sqrt{(-29)^2 - 4(3.8)(-233)}}{2(3.8)}$$

Choosing the positive solution, you find that

$$t = \frac{29 + \sqrt{(-29)^2 - 4(3.8)(-233)}}{2(3.8)} \approx 12.53 \approx 13.$$

Because $t = 5$ corresponds to 1985, it follows that $t = 13$ must correspond to 1993. Thus, the number of hours spent annually per person watching basic cable should have reached 400 by 1993.

A fourth type of application that often involves a quadratic equation is one dealing with the hypotenuse of a right triangle. These types of applications often use the Pythagorean Theorem, which states that

$$a^2 + b^2 = c^2 \qquad \text{Pythagorean Theorem}$$

where a and b are the legs of a right triangle and c is the hypotenuse.

EXAMPLE 8   *Cutting Across the Lawn*

Real Life

Your house is on a large corner lot. Several of the children in the neighborhood cut across your lawn, as shown in Figure 1.24. The distance across the lawn is 32 feet. How many feet does a person save by walking across the lawn instead of walking on the sidewalk?

Solution

In Figure 1.24, let x represent the length of the shorter part of the sidewalk. Using a ruler, you can find that the length of the longer part of the sidewalk is twice the shorter, so we represent its length by $2x$. Now, using the Pythagorean Theorem, you have

$$x^2 + (2x)^2 = 32^2 \qquad \text{Pythagorean Theorem}$$
$$5x^2 = 1024 \qquad \text{Combine like terms.}$$
$$x^2 = 204.8 \qquad \text{Divide both sides by 5.}$$
$$x = \sqrt{204.8} \qquad \text{Extract positive square root.}$$

The total distance on the sidewalk is

$$x + 2x = 3x = 3\sqrt{204.8} \approx 42.9 \text{ feet.}$$

Cutting across the lawn saves a person about $42.9 - 32$ or 10.9 feet.

2x

32 ft

x

FIGURE 1.24

GROUP ACTIVITY

COMPARING SOLUTION METHODS

In this section, you studied four *algebraic* methods for solving quadratic equations and one *graphical* method. Solve each of the quadratic equations below in several different ways. Discuss which method(s) you prefer. Does it depend on the equation? Explain your reasoning.

a. $x^2 - 4x - 5 = 0$ **b.** $x^2 - 4x = 0$

c. $x^2 - 4x - 3 = 0$ **d.** $x^2 - 4x - 6 = 0$

Factor the algebraic expression.

1. $3x^2 + 7x$
2. $4x^2 - 25$
3. $16 - (x - 11)^2$
4. $x^2 + 7x - 18$
5. $10x^2 + 13x - 3$
6. $6x^2 - 73x + 12$

Simplify the expression.

7. $\sqrt{9 - 4(3)(-12)}$
8. $\sqrt{36 - 4(2)(3)}$
9. $\sqrt{12^2 - 4(3)(4)}$
10. $\sqrt{15^2 + 4(9)(12)}$

1.4 Exercises

In Exercises 1–6, write the quadratic equation in standard form.

1. $2x^2 = 3 - 5x$
2. $x^2 = 25x$
3. $(x - 3)^2 = 2$
4. $12 - 3(x + 7)^2 = 0$
5. $\frac{1}{5}(3x^2 - 10) = 12x$
6. $x(x + 2) = 3x^2 + 1$

In Exercises 7–18, solve the quadratic equation for x by factoring.

7. $6x^2 + 3x = 0$

8. $9x^2 - 1 = 0$

9. $x^2 - 2x - 8 = 0$

10. $x^2 - 10x + 9 = 0$

11. $x^2 + 10x + 25 = 0$

12. $16x^2 + 56x + 49 = 0$

13. $3 + 5x - 2x^2 = 0$

14. $2x^2 = 19x + 33$

15. $x^2 + 4x = 12$

16. $-x^2 + 8x = 12$

17. $x^2 + 2ax + a^2 = 0$

18. $(x + a)^2 - b^2 = 0$

In Exercises 19–30, solve the equation by extracting square roots. List both the exact solution *and* the decimal solution rounded to two decimal places.

19. $x^2 = 16$
20. $x^2 = 144$

21. $x^2 = 7$
22. $x^2 = 27$

23. $3x^2 = 36$
24. $9x^2 = 25$

25. $(x - 12)^2 = 18$
26. $(x + 13)^2 = 21$

27. $(x + 2)^2 = 12$
28. $(x - 5)^2 = 20$

29. $(x - 7)^2 = (x + 3)^2$
30. $(x + 5)^2 = (x + 4)^2$

In Exercises 31–40, solve the quadratic equation by completing the square.

31. $x^2 - 2x = 0$

32. $x^2 + 4x = 0$

33. $x^2 + 4x - 32 = 0$

34. $x^2 - 2x - 3 = 0$

35. $x^2 + 6x + 2 = 0$

36. $x^2 + 8x + 14 = 0$

37. $9x^2 - 18x = -3$

38. $9x^2 - 12x = 14$

39. $8 + 4x - x^2 = 0$

40. $4x^2 - 4x - 99 = 0$

In Exercises 41–44, complete the square for the quadratic portion of the algebraic expression.

41. $\dfrac{1}{x^2 + 2x + 5}$

42. $\dfrac{4}{4x^2 + 4x - 3}$

43. $\dfrac{1}{\sqrt{6x - x^2}}$

44. $\dfrac{1}{\sqrt{16 - 6x - x^2}}$

Graphical Analysis **In Exercises 45–48, use a graphing utility to graph the equation. Use the graph to approximate any *x*-intercepts of the graph. Set *y* = 0 and solve the resulting equation. Compare the result with the *x*-intercepts of the graph.**

45. $y = (x + 3)^2 - 4$

46. $y = 1 - (x - 2)^2$

47. $y = -4x^2 + 4x + 3$

48. $y = x^2 + 3x - 4$

In Exercises 49–52, use the discriminant to determine the number of real solutions of the quadratic equation.

49. $2x^2 - 5x + 5 = 0$

50. $2x^2 - x - 1 = 0$

51. $\frac{1}{5}x^2 + \frac{6}{5}x - 8 = 0$

52. $\frac{1}{3}x^2 - 5x + 25 = 0$

In Exercises 53–74, use the Quadratic Formula to solve the equation.

53. $2x^2 + x - 1 = 0$

54. $2x^2 - x - 1 = 0$

55. $16x^2 + 8x - 3 = 0$

56. $25x^2 - 20x + 3 = 0$

57. $2 + 2x - x^2 = 0$

58. $x^2 - 10x + 22 = 0$

59. $x^2 + 14x + 44 = 0$

60. $6x = 4 - x^2$

61. $x^2 + 8x - 4 = 0$

62. $4x^2 - 4x - 4 = 0$

63. $12x - 9x^2 = -3$

64. $16x^2 + 22 = 40x$

65. $3x + x^2 - 1 = 0$

66. $36x^2 + 24x - 7 = 0$

67. $4x^2 + 4x = 7$

68. $16x^2 - 40x + 5 = 0$

69. $28x - 49x^2 = 4$

70. $9x^2 + 24x + 16 = 0$

71. $8t = 5 + 2t^2$

72. $25h^2 + 80h + 61 = 0$

73. $(y - 5)^2 = 2y$

74. $(z + 6)^2 = -2z$

In Exercises 75–78, use the Quadratic Formula to solve the equation. Round your solutions to three decimal places.

75. $5.1x^2 - 1.7x - 3.2 = 0$

76. $-0.005x^2 + 0.101x - 0.193 = 0$

77. $422x^2 - 506x - 347 = 0$

78. $2x^2 - 2.50x - 0.42 = 0$

In Exercises 79–88, solve the equation for *x* by any convenient method.

79. $x^2 - 2x - 1 = 0$

80. $11x^2 + 33x = 0$

81. $(x + 3)^2 = 81$

82. $x^2 - 14x + 49 = 0$

83. $x^2 - x - \frac{11}{4} = 0$

84. $x^2 + 3x - \frac{3}{4} = 0$

85. $(x + 1)^2 = x^2$

86. $a^2 x^2 - b^2 = 0$

87. $3x + 4 = 2x - 7$

88. $4x^2 + 2x + 4 = 2x + 8$

89. *True or False?* If $(2x - 3)(x + 5) = 8$, then $2x - 3 = 8$ or $x + 5 = 8$. Explain.

90. *Exploration* Solve $3(x + 4)^2 + (x + 4) - 2 = 0$ in two ways.

(a) Let $u = x + 4$, and solve the resulting equation for *u*. Then solve the *u*-solution for *x*.

(b) Expand and collect like terms in the equation, and solve the resulting equation for *x*.

(c) Which method is easier? Explain.

91. *Think About It* Write a quadratic equation whose solutions are -4 and 6. (There are many correct answers.)

92. *Exploration* Solve the equations, given that *a* and *b* are not zero.

a. $ax^2 + bx = 0$ **b.** $ax^2 - ax = 0$

93. *Dimensions of a Building* The floor of a one-story building is 14 feet longer than it is wide. The building has 1632 square feet of floor space.

(a) Draw a rectangle that gives a visual representation of the floor space. Represent the width as *w* and show the length in terms of *w*.

(b) Write a quadratic equation in terms of *w*.

(c) Find the length and width of the building floor.

94. *Dimensions of a Corral*   A rancher has 100 meters of fencing to enclose two adjacent rectangular corrals (see figure). Find the dimensions such that the enclosed area will be 350 square meters.

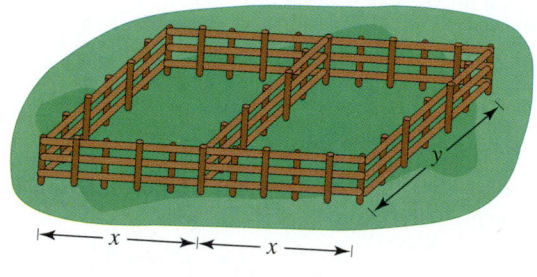

$$4x + 3y = 100$$

95. *Dimensions of a Box*   An open box with a square base is to be constructed from 84 square inches of material (see figure). What should be the dimensions of the base if the height of the box is to be 2 inches? (*Hint:* The surface area is given by $S = x^2 + 4xh$.)

$$84 = x^2 + 4x2$$

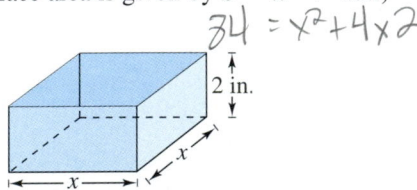

2 in.

96. *Dimensions of a Box*   An open box is to be made from a square piece of material by cutting 2-centimeter squares from each corner and turning up the sides (see figure). The volume of the finished box is to be 200 cubic centimeters. Find the size of the original piece of material.

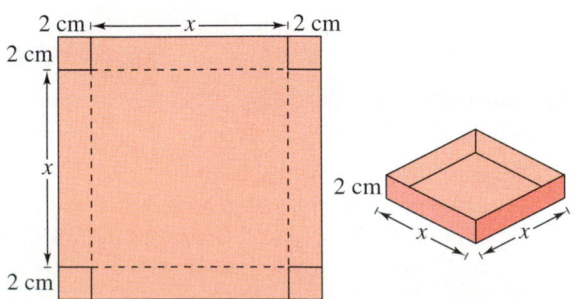

2 cm   ⟵—— x ——⟶   2 cm

2 cm

2 cm

2 cm

x

97. *Lawn Mowing*   Two people must mow a rectangular lawn 100 feet by 200 feet. Each wants to mow no more than half of the lawn. The first starts by mowing around the outside of the lawn. How wide a strip must the person mow on each of the four sides? If the mower has a 24-inch cut, approximate the required number of trips around the lawn.

98. *Seating Capacity*   A rectangular classroom seats 72 students. If the seats were rearranged with three more seats in each row, the classroom would have two fewer rows. Find the original number of seats in each row.

In Exercises 99 and 100, use the position equation given in Example 6 as the model for the problem.

99. *CN Tower*   At 1821 feet tall, the CN Tower in Toronto, Ontario is the world's tallest self-supporting structure. Suppose an object were dropped from the top of the tower.

 (a) Find the position equation $s = -16t^2 + v_0 t + s_0$.

 (b) Complete the table.

t	0	2	4	6	8	10
s						

 (c) From the table of part (b) you know the time until the object reaches ground level is greater than how many seconds? Find the time analytically.

100. *Warfare*   A bomber flying at 32,000 feet over level terrain drops a 500-pound bomb.

 (a) How long will it take until the bomb strikes the ground?

 (b) If the plane is flying at 600 miles per hour, how far will the bomb travel horizontally during its descent?

101. *An Isosceles Right Triangle*   The hypotenuse of an isosceles right triangle is 5 centimeters long. How long are its sides?

102. *An Equilateral Triangle*   An equilateral triangle has a height of 10 inches. How long are each of its sides? (*Hint:* Use the height of the triangle to partition the triangle into two congruent right triangles.)

103. *Flying Speed* Two planes leave simultaneously from the same airport, one flying due north and the other due east (see figure). The northbound plane is flying 50 miles per hour faster than the eastbound plane. After 3 hours the planes are 2440 miles apart. Find the speed of each plane.

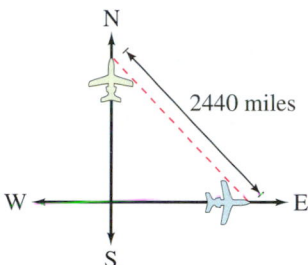

104. *Distance from the Dock* A windlass is used to tow a boat to the dock. The rope is attached to the boat at a point 15 feet below the level of the windlass.

(a) Draw a figure that gives a visual representation of the problem. Let *l* be the length of the rope and let *x* be the distance between the boat and the dock.

(b) Use the Pythagorean Theorem to write an equation giving the relationship between *l* and *x*. Find the distance from the boat to the dock when there is 75 feet of rope out.

105. *Total Revenue* The demand equation for a certain product is $p = 20 - 0.0002x$, where *p* is the price per unit and *x* is the number of units sold. The total revenue for selling *x* units is given by

$$\text{Revenue} = xp = x(20 - 0.0002x).$$

How many units must be sold to produce a revenue of $500,000?

106. *Total Revenue* The demand equation for a certain product is $p = 60 - 0.0004x$, where *p* is the price per unit and *x* is the number of units sold. The total revenue for selling *x* units is given by

$$\text{Revenue} = xp = x(60 - 0.0004x).$$

How many units must be sold to produce a revenue of $220,000?

Cost **In Exercises 107–110, use the cost equation to find the number of units *x* that a manufacturer can produce for the given cost *C*. Round your answer to the nearest positive integer.**

107. $C = 0.125x^2 + 20x + 500$ $C = \$14,000$

108. $C = 0.5x^2 + 15x + 5000$ $C = \$11,500$

109. $C = 800 + 0.04x + 0.002x^2$ $C = \$1680$

110. $C = 800 - 10x + \dfrac{x^2}{4}$ $C = \$896$

111. *Population of the U.S.* The population of the United States from 1800 to 1890 can be approximated by the model

$$\text{Population} = 0.6942t^2 + 6.183$$

where the population is given in millions of people and the time *t* represents the calendar year with $t = 0$ corresponding to 1800, $t = 1$ corresponding to 1810, and so on. If this model had continued to be valid up through the present time, when would the population of the United States have reached 250 million? Judging from the figure, would you say this model was a good representation of the population through 1890? Through 1990? (Source: U.S. Bureau of Census)

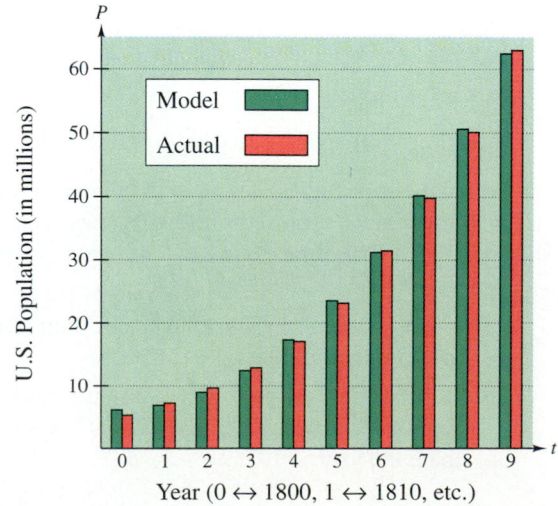

112. *Oxygen Consumption*   The metabolic rate of ectothermic organisms increases with increasing temperature within a certain range. The figure shows experimental data for the oxygen consumption C (in microliters per gram per hour) of a beetle for certain temperatures. This data can be approximated by the model

$$C = 0.45x^2 - 1.65x + 50.75, \qquad 10 \le x \le 25$$

where x is the air temperature in degrees Celsius.

(a) Find the air temperature if the oxygen consumption is 150 microliters per gram per hour.

(b) If the temperature is increased from $10°$ to $20°$, the oxygen consumption is increased by approximately what factor?

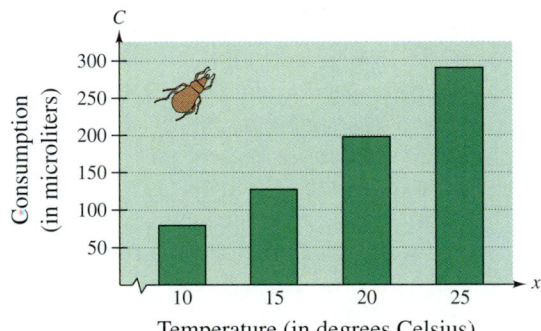

Temperature (in degrees Celsius)

113. *Pleasure Boats*   The total number of dollars spent on pleasure boats in the United States from 1988 through 1993 can be approximated by the model

$$\text{Spending} = 0.157t^2 - 1.041t + 7.385$$

where the spending is measured in billions of dollars and the time t represents the calendar year, with $t = 0$ corresponding to 1990. (Source: National Sporting Goods Association)

(a) Use a graphing utility to sketch a graph of the model over the interval $-2 \le t \le 3$.

(b) If the model is used to forecast future sales, will sales ever exceed 7 billion dollars? If so, estimate the year.

114. *Flying Distance*   A small commuter airline flies to three cities whose locations form the vertices of a right triangle (see figure). The total flight distance (from City A to City B to City C and back to City A) is 1400 kilometers. It is 600 kilometers between the two cities that are farthest apart. Approximate the other two distances between cities.

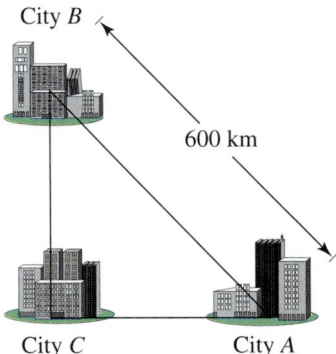

City B

600 km

City C   City A

Review   **Solve Exercises 115–123 as a review of the skills and problem-solving techniques you learned in previous sections.**

115. Identify the rule of algebra: $(10x)y = 10(xy)$

116. Identify the rule of algebra: $-4(x - 3) = -4x + 12$

117. Simplify: $\left(\dfrac{6u^2}{5v^{-3}} \right)^{-1}$

118. Multiply: $(3.5 \times 10^8)(2 \times 10^{-5})$

119. Factor completely: $x^5 - 27x^2$

120. Factor completely: $5(x + 5)x^{1/3} + 4x^{4/3}$

121. Simplify: $\dfrac{x^2 - 100}{10 - x}$

122. Add and simplify: $\dfrac{3}{2 - x} + \dfrac{5}{x} + \dfrac{1}{(x - 2)^2}$

123. Describe the viewing rectangle.

$$y = x^3 - 4x^2 + 10$$

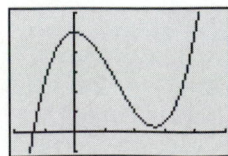

1.5 *Complex Numbers*

See Exercises 73–76 on page 140 for examples of how a graphing utility can be used to discover whether a quadratic equation has imaginary solutions.

The Imaginary Unit i ❑ *Operations with Complex Numbers* ❑
Complex Conjugates and Division ❑ *Complex Solutions of Quadratic Equations*

The Imaginary Unit *i*

In Section 1.4, you learned that some quadratic equations have no real solutions. For instance, the quadratic equation

$$x^2 + 1 = 0 \qquad\qquad \text{Equation with no real solution}$$

has no real solution because there is no real number x that can be squared to produce -1. To overcome this deficiency, mathematicians created an expanded system of numbers using the **imaginary unit *i*,** defined as

$$i = \sqrt{-1} \qquad\qquad \text{Imaginary unit}$$

where $i^2 = -1$. By adding real numbers to real multiples of this imaginary unit, we obtain the set of **complex numbers.** Each complex number can be written in the **standard form, *a* + *bi*.**

> **DEFINITION OF A COMPLEX NUMBER**
> For real numbers a and b, the number
>
> $a + bi$
>
> is a **complex number.** If $a = 0$ and $b \neq 0$, the complex number bi is an **imaginary number.**

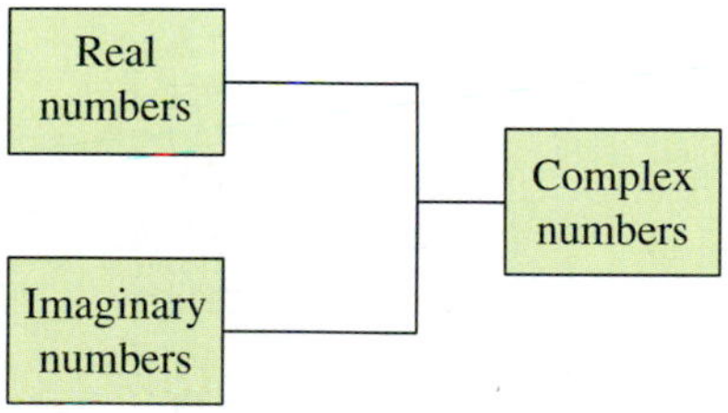

The set of real numbers is a subset of the set of complex numbers, as shown in Figure 1.25. This is true because every real number a can be written as a complex number using $b = 0$. That is, for every real number a, we can write $a = a + 0i$.

> **EQUALITY OF COMPLEX NUMBERS**
> Two complex numbers $a + bi$ and $c + di$, written in standard form, are **equal** to each other
>
> $a + bi = c + di \qquad\qquad$ Equality of two complex numbers
>
> if and only if $a = c$ and $b = d$.

```
┌──────────────┐
│ Real         │
│ numbers      │──┐
└──────────────┘  │   ┌──────────────┐
                  ├───│ Complex      │
┌──────────────┐  │   │ numbers      │
│ Imaginary    │  │   └──────────────┘
│ numbers      │──┘
└──────────────┘
```

FIGURE 1.25

Operations with Complex Numbers

To add (or subtract) two complex numbers, you add (or subtract) the real and imaginary parts of the numbers separately.

■ ADDITION AND SUBTRACTION OF COMPLEX NUMBERS

If $a + bi$ and $c + di$ are two complex numbers written in standard form, their sum and difference are defined as follows.

$$\text{Sum:} \quad (a + bi) + (c + di) = (a + c) + (b + d)i$$

$$\text{Difference:} \quad (a + bi) - (c + di) = (a - c) + (b - d)i$$

The **additive identity** in the complex number system is zero (the same as in the real number system). Furthermore, the **additive inverse** of the complex number $a + bi$ is

$$-(a + bi) = -a - bi. \qquad \text{\color{red}Additive inverse}$$

Thus, you have

$$(a + bi) + (-a - bi) = 0 + 0i = 0.$$

The *Interactive* CD-ROM shows every example with its solution; clicking on the *Try It!* button brings up similar problems. Guided Examples and Integrated Examples show step-by-step solutions to additional examples. Integrated Examples are related to several concepts in the section.

EXAMPLE 1   **Adding and Subtracting Complex Numbers**

a. $(3 - i) + (2 + 3i) = 3 - i + 2 + 3i$ {\color{red}Remove parentheses.}

$$= 3 + 2 - i + 3i \qquad \text{\color{red}Group like terms.}$$

$$= (3 + 2) + (-1 + 3)i$$

$$= 5 + 2i \qquad \text{\color{red}Standard form}$$

b. $2i + (-4 - 2i) = 2i - 4 - 2i$ {\color{red}Remove parentheses.}

$$= -4 + 2i - 2i \qquad \text{\color{red}Group like terms.}$$

$$= -4 \qquad \text{\color{red}Standard form}$$

c. $3 - (-2 + 3i) + (-5 + i) = 3 + 2 - 3i - 5 + i$

$$= 3 + 2 - 5 - 3i + i$$

$$= 0 - 2i$$

$$= -2i$$

NOTE   Note in Example 1(b) that the sum of two complex numbers can be a real number. ■■

Exploration

Complete the table:

$i^1 = i$ $i^7 = \boxed{}$
$i^2 = -1$ $i^8 = \boxed{}$
$i^3 = -i$ $i^9 = \boxed{}$
$i^4 = 1$ $i^{10} = \boxed{}$
$i^5 = \boxed{}$ $i^{11} = \boxed{}$
$i^6 = \boxed{}$ $i^{12} = \boxed{}$

What pattern do you see? Write a brief description of how you would find i raised to any positive integer power.

Many of the properties of real numbers are valid for complex numbers as well. Here are some examples.

Associative Property of Addition and Multiplication

Commutative Property of Addition and Multiplication

Distributive Property of Multiplication Over Addition

Notice below how these properties are used when two complex numbers are multiplied.

$$
\begin{aligned}
(a + bi)(c + di) &= a(c + di) + bi(c + di) && \text{Distributive}\\
&= ac + (ad)i + (bc)i + (bd)i^2 && \text{Distributive}\\
&= ac + (ad)i + (bc)i + (bd)(-1) && \text{Definition of } i\\
&= ac - bd + (ad)i + (bc)i && \text{Commutative}\\
&= (ac - bd) + (ad + bc)i && \text{Associative}
\end{aligned}
$$

Rather than trying to memorize this multiplication rule, we suggest that you simply remember how the distributive property is used to multiply two complex numbers. The procedure is similar to multiplying two polynomials and combining like terms (as in the FOIL Method).

EXAMPLE 2   Multiplying Complex Numbers

a. $(i)(-3i) = -3i^2$ Multiply.
$ = -3(-1)$ $i^2 = -1$
$ = 3$ Simplify.

b. $(2 - i)(4 + 3i) = 8 + 6i - 4i - 3i^2$ Product of binomials
$ = 8 + 6i - 4i - 3(-1)$ $i^2 = -1$
$ = 8 + 3 + 6i - 4i$ Group like terms.
$ = 11 + 2i$ Standard form

c. $(3 + 2i)(3 - 2i) = 9 - 6i + 6i - 4i^2$ Product of binomials
$ = 9 - 4(-1)$ $i^2 = -1$
$ = 9 + 4$ Simplify.
$ = 13$ Standard form

d. $(3 + 2i)^2 = 9 + 6i + 6i + 4i^2$ Product of binomials
$ = 9 + 4(-1) + 12i$ $i^2 = -1$
$ = 9 - 4 + 12i$ Simplify.
$ = 5 + 12i$ Standard form

Complex Conjugates and Division

Notice in Example 2(c) that the product of two complex numbers can be a real number. This occurs with pairs of complex numbers of the form $a + bi$ and $a - bi$, called **complex conjugates**.

$$
\begin{aligned}
(a + bi)(a - bi) &= a^2 - abi + abi - b^2 i^2 \\
&= a^2 - b^2(-1) \\
&= a^2 + b^2
\end{aligned}
$$

To find the quotient of $a + bi$ and $c + di$ where c and d are not both zero, multiply the numerator and denominator by the conjugate of the denominator to obtain

$$
\frac{a + bi}{c + di} = \frac{a + bi}{c + di}\left(\frac{c - di}{c - di}\right) = \frac{(ac + bd) + (bc - ad)i}{c^2 + d^2}.
$$

EXAMPLE 3   Dividing Complex Numbers

$$
\begin{aligned}
\frac{1}{1 + i} &= \frac{1}{1 + i}\left(\frac{1 - i}{1 - i}\right) && \text{Multiply by conjugate.} \\
&= \frac{1 - i}{1^2 - i^2} && \text{Expand.} \\
&= \frac{1 - i}{1 - (-1)} && i^2 = -1 \\
&= \frac{1 - i}{2} && \text{Simplify.} \\
&= \frac{1}{2} - \frac{1}{2}i && \text{Standard form}
\end{aligned}
$$

Some graphing utilities, such as the *TI–92* and the *TI-83* from Texas Instruments, can perform operations with complex numbers. For instance, to divide $2 + 3i$ by $4 - 2i$, enter

$\boxed{(}\ \boxed{2}\ \boxed{+}\ \boxed{3}\ \boxed{\text{2nd}}\ \boxed{i}\ \boxed{)}\ \boxed{\div}$

$\boxed{(}\ \boxed{4}\ \boxed{-}\ \boxed{2}\ \boxed{\text{2nd}}\ \boxed{i}\ \boxed{)}\ \boxed{\text{ENTER}}$.

The display is $1/10 + 4/5i$.

(Photo: Courtesy of Texas Instruments)

EXAMPLE 4   Dividing Complex Numbers

$$
\begin{aligned}
\frac{2 + 3i}{4 - 2i} &= \frac{2 + 3i}{4 - 2i}\left(\frac{4 + 2i}{4 + 2i}\right) && \text{Multiply by conjugate.} \\
&= \frac{8 + 4i + 12i + 6i^2}{16 - 4i^2} && \text{Expand.} \\
&= \frac{8 - 6 + 16i}{16 + 4} && i^2 = -1 \\
&= \frac{1}{20}(2 + 16i) && \text{Simplify.} \\
&= \frac{1}{10} + \frac{4}{5}i && \text{Standard form}
\end{aligned}
$$

Complex Solutions of Quadratic Equations

When using the Quadratic Formula to solve a quadratic equation, you often obtain a result such as $\sqrt{-3}$, which you know is not a real number. By factoring out $i = \sqrt{-1}$, you can write this number in standard form.

$$\sqrt{-3} = \sqrt{3(-1)} = \sqrt{3}\sqrt{-1} = \sqrt{3}i$$

The number $\sqrt{3}i$ is called the principal square root of -3.

Study Tip

The definition of principal square root uses the rule

$$\sqrt{ab} = \sqrt{a}\sqrt{b}$$

for $a > 0$ and $b < 0$. This rule is not valid if *both* a and b are negative. For example,

$$\sqrt{-5}\sqrt{-5} = \sqrt{5}i\sqrt{5}i$$
$$= \sqrt{25}i^2$$
$$= 5i^2 = -5$$

whereas

$$\sqrt{(-5)(-5)} = \sqrt{25} = 5.$$

To avoid problems with multiplying square roots of negative numbers, be sure to convert to standard form *before* multiplying.

PRINCIPAL SQUARE ROOT OF A NEGATIVE NUMBER

If a is a positive number, the **principal square root** of the negative number $-a$ is defined as

$$\sqrt{-a} = \sqrt{a}\,i.$$

EXAMPLE 5 *Writing Complex Numbers in Standard Form*

a. $\sqrt{-3}\sqrt{-12} = \sqrt{3}i\sqrt{12}i = \sqrt{36}i^2 = 6(-1) = -6$

b. $\sqrt{-48} - \sqrt{-27} = \sqrt{48}i - \sqrt{27}i = 4\sqrt{3}i - 3\sqrt{3}i = \sqrt{3}i$

c. $\left(-1 + \sqrt{-3}\right)^2 = \left(-1 + \sqrt{3}i\right)^2$
$$= (-1)^2 - 2\sqrt{3}i + \left(\sqrt{3}\right)^2(i^2)$$
$$= 1 - 2\sqrt{3}i + 3(-1)$$
$$= -2 - 2\sqrt{3}i$$

EXAMPLE 6 *Complex Solutions of a Quadratic Equation*

Solve $3x^2 - 2x + 5 = 0$.

Solution

$$x = \frac{-(-2) \pm \sqrt{(-2)^2 - 4(3)(5)}}{2(3)} \qquad \text{Quadratic Formula}$$

$$= \frac{2 \pm \sqrt{-56}}{6} \qquad \text{Simplify.}$$

$$= \frac{2 \pm 2\sqrt{14}i}{6} \qquad \text{Write in } i\text{-form.}$$

$$= \frac{1}{3} \pm \frac{\sqrt{14}}{3}i \qquad \text{Standard form}$$

GROUP ACTIVITY

ERROR ANALYSIS

Suppose you are a math instructor, and one of your students has handed in the following quiz. Find the error(s) in each solution and discuss how to explain each error to your student.

1. Write $\dfrac{5}{3 - 2i}$ in standard form.

$$\cancel{\dfrac{5}{3 - 2i} \cdot \dfrac{3 + 2i}{3 + 2i} = \dfrac{15 + 10i}{9 - 4} = 3 + 2i}$$

2. Multiply $\left(\sqrt{-4} + 3\right)\left(i - \sqrt{-3}\right)$.

$$\cancel{\begin{aligned}\left(\sqrt{-4} + 3\right)\left(i - \sqrt{-3}\right) &= i\sqrt{-4} - \sqrt{-4}\sqrt{-3} + 3i - 3\sqrt{-3} \\ &= -2i - \sqrt{12} + 3i - 3i\sqrt{3} \\ &= \left(1 - 3\sqrt{3}\right)i - 2\sqrt{3}\end{aligned}}$$

3. Sketch the graph of $y = -x^2 + 2$.

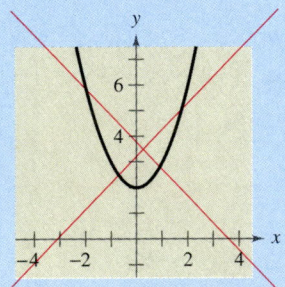

Simplify the expression.

1. $\sqrt{12}$ **2.** $\sqrt{500}$

3. $\sqrt{20} - \sqrt{5}$ **4.** $\sqrt{27} - \sqrt{243}$

5. $\sqrt{24}\sqrt{6}$ **6.** $2\sqrt{18}\sqrt{32}$

7. $\dfrac{1}{\sqrt{3}}$ **8.** $\dfrac{2}{\sqrt{2}}$

Solve the quadratic equation.

9. $x^2 + x - 1 = 0$ **10.** $x^2 + 2x - 1 = 0$

1.5 Exercises

In Exercises 1–4, find real numbers a and b so that the equation is true.

1. $a + bi = -10 + 6i$

2. $a + bi = 13 + 4i$

3. $(a - 1) + (b + 3)i = 5 + 8i$

4. $(a + 6) + 2bi = 6 - 5i$

In Exercises 5–16, write the complex number in standard form.

5. $4 + \sqrt{-9}$

6. $3 + \sqrt{-16}$

7. $2 - \sqrt{-27}$

8. $1 + \sqrt{-8}$

9. $\sqrt{-75}$

10. 45

11. $-6i + i^2$

12. $-4i^2 + 2i$

13. 8

14. $\left(\sqrt{-4}\right)^2 - 5$

15. $\sqrt{-0.09}$

16. $\sqrt{-0.0004}$

In Exercises 17–26, perform the addition or subtraction and write the result in standard form.

17. $(5 + i) + (6 - 2i)$

18. $(13 - 2i) + (-5 + 6i)$

19. $(8 - i) - (4 - i)$

20. $(3 + 2i) - (6 + 13i)$

21. $\left(-2 + \sqrt{-8}\right) + \left(5 - \sqrt{-50}\right)$

22. $\left(8 + \sqrt{-18}\right) - \left(4 + 3\sqrt{2}i\right)$

23. $13i - (14 - 7i)$

24. $22 + (-5 + 8i) + 10i$

25. $-\left(\frac{3}{2} + \frac{5}{2}i\right) + \left(\frac{5}{3} + \frac{11}{3}i\right)$

26. $(1.6 + 3.2i) + (-5.8 + 4.3i)$

In Exercises 27–40, perform the operation and write the result in standard form.

27. $\sqrt{-6} \cdot \sqrt{-2}$

28. $\sqrt{-5} \cdot \sqrt{-10}$

29. $\left(\sqrt{-10}\right)^2$

30. $\left(\sqrt{-75}\right)^2$

31. $(1 + i)(3 - 2i)$

32. $(6 - 2i)(2 - 3i)$

33. $6i(5 - 2i)$

34. $-8i(9 + 4i)$

35. $\left(\sqrt{14} + \sqrt{10}i\right)\left(\sqrt{14} - \sqrt{10}i\right)$

36. $\left(3 + \sqrt{-5}\right)\left(7 - \sqrt{-10}\right)$

37. $(4 + 5i)^2$

38. $(2 - 3i)^2$

39. $(2 + 3i)^2 + (2 - 3i)^2$

40. $(1 - 2i)^2 - (1 + 2i)^2$

41. *Error Analysis*   Describe the error.
$$\cancel{\sqrt{-6}\sqrt{-6} = \sqrt{(-6)(-6)} = \sqrt{36} = 6}$$

42. *Think About It*   **True or False?** There is no complex number that is equal to its conjugate. Explain.

In Exercises 43–50, write the conjugate of the complex number. Multiply the number and its conjugate.

43. $5 + 3i$

44. $9 - 12i$

45. $-2 - \sqrt{5}i$

46. $-4 + \sqrt{2}i$

47. $20i$

48. $\sqrt{-15}$

49. $\sqrt{8}$

50. $1 + \sqrt{8}$

In Exercises 51–64, perform the operation and write the result in standard form.

51. $\dfrac{6}{i}$

52. $-\dfrac{10}{2i}$

53. $\dfrac{4}{4 - 5i}$

54. $\dfrac{3}{1 - i}$

55. $\dfrac{2 + i}{2 - i}$

56. $\dfrac{8 - 7i}{1 - 2i}$

57. $\dfrac{6 - 7i}{i}$

58. $\dfrac{8 + 20i}{2i}$

59. $\dfrac{1}{(4 - 5i)^2}$

60. $\dfrac{(2 - 3i)(5i)}{2 + 3i}$

61. $\dfrac{2}{1 + i} - \dfrac{3}{1 - i}$

62. $\dfrac{2i}{2 + i} + \dfrac{5}{2 - i}$

63. $\dfrac{i}{3 - 2i} + \dfrac{2i}{3 + 8i}$

64. $\dfrac{1 + i}{i} - \dfrac{3}{4 - i}$

In Exercises 65–72, use the Quadratic Formula to solve the quadratic equation.

65. $x^2 - 2x + 2 = 0$

66. $x^2 + 6x + 10 = 0$

67. $4x^2 + 16x + 17 = 0$

68. $9x^2 - 6x + 37 = 0$

69. $4x^2 + 16x + 15 = 0$

70. $9x^2 - 6x - 35 = 0$

71. $16t^2 - 4t + 3 = 0$

72. $5s^2 + 6s + 3 = 0$

Graphical Reasoning In Exercises 73–76, use a graphing utility to graph the equation. Use the graph to approximate any *x*-intercepts of the graph. Set *y* = 0 and solve the resulting equation. Compare the result with the *x*-intercepts of the graph.

73. $y = \frac{1}{4}(4x^2 - 20x + 25)$

74. $y = -(x^2 - 4x + 3)$

75. $y = -(x^2 - 4x + 5)$

76. $y = \frac{1}{4}(x^2 - 2x + 9)$

77. **Essay** Use the results of Exercises 73–76 to describe the relationship between the number of *x*-intercepts of the graph of $y = ax^2 + bx + c$ and the solutions of the equation $ax^2 + bx + c = 0$.

78. Express each of the following powers of *i* as *i*, –*i*, 1, or –1.

(a) i^{40} (b) i^{25}

(c) i^{50} (d) i^{67}

In Exercises 79–86, simplify the complex number and write it in standard form.

79. $-6i^3 + i^2$

80. $4i^2 - 2i^3$

81. $-5i^5$

82. $(-i)^3$

83. $\left(\sqrt{-75}\right)^3$

84. $\left(\sqrt{-2}\right)^6$

85. $\dfrac{1}{i^3}$

86. $\dfrac{1}{(2i)^3}$

87. Cube the complex numbers.

$$2, \qquad -1 + \sqrt{3}\,i, \qquad -1 - \sqrt{3}\,i$$

88. Raise the numbers to the fourth power.

$$2, \qquad -2, \qquad 2i, \qquad -2i$$

89. Prove that the sum of a complex number $a + bi$ and its conjugate is a real number.

90. Prove that the difference of a complex number $a + bi$ and its conjugate is an imaginary number.

91. Prove that the product of a complex number $a + bi$ and its conjugate is a real number.

92. Prove that the conjugate of the product of two complex numbers $a_1 + b_1 i$ and $a_2 + b_2 i$ is the product of their conjugates.

93. Prove that the conjugate of the sum of two complex numbers $a_1 + b_1 i$ and $a_2 + b_2 i$ is the sum of their conjugates.

Review Solve Exercises 94–102 as a review of the skills and problem-solving techniques you learned in previous sections.

94. Subtract: $(x^3 - 3x^2) - (6 - 2x - 4x^2)$

95. Add: $(4 + 3x) + (8 - 6x - x^2)$

96. Multiply: $\left(3x - \frac{1}{2}\right)(x + 4)$

97. Expand: $(2x - 5)^2$

98. Expand: $[(x + y) + 3]^2$

99. **Volume of an Oblate Spheroid**

Solve for *a*: $V = \frac{4}{3}\pi a^2 b$

100. **Newton's Law of Universal Gravitation**

Solve for *r*: $F = \alpha\dfrac{m_1 m_2}{r^2}$

101. **Mixture Problem** A 5-liter container contains a mixture with a concentration of 50%. How much of this mixture must be withdrawn and replaced by 100% concentrate to bring the mixture up to 60% concentration?

102. **Average Speed** A business executive traveled at an average speed of 100 kilometers per hour on a 200-kilometer trip. Because of heavy traffic, the average speed on the return trip was 80 kilometers per hour. Find the average speed for the round trip.

1.6 *Other Types of Equations*

See Exercise 90 on page 151 for an example of how a radical equation can be used to model the temperature of saturated steam.

Polynomial Equations ▫ *Equations Involving Radicals* ▫
Equations with Fractions or Absolute Values ▫ *Applications*

Polynomial Equations

In this section you will extend the techniques for solving equations to nonlinear and nonquadratic equations. At this point in the text, you have only four basic methods for solving nonlinear equations—*factoring, extracting roots, completing the square,* and the *Quadratic Formula.* So the main goal of this section is to learn to *rewrite* nonlinear equations in a form to which you can apply one of these methods.

EXAMPLE 1 *Solving a Polynomial Equation by Factoring*

Solve $3x^4 = 48x^2$.

Solution

First write the polynomial equation in standard form with zero on one side, factor the other side, and then set each factor equal to zero.

$3x^4 = 48x^2$		Original equation
$3x^4 - 48x^2 = 0$		Standard form
$3x^2(x^2 - 16) = 0$		Factor
$3x^2(x + 4)(x - 4) = 0$		Factored form
$3x^2 = 0$	➡ $x = 0$	Set 1st factor equal to 0.
$x + 4 = 0$	➡ $x = -4$	Set 2nd factor equal to 0.
$x - 4 = 0$	➡ $x = 4$	Set 3rd factor equal to 0.

You can check these solutions by substituting in the original equation, as follows.

Check:

$3x^4 = 48x^2$	Original equation
$3(0)^4 = 48(0)^2$	0 checks. ✔
$3(-4)^4 = 48(-4)^2$	-4 checks. ✔
$3(4)^4 = 48(4)^2$	4 checks. ✔

After checking, you can conclude that the solutions are 0, -4, and 4.

Study Tip

A common mistake that is made in solving an equation such as that in Example 1 is dividing both sides of the equation by the variable factor x^2. This loses the solution $x = 0$. When using factoring to solve an equation, be sure to set each factor equal to zero. Don't divide both sides of an equation by a variable factor in an attempt to simplify the equation.

TECHNOLOGY

You can use a graphing utility to check graphically the solutions of the equation in Example 2. To do this, sketch the graph of

$$y = x^3 - 3x^2 - 3x + 9.$$

As shown below, the x-intercepts of the graph occur when $x = -\sqrt{3}$, $x = \sqrt{3}$, and $x = 3$, confirming the result found in Example 2.

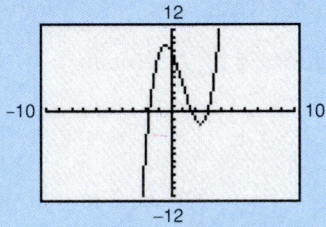

Try using a graphing utility to check the solutions found in Example 3.

EXAMPLE 2   *Solving a Polynomial Equation by Factoring*

Solve $x^3 - 3x^2 - 3x + 9 = 0$.

Solution

$x^3 - 3x^2 - 3x + 9 = 0$	Original equation
$x^2(x - 3) - 3(x - 3) = 0$	Factor by grouping.
$(x - 3)(x^2 - 3) = 0$	Distributive Property
$x - 3 = 0$ ⟹ $x = 3$	Set 1st factor equal to 0.
$x^2 - 3 = 0$ ⟹ $x = \pm\sqrt{3}$	Set 2nd factor equal to 0.

The solutions are 3, $\sqrt{3}$, and $-\sqrt{3}$. Check these in the original equation.

Occasionally, mathematical models involve equations that are of **quadratic type.** In general, an equation is of quadratic type if it can be written in the form

$$au^2 + bu + c = 0$$

where $a \neq 0$ and u is an algebraic expression.

EXAMPLE 3   *Solving an Equation of Quadratic Type*

Solve $x^4 - 3x^2 + 2 = 0$.

Solution

This equation is of quadratic type with $u = x^2$.

$$(x^2)^2 - 3(x^2) + 2 = 0$$

To solve this equation, you can factor the left side of the equation as the product of two second-degree polynomials.

$x^4 - 3x^2 + 2 = 0$	Original equation
$(x^2 - 1)(x^2 - 2) = 0$	Partially factor.
$(x + 1)(x - 1)(x^2 - 2) = 0$	Factor.
$x + 1 = 0$ ⟹ $x = -1$	Set 1st factor equal to 0.
$x - 1 = 0$ ⟹ $x = 1$	Set 2nd factor equal to 0.
$x^2 - 2 = 0$ ⟹ $x = \pm\sqrt{2}$	Set 3rd factor equal to 0.

The solutions are -1, 1, $\sqrt{2}$, and $-\sqrt{2}$. Check these in the original equation.

Equations Involving Radicals

The steps involved in solving the remaining equations in this section will often introduce *extraneous solutions,* as discussed in Section 1.2. Operations such as squaring both sides of an equation, raising both sides of an equation to a rational power, or multiplying both sides by a variable quantity all have this potential danger. Thus, when you use any of these operations, a check is crucial.

EXAMPLE 4 *Solving an Equation Involving a Rational Exponent*

Solve $4x^{3/2} - 8 = 0$.

Solution

$4x^{3/2} - 8 = 0$	Original equation
$4x^{3/2} = 8$	Add 8 to both sides.
$x^{3/2} = 2$	Isolate $x^{3/2}$.
$x = 2^{2/3}$	Raise both sides to $\frac{2}{3}$ power.
$x \approx 1.587$	Round to three decimal places.

The solution appears to be $2^{2/3}$. You can check this as follows.

NOTE The essential operations in Example 4 are isolating the factor with the rational exponent and raising both sides to the *reciprocal power.* In Example 5, this is equivalent to isolating the square root and squaring both sides. ∎∎

Check: $4x^{3/2} - 8 = 0$	Original equation
$4(2^{2/3})^{3/2} \overset{?}{=} 8$	Substitute $2^{2/3}$ for x.
$4(2) \overset{?}{=} 8$	Property of exponents
$8 = 8$	Solution checks. ✔

EXAMPLE 5 *Solving an Equation Involving a Radical*

$\sqrt{2x + 7} - x = 2$	Original equation
$\sqrt{2x + 7} = x + 2$	Isolate the square root.
$2x + 7 = x^2 + 4x + 4$	Square both sides.
$0 = x^2 + 2x - 3$	Standard form
$0 = (x + 3)(x - 1)$	Factored form
$x + 3 = 0 \implies x = -3$	Set 1st factor equal to 0.
$x - 1 = 0 \implies x = 1$	Set 2nd factor equal to 0.

By checking these values, you can determine that the only solution is 1.

Equations with Fractions or Absolute Values

To solve an equation involving fractions, multiply both sides of the equation by the least common denominator of all terms in the equation. This procedure will "clear the equation of fractions." For instance, in the equation

$$\frac{2}{x^2 + 1} + \frac{1}{x} = \frac{2}{x}$$

you can multiply both sides of the equation by $x(x^2 + 1)$. Try doing this and solve the resulting equation. You should obtain one solution: $x = 1$.

EXAMPLE 6   *Solving an Equation Involving Fractions*

Solve $\dfrac{2}{x} = \dfrac{3}{x - 2} - 1$.

Solution

For this equation, the least common denominator of the three terms is $x(x - 2)$, so you begin by multiplying each term in the equation by this expression.

$$\frac{2}{x} = \frac{3}{x - 2} - 1$$

$$x(x - 2)\frac{2}{x} = x(x - 2)\frac{3}{x - 2} - x(x - 2)(1)$$

$$2(x - 2) = 3x - x(x - 2)$$

$$2x - 4 = -x^2 + 5x$$

$$x^2 - 3x - 4 = 0$$

$$(x - 4)(x + 1) = 0$$

$$x - 4 = 0 \implies x = 4$$

$$x + 1 = 0 \implies x = -1$$

Check:

$$\frac{2}{x} = \frac{3}{x - 2} - 1 \qquad \text{Original equation}$$

$$\frac{2}{4} \overset{?}{=} \frac{3}{4 - 2} - 1 \qquad \text{Substitute 4 for } x.$$

$$\frac{1}{2} = \frac{3}{2} - 1 \qquad \text{4 checks. } \checkmark$$

$$\frac{2}{-1} \overset{?}{=} \frac{3}{-1 - 2} - 1 \qquad \text{Substitute } -1 \text{ for } x.$$

$$-2 = -1 - 1 \qquad -1 \text{ checks. } \checkmark$$

The solutions are 4 and -1.

To solve an equation involving an absolute value, remember that the expression inside the absolute value signs can be positive or negative. This results in *two* separate equations, each of which must be solved. For instance, the equation

$$|x - 2| = 3$$

results in the two equations

$$x - 2 = 3 \quad \text{and} \quad -(x - 2) = 3$$

which implies that the equation has two solutions: 5 and -1.

EXAMPLE 7   *Solving an Equation Involving Absolute Value*

Solve $|x^2 - 3x| = -4x + 6$.

Solution

Because the variable expression inside the absolute value signs can be positive or negative, you must solve the following two equations.

First Equation

$x^2 - 3x = -4x + 6$	Use positive expression.
$x^2 + x - 6 = 0$	Standard form
$(x + 3)(x - 2) = 0$	Factored form
$x + 3 = 0 \implies x = -3$	Set 1st factor equal to 0.
$x - 2 = 0 \implies x = 2$	Set 2nd factor equal to 0.

Second Equation

$-(x^2 - 3x) = -4x + 6$	Use negative expression.
$x^2 - 7x + 6 = 0$	Standard form
$(x - 1)(x - 6) = 0$	Factored form
$x - 1 = 0 \implies x = 1$	Set 1st factor equal to 0.
$x - 6 = 0 \implies x = 6$	Set 2nd factor equal to 0.

Check:

$|(-3)^2 - 3(-3)| = -4(-3) + 6$	-3 checks. ✔
$|(2)^2 - 3(2)| \neq -4(2) + 6$	2 does not check.
$|(1)^2 - 3(1)| = -4(1) + 6$	1 checks. ✔
$|(6)^2 - 3(6)| \neq -4(6) + 6$	6 does not check.

The solutions are -3 and 1.

Applications

It would be impossible to categorize the many different types of applications that involve nonlinear and nonquadratic models. However, from the few examples and exercises that are given, we hope that you will gain some appreciation for the variety of applications that can occur.

EXAMPLE 8   *Reduced Rates*

Real Life

A ski club chartered a bus for a ski trip at a cost of $480. In an attempt to lower the bus fare per skier, the club invited nonmembers to go along. After five nonmembers joined the trip, the fare per skier decreased by $4.80. How many club members are going on the trip?

Solution

Begin the solution by creating a verbal model and assigning labels, as follows.

Verbal Model:

$$\boxed{\text{Cost per skier}} \cdot \boxed{\text{Number of skiers}} = \boxed{\text{Cost of trip}}$$

Labels:

Cost of trip $= 480$ *(dollars)*

Number of ski club members $= x$ *(people)*

Number of skiers $= x + 5$ *(people)*

Original cost per member $= \dfrac{480}{x}$ *(dollars per person)*

Cost per skier $= \dfrac{480}{x} - 4.80$ *(dollars per person)*

Equation:

$$\left(\frac{480}{x} - 4.80 \right)(x + 5) = 480$$

$$\left(\frac{480 - 4.8x}{x} \right)(x + 5) = 480$$

$$(480 - 4.8x)(x + 5) = 480x$$

$$480x - 4.8x^2 - 24x + 2400 = 480x$$

$$-4.8x^2 - 24x + 2400 = 0$$

$$x^2 + 5x - 500 = 0$$

$$(x + 25)(x - 20) = 0$$

$$x + 25 = 0 \implies x = -25$$

$$x - 20 = 0 \implies x = 20$$

Choosing the positive value of x, you can conclude that 20 ski club members are going on the trip. Check this in the original statement of the problem.

Interest in a savings account is calculated by one of three basic methods: simple interest, interest compounded n times per year, and interest compounded continuously. The next example uses the formula for interest that is compounded n times per year.

$$A = P\left(1 + \frac{r}{n}\right)^{nt}$$

In this formula, A is the balance in the account, P is the principal (or original deposit), r is the annual interest rate (in decimal form), n is the number of compoundings per year, and t is the time in years. Later, in Chapter 5, you will study the derivation of this formula for compound interest.

Real Life

EXAMPLE 9 *Compound Interest*

Suppose that when you were born your grandparents deposited $5000 in a long-term investment in which the interest was compounded quarterly. On your 25th birthday the value of the investment is $25,062.59. What was the annual interest rate for this investment?

Solution

Formula: $A = P\left(1 + \frac{r}{n}\right)^{nt}$

Labels: Balance $= A = 25{,}062.59$ *(dollars)*

Principal $= P = 5000$ *(dollars)*

Time $= t = 25$ *(years)*

Compoundings per year $= n = 4$ *(compoundings)*

Annual interest rate $= r$ *(percent in decimal form)*

Equation: $25{,}062.59 = 5000\left(1 + \dfrac{r}{4}\right)^{4(25)}$

$$\frac{25{,}062.59}{5000} = \left(1 + \frac{r}{4}\right)^{100}$$

$$5.0125 \approx \left(1 + \frac{r}{4}\right)^{100}$$

$$(5.0125)^{1/100} \approx 1 + \frac{r}{4}$$

$$1.01625 \approx 1 + \frac{r}{4}$$

$$0.01625 \approx \frac{r}{4}$$

$$0.065 \approx r$$

The annual interest rate is 0.065 or 6.5%. Check this in the original statement of the problem.

EXAMPLE 10   *Market Research*

The marketing department at a publishing firm is asked to determine the price of a book. The department determines that the demand for the book depends on the price of the book according to the formula

$$p = 40 - \sqrt{0.0001x + 1}, \qquad 0 \leq x$$

where p is the price per book in dollars and x is the number of books sold at the given price. For instance, in Figure 1.26 note that if the price were \$39, then (according to the model) no one would be willing to buy the book. On the other hand, if the price were \$17.60, 5 million copies could be sold. If the publisher set the price at \$12.95, how many copies would be sold?

Solution

$p = 40 - \sqrt{0.0001x + 1}$	Given model
$12.95 = 40 - \sqrt{0.0001x + 1}$	Set price at 12.95.
$\sqrt{0.0001x + 1} = 27.05$	Isolate the square root.
$0.0001x + 1 = 731.7025$	Square both sides.
$0.0001x = 730.7025$	Subtract 1 from both sides.
$x = 7{,}307{,}025$	Divide both sides by 0.0001.

Thus, by setting the book's price at \$12.95, the publisher can expect to sell about 7.3 million copies.

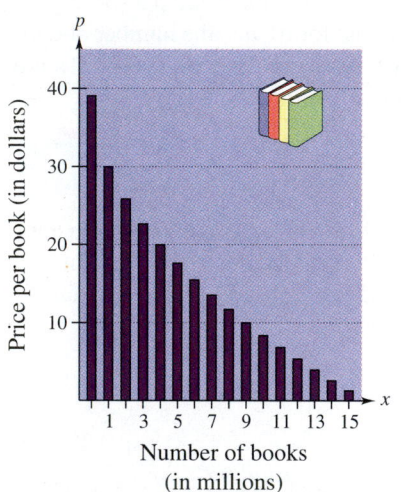

FIGURE 1.26

(Price per book (in dollars) vs. Number of books (in millions))

GROUP ACTIVITY

EXTENDING THE EXAMPLE

In Example 10, suppose the cost of producing the book is \$150,000 plus \$5.50 per book. Then the cost equation is

$$C = 5.5x + 150{,}000,$$

where C is measured in dollars and x represents the number of books produced. From Example 10, the total revenue given by

$$R = xp = x\left(40 - \sqrt{0.0001x + 1}\right).$$

Use the equations for R and C to find an equation for the profit P in terms of x. Construct a table that gives the profits for values of x ranging from 4.0 million to 6.5 million books in increments of 0.25 million. At what point does the profit appear to be maximized? Discuss ways in which you could refine or verify this estimate.

WARM UP

WARM UP

Find the real solution(s) of the equation.

1. $x^2 - 22x + 121 = 0$

2. $x(x - 20) + 3(x - 20) = 0$

3. $(x + 20)^2 = 625$

4. $5x^2 + x = 0$

5. $3x^2 + 4x - 4 = 0$

6. $12x^2 + 8x - 55 = 0$

7. $x^2 + 4x - 5 = 0$

8. $4x^2 + 4x - 15 = 0$

9. $x^2 - 3x + 1 = 0$

10. $x^2 - 4x + 2 = 0$

1.6 Exercises

In Exercises 1–22, find all solutions of the equation. Check your solutions in the original equation.

1. $4x^4 - 18x^2 = 0$

2. $20x^3 - 125x = 0$

3. $x^4 - 81 = 0$

4. $x^6 - 64 = 0$

5. $5x^3 + 30x^2 + 45x = 0$

6. $9x^4 - 24x^3 + 16x^2 = 0$

7. $x^3 - 3x^2 - x + 3 = 0$

8. $x^3 + 2x^2 + 3x + 6 = 0$

9. $x^4 - x^3 + x - 1 = 0$

10. $x^4 + 2x^3 - 8x - 16 = 0$

11. $x^4 - 4x^2 + 3 = 0$

12. $x^4 + 5x^2 - 36 = 0$

13. $4x^4 - 65x^2 + 16 = 0$

14. $36t^4 + 29t^2 - 7 = 0$

15. $x^6 + 7x^3 - 8 = 0$

16. $x^6 + 3x^3 + 2 = 0$

17. $\dfrac{1}{t^2} + \dfrac{8}{t} + 15 = 0$

18. $6\left(\dfrac{s}{s+1}\right)^2 + 5\left(\dfrac{s}{s+1}\right) - 6 = 0$

19. $2x + 9\sqrt{x} = 5$

20. $6x - 7\sqrt{x} - 3 = 0$

21. $3x^{1/3} + 2x^{2/3} = 5$

22. $9t^{2/3} + 24t^{1/3} + 16 = 0$

Graphical Analysis **In Exercises 23–26, use a graphing utility to graph the equation. Use the graph to approximate any x-intercepts of the graph. Set $y = 0$ and solve the resulting equation. Compare the result with the x-intercepts of the graph.**

23. $y = x^3 - 2x^2 - 3x$

24. $y = 2x^4 - 15x^3 + 18x^2$

25. $y = x^4 - 10x^2 + 9$

26. $y = x^4 - 29x^2 + 100$

In Exercises 27–48, find all solutions of the equation. Check your solutions in the original equation.

27. $\sqrt{2x} - 10 = 0$

28. $4\sqrt{x} - 3 = 0$

29. $\sqrt{x - 10} - 4 = 0$

30. $\sqrt{5 - x} - 3 = 0$

31. $\sqrt[3]{2x + 5} + 3 = 0$

32. $\sqrt[3]{3x + 1} - 5 = 0$

33. $-\sqrt{26 - 11x} + 4 = x$

34. $x + \sqrt{31 - 9x} = 5$

35. $\sqrt{x + 1} - 3x = 1$

36. $\sqrt{x + 5} = \sqrt{x - 5}$

37. $\sqrt{x} - \sqrt{x-5} = 1$

38. $\sqrt{x} + \sqrt{x-20} = 10$

39. $\sqrt{x+5} + \sqrt{x-5} = 10$

40. $2\sqrt{x+1} - \sqrt{2x+3} = 1$

41. $(x-5)^{2/3} = 16$

42. $(x+3)^{4/3} = 16$

43. $(x+3)^{2/3} = 8$

44. $(x^2+2)^{2/3} = 9$

45. $(x^2-5)^{2/3} = 16$

46. $(x^2-x-22)^{4/3} = 16$

47. $3x(x-1)^{1/2} + 2(x-1)^{3/2} = 0$

48. $4x^2(x-1)^{1/3} + 6x(x-1)^{4/3} = 0$

Graphical Analysis   In Exercises 49–52, use a graphing utility to graph an equation. Use the graph to approximate any *x*-intercepts of the graph. Set *y* = 0 and solve the resulting equation. Compare the result with the *x*-intercepts of the graph.

49. $y = \sqrt{11x-30} - x$

50. $y = 2x - \sqrt{15-4x}$

51. $y = \sqrt{7x+36} - \sqrt{5x+16} - 2$

52. $y = 3\sqrt{x} - \dfrac{4}{\sqrt{x}} - 4$

In Exercises 53–66, find all solutions of the equation. Check your solutions in the original equation.

53. $\dfrac{20-x}{x} = x$

54. $\dfrac{4}{x} - \dfrac{5}{3} = \dfrac{x}{6}$

55. $\dfrac{1}{x} - \dfrac{1}{x+1} = 3$

56. $\dfrac{x}{x^2-4} + \dfrac{1}{x+2} = 3$

57. $x = \dfrac{3}{x} + \dfrac{1}{2}$

58. $4x + 1 = \dfrac{3}{x}$

59. $\dfrac{4}{x+1} - \dfrac{3}{x+2} = 1$

60. $\dfrac{x+1}{3} - \dfrac{x+1}{x+2} = 0$

61. $|2x-1| = 5$

62. $|3x+2| = 7$

63. $|x| = x^2 + x - 3$

64. $|x^2+6x| = 3x + 18$

65. $|x+1| = x^2 - 5$

66. $|x-10| = x^2 - 10x$

Graphical Analysis   In Exercises 67–70, use a graphing utility to graph the equation. Use the graph to approximate any *x*-intercepts of the graph. Set *y* = 0 and solve the resulting equation. Compare the result with the *x*-intercepts of the graph.

67. $y = \dfrac{1}{x} - \dfrac{4}{x-1} - 1$

68. $y = x + \dfrac{9}{x+1} - 5$

69. $y = |x+1| - 2$

70. $y = |x-2| - 3$

In Exercises 71–74, find the real solutions of the equation analytically. Round to three decimal places.

71. $3.2x^4 - 1.5x^2 - 2.1 = 0$

72. $7.08x^6 + 4.15x^3 - 9.6 = 0$

73. $1.8x - 6\sqrt{x} - 5.6 = 0$

74. $4x^{2/3} + 8x^{1/3} + 3.6 = 0$

Think About It   In Exercises 75–78, find an equation having the given solutions. (There are many correct answers.)

75. $-3, 5$

76. $0, 2, \frac{5}{2}$

77. $\sqrt{2}, -\sqrt{2}, 4$

78. $-2, 2, i, -i$

79. **Sharing the Cost**   A college charters a bus for $1700 to take a group to a museum. When six more students join the trip, the cost per student drops by $7.50. How many students were in the original group?

80. **Sharing the Cost**   Three students are planning to rent an apartment for a year and share equally in the rent. By adding a fourth person, each person could save $75 a month. How much is the monthly rent?

81. **Airspeed**   An airline runs a commuter flight between two cities that are 720 miles apart. If the average speed of the plane could be increased by 40 miles per hour, the travel time would be decreased by 12 minutes. What airspeed is required to obtain this decrease in travel time?

82. Average Speed   A family drove 1080 miles to their vacation lodge. Because of increased traffic density, their average speed on the return trip was decreased by 6 miles per hour and the trip took $2\frac{1}{2}$ hours longer. Determine their average speed on the way to the lodge.

83. Compound Interest   A deposit of $2500 in a mutual fund reaches a balance of $3544.06 after 5 years. What annual interest rate on a certificate of deposit compounded monthly would yield an equivalent return?

84. Compound Interest   A sales representative for a mutual fund company describes a "guaranteed investment fund" that the company is offering to new investors. You are told that if you deposit $10,000 in the fund you will be guaranteed a return of at least $25,000 after 20 years. (Assume the interest is compounded quarterly.)

(a) What is the annual interest rate if the investment only meets the minimum guaranteed amount?

(b) If after 20 years you receive $35,000, what is the annual interest rate?

Exploration   **In Exercises 85 and 86, find x so that the distance between the points is 13.**

85. $(1, 2), (x, -10)$ **86.** $(-8, 0), (x, 5)$

Exploration   **In Exercises 87 and 88, find y so that the distance between the points is 17.**

87. $(0, 0), (8, y)$ **88.** $(-8, 4), (7, y)$

89. Airline Passengers   An airline offers daily flights between Chicago and Denver. The total monthly cost of these flights is given by

$$C = \sqrt{0.2x + 1}$$

where C is measured in millions of dollars and x is measured in thousands of passengers. The total cost of the flights for a certain month is 2.5 million dollars. How many passengers flew that month?

90. Saturated Steam   The temperature T (in degrees Fahrenheit) of saturated steam increases as pressure increases (see figure). This relationship is approximated by the model

$$T = 75.82 - 2.11x + 43.51\sqrt{x}, \qquad 5 \le x \le 40$$

where x is the absolute pressure in pounds per square inch.

(a) The temperature of steam at sea level ($x = 14.696$) is 212°F. Evaluate the model above at this pressure.

(b) Use the model to approximate the pressure if the temperature of the steam is 240°F.

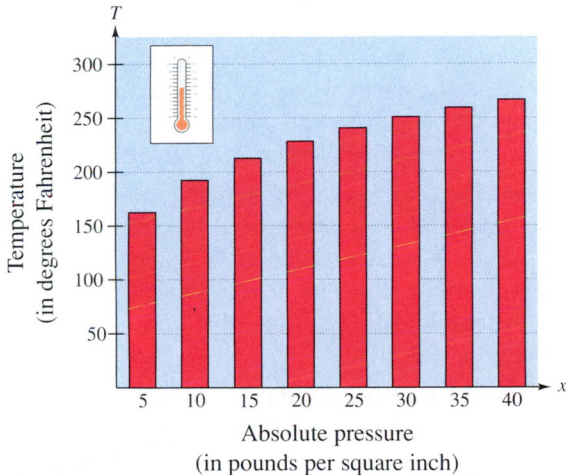

Absolute pressure
(in pounds per square inch)

91. Market Research   The demand equation for a certain product is modeled by

$$p = 40 - \sqrt{0.01x + 1}$$

where x is the number of units demanded per day and p is the price per unit. Approximate the demand if the price is $37.55.

92. Market Research   The demand equation for a certain product is modeled by

$$p = 30 - \sqrt{0.0001x + 1}$$

where x is the number of units demanded per day and p is the price per unit. Approximate the demand if the price is $34.70.

93. *Power Line* A power station is on one side of a river that is $\frac{3}{4}$ mile wide. A factory is 8 miles downstream on the other side of the river. It costs $24 per foot to run power lines overland and $30 per foot to run them underwater. The total cost of the project is $1,098,662.40. Find the length x as labeled in the figure.

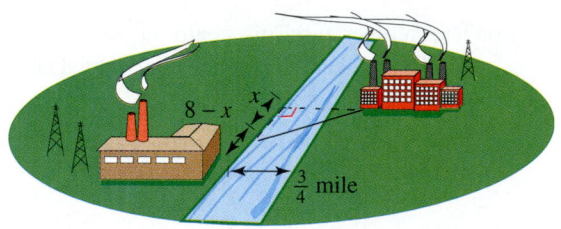

94. *Baseball Diamond* A baseball diamond has the shape of a square where the distance from home plate to second base is approximately $127\frac{1}{2}$ feet. Approximate the distance between each of the bases.

95. *Solving Graphically, Numerically, and Algebraically* A meteorologist is positioned 100 feet from the point where a weather balloon is launched. When the balloon is at height h, the distance between the meteorologist and the balloon is given by $d = \sqrt{100^2 + h^2}$.

(a) Use a graphing utility to graph the equation. Use the trace feature to approximate the value of h when $d = 200$.

(b) Complete the table. Use the table to approximate the value of h when $d = 200$.

h	160	165	170	175	180	185
d						

(c) Find h algebraically when $d = 200$.

(d) Compare the results of each method. In each case, what information did you gain that wasn't apparent in another solution method?

96. *Work Rate* Working together, two people can complete a task in 8 hours. Working alone, how long would it take each to do the task if one person takes 2 hours longer than the other?

In Exercises 97 and 98, solve for the indicated variable.

97. *Surface Area of a Cone*

Solve for h: $S = \pi r \sqrt{r^2 + h^2}$

98. *Inductance*

Solve for Q: $i = \pm \sqrt{\dfrac{1}{LC}} \sqrt{Q^2 - q}$

In Exercises 99 and 100, consider an equation of the form $x + \sqrt{x - a} = b$ where a and b are constants.

99. *Exploration* Find a and b if the solution to the equation is $x = 20$. (There are many correct answers.)

100. *Essay* Write a short paragraph listing the steps required in solving this equation involving radicals.

Review **Solve Exercises 101–108 as a review of the skills and problem-solving techniques you learned in previous sections. Perform the operations and simplify the result.**

101. Divide: $25y^2 \div \dfrac{xy}{5}$

102. Divide: $\dfrac{\left[\dfrac{24 - 18x}{(2 - x)^2}\right]}{\left(\dfrac{60 - 45x}{x^2 - 4x - 4}\right)}$

103. Multiply: $x^2 \cdot \dfrac{x + 1}{x^2 - x} \cdot \dfrac{(5x - 5)^2}{x^2 + 6x + 5}$

104. Multiply: $\dfrac{u^2 - 9v^2}{8rs} \cdot \dfrac{6r}{u^2 - 6uv + 9v^2}$

105. Add: $\dfrac{8}{3t} + \dfrac{3}{2t}$

106. Add: $(y - 2)^2 + \dfrac{3}{(y + 2)^2}$

107. Subtract: $\dfrac{2}{x^2 - 4} - \dfrac{1}{x^2 - 3x + 2}$

108. Subtract: $\dfrac{2}{z + 2} - \left(3 - \dfrac{2}{z}\right)$

1.7 *Linear Inequalities*

See Exercise 89 on page 162 for an example of how a linear inequality can be used to analyze data about grade-point averages of college students.

Introduction ❑ *Properties of Inequalities* ❑ *Solving a Linear Inequality* ❑ *Inequalities Involving Absolute Value* ❑ *Applications*

Introduction

Simple inequalities were reviewed in Section P.1. There, you used inequality symbols $<, \leq, >$, and $\geq$ to compare two numbers and to denote subsets of real numbers. For instance, the simple inequality

$$x \geq 3$$

denotes all real numbers x that are greater than or equal to 3.

In this section you will expand your work with inequalities to include more involved statements such as

$$5x - 7 < 3x + 9 \qquad \text{and} \qquad -3 \leq 6x - 1 < 3.$$

As with an equation, you **solve an inequality** in the variable x by finding all values of x for which the inequality is true. Such values are **solutions** and are said to **satisfy** the inequality. The set of all real numbers that are solutions of an inequality is the **solution set** of the inequality. For instance, the solution set of $x + 1 < 4$ is all real numbers that are less than 3.

The set of all points on the real number line that represent the solution set is the **graph** of the inequality. Graphs of many types of inequalities consist of intervals on the real number line. You can review the nine basic types of intervals on the real number line by turning to pages 3 and 4 in Section P.1. On those pages, note that each type of interval can be classified as *bounded* or *unbounded*.

EXAMPLE 1 *Intervals and Inequalities*

Write an inequality to represent each interval and state whether the interval is bounded or unbounded.

a. $(-3, 5]$ **b.** $(-3, \infty)$ **c.** $[0, 2]$

Solution

a. $(-3, 5]$ corresponds to $-3 < x \leq 5$.   Bounded

b. $(-3, \infty)$ corresponds to $-3 < x$.   Unbounded

c. $[0, 2]$ corresponds to $0 \leq x \leq 2$.   Bounded

Properties of Inequalities

The procedures for solving linear inequalities in one variable are much like those for solving linear equations. To isolate the variable you can make use of the **properties of inequalities.** These properties are similar to the properties of equality, but there are two important exceptions. When both sides of an inequality are multiplied or divided by a negative number, the direction of the inequality symbol must be reversed. Here is an example.

$$-2 < 5 \qquad \text{Original inequality}$$
$$(-3)(-2) > (-3)(5) \qquad \text{Multiply both sides by } -3 \text{ and reverse inequality.}$$
$$6 > -15$$

Two inequalities that have the same solution set are **equivalent.** For instance, the inequalities

$$x + 2 < 5 \qquad \text{and} \qquad x < 3$$

are equivalent. To obtain the second inequality from the first, you can subtract 2 from each side of the inequality. The following list describes the operations that can be used to create equivalent inequalities.

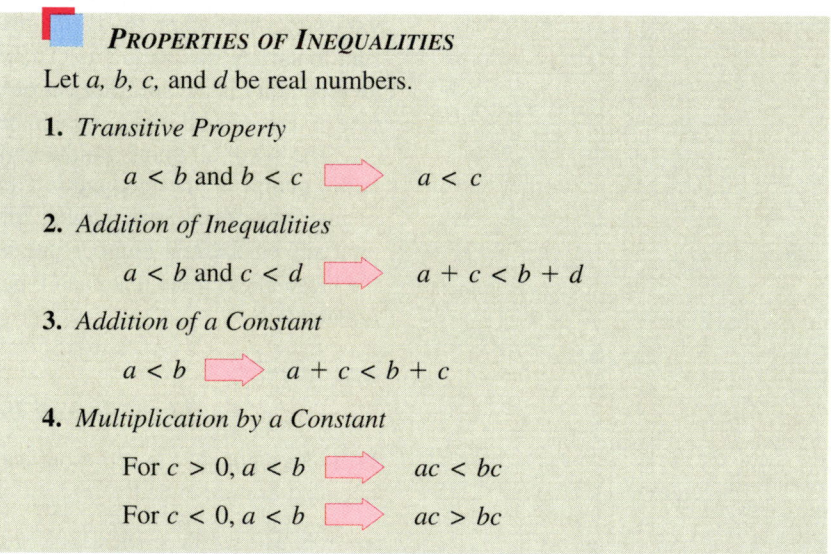

PROPERTIES OF INEQUALITIES

Let a, b, c, and d be real numbers.

1. *Transitive Property*

$$a < b \text{ and } b < c \implies a < c$$

2. *Addition of Inequalities*

$$a < b \text{ and } c < d \implies a + c < b + d$$

3. *Addition of a Constant*

$$a < b \implies a + c < b + c$$

4. *Multiplication by a Constant*

$$\text{For } c > 0, a < b \implies ac < bc$$
$$\text{For } c < 0, a < b \implies ac > bc$$

NOTE Each of the properties above is true if the symbol $<$ is replaced by $\leq$ and $>$ is replaced by $\geq$. For instance, another form of the multiplication property would be as follows.

$$\text{For } c > 0, a \leq b \implies ac \leq bc$$
$$\text{For } c < 0, a \leq b \implies ac \geq bc \quad \blacksquare\blacksquare$$

Solving a Linear Inequality

The simplest type of inequality is a **linear inequality** in a single variable. For instance, $2x + 3 > 4$ is a linear inequality in x.

In the following examples, pay special attention to the steps in which the inequality symbol is reversed. Remember that when you multiply or divide by a negative number, you must reverse the inequality symbol.

EXAMPLE 2   Solving a Linear Inequality

Solve $5x - 7 > 3x + 9$.

Solution

$5x - 7 > 3x + 9$	Original inequality
$5x > 3x + 16$	Add 7 to both sides.
$5x - 3x > 16$	Subtract $3x$ from both sides.
$2x > 16$	Combine like terms.
$x > 8$	Divide both sides by 2.

The solution set is all real numbers that are greater than 8, which is denoted by $(8, \infty)$. The graph is shown in Figure 1.27.

Checking the solution set of an inequality is not as simple as checking the solutions of an equation. You can, however, get an indication of the validity of a solution set by substituting a few convenient values of x.

EXAMPLE 3   Solving a Linear Inequality

Solve $1 - \dfrac{3x}{2} \geq x - 4$.

Solution

$1 - \dfrac{3x}{2} \geq x - 4$	Original inequality
$2 - 3x \geq 2x - 8$	Multiply both sides by 2.
$-3x \geq 2x - 10$	Subtract 2 from both sides.
$-5x \geq -10$	Subtract $2x$ from both sides.
$x \leq 2$	Divide both sides by -5 and reverse inequality.

The solution set is all real numbers that are less than or equal to 2, which is denoted by $(-\infty, 2]$. The graph is shown in Figure 1.28.

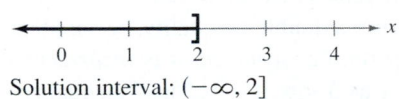

Solution interval: $(8, \infty)$

FIGURE 1.27

Solution interval: $(-\infty, 2]$

FIGURE 1.28

Sometimes it is possible to write two inequalities as a **double inequality.** For instance, you can write the two inequalities $-4 \leq 5x - 2$ and $5x - 2 < 7$ more simply as

$$-4 \leq 5x - 2 < 7.$$

This form allows you to solve the two inequalities together, as demonstrated in Example 4.

EXAMPLE 4   *Solving a Double Inequality*

To solve the double inequality, you can isolate x as the middle term.

$-3 \leq 6x - 1 < 3$ Original inequality

$-2 \leq 6x < 4$ Add 1 to all three parts.

$-\dfrac{1}{3} \leq x < \dfrac{2}{3}$ Divide all three parts by 6 and reduce.

The solution set is all real numbers that are greater than or equal to $-\frac{1}{3}$ and less than $\frac{2}{3}$, which is denoted by $\left[-\frac{1}{3}, \frac{2}{3}\right)$. The graph of this solution set is shown in Figure 1.29.

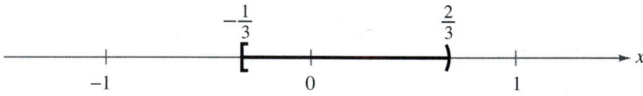

Solution interval: $\left[-\frac{1}{3}, \frac{2}{3}\right)$

FIGURE 1.29

The double inequality in Example 4 could have been solved in two parts as follows.

$$-3 \leq 6x - 1 \qquad \text{and} \qquad 6x - 1 < 3$$
$$-2 \leq 6x \qquad\qquad\qquad\quad 6x < 4$$
$$-\frac{1}{3} \leq x \qquad\qquad\qquad\quad x < \frac{2}{3}$$

The solution set consists of all real numbers that satisfy *both* inequalities. In other words, the solution set is the set of all values of x for which $-\frac{1}{3} \leq x < \frac{2}{3}$.

When combining two inequalities to form a double inequality, be sure that the inequalities satisfy the Transitive Property. For instance, it is *incorrect* to combine the inequalities $3 < x$ and $x \leq -1$ as $3 < x \leq -1$. This "inequality" is obviously wrong because 3 is not less than -1.

Inequalities Involving Absolute Value

TECHNOLOGY

A graphing utility can be used to give a rough indication of the graph of an inequality. For instance, on a *TI-83* or a *TI-82*, you can graph $|x - 5| < 2$ (see Example 5) by entering

$Y_1 = \text{abs}(X - 5) < 2$

and pressing the graph key. With a standard setting, the graph should look like that shown below.

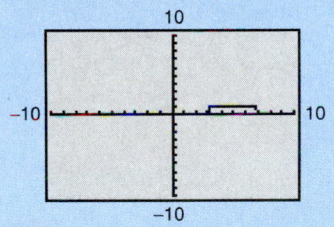

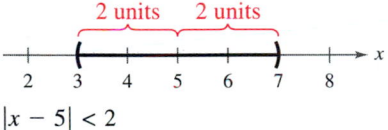

$|x - 5| < 2$

FIGURE 1.30

SOLVING AN ABSOLUTE VALUE INEQUALITY

Let x be a variable of an algebraic expression and let a be a real number such that $a \geq 0$.

1. The solutions of $|x| < a$ are all values of x that lie between $-a$ and a.

$$|x| < a \qquad \text{if and only if} \qquad -a < x < a.$$

2. The solutions of $|x| > a$ are all values of x that are less than $-a$ or greater than a.

$$|x| > a \qquad \text{if and only if} \qquad x < -a \quad \text{or} \quad x > a.$$

These rules are also valid if $<$ is replaced by $\leq$ and $>$ is replaced by $\geq$.

EXAMPLE 5   *Solving an Absolute Value Inequality*

Solve $|x - 5| < 2$.

Solution

$|x - 5| < 2$	Original inequality
$-2 < x - 5 < 2$	Equivalent inequalities
$-2 + 5 < x - 5 + 5 < 2 + 5$	Add 5 to all three parts.
$3 < x < 7$	Simplify.

The solution set is all real numbers that are greater than 3 and less than 7, which is denoted by $(3, 7)$. The graph is shown in Figure 1.30.

EXAMPLE 6   *Solving an Absolute Value Inequality*

Solve $|x + 3| \geq 7$.

Solution

$|x + 3| \geq 7$			Original inequality
$x + 3 \leq -7$	or	$x + 3 \geq 7$	Equivalent inequalities
$x + 3 - 3 \leq -7 - 3$		$x + 3 - 3 \geq 7 - 3$	Subtract 3 from both sides.
$x \leq -10$		$x \geq 4$	Simplify.

The solution set is all real numbers that are less than or equal to -10 *or* greater than or equal to 4, which is denoted by $(-\infty, -10] \cup [4, \infty)$. The graph is shown in Figure 1.31.

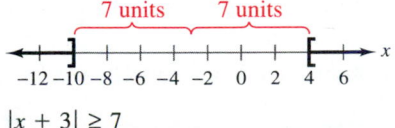

$|x + 3| \geq 7$

FIGURE 1.31

The smallest registered street-legal car in the U.S. was built by Arlis Sluder and is now owned by Jeff Gibson. Its overall length is $88\frac{3}{4}$ inches and its width is $40\frac{1}{2}$ inches.
(Photo: Ron McPherson)

Applications

EXAMPLE 7   *Comparative Shopping*

Real Life

A subcompact car can be rented from Company A for \$180 per week with no extra charge for mileage. A similar car can be rented from Company B for \$100 per week, plus 20 cents for each mile driven. How many miles must you drive in a week to make the rental fee for Company B more than that for Company A?

Solution

Verbal Model:	Weekly cost for Company B	>	Weekly cost for Company A

Labels: Miles driven in one week $= m$ *(miles)*
Weekly cost for Company A $= 180$ *(dollars)*
Weekly cost for Company B $= 100 + 0.20m$ *(dollars)*

Inequality: $100 + 0.2m > 180$
$0.2m > 80$
$m > 400 \text{ miles}$

If you drive more than 400 miles in a week, Company B costs more.

EXAMPLE 8   *Exercise Program*

Real Life

A man begins an exercise and diet program that is designed to reduce his weight by at least 2 pounds per week. At the beginning of the diet the man weighs 225 pounds. Find the maximum number of weeks before the man's weight will reach (or fall below) his goal of 192 pounds.

Solution

Verbal Model:	Desired weight	≤	Current weight	−	2 pounds	·	Number of weeks

Labels: Desired weight $= 192$ *(pounds)*
Current weight $= 225$ *(pounds)*
Number of weeks $= x$ *(weeks)*

Inequality: $192 \leq 225 - 2x$
$-33 \leq -2x$
$16.5 \geq x$

Losing at least 2 pounds a week, he will reach his goal in at most $16\frac{1}{2}$ weeks.

EXAMPLE 9 **Accuracy of a Measurement**

You go to a candy store to buy chocolates that cost $9.89 per pound. The scale that is used in the store has a state seal of approval that indicates the scale is accurate to within half an ounce. According to the scale, your purchase weighs one-half pound and costs $4.95. How much might you have been undercharged or overcharged due to an error in the scale?

Solution

Let x represent the *true* weight of the candy. Because the scale is accurate to within half an ounce (or $\frac{1}{32}$ of a pound), the difference between the exact weight (x) and the scale weight $\left(\frac{1}{2}\right)$ is less than or equal to $\frac{1}{32}$ of a pound. That is,

$$\left| x - \tfrac{1}{2} \right| \le \tfrac{1}{32}.$$

You can solve this inequality as follows.

$$-\tfrac{1}{32} \le x - \tfrac{1}{2} \le \tfrac{1}{32}$$
$$\tfrac{15}{32} \le x \le \tfrac{17}{32}$$
$$0.46875 \le x \le 0.53125$$

In other words, your "one-half pound" of candy could have weighed as little as 0.46875 pound (which would have cost $4.64) or as much as 0.53125 pound (which would have cost $5.25). Thus, you could have been undercharged by as much as $0.30 or overcharged by as much as $0.31.

GROUP ACTIVITY

COMMUNICATING MATHEMATICALLY

Some people find that it is easier to remember a verbal statement, a numerical example, or a picture than it is to remember a mathematical formula. For instance, you can remember the factoring formula $(u + v)(u - v) = u^2 - v^2$ as "the product of the sum and difference of two terms is the difference of each term squared."

Four different properties of inequalities are listed on page 154. For each property, (a) translate the mathematical statement into a *verbal statement*, (b) compile a list of several numerical examples that demonstrate the property, and (c) construct a number line or series of number lines that graphically illustrates the property.

WARM UP

Determine which of the two numbers is larger.

1. $-\frac{1}{2}, -7$

2. $-\frac{1}{3}, -\frac{1}{6}$

3. $-\pi, -3$

4. $-6, \frac{13}{2}$

Use inequality notation to describe the statement.

5. x is nonnegative.

6. z is strictly between -3 and 10.

7. P is no more than 2.

8. W is at least 200.

Evaluate the expression for the given values of x.

9. $|x - 10|$, $x = 12, x = 3$

10. $|2x - 3|$, $x = \frac{3}{2}, x = 1$

1.7 Exercises

In Exercises 1 and 2, write a verbal description of the inequality. Do not use the variable.

1. $x \le 25$

2. $-10 \le s \le 10$

In Exercises 3–6, write an inequality to represent the interval and state whether the interval is bounded or unbounded.

3. $[-1, 3]$

4. $(4, 10]$

5. $(10, \infty)$

6. $[-6, \infty)$

In Exercises 7–14, match the inequality with its graph. [The graphs are labeled (a), (b), (c), (d), (e), (f), (g), and (h).]

7. $x < 3$

8. $x \ge 5$

9. $-3 < x \le 4$

10. $0 \le x \le \frac{9}{2}$

11. $|x| < 3$

12. $|x| > 4$

13. $|x - 4| > 2$

14. $|x + 5| < 3$

(a)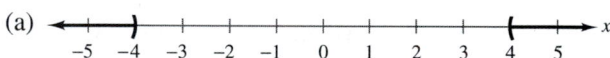

(b)

(c)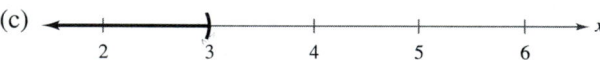

(d)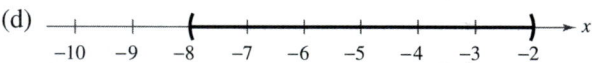

(e)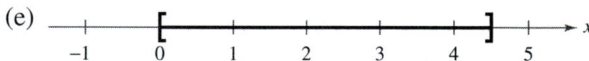

(f)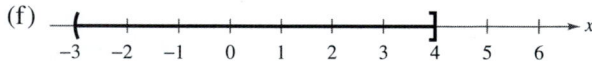

(g)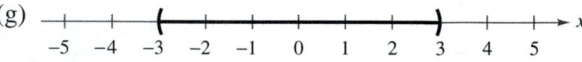

(h)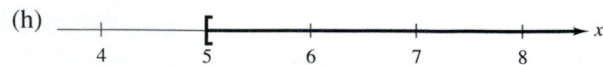

In Exercises 15–20, determine whether the values of x are solutions of the inequality.

Inequality	Values	
15. $5x - 12 > 0$	(a) $x = 3$	(b) $x = -3$
	(c) $x = \frac{5}{2}$	(d) $x = \frac{3}{2}$
16. $x + 1 < \dfrac{2x}{3}$	(a) $x = 0$	(b) $x = 4$
	(c) $x = -4$	(d) $x = -3$
17. $0 < \dfrac{x-2}{4} < 2$	(a) $x = 4$	(b) $x = 10$
	(c) $x = 0$	(d) $x = \frac{7}{2}$
18. $-1 < \dfrac{3-x}{2} \le 1$	(a) $x = 0$	(b) $x = \sqrt{5}$
	(c) $x = 1$	(d) $x = 5$
19. $|x - 10| \ge 3$	(a) $x = 13$	(b) $x = -1$
	(c) $x = 14$	(d) $x = 9$
20. $|2x - 3| < 15$	(a) $x = -6$	(b) $x = 0$
	(c) $x = 12$	(d) $x = 7$

In Exercises 21–38, solve the inequality and sketch the solution on the real number line.

21. $4x < 12$ **22.** $2x > 3$

23. $-10x < 40$ **24.** $-6x > 15$

25. $x - 5 \ge 7$ **26.** $x + 7 \le 12$

27. $2x + 7 < 3$ **28.** $4(x + 1) < 2x + 3$

29. $2x - 1 \ge 0$ **30.** $3x + 1 \ge 2$

31. $4 - 2x < 3$ **32.** $6x - 4 \le 2$

33. $1 < 2x + 3 < 9$

34. $-8 \le 1 - 3(x - 2) < 13$

35. $-4 < \dfrac{2x - 3}{3} < 4$ **36.** $0 \le \dfrac{x + 3}{2} < 5$

37. $\dfrac{3}{4} > x + 1 > \dfrac{1}{4}$ **38.** $-1 < -\dfrac{x}{3} < 1$

Graphical Analysis   In Exercises 39–44, use a graphing utility to graph the inequality.

39. $6x > 12$ **40.** $3x - 1 \le 5$

41. $5 - 2x \ge 1$ **42.** $3(x + 1) < x + 7$

43. $0 \le 2(x + 4) < 20$ **44.** $-2 < 3x + 1 < 10$

Graphical Analysis   In Exercises 45–48, use a graphing utility to graph the equation. Use the graph to approximate the values of x satisfying the specified inequalities.

Equation	Inequalities	
45. $y = 2x - 3$	(a) $y \ge 1$	(b) $y \le 0$
46. $y = \frac{2}{3}x + 1$	(a) $y \le 5$	(b) $y \ge 0$
47. $y = -\frac{1}{2}x + 2$	(a) $0 \le y \le 3$	(b) $y \ge 0$
48. $y = -3x + 8$	(a) $-1 \le y \le 3$	(b) $y \le 0$

In Exercises 49–54, find the interval(s) on the real number line for which the radicand is nonnegative (greater than or equal to zero).

49. $\sqrt{x - 5}$ **50.** $\sqrt{x - 10}$

51. $\sqrt{x + 3}$ **52.** $\sqrt[4]{6x + 15}$

53. $\sqrt[4]{7 - 2x}$ **54.** $\sqrt{3 - x}$

In Exercises 55–70, solve the inequality and sketch the solution on the real number line.

55. $|x| < 5$ **56.** $|2x| < 6$

57. $\left|\dfrac{x}{2}\right| > 3$ **58.** $|5x| > 10$

59. $|x - 20| \le 4$ **60.** $|x - 7| < 6$

61. $|x - 20| \ge 4$ **62.** $|x + 14| + 3 > 17$

63. $\left|\dfrac{x - 3}{2}\right| \ge 5$ **64.** $|1 - 2x| < 5$

65. $|9 - 2x| - 2 < -1$ **66.** $\left|1 - \dfrac{2x}{3}\right| < 1$

67. $2|x + 10| \ge 9$ **68.** $3|4 - 5x| \le 9$

69. $|x - 5| < 0$ **70.** $|x - 5| \ge 0$

Graphical Analysis   In Exercises 71 and 72, graph the equation, and use it to approximate the values of x satisfying the specified inequalities.

Equation	Inequalities	
71. $y = |x - 3|$	(a) $y \le 2$	(b) $y \ge 4$
72. $y = \left|\frac{1}{2}x + 1\right|$	(a) $y \le 4$	(b) $y \ge 1$

In Exercises 73–80, use absolute value notation to define each interval (or pair of intervals) on the real number line.

73.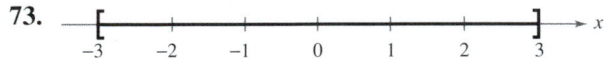

74.

75.

76.

77. All real numbers within 10 units of 12

78. All real numbers at least five units from 8

79. All real numbers more than five units from −3

80. All real numbers no more than seven units from −6

81. **Think About It**   The graph of $|x - 5| < 3$ can be described as *all real numbers within three units of 5*. Give a similar description of $|x - 10| < 8$.

82. **Think About It**   The graph of $|x - 2| > 5$ can be described as *all real numbers more than five units from 2*. Give a similar description of $|x - 8| > 4$.

83. **Comparative Shopping**   You can rent a midsize car from Company A for $250 per week with unlimited mileage. A similar car can be rented from Company B for $150 per week, plus 25 cents for each mile driven. How many miles must you drive in a week to make the rental fee for Company B *greater than* that for Company A?

84. **Comparative Shopping**   Your department sends its copying to the photocopy center of your company. The center bills your department $0.10 per page. You have investigated the possibility of buying a departmental copier for $3000. With your own copier the cost per page would be $0.03. The expected life of the copier is 4 years. How many copies must you make in the 4-year period to justify buying the copier?

85. **Simple Interest**   In order for an investment of $1000 to grow to *more than* $1250 in 2 years, what must the annual interest rate be? $[A = P(1 + rt)]$

86. **Simple Interest**   In order for an investment of $750 to grow to *more than* $1050 in 2 years, what must the annual interest rate be? $[A = P(1 + rt)]$

87. **Break-Even Analysis**   The revenue for selling x units of a product is $R = 115.95x$. The cost of producing x units is $C = 95x + 750$. To obtain a profit, the revenue must be *greater than* the cost. For what values of x will this product return a profit?

88. **Annual Operating Cost**   A utility company has a fleet of vans. The annual operating cost per van is

$$C = 0.32m + 2300$$

where m is the number of miles traveled by a van in a year. What number of miles will yield an annual operating cost that is less than $10,000?

89. **Data Analysis**   The admissions office of a college wants to determine if there is a relationship between IQ scores x and grade-point averages y after the first year. A sample of 12 students yielded the following data.

x	118	131	125	123	133	136
y	2.2	2.4	3.2	2.4	3.5	3.0

x	128	124	116	120	134	131
y	3.0	2.8	2.2	1.8	3.4	3.6

(a) Use a graphing utility to plot the data.

(b) A model for this data is given by

$$y = 0.067x - 5.638.$$

Use a graphing utility to graph the equation on the same display used in part (a).

(c) Which values of x predict a grade-point average of at least 3.0?

(d) Use the graph to write a statement about the accuracy of the model. If you think the graph indicates that IQ scores are not particularly good predictors of grade-point average, list other factors that may influence college performance.

90. Teachers' Salaries The average salary S (in thousands of dollars) for elementary and secondary teachers in the United States from 1984 to 1993 is approximated by the model

$$S = 15.812 + 1.472t$$

where $t = 4$ represents 1984 (see figure). According to this model, when will the average salary *exceed* $40,000? (Source: National Education Association)

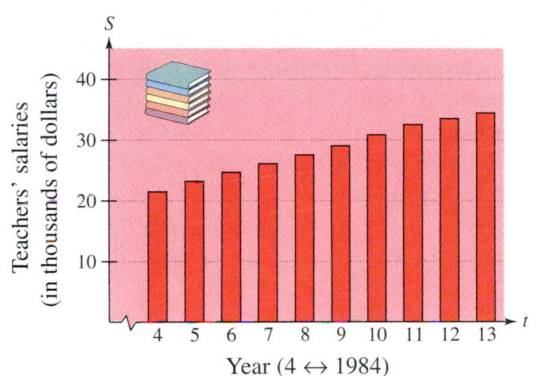

Year (4 ↔ 1984)

91. Accuracy of Measurement The side of a square is measured as 10.4 inches with a possible error of $\frac{1}{16}$ inch. Using these measurements, determine the interval containing the area of the square.

92. Accuracy of Measurement You buy a bag of oranges for $0.95 per pound. The weight that is listed on the bag is 4.65 pounds. The scale that weighed the bag is accurate to within 1 ounce. How much might you have been undercharged or overcharged?

93. Height The heights h of two-thirds of the members of a certain population satisfy the inequality

$$\left| \frac{h - 68.5}{2.7} \right| \leq 1$$

where h is measured in inches. Determine the interval on the real number line in which these heights lie.

94. Relative Humidity A certain electronic device is to be operated in an environment with relative humidity h in the interval defined by

$$|h - 50| \leq 30.$$

What are the minimum and maximum relative humidities for the operation of this device?

95. Exploration Find sets of values of a, b, and c so that $0 \leq x \leq 10$ is a solution of the inequality $|ax - b| \leq c$.

96. Given two real numbers a and b, such that $0 < a < b$, prove that

$$\frac{1}{a} > \frac{1}{b}.$$

Chapter Opener In Exercises 97–100, use the information given in the chapter opener on page 81.

97. Estimate the frequency if the plate thickness is 2 millimeters.

98. Estimate the plate thickness if the frequency is 600 vibrations per second.

99. Approximate the interval for the plate thickness if the frequency is between 200 and 400 vibrations per second.

100. Approximate the interval for the frequency if the plate thickness is less than 3 millimeters.

Review Solve Exercises 101–104 as a review of the skills and problem-solving techniques you learned in previous sections.

101. Find the coordinates of the point located three units to the left of the y-axis and 10 units above the x-axis.

102. Determine the quadrants in which the point (x, y) is located if $y > 0$.

103. Find the distance between the points $(-4, 2)$ and $(1, 12)$.

104. Find the coordinates of the midpoint of the line segment joining the points $(0, 3)$ and $(18, 12)$.

1.8 *Other Types of Inequalities*

See Exercise 75 on page 173 for an example of how a quadratic inequality can be used to model the percent of the American population that are college graduates.

Polynomial Inequalities ◻ *Rational Inequalities* ◻ *Applications*

Polynomial Inequalities

To solve a polynomial inequality such as

$$x^2 - 2x - 3 < 0$$

you can use the fact that a polynomial can change signs only at its zeros (the x-values that make the polynomial equal to zero). Between two consecutive zeros a polynomial must be entirely positive or entirely negative. This means that when the real zeros of a polynomial are put in order, they divide the real number line into intervals in which the polynomial has no sign changes. These zeros are the **critical numbers** of the inequality, and the resulting intervals are the **test intervals** for the inequality. For instance, the polynomial

$$x^2 - 2x - 3 = (x + 1)(x - 3)$$

has two zeros, $x = -1$ and $x = 3$, and these zeros divide the real number line into three test intervals:

$$(-\infty, -1), \quad (-1, 3), \quad \text{and} \quad (3, \infty).$$

Thus, to solve the inequality $x^2 - 2x - 3 < 0$, you need only test one value from each of these test intervals. You can use the same basic approach to determine the test intervals for any polynomial.

FINDING TEST INTERVALS FOR A POLYNOMIAL

To determine the intervals on which the values of a polynomial are entirely negative or entirely positive, use the following steps.

1. Find all real zeros of the polynomial, and arrange the zeros in increasing order (from smallest to largest). These zeros are the **critical numbers** of the polynomial.
2. Use the critical numbers of the polynomial to determine its **test intervals.**
3. Choose one representative x-value in each test interval and evaluate the polynomial at that value. If the value of the polynomial is negative, the polynomial will have negative values for *every* x-value in the interval. If the value of the polynomial is positive, the polynomial will have positive values for *every* x-value in the interval.

EXAMPLE 1   *Solving a Polynomial Inequality*

Solve $x^2 - x - 6 < 0$.

Solution

By factoring the quadratic as

$$x^2 - x - 6 = (x + 2)(x - 3)$$

you can see that the critical numbers are $x = -2$ and $x = 3$. Thus, the polynomial's test intervals are

$$(-\infty, -2), \quad (-2, 3), \quad \text{and} \quad (3, \infty). \qquad \textcolor{red}{\text{Test intervals}}$$

In each test interval, choose a representative x-value and evaluate the polynomial.

Interval	x-Value	Polynomial Value	Conclusion
$(-\infty, -2)$	$x = -3$	$(-3)^2 - (-3) - 6 = 6$	Positive
$(-2, 3)$	$x = 0$	$(0)^2 - (0) - 6 = -6$	Negative
$(3, \infty)$	$x = 4$	$(4)^2 - (4) - 6 = 6$	Positive

From this you can conclude that the polynomial is positive for all x-values in $(-\infty, -2)$ and $(3, \infty)$, and is negative for all x-values in $(-2, 3)$. This implies that the solution of the inequality $x^2 - x - 6 < 0$ is the interval $(-2, 3)$, as shown in Figure 1.32.

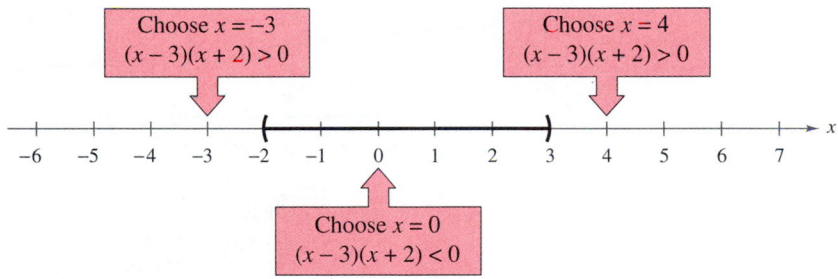

FIGURE 1.32

As with linear inequalities, you can check the reasonableness of a solution by substituting x-values into the original inequality. For instance, to check the solution found in Example 1, try substituting several x-values from the interval $(-2, 3)$ into the inequality

$$x^2 - x - 6 < 0.$$

Regardless of which x-values you choose, the inequality should be satisfied.

In Example 1, the polynomial inequality was given in standard form. Whenever this is not the case, you should begin the solution process by writing the inequality in standard form—with the polynomial on one side and zero on the other.

EXAMPLE 2   *Solving a Polynomial Inequality*

Solve $2x^2 + 5x > 12$.

Solution

Begin by writing the inequality in standard form.

$$2x^2 + 5x > 12 \qquad \text{Original inequality}$$

$$2x^2 + 5x - 12 > 0 \qquad \text{Standard form}$$

$$(x + 4)(2x - 3) > 0 \qquad \text{Factored form}$$

Critical numbers: $x = -4, x = \frac{3}{2}$

Test intervals: $\left(-\infty, -4\right), \left(-4, \frac{3}{2}\right), \left(\frac{3}{2}, \infty\right)$

Test: Is $(x + 4)(2x - 3) > 0$?

After testing these intervals, as shown in Figure 1.33, you can see that the polynomial $2x^2 + 5x - 12$ is positive on the open intervals $(-\infty, -4)$ and $\left(\frac{3}{2}, \infty\right)$. Therefore, the solution set consists of all real numbers in the intervals $(-\infty, -4)$ and $\left(\frac{3}{2}, \infty\right)$.

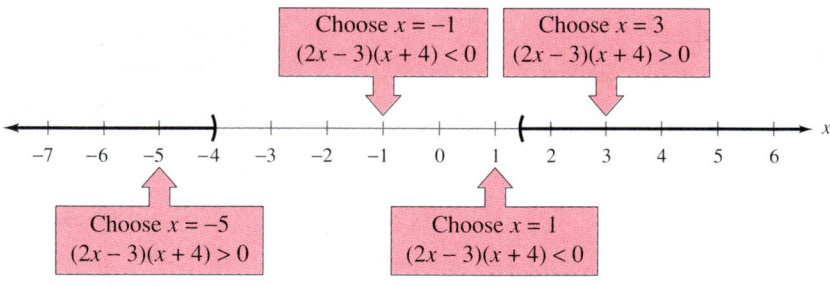

FIGURE 1.33

When solving a quadratic inequality, be sure you have accounted for the particular type of inequality symbol given in the inequality. For instance, in Example 2, note that the solution consisted of two open intervals because the original inequality contained a "greater than" symbol. If the original inequality had been $2x^2 + 5x \geq 12$, the solution would have consisted of the two half-open intervals $(-\infty, -4]$ and $\left[\frac{3}{2}, \infty\right)$.

Each of the polynomial inequalities in Examples 1 and 2 has a solution set that consists of a single interval or the union of two intervals. When solving the exercises for this section, watch for unusual solution sets, as illustrated in Example 3.

EXAMPLE 3 *Unusual Solution Sets*

a. The solution set of the following inequality consists of the entire set of real numbers, $(-\infty, \infty)$.

$$x^2 + 2x + 4 > 0$$

b. The solution set of the following inequality consists of the single real number $\{-1\}$.

$$x^2 + 2x + 1 \le 0$$

c. The solution set of the following inequality is empty.

$$x^2 + 3x + 5 < 0$$

d. The solution set of the following inequality consists of all real numbers except the number 2.

$$x^2 - 4x + 4 > 0$$

Exploration

You can use a graphing utility to verify the results given in Example 3. For instance, the graph of $y = x^2 + 2x + 4$ is shown below. Notice that the y-values are greater than 0 for *all* values of x, as stated in Example 3(a). Use the graphing utility to graph the following:

$$y = x^2 + 2x + 1 \qquad y = x^2 + 3x + 5 \qquad y = x^2 - 4x + 4$$

Explain how you can use the graphs to verify the results of parts (b), (c), and (d) of Example 3.

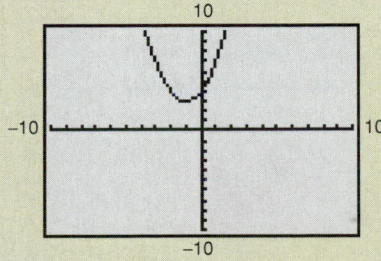

Rational Inequalities

The concepts of critical numbers and test intervals can be extended to rational inequalities. To do this, use the fact that the value of a rational expression can change sign only at its *zeros* (the x-values for which its numerator is zero) and its *undefined values* (the x-values for which its denominator is zero). These two types of numbers make up the **critical numbers** of a rational inequality.

EXAMPLE 4   Solving a Rational Inequality

Solve $\dfrac{2x - 7}{x - 5} \le 3$.

Solution

$$\frac{2x - 7}{x - 5} \le 3 \qquad \text{Original inequality}$$

$$\frac{2x - 7}{x - 5} - 3 \le 0 \qquad \text{Standard form}$$

$$\frac{2x - 7 - 3x + 15}{x - 5} \le 0 \qquad \text{Add fractions.}$$

$$\frac{-x + 8}{x - 5} \le 0 \qquad \text{Simplify.}$$

Critical numbers: $x = 5, x = 8$

Test intervals: $(-\infty, 5), (5, 8), (8, \infty)$

Test: Is $\dfrac{-x + 8}{x - 5} \le 0$?

After testing these intervals, as shown in Figure 1.34, you can see that the rational expression $(-x + 8)/(x - 5)$ is negative in the open intervals $(-\infty, 5)$ and $(8, \infty)$. Moreover, because $(-x + 8)/(x - 5) = 0$ when $x = 8$, you can conclude that the solution set consists of all real numbers in the intervals

$(-\infty, 5) \cup [8, \infty).$ Solution set

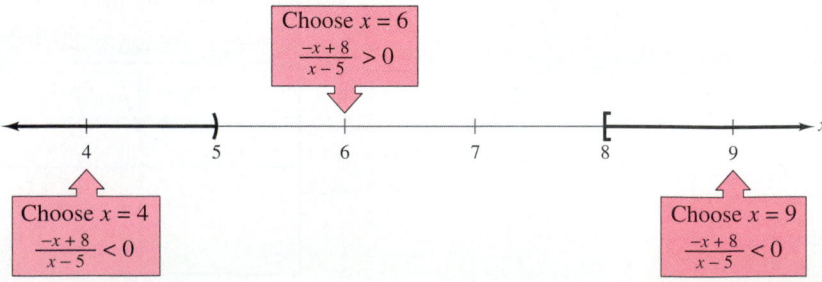

FIGURE 1.34

Applications

One common application of inequalities comes from business and involves profit, revenue, and cost. The formula that relates these three quantities is

| Profit | = | Revenue | − | Cost |

$$P = R - C.$$

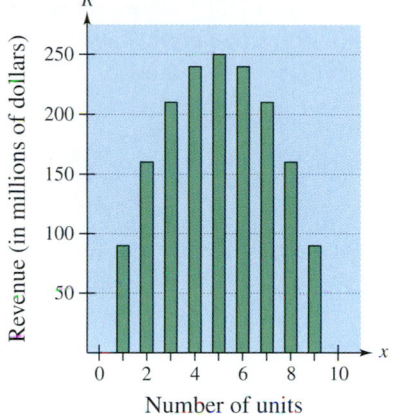

FIGURE 1.35

EXAMPLE 5   *Increasing the Profit for a Product*

Real Life

The marketing department of a calculator manufacturer has determined that the demand for a new model of calculator is given by

$$p = 100 - 0.00001x, \qquad 0 \le x \le 10{,}000{,}000 \qquad \text{Demand equation}$$

where the price per calculator p is in dollars and x represents the number of calculators sold. (If this model is accurate, no one would be willing to pay $100 for the calculator. At the other extreme, the company couldn't *give* away more than 10 million calculators.) The revenue for selling x calculators is given by

$$R = xp = x(100 - 0.00001x) \qquad \text{Revenue equation}$$

as shown in Figure 1.35. The total cost of producing x calculators is $10 per calculator plus a development cost of $2,500,000. Thus, the total cost is

$$C = 10x + 2{,}500{,}000. \qquad \text{Cost equation}$$

What price should the company charge per calculator to obtain a profit of at least $190,000,000?

Solution

Verbal Model:

| Profit | = | Revenue | − | Cost |

Equation: $\quad P = R - C$

$$P = 100x - 0.00001x^2 - (10x + 2{,}500{,}000)$$

$$P = -0.00001x^2 + 90x - 2{,}500{,}000$$

To answer the question, you must now solve the inequality

$$-0.00001x^2 + 90x - 2{,}500{,}000 \ge 190{,}000{,}000.$$

Using the techniques described in this section, you can find the solution to be $3{,}500{,}000 \le x \le 5{,}500{,}000$, as shown in Figure 1.36. The prices that correspond to these x values are given by

$$\$45.00 \le p \le \$65.00.$$

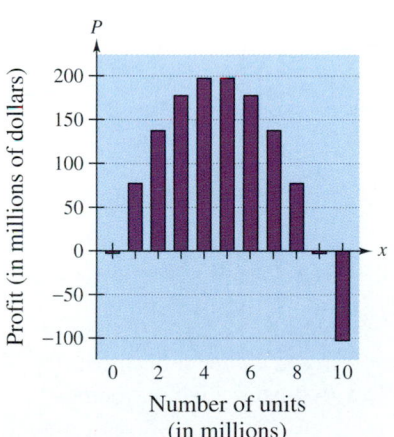

FIGURE 1.36

Another common application of inequalities is finding the domain of an expression that involves a square root, as shown in Example 6.

EXAMPLE 6   *Finding the Domain of an Expression*

Find the domain of $\sqrt{64 - 4x^2}$.

Solution

Remember that the domain of an expression is the set of all x-values for which the expression is defined. Because $\sqrt{64 - 4x^2}$ is defined (has real values) only if $64 - 4x^2$ is nonnegative, the domain is given by $64 - 4x^2 \geq 0$.

$64 - 4x^2 \geq 0$	Standard form
$16 - x^2 \geq 0$	Divide both sides by 4.
$(4 - x)(4 + x) \geq 0$	Factored form

Thus, the inequality has two critical numbers: -4 and 4. You can use these two numbers to test the inequality as follows.

Critical numbers: $x = -4, x = 4$

Test intervals: $(-\infty, -4), (-4, 4), (4, \infty)$

Test: Is $(4 - x)(4 + x) \geq 0$?

A test shows that $64 - 4x^2$ is greater than or equal to 0 in the *closed interval* $[-4, 4]$. Thus, the domain of the expression $\sqrt{64 - 4x^2}$ is the interval $[-4, 4]$, as shown in Figure 1.37.

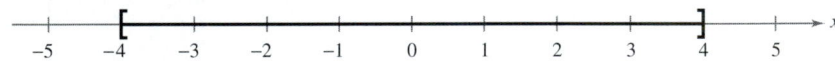

FIGURE 1.37

GROUP ACTIVITY

PROFIT ANALYSIS

Consider the relationship $P = R - C$ described on page 169. Discuss why it might be beneficial to solve $P < 0$ if you owned a business. Use the situation described in Example 5 to illustrate the reasoning.

Solve the inequality.

1. $-\dfrac{y}{3} > 2$

2. $-6z < 27$

3. $-3 \le 2x + 3 < 5$

4. $-3x + 5 \ge 20$

5. $10 > 4 - 3(x + 1)$

6. $3 < 1 + 2(x - 4) < 7$

7. $2|x| \le 7$

8. $|x - 3| > 1$

9. $|x + 4| > 2$

10. $|2 - x| \le 4$

1.8 Exercises

In Exercises 1–4, determine whether the values of x are solutions of the inequality.

Inequality	Values	
1. $x^2 - 3 < 0$	(a) $x = 3$	(b) $x = 0$
	(c) $x = \frac{3}{2}$	(d) $x = -5$
2. $x^2 - x - 12 \ge 0$	(a) $x = 5$	(b) $x = 0$
	(c) $x = -4$	(d) $x = -3$
3. $\dfrac{x + 2}{x - 4} \ge 3$	(a) $x = 5$	(b) $x = 4$
	(c) $x = -\frac{9}{2}$	(d) $x = \frac{9}{2}$
4. $\dfrac{3x^2}{x^2 + 4} < 1$	(a) $x = -2$	(b) $x = -1$
	(c) $x = 0$	(d) $x = 3$

In Exercises 5–8, find the critical numbers.

5. $2x^2 - x - 6$

6. $9x^3 - 25x^2$

7. $2 + \dfrac{3}{x - 5}$

8. $\dfrac{x}{x + 2} - \dfrac{2}{x - 1}$

In Exercises 9–28, solve the inequality and graph the solution on the real number line.

9. $x^2 \le 9$

10. $x^2 < 5$

11. $x^2 > 4$

12. $(x - 3)^2 \ge 1$

13. $(x + 2)^2 < 25$

14. $(x + 6)^2 \le 8$

15. $x^2 + 4x + 4 \ge 9$

16. $x^2 - 6x + 9 < 16$

17. $x^2 + x < 6$

18. $x^2 + 2x > 3$

19. $3(x - 1)(x + 1) > 0$

20. $6(x + 2)(x - 1) < 0$

21. $x^2 + 2x - 3 < 0$

22. $x^2 - 4x - 1 > 0$

23. $4x^3 - 6x^2 < 0$

24. $4x^3 - 12x^2 > 0$

25. $x^3 - 4x \ge 0$

26. $2x^3 - x^4 \le 0$

27. $(x - 1)^2(x + 2)^3 \ge 0$

28. $x^4(x - 3) \le 0$

Graphical Analysis   **In Exercises 29–32, use a graphing utility to graph the equation. Use the graph to approximate the values of x satisfying the specified inequalities.**

Equation	Inequalities	
29. $y = -x^2 + 2x + 3$	(a) $y \le 0$	(b) $y \ge 3$
30. $y = \frac{1}{2}x^2 - 2x + 1$	(a) $y \le 1$	(b) $y \ge 7$
31. $y = \frac{1}{8}x^3 - \frac{1}{2}x$	(a) $y \ge 0$	(b) $y \le 6$
32. $y = x^3 - x^2 - 16x + 16$	(a) $y \le 0$	(b) $y \ge 36$

In Exercises 33–46, solve the inequality and graph the solution on the real number line.

33. $\dfrac{1}{x} - x > 0$

34. $\dfrac{1}{x} - 4 < 0$

35. $\dfrac{x + 6}{x + 1} - 2 < 0$

36. $\dfrac{x + 12}{x + 2} - 3 \geq 0$

37. $\dfrac{3x - 5}{x - 5} > 4$

38. $\dfrac{5 + 7x}{1 + 2x} < 4$

39. $\dfrac{4}{x + 5} > \dfrac{1}{2x + 3}$

40. $\dfrac{5}{x - 6} > \dfrac{3}{x + 2}$

41. $\dfrac{1}{x - 3} \leq \dfrac{9}{4x + 3}$

42. $\dfrac{1}{x} \geq \dfrac{1}{x + 3}$

43. $\dfrac{x^2 + 2x}{x^2 - 9} \leq 0$

44. $\dfrac{x^2 + x - 6}{x} \geq 0$

45. $\dfrac{3}{x - 1} - \dfrac{2}{x + 1} < 1$

46. $\dfrac{3x}{x - 1} \leq \dfrac{x}{x + 4}$

Graphical Analysis   In Exercises 47–50, use a graphing utility to graph the equation. Use the graph to approximate the values of x satisfying the specified inequalities.

Equation	Inequalities	
47. $y = \dfrac{3x}{x - 2}$	(a) $y \leq 0$	(b) $y \geq 6$
48. $y = \dfrac{2(x - 2)}{x + 1}$	(a) $y \leq 0$	(b) $y \geq 8$
49. $y = \dfrac{2x^2}{x^2 + 4}$	(a) $y \geq 1$	(b) $y \leq 2$
50. $y = \dfrac{5x}{x^2 + 4}$	(a) $y \geq 1$	(b) $y \leq 0$

In Exercises 51–56, find the domain of x in the expression.

51. $\sqrt[4]{4 - x^2}$

52. $\sqrt{x^2 - 4}$

53. $\sqrt{x^2 - 7x + 12}$

54. $\sqrt{144 - 9x^2}$

55. $\sqrt{12 - x - x^2}$

56. $\sqrt{\dfrac{x}{x^2 - 9}}$

In Exercises 57–62, solve the inequality. Round each number in your solution to two decimal places.

57. $0.4x^2 + 5.26 < 10.2$

58. $-1.3x^2 + 3.78 > 2.12$

59. $-0.5x^2 + 12.5x + 1.6 > 0$

60. $1.2x^2 + 4.8x + 3.1 < 5.3$

61. $\dfrac{1}{2.3x - 5.2} > 3.4$

62. $\dfrac{2}{3.1x - 3.7} > 5.8$

Exploration   In Exercises 63–66, find the interval for b such that the equation has at least one real solution.

63. $x^2 + bx + 4 = 0$

64. $x^2 + bx - 4 = 0$

65. $3x^2 + bx + 10 = 0$

66. $2x^2 + bx + 5 = 0$

67. *Conjecture*   Write a conjecture about the interval for b in Exercises 63–66. Explain your reasoning.

68. *Think About It*   What is the center of the interval for b in Exercises 63–66?

69. *Height of a Projectile*   A projectile is fired straight upward from ground level with an initial velocity of 160 feet per second.

(a) At what instant will it be back at ground level?

(b) When will the height exceed 384 feet?

70. *Height of a Projectile*   A projectile is fired straight upward from ground level with an initial velocity of 128 feet per second.

(a) At what instant will it be back at ground level?

(b) When will the height be less than 128 feet?

71. *Dimensions of a Field*   A rectangular playing field with a perimeter of 100 meters is to have an area of at least 500 square meters. Within what bounds must the length of the rectangle lie?

72. *Compound Interest* P dollars, invested at interest rate r compounded annually, increases to an amount

$$A = P(1 + r)^2$$

in 2 years. If an investment of $1000 is to increase to an amount greater than $1200 in 2 years, then the interest rate must be greater than what percent?

73. Resistors When two resistors of resistance R_1 and R_2 are connected in parallel (see figure), the total resistance R satisfies the equation

$$\frac{1}{R} = \frac{1}{R_1} + \frac{1}{R_2}.$$

Find R_1 for a parallel circuit in which $R_2 = 2$ ohms and R must be at least 1 ohm.

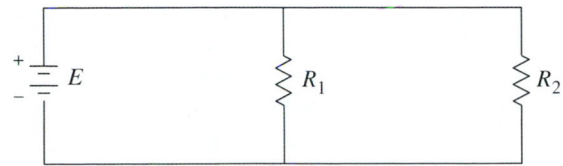

74. Company Profits The revenue and cost equations for a product are given by

$$R = x(50 - 0.0002x)$$
$$C = 12x + 150{,}000$$

where R and C are measured in dollars and x represents the number of units sold. How many units must be sold to obtain a profit of at least $1,650,000?

75. Percent of College Graduates The percent P of the American population that graduated from college from 1960 to 1993 is approximated by the model

$$P = 7.34 + 0.41t + 0.002t^2$$

where the time t represents the calendar year, with $t = 0$ corresponding to 1960 (see figure). According to this model, when will the percent of college graduates exceed 25% of the population? (Source: U.S. Bureau of the Census)

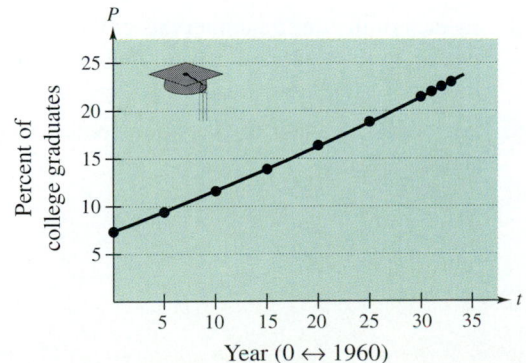

Year (0 ↔ 1960)

76. Safe Beam Loads The maximum safe load uniformly distributed over a 1-foot section of a 2-inch-wide wooden beam is approximated by the model

$$\text{Load} = 168.5d^2 - 472.1$$

where d is the depth of the beam.

(a) Evaluate the model for $d = 4, d = 6, d = 8$, $d = 10$, and $d = 12$. Use the results to create a bar graph.

(b) Determine the minimum depth of the beam that will safely support a load of 2000 pounds.

77. Power Supply Two factories are located at the coordinates $(-4, 0)$ and $(4, 0)$ with their power supply located at the point $(0, 6)$ (see figure).

(a) Write an equation, in terms of y, giving the amount L of power line required to supply both factories.

(b) Determine the interval of values for y in the context of the problem. Determine L for the two endpoints of the interval. Will L increase or decrease for values of y not at the endpoints of the interval?

(c) Use a graphing utility to graph the equation of part (a) and use the graph to verify your answers to part (b).

(d) Find the values of y such that $L < 13$.

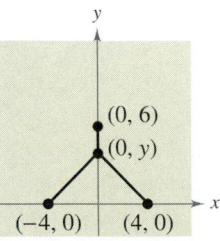

Review **Solve Exercises 78–81, as a review of the skills and problem-solving techniques you learned in previous sections. Factor the expression.**

78. $4x^2 + 20x + 25$

79. $(x + 3)^2 - 16$

80. $x^2(x + 3) - 4(x + 3)$

81. $2x^4 - 54x$

FOCUS ON CONCEPTS

In this chapter, you studied algebraic and graphical methods related to solving equations and inequalities. You can use the following questions to check your understanding of several of these basic methods. The answers to these questions are given in the back of the book.

Classifying Common Graphs **In Exercises 1–4, transformations of the linear model, quadratic model, square root model, and absolute value model are given. Match each model with the corresponding graph. [The graphs are labeled (a), (b), (c), and (d).]**

1. $y = \frac{1}{2}x - 2$

2. $y = x^2 - 2$

3. $y = \sqrt{x - 2}$

4. $y = |x| - 2$

(a)
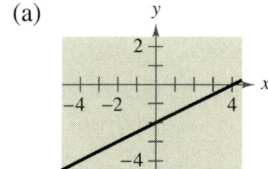

(b)
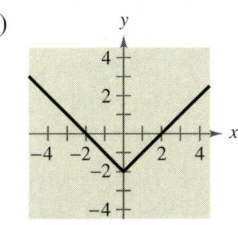

(c)
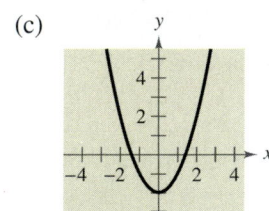

(d)
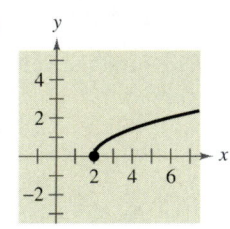

5. In your own words, explain how the display of a graphing utility changes if Xmax is changed from 10 to 20.

6. In your own words, explain what is meant by "equivalent equations." Describe the steps used to transform an equation into an equivalent equation.

7. Consider the linear equation $ax + b = 0$.

   (a) What is the sign of the solution if $ab > 0$?

   (b) What is the sign of the solution if $ab < 0$?

   In each case, explain your reasoning.

8. The graphs show the solution(s) of equations plotted on the real number line. In each case determine if the solution(s) is (are) for a linear equation, a quadratic equation, both, or neither. Explain.

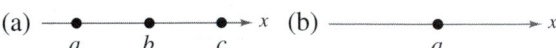

9. To solve the equation

$$2x^2 + 3x = 15x$$

a student divides both sides by x and solves the equation $2x + 3 = 15$. The resulting solution ($x = 6$) satisfies the given equation. Is there an error? Explain.

10. Identify the solution of the inequality $|x - a| \geq 2$.

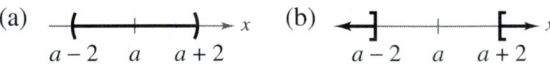

11. Consider the polynomial $(x - a)(x - b)$ and the real number line.

   (a) Identify the points on the line at which the polynomial is zero.

   (b) In each of the three subintervals of the line, write the sign of each factor and the sign of the product.

   (c) For what x-values does a polynomial change signs?

Review Exercises

In Exercises 1 and 2, complete a table of values. Use the solution points to sketch the graph of the equation.

1. $y = -\frac{1}{2}x + 2$ **2.** $y = x^2 - 3x$

In Exercises 3–12, sketch the graph *by hand.*

3. $y - 2x - 3 = 0$ **4.** $3x + 2y + 6 = 0$
5. $x - 5 = 0$ **6.** $y = 8 - |x|$
7. $y = \sqrt{5 - x}$ **8.** $y = \sqrt{x + 2}$
9. $y + 2x^2 = 0$ **10.** $y = x^2 - 4x$
11. $y = \sqrt{25 - x^2}$ **12.** $x^2 + y^2 = 10$

In Exercises 13–20, use a graphing utility to graph the equation. Approximate any intercepts.

13. $y = \frac{1}{4}(x + 1)^3$ **14.** $y = 4 - (x - 4)^2$
15. $y = \frac{1}{4}x^4 - 2x^2$ **16.** $y = \frac{1}{4}x^3 - 3x$
17. $y = x\sqrt{9 - x^2}$ **18.** $y = x\sqrt{x + 3}$
19. $y = |x - 4| - 4$ **20.** $y = |x + 2| + |3 - x|$

In Exercises 21 and 22, solve for y and sketch the graphs of the resulting equations.

21. $y^2 = 25 - x^2$ **22.** $(x + 2)^2 + y^2 = 16$

In Exercises 23 and 24, find a setting on a graphing utility so that the equation's graph matches the given graph.

23. $y = 10x^3 - 21x^2$ **24.** $y = 0.002x^2 - 0.06x - 1$

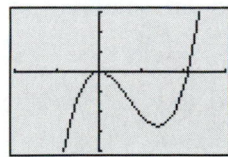   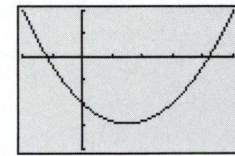

25. Find the center and radius of the circle given by $(x - 3)^2 + (y + 1)^2 = 9$.
Sketch the graph of the circle.

26. Find the standard form of the equation of the circle for which the endpoints of a diameter are $(0, 0)$ and $(4, -6)$.

In Exercises 27 and 28, determine whether the equation is an identity or a conditional equation.

27. $6 - (x - 2)^2 = 2 + 4x - x^2$
28. $3(x - 2) + 2x = 2(x + 3)$

In Exercises 29 and 30, determine whether the values of x are solutions of the equation.

Equation	*Values*
29. $3x^2 + 7x = x^2 + 4$	(a) $x = 0$ (b) $x = -4$
	(c) $x = \frac{1}{2}$ (d) $x = -1$
30. $6 + \dfrac{3}{x - 4} = 5$	(a) $x = 5$ (b) $x = 0$
	(c) $x = -2$ (d) $x = 7$

In Exercises 31–60, solve the equation (if possible) and check your solution.

31. $3x - 2(x + 5) = 10$
32. $4x + 2(7 - x) = 5$
33. $4(x + 3) - 3 = 2(4 - 3x) - 4$
34. $\frac{1}{2}(x - 3) - 2(x + 1) = 5$
35. $3\left(1 - \dfrac{1}{5t}\right) = 0$
36. $\dfrac{1}{x - 2} = 3$

37. $6x = 3x^2$ **38.** $15 + x - 2x^2 = 0$
39. $(x + 4)^2 = 18$ **40.** $16x^2 = 25$
41. $x^2 - 12x + 30 = 0$ **42.** $x^2 + 6x - 3 = 0$
43. $5x^4 - 12x^3 = 0$ **44.** $4x^3 - 6x^2 = 0$
45. $\dfrac{4}{(x - 4)^2} = 1$ **46.** $\dfrac{1}{(t + 1)^2} = 1$
47. $\sqrt{x + 4} = 3$ **48.** $\sqrt{x - 2} - 8 = 0$

49. $2\sqrt{x} - 5 = 0$

50. $\sqrt{3x - 2} = 4 - x$

51. $\sqrt{2x + 3} + \sqrt{x - 2} = 2$

52. $5\sqrt{x} - \sqrt{x - 1} = 6$

53. $(x - 1)^{2/3} - 25 = 0$

54. $(x + 2)^{3/4} = 27$

55. $(x + 4)^{1/2} + 5x(x + 4)^{3/2} = 0$

56. $8x^2(x^2 - 4)^{1/3} + (x^2 - 4)^{4/3} = 0$

57. $|x - 5| = 10$ **58.** $|2x + 3| = 7$

59. $|x^2 - 3| = 2x$ **60.** $|x^2 - 6| = x$

In Exercises 61–64, use a graphing utility to graph the equation. Use the graph to approximate any x-intercepts of the graph. Set $y = 0$ and solve the resulting equation. Compare the result with the x-intercepts of the graph.

61. $y = 4x^3 - 12x^2 + 8x$

62. $y = 12x^3 - 84x^2 + 120x$

63. $y = \dfrac{1}{x} + \dfrac{1}{x + 1} - 2$

64. $y = \dfrac{4}{x - 3} - \dfrac{4}{x} - 1$

In Exercises 65–68, solve the equation for the indicated variable.

65. Solve for r: $V = \frac{1}{3}\pi r^2 h$

66. Solve for X: $Z = \sqrt{R^2 - X^2}$

67. Solve for p: $L = \dfrac{k}{3\pi r^2 p}$

68. Solve for v: $E = 2kw\left(\dfrac{v}{2}\right)^2$

In Exercises 69 and 70, find the constant C such that the ordered pair is a solution point of the equation.

69. $y = C\sqrt{x + 1}$, $(3, 8)$

70. $x + C(y + 2) = 0$, $(4, 3)$

In Exercises 71–78, solve the inequality.

71. $x^2 - 2x \geq 3$ **72.** $\frac{1}{2}(3 - x) > \frac{1}{3}(2 - 3x)$

73. $\dfrac{x - 5}{3 - x} < 0$ **74.** $\dfrac{2}{x + 1} \leq \dfrac{3}{x - 1}$

75. $|x - 2| < 1$ **76.** $|x| \leq 4$

77. $\left|x - \frac{3}{2}\right| \geq \frac{3}{2}$ **78.** $|x - 3| > 4$

In Exercises 79–82, solve the inequality.

79. $\dfrac{x}{5} - 6 \leq -\dfrac{x}{2} + 6$ **80.** $2x^2 + x \geq 15$

81. $(x - 4)|x| > 0$ **82.** $|x(x - 6)| < 5$

In Exercises 83 and 84, find the domain of the expression by finding the interval(s) on the real number line for which the radicand is nonnegative.

83. $\sqrt{2x - 10}$ **84.** $\sqrt{x(x - 4)}$

In Exercises 85–94, perform the operations and write the result in standard form.

85. $(7 + 5i) + (-4 + 2i)$

86. $\left(\dfrac{\sqrt{2}}{2} - \dfrac{\sqrt{2}}{2}i\right) - \left(\dfrac{\sqrt{2}}{2} + \dfrac{\sqrt{2}}{2}i\right)$

87. $5i(13 - 8i)$ **88.** $(1 + 6i)(5 - 2i)$

89. $(10 - 8i)(2 - 3i)$ **90.** $i(6 + i)(3 - 2i)$

91. $\dfrac{6 + i}{i}$ **92.** $\dfrac{3 + 2i}{5 + i}$

93. $\dfrac{4}{-3i}$ **94.** $\dfrac{1}{(2 + i)^4}$

In Exercises 95 and 96, find all solutions of the equation.

95. $3x^2 + 1 = 0$ **96.** $2 + 8x^{-2} = 0$

97. *Monthly Profit*   In October, a company's total profit was 12% more than it was in September. The total profit for the two months was $689,000. Find the profit for each month.

98. Discount Rate   The price of a television set has been discounted $85. The sale price is $340. What was the percent discount?

99. Mixture Problem   A car radiator contains 10 liters of a 30% antifreeze solution. How many liters will have to be replaced with pure antifreeze if the resulting solution is to be 50% antifreeze?

100. Running Track   A fitness center has two running tracks around a rectangular playing floor (see figure). The tracks are 1 meter wide and form semicircles at the narrow ends of the floor. How much longer is the running distance on the outer track than on the inner track?

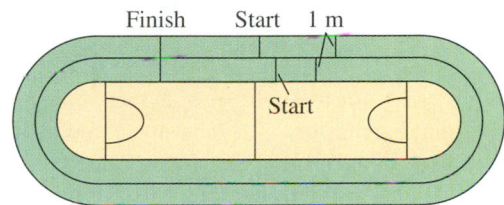

101. Cost Sharing   A group agrees to share equally in the cost of a $48,000 piece of machinery. If they can find two more group members, each member's share will decrease by $4000. How many are presently in the group?

102. Venture Capital   You are planning to start a small business that will require an investment of $90,000. You have found some people who are willing to share equally in the venture. If you can find three more people, each person's share will decrease by $2500. How many people have you found so far?

103. Average Speed   You commute 56 miles one way to work. The trip to work takes 10 minutes longer than the trip home. Your average speed on the trip home is 8 miles per hour faster. What is your average speed on the trip home?

104. Market Research   The demand equation for a product is given by

$$p = 42 - \sqrt{0.001x + 2}$$

where x is the number of units demanded per day and p is the price per unit. Find the demand if the price is set at $29.95.

105. Simply Supported Beam   A simply supported 20-foot beam supports a uniformly distributed load of 1000 pounds per foot. The bending moment M (in foot-pounds) x feet from one end of the beam is given by $M = 500x(20 - x)$.

(a) Use a graphing utility to graph the equation.

(b) Where is the bending moment zero?

(c) Use the graph to determine the point on the beam where the bending moment is the greatest.

(d) Determine the positions where the bending moment is less than 40,000 foot-pounds.

106. Break-Even Analysis   The revenue for selling x units of a product is $R = 125.95x$. The cost of producing x units is $C = 92x + 1200$. To obtain a profit, the revenue must be greater than the cost. For what values of x will this product return a profit?

107. Pendulum   The period of a pendulum is given by

$$T = 2\pi\sqrt{\frac{L}{32}}$$

where T is the time in seconds and L is the length of the pendulum in feet. If the period is to be at least 2 seconds, determine the minimum length of the pendulum.

108. Data Analysis   The total sales S (in billions of dollars) of sporting goods in the United States from 1981 to 1992 can be approximated by the model

$$S = 16.8091 + 0.7151t^2 - 0.0446t^3$$

where $t = 1$ represents 1981. The actual sales are given in the table. (Source: National Sporting Goods Association)

Year	1981	1982	1983	1984	1985	1986
Sales	18.7	18.7	23.1	26.4	27.4	30.6

Year	1987	1988	1989	1990	1991	1992
Sales	33.9	42.1	45.2	44.1	42.8	42.4

(a) Use a graphing utility to compare the data with the model.

(b) Use the model to estimate total sales in 1994.

(c) Explain why the model may not be accurate in predicting future sales.

CHAPTER PROJECT:

A Numerical Approach to Finding Falling Times

In this chapter, you studied two basic approaches for solving equations: an *algebraic* approach and a *graphical* approach. For many problems, you can gain insight by using a *numerical* approach instead of, or in addition to, the other two approaches.

Real Life

EXAMPLE 1   *Finding the Falling Time for an Object*

At 9:55 A.M. on Saturday, July 28, 1945, a terrible airplane accident occurred. A B-25 bomber crashed into the 78th and 79th floors of the Empire State Building in New York City. Hundreds of pieces of debris fell 975 feet to the streets below. How much time after hearing the crash did the people on the street have to get out of the way?

Solution

Assume that the debris *dropped* from a height of 975 feet. Using an initial velocity of $v_0 = 0$ and an initial height of $s_0 = 975$, you can model the height (in feet) of the falling debris as $s = -16t^2 + 975$, where t is the falling time in seconds. The table gives the heights at different times.

Time, t	0	1	2	3	4	5	6	7	7.8
Height, s	975	959	911	831	719	575	399	191	1.56

From the table, you can see that the debris took about 8 seconds to hit the ground. Because it would have taken about 1 second for the sound of the crash to reach the ground (sound travels at about 1100 feet per second), you can conclude that people had about 7 seconds to get out of the way. This conclusion was obtained numerically with a table. You can obtain the same conclusion algebraically by solving the equation $s = -16t^2 + 975$ for the time t that corresponds to a height of $s = 0$.

$$-16t^2 + 975 = s \qquad \text{Falling object model}$$

$$-16t^2 + 975 = 0 \qquad \text{Substitute 0 for } s.$$

$$975 = 16t^2 \qquad \text{Add } 16t^2 \text{ to both sides.}$$

$$\frac{975}{16} = t^2 \qquad \text{Divide both sides by 16.}$$

$$\sqrt{\frac{975}{16}} = t \qquad \text{Extract positive square root.}$$

$$7.8 \approx t \qquad \text{Use a calculator.}$$

Fourteen people were killed in this accident. The toll would have been greater, but it was a drizzly Saturday morning—most of the building's 15,000 office workers were at home.
(Photo: AP/World Wide Photos)

■ **E**XAMPLE **2**   *Adding Sophistication to a Model*   *Real Life*

In Example 1, because the airplane accident was accompanied by an explosion, some of the debris was most likely *propelled* downward. How does this consideration affect the time the people on the street had to get out of the way of the falling debris?

Solution

You could start by assuming that some of the debris was propelled straight down with an initial speed of about 100 feet per second. Using $v_0 = -100$ and $s_0 = 975$, you can obtain the model $s = -16t^2 - 100t + 975$. The table gives the heights of the debris at several times.

Time, t	0	1	2	3	4	5	5.3
Height, s	975	859	711	531	319	75	-4.44

With this model, note that the falling time has decreased considerably. Again figuring that it took about 1 second to hear the crash, you can conclude that the people on the street had only about 4.3 seconds to get out of the way.

CHAPTER PROJECT INVESTIGATIONS

1. *Real or False Accuracy*   In Example 2, suppose you solved the equation $0 = -16t^2 - 100t + 975$ algebraically to obtain

$$t = \frac{25 - 5\sqrt{181}}{-8} \approx 5.284.$$

Because sound traveling at 1100 feet per second takes about 0.886 second to travel 975 feet, you could reason that the people had $5.284 - 0.886$ or 4.398 seconds to get out of the way. This solution appears to be more accurate than that in Example 2. Is it?

2. *Exploration*   In Example 2, how long would the people on the street have had to get out of the way if some of the debris had been propelled downward with an initial velocity of 200 feet per second?

3. *Working Backward*   In the actual accident, all of the casualties were people who were in the building. What does this suggest about the initial velocity of any of the debris that was propelled downward?

4. *Speed*   Use the model and table in Example 1.
   (a) What was the average speed of the debris during its first second of fall?
   (b) Was the debris falling faster during its next second of fall? Explain.
   (c) Find the height of the debris after 7.7 seconds.
   (d) Use the result of part (c) to approximate the terminal speed of the debris.

5. Use the model given in Example 2 to approximate the terminal speed of the debris. Is the terminal speed given by this model greater than the terminal speed given by the model in Example 1?

6. In Example 2 it is assumed that some of the debris was propelled straight down with an initial speed of 100 feet per second. With this assumption, it seems reasonable that other parts of the debris would have been propelled straight up with the same initial speed. Under these conditions, approximate the duration during which debris continued to hit the street.

Chapter Test

Take this test as you would take a test in class. After you are done, check your work against the answers given in the back of the book.

The *Interactive* CD-ROM provides answers to the Chapter Tests and Cumulative Tests. It also offers Chapter Pre-Tests (which test key skills and concepts covered in previous chapters) and Chapter Post-Tests, both of which have randomly generated exercises with diagnostic capabilities.

In Exercises 1–6, use a graphing utility to graph the equation. Check for symmetry and identify any *x*- or *y*-intercepts.

1. $y = 4 - \frac{3}{4}x$ **2.** $y = 4 - \frac{3}{4}|x|$ **3.** $y = 4 - (x - 2)^2$

4. $y = x - x^3$ **5.** $y = \sqrt{3 - x}$ **6.** $(x - 3)^2 + y^2 = 9$

In Exercises 7–12, solve the equation (if possible).

7. $\frac{2}{3}(x - 1) + \frac{1}{4}x = 10$ **8.** $(x - 3)(x + 2) = 14$

9. $\dfrac{x - 2}{x + 2} + \dfrac{4}{x + 2} + 4 = 0$ **10.** $x^4 + x^2 - 6 = 0$

11. $2\sqrt{x} - \sqrt{2x + 1} = 1$ **12.** $|3x - 1| = 7$

In Exercises 13 and 14, solve the inequality and sketch the solution on the real number line.

13. $-3 \le 2(x + 4) < 14$ **14.** $\dfrac{2}{x} > \dfrac{5}{x + 6}$

15. On the first part of a 350-kilometer trip, a salesperson traveled 2 hours and 15 minutes at an average speed of 100 kilometers per hour. Find the average speed required for the remainder of the trip if the salesperson needs to arrive at the destination in another hour and 20 minutes.

16. The area of the ellipse in the figure is $A = \pi ab$. If a and b satisfy the constraint $a + b = 100$, find a and b if the area of the ellipse equals the area of the circle.

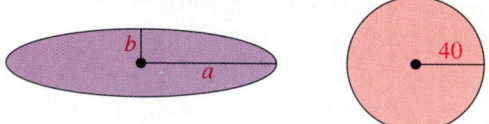

17. Perform the operations and write the result in standard form.

(a) $10i - \left(3 + \sqrt{-25}\right)$ (b) $\left(2 + \sqrt{3}i\right)\left(2 - \sqrt{3}i\right)$ (c) $\dfrac{5}{2 + i}$

2 Functions and Their Graphs

2.1 **Lines in the Plane and Slope**
2.2 **Functions**
2.3 **Analyzing Graphs of Functions**
2.4 **Translations and Combinations**
2.5 **Inverse Functions**

Many wildlife populations follow a cyclical "predator-prey" pattern. One example is the populations of snowshoe hare and lynx in the Yukon Territory. The researchers shown in the photo kept track of the lynx and hare populations from 1988 through 1995. Lynx numbers in a 350-square-kilometer region of the Yukon are shown below.

1988 (10)   1989 (16)
1990 (50)   1991 (60)
1992 (28)   1993 (15)
1994 (9)   1995 (8)

The hare population was low in 1986, increased to a high in 1990, and then decreased to a low again in 1992.

The numbers of lynx and snowshoe hare are functions of the year. For instance, the graph below depicts the number of lynx as a function of the year.

The data was supplied by Mark O'Donoghue, as part of the Kluane Boreal Forest Ecosystem Project.

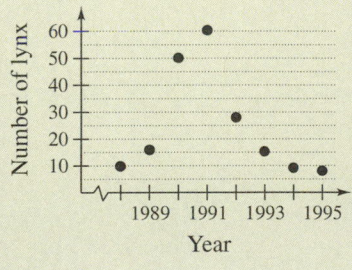

See Exercise 90 on page 210.

Photo: Alejandro Frid / Biological Photo Service

Husband and wife researchers Elizabeth Hofer and Peter Upton are measuring a lynx that has been trapped and sedated. The researchers work in the Yukon Territory, Canada.

▶ **2.1**   # *Lines in the Plane and Slope*

See Exercise 31 on page 192 for an example of how a linear equation can be used to model the earnings per share for General Mills stock from 1987 through 1994.

Using Slope ◻ *Finding the Slope of a Line* ◻ *Writing Linear Equations* ◻ *Parallel and Perpendicular Lines* ◻ *Application*

Using Slope

The simplest mathematical model for relating two variables is the **linear equation** $y = mx + b$. The equation is called *linear* because its graph is a line. (In mathematics, the term *line* means *straight line*.) By letting $x = 0$, you can see that the line crosses the y-axis at $y = b$, as shown in Figure 2.1. In other words, the y-intercept is $(0, b)$. The steepness or slope of the line is m.

$$y = mx + b$$

Slope ⌐   ⌐ y-intercept

The **slope** of a nonvertical line is the number of units the line rises (or falls) vertically for each unit of horizontal change from left to right, as shown in Figure 2.1.

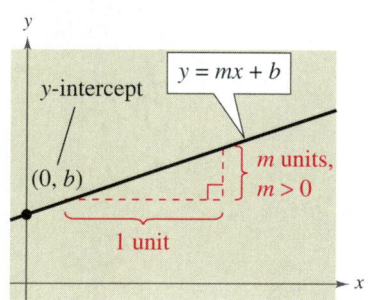

Positive slope, line rises.

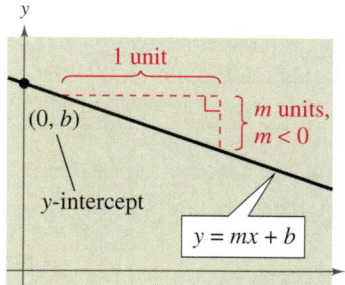

Negative slope, line falls.

FIGURE 2.1

A linear equation that is written in the form $y = mx + b$ is said to be written in **slope-intercept form.**

 THE SLOPE-INTERCEPT FORM OF THE EQUATION OF A LINE

The graph of the equation

$$y = mx + b$$

is a line whose slope is m and whose y-intercept is $(0, b)$.

▶ *Exploration*

Use a graphing utility to compare the slopes of the lines $y = mx$ where $m = 0.5$, 1, 2, and 4. Which line rises most quickly? Now, let $m = -0.5$, -1, -2, and -4. Which line falls most quickly? Use a square setting to obtain a true geometric perspective.

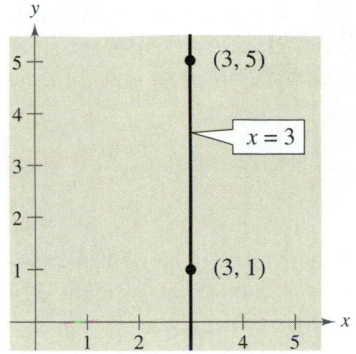

FIGURE 2.2   Slope is undefined.

Once you have determined the slope and the y-intercept of a line, it is a relatively simple matter to sketch its graph. In the following example, note that none of the lines is vertical. A vertical line has an equation of the form

$$x = a.$$ Vertical line

The equation of a vertical line cannot be written in the form $y = mx + b$ because the slope of a vertical line is undefined, as indicated in Figure 2.2.

EXAMPLE 1   *Graphing a Linear Equation*

Sketch the graph of each linear equation.

a. $y = 2x + 1$ **b.** $y = 2$ **c.** $x + y = 2$

Solution

a. Because $b = 1$, the y-intercept is $(0, 1)$. Moreover, because the slope is $m = 2$, the line *rises* two units for each unit the line moves to the right, as shown in Figure 2.3(a).

b. By writing this equation in the form $y = (0)x + 2$, you can see that the y-intercept is $(0, 2)$ and the slope is zero. A zero slope implies that the line is horizontal—that is, it doesn't rise *or* fall, as shown in Figure 2.3(b).

c. By writing this equation in slope-intercept form

$$x + y = 2$$ Original equation
$$y = -x + 2$$ Subtract x from both sides.
$$y = (-1)x + 2$$ Slope-intercept form

you can see that the y-intercept is $(0, 2)$. Moreover, because the slope is $m = -1$, this line *falls* one unit for each unit the line moves to the right, as shown in Figure 2.3(c).

(a) When m is positive, the line rises.

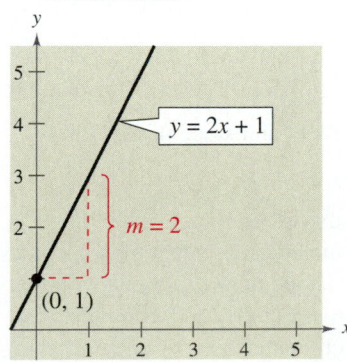

(b) When m is 0, the line is horizontal.

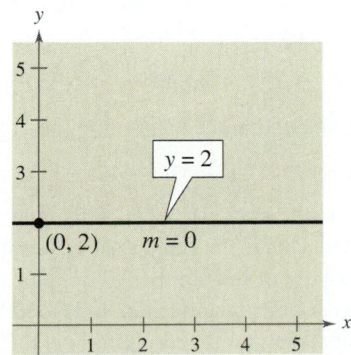

(c) When m is negative, the line falls.

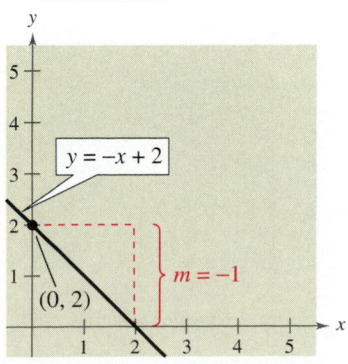

FIGURE 2.3

In real-life problems, the slope of a line can be interpreted as either a *ratio* or a *rate*. If the *x*-axis and *y*-axis have the same unit of measure, then the slope has no units and is a **ratio.** If the *x*-axis and *y*-axis have different units of measure, then the slope is a **rate** or **rate of change.**

EXAMPLE 2   *Using Slope as a Ratio*

The maximum recommended slope of a wheelchair ramp is $\frac{1}{12}$. A business is installing a wheelchair ramp that rises 22 inches over a horizontal length of 24 feet. Is the ramp steeper than recommended?   (Source: *American Disabilities Act Handbook*)

Solution

The horizontal length of the ramp is 12(24) or 288 inches, as shown in Figure 2.4. Thus, the slope of the ramp is

$$\text{Slope} = \frac{\text{vertical change}}{\text{horizontal change}} = \frac{22 \text{ in.}}{288 \text{ in.}} \approx 0.076 < 0.08\overline{3} = \frac{1}{12}.$$

Thus, the slope is not steeper than recommended.

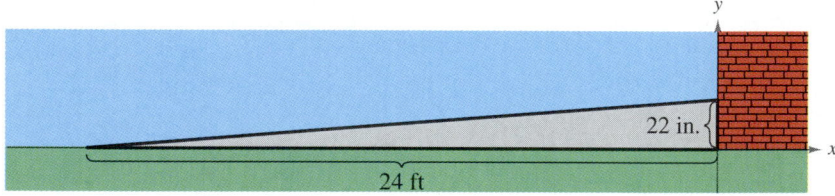

FIGURE 2.4

EXAMPLE 3   *Using Slope as a Rate of Change*

A manufacturing company determines that the total cost in dollars of producing *x* units of a product is

$$C = 25x + 3500. \qquad\qquad \text{Cost equation}$$

Describe the practical significance of the *y*-intercept and slope of this line.

Solution

The *y*-intercept $(0, 3500)$ tells you that the cost of producing zero units is $3500. This is the **fixed cost** of production—it includes costs that must be paid regardless of the number of units produced. The slope of $m = 25$ tells you that the cost of producing each unit is $25, as shown in Figure 2.5. Economists call the cost per unit the **marginal cost.** If the production increases by one unit, then the "margin" or extra amount of cost is $25.

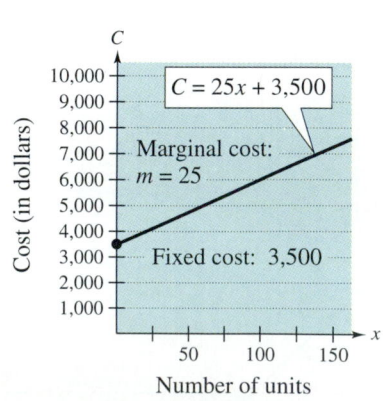

FIGURE 2.5   Production cost

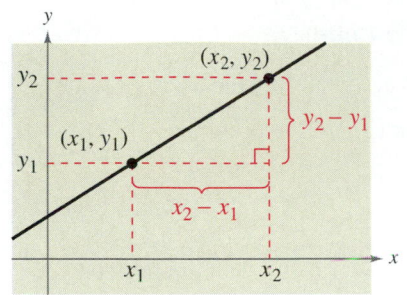

FIGURE 2.6

Finding the Slope of a Line

Given an equation of a line, you can find its slope by writing the equation in slope-intercept form. If you are not given an equation, you can still find the slope of a line. For instance, suppose you want to find the slope of the line passing through the points (x_1, y_1) and (x_2, y_2), as shown in Figure 2.6. As you move from left to right along this line, a change of $(y_2 - y_1)$ units in the vertical direction corresponds to a change of $(x_2 - x_1)$ units in the horizontal direction.

$$y_2 - y_1 = \text{the change in } y = \text{rise}$$

and

$$x_2 - x_1 = \text{the change in } x = \text{run}$$

The ratio of $(y_2 - y_1)$ to $(x_2 - x_1)$ represents the slope of the line that passes through the points (x_1, y_1) and (x_2, y_2).

$$\text{Slope} = \frac{\text{change in } y}{\text{change in } x}$$
$$= \frac{y_2 - y_1}{x_2 - x_1}$$

NOTE The French verb meaning to mount, to climb, or to rise is *monter*. Because Descartes was largely responsible for the development of analytical geometry, his use of *m*—short for *monter*—to indicate the slope became the accepted term among European mathematicians. ∎∎

THE SLOPE OF A LINE PASSING THROUGH TWO POINTS

The **slope** m of the nonvertical line through (x_1, y_1) and (x_2, y_2) is

$$m = \frac{y_2 - y_1}{x_2 - x_1}$$

where $x_1 \neq x_2$.

When this formula is used for slope, the *order of subtraction* is important. Given two points on a line, you are free to label either one of them as (x_1, y_1) and the other as (x_2, y_2). However, once you have done this, you must form the numerator and denominator using the same order of subtraction.

$$m = \frac{y_2 - y_1}{x_2 - x_1} \qquad m = \frac{y_1 - y_2}{x_1 - x_2} \qquad m = \frac{y_2 - y_1}{x_1 - x_2}$$

Correct ⎵   Correct ⎵   Incorrect

For instance, the slope of the line passing through the points $(3, 4)$ and $(5, 7)$ can be calculated as

$$m = \frac{7 - 4}{5 - 3} = \frac{3}{2}$$

or

$$m = \frac{4 - 7}{3 - 5} = \frac{-3}{-2} = \frac{3}{2}.$$

The *Interactive* CD-ROM shows every example with its solution; clicking on the *Try It!* button brings up similar problems. Guided Examples and Integrated Examples show step-by-step solutions to additional examples. Integrated Examples are related to several concepts in the section.

EXAMPLE 4 *Finding the Slope of a Line*

Find the slope of the line passing through each pair of points. (See Figure 2.7.)

a. $(-2, 0)$ and $(3, 1)$ **b.** $(-1, 2)$ and $(2, 2)$

c. $(0, 4)$ and $(1, -1)$ **d.** $(3, 4)$ and $(3, 1)$

Solution

a. Letting $(x_1, y_1) = (-2, 0)$ and $(x_2, y_2) = (3, 1)$, you obtain a slope of

$$m = \frac{y_2 - y_1}{x_2 - x_1} = \frac{1 - 0}{3 - (-2)} = \frac{1}{5}.$$

b. The slope of the line passing through $(-1, 2)$ and $(2, 2)$ is

$$m = \frac{2 - 2}{2 - (-1)} = \frac{0}{3} = 0.$$

c. The slope of the line passing through $(0, 4)$ and $(1, -1)$ is

$$m = \frac{-1 - 4}{1 - 0} = \frac{-5}{1} = -5.$$

d. The slope of the vertical line passing through $(3, 4)$ and $(3, 1)$ is not defined because division by zero is undefined.

NOTE   In Figure 2.7, note the following relationships between slope and the description of the line.

a. Positive slope; line rises from left to right

b. Zero slope; line is horizontal

c. Negative slope; line falls from left to right

d. Vertical line; undefined slope ▪▪

(a)

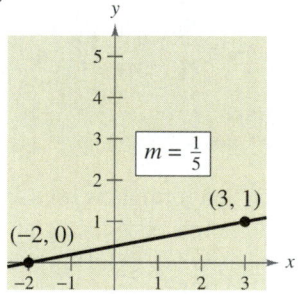

(b)

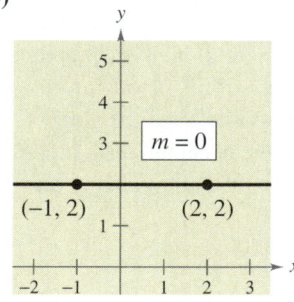

(c)

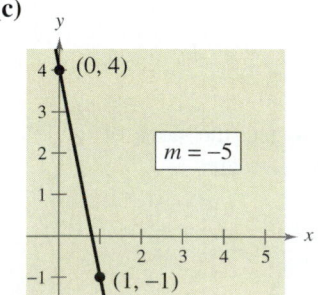

(d)

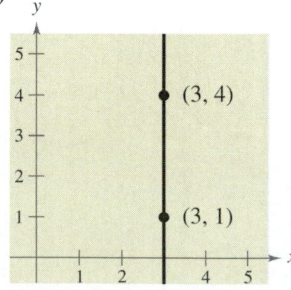

FIGURE 2.7

Writing Linear Equations

If (x_1, y_1) is a point lying on a line of slope m and (x, y) is *any other* point on the line, then

$$\frac{y - y_1}{x - x_1} = m.$$

This equation, involving the variables x and y, can be rewritten in the form $y - y_1 = m(x - x_1)$, which is the **point-slope form** of the equation of a line.

> ### POINT-SLOPE FORM OF THE EQUATION OF A LINE
>
> The equation of the line with slope m passing through the point (x_1, y_1) is
>
> $$y - y_1 = m(x - x_1).$$

The point-slope form is most useful for *finding* the equation of a line. You should remember this formula.

EXAMPLE 5   Using the Point-Slope Form

Find the equation of the line that has a slope of 3 and passes through the point $(1, -2)$, as shown in Figure 2.8.

Solution

Use the point-slope form with $m = 3$ and $(x_1, y_1) = (1, -2)$.

$y - y_1 = m(x - x_1)$	Point-slope form
$y - (-2) = 3(x - 1)$	Substitute for m, x_1, and y_1.
$y + 2 = 3x - 3$	Simplify.
$y = 3x - 5$	Slope-intercept form

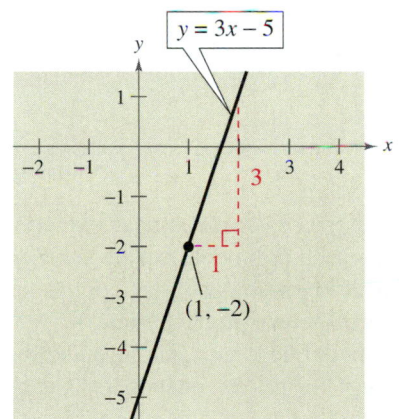

FIGURE 2.8

The point-slope form can be used to find an equation of the line passing through points (x_1, y_1) and (x_2, y_2). To do this, first find the slope of the line

$$m = \frac{y_2 - y_1}{x_2 - x_1}, \qquad x_1 \neq x_2,$$

and then use the point-slope form to obtain the equation

$$y - y_1 = \frac{y_2 - y_1}{x_2 - x_1}(x - x_1). \qquad \text{Two-point form}$$

This is sometimes called the **two-point form** of the equation of a line.

Bausch & Lomb, Inc.
Cash Flow

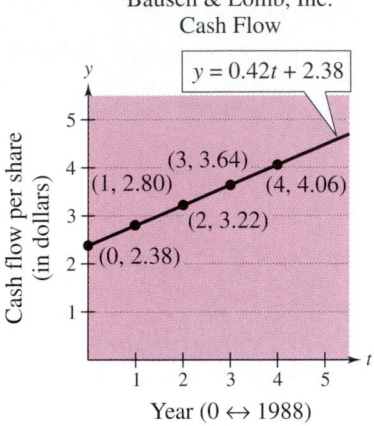

$y = 0.42t + 2.38$

FIGURE 2.9

$\blacksquare$

Real Life

EXAMPLE 6   *Predicting Cash Flow Per Share*

The cash flow per share for Bausch & Lomb, Inc. was \$2.38 in 1988 and \$2.80 in 1989. Using only this information, write a linear equation that gives the cash flow per share in terms of the year.   (Source: Bausch & Lomb, Inc.)

Solution

Let $t = 0$ represent 1988. Then the two given values are represented by the ordered pairs (0, 2.38) and (1, 2.80). The slope of the line passing through these points is

$$m = \frac{2.80 - 2.38}{1 - 0} = 0.42.$$

Using the point-slope form, you can find the equation that relates the cash flow C and the year t to be

$$y = 0.42t + 2.38.$$

You can use this model to predict future cash flows. For instance, it predicts the cash flows in 1990, 1991, and 1992 to be \$3.22, \$3.64, and \$4.06, respectively, as shown in Figure 2.9. (In this case, the predictions are quite good—the actual cash flows in 1990, 1991, and 1992 were \$3.38, \$3.65, and \$4.16, respectively.)

(a) Linear Extrapolation

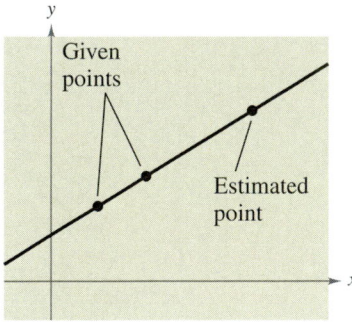

(b) Linear Interpolation

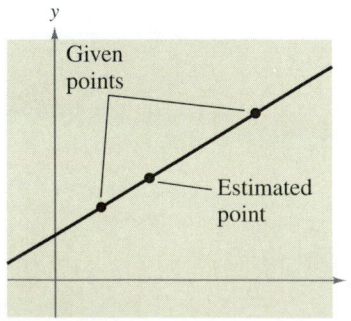

FIGURE 2.10

The prediction method illustrated in Example 6 is called **linear extrapolation.** Note in Figure 2.10(a) that an extrapolated point does not lie between the given points. When the estimated point lies between two given points, as shown in Figure 2.10(b), the procedure is called **linear interpolation.**

Because the slope of a vertical line is not defined, its equation cannot be written in slope-intercept form. However, every line has an equation that can be written in the **general form**

$$Ax + By + C = 0 \qquad\qquad \text{General form}$$

where A and B are not both zero. For instance, the vertical line given by $x = a$ can be represented by the general form $x - a = 0$.

◼ EQUATIONS OF LINES

1. General form: $Ax + By + C = 0$
2. Vertical line: $x = a$
3. Horizontal line: $y = b$
4. Slope-intercept form: $y = mx + b$
5. Point-slope form: $y - y_1 = m(x - x_1)$

Parallel and Perpendicular Lines

> **PARALLEL AND PERPENDICULAR LINES**
>
> 1. Two distinct nonvertical lines are **parallel** if and only if their slopes are equal. That is, $m_1 = m_2$.
> 2. Two nonvertical lines are **perpendicular** if and only if their slopes are negative reciprocals of each other. That is, $m_1 = -1/m_2$.

TECHNOLOGY

On a graphing utility, lines will not appear to have the correct slope unless you use a viewing rectangle that has a square setting. For instance, try graphing the lines in Example 7 using the standard setting $-10 \le x \le 10$ and $-10 \le y \le 10$. Then reset the viewing rectangle with the square setting $-9 \le x \le 9$ and $-6 \le y \le 6$. On which setting do the lines $y = \frac{2}{3}x - \frac{5}{3}$ and $y = -\frac{3}{2}x + 2$ appear perpendicular?

EXAMPLE 7   Finding Parallel and Perpendicular Lines

Find an equation of the line that passes through the point $(2, -1)$ and is (a) parallel to and (b) perpendicular to the line $2x - 3y = 5$.

Solution

By writing the equation in slope-intercept form

$2x - 3y = 5$	Original equation
$-3y = -2x + 5$	Subtract $2x$ from both sides.
$y = \frac{2}{3}x - \frac{5}{3}$	Slope-intercept form

you can see that it has a slope of $m = \frac{2}{3}$, as shown in Figure 2.11.

a. Any line parallel to the given line must also have a slope of $\frac{2}{3}$. Thus, the line through $(2, -1)$ that is parallel to the given line has the following equation.

$y - (-1) = \frac{2}{3}(x - 2)$	Point-slope form
$3(y + 1) = 2(x - 2)$	Multiply both sides by 3.
$3y + 3 = 2x - 4$	Distributive Property
$2x - 3y - 7 = 0$	General form
$y = \frac{2}{3}x - \frac{7}{3}$	Slope-intercept form

b. Any line perpendicular to the given line must have a slope of $-1/(2/3)$ or $-\frac{3}{2}$. Thus, the line through $(2, -1)$ that is perpendicular to the given line has the following equation.

$y - (-1) = -\frac{3}{2}(x - 2)$	Point-slope form
$2(y + 1) = -3(x - 2)$	Multiply both sides by 2.
$3x + 2y - 4 = 0$	General form
$y = -\frac{3}{2}x + 2$	Slope-intercept form

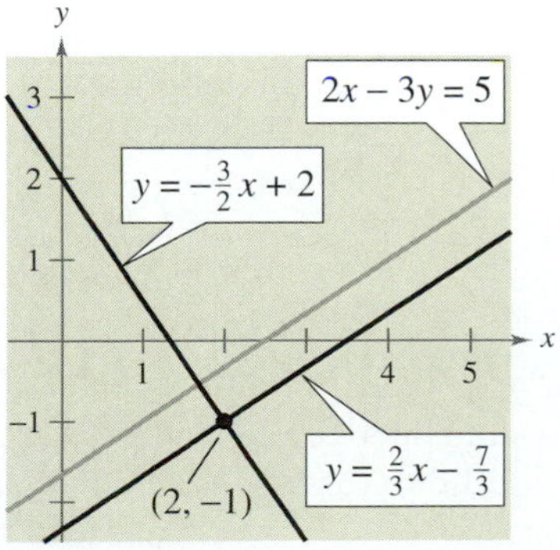

FIGURE 2.11

NOTE Most business expenses can be deducted the same year they occur. One exception is the cost of property that has a useful life of more than 1 year. Such costs must be **depreciated** over the useful life of the property. If the *same amount* is depreciated each year, the procedure is called **linear depreciation.** The **book value** is the difference between the original value and the total amount of depreciation accumulated to date. ∎∎

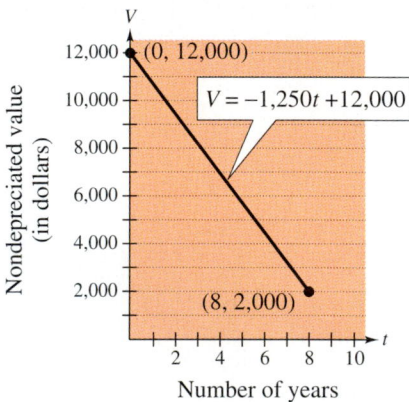

FIGURE 2.12   Straight-Line Depreciation

Application

Real Life

EXAMPLE 8   Depreciating Equipment

Your company has purchased a $12,000 machine that has a useful life of 8 years. The salvage value at the end of 8 years is $2000. Write a linear equation that describes the book value of the machine each year.

Solution

Let V represent the value of the machine at the end of year t. You can represent the initial value of the machine by the ordered pair (0, 12,000) and the salvage value of the machine by the ordered pair (8, 2000). The slope of the line is

$$m = \frac{2000 - 12,000}{8 - 0} = -\$1250$$

which represents the annual depreciation in *dollars per year.* Using the point-slope form, you can write the equation of the line as follows.

$$V - 12,000 = -1250(t - 0) \qquad \text{Point-slope form}$$
$$V = -1250t + 12,000 \qquad \text{Slope-intercept form}$$

The table shows the book value at the end of each year, and the graph of the equation is shown in Figure 2.12.

t	0	1	2	3	4	5	6	7	8
V	12,000	10,750	9500	8250	7000	5750	4500	3250	2000

GROUP ACTIVITY

MODELING LINEAR DATA

x	y	x	y
3	83.3	9	90.4
4	83.8	10	92.1
5	84.9	11	93.1
6	85.9	12	92.1
7	87.4	13	93.1
8	88.6	14	94.2

The table at the left shows the total number y (in millions) of households in the United States that owned at least one TV set during each year x from 1983 through 1994, where $x = 3$ represents 1983.   (Source: Nielsen Media Research)

Sketch a scatter plot of the data, and use a straight-edge to sketch the best-fitting line through the points. Find the equation of the line. Interpret the slope and y-intercept in the context of the data. Compare your model to those obtained by other students. Are all the different models valid? Explain. Use your model to estimate the number of TV households in 1997, and compare your estimate with those of other students.

The *Interactive* CD-ROM provides additional help with Warm-Up exercises by providing a hypertext link to the section in which the concept was introduced.

2.1 Exercises

In Exercises 1 and 2, identify the line that has the specified slope.

1. (a) $m = \frac{2}{3}$ (b) m is undefined.   (c) $m = -2$

2. (a) $m = 0$ (b) $m = -\frac{3}{4}$ (c) $m = 1$

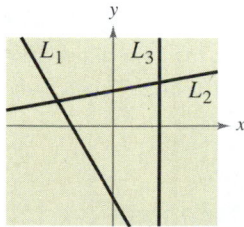   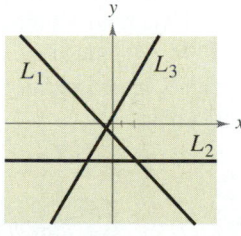

FIGURE FOR 1   FIGURE FOR 2

In Exercises 3 and 4, sketch the graphs of the lines through the given point with the indicated slopes. Make the sketches on the same set of coordinate axes.

Point	Slopes			
3. $(2, 3)$	(a) 0	(b) 1	(c) 2	(d) -3
4. $(-4, 1)$	(a) 3	(b) -3	(c) $\frac{1}{2}$	(d) undefined

In Exercises 5–10, estimate the slope of the line.

5.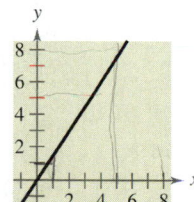

6.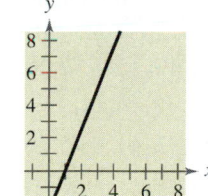

7.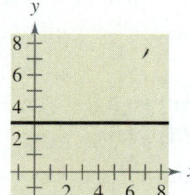

8.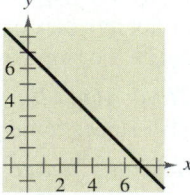

9.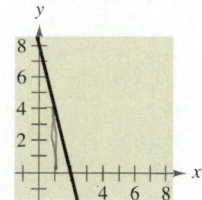

10.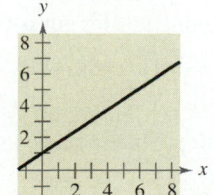

In Exercises 11–16, plot the points and find the slope of the line passing through the pair of points.

11. $(-3, -2), (1, 6)$ **12.** $(2, 4), (4, -4)$

13. $(-6, -1), (-6, 4)$ **14.** $(0, -10), (-4, 0)$

15. $(1, 2), (-2, -2)$ **16.** $\left(\frac{7}{8}, \frac{3}{4}\right), \left(\frac{5}{4}, -\frac{1}{4}\right)$

In Exercises 17–22, use the point on the line and the slope of the line to find three additional points through which the line passes. (The solution is not unique.)

	Point	*Slope*
17.	$(2, 1)$	$m = 0$
18.	$(-4, 1)$	m is undefined.
19.	$(5, -6)$	$m = 1$
20.	$(10, -6)$	$m = -1$
21.	$(-8, 1)$	m is undefined.
22.	$(-3, -1)$	$m = 0$

In Exercises 23–26, determine if the lines L_1 and L_2 passing through the pairs of points are parallel, perpendicular, or neither.

23. L_1: $(0, -1), (5, 9)$ **24.** L_1: $(-2, -1), (1, 5)$
  L_2: $(0, 3), (4, 1)$ L_2: $(1, 3), (5, -5)$

25. L_1: $(3, 6), (-6, 0)$ **26.** L_1: $(4, 8), (-4, 2)$
  L_2: $(0, -1), \left(5, \frac{7}{3}\right)$ L_2: $(3, -5), \left(-1, \frac{1}{3}\right)$

27. *Essay*   Write a brief paragraph explaining whether or not any pair of points on a line can be used to calculate the slope of the line.

28. *Think About It*   Is it possible for two lines with positive slopes to be perpendicular? Explain.

29. *Rate of Change*   The following are the slopes of lines representing annual sales y in terms of time x in years. Use the slopes to interpret any change in annual sales for a 1-year increase in time.

(a) The line has a slope of $m = 135$.

(b) The line has a slope of $m = 0$.

(c) The line has a slope of $m = -40$.

30. *Rate of Change*   The following are the slopes of lines representing daily revenues y in terms of time x in days. Use the slopes to interpret any change in daily revenues for a 1-day increase in time.

(a) The line has a slope of $m = 400$.

(b) The line has a slope of $m = 100$.

(c) The line has a slope of $m = 0$.

31. *Earnings per Share*   The graph gives the earnings per share of common stock for General Mills for the years 1987 through 1994. Use the slope to determine the years when earnings (a) decreased most rapidly and (b) increased most rapidly.   (Source: General Mills)

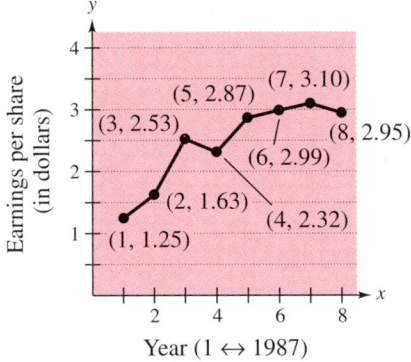

32. *Dividends per Share*   The graph gives the declared dividend per share of common stock for the Procter and Gamble Company for the years 1987 through 1994. Use the slope to determine the year when dividends increased most rapidly.   (Source: Procter and Gamble)

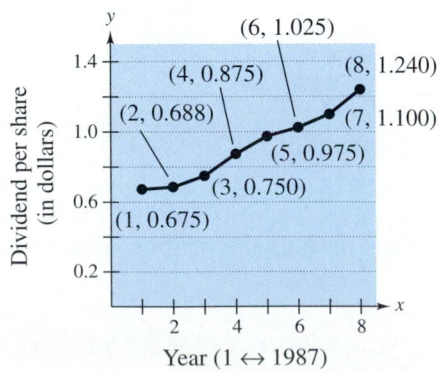

33. *Mountain Driving* When driving down a mountain road, you notice warning signs indicating that it is a "12% grade." This means that the slope of the road is $-\frac{12}{100}$. Approximate the amount of horizontal change in your position if you note from elevation markers that you have descended 2000 feet vertically.

34. *Attic Height* The "rise to run" in determining the steepness of the roof on a house is 3 to 4. Determine the maximum height in the attic of the house if the house is 32 feet wide (see figure).

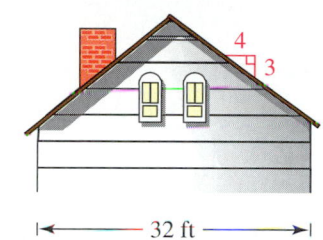

|← 32 ft →|

In Exercises 35–40, find the slope and *y*-intercept (if possible) of the equation of the line. Sketch a graph of the line.

35. $5x - y + 3 = 0$ **36.** $2x + 3y - 9 = 0$

37. $5x - 2 = 0$ **38.** $3y + 5 = 0$

39. $7x + 6y - 30 = 0$ **40.** $x - y - 10 = 0$

In Exercises 41–48, find an equation of the line passing through the points and sketch a graph of the line.

41. $(5, -1), (-5, 5)$ **42.** $(4, 3), (-4, -4)$

43. $\left(2, \frac{1}{2}\right), \left(\frac{1}{2}, \frac{5}{4}\right)$ **44.** $(-1, 4), (6, 4)$

45. $(-8, 1), (-8, 7)$ **46.** $(1, 1), \left(6, -\frac{2}{3}\right)$

47. $(1, 0.6), (-2, -0.6)$

48. $(-8, 0.6), (2, -2.4)$

In Exercises 49–58, find an equation of the line that passes through the given point and has the indicated slope. Sketch a graph of the line.

	Point	Slope
49.	$(0, -2)$	$m = 3$
50.	$(0, 10)$	$m = -1$
51.	$(-3, 6)$	$m = -2$
52.	$(0, 0)$	$m = 4$
53.	$(4, 0)$	$m = -\frac{1}{3}$
54.	$(-2, -5)$	$m = \frac{3}{4}$
55.	$(6, -1)$	m is undefined.
56.	$(-10, 4)$	$m = 0$
57.	$\left(4, \frac{5}{2}\right)$	$m = \frac{4}{3}$
58.	$\left(-\frac{1}{2}, \frac{3}{2}\right)$	$m = -3$

In Exercises 59–64, use the intercept form to find the equation of the line with the given intercepts. The intercept form of the equation of a line with intercepts $(a, 0)$ and $(0, b)$ is

$$\frac{x}{a} + \frac{y}{b} = 1, \qquad a \neq 0, b \neq 0.$$

59. *x*-intercept: $(2, 0)$ **60.** *x*-intercept: $(-3, 0)$
  y-intercept: $(0, 3)$ *y*-intercept: $(0, 4)$

61. *x*-intercept: $\left(-\frac{1}{6}, 0\right)$ **62.** *x*-intercept: $\left(\frac{2}{3}, 0\right)$
  y-intercept: $\left(0, -\frac{2}{3}\right)$ *y*-intercept: $(0, -2)$

63. Point on line: $(1, 2)$
  x-intercept: $(a, 0)$
  y-intercept: $(0, a)$, $a \neq 0$

64. Point on line: $(-3, 4)$
  x-intercept: $(a, 0)$
  y-intercept: $(0, a)$, $a \neq 0$

In Exercises 65–70, write equations of the lines through the given point (a) parallel to the given line and (b) perpendicular to the given line.

	Point	Line
65.	$(2, 1)$	$4x - 2y = 3$
66.	$(-3, 2)$	$x + y = 7$
67.	$(-6, 4)$	$3x + 4y = 7$
68.	$\left(\frac{7}{8}, \frac{3}{4}\right)$	$5x + 3y = 0$
69.	$(-1, 0)$	$y = -3$
70.	$(2, 5)$	$x = 4$

In Exercises 71–76, use a graphing utility to graph the pair of equations on the same viewing rectangle. Are the lines parallel, perpendicular, or neither? Use a square setting.

71. $L_1: y = \frac{1}{3}x - 2$
    $L_2: y = \frac{1}{3}x + 3$

72. $L_1: y = 2x - 1$
    $L_2: y = 2x + 1$

73. $L_1: y = \frac{1}{2}x - 3$
    $L_2: y = -\frac{1}{2}x + 1$

74. $L_1: y = -\frac{4}{5}x - 5$
    $L_2: y = \frac{5}{4}x + 1$

75. $L_1: y = \frac{2}{3}x - 3$
    $L_2: y = -\frac{3}{2}x + 2$

76. $L_1: y = -1.8x + 3.1$
    $L_2: y = 2.8x - 4.5$

Graphical Interpretation   In Exercises 77 and 78, use a graphing utility to graph the equation on each viewing rectangle. Which viewing rectangle is better? Explain your reasoning.

77. $y = 0.5x - 3$

Xmin = -5
Xmax = 10
Xscl = 1
Ymin = -1
Ymax = 10
Yscl = 1

Xmin = -2
Xmax = 10
Xscl = 1
Ymin = -4
Ymax = 1
Yscl = 1

78. $y = -8x + 5$

Xmin = -5
Xmax = 5
Xscl = 1
Ymin = -10
Ymax = 10
Yscl = 1

Xmin = -5
Xmax = 10
Xscl = 1
Ymin = -80
Ymax = 80
Yscl = 20

Graphical Interpretation   In Exercises 79–82, use a graphing utility to graph the three equations on the same viewing rectangle. Adjust the viewing rectangle so the slope appears visually correct. Identify any relationships that exist among the lines.

79. (a) $y = 2x$ (b) $y = -2x$ (c) $y = \frac{1}{2}x$

80. (a) $y = \frac{2}{3}x$ (b) $y = -\frac{3}{2}x$ (c) $y = \frac{2}{3}x + 2$

81. (a) $y = -\frac{1}{2}x$ (b) $y = -\frac{1}{2}x + 3$ (c) $y = 2x - 4$

82. (a) $y = x - 8$ (b) $y = x + 1$ (c) $y = -x + 3$

Rate of Change   In Exercises 83 and 84, you are given the dollar value of a product in 1996 *and* the rate at which the value of the item is expected to change during the next 5 years. Use this information to write a linear equation that gives the dollar value V of the product in terms of the year t. (Let $t = 6$ represent 1996.)

	1996 Value	*Rate*
83.	$2540	$125 increase per year
84.	$156	$4.50 increase per year

Graphical Interpretation   In Exercises 85–88, match the description with its graph. Also determine the slope and how it is interpreted in the situation. [The graphs are labeled (a), (b), (c), and (d).]

85. A person is paying $20 per week to a friend to repay a $200 loan.

86. An employee is paid $8.50 per hour plus $2 for each unit produced per hour.

87. A sales representative receives $30 per day for food plus $0.32 for each mile traveled.

88. A word processor that was purchased for $750 depreciates $100 per year.

(a)
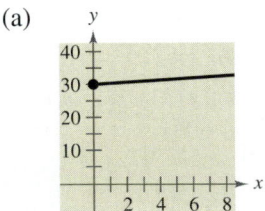

(b)
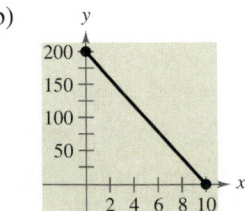

(c)
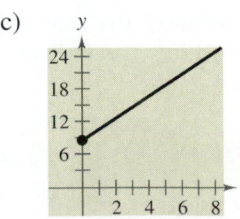

(d)
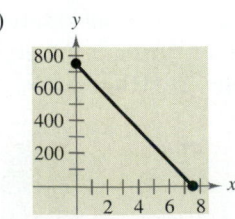

In Exercises 89 and 90, find a relationship between x and y so that (x, y) is equidistant from the two points.

89. $(4, -1), (-2, 3)$

90. $\left(3, \frac{5}{2}\right), (-7, 1)$

91. *Temperature* Find the equation of the line giving the relationship between the temperature in degrees Celsius C and degrees Fahrenheit F. Remember that water freezes at $0°$ Celsius ($32°$ Fahrenheit) and boils at $100°$ Celsius ($212°$ Fahrenheit).

92. *Temperature* Use the result of Exercise 91 to complete the table.

C		$-10°$	$10°$			$177°$
F	$0°$			$68°$	$90°$	

93. *Annual Salary* Your salary was \$28,500 in 1994 and \$32,900 in 1996. If your salary follows a linear growth pattern, what will your salary be in 1999?

94. *College Enrollment* A small college had 2546 students in 1994 and 2702 students in 1996. If the enrollment follows a linear growth pattern, how many students will the college have in 2000?

95. *Straight-Line Depreciation* A small business purchases a piece of equipment for \$875. After 5 years the equipment will be outdated and have no value. Write a linear equation giving the value V of the equipment during the 5 years it will be used.

96. *Straight-Line Depreciation* A small business purchases a piece of equipment for \$25,000. After 10 years the equipment will have to be replaced. Its value at that time is expected to be \$2000. Write a linear equation giving the value V of the equipment during the 10 years it will be used.

97. *Discount* A store is offering a 15% discount on all items. Write a linear equation giving the sale price S for an item with a list price L.

98. *Hourly Wages* A manufacturer pays its assembly line workers \$11.50 per hour. In addition, workers receive a piecework rate of \$0.75 per unit produced. Write a linear equation for the hourly wages W in terms of the number of units x produced per hour.

99. *Contracting Purchase* A contractor purchases a piece of equipment for \$36,500. The equipment requires an average expenditure of \$5.25 per hour for fuel and maintenance, and the operator is paid \$11.50 per hour.

(a) Write a linear equation giving the total cost C of operating this equipment for t hours. (Include the purchase cost of the equipment.)

(b) Assuming that customers are charged \$27 per hour of machine use, write an equation for the revenue R derived from t hours of use.

(c) Use the formula for profit $(P = R - C)$ to write an equation for the profit derived from t hours of use.

(d) **Break-Even Point** Use the result of part (c) to find the number of hours this equipment must be used to yield a profit of 0 dollars.

100. *Real Estate Purchase* A real estate office handles an apartment complex with 50 units. When the rent per unit is \$580 per month, all 50 units are occupied. However, when the rent is \$625 per month, the average number of occupied units drops to 47. Assume that the relationship between the monthly rent p and the demand x is linear.

(a) Write the equation of the line giving the demand x in terms of the rent p.

(b) Use this equation to predict the number of units occupied if the rent is \$655.

(c) Predict the number of units occupied if the rent is \$595.

101. *Perimeter* The length and width of a rectangular garden are 15 meters and 10 meters, respectively. A walkway of width x surrounds the garden.

(a) Draw a figure that gives a visual representation of the problem.

(b) Write the equation for the perimeter y of the walkway in terms of x.

(c) Use a graphing utility to graph the equation for the perimeter.

(d) Determine the slope of the graph in part (c). For each additional 1-meter increase in the width of the walkway, determine the increase in its perimeter.

102. Sales Commission A salesperson receives a monthly salary of $2500 plus a commission of 7% of sales. Write a linear equation for the salesperson's monthly wage W in terms of monthly sales S.

103. Daily Cost A sales representative of a company using a personal car receives $120 per day for lodging and meals plus $0.26 per mile driven. Write a linear equation giving the daily cost C to the company in terms of x, the number of miles driven.

104. Simple Interest An inheritance of $12,000 is invested in two different mutual funds. One fund pays $5\frac{1}{2}\%$ simple interest and the other pays 8% simple interest.

(a) If x dollars is invested in the fund paying $5\frac{1}{2}\%$, how much is invested in the fund paying 8%?

(b) Write the annual interest y in terms of x.

(c) Use a graphing utility to graph the function in part (b) over the interval $0 \le x \le 12,000$.

(d) Explain why the slope of the line in part (c) is negative.

105. Baseball Salaries The average annual salaries of major league baseball players (in thousands of dollars) from 1984 to 1994 are shown in the scatter plot. Find the equation of the line that you think best fits this data. (Let y represent the average salary and let t represent the year, with $t = 0$ corresponding to 1984.) (Source: Major League Baseball Players Association)

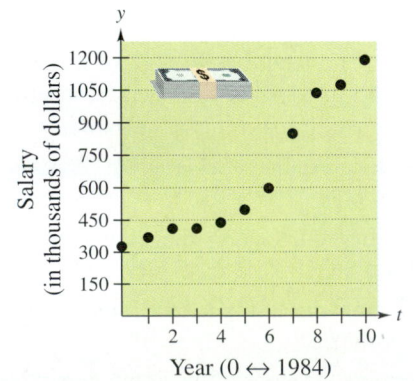

Year (0 ↔ 1984)

106. Data Analysis An instructor gives regular 20-point quizzes and 100-point exams in a mathematics course. Average scores for six students, given as ordered pairs (x, y) where x is the average quiz score and y is the average test score, are (18, 87), (10, 55), (19, 96), (16, 79), (13, 76), and (15, 82). [*Note*: The answers are not unique for parts (b)–(d).]

(a) Sketch a scatter plot of the data.

(b) Use a straight edge to sketch the "best-fitting" line through the points.

(c) Find an equation for the line sketched in part (b).

(d) Use the equation of part (c) to estimate the average test score for a person with an average quiz score of 17.

(e) If the instructor added 4 points to the average test score of everyone in the class, describe the changes in the positions of the plotted points and the change in the equation of the line.

Review Solve Exercises 107–110 as a review of the skills and problem-solving techniques you learned in previous sections. Match the equation with its graph. [The graphs are labeled (a), (b), (c), and (d).]

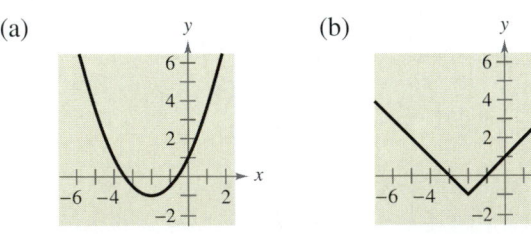

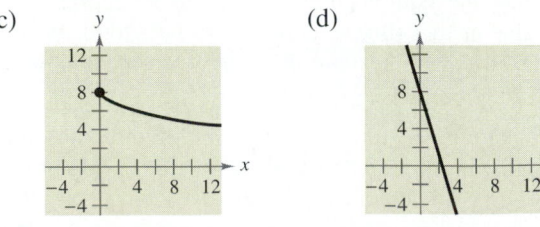

107. $y = 8 - 3x$

108. $y = 8 - \sqrt{x}$

109. $y = \frac{1}{2}x^2 + 2x + 1$

110. $y = |x + 2| - 1$

2.2   *Functions*

See Exercise 84 on page 209 for an example of how a piecewise-defined function can be used to model the price of mobile homes from 1974 through 1993.

Introduction to Functions ◻ *Function Notation* ◻ *The Domain of a Function* ◻ *Applications*

Introduction to Functions

Many everyday phenomena involve two quantities that are related to each other by some rule of correspondence. Here are some examples.

1. The simple interest I earned on $1000 for 1 year is related to the annual interest rate r by the formula $I = 1000r$.
2. The distance d traveled on a bicycle in 2 hours is related to the speed s of the bicycle by the formula $d = 2s$.
3. The area A of a circle is related to its radius r by the formula $A = \pi r^2$.

Not all correspondences between two quantities have simple mathematical formulas. For instance, people commonly match up NFL starting quarterbacks with touchdown passes and hours of the day with temperature. In each of these cases, however, there is some rule of correspondence that matches each item from one set with exactly one item from a different set. Such a rule of correspondence is called a **function.**

DEFINITION OF A FUNCTION

A **function** f from a set A to a set B is a rule of correspondence that assigns to each element x in the set A exactly one element y in the set B. The set A is the **domain** (or set of inputs) of the function f, and the set B contains the **range** (or set of outputs).

To help understand this definition, look at the function illustrated in Figure 2.13. This function can be represented by the following ordered pairs.

$$\{(1, 9°), (2, 13°), (3, 15°), (4, 15°), (5, 12°), (6, 10°)\}$$

In each ordered pair, the first coordinate is the input and the second coordinate is the output. In this example, note the following characteristics of a function.

1. Each element in A must be matched with an element of B.
2. Some elements in B may not be matched with any element in A.
3. Two or more elements of A may be matched with the same element of B.

The converse of the third statement is not true. That is, an element of A (the domain) cannot be matched with two different elements of B.

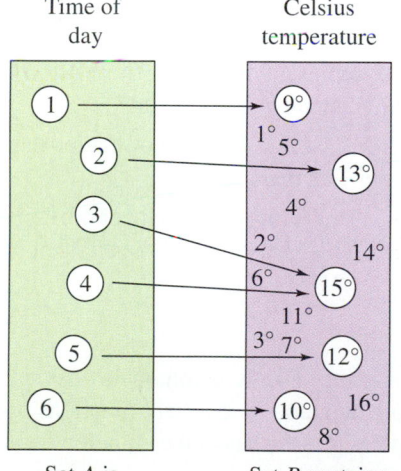

Time of day

Celsius temperature

Set A is the domain.
Inputs : 1, 2, 3, 4, 5, 6

Set B contains the range.
Outputs : 9°, 10°, 12°, 13°, 15°

FIGURE 2.13

In the following example, you are asked to decide whether different correspondences are functions. To do this, you must decide whether each element in the domain A is matched with exactly one element in the range B. If any element in A is matched with two or more elements in B, the correspondence is not a function.

■

EXAMPLE 1   *Testing for Functions*

Let $A = \{a, b, c\}$ and $B = \{1, 2, 3, 4, 5\}$. Which of the following sets of ordered pairs or figures represent functions from set A to set B?

a. $\{(a, 2), (b, 3), (c, 4)\}$ **b.** $\{(a, 4), (b, 5)\}$

c. 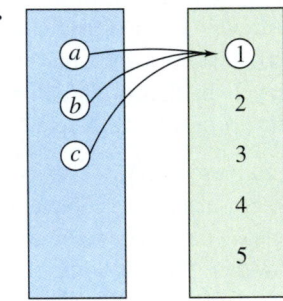   **d.**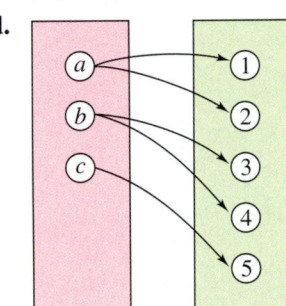

Solution

a. This collection of ordered pairs *does* represent a function from A to B. Each element of A is matched with exactly one element of B.

b. This collection of ordered pairs *does not* represent a function from A to B. Not every element of A is matched with an element of B.

c. This figure *does* represent a function from A to B. It does not matter that each element of A is matched with the same element of B.

d. This figure *does not* represent a function from A to B. The element a in A is matched with *two* elements, 1 and 2, of B. This is also true of the element b.

■

Representing functions by sets of ordered pairs is common in *discrete mathematics*. In algebra, however, it is more common to represent functions by equations or formulas involving two variables. For instance, the equation

$$y = x^2 \qquad \textit{y is a function of x.}$$

represents the variable y as a function of the variable x. In this equation, x is the **independent variable** and y is the **dependent variable.** The domain of the function is the set of all values taken on by the independent variable x, and the range of the function is the set of all values taken on by the dependent variable y.

Leonhard Euler (1707–1783), a Swiss mathematician, is considered to have been the most prolific and productive mathematician in history. One of his greatest influences on mathematics was his use of symbols, or notation. The function notation $y = f(x)$ was introduced by Euler.

EXAMPLE 2 *Testing for Functions Represented by Equations*

Which of the equations represent(s) y as a function of x?

a. $x^2 + y = 1$ **b.** $-x + y^2 = 1$

Solution

To determine whether y is a function of x, try to solve for y in terms of x.

a. Solving for y yields the following.

$$x^2 + y = 1 \qquad \text{Original equation}$$
$$y = 1 - x^2 \qquad \text{Solve for } y.$$

To each value of x there corresponds exactly one value of y. Thus, y *is a function of x.*

b. Solving for y yields the following.

$$-x + y^2 = 1 \qquad \text{Original equation}$$
$$y^2 = 1 + x \qquad \text{Add } x \text{ to both sides.}$$
$$y = \pm\sqrt{1 + x} \qquad \text{Solve for } y.$$

The $\pm$ indicates that to a given value of x there correspond two values of y. Thus, y *is not* a function of x.

Function Notation

When an equation is used to represent a function, it is convenient to name the function so that it can be referenced easily. For example, you know that the equation $y = 1 - x^2$ describes y as a function of x. Suppose you give this function the name "f." Then you can use the following **function notation.**

Input	*Output*	*Equation*
x	$f(x)$	$f(x) = 1 - x^2$

The symbol $f(x)$ is read as **the value of f at x** or simply **f of x.** The symbol $f(x)$ corresponds to the y-value for a given x. Thus, you can write $y = f(x)$. Keep in mind that f is the *name* of the function, whereas $f(x)$ is the *value* of the function at x. For instance, the function given by

$$f(x) = 3 - 2x$$

has *function values* denoted by $f(-1), f(0), f(2)$, and so on. To find these values, substitute the specified input values into the given equation.

For $x = -1$, $\quad f(-1) = 3 - 2(-1) = 3 + 2 = 5.$
For $x = 0$, $\quad f(0) = 3 - 2(0) = 3 - 0 = 3.$
For $x = 2$, $\quad f(2) = 3 - 2(2) = 3 - 4 = -1.$

Although f is often used as a convenient function name and x is often used as the independent variable, you can use other letters. For instance,

$$f(x) = x^2 - 4x + 7, \quad f(t) = t^2 - 4t + 7, \quad \text{and} \quad g(s) = s^2 - 4s + 7$$

all define the same function. In fact, the role of the independent variable is that of a "placeholder." Consequently, the function could be described by

$$f(\boxed{}) = (\boxed{})^2 - 4(\boxed{}) + 7.$$

NOTE In Example 3, note that $g(x + 2)$ is not equal to $g(x) + g(2)$. In general, $g(u + v) \neq g(u) + g(v)$. ∎∎

EXAMPLE 3   *Evaluating a Function*

Let $g(x) = -x^2 + 4x + 1$ and find the following.

a. $g(2)$ **b.** $g(t)$ **c.** $g(x + 2)$

Solution

a. Replacing x with 2 in $g(x) = -x^2 + 4x + 1$ yields the following.

$$g(2) = -(2)^2 + 4(2) + 1 = -4 + 8 + 1 = 5$$

b. Replacing x with t yields the following.

$$g(t) = -(t)^2 + 4(t) + 1 = -t^2 + 4t + 1$$

c. Replacing x with $x + 2$ yields the following.

$$\begin{aligned} g(x + 2) &= -(x + 2)^2 + 4(x + 2) + 1 \\ &= -(x^2 + 4x + 4) + 4x + 8 + 1 \\ &= -x^2 - 4x - 4 + 4x + 8 + 1 \\ &= -x^2 + 5 \end{aligned}$$

EXAMPLE 4   *A Piecewise-Defined Function*

Evaluate the function when $x = -1, 0,$ and 1.

$$f(x) = \begin{cases} x^2 + 1, & x < 0 \\ x - 1, & x \geq 0 \end{cases}$$

Solution

Because $x = -1$ is less than 0, use $f(x) = x^2 + 1$ to obtain

$$f(-1) = (-1)^2 + 1 = 2.$$

For $x = 0$, use $f(x) = x - 1$ to obtain

$$f(0) = (0) - 1 = -1.$$

For $x = 1$, use $f(x) = x - 1$ to obtain

$$f(1) = (1) - 1 = 0.$$

NOTE A function defined by two or more equations over a specified domain is called a **piecewise-defined** function. ∎∎

The Domain of a Function

The domain of a function can be described explicitly or it can be *implied* by the expression used to define the function. The **implied domain** is the set of all real numbers for which the expression is defined. For instance, the function given by

$$f(x) = \frac{1}{x^2 - 4}$$

has an implied domain that consists of all real x other than $x = \pm 2$. These two values are excluded from the domain because division by zero is undefined. Another common type of implied domain is that used to avoid even roots of negative numbers. For example, the function given by

$$f(x) = \sqrt{x}$$

is defined only for $x \geq 0$. Hence, its implied domain is the interval $[0, \infty)$. In general, the domain of a function *excludes* values that would cause division by zero *or* result in the even root of a negative number.

EXAMPLE 5   Finding the Domain of a Function

Find the domain of each function.

a. f: $\{(-3, 0), (-1, 4), (0, 2), (2, 2), (4, -1)\}$ **b.** $g(x) = \dfrac{1}{x + 5}$

c. Volume of a sphere: $V = \frac{4}{3}\pi r^3$ **d.** $h(x) = \sqrt{4 - x^2}$

Solution

a. The domain of f consists of all first coordinates in the set of ordered pairs.

Domain $= \{-3, -1, 0, 2, 4\}$

b. Excluding x-values that yield zero in the denominator, the domain of g is the set of all real numbers $x \neq -5$.

c. Because this function represents the volume of a sphere, the values of the radius r must be positive. Thus, the domain is the set of all real numbers r such that $r > 0$.

d. This function is defined only for x-values for which $4 - x^2 \geq 0$. Using the methods described in Section 1.8, you can conclude that $-2 \leq x \leq 2$. Thus, the domain is the interval $[-2, 2]$.

NOTE   In Example 5(c), note that the domain of a function may be implied by the physical context. For instance, from the equation $V = \frac{4}{3}\pi r^3$, you would have no reason to restrict r to positive values, but the physical context implies that a sphere cannot have negative radius. ■■

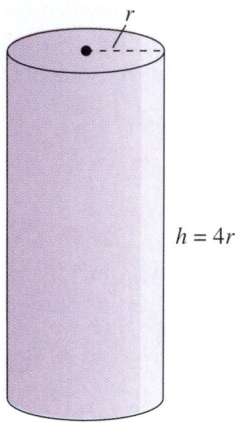

$h = 4r$

FIGURE 2.14

Applications

EXAMPLE 6   *The Dimensions of a Container*

You work in the marketing department of a soft-drink company and are experimenting with a new soft-drink can that is slightly narrower and taller than a standard can. For your experimental can, the ratio of the height to the radius is 4, as shown in Figure 2.14.

a. Express the volume of the can as a function of the radius r.

b. Express the volume of the can as a function of the height h.

Solution

a. $V = \pi r^2 h = \pi r^2 (4r) = 4\pi r^3$ *V as a function of r*

b. $V = \pi \left(\dfrac{h}{4}\right)^2 h = \dfrac{\pi h^3}{16}$ *V as a function of h*

EXAMPLE 7   *The Path of a Baseball*

A baseball is hit at a point 3 feet above ground at a velocity of 100 feet per second and an angle of 45°. The path of the baseball is given by the function

$$y = -0.0032x^2 + x + 3$$

where y and x are measured in feet, as shown in Figure 2.15. Will the baseball clear a 10-foot fence located 300 feet from home plate?

Solution

When $x = 300$, the height of the baseball is given by

$$y = -0.0032(300)^2 + 300 + 3 = 15 \text{ feet.}$$

Thus, the ball will clear the fence.

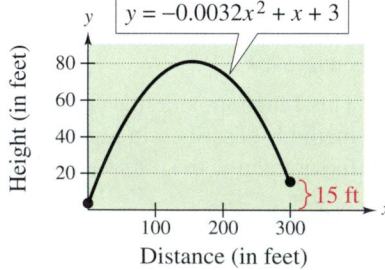

FIGURE 2.15

NOTE   In the equation in Example 7, the height of the baseball is a function of the distance from home plate. ▪▪

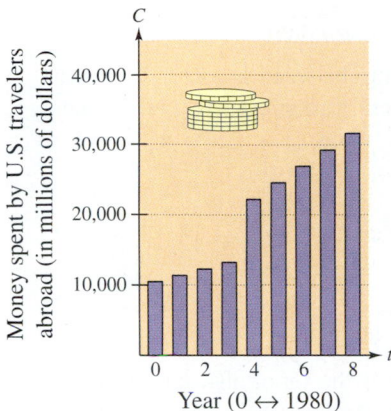

C

Money spent by U.S. travelers abroad (in millions of dollars)

40,000

30,000

20,000

10,000

0   2   4   6   8   t

Year (0 ↔ 1980)

FIGURE 2.16

EXAMPLE 8   *U.S. Travelers Abroad*

Real Life

The money C (in millions of dollars) spent by U.S. travelers in other countries increased in a linear pattern from 1980 to 1983, as shown in Figure 2.16. Then, in 1984, the money spent took a sharp jump and until 1988 increased in a *different* linear pattern. These two patterns can be approximated by the function

$$C = \begin{cases} 10{,}479 + 917.1t, & 0 \le t \le 3 \\ 12{,}808 + 2350.4t, & 4 \le t \le 8 \end{cases}$$

where $t = 0$ represents 1980. Use this function to approximate the total amount spent by U.S. travelers abroad between 1980 and 1988.   (Source: U.S. Bureau of Economic Analysis)

Solution

From 1980 to 1983, use the formula $C = 10{,}479 + 917.1t$.

$\underbrace{\$10{,}479}_{1980}, \ \underbrace{\$11{,}396}_{1981}, \ \underbrace{\$12{,}313}_{1982}, \ \underbrace{\$13{,}230}_{1983}$

From 1984 to 1988, use the formula $C = 12{,}808 + 2350.4t$.

$\underbrace{\$22{,}210}_{1984}, \ \underbrace{\$24{,}560}_{1985}, \ \underbrace{\$26{,}910}_{1986}, \ \underbrace{\$29{,}261}_{1987}, \ \underbrace{\$31{,}611}_{1988}$

The total of these nine amounts is $181,970, which implies that the total amount spent was approximately $181,970,000,000.

EXAMPLE 9   *From Calculus: Evaluating a Difference Quotient*

For $f(x) = x^2 - 4x + 7$, find $\dfrac{f(x + h) - f(x)}{h}$.

Solution

$$\frac{f(x + h) - f(x)}{h} = \frac{[(x + h)^2 - 4(x + h) + 7] - (x^2 - 4x + 7)}{h}$$

$$= \frac{x^2 + 2xh + h^2 - 4x - 4h + 7 - x^2 + 4x - 7}{h}$$

$$= \frac{2xh + h^2 - 4h}{h}$$

$$= \frac{h(2x + h - 4)}{h}$$

$$= 2x + h - 4, \qquad h \ne 0$$

NOTE   One of the basic definitions in calculus employs the ratio

$$\frac{f(x + h) - f(x)}{h}, \qquad h \ne 0$$

called a **difference quotient,** as illustrated in Example 9. ■■

SUMMARY OF FUNCTION TERMINOLOGY

Function: A **function** is a relationship between two variables such that to each value of the independent variable there corresponds exactly one value of the dependent variable.

Function Notation: $y = f(x)$

  f is the **name** of the function.
  y is the **dependent variable.**
  x is the **independent variable.**
  $f(x)$ is the **value of the function at x.**

Domain: The **domain** of a function is the set of all values (inputs) of the independent variable for which the function is defined. If x is in the domain of f, we say that f is **defined** at x. If x is not in the domain of f, we say that f is **undefined** at x.

Range: The **range** of a function is the set of all values (outputs) assumed by the dependent variable (that is, the set of all function values).

Implied Domain: If f is defined by an algebraic expression and the domain is not specified, the **implied domain** consists of all real numbers for which the expression is defined.

GROUP ACTIVITY

MODELING WITH PIECEWISE-DEFINED FUNCTIONS

The table at the right shows the monthly revenue (in thousands of dollars) for one year of a landscaping business, with $x = 1$ representing January.

A mathematical model that represents this data is:

$$f(x) = \begin{cases} -1.97x + 26.33 \\ 0.5|x^2 - 1.47x + 6.3|. \end{cases}$$

For what values of x is each part of the piecewise-defined function defined? How can you tell? Explain your reasoning.

Find $f(5)$ and $f(11)$, and interpret your results in the context of the problem. How do these model values compare with the actual data values?

x	y	x	y
1	5.2	7	12.8
2	5.6	8	10.1
3	6.6	9	8.6
4	8.3	10	6.9
5	11.5	11	4.5
6	15.8	12	2.7

WARM UP

Simplify the expression.

1. $2(-3)^3 + 4(-3) - 7$
2. $4(-1)^2 - 5(-1) + 4$
3. $(x + 1)^2 + 3(x + 1) - 4 - (x^2 + 3x - 4)$
4. $(x - 2)^2 - 4(x - 2) - (x^2 - 4)$

Solve for y in terms of x.

5. $2x + 5y - 7 = 0$ 6. $y^2 = x^2$

Solve the inequality.

7. $9 - 2x \geq 0$ 8. $3x + 2 \geq 0$
9. $9 - x^2 \geq 0$ 10. $x^2 - 3x + 2 \geq 0$

The *Interactive* CD-ROM contains step-by-step solutions to all odd-numbered Section and Review Exercises. It also provides Tutorial Exercises, which link to Guided Examples for additional help.

2.2 Exercises

In Exercises 1–4, is the relationship a function?

1.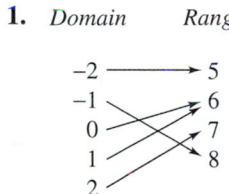

2.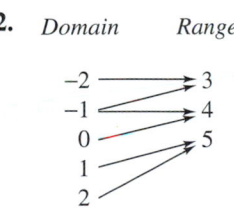

3.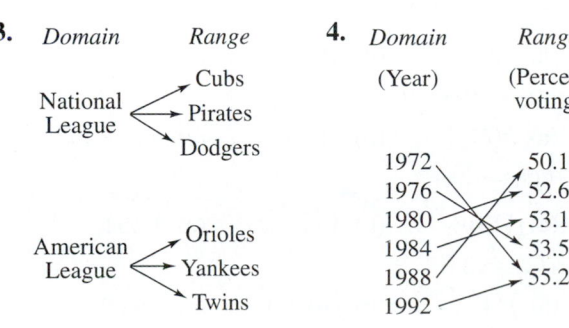

4.

In Exercises 5–8, does the table describe a function? Explain your reasoning.

5.

Input Value	−2	−1	0	1	2
Output Value	−8	−1	0	1	8

6.

Input Value	0	1	2	1	0
Output Value	−4	−2	0	2	4

7.

Input Value	10	7	4	7	10
Output Value	3	6	9	12	15

8.

Input Value	0	3	9	12	15
Output Value	3	3	3	3	3

In Exercises 9 and 10, which sets of ordered pairs represent function(s) from A to B? Explain.

9. $A = \{0, 1, 2, 3\}$ and $B = \{-2, -1, 0, 1, 2\}$
    (a) $\{(0, 1), (1, -2), (2, 0), (3, 2)\}$
    (b) $\{(0, -1), (2, 2), (1, -2), (3, 0), (1, 1)\}$
    (c) $\{(0, 0), (1, 0), (2, 0), (3, 0)\}$
    (d) $\{(0, 2), (3, 0), (1, 1)\}$
10. $A = \{a, b, c\}$ and $B = \{0, 1, 2, 3\}$
    (a) $\{(a, 1), (c, 2), (c, 3), (b, 3)\}$
    (b) $\{(a, 1), (b, 2), (c, 3)\}$
    (c) $\{(1, a), (0, a), (2, c), (3, b)\}$
    (d) $\{(c, 0), (b, 0), (a, 3)\}$

Circulation of Newspapers **In Exercises 11 and 12, use the graph, which shows the circulation (in millions) of daily newspapers in the United States.** (Source: Editor & Publisher Company)

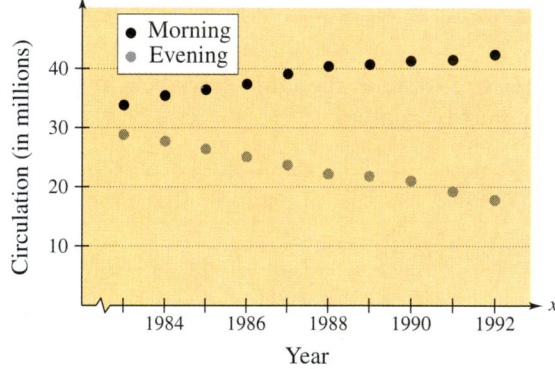

11. Is the circulation of morning newspapers a function of the year? Is the circulation of evening newspapers a function of the year? Explain.

12. Let $f(x)$ represent the circulation of evening newspapers in year x. Find $f(1988)$.

In Exercises 13–22, determine if the equation represents y as a function of x.

13. $x^2 + y^2 = 4$
14. $x = y^2$
15. $x^2 + y = 4$
16. $x + y^2 = 4$
17. $2x + 3y = 4$
18. $(x - 2)^2 + y^2 = 4$
19. $y^2 = x^2 - 1$
20. $y = \sqrt{x + 5}$
21. $y = |4 - x|$
22. $|y| = 4 - x$

In Exercises 23 and 24, fill in the blanks using the specified function and the given values of the independent variable.

23. $f(s) = \dfrac{1}{s + 1}$
    (a) $f(4) = \dfrac{1}{(\boxed{}) + 1}$
    (b) $f(0) = \dfrac{1}{(\boxed{}) + 1}$
    (c) $f(4x) = \dfrac{1}{(\boxed{}) + 1}$
    (d) $f(x + c) = \dfrac{1}{(\boxed{}) + 1}$

24. $g(x) = x^2 - 2x$
    (a) $g(2) = (\boxed{})^2 - 2(\boxed{})$
    (b) $g(-3) = (\boxed{})^2 - 2(\boxed{})$
    (c) $g(t + 1) = (\boxed{})^2 - 2(\boxed{})$
    (d) $g(x + c) = (\boxed{})^2 - 2(\boxed{})$

In Exercises 25–36, evaluate the function at the specified values of the independent variable and simplify.

25. $f(x) = 2x - 3$
    (a) $f(1)$ (b) $f(-3)$ (c) $f(x - 1)$
26. $g(y) = 7 - 3y$
    (a) $g(0)$ (b) $g(\frac{7}{3})$ (c) $g(s + 2)$
27. $h(t) = t^2 - 2t$
    (a) $h(2)$ (b) $h(1.5)$ (c) $h(x + 2)$
28. $V(r) = \frac{4}{3}\pi r^3$
    (a) $V(3)$ (b) $V(\frac{3}{2})$ (c) $V(2r)$
29. $f(y) = 3 - \sqrt{y}$
    (a) $f(4)$ (b) $f(0.25)$ (c) $f(4x^2)$
30. $f(x) = \sqrt{x + 8} + 2$
    (a) $f(-8)$ (b) $f(1)$ (c) $f(x - 8)$

31. $q(x) = \dfrac{1}{x^2 - 9}$

  (a) $q(0)$ (b) $q(3)$ (c) $q(y + 3)$

32. $q(t) = \dfrac{2t^2 + 3}{t^2}$

  (a) $q(2)$ (b) $q(0)$ (c) $q(-x)$

33. $f(x) = \dfrac{|x|}{x}$

  (a) $f(2)$ (b) $f(-2)$ (c) $f(x - 1)$

34. $f(x) = |x| + 4$

  (a) $f(2)$ (b) $f(-2)$ (c) $f(x^2)$

35. $f(x) = \begin{cases} 2x + 1, & x < 0 \\ 2x + 2, & x \geq 0 \end{cases}$

  (a) $f(-1)$ (b) $f(0)$ (c) $f(2)$

36. $f(x) = \begin{cases} x^2 + 2, & x \leq 1 \\ 2x^2 + 2, & x > 1 \end{cases}$

  (a) $f(-2)$ (b) $f(1)$ (c) $f(2)$

In Exercises 37–42, complete the table.

37. $f(x) = x^2 - 3$

x	-2	-1	0	1	2
$f(x)$					

38. $g(x) = \sqrt{x - 3}$

x	3	4	5	6	7
$g(x)$					

39. $h(t) = \frac{1}{2}|t + 3|$

t	-5	-4	-3	-2	-1
$h(t)$					

40. $f(s) = \dfrac{|s - 2|}{s - 2}$

s	0	1	$\frac{3}{2}$	$\frac{5}{2}$	4
$f(s)$					

41. $f(x) = \begin{cases} -\frac{1}{2}x + 4, & x \leq 0 \\ (x - 2)^2, & x > 0 \end{cases}$

x	-2	-1	0	1	2
$f(x)$					

42. $h(x) = \begin{cases} 9 - x^2, & x < 3 \\ x - 3, & x \geq 3 \end{cases}$

x	1	2	3	4	5
$h(x)$					

In Exercises 43–46, find all real values of x such that $f(x) = 0$.

43. $f(x) = 15 - 3x$

44. $f(x) = \dfrac{3x - 4}{5}$

45. $f(x) = x^2 - 9$

46. $f(x) = x^3 - x$

In Exercises 47–50, find the value(s) of x for which $f(x) = g(x)$.

47. $f(x) = x^2, \quad g(x) = x + 2$

48. $f(x) = x^2 + 2x + 1, \quad g(x) = 3x + 3$

49. $f(x) = \sqrt{3x} + 1, \quad g(x) = x + 1$

50. $f(x) = x^4 - 2x^2, \quad g(x) = 2x^2$

In Exercises 51–60, find the domain of the function.

51. $f(x) = 5x^2 + 2x - 1$

52. $g(x) = 1 - 2x^2$

53. $h(t) = \dfrac{4}{t}$

54. $s(y) = \dfrac{3y}{y + 5}$

55. $g(y) = \sqrt{y - 10}$

56. $f(t) = \sqrt[3]{t + 4}$

57. $f(x) = \sqrt[4]{1 - x^2}$

58. $h(x) = \dfrac{10}{x^2 - 2x}$

59. $g(x) = \dfrac{1}{x} - \dfrac{3}{x + 2}$

60. $f(s) = \dfrac{\sqrt{s - 1}}{s - 4}$

In Exercises 61–64, assume that the domain of f is the set $A = \{-2, -1, 0, 1, 2\}$. Determine the set of ordered pairs representing the function f.

61. $f(x) = x^2$

62. $f(x) = \dfrac{2x}{x^2 + 1}$

63. $f(x) = \sqrt{x + 2}$

64. $f(x) = |x + 1|$

65. Think About It   In your own words, explain the meanings of *domain* and *range*.

66. Think About It   Describe an advantage of function notation.

Exploration   In Exercises 67–70, select a function from $f(x) = cx$, $g(x) = cx^2$, $h(x) = c\sqrt{|x|}$, and $r(x) = c/x$ and determine the value of the constant c such that the function fits the data given in the table.

67.

x	-4	-1	0	1	4
y	-32	-2	0	-2	-32

68.

x	-4	-1	0	1	4
y	-1	$-\frac{1}{4}$	0	$\frac{1}{4}$	1

69.

x	-4	-1	0	1	4
y	-8	-32	undef.	32	8

70.

x	-4	-1	0	1	4
y	6	3	0	3	6

In Exercises 71–76, find the difference quotient and simplify your answer.

71. $f(x) = x^2 - x + 1$, $\dfrac{f(2 + h) - f(2)}{h}, h \neq 0$

72. $f(x) = 5x - x^2$, $\dfrac{f(5 + h) - f(5)}{h}, h \neq 0$

73. $f(x) = x^3$, $\dfrac{f(x + c) - f(x)}{c}, c \neq 0$

74. $f(x) = 2x$, $\dfrac{f(x + c) - f(x)}{c}, c \neq 0$

75. $g(x) = 3x - 1$, $\dfrac{g(x) - g(3)}{x - 3}, x \neq 3$

76. $f(t) = \dfrac{1}{t}$, $\dfrac{f(t) - f(1)}{t - 1}, t \neq 0$

77. Area of a Circle   Express the area A of a circle as a function of its circumference C.

78. Area of a Triangle   Express the area A of an equilateral triangle as a function of the length s of its sides.

79. Exploration   An open box of maximum volume is to be made from a square piece of material, 24 centimeters on a side, by cutting equal squares from the corners, and turning up the sides (see figure).

(a) Complete six rows of a table. (The first two rows are shown.) Use the result to guess the maximum volume.

Height x	Width	Volume V
1	$24 - 2(1)$	$1[24 - 2(1)]^2 = 484$
2	$24 - 2(2)$	$2[24 - 2(2)]^2 = 800$

(b) Plot the points (x, V). Is V a function of x?

(c) If V is a function of x, write the function and determine its domain.

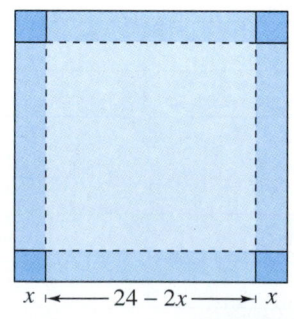

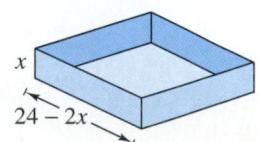

80. *Exploration* The cost per unit in the production of a certain radio model is $60. The manufacturer charges $90 per unit for orders of 100 or less. To encourage large orders, the manufacturer reduces the charge by $0.15 per radio for each unit ordered in excess of 100 (for example, there would be a charge of $87 per radio for an order size of 120).

(a) Complete six rows of the table. Use the result to estimate the maximum profit.

Units x	Price p	Profit P
102	$90 - 2(0.15)$	$xp - 102(60)$
104	$90 - 4(0.15)$	$xp - 104(60)$

(b) Plot the points (x, P). Is P a function of x?

(c) If P is a function of x, write the function and determine its domain.

81. *Area of a Triangle* A right triangle is formed in the first quadrant by the x- and y-axes and a line through the point $(2, 1)$ (see figure). Write the area of the triangle as a function of x, and determine the domain of the function.

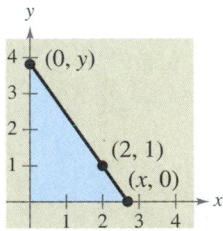

82. *Area of a Rectangle* A rectangle is bounded by the x-axis and the semicircle $y = \sqrt{36 - x^2}$ (see figure). Write the area of the rectangle as a function of x, and determine the domain of the function.

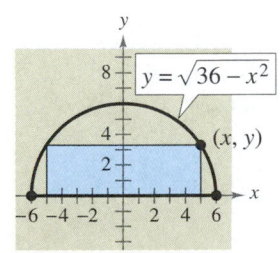

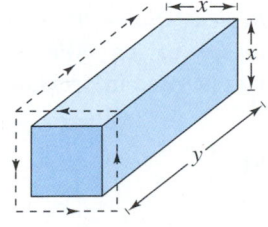

FIGURE FOR 82   **FIGURE FOR 83**

83. *Volume of a Package* A rectangular package to be sent by a postal service can have a maximum combined length and girth (perimeter of a cross section) of 108 inches (see figure). Write the volume of the package as a function of x. What is the domain of the function?

84. *Price of Mobile Homes* The average price p (in thousands of dollars) of a new mobile home in the United States from 1974 to 1993 can be approximated by the model

$$p(t) = \begin{cases} 19.247 + 1.694t, & -6 \le t \le -1 \\ 19.305 + 0.427t + 0.033t^2, & 0 \le t \le 13 \end{cases}$$

where $t = 0$ represents 1980 (see figure). Use this model to find the average price of a mobile home in 1978, 1988, and 1993. (Source: U.S. Bureau of Census, Construction Reports)

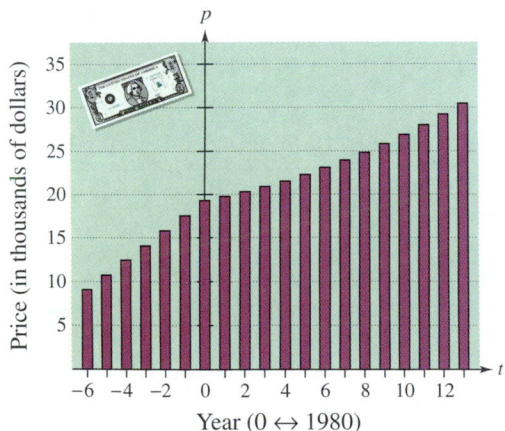

Year (0 ↔ 1980)

85. *Cost, Revenue and Profit* A company produces a product for which the variable cost is $12.30 per unit and the fixed costs are $98,000. The product sells for $17.98. Let x be the number of units produced and sold.

(a) Write the total cost C as a function of the number of units produced.

(b) Write the revenue R as a function of the number of units sold.

(c) Write the profit P as a function of the number of the units sold. [*Note*: $P = R - C$.]

86. *Cost Analysis*   The inventor of a new game believes that the variable cost for producing the game is $0.95 per unit and the fixed costs are $6000. The inventor sells each game for $1.69. Let x be the number of games sold.

 (a) Write the total cost C as a function of the number of games sold.

 (b) Write the average cost per unit $\overline{C} = C/x$ as a function of x.

87. *Charter Bus Fares*   For groups of 80 or more people, a charter bus company determines the rate per person according to the formula

$$\text{Rate} = 8 - 0.05(n - 80), \qquad n \geq 80$$

where the rate is given in dollars and n is the number of people.

 (a) Express the revenue R for the bus company as a function of n.

 (b) Use the function from part (a) to complete the table. What can you conclude?

n	90	100	110	120	130	140	150
$R(n)$							

88. *Fluid Force*   The force F (in tons) of water against the face of a dam is a function given by

$$F(y) = 149.76\sqrt{10}y^{5/2}$$

where y is the depth of the water in feet. Complete the table.

y	5	10	20	30	40
$F(y)$					

 (a) What can you conclude from the table?

 (b) Use the table to approximate the depth at which the force against the dam is 1,000,000 tons. How could you find a better estimate?

89. *Height of a Balloon*   A balloon carrying a transmitter ascends vertically from a point 3000 feet from the receiving station.

 (a) Draw a figure that gives a visual representation of the problem. Let h represent the height of the balloon and let d represent the distance between the balloon and the receiving station.

 (b) Express the height of the balloon as a function of d. What is the domain of the function?

90. *Chapter Opener*   Use the data on the opening page of this chapter to answer each question. Let $f(t)$ represent the number of lynx in year t.

 (a) Find $f(1992)$.

 (b) Find

$$\frac{f(1994) - f(1991)}{1994 - 1991}$$

and interpret the result in the context of the problem.

 (c) An approximate formula for the function is

$$N(t) = \frac{434t + 4387}{45t^2 - 55t + 100}$$

where N is the number of lynx and t is time in years, with $t = 0$ corresponding to 1990. Complete the table and compare the result with the data.

t	1988	1989	1990	1991
N				

t	1992	1993	1994	1995
N				

Review   **Solve the equations in Exercises 91–94 as a review of the skills and problem-solving techniques you learned in previous sections.**

91. $\dfrac{t}{3} + \dfrac{t}{5} = 1$

92. $\dfrac{3}{t} + \dfrac{5}{t} = 1$

93. $\dfrac{3}{x(x + 1)} - \dfrac{4}{x} = \dfrac{1}{x + 1}$

94. $\dfrac{12}{x} - 3 = \dfrac{4}{x} + 9$

2.3 *Analyzing Graphs of Functions*

See Exercise 67 on page 221 for an example of how a step function can be used to model the cost of a telephone call.

The Graph of a Function ◻ *Increasing and Decreasing Functions* ◻ *Step Functions* ◻ *Even and Odd Functions* ◻ *Summary of Graphs of Common Functions*

The Graph of a Function

In Section 2.2 you studied functions from an algebraic point of view. In this section, you will study functions from a geometric perspective. The **graph of a function** f is the collection of ordered pairs $(x, f(x))$ such that x is in the domain of f. As you study this section, remember that

$$x = \text{the directed distance from the } y\text{-axis}$$

$$f(x) = \text{the directed distance from the } x\text{-axis}$$

as shown in Figure 2.17. If the graph of a function has an x-intercept at $(a, 0)$, then a is a **zero** of the function. In other words, the zeros of a function are the values of x for which $f(x) = 0$. For instance, the function given by $f(x) = x^2 - 4$ has two zeros: -2 and 2.

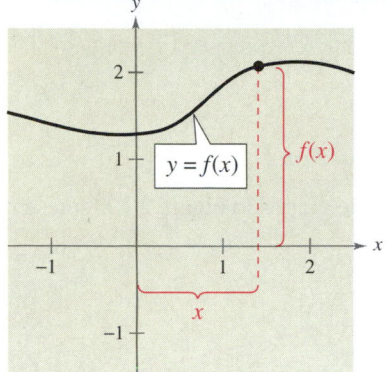

FIGURE 2.17

EXAMPLE 1 *Finding the Domain and Range of a Function*

Use the graph of the function f, shown in Figure 2.18, to find (a) the domain of f, (b) the function values $f(-1)$ and $f(2)$, and (c) the range of f.

Solution

a. A closed dot (on the left) indicates that $x = -1$ is in the domain of f, whereas the open dot (on the right) indicates $x = 4$ is not in the domain. Thus, the domain of f is all x in the interval $[-1, 4)$.

b. Because $(-1, -5)$ is a point on the graph of f, it follows that

$$f(-1) = -5.$$

Similarly, because $(2, 4)$ is a point on the graph of f, it follows that

$$f(2) = 4.$$

c. Because the graph does not extend below $f(-1) = -5$ or above $f(2) = 4$, the range of f is the interval $[-5, 4]$.

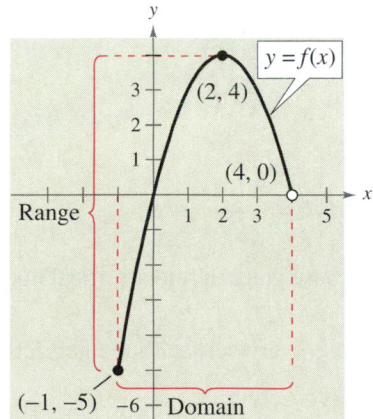

FIGURE 2.18

NOTE The use of dots (open or closed) at the extreme left and right points of a graph indicates that the graph does not extend beyond these points. If no such dots are shown, assume that the graph extends beyond these points. ▪▪

By the definition of a function, at most one *y*-value corresponds to a given *x*-value. It follows, then, that a vertical line can intersect the graph of a function at most once. This observation provides a convenient visual test called the **Vertical Line Test** for functions.

> ### VERTICAL LINE TEST FOR FUNCTIONS
>
> A set of points in a coordinate plane is the graph of *y* as a function of *x* if and only if no vertical line intersects the graph at more than one point.

EXAMPLE 2   *Vertical Line Test for Functions*

Use the Vertical Line Test to decide whether the graphs in Figure 2.19 represent *y* as a function of *x*.

(a)

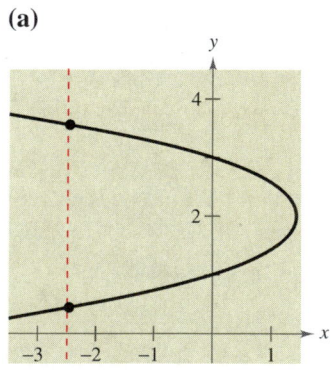

(b)

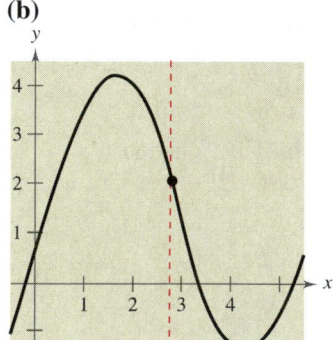

(c)

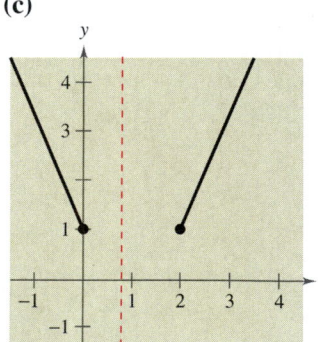

FIGURE 2.19

Solution

a. This *is not* a graph of *y* as a function of *x* because you can find a vertical line that intersects the graph twice.

b. This *is* a graph of *y* as a function of *x* because every vertical line intersects the graph at most once.

c. This *is* a graph of *y* as a function of *x*. (Note that if a vertical line does not intersect the graph, it simply means that the function is undefined for that particular value of *x*.)

Increasing and Decreasing Functions

The more you know about the graph of a function, the more you know about the function itself. Consider the graph shown in Figure 2.20. As you move from *left to right,* this graph decreases, then is constant, and then increases.

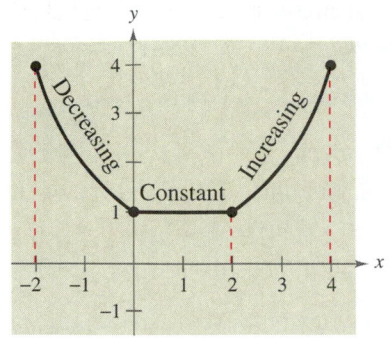

FIGURE 2.20

INCREASING, DECREASING, AND CONSTANT FUNCTIONS

A function f is **increasing** on an interval if, for any x_1 and x_2 in the interval, $x_1 < x_2$ implies $f(x_1) < f(x_2)$.

A function f is **decreasing** on an interval if, for any x_1 and x_2 in the interval, $x_1 < x_2$ implies $f(x_1) > f(x_2)$.

A function f is **constant** on an interval if, for any x_1 and x_2 in the interval, $f(x_1) = f(x_2)$.

EXAMPLE 3 *Increasing and Decreasing Functions*

In Figure 2.21, describe the increasing or decreasing behavior of the function.

Solution

a. This function is increasing over the entire real line.

b. This function is increasing on the interval $(-\infty, -1)$, decreasing on the interval $(-1, 1)$, and increasing on the interval $(1, \infty)$.

c. This function is increasing on the interval $(-\infty, 0)$, constant on the interval $(0, 2)$, and decreasing on the interval $(2, \infty)$.

(a)

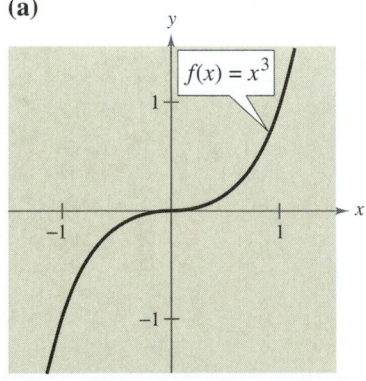

(b)

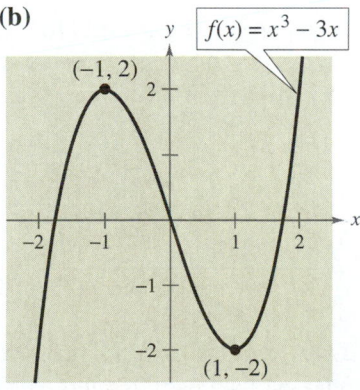

(c)

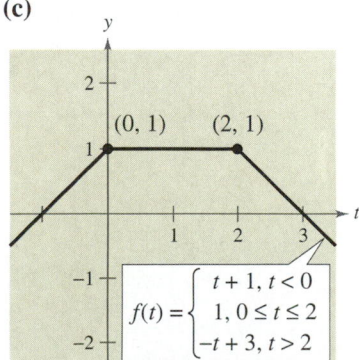

FIGURE 2.21

When analyzing the graph of a function, you want to find the points at which the function changes its increasing, decreasing, or constant behavior. These points often identify *maximum* or *minimum* values of the function.

EXAMPLE 4   *The Price of Diamonds*

Real Life

During the 1980s, the average price of a 1-carat polished diamond decreased and then increased according to the model

$$C = -0.7t^3 + 16.25t^2 - 106t + 388, \qquad 2 \leq t \leq 10$$

where C is the average price in dollars (on the Antwerp Index) and t represents the calendar year, with $t = 2$ corresponding to 1982. According to this model, during which years was the price of diamonds decreasing? During which years was the price of diamonds increasing? Approximate the minimum price of a 1-carat diamond between 1982 and 1990.   (Source: Diamond High Council)

Solution

To solve this problem, sketch an accurate graph of the function, as shown in Figure 2.22. From the graph, you can see that the price of diamonds decreased from 1982 until late 1984. Then, from late 1984 to 1990, the price increased. The minimum price during the 8-year period was approximately $175.

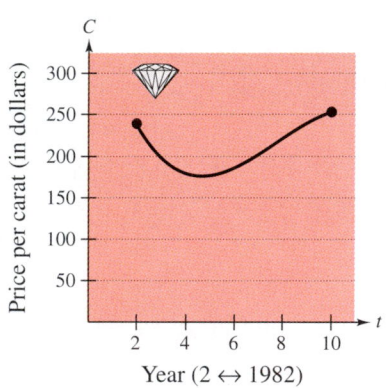

Year (2 ↔ 1982)

FIGURE 2.22

TECHNOLOGY

A graphing utility is useful for determining the minimum and maximum values of a function over a closed interval. For instance, graph

$$C = -0.7t^3 + 16.25t^2 - 106t + 388, \qquad 2 \leq t \leq 10$$

as shown below. By using the trace feature, you can determine that the minimum value occurs when $x \approx 4.7$.

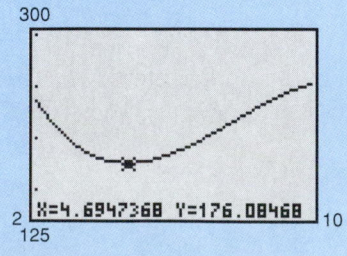

Step Functions

EXAMPLE 5   *The Greatest Integer Function*

The **greatest integer function** is denoted by $[\![x]\!]$ and is defined by

$[\![x]\!]$ = the greatest integer less than or equal to x.

The graph of this function is shown in Figure 2.23. Note that the graph of the greatest integer function jumps vertically one unit at each integer and is constant (a horizontal line segment) between each pair of consecutive integers. The greatest integer function is an example of a category of functions called **step functions.** Some values of the greatest integer function are as follows.

$$[\![-1]\!] = -1 \qquad\qquad [\![-0.5]\!] = -1$$
$$[\![0]\!] = 0 \qquad\qquad\quad [\![0.5]\!] = 0$$
$$[\![1]\!] = 1 \qquad\qquad\quad [\![1.5]\!] = 1$$

The range of the greatest integer function is the set of all integers.

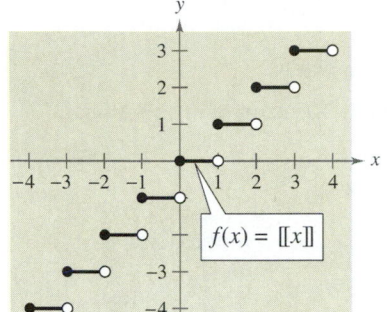

FIGURE 2.23

NOTE   If you use a graphing utility to sketch a step function, you should set the utility to *Dot* mode rather than *Connected* mode. ▪▪

EXAMPLE 6   *The Price of a Telephone Call*

Real Life

The cost of a telephone call between Los Angeles and San Francisco is $0.50 for the first minute and $0.36 for each additional minute (or portion of a minute). The greatest integer function can be used to create a model for the cost of this call, as follows.

$$C = 0.50 + 0.36 [\![t]\!], \qquad t > 0$$

where C is the total cost of the call in dollars and t is the length of the call in minutes. Sketch the graph of this function.

Solution

For calls up to 1 minute, the cost is $0.50. For calls between 1 and 2 minutes, the cost is $0.86, and so on.

Length of Call	$0 < t < 1$	$1 \le t < 2$	$2 \le t < 3$	$3 \le t < 4$	$4 \le t < 5$
Cost of Call	$0.50	$0.86	$1.22	$1.58	$1.94

Using these values, you can sketch the graph shown in Figure 2.24.

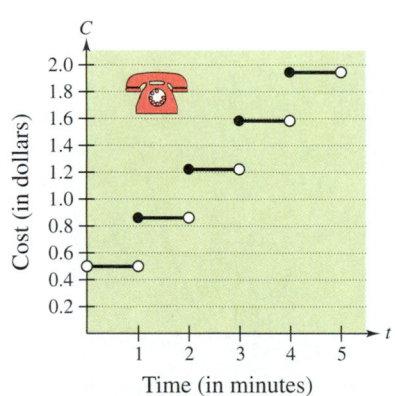

FIGURE 2.24

Even and Odd Functions

In Section 1.1, you studied different types of symmetry of a graph. In the terminology of functions, a function is said to be **even** if its graph is symmetric with respect to the y-axis and to be **odd** if its graph is symmetric with respect to the origin. The symmetry tests in Section 1.1 yield the following tests for even and odd functions.

TESTS FOR EVEN AND ODD FUNCTIONS

A function given by $y = f(x)$ is **even** if, for each x in the domain of f,

$$f(-x) = f(x).$$

A function given by $y = f(x)$ is **odd** if, for each x in the domain of f,

$$f(-x) = -f(x).$$

EXAMPLE 7   *Even and Odd Functions*

a. The function $g(x) = x^3 - x$ is odd because

$$g(-x) = (-x)^3 - (-x) = -x^3 + x = -(x^3 - x) = -g(x).$$

b. The function $h(x) = x^2 + 1$ is even because

$$h(-x) = (-x)^2 + 1 = x^2 + 1 = h(x).$$

The graphs of the two functions are shown in Figure 2.25.

(a) Symmetric to Origin

(b) Symmetric to y-Axis

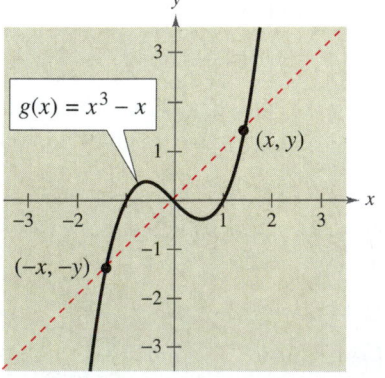
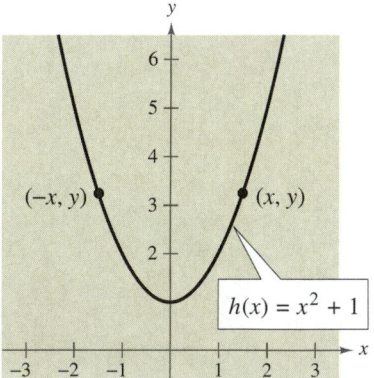

FIGURE 2.25

Summary of Graphs of Common Functions

Figure 2.26 shows the graphs of six common functions. You need to be familiar with these graphs.

(a) Constant Function

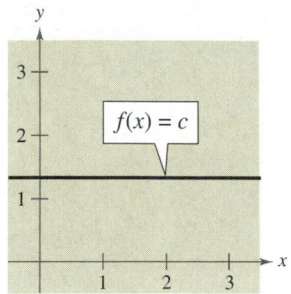

(b) Identity Function

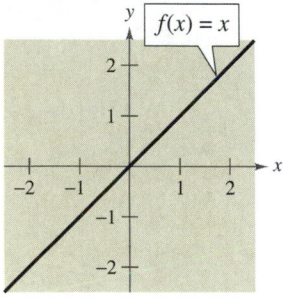

(c) Absolute Value

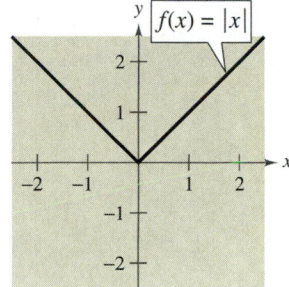

(d) Square Root

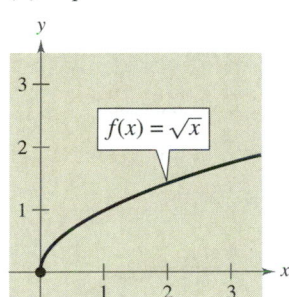

(e) Squaring Function

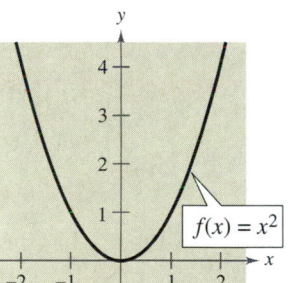

(f) Cubing Function

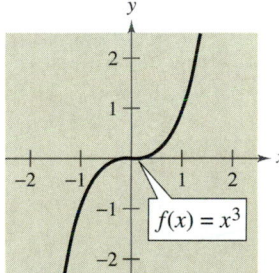

FIGURE 2.26

GROUP ACTIVITY

IDENTIFYING FUNCTIONS

The table gives the circulation y (in millions) and the annual acquisitions expenditures x (in millions of dollars) for top public libraries in the United States in 1993. Discuss whether the data itself represents a function. Explain your reasoning. Sketch and discuss the scatter plot of the data. On your scatter plot, draw what you think is the best-fitting model. Is the model that best represents the data a function? Defend your position. (Source: Public Library Association)

x	y	x	y
3.0	15.9	5.0	7.8
5.7	10.1	1.3	3.1
5.3	9.3	5.7	13.2
1.2	1.5	4.2	6.5
2.8	6.3	2.0	5.4

WARM UP

1. Find $f(2)$ for $f(x) = -x^3$.

2. Find $f(6)$ for $f(x) = x^2 - 6x$.

3. Find $f(-x)$ for $f(x) = \dfrac{3}{x}$.

4. Find $f(-x)$ for $f(x) = x^2 + 3$.

Solve for x.

5. $x^3 - 16x = 0$

6. $2x^2 - 3x + 1 = 0$

Find the domain of the function.

7. $g(x) = \dfrac{4}{x - 4}$

8. $f(x) = \dfrac{2x}{x^2 - 9x + 20}$

9. $h(t) = \sqrt[4]{5 - 3t}$

10. $f(t) = t^3 + 3t - 5$

2.3 Exercises

In Exercises 1–6, find the domain and range of the function.

1. $f(x) = 1 - x^2$

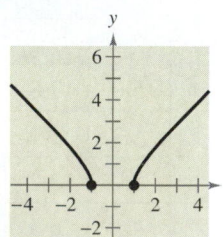

2. $f(x) = \sqrt{x - 1}$

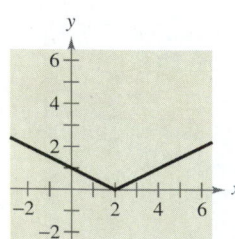

3. $f(x) = \sqrt{x^2 - 1}$

4. $f(x) = \frac{1}{2}|x - 2|$

5. $h(x) = \sqrt{16 - x^2}$

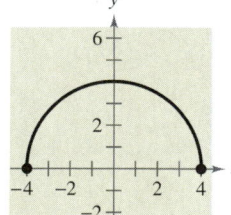

6. $g(x) = \dfrac{|x - 1|}{x - 1}$

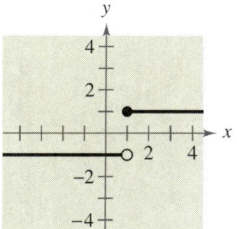

In Exercises 7–12, use the Vertical Line Test to determine whether y is a function of x.

7. $y = \frac{1}{2}x^2$

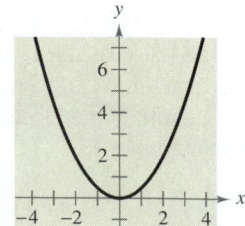

8. $y = \frac{1}{4}x^3$

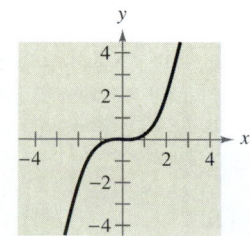

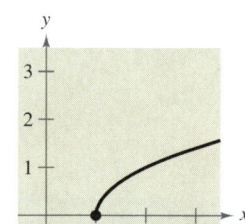
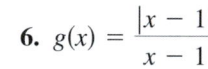

9. $x - y^2 = 1$

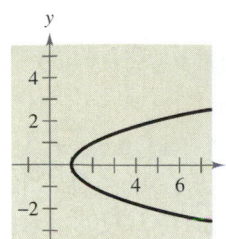

10. $x^2 + y^2 = 25$

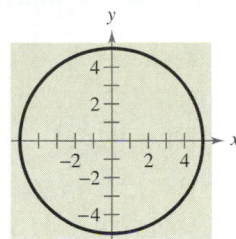

11. $x^2 = 2xy - 1$

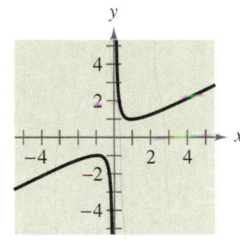

12. $x = |y + 2|$

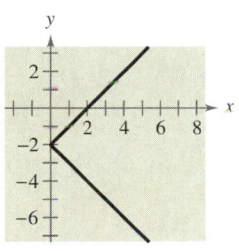

13. *Think About It*   Does the graph in Exercise 9 represent x as a function of y? Explain.

14. *Think About It*   Does the graph in Exercise 10 represent x as a function of y? Explain.

In Exercises 15–18, select the viewing rectangle that shows the most complete graph of the function.

15. $f(x) = -0.2x^2 + 3x + 32$

Xmin = -2	Xmin = -10	Xmin = 0
Xmax = 20	Xmax = 30	Xmax = 10
Xscl = 1	Xscl = 5	Xscl = 0.5
Ymin = -10	Ymin = -5	Ymin = 0
Ymax = 30	Ymax = 50	Ymax = 200
Yscl = 4	Yscl = 5	Yscl = 25

16. $f(x) = 6[x - (0.1x)^5]$

Xmin = -500	Xmin = -25	Xmin = -20
Xmax = 5000	Xmax = 25	Xmax = 20
Xscl = 50	Xscl = 5	Xscl = 5
Ymin = -500	Ymin = -25	Ymin = -100
Ymax = 500	Ymax = 25	Ymax = 100
Yscl = 50	Yscl = 5	Yscl = 20

17. $f(x) = 4x^3 - x^4$

Xmin = -2	Xmin = -50	Xmin = 0
Xmax = 6	Xmax = 50	Xmax = 2
Xscl = 1	Xscl = 5	Xscl = 0.2
Ymin = -10	Ymin = -50	Ymin = -2
Ymax = 30	Ymax = 50	Ymax = 2
Yscl = 4	Yscl = 5	Yscl = 0.5

18. $f(x) = 10x\sqrt{400 - x^2}$

Xmin = -5	Xmin = -20	Xmin = -25
Xmax = 50	Xmax = 20	Xmax = 25
Xscl = 5	Xscl = 2	Xscl = 5
Ymin = -5000	Ymin = -500	Ymin = -2000
Ymax = 5000	Ymax = 500	Ymax = 2000
Yscl = 500	Yscl = 50	Yscl = 200

In Exercises 19–22, (a) determine the intervals over which the function is increasing, decreasing, or constant, and (b) determine if the function is even, odd, or neither.

19. $f(x) = \frac{3}{2}x$

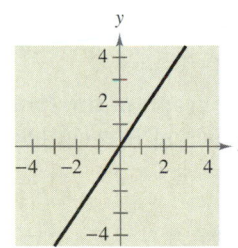

20. $f(x) = x^2 - 4x$

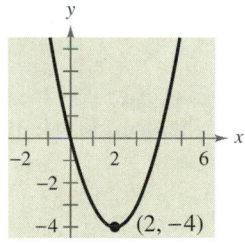

21. $f(x) = x^3 - 3x^2 + 2$

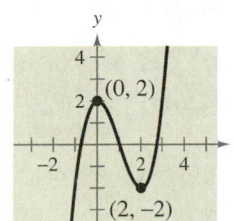

22. $f(x) = \sqrt{x^2 - 1}$

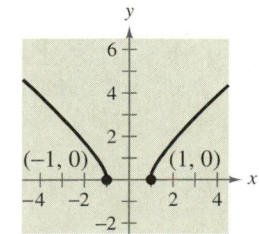

In Exercises 23–26, (a) use a graphing utility to graph the function, (b) determine the intervals over which the function is increasing, decreasing, or constant, and (c) determine if the function is even, odd, or neither.

23. $f(x) = 3x^4 - 6x^2$

24. $f(x) = x^{2/3}$

25. $f(x) = x\sqrt{x + 3}$

26. $f(x) = |x + 1| + |x - 1|$

In Exercises 27–32, determine whether the function is even, odd, or neither.

27. $f(x) = x^6 - 2x^2 + 3$ **28.** $h(x) = x^3 - 5$

29. $g(x) = x^3 - 5x$ **30.** $f(x) = x\sqrt{1 - x^2}$

31. $f(t) = t^2 + 2t - 3$ **32.** $g(s) = 4s^{2/3}$

Think About It In Exercises 33 and 34, find the coordinates of a second point on the graph of a function f if the given point is on the graph and the function is (a) even and (b) odd.

33. $\left(-\frac{3}{2}, 4\right)$

34. $(4, 9)$

In Exercises 35–46, sketch the graph of the function and determine whether the function is even, odd, or neither.

35. $f(x) = 3$ **36.** $g(x) = x$

37. $f(x) = 5 - 3x$ **38.** $h(x) = x^2 - 4$

39. $g(s) = \dfrac{s^2}{4}$ **40.** $f(t) = -t^4$

41. $f(x) = \sqrt{1 - x}$ **42.** $f(x) = x^{3/2}$

43. $g(t) = \sqrt[3]{t - 1}$ **44.** $f(x) = |x + 2|$

45. $f(x) = \begin{cases} x + 3, & x \le 0 \\ 3, & 0 < x \le 2 \\ 2x - 1, & x > 2 \end{cases}$

46. $f(x) = \begin{cases} 2x + 1, & x \le -1 \\ x^2 - 2, & x > -1 \end{cases}$

In Exercises 47–56, graph the function and determine the intervals for which $f(x) \ge 0$.

47. $f(x) = 4 - x$ **48.** $f(x) = 4x + 2$

49. $f(x) = x^2 - 9$ **50.** $f(x) = x^2 - 4x$

51. $f(x) = 1 - x^4$ **52.** $f(x) = \sqrt{x + 2}$

53. $f(x) = x^2 + 1$ **54.** $f(x) = -(1 + |x|)$

55. $f(x) = -5$ **56.** $f(x) = \frac{1}{2}(2 + |x|)$

In Exercises 57–60, graph the function.

57. $f(x) = \begin{cases} 2x + 3, & x < 0 \\ 3 - x, & x \ge 0 \end{cases}$

58. $f(x) = \begin{cases} \sqrt{4 + x}, & x < 0 \\ \sqrt{4 - x}, & x \ge 0 \end{cases}$

59. $f(x) = \begin{cases} x^2 + 5, & x \le 1 \\ -x^2 + 4x + 3, & x > 1 \end{cases}$

60. $f(x) = \begin{cases} 1 - (x - 1)^2, & x \le 2 \\ \sqrt{x - 2}, & x > 2 \end{cases}$

In Exercises 61 and 62, use a graphing utility to graph the function. State the domain and range of the function.

61. $f(x) = |x + 3|$

62. $h(t) = \sqrt{4 - t^2}$

In Exercises 63 and 64, use a graphing utility to graph the function. State the domain and range of the function. Describe the pattern of the graph.

63. $s(x) = 2\left(\frac{1}{4}x - \left[\!\left[\frac{1}{4}x\right]\!\right]\right)$

64. $g(x) = 2\left(\frac{1}{4}x - \left[\!\left[\frac{1}{4}x\right]\!\right]\right)^2$

65. *Essay* Use a graphing utility to graph each function. Write a paragraph describing any similarities and differences you observe among the graphs.

  (a) $y = x$ (b) $y = x^2$

  (c) $y = x^3$ (d) $y = x^4$

  (e) $y = x^5$ (f) $y = x^6$

66. *Conjecture*   Use the results of Exercise 65 to make a conjecture about the graphs of $y = x^7$ and $y = x^8$. Use a graphing utility to graph the functions and compare the results with the graph drawn by hand.

67. *Comparing Models*   The cost of a telephone call between two cities is \$0.65 for the first minute and \$0.40 for each additional minute.

   (a) It is required that a model be created for the cost C of a telephone call between the two cities lasting t minutes. Which of the following is the appropriate model? Explain.

$$C_1(t) = 0.65 + 0.4[\![t - 1]\!]$$
$$C_2(t) = 0.65 - 0.4[\![-(t - 1)]\!]$$

   (b) Graph the appropriate model. Determine the cost of a call lasting 18 minutes and 45 seconds.

68. *Cost of Overnight Delivery*   Suppose that the cost of sending an overnight package from New York to Atlanta is \$9.80 for under one pound and \$2.50 for each additional pound. Use the greatest integer function to create a model for the cost C of overnight delivery of a package weighing x pounds, $x > 0$. Sketch the graph of the function.

69. *Maximum Profit*   The marketing department of a company estimates that the demand for a product is given by $p = 100 - 0.0001x$, where p is the price per unit and x is the number of units. The cost of producing x units is given by $C = 350,000 + 30x$, and the profit for producing and selling x units is given by

$$P = R - C = xp - C.$$

Use a graphing utility to graph the profit function and estimate the number of units that would produce a maximum profit.

70. *Fluorescent Lamp*   The number of lumens (time rate of flow of light) L from a fluorescent lamp can be approximated by the model

$$L = -0.294x^2 + 97.744x - 664.875, \qquad 20 \le x \le 90$$

where x is the wattage of the lamp. Use a graphing utility to graph the function and estimate the wattage of a bulb necessary to obtain 2000 lumens.

In Exercises 71–74, write the height h of the rectangle as a function of x.

71.
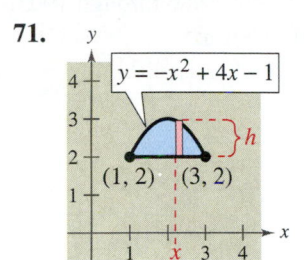

72.
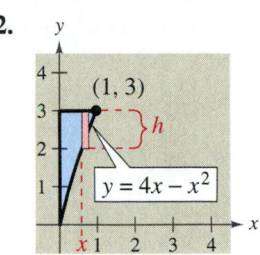

73.
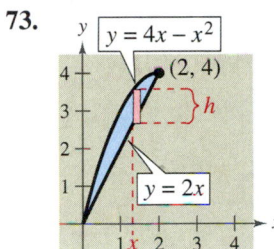

74.
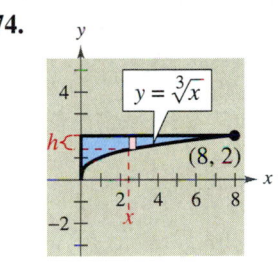

In Exercises 75–78, write the length L of the rectangle as a function of y.

75.
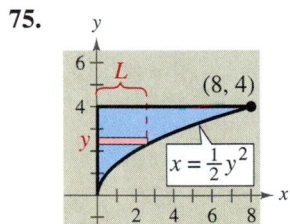

76.
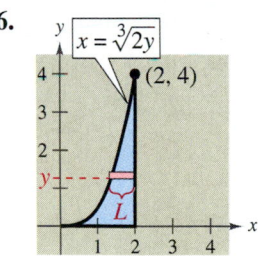

77.
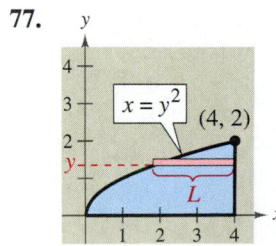

78.
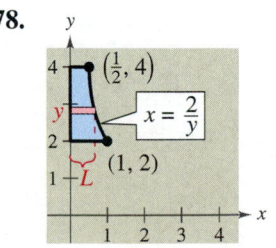

79. *Data Analysis* The table gives the amounts y (in billions of dollars) of the merchandise trade balance of the United States for the years 1986 through 1993. (Source: U.S. Bureau of the Census)

Year	1986	1987	1988	1989
y	−152.7	−152.1	−118.6	−109.6

Year	1990	1991	1992	1993
y	−101.7	−65.4	−84.5	−115.8

A model for this data is given by

$$y = -87.49 + 16.28t - 4.82t^2 - 1.17t^3$$

where $t = 0$ represents 1990.

(a) What is the domain of the model?

(b) Use a graphing utility to graph the data and the model on the same viewing rectangle.

(c) For which year does the model most accurately estimate the actual data? During which year is it least accurate?

(d) Why would economists be concerned if this model remained valid in the future?

80. *Geometry* Corners of equal size are cut from a square with sides of length 8 meters (see figure).

(a) Write the area A of the resulting figure as a function of x. Determine the domain of the function.

(b) Use a graphing utility to graph the area function over its domain. Use the graph to find the range of the function.

(c) Identify the resulting figure for the maximum value of x in the domain of the function. What is the length of each side of the figure?

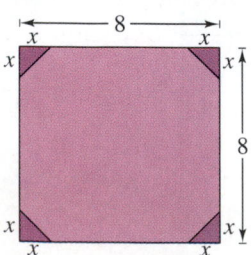

81. *Coordinate Axis Scale* It is necessary to graph the function $f(t)$, which models the specified data for the years 1980 through 1996, with $t = 0$ corresponding to 1980. State a possible scale for the vertical axis (e.g., hundreds, thousands, millions, etc.) of the graph and give a reason for your answer.

(a) $f(t)$ represents the average salary of college professors.

(b) $f(t)$ represents the U.S. population.

(c) $f(t)$ represents the percent of the civilian work force that is unemployed.

82. *Fluid Flow* The intake pipe of a 100-gallon tank has a flow rate of 10 gallons per minute, and two drainpipes have flow rates of 5 gallons per minute each. The figure shows the volume V of fluid in the tank as a function of time t. Determine the pipes in which the fluid is flowing in specific subintervals of the 1 hour of time shown on the graph. (There is more than one correct answer.)

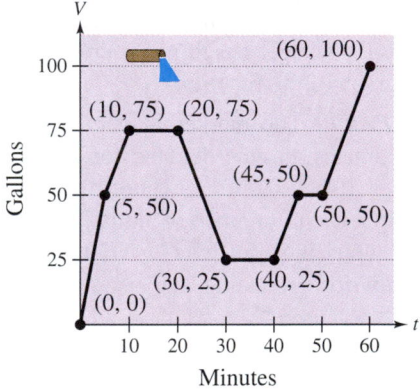

83. Prove that a function of the following form is odd.

$$y = a_{2n+1}x^{2n+1} + a_{2n-1}x^{2n-1} + \cdots + a_3x^3 + a_1x$$

84. Prove that a function of the following form is even.

$$y = a_{2n}x^{2n} + a_{2n-2}x^{2n-2} + \cdots + a_2x^2 + a_0$$

Review Solve the equations in Exercises 85–88 as a review of the skills and problem-solving techniques you learned in previous sections.

85. $x^2 - 10x = 0$

86. $100 - (x - 5)^2 = 0$

87. $x^3 + x = 0$

88. $16x^2 - 40x + 25 = 0$

2.4   *Translations and Combinations*

See Exercises 47 and 48 on pages 234 and 235 for examples of how combinations of functions can be used to analyze the costs of automobile upkeep from 1985 through 1991.

Shifting, Reflecting, and Stretching Graphs ❑ *Arithmetic Combinations of Functions* ❑ *Composition of Functions* ❑ *Applications*

Shifting, Reflecting, and Stretching Graphs

A computer animation of this concept appears in the *Interactive* CD-ROM.

Many functions have graphs that are simple transformations of the common graphs summarized on page 217. For example, you can obtain the graph of $h(x) = x^2 + 2$ by shifting the graph of $f(x) = x^2$ *up* two units, as shown in Figure 2.27. In function notation, h and f are related as follows.

$$h(x) = x^2 + 2 = f(x) + 2 \qquad\qquad \text{Upward shift of 2}$$

Similarly, you can obtain the graph of $g(x) = (x - 2)^2$ by shifting the graph of $f(x) = x^2$ to the *right* two units, as shown in Figure 2.28. In this case, the functions g and f have the following relationship.

$$g(x) = (x - 2)^2 = f(x - 2) \qquad\qquad \text{Right shift of 2}$$

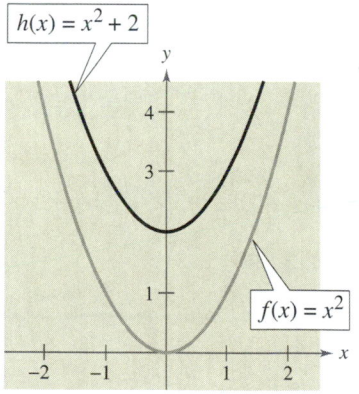

FIGURE 2.27

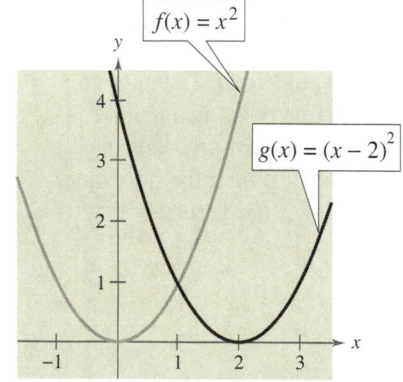

FIGURE 2.28

 VERTICAL AND HORIZONTAL SHIFTS

Let c be a positive real number. **Vertical and horizontal shifts** in the graph of $y = f(x)$ are represented as follows.

1. Vertical shift c units **upward:** $h(x) = f(x) + c$
2. Vertical shift c units **downward:** $h(x) = f(x) - c$
3. Horizontal shift c units to the **right:** $h(x) = f(x - c)$
4. Horizontal shift c units to the **left:** $h(x) = f(x + c)$

NOTE In items 3 and 4, be sure you see that $h(x) = f(x - c)$ corresponds to a *right* shift and $h(x) = f(x + c)$ corresponds to a *left* shift for $c > 0$. ▪▪

Some graphs can be obtained from a combination of vertical and horizontal shifts. This is demonstrated in Example 1(b).

NOTE   Vertical and horizontal shifts such as those shown in Example 1 generate a *family of functions*, each with the same shape but at different locations in the plane. ■ ■

EXAMPLE 1   Shifts in the Graph of a Function

Use the graph of $f(x) = x^3$ to sketch the graph of each function.

a. $g(x) = x^3 + 1$

b. $h(x) = (x + 2)^3 + 1$

Solution

Relative to the graph of $f(x) = x^3$, the graph of $g(x) = x^3 + 1$ is an upward shift of one unit, and the graph of $h(x) = (x + 2)^3 + 1$ involves a left shift of two units *and* an upward shift of one unit. The graphs of both functions are compared with the graph of $f(x) = x^3$ in Figure 2.29.

(a) 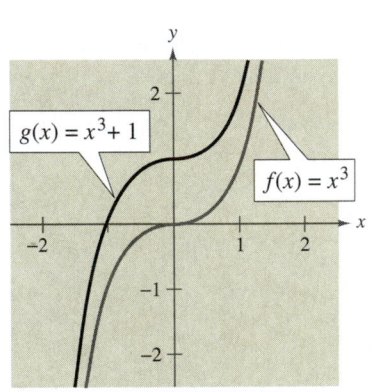   **(b)**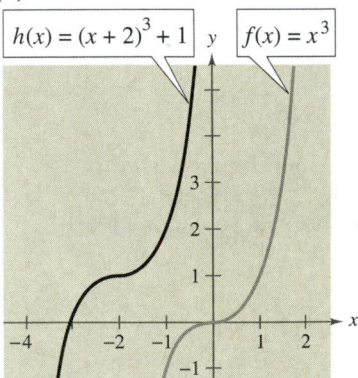

Study Tip

In part (b) of Figure 2.29, notice that the same result is obtained if the vertical shift precedes the horizontal shift *or* if the horizontal shift precedes the vertical shift.

FIGURE 2.29

Exploration

Graphing utilities are ideal tools for exploring translations of functions. Graph f, g, and h on the same screen. Before looking at the graphs, try to predict how the graphs of g and h relate to the graph of f.

a. $f(x) = x^2$, $g(x) = (x - 4)^2$, $h(x) = (x - 4)^2 + 3$

b. $f(x) = x^2$, $g(x) = (x + 1)^2$, $h(x) = (x + 1)^2 - 2$

c. $f(x) = x^2$, $g(x) = (x + 4)^2$, $h(x) = (x + 4)^2 + 2$

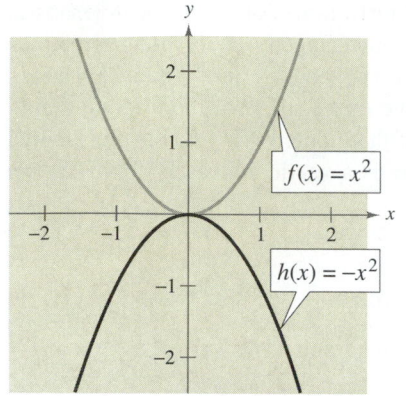

FIGURE 2.30

 A computer animation of this concept appears in the *Interactive* CD-ROM.

TECHNOLOGY

In the appendix, you will find programs for a variety of graphing calculator models that will give you practice working with reflections, horizontal shifts, and vertical shifts. These programs will sketch a graph of the function

$$y = R(x + H)^2 + V$$

where $R = \pm 1$, H is an integer between -6 and 6, and V is an integer between -3 and 3. Each time you run the program, different values of R, H, and V are possible. From the graph, you should be able to determine the values of R, H, and V. After you have determined the values, press the enter key to see the answer. (To look at the graph again, press the graph key.)

The second common type of transformation is a **reflection.** For instance, if you consider the x-axis to be a mirror, the graph of

$$h(x) = -x^2$$

is the mirror image (or reflection) of the graph of $f(x) = x^2$, as shown in Figure 2.30.

REFLECTIONS IN THE COORDINATE AXES

Reflections in the coordinate axes of the graph of $y = f(x)$ are represented as follows.

1. **Reflection in the x-axis:** $h(x) = -f(x)$
2. **Reflection in the y-axis:** $h(x) = f(-x)$

EXAMPLE 2 *Reflections and Shifts*

Compare the graphs of each function with the graph of $f(x) = \sqrt{x}$.

a. $g(x) = -\sqrt{x}$

b. $h(x) = \sqrt{-x}$

Solution

a. The graph of g is a reflection of the graph of f in the x-axis because

$$g(x) = -\sqrt{x} = -f(x).$$

b. The graph of h is a reflection of the graph of f in the y-axis because

$$h(x) = \sqrt{-x} = f(-x).$$

The graphs of both functions are shown in Figure 2.31.

(a)

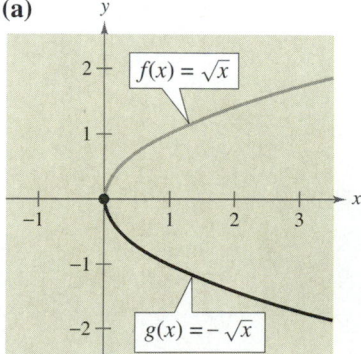

(b)

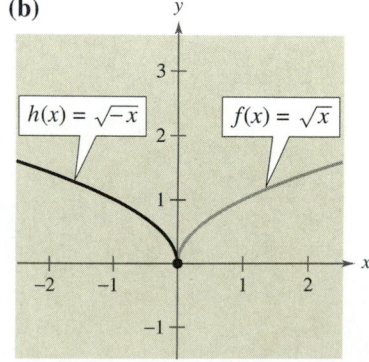

FIGURE 2.31

Horizontal shifts, vertical shifts, and reflections are **rigid** transformations because the basic shape of the graph is unchanged. These transformations change only the *position* of the graph in the *xy*-plane. **Nonrigid** transformations are those that cause a *distortion*—a change in the shape of the original graph. For instance, a nonrigid transformation of the graph of $y = f(x)$ is represented by $g(x) = cf(x)$, where the transformation is a **vertical stretch** if $c > 1$ and a **vertical shrink** if $0 < c < 1$.

Exploration

Sketch the graph of $f(x) = 2x^2$. Compare this graph with the graph of $h(x) = x^2$. Describe the effect of multiplying x^2 by a number greater than 1. Then graph $g(x) = \frac{1}{2}x^2$. Compare this with the graph of $h(x) = x^2$. Describe the effect of multiplying x^2 by a number less than 1. Can you think of an easy way to remember this generalization?

EXAMPLE 3 *Nonrigid Transformations*

Compare the graph of each function with the graph of $f(x) = |x|$.

a. $h(x) = 3|x|$

b. $g(x) = \frac{1}{3}|x|$

Solution

a. Relative to the graph of $f(x) = |x|$, the graph of

$$h(x) = 3|x| = 3f(x)$$

is a vertical stretch (multiply each *y*-value by 3) of the graph of *f*.

b. Similarly, the function

$$g(x) = \frac{1}{3}|x| = \frac{1}{3}f(x)$$

indicates that the graph of *g* is a vertical shrink of the graph of *f*.

The graphs of both functions are shown in Figure 2.32.

(a)

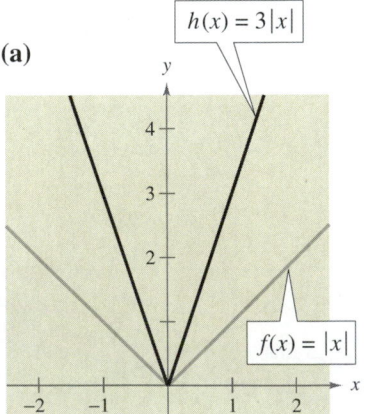

(b)

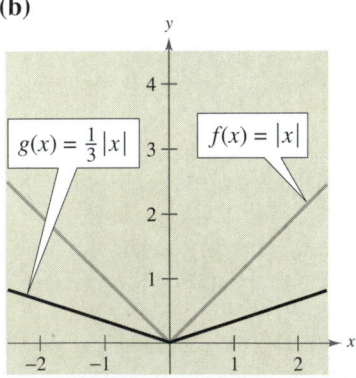

FIGURE 2.32

Arithmetic Combinations of Functions

Just as two real numbers can be combined by the operations of addition, subtraction, multiplication, and division to form other real numbers, two *functions* can be combined to create new functions. For example, the functions

$$f(x) = 2x - 3 \quad \text{and} \quad g(x) = x^2 - 1$$

can be combined to form the sum, difference, product, and quotient of f and g as follows.

$$f(x) + g(x) = (2x - 3) + (x^2 - 1) = x^2 + 2x - 4 \qquad \text{Sum}$$
$$f(x) - g(x) = (2x - 3) - (x^2 - 1) = -x^2 + 2x - 2 \qquad \text{Difference}$$
$$f(x)g(x) = (2x - 3)(x^2 - 1) = 2x^3 - 3x^2 - 2x + 3 \qquad \text{Product}$$
$$\frac{f(x)}{g(x)} = \frac{2x - 3}{x^2 - 1}, \quad x \neq \pm 1 \qquad \text{Quotient}$$

The domain of an arithmetic combination of functions f and g consists of all real numbers that are common to the domains of f and g. In the case of the quotient $f(x)/g(x)$, there is the further restriction that $g(x) \neq 0$.

SUM, DIFFERENCE, PRODUCT, AND QUOTIENT OF FUNCTIONS

Let f and g be two functions with overlapping domains. Then, for all x common to both domains, the **sum, difference, product,** and **quotient** of f and g are defined as follows.

1. *Sum:* $\qquad\qquad (f + g)(x) = f(x) + g(x)$
2. *Difference:* $\qquad (f - g)(x) = f(x) - g(x)$
3. *Product:* $\qquad\quad (fg)(x) = f(x) \cdot g(x)$
4. *Quotient:* $\qquad\quad \left(\dfrac{f}{g}\right)(x) = \dfrac{f(x)}{g(x)}, \qquad g(x) \neq 0$

EXAMPLE 4   *Finding the Sum of Two Functions*

Given $f(x) = 2x + 1$ and $g(x) = x^2 + 2x - 1$, find $(f + g)(x)$.

Solution

$$\begin{aligned}
(f + g)(x) &= f(x) + g(x) \\
&= (2x + 1) + (x^2 + 2x - 1) \\
&= x^2 + 4x
\end{aligned}$$

EXAMPLE 5 *Finding the Difference of Two Functions*

Given $f(x) = 2x + 1$ and $g(x) = x^2 + 2x - 1$, find $(f - g)(x)$. Then evaluate the difference when $x = 2$.

Solution

The difference of f and g is given by

$$
\begin{aligned}
(f - g)(x) &= f(x) - g(x) \\
&= (2x + 1) - (x^2 + 2x - 1) \\
&= -x^2 + 2.
\end{aligned}
$$

When $x = 2$, the value of this difference is

$$
(f - g)(2) = -(2)^2 + 2 = -2.
$$

In Examples 4 and 5, both f and g have domains that consist of all real numbers. Thus, the domains of $(f + g)$ and $(f - g)$ are also the set of all real numbers. Remember that any restrictions on the domains of f and g must be considered when forming the sum, difference, product, or quotient of f and g.

EXAMPLE 6 *Finding the Quotient of Two Functions*

Find the domains of $(f/g)(x)$ and $(g/f)(x)$ for the functions

$$
f(x) = \sqrt{x} \qquad \text{and} \qquad g(x) = \sqrt{4 - x^2}.
$$

Solution

The quotient of f and g is given by

$$
\left(\frac{f}{g}\right)(x) = \frac{f(x)}{g(x)} = \frac{\sqrt{x}}{\sqrt{4 - x^2}}
$$

and the quotient of g and f is given by

$$
\left(\frac{g}{f}\right)(x) = \frac{g(x)}{f(x)} = \frac{\sqrt{4 - x^2}}{\sqrt{x}}.
$$

The domain of f is $[0, \infty)$ and the domain of g is $[-2, 2]$. The interseetion of these domains is $[0, 2]$. Thus, the domains of f/g and g/f are as follows.

$$
\text{Domain of } \frac{f}{g}: [0, 2) \qquad \text{Domain of } \frac{g}{f}: (0, 2]
$$

Can you see why these two domains differ slightly?

Composition of Functions

Another way of combining two functions is to form the **composition** of one with the other. For instance, if $f(x) = x^2$ and $g(x) = x + 1$, the composition of f with g is given by

$$f(g(x)) = f(x + 1) = (x + 1)^2.$$

This composition is denoted as $f \circ g$.

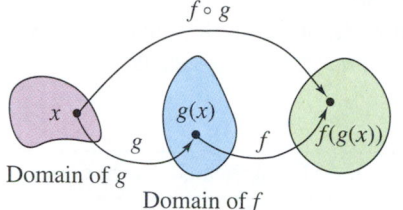

$f \circ g$

Domain of g

Domain of f

FIGURE 2.33

> ■ **DEFINITION OF COMPOSITION OF TWO FUNCTIONS**
> The **composition** of the function f with the function g is given by
>
> $$(f \circ g)(x) = f(g(x)).$$
>
> The domain of $(f \circ g)$ is the set of all x in the domain of g such that $g(x)$ is in the domain of f. (See Figure 2.33.)

EXAMPLE 7 *Composition of Functions*

Given $f(x) = x + 2$ and $g(x) = 4 - x^2$, find the following.

a. $(f \circ g)(x)$

b. $(g \circ f)(x)$

Solution

a. The composition of f with g is as follows.

$$
\begin{aligned}
(f \circ g)(x) &= f(g(x)) && \text{Definition of } f \circ g \\
&= f(4 - x^2) && \text{Definition of } g(x) \\
&= (4 - x^2) + 2 && \text{Definition of } f(x) \\
&= -x^2 + 6 && \text{Simplify.}
\end{aligned}
$$

b. The composition of g with f is as follows.

$$
\begin{aligned}
(g \circ f)(x) &= g(f(x)) && \text{Definition of } g \circ f \\
&= g(x + 2) && \text{Definition of } f(x) \\
&= 4 - (x + 2)^2 && \text{Definition of } g(x) \\
&= 4 - (x^2 + 4x + 4) && \text{Expand.} \\
&= -x^2 - 4x && \text{Simplify.}
\end{aligned}
$$

Note that, in this case, $(f \circ g)(x) \neq (g \circ f)(x)$.

NOTE The following tables of values help illustrate the composition of the functions f and g given in Example 7.

x	0	1	2	3
$g(x)$	4	3	0	−5

$g(x)$	4	3	0	−5
$f(g(x))$	6	5	2	−3

x	0	1	2	3
$f(g(x))$	6	5	2	−3

Note that the first two tables can be combined (or "composed") to produce the values given in the third table. ■■

Applications

EXAMPLE 8   *Political Makeup of the U.S. Senate*

Real Life

Consider three functions R, D, and I that represent the numbers of Republicans, Democrats, and Independents in the U.S. Senate from 1965 to 1995. Sketch the graphs of R, D, and I, and the sum of R, D, and I, in the same coordinate plane. The numbers of Republicans and Democrats in the Senate are shown below.   (Source: Secretary of the Senate)

Year	Republicans	Democrats
1965	32	68
1967	36	64
1969	43	57
1971	44	54
1973	42	56
1975	37	60
1977	38	61
1979	41	58

Year	Republicans	Democrats
1981	53	46
1983	54	46
1985	53	47
1987	45	55
1989	45	55
1991	44	56
1993	43	57
1995	53	47

Solution

The graphs of R, D, and I are shown in Figure 2.34. Note that the sum of R, D, and I is the constant function $R + D + I = 100$. This follows from the fact that the number of senators in the United States is 100 (two from each state).

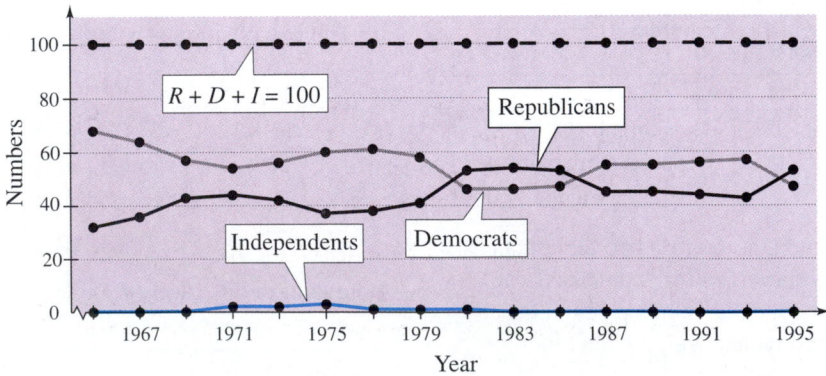

FIGURE 2.34

■

Real Life

EXAMPLE 9 *Bacteria Count*

The number of bacteria in a refrigerated food is given by

$$N(T) = 20T^2 - 80T + 500, \qquad 2 \le T \le 14$$

where T is the temperature of the food. When the food is removed from refrigeration, the temperature is given by

$$T(t) = 4t + 2, \qquad 0 \le t \le 3$$

where t is the time in hours. Find (a) the composite $N(T(t))$ and interpret its meaning in context and (b) the time when the bacteria count reaches 2000.

Solution

a. $\begin{aligned} N(T(t)) &= 20(4t + 2)^2 - 80(4t + 2) + 500 \\ &= 20(16t^2 + 16t + 4) - 320t - 160 + 500 \\ &= 320t^2 + 320t + 80 - 320t - 160 + 500 \\ &= 320t^2 + 420 \end{aligned}$

NOTE If you solve the equation in part (b) of Example 9, you will find that $t \approx \pm 2.2$. However, the negative value is rejected because it is not in the domain of the composite function. ■■

The composite function represents the number of bacteria in the food as a function of time.

b. The bacteria count will reach 2000 when $320t^2 + 420 = 2000$. Solve this equation to find that the count will reach 2000 when $t \approx 2.2$ hours.

■

GROUP ACTIVITY

ANALYZING COMBINATIONS OF FUNCTIONS

a. Use the graphs of f and $f + g$ to make a table showing the values of $g(x)$ when $x = 1, 2, 3, 4, 5,$ and 6. Explain your reasoning.

b. Use the graphs of f and $f - h$ to make a table showing the values of $h(x)$ when $x = 1, 2, 3, 4, 5,$ and 6. Explain your reasoning.

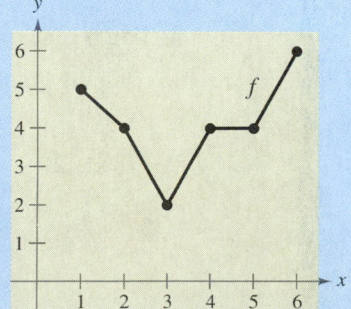

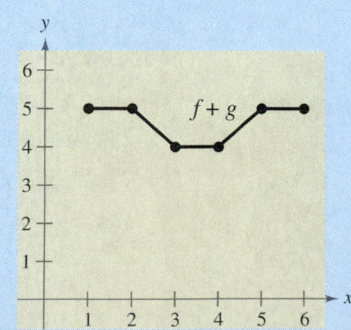

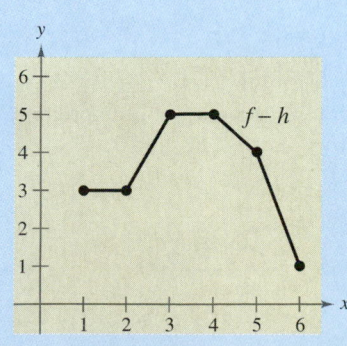

WARM UP

Perform the operations and simplify.

1. $\dfrac{1}{x} + \dfrac{1}{1-x}$

2. $\dfrac{2}{x+3} - \dfrac{2}{x-3}$

3. $\dfrac{3}{x-2} - \dfrac{2}{x(x-2)}$

4. $\dfrac{x}{x-5} + \dfrac{1}{3}$

5. $(x-1)\left(\dfrac{1}{\sqrt{x^2-1}}\right)$

6. $\left(\dfrac{x}{x^2-4}\right)\left(\dfrac{x^2-x-2}{x^2}\right)$

7. $(x^2-4) \div \left(\dfrac{x+2}{5}\right)$

8. $\left(\dfrac{x}{x^2+3x-10}\right) \div \left(\dfrac{x^2+3x}{x^2+6x+5}\right)$

9. $\dfrac{(1/x)+5}{3-(1/x)}$

10. $\dfrac{(x/4)-(4/x)}{x-4}$

2.4 Exercises

1. Sketch (on the same set of coordinate axes) a graph of f for $c = -2, 0,$ and 2.

   (a) $f(x) = x^3 + c$

   (b) $f(x) = (x-c)^3$

2. Sketch (on the same set of coordinate axes) a graph of f for $c = -2, 0,$ and 2.

   (a) $f(x) = x^2 + c$

   (b) $f(x) = (x-c)^2$

3. Sketch (on the same set of coordinate axes) a graph of f for $c = -1, 1,$ and 3.

   (a) $f(x) = |x| + c$

   (b) $f(x) = |x-c|$

   (c) $f(x) = |x+4| + c$

4. Sketch (on the same set of coordinate axes) a graph of f for $c = -3, -1, 1,$ and 3.

   (a) $f(x) = \sqrt{x} + c$

   (b) $f(x) = \sqrt{x-c}$

   (c) $f(x) = \sqrt{x-3} + c$

5. Use the graph of f (see figure) to sketch the graphs.

   (a) $y = f(x) + 2$ 　　　(b) $y = -f(x)$

   (c) $y = f(x-2)$ 　　　(d) $y = f(x+3)$

   (e) $y = f(2x)$ 　　　(f) $y = f(-x)$

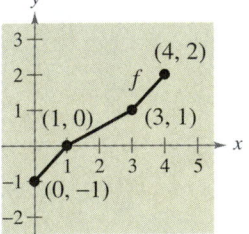 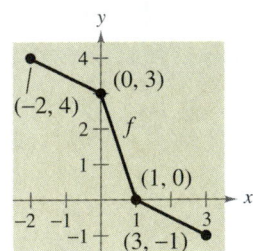

FIGURE FOR 5 　　　　FIGURE FOR 6

6. Use the graph of f (see figure) to sketch the graphs.

   (a) $y = f(x) - 1$ 　　　(b) $y = f(x+1)$

   (c) $y = f(x-1)$ 　　　(d) $y = -f(x-2)$

   (e) $y = f(-x)$ 　　　(f) $y = \frac{1}{2}f(x)$

7. Use the graph of $f(x) = x^2$ to write formulas for the functions whose graphs are shown below.

(a)

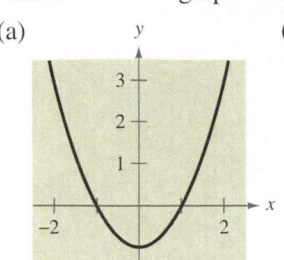

(b)

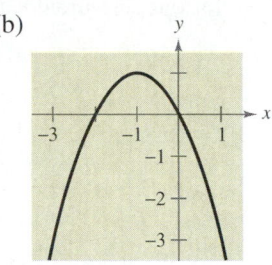

8. Use the graph of $f(x) = x^3$ to write formulas for the functions whose graphs are shown below.

(a)

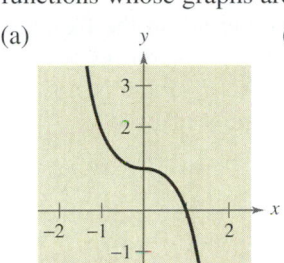

(b)

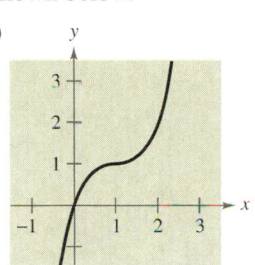

In Exercises 9–14, identify the common function and the transformation shown in the graph. Write the formula for the graphed function.

9.

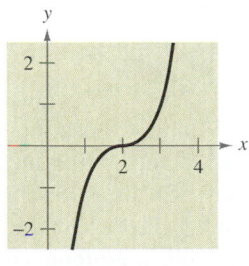

10.

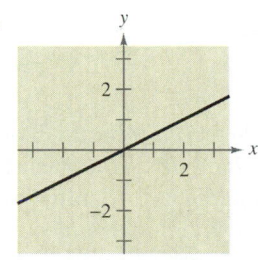

11.

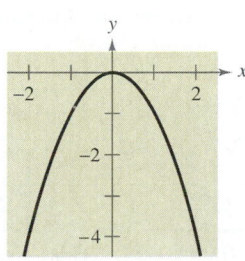

12.

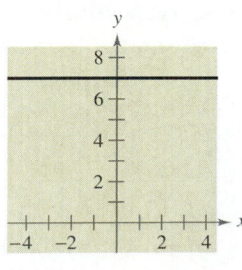

13.

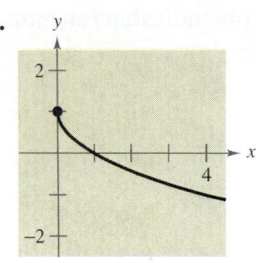

14.

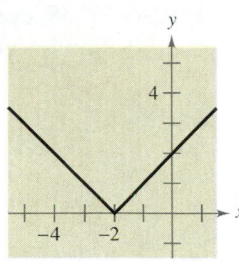

In Exercises 15–18, use the graphs of f and g to graph $h(x) = (f + g)(x)$.

15.

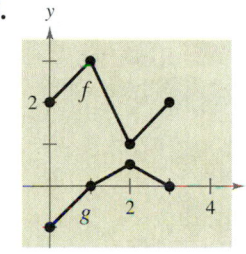

16.

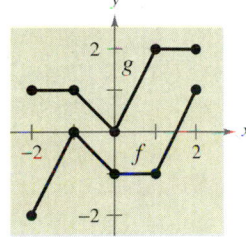

17.

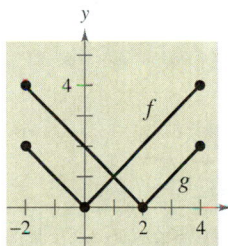

18.

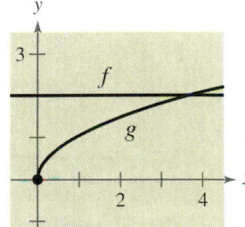

In Exercises 19–26, find (a) $(f + g)(x)$, (b) $(f - g)(x)$, (c) $(fg)(x)$, and (d) $(f / g)(x)$. What is the domain of f / g?

19. $f(x) = x + 1$, $g(x) = x - 1$

20. $f(x) = 2x - 5$, $g(x) = 1 - x$

21. $f(x) = x^2$, $g(x) = 1 - x$

22. $f(x) = 2x - 5$, $g(x) = 5$

23. $f(x) = x^2 + 5$, $g(x) = \sqrt{1 - x}$

24. $f(x) = \sqrt{x^2 - 4}$, $g(x) = \dfrac{x^2}{x^2 + 1}$

25. $f(x) = \dfrac{1}{x}$, $g(x) = \dfrac{1}{x^2}$

26. $f(x) = \dfrac{x}{x + 1}$, $g(x) = x^3$

In Exercises 27–38, evaluate the indicated function for $f(x) = x^2 + 1$ and $g(x) = x - 4$.

27. $(f + g)(3)$

28. $(f - g)(-2)$

29. $(f - g)(0)$

30. $(f + g)(1)$

31. $(f - g)(2t)$

32. $(f + g)(t - 1)$

33. $(fg)(4)$

34. $(fg)(-6)$

35. $\left(\dfrac{f}{g}\right)(5)$

36. $\left(\dfrac{f}{g}\right)(0)$

37. $\left(\dfrac{f}{g}\right)(-1) - g(3)$

38. $(2f)(5)$

In Exercises 39–42, graph the functions f, g, and $f + g$ on the same set of coordinate axes.

39. $f(x) = \dfrac{1}{2}x$, $g(x) = x - 1$

40. $f(x) = \dfrac{1}{3}x$, $g(x) = -x + 4$

41. $f(x) = x^2$, $g(x) = -2x$

42. $f(x) = 4 - x^2$, $g(x) = x$

Graphical Reasoning   In Exercises 43 and 44, use a graphing utility to sketch the graphs of f, g, and $f + g$ on the same viewing rectangle. Which function contributes most to the magnitude of the sum when $0 \le x \le 2$? Which function contributes most to the magnitude of the sum when $x > 5$?

43. $f(x) = 3x$, $g(x) = -\dfrac{x^3}{10}$

44. $f(x) = \dfrac{x}{2}$, $g(x) = \sqrt{x}$

45. *Stopping Distance*   While traveling in a car at x miles per hour, you are required to stop quickly to avoid an accident. The distance the car travels during your reaction time is given by $R(x) = \frac{3}{4}x$. The distance traveled while you are braking is given by

$$B(x) = \tfrac{1}{15}x^2.$$

Find the function giving total stopping distance T. Graph the functions R, B, and T on the same set of coordinate axes for $0 \le x \le 60$.

46. *Comparing Sales*   You own two restaurants. From 1990 to 1995, the sales R_1 (in thousands of dollars) for one restaurant can be modeled by

$$R_1 = 480 - 8t - 0.8t^2, \qquad t = 0, 1, 2, 3, 4, 5$$

where $t = 0$ represents 1990. During the same 6-year period, the sales R_2 (in thousands of dollars) for the second restaurant can be modeled by

$$R_2 = 254 + 0.78t, \qquad t = 0, 1, 2, 3, 4, 5.$$

(a) Write a function that represents the total sales for the two restaurants. Use a graphing utility to graph the total sales function.

(b) Use the *stacked bar graph* in the figure, which represents the total sales during the 6-year period, to determine whether the total sales have been increasing or decreasing.

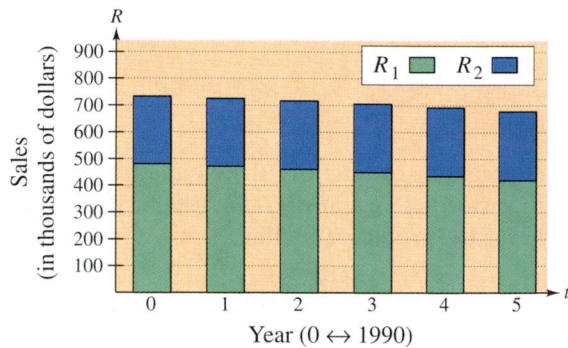

Data Analysis   In Exercises 47 and 48, use the table, which gives the variable costs for operating an automobile in the United States for the years 1985 through 1991. The variables y_1, y_2, and y_3 represent the costs in cents per mile for gas and oil, maintenance, and tires, respectively.   (Source: American Automobile Manufacturers Association)

Year	1985	1986	1987	1988	1989	1990	1991
y_1	6.16	4.48	4.80	5.20	5.20	5.40	6.70
y_2	1.23	1.37	1.60	1.60	1.90	2.10	2.20
y_3	0.65	0.67	0.80	0.80	0.80	0.90	0.90

47. Create a stacked bar graph for the data.

48. Mathematical models for the data are given by

$$y_1 = 0.16t^2 - 2.43t + 13.96$$

$$y_2 = 0.17t + 0.38$$

$$y_3 = 0.04t + 0.44$$

where $t = 5$ represents 1985. Use a graphing utility to graph y_1, y_2, y_3, and $y_1 + y_2 + y_3$ on the same viewing rectangle. Use the model to estimate the total variable cost per mile in 1995.

49. *Graphical Reasoning* An electronically controlled thermostat in a home is programmed to automatically lower the temperature during the night (see figure). The temperature in the house T, in degrees Fahrenheit, is given in terms of t, the time in hours on a 24-hour clock.

(a) Explain why T is a function of t.

(b) Approximate $T(4)$ and $T(15)$.

(c) Suppose the thermostat were reprogrammed to produce a temperature H where $H(t) = T(t - 1)$. How would this change the temperature?

(d) Suppose the thermostat were reprogrammed to produce a temperature H where $H(t) = T(t) - 1$. How would this change the temperature in the house?

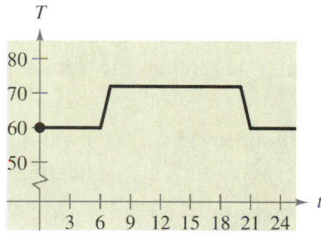

50. *Think About It* Write a piecewise-defined function that represents the graph in Exercise 49.

In Exercises 51–54, find (a) $f \circ g$, (b) $g \circ f$, and (c) $f \circ f$.

51. $f(x) = x^2$, $g(x) = x - 1$

52. $f(x) = \sqrt[3]{x - 1}$, $g(x) = x^3 + 1$

53. $f(x) = 3x + 5$, $g(x) = 5 - x$

54. $f(x) = x^3$, $g(x) = \dfrac{1}{x}$

In Exercises 55–62, find (a) $f \circ g$ and (b) $g \circ f$.

55. $f(x) = \sqrt{x + 4}$, $g(x) = x^2$

56. $f(x) = \sqrt[3]{x - 1}$, $g(x) = x^3 + 1$

57. $f(x) = \frac{1}{3}x - 3$, $g(x) = 3x + 1$

58. $f(x) = x^4$, $g(x) = x^4$

59. $f(x) = \sqrt{x}$, $g(x) = \sqrt{x}$

60. $f(x) = 2x - 3$, $g(x) = 2x - 3$

61. $f(x) = |x|$, $g(x) = x + 6$

62. $f(x) = x^{2/3}$, $g(x) = x^6$

In Exercises 63–66, use the graphs of f and g (see figures) to evaluate the functions.

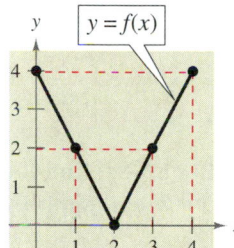
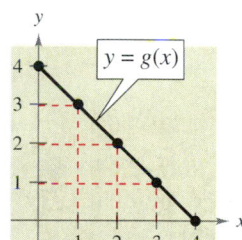

63. (a) $(f + g)(3)$ (b) $\left(\dfrac{f}{g}\right)(2)$

64. (a) $(f - g)(1)$ (b) $(fg)(4)$

65. (a) $(f \circ g)(2)$ (b) $(g \circ f)(2)$

66. (a) $(f \circ g)(1)$ (b) $(g \circ f)(3)$

Graphical Reasoning In Exercises 67–70, use the graph of f in the figure to sketch the graph of g.

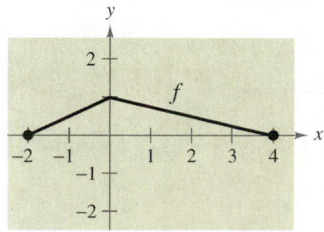

67. $g(x) = f(x) + 2$ **68.** $g(x) = f(x) - 1$

69. $g(x) = f(-x)$ **70.** $g(x) = -2f(x)$

In Exercises 71–76, find two functions f and g such that $(f \circ g)(x) = h(x)$. (There are many correct answers.)

71. $h(x) = (2x + 1)^2$ **72.** $h(x) = (1 - x)^3$

73. $h(x) = \sqrt[3]{x^2 - 4}$ **74.** $h(x) = \sqrt{9 - x}$

75. $h(x) = \dfrac{1}{x + 2}$ **76.** $h(x) = \dfrac{4}{(5x + 2)^2}$

In Exercises 77–80, determine the domain of (a) f, (b) g, and (c) $f \circ g$.

77. $f(x) = \sqrt{x}$, $g(x) = x^2 + 1$

78. $f(x) = \dfrac{1}{x}$, $g(x) = x + 3$

79. $f(x) = \dfrac{3}{x^2 - 1}$, $g(x) = x + 1$

80. $f(x) = 2x + 3$, $g(x) = \dfrac{x}{2}$

Average Rate of Change **In Exercises 81–84, find the difference quotient $[f(x + h) - f(x)]/h$ and simplify your answer.**

81. $f(x) = 3x - 4$ **82.** $f(x) = 1 - x^2$

83. $f(x) = \dfrac{4}{x}$ **84.** $f(x) = \sqrt{2x + 1}$

85. *Area* A square concrete foundation was prepared as a base for a cylindrical tank (see figure).

(a) Express the radius r of the tank as a function of the length x of the sides of the square.

(b) Express the area A of the circular base of the tank as a function of the radius r.

(c) Find and interpret $(A \circ r)(x)$.

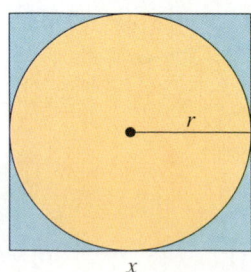

86. *Ripples* A pebble is dropped into a calm pond, causing ripples in the form of concentric circles. The radius (in feet) of the outer ripple is given by $r(t) = 0.6t$, where t is the time in seconds after the pebble strikes the water. The area of the circle is given by the function $A(r) = \pi r^2$. Find and interpret $(A \circ r)(t)$.

87. *Cost* The weekly cost of producing x units in a manufacturing process is given by the function

$$C(x) = 60x + 750.$$

The number of units produced in t hours is given by $x(t) = 50t$. Find and interpret $(C \circ x)(t)$.

88. *Think About It* You are a sales representative for an automobile manufacturer. You are paid an annual salary plus a bonus of 3% of your sales over $500,000. Consider the two functions

$$f(x) = x - 500{,}000 \quad \text{and} \quad g(x) = 0.03x.$$

If x is greater than $500,000, which of the following represents your bonus? Explain your reasoning.

(a) $f(g(x))$

(b) $g(f(x))$

89. *Exploration* The suggested retail price of a new car is p dollars. The dealership advertised a factory rebate of $1200 and an 8% discount.

(a) Write a function R in terms of p, giving the cost of the car after receiving the rebate from the factory.

(b) Write a function S in terms of p, giving the cost of the car after receiving the dealership discount.

(c) Form the composite functions $(R \circ S)(p)$ and $(S \circ R)(p)$ and interpret each.

(d) Find $(R \circ S)(18{,}400)$ and $(S \circ R)(18{,}400)$. Which yields the smaller cost for the car? Explain.

90. Prove that the product of two odd functions is an even function, and that the product of two even functions is an even function.

91. *Conjecture* Use examples to hypothesize whether the product of an odd function and an even function is even or odd. Then prove your hypothesis.

2.5 *Inverse Functions*

See Exercise 81 on page 247 for an example of how the inverse of a function can be used to analyze the behavior of a diesel engine.

See Exercise 81 on page 247 for an example of how the inverse of a function can be used to analyze the behavior of a diesel engine.

The Inverse of a Function ▫ *Finding the Inverse of a Function* ▫ *The Graph of the Inverse of a Function*

The Inverse of a Function

Recall from Section 2.2 that a function can be represented by a set of ordered pairs. For instance, the function $f(x) = x + 4$ from the set $A = \{1, 2, 3, 4\}$ to the set $B = \{5, 6, 7, 8\}$ can be written as follows.

$$f(x) = x + 4 : \{(1, 5), (2, 6), (3, 7), (4, 8)\}$$

By interchanging the first and second coordinates of each of these ordered pairs, you can form the **inverse function** of f, which is denoted by f^{-1}. It is a function from the set B to the set A, and can be written as follows.

$$f^{-1}(x) = x - 4 : \{(5, 1), (6, 2), (7, 3), (8, 4)\}$$

Note that the domain of f is equal to the range of f^{-1}, and vice versa, as shown in Figure 2.35. Also note that the functions f and f^{-1} have the effect of "undoing" each other. In other words, when you form the composition of f with f^{-1} or the composition of f^{-1} with f, you obtain the identity function.

$$f(f^{-1}(x)) = f(x - 4) = (x - 4) + 4 = x$$
$$f^{-1}(f(x)) = f^{-1}(x + 4) = (x + 4) - 4 = x$$

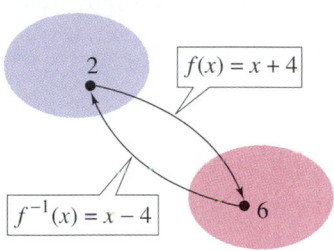

$f(x) = x + 4$

$f^{-1}(x) = x - 4$

FIGURE 2.35

NOTE A table of values can help you understand inverse functions. For instance, the following table shows several values of the function f in Example 1.

x	-2	-1	0	1	2
$f(x)$	-8	-4	0	4	8

By interchanging the table's rows, you obtain values of the inverse function f^{-1}.

x	-8	-4	0	4	8
$f^{-1}(x)$	-2	-1	0	1	2

In the first table, each output is 4 times the input, and in the second table, each output is $\frac{1}{4}$ the input. ■■

EXAMPLE 1 *Finding Inverse Functions Informally*

Find the inverse of $f(x) = 4x$. The verify that both $f(f^{-1}(x))$ and $f^{-1}(f(x))$ are equal to the identity function.

Solution

The given function *multiplies* each input by 4. To "undo" this function, you need to *divide* each input by 4. Thus, the inverse function of $f(x) = 4x$ is

$$f^{-1}(x) = \frac{x}{4}.$$

You can verify that both $f(f^{-1}(x))$ and $f^{-1}(f(x))$ are equal to the identity function as follows.

$$f(f^{-1}(x)) = f\left(\frac{x}{4}\right) = 4\left(\frac{x}{4}\right) = x$$

$$f^{-1}(f(x)) = f^{-1}(4x) = \frac{4x}{4} = x$$

EXAMPLE 2   *Finding Inverse Functions Informally*

Find the inverse of $f(x) = x - 6$. Then verify that both $f(f^{-1}(x))$ and $f^{-1}(f(x))$ are equal to the identity function.

Solution

The given function *subtracts* 6 from each input. To "undo" this function, you need to *add* 6 to each input. Thus, the inverse function of $f(x) = x - 6$ is

$$f^{-1}(x) = x + 6.$$

You can verify that both $f(f^{-1}(x))$ and $f^{-1}(f(x))$ are equal to the identity function as follows.

$$f(f^{-1}(x)) = f(x + 6) = (x + 6) - 6 = x$$
$$f^{-1}(f(x)) = f^{-1}(x - 6) = (x - 6) + 6 = x$$

The formal definition of the inverse of a function is as follows.

DEFINITION OF THE INVERSE OF A FUNCTION

Let f and g be two functions such that

$$f(g(x)) = x \qquad \text{for every } x \text{ in the domain of } g$$

and

$$g(f(x)) = x \qquad \text{for every } x \text{ in the domain of } f.$$

Under these conditions, the function g is the **inverse** of the function f. The function g is denoted by f^{-1} (read "f-inverse"). Thus,

$$f(f^{-1}(x)) = x \qquad \text{and} \qquad f^{-1}(f(x)) = x.$$

The domain of f must be equal to the range of f^{-1}, and the range of f must be equal to the domain of f^{-1}.

NOTE   Don't be confused by the use of -1 to denote the inverse function f^{-1}. In this text, whenever we write f^{-1}, we will *always* be referring to the inverse of the function f and *not* to the reciprocal of $f(x)$. ■■

    If the function g is the inverse of the function f, it must also be true that the function f is the inverse of the function g. For this reason, you can say that the functions f and g are *inverses of each other*.

EXAMPLE 3 **Verifying Inverse Functions**

Show that the functions are inverses of each other.

$$f(x) = 2x^3 - 1 \quad \text{and} \quad g(x) = \sqrt[3]{\frac{x+1}{2}}$$

Solution

$$f(g(x)) = f\left(\sqrt[3]{\frac{x+1}{2}}\right) = 2\left(\sqrt[3]{\frac{x+1}{2}}\right)^3 - 1$$

$$= 2\left(\frac{x+1}{2}\right) - 1$$

$$= x + 1 - 1$$

$$= x$$

$$g(f(x)) = g(2x^3 - 1) = \sqrt[3]{\frac{(2x^3 - 1) + 1}{2}}$$

$$= \sqrt[3]{\frac{2x^3}{2}}$$

$$= \sqrt[3]{x^3}$$

$$= x$$

EXAMPLE 4 **Verifying Inverse Functions**

Which of the functions is the inverse of $f(x) = \dfrac{5}{x-2}$?

$$g(x) = \frac{x-2}{5} \quad \text{and} \quad h(x) = \frac{5}{x} + 2$$

Solution

By forming the composition of f with g, you have

$$f(g(x)) = f\left(\frac{x-2}{5}\right) = \frac{5}{[(x-2)/5] - 2} = \frac{25}{x - 12} \neq x.$$

Because this composition is not equal to the identity function x, it follows that g *is not* the inverse of f. By forming the composition of f with h, you have

$$f(h(x)) = f\left(\frac{5}{x} + 2\right) = \frac{5}{(5/x) + 2 - 2} = \frac{5}{5/x} = x.$$

Thus, it appears that h *is* the inverse of f. You can confirm this by showing that the composition of h with f is also equal to the identity function. (Try doing this.)

Finding the Inverse of a Function

For simple functions (such as the ones in Examples 1 and 2) you can find inverse functions by inspection. For more complicated functions, however, it is best to use the following guidelines. The key step in these guidelines is Step 2—interchanging the roles of x and y. This step corresponds to the fact that inverse functions have ordered pairs with the coordinates reversed.

Study Tip

Note in Step 3 of the guidelines for finding the inverse of a function that it is possible that a function has no inverse. For instance, the function $f(x) = x^2$ has no inverse function.

> **FINDING THE INVERSE OF A FUNCTION**
>
> 1. In the equation for $f(x)$, replace $f(x)$ by y.
> 2. Interchange the roles of x and y.
> 3. If the new equation does not represent y as a function of x, the function f does not have an inverse function. If the new equation does represent y as a function of x, solve the new equation for y.
> 4. Replace y by $f^{-1}(x)$.
> 5. Verify that f and f^{-1} are inverses of each other by showing that the domain of f is equal to the range of f^{-1}, the range of f is equal to the domain of f^{-1}, and $f(f^{-1}(x)) = x = f^{-1}(f(x))$.

EXAMPLE 5   *Finding the Inverse of a Function*

Find the inverse of $f(x) = \dfrac{5 - 3x}{2}$.

Solution

$$f(x) = \frac{5 - 3x}{2} \qquad \text{Given function}$$

$$y = \frac{5 - 3x}{2} \qquad \text{Replace } f(x) \text{ by } y.$$

$$x = \frac{5 - 3y}{2} \qquad \text{Interchange } x \text{ and } y.$$

$$2x = 5 - 3y \qquad \text{Multiply both sides by 2.}$$

$$3y = 5 - 2x \qquad \text{Isolate the } y\text{-term.}$$

$$y = \frac{5 - 2x}{3} \qquad \text{Solve for } y.$$

$$f^{-1}(x) = \frac{5 - 2x}{3} \qquad \text{Replace } y \text{ by } f^{-1}(x).$$

Note that both f and f^{-1} have domains and ranges that consist of the entire set of real numbers. Check that $f(f^{-1}(x)) = x$ and $f^{-1}(f(x)) = x$.

The Graph of the Inverse of a Function

The graphs of a function f and its inverse f^{-1} are related to each other in the following way. If the point (a, b) lies on the graph of f, then the point (b, a) must lie on the graph of f^{-1}, and vice versa. This means that the graph of f^{-1} is a *reflection* of the graph of f in the line $y = x$, as shown in Figure 2.36.

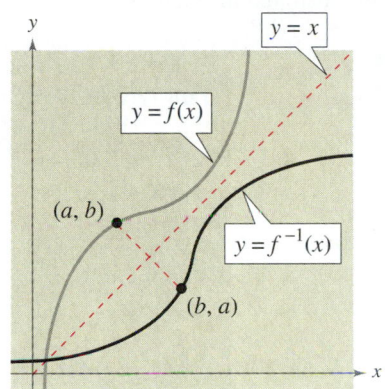

FIGURE 2.36

EXAMPLE 6   *The Graphs of f and f^{-1}*

Sketch the graphs of the inverse functions

$$f(x) = 2x - 3 \qquad \text{and} \qquad f^{-1}(x) = \tfrac{1}{2}(x + 3)$$

on the same rectangular coordinate system and show that the graphs are reflections of each other in the line $y = x$.

Solution

The graphs of f and f^{-1} are shown in Figure 2.37. Visually, it appears that the graphs are reflections of each other in the line $y = x$. You can further verify this reflective property by testing a few points on each graph. Note in the following list that if the point (a, b) is on the graph of f, the point (b, a) is on the graph of f^{-1}.

Graph of $f(x) = 2x - 3$	*Graph of $f^{-1}(x) = \tfrac{1}{2}(x + 3)$*
$(-1, -5)$	$(-5, -1)$
$(0, -3)$	$(-3, 0)$
$(1, -1)$	$(-1, 1)$
$(2, 1)$	$(1, 2)$
$(3, 3)$	$(3, 3)$

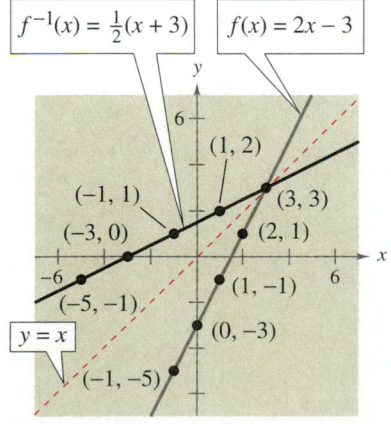

FIGURE 2.37

In the study tip on page 240, we mentioned that the function

$$f(x) = x^2$$

has no inverse. What we really meant is that *assuming the domain of f is the entire real line,* the function $f(x) = x^2$ has no inverse. If, however, you restrict the domain of f to the nonnegative real numbers, f does have an inverse, as demonstrated in Example 7.

EXAMPLE 7   *The Graphs of f and f^{-1}*

Sketch the graphs of the inverse functions

$$f(x) = x^2, \quad x \geq 0, \quad \text{and} \quad f^{-1}(x) = \sqrt{x}$$

on the same rectangular coordinate system and show that the graphs are reflections of each other in the line $y = x$.

Solution

The graphs of f and f^{-1} are shown in Figure 2.38. Visually, it appears that the graphs are reflections of each other in the line $y = x$. You can further verify this reflective property by testing a few points on each graph. Note in the following list that if the point (a, b) is on the graph of f, the point (b, a) is on the graph of f^{-1}.

Graph of $f(x) = x^2, \quad x \geq 0$	*Graph of* $f^{-1}(x) = \sqrt{x}$
$(0, 0)$	$(0, 0)$
$(1, 1)$	$(1, 1)$
$(2, 4)$	$(4, 2)$
$(3, 9)$	$(9, 3)$

Try showing that $f(f^{-1}(x)) = x$ and $f^{-1}(f(x)) = x$.

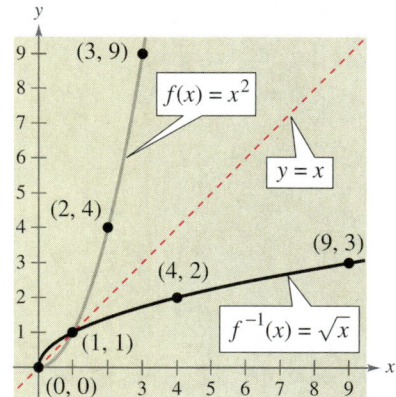

FIGURE 2.38

The guidelines for finding the inverse of a function, on page 240, include an *algebraic* test for determining whether a function has an inverse. The reflective property of the graphs of inverse functions gives you a nice *geometric* test for determining whether a function has an inverse. This test is called the **Horizontal Line Test** for inverse functions.

> ### HORIZONTAL LINE TEST FOR INVERSE FUNCTIONS
>
> A function f has an inverse function if and only if no *horizontal* line intersects the graph of f at more than one point.

EXAMPLE 8 *Applying the Horizontal Line Test*

a. The graph of the function $f(x) = x^3 - 1$ is shown in Figure 2.39(a). Because no horizontal line intersects the graph of f at more than one point, you can conclude that f *does* possess an inverse function.

b. The graph of the function $f(x) = x^2 - 1$ is shown in Figure 2.39(b). Because it is possible to find a horizontal line that intersects the graph of f at more than one point, you can conclude that f *does not* possess an inverse function.

(a)
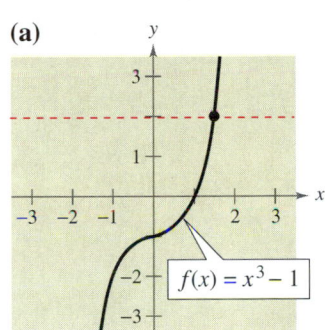

(b)
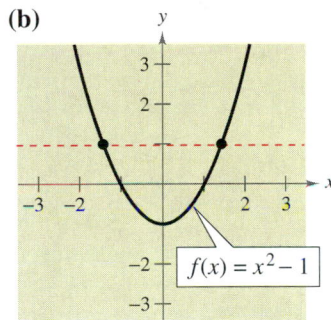

FIGURE 2.39

GROUP ACTIVITY

ERROR ANALYSIS

Suppose you are a math instructor, and one of your students has handed in the following quiz. Find the error(s) in each solution and discuss how to explain each error to your student.

1. Find the inverse f^{-1} of $y = \sqrt{2x - 5}$. $f(x) = \sqrt{2x - 5}$, so

$$f^{-1}(x) = \frac{1}{\sqrt{2x - 5}}$$

2. Find the inverse f^{-1} of $y = \frac{3}{5}x + \frac{1}{3}$. $f(x) = \frac{3}{5}x + \frac{1}{3}$, so

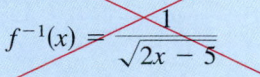
$$f^{-1}(x) = \frac{5}{3}x - 3$$

WARM UP

Find the domain of the function.

1. $f(x) = \sqrt[3]{x+1}$

2. $f(x) = \sqrt{x+1}$

3. $g(x) = \dfrac{2}{x^2 - 2x}$

4. $h(x) = \dfrac{x}{3x+5}$

Simplify the expression.

5. $2\left(\dfrac{x+5}{2}\right) - 5$

6. $7 - 10\left(\dfrac{7-x}{10}\right)$

7. $\sqrt[3]{2\left(\dfrac{x^3}{2} - 2\right) + 4}$

8. $\left(\sqrt[5]{x+2}\right)^5 - 2$

Solve for x in terms of y.

9. $y = \dfrac{2x-6}{3}$

10. $y = \sqrt[3]{2x-4}$

2.5 Exercises

In Exercises 1–4, match the graph of the function with the graph of its inverse. [The graphs of the inverse functions are labeled (a), (b), (c), and (d).]

(a)
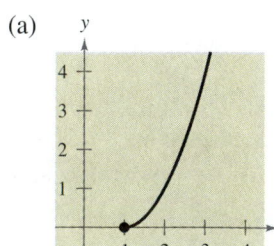

(b)
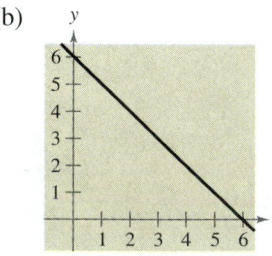

(c)
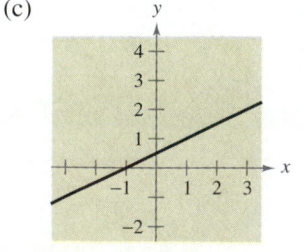

(d)
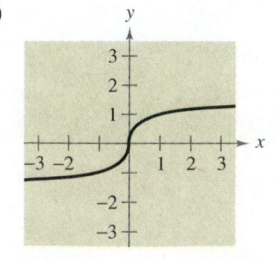

1.
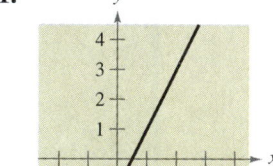

2.
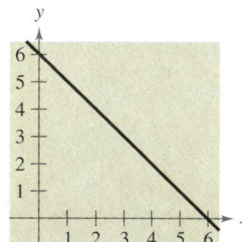

3.
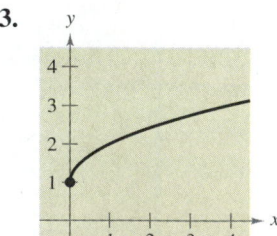

4.
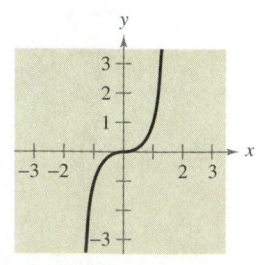

In Exercises 5–10, find the inverse of f informally. Verify that $f(f^{-1}(x)) = x$ and $f^{-1}(f(x)) = x$.

5. $f(x) = 8x$

6. $f(x) = \dfrac{1}{5}x$

7. $f(x) = x + 10$

8. $f(x) = x - 5$

9. $f(x) = \sqrt[3]{x}$

10. $f(x) = x^5$

In Exercises 11–20, show that f and g are inverse functions (a) algebraically and (b) graphically.

11. $f(x) = 2x, \quad g(x) = \dfrac{x}{2}$

12. $f(x) = x - 5, \quad g(x) = x + 5$

13. $f(x) = 5x + 1, \quad g(x) = \dfrac{x - 1}{5}$

14. $f(x) = 3 - 4x, \quad g(x) = \dfrac{3 - x}{4}$

15. $f(x) = x^3, \quad g(x) = \sqrt[3]{x}$

16. $f(x) = \dfrac{1}{x}, \quad g(x) = \dfrac{1}{x}$

17. $f(x) = \sqrt{x - 4}, \quad g(x) = x^2 + 4, \quad x \geq 0$

18. $f(x) = 1 - x^3, \quad g(x) = \sqrt[3]{1 - x}$

19. $f(x) = 9 - x^2, \quad x \geq 0, \quad g(x) = \sqrt{9 - x}, \quad x \leq 9$

20. $f(x) = \dfrac{1}{1 + x}, \quad x \geq 0$

$g(x) = \dfrac{1 - x}{x}, \quad 0 < x \leq 1$

In Exercises 21 and 22, does the function have an inverse?

21.

x	-1	0	1	2	3	4
$f(x)$	-2	1	2	1	-2	-6

22.

x	-3	-2	-1	0	2	2
$f(x)$	10	6	4	1	-3	-10

In Exercises 23–26, does the function have an inverse?

23.

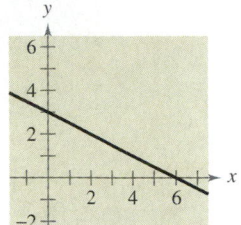

24.

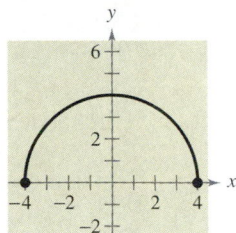

25.

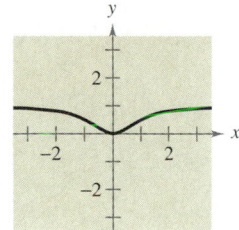

26.

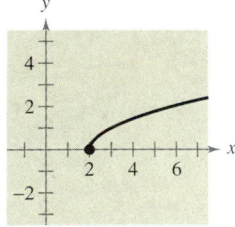

In Exercises 27–32, use a graphing utility to graph the function and use the Horizontal Line Test to determine whether the function has an inverse.

27. $g(x) = \dfrac{4 - x}{6}$

28. $f(x) = 10$

29. $h(x) = |x + 4| - |x - 4|$

30. $g(x) = (x + 5)^3$

31. $f(x) = -2x\sqrt{16 - x^2}$

32. $f(x) = \frac{1}{8}(x + 2)^2 - 1$

In Exercises 33–42, find the inverse of the function f. Then graph both f and f^{-1} on the same coordinate system.

33. $f(x) = 2x - 3$

34. $f(x) = 3x$

35. $f(x) = x^5$

36. $f(x) = x^3 + 1$

37. $f(x) = \sqrt{x}$

38. $f(x) = x^2, \quad x \geq 0$

39. $f(x) = \sqrt{4 - x^2}, \quad 0 \leq x \leq 2$

40. $f(x) = \dfrac{4}{x}$

41. $f(x) = \sqrt[3]{x - 1}$

42. $f(x) = x^{3/5}$

In Exercises 43–58, determine whether the function has an inverse. If it does, find its inverse.

43. $f(x) = x^4$

44. $f(x) = \dfrac{1}{x^2}$

45. $g(x) = \dfrac{x}{8}$

46. $f(x) = 3x + 5$

47. $p(x) = -4$

48. $f(x) = \dfrac{3x + 4}{5}$

49. $f(x) = (x + 3)^2, \quad x \geq -3$

50. $q(x) = (x - 5)^2$

51. $h(x) = \dfrac{1}{x}$

52. $f(x) = |x - 2|, \quad x \leq 2$

53. $f(x) = \sqrt{2x + 3}$

54. $f(x) = \sqrt{x - 2}$

55. $g(x) = x^2 - x^4$

56. $f(x) = \dfrac{x^2}{x^2 + 1}$

57. $f(x) = 25 - x^2, \quad x \leq 0$

58. $f(x) = ax + b, \quad a \neq 0$

In Exercises 59–62, delete part of the graph of the function so that the part that remains has an inverse. Find the inverse and give its domain. (There is more than one correct answer.)

59. $f(x) = (x - 2)^2$

60. $f(x) = 1 - x^4$

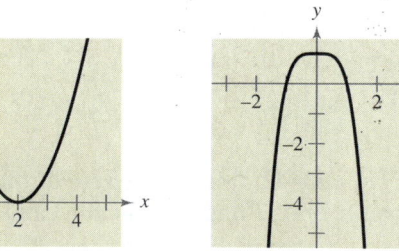

61. $f(x) = |x + 2|$

62. $f(x) = |x - 2|$

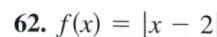

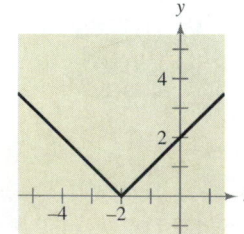
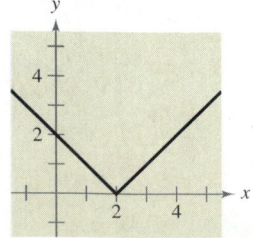

In Exercises 63 and 64, use the graph of the function f to complete the table and sketch the graph of f^{-1}.

63.

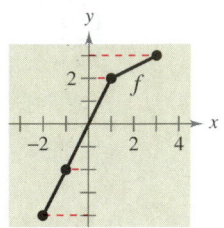

x	$f^{-1}(x)$
-4	
-2	
2	
3	

64.

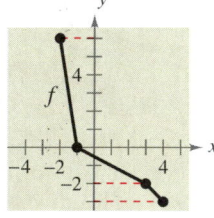

x	$f^{-1}(x)$
-3	
-2	
0	
6	

True or False? In Exercises 65–68, determine if the statement is true or false. If it is false, give an example that shows why it is false.

65. If f is an even function, f^{-1} exists.

66. If the inverse of f exists, the y-intercept of f is an x-intercept of f^{-1}.

67. If $f(x) = x^n$ where n is odd, f^{-1} exists.

68. There exists no function f such that $f = f^{-1}$.

In Exercises 69–74, use the functions $f(x) = \frac{1}{8}x - 3$ and $g(x) = x^3$ to find the indicated value or function.

69. $(f^{-1} \circ g^{-1})(1)$

70. $(g^{-1} \circ f^{-1})(-3)$

71. $(f^{-1} \circ f^{-1})(6)$

72. $(g^{-1} \circ g^{-1})(-4)$

73. $(f \circ g)^{-1}$

74. $g^{-1} \circ f^{-1}$

In Exercises 75–78, use the functions $f(x) = x + 4$ and $g(x) = 2x - 5$ to find the specified function.

75. $g^{-1} \circ f^{-1}$

76. $f^{-1} \circ g^{-1}$

77. $(f \circ g)^{-1}$

78. $(g \circ f)^{-1}$

79. *Hourly Wage* Your wage is $8.00 per hour plus $0.75 for each unit produced per hour. Thus, your hourly wage y in terms of the number of units produced is given by

$$y = 8 + 0.75x.$$

(a) Find the inverse of the function.

(b) What does each variable represent in the inverse function?

(c) Determine the number of units produced when your hourly wage averages $22.25.

80. *Cost* Suppose you need a total of 50 pounds of two commodities costing $1.25 and $1.60 per pound, respectively.

(a) Verify that the total cost is

$$y = 1.25x + 1.60(50 - x)$$

where x is the number of pounds of the less expensive commodity.

(b) Find the inverse of the cost function. What does each variable represent in the inverse function?

(c) Use the context of the problem to determine the domain of the inverse function.

(d) Determine the number of pounds of the less expensive commodity purchased if the total cost is $73.

81. *Diesel Engine* The function

$$y = 0.03x^2 + 245.50, \qquad 0 < x < 100$$

approximates the exhaust temperature y in degrees Fahrenheit where x is the percent load for a diesel engine.

(a) Find the inverse of the function. What does each variable represent in the inverse function?

(b) Use a graphing utility to graph the inverse function.

(c) Determine the percent load interval if the exhaust temperature of the engine must not exceed 500 degrees Fahrenheit.

82. *Think About It* The function

$$f(x) = k(2 - x - x^3)$$

has an inverse, and $f^{-1}(3) = -2$. Find k.

83. *Average Miles Per Gallon* The average miles per gallon f for cars in the United States for 1986 through 1991 is given in the table. The time in years is given by t, with $t = 6$ corresponding to 1986. (Source: U.S. Federal Highway Administration)

t	6	7	8	9	10	11
$f(t)$	18.27	19.20	19.95	20.40	21.02	21.68

(a) Does f^{-1} exist?

(b) If f^{-1} exists, what does it mean in the context of the problem?

(c) If f^{-1} exists, find $f^{-1}(19.95)$.

84. *Average Miles Per Gallon* If the table in Exercise 83 were extended to 1992 and if the average number of miles per gallon for that year were 21.02, would f^{-1} exist? Explain.

In Exercises 85–92, solve the equation by any convenient method.

85. $x^2 = 64$

86. $(x - 5)^2 = 8$

87. $4x^2 - 12x + 9 = 0$

88. $9x^2 + 12x + 3 = 0$

89. $x^2 - 6x + 4 = 0$

90. $2x^2 - 4x - 6 = 0$

91. $50 + 5x = 3x^2$

92. $2x^2 + 4x - 9 = 2(x - 1)^2$

93. *Consecutive Even Integers* Find two consecutive positive even integers whose product is 288.

94. *Lawn Mowing* Two people must mow a rectangular lawn measuring 100 feet by 200 feet. The first person agrees to mow three-fourths of the lawn and starts by mowing around the outside. How wide a strip must the person mow on each of the four sides? If the mower has a 24-inch cut, approximate the required number of trips around the lawn.

95. *Dimensions of a Triangular Sign* A triangular sign has a height that is equal to its base. The area of the sign is 10 square feet. Find the base and height of the sign.

96. *Dimensions of a Triangular Sign* A triangular sign has a height that is twice its base. The area of the sign is 10 square feet. Find the base and height of the sign.

FOCUS ON CONCEPTS

In this chapter, you studied functions and their graphs. You can use the following questions to check your understanding of several of the basic concepts discussed in this chapter. The answers to these questions are given in the back of the book.

1. With the information given in the graphs, is it possible to determine the slope of the two lines? Is it possible they could have the same slope? Explain.

(a) 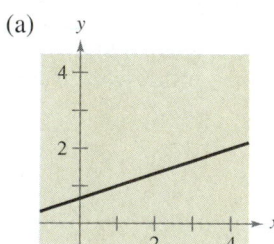   (b)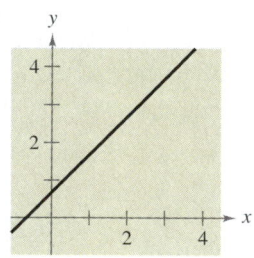

2. The slopes of two lines are -4 and $\frac{5}{2}$. Which is steeper? Explain.

3. The value V of a machine t years after it is purchased is $V = -4000t + 58,500, \ 0 \le t \le 5$. Explain what the V-intercept and slope measure.

4. Does the relationship shown in the figure represent a function from set A to set B? Explain.

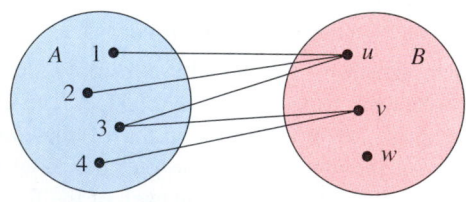

5. Select viewing rectangles using a graphing utility that would show these graphs.

(a) 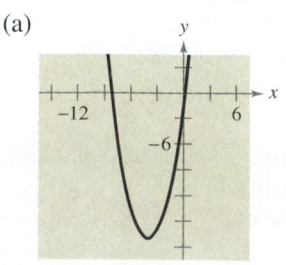   (b)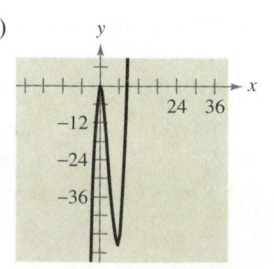

6. If f is an even function, determine if g is even, odd, or neither. Explain.

(a) $g(x) = -f(x)$ (b) $g(x) = f(-x)$

(c) $g(x) = f(x) - 2$ (d) $g(x) = f(x - 2)$

7. Management originally predicted that the profits from the sales of a new product would be approximated by the graph of the function f in the figure. The actual profits are shown by the function g along with a verbal description. Use the concepts of transformations of graphs to write g in terms of f.

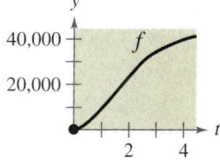

(a) The profits were only three-fourths as large as expected.

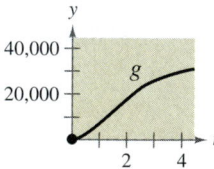

(b) The profits were consistently $10,000 greater than predicted.

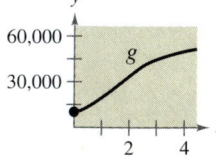

(c) There was a 2-year delay in the introduction of the product. After sales began, profits were as expected.

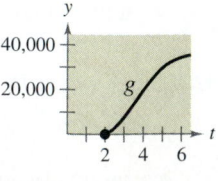

Review Exercises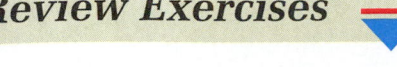

In Exercises 1 and 2, identify the line that has the specified slope.

1. (a) $m = \frac{3}{2}$ (b) $m = 0$ (c) $m = -3$

2. (a) m is undefined.   (b) $m = -1$ (c) $m = \frac{5}{2}$

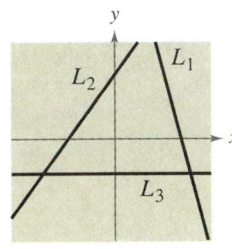

| FIGURE FOR **1** | FIGURE FOR **2** |

In Exercises 3 and 4, plot the points and find the slope of the line passing through the pair of points.

3. $(-4.5, 6), (2.1, 3)$ **4.** $(-3, 2), (8, 2)$

In Exercises 5 and 6, use the concept of slope to find t so that the three points are collinear.

5. $(-2, 5), (0, t), (1, 1)$ **6.** $(-6, 1), (1, t), (10, 5)$

In Exercises 7 and 8, use the point on the line and the slope of the line to find three additional points through which the line passes. (The solution is not unique.)

Point	Slope
7. $(2, -1)$	$m = \frac{1}{4}$
8. $(-3, 5)$	$m = -\frac{3}{2}$

In Exercises 9 and 10, find an equation of the line that passes through the points.

9. $(0, 0), (0, 10)$

10. $(-1, 4), (2, 0)$

In Exercises 11 and 12, find an equation of the line that passes through the given point and has the specified slope. Sketch the graph of the line.

Point	Slope
11. $(0, -5)$	$m = \frac{3}{2}$
12. $(-2, 6)$	$m = 0$

In Exercises 13 and 14, write an equation of the line through the point (a) parallel to the given line and (b) perpendicular to the given line.

Point	Line
13. $(3, -2)$	$5x - 4y = 8$
14. $(-8, 3)$	$2x + 3y = 5$

Rate of Change   In Exercises 15 and 16, you are given the dollar value of a product in 1996 *and* the rate at which the value of the item is expected to change during the next 5 years. Write a linear equation that gives the dollar value V of the product in terms of the year t. (Let $t = 6$ represent 1996.)

1996 Value	Rate
15. $12,500	$850 increase per year
16. $72.95	$5.15 increase per year

Exploration   In Exercises 17 and 18, find a relationship between x and y so that (x, y) is equidistant from the two points.

17. $(-2, -5), (6, 3)$ **18.** $\left(1, \frac{7}{2}\right), (5, 0)$

19. *Fourth-Quarter Sales*   During the second and third quarters of the year, a business had sales of $160,000 and $185,000, respectively. If the growth of sales follows a linear pattern, estimate sales during the fourth quarter.

20. *Dollar Value*   The dollar value of a product in 1995 is $85, and the product will increase in value at an expected rate of $3.75 per year.

  (a) Write a linear equation that gives a dollar value V of the product in terms of the year t. (Let $t = 5$ represent 1995.)

  (b) Use a graphing utility to graph the sales equation.

  (c) Move the cursor along the graph of the sales model to estimate the dollar value of the product in 2000.

In Exercises 21 and 22, determine which of the sets of ordered pairs represents a function from A to B. Give reasons for your answers.

21. $A = \{10, 20, 30, 40\}$ and $B = \{0, 2, 4, 6\}$

  (a) $\{(20, 4), (40, 0), (20, 6), (30, 2)\}$

  (b) $\{(10, 4), (20, 4), (30, 4), (40, 4)\}$

  (c) $\{(40, 0), (30, 2), (20, 4), (10, 6)\}$

  (d) $\{(20, 2), (10, 0), (40, 4)\}$

22. $A = \{u, v, w\}$ and $B = \{-2, -1, 0, 1, 2\}$

  (a) $\{(v, -1), (u, 2), (w, 0), (u, -2)\}$

  (b) $\{(u, -2), (v, 2), (w, 1)\}$

  (c) $\{(u, 2), (v, 2), (w, 1), (w, 1)\}$

  (d) $\{(w, -2), (v, 0), (w, 2)\}$

In Exercises 23–26, determine if the equation represents y as a function of x.

23. $16x - y^4 = 0$ **24.** $2x - y - 3 = 0$

25. $y = \sqrt{1 - x}$ **26.** $|y| = x + 2$

In Exercises 27 and 28, evaluate the function as indicated. Simplify your answers.

27. $f(x) = x^2 + 1$

  (a) $f(2)$ (b) $f(-4)$

  (c) $f(t^2)$ (d) $-f(x)$

28. $g(x) = x^{4/3}$

  (a) $g(8)$ (b) $g(t + 1)$

  (c) $\dfrac{g(8) - g(1)}{8 - 1}$ (d) $g(-x)$

In Exercises 29–34, determine the domain of the function. Verify your result with a graph.

29. $f(x) = \sqrt{25 - x^2}$ **30.** $f(x) = 3x + 4$

31. $g(s) = \dfrac{5}{3s - 9}$ **32.** $f(x) = \sqrt{x^2 + 8x}$

33. $h(x) = \dfrac{x}{x^2 - x - 6}$ **34.** $h(t) = |t + 1|$

In Exercises 35 and 36, select the viewing rectangle on a graphing utility that shows the most complete graph of the function.

35. $f(x) = \dfrac{3x}{2(3 - x)}$

Xmin = -4	Xmin = -5	Xmin = 0
Xmax = 4	Xmax = 10	Xmax = 20
Xscl = 1	Xscl = 1	Xscl = 2
Ymin = -3	Ymin = -8	Ymin = 0
Ymax = 3	Ymax = 6	Ymax = 10
Yscl = 1	Yscl = 1	Yscl = 2

36. $f(x) = 4[(0.3x)^3 - 5x]$

Xmin = -200	Xmin = -10	Xmin = -15
Xmax = 200	Xmax = 10	Xmax = 15
Xscl = 50	Xscl = 2	Xscl = 5
Ymin = -500	Ymin = -20	Ymin = -150
Ymax = 500	Ymax = 20	Ymax = 150
Yscl = 50	Yscl = 4	Yscl = 50

Graphical Analysis   **In Exercises 37–40, use a graphing utility to graph the function to (a) approximate the intervals in which the function is increasing, decreasing, or constant; and (b) determine if the function is even, odd, or neither.**

37. $g(x) = |x + 2| - |x - 2|$

38. $f(x) = (x^2 - 4)^2$

39. $h(x) = 4x^3 - x^4$

40. $g(x) = \sqrt[3]{x(x + 3)^2}$

41. *Vertical Motion* The velocity of a ball thrown vertically upward from ground level is given by $v(t) = -32t + 48$, where t is the time in seconds and v is the velocity in feet per second.

(a) Find the velocity when $t = 1$.

(b) Find the time when the ball reaches its maximum height. [*Hint:* Find the time when $v(t) = 0$.]

(c) Find the velocity when $t = 2$.

42. *Cost and Profit* A company produces a product for which the variable cost is $5.35 per unit and the fixed costs are $16,000. The company sells the product for $8.20 and can sell all that it produces.

(a) Find the total cost as a function of x, the number of units produced.

(b) Find the profit as a function of x.

43. *Dimensions of a Rectangle* A wire 24 inches long is to be cut into four pieces to form a rectangle with one side of length x.

(a) Express the area A of the rectangle as a function of x.

(b) Determine the domain of the function and use a graphing utility to graph the function over that domain.

(c) Use the graph of the function to approximate the maximum area of the rectangle. Make a conjecture about the dimensions of the rectangle.

44. *Mixture Problem* From a full 50-liter container of a 40% concentration of acid, x liters is removed and replaced with 100% acid.

(a) Write the amount of acid in the final mixture as a function of x.

(b) Determine the domain and range of the function.

(c) Determine x if the final mixture is 50% acid.

45. *Maximum Revenue* For groups of 80 or more, a charter bus company determines the rate per person according to the formula

$$\text{Rate} = \$8.00 - \$0.05(n - 80), \qquad n \geq 80.$$

(a) Determine the revenue as a function of n.

(b) Use a graphing utility to graph the revenue function. Move the cursor along the function to estimate the number of passengers that will maximize the revenue.

46. *Navigation* At noon, ship A is 160 kilometers due east of ship B. Ship A is sailing west at 20 kilometers per hour, and ship B is sailing south at 16 kilometers per hour (see figure).

(a) Verify that the distance between the ships is

$$d = \sqrt{(160 - 20t)^2 + (16t)^2}$$

where t is the time in hours, with $t = 0$ corresponding to noon.

(b) Use a graphing utility to graph the distance function. Move the cursor along the function to estimate (to two-decimal-place accuracy) the minimum distance between the ships. At what time will this occur?

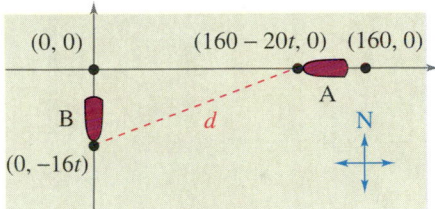

In Exercises 47–50, (a) find f^{-1}, (b) sketch the graphs of f and f^{-1} on the same coordinate system, and (c) verify that $f^{-1}(f(x)) = x = f(f^{-1}(x))$.

47. $f(x) = \frac{1}{2}x - 3$ **48.** $f(x) = 5x - 7$

49. $f(x) = \sqrt{x + 1}$ **50.** $f(x) = x^3 + 2$

In Exercises 51 and 52, restrict the domain of the function f to an interval over which the function is increasing and determine f^{-1} over that interval.

51. $f(x) = 2(x - 4)^2$ **52.** $f(x) = |x - 2|$

In Exercises 53–58, let $f(x) = 3 - 2x$, $g(x) = \sqrt{x}$, and $h(x) = 3x^2 + 2$. Find the indicated value.

53. $(f - g)(4)$ **54.** $(fh)(1)$

55. $(h \circ g)(7)$ **56.** $(g \circ f)(-2)$

57. $(f \circ h)(3)$ **58.** $(f + g)(4)$

CHAPTER PROJECT:

A Graphical Approach to Maximization

In business, a **demand function** gives the price per unit p in terms of the number of units sold x. The demand function whose graph is shown below is

$$p = 40 - 5x^2, \qquad 0 \le x \le \sqrt{8} \qquad\qquad \text{Demand function}$$

where x is measured in millions of units. Note that as the price decreases, the number of units sold increases. The **revenue** R (in millions of dollars) is determined by multiplying the number of units sold by the price per unit. Thus,

$$R = xp = x(40 - 5x^2), \qquad 0 \le x \le \sqrt{8} \qquad\qquad \text{Revenue function}$$

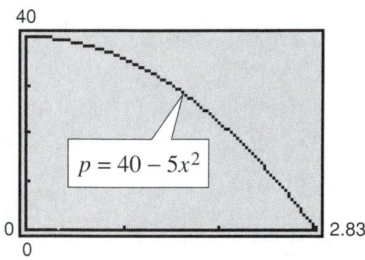

EXAMPLE 1   *Finding the Maximum Revenue*

Real Life

Use a graphing utility to sketch the graph of the revenue function

$$R = 40x - 5x^3, \qquad 0 \le x \le \sqrt{8}.$$

How many units should be sold to obtain maximum revenue? What price per unit should be charged to obtain maximum revenue?

Solution

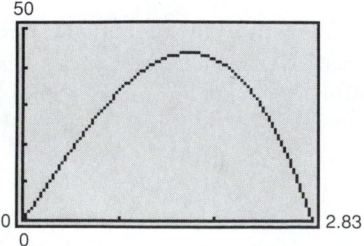

To begin, you need to determine a viewing rectangle that will display the part of the graph that is important to this problem. The domain is given, so you can set the x-boundaries of the graph between 0 and $\sqrt{8}$. To determine the y-boundaries, however, you need to experiment a little. After calculating several values of R, you could decide to use y-boundaries between 0 and 50, as shown in the graph at the left. Next you can use the trace key to find that the maximum revenue of about \$43.5 million occurs when x is approximately 1.64 million units. To find the price per unit that corresponds to this maximum revenue, you can substitute $x = 1.64$ into the demand function to obtain

$$p = 40 - 5(1.64)^2 \approx \$26.55.$$

EXAMPLE 2 *Finding the Maximum Profit*

For the revenue function discussed in Example 1, the cost of producing each unit is $15. How many units should be sold to obtain maximum profit? What price per unit should be charged to obtain maximum profit?

Solution

The total cost C (in millions of dollars) of producing x million units is $C = 15x$. This implies that the profit P (in millions of dollars) obtained by selling x million units is

$$
\begin{aligned}
P &= R - C \\
&= (40x - 5x^3) - 15x \\
&= -5x^3 + 25x.
\end{aligned}
$$

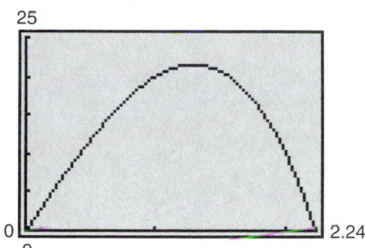

As in Example 1, you can use a graphing utility to graph this function. From the graph, you can approximate that the maximum profit of about $21.5 million occurs when x is approximately 1.28 million units. The price per unit that corresponds to this maximum profit is

$$p = 40 - 5(1.28)^2 \approx \$31.81.$$

CHAPTER PROJECT INVESTIGATIONS

1. *Can't Give It Away!* For the demand function $p = 40 - 5x^2$, match the points $(0, 40)$ and $\left(\sqrt{8}, 0\right)$ with statement (a) or (b). Explain your reasoning.

(a) No one will buy the product at this price.

(b) You can't *give* more than this number away.

2. *Exploration* Use a graphing utility to zoom in on the maximum point of the revenue function in Example 1. (Use a setting of $1.62 \le x \le 1.65$ and $43.5 \le y \le 43.6$.) Use the trace feature to improve the accuracy of the approximation obtained in Example 1. Do you think this improved accuracy is appropriate in the context of this particular problem? Does it change the price?

3. *Exploration* Find a setting that allows you to improve graphically the accuracy of the solution in Example 2.

4. *Exploration* In Example 2, suppose that, in addition to the cost of $15 per unit, there is an initial cost of $250,000. How does this change the profit function? Does this affect the *price* that corresponds to a maximum profit? Does this affect the *amount* of the maximum profit? Explain.

5. *Maximum Volume of a Box* In the Chapter Project on page 79, you were asked to maximize the volume of a box with a square base and a surface area of 216 square inches. In that problem, x represents the length (in inches) of each side of the base and

$$V = 54x - \tfrac{1}{2}x^3$$

represents the volume (in cubic inches). Use a graphing utility to approximate the dimensions that produce a box of maximum volume. (Use a graph that yields an accuracy of 0.0001.)

Cumulative Test for Chapters P–2

Take this test as you would take a test in class. After you are done, check your work against the answers in the back of the book.

The *Interactive* CD-ROM provides answers to the Chapter Tests and Cumulative Tests. It also offers Chapter Pre-Tests (which test key skills and concepts covered in previous chapters) and Chapter Post-Tests, both of which have randomly generated exercises with diagnostic capabilities.

In Exercises 1 and 2, simplify the expression.

1. $\dfrac{8x^2y^{-3}}{30x^{-1}y^2}$ **2.** $\sqrt{24x^4y^3}$

In Exercises 3–5, perform the operations and simplify the result.

3. $4x - [2x + 3(2 - x)]$ **4.** $(x - 2)(x^2 + x - 3)$ **5.** $\dfrac{2}{s + 3} - \dfrac{1}{s + 1}$

In Exercises 6–8, factor the expression completely.

6. $25 - (x - 2)^2$ **7.** $x - 5x^2 - 6x^3$ **8.** $54 - 16x^3$

In Exercises 9–11, graph the equation without the aid of a graphing utility.

9. $x - 3y + 12 = 0$ **10.** $y = x^2 - 9$ **11.** $y = \sqrt{4 - x}$

In Exercises 12–14, solve (if possible) the equation.

12. $2x - 3(x - 4) = 5$ **13.** $3y^2 + 6y + 2 = 0$ **14.** $\sqrt{x + 10} = x - 2$

15. Solve: $|4(x - 2)| < 28$

16. Find an equation for the line passing through $\left(-\frac{1}{2}, 1\right)$ and $(3, 8)$.

17. Explain why the graph at the right does not represent y as a function of x.

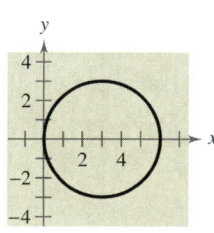

FIGURE FOR 17

18. Evaluate (if possible) the function $f(x) = \dfrac{x}{x - 2}$ for the given values.

(a) $f(6)$ (b) $f(2)$ (c) $f(s + 2)$

19. Describe how the graph of each function would differ from the graph of $y = \sqrt[3]{x}$. (*Note:* It is not necessary to sketch the graphs.)

(a) $r(x) = \frac{1}{2}\sqrt[3]{x}$ (b) $h(x) = \sqrt[3]{x} + 2$ (c) $g(x) = \sqrt[3]{x + 2}$

20. Determine whether $h(x) = 5x - 2$ has an inverse. If so, find it.

21. A group of n people decide to buy a $36,000 minibus. Each person will pay an equal share of the cost. If three additional people join the group, the cost per person will decrease by $1000. Find n.

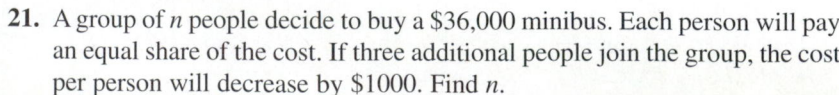

3 Zeros of Polynomial Functions

The Fundamental Theorem of Algebra implies that an nth-degree polynomial equation has precisely n solutions. This result, however, is true only if repeated and complex solutions are counted. For example, the equation

$$x^4 - 2x^3 + 2x^2 - 2x + 1 = 0$$

has solutions of 1, 1, i, and $-i$.

When first developed, complex numbers were used primarily for theoretical results such as this. Today, however, complex numbers have several other uses.

One use is in creating fractals like that shown in the photograph. To program the fractal with a computer, you plot complex numbers in the complex plane.

In the complex plane, the point (a, b) represents the complex number $a + bi$. For example, the number $2 + 3i$ is plotted below.

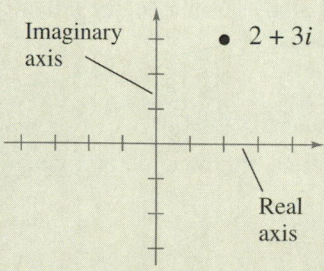

The Complex Plane

See Exercises 67 and 68 on page 310.

Photos: Jim Zuckerman; Rondi Ballard (inset)

This fractal was created on a computer by photographer Jim Zuckerman. Zuckerman owns and operates a photography and digital imaging company in Northridge, California.

3.1   *Quadratic Functions*

See Exercise 73 on page 267 for an example of how a quadratic function can be used in forestry to model the number of board feet yielded by a fallen log.

The Graph of a Quadratic Function   □   *The Standard Form of a Quadratic Function*   □   *Applications*

The Graph of a Quadratic Function

In this and the next section, you will study the graphs of polynomial functions.

> **DEFINITION OF POLYNOMIAL FUNCTION**
>
> Let n be a nonnegative integer and let a_n, a_{n-1}, . . . , a_2, a_1, a_0 be real numbers with $a_n \neq 0$. The function given by
>
> $$f(x) = a_n x^n + a_{n-1} x^{n-1} + \cdots + a_2 x^2 + a_1 x + a_0$$
>
> is called a **polynomial function of x with degree n.**

Polynomial functions are classified by degree. For instance, the polynomial function

$$f(x) = a, \qquad a \neq 0 \qquad \textcolor{red}{\text{Constant function}}$$

has degree 0 and is called a **constant function.** In Chapter 2, you learned that the graph of this type of function is a horizontal line. The polynomial function

$$f(x) = ax + b, \qquad a \neq 0 \qquad \textcolor{red}{\text{Linear function}}$$

has degree 1 and is called a **linear function.** In Chapter 2, you learned that the graph of the linear function $f(x) = ax + b$ is a line whose slope is a and whose y-intercept is $(0, b)$. In this section you will study second-degree polynomial functions, which are called **quadratic functions.**

> **DEFINITION OF QUADRATIC FUNCTION**
>
> Let a, b, and c be real numbers with $a \neq 0$. The function of x given by
>
> $$f(x) = ax^2 + bx + c \qquad \textcolor{red}{\text{Quadratic function}}$$
>
> is called a **quadratic function.**

NOTE   The graph of a quadratic function is a "U"-shaped curve that is called a **parabola.** ■■

All parabolas are symmetric with respect to a line called the **axis of symmetry,** or simply the **axis** of the parabola. The point where the axis intersects the parabola is the **vertex** of the parabola, as shown in Figure 3.1. If the leading coefficient is positive, the graph of $f(x) = ax^2 + bx + c$ is a parabola that opens upward, and if the leading coefficient is negative, the graph of $f(x) = ax^2 + bx + c$ is a parabola that opens downward.

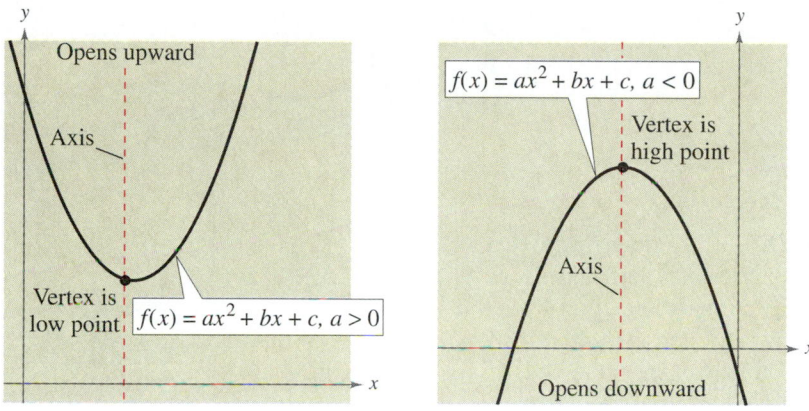

FIGURE 3.1

The simplest type of quadratic function is $f(x) = ax^2$. Its graph is a parabola whose vertex is $(0, 0)$. If $a > 0$, the vertex is the *minimum* point on the graph, and if $a < 0$, the vertex is the *maximum* point on the graph, as shown in Figure 3.2.

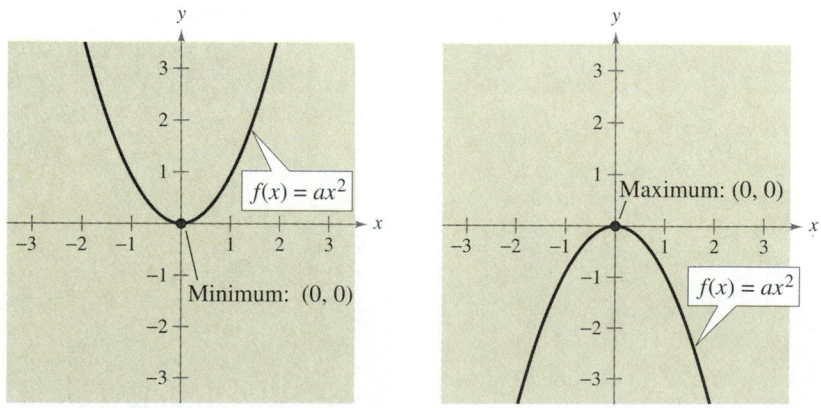

FIGURE 3.2

When sketching the graph of $f(x) = ax^2$, it is helpful to use the graph of $y = x^2$ as a reference, as discussed in Section 2.4.

EXAMPLE 1   *Sketching the Graph of a Quadratic Function*

a. Compared with $y = x^2$, each output of $f(x) = \frac{1}{3}x^2$ "shrinks" by a factor of $\frac{1}{3}$, creating the broader parabola shown in Figure 3.3(a).

b. Compared with $y = x^2$, each output of $g(x) = 2x^2$ "stretches" by a factor of 2, creating the narrower parabola shown in Figure 3.3(b).

NOTE   In Example 1, note that the coefficient a determines how widely the parabola given by $f(x) = ax^2$ opens. If $|a|$ is small, the parabola opens more widely than if $|a|$ is large. ▪▪

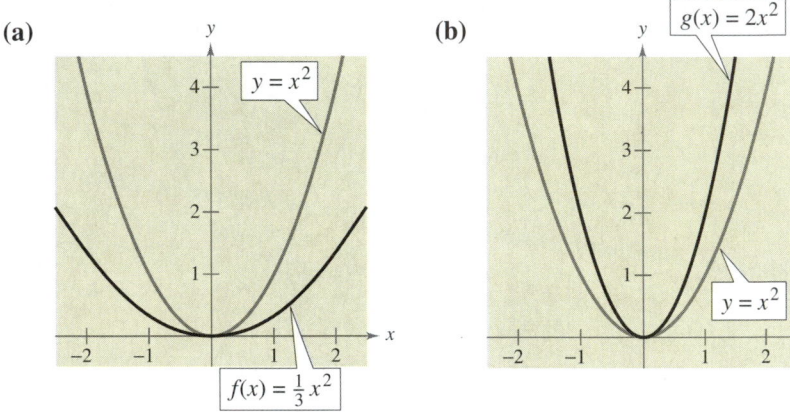

(a)   **(b)**

FIGURE 3.3

▶ *Exploration*

Graph $y = ax^2$ for $a = -2$, $-1, -0.5, 0.5, 1$, and 2. How does the value of a affect the graph?

Graph $y = (x - h)^2$ for $h = -4, -2, 2$, and 4. How does the value of h affect the graph?

Graph $y = x^2 + k$ for $k = -4, -2, 2$, and 4. How does the value of k affect the graph?

Recall from Section 2.4 that the graphs of $y = f(x \pm c)$, $y = f(x) \pm c$, $y = f(-x)$, and $y = -f(x)$ are rigid transformations of the graph of $y = f(x)$. For instance, in Figure 3.4, notice how the graph of $y = x^2$ can be transformed to produce the graphs of $f(x) = -x^2 + 1$ and $g(x) = (x + 2)^2 - 3$.

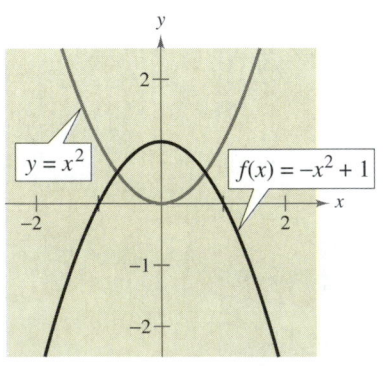 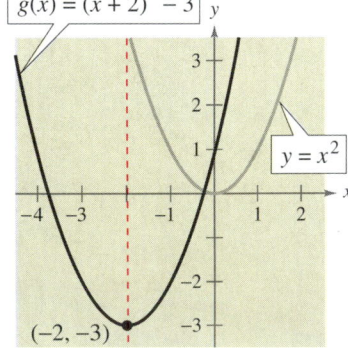

FIGURE 3.4

The Standard Form of a Quadratic Function

The **standard form** of a quadratic function is

$$f(x) = a(x - h)^2 + k.$$

This form is especially convenient for sketching a parabola because it identifies the vertex of the parabola.

STANDARD FORM OF A QUADRATIC FUNCTION

The quadratic function

$$f(x) = a(x - h)^2 + k, \qquad a \neq 0$$

is in **standard form.** The graph of f is a parabola whose axis is the vertical line $x = h$ and whose vertex is the point (h, k). If $a > 0$, the parabola opens upward, and if $a < 0$, the parabola opens downward.

To write a quadratic function in standard form, you can use the process of *completing the square,* as illustrated in Example 2.

EXAMPLE 2 *Writing a Quadratic Function in Standard Form*

Sketch the graph of $f(x) = 2x^2 + 8x + 7$ and identify the vertex.

Solution

Begin by writing the quadratic function in standard form. Notice that the first step in completing the square is to factor out any coefficient of x^2 that is not 1.

$f(x) = 2x^2 + 8x + 7$	Original function
$= 2(x^2 + 4x) + 7$	Factor 2 out of x terms.
$= 2(x^2 + 4x + 4 - 4) + 7$	Add and subtract 4 within parentheses.

$$2^2$$

$= 2(x^2 + 4x + 4) - 2(4) + 7$	Regroup terms.
$= 2(x^2 + 4x + 4) - 8 + 7$	Simplify.
$= 2(x + 2)^2 - 1$	Standard form

From this form, you can see that the graph of f is a parabola that opens upward and has its vertex at $(-2, -1)$. This corresponds to a left shift of two units and a downward shift of one unit relative to the graph of $y = 2x^2$, as shown in Figure 3.5.

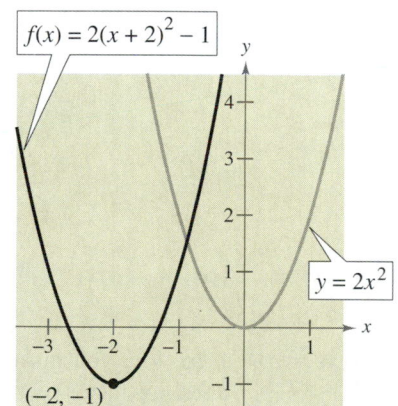

$f(x) = 2(x + 2)^2 - 1$

$y = 2x^2$

$(-2, -1)$

FIGURE 3.5

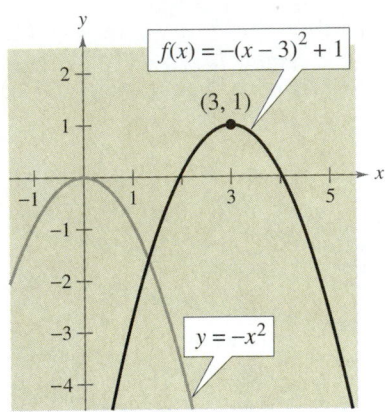

FIGURE 3.6

EXAMPLE 3 *Writing a Quadratic Function in Standard Form*

Sketch the graph of $f(x) = -x^2 + 6x - 8$ and identify the vertex.

Solution

As in Example 2, begin by writing the quadratic function in standard form.

$$f(x) = -x^2 + 6x - 8 \qquad \text{Original function}$$
$$= -(x^2 - 6x) - 8 \qquad \text{Factor } -1 \text{ out of } x \text{ terms.}$$
$$= -(x^2 - 6x + 9 - 9) - 8 \qquad \text{Add and subtract 9 within parentheses.}$$

$$3^2$$

$$= -(x^2 - 6x + 9) - (-9) - 8 \qquad \text{Regroup terms.}$$
$$= -(x - 3)^2 + 1 \qquad \text{Standard form}$$

Thus, the graph of f is a parabola that opens downward with vertex at $(3, 1)$, as shown in Figure 3.6.

EXAMPLE 4 *Finding the Equation of a Parabola*

Find an equation for the parabola whose vertex is $(1, 2)$ and that passes through the point $(0, 0)$, as shown in Figure 3.7.

Solution

Because the parabola has a vertex at $(h, k) = (1, 2)$, the equation must have the form

$$f(x) = a(x - 1)^2 + 2. \qquad \text{Standard form}$$

Because the parabola passes through the point $(0, 0)$, it follows that $f(0) = 0$. Thus, you obtain

$$0 = a(0 - 1)^2 + 2 \qquad \Longrightarrow \qquad a = -2$$

which implies that the equation is

$$f(x) = -2(x - 1)^2 + 2 = -2x^2 + 4x.$$

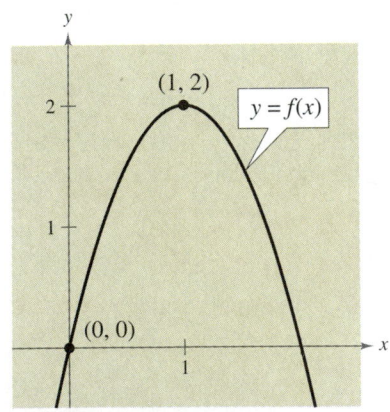

FIGURE 3.7

To find the x-intercepts of the graph of $f(x) = ax^2 + bx + c$, you must solve the equation $ax^2 + bx + c = 0$. If $ax^2 + bx + c$ does not factor, you can use the Quadratic Formula to find the x-intercepts. Remember, however, that a parabola may have no x-intercepts.

Applications

Many applications involve finding the maximum or minimum value of a quadratic function. By writing the quadratic function $f(x) = ax^2 + bx + c$ in standard form, you can determine that the vertex occurs when $x = -b/2a$.

EXAMPLE 5   *The Maximum Height of a Baseball*

A baseball is hit at a point 3 feet above the ground at a velocity of 100 feet per second and at an angle of 45° with respect to the ground. The path of the baseball is given by the function

$$f(x) = -0.0032x^2 + x + 3$$

where $f(x)$ is the height of the baseball (in feet) and x is the distance from home plate (in feet). What is the maximum height reached by the baseball?

Solution

For this quadratic function, you have

$$f(x) = ax^2 + bx + c = -0.0032x^2 + x + 3.$$

Thus, $a = -0.0032$ and $b = 1$. Because the function has a maximum at $x = -b/2a$, you can conclude that the baseball reaches its maximum height when it is

$$x = -\frac{b}{2a} = -\frac{1}{2(-0.0032)} = 156.25 \text{ feet}$$

from home plate. At this distance, the maximum height is

$$f(156.25) = -0.0032(156.25)^2 + 156.25 + 3 = 81.125 \text{ feet}.$$

The path of the baseball is shown in Figure 3.8.

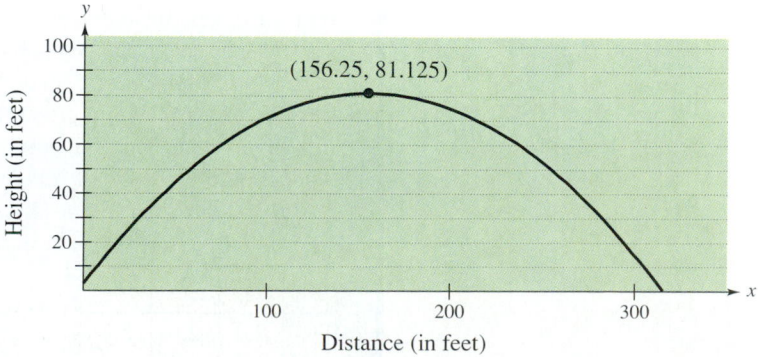

FIGURE 3.8

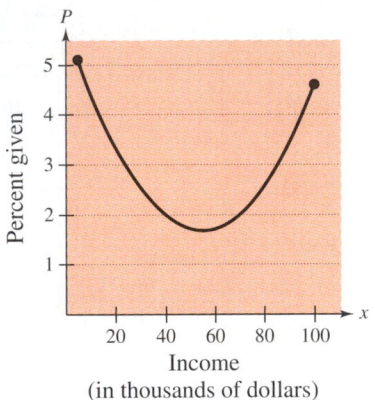

FIGURE 3.9

In Erie, Pennsylvania, over 4000 walkers raised over $220,000 for the Campaign for Healthier Babies. *(Photo: Ed Bernik)*

EXAMPLE 6   *Charitable Contributions*

According to a survey conducted by *Independent Sector*, the percent of income that Americans give to charities is related to the amount of income. For families with annual incomes of $100,000 or less, the percent can be modeled by

$$P = 0.0014x^2 - 0.1529x + 5.855, \qquad 5 \le x \le 100$$

where x is the annual income in thousands of dollars. According to this model, what income level corresponds to the minimum percent of charitable contributions?

Solution

There are two ways to answer this question. One is to sketch the graph of the quadratic function, as shown in Figure 3.9. From this graph, it appears that the minimum percent corresponds to an income level of about $55,000. The other way to answer the question is to use the fact that the minimum point of the parabola occurs when $x = -b/2a$.

$$x = -\frac{b}{2a} = -\frac{-0.1529}{2(0.0014)} \approx 54.6.$$

From this x-value, you can conclude that the minimum percent corresponds to an income level of about $54,600.

GROUP ACTIVITY

QUADRATIC MODELING

The tables below give the United States' annual wheat production y_1 (in billions of bushels) and annual grapefruit production y_2 (in millions of boxes) for the year x, where $x = 13$ represents 1993.

Create a scatter plot of each data set. Find the quadratic model that fits each set of data and explain how you found each model. What factors do you think would affect the production of wheat and grapefruit in a given year? Do you think your models are valid for estimating values before 1985 or in the future? Explain your reasoning. *(Source: U.S. Department of Agriculture)*

x	5	9	13
y_1	2.4	2.0	2.4

x	5	8	11
y_2	56	69	56

The *Interactive* CD-ROM provides additional help with Warm-Up exercises by providing a hypertext link to the section in which the concept was introduced.

WARM UP

Solve the equation by factoring.

1. $2x^2 + 11x - 6 = 0$ **2.** $5x^2 - 12x - 9 = 0$

3. $3 + x - 2x^2 = 0$ **4.** $x^2 + 20x + 100 = 0$

Solve the equation by completing the square.

5. $x^2 - 6x + 4 = 0$ **6.** $x^2 + 4x + 1 = 0$

7. $2x^2 - 16x + 25 = 0$ **8.** $3x^2 + 30x + 74 = 0$

Use the Quadratic Formula to solve the equation.

9. $x^2 + 3x + 3 = 0$ **10.** $x^2 + 3x - 3 = 0$

The *Interactive* CD-ROM contains step-by-step solutions to all odd-numbered Section and Review Exercises. It also provides Tutorial Exercises, which link to Guided Examples for additional help.

3.1 Exercises

In Exercises 1–8, match the quadratic function with its graph. [The graphs are labeled (a), (b), (c), (d), (e), (f), (g), and (h).]

(a)
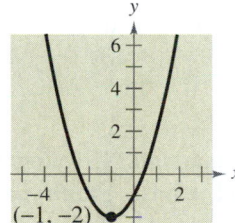

(b)
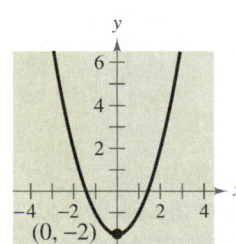

(c)
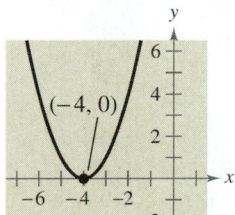

(d)
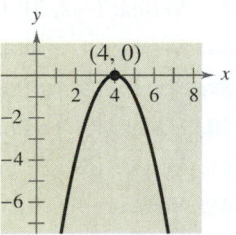

(e)
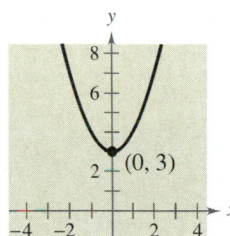

(f)

(g)
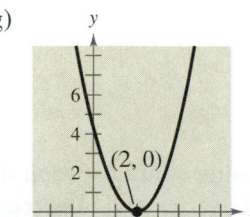

(h)
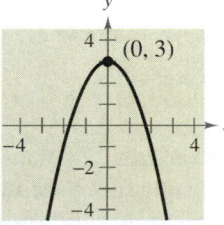

1. $f(x) = (x - 2)^2$ **2.** $f(x) = (x + 4)^2$

3. $f(x) = x^2 - 2$ **4.** $f(x) = 3 - x^2$

5. $f(x) = 4 - (x - 2)^2$ **6.** $f(x) = (x + 1)^2 - 2$

7. $f(x) = x^2 + 3$ **8.** $f(x) = -(x - 4)^2$

Exploration   In Exercises 9–12, graph each equation. Describe how each differs from the graph of $y = x^2$.

9. (a) $y = \frac{1}{2}x^2$ (b) $y = -\frac{1}{8}x^2$
   (c) $y = \frac{3}{2}x^2$ (d) $y = -3x^2$

10. (a) $y = x^2 + 1$ (b) $y = x^2 - 1$
    (c) $y = x^2 + 3$ (d) $y = x^2 - 3$

11. (a) $y = (x - 1)^2$ (b) $y = (x + 1)^2$
    (c) $y = (x - 3)^2$ (d) $y = (x + 3)^2$

12. (a) $y = -\frac{1}{2}(x - 2)^2 + 1$ (b) $y = \frac{1}{2}(x - 2)^2 + 1$

In Exercises 13–26, sketch the graph of the quadratic function without the aid of a graphing utility. Identify the vertex and intercepts.

13. $f(x) = x^2 - 5$
14. $f(x) = \frac{1}{2}x^2 - 4$
15. $f(x) = 16 - x^2$
16. $h(x) = 25 - x^2$
17. $f(x) = (x + 5)^2 - 6$
18. $f(x) = (x - 6)^2 + 3$
19. $h(x) = x^2 - 8x + 16$
20. $g(x) = x^2 + 2x + 1$
21. $f(x) = x^2 - x + \frac{5}{4}$
22. $f(x) = x^2 + 3x + \frac{1}{4}$
23. $f(x) = -x^2 + 2x + 5$
24. $f(x) = -x^2 - 4x + 1$
25. $h(x) = 4x^2 - 4x + 21$
26. $f(x) = 2x^2 - x + 1$

In Exercises 27–30, use a graphing utility to graph the quadratic function. Identify the vertex and intercepts. Then check your results algebraically by completing the square.

27. $f(x) = -(x^2 + 2x - 3)$
28. $g(x) = x^2 + 8x + 11$
29. $f(x) = 2x^2 - 16x + 31$
30. $g(x) = \frac{1}{2}(x^2 + 4x - 2)$

In Exercises 31–36, find an equation for the parabola.

31.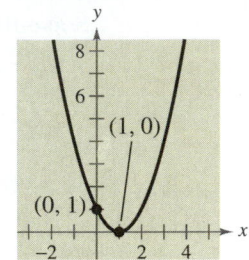

32.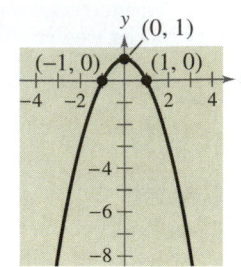

33.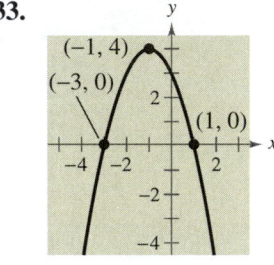

34.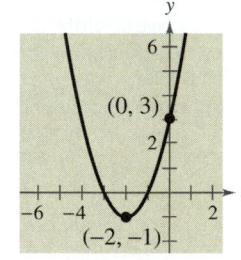

35.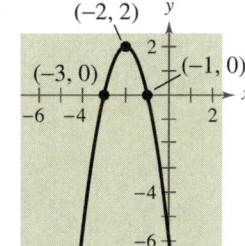

36.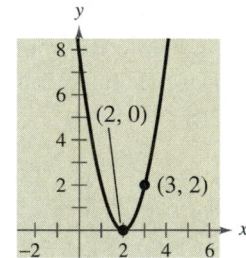

In Exercises 37–42, find the quadratic function that has the indicated vertex and whose graph passes through the given point.

37. Vertex: $(-2, 5)$; Point: $(0, 9)$
38. Vertex: $(4, -1)$; Point: $(2, 3)$
39. Vertex: $(3, 4)$; Point: $(1, 2)$
40. Vertex: $(2, 3)$; Point: $(0, 2)$
41. Vertex: $(5, 12)$; Point: $(7, 15)$
42. Vertex: $(-2, -2)$; Point: $(-1, 0)$

Graphical Reasoning In Exercises 43–46, determine the *x*-intercepts of the graph visually. How do the *x*-intercepts correspond to the solutions of the quadratic equation when *y* = 0?

43. $y = x^2 - 16$

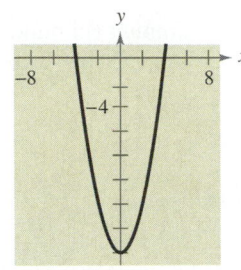

44. $y = x^2 - 6x + 9$

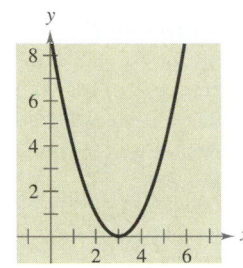

45. $y = x^2 - 4x - 5$

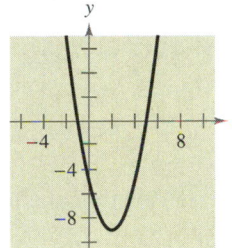

46. $y = 2x^2 + 5x - 3$

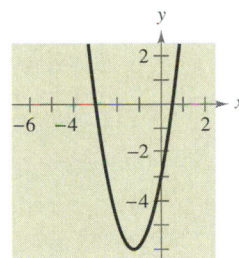

In Exercises 47–50, use a graphing utility to graph the quadratic function. Find the *x*-intercepts of the graph and compare them with the solutions of the corresponding quadratic equation when *y* = 0.

47. $y = x^2 - 4x$

48. $y = x^2 - 9x + 18$

49. $y = 2x^2 - 7x - 30$

50. $y = -\frac{1}{2}(x^2 - 6x - 7)$

In Exercises 51–56, find two quadratic functions whose graphs have the given *x*-intercepts and of which one opens upward and the other downward. (The answers are not unique.)

51. $(-1, 0), (3, 0)$

52. $\left(-\frac{5}{2}, 0\right), (2, 0)$

53. $(0, 0), (10, 0)$

54. $(4, 0), (8, 0)$

55. $(-3, 0), \left(-\frac{1}{2}, 0\right)$

56. $(-5, 0), (5, 0)$

In Exercises 57–60, find two positive real numbers whose product is a maximum.

57. The sum is 110.

58. The sum is *S*.

59. The sum of the first and twice the second is 24.

60. The sum is 50.

Maximum Area In Exercises 61 and 62, consider a rectangle of length *x* and perimeter *P*. (a) Express the area *A* as a function of *x* and determine the domain of the function. (b) Graph the area function. (c) Find the length and width of the rectangle of maximum area.

61. $P = 100$ feet   **62.** $P = 36$ meters

63. *Numerical, Graphical, and Analytical Analysis* A rancher has 200 feet of fencing to enclose two adjacent rectangular corrals (see figure).

(a) Complete six rows of a table such as the one below.

x	y	Area
2	$\frac{1}{3}[200 - 4(2)]$	$2xy = 256$
4	$\frac{1}{3}[200 - 4(4)]$	$2xy \approx 491$

(b) Use a graphing utility to generate additional rows of the table. Use the table to estimate the dimensions that will enclose the maximum area.

(c) Write the area *A* as a function of *x*.

(d) Use a graphing utility to graph the area function. Use the graph to approximate the dimensions that will produce the maximum enclosed area.

(e) Write the area function in standard form to find analytically the dimensions that will produce the maximum area.

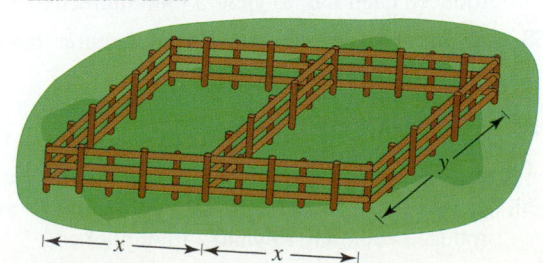

64. **Maximum Area**   An indoor physical fitness room consists of a rectangular region with a semicircle on each end. The perimeter of the room is to be a 200-meter running track.

   (a) Draw a figure that visually represents the problem. Let x and y represent the length and width of the rectangular region, respectively.

   (b) Determine the radius of the semicircular ends of the track. Determine the distance, in terms of y, around the two semicircular parts of the track.

   (c) Use the result of part (b) to write an equation, in terms of x and y, for the distance traveled in one lap around the track. Solve for y.

   (d) Use the result of part (c) to write the area A of the rectangular region as a function of x. What dimensions will produce a maximum area of the rectangle?

65. **Maximum Revenue**   Find the number of units that produce a maximum revenue

   $$R = 900x - 0.1x^2$$

   where R is the total revenue in dollars and x is the number of units sold.

66. **Maximum Revenue**   Find the number of units that produce a maximum revenue

   $$R = 100x - 0.0002x^2$$

   where R is the total revenue in dollars and x is the number of units sold.

67. **Minimum Cost**   A manufacturer of lighting fixtures has daily production costs of

   $$C = 800 - 10x + 0.25x^2$$

   where C is the total cost in dollars and x is the number of units produced. How many fixtures should be produced each day to yield a minimum cost?

68. **Minimum Cost**   A textile manufacturer has daily production costs of

   $$C = 10{,}000 - 110x + 0.045x^2$$

   where C is the total cost in dollars and x is the number of units produced. How many units should be produced each day to yield a minimum cost?

69. **Maximum Profit**   The profit for a company is given by

   $$P = -0.0002x^2 + 140x - 250{,}000$$

   where x is the number of units sold. What sales level will yield a maximum profit?

70. **Maximum Profit**   Let x be the amount (in hundreds of dollars) a company spends on advertising, and let P be the profit, where

   $$P = 230 + 20x - 0.5x^2.$$

   What expenditure for advertising gives the maximum profit?

71. **Trajectory of a Ball**   The height y (in feet) of a ball thrown by a child is given by

   $$y = -\frac{1}{12}x^2 + 2x + 4$$

   where x is the horizontal distance (in feet) from where the ball is thrown (see figure).

   (a) How high is the ball when it leaves the child's hand? (*Note:* Find y when $x = 0$.)

   (b) How high is the ball when it is at its maximum height?

   (c) How far from the child does the ball strike the ground?

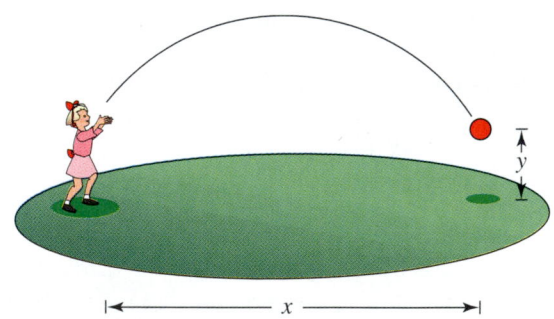

72. **Maximum Height of a Dive**   The path of a diver is given by

   $$y = -\frac{4}{9}x^2 + \frac{24}{9}x + 12$$

   where y is the height in feet and x is the horizontal distance from the end of the diving board in feet. What is the maximum height of the dive?

73. Forestry The number of board feet in a 16-foot log is approximated by the model

$$V = 0.77x^2 - 1.32x - 9.31, \qquad 5 \le x \le 40$$

where V is the number of board feet and x is the diameter (in inches) of the log at the small end. (One board foot is a measure of volume equivalent to a board that is 12 inches wide, 12 inches long, and 1 inch thick.)

(a) Sketch a graph of the function.

(b) Estimate the number of board feet in a 16-foot log with a diameter of 16 inches.

(c) Estimate the diameter of a 16-foot log that scaled 500 board feet when the lumber was sold.

74. Automobile Aerodynamics The number of horsepower y required to overcome wind drag on a certain automobile is approximated by

$$y = 0.002s^2 + 0.005s - 0.029, \qquad 0 \le s \le 100$$

where s is the speed of the car in miles per hour.

(a) Use a graphing utility to graph the function.

(b) Graphically estimate the maximum speed of the car if the power required to overcome wind drag is not to exceed 10 horsepower. Verify analytically.

75. Graphical Analysis From 1950 to 1990, the average annual consumption C of cigarettes by Americans (18 and older) can be modeled by

$$C = 4024.5 + 51.4t - 3.1t^2, \qquad -10 \le t \le 30$$

where t is the year, with $t = 0$ corresponding to 1960. (Source: U.S. Center for Disease Control)

(a) Use a graphing utility to graph the model.

(b) Use the graph of the model to approximate the maximum average annual consumption. Beginning in 1966, all cigarette packages were required by law to carry a health warning. Do you think the warning had any effect? Explain.

(c) In 1960, the U.S. population (18 and over) was 116,530,000. Of those, about 48,500,000 were smokers. What was the average annual cigarette consumption *per smoker* in 1960? What was the average daily cigarette consumption *per smoker?*

76. Data Analysis The number y (in millions) of VCRs in use in the United States for the years 1984 through 1993 are given in the table. The variable t represents time in years, where $t = 4$ represents 1984. (Source: Television Bureau of Advertising, Inc.)

t	4	5	6	7	8	9	10	11	12	13
y	9	18	31	43	51	58	63	67	69	72

(a) Use a graphing utility to sketch a scatter plot of the data.

(b) A model for this data is given by

$$y = -0.73t^2 + 19.46t - 58.77.$$

Use a graphing utility to graph the model on the same viewing rectangle as the scatter plot.

(c) Do you think the model can be used to predict VCR utilization in 2000? Explain.

77. Assume that the function $f(x) = ax^2 + bx + c$ $(a \ne 0)$ has two real zeros. Show that the x-coordinate of the vertex of the graph is the average of the zeros of f. (*Hint:* Use the Quadratic Formula.)

78. Use a graphing utility to demonstrate the result of Exercise 77 for each of the following functions.

(a) $f(x) = \frac{1}{2}(x - 3)^2 - 2$

(b) $f(x) = 6 - \frac{2}{3}(x + 1)^2$

Review **Solve Exercises 79–82 as a review of the skills and problem-solving techniques you learned in previous sections.**

79. Find an equation of the line through the points $(-4, 3)$ and $(2, 1)$.

80. Find an equation of the line through the point $\left(\frac{7}{2}, 2\right)$ with the slope $m = \frac{3}{2}$.

81. Write an equation of the line through the point $(0, 3)$ perpendicular to the line $4x + 5y = 10$.

82. Describe the relationship between the graphs of the lines $2x - 3y = 4$ and $-4x + 6y = 15$.

3.2 *Polynomial Functions of Higher Degree*

See Exercises 78 and 80 on pages 279 and 280 for an example of how a cubic polynomial can be used to model the volume of a box.

Graphs of Polynomial Functions ▫ *The Leading Coefficient Test* ▫
Zeros of Polynomial Functions ▫ *The Intermediate Value Theorem*

Graphs of Polynomial Functions

In this section, you will study basic features of the graphs of polynomial functions. The first feature is that the graph of a polynomial function is **continuous.** Essentially, this means that the graph of a polynomial function has no breaks, as shown in Figure 3.10(a).

(a) Continuous

(b) Discontinuous

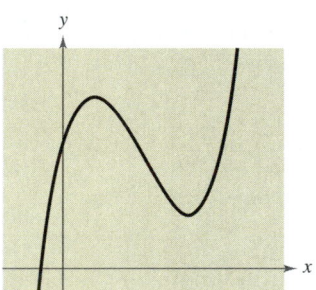
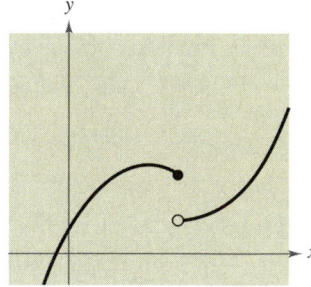

FIGURE 3.10

Study Tip

The graphs of polynomial functions of degree greater than 2 are more difficult to analyze than graphs of polynomials of degree 0, 1, or 2. However, using the features presented in this section, together with point plotting, intercepts, and symmetry, you should be able to make reasonably accurate sketches *by hand.* Of course, if you have a graphing utility, the task is easier.

The second feature is that the graph of a polynomial function has only smooth, rounded turns, as shown in Figure 3.11(a). A polynomial function cannot have a sharp turn. For instance, the function $f(x) = |x|$, which has a sharp turn at the point $(0, 0)$, as shown in Figure 3.11(b), is not a polynomial function.

(a) Polynomial functions have rounded graphs.

(b) Functions with sharp turns are not polynomial functions.

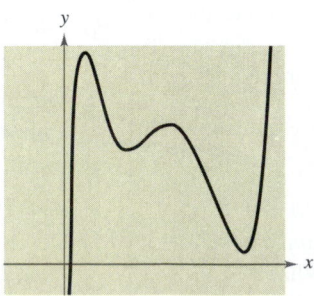
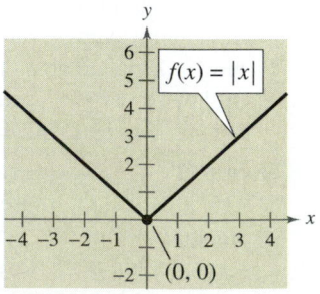

FIGURE 3.11

The polynomial functions that have the simplest graphs are monomials of the form $f(x) = x^n$, where n is an integer greater than zero. From Figure 3.12, you can see that when n is *even* the graph is similar to the graph of $f(x) = x^2$ and when n is *odd* the graph is similar to the graph of $f(x) = x^3$. Moreover, the greater the value of n, the flatter the graph near the origin.

(a) If n is even, the graph of $y = x^n$ *touches* the axis at the x-intercept.

(b) If n is odd, the graph of $y = x^n$ *crosses* the axis at the x-intercept.

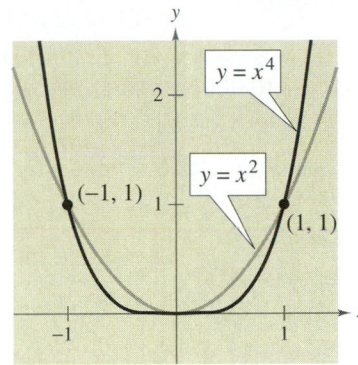

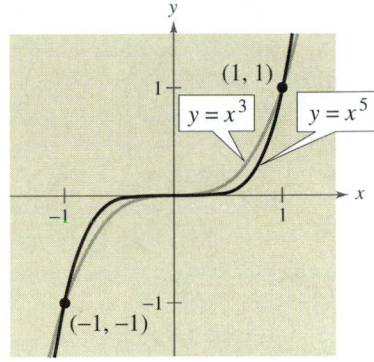

FIGURE 3.12

EXAMPLE 1 *Sketching Transformations of Monomial Functions*

a. Because the degree of $f(x) = -x^5$ is odd, its graph is similar to the graph of $y = x^3$. In Figure 3.13(a), note that the negative coefficient has the effect of reflecting the graph about the x-axis.

b. The graph of $h(x) = (x + 1)^4$, as shown in Figure 3.13(b), is a left shift, by one unit, of the graph of $y = x^4$.

(a)

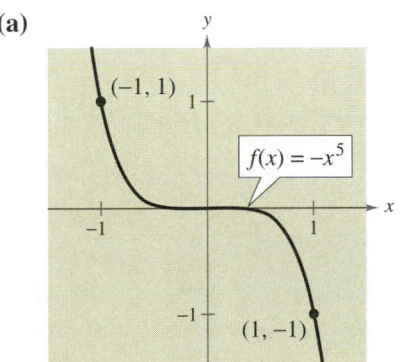

(b)

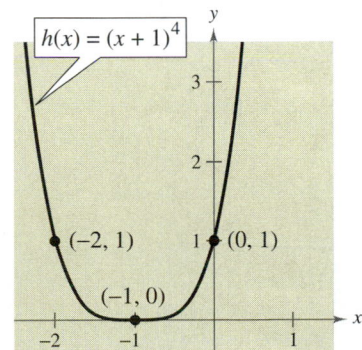

FIGURE 3.13

The Leading Coefficient Test

In Example 1, note that both graphs eventually rise or fall without bound as x moves to the right. Whether the graph of a polynomial eventually rises or falls can be determined by the function's degree (even or odd) and by its leading coefficient, as indicated in the **Leading Coefficient Test.**

LEADING COEFFICIENT TEST

As x moves without bound to the left or to the right, the graph of the polynomial function $f(x) = a_n x^n + \cdots + a_1 x + a_0$ eventually rises or falls in the following manner.

1. When n is *odd:*

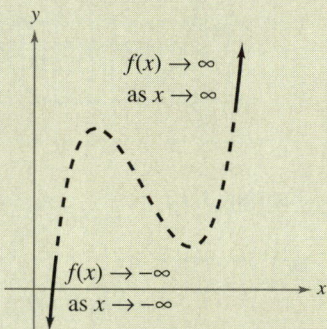 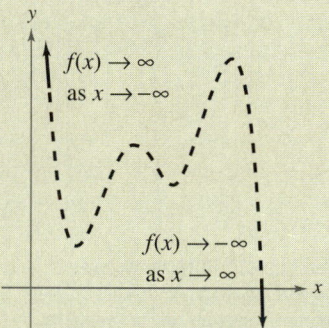

If the leading coefficient is positive ($a_n > 0$), the graph falls to the left and rises to the right.

If the leading coefficient is negative ($a_n < 0$), the graph rises to the left and falls to the right.

2. When n is *even:*

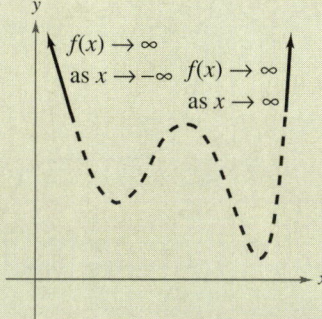 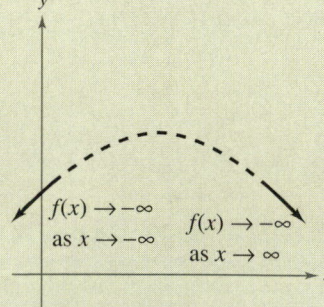

If the leading coefficient is positive ($a_n > 0$), the graph rises to the left and right.

If the leading coefficient is negative ($a_n < 0$), the graph falls to the left and right.

NOTE   The dashed portions of the graphs indicate that the test determines *only* the right and left behavior of the graph. ▪▪

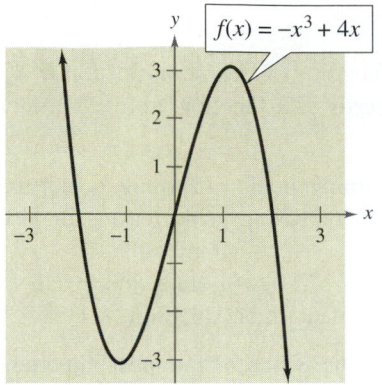

$f(x) = -x^3 + 4x$

FIGURE 3.14

EXAMPLE 2 *Applying the Leading Coefficient Test*

Describe the right and left behavior of the graph of

$$f(x) = -x^3 + 4x.$$

Solution

Because the degree is odd and the leading coefficient is negative, the graph rises to the left and falls to the right, as shown in Figure 3.14.

In Example 2, note that the Leading Coefficient Test only tells you whether the graph *eventually* rises or falls to the right or left. Other characteristics of the graph, such as intercepts and minimum and maximum points, must be determined by means of other tests.

EXAMPLE 3 *Applying the Leading Coefficient Test*

Describe the right and left behavior of the graph of each function.

a. $f(x) = x^4 - 5x^2 + 4$

b. $f(x) = x^5 - x$

Solution

a. Because the degree is even and the leading coefficient is positive, the graph rises to the left and right, as shown in Figure 3.15(a).

b. Because the degree is odd and the leading coefficient is positive, the graph falls to the left and rises to the right, as shown in Figure 3.15(b).

(a)

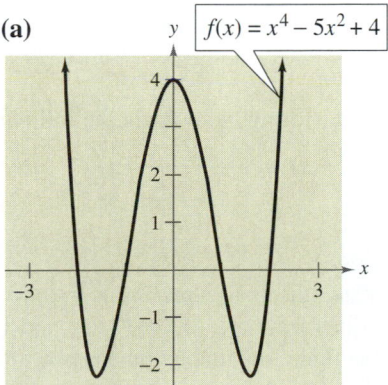

$f(x) = x^4 - 5x^2 + 4$

(b)

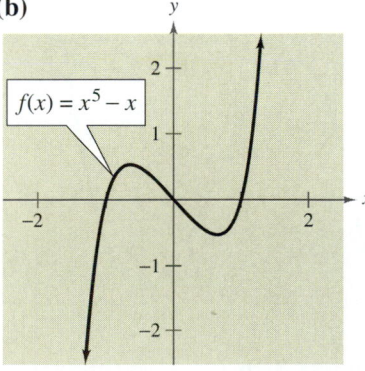

$f(x) = x^5 - x$

FIGURE 3.15

Zeros of Polynomial Functions

It can be shown that for a polynomial function f of degree n, the following statements are true. (Remember that the **zeros** of a function are the x-values for which the function is zero.)

1. The graph of f has, at most, $n - 1$ turning points. (Turning points are points at which the graph changes from increasing to decreasing or vice versa.)
2. The function f has, at most, n real zeros. (You will study this result in detail in Section 3.5 on the Fundamental Theorem of Algebra.)

Finding the zeros of polynomial functions is one of the most important problems in algebra. There is a strong interplay between graphical and algebraic approaches to this problem. Sometimes you can use information about the graph of a function to help find its zeros, and in other cases you can use information about the zeros of a function to help sketch its graph.

NOTE In the equivalent statements at the right, notice that finding zeros of polynomial functions is closely related to factoring and finding x-intercepts. ■■

REAL ZEROS OF POLYNOMIAL FUNCTIONS

If f is a polynomial function and a is a real number, the following statements are equivalent.

1. $x = a$ is a *zero* of the function f.
2. $x = a$ is a *solution* of the polynomial equation $f(x) = 0$.
3. $(x - a)$ is a *factor* of the polynomial $f(x)$.
4. $(a, 0)$ is an *x-intercept* of the graph of f.

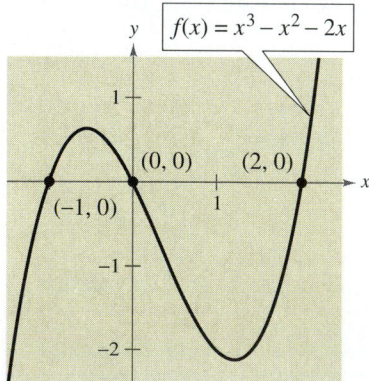

$f(x) = x^3 - x^2 - 2x$

FIGURE 3.16

*E*XAMPLE 4   *Finding Zeros of a Polynomial Function*

Find all real zeros of $f(x) = x^3 - x^2 - 2x$.

Solution

By factoring, you obtain the following.

$$f(x) = x^3 - x^2 - 2x \qquad \text{Original function}$$
$$= x(x^2 - x - 2) \qquad \text{Remove common monomial factor.}$$
$$= x(x - 2)(x + 1) \qquad \text{Factor completely.}$$

Thus, the real zeros are $x = 0$, $x = 2$, and $x = -1$, and the corresponding x-intercepts are $(0, 0)$, $(2, 0)$, and $(-1, 0)$, as shown in Figure 3.16. Note in the figure that the graph has two turning points. This is consistent with the fact that a third-degree polynomial can have *at most* two turning points.

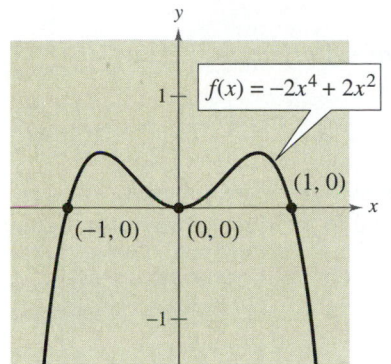

FIGURE 3.17

EXAMPLE 5 *Finding Zeros of a Polynomial Function*

Find all real zeros of $f(x) = -2x^4 + 2x^2$.

Solution

In this case, the polynomial factors as follows.

$$\begin{aligned} f(x) &= -2x^4 + 2x^2 & \text{Original function} \\ &= -2x^2(x^2 - 1) & \text{Remove common monomial factor.} \\ &= -2x^2(x - 1)(x + 1) & \text{Factor completely.} \end{aligned}$$

Thus, the real zeros are $x = 0$, $x = 1$, and $x = -1$, and the corresponding x-intercepts are $(0, 0)$, $(1, 0)$, and $(-1, 0)$, as shown in Figure 3.17. Note in the figure that the graph has three turning points. This is consistent with the fact that a fourth-degree polynomial can have *at most* three turning points.

In Example 5, the real zero arising from $-2x^2 = 0$ is called a **repeated zero.** In general, a factor $(x - a)^k$ yields a repeated zero $x = a$ of **multiplicity** k. If k is odd, the graph *crosses* the x-axis at $x = a$. If k is even, the graph *touches* the x-axis (but does not cross the x-axis) at $x = a$. Note how this occurs in Figure 3.17.

EXAMPLE 6 *Charitable Contributions Revisited*

Real Life

Example 6 in Section 3.1 discusses the model

$$P = 0.0014x^2 - 0.1529x + 5.855, \qquad 5 \le x \le 100$$

where P is the percent of annual income given and x is the annual income in thousands of dollars. Note that this model gives the charitable contributions as a *percent* of annual income. To find the average *amount* that a family gives to charity, you can multiply the given model by the income $1000x$ (and divide by 100 to change from percent to decimal form) to obtain

$$A = 0.014x^3 - 1.529x^2 + 58.55x, \qquad 5 \le x \le 100$$

where A represents the amount of charitable contributions in dollars. Sketch the graph of this function and use the graph to estimate the annual income for a family that gives $1000 a year to charities.

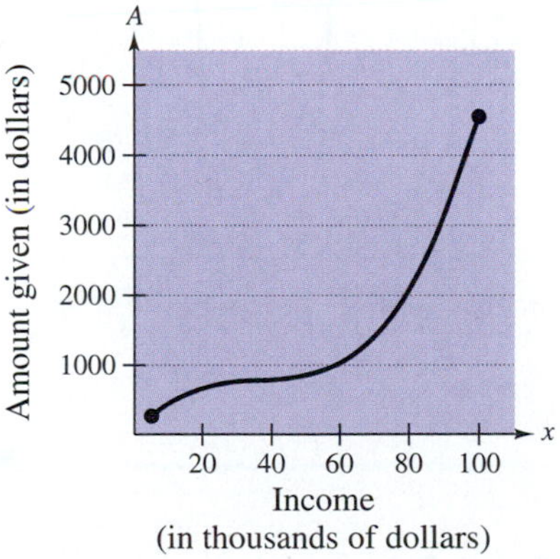

Solution

The graph of this function is shown in Figure 3.18. From the graph you see that an average contribution of $1000 corresponds to an annual income of about $59,000.

Amount given (in dollars)

Income
(in thousands of dollars)

FIGURE 3.18

Example 7 uses an "algebraic approach" to describe the graph of the function. A graphing utility is a comple-ment to this approach. Remember that the most important aspect of using a graphing utility is to find a viewing window that shows all significant features of the graph. For instance, which of the graphs below shows all of the significant features of the function in Example 7?

(a)

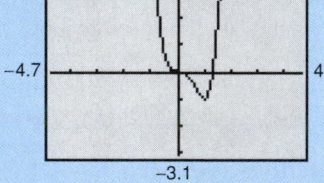

(b)

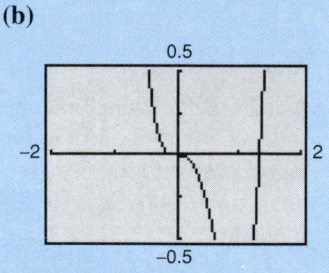

Example 7 shows how the Leading Coefficient Test and zeros of polyno-mial functions can be used as sketching aids.

EXAMPLE 7   *Sketching the Graph of a Polynomial Function*

Sketch the graph of $f(x) = 3x^4 - 4x^3$.

Solution

Because the leading coefficient is positive and the degree is even, you know that the graph eventually rises to the left and to the right, as shown in Figure 3.19(a). By factoring

$$f(x) = 3x^4 - 4x^3$$
$$= x^3(3x - 4)$$

you can see that the zeros of f are $x = 0$ and $x = \frac{4}{3}$ (both of odd multiplicity). Thus, the x-intercepts occur at $(0, 0)$ and $\left(\frac{4}{3}, 0\right)$. To sketch the graph by hand, find a few additional points, as shown in the table. Then plot the points and complete the graph, as shown in Figure 3.19(b).

x	-1	0.5	1	1.5
$f(x)$	7	-0.3125	-1	1.6875

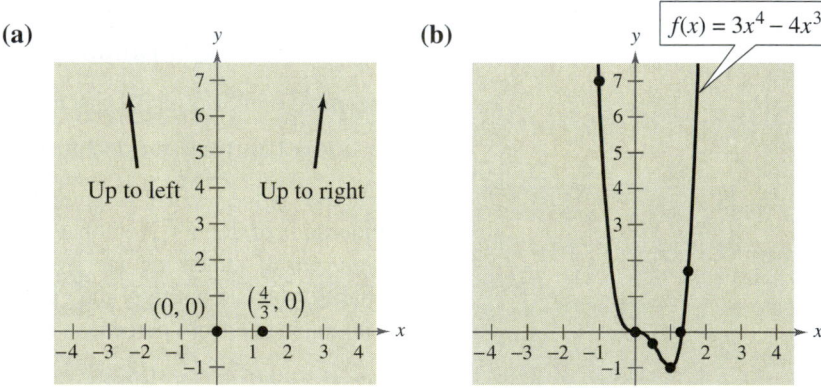

FIGURE 3.19

NOTE   Before applying the Leading Coefficient Test to a polynomial function, it is a good idea to check that the function is written in standard form. For instance, if the function in Example 7 had been given as $f(x) = -4x^3 + 3x^4$, it might have appeared that the leading coefficient was negative. ■■

The Intermediate Value Theorem

The next theorem, called the **Intermediate Value Theorem,** tells you of the existence of real zeros of polynomial functions. The theorem implies that if $(a, f(a))$ and $(b, f(b))$ are two points on the graph of a polynomial function such that $f(a) \neq f(b)$, then for any number d between $f(a)$ and $f(b)$ there must be a number c between a and b such that $f(c) = d$. (See Figure 3.20.)

> ### INTERMEDIATE VALUE THEOREM
>
> Let a and b be real numbers such that $a < b$. If f is a polynomial function such that $f(a) \neq f(b)$, then, in the interval $[a, b]$, f takes on every value between $f(a)$ and $f(b)$.

The Intermediate Value Theorem helps you locate the real zeros of a polynomial function in the following way. If you can find a value $x = a$ where a polynomial function is positive, and another value $x = b$ where it is negative, you can conclude that the function has at least one real zero between these two values. For example, the function

$$f(x) = x^3 + x^2 + 1$$

is negative when $x = -2$ and positive when $x = -1$. Therefore, it follows from the Intermediate Value Theorem that f must have a real zero somewhere between -2 and -1, as shown in Figure 3.21.

By continuing this line of reasoning, you can approximate any real zeros of a polynomial function to any desired accuracy. This concept is further demonstrated in Example 8.

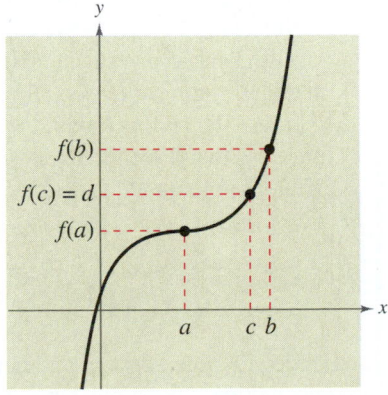

FIGURE 3.20

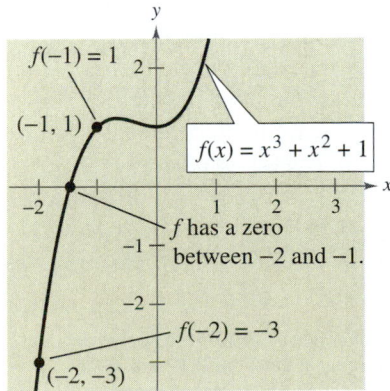

FIGURE 3.21

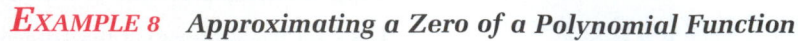

EXAMPLE 8   *Approximating a Zero of a Polynomial Function*

Use the Intermediate Value Theorem to approximate the real zero of

$$f(x) = x^3 - x^2 + 1.$$

Solution

Begin by computing a few function values, as follows.

x	-2	-1	0	1
$f(x)$	-11	-1	1	1

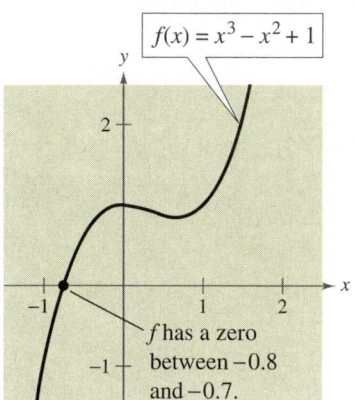

$f(x) = x^3 - x^2 + 1$

f has a zero between -0.8 and -0.7.

FIGURE 3.22

Because $f(-1)$ is negative and $f(0)$ is positive, you can apply the Intermediate Value Theorem to conclude that the function has a zero between -1 and 0. To pinpoint this zero more closely, divide the interval $[-1, 0]$ into tenths and evaluate the function at each point. When you do this, you will find that

$$f(-0.8) = -0.152 \qquad \text{and} \qquad f(-0.7) = 0.167.$$

Thus, f must have a zero between -0.8 and -0.7, as shown in Figure 3.22. By continuing this process you can approximate this zero to any desired accuracy.

TECHNOLOGY

The approximation process in Example 8 adapts very well to a graphing utility. By repeatedly using the zoom and trace features, you can find that the real zero of $f(x) = x^3 - x^2 + 1$ occurs between -0.755 and -0.754.

The *Interactive* CD-ROM offers graphing utility emulators of the *TI-82* and *TI-83*, which can be used with the Examples, Explorations, Technology notes, and Exercises.

GROUP ACTIVITY

CREATING POLYNOMIAL FUNCTIONS

Suppose you are a math instructor and are writing a quiz for your algebra class. You want to make up several polynomial functions for your students to investigate. Discuss how you could find polynomial functions that have reasonably simple zeros. Justify your reasoning. Then use the methods you have discussed to find polynomial functions that have the following zeros. For (a) and (c), find two different polynomial functions having the specified zeros.

a. $-5, \frac{1}{2}$ **b.** $-\frac{1}{6}, 7$

c. $2, 4, -3$ **d.** $3, -\frac{2}{3}, \frac{3}{4}$

e. $-3, -3, -3, -3$ **f.** $1, -2, 3, -4$

Trade polynomials with another person in your class. Verify that one another's functions have the specified zeros.

WARM UP

Factor the expression completely.

1. $12x^2 + 7x - 10$ **2.** $25x^3 - 60x^2 + 36x$

3. $12z^4 + 17z^3 + 5z^2$ **4.** $y^3 + 125$

5. $x^3 + 3x^2 - 4x - 12$ **6.** $x^3 + 2x^2 + 3x + 6$

Find all real solutions of the equation.

7. $5x^2 + 8 = 0$ **8.** $x^2 - 6x + 4 = 0$

9. $4x^2 + 4x - 11 = 0$ **10.** $x^4 - 18x^2 + 81 = 0$

3.2 Exercises

In Exercises 1–8, match the polynomial function with its graph. [The graphs are labeled (a) through (h).]

(a)
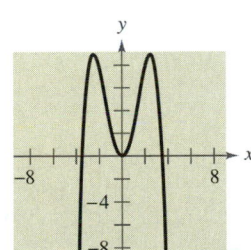

(b)
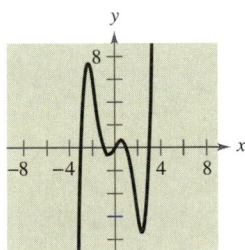

(g)
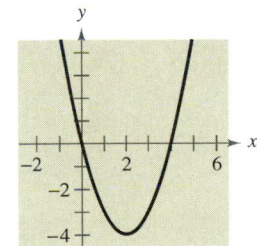

(h)
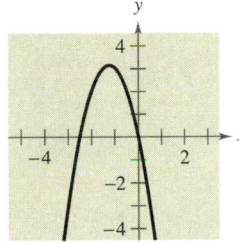

(c)
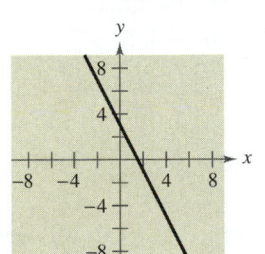

(d)
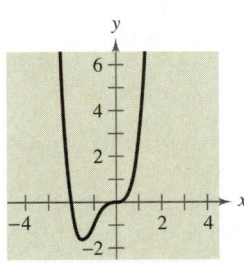

1. $f(x) = -2x + 3$ **2.** $f(x) = x^2 - 4x$

3. $f(x) = -2x^2 - 5x$ **4.** $f(x) = 2x^3 - 3x + 1$

5. $f(x) = -\frac{1}{4}x^4 + 3x^2$ **6.** $f(x) = -\frac{1}{3}x^3 + x^2 - \frac{4}{3}$

7. $f(x) = x^4 + 2x^3$ **8.** $f(x) = \frac{1}{5}x^5 - 2x^3 + \frac{9}{5}x$

In Exercises 9–12, sketch the graph of $y = x^n$ and the specified transformations.

9. $y = x^3$

    (a) $f(x) = (x - 2)^3$ (b) $f(x) = x^3 - 2$

    (c) $f(x) = -\frac{1}{2}x^3$ (d) $f(x) = (x - 2)^3 - 2$

10. $y = x^5$

    (a) $f(x) = (x + 1)^5$ (b) $f(x) = x^5 + 1$

    (c) $f(x) = 1 - \frac{1}{2}x^5$ (d) $f(x) = -\frac{1}{2}(x + 1)^5$

(e)
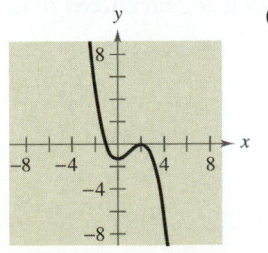

(f)
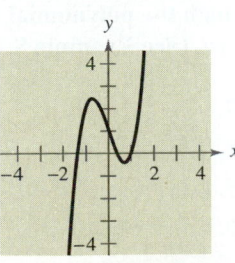

11. $y = x^4$

(a) $f(x) = (x + 3)^4$ (b) $f(x) = x^4 - 3$

(c) $f(x) = 4 - x^4$ (d) $f(x) = \frac{1}{2}(x - 1)^4$

12. $y = x^6$

(a) $f(x) = -\frac{1}{8}x^6$ (b) $f(x) = (x + 2)^6 - 4$

(c) $f(x) = x^6 - 4$ (d) $f(x) = -\frac{1}{4}x^6 + 1$

In Exercises 13–22, determine the right-hand and left-hand behavior of the graph of the polynomial function.

13. $f(x) = \frac{1}{3}x^3 + 5x$

14. $f(x) = 2x^2 - 3x + 1$

15. $g(x) = 5 - \frac{7}{2}x - 3x^2$

16. $h(x) = 1 - x^6$

17. $f(x) = -2.1x^5 + 4x^3 - 2$

18. $f(x) = 2x^5 - 5x + 7.5$

19. $f(x) = 6 - 2x + 4x^2 - 5x^3$

20. $f(x) = \dfrac{3x^4 - 2x + 5}{4}$

21. $h(t) = -\frac{2}{3}(t^2 - 5t + 3)$

22. $f(s) = -\frac{7}{8}(s^3 + 5s^2 - 7s + 1)$

▦ *Graphical Analysis*   **In Exercises 23–26, use a graphing utility to graph the functions f and g on the same viewing rectangle. Zoom out sufficiently far to show that the right-hand and left-hand behavior of f and g appear identical.**

23. $f(x) = 3x^3 - 9x + 1,$ $g(x) = 3x^3$

24. $f(x) = -\frac{1}{3}(x^3 - 3x + 2),$ $g(x) = -\frac{1}{3}x^3$

25. $f(x) = -(x^4 - 4x^3 + 16x),$ $g(x) = -x^4$

26. $f(x) = 3x^4 - 6x^2,$ $g(x) = 3x^4$

In Exercises 27–42, find all the real zeros of the polynomial function.

27. $f(x) = x^2 - 25$ **28.** $f(x) = 49 - x^2$

29. $h(t) = t^2 - 6t + 9$ **30.** $f(x) = x^2 + 10x + 25$

31. $f(x) = x^2 + x - 2$ **32.** $f(x) = \frac{1}{2}x^2 + \frac{5}{2}x - \frac{3}{2}$

33. $f(x) = 3x^2 - 12x + 3$

34. $g(x) = 5(x^2 - 2x - 1)$

35. $f(t) = t^3 - 4t^2 + 4t$

36. $f(x) = x^4 - x^3 - 20x^2$

37. $g(t) = \frac{1}{2}t^4 - \frac{1}{2}$

38. $f(x) = x^5 + x^3 - 6x$

39. $f(x) = 2x^4 - 2x^2 - 40$

40. $g(t) = t^5 - 6t^3 + 9t$

41. $f(x) = 5x^4 + 15x^2 + 10$

42. $f(x) = x^3 - 4x^2 - 25x + 100$

▦ *Graphical Analysis*   **In Exercises 43–46, use a graphing utility to graph the function. Use the graph to approximate any x-intercepts of the graph. Set $y = 0$ and solve the resulting equation. Compare the result with any x-intercepts of the graph.**

43. $y = 4x^3 - 20x^2 + 25x$

44. $y = 4x^3 + 4x^2 - 7x + 2$

45. $y = x^5 - 5x^3 + 4x$

46. $y = \frac{1}{4}x^3(x^2 - 9)$

In Exercises 47–56, find a polynomial function that has the given zeros. (There are many correct answers.)

47. $0, 10$ **48.** $0, -3$

49. $2, -6$ **50.** $-4, 5$

51. $0, -2, -3$ **52.** $0, 2, 5$

53. $4, -3, 3, 0$ **54.** $-2, -1, 0, 1, 2$

55. $1 + \sqrt{3}, 1 - \sqrt{3}$ **56.** $2, 4 + \sqrt{5}, 4 - \sqrt{5}$

▦ **Exercises 57–60, use the Intermediate Value Theorem and a graphing utility to find intervals of length 1 in which the polynomial function is guaranteed to have a zero. (See Example 8.)**

57. $f(x) = x^3 - 3x^2 + 3$

58. $f(x) = 0.11x^3 - 2.07x^2 + 9.81x - 6.88$

59. $g(x) = 3x^4 + 4x^3 - 3$

60. $h(x) = x^4 - 10x^2 + 3$

In Exercises 61–72, sketch the graph of the function.

61. $f(x) = -\frac{3}{2}$

62. $h(x) = \frac{1}{3}x - 3$

63. $f(t) = \frac{1}{4}(t^2 - 2t + 15)$

64. $g(x) = -x^2 + 10x - 16$

65. $f(x) = x^3 - 3x^2$

66. $f(x) = 1 - x^3$

67. $g(t) = -\frac{1}{4}(t - 2)^2(t + 2)^2$

68. $f(x) = x^2(x - 4)$

69. $h(x) = \frac{1}{3}x^3(x - 4)^2$

70. $g(x) = \frac{1}{10}(x + 1)^2(x - 3)^3$

71. $f(x) = 1 - x^6$

72. $g(x) = 1 - (x + 1)^6$

In Exercises 73–76, use a graphing utility to graph the function.

73. $f(x) = x^3 - 4x$

74. $f(x) = \frac{1}{4}x^4 - 2x^2$

75. $g(x) = \frac{1}{5}(x + 1)^2(x - 3)(2x - 9)$

76. $h(x) = \frac{1}{5}(x + 2)^2(3x - 5)^2$

77. *Exploration* Explore the transformations of the form $g(x) = a(x - h)^5 + k$.

(a) Use a graphing utility to graph the functions

$$y_1 = -\frac{1}{3}(x - 2)^5 + 1$$

and

$$y_2 = \frac{3}{5}(x + 2)^5 - 3.$$

Determine whether the graphs are increasing or decreasing. Explain.

(b) Will the graph of g always be increasing or decreasing? If so, is this behavior determined by a, h, or k? Explain.

(c) Use a graphing utility to graph the function

$$H(x) = x^5 - 3x^3 + 2x + 1.$$

Use the graph and the result of part (b) to determine whether H can be written in the form $H(x) = a(x - h)^5 + k$. Explain.

(d) Determine the natural number exponents so that the results of part (b) are true.

78. *Numerical and Graphical Analysis* An open box is to be made from a square piece of material, 36 centimeters on a side, by cutting equal squares from the corners and turning up the sides (see figure).

(a) Complete four rows of a table such as the one below.

Height	Width	Volume
1	$36 - 2(1)$	$1[36 - 2(1)]^2 = 1156$
2	$36 - 2(2)$	$2[36 - 2(2)]^2 = 2048$

(b) Use a graphing utility to generate additional rows of the table. Use the table to estimate the dimensions that will produce a maximum volume.

(c) Verify that the volume of the box is given by

$$V(x) = x(36 - 2x)^2$$

Determine the domain of the function.

(d) Use a graphing utility to graph V and use the graph to estimate the value of x for which $V(x)$ is maximum. Compare your result with that of part (b).

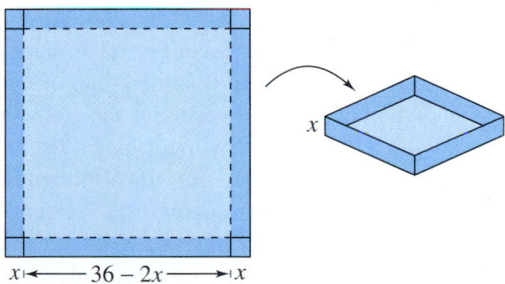

79. *Graphical Reasoning* Sketch a graph of the function $f(x) = x^4$. Explain how the graph of g differs (if it does) from the graph of f. Determine whether g is odd, even, or neither.

(a) $g(x) = f(x) + 2$

(b) $g(x) = f(x + 2)$

(c) $g(x) = f(-x)$

(d) $g(x) = -f(x)$

(e) $g(x) = f\left(\frac{1}{2}x\right)$

(f) $g(x) = \frac{1}{2}f(x)$

(g) $g(x) = f\left(x^{3/4}\right)$

(h) $g(x) = (f \circ f)(x)$

80. *Volume of a Box*   An open box with locking tabs is to be made from a square piece of material 24 inches on a side. This is done by cutting equal squares from the corners and folding along the dashed lines shown in the figure.

(a) Verify that the volume of the box is given by $V(x) = 8x(6 - x)(12 - x)$.

(b) Determine the domain of the function V.

(c) Sketch the graph of the function and estimate the value of x for which $V(x)$ is maximum.

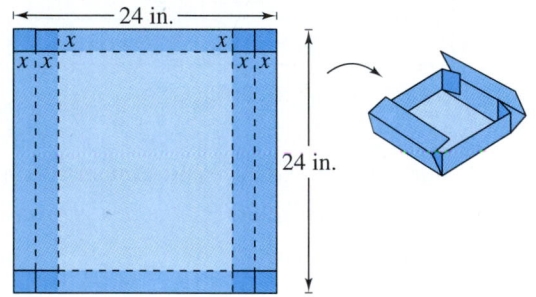

81. *Advertising Expenses*   The total revenue R (in millions of dollars) for a company is related to its advertising expense by the function

$$R = \frac{1}{100,000}(-x^3 + 600x^2), \qquad 0 \le x \le 400$$

where x is the amount spent on advertising (in tens of thousands of dollars). Use the graph of this function, shown in the figure, to estimate the point on the graph at which the function is increasing most rapidly. This point is called the **point of diminishing returns** because any expense above this amount will yield less return per dollar invested in advertising.

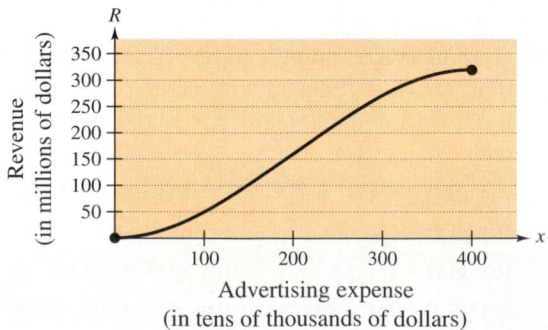

Advertising expense
(in tens of thousands of dollars)

82. *Tree Growth*   The growth of a red oak tree is approximated by the function

$$G = -0.003t^3 + 0.137t^2 + 0.458t - 0.839$$

where G is the height of the tree in feet and t ($2 \le t \le 34$) is its age in years. Use a graphing utility to graph the function and estimate the age of the tree when it is growing most rapidly. This point is called the **point of diminishing returns** because the increase in yield will be less with each additional year. (*Hint:* Use a viewing rectangle in which $-10 \le x \le 45$ and $-5 \le y \le 60$.)

83. Use the graphs of the function f and g to answer each question.

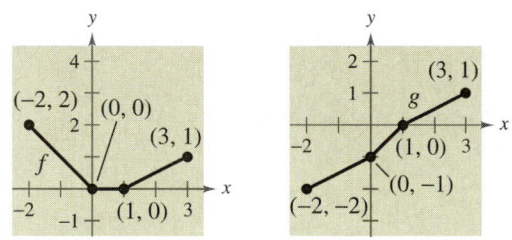

(a) Find $(f \circ g)(3)$.

(b) Explain why f does not have an inverse.

(c) Find $g^{-1}(1)$.

(d) Find $(g^{-1} \circ f)(0)$.

Review   **Solve Exercises 84–87 as a review of the skills and problem-solving techniques you learned in previous sections.**

84. Does the equation $3x - y^2 = 4$ determine y as a function of x? Explain.

85. Does the equation $3x^2 - y = 4$ determine y as a function of x? Explain.

86. Determine the domain of $f(x) = \dfrac{x}{\sqrt{4 - x}}$.

87. Find and simplify the difference quotient $\dfrac{g(x) - g(4)}{x - 4}$ where $g(x) = x^2 - 1$.

3.3 *Polynomial and Synthetic Division*

See Exercise 76 on page 291 for an example of how synthetic division can be used to evaluate a model for the average monthly cable television rates in the United States.

Long Division of Polynomials □ *Synthetic Division* □ *The Remainder and Factor Theorems* □ *Application*

Long Division of Polynomials

In this section, you will study two procedures for *dividing* polynomials. These procedures are especially valuable in factoring and finding the zeros of polynomial functions. To begin, suppose you are given the graph of

$$f(x) = 6x^3 - 19x^2 + 16x - 4.$$

Notice that a zero of f occurs at $x = 2$, as shown in Figure 3.23. Because $x = 2$ is a zero of f, you know that $(x - 2)$ is a factor of $f(x)$. This means that there exists a second-degree polynomial $q(x)$ such that

$$f(x) = (x - 2) \cdot q(x).$$

To find $q(x)$, you can use **long division,** as illustrated in Example 1.

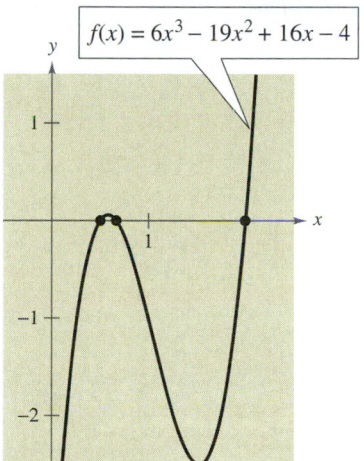

$f(x) = 6x^3 - 19x^2 + 16x - 4$

FIGURE 3.23

EXAMPLE 1   *Long Division of Polynomials*

Divide $6x^3 - 19x^2 + 16x - 4$ by $x - 2$, and use the result to factor the polynomial completely.

Solution

Partial quotients

$$
\begin{array}{r}
6x^2 - 7x + 2 \\
x - 2{\overline{\smash{\big)}\,6x^3 - 19x^2 + 16x - 4}} \\
\underline{6x^3 - 12x^2} \\
-7x^2 + 16x \\
\underline{-7x^2 + 14x} \\
2x - 4 \\
\underline{2x - 4} \\
0
\end{array}
$$

Multiply: $6x^2(x - 2)$.
Subtract.
Multiply: $-7x(x - 2)$.
Subtract.
Multiply: $2(x - 2)$.
Subtract.

From this division, you can conclude that

$$6x^3 - 19x^2 + 16x - 4 = (x - 2)(6x^2 - 7x + 2)$$

and by factoring the quadratic $6x^2 - 7x + 2$, you have

$$6x^3 - 19x^2 + 16x - 4 = (x - 2)(2x - 1)(3x - 2).$$

NOTE   Note that the factorization shown in Example 1 agrees with the graph shown in Figure 3.23 in that the three x-intercepts occur at $x = 2$, $x = \frac{1}{2}$, and $x = \frac{2}{3}$. ■■

In Example 1, $x - 2$ is a factor of the polynomial $6x^3 - 19x^2 + 16x - 4$, and the long division process produces a remainder of zero. Often, long division will produce a nonzero remainder. For instance, if you divide $x^2 + 3x + 5$ by $x + 1$, you obtain the following.

$$
\begin{array}{r}
x + 2 \quad\longleftarrow \text{ Quotient}\\
x + 1\overline{)\,x^2 + 3x + 5} \quad\longleftarrow \text{ Dividend}\\
\underline{x^2 + x}\\
2x + 5\\
\underline{2x + 2}\\
3 \quad\longleftarrow \text{ Remainder}
\end{array}
$$

Divisor $\longrightarrow$

In fractional form, you can write this result as follows.

$$
\underbrace{\frac{\overbrace{x^2 + 3x + 5}^{\text{Dividend}}}{\underbrace{x + 1}_{\text{Divisor}}}} = \overbrace{x + 2}^{\text{Quotient}} + \frac{\overset{\text{Remainder}}{3}}{\underbrace{x + 1}_{\text{Divisor}}}
$$

This implies that $x^2 + 3x + 5 = (x + 1)(x + 2) + 3$, which illustrates the following well-known theorem called the **Division Algorithm.**

  ### THE DIVISION ALGORITHM

If $f(x)$ and $d(x)$ are polynomials such that $d(x) \neq 0$, and the degree of $d(x)$ is less than or equal to the degree of $f(x)$, there exist unique polynomials $q(x)$ and $r(x)$ such that

$$
f(x) = d(x)q(x) + r(x)
$$

Dividend — Quotient

Divisor — Remainder

where $r(x) = 0$ *or* the degree of $r(x)$ is less than the degree of $d(x)$. If the remainder $r(x)$ is zero, $d(x)$ **divides evenly** into $f(x)$.

The Division Algorithm can also be written as

$$
\frac{f(x)}{d(x)} = q(x) + \frac{r(x)}{d(x)}.
$$

In the Division Algorithm, the rational expression $f(x)/d(x)$ is **improper** because the degree of $f(x)$ is greater than or equal to the degree of $d(x)$. On the other hand, the rational expression $r(x)/d(x)$ is **proper** because the degree of $r(x)$ is less than the degree of $d(x)$.

EXAMPLE 2 *Long Division of Polynomials*

Divide $x^3 - 1$ by $x - 1$.

Solution

Because there is no x^2-term or x-term in the dividend, you need to line up the subtraction by using zero coefficients (or leaving spaces) for the missing terms.

$$
\require{enclose}
\begin{array}{r}
x^2 + x + 1 \\
x - 1 \enclose{longdiv}{x^3 + 0x^2 + 0x - 1} \\
\underline{x^3 - x^2} \\
x^2 \\
\underline{x^2 - x} \\
x - 1 \\
\underline{x - 1} \\
0
\end{array}
$$

Thus, $x - 1$ divides evenly into $x^3 - 1$ and you can write

$$\frac{x^3 - 1}{x - 1} = x^2 + x + 1, \quad x \neq 1.$$

NOTE You can check the result of a division problem by multiplying. For instance, in Example 2, try checking that $(x - 1)(x^2 + x + 1) = x^3 - 1$. ■■

EXAMPLE 3 *Long Division of Polynomials*

Divide $2x^4 + 4x^3 - 5x^2 + 3x - 2$ by $x^2 + 2x - 3$.

Solution

$$
\require{enclose}
\begin{array}{r}
2x^2 + 1 \\
x^2 + 2x - 3 \enclose{longdiv}{2x^4 + 4x^3 - 5x^2 + 3x - 2} \\
\underline{2x^4 + 4x^3 - 6x^2} \\
x^2 + 3x - 2 \\
\underline{x^2 + 2x - 3} \\
x + 1
\end{array}
$$

Note that the first subtraction eliminated two terms from the dividend. When this happens, the quotient skips a term. Thus, you can write

$$\frac{2x^4 + 4x^3 - 5x^2 + 3x - 2}{x^2 + 2x - 3} = 2x^2 + 1 + \frac{x + 1}{x^2 + 2x - 3}.$$

Synthetic Division

There is a nice shortcut for long division of polynomials by divisors of the form $x - k$. The shortcut is called **synthetic division.** We summarize the pattern for synthetic division of a cubic polynomial as follows. (The pattern for higher-degree polynomials is similar.)

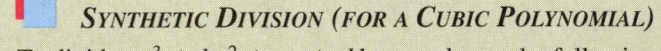

SYNTHETIC DIVISION (FOR A CUBIC POLYNOMIAL)

To divide $ax^3 + bx^2 + cx + d$ by $x - k$, use the following pattern.

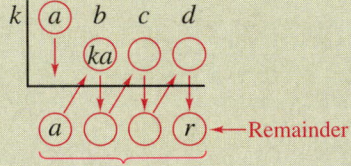

Remainder

Coefficients of quotient

Vertical pattern: Add terms.
Diagonal pattern: Multiply by k.

NOTE Synthetic division works *only* for divisors of the form $x - k$. [Remember that $x + k = x - (-k)$.] You cannot use synthetic division to divide a polynomial by a quadratic such as $x^2 - 3$. ▪▪

EXAMPLE 4 *Using Synthetic Division*

Use synthetic division to divide $x^4 - 10x^2 - 2x + 4$ by $x + 3$.

Solution

You should set up the array as follows. Note that a zero is included for each missing term in the dividend.

Divisor: $x + 3$ Dividend: $x^4 - 10x^2 - 2x + 4$

$$
\begin{array}{r|rrrrr}
-3 & 1 & 0 & -10 & -2 & 4 \\
   &   & -3 & 9 & 3 & -3 \\
\hline
   & 1 & -3 & -1 & 1 & 1
\end{array}
$$

←Remainder: 1

Quotient: $x^3 - 3x^2 - x + 1$

Thus, you have

$$\frac{x^4 - 10x^2 - 2x + 4}{x + 3} = x^3 - 3x^2 - x + 1 + \frac{1}{x + 3}.$$

TECHNOLOGY

With the *TI-83* or the *TI-82*, you can evaluate a function by using the **value** feature. For instance, to evaluate the function in Example 5 (on page 285) at $x = -2$, enter the function in Y₁. Then use the following keystrokes.

| CALC | (1: value)
| ENTER |
| (-) | 2 | ENTER |

The display should be -9.

The Remainder and Factor Theorems

The remainder obtained in the synthetic division process has an important interpretation, as described in the **Remainder Theorem.**

THINK ABOUT THE PROOF

To prove the Remainder Theorem, you can use the Division Algorithm to write $f(x)$ as

$f(x) = (x - k)q(x) + r(x).$

By the Division Algorithm, you know that either $r(x) = 0$ or the degree of $r(x)$ is less than the degree of $x - k$. How does this allow you to prove the theorem? The details of the proof are in the appendix.

THE REMAINDER THEOREM

If a polynomial $f(x)$ is divided by $x - k$, the remainder is

$r = f(k).$

The Remainder Theorem tells you that synthetic division can be used to evaluate a polynomial function. That is, to evaluate a polynomial function $f(x)$ when $x = k$, divide $f(x)$ by $x - k$. The remainder will be $f(k)$, as illustrated in Example 5.

EXAMPLE 5   Using the Remainder Theorem

Use the Remainder Theorem to evaluate the following function at $x = -2$.

$f(x) = 3x^3 + 8x^2 + 5x - 7$

Solution

Using synthetic division, you obtain the following.

$$
\begin{array}{r|rrrr}
-2 & 3 & 8 & 5 & -7 \\
   &   & -6 & -4 & -2 \\
\hline
   & 3 & 2 & 1 & -9
\end{array}
$$

Because the remainder is $r = -9$, you can conclude that

$f(-2) = -9.$

This means that $(-2, -9)$ is a point on the graph of f. Try checking this by substituting $x = -2$ in the original function.

THINK ABOUT THE PROOF

To prove the Factor Theorem, you can use the Division Algorithm to write $f(x)$ as

$f(x) = (x - k)q(x) + r(x)$

By the Remainder Theorem, you know that $r(x) = f(k)$. Thus,

$f(x) = (x - k)q(x) + f(k).$

where $q(x)$ is a polynomial of lesser degree than $f(x)$. How does this allow you to prove the theorem? The details of the proof are in the appendix.

Another important theorem is the **Factor Theorem,** which is stated below. This theorem states that you can test to see whether a polynomial has $(x - k)$ as a factor by evaluating the polynomial at $x = k$. If the result is 0, $(x - k)$ is a factor.

THE FACTOR THEOREM

A polynomial $f(x)$ has a factor $(x - k)$ if and only if $f(k) = 0$.

---

*E*XAMPLE 6   *Factoring a Polynomial*

Show that $(x - 2)$ and $(x + 3)$ are factors of

$$f(x) = 2x^4 + 7x^3 - 4x^2 - 27x - 18.$$

Then find the remaining factors of $f(x)$.

Solution

Using synthetic division with 2 and -3 *successively,* you obtain the following.

$$
\begin{array}{r|rrrrr}
2 & 2 & 7 & -4 & -27 & -18 \\
  &   & 4 & 22 & 36 & 18 \\
\hline
  & 2 & 11 & 18 & 9 & 0
\end{array}
$$

0 remainder

$\longrightarrow$ $(x - 2)$ is a factor.

$$
\begin{array}{r|rrrr}
-3 & 2 & 11 & 18 & 9 \\
   &   & -6 & -15 & -9 \\
\hline
   & 2 & 5 & 3 & 0
\end{array}
$$

0 remainder

$\longrightarrow$ $(x + 3)$ is a factor.

Because the resulting quadratic factors as

$$2x^2 + 5x + 3 = (2x + 3)(x + 1)$$

the complete factorization of $f(x)$ is

$$f(x) = (x - 2)(x + 3)(2x + 3)(x + 1).$$

Note that this factorization implies that f has four real zeros:

$$2, -3, -\tfrac{3}{2}, \text{ and } -1.$$

This is confirmed by the graph of f, which is shown in Figure 3.24.

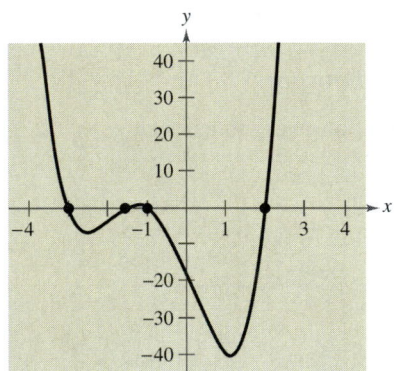

FIGURE 3.24

In summary, the remainder r, obtained in the synthetic division of $f(x)$ by $x - k$, provides the following information.

1. The remainder r gives the value of f at $x = k$. That is, $r = f(k)$.
2. If $r = 0$, $(x - k)$ is a factor of $f(x)$.
3. If $r = 0$, $(k, 0)$ is an x-intercept of the graph of f.

NOTE   Throughout this text, we have emphasized the importance of developing several problem-solving strategies. In the exercises for this section, try using more than one strategy to solve several of the exercises. For instance, if you find that $x - k$ divides evenly into $f(x)$, try sketching the graph of f. You should find that $(k, 0)$ is an x-intercept of the graph. ∎

Application

EXAMPLE 7   *Take-Home Pay*

The 1994 monthly take-home pay for an employee who was married and claimed four deductions is given by the function

$$y = -0.00002079x^2 + 0.86589x + 37.27, \qquad 500 \leq x \leq 5000$$

where y represents the take-home pay and x represents the monthly salary. Find a function that gives the take-home pay as a *percent* of the monthly salary. (Source: Hooper International Accounting Software, based on a state and local income tax rate of 3.1%)

Solution

Because the monthly salary is given by x and the take-home pay is given by y, the percent of monthly salary that the person takes home is

$$P = \frac{y}{x}$$

$$= \frac{-0.00002079x^2 + 0.86589x + 37.27}{x}$$

$$= -0.00002079x + 0.86589 + \frac{37.27}{x}.$$

The difference between gross pay and take-home pay results primarily from federal deductions for income tax, Social Security tax, and Medicare tax. Because the federal income tax is graduated, the percent that is deducted depends on a person's income. People who earn more do not simply pay more tax—they pay much more tax because the tax rate increases.

The graphs of these functions are shown in Figure 3.25(a) and (b). Note in Figure 3.25(b) that as a person's monthly salary increases, the percent that he or she takes home decreases.

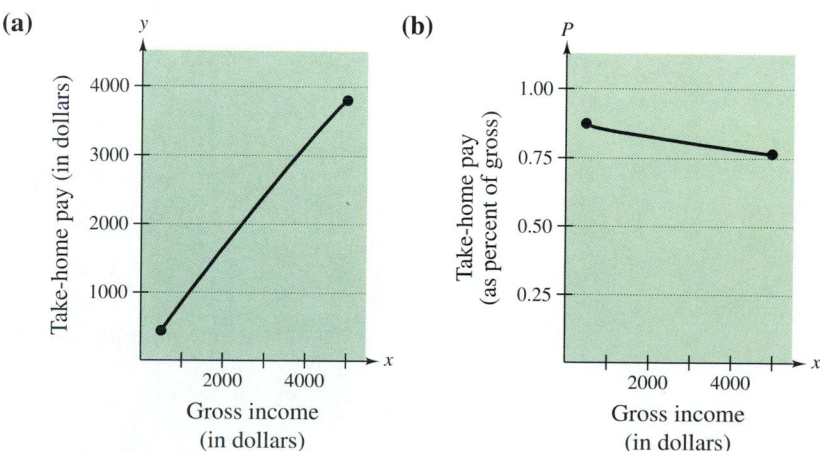

(a)   **(b)**

FIGURE 3.25

Gʀᴏᴜᴘ Aᴄᴛɪᴠɪᴛʏ

ANALYZING A SLANT ASYMPTOTE

By use of long division, the rational function given by $f(x) = (x^2 + 2x + 3)/(x - 1)$ can be written as

$$f(x) = \frac{x^2 + 2x + 3}{x - 1} = x + 3 + \frac{6}{x - 1}.$$

$f(x) = \dfrac{x^2 + 2x + 3}{x - 1}$

$y = x + 3$

As x approaches $-\infty$ or ∞, the value of the fraction $6/(x - 1)$ is insignificant and the graph of f is close to the graph of $y = x + 3$ (see figure). The line

$$y = x + 3 \qquad\qquad \text{Slant asymptote}$$

is a **slant asymptote** of the graph of f. Complete the table to show that the values of $f(x)$ and y are nearly the same for large values of x.

x	$-10{,}000$	-1000	-100	100	1000	$10{,}000$
$f(x)$						
y						

Use long division to show that $g(x) = (2x^3 - 5)/(x^2)$ also has a slant asymptote. What is the equation of this slant asymptote? Graph g and complete the table for g similar to the one above for f. What can you conclude?

WARM UP

Write the expression in standard polynomial form.

1. $(x - 1)(x^2 + 2) + 5$ **2.** $(x^2 - 3)(2x + 4) + 8$

3. $(x^2 + 1)(x^2 - 2x + 3) - 10$ **4.** $(x + 6)(2x^3 - 3x) - 5$

Factor the polynomial.

5. $x^2 - 4x + 3$ **6.** $4x^3 - 10x^2 + 6x$

Find a polynomial function that has the given zeros.

7. $0, 3, 4$ **8.** $-6, 1$

9. $-3, 1 + \sqrt{2}, 1 - \sqrt{2}$ **10.** $1, -2, 2 + \sqrt{3}, 2 - \sqrt{3}$

3.3 Exercises

Analytical Analysis In Exercises 1–4, find the sum or difference of the terms of y_2 to verify that $y_1 = y_2$.

1. $y_1 = \dfrac{4x}{x-1}$, $y_2 = 4 + \dfrac{4}{x-1}$

2. $y_1 = \dfrac{3x-5}{x-3}$, $y_2 = 3 + \dfrac{4}{x-3}$

3. $y_1 = \dfrac{x^2}{x+2}$, $y_2 = x - 2 + \dfrac{4}{x+2}$

4. $y_1 = \dfrac{x^4 - 3x^2 - 1}{x^2 + 5}$, $y_2 = x^2 - 8 + \dfrac{39}{x^2+5}$

Graphical Analysis In Exercises 5 and 6, use a graphing utility to graph the two equations on the same viewing rectangle. Use the graphs to verify that the expressions are equivalent. Verify the results algebraically.

5. $y_1 = \dfrac{x^5 - 3x^3}{x^2 + 1}$, $y_2 = x^3 - 4x + \dfrac{4x}{x^2+1}$

6. $y_1 = \dfrac{x^3 - 2x^2 + 5}{x^2 + x + 1}$, $y_2 = x - 3 + \dfrac{2(x+4)}{x^2+x+1}$

In Exercises 7–20, divide by long division.

7. Divide $2x^2 + 10x + 12$ by $x + 3$.

8. Divide $5x^2 - 17x - 12$ by $x - 4$.

9. Divide $4x^3 - 7x^2 - 11x + 5$ by $4x + 5$.

10. Divide $6x^3 - 16x^2 + 17x - 6$ by $3x - 2$.

11. Divide $x^4 + 5x^3 + 6x^2 - x - 2$ by $x + 2$.

12. Divide $x^3 + 4x^2 - 3x - 12$ by $x^2 - 3$.

13. Divide $7x + 3$ by $x + 2$.

14. Divide $8x - 5$ by $2x + 1$.

15. $(6x^3 + 10x^2 + x + 8) \div (2x^2 + 1)$

16. $(x^3 - 9) \div (x^2 + 1)$

17. $\dfrac{x^4 + 3x^2 + 1}{x^2 - 2x + 3}$

18. $\dfrac{x^5 + 7}{x^3 - 1}$

19. $\dfrac{x^4}{(x-1)^3}$

20. $\dfrac{2x^3 - 4x^2 - 15x + 5}{(x-1)^2}$

Think About It In Exercises 21–22, perform the division by assuming that n is a positive integer.

21. $\dfrac{x^{3n} + 9x^{2n} + 27x^n + 27}{x^n + 3}$

22. $\dfrac{x^{3n} - 3x^{2n} + 5x^n - 6}{x^n - 2}$

In Exercises 23–40, divide by synthetic division.

23. $(3x^3 - 17x^2 + 15x - 25) \div (x - 5)$

24. $(5x^3 + 18x^2 + 7x - 6) \div (x + 3)$

25. $(4x^3 - 9x + 8x^2 - 18) \div (x + 2)$

26. $(9x^3 - 16x - 18x^2 + 32) \div (x - 2)$

27. $(-x^3 + 75x - 250) \div (x + 10)$

28. $(3x^3 - 16x^2 - 72) \div (x - 6)$

29. $(5x^3 - 6x^2 + 8) \div (x - 4)$

30. $(5x^3 + 6x + 8) \div (x + 2)$

31. $\dfrac{10x^4 - 50x^3 - 800}{x - 6}$

32. $\dfrac{x^5 - 13x^4 - 120x + 80}{x + 3}$

33. $\dfrac{x^3 + 512}{x + 8}$

34. $\dfrac{5x^3}{x + 3}$

35. $\dfrac{-3x^4}{x - 2}$

36. $\dfrac{-3x^4}{x + 2}$

37. $\dfrac{180x - x^4}{x - 6}$

38. $\dfrac{5 - 3x + 2x^2 - x^3}{x + 1}$

39. $\dfrac{4x^3 + 16x^2 - 23x - 15}{x + \frac{1}{2}}$

40. $\dfrac{3x^3 - 4x^2 + 5}{x - \frac{3}{2}}$

41. *Essay* Briefly explain what it means for a divisor to *divide evenly* into a dividend.

42. *Essay* Briefly explain how to check polynomial division, and justify your reasoning. Give an example.

Exploration In Exercises 43 and 44, find the constant c such that the denominator will divide evenly into the numerator.

43. $\dfrac{x^3 + 4x^2 - 3x + c}{x - 5}$ 44. $\dfrac{x^5 - 2x^2 + x + c}{x + 2}$

In Exercises 45–48, express the function in the form $f(x) = (x - k)q(x) + r$ for the given value of k, and demonstrate that $f(k) = r$.

Function	*Value of k*
45. $f(x) = x^3 - x^2 - 14x + 11$,	$k = 4$
46. $f(x) = 15x^4 + 10x^3 - 6x^2 + 14$,	$k = -\frac{2}{3}$
47. $f(x) = x^3 + 3x^2 - 2x - 14$,	$k = \sqrt{2}$
48. $f(x) = 4x^3 - 6x^2 - 12x - 4$,	$k = 1 - \sqrt{3}$

In Exercises 49–52, use synthetic division to find the function values. Verify with another method.

49. $f(x) = 4x^3 - 13x + 10$
   (a) $f(1)$ (b) $f(-2)$
   (c) $f\left(\frac{1}{2}\right)$ (d) $f(8)$

50. $g(x) = x^6 - 4x^4 + 3x^2 + 2$
   (a) $g(2)$ (b) $g(-4)$
   (c) $g(3)$ (d) $g(-1)$

51. $h(x) = 3x^3 + 5x^2 - 10x + 1$
   (a) $h(3)$ (b) $h\left(\frac{1}{3}\right)$
   (c) $h(-2)$ (d) $h(-5)$

52. $f(x) = 0.4x^4 - 1.6x^3 + 0.7x^2 - 2$
   (a) $f(1)$ (b) $f(-2)$
   (c) $f(5)$ (d) $f(-10)$

Think About It In Exercises 53 and 54, answer the questions about the division $f(x)/(x - k)$ where

$$f(x) = (x + 3)^2(x - 3)(x + 1)^3.$$

53. What is the remainder when $k = -3$? Explain.

54. If it is necessary to find $f(2)$, is it easier to evaluate it directly or use synthetic division? Explain.

In Exercises 55–62, use synthetic division to show that x is a solution of the third-degree polynomial equation, and use the result to factor the polynomial completely. List all the real zeros of the function.

Polynomial Equation	*Value of x*
55. $x^3 - 7x + 6 = 0$,	$x = 2$
56. $x^3 - 28x - 48 = 0$,	$x = -4$
57. $2x^3 - 15x^2 + 27x - 10 = 0$,	$x = \frac{1}{2}$
58. $48x^3 - 80x^2 + 41x - 6 = 0$,	$x = \frac{2}{3}$
59. $x^3 + 2x^2 - 3x - 6 = 0$,	$x = \sqrt{3}$
60. $x^3 + 2x^2 - 2x - 4 = 0$,	$x = \sqrt{2}$
61. $x^3 - 3x^2 + 2 = 0$,	$x = 1 + \sqrt{3}$
62. $x^3 - x^2 - 13x - 3 = 0$,	$x = 2 - \sqrt{5}$

Graphical Analysis In Exercises 63–66, (a) use the root-finding capabilities of a graphing utility to approximate the zeros of the function accurate to three decimal places. (b) Determine one of the exact zeros and use synthetic division to verify your result, and then factor the polynomial completely.

63. $f(x) = x^3 - 2x^2 - 5x + 10$

64. $g(x) = x^3 - 4x^2 - 2x + 8$

65. $h(t) = t^3 - 2t^2 - 7t + 2$

66. $f(s) = s^3 - 12s^2 + 40s - 24$

In Exercises 67–72, simplify the rational expression.

67. $\dfrac{4x^3 - 8x^2 + x + 3}{2x - 3}$

68. $\dfrac{x^3 + x^2 - 64x - 64}{x + 8}$

69. $\dfrac{x^3 + 3x^2 - x - 3}{x + 1}$

70. $\dfrac{2x^3 + 3x^2 - 3x - 2}{x - 1}$

71. $\dfrac{x^4 + 6x^3 + 11x^2 + 6x}{x^2 + 3x + 2}$

72. $\dfrac{x^4 + 9x^3 - 5x^2 - 36x + 4}{x^2 - 4}$

73. *Power of an Engine* The horsepower y developed by a compact car engine is approximated by the model

$$y = -1.42x^3 + 5.04x^2 + 32.45x - 0.75, \quad 1 \leq x \leq 5$$

where x is the engine speed in thousands of revolutions per minute. Note in the graph that there are two engine speeds that develop 110 horsepower, one of which is 5000 rpm. Approximate the second engine speed. Verify with another method.

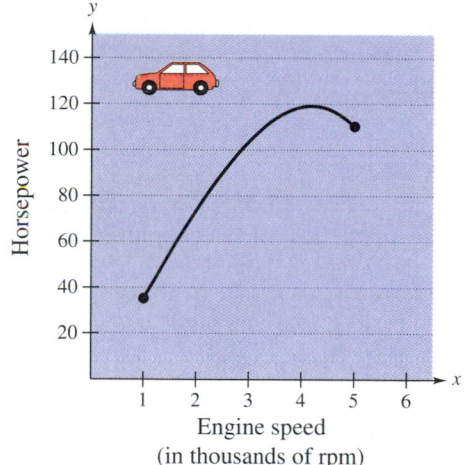

Engine speed
(in thousands of rpm)

74. *Automobile Emissions* The number of parts per million of nitric oxide emissions y from a certain car engine is approximated by the model

$$y = -5.05x^3 + 3857x - 38411.25, \quad 13 \leq x \leq 18$$

where x is the air-fuel ratio. Use a graphing utility to graph the model. Observe that there are two air-fuel ratios that produce 2400 parts per million of nitric oxide, one of which is 15. Approximate the second air-fuel ratio.

75. *Exploration* Use the form

$$f(x) = (x - k)q(x) + r$$

to create a cubic function that (a) passes through the point $(2, 5)$ and rises to the right, and (b) passes through the point $(-3, 1)$ and falls to the right. (The answers are not unique.)

76. *Data Analysis* The average monthly basic rate R for cable television in the United States is given for the years 1985 through 1993 in the table. The variable t represents the year, with $t = 0$ corresponding to 1990. (Source: *The Cable TV Financial Databook*)

t	-5	-4	-3	-2	-1
R	9.73	10.67	12.18	13.86	15.21

t	0	1	2	3
R	16.78	18.10	19.08	19.39

(a) Use a graphing utility to sketch a scatter plot of the data.

(b) A model for the data is given by

$$R = 16.823 + 1.415t - 0.115t^2 - 0.023t^3.$$

Use a graphing utility to graph the model on the same viewing rectangle as the scatter plot. Compare the model with the data.

(c) Use the model to create a table of estimated values of R. Compare the estimated values with the actual data.

(d) Use synthetic division to evaluate the model for the year 1996. Even though the model is relatively accurate for estimating the given data, do you think it is accurate in predicting future cable rates? Explain.

Review **Solve Exercises 77–80 as a review of the skills and problem-solving techniques you learned in previous sections.**

77. Find the inverse of $f(x) = \sqrt[3]{x} + 2$. Use a graphing utility to graph f and f^{-1}.

78. Verify algebraically that $f(x) = 2 + (1/x)$ is the inverse of $g(x) = 1/(x - 2)$.

79. Determine the domain of $f \circ g$ if $f(x) = \sqrt{x}$ and $g(x) = x^2 - 25$.

80. Determine if $f(x) = x^3 + 3x^2$ has an inverse.

See Exercise 72 on page 301 for an example of how the zeros of a polynomial function can help analyze data dealing with the value of imported goods in the United States from 1984 through 1993.

3.4   *Real Zeros of Polynomial Functions*

Descartes's Rule of Signs   □   *The Rational Zero Test*   □
Bounds for Real Zeros of Polynomial Functions

Descartes's Rule of Signs

In Section 3.2, you learned that an *n*th-degree polynomial function can have *at most n* real zeros. Of course, many *n*th-degree polynomials do not have that many zeros. For instance, $f(x) = x^2 + 1$ has no real zeros, and $f(x) = x^3 + 1$ has only one real zero. The following theorem, called **Descartes's Rule of Signs,** sheds more light on the number of real zeros of a polynomial.

NOTE   When there is only one variation in sign, Descartes's Rule of Signs guarantees the existence of exactly one positive (or negative) real zero. ▪▪

> **DESCARTES'S RULE OF SIGNS**
>
> Let $f(x) = a_n x^n + a_{n-1} x^{n-1} + \cdots + a_2 x^2 + a_1 x + a_0$ be a polynomial with real coefficients and $a_0 \neq 0$.
>
> 1. The number of *positive real zeros* of f is either equal to the number of variations in sign of $f(x)$ *or* less than that number by an even integer.
> 2. The number of *negative real zeros* of f is either equal to the number of variations in sign of $f(-x)$ *or* less than that number by an even integer.

A **variation in sign** means that two consecutive coefficients have opposite signs. For example, the polynomial

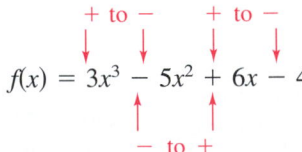

$$f(x) = 3x^3 - 5x^2 + 6x - 4$$

has *three* variations in sign, whereas

$$f(-x) = 3(-x)^3 - 5(-x)^2 + 6(-x) - 4$$
$$= -3x^3 - 5x^2 - 6x - 4$$

has no variations in sign. Thus, from Descartes's Rule of Signs, the polynomial $f(x) = 3x^3 - 5x^2 + 6x - 4$ has either three positive real zeros or one positive real zero, and has no negative real zeros. From the graph in Figure 3.26, you can see that the function has only one real zero (it is a positive number, near 1).

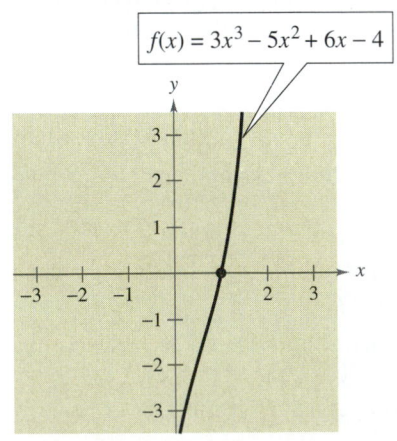

$f(x) = 3x^3 - 5x^2 + 6x - 4$

FIGURE 3.26

The Rational Zero Test

The **Rational Zero Test** relates the possible rational zeros of a polynomial (having integer coefficients) to the leading coefficient and to the constant term of the polynomial.

> ### THE RATIONAL ZERO TEST
>
> If the polynomial $f(x) = a_n x^n + a_{n-1} x^{n-1} + \cdots + a_2 x^2 + a_1 x + a_0$ has *integer* coefficients, every rational zero of f has the form
>
> $$\text{Rational zero} = \frac{p}{q}$$
>
> where p and q have no common factors other than 1, and
>
> p = a factor of the constant term a_0
>
> q = a factor of the leading coefficient a_n.

NOTE When the leading coefficient is 1, the possible rational zeros are simply the factors of the constant term. ∎∎

To use the Rational Zero Test, you should first list all rational numbers whose numerators are factors of the constant term and whose denominators are factors of the leading coefficient.

$$\text{Possible rational zeros} = \frac{\text{factors of constant term}}{\text{factors of leading coefficient}}$$

Having formed this list of *possible rational zeros,* use a trial-and-error method to determine which, if any, are actual zeros of the polynomial.

EXAMPLE 1 *Rational Zero Test with Leading Coefficient of 1*

Find the rational zeros of $f(x) = x^3 + x + 1$.

Solution

Because the leading coefficient is 1, the possible rational zeros are ± 1, the factors of the constant term. By testing these possible zeros, you can see that neither works.

$$f(1) = (1)^3 + 1 + 1 = 3$$
$$f(-1) = (-1)^3 + (-1) + 1 = -1$$

Thus, you can conclude that the given polynomial has *no* rational zeros. Note from the graph of f in Figure 3.27 that f does have one real zero (between -1 and 0). However, by the Rational Zero Test, you know that this real zero is *not* a rational number.

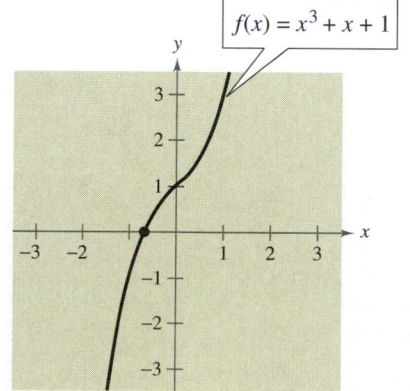

$f(x) = x^3 + x + 1$

FIGURE 3.27

EXAMPLE 2 *Rational Zero Test with Leading Coefficient of 1*

Find the rational zeros of

$$f(x) = x^4 - x^3 + x^2 - 3x - 6.$$

Solution

Because the leading coefficient is 1, the possible rational zeros are the factors of the constant term.

Possible rational zeros: $\pm 1, \pm 2, \pm 3, \pm 6$

A test of these possible zeros shows that $x = -1$ and $x = 2$ are the only two that work. Check the others to be sure.

If the leading coefficient of a polynomial is not 1, the list of possible rational zeros can increase dramatically. In such cases the search can be shortened in several ways: (1) a programmable calculator can be used to speed up the calculations; (2) a graph, drawn either by hand or with a graphing utility, can give a good estimate of the locations of the zeros; and (3) synthetic division can be used to test the possible rational zeros.

To see how to use synthetic division to test the possible rational zeros, let's take another look at the function

$$f(x) = x^4 - x^3 + x^2 - 3x - 6$$

given in Example 2. To test that $x = -1$ and $x = 2$ are zeros of f, you can apply synthetic division successively, as follows.

$$
\begin{array}{r|rrrrr}
-1 & 1 & -1 & 1 & -3 & -6 \\
   &   & -1 & 2 & -3 & 6 \\
\hline
   & 1 & -2 & 3 & -6 & 0
\end{array}
$$

$$
\begin{array}{r|rrrr}
2 & 1 & -2 & 3 & -6 \\
  &   & 2 & 0 & 6 \\
\hline
  & 1 & 0 & 3 & 0
\end{array}
$$

Thus, you have

$$f(x) = (x + 1)(x - 2)(x^2 + 3).$$

Because the factor $(x^2 + 3)$ produces no real zeros, you can conclude that $x = -1$ and $x = 2$ are the only *real* zeros of f.

Finding the first zero is often the hardest part. After that, the search is simplified by using the lower-degree polynomial obtained in synthetic division.

EXAMPLE 3 *Using the Rational Zero Test*

Find the rational zeros of $f(x) = 2x^3 + 3x^2 - 8x + 3$.

Solution

The leading coefficient is 2 and the constant term is 3.

$$\textit{Possible rational zeros: } \frac{\text{Factors of 3}}{\text{Factors of 2}} = \frac{\pm 1, \pm 3}{\pm 1, \pm 2} = \pm 1, \pm 3, \pm \frac{1}{2}, \pm \frac{3}{2}$$

By synthetic division, you can determine that $x = 1$ is a zero.

$$
\begin{array}{r|rrrr}
1 & 2 & 3 & -8 & 3 \\
  &   & 2 & 5 & -3 \\
\hline
  & 2 & 5 & -3 & 0
\end{array}
$$

Thus, $f(x)$ factors as

$$f(x) = (x - 1)(2x^2 + 5x - 3) = (x - 1)(2x - 1)(x + 3)$$

and you can conclude that the rational zeros of f are $x = 1$, $x = \frac{1}{2}$, and $x = -3$.

EXAMPLE 4 *Using the Rational Zero Test*

Find all the real zeros of $f(x) = 10x^3 - 15x^2 - 16x + 12$.

Solution

The leading coefficient is 10 and the constant term is 12.

$$\textit{Possible rational zeros: } \frac{\text{Factors of 12}}{\text{Factors of 10}} = \frac{\pm 1, \pm 2, \pm 3, \pm 4, \pm 6, \pm 12}{\pm 1, \pm 2, \pm 5, \pm 10}$$

With so many possibilities (32, in fact), it is worth your time to stop and sketch a graph. From Figure 3.28, it looks like three reasonable choices would be $x = -\frac{6}{5}$, $x = \frac{1}{2}$, and $x = 2$. Testing these by synthetic division shows that only $x = 2$ works. Thus, you have

$$f(x) = (x - 2)(10x^2 + 5x - 6).$$

Using the Quadratic Formula, you find that the two additional zeros are irrational numbers.

$$x = \frac{-5 + \sqrt{265}}{20} \approx 0.5639 \quad \text{and} \quad x = \frac{-5 - \sqrt{265}}{20} \approx -1.0639$$

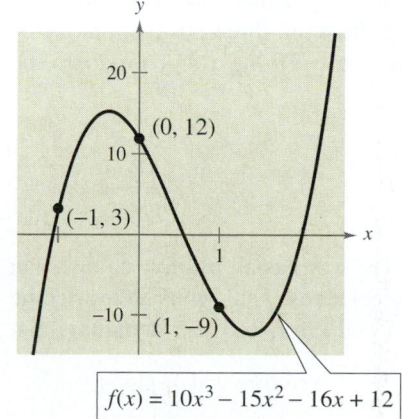

$$f(x) = 10x^3 - 15x^2 - 16x + 12$$

FIGURE 3.28

Bounds for Real Zeros of Polynomial Functions

The third test for zeros of a polynomial function is related to the sign pattern in the last row of the synthetic division tableau. This test can give you an upper or lower bound of the real zeros of f.

NOTE   A real number b is an **upper bound** for the real zeros of f if no zeros are greater than b. Similarly, b is a **lower bound** if no real zeros of f are less than b. ▪▪

UPPER AND LOWER BOUND RULE

Let $f(x)$ be a polynomial with real coefficients and a positive leading coefficient. Suppose $f(x)$ is divided by $x - c$, using synthetic division.

1. If $c > 0$ and each number in the last row is either positive or zero, c is an *upper bound* for the real zeros of f.
2. If $c < 0$ and the numbers in the last row are alternately positive and negative (zero entries count as positive or negative), c is a *lower bound* for the real zeros of f.

▶ *Exploration*

Graph

$f(x) = 6x^3 - 4x^2 + 3x - 2.$

Notice that the graph intersects the x-axis at the point $\left(\frac{2}{3}, 0\right)$. How does this relate to the real zero found in Example 5? Graph

$g(x) = x^4 - 5x^3 + 3x^2 + x.$

How many times does the graph intersect the x-axis? How many real zeros does g have?

EXAMPLE 5   *Finding the Zeros of a Polynomial Function*

Find the real zeros of $f(x) = 6x^3 - 4x^2 + 3x - 2$.

Solution

The possible real zeros are as follows.

$$\frac{\text{Factors of } 2}{\text{Factors of } 6} = \frac{\pm 1, \pm 2}{\pm 1, \pm 2, \pm 3, \pm 6} = \pm 1, \pm\frac{1}{2}, \pm\frac{1}{3}, \pm\frac{1}{6}, \pm\frac{2}{3}, \pm 2$$

Because $f(x)$ has three variations in sign and $f(-x)$ has none, you can apply Descartes's Rule of Signs to conclude that there are three positive real zeros or one positive real zero, and no negative zeros. Trying $x = 1$ produces the following.

$$
\begin{array}{r|rrrr}
1 & 6 & -4 & 3 & -2 \\
  &   & 6 & 2 & 5 \\
\hline
  & 6 & 2 & 5 & 3
\end{array}
$$

Thus, $x = 1$ is not a zero, but because the last row has all positive entries, you know that $x = 1$ is an upper bound for the real zeros. Thus, you can restrict the search to zeros between 0 and 1. By trial and error, you can determine that $x = \frac{2}{3}$ is a zero. Thus,

$$f(x) = \left(x - \tfrac{2}{3}\right)(6x^2 + 3).$$

Because $6x^2 + 3$ has no real zeros, if follows that $x = \frac{2}{3}$ is the only real zero.

Before concluding this section, we list two additional hints that can help you find the real zeros of a polynomial.

1. If the terms of $f(x)$ have a common monomial factor, it should be factored out before applying the tests in this section. For instance, by writing

$$f(x) = x^4 - 5x^3 + 3x^2 + x$$
$$= x(x^3 - 5x^2 + 3x + 1)$$

you can see that $x = 0$ is a zero of f and that the remaining zeros can be obtained by analyzing the cubic factor.

2. If you are able to find all but two zeros of $f(x)$, you are home free because you can always use the Quadratic Formula on the remaining quadratic factor. For instance, if you succeeded in writing

$$f(x) = x^4 - 5x^3 + 3x^2 + x$$
$$= x(x - 1)(x^2 - 4x - 1)$$

you can apply the Quadratic Formula to $x^2 - 4x - 1$ to conclude that the two remaining zeros are

$$x = 2 + \sqrt{5} \quad \text{and} \quad x = 2 - \sqrt{5}.$$

GROUP ACTIVITY

COMPARING REAL ZEROS AND RATIONAL ZEROS

Discuss the meanings of *real zeros* and *rational zeros* of a polynomial function and compare these two types of zeros. Then answer the following questions, explain your reasoning, and construct a table summarizing your results.

a. Is it possible for a polynomial function to have no rational zeros but to have real zeros? If so, give an example.

b. If a polynomial function has three real zeros, and only one of them is a rational number, must the other two zeros be irrational numbers? If so, give an example.

c. Consider a cubic polynomial function

$$f(x) = ax^3 + bx^2 + cx + d$$

where $a \neq 0$. Is it possible that f has no real zeros? If so, give an example. Is is possible that f has no rational zeros? If so, give an example.

d. Is it possible that a second-degree polynomial function with integer coefficients has one rational zero and one irrational zero? If so, give an example.

Find a polynomial function with integer coefficients having the given zeros.

1. $-1, \frac{2}{3}, 3$ 2. $-2, 0, \frac{3}{4}, 2$

Divide by synthetic division.

3. $\dfrac{x^5 - 9x^3 + 5x + 18}{x + 3}$ 4. $\dfrac{3x^4 + 17x^3 + 10x^2 - 9x - 8}{x + (2/3)}$

Use the given zero to find all the real zeros of f.

5. $f(x) = 2x^3 + 11x^2 + 2x - 4, \quad x = \frac{1}{2}$

6. $f(x) = 6x^3 - 47x^2 - 124x - 60, \quad x = 10$

7. $f(x) = 4x^3 - 13x^2 - 4x + 6, \quad x = -\frac{3}{4}$

8. $f(x) = 10x^3 + 51x^2 + 48x - 28, \quad x = \frac{2}{5}$

Find all real solutions of the equation.

9. $x^4 - 3x^2 + 2 = 0$ 10. $x^4 - 7x^2 + 12 = 0$

3.4 Exercises

In Exercises 1–10, use Descartes's Rule of Signs to determine the possible number of positive and negative zeros of the function.

1. $f(x) = x^3 + 3$

2. $g(x) = x^3 + 3x^2$

3. $g(x) = 5x^5 + 10x$

4. $h(x) = 4x^2 - 8x + 3$

5. $h(x) = 3x^4 + 2x^2 + 1$

6. $h(x) = 2x^4 - 3x + 2$

7. $g(x) = 2x^3 - 3x^2 - 3$

8. $f(x) = 4x^3 - 3x^2 + 2x - 1$

9. $f(x) = -5x^3 + x^2 - x + 5$

10. $f(x) = 3x^3 + 2x^2 + x + 3$

In Exercises 11–16, use the Rational Zero Test to list all possible rational zeros of f. Verify that the zeros of f shown on the graph are contained in the list.

11. $f(x) = x^3 + 3x^2 - x - 3$

12. $f(x) = x^3 - 4x^2 - 4x + 16$

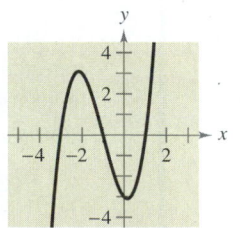

FIGURE FOR 11

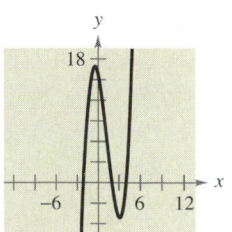

FIGURE FOR 12

13. $f(x) = 2x^4 - 17x^3 + 35x^2 + 9x - 45$

14. $f(x) = 6x^3 - 71x^2 - 13x + 12$

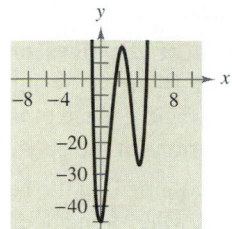 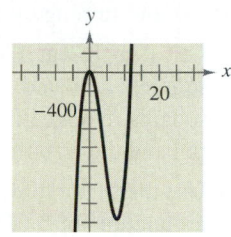

FIGURE FOR 13   FIGURE FOR 14

15. $f(x) = 20x^4 + 144x^3 - 253x^2 - 900x + 800$

16. $f(x) = 4x^5 - 8x^4 - 5x^3 + 10x^2 + x - 2$

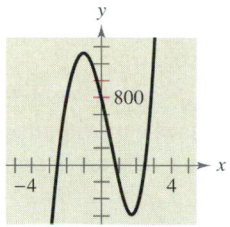 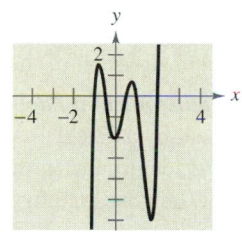

FIGURE FOR 15   FIGURE FOR 16

In Exercises 17–28, find all the real zeros of the function.

17. $f(x) = x^3 - 6x^2 + 11x - 6$

18. $f(x) = x^3 - 7x - 6$

19. $g(x) = x^3 - 4x^2 - x + 4$

20. $h(x) = x^3 - 9x^2 + 20x - 12$

21. $h(t) = t^3 + 12t^2 + 21t + 10$

22. $f(x) = x^3 + 6x^2 + 12x + 8$

23. $f(x) = x^3 - 4x^2 + 5x - 2$

24. $p(x) = x^3 - 9x^2 + 27x - 27$

25. $C(x) = 2x^3 + 3x^2 - 1$

26. $f(x) = 3x^3 - 19x^2 + 33x - 9$

27. $f(x) = 9x^4 - 9x^3 - 58x^2 + 4x + 24$

28. $f(x) = 2x^4 - 15x^3 + 23x^2 + 15x - 25$

In Exercises 29–32, find all real solutions of the polynomial equation.

29. $z^4 - z^3 - 2z - 4 = 0$

30. $x^4 - 13x^2 - 12x = 0$

31. $2y^4 + 7y^3 - 26y^2 + 23y - 6 = 0$

32. $x^5 - x^4 - 3x^3 + 5x^2 - 2x = 0$

In Exercises 33–36, (a) list the possible rational zeros of f, (b) sketch the graph of f so that some of the possible zeros in part (a) can be disregarded, and then (c) determine all real zeros of f.

33. $f(x) = x^3 + x^2 - 4x - 4$

34. $f(x) = -3x^3 + 20x^2 - 36x + 16$

35. $f(x) = -4x^3 + 15x^2 - 8x - 3$

36. $f(x) = 4x^3 - 12x^2 - x + 15$

In Exercises 37–40, (a) list the possible rational zeros of f, (b) use a graphing utility to graph f so that some of the possible zeros in part (a) can be disregarded, and then (c) determine all real zeros of f.

37. $f(x) = -2x^4 + 13x^3 - 21x^2 + 2x + 8$

38. $f(x) = 4x^4 - 17x^2 + 4$

39. $f(x) = 32x^3 - 52x^2 + 17x + 3$

40. $f(x) = 4x^3 + 7x^2 - 11x - 18$

Graphical Analysis **In Exercises 41–44, (a) use the root-finding capabilities of a graphing utility to approximate the zeros of the function accurate to three decimal places. (b) Determine one of the exact zeros and use synthetic division to verify your result, and then factor the polynomial completely.**

41. $f(x) = x^4 - 3x^2 + 2$

42. $P(t) = t^4 - 7t^2 + 12$

43. $h(x) = x^5 - 7x^4 + 10x^3 + 14x^2 - 24x$

44. $g(x) = 6x^4 - 11x^3 - 51x^2 + 99x - 27$

In Exercises 45–48, use synthetic division to verify the upper and lower bounds of the zeros of f.

45. $f(x) = x^4 - 4x^3 + 15$

    (a) Upper: $x = 4$ (b) Lower: $x = -1$

46. $f(x) = 2x^3 - 3x^2 - 12x + 8$

    (a) Upper: $x = 4$ (b) Lower: $x = -3$

47. $f(x) = x^4 - 4x^3 + 16x - 16$

    (a) Upper: $x = 5$ (b) Lower: $x = -3$

48. $f(x) = 2x^4 - 8x + 3$

    (a) Upper: $x = 3$ (b) Lower: $x = -4$

In Exercises 49–52, find all the real zeros of the function.

49. $f(x) = 4x^3 - 3x - 1$

50. $f(z) = 12z^3 - 4z^2 - 27z + 9$

51. $f(y) = 4y^3 + 3y^2 + 8y + 6$

52. $g(x) = 3x^3 - 2x^2 + 15x - 10$

In Exercises 53–56, find all the rational zeros of the polynomial function.

53. $P(x) = x^4 - \frac{25}{4}x^2 + 9 = \frac{1}{4}(4x^4 - 25x^2 + 36)$

54. $f(x) = x^3 - \frac{3}{2}x^2 - \frac{23}{2}x + 6 = \frac{1}{2}(2x^3 - 3x^2 - 23x + 12)$

55. $f(x) = x^3 - \frac{1}{4}x^2 - x + \frac{1}{4} = \frac{1}{4}(4x^3 - x^2 - 4x + 1)$

56. $f(z) = z^3 + \frac{11}{6}z^2 - \frac{1}{2}z - \frac{1}{3} = \frac{1}{6}(6z^3 + 11z^2 - 3z - 2)$

In Exercises 57–60, match the cubic function with the numbers of rational and irrational zeros.

    (a) Rational zeros: 0;   Irrational zeros: 1

    (b) Rational zeros: 3;   Irrational zeros: 0

    (c) Rational zeros: 1;   Irrational zeros: 2

    (d) Rational zeros: 1;   Irrational zeros: 0

57. $f(x) = x^3 - 1$ **58.** $f(x) = x^3 - 2$

59. $f(x) = x^3 - x$ **60.** $f(x) = x^3 - 2x$

61. *Dimensions of a Box*   An open box is to be made from a rectangular piece of material, 15 centimeters by 9 centimeters, by cutting equal squares from the corners and turning up the sides.

    (a) Let x represent the length of the sides of the squares removed. Draw a figure showing the squares removed from the original piece of material and the resulting dimensions of the open box.

    (b) Use the figure to write the volume V of the box as a function of x. Determine the domain of the function.

    (c) Sketch the graph of the function and approximate the dimensions of the box that yield a maximum volume.

    (d) Find values of x such that $V = 56$. Which of these values is a physical impossibility in the construction of the box? Explain.

62. *Dimensions of a Package*   A rectangular package to be sent by a postal service can have a maximum combined length and girth (perimeter of a cross section) of 120 inches (see figure).

    (a) Show that the volume of the box is given by
$$V(x) = 4x^2(30 - x).$$

    (b) Use a graphing utility to graph the function and approximate the dimensions of the box that yield a maximum volume.

    (c) Find values of x such that $V = 13{,}500$. Which of these values is a physical impossibility in the construction of the package? Explain.

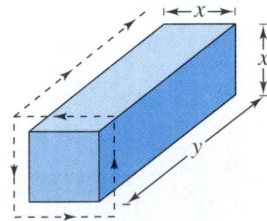

Think About It In Exercises 63–68, determine (if possible) the zeros of the function g if the function f has zeros at $x = r_1$, $x = r_2$, and $x = r_3$.

63. $g(x) = -f(x)$

64. $g(x) = 3f(x)$

65. $g(x) = f(x - 5)$

66. $g(x) = f(2x)$

67. $g(x) = 3 + f(x)$

68. $g(x) = f(-x)$

69. *Advertising Costs* A company that produces portable cassette players estimates that the profit for selling a particular model is given by

$$P = -76x^3 + 4830x^2 - 320{,}000, \quad 0 \le x \le 60$$

where P is the profit in dollars and x is the advertising expense in tens of thousands of dollars. Using this model, find the smaller of two advertising amounts that yields a profit of $2,500,000.

70. *Advertising Costs* A company that manufactures bicycles estimates that the profit for selling a particular model is given by

$$P = -45x^3 + 2500x^2 - 275{,}000, \quad 0 \le x \le 50$$

where P is the profit in dollars and x is the advertising expense in tens of thousands of dollars. Using this model, find the smaller of two advertising amounts that yields a profit of $800,000.

71. *Ordering and Transportation Cost* The ordering and transportation cost C for the components used in manufacturing a certain product is given by

$$C = 100\left(\frac{200}{x^2} + \frac{x}{x + 30}\right), \quad 1 \le x$$

where C is measured in thousands of dollars and x is the order size in hundreds. In calculus, it can be shown that the cost is a minimum when

$$3x^3 - 40x^2 - 2400x - 36{,}000 = 0.$$

Use a calculator to approximate the optimal order size to the nearest hundred units.

72. *Imports into the United States* The values (in billions of dollars) of goods imported into the United States for the years 1984 through 1993 are given in the table. (Source: *U.S. General Imports and Imports for Consumption*)

Year	1984	1985	1986	1987	1988
Imports	325.7	345.3	370.0	406.2	441.0

Year	1989	1990	1991	1992	1993
Imports	473.4	495.3	487.1	532.7	580.5

(a) A model for this data is given by

$$I = 0.161t^3 + 0.626t^2 + 25.646t + 484.137$$

where I is the annual value of goods imported (in billions of dollars) and t represents the calendar year, with $t = 0$ corresponding to 1990. Use a graphing utility to sketch a scatter plot of the data and the model on the same viewing rectangle. How do they compare?

(b) According to this model, when did the annual value of imports reach 525 billion dollars?

(c) According to the right-hand behavior of the model, will the value of imports continue to increase? Explain.

Review Solve Exercises 73–78 as a review of the skills and problem-solving techniques you learned in previous sections. Use the given graph to graph the function g.

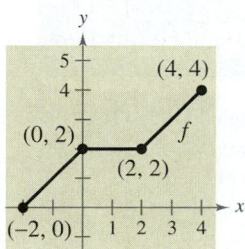

73. $g(x) = f(x - 2)$

74. $g(x) = f(x) - 2$

75. $g(x) = 2f(x)$

76. $g(x) = f(-x)$

77. $g(x) = f(2x)$

78. $g(x) = f\left(\frac{1}{2}x\right)$

3.5   *The Fundamental Theorem of Algebra*

See Exercise 64 on page 310 for an example of how the zeros of a polynomial function can help you analyze the profit function for a business.

The Fundamental Theorem of Algebra ◻ *Conjugate Pairs* ◻
Factoring a Polynomial

The Fundamental Theorem of Algebra

You have been using the fact that an nth-degree polynomial can have at most n real zeros. In the complex number system, this statement can be improved. That is, in the complex number system, every nth-degree polynomial function has *precisely* n zeros. This important result is derived from the **Fundamental Theorem of Algebra,** first proved by the famous German mathematician Carl Friedrich Gauss (1777–1855).

> ### THE FUNDAMENTAL THEOREM OF ALGEBRA
>
> If $f(x)$ is a polynomial of degree n, where $n > 0$, then f has at least one zero in the complex number system.

Using the Fundamental Theorem of Algebra and the equivalence of zeros and factors, you obtain the following theorem.

> ### LINEAR FACTORIZATION THEOREM
>
> If $f(x)$ is a polynomial of degree n
>
> $$f(x) = a_n x^n + a_{n-1}x^{n-1} + \cdots + a_1 x + a_0$$
>
> where $n > 0$, then f has precisely n linear factors
>
> $$f(x) = a_n(x - c_1)(x - c_2) \cdots (x - c_n)$$
>
> where $c_1, c_2, \ldots, c_n$ are complex numbers and a_n is the leading coefficient of $f(x)$.

Jean Le Rond d'Alembert (1717–1783) worked independently of Carl Gauss in trying to prove the Fundamental Theorem of Algebra. His efforts were such that, in France, the Fundamental Theorem of Algebra is frequently known as the Theorem of d'Alembert.
(Photo: The Fogg Art Museum)

Note that neither the Fundamental Theorem of Algebra nor the Linear Factorization Theorem tells you *how* to find the zeros or factors of a polynomial. Such theorems are called **existence theorems.** To find the zeros of a polynomial function, you still must rely on the techniques developed earlier in the text.

Remember that the n zeros of a polynomial function can be real or complex, and they may be repeated. Example 1 illustrates several cases.

THINK ABOUT THE PROOF

To prove the Linear Factorization Theorem, you can use the Fundamental Theorem of Algebra to conclude that f must have at least one zero, c_1. Thus, $(x - c_1)$ is a factor of $f(x)$, and by the Factor Theorem it follows that

$$f(x) = (x - c_1)f_1(x).$$

If the degree of $f_1(x)$ is greater than zero, you can apply the Fundamental Theorem of Algebra again to conclude that f_1 has a zero, c_2. How can you continue this reasoning to complete the proof? The details of the proof are in the appendix.

EXAMPLE 1 *Zeros of Polynomial Functions*

a. The first-degree polynomial $f(x) = x - 2$ has exactly *one* zero: $x = 2$.

b. Counting multiplicity, the second-degree polynomial function

$$f(x) = x^2 - 6x + 9 = (x - 3)(x - 3)$$

has exactly *two* zeros: $x = 3$ and $x = 3$. (This is called a **repeated zero.**)

c. The third-degree polynomial function

$$f(x) = x^3 + 4x = x(x^2 + 4) = x(x - 2i)(x + 2i)$$

has exactly *three* zeros: $x = 0$, $x = 2i$, and $x = -2i$.

d. The fourth-degree polynomial function

$$f(x) = x^4 - 1 = (x - 1)(x + 1)(x - i)(x + i)$$

has exactly *four* zeros: $x = 1$, $x = -1$, $x = i$, and $x = -i$.

TECHNOLOGY

When a graphing utility is used to locate the zeros of a function, the only zeros that appear as x-intercepts are the *real zeros*. Compare the graphs below with the four polynomial functions in Example 1. Which zeros appear on the graphs?

(a)
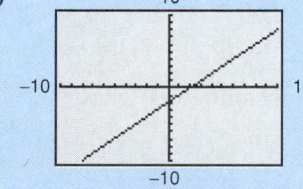

(b)
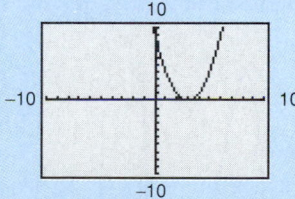

(c)
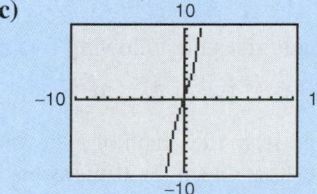

(d)
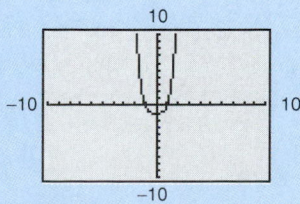

Example 2 shows how to use the methods described in Sections 3.3 and 3.4 (Descarte's Rule of Signs, the Rational Zero Test, synthetic division, and factoring) to find all the zeros of a polynomial function, including complex zeros.

EXAMPLE 2   *Finding the Zeros of a Polynomial Function*

Write $f(x) = x^5 + x^3 + 2x^2 - 12x + 8$ as the product of linear factors, and list all of its zeros.

Solution

Descartes's Rule of Signs indicates two or no positive real zeros and one negative real zero. Moreover, the possible rational zeros are $\pm 1, \pm 2, \pm 4,$ and ± 8. Synthetic division produces the following.

$$
\begin{array}{r|rrrrrr}
1 & 1 & 0 & 1 & 2 & -12 & 8 \\
  &   & 1 & 1 & 2 & 4 & -8 \\
\hline
  & 1 & 1 & 2 & 4 & -8 & 0
\end{array}
\longrightarrow \quad \text{1 is a zero.}
$$

$$
\begin{array}{r|rrrrr}
1 & 1 & 1 & 2 & 4 & -8 \\
  &   & 1 & 2 & 4 & 8 \\
\hline
  & 1 & 2 & 4 & 8 & 0
\end{array}
\longrightarrow \quad \text{1 is a repeated zero.}
$$

$$
\begin{array}{r|rrrr}
-2 & 1 & 2 & 4 & 8 \\
   &   & -2 & 0 & -8 \\
\hline
   & 1 & 0 & 4 & 0
\end{array}
\longrightarrow \quad \text{--2 is a zero.}
$$

Thus, you have

$$f(x) = x^5 + x^3 + 2x^2 - 12x + 8$$
$$= (x - 1)(x - 1)(x + 2)(x^2 + 4).$$

By factoring $x^2 + 4$ as

$$x^2 - (-4) = \left(x - \sqrt{-4}\right)\left(x + \sqrt{-4}\right) = (x - 2i)(x + 2i)$$

you obtain

$$f(x) = (x - 1)(x - 1)(x + 2)(x - 2i)(x + 2i)$$

which gives the following five zeros of f.

$$1, \quad 1, \quad -2, \quad 2i, \quad \text{and} \quad -2i$$

Note from the graph of f shown in Figure 3.29 that the *real* zeros are the only ones that appear as x-intercepts.

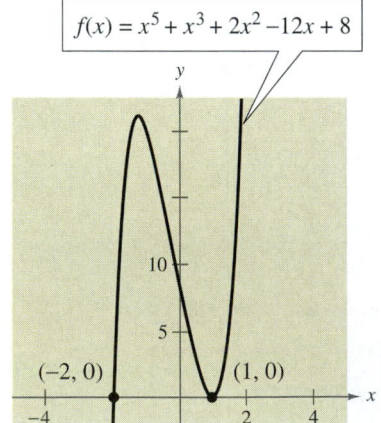

$f(x) = x^5 + x^3 + 2x^2 - 12x + 8$

FIGURE 3.29

Conjugate Pairs

In Example 2, note that the two complex zeros are **conjugates.** That is, they are of the form $a + bi$ and $a - bi$.

> ### COMPLEX ZEROS OCCUR IN CONJUGATE PAIRS
>
> Let $f(x)$ be a polynomial function that has *real coefficients.* If $a + bi$, where $b \neq 0$, is a zero of the function, the conjugate $a - bi$ is also a zero of the function.

NOTE Be sure you see that this result is true only if the polynomial function has *real coefficients.* For instance, the result applies to the function $f(x) = x^2 + 1$, but not to the function $g(x) = x - i$. ■■

EXAMPLE 3 *Finding a Polynomial with Given Zeros*

Find a fourth-degree polynomial function, with real coefficients, that has -1, -1, and $3i$ as zeros.

Solution

Because $3i$ is a zero *and* the polynomial is stated to have real coefficients, you know that the conjugate $-3i$ must also be a zero. Thus, from the Linear Factorization Theorem, $f(x)$ can be written as

$$f(x) = a(x + 1)(x + 1)(x - 3i)(x + 3i).$$

For simplicity, let $a = 1$, and obtain

$$f(x) = (x^2 + 2x + 1)(x^2 + 9)$$
$$= x^4 + 2x^3 + 10x^2 + 18x + 9.$$

Factoring a Polynomial

The Linear Factorization Theorem shows that you can write any nth-degree polynomial as the product of n linear factors.

$$f(x) = a(x - c_1)(x - c_2)(x - c_3) \cdots (x - c_n)$$

However, this result includes the possibility that some of the values of c_i are complex. The boxed statement at the top of page 306 implies that even if you do not want to get involved with "complex factors," you can still write $f(x)$ as the product of linear and/or quadratic factors.

THINK ABOUT THE PROOF

To prove the theorem at the right, begin by using the Linear Factorization Theorem to write the polynomial function as

$$f(x) = a(x - c_1) \cdots (x - c_n).$$

If each c_i is real, there is nothing to prove. If any of the c_i's are complex, how can you use the theorem on page 305 to conclude that $f(x)$ has a quadratic factor with real coefficients? The details of the proof are given in the appendix.

FACTORS OF A POLYNOMIAL

Every polynomial of degree $n > 0$ with real coefficients can be written as the product of linear and quadratic factors with real coefficients, where the quadratic factors have no real zeros.

A quadratic factor with no real zeros is said to be **irreducible over the reals.** Be sure you see that this is not the same as being *irreducible over the rationals.* For example, the quadratic

$$x^2 + 1 = (x - i)(x + i)$$

is irreducible over the reals (and therefore over the rationals). On the other hand, the quadratic

$$x^2 - 2 = (x - \sqrt{2})(x + \sqrt{2})$$

is irreducible over the rationals but *reducible* over the reals.

EXAMPLE 4   *Factoring a Polynomial*

Write the polynomial $f(x) = x^4 - x^2 - 20$:

a. as the product of factors that are irreducible over the *rationals,*

b. as the product of linear factors and quadratic factors that are irreducible over the *reals,* and

c. in completely factored form.

Solution

a. Begin by factoring the polynomial into the product of two quadratic polynomials.

$$x^4 - x^2 - 20 = (x^2 - 5)(x^2 + 4)$$

Both of these factors are irreducible over the rationals.

b. By factoring over the reals, you have

$$x^4 - x^2 - 20 = (x + \sqrt{5})(x - \sqrt{5})(x^2 + 4)$$

where the quadratic factor is irreducible over the reals.

c. In completely factored form, you have

$$x^4 - x^2 - 20 = (x + \sqrt{5})(x - \sqrt{5})(x - 2i)(x + 2i).$$

NOTE   In Example 4, notice from the completely factored form that the 4th-degree polynomial has four zeros. ■ ■

EXAMPLE 5   *Finding the Zeros of a Polynomial Function*

Find all the zeros of $f(x) = x^4 - 3x^3 + 6x^2 + 2x - 60$, given that $1 + 3i$ is a zero of f.

Solution

Because complex zeros occur in conjugate pairs, you know that $1 - 3i$ is also a zero of f. This means that both

$$[x - (1 + 3i)] \quad \text{and} \quad [x - (1 - 3i)]$$

are factors of $f(x)$. Multiplying these two factors produces

$$[x - (1 + 3i)][x - (1 - 3i)] = [(x - 1) - 3i][(x - 1) + 3i]$$
$$= (x - 1)^2 - 9i^2$$
$$= x^2 - 2x + 10.$$

Using long division, you can divide $x^2 - 2x + 10$ into $f(x)$ to obtain the following.

$$
\require{enclose}
\begin{array}{r}
x^2 - x - 6 \\
x^2 - 2x + 10 \enclose{longdiv}{x^4 - 3x^3 + 6x^2 + 2x - 60} \\
\underline{x^4 - 2x^3 + 10x^2 } \\
-x^3 - 4x^2 + 2x \\
\underline{-x^3 + 2x^2 - 10x } \\
-6x^2 + 12x - 60 \\
\underline{-6x^2 + 12x - 60} \\
0
\end{array}
$$

Therefore, you have

$$f(x) = (x^2 - 2x + 10)(x^2 - x - 6) = (x^2 - 2x + 10)(x - 3)(x + 2)$$

and you can conclude that the zeros of f are $1 + 3i$, $1 - 3i$, 3, and -2.

GROUP ACTIVITY

FACTORING A POLYNOMIAL

Compile a list of all the various techniques for factoring a polynomial that have been covered thus far in the text. Give an example illustrating each technique, and discuss when the use of each technique is appropriate.

WARM UP

Write the complex number in standard form and give its complex conjugate.

1. $4 - \sqrt{-29}$

2. $-5 - \sqrt{-144}$

3. $-1 + \sqrt{-32}$

4. $6 + \sqrt{-1/4}$

Perform the operations.

5. $(-3 + 6i) - (10 - 3i)$

6. $(12 - 4i) + 20i$

7. $(4 - 2i)(3 + 7i)$

8. $(2 - 5i)(2 + 5i)$

9. $\dfrac{1 + i}{1 - i}$

10. $(3 + 2i)^3$

3.5 Exercises

In Exercises 1–4, find all the zeros of the function.

1. $f(x) = x(x - 6)^2$

2. $g(x) = (x - 2)(x + 4)^3$

3. $h(t) = (t - 3)(t - 2)(t - 3i)(t + 3i)$

4. $h(m) = (m - 4)^2(m - 2 + 4i)(m - 2 - 4i)$

Graphical and Analytical Analysis   In Exercises 5–8, describe the relationship between the real zeros of the function and the x-intercepts of the graph.

5. $f(x) = x^3 - 4x^2 + x - 4$

6. $f(x) = x^3 - 4x^2 - 4x + 16$

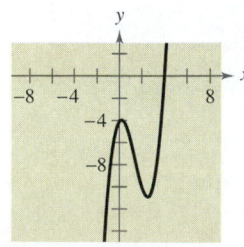

FIGURE FOR 5   **FIGURE FOR 6**

7. $f(x) = x^4 + 4x^2 + 4$

8. $f(x) = x^4 - 3x^2 - 4$

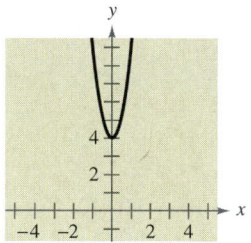
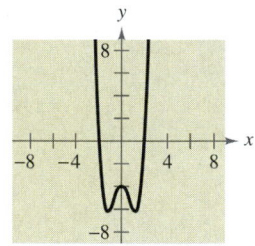

FIGURE FOR 7   **FIGURE FOR 8**

In Exercises 9–28, find all the zeros of the function and write the polynomial as a product of linear factors.

9. $f(x) = x^2 + 25$

10. $f(x) = x^2 - x + 56$

11. $h(x) = x^2 - 4x + 1$

12. $g(x) = x^2 + 10x + 23$

13. $f(x) = x^4 - 81$

14. $f(y) = y^4 - 625$

15. $f(z) = z^2 - 2z + 2$

16. $h(x) = x^3 - 3x^2 + 4x - 2$

17. $g(x) = x^3 - 6x^2 + 13x - 10$

18. $f(x) = x^3 - 2x^2 - 11x + 52$

19. $f(t) = t^3 - 3t^2 - 15t + 125$

20. $f(x) = x^3 + 11x^2 + 39x + 29$

21. $h(x) = x^3 - x + 6$

22. $h(x) = x^3 + 9x^2 + 27x + 35$

23. $f(x) = 5x^3 - 9x^2 + 28x + 6$

24. $g(x) = 3x^3 - 4x^2 + 8x + 8$

25. $g(x) = x^4 - 4x^3 + 8x^2 - 16x + 16$

26. $h(x) = x^4 + 6x^3 + 10x^2 + 6x + 9$

27. $f(x) = x^4 + 10x^2 + 9$

28. $f(x) = x^4 + 29x^2 + 100$

In Exercises 29–34, find all the zeros of the function. When there is an extended list of possible rational zeros, use a graphing utility to graph the function in order to discard any rational zeros that are obviously not zeros of the function.

29. $f(x) = x^3 + 24x^2 + 214x + 740$

30. $f(s) = 2s^3 - 5s^2 + 12s - 5$

31. $f(x) = 16x^3 - 20x^2 - 4x + 15$

32. $f(x) = 9x^3 - 15x^2 + 11x - 5$

33. $f(x) = 2x^4 + 5x^3 + 4x^2 + 5x + 2$

34. $g(x) = x^5 - 8x^4 + 28x^3 - 56x^2 + 64x - 32$

In Exercises 35–44, find a polynomial function with integer coefficients that has the given zeros.

35. $1, 5i, -5i$

36. $4, 3i, -3i$

37. $6, -5 + 2i, -5 - 2i$

38. $2, 4 + i, 4 - i$

39. $i, -i, 6i, -6i$

40. $2, 2, 2, 4i, -4i$

41. $\frac{2}{3}, -1, 3 + \sqrt{2}i$

42. $-5, -5, 1 + \sqrt{3}i$

43. $\frac{3}{4}, -2, -\frac{1}{2} + i$

44. $0, 0, 4, 1 + i$

In Exercises 45–48, write the polynomial (a) as the product of factors that are irreducible over the *rationals*, (b) as the product of linear and quadratic factors that are irreducible over the *reals*, and (c) in completely factored form.

45. $f(x) = x^4 + 6x^2 - 27$

46. $f(x) = x^4 - 2x^3 - 3x^2 + 12x - 18$
(*Hint:* One factor is $x^2 - 6$.)

47. $f(x) = x^4 - 4x^3 + 5x^2 - 2x - 6$
(*Hint:* One factor is $x^2 - 2x - 2$.)

48. $f(x) = x^4 - 3x^3 - x^2 - 12x - 20$
(*Hint:* One factor is $x^2 + 4$.)

In Exercises 49–58, use the given zero to find all the zeros of the function.

Function	Zero
49. $f(x) = 2x^3 + 3x^2 + 50x + 75,$	$5i$
50. $f(x) = x^3 + x^2 + 9x + 9,$	$3i$
51. $f(x) = 2x^4 - x^3 + 7x^2 - 4x - 4,$	$2i$
52. $g(x) = x^3 - 7x^2 - x + 87,$	$5 + 2i$
53. $g(x) = 4x^3 + 23x^2 + 34x - 10,$	$-3 + i$
54. $h(x) = 3x^3 - 4x^2 + 8x + 8,$	$1 - \sqrt{3}i$
55. $f(x) = x^4 + 3x^3 - 5x^2 - 21x + 22,$	$-3 + \sqrt{2}i$
56. $f(x) = x^3 + 4x^2 + 14x + 20,$	$-1 - 3i$
57. $h(x) = 8x^3 - 14x^2 + 18x - 9,$	$\frac{1}{2}(1 - \sqrt{5}i)$
58. $f(x) = 25x^3 - 55x^2 - 54x - 18,$	$\frac{1}{5}(-2 + \sqrt{2}i)$

59. *Think About It* A zero of the function $f(x) = x^3 + ix^2 + ix - 1$ is $x = -i$.

 (a) Show that the conjugate $x = i$ is not a zero of f.

 (b) Does the result of part (a) contradict the theorem that complex zeros occur in conjugate pairs? Explain.

60. *Essay* Explain whether or not there exists a third-degree polynomial function with integer coefficients that has no real zeros.

61. *Exploration* Use a graphing utility to graph the function $f(x) = x^4 - 4x^2 + k$ for different values of k. Find values of k such that the zeros of f satisfy the specified characteristics. (Some parts do not have unique answers.)

(a) Four real zeros

(b) Two real zeros each of multiplicity 2

(c) Two real zeros and two complex roots

(d) Four complex zeros

62. *Think About It* Will the answers to Exercise 61 change for the function g?

(a) $g(x) = f(x - 2)$

(b) $g(x) = f(2x)$

63. *Maximum Height* A baseball is thrown upward from ground level with an initial velocity of 48 feet per second, and its height h in feet is given by

$$h = -16t^2 + 48t, \quad 0 \le t \le 3$$

where t is the time in seconds. Suppose you are told the ball reaches a height of 64 feet. Is this possible?

64. *Profit* The demand equation for a certain product is given by $p = 140 - 0.0001x$, where p is the unit price (in dollars) of the product and x is the number of units produced and sold. The cost equation for the product is $C = 80x + 150{,}000$, where C is the total cost (in dollars) and x is the number of units produced. The total profit obtained by producing and selling x units is given by

$$P = R - C = xp - C.$$

You are working in the marketing department of the company that produces this product, and you are asked to determine a price p that will yield a profit of 9 million dollars. Is this possible? Explain.

65. (a) Find a quadratic function f (with integer coefficients) that has $\pm \sqrt{b}\,i$ as zeros. Assume that b is a positive integer.

(b) Find a quadratic function f (with integer coefficients) that has $a \pm bi$ as zeros. Assume that b is a positive integer.

66. *Graphical Reasoning* The graph of one of the following functions is given in the figure. Identify the function. Explain why each of the others is not the correct function. Use a graphing utility to verify your result.

(a) $f(x) = x^2(x + 2)(x - 3.5)$

(b) $g(x) = (x + 2)(x - 3.5)$

(c) $h(x) = (x + 2)(x - 3.5)(x^2 + 1)$

(d) $k(x) = (x + 1)(x + 2)(x - 3.5)$

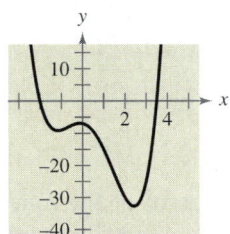

Chapter Opener **In Exercises 67 and 68 plot the complex numbers.**

67. $-4 + 2i$ **68.** $\frac{3}{2} + \frac{11}{2}i$

Review **Solve Exercises 69–72 as a review of the skills and problem-solving techniques you learned in previous sections.**

69. What is 150% of 420?

70. Two cars start at a given point and travel in the same direction at average speeds of 84 kilometers per hour and 100 kilometers per hour. How much time must elapse before the two cars are 12 kilometers apart?

71. To obtain the height of a flagpole you measure the flagpole's shadow and find that it is 10 feet long. You also measure the shadow of a yardstick and find that it is 15 inches long. How tall is the flagpole?

72. A baseball diamond has the shape of a square with sides 90 feet long. A player running from second base to third base is x feet from third. Let d be the distance from home plate to the runner.

(a) Write an equation giving the relationship between x and d.

(b) Find d if $x = 20$.

3.6   *Mathematical Modeling*

See Exercise 87 on page 323 for an example of how a linear function can be used to model the salaries of professional football players from 1987 through 1992.

Introduction ◻ *Direct Variation* ◻ *Direct Variation as nth Power* ◻
Inverse Variation ◻ *Joint Variation* ◻ *Least Squares Regression*

Introduction

You have already studied some techniques for fitting models to data. For instance, in Section 2.1, you learned how to find the equation of a line that passes through two points. In this section, you will study other techniques for fitting models to data: *direct and inverse variation* and *least squares regression.* The resulting models are either polynomial functions or rational functions. (Rational functions will be studied in Chapter 4.)

Real Life

EXAMPLE 1   *A Mathematical Model*

The annual amounts of advertising expenses y (in billions of dollars) in the United States from 1983 to 1992 are given in the table.   (Source: McCann Erickson)

Year	1983	1984	1985	1986	1987	1988	1989	1990	1991	1992
y	75.9	88.1	94.8	102.1	109.8	118.1	125.6	128.6	126.4	131.7

A linear model that approximates this data is

$$y = 6.1703t + 63.8327, \quad 3 \le t \le 12$$

where $t = 3$ corresponds to 1983. Plot the actual data *and* the model on the same graph. How closely does the model represent the data?

Solution

The actual data is plotted in Figure 3.30, along with the graph of the linear model. From the figure, it appears that the model is a "good fit" for the actual data. You can see how well the model fits by comparing the actual values of y with the values of y given by the model (the values given by the model are labeled y^* in the table below).

t	3	4	5	6	7	8	9	10	11	12
y	75.9	88.1	94.8	102.1	109.8	118.1	125.6	128.6	126.4	131.7
y^*	82.3	88.5	94.7	100.9	107.0	113.2	119.4	125.5	131.7	137.9

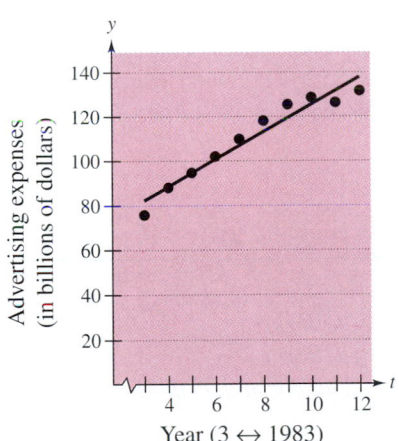

FIGURE 3.30

Direct Variation

There are two basic types of linear models. The more general model has a *y*-intercept that is nonzero: $y = mx + b$, $b \neq 0$. The simpler one, $y = mx$, has a *y*-intercept that is zero. In the simpler model, *y* is said to **vary directly as *x*,** or to be **directly proportional to *x*.**

DIRECT VARIATION

The following statements are equivalent.

1. *y* **varies directly** as *x*.
2. *y* is **directly proportional** to *x*.
3. $y = mx$ for some nonzero constant *m*.

m is the **constant of variation** or the **constant of proportionality.**

EXAMPLE 2   *Direct Variation*

In Pennsylvania, the state income tax is directly proportional to *gross income*. Suppose you were working in Pennsylvania and your state income tax deduction was $31.50 for a gross monthly income of $1500.00. Find a mathematical model that gives the Pennsylvania state income tax in terms of the gross income.

Solution

| *Verbal Model:* | State income tax | = | *m* | · | Gross income |

Labels: State income tax = *y*   *(dollars)*
             Gross income = *x*   *(dollars)*
             Income tax rate = *m*   *(percent in decimal form)*

Equation: $y = mx$

To solve for *m*, substitute the given information into the equation $y = mx$, and then solve for *m*.

$$y = mx \qquad\qquad \text{Direct variation model}$$
$$31.50 = m(1500) \qquad \text{Substitute } y = 31.50 \text{ and } x = 1500.$$
$$0.021 = m \qquad\qquad \text{Income tax rate}$$

Thus, the equation (or model) for state income tax in Pennsylvania is $y = 0.021x$. In other words, Pennsylvania has a state income tax rate of 2.1% of the gross income. The graph of this equation is shown in Figure 3.31.

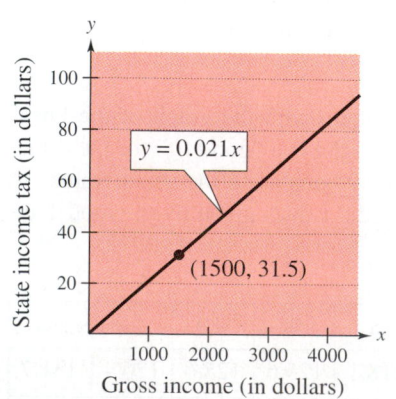

FIGURE 3.31

Direct Variation as *n*th Power

Another type of direct variation relates one variable to a *power* of another variable. For example, in the formula for the area of a circle, $A = \pi r^2$, the area A is directly proportional to the square of the radius r. Note that for this formula, π is the constant of proportionality.

DIRECT VARIATION AS NTH POWER

The following statements are equivalent.

1. y **varies directly as the *n*th power** of x.
2. y is **directly proportional to the *n*th power** of x.
3. $y = kx^n$ for some constant k.

Real Life

EXAMPLE 3 *Direct Variation as nth Power*

The distance a ball rolls down an inclined plane is directly proportional to the square of the time it rolls. During the first second the ball rolls 8 feet. (See Figure 3.32.)

a. Write an equation relating the distance traveled to the time.

b. How far will the ball roll during the first 3 seconds?

$t = 0$ sec
$t = 1$ sec
10 20 30 40 50 60 70
$t = 3$ sec

FIGURE 3.32

Solution

a. Letting d be the distance (in feet) the ball rolls and letting t be the time (in seconds),

$$d = kt^2.$$

Now, because $d = 8$ when $t = 1$, you can see that $k = 8$. Thus, the equation relating distance to time is

$$d = 8t^2.$$

b. When $t = 3$, the distance traveled is $d = 8(3^2) = 8(9) = 72$ feet.

In Examples 2 and 3, the direct variations are such that an *increase* in one variable corresponds to an *increase* in the other variable. This is also true in the model $d = \frac{1}{5}F$, $F > 0$, where an increase in F results in an increase in d. You should not, however, assume that this always occurs with direct variation. For example, in the model $y = -3x$, an increase in x results in a *decrease* in y, and yet we say that y varies directly as x.

Inverse Variation

NOTE   If x and y are related by an equation of the form

$$y = \frac{k}{x^n}$$

then y varies inversely as the nth power of x (or y is inversely proportional to the nth power of x).

▪▪

   INVERSE VARIATION

The following statements are equivalent.

1. y **varies inversely** as x.
2. y is **inversely proportional** to x.
3. $y = \dfrac{k}{x}$ for some constant k.

EXAMPLE 4   *Inverse Variation*

Real Life

A gas law states that the volume of an enclosed gas varies directly as the temperature *and* inversely as the pressure, as shown in Figure 3.33. The pressure of a gas is 0.75 kilogram per square centimeter when the temperature is 294°K and the volume is 8000 cubic centimeters.

a. Write an equation relating the pressure, temperature, and volume.

b. Find the pressure when the temperature is 300°K and the volume is 7000 cubic centimeters.

Solution

a. Let V be the volume (in cubic centimeters), P the pressure (in kilograms per square centimeter), and T the temperature (in degrees Kelvin). Because V varies directly as T *and* inversely as P,

$$V = \frac{kT}{P}.$$

Now, because $P = 0.75$ when $T = 294$ and $V = 8000$,

$$8000 = \frac{k(294)}{0.75}$$

$$\frac{8000(0.75)}{294} = k$$

$$k = \frac{6000}{294} = \frac{1000}{49}.$$

Thus, the equation relating pressure, temperature, and volume is

$$V = \frac{1000}{49}\left(\frac{T}{P}\right).$$

b. When $T = 300$ and $V = 7000$, the pressure is

$$P = \frac{1000}{49}\left(\frac{300}{7000}\right) = \frac{300}{343} \approx 0.87 \text{ kilogram per square centimeter.}$$

P_1 P_1

V_1

$P_2 > P_1$
then
$V_2 < V_1$

P_2 P_2

V_2

FIGURE 3.33   If the temperature is held constant and pressure increases, volume decreases.

Joint Variation

In Example 4, note that when a direct and inverse variation occur in the same statement, we couple them with the word "and." To describe two different *direct* variations in the same statement, we use the word **jointly.**

NOTE If *x*, *y*, and *z* are related by an equation of the form

$$z = kx^n y^m$$

then *z* varies jointly as the *n*th power of *x* and the *m*th power of *y*. ∎∎

>
> **JOINT VARIATION**
>
> The following statements are equivalent.
>
> **1.** *z* **varies jointly** as *x* and *y*.
> **2.** *z* is **jointly proportional** to *x* and *y*.
> **3.** $z = kxy$ for some constant *k*.

Real
Life

EXAMPLE 5 *Joint Variation*

The *simple* interest for a certain savings account is jointly proportional to the time and the principal. After one quarter (three months), the interest on a principal of $5000 is $106.25.

a. Write an equation relating the interest, principal, and time.

b. Find the interest after three quarters.

Solution

a. Let I = interest (in dollars), P = principal (in dollars), and t = time (in years). Because I is jointly proportional to P and t,

$$I = kPt.$$

For $I = 106.25$, $P = 5000$, and $t = \frac{1}{4}$,

$$106.25 = k(5000)\left(\frac{1}{4}\right)$$

which implies that $k = 4(106.25)/5000 = 0.085$. Thus, the equation relating interest, principal, and time is

$$I = 0.085Pt$$

which is the familiar equation for simple interest where the constant of proportionality, 0.085, represents an annual interest rate of 8.5%.

b. When $P = \$5000$ and $t = \frac{3}{4}$, the interest is

$$I = (0.085)(5000)\left(\frac{3}{4}\right) = \$318.75.$$

Most graphing utilities have built-in programs that will calculate the equation of the "best-fitting" line for a collection of points. For instance, to find the "best-fitting" line on a *TI-83* or a *TI-82*, enter the *x*- and *y*-values with the STAT (Edit) menu so that the *x*-values are in L_1 and the *y*-values are in L_2. Then, from the STAT (Calc) menu, choose "Lin-Reg(ax+b)." The graph below shows the "best-fiting" line for the data in Example 6.

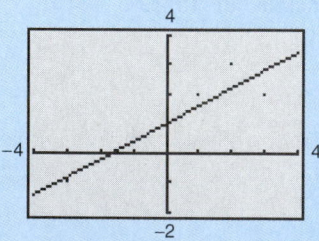

The *TI-83* and the *TI-82* are also capable of finding "best-fitting" quadratic, cubic, quartic, and direct variation as *n*th-power models. The *TI-83* can also find logistic and sinusoidal models.

x_i	y_i	$x_i y_i$	x_i^2
-3	-1	3	9
-2	0	0	4
$\vdots$	$\vdots$	$\vdots$	$\vdots$
3	2	6	9
0	7	17	28

Least Squares Regression

So far in the text, you have worked with many different types of mathematical models that approximate real-life data. For instance, in Example 1 on page 311 you analyzed a model for data on the amount of money spent on advertising in the United States.

To find such a model, statisticians use a measure called the **sum of square differences,** which is the sum of the squares of the differences between the actual data values and model values. The "best-fitting" linear model is the one with the least sum of square differences. This best-fitting linear model is called the **least squares regression line.**

> ■
> ### LEAST SQUARES REGRESSION LINE
> The least squares regression line, $y = ax + b$, for the points (x_1, y_1), $(x_2, y_2), (x_3, y_3), \ldots, (x_n, y_n)$ is
> $$a = \frac{n\sum_{i=1}^{n} x_i y_i \sum_{i=1}^{n} x_i \sum_{i=1}^{n} y_i}{n\sum_{i=1}^{n} x_i^2 - \left(\sum_{i=1}^{n} x_i\right)^2} \quad \text{and} \quad b = \frac{1}{n}\left(\sum_{i=1}^{n} y_i - a\sum_{i=1}^{n} x_i\right).$$

■
EXAMPLE 6 *Finding a Least Squares Regression Line*

Find the least squares regression line for the data $(-3, -1)$, $(-2, 0)$, $(-1, 0)$, $(0, 1)$, $(1, 2)$, $(2, 3)$, and $(3, 2)$.

Solution

Begin by constructing a table such as the one at the left. Applying the formulas for the least squares regression line with $n = 7$ produces

$$a = \frac{n\sum_{i=1}^{n} x_i y_i \sum_{i=1}^{n} x_i \sum_{i=1}^{n} y_i}{n\sum_{i=1}^{n} x_i^2 - \left(\sum_{i=1}^{n} x_i\right)^2} = \frac{7(17) - (0)(7)}{7(28) - (0)^2} = \frac{17}{28}$$

and

$$b = \frac{1}{n}\left(\sum_{i=1}^{n} y_i - a\sum_{i=1}^{n} x_i\right) = \frac{1}{7}\left[7 - \frac{17}{28}(0)\right] = 1.$$

Thus, the least squares regression line is $y = \frac{17}{28}x + 1$.

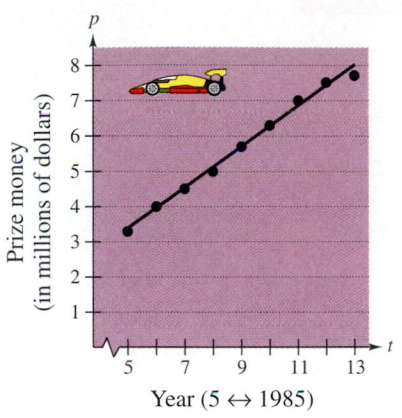

FIGURE 3.34

Prize money (in millions of dollars) — axis labels: 1, 2, 3, 4, 5, 6, 7, 8; Year (5 ↔ 1985): 5, 7, 9, 11, 13

Real Life

EXAMPLE 7 *Prize Money at the Indianapolis 500*

The total prize money p (in millions of dollars) awarded at the Indianapolis 500 race from 1985 to 1993 is given in the table. Construct a scatter plot that represents the data and find a linear model that approximates the data. (Source: Indianapolis Motor Speedway Hall of Fame)

Year	1985	1986	1987	1988	1989	1990	1991	1992	1993
p	3.27	4.00	4.49	5.03	5.72	6.33	7.01	7.53	7.68

Solution

Let $t = 5$ represent 1985. The scatter plot for the points is shown in Figure 3.34. For this data, $n = 9$. Using the least squares regression line formulas, you can determine

$$a = \frac{9(494.11) - (81)(51.05)}{9(789) - (81)^2} \approx 0.58$$

and

$$b = \frac{1}{9}[51.06 - (0.58)(81)] \approx 0.49.$$

The equation of the line is $p = 0.58t + 0.49$. To check this model, compare the actual p-values with the p-values given by the model (the values given by the model are labeled $p*$ below).

t	5	6	7	8	9	10	11	12	13
p	3.27	4.00	4.49	5.03	5.72	6.33	7.01	7.53	7.68
$p*$	3.39	3.97	4.55	5.13	5.71	6.29	6.87	7.45	8.03

GROUP ACTIVITY

RESEARCH PROJECT

Use your school's library, or some other reference source, to locate some data that you think describes a linear relationship. Create a scatter plot of the data, and find the least squares regression line that represents the points. Interpret the slope and y-intercept in the context of the data.

WARM UP

Solve for k.

1. $15 = k(45)$

2. $9 = k(4^2)$

3. $20 = \dfrac{k(15)}{32}$

4. $30 = \dfrac{k(0.2)}{0.5}$

5. $110 = k(27)(0.4)$

6. $210 = k(4^2)(16)$

Find the indicated value.

7. Let $d = 2.7r$. Find d when $r = 10$.

8. Let $s = 3tp^3$. Find s when $t = 2$ and $p = 1/3$.

9. Let $R = \dfrac{4t}{h}$. Find R when $t = 7$ and $h = 13$.

10. Let $M = 14rst$. Find M when $r = 0.01$, $s = 150$, and $t = 7.5$.

3.6 Exercises

1. *Employment*   The total numbers of employees (in thousands) in the United States from 1987 to 1993 are given by the following ordered pairs.

(1987, 121,602)
(1988, 123,378)
(1989, 125,557)
(1990, 126,424)
(1991, 126,867)
(1992, 128,548)
(1993, 129,525)

A linear model that approximates this data is

$$y = 113{,}336.2 + 1265.0t, \qquad 7 \le t \le 13$$

where y represents the number of employed (in thousands) and $t = 0$ represents 1980. Plot the actual data *and* the model on the same graph. How closely does the model represent the data?   (Source: U.S. Bureau of Labor Statistics)

2. *Olympic Swimming*   The winning times (in minutes) in the women's 400-meter freestyle swimming event in the Olympics from 1948 to 1992 are given by the following ordered pairs.

(1948, 5.30), (1952, 5.20)
(1956, 4.91), (1960, 4.84)
(1964, 4.72), (1968, 4.53)
(1972, 4.32), (1976, 4.16)
(1980, 4.15), (1984, 4.12)
(1988, 4.06), (1992, 4.12)

A linear model that approximates this data is

$$y = 5.4 - 0.03t, \qquad 8 \le t \le 52$$

where y represents the winning time in minutes and $t = 0$ represents 1940. Plot the actual data *and* the model on the same graph. How closely does the model represent the data?   (Source: Olympic Committee)

Think About It   In Exercises 3 and 4, use the graph to determine whether y varies directly as some power of x or inversely as some power of x. Explain.

3.
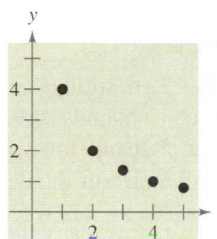

4.
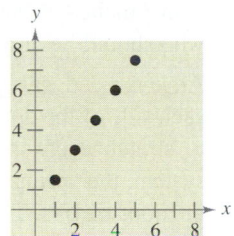

In Exercises 5–8, use the given value of k to complete the table for the direct variation model $y = kx^2$. Plot the points of a rectangular coordinate system.

x	2	4	6	8	10
$y = kx^2$					

5. $k = 1$ **6.** $k = 2$

7. $k = \frac{1}{2}$ **8.** $k = \frac{1}{4}$

In Exercises 9–12, use the given value of k to complete the table for the inverse variation model $y = k/x^2$. Plot the points on a rectangular coordinate system.

x	2	4	6	8	10
$y = \dfrac{k}{x^2}$					

9. $k = 2$ **10.** $k = 5$

11. $k = 10$ **12.** $k = 20$

In Exercises 13–16, determine whether the variation model is of the form $y = kx$ or $y = k/x$ and find k.

13.

x	5	10	15	20	25
y	1	$\frac{1}{2}$	$\frac{1}{3}$	$\frac{1}{4}$	$\frac{1}{5}$

14.

x	5	10	15	20	25
y	2	4	6	8	10

15.

x	5	10	15	20	25
y	−3.5	−7	−10.5	−14	−17.5

16.

x	5	10	15	20	25
y	24	12	8	6	$\frac{24}{5}$

Direct Variation   In Exercises 17–20, assume that y is proportional to x. Use the given x-value and y-value to find a linear model that relates y and x.

x-Value	y-Value		x-Value	y-Value
17. $x = 5$	$y = 12$	**18.**	$x = 2$	$y = 14$
19. $x = 10$	$y = 2050$	**20.**	$x = 6$	$y = 580$

21. *Simple Interest*   The simple interest on an investment is directly proportional to the amount of the investment. By investing \$2500 in a certain bond issue, you obtained an interest payment of \$187.50 at the end of 1 year. Find a mathematical model that gives the interest I for this bond issue at the end of 1 year in terms of the amount invested P.

22. *Simple Interest*   The simple interest on an investment is directly proportional to the amount of the investment. By investing \$5000 in a municipal bond, you obtained an interest payment of \$337.50 at the end of 1 year. Find a mathematical model that gives the interest I for this municipal bond at the end of 1 year in terms of the amount invested P.

23. *Property Tax*   Property tax is based on the assessed value of the property. A house that has an assessed value of \$50,000 has a property tax of \$1840. Find a mathematical model that gives the amount of property tax y in terms of the assessed value x of the property. Use the model to find the property tax on a house that has an assessed value of \$85,000.

24. *State Sales Tax*   State sales tax is based on the retail price. An item that sells for \$145.99 has a sales tax of \$10.22. Find a mathematical model that gives the amount of sales tax y in terms of the retail price x. Use the model to find the sales tax on a \$540.50 purchase.

25. *Centimeters and Inches*   On a yardstick with scales in inches and centimeters, you notice that 13 inches is approximately the same length as 33 centimeters. Use this information to find a mathematical model that relates centimeters to inches. Then use the model to complete the table.

Inches	5	10	20	25	30
Centimeters					

26. *Liters and Gallons*   When buying gasoline, you notice that 14 gallons of gasoline is approximately the same as 53 liters. Use this information to find a linear model that relates gallons to liters. Use the model to complete the table.

Gallons	5	10	20	25	30
Liters					

Hooke's Law   In Exercises 27–30, use Hooke's Law for springs, which states that the distance a spring is stretched (or compressed) varies directly as the force on the spring (see figure).

27. A force of 265 newtons stretches a spring 0.15 meter.

(a) How far will a force of 90 newtons stretch the spring?

(b) What force is required to stretch the spring 0.1 meter?

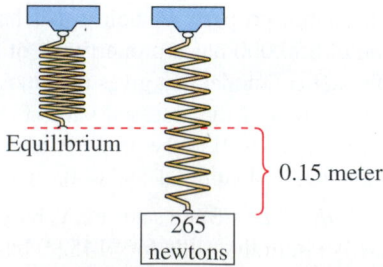

28. A force of 220 newtons stretches a spring 0.12 meter. What force is required to stretch the spring 0.16 meter?

29. The coiled spring of a toy supports the weight of a child. The spring compresses a distance of 1.9 inches under the weight of a 25-pound child. The toy will not work properly if its spring is compressed more than 3 inches. What is the weight of the heaviest child who should be allowed to use the toy?

30. An overhead garage door has two springs, one on each side of the door. A force of 15 pounds is required to stretch each spring 1 foot. Because of a pulley system, the springs stretch only one half the distance the door travels. The door moves a total of 8 feet and the springs are at their natural length when the door is open. Find the combined lifting force applied to the door by the springs when the door is closed.

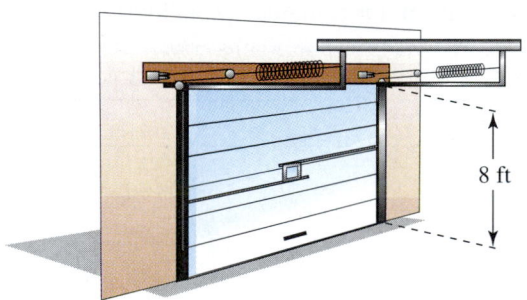

In Exercises 31–44, find a mathematical model for the verbal statement.

31. A varies directly as the square of r.

32. V varies directly as the cube of e.

33. y varies inversely as the square of x.

34. h varies inversely as the square root of s.

35. z is proportional to the cube root of u.

36. x is inversely proportional to $t + 1$.

37. z varies jointly as u and v.

38. V varies jointly as l, w, and h.

39. F varies directly as g and inversely as r^2.

40. z is jointly proportional to the square of x and y^3.

41. *Boyle's Law*   For a constant temperature, the pressure P of a gas is inversely proportional to the volume V of the gas.

42. *Newton's Law of Cooling*   The rate of change R of the temperature of an object is proportional to the difference between the temperature T of the object and the temperature T_e of the environment in which the object is placed.

43. *Newton's Law of Universal Gravitation*   The gravitational attraction F between two objects of masses m_1 and m_2 is proportional to the product of the masses and inversely proportional to the square of the distance r between the objects.

44. *Logistics Growth*   The rate of growth R of a population is jointly proportional to the size S of the population and the difference between S and the maximum size L that the environment can support.

In Exercises 45–50, write a sentence using the variation terminology of this section to describe the formula.

45. Area of a Triangle: $A = \frac{1}{2}bh$

46. Surface Area of a Sphere: $S = 4\pi r^2$

47. Volume of a Sphere: $V = \frac{4}{3}\pi r^3$

48. Volume of a Right Circular Cylinder: $V = \pi r^2 h$

49. Average Speed: $r = \dfrac{d}{t}$

50. Free Vibrations: $\omega = \sqrt{\dfrac{kg}{W}}$

In Exercises 51–64, find a mathematical model representing the statement. (In each case, determine the constant of proportionality.)

51. A varies directly as r^2. ($A = 9\pi$ when $r = 3$.)

52. s is directly proportional to t^2. ($s = 64$ when $t = 2$.)

53. y varies inversely as x. ($y = 3$ when $x = 25$.)

54. y is inversely proportional to x. ($y = 7$ when $x = 4$.)

55. h is inversely proportional to the third power of t. ($h = \frac{3}{16}$ when $t = 4$.)

56. z is jointly proportional to x and y. ($z = 32$ when $x = 10$ and $y = 16$.)

57. z varies jointly as x and y. ($z = 64$ when $x = 4$ and $y = 8$.)

58. R varies inversely as s^2. ($R = 80$ when $s = \frac{1}{5}$.)

59. F is jointly proportional to r and the third power of s. ($F = 4158$ when $r = 11$ and $s = 3$.)

60. P varies directly as x and inversely as the square of y. ($P = \frac{28}{3}$ when $x = 42$ and $y = 9$.)

61. z varies directly as the square of x and inversely as y. ($z = 6$ when $x = 6$ and $y = 4$.)

62. v varies jointly as p and q and inversely as the square of s. ($v = 1.5$ when $p = 4.1$, $q = 6.3$, and $s = 1.2$.)

63. S varies directly as L and inversely as $L - S$. ($S = 4$ when $L = 6$.)

64. P is jointly proportional to S and $L - S$. ($P = 10$ when $S = 4$ and $L = 6$.)

Erosion   **In Exercises 65 and 66, use the fact that the diameter of the largest particle that can be moved by a stream varies approximately as the square of the velocity of the stream.**

65. A stream with a velocity of 1/4 mile per hour can move coarse sand particles about 0.02 inch in diameter. Approximate the velocity required to carry particles 0.12 inch in diameter.

66. A stream of velocity v can move particles of diameter d or less. By what factor does d increase when the velocity is doubled?

Electrical Resistance   **In Exercises 67 and 68, use the fact that the resistance of a wire carrying an electrical current is directly proportional to its length and inversely proportional to its cross-sectional area.**

67. If #28 copper wire (which has a diameter of 0.0126 inch) has a resistance of 66.17 ohms per thousand feet, what length of #28 copper wire will produce a resistance of 33.5 ohms?

68. A 14-foot piece of copper wire produces a resistance of 0.05 ohm. Use the constant of proportionality of Exercise 67 to find the diameter of the wire.

69. *Free Fall*   Neglecting air resistance, the distance s an object falls varies directly as the square of the duration t of the fall. An object falls a distance of 144 feet in 3 seconds. How far will it fall in 5 seconds?

70. Stopping Distance   The stopping distance d of an automobile is directly proportional to the square of its speed s. A car required 75 feet to stop when its speed was 30 miles per hour. Estimate the stopping distance if the brakes are applied when the car is traveling at 50 miles per hour.

71. Comparative Shopping   The prices of three sizes of pizza at a pizza shop are as follows.

9-inch: $8.78
12-inch: $11.78
15-inch: $14.18

One would expect that the price of a certain size pizza would be directly proportional to its surface area. Is that the case for this pizza shop? If not, which size pizza is the best buy?

72. Demand for a Product   A company has found that the demand for its product varies inversely as the price of the product. When the price is $3.75, the demand is 500 units. Approximate the demand when the price is $4.25.

73. Fluid Flow   The velocity v of a fluid flowing in a conduit is inversely proportional to the cross-sectional area of the conduit. (Assume that the volume of the flow per unit of time is held constant.)

(a) Determine the change of velocity of water flowing from a hose when a person places a finger over the end of the hose to decrease its cross-sectional area by 25%.

(b) Use the fluid velocity model of part (a) to determine the effect on the velocity of a stream when it is dredged to increase its cross-sectional area by one-third.

74. Safe Load of a Beam   The maximum load that can be safely supported by a horizontal beam varies jointly as the width of the beam and the square of its depth and inversely as the length of the beam. Determine the change in the maximum safe load under the following conditions.

(a) The width and length of the beam are doubled.

(b) The width and depth of the beam are doubled.

(c) All three of the dimensions are doubled.

(d) The depth of the beam is halved.

75. Data Analysis   An experiment in a physics lab requires a student to measure the compressed length x (in centimeters) of a spring when a force of F grams is applied. The data is given in the table.

F	0	50	100	150	200	250	300
x	12.00	10.85	9.72	8.49	7.35	6.10	4.84

(a) Sketch a scatter plot of the data.

(b) Does it appear that the data can be modeled by Hooke's Law? If so, estimate k. (See Exercises 27–30.)

(c) Use the model of part (b) to approximate the force required to compress the spring 9 centimeters.

76. Data Analysis   An oceanographer took readings of the water temperature C (in degrees Celsius) at depth d (in meters). The data collected is shown in the table.

d	1000	2000	3000	4000	5000
C	4.2°	1.9°	1.4°	1.2°	0.9°

(a) Sketch a scatter plot of the data.

(b) Does it appear that the data can be modeled by the inverse proportion model $C = k/d$? If so, estimate k.

(c) Use a graphing utility to plot the data points and the inverse model of part (b).

(d) Use the model to approximate the depth at which the water temperature is 3°C.

77. Data Analysis   A light probe is located x centimeters from a light source, and the intensity y (in microwatts per square centimeter) of the light is measured. The results are given in the table.

x	30	34	38	42	46	50
y	0.1881	0.1543	0.1172	0.0998	0.0775	0.0645

A model for the data is given by $y = 262.76/x^{2.12}$.

(a) Use a graphing utility to plot the data points and the model on the same viewing rectangle.

(b) Use the model to approximate the light intensity 25 centimeters from the light source.

78. *Illumination of a Light* The illumination from a light source varies inversely as the square of the distance from the light source. When the distance from a light source is doubled, how does the illumination change? Discuss this model in terms of the data given in Exercise 77. Give a possible explanation of the difference.

In Exercises 79–82, how well can the data shown in the scatter plot be approximated by a linear model?

79.
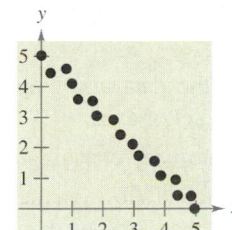

80.
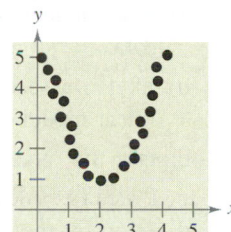

81.
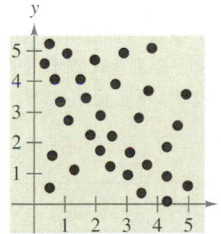

82.
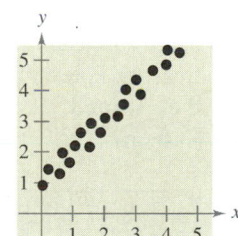

In Exercises 83–86, sketch the line that you think best approximates the data in the scatter plot. Then find an equation of the line.

83.
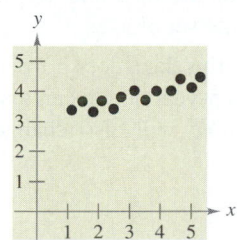

84.
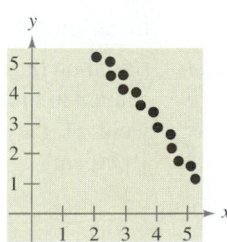

85.
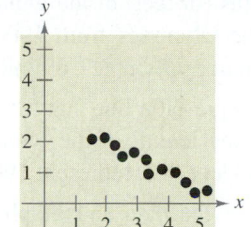

86.
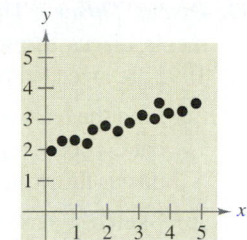

87. *Football Salaries* The average annual salaries of professional football players (in thousands of dollars) from 1987 to 1992 are given in the table. (Source: National Football League Players Association)

(a) Find the least squares regression line that fits this data. (Let y represent the average salary and let $t = 7$ represent 1987.)

(b) Sketch a scatter plot of the data and graph the linear model you found in part (a) on the same set of axes.

(c) Use the model to estimate the average salaries in 1994, 1995, and 1996.

(d) Use your school's library or some other reference source to analyze the accuracy of the salary estimates in part (c).

Year	1987	1988	1989	1990	1991	1992
Salary	203	239	295	352	415	488

88. *Essay* A linear mathematical model for predicting prize winnings in a race is based on data for only 3 years. Write a paragraph discussing the potential accuracy and inaccuracy of such a model.

89. *Discus Throw* The lengths (in feet) of the winning men's discus throws in the Olympics from 1900 to 1992 are listed below. (Source: Olympic Committee)

(a) Find the least squares regression line that fits this data. (Let y represent the length of the winning discus throw in feet and let $t = 0$ represent 1900.)

(b) Sketch a scatter plot of the data and graph the linear model you found in part (a) on the same set of axes.

(c) Use the model to estimate the winning men's discus throw in 1996.

(d) Use your school's library or some other reference source to analyze the accuracy of the estimate in part (c).

1900	118.2	1932	162.4	1968	212.5
1904	128.9	1936	165.6	1972	211.3
1908	134.2	1948	173.2	1976	221.5
1912	145.0	1952	180.5	1980	218.7
1920	146.6	1956	184.9	1984	218.5
1924	151.4	1960	194.2	1988	225.8
1928	155.2	1964	200.1	1992	213.7

90. *Total Assets* The total assets for Boatmen's Bancshares, Inc. from 1985 to 1993 (in millions of dollars) are listed below. (Source: Standard OTC Stock Reports)

(a) Find the least squares regression line that fits this data. (Let y represent the total assets and let $t = 5$ represent 1985.)

(b) Use a graphing utility to sketch a scatter plot of the data and the graph of the model on the same viewing rectangle.

(c) Use the model to estimate the assets of Boatmen's Bancshares in 1996.

(d) Use your school's library or some other reference source to analyze the accuracy of the estimate.

1985	7054.6	1988	14676.0	1991	17635.0
1986	9919.0	1989	14542.0	1992	23387.0
1987	9884.9	1990	17469.0	1993	26654.0

91. *Meat Consumption* The table gives the per capita consumption x (in pounds) of meat from poultry products and the per capita consumption y (in pounds) of beef for the years 1987 through 1992 in the United States. (Source: U.S. Department of Agriculture)

x	50.7	51.7	53.6	55.9	58.1	60.1
y	69.5	68.6	65.4	63.9	63.1	62.8

(a) Find the least squares regression line that fits this data.

(b) Sketch a scatter plot of the data and graph the linear model on the same set of axes.

(c) Use the model to estimate beef consumption if poultry consumption is 62 pounds.

(d) Interpret the meaning of the slope of the linear model in the context of the problem.

92. *U.S. Treasury Notes* The table gives the yield of 5-year treasury notes x and the yield of 10-year treasury notes y for the years 1987 through 1992 in the United States. (Source: Federal Reserve System)

x	8.47	8.50	8.37	7.37	6.19	5.14
y	8.85	8.49	8.55	7.86	7.01	5.87

(a) Find the least squares regression line that fits this data.

(b) Sketch a scatter plot of the data and graph the linear model on the same set of axes.

(c) Based on the model for this data, does a 1 percent change in the yield of a 5-year note mean that the yield of the 10-year note will also change by 1 percent?

FOCUS ON CONCEPTS

In this chapter, you studied several concepts related to the graphs and zeros of polynomial functions. Answer the following questions to check your understanding of several of these basic concepts discussed in this chapter. The answers to these questions are given in the back of the book.

1. *Modeling Company Profits*   The profits P (in millions of dollars) for a company are modeled by a quadratic function of the form

$$P = at^2 + bt + c$$

where t represents the year. If you were president of the company, which of the models would you prefer? Explain your reasoning.

(a) a is positive and $-b/(2a) \le t$.

(b) a is positive and $t \le -b/(2a)$.

(c) a is negative and $-b/(2a) \le t$.

(d) a is negative and $t \le -b/(2a)$.

2. *Graphs of Polynomial Functions*   Describe a polynomial function that could represent the graph. (Indicate the degree of the function and the sign of its leading coefficient.)

(a)
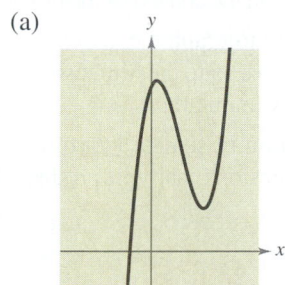

(b)
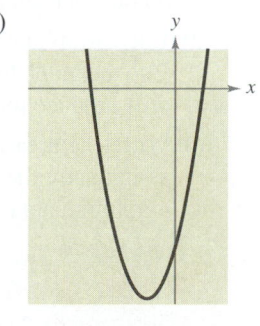

(c)
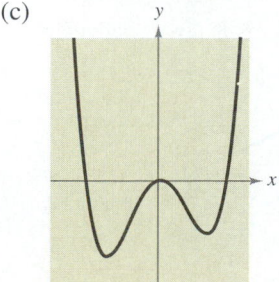

(d)
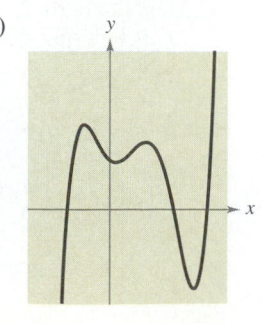

3. *Zeros of Polynomial Functions*   The real line shows all of the real zeros of a polynomial function.

(a) Could the function be a second-degree polynomial function? If so, sketch its graph.

(b) Could the function be a third-degree polynomial function? If so, sketch its graph.

(c) Could the function be a fourth-degree polynomial function? If so, sketch its graph.

(d) Could the function be a fifth-degree polynomial function? If so, sketch its graph.

4. *Mathematical Modeling*   The scatter plots show data taken from real-life situations. Decide whether the data would be better modeled by a first-degree or second-degree polynomial function. Also state whether the leading coefficient of the function is positive or negative.

(a)
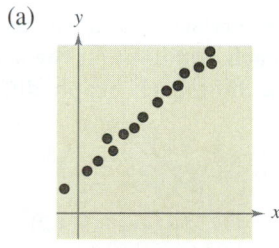

(b)
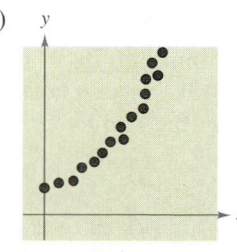

(c)
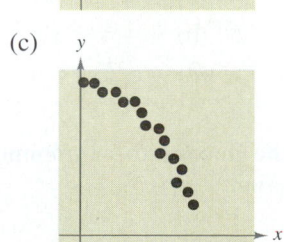

(d)
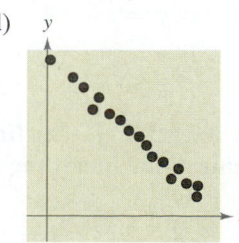

Review Exercises

In Exercises 1–4, sketch the graph of the quadratic function. Identify the vertex and the intercepts.

1. $f(x) = \left(x + \frac{3}{2}\right)^2 + 1$
2. $f(x) = (x - 4)^2 - 4$
3. $f(x) = \frac{1}{3}(x^2 + 5x - 4)$
4. $f(x) = 3x^2 - 12x + 11$

In Exercises 5–8, find the quadratic function that has the indicated vertex and whose graph passes through the given point.

5.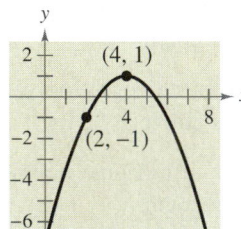

6.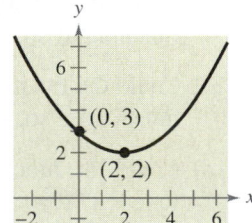

7. Vertex: $(1, -4)$; Point: $(2, -3)$
8. Vertex: $(2, 3)$; Point: $(-1, 6)$

Graphical Reasoning In Exercises 9 and 10, use a graphing utility to graph each equation on the same viewing rectangle. Describe how each graph differs from the graph of $y = x^2$.

9. (a) $y = 2x^2$
    (b) $y = -2x^2$
    (c) $y = x^2 + 2$
    (d) $y = (x + 2)^2$
10. (a) $y = x^2 - 4$
    (b) $y = 4 - x^2$
    (c) $y = (x - 3)^2$
    (d) $y = \frac{1}{2}x^2 - 1$

In Exercises 11–18, find the maximum or minimum value of the quadratic function.

11. $g(x) = x^2 - 2x$
12. $f(x) = x^2 + 8x + 10$
13. $f(x) = 6x - x^2$
14. $h(x) = 3 + 4x - x^2$
15. $f(t) = -2t^2 + 4t + 1$
16. $h(x) = 4x^2 + 4x + 13$
17. $h(x) = x^2 + 5x - 4$
18. $f(x) = 4x^2 + 4x + 5$

19. *Numerical, Graphical, and Analytical Analysis*   A rectangle is inscribed in the region bounded by the x-axis, the y-axis, and the graph of $x + 2y - 8 = 0$ (see figure).

(a) Complete six rows of a table such as the one below.

x	y	Area
1	$4 - \frac{1}{2}(1)$	$(1)[4 - \frac{1}{2}(1)] = \frac{7}{2}$
2	$4 - \frac{1}{2}(2)$	$(2)[4 - \frac{1}{2}(2)] = 6$

(b) Use a graphing utility to generate additional rows of the table. Use the table to estimate the dimensions that will produce the maximum area.

(c) Write the area A as a function of x. Determine the domain of the function in the context of the problem.

(d) Use a graphing utility to graph the area function. Use the graph to approximate the dimensions that will produce the maximum area.

(e) Write the area function in standard form to find analytically the dimensions that will produce the maximum area.

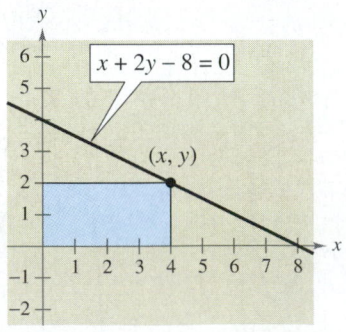

20. *Maximum Area* The perimeter of a rectangle is 200 meters.

(a) Draw a rectangle that gives a visual representation of the problem. Label the length and width in terms of x and y, respectively.

(b) Write y as a function of x. Use the result to write the area as a function of x.

(c) Of all possible rectangles with perimeters of 200 meters, find the dimensions of the one with the maximum area.

21. *Maximum Profit* A real estate office handles 50 apartment units. When the rent is $540 per month, all units are occupied. However, for each $30 increase in rent, one unit becomes vacant. Each occupied unit requires an average of $18 per month for service and repairs. What rent should be charged to obtain the most profit?

22. *Minimum Cost* A manufacturer has daily production costs of

$$C = 20{,}000 - 120x + 0.055x^2$$

where C is the total cost in dollars and x is the number of units produced. How many units should be produced each day to yield a minimum cost?

In Exercises 23–26, determine the right-hand and left-hand behavior of the graph of the polynomial function.

23. $f(x) = -x^2 + 6x + 9$

24. $f(x) = \frac{1}{2}x^3 + 2x$

25. $g(x) = \frac{3}{4}(x^4 + 3x^2 + 2)$

26. $h(x) = -x^5 - 7x^2 + 10x$

Graphical Analysis **In Exercises 27–30, use a graphing utility to graph the functions f and g on the same viewing rectangle. Zoom out sufficiently far to show that the right-hand and left-hand behavior of f and g appear identical.**

27. $f(x) = \frac{1}{2}x^3 - 2x + 1, \quad g(x) = \frac{1}{2}x^3$

28. $f(x) = -\frac{1}{5}(x^3 - 6x^2 + 15), \quad g(x) = -\frac{1}{5}x^3$

29. $f(x) = -x^4 + 2x^3, \quad g(x) = -x^4$

30. $f(x) = x^5 - 5x, \quad g(x) = x^5$

In Exercises 31–34, sketch the graph of the function.

31. $f(x) = -(x - 2)^3$

32. $g(x) = x^4 - x^3 - 2x^2$

33. $f(t) = t^3 - 3t$

34. $f(x) = x(x + 3)^2$

35. *Volume* A rectangular package can have a maximum combined length and girth (perimeter of a cross section) of 216 centimeters (see figure).

(a) Write the volume V as a function of x.

(b) Use a graphing utility to graph the volume function, and use the graph to estimate the dimensions of the package of maximum volume.

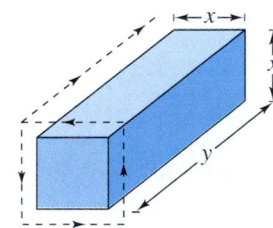

36. *Volume* Rework Exercise 35 for a cylindrical package. (The cross sections are circular.)

Graphical Analysis **In Exercises 37 and 38, use a graphing utility to graph the two equations on the same viewing rectangle. Use the graphs to verify that the expressions are equivalent. Verify the results analytically.**

37. $y_1 = \dfrac{x^2}{x - 2}, \quad y_2 = x + 2 + \dfrac{4}{x - 2}$

38. $y_1 = \dfrac{x^4 + 1}{x^2 + 2}, \quad y_2 = x^2 - 2 + \dfrac{5}{x^2 + 2}$

In Exercises 39–44, perform the division.

39. $\dfrac{24x^2 - x - 8}{3x - 2}$

40. $\dfrac{4x + 7}{3x - 2}$

41. $\dfrac{5x^3 - 13x^2 - x + 2}{x^2 - 3x + 1}$

42. $\dfrac{3x^4}{x^2 - 1}$

43. $\dfrac{x^4 - 3x^3 + 4x^2 - 6x + 3}{x^2 + 2}$

44. $\dfrac{6x^4 + 10x^3 + 13x^2 - 5x + 2}{2x^2 - 1}$

In Exercises 45–48, use synthetic division to perform the division.

45. $\dfrac{6x^4 - 4x^3 - 27x^2 + 18x}{x - \left(\frac{2}{3}\right)}$

46. $\dfrac{0.1x^3 + 0.3x^2 - 0.5}{x - 5}$

47. $\dfrac{2x^3 - 5x^2 + 12x - 5}{x - (1 + 2i)}$

48. $\dfrac{9x^3 - 15x^2 + 11x - 5}{x - \left[\left(\frac{1}{3}\right) + \left(\frac{2}{3}\right)i\right]}$

In Exercises 49 and 50, use synthetic division to determine whether the given values of x are zeros of the function.

49. $f(x) = 20x^4 + 9x^3 - 14x^2 - 3x$

(a) $x = -1$ (b) $x = \frac{3}{4}$

(c) $x = 0$ (d) $x = 1$

50. $f(x) = 3x^3 - 26x^2 + 364x - 232$

(a) $x = 4 - 10i$ (b) $x = 4$

(c) $x = \frac{2}{3}$ (d) $x = -1$

In Exercises 51 and 52, use synthetic division to find the specified value of the function.

51. $f(x) = x^4 + 10x^3 - 24x^2 + 20x + 44$

(a) $f(-3)$ (b) $f(\sqrt{2}i)$

52. $g(t) = 2t^5 - 5t^4 - 8t + 20$

(a) $g(-4)$ (b) $g(\sqrt{2})$

In Exercises 53–56, find a polynomial with integer coefficients that has the given zeros.

53. $-1, -1, \frac{1}{3}, -\frac{1}{2}$

54. $5, 1 - \sqrt{2}, 1 + \sqrt{2}$

55. $\frac{2}{3}, 4, \sqrt{3}i, -\sqrt{3}i$

56. $2, -3, 1 - 2i, 1 + 2i$

In Exercises 57 and 58, use Descartes's Rule of Signs to determine the possible numbers of positive and negative zeros of the function.

57. $g(x) = 5x^3 + 3x^2 - 6x + 9$

58. $h(x) = -2x^5 + 4x^3 - 2x^2 + 5$

In Exercises 59 and 60, use the Rational Zero Test to list all possible rational zeros of f.

59. $f(x) = -4x^3 + 8x^2 - 3x + 15$

60. $f(x) = 3x^4 + 4x^3 - 5x^2 - 8$

In Exercises 61–64, find all the zeros of the function.

61. $f(x) = 4x^3 - 11x^2 + 10x - 3$

62. $f(x) = x^3 - 1.3x^2 - 1.7x + 0.6$

63. $f(x) = 6x^4 - 25x^3 + 14x^2 + 27x - 18$

64. $f(x) = 5x^4 + 126x^3 + 25$

In Exercises 65–68, use a graphing utility to (a) graph the function, (b) determine the number of real zeros of the function, and (c) approximate the real zeros of the function to the nearest hundredth.

65. $f(x) = x^4 + 2x + 1$

66. $g(x) = x^3 - 3x^2 + 3x + 2$

67. $h(x) = x^3 - 6x^2 + 12x - 10$

68. $f(x) = x^5 + 2x^3 - 3x - 20$

69. *Data Analysis* Sales (in billions of dollars) of recreational vehicles in the United States for the years 1980 through 1993 are given in the table. (Source: National Sporting Goods Association)

Year	0	1	2	3	4	5	6
Sales	1.2	1.8	1.7	3.4	4.1	3.5	3.9

Year	7	8	9	10	11	12	13
Sales	4.5	4.8	4.5	4.1	3.6	4.4	4.8

A model for the data is given by $S = 1.2087 + 0.2896t + 0.1762t^2 - 0.0309t^3 + 0.0013t^4$, where S is sales in billions of dollars and t is the time in years, with $t = 0$ corresponding to 1980.

(a) Use a graphing utility to sketch a scatter plot of the data and the model on the same viewing rectangle. How do they compare?

(b) The table shows that sales were down from 1989 through 1991. Give a possible explanation. Does the model show the downturn in sales?

(c) Use a graphing utility to approximate the magnitude of the decrease in sales during the slump described in part (b). Was the actual decrease more or less than indicated by the model?

(d) Use the model to estimate sales in 1995.

70. *Age of the Groom* The average age of the groom at a wedding for a given age of the bride can be approximated by the model $y = -0.00428x^2 + 1.442x - 3.136$, $20 \le x \le 55$ where y is the age of the groom and x is the age of the bride. For what age of the bride is the average age of the groom 30? (Source: U.S. National Center for Health Statistics)

In Exercises 71 and 72, find a mathematical model representing the statement. (In each case, determine the constant of proportionality.)

71. F is jointly proportional to x and the square root of y. ($F = 6$ when $x = 9$ and $y = 4$.)

72. w varies jointly as x and y and inversely as the cube of z. ($w = 44/9$ when $x = 12$, $y = 11$, and $z = 6$.)

73. *Wind Power* The power P produced by a wind turbine is proportional to the cube of the wind speed S. A wind speed of 27 miles per hour produces a power output of 750 kilowatts. Find the output for a wind speed of 40 miles per hour.

74. *Frictional Force* The frictional force F between the tires and the road required to keep a car on a curved section of a highway is directly proportional to the square of the speed s of the car. If the speed of the car is doubled, the force will change by what factor?

75. *Kilometers and Miles* Suppose you are driving on the highway and you notice that a billboard indicates that it is 2.5 miles or 4 kilometers to the next restaurant of a national fast-food chain. Use this information to find a linear model that relates miles to kilometers. Use the model to complete the table.

Miles	2	5	10	12
Kilometers				

76. *Average Hourly Wage* The table gives the average hourly wages (y_1) for workers in the mining industry and the average hourly wages (y_2) for workers in the construction industry in the United States for the years 1990 through 1993. The time in years is given by t, where $t = 0$ represents 1990. (Source: U.S. Bureau of Labor Statistics)

t	0	1	2	3
y_1	$13.68	$14.18	$14.51	$14.60
y_2	$13.77	$13.99	$14.11	$14.35

(a) Find the least squares regression lines for mining wages versus time and construction wages versus time.

(b) Sketch a scatter plot of the data and graph the linear models you found in part (a) on the same set of axes.

(c) Interpret the slope of each model in the context of the problem.

(d) Use the models to estimate the wages in each industry for the year 2000.

CHAPTER PROJECT: *A Graphical Approach to Finding Zeros*

You can use the zoom feature of a graphing utility to approximate a function's real zeros (the *x*-intercepts of its graph) to any desired accuracy.

EXAMPLE 1   *Approximating a Zero of a Function*

Use a graphing utility to approximate the positive zero of $f(x) = x^2 - 5$. Compare your approximation with the actual zero.

Solution

(a)
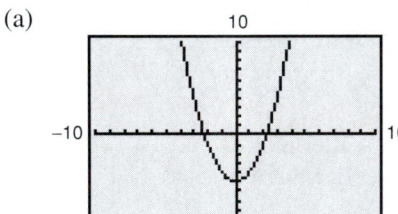
Graph $y = x^2 - 5$.

(b)
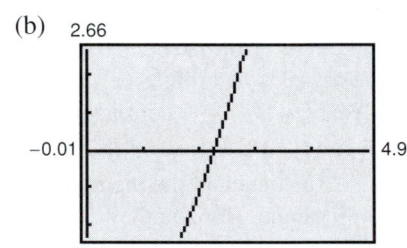
Zoom once to get a closer view of the positive *x*-intercept.

(c)
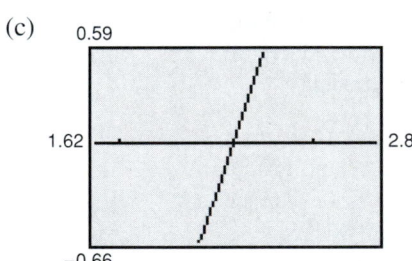
Zoom a second time to get an even better view. Use the cursor keys to determine that the *x*-intercept is about 2.2

(d)
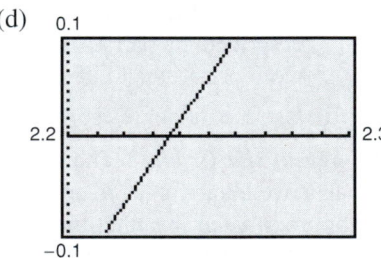
Set the *x*-values to vary from 2.2 to 2.3, with an *x*-scale of 0.01. Use the trace key to approximate the *x*-intercept to be 2.236.

To find the actual zero, you can set $f(x)$ equal to zero and solve the resulting equation.

$$x = \sqrt{5} \approx 2.236068 \qquad \text{Positive solution of } x^2 - 5 = 0$$

From this result, you can see that the *graphical* approach agrees with the *analytical* approach with an accuracy of 0.001.

EXAMPLE 2   *Finding Points of Intersection of Two Graphs*

Find the points of intersection of the circle and parabola given by

$$x^2 + y^2 - 3x + 5y - 11 = 0 \quad \text{and} \quad y = x^2 - 4x + 5.$$

Solution

Begin by writing the circle as the union of two functions.

$$y = \tfrac{1}{2}\left(-5 + \sqrt{69 + 12x - 4x^2}\right) \qquad \text{Top half of circle}$$
$$y = \tfrac{1}{2}\left(-5 - \sqrt{69 + 12x - 4x^2}\right) \qquad \text{Bottom half of circle}$$
$$y = x^2 - 4x + 5 \qquad \text{Parabola}$$

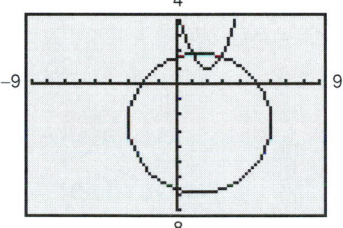

Next, graph all three functions on the same viewing rectangle, as shown at the left. From the graphs, you can see that the points of intersection are roughly (1, 1.9) and (2.8, 1.7). To obtain a more accurate approximation of the points of intersection, you can use the zoom feature to obtain a new viewing rectangle on your graphing utility (see Question 1 below). Another approach to finding the points of intersection is to substitute $x^2 - 4x + 5$ for y in the equation of the circle. This produces a fourth-degree polynomial equation that can be solved for x.

$$x^2 + y^2 - 3x + 5y - 11 = 0$$
$$x^2 + (x^2 - 4x + 5)^2 - 3x + 5(x^2 - 4x + 5) - 11 = 0$$
$$x^4 - 8x^3 + 32x^2 - 63x + 39 = 0$$

Using a graphing utility, you can approximate the solutions of this equation to be $x \approx 1.055$ and $x \approx 2.841$.

CHAPTER PROJECT INVESTIGATIONS

1. *Exploration*   Using a setting of $1.05 \le x \le 1.06$ and $1.89 \le y \le 1.90$, graph the top half of the circle and the parabola in Example 2 on the same screen. Then use the trace feature to approximate (accurate to three decimal places) the y-coordinate of the point of intersection that is shown on the screen.

2. *Exploration*   Graph the fourth-degree polynomial function given in Example 2.

    (a) Find a setting that allows you to approximate the solution $x \approx 1.055$ to two more decimal places.

    (b) Find a setting that allows you to approximate the solution $x \approx 2.841$ to two more decimal places.

3. *Finding Zeros of Functions*   Use a graphing utility to approximate the real zeros of the function

$$f(x) = x^2 + x + 1.$$

Then find the zeros analytically and compare your results with those obtained graphically.

4. *Finding Points of Intersection*   Use a graphing utility to find the points of intersection of the circle and the parabola given by

$$x^2 + y^2 - 5x + 4y - 13 = 0$$
$$y = x^2 - 3x + 2.$$

Chapter Test

Take this test as you would take a test in class. After you are done, check your work against the answers given in the back of the book.

The *Interactive* CD-ROM provides answers to the Chapter Tests and Cumulative Tests. It also offers Chapter Pre-Tests (which test key skills and concepts covered in previous chapters) and Chapter Post-Tests, both of which have randomly generated exercises with diagnostic capabilities.

1. Describe how the graph of g differs from the graph of $f(x) = x^2$.

   (a) $g(x) = 2 - x^2$ (b) $g(x) = \left(x - \frac{3}{2}\right)^2$

2. Identify the vertex and intercepts of the graph of $y = x^2 + 4x + 3$.

3. Find an equation of the parabola shown in the figure.

4. The path of a ball is given by $y = -\frac{1}{20}x^2 + 3x + 5$, where y is the height in feet and x is the horizontal distance (in feet) from where the ball was thrown.

   (a) Find the maximum height of the ball.

   (b) Which constant determines the height at which the ball was thrown? Does changing this constant change the coordinates of the maximum height of the ball? Explain.

5. Determine the right-hand and left-hand behavior of the graph of the function $h(t) = -\frac{3}{4}t^5 + 2t^2$.

6. Divide by long division.

   $$\dfrac{3x^3 + 4x - 1}{x^2 + 1}$$

7. Divide by synthetic division.

   $$\dfrac{2x^4 - 5x^2 - 3}{x - 2}$$

8. Use synthetic division to show that $x = \sqrt{3}$ is a solution of the equation $4x^3 - x^2 - 12x + 3 = 0$. Use the result to factor the polynomial completely and list all the real solutions of the equation.

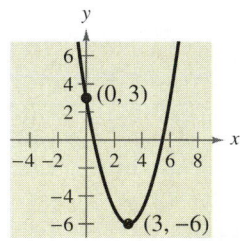

FIGURE FOR 3

In Exercises 9 and 10, list all the possible rational zeros of the function. Use a graphing utility to graph the function and find all the rational zeros.

9. $g(t) = 2t^4 - 3t^3 + 16t - 24$ 10. $h(x) = 3x^5 + 2x^4 - 3x - 2$

In Exercises 11 and 12, use the root-finding capabilities of a graphing utility to approximate the real zeros of the function accurate to three decimal places.

11. $f(x) = x^4 - x^3 - 1$ 12. $f(x) = 3x^5 + 2x^4 - 12x - 8$

In Exercises 13 and 14, find a polynomial function with integer coefficients that has the given zeros.

13. $0, 3, 3 + i, 3 - i$ 14. $1 + \sqrt{3}i, 1 - \sqrt{3}i, 2, 2$

On July 16, 1994, the comet Shoemaker-Levy 9 collided with Jupiter, the largest planet in our solar system. The dark spots in the photo of Jupiter, below, show the results of this collision. The impact was filmed from the space probe Galileo, on its way to Jupiter.

On July 16, 1994, the comet Shoemaker-Levy 9 collided with Jupiter, the largest planet in our solar system. The dark spots in the photo of Jupiter, below, show the results of this collision. The impact was filmed from the space probe Galileo, on its way to Jupiter.

All the planets in our solar system travel in elliptical paths around the Sun. Comets, on the other hand, can have varying paths. Some are elliptical, some are parabolic, and some are hyperbolic.

Since the solar system's formation, thousands of comets have collided with the planets and their moons. For evidence of these collisions, all you need do is to look at the moon's surface through a telescope.

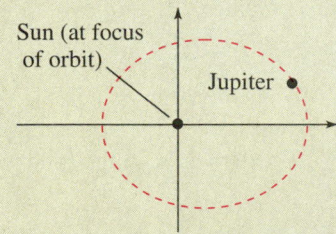

Sun (at focus of orbit)

Jupiter

See Exercises 91 and 92 on page 384.

Rational Functions and Conics

4

4.1 *Rational Functions and Asymptotes*

4.2 *Graphs of Rational Functions*

4.3 *Partial Fractions*

4.4 *Conics*

4.5 *Translations of Conics*

Photos: Alan Levenson; Courtesy of the Space Telescope Science Institute/NASA (inset)

Astronomers David Levy, Carolyn Shoemaker, and Eugene Shoemaker (from left to right) are shown with the 18-inch Schmidt telescope they used to discover the comet that is now named after them.

4.1 *Rational Functions and Asymptotes*

See Exercise 39 on page 341 for an example of how a rational function can be used to model the cost of removing pollutants from a river.

Introduction ❏ *Horizontal and Vertical Asymptotes* ❏ *Applications*

Introduction

A **rational function** can be written in the form

$$f(x) = \frac{p(x)}{q(x)}$$

where $p(x)$ and $q(x)$ are polynomials and $q(x)$ is not the zero polynomial. In this section we assume that $p(x)$ and $q(x)$ have no common factors.

In general, the *domain* of a rational function of x includes all real numbers except x-values that make the denominator zero. Much of our discussion of rational functions will focus on their graphical behavior near these x-values.

EXAMPLE 1   *Finding the Domain of a Rational Function*

Find the domain of

$$f(x) = \frac{1}{x}$$

and discuss the behavior of f near any excluded x-values.

Solution

Because the denominator is zero when $x = 0$, the domain of f is all real numbers except $x = 0$. To determine the behavior of f near this excluded value, evaluate $f(x)$ to the left and right of $x = 0$, as indicated in the following two tables.

x	-1	-0.5	-0.1	-0.01	-0.001	⟶ 0
$f(x)$	-1	-2	-10	-100	-1000	⟶ $-\infty$

x	0 ⟵	0.001	0.01	0.1	0.5	1
$f(x)$	∞ ⟵	1000	100	10	2	1

Note that as x approaches 0 *from the left*, $f(x)$ decreases without bound, whereas as x approaches 0 *from the right*, $f(x)$ increases without bound. The graph of f is shown in Figure 4.1.

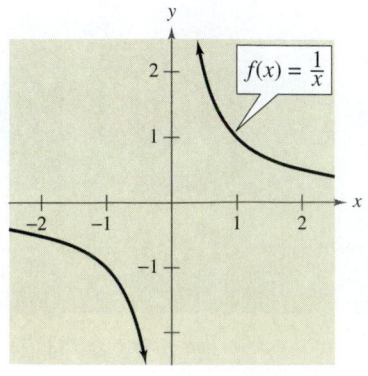

FIGURE 4.1

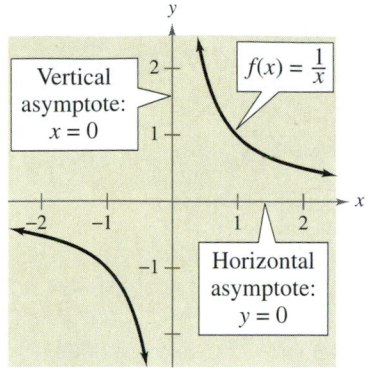

FIGURE 4.2

NOTE The graphs of

$$f(x) = \frac{1}{x} \qquad \text{Figure 4.2}$$

and

$$f(x) = \frac{2x + 1}{x + 1} \qquad \text{Figure 4.3(a)}$$

are **hyperbolas.** You will study hyperbolas in Sections 4.4 and 4.5. ■■

Horizontal and Vertical Asymptotes

In Example 1, the behavior of f near $x = 0$ is denoted as follows.

$$f(x) \longrightarrow -\infty \text{ as } x \longrightarrow 0^- \qquad f(x) \longrightarrow \infty \text{ as } x \longrightarrow 0^+$$

$f(x)$ decreases without bound as x approaches 0 from the left. $f(x)$ increases without bound as x approaches 0 from the right.

The line $x = 0$ is a **vertical asymptote** of the graph of f, as shown in Figure 4.2. From this figure, you can see that the graph of f also has a **horizontal asymptote**—the line $y = 0$. This means that the values of $f(x) = 1/x$ approach zero as x increases or decreases without bound.

$$f(x) \longrightarrow 0 \text{ as } x \longrightarrow -\infty \qquad f(x) \longrightarrow 0 \text{ as } x \longrightarrow \infty$$

$f(x)$ approaches 0 as x decreases without bound. $f(x)$ approaches 0 as x increases without bound.

DEFINITION OF VERTICAL AND HORIZONTAL ASYMPTOTES

1. The line $x = a$ is a **vertical asymptote** of the graph of f if

$$f(x) \longrightarrow \infty \qquad \text{or} \qquad f(x) \longrightarrow -\infty$$

as $x \longrightarrow a$, either from the right or from the left.

2. The line $y = b$ is a **horizontal asymptote** of the graph of f if

$$f(x) \longrightarrow b$$

as $x \longrightarrow \infty$ or $x \longrightarrow -\infty$.

Eventually (as $x \longrightarrow \infty$ or $x \longrightarrow -\infty$) the distance between the horizontal asymptote and the points on the graph must approach zero. Figure 4.3 shows the horizontal and vertical asymptotes of the graphs of three rational functions.

(a)

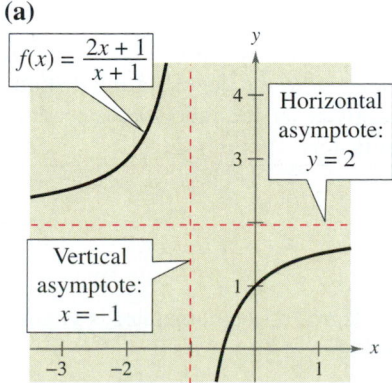

(b)   **(c)**

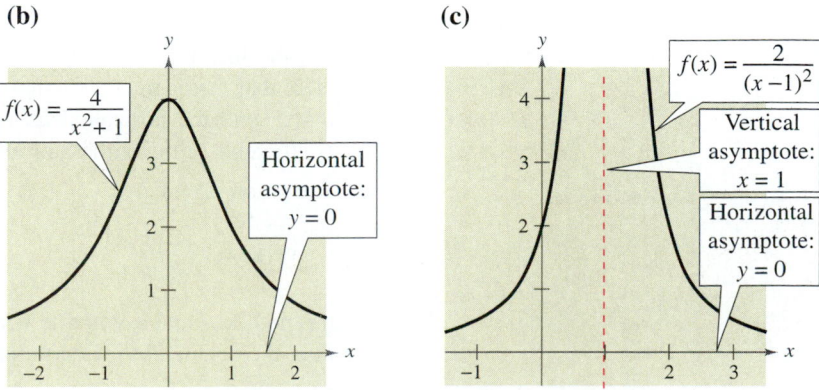

FIGURE 4.3

Let f be the rational function given by

$$f(x) = \frac{p(x)}{q(x)} = \frac{a_n x^n + a_{n-1} x^{n-1} + \cdots + a_1 x + a_0}{b_m x^m + b_{m-1} x^{m-1} + \cdots + b_1 x + b_0}$$

where $p(x)$ and $q(x)$ have no common factors.

1. The graph of f has *vertical* asymptotes at the zeros of $q(x)$.
2. The graph of f has one or no *horizontal* asymptotes determined as follows.
   a. If $n < m$, the graph of f has the x-axis ($y = 0$) as a horizontal asymptote.
   b. If $n = m$, the graph of f has the line $y = a_n/b_m$ as a horizontal asymptote.
   c. If $n > m$, the graph of f has no horizontal asymptote.

You can apply this theorem by comparing the degrees of the numerator and denominator.

EXAMPLE 2 *Finding Horizontal Asymptotes*

a. The graph of

$$f(x) = \frac{2x}{3x^2 + 1}$$

has the x-axis as a horizontal asymptote, as shown in Figure 4.4(a). Note that the degree of the numerator is *less than* the degree of the denominator.

b. The graph of

$$f(x) = \frac{2x^2}{3x^2 + 1}$$

has the line $y = \frac{2}{3}$ as a horizontal asymptote, as shown in Figure 4.4(b). Note that the degree of the numerator is *equal to* the degree of the denominator, and the horizontal asymptote is given by the ratio of the leading coefficients of the numerator and denominator.

c. The graph of

$$f(x) = \frac{2x^3}{3x^2 + 1}$$

has no horizontal asymptote because the degree of the numerator is *greater than* the degree of the denominator. See Figure 4.4(c).

(a)

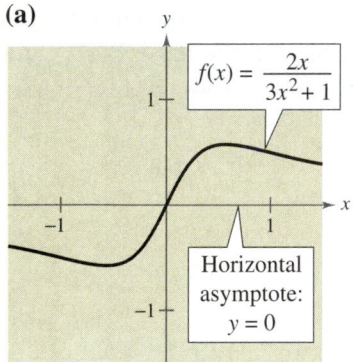

$f(x) = \dfrac{2x}{3x^2 + 1}$

Horizontal asymptote: $y = 0$

(b)

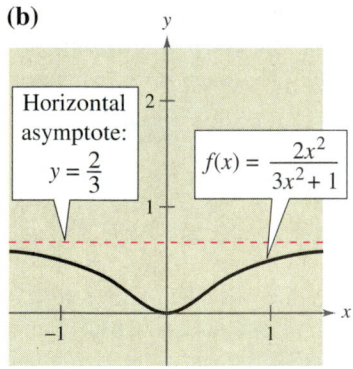

Horizontal asymptote: $y = \dfrac{2}{3}$

$f(x) = \dfrac{2x^2}{3x^2 + 1}$

(c)

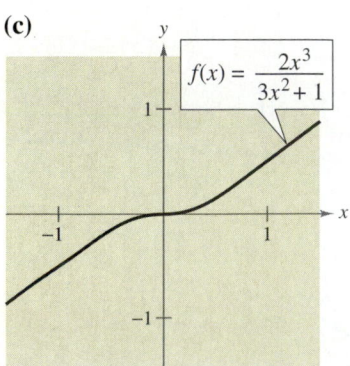

$f(x) = \dfrac{2x^3}{3x^2 + 1}$

FIGURE 4.4

Hundreds of millions of tons of gases and particulates poured into the atmosphere turn clear, odorless air into hazy, smelly air. *(Photo:"Images © 1995 PhotoDisc, Inc.")*

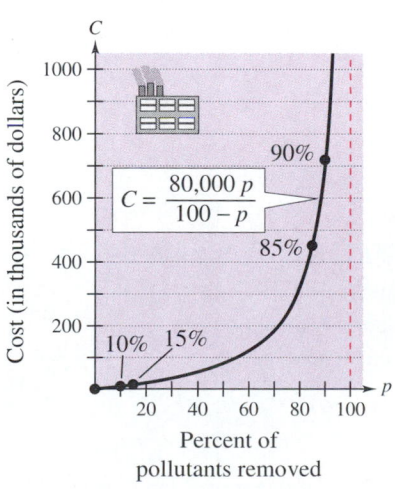

FIGURE 4.5

Applications

There are many examples of asymptotic behavior in real life. For instance, Example 3 shows how a vertical asymptote can be used to analyze the cost of removing pollutants from smokestack emissions.

EXAMPLE 3   *Cost-Benefit Model*

A utility company burns coal to generate electricity. The cost of removing a certain *percent* of the pollutants from the smokestack emission is typically not a linear function. That is, if it costs C dollars to remove 25% of the pollutants, it would cost more than 2C dollars to remove 50% of the pollutants. As the percent of removed pollutants approaches 100%, the cost tends to become prohibitive. Suppose that the cost C of removing $p\%$ of the smokestack pollutants is given by

$$C = \frac{80{,}000p}{100 - p}, \qquad 0 \le p < 100.$$

Sketch the graph of this function. Suppose you are a member of a state legislature that is considering a law that would require utility companies to remove 90% of the pollutants from their smokestack emissions. If the current law requires 85% removal, how much additional expense would the new law ask the utility company to spend?

Solution

The graph of this function is shown in Figure 4.5. Note that the graph has a vertical asymptote at $p = 100$. Because the current law requires 85% removal, the current cost to the utility company is

$$C = \frac{80{,}000(85)}{100 - 85} \qquad \text{Evaluate } C \text{ when } p = 85.$$

$$\approx \$453{,}333.$$

If the new law increases the percent removal to 90%, the cost to the utility company will be

$$C = \frac{80{,}000(90)}{100 - 90} \qquad \text{Evaluate } C \text{ when } p = 90.$$

$$= \$720{,}000.$$

Therefore, the new law requires the utility company to spend an additional

$$720{,}000 - 453{,}333 = \$266{,}667.$$

EXAMPLE 4   *Average Cost of Producing a Product*

Real Life

A business has a cost function of $C = 0.5x + 5000$, where C is measured in dollars and x is the number of units produced. The *average cost per unit* is

$$\overline{C} = \frac{C}{x} = \frac{0.5x + 5000}{x}.$$

Find the average cost per unit when $x = 1000$, 5000, $10{,}000$, and $100{,}000$. What is the horizontal asymptote for this function, and what does it represent?

Solution

When $x = 1000$, $\overline{C} = \dfrac{0.5(1000) + 5000}{1000} = \$5.50$.

When $x = 5000$, $\overline{C} = \dfrac{0.5(5000) + 5000}{5000} = \$1.50$.

When $x = 10{,}000$, $\overline{C} = \dfrac{0.5(10{,}000) + 5000}{10{,}000} = \$1.00$.

When $x = 100{,}000$, $\overline{C} = \dfrac{0.5(100{,}000) + 5000}{100{,}000} = \$0.55$.

As shown in Figure 4.6, the horizontal asymptote is given by the line $\overline{C} = 0.50$. This line represents the least possible unit cost for the product.

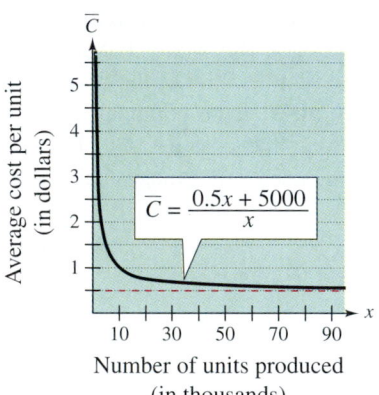

$$\overline{C} = \frac{0.5x + 5000}{x}$$

Average cost per unit (in dollars)

Number of units produced (in thousands)

FIGURE 4.6

NOTE   Note that this example points out one of the major problems of a small business. That is, it is difficult to have competitively low prices when the production level is low. ■■

GROUP ACTIVITY

COMMON FACTORS IN THE NUMERATOR AND DENOMINATOR

When sketching the graph of a rational function, be sure that the rational function has no factor that is common to its numerator and denominator. To see why, consider the function given by

$$f(x) = \frac{2x^2 + x - 1}{x + 1} = \frac{(x + 1)(2x - 1)}{x + 1}$$

which has a common factor of $x + 1$ in the numerator and denominator. Sketch the graph of this function. Does it have a vertical asymptote at $x = -1$?

Factor the polynomial.

1. $x^2 - 3x - 10$ **2.** $x^2 - 7x + 10$

3. $x^3 + 4x^2 + 3x$ **4.** $x^3 - 4x^2 - 2x + 8$

Sketch the graph of the equation.

5. $y = 2$ **6.** $x = -1$

7. $y = x - 2$ **8.** $y = -x + 1$

Divide using long division.

9. $(x^2 + 5x + 6) \div (x - 4)$ **10.** $(x^2 + 5x + 6) \div (x + 4)$

4.1 Exercises

In Exercises 1–6, (a) complete each table, (b) determine the vertical and horizontal asymptotes of the function, and (c) find the domain of the function.

x	$f(x)$
0.5	
0.9	
0.99	
0.999	

x	$f(x)$
1.5	
1.1	
1.01	
1.001	

x	$f(x)$
5	
10	
100	
1000	

1. $f(x) = \dfrac{1}{x - 1}$

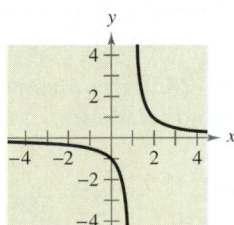

2. $f(x) = \dfrac{5x}{x - 1}$

3. $f(x) = \dfrac{3x}{|x - 1|}$

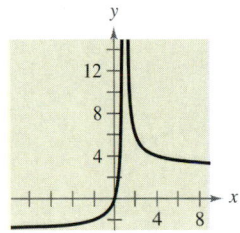

4. $f(x) = \dfrac{3}{|x - 1|}$

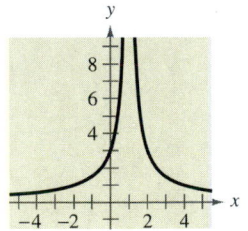

5. $f(x) = \dfrac{3x^2}{x^2 - 1}$

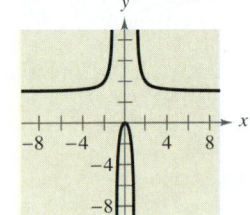

6. $f(x) = \dfrac{4x}{x^2 - 1}$

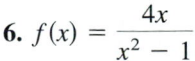
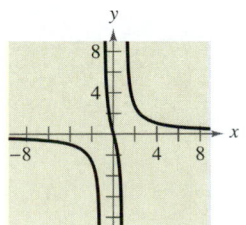

In Exercises 7–14, find the domain of the function and identify any horizontal and vertical asymptotes.

7. $f(x) = \dfrac{1}{x^2}$

8. $f(x) = \dfrac{4}{(x-2)^3}$

9. $f(x) = \dfrac{2+x}{2-x}$

10. $f(x) = \dfrac{1-5x}{1+2x}$

11. $f(x) = \dfrac{x^3}{x^2-1}$

12. $f(x) = \dfrac{2x^2}{x+1}$

13. $f(x) = \dfrac{3x^2+1}{x^2+x+9}$

14. $f(x) = \dfrac{3x^2+x-5}{x^2+1}$

In Exercises 15–20, match the rational function with its graph. [The graphs are labeled (a) through (f).]

(a)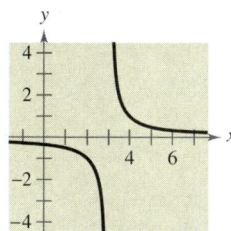

(b)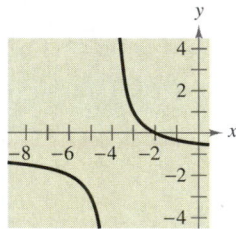

(c)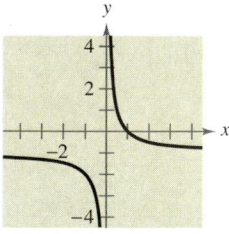

(d)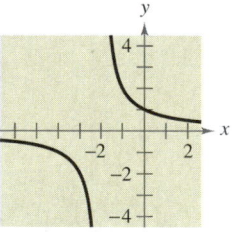

(e)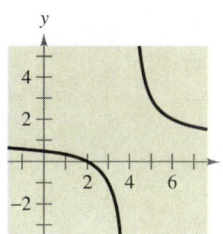

(f)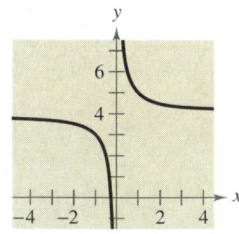

15. $f(x) = \dfrac{2}{x+2}$

16. $f(x) = \dfrac{1}{x-3}$

17. $f(x) = \dfrac{4x+1}{x}$

18. $f(x) = \dfrac{1-x}{x}$

19. $f(x) = \dfrac{x-2}{x-4}$

20. $f(x) = \dfrac{x+2}{x+4}$

Analytical and Numerical Analysis **In Exercises 21–24, (a) determine the domain of f and g, (b) find any vertical asymptotes of f, (c) complete the table, and (d) explain how the two functions differ.**

21. $f(x) = \dfrac{x^2-4}{x+2}$, $g(x) = x-2$

x	-4	-3	-2.5	-2	-1.5	-1	0
$f(x)$							
$g(x)$							

22. $f(x) = \dfrac{x^2(x-3)}{x^2-3x}$, $g(x) = x$

x	-1	0	1	2	3	3.5	4
$f(x)$							
$g(x)$							

23. $f(x) = \dfrac{x-3}{x^2-3x}$, $g(x) = \dfrac{1}{x}$

x	-1	-0.5	0	0.5	2	3	4
$f(x)$							
$g(x)$							

24. $f(x) = \dfrac{2x-8}{x^2-9x+20}$, $g(x) = \dfrac{2}{x-5}$

x	0	1	2	3	4	5	6
$f(x)$							
$g(x)$							

Think About It **In Exercises 25–28, write a rational function f having the specified characteristics.**

25. Vertical asymptotes: $x = -2$, $x = 1$

26. Vertical asymptote: None
    Horizontal asymptote: $y = 0$

27. Vertical asymptote: None
Horizontal asymptote: $y = 2$

28. Vertical asymptotes: $x = 0, x = \frac{5}{2}$
Horizontal asymptote: $y = -3$

Exploration In Exercises 29–32, (a) determine the value the function f approaches as the magnitude of x increases. Is $f(x)$ greater than or less than this functional value when (b) x is positive and large in magnitude and (c) x is negative and large in magnitude?

29. $f(x) = 4 - \dfrac{1}{x}$

30. $f(x) = 2 + \dfrac{1}{x - 3}$

31. $f(x) = \dfrac{2x - 1}{x - 3}$

32. $f(x) = \dfrac{2x - 1}{x^2 + 1}$

In Exercises 33–36, find the zeros (if any) of the rational function.

33. $g(x) = \dfrac{x^2 - 4}{x + 1}$

34. $h(x) = 4 + \dfrac{5}{x^2 + 2}$

35. $f(x) = 1 - \dfrac{2}{x - 3}$

36. $g(x) = \dfrac{x^3 - 8}{x^2 + 1}$

Data Analysis In Exercises 37 and 38, consider a physics laboratory experiment designed to determine an unknown mass. A flexible metal meter stick was clamped to a table with 50 centimeters overhanging the edge (see figure). Known masses M ranging from 200 grams to 2000 grams were attached to the end of the meter stick. For each mass the meter stick was displaced vertically and then allowed to oscillate. The average duration t in seconds of one oscillation for each mass was recorded in the table.

M	200	400	600	800	1000
t	0.450	0.597	0.721	0.831	0.906

M	1200	1400	1600	1800	2000
t	1.003	1.088	1.168	1.218	1.338

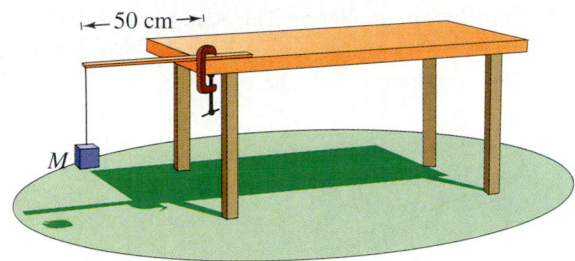

FIGURE FOR 37 AND 38

37. A model for the data is given by

$$t = \frac{38M + 16{,}965}{10(M + 5000)}.$$

Use this model to create a table showing the predicted time for each of the masses shown in the table. What can you conclude?

38. Use the model to approximate the mass of an object for which $t = 1.056$ seconds.

39. *Cost of Clean Water* The cost in millions of dollars for removing $p\%$ of the industrial and municipal pollutants discharged into a river is given by

$$C = \frac{255p}{100 - p}, \qquad 0 \le p < 100.$$

(a) Find the cost of removing 10% of the pollutants.

(b) Find the cost of removing 40% of the pollutants.

(c) Find the cost of removing 75% of the pollutants.

(d) According to this model, would it be possible to remove 100% of the pollutants? Explain.

40. *Recycling Costs* In a pilot project, a rural township is given recycling bins for separating and storing recyclable products. The cost in dollars for supplying bins to $p\%$ of the population is given by

$$C = \frac{25{,}000p}{100 - p}, \qquad 0 \le p < 100.$$

(a) Find the cost if 15% of the population gets bins.

(b) Find the cost if 50% of the population gets bins.

(c) Find the cost if 90% of the population gets bins.

(d) According to this model, would it be possible to supply bins to 100% of the residents? Explain.

41. *Population of Deer*   The game commission introduces 100 deer into newly acquired state game lands. The population of the herd is given by

$$N = \frac{20(5 + 3t)}{1 + 0.04t}, \qquad 0 \le t$$

where t is the time in years (see figure).

(a) Find the population when t is 5, 10, and 25.

(b) What is the limiting size of the herd as time increases?

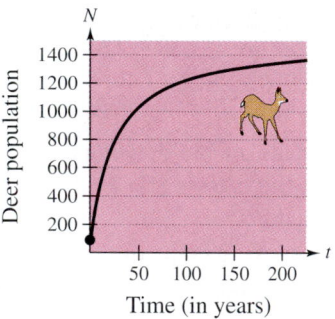

42. *Food Consumption*   A biology class performs an experiment comparing the quantity of food consumed by a certain kind of moth with the quantity supplied (see figure). The model for their experimental data is given by

$$y = \frac{1.568x - 0.001}{6.360x + 1}, \qquad 0 < x$$

where x is the quantity (in milligrams) of food supplied and y is the quantity (in milligrams) eaten. At what level of consumption will the moth become satiated?

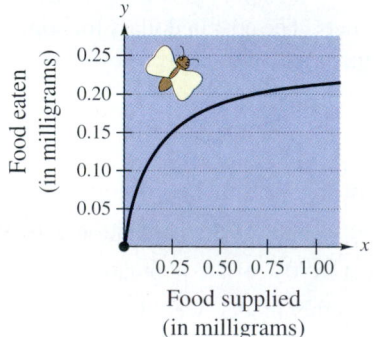

43. *Human Memory Model*   Psychologists have developed mathematical models to predict performance as a function of the number of trials n for a certain task. Consider the learning curve given by

$$P = \frac{0.5 + 0.9(n - 1)}{1 + 0.9(n - 1)}, \qquad 0 < n$$

where P is the percent of correct responses after n trials.

(a) Complete the table for this model. What does it suggest?

n	1	2	3	4	5	6	7	8	9	10
P										

(b) According to this model, what is the limiting percent of correct responses as n increases?

44. *Human Memory Model*   How would the limiting percent of correct responses change if the human memory model of Exercise 43 were changed to

$$P = \frac{0.5 + 0.6(n - 1)}{1 + 0.8(n - 1)}, \qquad 0 < n?$$

45. The average annual percent yields on 5-year U.S. Treasury notes for the years 1986 through 1993 are given in the table. (Source: *Federal Reserve Bulletin*)

Year	1986	1987	1988	1989
Yield	7.30	7.94	8.47	8.50

Year	1990	1991	1992	1993
Yield	8.37	7.37	6.19	5.14

A model for this data is $y = (8216 - 585t)/(100 + 3t + 29t^2)$ where t is time in years, with $t = 0$ corresponding to 1990.

(a) Use a graphing utility to plot the data and graph the model on the same viewing rectangle.

(b) Use the model to estimate the yield in 1994.

(c) Would this model be good for estimating yields for future years? Explain.

4.2 *Graphs of Rational Functions*

See Exercise 82 on page 352 for an example of how a rational function can be used to model average speed.

Sketching the Graph of a Rational Function □ *Slant Asymptotes* □ *Application*

Sketching the Graph of a Rational Function

GUIDELINES FOR GRAPHING RATIONAL FUNCTIONS

Let $f(x) = p(x)/q(x)$, where $p(x)$ and $q(x)$ are polynomials with no common factors.

1. Find and plot the y-intercept (if any) by evaluating $f(0)$.
2. Find the zeros of the numerator (if any) by solving the equation $p(x) = 0$. Then plot the corresponding x-intercepts.
3. Find the zeros of the denominator (if any) by solving the equation $q(x) = 0$. Then sketch the corresponding vertical asymptotes.
4. Find and sketch the horizontal asymptote (if any) by using the rule for finding the horizontal asymptote of a rational function.
5. Plot at least one point *between* and one point *beyond* each x-intercept and vertical asymptote.
6. Use smooth curves to complete the graph between and beyond the vertical asymptotes.

NOTE Testing for symmetry can be useful, especially for simple rational functions. For example, the graph of $f(x) = 1/x$ is symmetrical with respect to the origin, and the graph of $g(x) = 1/x^2$ is symmetrical with respect to the y-axis. ■■

TECHNOLOGY

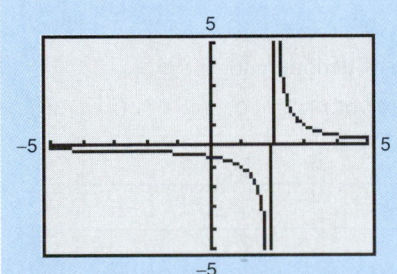

Graphing utilities have difficulty sketching graphs of rational functions that have vertical asymptotes. Often, the utility will connect parts of the graph that are not supposed to be connected. For instance, the screen at the left shows the graph of

$$f(x) = \frac{1}{x-2}.$$

Notice that the graph should consist of two *separated* portions—one to the left of $x = 2$ and the other to the right of $x = 2$. To eliminate this problem, you can try changing the *mode* of the graphing utility to *dot mode*. The problem with this is that the graph is then represented as a collection of dots rather than as a smooth curve.

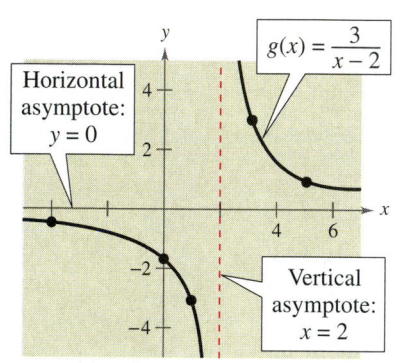

FIGURE 4.7

EXAMPLE 1 *Sketching the Graph of a Rational Function*

Sketch the graph of $g(x) = \dfrac{3}{x - 2}$.

Solution

y-intercept:	$\left(0, -\frac{3}{2}\right)$, from $g(0) = -\frac{3}{2}$.
x-intercept:	None, because $3 \neq 0$.
Vertical Asymptote:	$x = 2$, zero of denominator
Horizontal Asymptote:	$y = 0$, degree of $p(x) <$ degree of $q(x)$
Additional Points:	

x	-4	1	3	5
$g(x)$	-0.5	-3	3	1

By plotting the intercepts, asymptotes, and a few additional points, you can obtain the graph shown in Figure 4.7.

NOTE The graph of g is a vertical stretch and a right shift of the graph of $f(x) = 1/x$ because

$$g(x) = \frac{3}{x - 2} = 3\left(\frac{1}{x - 2}\right) = 3f(x - 2).$$

EXAMPLE 2 *Sketching the Graph of a Rational Function*

Sketch the graph of $f(x) = \dfrac{2x - 1}{x}$.

Solution

y-intercept:	None, because $x = 0$ is not in the domain.
x-intercept:	$\left(\frac{1}{2}, 0\right)$, from $2x - 1 = 0$.
Vertical Asymptote:	$x = 0$, zero of denominator
Horizontal Asymptote:	$y = 2$, degree of $p(x) =$ degree of $q(x)$
Additional Points:	

x	-4	-1	$\frac{1}{4}$	4
$f(x)$	2.25	3	-2	1.75

By plotting the intercepts, asymptotes, and a few additional points, you can obtain the graph shown in Figure 4.8.

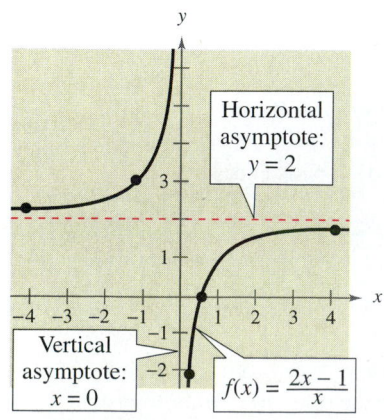

FIGURE 4.8

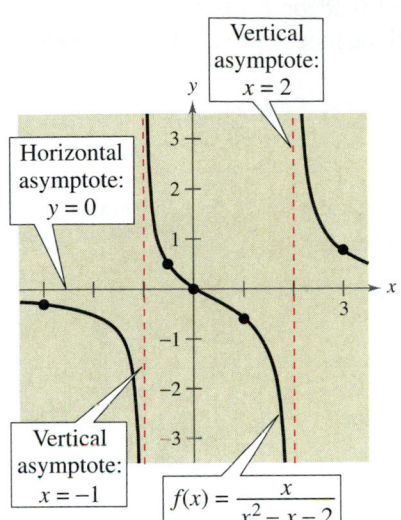

FIGURE 4.9

EXAMPLE 3 *Sketching the Graph of a Rational Function*

Sketch the graph of $f(x) = \dfrac{x}{x^2 - x - 2}$.

Solution

By factoring the denominator, you have

$$f(x) = \frac{x}{x^2 - x - 2} = \frac{x}{(x + 1)(x - 2)}.$$

y-intercept:	$(0, 0)$, because $f(0) = 0$.
x-intercept:	$(0, 0)$
Vertical Asymptotes:	$x = -1$, $x = 2$, zero of denominator
Horizontal Asymptote:	$y = 0$, degree of $p(x) <$ degree of $q(x)$
Additional Points:	

x	-3	-0.5	1	3
$f(x)$	-0.3	0.4	-0.5	0.75

The graph is shown in Figure 4.9.

EXAMPLE 4 *Sketching the Graph of a Rational Function*

Sketch the graph of $f(x) = \dfrac{2(x^2 - 9)}{x^2 - 4}$.

Solution

By factoring the numerator and denominator, you have

$$f(x) = \frac{2(x^2 - 9)}{x^2 - 4} = \frac{2(x - 3)(x + 3)}{(x - 2)(x + 2)}.$$

y-intercept:	$\left(0, \frac{9}{2}\right)$, because $f(0) = \frac{9}{2}$.
x-intercepts:	$(-3, 0)$ and $(3, 0)$
Vertical Asymptotes:	$x = -2$, $x = 2$, zero of denominator
Horizontal Asymptote:	$y = 2$, degree of $p(x) =$ degree of $q(x)$
Symmetry:	With respect to y-axis, because $f(-x) = f(x)$.
Additional Points:	

x	0.5	2.5	6
$f(x)$	4.67	-2.44	1.69

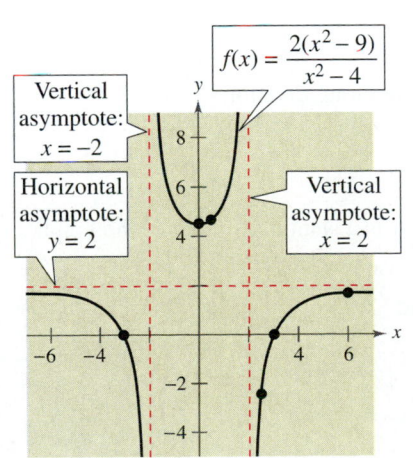

FIGURE 4.10

The graph is shown in Figure 4.10.

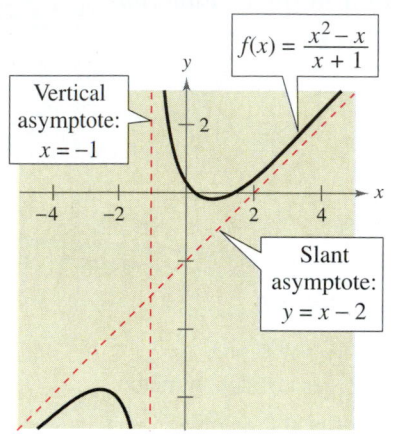

FIGURE 4.11

Slant Asymptotes

If the degree of the numerator of a rational function is exactly *one more* than the degree of the denominator, the graph of the function has a **slant asymptote.** For example, the graph of

$$f(x) = \frac{x^2 - x}{x + 1}$$

has a slant asymptote, as shown in Figure 4.11. To find the equation of a slant asymptote, use long division. For instance, by dividing $x + 1$ into $x^2 - x$, you have

$$f(x) = \frac{x^2 - x}{x + 1} = \underbrace{x - 2}_{} + \frac{2}{x + 1}.$$

Slant asymptote
$(y = x - 2)$

In Figure 4.11, notice that the graph of f approaches the line $y = x - 2$ as x moves to the right or left.

EXAMPLE 5   *A Rational Function with a Slant Asymptote*

Sketch the graph of $f(x) = \dfrac{x^2 - x - 2}{x - 1}$.

Solution

First write $f(x)$ in two different ways. Factoring the numerator

$$f(x) = \frac{x^2 - x - 2}{x - 1} = \frac{(x - 2)(x + 1)}{x - 1}$$

allows you to recognize the x-intercepts, and long division

$$f(x) = \frac{x^2 - x - 2}{x - 1} = x - \frac{2}{x - 1}$$

allows you to recognize that the line $y = x$ is a slant asymptote of the graph.

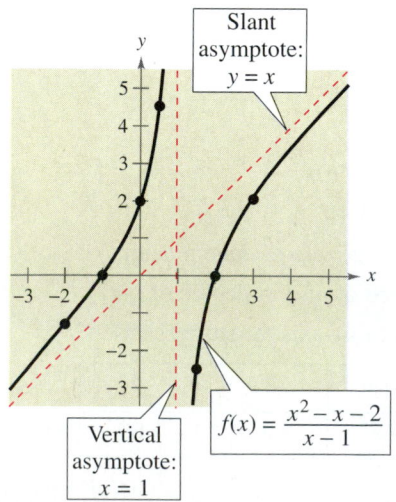

FIGURE 4.12

y-intercept:	$(0, 2)$, because $f(0) = 2$.
x-intercepts:	$(-1, 0)$ and $(2, 0)$
Vertical Asymptote:	$x = 1$
Slant Asymptote:	$y = x$
Additional Points:	

x	-2	0.5	1.5	3
$f(x)$	-1.33	4.5	-2.5	2

The graph is shown in Figure 4.12.

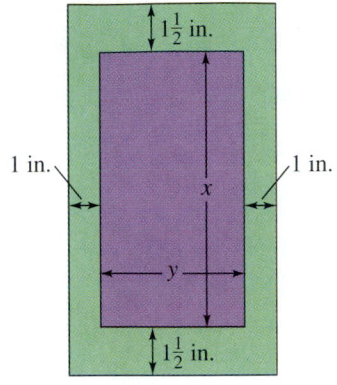

$1\frac{1}{2}$ in.

1 in.   1 in.

x

y

$1\frac{1}{2}$ in.

FIGURE 4.13

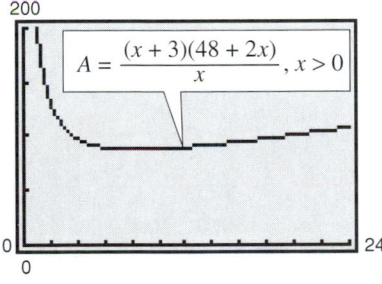

200

$A = \dfrac{(x+3)(48+2x)}{x}, \; x > 0$

0

0   24

FIGURE 4.14

The *Interactive* CD-ROM offers graphing utility emulators of the *TI-82* and *TI-83*, which can be used with the Examples, Explorations, Technology notes, and Exercises.

Application

Real Life

EXAMPLE 6   Finding a Minimum Area

A rectangular page is to contain 48 square inches of print. The margins at the top and bottom of the page are each $1\frac{1}{2}$ inches deep. The margins on each side are 1 inch wide. What should the dimensions of the page be so that the least amount of paper is used?

Solution

Let A be the area to be minimized. From Figure 4.13, you can write

$$A = (x + 3)(y + 2).$$

The printed area inside the margins is given by $48 = xy$ or $y = 48/x$. To find the minimum area, rewrite the equation for A in terms of just one variable by substituting $48/x$ for y.

$$A = (x + 3)\left(\frac{48}{x} + 2\right) = \frac{(x + 3)(48 + 2x)}{x}, \qquad x > 0$$

The graph of this rational function is shown in Figure 4.14. Because x represents the height of the printed area, you need consider only the portion of the graph for which x is positive. Using the zoom and trace features of a graphing utility, you can approximate the minimum value of A to occur when $x \approx 8.5$ inches. The corresponding value of y is $48/8.5 \approx 5.6$ inches. Thus, the dimensions should be

$$x + 3 \approx 11.5 \text{ inches} \qquad \text{by} \qquad y + 2 \approx 7.6 \text{ inches.}$$

If you go on to take a course in calculus, you will learn a technique for finding the exact value of x that produces a minimum area. In this case, that value is $x = 6\sqrt{2} \approx 8.485$.

GROUP ACTIVITY

ASYMPTOTES OF GRAPHS OF RATIONAL FUNCTIONS

Discuss whether or not it is possible for the graph of a rational function to cross its horizontal asymptote or its slant asymptote. Use the graphs of the following functions to investigate these questions. What can you conclude?

$$f(x) = \frac{x}{x^2 + 1} \qquad \text{and} \qquad g(x) = \frac{x^3}{x^2 + 1}$$

The *Interactive* CD-ROM provides additional help with Warm-Up exercises by providing a hypertext link to the section in which the concept was introduced.

WARM UP

Find any intercepts of the graph.

1. $2x - xy - y + 6 = 0$ **2.** $4x - 5y + 12 = 0$

Is the function even, odd, or neither?

3. $g(x) = x(x^2 - 5)$ **4.** $h(x) = \frac{1}{8}x^4 - 3x^2 - 4$

Find the domain of the function and identify any horizontal or vertical asymptotes.

5. $f(x) = \dfrac{6}{x - 8}$ **6.** $f(x) = \dfrac{3x - 1}{4x + 1}$

7. $h(x) = \dfrac{2x^2 + 5}{x^2 - 9}$ **8.** $g(x) = 4 - \dfrac{1}{x}$

Perform the specified division.

9. $\dfrac{4x^2 + 5x + 8}{2x - 1}$ **10.** $\dfrac{x^3 + 1}{x^2}$

The *Interactive* CD-ROM contains step-by-step solutions to all odd-numbered Section and Review Exercises. It also provides Tutorial Exercises, which link to Guided Examples for additional help.

4.2 Exercises

In Exercises 1–4, use the graph of $f(x) = 2/x$ to sketch the graph of g.

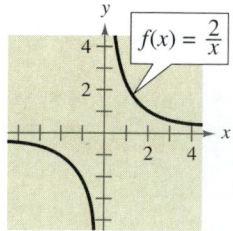

1. $g(x) = \dfrac{2}{x} + 1$ **2.** $g(x) = \dfrac{2}{x - 1}$

3. $g(x) = -\dfrac{2}{x}$ **4.** $g(x) = \dfrac{1}{x + 2}$

In Exercises 5–8, use the graph of $f(x) = 2/x^2$ to sketch the graph of g.

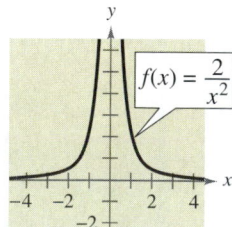

5. $g(x) = \dfrac{2}{x^2} - 2$ **6.** $g(x) = -\dfrac{2}{x^2}$

7. $g(x) = \dfrac{2}{(x - 2)^2}$ **8.** $g(x) = \dfrac{1}{2x^2}$

In Exercises 9–12, use the graph of $f(x) = 4/x^3$ to sketch the graph of g.

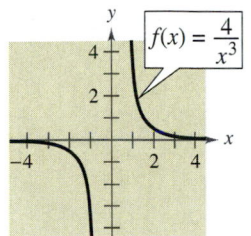

9. $g(x) = \dfrac{4}{(x + 2)^3}$

10. $g(x) = \dfrac{4}{x^3} + 1$

11. $g(x) = -\dfrac{4}{x^3}$

12. $g(x) = \dfrac{1}{x^3}$

In Exercises 13–34, sketch the graph of the rational function. As sketching aids, check for intercepts, symmetry, vertical asymptotes, and horizontal asymptotes.

13. $f(x) = \dfrac{1}{x + 2}$

14. $f(x) = \dfrac{1}{x - 3}$

15. $h(x) = \dfrac{-1}{x + 2}$

16. $g(x) = \dfrac{1}{3 - x}$

17. $C(x) = \dfrac{5 + 2x}{1 + x}$

18. $P(x) = \dfrac{1 - 3x}{1 - x}$

19. $g(x) = \dfrac{1}{x + 2} + 2$

20. $f(t) = \dfrac{1 - 2t}{t}$

21. $f(x) = \dfrac{x^2}{x^2 + 9}$

22. $f(x) = 2 - \dfrac{3}{x^2}$

23. $h(x) = \dfrac{x^2}{x^2 - 9}$

24. $g(x) = \dfrac{x}{x^2 - 9}$

25. $g(s) = \dfrac{s}{s^2 + 1}$

26. $f(x) = -\dfrac{1}{(x - 2)^2}$

27. $g(x) = \dfrac{4(x + 1)}{x(x - 4)}$

28. $h(x) = \dfrac{2}{x^2(x - 2)}$

29. $f(x) = \dfrac{3x}{x^2 - x - 2}$

30. $f(x) = \dfrac{2x}{x^2 + x - 2}$

31. $f(x) = \dfrac{2 + x}{1 - x}$

32. $f(x) = \dfrac{3 - x}{2 - x}$

33. $f(t) = \dfrac{3t + 1}{t}$

34. $h(x) = \dfrac{1}{x - 3} + 1$

Analytical, Numerical, and Graphical Analysis In Exercises 35–38, (a) determine the domains of f and g, (b) find any vertical asymptotes of f, (c) compare the functions by completing the table, (d) use a graphing utility to graph f and g on the same viewing rectangle, and (e) explain why the graphing utility may not show the difference in the domains of f and g.

35. $f(x) = \dfrac{x^2 - 1}{x + 1}, \qquad g(x) = x - 1$

x	-3	-2	-1.5	-1	-0.5	0	1
$f(x)$							
$g(x)$							

36. $f(x) = \dfrac{x^2(x - 2)}{x^2 - 2x}, \qquad g(x) = x$

x	-1	0	1	1.5	2	2.5	3
$f(x)$							
$g(x)$							

37. $f(x) = \dfrac{x - 2}{x^2 - 2x}, \qquad g(x) = \dfrac{1}{x}$

x	-0.5	0	0.5	1	1.5	2	3
$f(x)$							
$g(x)$							

38. $f(x) = \dfrac{2x - 6}{x^2 - 7x + 12}, \qquad g(x) = \dfrac{2}{x - 4}$

x	0	1	2	3	4	5	6
$f(x)$							
$g(x)$							

In Exercises 39–44, sketch a graph of the function. Give the domain of the function and identify any vertical or horizontal asymptotes.

39. $h(t) = \dfrac{4}{t^2 + 1}$

40. $g(x) = -\dfrac{x}{(x-2)^2}$

41. $f(t) = \dfrac{2t^2}{t^2 - 4}$

42. $f(x) = \dfrac{x + 4}{x^2 + x - 6}$

43. $f(x) = \dfrac{20x}{x^2 + 1} - \dfrac{1}{x}$

44. $f(x) = 5\left(\dfrac{1}{x - 4} - \dfrac{1}{x + 2}\right)$

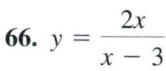 *Exploration*   In Exercises 45 and 46, use a graphing utility to obtain the graph of the function and note that a graph may have two horizontal asymptotes.

45. $h(x) = \dfrac{6x}{\sqrt{x^2 + 1}}$

46. $g(x) = \dfrac{4|x - 2|}{x + 1}$

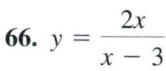 *Exploration*   In Exercises 47 and 48, use a graphing utility to obtain the graph of the function and note that the graph may cross its horizontal asymptote.

47. $f(x) = \dfrac{4(x - 1)^2}{x^2 - 4x + 5}$

48. $g(x) = \dfrac{3x^4 - 5x + 3}{x^4 + 1}$

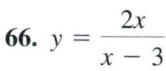 *Think About It*   In Exercises 49 and 50, use a graphing utility to obtain the graph of the function. Explain why there is no vertical asymptote when a superficial examination of the function may indicate that there should be one.

49. $h(x) = \dfrac{6 - 2x}{3 - x}$

50. $g(x) = \dfrac{x^2 + x - 2}{x - 1}$

51. *True or False?*   If the graph of a rational function f has a vertical asymptote at $x = 5$, it is possible to sketch the graph without lifting your pencil from the paper. Explain.

52. *Essay*   Write a paragraph discussing whether every rational function has a vertical asymptote.

In Exercises 53–60, sketch the graph of the rational function. As sketching aids, check for intercepts, symmetry, vertical asymptotes, and slant asymptotes.

53. $f(x) = \dfrac{2x^2 + 1}{x}$

54. $f(x) = \dfrac{1 - x^2}{x}$

55. $g(x) = \dfrac{x^2 + 1}{x}$

56. $h(x) = \dfrac{x^2}{x - 1}$

57. $f(x) = \dfrac{x^3}{x^2 - 1}$

58. $g(x) = \dfrac{x^3}{2x^2 - 8}$

59. $f(x) = \dfrac{x^2 - x + 1}{x - 1}$

60. $f(x) = \dfrac{2x^2 - 5x + 5}{x - 2}$

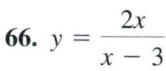 In Exercises 61–64, use a graphing utility to graph the rational function. Give the domain of the function and identify any asymptotes. Then zoom out sufficiently far so the graph appears as a line. Identify the line.

61. $f(x) = \dfrac{x^2 + 5x + 8}{x + 3}$

62. $f(x) = \dfrac{2x^2 + x}{x + 1}$

63. $g(x) = \dfrac{1 + 3x^2 - x^3}{x^2}$

64. $h(x) = \dfrac{12 - 2x - x^2}{2(4 + x)}$

Graphical Reasoning   In Exercises 65–68, (a) use the graph to determine any x-intercepts of the rational function, and (b) set $y = 0$ and solve the resulting equation to confirm your result in part (a).

65. $y = \dfrac{x + 1}{x - 3}$

66. $y = \dfrac{2x}{x - 3}$

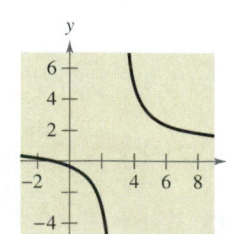

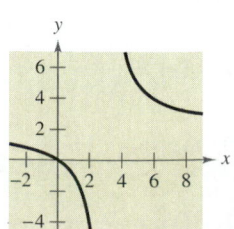

67. $y = \dfrac{1}{x} - x$

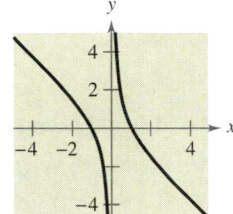

68. $y = x - 3 + \dfrac{2}{x}$

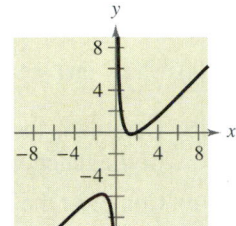

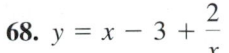 *Graphical Reasoning* In Exercises 69–72, **(a) use a graphing utility to graph the function and determine any x-intercepts, and (b) set y = 0 and solve the resulting equation to confirm your result in part (a).**

69. $y = \dfrac{1}{x + 5} + \dfrac{4}{x}$

70. $y = 20\left(\dfrac{2}{x + 1} - \dfrac{3}{x}\right)$

71. $y = x - \dfrac{6}{x - 1}$

72. $y = x - \dfrac{9}{x}$

73. *Concentration of a Mixture* A 1000-liter tank contains 50 liters of a 25% brine solution. You add x liters of a 75% brine solution to the tank.

(a) Show that the concentration C of the final mixture is given by

$$C = \dfrac{3x + 50}{4(x + 50)}.$$

(b) Determine the domain of the function based on the physical constraints of the problem.

(c) Graph the concentration function. As the tank is filled, what happens to the rate at which the concentration of brine is increasing? What does the concentration of brine appear to approach?

74. *Dimensions of a Rectangle* A rectangular region of length x and width y has an area of 500 square meters.

(a) Express the width y as a function of x.

(b) Determine the domain of the function based on the physical constraints of the problem.

(c) Sketch a graph of the function and determine the width of the rectangle if $x = 30$ meters.

75. *Page Design* A page that is x inches wide and y inches high contains 30 square inches of print. The top and bottom margins are 1 inch deep and the margins on each side are 2 inches wide (see figure).

(a) Show that the total area A on the page is given by

$$A = \dfrac{2x(x + 11)}{x - 4}.$$

(b) Determine the domain of the function based on the physical constraints of the problem.

(c) Use a graphing utility to obtain a graph of the area function and approximate the page size for which the least amount of paper will be used.

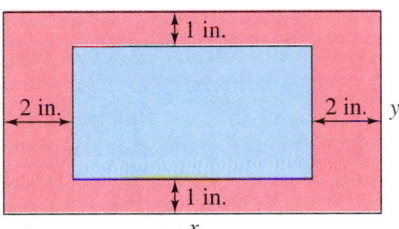

76. *Minimum Area* A right triangle is formed in the first quadrant by the x-axis, the y-axis, and a line segment through the point $(3, 2)$ (see figure).

(a) Show that an equation of the line segment is

$$y = \dfrac{2(a - x)}{a - 3}, \qquad 0 \le x \le a.$$

(b) Show that the area of the triangle is

$$A = \dfrac{a^2}{a - 3}.$$

(c) Graph the area function and, from the graph, estimate the value of a that yields a minimum area. Estimate the minimum area.

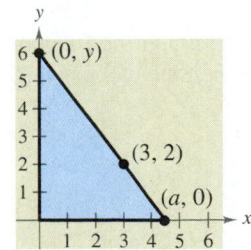

In Exercises 77 and 78, use a graphing utility to obtain the graph of the function and locate any relative maximum or minimum points on the graph.

77. $f(x) = \dfrac{3(x + 1)}{x^2 + x + 1}$

78. $C(x) = x + \dfrac{32}{x}$

79. *Ordering and Transportation Cost*   The ordering and transportation cost C for the components used in manufacturing a certain product is given by

$$C = 100\left(\dfrac{200}{x^2} + \dfrac{x}{x + 30}\right), \qquad 1 \le x$$

where C is measured in thousands of dollars and x is the order size in hundreds. Use a graphing utility to obtain the graph of the cost function and, from the graph, estimate the order size that minimizes cost.

80. *Average Cost*   The cost of producing x units is $C = 0.2x^2 + 10x + 5$, and therefore the average cost per unit is

$$\overline{C} = \dfrac{C}{x} = \dfrac{0.2x^2 + 10x + 5}{x}, \qquad 0 < x.$$

Sketch the graph of the average cost function, and estimate the number of units that should be produced to minimize the average cost per unit.

81. *Medicine*   The concentration of a certain chemical in the bloodstream t hours after injection into muscle tissue is given by

$$C = \dfrac{3t^2 + t}{t^3 + 50}, \qquad 0 \le t.$$

(a) Determine the horizontal asymptote of the function and interpret its meaning in the context of the problem.

(b) Use a graphing utility to graph the function and approximate the time when the bloodstream concentration is greatest.

82. *Average Speed*   A driver averaged 50 miles per hour on the round trip between home and a city 100 miles away. The average speeds for going and returning were x and y miles per hour, respectively.

(a) Show that $y = \dfrac{25x}{x - 25}$.

(b) Determine the vertical and horizontal asymptotes of the function.

(c) Complete the table.

x	30	35	40	45	50	55	60
y							

(d) Are the results in the table unexpected? Explain.

(e) Is it possible to average 20 miles per hour in one direction and still average 50 miles per hour on the round trip? Explain.

83. *Think About It*   Write a rational function satisfying the following criteria.

Vertical asymptote: $x = 2$
Slant asymptote: $y = x + 1$
Zero of the function: $x = -2$

84. *Think About It*   Write a rational function satisfying the following criteria.

Vertical asymptote: $x = -1$
Horizontal asymptote: $y = 2$
Zero of the function: $x = 3$

Review   **Solve Exercises 85–88 as a review of the skills and problem-solving techniques you learned in previous sections. Solve the inequality and show the solution on the real number line.**

85. $10 - 3x \le 0$

86. $5 - 2x > 5(x + 1)$

87. $|4(x - 2)| < 20$

88. $\frac{1}{2}|2x + 3| \ge 5$

▶ 4.3 *Partial Fractions*

See Exercise 47 on page 360 for an example of how partial fractions can help analyze the exhaust temperatures of a diesel engine.

Introduction □ *Partial Fraction Decomposition*

Introduction

In this section, you will learn to write a rational expression as the sum of two or more simpler rational expressions. For example, the rational expression $(x + 7)/(x^2 - x - 6)$ can be written as the sum of two fractions with first-degree denominators. That is,

$$\frac{x + 7}{x^2 - x - 6} = \frac{2}{x - 3} + \frac{-1}{x + 2}.$$

Each fraction on the right side of the equation is a **partial fraction,** and together they make up the **partial fraction decomposition** of the left side.

NOTE In step 1 at right, $N_1(x)$ is the remainder from the division of $N(x)$ by $D(x)$. ▪▪

DECOMPOSITION OF $N(x)/D(x)$ INTO PARTIAL FRACTIONS

1. *Divide if improper:* If $N(x)/D(x)$ is an improper fraction, divide the denominator into the numerator to obtain

$$\frac{N(x)}{D(x)} = (\text{polynomial}) + \frac{N_1(x)}{D(x)}$$

and apply Steps 2, 3, and 4 (below) to the proper rational expression $N_1(x)/D(x)$.

2. *Factor denominator:* Completely factor the denominator into factors of the form

$$(px + q)^m \qquad \text{and} \qquad (ax^2 + bx + c)^n$$

where $(ax^2 + bx + c)$ is irreducible.

3. *Linear factors:* For *each* factor of the form $(px + q)^m$, the partial fraction decomposition must include the following sum of m fractions.

$$\frac{A_1}{(px + q)} + \frac{A_2}{(px + q)^2} + \cdots + \frac{A_m}{(px + q)^m}$$

4. *Quadratic factors:* For *each* factor of the form $(ax^2 + bx + c)^n$, the partial fraction decomposition must include the following sum of n fractions.

$$\frac{B_1 x + C_1}{ax^2 + bx + c} + \frac{B_2 x + C_2}{(ax^2 + bx + c)^2} + \cdots + \frac{B_n x + C_n}{(ax^2 + bx + c)^n}$$

Partial Fraction Decomposition

Algebraic techniques for determining the constants in the numerators of partial fractions are demonstrated in the examples that follow. Note that the techniques vary slightly, depending on the type of factors of the denominator: linear or quadratic, distinct or repeated.

■

EXAMPLE 1   *Distinct Linear Factors*

Write the partial fraction decomposition for

$$\frac{x + 7}{x^2 - x - 6}.$$

Solution

Because $x^2 - x - 6 = (x - 3)(x + 2)$, you should include one partial fraction with a constant numerator for each linear factor of the denominator and write

$$\frac{x + 7}{x^2 - x - 6} = \frac{A}{x - 3} + \frac{B}{x + 2}.$$

Multiplying both sides of this equation by the least common denominator, $(x - 3)(x + 2)$, leads to the **basic equation**

$$x + 7 = A(x + 2) + B(x - 3). \qquad \text{\color{red}Basic equation}$$

Because this equation is true for all x, you can substitute any *convenient* values of x that will help determine the constants A and B. Values of x that are especially convenient are ones that make the factors $(x + 2)$ and $(x - 3)$ equal to zero. For instance, let $x = -2$. Then

$$
\begin{aligned}
-2 + 7 &= A(0) + B(-5) && \text{\color{red}Substitute } -2 \text{ for } x. \\
5 &= -5B \\
-1 &= B.
\end{aligned}
$$

To solve for A, let $x = 3$ and obtain

$$
\begin{aligned}
3 + 7 &= A(5) + B(0) && \text{\color{red}Substitute 3 for } x. \\
10 &= 5A \\
2 &= A.
\end{aligned}
$$

Therefore, the decomposition is

$$\frac{x + 7}{x^2 - x - 6} = \frac{2}{x - 3} + \frac{-1}{x + 2}$$

as indicated at the beginning of this section. Check this result by combining the two partial fractions on the right side of the equation.

■

TECHNOLOGY

You can use a graphing utility to check *graphically* the decomposition found in Example 1. To do this, graph

$$y_1 = \frac{x + 7}{x^2 - x - 6}$$

and

$$y_2 = \frac{2}{x - 3} + \frac{-1}{x + 2}$$

on the same viewing rectangle. Their graphs should be identical, as shown below.

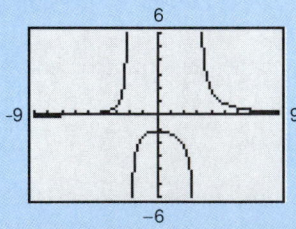

The next example shows how to find the partial fraction decomposition for a rational function whose denominator has a repeated linear factor.

EXAMPLE 2   *Repeated Linear Factors*

Write the partial fraction decomposition for

$$\frac{5x^2 + 20x + 6}{x^3 + 2x^2 + x}.$$

Solution

Because the denominator factors as

$$x^3 + 2x^2 + x = x(x^2 + 2x + 1) = x(x + 1)^2$$

you should include one partial fraction with a constant numerator for each power of x and $(x + 1)$ and write

$$\frac{5x^2 + 20x + 6}{x(x + 1)^2} = \frac{A}{x} + \frac{B}{x + 1} + \frac{C}{(x + 1)^2}.$$

Multiplying by the LCD, $x(x + 1)^2$, leads to the basic equation

$$5x^2 + 20x + 6 = A(x + 1)^2 + Bx(x + 1) + Cx. \qquad \text{Basic equation}$$

Letting $x = -1$ eliminates the A and B terms and yields

$$5 - 20 + 6 = 0 + 0 - C \qquad \text{Substitute } -1 \text{ for } x.$$

$$C = 9.$$

Letting $x = 0$ eliminates the B and C terms and yields

$$6 = A(1) + 0 + 0 \qquad \text{Substitute } 0 \text{ for } x.$$

$$6 = A.$$

At this point, you have exhausted the most convenient choices for x, so to find the value of B, use *any other value* for x along with the known values of A and C. Thus, using $x = 1$, $A = 6$, and $C = 9$,

$$5 + 20 + 6 = A(4) + B(2) + C$$

$$31 = 6(4) + 2B + 9$$

$$-2 = 2B$$

$$-1 = B.$$

Therefore, the partial fraction decomposition is

$$\frac{5x^2 + 20x + 6}{x(x + 1)^2} = \frac{6}{x} + \frac{-1}{x + 1} + \frac{9}{(x + 1)^2}.$$

The procedure used to solve for the constants in Examples 1 and 2 works well when the factors of the denominator are linear. However, when the denominator contains irreducible quadratic factors, you should use a different procedure, which involves writing the right side of the basic equation in polynomial form and *equating the coefficients* of like terms.

EXAMPLE 3   *Distinct Linear and Quadratic Factors*

Write the partial fraction decomposition for $\dfrac{3x^2 + 4x + 4}{x^3 + 4x}$.

Solution

Because the denominator factors as

$$x^3 + 4x = x(x^2 + 4)$$

you should include one partial fraction with a constant numerator and one partial fraction with a linear numerator and write

$$\frac{3x^2 + 4x + 4}{x^3 + 4x} = \frac{A}{x} + \frac{Bx + C}{x^2 + 4}.$$

Multiplying by the LCD, $x(x^2 + 4)$, yields the basic equation

$$3x^2 + 4x + 4 = A(x^2 + 4) + (Bx + C)x. \qquad \text{\color{red}{Basic equation}}$$

Expanding this basic equation and collecting like terms produces

$$3x^2 + 4x + 4 = Ax^2 + 4A + Bx^2 + Cx$$
$$= (A + B)x^2 + Cx + 4A. \qquad \text{\color{red}{Polynomial form}}$$

Finally, because two polynomials are equal if and only if the coefficients of like terms are equal,

$$3x^2 + 4x + 4 = (A + B)x^2 + Cx + 4A \qquad \text{\color{red}{Equate coefficients of like terms.}}$$

you obtain the following equations.

$$3 = A + B, \qquad 4 = C, \qquad \text{and} \qquad 4 = 4A$$

Thus, $A = 1$ and $C = 4$. Moreover, substituting $A = 1$ in the equation $3 = A + B$ yields

$$3 = 1 + B$$
$$2 = B.$$

Therefore, the partial fraction decomposition is

$$\frac{3x^2 + 4x + 4}{x^3 + 4x} = \frac{1}{x} + \frac{2x + 4}{x^2 + 4}.$$

The method of partial fractions was introduced by John Bernoulli (1667–1748), a Swiss mathematician who was instrumental in the early development of calculus. John Bernoulli was a professor at the University of Basel and taught many outstanding students, the most famous of whom was Leonhard Euler.

The next example shows how to find the partial fraction decomposition for a rational function whose denominator has a repeated quadratic factor.

EXAMPLE 4   *Repeated Quadratic Factors*

Write the partial fraction decomposition for

$$\frac{8x^3 + 13x}{(x^2 + 2)^2}.$$

Solution

We include one partial fraction with a linear numerator for each power of $(x^2 + 2)$, and write

$$\frac{8x^3 + 13x}{(x^2 + 2)^2} = \frac{Ax + B}{x^2 + 2} + \frac{Cx + D}{(x^2 + 2)^2}.$$

Multiplying by the LCD, $(x^2 + 2)^2$, yields the basic equation

$$8x^3 + 13x = (Ax + B)(x^2 + 2) + Cx + D \qquad \text{Basic equation}$$
$$= Ax^3 + 2Ax + Bx^2 + 2B + Cx + D$$
$$= Ax^3 + Bx^2 + (2A + C)x + (2B + D). \qquad \text{Polynomial form}$$

Equating coefficients of like terms

$$8x^3 + 0x^2 + 13x + 0 = Ax^3 + Bx^2 + (2A + C)x + (2B + D)$$

produces

$$8 = A, 0 = B, 13 = 2A + C, \quad \text{and} \quad 0 = 2B + D. \qquad \text{Equate coefficients.}$$

Finally, use the values $A = 8$ and $B = 0$ to obtain the following.

$$13 = 2A + C = 2(8) + C \qquad\qquad 0 = 2B + D = 2(0) + D$$
$$-3 = C \qquad\qquad\qquad\qquad\quad 0 = D$$

Therefore,

$$\frac{8x^3 + 13x}{(x^2 + 2)^2} = \frac{8x}{x^2 + 2} + \frac{-3x}{(x^2 + 2)^2}.$$

NOTE   By equating coefficients of like terms in Examples 3 and 4, you obtained several equations involving A, B, C, and D, which were solved by *substitution*. In a later chapter you will study a more general method for solving systems of equations. ∎

GUIDELINES FOR SOLVING THE BASIC EQUATION

Linear Factors

1. Substitute the *zeros* of the distinct linear factors into the basic equation.
2. For repeated linear factors, use the coefficients determined above to rewrite the basic equation. Then substitute *other* convenient values for x and solve for the remaining coefficients.

Quadratic Factors

1. Expand the basic equation.
2. Collect terms according to powers of x.
3. Equate the coefficients of like terms to obtain equations involving A, B, C, and so on.
4. Use substitution to solve for A, B, C, ...

Keep in mind that for *improper* rational expressions such as

$$\frac{N(x)}{D(x)} = \frac{2x^3 + x^2 - 7x + 7}{x^2 + x - 2}$$

you must first divide before applying partial fraction decomposition.

GROUP ACTIVITY

ERROR ANALYSIS

Suppose you are tutoring a student in algebra. In trying to find a partial fraction decomposition, your student wrote the following.

$$\frac{x^2 + 1}{x(x - 1)} = \frac{A}{x} + \frac{B}{x - 1}$$

$$\frac{x^2 + 1}{x(x - 1)} = \frac{A(x - 1)}{x(x - 1)} + \frac{Bx}{x(x - 1)}$$

$$x^2 + 1 = A(x - 1) + Bx \qquad \text{Basic equation}$$

By substituting $x = 0$ and $x = 1$ into the basic equation, your student concluded that $A = -1$ and $B = 2$. However, in checking this solution, your student obtained the following. What went wrong?

$$\frac{-1}{x} + \frac{2}{x - 1} = \frac{(-1)(x - 1) + 2(x)}{x(x - 1)} = \frac{x + 1}{x(x - 1)} \neq \frac{x^2 + 1}{x(x - 1)}.$$

Find the sum and simplify.

1. $\dfrac{2}{x} + \dfrac{3}{x + 1}$

2. $\dfrac{5}{x + 2} + \dfrac{3}{x}$

3. $\dfrac{7}{x - 2} - \dfrac{3}{2x - 1}$

4. $\dfrac{2}{x + 5} - \dfrac{5}{x + 12}$

5. $\dfrac{-5}{x + 2} + \dfrac{4}{(x + 2)^2}$

6. $\dfrac{1}{x - 3} + \dfrac{3}{(x - 3)^2} - \dfrac{5}{(x - 3)^3}$

7. $\dfrac{-3}{x} + \dfrac{3x - 1}{x^2 + 3}$

8. $\dfrac{5}{x + 1} - \dfrac{x - 6}{x^2 + 5}$

9. $\dfrac{3}{x^2 + 1} + \dfrac{x - 3}{(x^2 + 1)^2}$

10. $\dfrac{x}{x^2 + x + 1} - \dfrac{x - 1}{(x^2 + x + 1)^2}$

4.3 Exercises

In Exercises 1–6, write the form of the partial fraction decomposition of the rational expression. Do not solve for the constants.

1. $\dfrac{7}{x^2 - 14x}$

2. $\dfrac{x - 2}{x^2 + 4x + 3}$

3. $\dfrac{12}{x^3 - 10x^2}$

4. $\dfrac{4x^2 + 3}{(x - 5)^3}$

5. $\dfrac{2x - 3}{x^3 + 10x}$

6. $\dfrac{x - 1}{x(x^2 + 1)^2}$

In Exercises 7–30, write the partial fraction decomposition for the rational expression. Check your result algebraically.

7. $\dfrac{1}{x^2 - 1}$

8. $\dfrac{1}{4x^2 - 9}$

9. $\dfrac{1}{x^2 + x}$

10. $\dfrac{3}{x^2 - 3x}$

11. $\dfrac{1}{2x^2 + x}$

12. $\dfrac{5}{x^2 + x - 6}$

13. $\dfrac{3}{x^2 + x - 2}$

14. $\dfrac{x + 1}{x^2 + 4x + 3}$

15. $\dfrac{x^2 + 12x + 12}{x^3 - 4x}$

16. $\dfrac{x + 2}{x(x - 4)}$

17. $\dfrac{4x^2 + 2x - 1}{x^2(x + 1)}$

18. $\dfrac{2x - 3}{(x - 1)^2}$

19. $\dfrac{3x}{(x - 3)^2}$

20. $\dfrac{6x^2 + 1}{x^2(x - 1)^3}$

21. $\dfrac{x^2 - 1}{x(x^2 + 1)}$

22. $\dfrac{x}{(x - 1)(x^2 + x + 1)}$

23. $\dfrac{x^2}{x^4 - 2x^2 - 8}$

24. $\dfrac{2x^2 + x + 8}{(x^2 + 4)^2}$

25. $\dfrac{x}{16x^4 - 1}$

26. $\dfrac{x^2 - 4x + 7}{(x + 1)(x^4 - 2x + 3)}$

27. $\dfrac{x^2 + 5}{(x + 1)(x^2 - 2x + 3)}$

28. $\dfrac{x + 1}{x^3 + x}$

29. $\dfrac{x^4}{(x - 1)^3}$

30. $\dfrac{x^2 - x}{x^2 + x + 1}$

In Exercises 31–38, write the partial fraction decomposition for the rational expression. Use a graphing utility to check your result graphically.

31. $\dfrac{5 - x}{2x^2 + x - 1}$

32. $\dfrac{3x^2 - 7x - 2}{x^3 - x}$

33. $\dfrac{x - 1}{x^3 + x^2}$

34. $\dfrac{4x^2 - 1}{2x(x + 1)^2}$

35. $\dfrac{x^2 + x + 2}{(x^2 + 2)^2}$

36. $\dfrac{x^3}{(x + 2)^2(x - 2)^2}$

37. $\dfrac{2x^3 - 4x^2 - 15x + 5}{x^2 - 2x - 8}$

38. $\dfrac{x^3 - x + 3}{x^2 + x - 2}$

In Exercises 39–42, write the partial fraction decomposition for the rational expression. Check your result algebraically. Then assign a value to the constant a to check the result graphically.

39. $\dfrac{1}{a^2 - x^2}$

40. $\dfrac{1}{x(x + a)}$

41. $\dfrac{1}{y(a - y)}$

42. $\dfrac{1}{(x + 1)(a - x)}$

Graphical Analysis In Exercises 43–46, write the partial fraction decomposition for the rational function. Identify the graph of the rational function and the graph of each term of its decomposition. State any relationship between the vertical asymptotes of the rational function and the vertical asymptotes of the terms of decomposition.

43. $y = \dfrac{x - 12}{x(x - 4)}$

44. $y = \dfrac{2(x + 1)^2}{x(x^2 + 1)}$

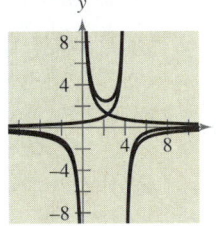

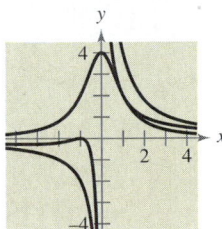

45. $y = \dfrac{2(4x - 3)}{x^2 - 9}$

46. $y = \dfrac{2(4x^2 - 15x + 39)}{x^2(x^2 - 10x + 26)}$

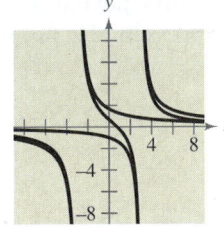

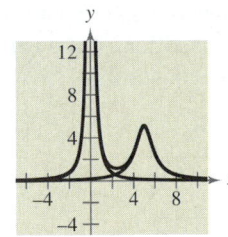

47. *Exhaust Temperatures* The magnitude of the range of exhaust temperatures in degrees Fahrenheit in an experimental Diesel engine is approximated by

$$R = \dfrac{2000(4 - 3x)}{(11 - 7x)(7 - 4x)}, \qquad 0 \le x \le 1$$

where x is the relative load.

(a) Write the partial fraction decomposition of the expression.

(b) The decomposition of part (a) is the difference of two fractions. The absolute values of the terms give the expected maximum and minimum temperatures of the exhaust gases. Use a graphing utility to graph each term.

4.4 *Conics*

See Exercise 30 on page 371 for an example of how a parabola can be used to model the cables of a suspension bridge.

Introduction ◻ *Parabolas* ◻ *Ellipses* ◻ *Hyperbolas*

Introduction

Conic sections were discovered during the classical Greek period, 600 to 300 B.C. This early Greek study was largely concerned with the geometrical properties of conics. It was not until the early 17th century that the broad applicability of conics became apparent and played a prominent role in the early development of calculus.

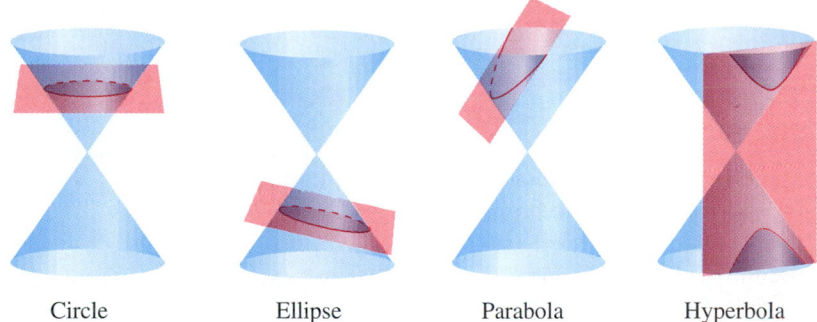

Circle   Ellipse   Parabola   Hyperbola

FIGURE 4.15   Basic Conics

A **conic section** (or simply conic) is the intersection of a plane and a double-napped cone. Notice from Figure 4.15 that in the formation of the four basic conics, the intersecting plane does not pass through the vertex of the cone. When the plane does pass through the vertex, the resulting figure is a **degenerate conic,** as shown in Figure 4.16.

There are several ways to approach the study of conics. You could begin by defining conics in terms of the intersections of planes and cones, as the Greeks did, or you could define them algebraically, in terms of the general second-degree equation

$$Ax^2 + Bxy + Cy^2 + Dx + Ey + F = 0.$$

However, we will use a third approach, in which each of the conics is defined as a *locus* (collection) of points satisfying a certain geometric property. For example, in Section 1.1 you saw how the definition of a circle as *the collection of all points* (x, y) *that are equidistant from a fixed point* (h, k) led easily to the standard equation of a circle

$$(x - h)^2 + (y - k)^2 = r^2. \qquad \text{\color{red}Equation of circle}$$

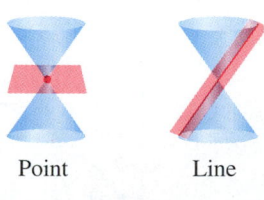

Point   Line

Two Intersecting Lines

FIGURE 4.16   Degenerate Conics

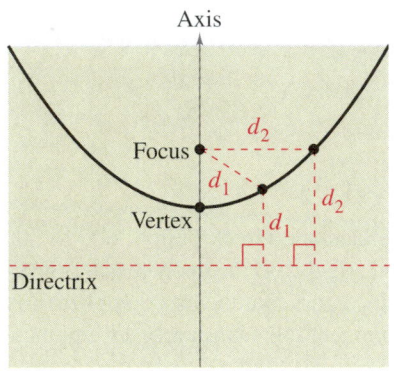

FIGURE 4.17 Parabola

THINK ABOUT THE PROOF

The proofs of the two cases at the right are similar. To prove the case with the vertical axis, use the diagram in Figure 4.18(a). Begin by assuming that the directrix, $y = -p$, is parallel to the x-axis. In Figure 4.18(a), assume that $p > 0$. Because p is the *directed* distance from the vertex to the focus, the focus must lie above the vertex. How does this information allow you to complete the proof? The details of the proof are given in the appendix.

Parabolas

In Section 3.1, you learned that the graph of the quadratic function $f(x) = ax^2 + bx + c$ is a parabola that opens upward or downward. The following definition of a parabola is more general in the sense that it is independent of the orientation of the parabola.

> ### DEFINITION OF A PARABOLA
>
> A **parabola** is the set of all points (x, y) that are equidistant from a fixed line, the **directrix,** and a fixed point, the **focus,** not on the line. See Figure 4.17.

The midpoint between the focus and the directrix is the **vertex,** and the line passing through the focus and the vertex is the **axis** of the parabola.

> ### STANDARD EQUATION OF A PARABOLA (VERTEX AT ORIGIN)
>
> The **standard form of the equation of a parabola** with vertex at $(0, 0)$ and directrix $y = -p$ is
>
> $$x^2 = 4py, \qquad p \neq 0. \qquad \text{Vertical axis}$$
>
> For directrix $x = -p$, the equation is
>
> $$y^2 = 4px, \qquad p \neq 0. \qquad \text{Horizontal axis}$$
>
> The focus is on the axis p units (directed distance) from the vertex.

Notice that a parabola can have a vertical or a horizontal axis. Examples of each are shown in Figure 4.18.

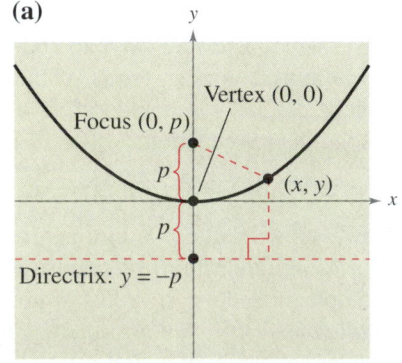

Parabola with vertical axis   Parabola with horizontal axis

FIGURE 4.18

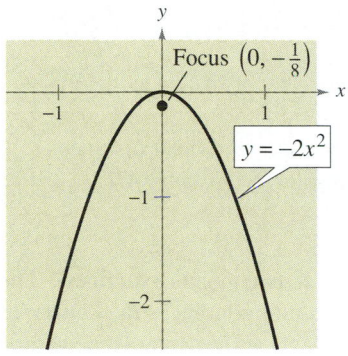

FIGURE 4.19

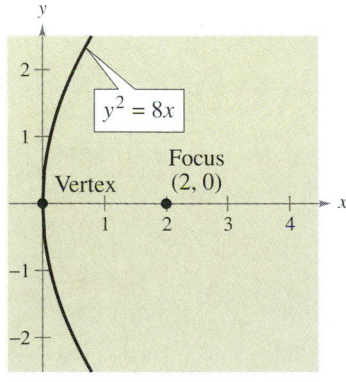

FIGURE 4.20

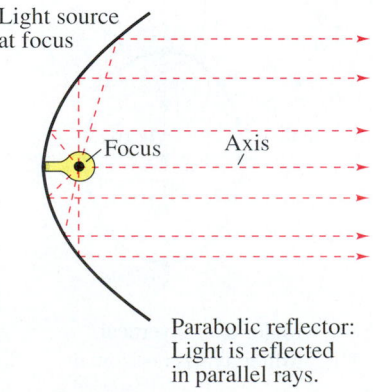

Parabolic reflector:
Light is reflected
in parallel rays.

FIGURE 4.21

EXAMPLE 1   *Finding the Focus of a Parabola*

Find the focus of the parabola whose equation is $y = -2x^2$.

Solution

Because the squared term in the equation involves x, you know that the axis is vertical, and the equation is of the form

$$x^2 = 4py.$$

You can write the given equation in this form as follows.

$$x^2 = -\frac{1}{2}y$$

$$x^2 = 4\left(-\frac{1}{8}\right)y \qquad \text{Standard form}$$

Thus, $p = -\frac{1}{8}$. Because p is negative, the parabola opens downward (see Figure 4.19), and the focus of the parabola is $(0, p) = \left(0, -\frac{1}{8}\right)$.

EXAMPLE 2   *A Parabola with a Horizontal Axis*

Write the standard form of the equation of the parabola with vertex at the origin and focus at $(2, 0)$.

Solution

The axis of the parabola is horizontal, passing through $(0, 0)$ and $(2, 0)$, as shown in Figure 4.20. Thus, the standard form is

$$y^2 = 4px.$$

Because the focus is $p = 2$ units from the vertex, the equation is

$$y^2 = 4(2)x$$
$$y^2 = 8x.$$

Parabolas occur in a wide variety of applications. For instance, a parabolic reflector can be formed by revolving a parabola about its axis. The resulting surface has the property that all incoming rays parallel to the axis are reflected through the focus of the parabola—this is the principle behind the construction of the parabolic mirrors used in reflecting telescopes. Conversely, the light rays emanating from the focus of a parabolic reflector used in a flashlight are all parallel to one another, as shown in Figure 4.21.

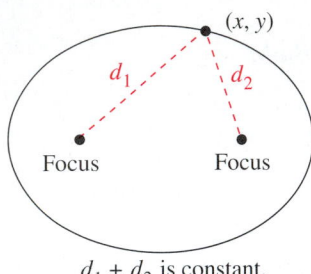

$d_1 + d_2$ is constant.

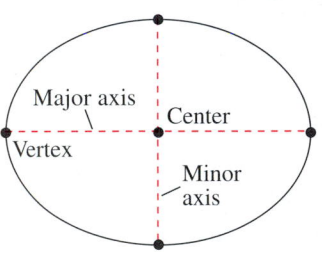

FIGURE 4.22

 A computer animation of this concept appears in the *Interactive* CD-ROM.

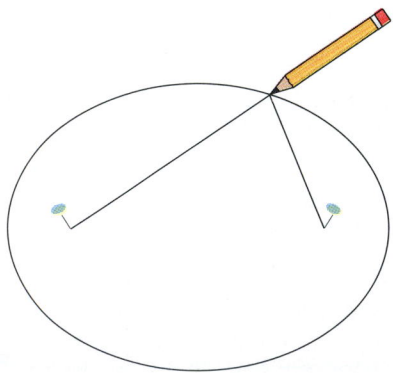

FIGURE 4.23

Ellipses

DEFINITION OF AN ELLIPSE

An **ellipse** is the set of all points (x, y) the sum of whose distances from two distinct fixed points (**foci**) is constant. See Figure 4.22.

The line through the foci intersects the ellipse at two points (**vertices).** The chord joining the vertices is the **major axis,** and its midpoint is the **center** of the ellipse. The chord perpendicular to the major axis at the center is the **minor axis** of the ellipse.

You can visualize the definition of an ellipse by imagining two thumbtacks placed at the foci, as shown in Figure 4.23. If the ends of a fixed length of string are fastened to the thumbtacks and the string is drawn taut with a pencil, the path traced by the pencil will be an ellipse.

The standard form of the equation of an ellipse takes one of two forms, depending upon whether the major axis is horizontal or vertical.

STANDARD EQUATION OF AN ELLIPSE (CENTER AT ORIGIN)

The **standard form of the equation of an ellipse** with center at the origin and major and minor axes of lengths $2a$ and $2b$ (where $0 < b < a$) is

$$\frac{x^2}{a^2} + \frac{y^2}{b^2} = 1 \qquad \text{or} \qquad \frac{x^2}{b^2} + \frac{y^2}{a^2} = 1.$$

The vertices and foci lie on the major axis, a and c units, respectively, from the center, as shown in Figure 4.24. Moreover, a, b, and c are related by the equation $c^2 = a^2 - b^2$.

(a)

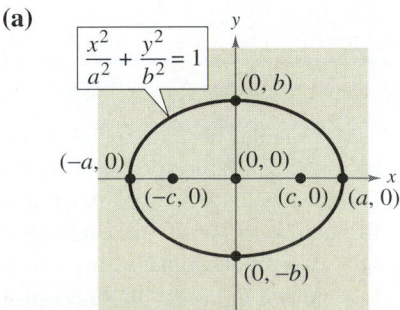

Major axis is horizontal.
Minor axis is vertical.

(b)

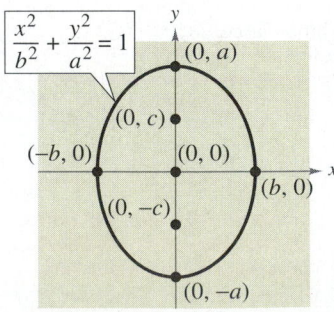

Major axis is vertical.
Minor axis is horizontal.

FIGURE 4.24

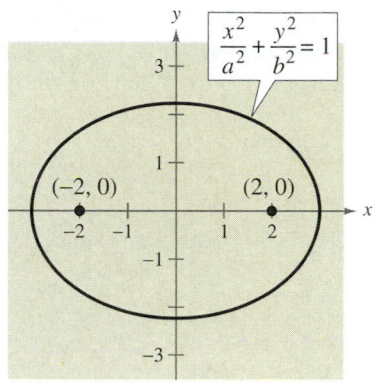

FIGURE 4.25

EXAMPLE 3   Finding the Standard Equation of an Ellipse

Find the standard form of the equation of the ellipse that has a major axis of length 6 and foci at $(-2, 0)$ and $(2, 0)$, as shown in Figure 4.25.

Solution

Because the foci occur at $(-2, 0)$ and $(2, 0)$, the center of the ellipse is $(0, 0)$, and the major axis is horizontal. Thus, the ellipse has an equation of the form

$$\frac{x^2}{a^2} + \frac{y^2}{b^2} = 1. \qquad \text{Standard form}$$

Because the length of the major axis is 6,

$$2a = 6 \qquad \text{Length of major axis}$$

which implies that $a = 3$. Moreover, the distance from the center to either focus is $c = 2$. Finally,

$$b^2 = a^2 - c^2 = 3^2 - 2^2 = 9 - 4 = 5$$

which yields the equation

$$\frac{x^2}{9} + \frac{y^2}{5} = 1.$$

EXAMPLE 4   Sketching an Ellipse

Sketch the ellipse given by $4x^2 + y^2 = 36$, and identify the vertices.

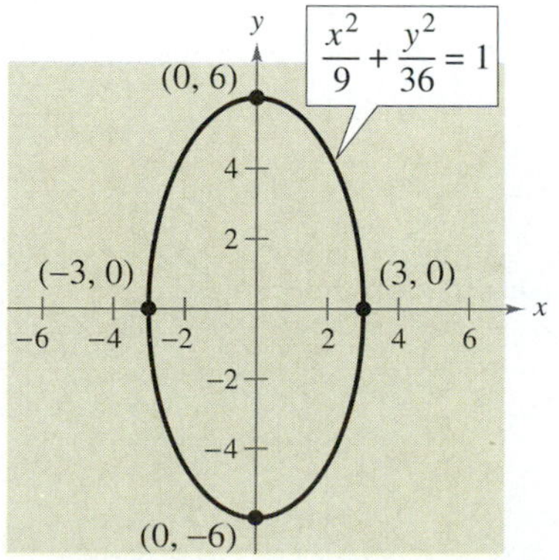

Solution

$$4x^2 + y^2 = 36 \qquad \text{Given equation}$$

$$\frac{4x^2}{36} + \frac{y^2}{36} = \frac{36}{36}$$

$$\frac{x^2}{3^2} + \frac{y^2}{6^2} = 1 \qquad \text{Standard form}$$

Because the denominator of the y^2-term is larger than the denominator of the x^2-term, you can conclude that the major axis is vertical. Moreover, because $a = 6$, the vertices are $(0, -6)$ and $(0, 6)$. Finally, because $b = 3$, the endpoints of the minor axis are $(-3, 0)$ and $(3, 0)$, as shown in Figure 4.26. Note that you can sketch the ellipse by locating the endpoints of the two axes. Because 3^2 is the denominator of the x^2-term, move three units to the *right and left* of the center to locate the endpoints of the horizontal axis. Similarly, because 6^2 is the denominator of the y^2-term, move six units *up and down* from the center to locate the endpoints of the vertical axis.

FIGURE 4.26

Hyperbolas

The definition of a **hyperbola** is similar to that of an ellipse. The difference is that, for an ellipse, the *sum* of the distances between the foci and a point on the ellipse is constant, while, for a hyperbola, the *difference* of the distances between the foci and a point on the hyperbola is constant.

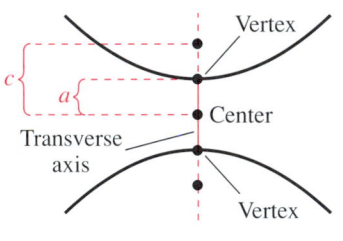

> **DEFINITION OF A HYPERBOLA**
>
> A **hyperbola** is the set of all points (x, y) the difference of whose distances from two distinct fixed points (**foci**) is constant. See Figure 4.27.

The graph of a hyperbola has two disconnected parts (**branches**). The line through the two foci intersects the hyperbola at two points (**vertices**). The line segment connecting the vertices is the **transverse axis,** and the midpoint of the transverse axis is the **center** of the hyperbola.

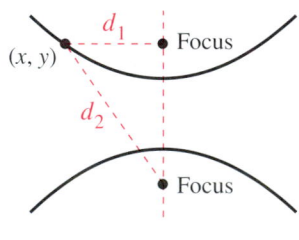

$d_2 - d_1$ is constant.

FIGURE 4.27

> **STANDARD EQUATION OF A HYPERBOLA (CENTER AT ORIGIN)**
>
> The **standard form of the equation of a hyperbola** with center at the origin (where $a \neq 0$ and $b \neq 0$), is
>
> $$\frac{x^2}{a^2} - \frac{y^2}{b^2} = 1 \qquad \text{or} \qquad \frac{y^2}{a^2} - \frac{x^2}{b^2} = 1.$$
>
> The vertices and foci are, respectively, a and c units from the center, and $b^2 = c^2 - a^2$. See Figure 4.28.

(a)

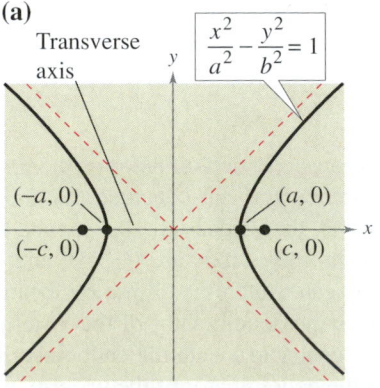

(b)

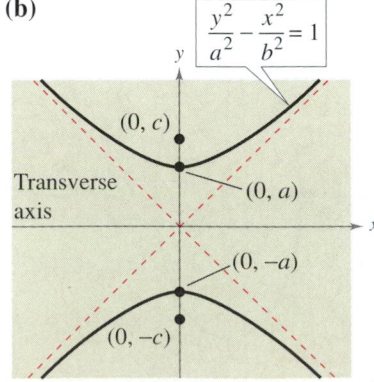

FIGURE 4.28

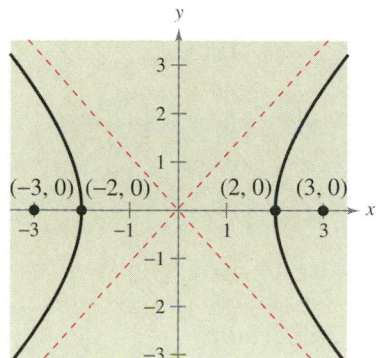

FIGURE 4.29

EXAMPLE 5   *Finding the Standard Equation of a Hyperbola*

Find the standard form of the equation of the hyperbola with foci at $(-3, 0)$ and $(3, 0)$ and vertices at $(-2, 0)$ and $(2, 0)$, as shown in Figure 4.29.

Solution

It can be seen that $c = 3$, because the foci are three units from the center. Moreover, $a = 2$ because the vertices are two units from the center. Thus, it follows that

$$b^2 = c^2 - a^2 = 3^2 - 2^2 = 9 - 4 = 5.$$

Because the transverse axis is horizontal, the standard form of the equation is

$$\frac{x^2}{a^2} - \frac{y^2}{b^2} = 1.$$

Finally, substitute $a^2 = 4$ and $b^2 = 5$ to obtain

$$\frac{x^2}{4} - \frac{y^2}{5} = 1. \qquad \text{Standard form}$$

An important aid in sketching the graph of a hyperbola is the determination of its **asymptotes,** as shown in Figure 4.30. Each hyperbola has two asymptotes that intersect at the center of the hyperbola. Furthermore, the asymptotes pass through the corners of a rectangle of dimensions $2a$ by $2b$. The line segment of length $2b$ joining $(0, b)$ and $(0, -b)$ [or $(-b, 0)$ and $(b, 0)$] is the **conjugate axis** of the hyperbola.

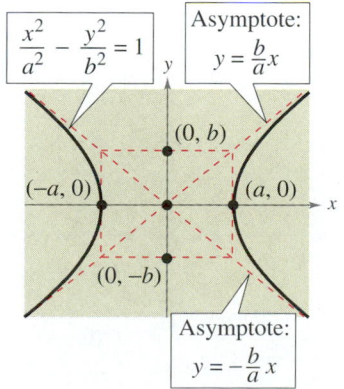

Transverse axis is horizontal.

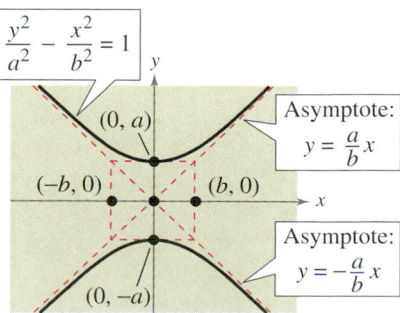

Transverse axis is vertical.

FIGURE 4.30

> **◼ ◻ ASYMPTOTES OF A HYPERBOLA (CENTER AT ORIGIN)**
>
> The **asymptotes of a hyperbola** with center at $(0, 0)$ are
>
> $$y = \frac{b}{a}x \quad \text{and} \quad y = -\frac{b}{a}x \qquad \text{Transverse axis is horizontal.}$$
>
> or
>
> $$y = \frac{a}{b}x \quad \text{and} \quad y = -\frac{a}{b}x. \qquad \text{Transverse axis is vertical.}$$

Exploration

Use a graphing utility to graph the hyperbola in Example 6. Does your graph look like that shown in Figure 4.31(b)? If not, what must you do to obtain both the upper and lower portions of the hyperbola? Explain your reasoning.

EXAMPLE 6   *Sketching the Graph of a Hyperbola*

Sketch the graph of the hyperbola whose equation is $4x^2 - y^2 = 16$.

Solution

$$4x^2 - y^2 = 16 \qquad \text{Original equation}$$

$$\frac{4x^2}{16} - \frac{y^2}{16} = \frac{16}{16}$$

$$\frac{x^2}{2^2} - \frac{y^2}{4^2} = 1 \qquad \text{Standard form}$$

Because the x^2-term is positive, you can conclude that the transverse axis is horizontal and the vertices occur at $(-2, 0)$ and $(2, 0)$. Moreover, the endpoints of the conjugate axis occur at $(0, -4)$ and $(0, 4)$, and you can sketch the rectangle shown in Figure 4.31(a). Finally, by drawing the asymptotes through the corners of this rectangle, you can complete the sketch shown in Figure 4.31(b).

(a)
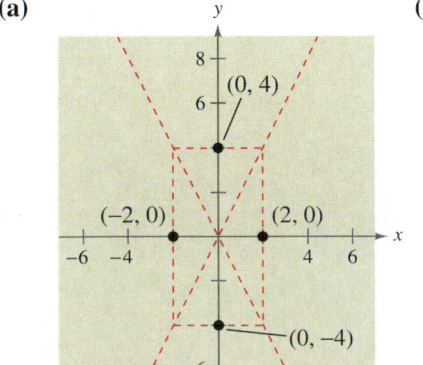

(b)
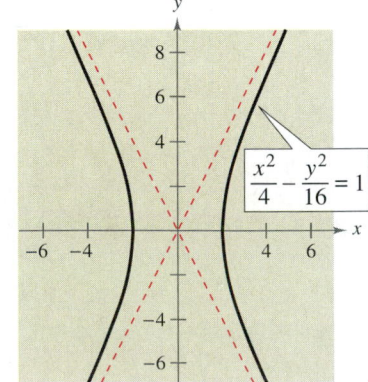

FIGURE 4.31

The *Interactive* CD-ROM shows every example with its solution; clicking on the *Try It!* button brings up similar problems. Guided Examples and Integrated Examples show step-by-step solutions to additional examples. Integrated Examples are related to several concepts in the section.

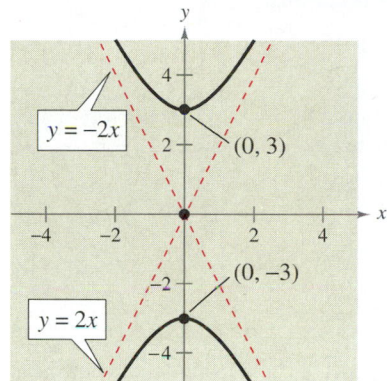

FIGURE 4.32

EXAMPLE 7 *Finding the Standard Equation of a Hyperbola*

Find the standard form of the equation of a hyperbola having vertices at $(0, -3)$ and $(0, 3)$ and with asymptotes $y = -2x$ and $y = 2x$, as shown in Figure 4.32.

Solution

Because the transverse axis is vertical, the asymptotes are of the form

$$y = \frac{a}{b}x \qquad \text{and} \qquad y = -\frac{a}{b}x.$$

Thus,

$$\frac{a}{b} = 2$$

and, because $a = 3$, you can determine that $b = \frac{3}{2}$. Finally, you can conclude that the hyperbola has the following equation.

$$\frac{y^2}{3^2} - \frac{x^2}{(3/2)^2} = 1 \qquad \qquad \text{Standard form}$$

GROUP ACTIVITY

HYPERBOLAS IN APPLICATION

At the beginning of this section, we mentioned that each type of conic section can be formed by the intersection of a plane and a double-napped cone. The figure below shows three examples of how such an intersection can occur in physical situations.

Identify the cone and hyperbola (or portion of a hyperbola) in each of the three situations. Can you think of other examples of physical situations in which hyperbolas are formed?

Rewrite the equation so that it has no fractions.

1. $\dfrac{x^2}{16} + \dfrac{y^2}{9} = 1$

2. $\dfrac{x^2}{32} + \dfrac{4y^2}{32} = \dfrac{32}{32}$

3. $\dfrac{x^2}{1/4} - \dfrac{y^2}{4} = 1$

4. $\dfrac{3x^2}{1/9} + \dfrac{4y^2}{9} = 1$

Solve for c. (Assume $c > 0$.)

5. $c^2 = 3^2 - 1^2$

6. $c^2 = 2^2 + 3^2$

7. $c^2 + 2^2 = 4^2$

8. $c^2 - 1^2 = 2^2$

Find the distance between the point and the origin.

9. $(0, -4)$

10. $(-2, 0)$

4.4 Exercises

In Exercises 1–8, match the equation with its graph. If the equation's graph is not shown, say so. [The graphs are labeled (a), (b), (c), (d), (e), and (f).]

(a)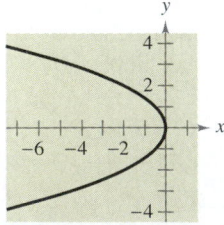

(b)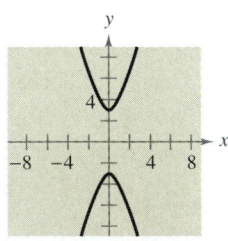

(c)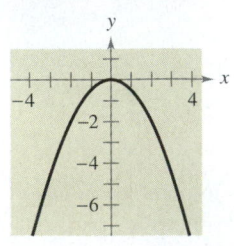

(d)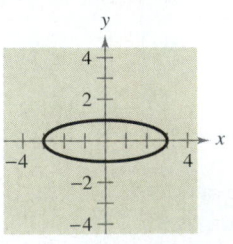

(e)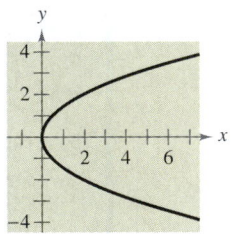

(f)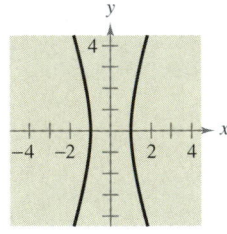

1. $x^2 = 2y$

2. $x^2 = -2y$

3. $y^2 = 2x$

4. $y^2 = -2x$

5. $9x^2 + y^2 = 9$

6. $x^2 + 9y^2 = 9$

7. $9x^2 - y^2 = 9$

8. $y^2 - 9x^2 = 9$

In Exercises 9–14, find the vertex and focus of the parabola and sketch its graph.

9. $y = \frac{1}{2}x^2$

10. $y = 2x^2$

11. $y^2 = -6x$

12. $y^2 = 3x$

13. $x^2 + 8y = 0$

14. $x + y^2 = 0$

In Exercises 15 and 16, use a graphing utility to graph the parabola and its tangent line. Identify the point of tangency.

Parabola	*Tangent Line*
15. $y^2 - 8x = 0$	$x - y + 2 = 0$
16. $x^2 + 12y = 0$	$x + y - 3 = 0$

In Exercises 17–26, find an equation of the parabola with vertex at the origin.

17. Focus: $\left(0, -\frac{3}{2}\right)$

18. Focus: $(2, 0)$

19. Focus: $(-2, 0)$

20. Focus: $(0, -2)$

21. Directrix: $y = -1$

22. Directrix: $x = 3$

23. Directrix: $y = 2$

24. Directrix: $x = -2$

25. Passes through the point $(4, 6)$; horizontal axis

26. Passes through the point $(-2, -2)$; vertical axis

In Exercises 27 and 28, find an equation of the parabola and determine the coordinates of the focus.

27.

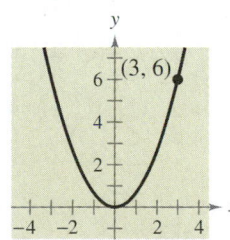

28.

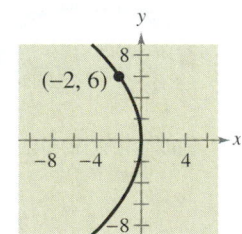

29. *Satellite Antenna* Write an equation for a cross section of the parabolic television dish antenna shown in the figure.

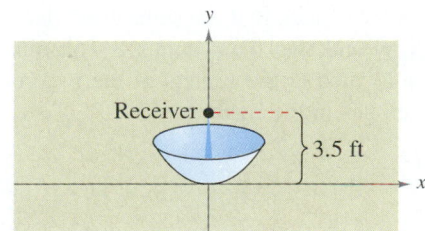

30. *Suspension Bridge* Each cable of a suspension bridge is suspended (in the shape of a parabola) between two towers that are 120 meters apart. The top of each tower is 20 meters above the roadway. The cables touch the roadway midway between the towers.

(a) Draw a sketch of the bridge. Locate the origin of a rectangular coordinate system at the center of the roadway. Label the coordinates of the known points.

(b) Write an equation that models the cables.

(c) Complete the table by finding the height y of the suspension cables over the roadway at a distance of x meters from the center of the bridge.

x	0	20	40	60
y				

31. *Beam Deflection* A simply supported beam is 64 feet long and has a load at the center (see figure). The deflection of the beam at its center is 1 inch. The shape of the deflected beam is parabolic.

(a) Find an equation of the parabola. (Assume that the origin is at the center of the beam.)

(b) How far from the center of the beam is the deflection one-half inch?

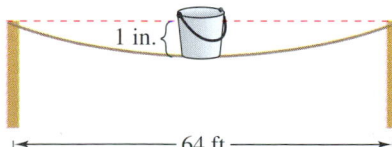

32. *Exploration* Consider the equation $x^2 = 4py$.

(a) Use a graphing utility to graph the parabolas for $p = 1, p = 2, p = 3$, and $p = 4$. Describe the effect on the graph when p increases.

(b) Locate the focus for each parabola in part (a).

(c) For each parabola in part (a), find the length of the chord passing through the focus parallel to the directrix. How can the length of this chord be determined directly from $x^2 = 4py$?

(d) Explain how the result of part (c) can be used as a sketching aid when graphing parabolas.

33. *True or False?* "It is possible for a parabola to intersect its directrix." Explain your reasoning.

34. *Exploration* Let (x_1, y_1) be the coordinates of a point on the parabola $x^2 = 4py$. The equation of the line tangent to the parabola at the point is

$$y - y_1 = \frac{x_1}{2p}(x - x_1).$$

(a) What is the slope of the tangent line?

(b) Find an equation of the tangent line at the endpoint of each of the chords for the parabolas in Exercise 32. Graph the tangent lines.

In Exercises 35–40, find the center and vertices of the ellipse and sketch its graph.

35. $\dfrac{x^2}{25} + \dfrac{y^2}{16} = 1$ **36.** $\dfrac{x^2}{144} + \dfrac{y^2}{169} = 1$

37. $\dfrac{x^2}{16} + \dfrac{y^2}{25} = 1$ **38.** $\dfrac{x^2}{169} + \dfrac{y^2}{144} = 1$

39. $\dfrac{x^2}{9} + \dfrac{y^2}{5} = 1$ **40.** $\dfrac{x^2}{28} + \dfrac{y^2}{64} = 1$

In Exercises 41 and 42, use a graphing utility to graph the ellipse. (*Hint:* Use two equations.)

41. $5x^2 + 3y^2 = 15$ **42.** $x^2 + 4y^2 = 4$

Think About It **In Exercises 43 and 44, which part of the ellipse $4x^2 + 9y^2 = 36$ is represented by the equation?**

43. $x = -\frac{3}{2}\sqrt{4 - y^2}$ **44.** $y = \frac{2}{3}\sqrt{9 - x^2}$

In Exercises 45–52, find an equation of the ellipse with center at the origin.

45.

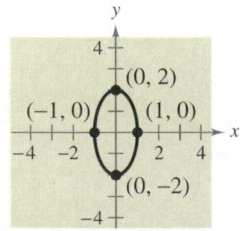

46.

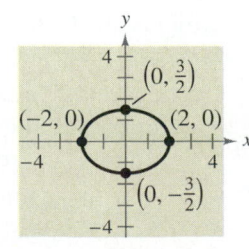

47. Vertices: $(\pm 5, 0)$; Foci: $(\pm 2, 0)$

48. Vertices: $(0, \pm 8)$; Foci: $(0, \pm 4)$

49. Foci: $(\pm 5, 0)$; Major axis of length 12

50. Foci: $(\pm 2, 0)$; Major axis of length 8

51. Vertices: $(0, \pm 5)$; Passes through the point $(4, 2)$

52. Major axis vertical; Passes through the points $(0, 4)$ and $(2, 0)$

53. *Think About It* On page 364 (see Figure 4.23), it is noted that an ellipse can be drawn using two thumbtacks, a string of fixed length (greater than the distance between the two tacks), and a pencil.

(a) What is the length of the string in terms of a?

(b) Explain why the path is an ellipse.

54. *Fireplace Arch* A fireplace arch is to be constructed in the shape of a semiellipse. The opening is to have a height of 2 feet at the center and a width of 6 feet along the base (see figure). The contractor draws the outline of the ellipse by the method in Exercise 53. Give the required positions of the tacks and the length of the string.

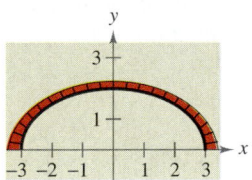

FIGURE FOR 54

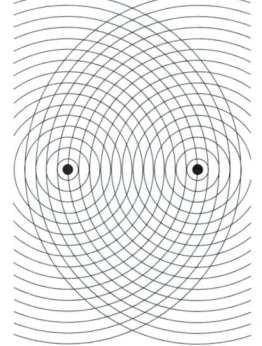

FIGURE FOR 55

55. *Geometry* Sketch a graph of the ellipse that consists of all points (x, y) such that the sum of the distances between (x, y) and two fixed points is 16 units and the foci are located at the centers of the two sets of concentric circles in the figure.

56. Mountain Tunnel   A semielliptical arch over a tunnel for a road through a mountain has a major axis of 100 feet and a height at the center of 30 feet.

(a) Make a sketch to solve the problem. Draw a rectangular coordinate system on the tunnel with the center of the road entering the tunnel at the origin. Identify the coordinates of the known points.

(b) Find an equation of the elliptical tunnel.

(c) Determine the height of the arch 5 feet from the edge of the tunnel.

57. Exploration   Consider the ellipse

$$\frac{x^2}{a^2} + \frac{y^2}{b^2} = 1, \quad a + b = 20.$$

(a) The area of the ellipse is given by $A = \pi ab$. Write the area of the ellipse as a function of a.

(b) Find the equation of an ellipse with an area of 264 square centimeters.

(c) Complete the table using your equation from part (a), and make a conjecture about the shape of the ellipse with maximum area.

a	8	9	10	11	12	13
A						

(d) Use a graphing utility to graph the area function and use the graph to make a conjecture about the shape of the ellipse that yields maximum area.

58. Geometry   The area of the ellipse in the figure is twice the area of the circle. What is the length of the major axis? (The area of an ellipse is $A = \pi ab$.)

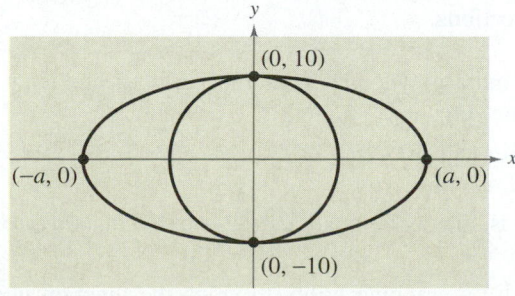

59. Think About It   Is the graph of $x^2 + 4y^4 = 4$ an ellipse? Explain.

60. Think About It   The graph of $x^2 + y^2 = 0$ is a degenerate conic. Sketch the graph of this equation.

61. Essay   Write a paragraph discussing the change in the shape and orientation of the graph of the ellipse

$$\frac{x^2}{a^2} + \frac{y^2}{4^2} = 1$$

as a increases from 1 to 8.

62. Geometry   A line segment through a focus of an ellipse with endpoints on the ellipse and perpendicular to the major axis is called a **latus rectum** of the ellipse. Therefore, an ellipse has two latera recta. Knowing the length of the latera recta is helpful in sketching an ellipse because it yields other points on the curve (see figure). Show that the length of each latus rectum is $2b^2/a$.

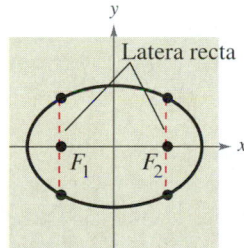

In Exercises 63–66, sketch the graph of the ellipse, making use of the latera recta (see Exercise 62).

63. $\dfrac{x^2}{4} + \dfrac{y^2}{1} = 1$

64. $\dfrac{x^2}{9} + \dfrac{y^2}{16} = 1$

65. $9x^2 + 4y^2 = 36$

66. $5x^2 + 3y^2 = 15$

In Exercises 67–72, find the center and vertices of the hyperbola and sketch its graph, using asymptotes as sketching aids.

67. $x^2 - y^2 = 1$

68. $\dfrac{x^2}{9} - \dfrac{y^2}{16} = 1$

69. $\dfrac{y^2}{1} - \dfrac{x^2}{4} = 1$

70. $\dfrac{y^2}{9} - \dfrac{x^2}{1} = 1$

71. $\dfrac{y^2}{25} - \dfrac{x^2}{144} = 1$

72. $\dfrac{x^2}{36} - \dfrac{y^2}{4} = 1$

In Exercises 73 and 74, use a graphing utility to graph the hyperbola and its asymptotes.

73. $2x^2 - 3y^2 = 6$ **74.** $3y^2 - 5x^2 = 15$

Think About It In Exercises 75 and 76, state the part of the graph of the hyperbola $4x^2 - 9y^2 = 36$ that satisfies the given equation.

75. $y = -\frac{2}{3}\sqrt{x^2 - 9}$ **76.** $x = \frac{3}{2}\sqrt{y^2 + 4}$

In Exercises 77–84, find an equation of the specified hyperbola with center at the origin.

77. Vertices: $(0, \pm 2)$; Foci: $(0, \pm 4)$

78. Vertices: $(\pm 3, 0)$; Foci: $(\pm 5, 0)$

79. Vertices: $(\pm 1, 0)$; Asymptotes: $y = \pm 3x$

80. Vertices: $(0, \pm 3)$; Asymptotes: $y = \pm 3x$

81. Foci: $(0, \pm 8)$; Asymptotes: $y = \pm 4x$

82. Foci: $(\pm 10, 0)$; Asymptotes: $y = \pm \frac{3}{4}x$

83.
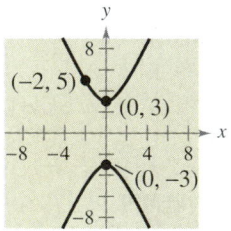

84.
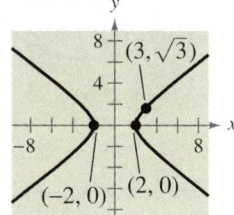

85. *Hyperbolic Mirror* A hyperbolic mirror (used in some telescopes) has the property that a light ray directed at the focus will be reflected to the other focus (see figure). The focus of a hyperbolic mirror has coordinates $(24, 0)$. Find the vertex of the mirror if its mount at the top edge of the mirror has coordinates $(24, 24)$.

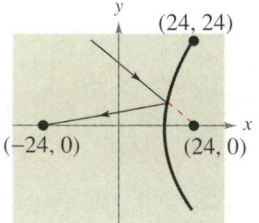

86. *LORAN* Long-distance radio navigation for aircraft and ships uses synchronized pulses transmitted by widely separated transmitting stations. These pulses travel at the speed of light (186,000 miles per second). The difference in the times of arrival of these pulses at an aircraft or ship is constant on a hyperbola having the transmitting stations as foci. Assume that two stations, 300 miles apart, are positioned on the rectangular coordinate system at points with coordinates $(-150, 0)$ and $(150, 0)$ and that a ship is traveling on a path with coordinates $(x, 75)$ (see figure). Find the x-coordinate of the position of the ship if the time difference between the pulses from the transmitting stations is 1000 microseconds (0.001 second).

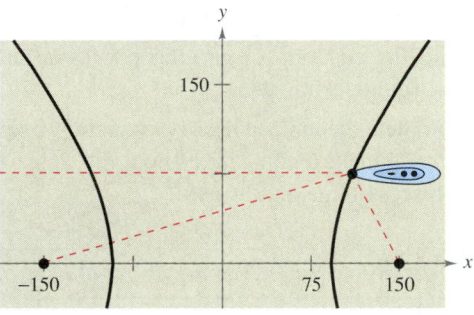

87. Use the definition of an ellipse to derive the standard form of the equation of an ellipse.

88. Use the definition of a hyperbola to derive the standard form of the equation of a hyperbola.

Review Solve Exercises 89–92 as a review of the skills and problem-solving techniques you learned in previous sections.

89. Find a polynomial with integer coefficients that has the zeros $3, 2 + i$, and $2 - i$.

90. Find all the zeros of $f(x) = 2x^3 - 3x^2 + 50x - 75$ if one of the zeros is $x = \frac{3}{2}$.

91. List the possible rational zeros of the function $g(x) = 6x^4 + 7x^3 - 29x^2 - 28x + 20$.

92. Use a graphing utility to graph the function $h(x) = 2x^4 + x^3 - 19x^2 - 9x + 9$. Use the graph and the Rational Zero Test to find the zeros of h.

4.5 *Translations of Conics*

See Exercise 55 on page 382 for an example of how an ellipse can be used to model the equation of a satellite's orbit around the earth.

Vertical and Horizontal Shifts of Conics in Standard Form   □   *Writing Equations of Conics*

Vertical and Horizontal Shifts of Conics

In Section 4.4 you looked at conic sections whose graphs were in *standard position*. In this section you will study the equations of conic sections that have been shifted vertically or horizontally in the plane.

STANDARD FORMS OF EQUATIONS OF CONICS

Circle: Center $= (h, k)$, Radius $= r$

$$(x - h)^2 + (y - k)^2 = r^2$$

Ellipse: Center $= (h, k)$

Major axis length $= 2a$

Minor axis length $= 2b$

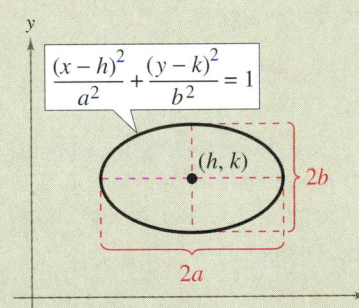

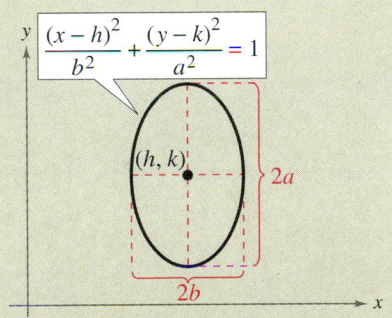

Hyperbola: Center $= (h, k)$

Transverse axis length $= 2a$

Conjugate axis length $= 2b$

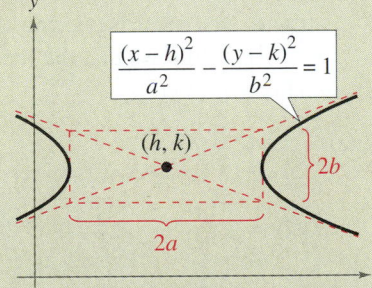

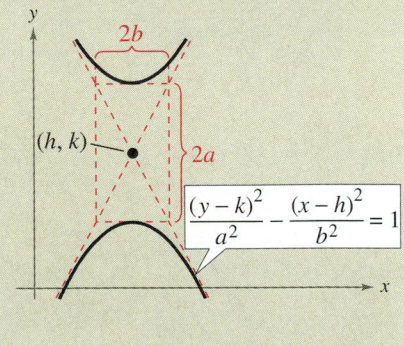

Parabola: Vertex $= (h, k)$

Directed distance from vertex to focus $= p$

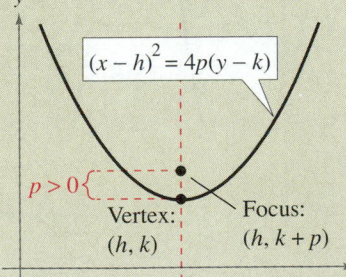

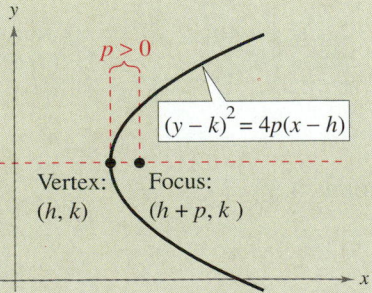

EXAMPLE 1   *Equations of Conic Sections*

a. The graph of

$$(x - 1)^2 + (y + 2)^2 = 3^2$$

is a circle whose center is the point $(1, -2)$ and whose radius is 3, as shown in Figure 4.33(a).

b. The graph of

$$\frac{(x - 2)^2}{3^2} + \frac{(y - 1)^2}{2^2} = 1$$

is an ellipse whose center is the point $(2, 1)$. The major axis of the ellipse is horizontal and of length $2(3) = 6$. The minor axis of the ellipse is vertical and of length $2(2) = 4$, as shown in Figure 4.33(b).

c. The graph of

$$\frac{(x - 3)^2}{1^2} - \frac{(y - 2)^2}{3^2} = 1$$

is a hyperbola whose center is the point $(3, 2)$. The transverse axis is horizontal and of length $2(1) = 2$. The conjugate axis is vertical and of length $2(3) = 6$, as shown in Figure 4.33(c).

d. The graph of

$$(x - 2)^2 = 4(-1)(y - 3)$$

is a parabola whose vertex is the point $(2, 3)$. The axis of the parabola is vertical. The focus is one unit above or below the vertex. Moreover, because $p = -1$, it follows that the focus lies *below* the vertex, as shown in Figure 4.33(d).

(a)

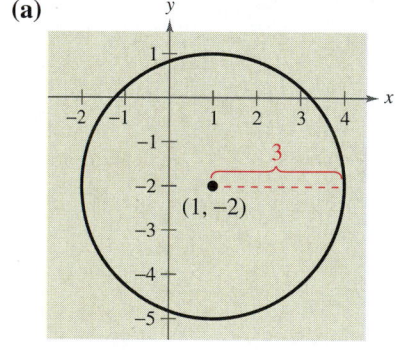

(b)

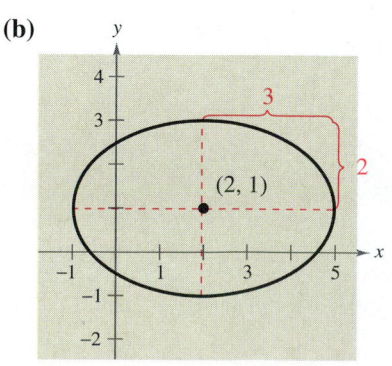

(c)

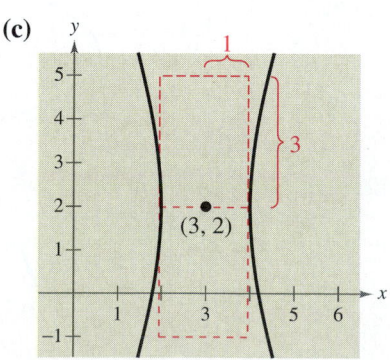

(d)

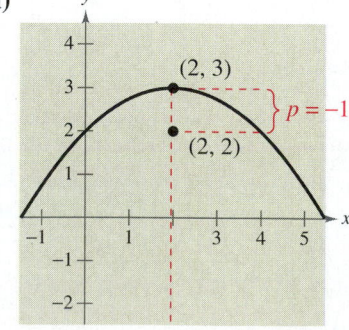

FIGURE 4.33

Writing Equations of Conics in Standard Form

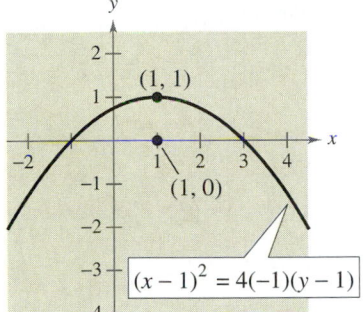

FIGURE 4.34

$(1, 1)$
$(1, 0)$
$(x - 1)^2 = 4(-1)(y - 1)$

NOTE Note in Example 2 that p is the *directed distance* from the vertex to the focus. Because the axis of the parabola is vertical and $p = -1$, the focus is one unit *below* the vertex, and the parabola opens downward. ■■

EXAMPLE 2 *Finding the Standard Form of a Parabola*

Find the vertex and focus of the parabola $x^2 - 2x + 4y - 3 = 0$.

Solution

Complete the square to write the equation in standard form.

$x^2 - 2x + 4y - 3 = 0$	Original equation
$x^2 - 2x = -4y + 3$	Group terms.
$x^2 - 2x + 1 = -4y + 3 + 1$	Add 1 to both sides.
$(x - 1)^2 = -4y + 4$	Completed square form
$(x - 1)^2 = 4(-1)(y - 1)$	$(x - h)^2 = 4p(y - k)$

From this standard form, it follows that $h = 1$, $k = 1$, and $p = -1$. Because the axis is vertical and p is negative, the parabola opens downward. The vertex is $(h, k) = (1, 1)$ and the focus is $(h, k + p) = (1, 0)$. The graph is shown in Figure 4.34.

EXAMPLE 3 *Sketching an Ellipse*

Sketch the graph of the ellipse $x^2 + 4y^2 + 6x - 8y + 9 = 0$.

Solution

Complete the square to write the equation in standard form.

$x^2 + 4y^2 + 6x - 8y + 9 = 0$	Original equation
$(x^2 + 6x + \blacksquare) + (4y^2 - 8y + \blacksquare) = -9$	Group terms.
$(x^2 + 6x + \blacksquare) + 4(y^2 - 2y + \blacksquare) = -9$	Factor 4 out of y-terms.
$(x^2 + 6x + 9) + 4(y^2 - 2y + 1) = -9 + 9 + 4(1)$	Add 9 and 4 to each side.
$(x + 3)^2 + 4(y - 1)^2 = 4$	Completed square form
$\dfrac{(x + 3)^2}{4} + \dfrac{(y - 1)^2}{1} = 1$	$\dfrac{(x - h)^2}{a^2} + \dfrac{(y - k)^2}{b^2} = 1$

From this standard form, it follows that the center is $(h, k) = (-3, 1)$. Because the denominator of the x-term is $4 = a^2 = 2^2$, the endpoints of the major axis lie two units to the right and left of the center. Similarly, because the denominator of the y-term is $1 = b^2 = 1^2$, the endpoints of the minor axis lie one unit up and down from the center. The ellipse is shown in Figure 4.35.

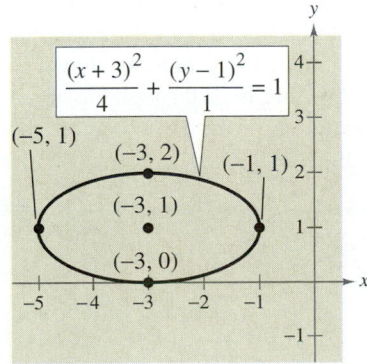

FIGURE 4.35

$\dfrac{(x + 3)^2}{4} + \dfrac{(y - 1)^2}{1} = 1$
$(-5, 1)$
$(-3, 2)$
$(-1, 1)$
$(-3, 1)$
$(-3, 0)$

EXAMPLE 4   *Sketching a Hyperbola*

Sketch the graph of the hyperbola given by the equation

$$y^2 - 4x^2 + 4y + 24x - 41 = 0.$$

Solution

Complete the square to write the equation in standard form.

$$y^2 - 4x^2 + 4y + 24x - 41 = 0 \qquad \text{Original equation}$$

$$(y^2 + 4y + \blacksquare) - (4x^2 - 24x + \blacksquare) = 41 \qquad \text{Group terms.}$$

$$(y^2 + 4y + \blacksquare) - 4(x^2 - 6x + \blacksquare) = 41 \qquad \text{Factor 4 out of } x\text{-terms.}$$

$$(y^2 + 4y + 4) - 4(x^2 - 6x + 9) = 41 + 4 - 4(9) \quad \text{Add 4, subtract 36.}$$

$$(y + 2)^2 - 4(x - 3)^2 = 9 \qquad \text{Completed square form}$$

$$\frac{(y + 2)^2}{9} - \frac{4(x - 3)^2}{9} = 1 \qquad \text{Divide both sides by 9.}$$

$$\frac{(y + 2)^2}{9} - \frac{(x - 3)^2}{9/4} = 1 \qquad \text{Change 4 to } \frac{1}{1/4}.$$

$$\frac{(y + 2)^2}{3^2} - \frac{(x - 3)^2}{(3/2)^2} = 1 \qquad \frac{(y - k)^2}{a^2} - \frac{(x - h)^2}{b^2} = 1$$

From this standard form, it follows that the transverse axis is vertical and the center lies at $(h, k) = (3, -2)$. Because the denominator of the y-term is $a^2 = 3^2$, you know that the vertices occur three units above and below the center.

$$(3, -5) \qquad \text{and} \qquad (3, 1) \qquad \text{Vertices}$$

To sketch the hyperbola, draw a rectangle whose top and bottom pass through the vertices. Because the denominator of the x-term is $b^2 = (3/2)^2$, locate the sides of the rectangle $\frac{3}{2}$ units to the right and left of the center, as shown in Figure 4.36. Finally, sketch the asymptotes by drawing lines through the opposite corners of the rectangle. Using these asymptotes, you can complete the graph of the hyperbola, as shown in Figure 4.36.

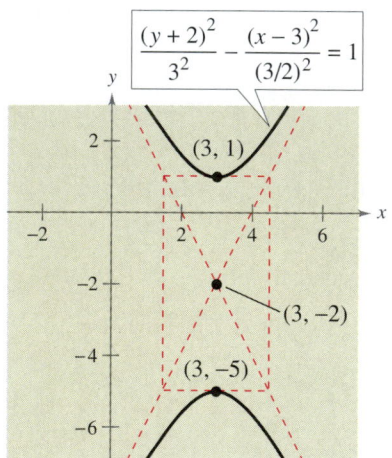

$$\frac{(y + 2)^2}{3^2} - \frac{(x - 3)^2}{(3/2)^2} = 1$$

FIGURE 4.36

To find the foci in Example 4, first find c.

$$c^2 = a^2 + b^2 = 9 + \frac{9}{4} = \frac{45}{4} \qquad \Longrightarrow \qquad c = \frac{3\sqrt{5}}{2}$$

Because the transverse axis is vertical, the foci lie c units above and below the center.

$$\left(3, -2 + \tfrac{3}{2}\sqrt{5}\right) \qquad \text{and} \qquad \left(3, -2 - \tfrac{3}{2}\sqrt{5}\right) \qquad \text{Foci}$$

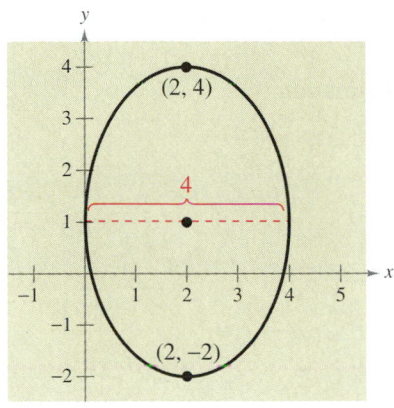

FIGURE 4.37

EXAMPLE 5 *Writing the Equation of an Ellipse*

Write the standard form of the equation of the ellipse whose vertices are $(2, -2)$ and $(2, 4)$. The length of the minor axis of the ellipse is 4, as shown in Figure 4.37.

Solution

The center of the ellipse lies at the midpoint of its vertices. Thus, the center is

$$(h, k) = (2, 1). \qquad \textcolor{red}{\text{Center}}$$

Because the vertices lie on a vertical line and are six units apart, it follows that the major axis is vertical and has a length of $2a = 6$. Thus, $a = 3$. Moreover, because the minor axis has a length of 4, it follows that $2b = 4$, which implies that $b = 2$. Therefore, the standard form of the ellipse is as follows.

$$\frac{(x - h)^2}{b^2} + \frac{(y - k)^2}{a^2} = 1 \qquad \textcolor{red}{\text{Major axis is vertical.}}$$

$$\frac{(x - 2)^2}{2^2} + \frac{(y - 1)^2}{3^2} = 1 \qquad \textcolor{red}{\text{Standard form}}$$

An interesting application of conic sections involves the orbits of comets in our solar system. Of the 610 comets identified prior to 1970, 245 have elliptical orbits, 295 have parabolic orbits, and 70 have hyperbolic orbits. For example, Halley's Comet has an elliptical orbit, and reappearance of this comet can be predicted every 76 years. The center of the sun is a focus of each of these orbits, and each orbit has a vertex at the point where the comet is closest to the sun, as shown in Figure 4.38.

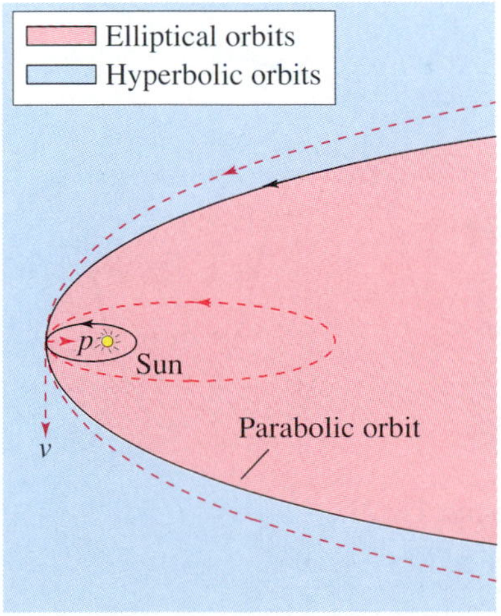

FIGURE 4.38

GROUP ACTIVITY

IDENTIFYING EQUATIONS OF CONICS

Make up an equation for one of the conics you studied in this section, and sketch its graph as accurately as possible on graph paper. Don't write the equation on your graph. Exchange graphs with another student. Try to reconstruct the equation of the conic that is represented by the graph you received. Compare results with the student who used your graph.

WARM UP

Identify the conic represented by the equation.

1. $\dfrac{x^2}{4} - \dfrac{y^2}{4} = 1$ 2. $\dfrac{x^2}{9} + \dfrac{y^2}{1} = 1$

3. $2x + y^2 = 0$ 4. $4x^2 + 4y^2 = 25$

5. $\dfrac{x^2}{4} + \dfrac{y^2}{16} = 1$ 6. $\dfrac{x^2}{9} - \dfrac{y^2}{4} = 1$

7. $3x - y^2 = 0$ 8. $x^2 - 6y = 0$

9. $\dfrac{y^2}{4} - \dfrac{x^2}{2} = 1$ 10. $\dfrac{x^2}{9/4} + \dfrac{y^2}{4} = 1$

4.5 Exercises

In Exercises 1–8, find the vertex, focus, and directrix of the parabola, and sketch its graph.

1. $(x - 1)^2 + 8(y + 2) = 0$
2. $(x + 3) + (y - 2)^2 = 0$
3. $\left(y + \frac{1}{2}\right)^2 = 2(x - 5)$
4. $\left(x + \frac{1}{2}\right)^2 = 4(y - 3)$
5. $y = \frac{1}{4}(x^2 - 2x + 5)$
6. $4x - y^2 - 2y - 33 = 0$
7. $y^2 + 6y + 8x + 25 = 0$
8. $y^2 - 4y - 4x = 0$

In Exercises 9–12, find the vertex, focus, and directrix of the parabola, and use a graphing utility to obtain its graph.

9. $y = -\frac{1}{6}(x^2 + 4x - 2)$
10. $x^2 - 2x + 8y + 9 = 0$
11. $y^2 + x + y = 0$
12. $y^2 - 4x - 4 = 0$

In Exercises 13–22, find an equation of the parabola.

13.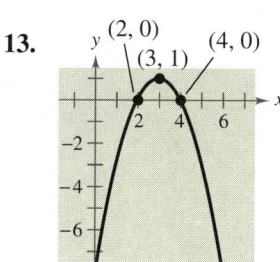

14.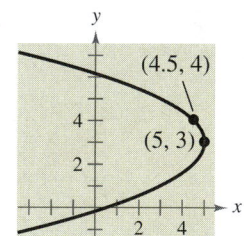

15.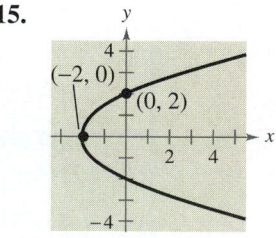

16.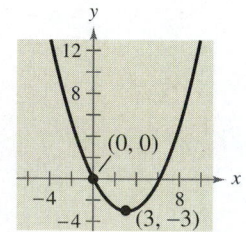

17. Vertex: $(3, 2)$; Focus: $(1, 2)$
18. Vertex: $(-1, 2)$; Focus: $(-1, 0)$
19. Vertex: $(0, 4)$; Directrix: $y = 2$

20. Vertex: $(-2, 1)$; Directrix: $x = 1$

21. Focus: $(2, 2)$; Directrix: $x = -2$

22. Focus: $(0, 0)$; Directrix: $y = 4$

Think About It In Exercises 23 and 24, change the equation so that its graph matches the description.

23. $(y - 3)^2 = 6(x + 1)$; Upper half of parabola

24. $(y + 1)^2 = 2(x - 2)$; Lower half of parabola

25. ***Satellite Orbit*** An earth satellite in a 100-mile-high circular orbit around the earth has a velocity of approximately 17,500 miles per hour. If this velocity is multiplied by $\sqrt{2}$, the satellite will have the minimum velocity necessary to escape the earth's gravity and it will follow a parabolic path with the center of the earth as the focus (see figure).

(a) Find the escape velocity of the satellite.

(b) Find an equation of its path (assume that the radius of the earth is 4000 miles).

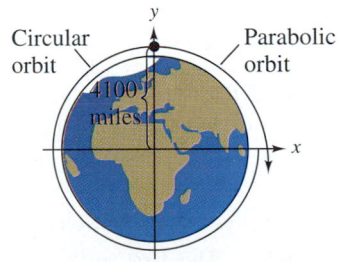

Circular orbit   Parabolic orbit
4100 miles

26. ***Projectile Motion*** A bomber is flying at an altitude of 30,000 feet and a speed of 540 miles per hour (792 feet per second). How many feet will a bomb dropped from the plane travel horizontally before it hits the target if the path of the bomb is modeled by

$$y = 30,000 - \frac{x^2}{39,204}?$$

27. ***Path of a Projectile*** The path of a softball is given by the equation

$$y = -0.08x^2 + x + 4.$$

The coordinates x and y are measured in feet, with $x = 0$ corresponding to the position from which the ball was thrown.

(a) Use a graphing utility to graph the trajectory of the softball.

(b) Move the cursor along the path to approximate the highest point and the range of the trajectory.

28. ***Revenue*** The revenue R generated by the sale of x units is given by $R = 375x - \frac{3}{2}x^2$.

(a) Use a graphing utility to graph the function.

(b) Use the trace feature of the graphing utility to approximate *graphically* the sales that will maximize the revenue.

(c) Use a table to approximate *numerically* the sales that will maximize the revenue.

(d) Find the coordinates of the vertex to determine *analytically* the sales that will maximize the revenue.

(e) Compare the results of parts (b)–(d). What did you learn by using all three approaches?

In Exercises 29–34, find the center, foci, and vertices of the ellipse, and sketch its graph.

29. $\dfrac{(x - 1)^2}{9} + \dfrac{(y - 5)^2}{25} = 1$

30. $(x + 2)^2 + \dfrac{(y + 4)^2}{1/4} = 1$

31. $9x^2 + 4y^2 + 36x - 24y + 36 = 0$

32. $9x^2 + 4y^2 - 36x + 8y + 31 = 0$

33. $16x^2 + 25y^2 - 32x + 50y + 16 = 0$

34. $9x^2 + 25y^2 - 36x - 50y + 61 = 0$

In Exercises 35 and 36, find the center, foci, and vertices of the ellipse, and graph the ellipse with the aid of a graphing utility.

35. $12x^2 + 20y^2 - 12x + 40y - 37 = 0$

36. $36x^2 + 9y^2 + 48x - 36y + 43 = 0$

In Exercises 37–48, find an equation for the ellipse.

37.

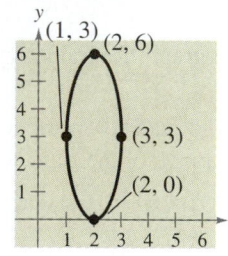

38.

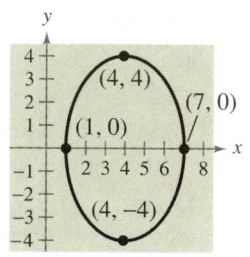

39.

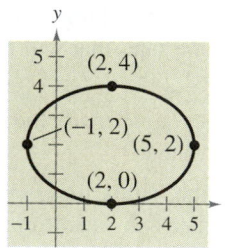

40.

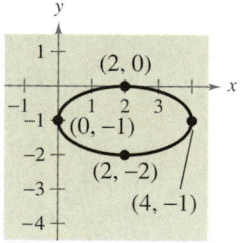

41. Vertices: $(0, 2)$, $(4, 2)$;   Minor axis of length 2

42. Foci: $(0, 0)$, $(4, 0)$;   Major axis of length 8

43. Foci: $(0, 0)$, $(0, 8)$;   Major axis of length 16

44. Center: $(2, -1)$;   Vertex: $\left(2, \frac{1}{2}\right)$;
    Minor axis of length 2

45. Vertices: $(3, 1)$, $(3, 9)$;   Minor axis of length 6

46. Center: $(3, 2)$; $a = 3c$;   Foci: $(1, 2)$, $(5, 2)$

47. Center: $(0, 4)$; $a = 2c$;   Vertices: $(-4, 4)$, $(4, 4)$

48. Vertices: $(5, 0)$, $(5, 12)$;
    Endpoints of the minor axis: $(0, 6)$, $(10, 6)$

Think About It   **In Exercises 49 and 50, change the equation so that its graph matches the description.**

49. $\dfrac{(x - 3)^2}{9} + \dfrac{y^2}{4} = 1$; Right half of ellipse

50. $\dfrac{(x + 1)^2}{16} + \dfrac{(y - 2)^2}{25} = 1$; Bottom half of ellipse

In Exercises 51–56, use or determine the eccentricity of the ellipse, which is defined by $e = c/a$. It measures the flatness of the ellipse.

51. Find an equation of the ellipse with vertices $(\pm 5, 0)$ and eccentricity $e = 3/5$.

52. Find an equation of the ellipse with vertices $(0, \pm 8)$ and eccentricity $e = 1/2$.

53. ***Planetary Motion***   The planet Pluto moves in an elliptical orbit with the sun at one of the foci (see figure). The length of half of the major axis is 3.666×10^9 miles and the eccentricity is 0.248. Find the smallest distance (*perihelion*) and the greatest distance (*aphelion*) of Pluto from the center of the sun.

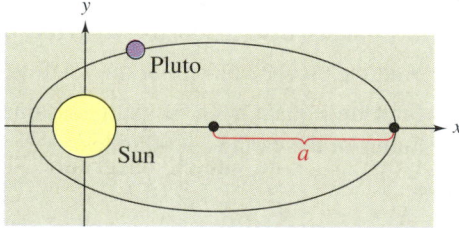

54. ***Orbit of Saturn***   Saturn moves in an elliptical orbit with the sun at one of the foci. The smallest distance and the greatest distance of the planet from the sun are 1.3495×10^9 and 1.5045×10^9 kilometers, respectively. Find the eccentricity of the orbit.

55. ***Satellite Orbit***   The first artificial satellite to orbit the earth was Sputnik I (launched by the former Soviet Union in 1957). Its highest point above the earth's surface was 938 kilometers, and its lowest point was 212 kilometers (see figure). The center of the earth is the focus of the elliptical orbit and the radius of the earth is 6378 kilometers. Find the eccentricity of the orbit.

56. *Exploration*   Use the definition of eccentricity in Exercises 51 and 52 and consider the ellipse given by

$$\frac{x^2}{a^2} + \frac{y^2}{b^2} = 1.$$

(a) Show that the equation of the ellipse can be written as

$$\frac{(x-h)^2}{a^2} + \frac{(y-k)^2}{a^2(1-e^2)} = 1.$$

(b) Use a graphing utility to graph the ellipse

$$\frac{(x-2)^2}{4} + \frac{(y-3)^2}{4(1-e^2)} = 1$$

for $e = 0.95, 0.75, 0.5, 0.25$, and 0.

(c) Use the results of part (b) to make a conjecture about the change in the shape of the ellipse as e approaches 0.

In Exercises 57–64, find the center, vertices, and foci of the hyperbola, and sketch its graph. Sketch the asymptotes as an aid in obtaining the graph of the hyperbola.

57. $\dfrac{(x-1)^2}{4} - \dfrac{(y+2)^2}{1} = 1$

58. $\dfrac{(x+1)^2}{144} - \dfrac{(y-4)^2}{25} = 1$

59. $(y+6)^2 - (x-2)^2 = 1$

60. $\dfrac{(y-1)^2}{1/4} - \dfrac{(x+3)^2}{1/9} = 1$

61. $9x^2 - y^2 - 36x - 6y + 18 = 0$

62. $x^2 - 9y^2 + 36y - 72 = 0$

63. $x^2 - 9y^2 + 2x - 54y - 80 = 0$

64. $16y^2 - x^2 + 2x + 64y + 63 = 0$

In Exercises 65 and 66, find the center, vertices and foci of the hyperbola, and graph the hyperbola and its asymptotes with the aid of a graphing utility.

65. $9y^2 - x^2 + 2x + 54y + 62 = 0$

66. $9x^2 - y^2 + 54x + 10y + 55 = 0$

In Exercises 67–78, find an equation for the hyperbola.

67.

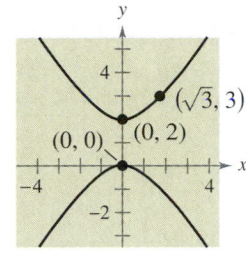

68.

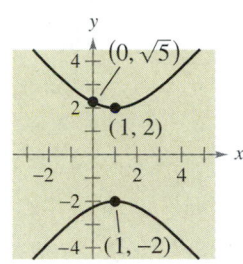

69.

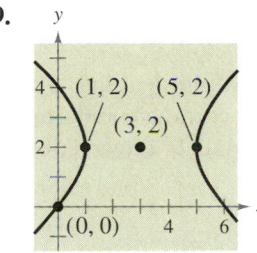

70.

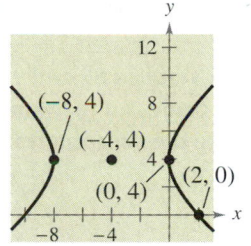

71. Vertices: $(2, 0), (6, 0)$; Foci: $(0, 0), (8, 0)$

72. Vertices: $(2, 3), (2, -3)$; Foci: $(2, 5), (2, -5)$

73. Vertices: $(4, 1), (4, 9)$; Foci: $(4, 0), (4, 10)$

74. Vertices: $(-2, 1), (2, 1)$; Foci: $(-3, 1), (3, 1)$

75. Vertices: $(2, 3), (2, -3)$;
Passes through the point $(0, 5)$

76. Vertices: $(-2, 1), (2, 1)$;
Passes through the point $(4, 3)$

77. Vertices: $(0, 2), (6, 2)$;
Asymptotes: $y = \frac{2}{3}x, y = 4 - \frac{2}{3}x$

78. Vertices: $(3, 0), (3, 4)$;
Asymptotes: $y = \frac{2}{3}x, y = 4 - \frac{2}{3}x$

Think About It   **In Exercises 79 and 80, describe the part of the hyperbola**

$$\frac{(x-3)^2}{4} - \frac{(y-1)^2}{9} = 1$$

given by the equation.

79. $x = 3 - \frac{2}{3}\sqrt{9 + (y-1)^2}$

80. $y = 1 + \frac{3}{2}\sqrt{(x-3)^2 - 4}$

In Exercises 81–88, classify the graph of the equation as a circle, a parabola, an ellipse, or a hyperbola.

81. $x^2 + y^2 - 6x + 4y + 9 = 0$

82. $x^2 + 4y^2 - 6x + 16y + 21 = 0$

83. $4x^2 - y^2 - 4x - 3 = 0$

84. $y^2 - 4y - 4x = 0$

85. $4x^2 + 3y^2 + 8x - 24y + 51 = 0$

86. $4y^2 - 2x^2 - 4y - 8x - 15 = 0$

87. $25x^2 - 10x - 200y - 119 = 0$

88. $4x^2 + 4y^2 - 16y + 15 = 0$

In Exercises 89 and 90, use a graphing utility to graph the conics on the same viewing rectangle and approximate the coordinates of any points of intersection.

89. $x^2 + 9y^2 = 9$, $y = x^2 - 4$

90. $x^2 + y^2 = 25$, $x^2 - y^2 = 1$

91. *Chapter Opener* Halley's Comet has an elliptical orbit with the sun at one focus. The eccentricity of the orbit is approximately 0.97. The length of the major axis of the orbit is approximately 36.23 astronomical units. (An astronomical unit is about 93 million miles.) Find an equation for the orbit. Place the center of the orbit at the origin, and place the major axis on the *x*-axis.

92. *Chapter Opener* The comet Encke has an elliptical orbit with the sun at one focus. Encke ranges from 0.34 to 4.08 astronomical units from the sun. Find an equation of the orbit. Place the center of the orbit at the origin, and place the major axis on the *x*-axis.

Review Solve Exercises 93–96 as a review of the skills and problem-solving techniques you learned in previous sections. Identify the rule of algebra illustrated.

93. $(a + 4) - (a + 4) = 0$

94. $u + (v + 10) = (u + v) + 10$

95. $(x + 3)(a + b) = x(a + b) + 3(a + b)$

96. $\dfrac{1}{z - 3}(z - 3) = 1$, $z \neq 3$

A comet's tail is made of two components. The dust tail, composed of dust particles, reflects sunlight and appears white or yellow. The plasma tail, composed of gas, appears blue because of its ionized carbon monoxide molecules. *(Photo: Courtesy of the Astronomical Society of the Pacific)*

FOCUS ON CONCEPTS

In this chapter, you studied rational functions and conics. Answer the following questions to check your understanding of several of the basic concepts discussed in this chapter. The answers to these questions are given in the back of the book.

1. Give an example of a rational function whose domain is the set of all real numbers. Give an example whose domain is the set of all real numbers except $x = 20$.

2. Describe what is meant by an asymptote of a graph.

3. Does every rational function have a vertical and/or horizontal asymptote? If not, give an example.

4. Let $f(x) = p(x)/q(x)$, where $p(x)$ and $q(x)$ are polynomials with no common factors.

   (a) How do you find the zeros of the function?

   (b) How do you find any vertical asymptotes?

   (c) What is the horizontal asymptote if the degree of p is less than the degree of q?

In Exercises 5–8, identify a characteristic of the rational function that enables you to identify its graph.

(a)

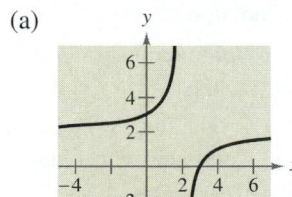

(b)

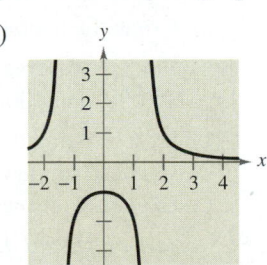

(c)

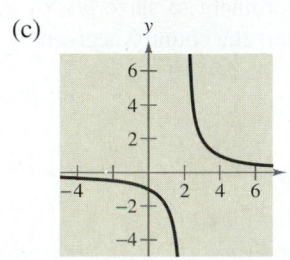

(d)

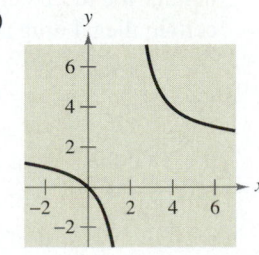

5. $y = \dfrac{2}{x-2}$

6. $y = \dfrac{2}{x^2-2}$

7. $y = \dfrac{2x}{x-2}$

8. $y = \dfrac{2(x-3)}{x-2}$

In Exercises 9–12, match the rational expression with the form of its decomposition. [The decompositions are labeled (a), (b), (c), and (d).]

(a) $\dfrac{A}{x} + \dfrac{B}{x+2} + \dfrac{C}{x-2}$

(b) $\dfrac{A}{x} + \dfrac{B}{x-4}$

(c) $\dfrac{A}{x} + \dfrac{B}{x^2} + \dfrac{C}{x-4}$

(d) $\dfrac{A}{x} + \dfrac{Bx+C}{x^2+4}$

9. $\dfrac{3x-1}{x(x-4)}$

10. $\dfrac{3x-1}{x^2(x-4)}$

11. $\dfrac{3x-1}{x(x^2+4)}$

12. $\dfrac{3x-1}{x(x^2-4)}$

In Exercises 13 and 14, an equation and four variations of it are given. In your own words, describe how the graph of each of the variations would differ from the graph of the original equation.

13. $y^2 = 8x$

   (a) $(y-2)^2 = 8x$ (b) $y^2 = 8(x+1)$

   (c) $y^2 = -8x$ (d) $y^2 = 4x$

14. $\dfrac{x^2}{4} + \dfrac{y^2}{9} = 1$

   (a) $\dfrac{x^2}{9} + \dfrac{y^2}{4} = 1$ (b) $\dfrac{x^2}{4} + \dfrac{y^2}{4} = 1$

   (c) $\dfrac{x^2}{4} + \dfrac{y^2}{25} = 1$ (d) $\dfrac{(x-3)^2}{4} + \dfrac{y^2}{9} = 1$

15. Explain how the central rectangle of a hyperbola can be used to sketch its asymptotes.

16. Consider an ellipse whose major axis is horizontal and 10 units in length. The value of b in the standard form of the ellipse must be less than what real number? Explain the change in the shape of the ellipse as b approaches this number.

Review Exercises

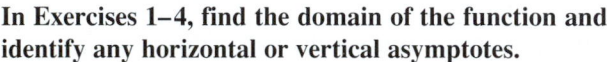

In Exercises 1–4, find the domain of the function and identify any horizontal or vertical asymptotes.

1. $f(x) = \dfrac{4}{x + 3}$

2. $f(x) = \dfrac{2x^2 + 5x - 3}{x^2 + 2}$

3. $g(x) = \dfrac{x^2}{x^2 - 4}$

4. $g(x) = \dfrac{1}{(x - 3)^2}$

In Exercises 5–16, sketch the graph of the rational function. As sketching aids, check for intercepts, symmetry, vertical asymptotes, horizontal asymptotes, and slant asymptotes.

5. $f(x) = \dfrac{-5}{x^2}$

6. $f(x) = \dfrac{4}{x}$

7. $g(x) = \dfrac{2 + x}{1 - x}$

8. $h(x) = \dfrac{x - 3}{x - 2}$

9. $p(x) = \dfrac{x^2}{x^2 + 1}$

10. $f(x) = \dfrac{2x}{x^2 + 4}$

11. $f(x) = \dfrac{x}{x^2 + 1}$

12. $h(x) = \dfrac{4}{(x - 1)^2}$

13. $f(x) = \dfrac{2x^2}{x^2 + 1}$

14. $y = \dfrac{2x^2}{x^2 - 4}$

15. $y = \dfrac{x}{x^2 - 1}$

16. $g(x) = \dfrac{-2}{(x + 3)^2}$

In Exercises 17–20, use a graphing utility to graph the function. Identify any vertical, horizontal, or slant asymptotes.

17. $s(x) = \dfrac{8x^2}{x^2 + 4}$

18. $y = \dfrac{5x}{x^2 - 4}$

19. $g(x) = \dfrac{x^2 + 1}{x + 1}$

20. $y = \dfrac{1}{x + 3} + 2$

Think About It **In Exercises 21 and 22, write a rational function f having the specified characteristics.**

21. Vertical asymptotes: $x = -3$, $x = 4$
Horizontal asymptote: $y = 2$

22. Vertical asymptote: $x = 5$
Slant asymptote: $y = 2x$

23. *Average Cost* A business has a cost of $C = 0.5x + 500$ for producing x units. The average cost per unit is

$$\overline{C} = \frac{C}{x} = \frac{0.5x + 500}{x}, \qquad 0 < x.$$

Determine the average cost per unit as x increases without bound. (Find the horizontal asymptote.)

24. *Average Cost* The cost of producing x units is C, and the average cost per unit is

$$\overline{C} = \frac{C}{x} = \frac{100{,}000 + 0.9x}{x}, \qquad 0 < x.$$

(a) Graph the average cost function.

(b) Find the average cost of producing $x = 1000$, 10,000, and 100,000 units.

(c) By increasing the level of production, what is the smallest average cost per unit you can obtain? Explain your reasoning.

25. *Seizure of Illegal Drugs* The cost in millions of dollars for the federal government to seize $p\%$ of a certain illegal drug as it enters the country is given by

$$C = \frac{528p}{100 - p}, \qquad 0 \le p \le 100.$$

(a) Find the cost of seizing 25%.

(b) Find the cost of seizing 50%.

(c) Find the cost of seizing 75%.

(d) According to this model, would it be possible to seize 100% of the drug?

26. *Capillary Attraction* The rise of distilled water in tubes of diameter x inches is approximated by the model

$$y = \left(\frac{0.80 - 0.54x}{1 + 2.72x}\right)^2, \qquad 0 < x$$

where y is measured in inches (see figure). Approximate the diameter of the tube that will cause the water to rise 0.1 inch.

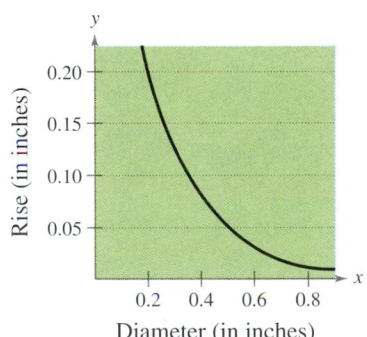

Diameter (in inches)

27. *Numerical and Graphical Analysis* A right triangle is formed in the first quadrant by the x- and y-axes and a line through the point $(2, 3)$.

(a) Create a figure that illustrates the problem. Label the known and unknown quantities.

(b) Verify that the area of the triangle is given by

$$A = \frac{3x^2}{2(x - 2)}, \qquad x > 2.$$

(c) Create a table that gives values of area for various values of x. Start the table with $x = 2.5$ and increment x in steps of 0.5. Continue until you can approximate the dimensions of the triangle of minimum area.

(d) Use a graphing utility to graph the area function. Use the graph to approximate the dimensions of the triangle of minimum area.

(e) Determine the slant asymptote of the area function. Explain its meaning.

28. *Population of Fish* The Parks and Wildlife Commission introduces 80,000 fish into a large human-made lake. The population of the fish in thousands is given by

$$N = \frac{20(4 + 3t)}{1 + 0.05t}, \qquad 0 \le t$$

where t is time in years.

(a) Find the populations when t is 5, 10, and 25.

(b) What is the limiting number of fish in the lake as time increases?

29. *Photosynthesis* The amount y of CO_2 uptake in milligrams per square decimeter per hour at optimal temperatures and with the natural supply of CO_2 is approximated by the model

$$y = \frac{18.47x - 2.96}{0.23x + 1}, \qquad 0 < x$$

where x is the light intensity in watts per square meter. Use a graphing utility to graph the function and determine the limiting amount of CO_2 uptake.

30. *Think About It* Use a graphing utility to graph the functions

$$f(x) = \frac{x^2 - 9}{x + 3} \qquad \text{and} \qquad g(x) = x - 3$$

on the same viewing rectangle. Does the graphing utility show the difference in the domains of the functions? Explain.

In Exercises 31–38, write the partial fraction decomposition for the rational expression.

31. $\dfrac{4 - x}{x^2 + 6x + 8}$

32. $\dfrac{-x}{x^2 + 3x + 2}$

33. $\dfrac{x^2}{x^2 + 2x - 15}$

34. $\dfrac{9}{x^2 - 9}$

35. $\dfrac{x^2 + 2x}{x^3 - x^2 + x - 1}$

36. $\dfrac{4x - 2}{3(x - 1)^2}$

37. $\dfrac{3x^2 + 4x}{(x^2 + 1)^2}$

38. $\dfrac{4x^2}{(x - 1)(x^2 + 1)}$

In Exercises 39–50, identify the conic and sketch its graph.

39. $4x - y^2 = 0$

40. $8y + x^2 = 0$

41. $x^2 - 6x + 2y + 9 = 0$

42. $y^2 - 12y - 8x + 20 = 0$

43. $x^2 + y^2 - 2x - 4y + 5 = 0$

44. $16x^2 + 16y^2 - 16x + 24y - 3 = 0$

45. $4x^2 + y^2 = 16$

46. $2x^2 + 6y^2 = 18$

47. $x^2 + 9y^2 + 10x - 18y + 25 = 0$

48. $4x^2 + y^2 - 16x + 15 = 0$

49. $5y^2 - 4x^2 = 20$

50. $x^2 - 9y^2 + 10x + 18y + 7 = 0$

Graphical Reasoning In Exercises 51 and 52, consider the general form of a conic whose axes are not parallel to either the *x*-axis or the *y*-axis (note the *xy*-term). Use the quadratic formula to solve for *y*, and use a graphing utility to graph the resulting equations. Identify the conic.

51. $x^2 - 10xy + y^2 + 1 = 0$

52. $40x^2 + 36xy + 25y^2 - 52 = 0$

In Exercises 53–58, find an equation of the specified parabola.

53.

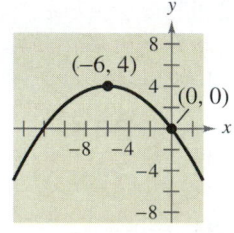

54.

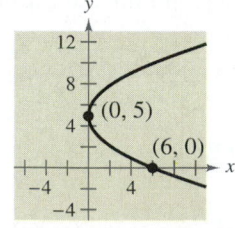

55. Vertex: $(4, 2)$; Focus: $(4, 0)$

56. Vertex: $(2, 0)$; Focus: $(0, 0)$

57. Vertex: $(0, 2)$; Horizontal axis; Passes through $(-1, 0)$

58. Vertex: $(2, 2)$; Directrix: $y = 0$

In Exercises 59–64, find an equation of the specified ellipse.

59.

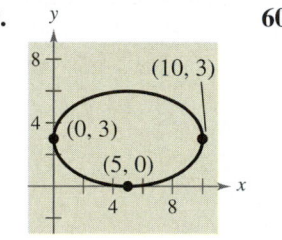

60.

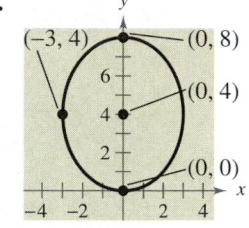

61. Vertices: $(-3, 0)$, $(7, 0)$; Foci: $(0, 0)$, $(4, 0)$

62. Vertices: $(2, 0)$, $(2, 4)$; Foci: $(2, 1)$, $(2, 3)$

63. Vertices: $(0, \pm 6)$; Passes through: $(2, 2)$

64. Vertices: $(0, 1)$, $(4, 1)$; Co-vertices: $(2, 0)$, $(2, 2)$

In Exercises 65–70, find an equation of the specified hyperbola.

65.

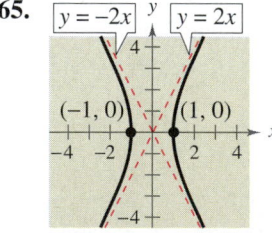

66.

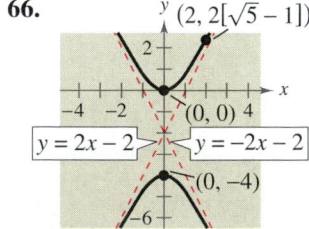

67. Vertices: $(0, \pm 1)$; Foci: $(0, \pm 3)$

68. Vertices: $(2, 2)$, $(-2, 2)$; Foci: $(4, 2)$, $(-2, 2)$

69. Foci: $(0, 0)$, $(8, 0)$; Asymptotes: $y = \pm 2(x - 4)$

70. Foci: $(3, \pm 2)$; Asymptotes: $y = \pm 2(x - 3)$

71. *Satellite Antenna* A cross section of a large parabolic antenna (see figure) is given by

$$y = \frac{x^2}{200}, \qquad -100 \le x \le 100.$$

The receiving and transmitting equipment is positioned at the focus. Find the coordinates of the focus.

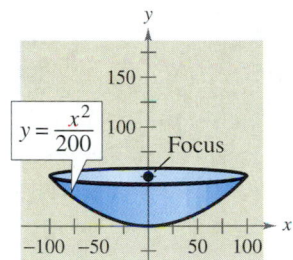

72. *Parabolic Archway* A parabolic archway is 12 meters high at the vertex. At a height of 10 meters, the width of the archway is 8 meters (see figure). How wide is the archway at ground level?

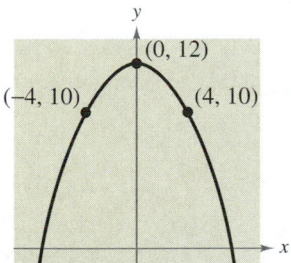

73. *Architecture* A church window is bounded above by a parabola and below by the arc of a circle (see figure).

(a) Find equations for the parabola and the circle.

(b) Complete the table by filling in the vertical distance d between the circle and parabola for each given value of x.

x	0	1	2	3	4
d					

74. *CompuServe* CompuServe is a part of H & R Block, Inc. The pretax earnings (in millions of dollars) for the years 1990 through 1994 are given in the table. (Source: *1994 H & R Block, Inc. Annual Report*)

Year	1990	1991	1992	1993	1994
Earnings	40.3	48.6	55.4	74.0	102.3

A model for this data is given by

$$y = 3.7t^2 + 0.14t + 41.64$$

where y is the earnings in millions of dollars and t is the time in years, with $t = 0$ corresponding to 1990.

(a) Use a graphing utility to plot the data in the table and graph the model on the same viewing rectangle.

(b) Use the model to estimate the earnings for 1995.

75. *Semielliptical Archway* A semielliptical archway is to be formed over the entrance to an estate. The arch is to be set on pillars that are 10 feet apart and is to have a height (atop the pillars) of 4 feet (see figure). Where should the foci be placed in order to sketch the arch?

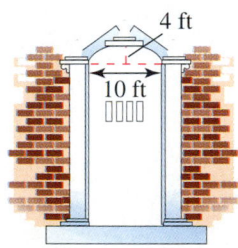

76. *Heating and Plumbing* Find the diameter d of the largest water pipe that can be placed behind a ventilation duct with a diameter of 60 centimeters as shown in the figure.

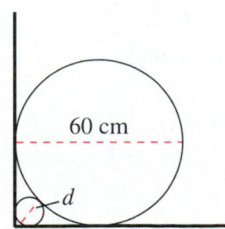

CHAPTER PROJECT: *A Graphical Approach to Finding Average Cost*

A graphing utility can be used to investigate asymptotic behavior. For instance, the example below shows how to use a graphing utility to visualize how the average cost of producing a product changes as the number of units produced increases.

Real Life

EXAMPLE 1   *Finding the Average Cost of a Product*

You have started a small business that presses and packages compact discs. Your initial investment is $250,000, and the cost of producing and packaging each disc is $0.32. Describe the *average cost per unit* of producing each disc as a function of *x* (the number of units produced).

Solution

The total cost *C* of producing *x* units is

$$C = \boxed{\text{Initial cost}} + \boxed{\text{Cost per unit}} \cdot \boxed{\text{Number of units}}$$

$$= 250{,}000 + 0.32x.$$

To obtain the average cost per unit $\overline{C}$ of producing *x* units, divide the total cost by *x*.

$$\overline{C} = \frac{C}{x} = \frac{250{,}000 + 0.32x}{x} = \frac{250{,}000}{x} + 0.32$$

The table shows the average costs per unit for several different levels of production.

x	1000	10,000	100,000	1,000,000	10,000,000
C	\$250,320	\$253,200	\$282,000	\$570,000	\$3,450,000
$\overline{C}$	\$250.32	\$25.32	\$2.82	\$0.57	\$0.345

The graph of the average cost function is shown at the left, together with the graph of the horizontal line $y = 0.32$. Notice that as the number of units produced increases, the average cost per unit gets closer and closer to the unit cost of $0.32.

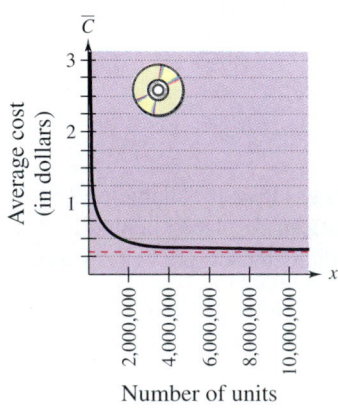

Average cost (in dollars)

Number of units

In Example 1, it is assumed that the business can continue to increase its production level without having to invest additional capital. Example 2 shows how the problem changes when additional capital investments are required.

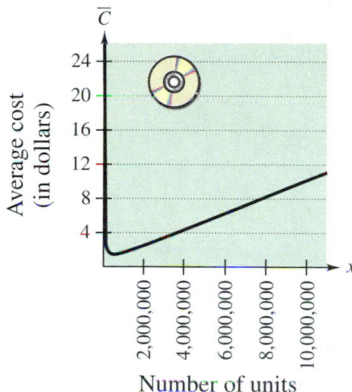

Average cost (in dollars) — vertical axis with values 4, 8, 12, 16, 20, 24 labeled $\overline{C}$

Number of units — horizontal axis with values 2,000,000; 4,000,000; 6,000,000; 8,000,000; 10,000,000 labeled x

Real Life

EXAMPLE 2   *Finding the Minimum Average Cost*

As the production level in Example 1 increases and your business grows, suppose you must invest additional capital to buy more equipment, advertise, build new offices, and so on. You have determined that the additional investments are proportional to the square of the number of units produced, and you have derived the following model for the total cost of producing x units.

$$C = 250,000 + 0.32x + 0.000001x^2$$

Describe the average cost per unit as a function of the number of units produced.

Solution

With this model of the total cost, the average cost function is

$$\overline{C} = \frac{250,000 + 0.32x + 0.000001x^2}{x} = \frac{250,000}{x} + 0.32 + 0.000001x.$$

The graph of the average cost function is shown at the left. Notice that the average cost decreases until the production level reaches about half a million units. Then, with increased production, the average cost per unit increases.

CHAPTER PROJECT INVESTIGATIONS

1. *Exploration*   In Example 1, would it be possible to produce enough units so that the average cost was exactly $0.32? Would it be possible to produce enough units so that the average cost was arbitrarily close to $0.32?

2. *Exploration*   In Example 1, how many units should be produced to obtain an average cost per unit of $0.33? To answer this question, did you use a numerical approach, an analytic approach, or a graphical approach? Explain why you chose the approach you used.

3. *Exploration*   In Example 2, use the zoom feature of a graphing utility to approximate the *minimum* average cost. At what production level does this minimum average cost occur?

4. *Slant Asymptote*   The graph shown in Example 2 has a slant asymptote. Find its equation, sketch its graph, and interpret it in the context of the problem.

5. *Economy of Scale*   In business, the expression *economy of scale* means that the average cost per unit tends to decrease as the production level increases. Judging from Examples 1 and 2 *and* your own knowledge of business, what factors could temper an economy of scale in business?

6. *Testing a Hypothesis*   Which graph has the x-axis as an asymptote? Which has a horizontal asymptote that is not the x-axis? Which has a slant asymptote?

(a) $y = \dfrac{3}{2x + 1}, \qquad 0 \leq x$

(b) $y = \dfrac{3x}{2x + 1}, \qquad 0 \leq x$

(c) $y = \dfrac{3x^2}{2x + 1}, \qquad 0 \leq x$

Form a hypothesis about the horizontal or slant asymptotes of the graph of $y = ax^n/(bx + 1)$. Test your hypothesis with examples.

Chapter Test

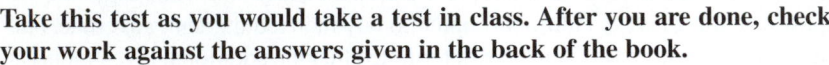

Take this test as you would take a test in class. After you are done, check your work against the answers given in the back of the book.

In Exercises 1–3, find the domain of the function and identify any asymptotes.

1. $y = \dfrac{3}{4 - x}$ **2.** $f(x) = \dfrac{2 - x^2}{2 + x^2}$ **3.** $g(x) = \dfrac{x^2 + 2x - 3}{x - 2}$

In Exercises 4 and 5, graph the function.

4. $h(x) = \dfrac{4}{x^2} - 1$ **5.** $g(x) = \dfrac{x^2 + 2}{x - 1}$

6. Find a rational function that has vertical asymptotes at $x = \pm 3$ and a horizontal asymptote at $y = 4$.

7. A triangle is formed by the coordinate axes and a line through the point $(2, 1)$ (see figure).

(a) Verify that $y = 1 + \dfrac{2}{x - 2}$.

(b) Write the area A of the triangle as a function of x. Determine the domain of the function in the context of the problem.

(c) Use a graphing utility to graph the area function. What is the minimum area of the triangle?

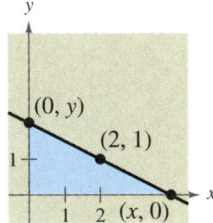

FIGURE FOR 7

In Exercises 8–11, write the partial fraction decomposition of the rational expression.

8. $\dfrac{2x + 5}{x^2 - x - 2}$ **9.** $\dfrac{3x^2 - 2x + 4}{x^2(2 - x)}$ **10.** $\dfrac{x^2 + 1}{x^3 - x}$ **11.** $\dfrac{x^2 - 1}{x^3 + x}$

In Exercises 12–14, graph the conic and identify any vertices and foci.

12. $y^2 - 4x + 4 = 0$ **13.** $x^2 - \dfrac{y^2}{4} = 1$ **14.** $x^2 - 4y^2 - 4x = 0$

15. Find an equation of the ellipse with vertices $(0, 2)$ and $(8, 2)$ and minor axis of length 4.

16. Find an equation of the hyperbola with vertices $(0, \pm 3)$ and asymptotes $y = \pm\frac{3}{2}x$.

17. Use a graphing utility to graph the conics $x^2 + y^2 = 36$ and $x^2 - (y^2/4) = 1$ on the same viewing rectangle and approximate the coordinates of any points of intersection.

Automobiles are designed with crumple zones that allow the occupants to move short distances when the automobiles come to abrupt stops. The greater the distance moved, the less g's the crash victims experience. (One g is equal to the acceleration due to gravity. For very short periods of time, humans have withstood as much as 40 g's.)

In crash tests with a vehicle moving at 90 kilometers per hour, analysts measured the number of g's that were experienced during deceleration by a crash dummy that was permitted to move x meters during impact.

x	0.2	0.4	0.6	0.8	1.0
g	158	80	53	40	32

A model for this data is

$$g = -3.00 + 11.88 \ln x + \frac{36.94}{x}.$$

The graph of this model is shown below.

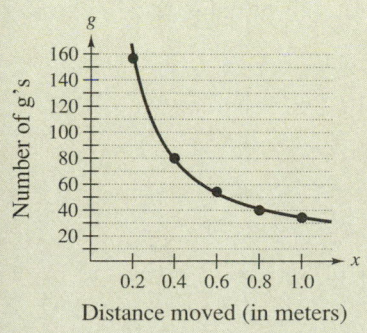

Distance moved (in meters)

See Exercise 94 on page 434.

Photo: Brad Trent

Exponential and Logarithmic Functions

5

5.1 *Exponential Functions and Their Graphs*
5.2 *Logarithmic Functions and Their Graphs*
5.3 *Properties of Logarithms*
5.4 *Exponential and Logarithmic Equations*
5.5 *Exponential and Logarithmic Models*

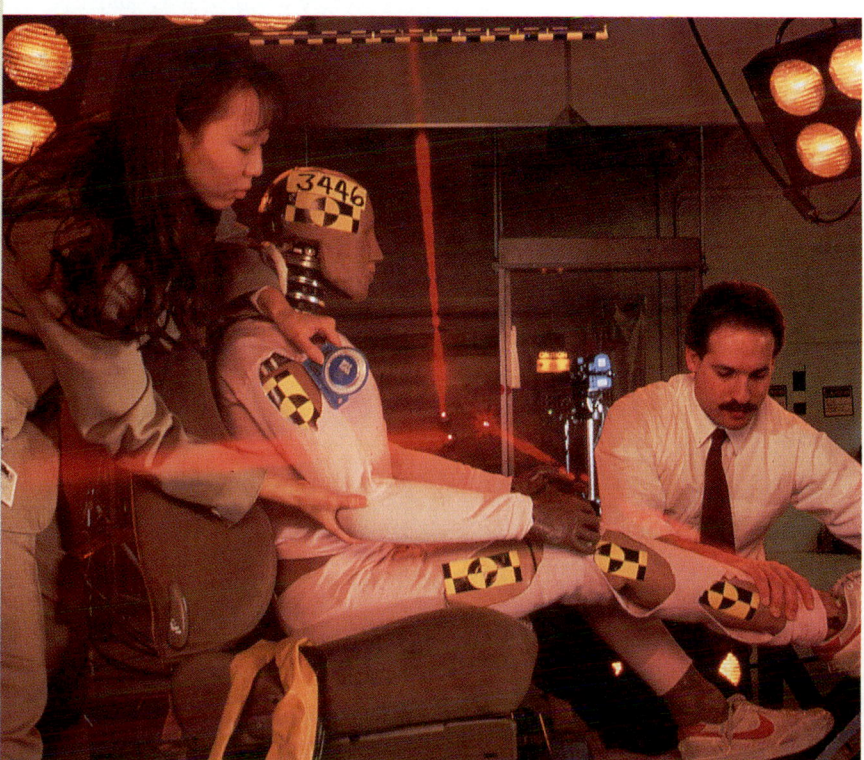

At a General Motors lab, engineer Bonnie Cheung and physicist Stephen Rouhana prepare a dummy for a simulated automobile crash. The laser (in red) helps position the dummy.

393

5.1   *Exponential Functions and Their Graphs*

See Exercise 63 on page 405 for an example of how an exponential function can be used to model the meteorological relationship between atmospheric pressure and altitude.

Exponential Functions   □   *Graphs of Exponential Functions*   □
The Natural Base e   □   *Applications*

Exponential Functions

So far, this book has dealt only with **algebraic functions,** which include polynomial functions and rational functions. In this chapter you will study two types of nonalgebraic functions—*exponential* functions and *logarithmic* functions. These functions are examples of **transcendental functions.**

NOTE   The base $a = 1$ is excluded because it yields

$$f(x) = 1^x = 1.$$

This is a constant function, not an exponential function. ▪▪

   DEFINITION OF EXPONENTIAL FUNCTION

The **exponential function** f with base a is denoted by

$$f(x) = a^x$$

where $a > 0$, $a \neq 1$, and x is any real number.

You already know how to evaluate a^x for integer and rational values of x. For example, you know that $4^3 = 64$ and $4^{1/2} = 2$. However, to evaluate 4^x for any real number x, you need to interpret forms with *irrational* exponents. For the purposes of this book, it is sufficient to think of

$$a^{\sqrt{2}} \text{ (where } \sqrt{2} \approx 1.414214)$$

as that value having the successively closer approximations

$$a^{1.4}, a^{1.41}, a^{1.414}, a^{1.4142}, a^{1.41421}, a^{1.414214}, \ldots .$$

Example 1 shows how to use a calculator to evaluate exponential expressions.

EXAMPLE 1   *Evaluating Exponential Expressions*

Use a calculator to evaluate each expression.
a. $2^{-3.1}$ **b.** $2^{-\pi}$

Solution

Number	Graphing Calculator Keystrokes	Display
a. $2^{-3.1}$	2 ∧ (-) 3.1 ENTER	0.1166291
b. $2^{-\pi}$	2 ∧ (-) π ENTER	0.1133147

Graphs of Exponential Functions

The graphs of all exponential functions have similar characteristics, as shown in Examples 2, 3, and 4.

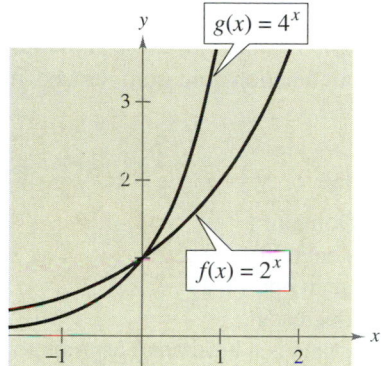

FIGURE 5.1

EXAMPLE 2 Graphs of $y = a^x$

In the same coordinate plane, sketch the graph of each function.

a. $f(x) = 2^x$

b. $g(x) = 4^x$

Solution

The table below lists some values for each function, and Figure 5.1 shows their graphs. Note that both graphs are increasing. Moreover, the graph of $g(x) = 4^x$ is increasing more rapidly than the graph of $f(x) = 2^x$.

x	-2	-1	0	1	2	3
2^x	$\frac{1}{4}$	$\frac{1}{2}$	1	2	4	8
4^x	$\frac{1}{16}$	$\frac{1}{4}$	1	4	16	64

EXAMPLE 3 Graphs of $y = a^{-x}$

In the same coordinate plane, sketch the graph of each function.

a. $F(x) = 2^{-x}$

b. $G(x) = 4^{-x}$

Solution

The table below lists some values for each function, and Figure 5.2 shows their graphs. Note that both graphs are decreasing. Moreover, the graph of $G(x) = 4^{-x}$ is decreasing more rapidly than the graph of $F(x) = 2^{-x}$.

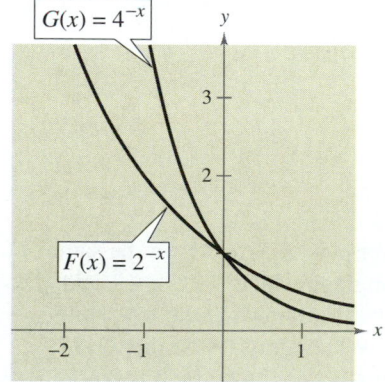

FIGURE 5.2

x	-3	-2	-1	0	1	2
2^{-x}	8	4	2	1	$\frac{1}{2}$	$\frac{1}{4}$
4^{-x}	64	16	4	1	$\frac{1}{4}$	$\frac{1}{16}$

NOTE The tables in Examples 2 and 3 were evaluated by hand. You could, of course, use a graphing utility to construct tables with even more values. ■■

The *Interactive* CD-ROM offers graphing utility emulators of the *TI-82* and *TI-83*, which can be used with the Examples, Explorations, Technology notes, and Exercises.

Comparing the functions in Examples 2 and 3, observe that

$$F(x) = 2^{-x} = f(-x) \qquad \text{and} \qquad G(x) = 4^{-x} = g(-x).$$

Consequently, the graph of F is a reflection (in the y-axis) of the graph of f. The graphs of G and g have the same relationship. The graphs in Figures 5.1 and 5.2 are typical of the exponential functions a^x and a^{-x}. They have one y-intercept and one horizontal asymptote (the x-axis), and they are continuous. The basic characteristics of these exponential functions are summarized in Figure 5.3.

Graph of $y = a^x$, $a > 1$

- Domain: $(-\infty, \infty)$
- Range: $(0, \infty)$
- Intercept: $(0, 1)$
- Increasing
- x-axis is a horizontal asymptote
  $(a^x \to 0$ as $x \to -\infty)$
- Continuous

Graph of $y = a^{-x}$, $a > 1$

- Domain: $(-\infty, \infty)$
- Range: $(0, \infty)$
- Intercept: $(0, 1)$
- Decreasing
- x-axis is a horizontal asymptote
  $(a^{-x} \to 0$ as $x \to \infty)$
- Continuous

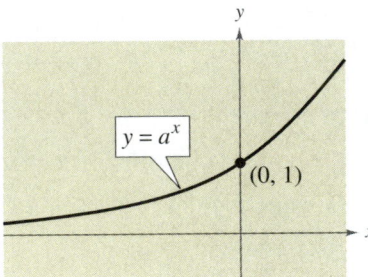

FIGURE 5.3

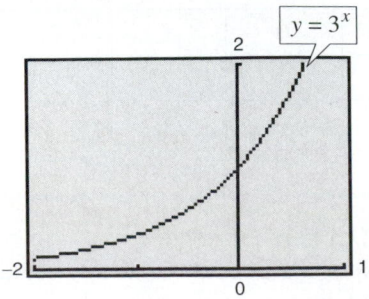

Exploration

Use a graphing utility to graph $y = a^x$ for $a = 3, 5,$ and 7 on the same viewing rectangle. (Use a viewing rectangle in which $-2 \le x \le 1$ and $0 \le y \le 2$.) For instance, the graph of $y = 3^x$ is shown at the left. How do the graphs compare with each other? Which graph is on the top in the interval $(-\infty, 0)$? Which is on the bottom? Which graph is on the top in the interval $(0, \infty)$? Which is on the bottom? Repeat this experiment with the graphs of $y = a^x$ for $a = \frac{1}{3}, \frac{1}{5},$ and $\frac{1}{7}$. (Use a viewing rectangle in which $-1 \le x \le 2$ and $0 \le y \le 2$.) What can you conclude about the relationship between the function's behavior and the value of a?

In the following example, notice how the graph of $y = a^x$ can be used to sketch the graphs of functions of the form $f(x) = b \pm a^{x+c}$.

EXAMPLE 4   Sketching Graphs of Exponential Functions

Each of the following graphs is a transformation of the graph of $f(x) = 3^x$, as shown in Figure 5.4.

a. Because $g(x) = 3^{x+1} = f(x + 1)$, the graph of g can be obtained by shifting the graph of f one unit to the left.

b. Because $h(x) = 3^x - 2 = f(x) - 2$, the graph of h can be obtained by shifting the graph of f down two units.

c. Because $k(x) = -3^x = -f(x)$, the graph of k can be obtained by reflecting the graph of f in the x-axis.

d. Because $j(x) = 3^{-x} = f(-x)$, the graph of j can be obtained by reflecting the graph of f in the y-axis.

NOTE   In Figure 5.4, notice that the transformations in parts (a), (c), and (d) keep the x-axis as a horizontal asymptote, but the transformation in part (b) yields a new horizontal asymptote of $y = -2$. Also, be sure to note how the y-intercept is affected by each transformation. ▪▪

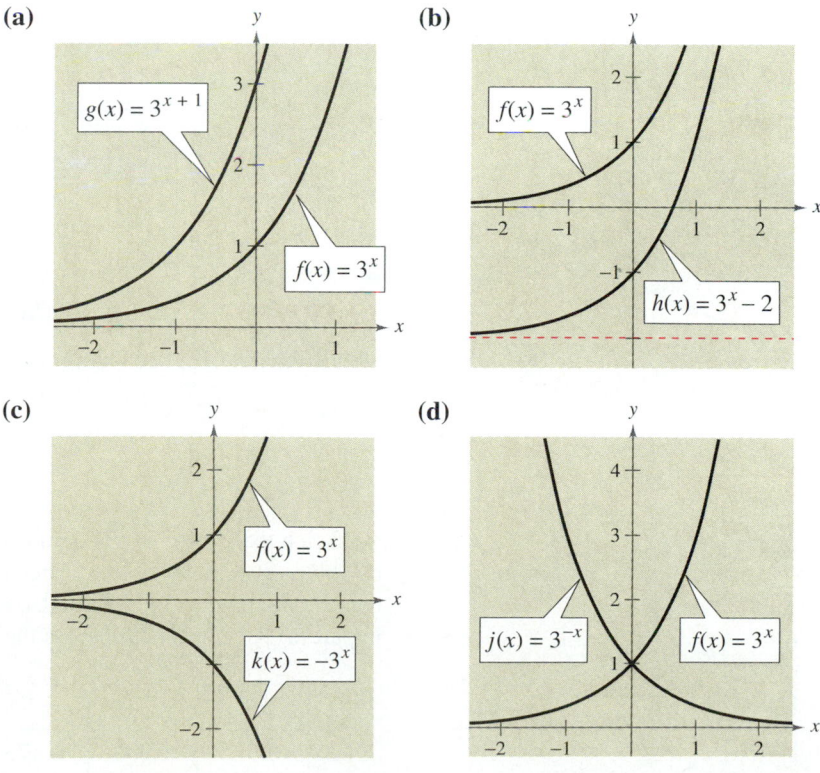

FIGURE 5.4

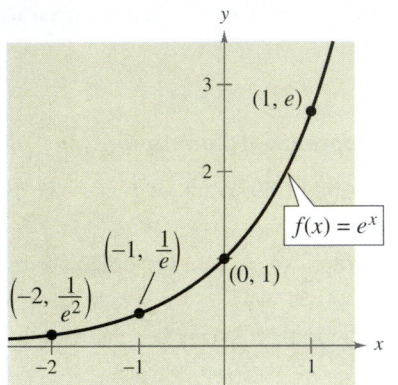

FIGURE 5.5

The Natural Base e

In many applications, the most convenient choice for a base is the irrational number

$$e \approx 2.71828 \ldots .$$

This number is called the **natural base.** The function $f(x) = e^x$ is called the **natural exponential function.** Its graph is shown in Figure 5.5. Be sure you see that for the exponential function $f(x) = e^x$, e is the constant $2.71828 \ldots$, whereas x is the variable.

EXAMPLE 5   *Evaluating the Natural Exponential Function*

Use a calculator to evaluate each expression.

a. e^{-2} **b.** e^{-1} **c.** e^1 **d.** e^2

Solution

	Number	Graphing Calculator Keystrokes	Display
a.	e^{-2}	e^x (-) 2 ENTER	0.1353353
b.	e^{-1}	e^x (-) 1 ENTER	0.3678794
c.	e^1	e^x 1 ENTER	2.7182818
d.	e^2	e^x 2 ENTER	7.3890561

EXAMPLE 6   *Graphing Natural Exponential Functions*

Sketch the graph of each natural exponential function.

a. $f(x) = 2e^{0.24x}$

b. $g(x) = \frac{1}{2}e^{-0.58x}$

Solution

To sketch these two graphs, you can use a graphing utility to construct a table of values, as shown below. After constructing the table, plot the points and connect them with smooth curves, as shown in Figure 5.6. Note that the graph in part (a) is increasing whereas the graph in part (b) is decreasing.

(a)

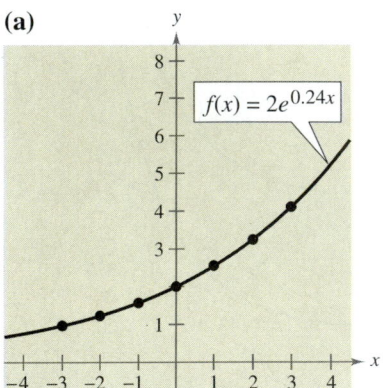

(b)

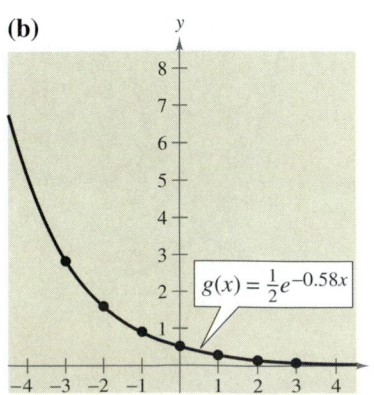

FIGURE 5.6

x	-3	-2	-1	0	1	2	3
$f(x)$	0.974	1.238	1.573	2.000	2.543	3.232	4.109
$g(x)$	2.849	1.595	0.893	0.500	0.280	0.157	0.088

Applications

One of the most familiar examples of exponential growth is that of an investment earning *continuously compounded interest*. On page 147 in Section 1.6, you were introduced to the formula for the balance in an account that is compounded annually. Using exponential functions, you can now *develop* that formula and show how it leads to continuous compounding.

Suppose a principal P is invested at an annual interest rate r, compounded once a year. If the interest is added to the principal at the end of the year, the balance is

$$P_1 = P + Pr = P(1 + r).$$

This pattern of multiplying the previous principal by $1 + r$ is then repeated each successive year, as shown below.

Year	Balance After Each Compounding
0	$P = P$
1	$P_1 = P(1 + r)$
2	$P_2 = P_1(1 + r) = P(1 + r)(1 + r) = P(1 + r)^2$
3	$P_3 = P_2(1 + r) = P(1 + r)^2(1 + r) = P(1 + r)^3$
$\vdots$	
t	$P_1 = P(1 + r)^t$

To accommodate more frequent (quarterly, monthly, or daily) compounding of interest, let n be the number of compoundings per year and let t be the number of years. Then the rate per compounding is r/n and the account balance after t years is

$$A = P\left(1 + \frac{r}{n}\right)^{nt}. \qquad \textcolor{red}{\text{Amount with } n \text{ compoundings per year}}$$

If you let the number of compoundings n increase without bound, the process approaches what is called **continuous compounding.** In the formula for n compoundings per year, let $m = n/r$. This produces

$$A = P\left(1 + \frac{r}{n}\right)^{nt}$$

$$= P\left(1 + \frac{1}{m}\right)^{mrt}$$

$$= P\left[\left(1 + \frac{1}{m}\right)^{m}\right]^{rt}.$$

As m increases without bound, it can be shown that $[1 + (1/m)]^m$ approaches e. (Try the values $m = 10$, $10,000$, and $10,000,000$.) From this, you can conclude that the formula for continuous compounding is $A = Pe^{rt}$.

▶ *Exploration*

Use the formula $A = P(1 + r/n)^{nt}$ to calculate the amount in an account when $P = \$3000$, $r = 6\%$, $t = 10$ years, and the number of compoundings is (1) by the day, (2) by the hour, (3) by the minute, and (4) by the second. Use these results to present an argument that increasing the number of compoundings does not mean unlimited growth of the amount in the account.

FORMULAS FOR COMPOUND INTEREST

After t years, the balance A in an account with principal P and annual interest rate r (in decimal form) is given by the following formulas.

1. For n compoundings per year: $A = P\left(1 + \dfrac{r}{n}\right)^{nt}$

2. For continuous compounding: $A = Pe^{rt}$

NOTE   Be sure you see that the annual interest rate must be expressed in decimal form. For instance, 6% should be expressed as 0.06. ▪▪

EXAMPLE 7   *Compounding n Times and Continuously*   Real Life

A total of $12,000 is invested at an annual interest rate of 9%. Find the balance after 5 years if it is compounded

a. quarterly.

b. continuously.

Solution

a. For quarterly compoundings, you have $n = 4$. Thus, in 5 years at 9%, the balance is

$$A = P\left(1 + \frac{r}{n}\right)^{nt} \qquad \text{Formula for compound interest}$$

$$= 12{,}000\left(1 + \frac{0.09}{4}\right)^{4(5)} \qquad \text{Substitute for } P, r, n, \text{ and } t.$$

$$= \$18{,}726.11. \qquad \text{Use a calculator.}$$

b. For continuous compounding, the balance is

$$A = Pe^{rt} \qquad \text{Formula for continuous compounding}$$

$$= 12{,}000e^{0.09(5)} \qquad \text{Substitute for } P, r, \text{ and } t.$$

$$= \$18{,}819.75. \qquad \text{Use a calculator.}$$

Note that continuous compounding yields

$$\$18{,}819.75 - \$18{,}726.11 = \$93.64$$

more than quarterly compounding. This is typical of the two types of compounding. That is, for a given principal, interest rate, and time, continuous compounding will always yield a larger balance than compounding n times a year.

EXAMPLE 8 *Radioactive Decay*

In 1986, a nuclear reactor accident occurred in Chernobyl in what was then the Soviet Union. The explosion spread radioactive chemicals over hundreds of square miles, and the government evacuated the city and the surrounding area. To see why the city is now uninhabited, consider the following model.

$$P = 10e^{-0.00002845t}$$

This model represents the amount of plutonium that remains (from an initial amount of 10 pounds) after t years. Sketch the graph of this function over the interval from $t = 0$ to $t = 100,000$. How much of the 10 pounds will remain after 100,000 years?

Solution

The graph of this function is shown in Figure 5.7. Note from this graph that plutonium has a *half-life* of about 24,360 years. That is, after 24,360 years, *half* of the original amount will remain. After another 24,360 years, one-quarter of the original amount will remain, and so on. After 100,000 years, there will still be

$$P = 10e^{-0.00002845(100,000)} = 10e^{-2.845} \approx 0.58 \text{ pounds}$$

of plutonium remaining.

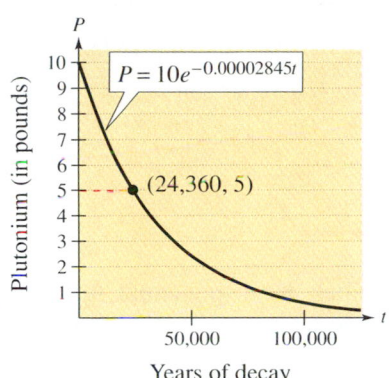

FIGURE 5.7

GROUP ACTIVITY

IDENTIFYING EXPONENTIAL FUNCTIONS

Which of the following functions generated the two tables below? Discuss how you were able to decide. What do these functions have in common? Are any the same? If so, explain why.

a. $f_1(x) = 2^{(x+3)}$ **b.** $f_2(x) = 8\left(\frac{1}{2}\right)^x$ **c.** $f_3(x) = \left(\frac{1}{2}\right)^{(x-3)}$

d. $f_4(x) = \left(\frac{1}{2}\right)^x + 7$ **e.** $f_5(x) = 7 + 2^x$ **f.** $f_6(x) = (8)2^x$

x	-1	0	1	2	3
$g(x)$	7.5	8	9	11	15

x	-2	-1	0	1	2
$h(x)$	32	16	8	4	2

Create two different exponential functions with y-intercepts of $(0, -3)$. Compare your functions with those of other students in your class.

The *Interactive* CD-ROM provides additional help with Warm-Up exercises by providing a hypertext link to the section in which the concept was introduced.

The *Interactive* CD-ROM contains step-by-step solutions to all odd-numbered Section and Review Exercises. It also provides Tutorial Exercises, which link to Guided Examples for additional help.

WARM UP

Use the properties of exponents to simplify the expression.

1. $5^{2x}(5^{-x})$

2. $3^{-x}(3^{3x})$

3. $\dfrac{4^{5x}}{4^{2x}}$

4. $\dfrac{10^{2x}}{10^{x}}$

5. $(4^{x})^2$

6. $(4^{2x})^5$

7. $\left(\dfrac{2^{x}}{3^3}\right)^{-1}$

8. $(4^{6x})^{1/2}$

9. $(2^{3x})^{-1/3}$

10. $(16^{x})^{1/4}$

5.1 Exercises

In Exercises 1–10, evaluate the expression. Round your result to three decimal places.

1. $(3.4)^{5.6}$

2. $5000(2^{-1.5})$

3. $(1.005)^{400}$

4. $8^{2\pi}$

5. $5^{-\pi}$

6. $\sqrt[3]{4395}$

7. $100^{\sqrt{2}}$

8. $e^{1/2}$

9. $e^{-3/4}$

10. $e^{3.2}$

Think About It   In Exercises 11–14, use properties of exponents to determine which functions (if any) are the same.

11. $f(x) = 3^{x-2}$
    $g(x) = 3^x - 9$
    $h(x) = \frac{1}{9}(3x)$

12. $f(x) = 4^x + 12$
    $g(x) = 2^{2x+6}$
    $h(x) = 64(4^x)$

13. $f(x) = 16(4^{-x})$
    $g(x) = \left(\frac{1}{4}\right)^{x-2}$
    $h(x) = 16(2^{-2x})$

14. $f(x) = 5^{-x} + 3$
    $g(x) = 5^{3-x}$
    $h(x) = -5^{x-3}$

In Exercises 15–18, match the exponential function with its graph. [The graphs are labeled (a), (b), (c), and (d).]

(a)
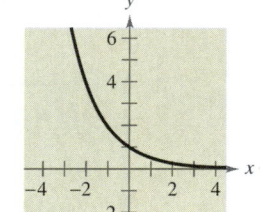

(b)
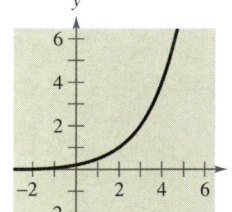

(c)
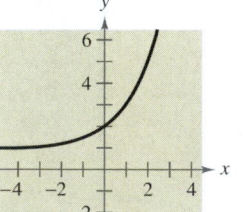

(d)
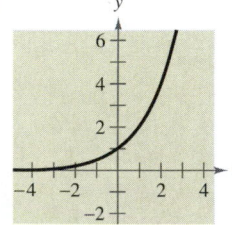

15. $f(x) = 2^x$

16. $f(x) = 2^x + 1$

17. $f(x) = 2^{-x}$

18. $f(x) = 2^{x-2}$

In Exercises 19–36, graph the exponential function.

19. $g(x) = 5^x$

20. $f(x) = \left(\frac{3}{2}\right)^x$

21. $f(x) = \left(\frac{1}{5}\right)^x = 5^{-x}$

22. $h(x) = \left(\frac{3}{2}\right)^{-x}$

23. $h(x) = 5^{x-2}$

24. $g(x) = \left(\frac{3}{2}\right)^{x+2}$

25. $g(x) = 5^{-x} - 3$

26. $f(x) = \left(\frac{3}{2}\right)^{-x} + 2$

27. $y = 2^{-x^2}$

28. $y = 3^{-|x|}$

29. $y = 3^{x-2} + 1$

30. $y = 4^{x+1} - 2$

31. $y = 1.08^{-5x}$

32. $y = 1.08^{5x}$

33. $s(t) = 2e^{0.12t}$

34. $s(t) = 3e^{-0.2t}$

35. $g(x) = 1 + e^{-x}$

36. $h(x) = e^{x-2}$

37. Graph the functions $y = 3^x$ and $y = 4^x$ and use the graphs to solve the inequalities.

(a) $4^x < 3^x$

(b) $4^x > 3^x$

38. Graph the functions $y = \left(\frac{1}{2}\right)^x$ and $y = \left(\frac{1}{4}\right)^x$ and use the graphs to solve the inequalities.

(a) $\left(\frac{1}{4}\right)^x < \left(\frac{1}{2}\right)^x$

(b) $\left(\frac{1}{4}\right)^x > \left(\frac{1}{2}\right)^x$

39. Use the graph of $f(x) = 3^x$ to graph each of the functions. Identify the transformation.

(a) $g(x) = f(x-2) = 3^{x-2}$

(b) $h(x) = -\frac{1}{2}f(x) = -\frac{1}{2}(3^x)$

(c) $q(x) = f(-x) + 3 = 3^{-x} + 3$

40. Use a graphing utility to graph each function. Use the graph to find any asymptotes of the function.

(a) $f(x) = \dfrac{8}{1 + e^{-0.5x}}$

(b) $g(x) = \dfrac{8}{1 + e^{-0.5/x}}$

41. Use a graphing utility to graph each function. Use the graph to find where the function is increasing and decreasing, and approximate any relative maximum or minimum values.

(a) $f(x) = x^2 e^{-x}$

(b) $g(x) = x2^{3-x}$

42. *Comparing Functions* Use a graphing utility to graph $y_1 = e^x$ and each of the functions $y_2 = x^2$, $y_3 = x^3$, $y_4 = \sqrt{x}$, and $y_5 = |x|$. Which function increases at the fastest rate for "large" values of x?

43. *Conjecture* Use the result of Exercise 42 to make a conjecture about the rate of growth of $y_1 = e^x$ and $y = x^n$ where n is a natural number and x is "large."

44. *Essay* Use the results of Exercises 42 and 43 to describe what is implied when it is stated that a quantity is growing exponentially.

45. *Graphical Analysis* Use a graphing utility to graph

$$f(x) = \left(1 + \frac{0.5}{x}\right)^x \quad \text{and} \quad g(x) = e^{0.5}$$

on the same viewing rectangle. What is the relationship between f and g as x increases without bound?

46. *Conjecture* Use the result of Exercise 45 to make a conjecture about the value of $[1 + (r/x)]^x$ as x increases without bound. Create a table that illustrates your conjecture for $r = 1$.

Compound Interest **In Exercises 47–50, complete the table to determine the balance A for P dollars invested at rate r for t years and compounded n times per year.**

n	1	2	4	12	365	Continuous
A						

47. $P = \$2500$, $r = 12\%$, $t = 10$ years

48. $P = \$1000$, $r = 10\%$, $t = 10$ years

49. $P = \$2500$, $r = 12\%$, $t = 20$ years

50. $P = \$1000$, $r = 10\%$, $t = 40$ years

Compound Interest **In Exercises 51 and 52, complete the table to determine the amount of money P that should be invested at rate r to produce a final balance of $\$100,000$ in t years.**

t	1	10	20	30	40	50
P						

51. $r = 9\%$, compounded continuously

52. $r = 12\%$, compounded continuously

53. Trust Fund   On the day of a child's birth, a deposit of $25,000 is made in a trust fund that pays 8.75% interest, compounded continuously. Determine the balance in this account on the child's 25th birthday.

54. Trust Fund   A deposit of $5000 is made in a trust fund that pays 7.5% interest, compounded continuously. It is specified that the balance will be given to the college from which the donor graduated after the money has earned interest for 50 years. How much will the college receive?

55. Graphical Reasoning   There are two options for investing $500. The first earns 7% compounded annually and the second earns 7% simple interest. The figure shows the growth of each investment over a 30-year period.

(a) Identify the two types of investments in the figure. Explain your reasoning.

(b) Verify your answer in part (a) by finding the equations that model the investment growth and graphing the models.

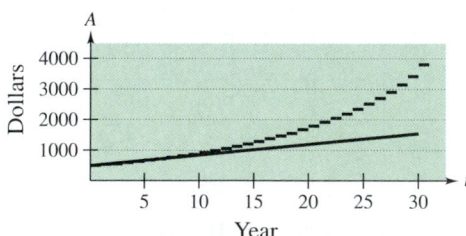

56. Depreciation   After t years, the value of a car that cost $20,000 is given by

$$V(t) = 20,000\left(\tfrac{3}{4}\right)^t.$$

Graph the function and determine the value of the car 2 years after it was purchased.

57. Inflation   If the annual rate of inflation averages 4% over the next 10 years, the approximate cost C of goods or services during any year in that decade will be given by

$$C(t) = P(1.04)^t$$

where t is the time in years and P is the present cost. If the price of an oil change for your car is presently $23.95, estimate the price 10 years from now.

58. Demand Function   The demand equation for a certain product is given by

$$p = 5000\left(1 - \frac{4}{4 + e^{-0.002x}}\right).$$

(a) Use a graphing utility to graph the demand function for $x > 0$ and $p > 0$.

(b) Find the price p for a demand of $x = 500$ units.

(c) Use the graph in part (a) to approximate the greatest price that will still yield a demand of at least 600 units.

59. Bacteria Growth   A certain type of bacteria increases according to the model

$$P(t) = 100e^{0.2197t}$$

where t is the time in hours. Find (a) $P(0)$, (b) $P(5)$, and (c) $P(10)$.

60. Population Growth   The population of a town increases according to the model

$$P(t) = 2500e^{0.0293t}$$

where t is the time in years, with $t = 0$ corresponding to 1990. Use the model to estimate the population in (a) 2000 and (b) 2010.

61. Radioactive Decay   Let Q represent a mass of radium (^{226}Ra), whose half-life is 1620 years. The quantity of radium present after t years is given by

$$Q = 25\left(\tfrac{1}{2}\right)^{t/1620}.$$

(a) Determine the initial quantity (when $t = 0$).

(b) Determine the quantity present after 1000 years.

(c) Use a graphing utility to graph the function over the interval $t = 0$ to $t = 5000$.

62. Radioactive Decay   Let Q represent a mass of carbon 14(^{14}C), whose half-life is 5730 years. The quantity of carbon 14 present after t years is given by

$$Q = 10\left(\tfrac{1}{2}\right)^{t/5730}.$$

(a) Determine the initial quantity (when $t = 0$).

(b) Determine the quantity present after 2000 years.

(c) Sketch the graph of this function over the interval $t = 0$ to $t = 10,000$.

63. *Data Analysis* A meteorologist measures the atmospheric pressure P (in kilograms per square meter) at altitude h (in kilometers). The data is shown in the table.

h	0	5	10	15	20
P	10,332	5583	2376	1240	517

A model for the data is given by

$P = 10,958e^{-0.15h}$.

(a) Sketch a scatter plot of the data and graph the model on the same set of axes.

(b) Create a table to compare the model with the sample data.

(c) Estimate the atmospheric pressure at a height of 8 kilometers.

(d) Use the graph in part (a) to estimate the altitude at which the atmospheric pressure is 2000 kilograms per square meter.

64. *Data Analysis* To estimate the amount of defoliation caused by the gypsy moth during a given year, a forester counts the number x of egg masses on $\frac{1}{40}$ of an acre (circle of radius 18.6 feet) the preceding fall. The percent of defoliation y the next spring is given in the table. (Source: USDA, Forest Service)

x	0	25	50	75	100
y	12	44	81	96	99

(a) A model for the data is given by

$$y = \frac{300}{3 + 17e^{-0.065x}}.$$

Use a graphing utility to create a scatter plot of the data and graph the model on the same viewing rectangle.

(b) Create a table to compare the model with the sample data.

(c) Estimate the percent of defoliation if 36 egg masses are counted on $\frac{1}{40}$ acre.

(d) Use the graph in part (a) to estimate the number of egg masses per $\frac{1}{40}$ acre if you observe that $\frac{2}{3}$ of a forest is defoliated the following spring.

65. *True or False?* $e = \dfrac{271,801}{99,990}$. Explain.

66. *Think About It* Which functions are exponential?

(a) $3x$ (b) $3x^2$

(c) 3^x (d) 2^{-x}

67. *Think About It* Without using a calculator, why do you know that $2^{\sqrt{2}}$ is greater than 2, but less than 4?

68. *Exploration* Use a graphing utility to compare the graph of the function $y = e^x$ with the graphs of the following functions.

(a) $y_1 = 1 + \dfrac{x}{1!}$

(b) $y_2 = 1 + \dfrac{x}{1!} + \dfrac{x^2}{2!}$

(c) $y_3 = 1 + \dfrac{x}{1!} + \dfrac{x^2}{2!} + \dfrac{x^3}{3!}$

69. *Finding a Pattern* Identify the pattern of successive polynomials given in Exercise 68. Extend the pattern one more term and compare the graph of the resulting polynomial function with the graph of $y = e^x$. What do you think this pattern implies?

70. Given the exponential function $f(x) = a^x$, show that

(a) $f(u + v) = f(u) \cdot f(v)$.

(b) $f(2x) = [f(x)]^2$.

Review Solve Exercises 71–74 as a review of the skills and problem-solving techniques you learned in previous sections. Solve for y.

71. $2x - 7y + 14 = 0$

72. $x^2 + 3y = 4$

73. $x^2 + y^2 = 25$

74. $x - |y| = 2$

5.2   *Logarithmic Functions and Their Graphs*

See Exercises 75 and 76 on page 415 for an example of how a logarithmic function can be used to model the minimum required ventilation rates in public school classrooms.

Logarithmic Functions   □   Graphs of Logarithmic Functions   □ The Natural Logarithmic Function   □   Application

Logarithmic Functions

In Section 2.5, you studied the concept of the inverse of a function. There, you learned that if a function has the property that no horizontal line intersects the graph of the function more than once, the function must have an inverse. By looking back at the graphs of the exponential functions introduced in Section 5.1, you will see that every function of the form $f(x) = a^x$ passes the "Horizontal Line Test" and therefore must have an inverse. This inverse function is called the **logarithmic function with base *a*.**

NOTE   The equations

$$y = \log_a x \qquad \text{and} \qquad x = a^y$$

are equivalent. The first equation is in logarithmic form and the second is in exponential form.   ■■

> **DEFINITION OF LOGARITHMIC FUNCTION**
> For $x > 0$ and $0 < a \neq 1$,
>
> $$y = \log_a x \text{ if and only if } x = a^y.$$
>
> The function given by
>
> $$f(x) = \log_a x$$
>
> is called the **logarithmic function with base *a*.**

When evaluating logarithms, remember that *a logarithm is an exponent.* This means that $\log_a x$ is the exponent to which a must be raised to obtain x. For instance, $\log_2 8 = 3$ because 2 must be raised to the third power to get 8.

EXAMPLE 1   *Evaluating Logarithms*

a. $\log_2 32 = 5$ because $2^5 = 32.$

b. $\log_3 27 = 3$ because $3^3 = 27.$

c. $\log_4 2 = \frac{1}{2}$ because $4^{1/2} = \sqrt{4} = 2.$

d. $\log_{10} \frac{1}{100} = -2$ because $10^{-2} = \frac{1}{10^2} = \frac{1}{100}.$

e. $\log_3 1 = 0$ because $3^0 = 1.$

f. $\log_2 2 = 1$ because $2^1 = 2.$

The logarithmic function with base 10 is called the **common logarithmic function.** On most calculators, this function is denoted by $\boxed{\text{LOG}}$.

EXAMPLE 2   *Evaluating Logarithms on a Calculator*

Use a calculator to evaluate each expression.

a. $\log_{10} 10$ **b.** $2\log_{10} 2.5$ **c.** $\log_{10}(-2)$

Solution

Number	Graphing Calculator Keystrokes	Display
a. $\log_{10} 10$	$\boxed{\text{LOG}}$ 10 $\boxed{\text{ENTER}}$	1
b. $2\log_{10} 2.5$	2 $\boxed{\times}$ $\boxed{\text{LOG}}$ 2.5 $\boxed{\text{ENTER}}$	0.7958800
c. $\log_{10}(-2)$	$\boxed{\text{LOG}}$ $\boxed{(\text{-})}$ 2 $\boxed{\text{ENTER}}$	ERROR

Note that the calculator displays an error message when you try to evaluate $\log_{10}(-2)$. The reason for this is that the domain of every logarithmic function is the set of *positive real numbers.*

The following properties follow directly from the definition of the logarithmic function with base a.

> **Properties of Logarithms**
>
> **1.** $\log_a 1 = 0$ because $a^0 = 1$.
> **2.** $\log_a a = 1$ because $a^1 = a$.
> **3.** $\log_a a^x = x$ because $a^x = a^x$.
> **4.** If $\log_a x = \log_a y$, then $x = y$.

EXAMPLE 3   *Using Properties of Logarithms*

Solve each equation for x.

a. $\log_2 x = \log_2 3$
b. $\log_4 4 = x$

Solution

a. Using Property 4, you can conclude that $x = 3$.
b. Using Property 2, you can conclude that $x = 1$.

Study Tip

Because $\log_a x$ is the inverse function of a^x, it follows that the domain of $\log_a x$ is the range of a^x, $(0, \infty)$. In other words, $\log_a x$ is defined only if x is positive.

Graphs of Logarithmic Functions

To sketch the graph of $y = \log_a x$, you can use the fact that the graphs of inverse functions are reflections of each other in the line $y = x$.

EXAMPLE 4   *Graphs of Exponential and Logarithmic Functions*

In the same coordinate plane, sketch the graph of each function.

a. $f(x) = 2^x$ **b.** $g(x) = \log_2 x$

Solution

a. For $f(x) = 2^x$, construct a table of values, as follows.

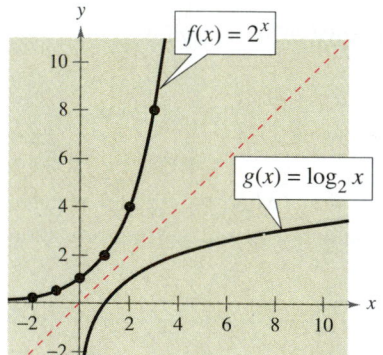

x	-2	-1	0	1	2	3
$f(x) = 2^x$	$\frac{1}{4}$	$\frac{1}{2}$	1	2	4	8

By plotting these points and connecting them with a smooth curve, you obtain the graph shown in Figure 5.8.

b. Because $g(x) = \log_2 x$ is the inverse of $f(x) = 2^x$, the graph of g is obtained by reflecting the graph of f in the line $y = x$, as shown in Figure 5.8.

FIGURE 5.8

Before you can confirm the result of Example 4 with a graphing utility, you need to know how to enter $\log_2 x$. A procedure using the change-of-base formula is discussed in Section 5.3.

EXAMPLE 5   *Sketching the Graph of a Logarithmic Function*

Sketch the graph of the common logarithmic function $f(x) = \log_{10} x$.

Solution

Begin by constructing a table of values. Note that some of the values can be obtained without a calculator, whereas others require a calculator. Next, plot the points and connect them with a smooth curve, as shown in Figure 5.9.

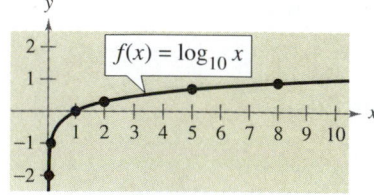

FIGURE 5.9

	Without Calculator				With Calculator		
x	$\frac{1}{100}$	$\frac{1}{10}$	1	10	2	5	8
$\log_{10} x$	-2	-1	0	1	0.301	0.699	0.903

The nature of the graph in Figure 5.9 is typical of functions of the form $f(x) = \log_a x, a > 1$. They have one x-intercept and one vertical asymptote. Notice how slowly the graph rises for $x > 1$. In Figure 5.9 you would need to move out to $x = 1000$ before the graph rose to $y = 3$. The basic characteristics of logarithmic graphs are summarized in Figure 5.10.

NOTE In the graph at the right, note that the vertical asymptote occurs at $x = 0$, where $\log_a x$ is *undefined*. ■■

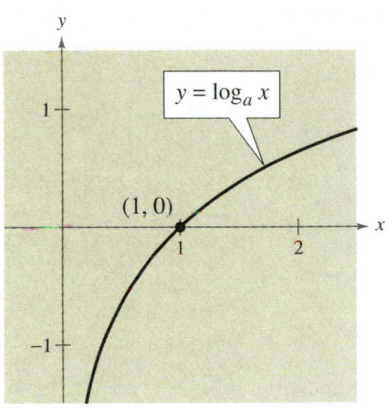

Graph of $y = \log_a x, a > 1$

- Domain: $(0, \infty)$
- Range: $(-\infty, \infty)$
- Intercept: $(1, 0)$
- Increasing
- y-axis is a vertical asymptote
  ($\log_a x \to -\infty$ as $x \to 0^+$)
- Continuous
- Reflection of graph of $y = a^x$
  about the line $y = x$

FIGURE 5.10

In the following example, the graph of $\log_a x$ is used to sketch the graphs of functions of the form $y = b \pm \log_a(x + c)$.

EXAMPLE 6 *Sketching the Graphs of Logarithmic Functions*

The graph of each of the following functions is similar to the graph of $f(x) = \log_{10} x$, as shown in Figure 5.11.

a. Because $g(x) = \log_{10}(x - 1) = f(x - 1)$, the graph of g can be obtained by shifting the graph of f one unit to the right.

b. Because $h(x) = 2 + \log_{10} x = 2 + f(x)$, the graph of h can be obtained by shifting the graph of f two units up.

NOTE In Figure 5.11, notice how each transformation affects the vertical asymptote. ■■

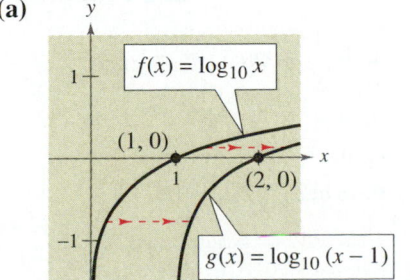

FIGURE 5.11

The Natural Logarithmic Function

As with exponential functions, the most widely used base for logarithmic functions is the number e, where

$$e \approx 2.718281828 \ldots .$$

The logarithmic function with base e is the **natural logarithmic function** and is denoted by the special symbol $\ln x$, read as "el en of x."

 THE NATURAL LOGARITHMIC FUNCTION

The function defined by

$$f(x) = \log_e x = \ln x, \quad x > 0$$

is called the **natural logarithmic function.**

The four properties of logarithms listed on page 407 are also valid for natural logarithms.

 PROPERTIES OF NATURAL LOGARITHMS

1. $\ln 1 = 0$ because $e^0 = 1$.
2. $\ln e = 1$ because $e^1 = e$.
3. $\ln e^x = x$ because $e^x = e^x$.
4. If $\ln x = \ln y$, then $x = y$.

The graph of the natural logarithmic function is shown in Figure 5.12. Try using a graphing utility to confirm this graph. What is the domain of the natural logarithmic function?

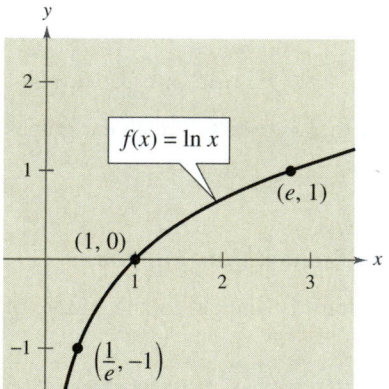

FIGURE 5.12

EXAMPLE 7   *Using Properties of Natural Logarithms*

a. $\ln \dfrac{1}{e} = \ln e^{-1} = -1$ Property 3

b. $\ln e^2 = 2$ Property 3

c. $\ln e^0 = 0$ Property 1

d. $2 \ln e = 2(1) = 2$ Property 2

On most calculators, the natural logarithm is denoted by $\boxed{\text{LN}}$, as illustrated in Example 8.

EXAMPLE 8 **Evaluating the Natural Logarithmic Function**

Use a calculator to evaluate each expression.

a. ln 2   **b.** ln 0.3   **c.** ln e^2 **d.** ln(−1)

Solution

Number	Graphing Calculator Keystrokes	Display
a. ln 2	$\boxed{\text{LN}}$ 2 $\boxed{\text{ENTER}}$	0.6931472
b. ln 0.3	$\boxed{\text{LN}}$ .3 $\boxed{\text{ENTER}}$	−1.2039728
c. ln e^2	$\boxed{\text{LN}}$ $\boxed{e^x}$ 2 $\boxed{\text{ENTER}}$	2
d. ln(−1)	$\boxed{\text{LN}}$ $\boxed{\text{(-)}}$ 1 $\boxed{\text{ENTER}}$	ERROR

In Example 8, be sure you see that ln(−1) gives an error message on most calculators. This occurs because the domain of ln x is the set of positive real numbers (see Figure 5.12). Hence, ln(−1) is undefined.

EXAMPLE 9 **Finding the Domains of Logarithmic Functions**

Find the domain of each function.

a. $f(x) = \ln(x - 2)$ **b.** $g(x) = \ln(2 - x)$ **c.** $h(x) = \ln x^2$

Solution

a. Because $\ln(x - 2)$ is defined only if $x - 2 > 0$, it follows that the domain of f is $(2, \infty)$.

b. Because $\ln(2 - x)$ is defined only if $2 - x > 0$, it follows that the domain of g is $(-\infty, 2)$. The graph of g is shown in Figure 5.13.

c. Because $\ln x^2$ is defined only if $x^2 > 0$, it follows that the domain of h is all real numbers except $x = 0$.

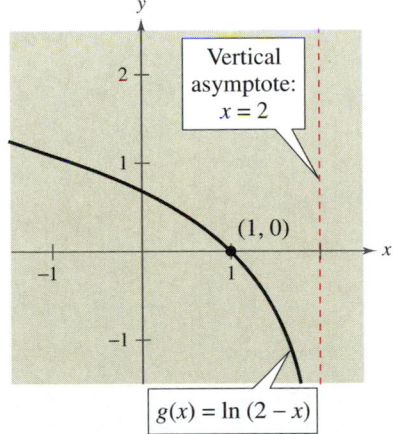

Vertical asymptote: $x = 2$

(1, 0)

$g(x) = \ln(2 - x)$

FIGURE 5.13

NOTE In Example 9, suppose you had been asked to analyze the function given by $h(x) = \ln |x - 2|$. How would the domain of this function compare with the domains of the functions given in parts (a) and (b) of the example? ▪▪

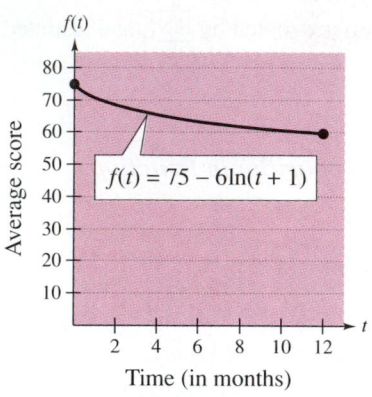

FIGURE 5.14

> ▶ *Exploration*
>
> Use a graphing utility to determine the time in months when the average score in Example 10 was 60. Explain your method of solving the problem. Describe another way that you can use a graphing utility to determine the answer.

x	$f_1(x)$	$f_2(x)$
−3	Undefined	Undefined
−2.99999	−11.513	−15.513
−2.9999	−9.210	−13.210
−2	0	−4
0	1.099	−2.901
3	1.792	−2.208
6	2.197	−1.803

Application

EXAMPLE 10   Human Memory Model

Students participating in a psychological experiment attended several lectures on a subject and were given an exam. Every month for a year after the exam, the students were retested to see how much of the material they remembered. The average scores for the group are given by the *human memory model*

$$f(t) = 75 - 6 \ln(t + 1), \quad 0 \le t \le 12$$

where t is the time in months.

a. What was the average score on the original ($t = 0$) exam?

b. What was the average score at the end of $t = 2$ months?

c. What was the average score at the end of $t = 6$ months?

Solution

a. The original average score was

$$f(0) = 75 - 6 \ln 1 = 75 - 6(0) = 75.$$

b. After 2 months, the average score was

$$f(2) = 75 - 6 \ln 3 \approx 75 - 6(1.0986) \approx 68.4.$$

c. After 6 months, the average score was

$$f(6) = 75 - 6 \ln 7 \approx 75 - 6(1.9459) \approx 63.3.$$

The graph of f is shown in Figure 5.14.

GROUP ACTIVITY

TRANSFORMING LOGARITHMIC FUNCTIONS

The table at left gives selected points for two natural logarithmic functions of the form $f(x) = b \pm \ln(x + c)$. Study the table. What can you infer? Compare the given data with data for $f(x) = \ln x$. Try to find a natural logarithmic function that fits each set of data. Explain the method you used. (*Hint:* You might find it easier to find the value of c first.)

5.2 Exercises

In Exercises 1–8, write the logarithmic equation in exponential form. For example, the exponential form of $\log_5 25 = 2$ is $5^2 = 25$.

1. $\log_4 64 = 3$ **2.** $\log_3 81 = 4$

3. $\log_7 \frac{1}{49} = -2$ **4.** $\log_{10} \frac{1}{1000} = -3$

5. $\log_{32} 4 = \frac{2}{5}$ **6.** $\log_{16} 8 = \frac{3}{4}$

7. $\ln 1 = 0$ **8.** $\ln 4 = 1.386 \ldots$

In Exercises 9–18, write the exponential equation in logarithmic form. For example, the logarithmic form of $2^3 = 8$ is $\log_2 8 = 3$.

9. $5^3 = 125$ **10.** $8^2 = 64$

11. $81^{1/4} = 3$ **12.** $9^{3/2} = 27$

13. $6^{-2} = \frac{1}{36}$ **14.** $10^{-3} = 0.001$

15. $e^3 = 20.0855 \ldots$ **16.** $e^0 = 1$

17. $e^x = 4$ **18.** $u^v = w$

In Exercises 19–30, evaluate the expression without using a calculator.

19. $\log_2 16$ **20.** $\log_2 \frac{1}{8}$

21. $\log_{16} 4$ **22.** $\log_{27} 9$

23. $\log_7 1$ **24.** $\log_{10} 1000$

25. $\log_{10} 0.01$ **26.** $\log_{10} 10$

27. $\log_8 32$ **28.** $\log_9 243$

29. $\ln e^3$ **30.** $\log_a a^2$

In Exercises 31–40, use a calculator to evaluate the logarithm. Round to three decimal places.

31. $\log_{10} 345$ **32.** $\log_{10} \frac{4}{5}$

33. $\log_{10} 145$ **34.** $\log_{10} 12.5$

35. $\ln 18.42$ **36.** $\ln \sqrt{42}$

37. $\ln(1 + \sqrt{3})$ **38.** $\ln(\sqrt{5} - 2)$

39. $\ln 0.32$ **40.** $\ln 0.75$

In Exercises 41–44, describe the relationship between the graphs of f and g. What is the relationship between the functions f and g?

41. $f(x) = 3^x$

$g(x) = \log_3 x$

42. $f(x) = 5^x$

$g(x) = \log_5 x$

43. $f(x) = e^x$

$g(x) = \ln x$

44. $f(x) = 10^x$

$g(x) = \log_{10} x$

In Exercises 45–50, use the graph of $y = \log_3 x$ to match the given function with its graph. [The graphs are labeled (a), (b), (c), (d), (e), and (f).]

(a)

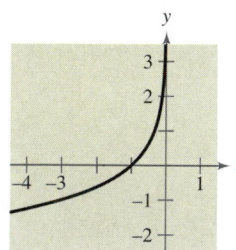

(b)

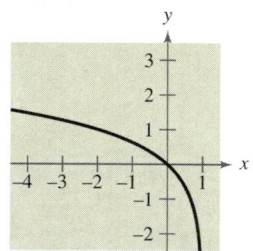

(c)

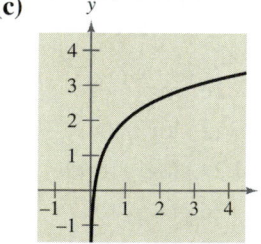

(d)

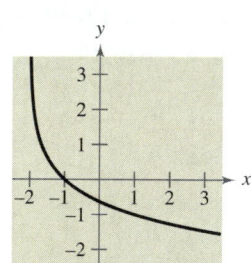

(e)

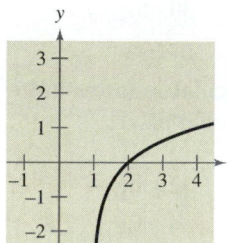

(f)

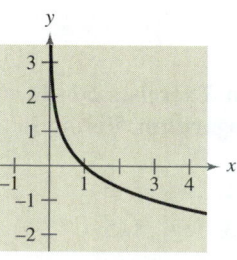

45. $f(x) = \log_3 x + 2$

46. $f(x) = -\log_3 x$

47. $f(x) = -\log_3(x + 2)$

48. $f(x) = \log_3(x - 1)$

49. $f(x) = \log_3(1 - x)$

50. $f(x) = -\log_3(-x)$

In Exercises 51–62, find the domain, vertical asymptote, and x-intercept of the logarithmic function and sketch its graph.

51. $f(x) = \log_4 x$

52. $g(x) = \log_6 x$

53. $y = -\log_3 x + 2$

54. $h(x) = \log_4(x - 3)$

55. $f(x) = -\log_6(x + 2)$

56. $y = \log_5(x - 1) + 4$

57. $y = \log_{10}\left(\dfrac{x}{5}\right)$

58. $y = \log_{10}(-x)$

59. $f(x) = \ln(x - 2)$

60. $h(x) = \ln(x + 1)$

61. $g(x) = \ln(-x)$

62. $f(x) = \ln(3 - x)$

In Exercises 63–66, use a graphing utility to graph the function. Use the graph to determine the intervals in which the function is increasing and decreasing and approximate any relative maximum or minimum values of the function.

63. $f(x) = |\ln x|$

64. $h(x) = \ln(x^2 + 1)$

65. $f(x) = \dfrac{x}{2} - \ln\dfrac{x}{4}$

66. $g(x) = \dfrac{12 \ln x}{x}$

67. *Graphical Analysis*   Use a graphing utility to graph f and g on the same viewing rectangle and determine which is increasing at the greater rate for "large" values of x. What can you conclude about the rate of growth of the natural logarithmic function?

(a) $f(x) = \ln x$, $g(x) = \sqrt{x}$

(b) $f(x) = \ln x$, $g(x) = \sqrt[4]{x}$

68. *Exploration*   The table of values was obtained from evaluating a function. Determine which of the statements may be true and which must be false.

x	1	2	8
y	0	1	3

(a) y is an exponential function of x.

(b) y is a logarithmic function of x.

(c) x is an exponential function of y.

(d) y is a linear function of x.

69. Exploration   Use a graphing utility to compare the graph of the function $y = \ln x$ with the graphs of the following functions.

(a) $y = x - 1$

(b) $y = (x - 1) - \frac{1}{2}(x - 1)^2$

(c) $y = (x - 1) - \frac{1}{2}(x - 1)^2 + \frac{1}{3}(x - 1)^3$

70. Finding a Pattern   Identify the pattern of successive polynomials given in Exercise 69. Extend the pattern one more term and compare the graph of the resulting polynomial function with the graph of $y = \ln x$. What do you think the pattern implies?

71. Human Memory Model   Students in a mathematics class were given an exam and then retested monthly with an equivalent exam. The average scores for the class are given by the human memory model

$$f(t) = 80 - 17 \log_{10}(t + 1), \qquad 0 \le t \le 12$$

where t is the time in months.

(a) What was the average score on the original exam $(t = 0)$?

(b) What was the average score after 4 months?

(c) What was the average score after 10 months?

72. Population Growth   The population of a town will double in

$$t = \frac{10 \ln 2}{\ln 67 - \ln 50} \text{ years.}$$

Find t.

73. World Population Growth   The time t in years for the world population to double if it is increasing at a continuous rate of r is given by

$$t = \frac{\ln 2}{r}.$$

(a) Complete the table.

r	0.005	0.01	0.015	0.02	0.025	0.03
t						

(b) Use a reference source to decide which value of r best approximates the actual rate of growth for the world population.

74. Investment Time   A principal P, invested at $9\frac{1}{2}\%$ and compounded continuously, increases to an amount K times the original principal after t years, where t is given by

$$t = \frac{\ln K}{0.095}.$$

(a) Complete the table and interpret your results.

K	1	2	4	6	8	10	12
t							

(b) Sketch a graph of the function.

Ventilation Rates   In Exercises 75 and 76, use the model

$$y = 80.4 - 11 \ln x, \qquad 100 \le x \le 1500$$

which approximates the minimum required ventilation rate in terms of the air space per child in a public school classroom. In the model, x is the air space per child in cubic feet and y is the ventilation rate in cubic feet per minute.

75. Use a graphing utility to graph the function and approximate the required ventilation rate if there is 300 cubic feet of air space per child.

76. A classroom is designed for 30 students. The air-conditioning system in the room has the capacity of moving 450 cubic feet of air per minute.

(a) Determine the ventilation rate per child, assuming that the room is filled to capacity.

(b) Use the graph of Exercise 75 to estimate the air space required per child.

(c) Determine the minimum number of square feet of floor space required for the room if the ceiling height is 30 feet.

77. Work   The work (in foot-pounds) done in compressing a volume of 9 cubic feet at a pressure of 15 pounds per square inch to a volume of 3 cubic feet is

$$W = 19,440(\ln 9 - \ln 3).$$

Find W.

78. *Sound Intensity*   The relationship between the number of decibels β and the intensity of a sound I in watts per square centimeter is given by

$$\beta = 10 \log_{10}\left(\frac{I}{10^{-16}}\right).$$

(a) Determine the number of decibels of a sound with an intensity of 10^{-4} watts per square centimeter.

(b) Determine the number of decibels of a sound with an intensity of 10^{-6} watts per square centimeter.

(c) The intensity of the sound in part (a) is 100 times as great as that in part (b). Is the number of decibels 100 times as great? Explain.

Monthly Payment   In Exercises 79–82, use the model

$$t = 10.042 \ln\left(\frac{x}{x - 1250}\right), \qquad 1250 < x$$

which approximates the length of a home mortgage of $150,000 at 10% in terms of the monthly payment. In the model, t is the length of the mortgage in years and x is the monthly payment in dollars (see figure).

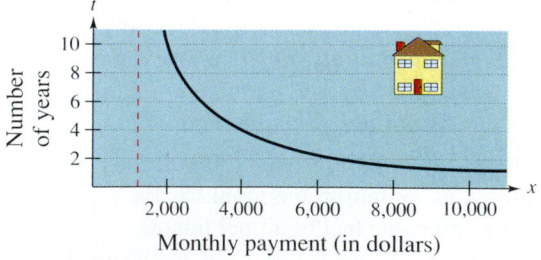

Monthly payment (in dollars)

79. Use the model to approximate the length of the mortgage (for $150,000 at 10%) if the monthly payment is $1316.35.

80. Use the model to approximate the length of the mortgage (for $150,000 at 10%) if the monthly payment is $1982.26.

81. Approximate the total amount paid over the term of the mortgage with a monthly payment of $1316.35. What is the total interest charge?

82. Approximate the total amount paid over the term of the mortgage with a monthly payment of $1982.26. What is the total interest charge?

83. (a) Complete the table for the function

$$f(x) = \frac{\ln x}{x}.$$

x	1	5	10	10^2	10^4	10^6
$f(x)$						

(b) Use the table in part (a) to determine what value $f(x)$ approaches as x increases without bound.

(c) Use a graphing utility to confirm the result of part (b).

84. *Exploration*   Answer the following questions for the function $f(x) = \log_{10} x$. Do not use a calculator.

(a) What is the domain of f?

(b) What is f^{-1}?

(c) If x is a real number between 1000 and 10,000, in which interval will $f(x)$ be found?

(d) In which interval will x be found if $f(x)$ is negative?

(e) If $f(x)$ is increased by one unit, x must have been increased by what factor?

(f) If $f(x_1) = 3n$ and $f(x_2) = n$, what is the ratio of x_1 to x_2?

Review   **Solve Exercises 85–88 as a review of the skills and problem-solving techniques you learned in previous sections. Translate the statement into an algebraic expression.**

85. The product of 8 and n is decreased by 3.

86. The total hourly wage for an employee is $9.25 per hour plus 75 cents for each of q units produced per hour.

87. The total cost for auto repairs if the cost of parts was $83.95 and there were t hours of labor at $37.50 per hour

88. The area of a rectangle if the length is 10 units more than the width w

5.3 *Properties of Logarithms*

See Exercise 89 on page 423 for an example of how a logarithmic function can be used as a human memory model.

Change of Base □ *Properties of Logarithms* □ *Rewriting Logarithmic Expressions* □ *Application*

Change of Base

Most calculators have only two types of log keys, one for common logarithms (base 10) and one for natural logarithms (base e). Although common logs and natural logs are the most frequently used, you may occasionally need to evaluate logarithms to other bases. To do this, you can use the following *change-of-base formula*.

> ### Change-of-Base Formula
>
> Let a, b, and x be positive real numbers such that $a \neq 1$ and $b \neq 1$. Then $\log_a x$ is given by
>
> $$\log_a x = \frac{\log_b x}{\log_b a}.$$

One way to look at the change-of-base formula is that logarithms to base a are simply *constant multiples* of logarithms to base b. The constant multiplier is $1/(\log_b a)$.

John Napier, a Scottish mathematician, developed logarithms as a way to simplify some of the tedious calculations of his day. Beginning in 1594, Napier worked about 20 years on the invention of logarithms. Napier was only partially successful in his quest to simplify tedious calculations. Nonetheless, the development of logarithms was a step forward and received immediate recognition.

EXAMPLE 1 *Changing Bases Using Common Logarithms*

a. $\log_4 30 = \dfrac{\log_{10} 30}{\log_{10} 4} \approx \dfrac{1.47712}{0.60206} \approx 2.4534$

b. $\log_2 14 = \dfrac{\log_{10} 14}{\log_{10} 2} \approx \dfrac{1.14613}{0.30103} \approx 3.8074$

EXAMPLE 2 *Changing Bases Using Natural Logarithms*

a. $\log_4 30 = \dfrac{\ln 30}{\ln 4} \approx \dfrac{3.40120}{1.38629} \approx 2.4534$

b. $\log_2 14 = \dfrac{\ln 14}{\ln 2} \approx \dfrac{2.63906}{0.693147} \approx 3.8074$

Properties of Logarithms

You know from the previous section that the logarithmic function with base a is the *inverse* of the exponential function with base a. Thus, it makes sense that the properties of exponents should have corresponding properties involving logarithms. For instance, the exponential property $a^0 = 1$ has the corresponding logarithmic property $\log_a 1 = 0$.

NOTE There is no general property that can be used to rewrite $\log_a(u \pm v)$. Specifically, $\log_a(x + y)$ is not equal to $\log_a x + \log_a y$. ■■

PROPERTIES OF LOGARITHMS

Let a be a positive number such that $a \neq 1$, and let n be a real number. If u and v are positive real numbers, the following properties are true.

1. $\log_a(uv) = \log_a u + \log_a v$ **1.** $\ln(uv) = \ln u + \ln v$

2. $\log_a \dfrac{u}{v} = \log_a u - \log_a v$ **2.** $\ln \dfrac{u}{v} = \ln u - \ln v$

3. $\log_a u^n = n \log_a u$ **3.** $\ln u^n = n \ln u$

EXAMPLE 3   *Using Properties of Logarithms*

Write the logarithm in terms of $\ln 2$ and $\ln 3$.

a. $\ln 6$ **b.** $\ln \dfrac{2}{27}$

Solution

a. $\ln 6 = \ln(2 \cdot 3)$ Rewrite 6 as $2 \cdot 3$.

$\qquad = \ln 2 + \ln 3$ Property 1

b. $\ln \dfrac{2}{27} = \ln 2 - \ln 27$ Property 2

$\qquad\qquad = \ln 2 - \ln 3^3$ Rewrite 27 as 3^3.

$\qquad\qquad = \ln 2 - 3 \ln 3$ Property 3

THINK ABOUT THE PROOF

To prove Property 1, let $x = \log_a u$ and $y = \log_a v$. The corresponding exponential forms of these two equations are $a^x = u$ and $a^y = v$. Multiplying u and v produces

$$uv = a^x a^y = a^{x+y}.$$

Can you see how to use this equation to complete the proof? The details of the proof are listed in the appendix.

EXAMPLE 4   *Using Properties of Logarithms*

Use the properties of logarithms to verify that $-\ln \frac{1}{2} = \ln 2$.

Solution

$$-\ln \tfrac{1}{2} = -\ln(2^{-1}) = -(-1) \ln 2 = \ln 2$$

Try checking this result on your calculator.

Rewriting Logarithmic Expressions

The properties of logarithms are useful for rewriting logarithmic expressions in forms that simplify the operations of algebra. This is true because they convert complicated products, quotients, and exponential forms into simpler sums, differences, and products, respectively.

EXAMPLE 5 *Rewriting the Logarithm of a Product*

$$\log_{10} 5x^3y = \log_{10} 5 + \log_{10} x^3y$$
$$= \log_{10} 5 + \log_{10} x^3 + \log_{10} y$$
$$= \log_{10} 5 + 3\log_{10} x + \log_{10} y$$

EXAMPLE 6 *Rewriting the Logarithm of a Quotient*

$$\ln \frac{\sqrt{3x-5}}{7} = \ln(3x-5)^{1/2} - \ln 7$$
$$= \frac{1}{2}\ln(3x-5) - \ln 7$$

In Examples 5 and 6, the properties of logarithms were used to *expand* logarithmic expressions. In Examples 7 and 8, this procedure is reversed and the properties of logarithms are used to *condense* logarithmic expressions.

EXAMPLE 7 *Condensing a Logarithmic Expression*

$$\tfrac{1}{2}\log_{10}x + 3\log_{10}(x+1) = \log_{10}x^{1/2} + \log_{10}(x+1)^3$$
$$= \log_{10}\left[\sqrt{x}\cdot(x+1)^3\right]$$

EXAMPLE 8 *Condensing a Logarithmic Expression*

$$2\ln(x+2) - \ln x = \ln(x+2)^2 - \ln x$$
$$= \ln\frac{(x+2)^2}{x}$$

Exploration

Use a graphing utility to graph the functions

$$y = \ln x - \ln(x-3)$$

and

$$y = \ln\frac{x}{x-3}$$

on the same viewing rectangle. Does the graphing utility show the functions with the same domain? If so, should it? Explain your reasoning.

Application

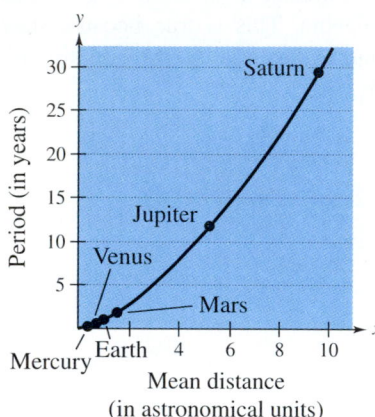

FIGURE 5.15

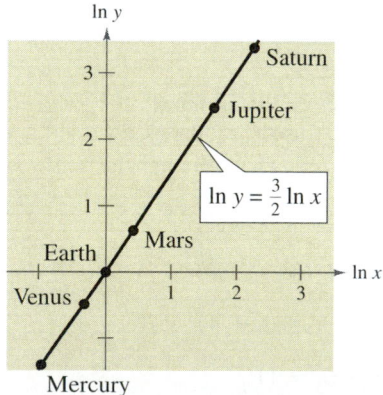

FIGURE 5.16

NOTE This Group Activity is related to the one on page 22 in Section P.2. If you didn't work on that activity, try both activities now. Which approach do you prefer? Explain your reasoning. ■■

EXAMPLE 9   *Finding a Mathematical Model*

Real Life

The table gives the mean distance x and the period y of the six planets that are closest to the sun. In the table, the mean distance is given in terms of astronomical units (where the earth's mean distance is defined as 1.0), and the period is given in terms of years. Find an equation that expresses y as a function of x.

Planet	Mercury	Venus	Earth	Mars	Jupiter	Saturn
Period, y	0.241	0.615	1.0	1.881	11.861	29.457
Mean Distance, x	0.387	0.723	1.0	1.523	5.203	9.541

Solution

The points in the table are plotted in Figure 5.15. From this figure it is not clear how to find an equation that relates y and x. To solve this problem, take the natural log of each of the x- and y-values given in the table. This produces the following results.

Planet	Mercury	Venus	Earth	Mars	Jupiter	Saturn
ln y	−1.423	−0.486	0.0	0.632	2.473	3.383
ln x	−0.949	−0.324	0.0	0.421	1.649	2.256

Now, by plotting the points in the second table, you can see that all six of the points appear to lie in a line (see Figure 5.16). You can use a graphical approach or the algebraic approach discussed in Section 3.6 to find that the slope of this line is $\frac{3}{2}$, and you can therefore conclude that $\ln y = \frac{3}{2} \ln x$. [Try to convert this to $y = f(x)$ form.]

GROUP ACTIVITY

KEPLER'S LAW

The relationship described in Example 9 was first discovered by Johannes Kepler. Use properties of logarithms to rewrite the relationship so that y is expressed as a function of x.

WARM UP

WARM UP

Evaluate the expression without using a calculator.

1. $\log_7 49$

2. $\log_2 \frac{1}{32}$

3. $\ln \frac{1}{e^2}$

4. $\log_1 0.001$

Simplify the expression.

5. $e^2 e^3$

6. $\dfrac{e^2}{e^3}$

7. $(e^2)^3$

8. $(e^2)^0$

Rewrite the expression in exponential form.

9. $\dfrac{1}{x^2}$

10. $\sqrt{x}$

5.3 Exercises

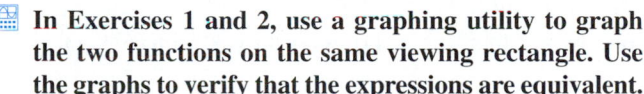

In Exercises 1 and 2, use a graphing utility to graph the two functions on the same viewing rectangle. Use the graphs to verify that the expressions are equivalent.

1. $f(x) = \log_{10} x$

  $g(x) = \dfrac{\ln x}{\ln 10}$

2. $f(x) = \ln x$

  $g(x) = \dfrac{\log_{10} x}{\log_{10} e}$

In Exercises 3–6, rewrite the logarithm as a multiple of a common logarithm.

3. $\log_3 5$

4. $\log_4 10$

5. $\log_2 x$

6. $\ln 5$

In Exercises 7–10, rewrite the logarithm as a multiple of a natural logarithm.

7. $\log_3 5$

8. $\log_4 10$

9. $\log_2 x$

10. $\log_{10} 5$

In Exercises 11–18, evaluate the logarithm using the change-of-base formula. Round your result to three decimal places.

11. $\log_3 7$

12. $\log_7 4$

13. $\log_{1/2} 4$

14. $\log_4 0.55$

15. $\log_9 0.4$

16. $\log_{20} 125$

17. $\log_{15} 1250$

18. $\log_{1/3} 0.015$

In Exercises 19–38, use the properties of logarithms to write the expression as a sum, difference, and/or constant multiple of logarithms. (Assume all variables are positive.)

19. $\log_{10} 5x$

20. $\log_{10} 10z$

21. $\log_{10} \dfrac{5}{x}$

22. $\log_{10} \dfrac{y}{2}$

23. $\log_8 x^4$

24. $\log_6 z^{-3}$

25. $\ln \sqrt{z}$

26. $\ln \sqrt[3]{t}$

27. $\ln xyz$

28. $\ln \dfrac{xy}{z}$

29. $\ln \sqrt{a-1}, \ a>1$

30. $\ln\left(\dfrac{x^2-1}{x^3}\right), \ x>1$

31. $\ln z(z-1)^2, \ z>1$

32. $\ln \sqrt{\dfrac{x^2}{y^3}}$

33. $\ln \sqrt[3]{\dfrac{x}{y}}$

34. $\ln \dfrac{x}{\sqrt{x^2+1}}$

35. $\ln \dfrac{x^4\sqrt{y}}{z^5}$

36. $\ln \sqrt{x^2(x+2)}$

37. $\log_b \dfrac{x^2}{y^2 z^3}$

38. $\log_b \dfrac{\sqrt{x}y^4}{z^4}$

Graphical Analysis In Exercises 39 and 40, use a graphing utility to graph the two equations on the same viewing rectangle. Use the graphs to verify that the expressions are equivalent.

39. $y_1 = \ln[x^3(x+4)]$
$y_2 = 3\ln x + \ln(x+4)$

40. $y_1 = \ln\left(\dfrac{\sqrt{x}}{x-2}\right)$
$y_2 = \tfrac{1}{2}\ln x - \ln(x-2)$

In Exercises 41–60, write the expression as the logarithm of a single quantity.

41. $\ln x + \ln 2$

42. $\ln y + \ln z$

43. $\log_4 z - \log_4 y$

44. $\log_5 8 - \log_5 t$

45. $2\log_2(x+4)$

46. $-4\log_6 2x$

47. $\tfrac{1}{3}\log_3 5x$

48. $\tfrac{3}{2}\log_7(z-2)$

49. $\ln x - 3\ln(x+1)$

50. $2\ln 8 + 5\ln z$

51. $\ln(x-2) - \ln(x+2)$

52. $3\ln x + 2\ln y - 4\ln z$

53. $\ln x - 2[\ln(x+2) + \ln(x-2)]$

54. $4[\ln z + \ln(z+5)] - 2\ln(z-5)$

55. $\tfrac{1}{3}[2\ln(x+3) + \ln x - \ln(x^2-1)]$

56. $2[\ln x - \ln(x+1) - \ln(x-1)]$

57. $\tfrac{1}{3}[\ln y + 2\ln(y+4)] - \ln(y-1)$

58. $\tfrac{1}{2}[\ln(x+1) + 2\ln(x-1)] + 3\ln x$

59. $2\ln 3 - \tfrac{1}{2}\ln(x^2+1)$

60. $\tfrac{3}{2}\ln 5t^6 - \tfrac{3}{4}\ln t^4$

Graphical Analysis In Exercises 61 and 62, use a graphing utility to graph the two equations on the same viewing rectangle. Use the graphs to verify that the expressions are equivalent.

61. $y_1 = 2[\ln 8 - \ln(x^2+1)], \quad y_2 = \ln\left[\dfrac{64}{(x^2+1)^2}\right]$

62. $y_1 = \ln x + \tfrac{1}{3}\ln(x+1), \quad y_2 = \ln(x\sqrt[3]{x+1})$

Think About It In Exercises 63 and 64, use a graphing utility to graph the two equations on the same viewing rectangle. Are the expressions equivalent? Explain.

63. $y_1 = \ln x^2, \ y_2 = 2\ln x$

64. $y_1 = \tfrac{1}{4}\ln[x^4(x^2+1)], \ y_2 = \ln x + \tfrac{1}{4}\ln(x^2+1)$

Comparing Logarithmic Quantities In Exercises 65 and 66, compare the logarithmic quantities. If two are equal, explain why.

65. $\dfrac{\log_2 32}{\log_2 4}, \quad \log_2 \dfrac{32}{4}, \quad \log_2 32 - \log_2 4$

66. $\log_7 \sqrt{70}, \quad \log_7 35, \quad \tfrac{1}{2} + \log_7 \sqrt{10}$

67. *Think About It* Graph

$$f(x) = \ln \dfrac{x}{2}, \quad g(x) = \dfrac{\ln x}{\ln 2}, \quad h(x) = \ln x - \ln 2$$

on the same set of axes. Which two functions have identical graphs? Explain why.

68. *Exploration* Approximate the natural logarithms of as many integers as possible between 1 and 20 given that $\ln 2 \approx 0.6931$, $\ln 3 \approx 1.0986$, and $\ln 5 \approx 1.6094$. (Do not use a calculator.)

In Exercises 69–82, find the exact value of the logarithm without using a calculator. (If this is not possible, state the reason.)

69. $\log_3 9$

70. $\log_6 \sqrt[3]{6}$

71. $\log_4 16^{1.2}$

72. $\log_5 \frac{1}{125}$

73. $\log_3 (-9)$

74. $\log_2 (-16)$

75. $\log_5 75 - \log_5 3$

76. $\log_4 2 + \log_4 32$

77. $\ln e^2 - \ln e^5$

78. $3 \ln e^4$

79. $\log_{10} 0$

80. $\ln 1$

81. $\ln e^{4.5}$

82. $\ln \sqrt[4]{e^3}$

In Exercises 83–88, use the properties of logarithms to simplify the logarithmic expression.

83. $\log_4 8$

84. $\log_2(4^2 \cdot 3^4)$

85. $\log_5 \frac{1}{250}$

86. $\log_{10} \frac{9}{300}$

87. $\ln(5e^6)$

88. $\ln \frac{6}{e^2}$

89. *Human Memory Model*   Students participating in a psychological experiment attended several lectures and were given an exam. Every month for a year after the exam, the students were retested to see how much of the material they remembered. The average scores for the group are given by the memory model

$$f(t) = 90 - 15 \log_{10}(t + 1), \qquad 0 \le t \le 12$$

where t is the time in months.

(a) What was the average score on the original exam $(t = 0)$?

(b) What was the average score after 6 months?

(c) What was the average score after 12 months?

(d) When will the average score decrease to 75?

(e) Use the properties of logarithms to write the function in another form.

(f) Sketch the graph of the function over the specified domain.

90. *Sound Intensity*   The relationship between the number of decibels β and the intensity of a sound I in watts per square centimeter is given by

$$\beta = 10 \log_{10}\left(\frac{I}{10^{-16}}\right).$$

Use the properties of logarithms to write the formula in simpler form, and determine the number of decibels of a sound with an intensity of 10^{-10} watts per square centimeter.

True or False?   **In Exercises 91–96, determine if the statement is true or false given that $f(x) = \ln x$. If the statement is false, state why or give an example to show that it is false.**

91. $f(0) = 0$

92. $f(ax) = f(a) + f(x), \qquad a > 0, x > 0$

93. $f(x - 2) = f(x) - f(2), \qquad x > 2$

94. $\sqrt{f(x)} = \frac{1}{2}f(x)$

95. If $f(u) = 2f(v)$, then $v = u^2$.

96. If $f(x) < 0$, then $0 < x < 1$.

97. Prove that $\log_b \dfrac{u}{v} = \log_b u - \log_b v$.

98. Prove that $\log_b u^n = n \log_b u$.

Review   **Solve Exercises 99–102 as a review of the skills and problem-solving techniques you learned in previous sections. Simplify the expressions.**

99. $\dfrac{24xy^{-2}}{16x^{-3}y}$

100. $\left(\dfrac{2x^2}{3y}\right)^{-3}$

101. $(18x^3y^4)^{-3}(18x^3y^4)^3$

102. $xy(x^{-1} + y^{-1})^{-1}$

5.4 *Exponential and Logarithmic Equations*

See Exercise 94 on page 434 for an example of how a logarithmic function can be used to model crumple zones for automobile crash tests.

Introduction ◻ *Solving Exponential Equations* ◻
Solving Logarithmic Equations ◻ *Applications*

Introduction

So far in this chapter, you have studied the definitions, graphs, and properties of exponential and logarithmic functions. In this section, you will study procedures for *solving equations* involving exponential and logarithmic functions. As a simple example, consider the exponential equation $2^x = 32$. One property of exponential functions states that $a^x = a^y$ if and only if $x = y$. You can solve $2^x = 32$ by rewriting the equation in the form $2^x = 2^5$, which implies that $x = 5$.

Although this method works in some cases, it does not work for an equation as simple as $e^x = 7$. In such a case, solution procedures are based on the fact that the exponential and logarithmic functions are inverses of each other. The following properties are the **inverse properties** of exponential and logarithmic functions.

	Base a	*Base e*
1.	$\log_a a^x = x$	$\ln e^x = x$
2.	$a^{\log_a x} = x$	$e^{\ln x} = x$

Now, to solve $e^x = 7$, you can take the natural logarithms of both sides to obtain

$e^x = 7$	Original equation
$\ln e^x = \ln 7$	Take logarithms of both sides.
$x = \ln 7$	$\ln e^x = x$ because $e^x = e^x$.

Here are some guidelines for solving exponential and logarithmic equations.

SOLVING EXPONENTIAL AND LOGARITHMIC EQUATIONS

1. *To solve an exponential equation,* first isolate the exponential expression, then take the logarithms of both sides and solve for the variable.

2. *To solve a logarithmic equation,* rewrite the equation in exponential form and solve for the variable.

Solving Exponential Equations

EXAMPLE 1   Solving an Exponential Equation

Solve $e^x = 72$.

Solution

$e^x = 72$	Original equation
$\ln e^x = \ln 72$	Take logarithms of both sides.
$x = \ln 72$	Inverse property of logs and exponents
$x \approx 4.277$	Use a calculator.

The solution is $\ln 72$. Check this in the original equation.

EXAMPLE 2   Solving an Exponential Equation

Solve $e^x + 5 = 60$.

Solution

$e^x + 5 = 60$	Original equation
$e^x = 55$	Subtract 5 from both sides.
$\ln e^x = \ln 55$	Take logarithms of both sides.
$x = \ln 55$	Inverse property of logs and exponents
$x \approx 4.007$	Use a calculator.

The solution is $\ln 55$. Check this in the original equation.

TECHNOLOGY

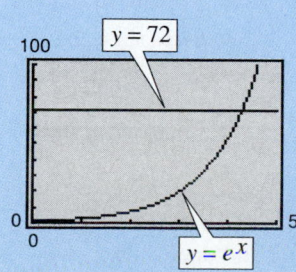

When solving an exponential or logarithmic equation, remember that you can check your solution graphically by "graphing the left and right sides separately" and estimating the x-coordinate of the point of intersection. For instance, to check the solution of the equation in Example 1, you can sketch the graphs of

$$y = e^x \quad \text{and} \quad y = 72$$

on the same viewing rectangle, as shown at the left. Notice that the graphs intersect when $x \approx 4.277$, which confirms the solution found in Example 1.

EXAMPLE 3   *Solving an Exponential Equation*

Solve $4e^{2x} = 5$.

Solution

$4e^{2x} = 5$	Original equation
$e^{2x} = \dfrac{5}{4}$	Divide both sides by 4.
$\ln e^{2x} = \ln \dfrac{5}{4}$	Take logarithms of both sides.
$2x = \ln \dfrac{5}{4}$	Inverse property of logs and exponents
$x = \dfrac{1}{2} \ln \dfrac{5}{4}$	Divide both sides by 2.
$x \approx 0.112$	Use a calculator.

The solution is $\frac{1}{2} \ln \frac{5}{4}$. Check this in the original equation.

When an equation involves two or more exponential expressions, you can still use a procedure similar to that demonstrated in the first three examples. However, the algebra is a bit more complicated.

EXAMPLE 4   *Solving an Exponential Equation*

Solve $e^{2x} - 3e^x + 2 = 0$.

Solution

$e^{2x} - 3e^x + 2 = 0$		Original equation
$(e^x)^2 - 3e^x + 2 = 0$		Quadratic form
$(e^x - 2)(e^x - 1) = 0$		Factor.
$e^x - 2 = 0$ ➡ $x = \ln 2$		Set 1st factor equal to 0.
$e^x - 1 = 0$ ➡ $x = 0$		Set 2nd factor equal to 0.

The solutions are $\ln 2$ and 0. Check these in the original equation.

NOTE In Example 4, use a graphing utility to graph $y = e^{2x} - 3e^x + 2$. It should have two x-intercepts: one when $x = \ln 2$ and one when $x = 0$. ▪▪

Solving Logarithmic Equations

To solve a logarithmic equation such as

$$\ln x = 3 \qquad \text{Logarithmic form}$$

write the equation in exponential form as follows.

$$e^{\ln x} = e^3 \qquad \text{Exponentiate both sides.}$$
$$x = e^3 \qquad \text{Exponential form}$$

This procedure is called *exponentiating* both sides of an equation.

EXAMPLE 5   *Solving a Logarithmic Equation*

Solve $\ln x = 2$.

Solution

$$\ln x = 2 \qquad \text{Original equation}$$
$$e^{\ln x} = e^2 \qquad \text{Exponentiate both sides.}$$
$$x = e^2 \qquad \text{Inverse property of exponents and logs}$$
$$x \approx 7.389 \qquad \text{Use a calculator.}$$

The solution is e^2. Check this in the original equation.

EXAMPLE 6   *Solving a Logarithmic Equation*

Solve $5 + 2 \ln x = 4$.

Solution

$$5 + 2 \ln x = 4 \qquad \text{Original equation}$$
$$2 \ln x = -1 \qquad \text{Subtract 5 from both sides.}$$
$$\ln x = -\frac{1}{2} \qquad \text{Divide both sides by 2.}$$
$$e^{\ln x} = e^{-1/2} \qquad \text{Exponentiate both sides.}$$
$$x = e^{-1/2} \qquad \text{Inverse property of exponents and logs}$$
$$x \approx 0.607 \qquad \text{Use a calculator.}$$

The solution is $e^{-1/2}$. To check this result graphically, you can sketch the graphs of $y = 5 + 2 \ln x$ and $y = 4$ in the same coordinate plane, as shown in Figure 5.17.

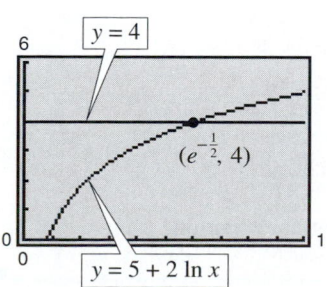

$y = 4$

$\left(e^{-\frac{1}{2}}, 4\right)$

$y = 5 + 2 \ln x$

FIGURE 5.17

EXAMPLE 7   *Solving a Logarithmic Equation*

Solve $2 \ln 3x = 4$.

Solution

$2 \ln 3x = 4$	Original equation
$\ln 3x = 2$	Divide both sides by 2.
$e^{\ln 3x} = e^2$	Exponentiate both sides.
$3x = e^2$	Inverse property of exponents and logs
$x = \dfrac{1}{3}e^2$	Divide both sides by 3.
$x \approx 2.463$	Use a calculator.

The solution is $\frac{1}{3}e^2$. Check this in the original equation.

EXAMPLE 8   *Solving a Logarithmic Equation*

Solve $\ln x - \ln(x - 1) = 1$.

Solution

$\ln x - \ln(x - 1) = 1$	Original equation
$\ln \dfrac{x}{x - 1} = 1$	Property 2 of logarithms
$\dfrac{x}{x - 1} = e^1$	Exponentiate both sides.
$x = ex - e$	Multiply both sides by $x - 1$.
$x - ex = -e$	Subtract ex from both sides.
$x(1 - e) = -e$	Factor.
$x = \dfrac{-e}{1 - e}$	Divide both sides by $1 - e$.
$x = \dfrac{e}{e - 1}$	Simplify.

The solution is $e/(e - 1)$. Check this in the original equation.

In solving exponential or logarithmic equations, the following properties are useful. Can you see where these properties were used in this section?

1. $x = y$ if and only if $\log_a x = \log_a y$.
2. $x = y$ if and only if $a^x = a^y$, $a > 0$, $a \neq 1$.

Applications

EXAMPLE 9 Doubling an Investment

Real Life

You have deposited $500 in an account that pays 6.75% interest, compounded continuously. How long will it take your money to double?

Solution

Using the formula for continuous compounding, you can find that the balance in the account is given by

$$A = Pe^{rt} = 500e^{0.0675t}.$$

To find the time required for the balance to double, let $A = 1000$, and solve the resulting equation for t.

$500e^{0.0675t} = 1000$	Let $A = 1000$.
$e^{0.0675t} = 2$	Divide both sides by 500.
$\ln e^{0.0675t} = \ln 2$	Take logarithms of both sides.
$0.0675t = \ln 2$	Inverse property of logs and exponents
$t = \dfrac{\ln 2}{0.0675}$	Divide both sides by 0.0675.
$t \approx 10.27$	Use a calculator.

The balance in the account will double after approximately 10.27 years. This result is graphically demonstrated in Figure 5.18.

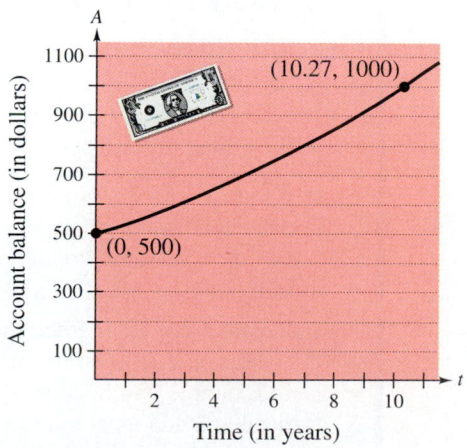

FIGURE 5.18

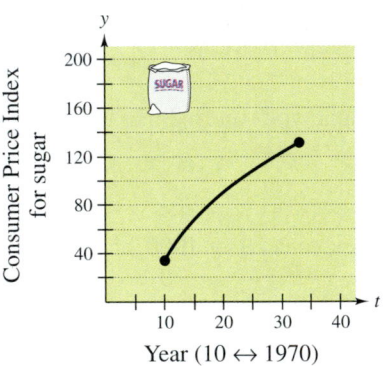

Consumer Price Index for sugar

Year (10 ↔ 1970)

FIGURE 5.19

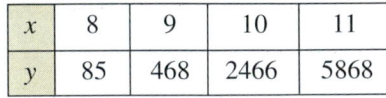

x	8	9	10	11
y	85	468	2466	5868

x	12	13	14	15
y	12,153	24,016	45,283	60,399

Real Life

EXAMPLE 10   Consumer Price Index for Sugar

From 1970 to 1993, the Consumer Price Index (CPI) value y for a fixed amount of sugar for the year t can be modeled by the equation

$$y = -169.8 + 86.8 \ln t$$

where $t = 10$ represents 1970 (see Figure 5.19). During which year did the price of sugar reach 4 times its 1970 price of 30.5 on the CPI?   (Source: U.S. Bureau of Labor Statistics)

Solution

$-169.8 + 86.8 \ln t = y$	Given model
$-169.8 + 86.8 \ln t = 122$	Let $y = (4)(30.5) = 122$.
$86.8 \ln t = 291.8$	Add 169.8 to both sides.
$\ln t \approx 3.362$	Divide both sides by 86.8.
$e^{\ln t} \approx e^{3.362}$	Exponentiate both sides.
$t \approx 29$	Inverse property of exponents and logs

The solution is $t \approx 29$ years. Because $t = 10$ represents 1970, it follows that the price of sugar reached 4 times its 1970 price in 1988.

GROUP ACTIVITY

COMPARING MATHEMATICAL MODELS

The table gives the monthly amount of traffic y (in millions of packets) on the NSFNET, the backbone of the Internet, for January of each year x from 1988 through 1995, where $x = 8$ represents 1988.   (Source: Network Information Center)

(a) Create a scatter plot of the data. Find a linear model for the data, and add its graph to your scatter plot. According to this model, when will monthly network traffic reach 150,000 millions of packets?

(b) Create a new table giving values for $\ln x$ and $\ln y$ and create a scatter plot of this transformed data. Use the method illustrated in Example 9 in Section 5.3 to find a model for the transformed data, and add its graph to your scatter plot. According to this model, when will monthly network traffic reach 150,000 millions of packets?

(c) Solve the model in part (b) for y, and add its graph to your scatter plot in part (a). Which model better fits the original data? Which model will better predict future traffic levels? Explain.

5.4 Exercises

In Exercises 1–6, determine whether the x-values are solutions of the equation.

1. $4^{2x-7} = 64$

  (a) $x = 5$

  (b) $x = 2$

2. $2^{3x+1} = 32$

  (a) $x = -1$

  (b) $x = 2$

3. $3e^{x+2} = 75$

  (a) $x = -2 + e^{25}$

  (b) $x = -2 + \ln 25$

  (c) $x \approx 1.2189$

4. $5^{2x+3} = 812$

  (a) $x = -1.5 + \log_5 \sqrt{812}$

  (b) $x \approx 0.5813$

  (c) $x = \dfrac{1}{2}\left(-3 + \dfrac{\ln 812}{\ln 5}\right)$

5. $\log_4(3x) = 3$

  (a) $x \approx 20.3560$

  (b) $x = -4$

  (c) $x = \frac{64}{3}$

6. $\ln(x - 1) = 3.8$

  (a) $x = 1 + e^{3.8}$

  (b) $x \approx 45.7012$

  (c) $x = 1 + \ln 3.8$

In Exercises 7–10, approximate the point of intersection of the graphs of f and g. Then solve the equation $f(x) = g(x)$ algebraically.

7. $f(x) = 2^x$

  $g(x) = 8$

8. $f(x) = 27^x$

  $g(x) = 9$

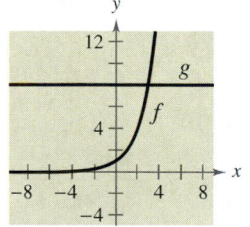
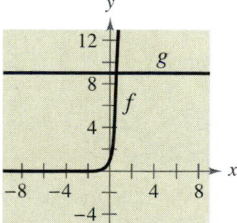

9. $f(x) = \log_3 x$

  $g(x) = 2$

10. $f(x) = \ln(x - 4)$

  $g(x) = 0$

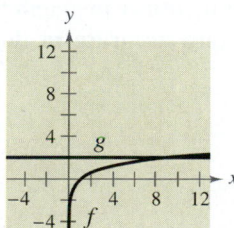
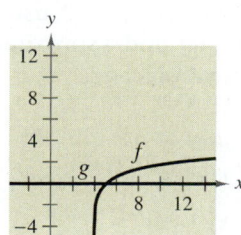

In Exercises 11–20, solve for x.

11. $4^x = 16$ 12. $3^x = 243$

13. $7^x = \frac{1}{49}$ 14. $8^x = 4$

15. $\left(\frac{3}{4}\right)^x = \frac{27}{64}$ 16. $3^{x-1} = 27$

17. $\log_4 x = 3$ 18. $\log_x 625 = 4$

19. $\log_{10} x = -1$ 20. $\ln(2x - 1) = 0$

In Exercises 21–26, apply the inverse properties of $\ln x$ and e^x to simplify the expression.

21. $\log_{10} 10^{x^2}$ 22. $\log_6 6^{2x-1}$

23. $e^{\ln(5x+2)}$ 24. $-1 + \ln e^{2x}$

25. $e^{\ln x^2}$ 26. $-8 + e^{\ln x^3}$

In Exercises 27–46, solve the exponential equation algebraically. Round the result to three decimal places.

27. $e^x = 10$ 28. $4e^x = 91$

29. $7 - 2e^x = 5$ 30. $-14 + 3e^x = 11$

31. $e^{3x} = 12$ 32. $e^{2x} = 50$

33. $500e^{-x} = 300$ 34. $1000e^{-4x} = 75$

35. $e^{2x} - 4e^x - 5 = 0$ 36. $e^{2x} - 5e^x + 6 = 0$

37. $20(100 - e^{x/2}) = 500$ 38. $\dfrac{400}{1 + e^{-x}} = 350$

39. $10^x = 42$ 40. $10^x = 570$

41. $3^{2x} = 80$ 42. $6^{5x} = 3000$

43. $5^{-t/2} = 0.20$ 44. $4^{-3t} = 0.10$

45. $2^{3-x} = 565$ 46. $\left(1 + \dfrac{0.10}{12}\right)^{12t} = 2$

In Exercises 47–50, use a graphing utility to graph the function and approximate its zero accurate to three decimal places.

47. $g(x) = 6e^{1-x} - 25$

48. $f(x) = 3e^{3x/2} - 962$

49. $g(t) = e^{0.09t} - 3$

50. $h(t) = e^{0.125t} - 8$

In Exercises 51–54, solve the exponential equation. Round the result to three decimal places.

51. $8(10^{3x}) = 12$

52. $3(5^{x-1}) = 21$

53. $\left(1 + \dfrac{0.065}{365}\right)^{365t} = 4$

54. $\dfrac{3000}{2 + e^{2x}} = 2$

In Exercises 55–70, solve the logarithmic equation algebraically. Round the result to three decimal places.

55. $\ln x = -3$

56. $\ln x = 2$

57. $\ln 2x = 2.4$

58. $3\ln 5x = 10$

59. $\ln\sqrt{x + 2} = 1$

60. $\ln(x + 1)^2 = 2$

61. $\log_{10}(z - 3) = 2$

62. $\log_{10} x^2 = 6$

63. $\ln x + \ln(x - 2) = 1$

64. $\ln x + \ln(x + 3) = 1$

65. $\log_{10}(x + 4) - \log_{10} x = \log_{10}(x + 2)$

66. $\log_4 x - \log_4(x - 1) = \frac{1}{2}$

67. $\log_3 x + \log_3(x^2 - 8) = \log_3 8x$

68. $\log_2 x + \log_2(x + 2) = \log_2(x + 6)$

69. $\ln(x + 5) = \ln(x - 1) - \ln(x + 1)$

70. $\ln(x + 1) - \ln(x - 2) = \ln x^2$

In Exercises 71–76, solve the logarithmic equation. Round the result to three decimal places.

71. $6\log_3(0.5x) = 11$

72. $5\log_{10}(x - 2) = 11$

73. $2\ln x = 7$

74. $\ln 4x = 1$

75. $\ln x + \ln(x^2 + 1) = 8$

76. $\log_{10} 8x - \log_{10}\left(1 + \sqrt{x}\right) = 2$

In Exercises 77–80, use a graphing utility to approximate the point of intersection of the graphs. Round the result to three decimal places.

77. $y_1 = 7$
$y_2 = 2^x$

78. $y_1 = 500$
$y_2 = 1500e^{-x/2}$

79. $y_1 = 3$
$y_2 = \ln x$

80. $y_1 = 10$
$y_2 = 4\ln(x - 2)$

Compound Interest **In Exercises 81 and 82, find the time required for a $1000 investment to double at interest rate r, compounded continuously.**

81. $r = 0.085$ **82.** $r = 0.12$

83. Think About It Are the times required for the investments of Exercises 81 and 82 to quadruple twice as long as the times for them to double? Give a reason for your answer and verify your answer algebraically.

84. Essay Write a paragraph explaining whether or not the time required for an investment to double is dependent on the size of the investment.

Compound Interest **In Exercises 85 and 86, find the time required for a $1000 investment to triple at interest rate r, compounded continuously.**

85. $r = 0.085$ **86.** $r = 0.12$

87. Demand Function The demand equation for a certain product is given by

$$p = 500 - 0.5(e^{0.004x}).$$

Find the demand x for a price of (a) $p = \$350$ and (b) $p = \$300$.

88. Demand Function The demand equation for a certain product is given by

$$p = 5000\left(1 - \frac{4}{4 + e^{-0.002x}}\right).$$

Find the demand x for a price of (a) $p = \$600$ and (b) $p = \$400$.

89. Forest Yield The yield V (in millions of cubic feet per acre) for a forest at age t years is given by

$$V = 6.7e^{-48.1/t}.$$

(a) Use a graphing utility to graph the function.

(b) Determine the horizontal asymptote of the function. Interpret its meaning in the context of the problem.

(c) Find the time necessary to obtain a yield of 1.3 million cubic feet.

90. Trees per Acre The number of trees per acre N of a certain species is approximated by the model

$$N = 68(10^{-0.04x}), \qquad 5 \le x \le 40$$

where x is the average diameter of the trees 3 feet above the ground. Use the model to approximate the average diameter of the trees in a test plot when $N = 21$.

91. Average Heights The percent of American males between the ages of 18 and 24 who are at least x inches tall is given by

$$m(x) = \frac{100}{1 + e^{-0.6114(x - 69.71)}}.$$

The percent of American females between the ages of 18 and 24 who are at least x inches tall is given by

$$f(x) = \frac{100}{1 + e^{-0.66607(x - 64.51)}}$$

where m and f are the percents and x is the height in inches. (Source: U.S. National Center for Health Statistics)

(a) Use the graph to determine any horizontal asymptotes of the functions. What do they mean?

(b) What is the median height of each sex?

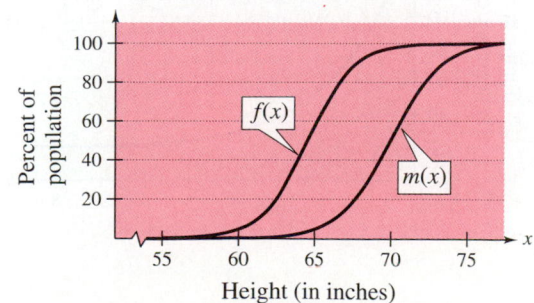

92. *Human Memory Model* In a group project in learning theory, a mathematical model for the proportion P of correct responses after n trials was found to be

$$P = \frac{0.83}{1 + e^{-0.2n}}.$$

(a) Use a graphing utility to graph the function.

(b) Use the graph to determine any horizontal asymptotes of the function. Interpret the meaning of the upper asymptote in the context of this problem.

(c) After how many trials will 60% of the responses be correct?

93. *Data Analysis* An object at a temperature of 160°C was removed from a furnace and placed in a room at 20°C. The temperature T of the object was measured each hour h and recorded in the table.

h	0	1	2	3	4	5
T	160°	90°	56°	38°	29°	24°

A model for this data is

$$T = 20[1 + 7(2^{-h})].$$

(a) The graph of this model is shown in the figure. Use the graph to identify the horizontal asymptote of the model and interpret the asymptote in the context of the problem.

(b) Approximate the time when the temperature of the object was 100°C.

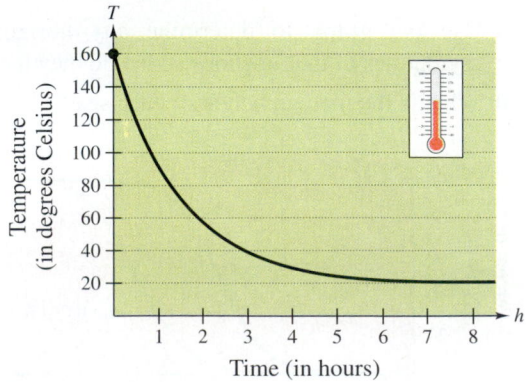
Time (in hours)

94. *Chapter Opener* Automobiles are designed with crumple zones that help protect their occupants in crashes. The crumple zones allow the occupants to move short distances when the automobiles come to abrupt stops. The greater the distance moved, the less g's the crash victims experience. (One g is equal to the acceleration due to gravity. For very short periods of time, humans have withstood as much as 40 g's.) In crash tests with vehicles moving at 90 kilometers per hour, analysts measured the numbers, y, of g's experienced during deceleration by crash dummies that were permitted to move x meters during impact. The data is shown in the table.

x	0.2	0.4	0.6	0.8	1
y	158	80	53	40	32

A model for this data is

$$y = -3.00 + 11.88 \ln x + \frac{36.94}{x}.$$

(a) Use a graphing utility to graph the data points and the model on the same viewing rectangle. How do they compare?

(b) Use the model to estimate the distance traveled during impact if the passenger deceleration must not exceed 30 g's.

(c) Do you think it is practical to lower the number of g's experienced during impact to less than 23? Explain your reasoning.

Review **Solve Exercises 95–98 as a review of the skills and problem-solving techniques you learned in previous sections. Simplify the expression.**

95. $\sqrt{48x^2y^5}$

96. $\sqrt{32} - 2\sqrt{25}$

97. $\sqrt[3]{25} \cdot \sqrt[3]{15}$

98. $\dfrac{3}{\sqrt{10} - 2}$

See Exercise 71 on page 447 for an example of comparing an exponential growth model, a linear model, a logarithmic model, and a quadratic model for the same set of data.

5.5   *Exponential and Logarithmic Models*

Introduction ▫ *Exponential Growth and Decay* ▫
Gaussian Models ▫ *Logistics Growth Models* ▫ *Logarithmic Models*

Introduction

The five most common types of mathematical models involving exponential functions and logarithmic functions are as follows.

1. Exponential growth model: $y = ae^{bx}, \quad b > 0$
2. Exponential decay model: $y = ae^{-bx}, \quad b > 0$
3. Gaussian model: $y = ae^{-(x-b)^2/c}$
4. Logistics growth model: $y = \dfrac{a}{1 + be^{-(x-c)/d}}$
5. Logarithmic models: $y = a + b \ln x, \quad y = a + b \log_{10} x$

The graphs of the basic forms of these functions are shown in Figure 5.20.

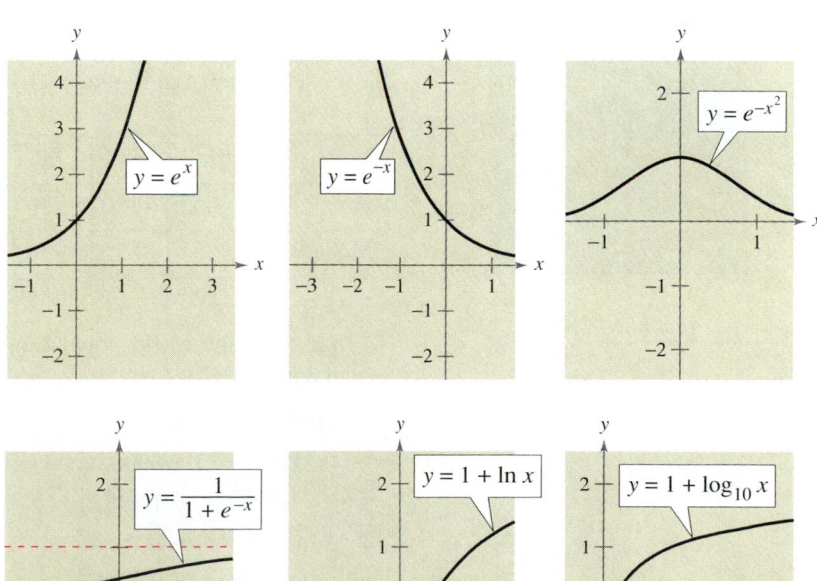

FIGURE 5.20

Study Tip

You can often gain quite a bit of insight into a situation modeled by an exponential or logarithmic function by identifying and interpreting the function's asymptotes. Use the graphs in Figure 5.20 to identify the asymptotes of each function.

Exponential Growth and Decay

EXAMPLE 1 *Population Increase*

Estimates of the world population (in millions) from 1980 through 1992 are shown in the table. The scatter plot of the data is shown in Figure 5.21(a). (Source: Statistical Office of the United Nations)

Year	1980	1985	1986	1987	1988	1989	1990	1991	1992
Population	4453	4850	4936	5024	5112	5202	5294	5384	5478

An exponential growth model that approximates this data is given by

$$P = 4451e^{0.017303t}, \quad 0 \le t \le 12$$

where P is the population (in millions) and $t = 0$ represents 1980. Compare the values given by the model with the estimates given by the United Nations. According to this model, when will the world population reach 6 billion?

Solution

The following table compares the two sets of population figures. The graph of the model is shown in Figure 5.21(b).

Year	1980	1985	1986	1987	1988	1989	1990	1991	1992
Population	4453	4850	4936	5024	5112	5202	5294	5384	5478
Model	4451	4853	4938	5024	5112	5201	5292	5384	5478

To find when the world population will reach 6 billion, let $P = 6000$ in the model and solve for t.

$4451e^{0.017303t} = P$	Given model
$4451e^{0.017303t} = 6000$	Let $P = 6000$.
$e^{0.017303t} \approx 1.348$	Divide both sides by 4451.
$\ln e^{0.017303t} \approx \ln 1.348$	Take logarithms of both sides.
$0.017303t \approx 0.2986$	Inverse property of logs and exponents
$t \approx 17.26$	Divide both sides by 0.017303.

According to the model, the world population will reach 6 billion in 1997.

NOTE An exponential model increases (or decreases) by the same percent each year. What is the annual percent increase for the model above?

(a)

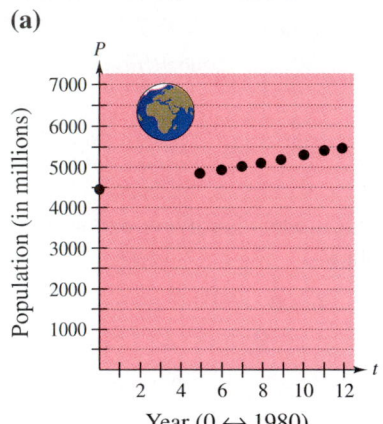

(b)

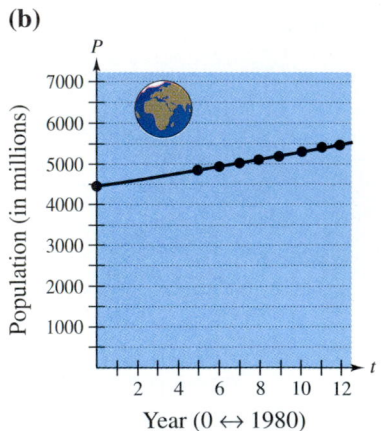

FIGURE 5.21

In Example 1, you were given the exponential growth model. But suppose this model were not given; how could you find such a model? One technique for doing this is demonstrated in Example 2.

■

EXAMPLE 2   *Finding an Exponential Growth Model*

Find an exponential growth model whose graph passes through the points $(0, 4453)$ and $(7, 5024)$, as shown in Figure 5.22(a).

Solution

The general form of the model is

$$y = ae^{bx}.$$

From the fact that the graph passes through the point $(0, 4453)$, you know that $y = 4453$ when $x = 0$. By substituting these values into the general form of the model, you have

$$4453 = ae^0 \qquad \Longrightarrow \qquad a = 4453.$$

In a similar way, from the fact that the graph passes through the point $(7, 5024)$, you know that $y = 5024$ when $x = 7$. By substituting these values into the model, you have

$$5024 = 4453e^{7b} \qquad \Longrightarrow \qquad b = \frac{1}{7}\ln\frac{5024}{4453} \approx 0.01724.$$

Thus, the exponential growth model is

$$y = 4453e^{0.01724x}.$$

The graph of the model is shown in Figure 5.22(b).

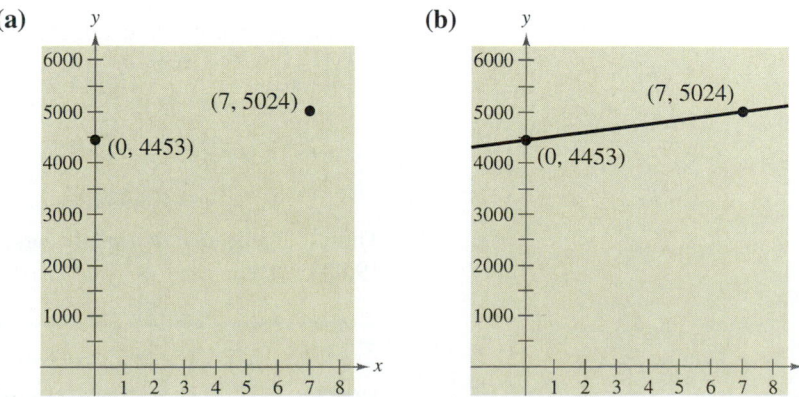

FIGURE 5.22

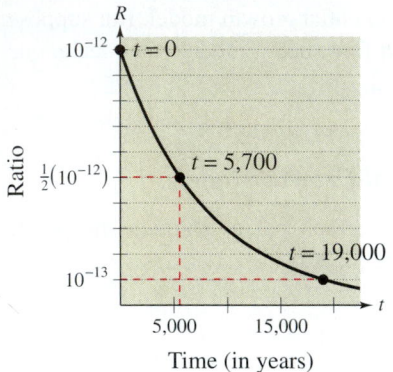

FIGURE 5.23

In living organic material, the ratio of the number of radioactive carbon isotopes (carbon 14) to the number of nonradioactive carbon isotopes (carbon 12) is about 1 to 10^{12}. When organic material dies, its carbon 12 content remains fixed, whereas its radioactive carbon 14 begins to decay with a half-life of about 5700 years. To estimate the age of dead organic material, scientists use the following formula, which denotes the ratio of carbon 14 to carbon 12 present at any time t (in years).

$$R = \frac{1}{10^{12}} e^{-t/8223}$$

The graph of R is shown in Figure 5.23. Note that R decreases as t increases.

EXAMPLE 3   *Carbon Dating*

Real Life

The ratio of carbon 14 to carbon 12 in a newly discovered fossil is

$$R = \frac{1}{10^{13}}.$$

Estimate the age of the fossil.

Solution

In the carbon dating model, substitute the given value of R to obtain the following.

$\dfrac{1}{10^{12}} e^{-t/8223} = R$	Given model
$\dfrac{e^{-t/8223}}{10^{12}} = \dfrac{1}{10^{13}}$	Let $R = \dfrac{1}{10^{13}}$.
$e^{-t/8223} = \dfrac{1}{10}$	Multiply both sides by 10^{12}.
$\ln e^{-t/8223} = \ln \dfrac{1}{10}$	Take logarithms of both sides.
$-\dfrac{t}{8223} \approx -2.3026$	Inverse property of logs and exponents
$t \approx 18{,}934$	Multiply both sides by -8223.

Thus, to the nearest thousand years, you can estimate the age of the fossil to be 19,000 years.

NOTE The carbon dating model in Example 3 assumed that the carbon 14/carbon 12 ratio was one part in 10,000,000,000,000. Suppose an error in measurement occurred and the actual ratio was only one part in 8,000,000,000,000. The fossil age corresponding to the actual ratio would then be approximately 17,000 years. Try checking this result. ■■

Gaussian Models

As mentioned at the beginning of this section, Gaussian models are of the form

$$y = ae^{-(x-b)^2/c}.$$

This type of model is commonly used in probability and statistics to represent populations that are **normally distributed.** For *standard* normal distributions, the model takes the form

$$y = \frac{1}{\sigma\sqrt{2\pi}}e^{-x^2/2\sigma^2}$$

where σ is the standard deviation (σ is the lowercase Greek letter sigma). The graph of a Gaussian model is called a **bell-shaped curve.** Try assigning a value to σ and sketching a normal distribution curve with a graphing utility. Can you see why it is called a bell-shaped curve?

The highest possible score for the Scholastic Aptitude Test (SAT) is 1600. The largest number of students ever attaining perfect scores in one year is 13, which has occurred twice, in 1987 and in 1992. *(Photo: Chuck Savage/The Stock Market)*

Real Life

EXAMPLE 4 *SAT Scores*

In 1993, the Scholastic Aptitude Test (SAT) scores for males roughly followed a normal distribution given by

$$y = 0.0026e^{-(x-500)^2/48,000}, \quad 200 \le x \le 800$$

where x is the SAT score for mathematics. Sketch the graph of this function. From the graph, estimate the average SAT score.   (Source: College Board)

Solution

The graph of the function is given in Figure 5.24. From the graph, you can see that the average mathematics score for males in 1993 was 500.

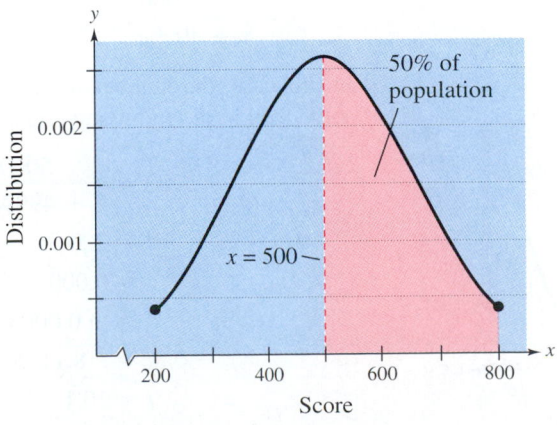

FIGURE 5.24

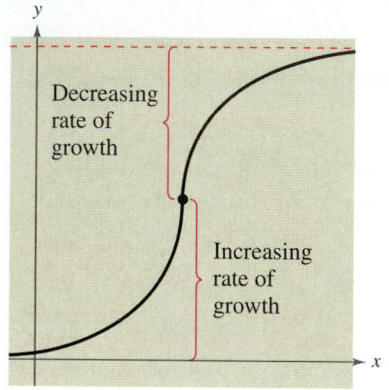

Decreasing
rate of
growth

Increasing
rate of
growth

FIGURE 5.25

Logistics Growth Models

Some populations initially have rapid growth, followed by a declining rate of growth, as indicated by the graph in Figure 5.25. One model for describing this type of growth pattern is the **logistics curve** given by the function

$$y = \frac{a}{1 + be^{-(x-c)/d}}$$

where y is the population size and x is the time. An example is a bacteria culture that is initially allowed to grow under ideal conditions, and then under less favorable conditions that inhibit growth. A logistics growth curve is also called a **sigmoidal curve.**

Real
Life

EXAMPLE 5   *Spread of a Virus*

On a college campus of 5000 students, one student returns from vacation with a contagious flu virus. The spread of the virus is modeled by

$$y = \frac{5000}{1 + 4999e^{-0.8t}}, \quad 0 \le t$$

where y is the total number infected after t days. The college will cancel classes when 40% or more of the students are ill.

a. How many are infected after 5 days?

b. After how many days will the college cancel classes?

Solution

a. After 5 days, the number infected is

$$y = \frac{5000}{1 + 4999e^{-0.8(5)}} = \frac{5000}{1 + 4999e^{-4}} \approx 54.$$

b. In this case, the number infected is $(0.40)(5000) = 2000$. Therefore, you solve for t in the following equation.

$$2000 = \frac{5000}{1 + 4999e^{-0.8t}}$$

$$1 + 4999e^{-0.8t} = 2.5$$

$$e^{-0.8t} \approx 0.0003$$

$$\ln e^{-0.8t} \approx \ln 0.0003$$

$$-0.8t \approx -8.1115$$

$$t \approx 10.1$$

Hence, after 10 days, at least 40% of the students will be infected, and the college will cancel classes. The graph of the function is shown in Figure 5.26.

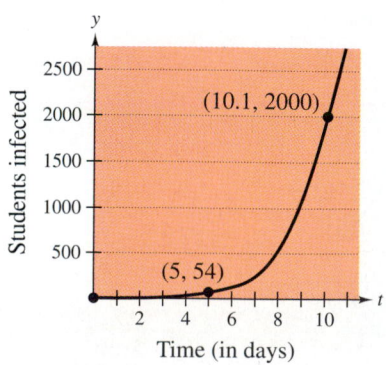

Students infected

2500
2000
1500
1000
500

(10.1, 2000)

(5, 54)

2   4   6   8   10

Time (in days)

FIGURE 5.26

Logarithmic Models

EXAMPLE 6   *Magnitude of Earthquakes*

On the Richter scale, the magnitude R of an earthquake of intensity I is given by

$$R = \log_{10} \frac{I}{I_0}$$

where $I_0 = 1$ is the minimum intensity used for comparison. Find the intensities per unit of area for the following earthquakes. (Intensity is a measure of the wave energy of an earthquake.)

a. Tokyo and Yokohama, Japan in 1923, $R = 8.3$.

b. Kobe, Japan in 1995, $R = 7.2$.

Solution

a. Because $I_0 = 1$ and $R = 8.3$, you have

$$8.3 = \log_{10} I$$
$$I = 10^{8.3} \approx 199,526,000.$$

b. For $R = 7.2$, you have $7.2 = \log_{10} I$, and

$$I = 10^{7.2} \approx 15,849,000.$$

Note that an increase of 1.1 units on the Richter scale (from 7.2 to 8.3) represents an increase in intensity by a factor of

$$\frac{199,526,000}{15,849,000} \approx 13.$$

In other words, the earthquake in 1923 had a magnitude about 13 times greater than that of the 1995 quake.

Twenty seconds of a 7.2 magnitude earthquake in Kobe, Japan, on January 17, 1995, left damage approaching $60 billion. *(Photo: Ben Simmons/The Stock Market)*

t	Year	Population
1	1810	7.23
3	1830	12.87
5	1850	23.19
7	1870	39.82
9	1890	62.95
11	1910	91.97
13	1930	122.78
15	1950	151.33
17	1970	203.30
19	1990	250.00

GROUP ACTIVITY

COMPARING POPULATION MODELS

The population (in millions) of the United States from 1810 to 1990 is given in the table. (Source: U.S. Bureau of Census) Least squares regression analysis gives the best quadratic model for this data as $P = 0.657t^2 + 0.305t + 6.118$ and the best exponential model for this data as $P = 8.325e^{0.195t}$. Which model better fits the data? Describe the method you used to reach your conclusion.

WARM UP

Sketch the graph of the equation.

1. $y = 2^{0.25x}$

2. $y = 2^{-0.25x}$

3. $y = 4 \log_2 x$

4. $y = \ln(x - 3)$

5. $y = e^{-x^2/5}$

6. $y = \dfrac{2}{1 + e^{-x}}$

Solve the equation for x. Round the result to three decimal places.

7. $3e^{2x} = 7$

8. $2e^{-0.2x} = 0.002$

9. $4 \ln 5x = 14$

10. $6 \ln 2x = 12$

5.5 Exercises

In Exercises 1–6, match the function with its graph. [The graphs are labeled (a) through (f).]

1. $y = 2e^{x/4}$

2. $y = 6e^{-x/4}$

3. $y = \frac{1}{16}(x^2 + 8x + 32)$

4. $y = \dfrac{12}{x + 4}$

5. $y = \ln(x + 1)$

6. $y = \sqrt{x}$

(a)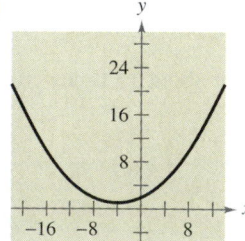

(b)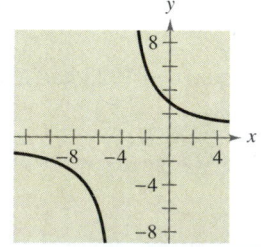

(c)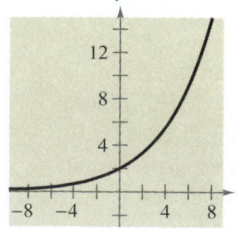

(d)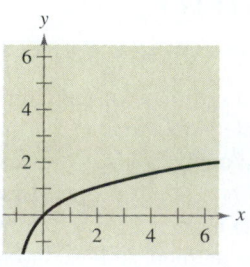

(e)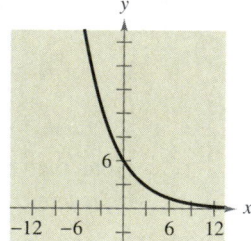

(f)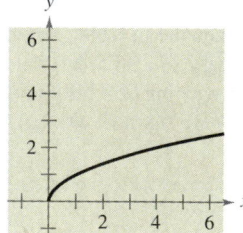

Compound Interest **In Exercises 7–14, complete the table for a savings account in which interest is compounded continuously.**

	Initial Investment	Annual % Rate	Time to Double	Amount After 10 Years
7.	$1000	12%		
8.	$20,000	$10\frac{1}{2}\%$		
9.	$750		$7\frac{3}{4}$ yr	
10.	$10,000		5 yr	
11.	$500			$1292.85
12.	$600			$19,205.00
13.		4.5%		$10,000.00
14.		8%		$20,000.00

Compound Interest In Exercises 15 and 16, determine the principal P that must be invested at rate r, compounded monthly, so that $500,000 will be available for retirement in t years.

15. $r = 7\frac{1}{2}\%, t = 20$ **16.** $r = 12\%, t = 40$

Compound Interest In Exercises 17 and 18, determine the time necessary for $1000 to double if it is invested at interest rate r compounded (a) annually, (b) monthly, (c) daily, and (d) continuously.

17. $r = 11\%$ **18.** $r = 10\frac{1}{2}\%$

19. *Compound Interest* Complete the table for the time t necessary for P dollars to triple if interest is compounded continuously at rate r.

r	2%	4%	6%	8%	10%	12%
t						

20. *Modeling Data* Draw a scatter plot of the data in Exercise 19. Use the curve-fitting capabilities of a graphing utility to find a model for the data.

21. *Compound Interest* Complete the table for the time t necessary for P dollars to triple if interest is compounded annually at rate r.

r	2%	4%	6%	8%	10%	12%
t						

22. *Modeling Data* Draw a scatter plot of the data in Exercise 21. Use the curve-fitting capabilities of a graphing utility to find a model for the data.

23. *Comparing Investments* If $1 is invested in an account over a 10-year period, the amount in the account, where t represents the time in years, is

$$A = 1 + 0.075[\![t]\!] \quad \text{or} \quad A = e^{0.07t}$$

depending on whether the account pays simple interest at $7\frac{1}{2}\%$ or continuous compound interest at 7%. Graph each function on the same set of axes. Which grows at the faster rate?

24. *Comparing Investments* If $1 is invested in an account over a 10-year period, the amount in the account, where t represents the time in years, is

$$A = 1 + 0.06[\![t]\!] \quad \text{or} \quad A = \left(1 + \frac{0.055}{365}\right)^{[\![365t]\!]}$$

depending on whether the account pays simple interest at 6% or compound interest at $5\frac{1}{2}\%$ compounded daily. Use a graphing utility to graph each function on the same viewing rectangle. Which grows at the faster rate?

In Exercises 25–30, complete the table for the given radioactive isotope.

Isotope	Half-life (years)	Initial Quantity	Amount After 1000 Years
25. ^{226}Ra	1620	10g	
26. ^{226}Ra	1620		1.5g
27. ^{14}C	5730		2g
28. ^{14}C	5730	3g	
29. ^{230}Pu	24,360		2.1g
30. ^{230}Pu	24,360		0.4g

In Exercises 31–34, find the exponential model $y = ae^{bx}$ that fits the points given in the graph or table.

31.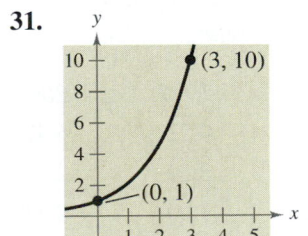
32.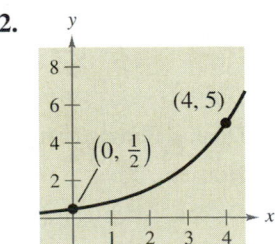

33.

x	0	3
y	1	$\frac{1}{4}$

34.

x	0	4
y	5	1

35. *Population*   The population P of a city is given by

$$P = 105,300e^{0.015t}$$

where $t = 0$ represents 1990. According to this model, when will the population reach 150,000?

36. *Population*   The population P of a city is given by

$$P = 240,360e^{0.012t}$$

where $t = 0$ represents 1990. According to this model, when will the population reach 275,000?

37. *Population*   The population P of a city is given by

$$P = 2500e^{kt}$$

where $t = 0$ represents 1990. In 1945, the population was 1350. Find the value of k, and use this result to predict the population in the year 2010.

38. *Population*   The population P of a city is given by

$$P = 140,500e^{kt}$$

where $t = 0$ represents 1990. In 1960, the population was 100,250. Find the value of k, and use this result to predict the population in the year 2000.

Population   **In Exercises 39–42, the table gives the population (in millions) of a city in 1990 and the projected population (in millions) for the year 2000. Find the exponential growth model $y = ae^{bt}$ for the population by letting $t = 0$ correspond to 1990. Use the model to predict the population of the city in 2010.**   (Source: U.S. Bureau of the Census, International Database)

City	1990	2000
39. Dhaka, Bangladesh	4.22	6.49
40. Houston, Texas	2.30	2.65
41. Detroit, Michigan	3.00	2.74
42. London, United Kingdom	9.17	8.57

43. *Think About It*   In Exercises 39 and 40, you can see that the populations of Dhaka and Houston are growing at different rates. What constant in the equation $y = ae^{bt}$ is determined by these different growth rates? Discuss the relationship between the different growth rates and the magnitude of the constant.

44. *Think About It*   In Exercises 39 and 41, you can see that one population is increasing whereas the other is decreasing. What constant in the equation $y = ae^{bt}$ reflects this difference? Explain.

45. *Bacteria Growth*   The number of bacteria N in a culture is given by the model

$$N = 100e^{kt}$$

where t is the time in hours. If $N = 300$ when $t = 5$, estimate the time required for the population to double in size.

46. *Bacteria Growth*   The number of bacteria N in a culture is given by the model

$$N = 250e^{kt}$$

where t is the time in hours. If $N = 280$ when $t = 10$, estimate the time required for the population to double in size.

47. *Radioactive Decay*   The half-life of radioactive radium (^{226}Ra) is 1620 years. What percent of a present amount of radioactive radium will remain after 100 years?

48. *Radioactive Decay*   Carbon 14 dating assumes that the carbon dioxide on earth today has the same radioactive content as it did centuries ago. If this is true, the amount of ^{14}C absorbed by a tree that grew several centuries ago should be the same as the amount of ^{14}C absorbed by a tree growing today. A piece of ancient charcoal contains only 15% as much radioactive carbon as a piece of modern charcoal. How long ago was the tree burned to make the ancient charcoal if the half-life of ^{14}C is 5730 years?

49. *Comparing Depreciation Models*   A car that cost $22,000 new has a book value of $13,000 after 2 years.

(a) Find the straight-line model $V = mt + b$.

(b) Find the exponential model $V = ae^{kt}$.

(c) Use a graphing utility to graph the two models on the same viewing rectangle. Which model depreciates faster in the first 2 years?

(d) Find the book values of the car after 1 year and after 3 years using each model.

(e) Interpret the slope of the straight-line model.

50. *Depreciation* A computer that cost $4600 new has a book value of $3000 after 2 years. Find the value of the computer after 3 years by using the exponential model $y = ae^{bt}$.

51. *Sales* The sales S (in thousands of units) of a new product after it has been on the market t years are given by

$$S(t) = 100(1 - e^{kt}).$$

Fifteen thousand units of the new product were sold the first year.

(a) Complete the model by solving for k.

(b) Sketch the graph of the model.

(c) Use the model to estimate the number of units sold after 5 years.

52. *Sales and Advertising* After discontinuing all advertising for a certain product in 1994, the manufacturer noted that sales began to drop according to the model

$$S = \frac{500,000}{1 + 0.6e^{kt}}$$

where S represents the number of units sold and $t = 0$ represents 1994. In 1996, the company sold 300,000 units.

(a) Complete the model by solving for k.

(b) Estimate sales in 1999.

53. *Sales and Advertising* The sales S (in thousands of units) of a product after x hundred dollars is spent on advertising is given by

$$S = 10(1 - e^{kx}).$$

When $500 is spent on advertising, 2500 units are sold.

(a) Complete the model by solving for k.

(b) Estimate the number of units that will be sold if advertising expenditures are raised to $700.

54. *Profits* Because of a slump in the economy, a company finds that its annual profits have dropped from $742,000 in 1994 to $632,000 in 1996. If the profit follows an exponential pattern of decline, what is the expected profit for 1997? (Let $t = 0$ represent 1994.)

55. *Learning Curve* The management at a factory has found that the maximum number of units a worker can produce in a day is 30. The learning curve for the number of units N produced per day after a new employee has worked t days is given by

$$N = 30(1 - e^{kt}).$$

After 20 days on the job, a new employee produces 19 units.

(a) Find the learning curve for this employee (first, find the value of k).

(b) How many days should pass before this employee is producing 25 units per day?

(c) Is the employee's production increasing at a linear rate? Explain your reasoning.

56. *Endangered Species* A conservation organization releases 100 animals of an endangered species into a game preserve. The organization believes that the preserve has a carrying capacity of 1000 animals and that the growth of the herd will be modeled by the logistics curve

$$p(t) = \frac{1000}{1 + 9e^{-0.1656t}}$$

where t is measured in months (see figure).

(a) Use a graphing utility to graph the function. Use the graph to determine the horizontal asymptotes, and interpret the meaning of the larger p-value in the context of the problem.

(b) Estimate the population after 5 months.

(c) After how many months will the population be 500?

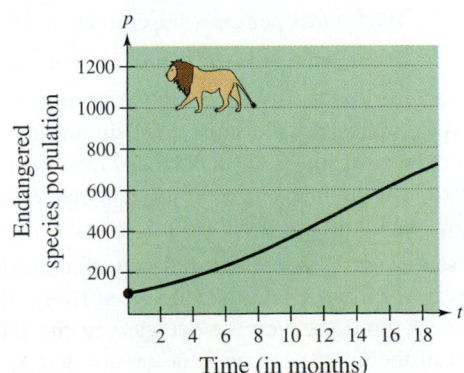

Time (in months)

Earthquake Magnitudes   In Exercises 57 and 58, use the Richter scale (see page 441) for measuring the magnitudes of earthquakes.

57. Find the magnitude R of an earthquake of intensity I (let $I_0 = 1$).

   (a) $I = 80,500,000$

   (b) $I = 48,275,000$

58. Find the intensity I of an earthquake measuring R on the Richter scale (let $I_0 = 1$).

   (a) Colombia in 1906, $R = 8.6$

   (b) Los Angeles in 1971, $R = 6.7$

Intensity of Sound   In Exercises 59–62, use the following information for determining sound intensity. The level of sound β, in decibels, with an intensity of I is given by

$$\beta(I) = 10 \log_{10} \frac{I}{I_0}$$

where I_0 is an intensity of 10^{-16} watts per square centimeter, corresponding roughly to the faintest sound that can be heard by the human ear.

59. (a) $I = 10^{-14}$ watts per cm² (faint whisper)

   (b) $I = 10^{-9}$ watts per cm² (busy street corner)

   (c) $I = 10^{-6.5}$ watts per cm² (air hammer)

   (d) $I = 10^{-4}$ watts per cm² (threshold of pain)

60. (a) $I = 10^{-13}$ watts per cm² (whisper)

   (b) $I = 10^{-7.5}$ watts per cm² (jet 4 miles from takeoff)

   (c) $I = 10^{-7}$ watts per cm² (diesel truck at 25 feet)

   (d) $I = 10^{-4.5}$ watts per cm² (auto horn at 3 feet)

61. *Noise Level*   Due to the installation of noise suppression materials, the noise level in an auditorium was reduced from 93 to 80 decibels. Find the percent decrease in the intensity level of the noise as a result of the installation of these materials.

62. *Noise Level*   Due to the installation of a muffler, the noise level in an engine was reduced from 88 to 72 decibels. Find the percent decrease in the intensity level of the noise as a result of the installation of the muffler.

Acidity   In Exercises 63–68, use the acidity model given by pH $= -\log_{10}[H^+]$, where acidity (pH) is a measure of the hydrogen ion concentration $[H^+]$ (measured in moles of hydrogen per liter) of a solution.

63. Find the pH if $[H^+] = 2.3 \times 10^{-5}$.

64. Find the pH if $[H^+] = 11.3 \times 10^{-6}$.

65. Compute $[H^+]$ for a solution in which pH $= 5.8$.

66. Compute $[H^+]$ for a solution in which pH $= 3.2$.

67. A certain fruit has a pH of 2.5 and an antacid tablet has a pH of 9.5. The hydrogen ion concentration of the fruit is how many times the concentration of the tablet?

68. If the pH of a solution is decreased by one unit, the hydrogen ion concentration is increased by what factor?

69. *Home Mortgage*   A $120,000 home mortgage for 35 years at $9\frac{1}{2}\%$ has a monthly payment of $985.93. Part of the monthly payment goes for the interest charge on the unpaid balance, and the remainder of the payment is used to reduce the principal. The amount that goes for interest is given by

$$u = M - \left(M - \frac{Pr}{12}\right)\left(1 + \frac{r}{12}\right)^{12t}$$

and the amount that goes toward reduction of the principal is given by

$$v = \left(M - \frac{Pr}{12}\right)\left(1 + \frac{r}{12}\right)^{12t}.$$

In these formulas, P is the size of the mortgage, r is the interest rate, M is the monthly payment, and t is the time in years.

   (a) Use a graphing utility to graph each function on the same viewing rectangle. (The viewing rectangle should show all 35 years of mortgage payments.)

   (b) In the early years of the mortgage, the larger part of the monthly payment goes for what purpose? Approximate the time when the monthly payment is evenly divided between interest and principal reduction.

   (c) Repeat parts (a) and (b) for a repayment period of 20 years ($M = 1118.56). What can you conclude?

70. *Home Mortgage* The total interest u paid on a home mortgage of P dollars at interest rate r for t years is given by

$$u = P \left[\frac{rt}{1 - \left(\dfrac{1}{1 + r/12} \right)^{12t}} - 1 \right].$$

Consider a $120,000 home mortgage at $9\frac{1}{2}\%$.

(a) Use a graphing utility to graph the total interest function.

(b) Approximate the length of the mortgage for which the total interest paid is the same as the size of the mortgage. Is it possible that some people are paying twice as much in interest charges as the size of the mortgage?

71. *Data Analysis* The time t (in seconds) required to attain a speed of s miles per hour from a standing start for a 1995 Dodge Avenger is given in the table. (Source: *Road & Track*, March 1995)

s	30	40	50	60	70	80	90
t	3.4	5.0	7.0	9.3	12.0	15.8	20.0

Two models for this data are

$t_1 = 40.757 + 0.556s - 15.817 \ln s$,

$t_2 = 1.2259 + 0.0023s^2$.

(a) Use a graphing utility to fit a linear model t_3 and an exponential model t_4 to the data.

(b) Use a graphing utility to graph the data points and each model.

(c) Create a table to compare the given data with estimates obtained from each model.

(d) Use the results of part (c) to find the sum of the absolute values of the differences between the data and estimated values given by each model. Based on the four sums, which model do you think better fits the data? Explain.

72. *Essay* Use your school's library or some other reference source to write a paper describing John Napier's work with logarithms.

73. *Essay* Before the development of electronic calculators and graphing utilities, some computations were done on slide rules. Use your school's library or some other reference source to write a paper describing the use of logarithmic scales on a slide rule.

74. *Estimating the Time of Death* At 8:30 A.M., a coroner was called to the home of a person who had died during the night. In order to estimate the time of death, the coroner took the person's temperature twice. At 9:00 A.M. the temperature was 85.7°F and at 9:30 A.M. the temperature was 82.8°F. From these two temperatures the coroner was able to determine that the time elapsed since death and the body temperature were related by the formula

$$t = -2.5 \ln \frac{T - 70}{98.6 - 70}$$

where t is the time in hours elapsed since the person died and T is the temperature (in degrees Fahrenheit) of the person's body. Assume the person had a normal body temperature of 98.6°F at death, and the room temperature was a constant 70°F. (This formula is derived from a general cooling principle called Newton's Law of Cooling.) Use the formula to estimate the time of death of the person.

Review **Solve Exercises 75–78 as a review of the skills and problem-solving techniques you learned in previous sections. Divide by synthetic division.**

75. $\dfrac{4x^3 + 4x^2 - 39x + 36}{x + 4}$

76. $\dfrac{8x^3 - 36x^2 + 54x - 27}{x - \frac{3}{2}}$

77. $(2x^3 - 8x^2 + 3x - 9) \div (x - 4)$

78. $(x^4 - 3x + 1) \div (x + 5)$

FOCUS ON CONCEPTS

In this chapter, you studied several concepts related to exponential and logarithmic functions. Answer the following questions to check your understanding of several of the basic concepts discussed in this chapter. The answers to these questions are given in the back of the book.

1. *Comparing Graphs* The graphs of $y = e^{kt}$ are shown for $k = a, b, c,$ and d. Use the graphs to order $a, b, c,$ and d. Which of the four values are negative? Which are positive?

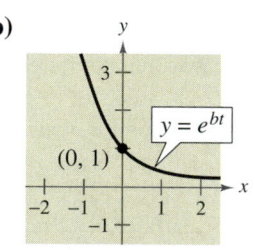

2. *True or False?* Rewrite each verbal statement as an equation. Then decide whether the statement is true or false. If it is false, give an example that shows it is false.

   (a) The logarithm of the product of two numbers is equal to the sum of the logarithms of the numbers.

   (b) The logarithm of the sum of two numbers is equal to the product of the logarithms of the numbers.

   (c) The logarithm of the difference of two numbers is equal to the difference of the logarithms of the numbers.

   (d) The logarithm of the quotient of two numbers is equal to the difference of the logarithms of the numbers.

3. *Investing Money* You are investing P dollars at an annual rate of r, compounded continuously, for t years. Which of the following would be most advantageous? Explain your reasoning.

   (a) Double the amount you invest.

   (b) Double your interest rate.

   (c) Double the number of years.

4. Identify the model as linear, logarithmic, exponential, logistic, or none of the above. Explain your reasoning.

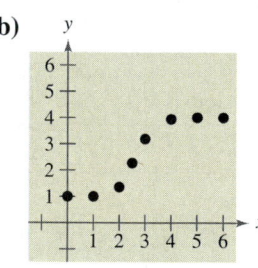

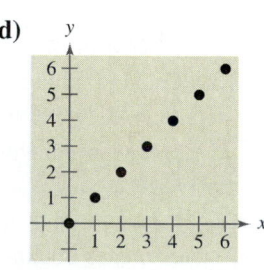

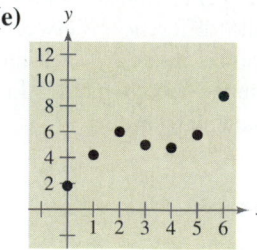

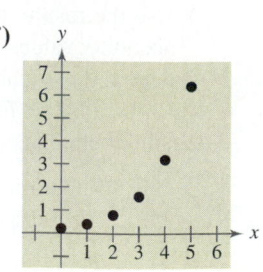

Review Exercises

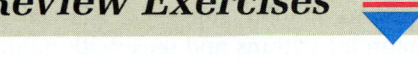

In Exercises 1–6, match the function with its graph. [The graphs are labeled (a) through (f).]

1. $f(x) = 4^x$

2. $f(x) = 4^{-x}$

3. $f(x) = -4^x$

4. $f(x) = 4^x + 1$

5. $f(x) = \log_4 x$

6. $f(x) = \log_4(x - 1)$

(a)
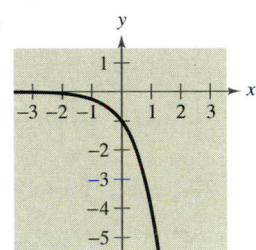

(b)
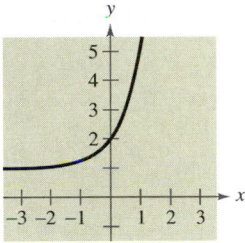

(c)
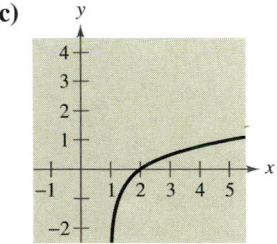

(d)
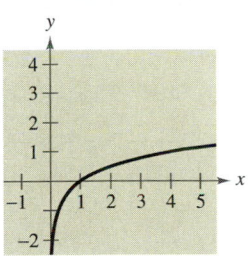

(e)
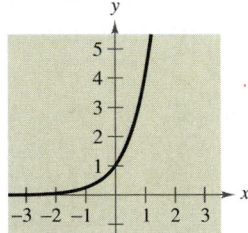

(f)
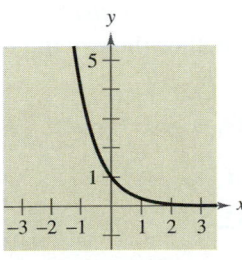

In Exercises 7–12, sketch the graph of the function.

7. $f(x) = 0.3^x$

8. $g(x) = 0.3^{-x}$

9. $h(x) = e^{-x/2}$

10. $h(x) = 2 - e^{-x/2}$

11. $f(x) = e^{x+2}$

12. $s(t) = 4e^{-2/t}, \quad t > 0$

In Exercises 13 and 14, use a graphing utility to graph the function. Identify any asymptotes.

13. $g(x) = 200e^{4/x}$

14. $f(x) = \dfrac{10}{1 + 2^{-0.05x}}$

In Exercises 15 and 16, complete the table to determine the balance A for P dollars invested at rate r for t years and compounded n times per year.

n	1	2	4	12	365	Continuous
A						

15. $P = \$3500, \quad r = 10.5\%, \quad t = 10$ years

16. $P = \$2000, \quad r = 12\%, \quad t = 30$ years

In Exercises 17 and 18, complete the table to determine the amount P that should be invested at rate r to produce a balance of \$200,000 in t years.

t	1	10	20	30	40	50
P						

17. $r = 8\%$, compounded continuously

18. $r = 10\%$, compounded monthly

19. *Waiting Times* The average time between incoming calls at a switchboard is 3 minutes. The probability of waiting less than t minutes until the next incoming call is approximated by the model

$$F(t) = 1 - e^{-t/3}.$$

If a call has just come in, find the probability that the next call will be within

(a) $\frac{1}{2}$ minute. (b) 2 minutes. (c) 5 minutes.

20. *Depreciation* After t years, the value of a car that cost \$14,000 is given by

$$V(t) = 14,000\left(\tfrac{3}{4}\right)^t.$$

(a) Use a graphing utility to graph the function.

(b) Find the value of the car 2 years after it was purchased.

(c) According to the model, when does the car depreciate most rapidly? Is this realistic? Explain.

21. *Trust Fund* On the day a person was born, a deposit of $50,000 was made in a trust fund that pays 8.75% interest, compounded continuously.

(a) Find the balance on the person's 35th birthday.

(b) How much longer would the person have to wait to get twice as much?

22. *Fuel Efficiency* A certain automobile gets 28 miles per gallon of gasoline for speeds up to 50 miles per hour. Over 50 miles per hour, the number of miles per gallon drops at a rate of 12% for each additional 10 miles per hour. If s is the speed and y is the number of miles per gallon, then

$$y = 28e^{0.6 - 0.012s}, \qquad s \geq 50.$$

Use this model to complete the table.

s	50	55	60	65	70
y					

In Exercises 23–28, sketch the graph of the function. Identify any asymptotes.

23. $g(x) = \log_2 x$

24. $g(x) = \log_5 x$

25. $f(x) = \ln x + 3$

26. $f(x) = \ln(x - 3)$

27. $h(x) = \ln(e^{x-1})$

28. $f(x) = \frac{1}{4} \ln x$

In Exercises 29 and 30, use a graphing utility to graph the function.

29. $y = \log_{10}(x^2 + 1)$

30. $y = \sqrt{x} \ln(x + 1)$

In Exercises 31 and 32, write the exponential equation in logarithmic form.

31. $4^3 = 64$

32. $25^{3/2} = 125$

In Exercises 33–36, evaluate the expression by hand.

33. $\log_{10} 1000$

34. $\log_9 3$

35. $\ln e^7$

36. $\log_a \dfrac{1}{a}$

In Exercises 37–40, evaluate the logarithm using the change-of-base formula. Do each problem twice, once with common logarithms and once with natural logarithms. Round the result to three decimal places.

37. $\log_4 9$

38. $\log_{1/2} 5$

39. $\log_{12} 200$

40. $\log_3 0.28$

In Exercises 41–44, use the properties of logarithms to write the expression as a sum, difference, and/or multiple of logarithms.

41. $\log_5 5x^2$

42. $\log_7 \dfrac{\sqrt{x}}{4}$

43. $\log_{10} \dfrac{5\sqrt{y}}{x^2}$

44. $\ln \left| \dfrac{x - 1}{x + 1} \right|$

In Exercises 45–48, write the expression as the logarithm of a single quantity.

45. $\log_2 5 + \log_2 x$

46. $\log_6 y - 2 \log_6 z$

47. $\frac{1}{2} \ln|2x - 1| - 2 \ln|x + 1|$

48. $5 \ln|x - 2| - \ln|x + 2| - 3 \ln|x|$

True or False? **In Exercises 49–54, determine whether the equation or statement is true or false.**

49. $\log_b b^{2x} = 2x$

50. $e^{x-1} = \dfrac{e^x}{e}$

51. $\ln(x + y) = \ln x + \ln y$

52. $\ln(x + y) = \ln(x \cdot y)$

53. $\log\left(\dfrac{10}{x}\right) = 1 - \log x$

54. The domain of the function $f(x) = \ln x$ is the set of all real numbers.

55. *Snow Removal* The number of miles s of roads cleared of snow is approximated by the model

$$s = 25 - \dfrac{13 \ln(h/12)}{\ln 3}, \qquad 2 \leq h \leq 15$$

where h is the depth of the snow in inches. Use this model to find s when $h = 10$ inches.

56. Climb Rate The time t, in minutes, for a small plane to climb to an altitude of h feet is given by

$$t = 50 \log_{10} \frac{18{,}000}{18{,}000 - h}$$

where 18,000 feet is the plane's absolute ceiling.

(a) Determine the domain of the function appropriate for the context of the problem.

(b) Use a graphing utility to graph the time function and identify any asymptotes.

(c) As the plane approaches its absolute ceiling, what can be said about the time required to further increase its altitude?

(d) Find the time for the plane to climb to an altitude of 4000 feet.

In Exercises 57–62, solve the exponential equation. Round your result to three decimal places.

57. $e^x = 12$

58. $e^{3x} = 25$

59. $3e^{-5x} = 132$

60. $14e^{3x+2} = 560$

61. $e^{2x} - 7e^x + 10 = 0$

62. $e^{2x} - 6e^x + 8 = 0$

In Exercises 63–68, solve the logarithmic equation. Round the result to three decimal places.

63. $\ln 3x = 8.2$

64. $2 \ln 4x = 15$

65. $\ln x - \ln 3 = 2$

66. $\ln \sqrt{x+1} = 2$

67. $\log(x - 1) = \log(x - 2) - \log(x + 2)$

68. $\log(1 - x) = -1$

In Exercises 69–72, use a graphing utility to solve the equation. Round the result to two decimal places.

69. $2^{0.6x} - 3x = 0$

70. $25e^{-0.3x} = 12$

71. $2 \ln(x + 3) + 3x = 8$

72. $6 \log_{10}(x^2 + 1) - x = 0$

In Exercises 73 and 74, find the exponential function $y = ae^{bx}$ that passes through the points.

73. $(0, 2), (4, 3)$

74. $\left(0, \frac{1}{2}\right), (5, 5)$

75. Demand Function The demand equation for a certain product is given by

$$p = 500 - 0.5e^{0.004x}.$$

Find the demand x for a price of (a) $p = \$450$ and (b) $p = \$400$.

76. Typing Speed In a typing class, the average number of words per minute typed after t weeks of lessons was found to be

$$N = \frac{157}{1 + 5.4e^{-0.12t}}.$$

Find the time necessary to type (a) 50 words per minute and (b) 75 words per minute.

77. Compound Interest A deposit of $10,000 is made in a savings account for which the interest is compounded continuously. The balance will double in 5 years.

(a) What is the annual interest rate for this account?

(b) Find the balance after 1 year.

78. Sound Intensity The relationship between the number of decibels β and the intensity of a sound I in watts per square centimeter is given by

$$\beta = 10 \log_{10}\left(\frac{I}{10^{-16}}\right).$$

Determine the intensity of a sound in watts per square centimeter if the decibel level is 125.

79. Earthquake Magnitudes On the Richter scale, the magnitude R of an earthquake of intensity I is given by

$$R = \log_{10}\frac{I}{I_0}$$

where $I_0 = 1$ is the minimum intensity used for comparison. Find the intensity per unit of area for the following values of R.

(a) $R = 8.4$

(b) $R = 6.85$

(c) $R = 9.1$

CHAPTER PROJECT: *A Graphical Approach to Compound Interest*

A graphing utility can be used to investigate the rates of growth of different types of compound interest.

*Real
Life*

EXAMPLE 1   *Comparing Balances*

You are depositing $1000 in a savings account. Which of the following will produce a larger balance?

a. 6% annual interest rate, compounded annually

b. 6% annual interest rate, compounded continuously

c. 6.25% annual interest rate, compounded quarterly

Solution

Option (b) is better than option (a) because, for a given interest rate, continuous compounding yields a larger balance than compounding n times per year. Distinguishing between the second and third options is not as straightforward— the higher interest rate favors option (c), but the "more frequent" compounding favors option (b). One way to compare all three options is to sketch their graphs on the same viewing rectangle.

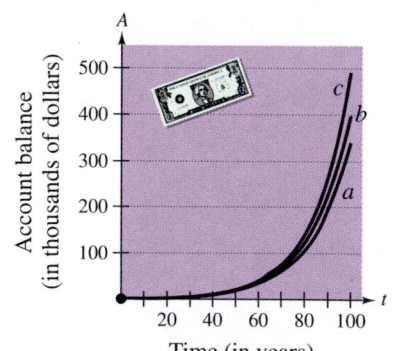

Time (in years)

Option (a)	*Option (b)*	*Option (c)*
$A = 1000(1 + 0.06)^t$	$A = 1000e^{0.06t}$	$A = 1000\left(1 + \dfrac{0.0625}{4}\right)^{4t}$

The graphs are shown at the left. On the graph, the t-values vary from 0 years through 100 years. From the graphs, you can conclude that option (c) is better than option (b), and option (b) is better than option (a). Note that for the first 50 years, there is little difference in the graphs. Between 50 and 100 years, however, the balances obtained begin to differ significantly. At the end of 100 years, the balances are (a) $339,302, (b) $403,429, and (c) $493,575.

To help distinguish among different rates and types of compounding, banks use the concept of *effective yield*. The **effective yield** of a savings plan is the percent increase in the balance after *one* year. For instance, in Example 1 the one-year balances are (a) $1060.00, (b) $1061.84, and (c) $1063.98.

Effective Yield (a)	*Effective Yield (b)*	*Effective Yield (c)*
6.000%	6.184%	6.398%

Because option (c) has the greatest effective yield, it is the best option and will yield the highest balance.

If you were to create a retirement plan with a regular savings account, the income tax on the interest would be due each year. With a *tax-deferred* retirement plan, the interest is allowed to build without being taxed until the account reaches maturity.

EXAMPLE 2   *To Defer or Not to Defer*

You deposit $25,000 in an account to accrue interest for 40 years. The account pays 8% compounded annually. Assume that the income tax on the earned interest is 30%. Which of the following plans produces a larger balance after all income tax is paid?

a. *Deferred* The income tax on the interest that is earned is paid in one lump sum at the end of 40 years.

b. *Not Deferred* The income tax on the interest that is earned each year is paid at the end of each year.

Solution

a. The untaxed balance at the end of 40 years is

$$A = 25,000(1 + 0.08)^{40} = \$543,113.04.$$

The income tax due is $0.3(518,113.04) = \$155,433.91$, so you are left with a balance of $\$387,679.13$.

b. You can reason that only 70% of the earned interest will remain in the account each year. The taxed balance at the end of 40 years is

$$A = 25,000[1 + 0.08(0.7)]^{40} = \$221,053.16.$$

Thus, the tax-deferred plan will produce a significantly greater balance at the end of 40 years. The balances are compared graphically at the left.

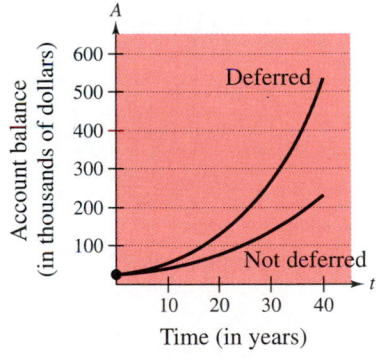

CHAPTER PROJECT INVESTIGATIONS

1. *Comparing Savings Plans* Which would produce a larger balance: an annual interest rate of 8.05% compounded monthly, or an annual interest rate of 8% compounded continuously? Explain.

2. *Exploration* You deposit $1000 in each of two savings accounts. The interest for the accounts is paid according to the two options described in Question 1. How long would it take for the balance in one of the accounts to exceed the balance in the other account by $100? By $100,000?

3. *Comparing Retirement Plans* No income tax is due on the interest earned in some types of investments. You deposit $25,000 into an account. Which of the following plans is better? Explain.

(a) *Tax-free* The account pays 5% compounded annually. There is no income tax due on the earned interest.

(b) *Tax-deferred* The account pays 7% compounded annually. At maturity, the earned interest is taxable at a rate of 40%.

Cumulative Test for Chapters 3–5

Take this test as you would take a test in class. After you are done, check your work against the answers given in the back of the book.

The *Interactive* CD-ROM provides answers to the Chapter Tests and Cumulative Tests. It also offers Chapter Pre-Tests (which test key skills and concepts covered in previous chapters) and Chapter Post-Tests, both of which have randomly generated exercises with diagnostic capabilities.

In Exercises 1–6, sketch the graph of the function without the aid of a graphing utility.

1. $h(x) = -(x^2 + 4x)$

2. $f(t) = \frac{1}{4}t(t-2)^2$

3. $g(s) = \dfrac{2s}{s-3}$

4. $g(s) = \dfrac{2s^2}{s-3}$

5. $f(x) = 6(2^{-x})$

6. $g(x) = \log_3 x$

7. Divide: $\dfrac{6x^3 - 4x^2}{2x^2 + 1}$.

8. Find all the zeros of $f(x) = x^3 + 2x^2 + 4x + 8$.

9. Use a graphing utility to approximate the real zero of the function $g(x) = x^3 + 3x^2 - 6$ to the nearest hundredth.

In Exercises 10 and 11, sketch a graph of the conic.

10. $6x - y^2 = 0$

11. $\dfrac{(x-2)^2}{4} + \dfrac{(y+1)^2}{9} = 1$

12. Find an equation of the parabola in the figure.

13. Find an equation of the hyperbola with foci $(0, 0)$ and $(0, 4)$ and asymptotes $y = \pm\frac{1}{2}x + 2$.

14. Write $2 \ln x - \frac{1}{2} \ln(x + 5)$ as a logarithm of a single quantity.

15. Use a graphing utility to graph $f(x) = \dfrac{1000}{1 + 4e^{-0.2x}}$ and determine the horizontal asymptotes.

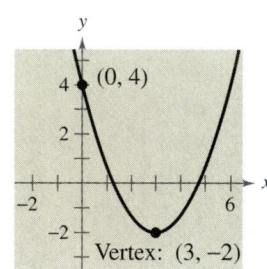

FIGURE FOR 12

In Exercises 16 and 17, solve the equation.

16. $6e^{2x} = 72$

17. $\log_2 x + \log_2 5 = 6$

18. Let x be the amount (in hundreds of dollars) that a company spends on advertising, and let P be the profit (in thousands of dollars), where

$$P = 230 + 20x - \tfrac{1}{2}x^2.$$

What amount of advertising will yield a maximum profit?

19. On the day a grandchild is born, a grandparent deposits $2500 into a fund earning 7.5%, compounded continuously. Determine the balance in the account at the time of the grandchild's 25th birthday.

In 1995, a huge geological project, called *Project Deep Probe*, set off explosions along a 2100-mile line from northern Canada to the Mexican border. The explosions were recorded by nearly 800 seismographs in Alberta, Montana, and Wyoming.

The goal of the project was to use the seismograph readings to obtain a geological profile of earth's mantle.

The seismographs were located 0.8 miles apart. To find the central angle between two adjacent seismographs, geologists used the formula $s = \theta r$, where θ is measured in radians. Using $s = 0.8$ miles and $r = 4000$ miles, you obtain an angle of 0.0002 radians or 0.0115°.

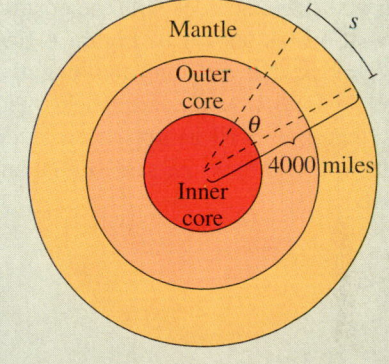

See Exercises 98 and 99 on page 466.

6 Trigonometry

Photos: © Michael Milstein

Project Deep Probe used about 800 portable seismographs from southern Alberta to central Wyoming. Geologists Holger and Reingard Mandler are shown checking a seismograph before burying it in Wyoming.

455

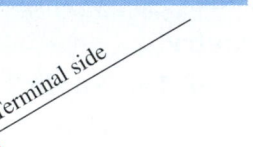

6.1 *Angles and Their Measure*

See Exercises 83–86 on page 465 for examples of how trigonometry can be used to find the distance between two cities of a given longitude.

Angles □ *Degree Measure* □ *Radian Measure* □ *Conversion of Angle Measure* □ *Applications*

Angles

As derived from the Greek language, the word **trigonometry** means "measurement of triangles." Initially, trigonometry dealt with relationships among the sides and angles of triangles and was used in the development of astronomy, navigation, and surveying. With the development of calculus and the physical sciences in the 17th century, a different perspective arose–one that viewed the classic trigonometric relationships as *functions* with the set of real numbers as their domains. Consequently, the applications of trigonometry expanded to include a vast number of physical phenomena involving rotations and vibrations. These phenomena include sound waves, light rays, planetary orbits, vibrating strings, pendulums, and orbits of atomic particles.

The approach in this text incorporates *both* perspectives, starting with angles and their measure.

An **angle** is determined by rotating a ray (half-line) about its endpoint. The starting position of the ray is the **initial side** of the angle, and the position after rotation is the **terminal side,** as shown in Figure 6.1(a). The endpoint of the ray is the **vertex** of the angle. This perception of an angle fits a coordinate system in which the origin is the vertex and the initial side coincides with the positive *x*-axis. Such an angle is in **standard position,** as shown in Figure 6.1(b). **Positive angles** are generated by counterclockwise rotation, and **negative angles** by clockwise rotation, as shown in Figure 6.2. Angles are labeled with Greek letters α (alpha), β (beta), and θ (theta), as well as uppercase letters *A, B,* and *C.* In Figure 6.3, note that angles α and β have the same initial and terminal sides. Such angles are **coterminal.**

(a)

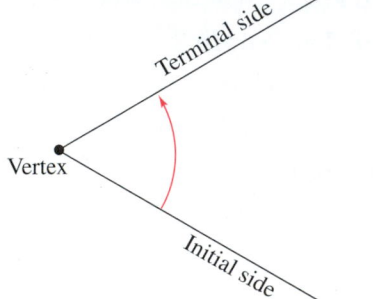

(b)

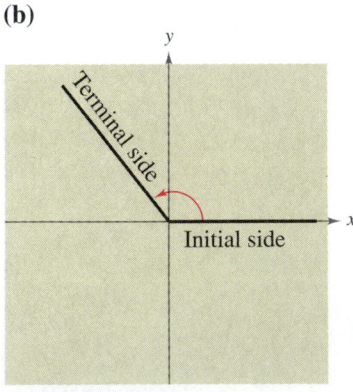

FIGURE 6.1

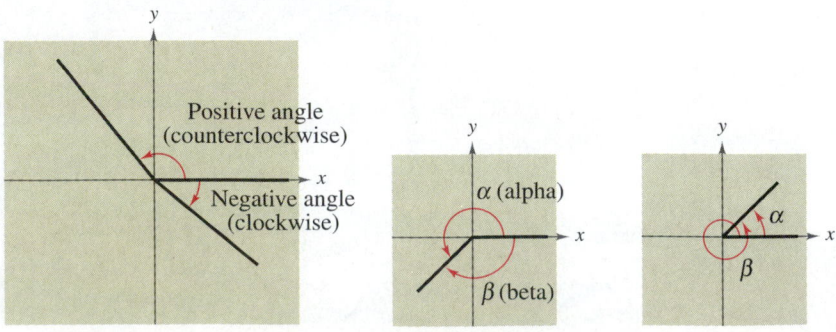

FIGURE 6.2   FIGURE 6.3

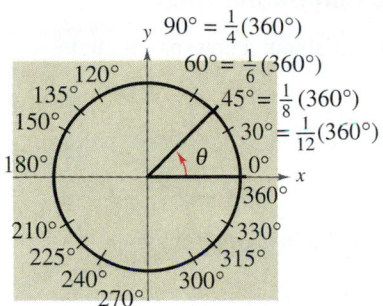

FIGURE 6.4

Degree Measure

The **measure of an angle** is determined by the amount of rotation from the initial to the terminal side. The most common unit of angle measure is the **degree,** denoted by the symbol °. A measure of **one degree (1°)** is equivalent to a rotation of $1/360$ of a complete revolution about the vertex. To measure angles, it is convenient to mark degrees on the circumference of a circle, as shown in Figure 6.4. Thus, a full revolution (counterclockwise) corresponds to 360°, a half revolution to 180°, a quarter revolution to 90°, and so on.

Recall that the four quadrants in a coordinate system are numbered I, II, III, and IV. Figure 6.5 shows which angles between 0° and 360° lie in each of the four quadrants. Figure 6.6 shows several common angles with their degree measures. Note that angles between 0° and 90° are **acute** and angles between 90° and 180° are **obtuse.**

NOTE The phrase "the terminal side of θ lies in a quadrant" is often abbreviated by simply saying that "θ lies in a quadrant." The terminal sides of the "quadrant angles" 0°, 90°, 180°, and 270° do not lie within quadrants. ■■

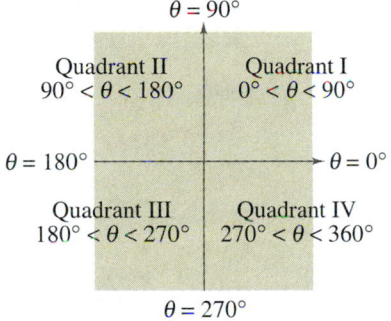

FIGURE 6.5

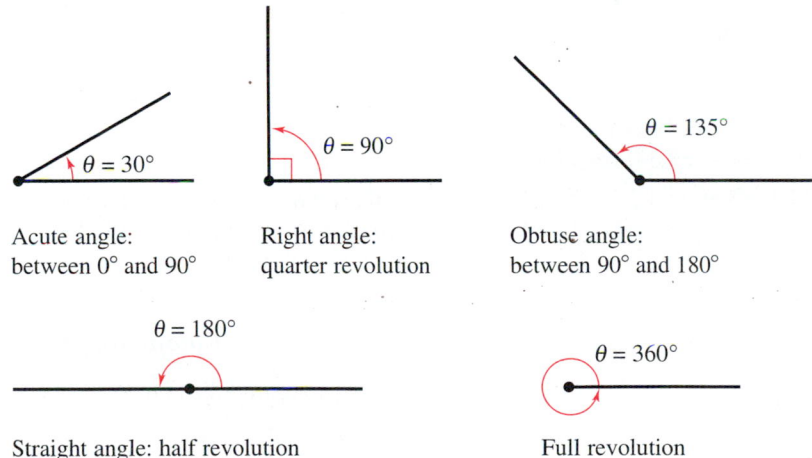

FIGURE 6.6

Two angles are coterminal if they have the same initial and terminal sides. For instance, the angles 0° and 360° are coterminal, as are the angles 30° and 390°. You can find an angle that is coterminal to a given angle θ by adding or subtracting 360° (one revolution), as demonstrated in Example 1. A given angle θ has many coterminal angles. For instance, $\theta = 30°$ is coterminal with

$$30° + n(360°)$$

where n is an integer.

With calculators it is convenient to use *decimal* degrees to denote fractional parts of degrees. Historically, however, fractional parts of degrees were expressed in *minutes* and *seconds*, using the prime (′) and double prime (″) notations, respectively. That is,

$1′ = $ one minute $= \frac{1}{60}(1°)$

$1″ = $ one second $= \frac{1}{3600}(1°)$.

Consequently, an angle of 64 degrees, 32 minutes, and 47 seconds, is represented by $\theta = 64° \, 32′ \, 47″$. Many calculators have special keys for converting an angle in degrees, minutes, and seconds (D° M′ S″) into decimal degree form, and vice versa.

EXAMPLE 1   *Sketching and Finding Coterminal Angles*

a. For the positive angle 390°, subtract 360° to obtain a coterminal angle.

$$390° - 360° = 30° \qquad \text{See Figure 6.7(a).}$$

b. For the positive angle 135°, subtract 360° to obtain a coterminal angle.

$$135° - 360° = -225° \qquad \text{See Figure 6.7(b).}$$

c. For the negative angle −120°, add 360° to obtain a coterminal angle.

$$-120° + 360° = 240° \qquad \text{See Figure 6.7(c).}$$

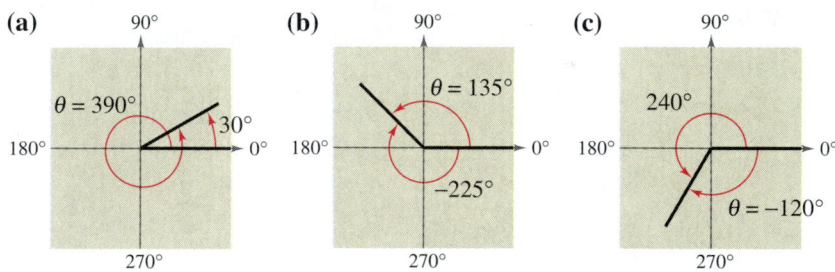

FIGURE 6.7

Two positive angles α and β are **complementary** (complements of each other) if their sum is 90°. Two positive angles are **supplementary** (supplements of each other) if their sum is 180°. See Figure 6.8.

EXAMPLE 2   *Complementary and Supplementary Angles*

If possible, find the complement and the supplement of (a) 72° and (b) 148°.

Solution

a. The complement of $\theta = 72°$ is

$$90° - \theta = 90° - 72° = 18°.$$

The supplement of $\theta = 72°$ is

$$180° - \theta = 180° - 72° = 108°.$$

b. Because $\theta = 148°$ is greater than 90°, it has no complement. (Remember to use only *positive* angles for complements.) The supplement is

$$180° - \theta = 180° - 148° = 32°.$$

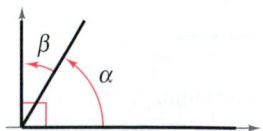

Complementary Angles

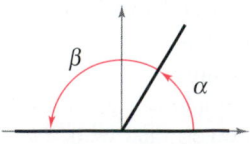

Supplementary Angles

FIGURE 6.8

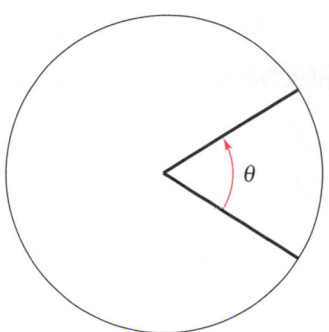

FIGURE 6.9

NOTE One revolution around a circle of radius r corresponds to an angle of 2π radians because

$$\frac{s}{r} = \frac{2\pi r}{r} = 2\pi \text{ radians.} \quad \blacksquare\blacksquare$$

(a)

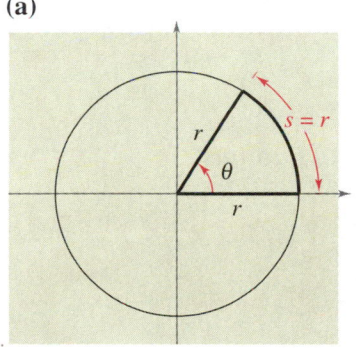

Arc length = radius when $\theta = 1$ radian

(b)

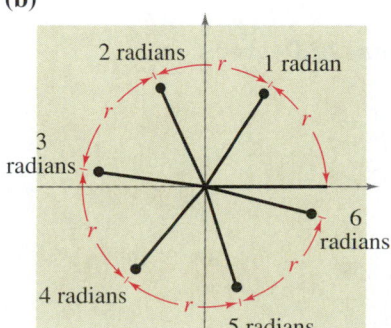

FIGURE 6.10

Radian Measure

A second way to measure angles is in radians. This type of measure is especially useful in calculus. To define a radian, you can use a **central angle** of a circle, one whose vertex is the center of the circle, as shown in Figure 6.9.

> ### DEFINITION OF A RADIAN
> One **radian** is the measure of a central angle θ that intercepts an arc s equal in length to the radius r of the circle. See Figure 6.10(a).

Because the circumference of a circle is $2\pi r$, it follows that a central angle of one full revolution (counterclockwise) corresponds to an arc length of $s = 2\pi r$. Moreover, because $2\pi \approx 6.28$, there are just over six radius lengths in a full circle, as shown in Figure 6.10(b). In general, the radian measure of a central angle θ is obtained by dividing the arc length s by r. That is, $s/r = \theta$, where θ *is measured in radians*. Because the units of measure for s and r are the same, this ratio is unitless–it is simply a real number.

EXAMPLE 3 *Complementary, Supplementary, and Coterminal Angles*

a. The complement of $\theta = \pi/12$ is

$$\pi/2 - \pi/12 = 6\pi/12 - \pi/12 = 5\pi/12. \qquad \text{See Figure 6.11(a).}$$

b. The supplement of $\theta = 5\pi/6$ is

$$\pi - 5\pi/6 = 6\pi/6 - 5\pi/6 = \pi/6. \qquad \text{See Figure 6.11(b).}$$

c. In radian measure, a coterminal angle is found by adding or subtracting 2π. For $\theta = 17\pi/6$, subtract 2π to obtain a coterminal angle.

$$17\pi/6 - 2\pi = 17\pi/6 - 12\pi/6 = 5\pi/6 \qquad \text{See Figure 6.11(c).}$$

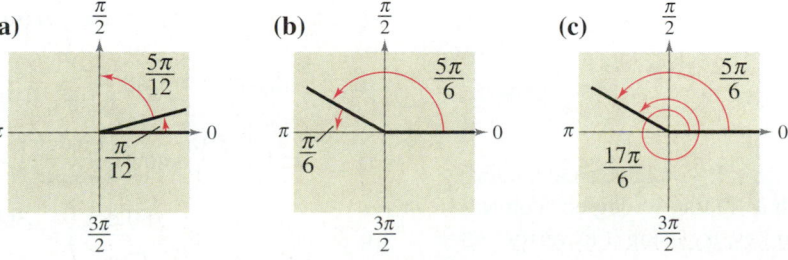

FIGURE 6.11

Conversion of Angle Measure

Because 2π radians corresponds to one complete revolution, degrees and radians are related by the equations

$$360° = 2\pi \text{ rad} \qquad \text{and} \qquad 180° = \pi \text{ rad.}$$

From the latter equation, you obtain

$$1° = \frac{\pi}{180} \text{ rad} \qquad \text{and} \qquad 1 \text{ rad} = \left(\frac{180}{\pi}\right)°$$

which lead to the following conversion rules.

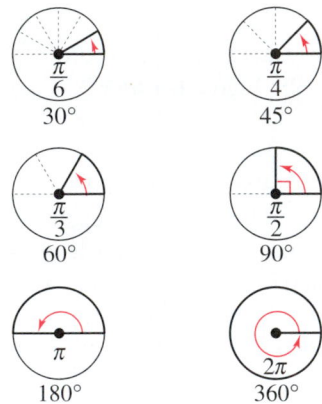

FIGURE 6.12

> ### CONVERSIONS BETWEEN DEGREES AND RADIANS
>
> **1.** To convert degrees to radians, multiply degrees by $\dfrac{\pi \text{ rad}}{180°}$.
>
> **2.** To convert radians to degrees, multiply radians by $\dfrac{180°}{\pi \text{ rad}}$.
>
> To apply these two conversion rules, use the basic relationship $\pi \text{ rad} = 180°$. (See Figure 6.12.)

NOTE Note that when no units of angle measure are specified, *radian measure is implied.* For instance, if you write $\theta = \pi$ or $\theta = 2$, you should mean $\theta = \pi$ radians or $\theta = 2$ radians. ■■

EXAMPLE 4 *Converting from Degrees to Radians*

a. $135° = (135 \text{ deg})\left(\dfrac{\pi \text{ rad}}{180 \text{ deg}}\right) = \dfrac{3\pi}{4} \text{ rad}$ Multiply by $\pi/180$.

b. $540° = (540 \text{ deg})\left(\dfrac{\pi \text{ rad}}{180 \text{ deg}}\right) = 3\pi \text{ rad}$ Multiply by $\pi/180$.

c. $-270° = (-270 \text{ deg})\left(\dfrac{\pi \text{ rad}}{180 \text{ deg}}\right) = -\dfrac{3\pi}{2} \text{ rad}$ Multiply by $\pi/180$.

NOTE If you have a calculator with a "radian-to-degree" conversion key, try using it to verify the result shown in part (c) of Example 5. ■■

EXAMPLE 5 *Converting from Radians to Degrees*

a. $-\dfrac{\pi}{2} \text{ rad} = \left(-\dfrac{\pi}{2} \text{ rad}\right)\left(\dfrac{180 \text{ deg}}{\pi \text{ rad}}\right) = -90°$ Multiply by $180/\pi$.

b. $\dfrac{9\pi}{2} \text{ rad} = \left(\dfrac{9\pi}{2} \text{ rad}\right)\left(\dfrac{180 \text{ deg}}{\pi \text{ rad}}\right) = 810°$ Multiply by $180/\pi$.

c. $2 \text{ rad} = (2 \text{ rad})\left(\dfrac{180 \text{ deg}}{\pi \text{ rad}}\right) = \dfrac{360}{\pi} \approx 114.59°$ Multiply by $180/\pi$.

Applications

The *radian measure* formula, $\theta = s/r$, can be used to measure arc length along a circle. Specifically, for a circle of radius r, a central angle θ intercepts an arc of length s given by

$$s = r\theta \qquad\qquad \text{Length of circular arc}$$

where θ is measured in radians.

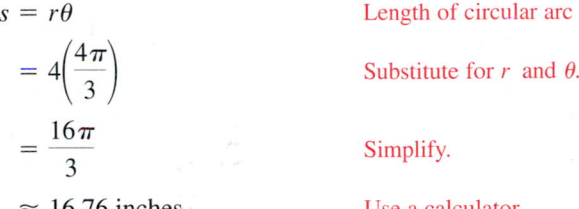

EXAMPLE 6   *Finding Arc Length*

A circle has a radius of 4 inches. Find the length of the arc intercepted by a central angle of 240°, as shown in Figure 6.13.

Solution

To use the formula $s = r\theta$, first convert 240° to radian measure.

$$240° = (240 \text{ deg})\left(\frac{\pi \text{ rad}}{180 \text{ deg}}\right) \qquad \text{Convert from degrees to radians.}$$

$$= \frac{4\pi}{3} \text{ rad} \qquad\qquad \text{Simplify.}$$

Then, using a radius of $r = 4$ inches, you can find the arc length to be

$$s = r\theta \qquad\qquad \text{Length of circular arc}$$

$$= 4\left(\frac{4\pi}{3}\right) \qquad\qquad \text{Substitute for } r \text{ and } \theta.$$

$$= \frac{16\pi}{3} \qquad\qquad \text{Simplify.}$$

$$\approx 16.76 \text{ inches} \qquad\qquad \text{Use a calculator.}$$

Note that the units for $r\theta$ are determined by the units for r because θ is given in radian measure and therefore has no units.

The formula for the length of a circular arc can be used to analyze the motion of a particle moving at a *constant speed* along a circular path. Consider a circle of radius r. If s is the length of the arc traveled in time t, the **speed** of the particle is

$$\text{Speed} = \frac{\text{distance}}{\text{time}} = \frac{s}{t}.$$

Moreover, if θ is the angle (in radian measure) corresponding to the arc length s, the **angular speed** of the particle is

$$\text{Angular speed} = \frac{\theta}{t}.$$

FIGURE 6.13

The *Interactive* CD-ROM shows every example with its solution; clicking on the *Try It!* button brings up similar problems. Guided Examples and Integrated Examples show step-by-step solutions to additional examples. Integrated Examples are related to several concepts in the section.

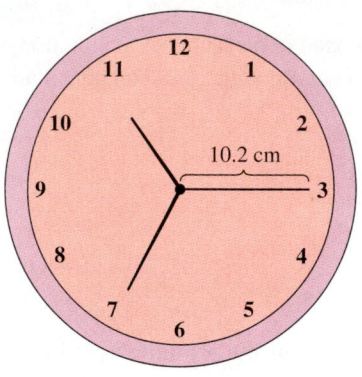

FIGURE 6.14

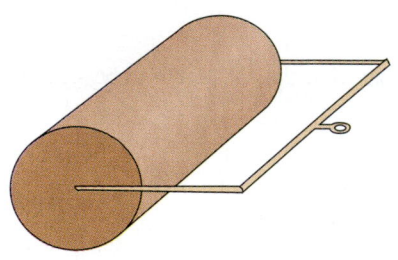

FIGURE 6.15

Real Life

EXAMPLE 7   *Finding the Speed of an Object*

The second hand of a clock is 10.2 centimeters long, as shown in Figure 6.14. Find the speed of the tip of this second hand.

Solution

The time required for the second hand to make one full revolution is

$$t = 60 \text{ seconds} = 1 \text{ minute.}$$

The distance traveled by the tip of the second hand in one revolution is

$$s = 2\pi \,(\text{radius}) = 2\pi(10.2) = 20.4\pi \text{ centimeters.}$$

Therefore, the speed of the tip of the second hand is

$$\text{Speed} = \frac{s}{t} = \frac{20.4\pi \text{ centimeters}}{60 \text{ seconds}} \approx 1.068 \text{ centimeters per second.}$$

Real Life

EXAMPLE 8   *Finding Angular Speed*

A lawn roller, as shown in Figure 6.15, makes 1.2 revolutions per second. Find the angular speed of the roller in radians per second.

Solution

Because each revolution generates 2π radians, it follows that the roller turns $(1.2)(2\pi) = 2.4\pi$ radians per second. In other words, the angular speed is

$$\text{Angular speed} = \frac{\theta}{t} = \frac{2.4\pi \text{ radians}}{1 \text{ second}} = 2.4\pi \text{ radians per second.}$$

GROUP ACTIVITY

DEGREE AND RADIAN MEASURE

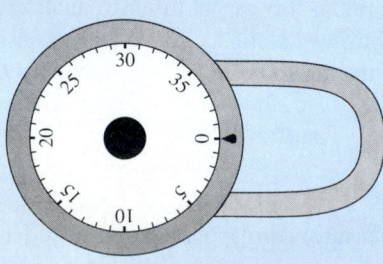

A standard combination lock has 40 numbers (0–39). Suppose the lock is positioned as shown in the figure, with its dial pointing to 0. Choose a three-number combination. Without revealing the combination, describe to your partner how to turn the dial *in terms of degree measure* to open the lock. Each angle measure should be given in standard position from the pointer. Remember that most locks follow a right-left-right pattern. Ask your partner to verify each number of the combination as you go. Switch roles and use radians to describe the combination.

The *Interactive* CD-ROM provides additional help with Warm-Up exercises by providing a hypertext link to the section in which the concept was introduced.

The *Interactive* CD-ROM contains step-by-step solutions to all odd-numbered Section and Review Exercises. It also provides Tutorial Exercises, which link to Guided Examples for additional help.

6.1 Exercises

In Exercises 1–4, estimate the number of degrees in the angle.

1.

2.

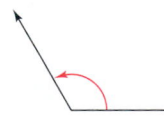

3.

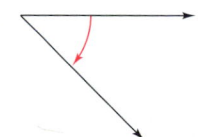

4.

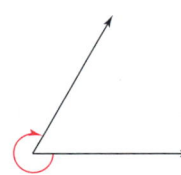

In Exercises 5–8, determine the quadrant in which the angle lies.

5. (a) $130°$ (b) $285°$

6. (a) $8.3°$ (b) $257° \, 30'$

7. (a) $-132° \, 50'$ (b) $-336°$

8. (a) $-260°$ (b) $-3.4°$

In Exercises 9–12, sketch the angle in standard position.

9. (a) $30°$ (b) $150°$

10. (a) $-270°$ (b) $-120°$

11. (a) $405°$ (b) $-480°$

12. (a) $750°$ (b) $-600°$

In Exercises 13–16, determine two coterminal angles (one positive and one negative) for the angle. Give the answers in degrees.

13. (a)   (b)

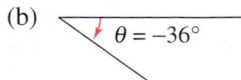

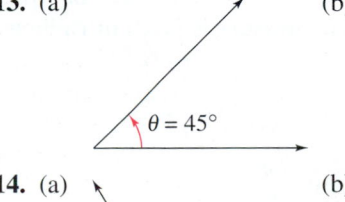

14. (a)   (b)

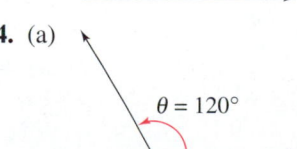
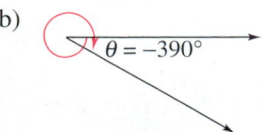

15. (a) $300°$ (b) $740°$

16. (a) $-420°$ (b) $230°$

In Exercises 17–20, convert the measure to decimal degree form.

17. (a) $54° 45'$ (b) $-128° 30'$

18. (a) $245° 10'$ (b) $2° 12'$

19. (a) $85° 18' 30''$ (b) $330° 25''$

20. (a) $-135° 36''$ (b) $-408° 16' 20''$

In Exercises 21–24, convert the measure to D° M′ S″ form.

21. (a) $240.6°$ (b) $-145.8°$

22. (a) $-345.12°$ (b) 0.45

23. (a) 2.5 (b) -3.58

24. (a) -0.355 (b) 0.7865

In Exercises 25–28, estimate the angle to the nearest one-half radian.

25. 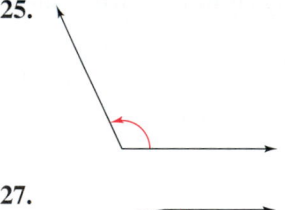   **26.**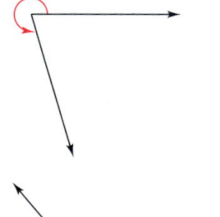

27.   **28.**

In Exercises 29–34, determine the quadrant in which the angle lies. (The angle measure is given in radians.)

29. (a) $\dfrac{\pi}{5}$ (b) $\dfrac{7\pi}{5}$

30. (a) $\dfrac{5\pi}{4}$ (b) $\dfrac{7\pi}{4}$

31. (a) $-\dfrac{\pi}{12}$ (b) $-\dfrac{11\pi}{9}$

32. (a) -1 (b) -2

33. (a) 3.5 (b) 2.25

34. (a) 5.63 (b) -2.25

In Exercises 35–38, sketch the angle in standard position.

35. (a) $\dfrac{5\pi}{4}$ (b) $\dfrac{2\pi}{3}$

36. (a) $-\dfrac{7\pi}{4}$ (b) $-\dfrac{5\pi}{2}$

37. (a) $\dfrac{11\pi}{6}$ (b) 7π

38. (a) 4 (b) -3

In Exercises 39–42, determine two coterminal angles (one positive and one negative) for the angle. Give the answers in radians.

39. (a) 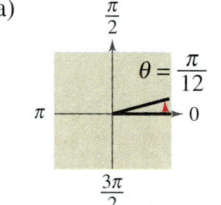   (b)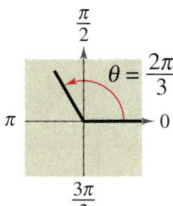

40. (a) 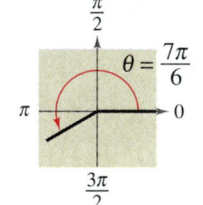   (b)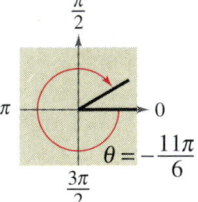

41. (a) $-\dfrac{9\pi}{4}$ (b) $-\dfrac{2\pi}{15}$ **42.** (a) $\dfrac{8\pi}{9}$ (b) $\dfrac{8\pi}{45}$

In Exercises 43–46, find (if possible) the complement and supplement of the angle.

43. (a) $18°$ (b) $115°$ **44.** (a) $79°$ (b) $150°$

45. (a) $\dfrac{\pi}{3}$ (b) $\dfrac{3\pi}{4}$ **46.** (a) 1 (b) 2

In Exercises 47–50, express the angle in radian measure as a multiple of π. (Do not use a calculator.)

47. (a) $30°$ (b) $150°$ **48.** (a) $315°$ (b) $120°$

49. (a) $-20°$ (b) $-240°$ **50.** (a) $-270°$ (b) $144°$

In Exercises 51–54, express the angle in degree measure. (Do not use a calculator.)

51. (a) $\dfrac{3\pi}{2}$ (b) $\dfrac{7\pi}{6}$

52. (a) $-\dfrac{7\pi}{12}$ (b) $\dfrac{\pi}{9}$

53. (a) $\dfrac{7\pi}{3}$ (b) $-\dfrac{11\pi}{30}$

54. (a) $\dfrac{11\pi}{6}$ (b) $\dfrac{34\pi}{15}$

In Exercises 55–62, convert the measure from degrees to radians. Round to three decimal places.

55. $115°$

56. $87.4°$

57. $-216.35°$

58. $-48.27°$

59. $532°$

60. $0.54°$

61. $-0.83°$

62. $345°$

In Exercises 63–70, convert the measure from radians to degrees. Round to three decimal places.

63. $\dfrac{\pi}{7}$

64. $\dfrac{5\pi}{11}$

65. $\dfrac{15\pi}{8}$

66. 6.5π

67. -4.2π

68. 4.8

69. -2

70. -0.57

In Exercises 71–74, find the angle in radians.

71.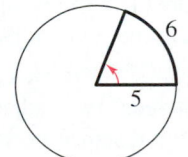

72.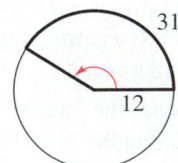

73.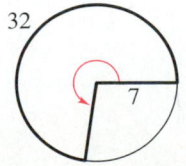

74.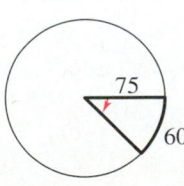

In Exercises 75–78, find the radian measure of the central angle of a circle of radius r that intercepts an arc of length s.

Radius	Arc Length
75. 15 inches	4 inches
76. 16 feet	10 feet
77. 14.5 centimeters	25 centimeters
78. 80 kilometers	160 kilometers

In Exercises 79–82, find the length of the arc on a circle of radius r intercepted by a central angle θ.

Radius	Central Angle
79. 15 inches	$180°$
80. 9 feet	$60°$
81. 6 meters	2 radians
82. 40 centimeters	$3\pi/4$ radians

Distance Between Cities   **In Exercises 83–86, find the distance between the cities. Assume that earth is a sphere of radius 4000 miles and the cities are on the same meridian (one city is due north of the other).**

City	Latitude
83. Dallas, Texas	$32°\ 47'\ 9''$N
Omaha, Nebraska	$41°\ 15'\ 42''$N
84. San Francisco, California	$37°\ 46'\ 39''$N
Seattle, Washington	$47°\ 36'\ 32''$N
85. Miami, Florida	$25°\ 46'\ 37''$N
Erie, Pennsylvania	$42°\ 7'\ 15''$N
86. Johannesburg, South Africa	$26°\ 10'$S
Jerusalem, Israel	$31°\ 47'$N

87. ***Difference in Latitudes***   Assuming that earth is a sphere of radius 6378 kilometers, what is the difference in latitude of two cities, one of which is 600 kilometers due north of the other?

88. ***Difference in Latitudes***   Assuming that earth is a sphere of radius 6378 kilometers, what is the difference in latitude of two cities, one of which is 800 kilometers due north of the other?

89. *Instrumentation*   The pointer on a voltmeter is 6 centimeters in length (see figure). Find the angle through which the pointer rotates when it moves 2.5 centimeters on the scale.

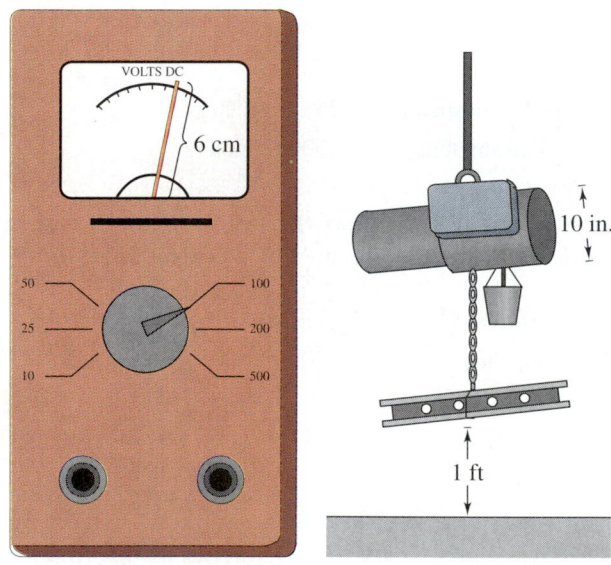

FIGURE FOR **89**   FIGURE FOR **90**

90. *Electric Hoist*   An electric hoist is being used to lift a piece of equipment (see figure). The diameter of the drum on the hoist is 10 inches, and the equipment must be raised 1 foot. Find the number of degrees through which the drum must rotate.

91. *Angular Speed*   A car is moving at a rate of 50 miles per hour, and the diameter of its wheels is 2.5 feet.

   (a) Find the number of revolutions per minute the wheels are rotating.

   (b) Find the angular speed of the wheels in radians per minute.

92. *Angular Speed*   A 2-inch-diameter pulley on an electric motor that runs at 1700 revolutions per minute is connected by a belt to a 4-inch-diameter pulley on a saw arbor.

   (a) Find the angular speed (in radians per minute) of each pulley.

   (b) Find the revolutions per minute of the saw.

93. *Think About It*   Is a degree or a radian the larger unit of measure? Explain.

94. *Essay*   If the radius of a circle is increasing and the magnitude of a central angle is held constant, how is the length of the intercepted arc changing? Explain your reasoning.

95. *Floppy Disk*   The radius of the magnetic disk in a 3.5-inch diskette is 1.68 inches. Find the linear speed of a point on the circumference of the disk if it is rotating at a speed of 360 revolutions per minute.

96. *Speed of a Bicycle*   The radii of the sprocket assemblies and the wheel of the bicycle in the figure are 4 inches, 2 inches, and 14 inches, respectively. If the cyclist is pedaling at a rate of 1 revolution per second, find the speed of the bicycle in (a) feet per second and (b) miles per hour.

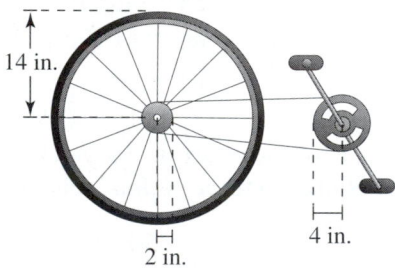

97. *Geometry*   Prove that the area of a circular sector of radius r with central angle θ is $A = \frac{1}{2}\theta r^2$, where θ is measured in radians.

Chapter Opener   In Exercises 98 and 99, use the information given in the chapter opener on page 455.

98. Suppose the seismographs were 1.2 miles apart. Find the central angle between the two adjacent seismographs.

99. Suppose the central angle between two adjacent seismographs is 0.031°. Find the distance between the seismographs.

6.2   *Right Triangle Trigonometry*

*See Exercises 71 and 72
on page 477 for examples of
how trigonometry can help find
dimensions of machined parts.*

*The Six Trigonometric Functions   □   Trigonometric Identities   □
Evaluating Trigonometric Functions with a Calculator   □
Applications Involving Right Triangles*

The Six Trigonometric Functions

Our first look at the trigonometric functions is from a *right triangle* perspective.
Consider a right triangle, one of whose acute angles is labeled θ, as shown in
Figure 6.16. The three sides of the triangle are the **hypotenuse,** the **opposite
side** (the side opposite the angle θ), and the **adjacent side** (the side adjacent to
the angle θ). Using the lengths of these three sides, you can form six ratios that
define the six trigonometric functions of the acute angle θ.

sine   cosecant   cosine   secant   tangent   cotangent

These six functions are normally abbreviated as sin, csc, cos, sec, tan, and cot,
respectively. In the following definition it is important to see that $0° < \theta < 90°$
and that for such angles the value of each trigonometric function is *positive*.

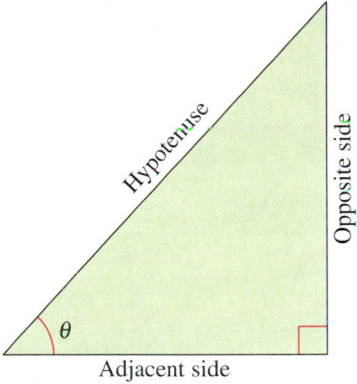

FIGURE 6.16

> ### RIGHT TRIANGLE DEFINITIONS OF TRIGONOMETRIC FUNCTIONS
>
> Let θ be an *acute* angle of a right triangle. The six trigonometric
> functions *of the angle* θ are defined as follows.
>
> $$\sin \theta = \frac{\text{opp}}{\text{hyp}} \qquad \csc \theta = \frac{\text{hyp}}{\text{opp}}$$
>
> $$\cos \theta = \frac{\text{adj}}{\text{hyp}} \qquad \sec \theta = \frac{\text{hyp}}{\text{adj}}$$
>
> $$\tan \theta = \frac{\text{opp}}{\text{adj}} \qquad \cot \theta = \frac{\text{adj}}{\text{opp}}$$
>
> The abbreviation opp, adj, and hyp represent the lengths of the three
> sides of a right triangle.
>
> opp = the length of the side *opposite* θ
>
> adj = the length of the side *adjacent* θ
>
> hyp = the length of the *hypotenuse*

The leading Teutonic mathematical
astronomer of the 16th century was
Georg Joachim Rhaeticus (1514–1576).
He was the first to define the trigonometric functions as ratios of the sides of
a right triangle.

NOTE   The functions in the second column above are the *reciprocals* of the
corresponding functions in the first column. ■■

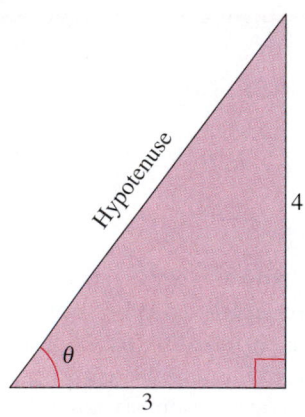

FIGURE **6.17**

EXAMPLE 1 *Evaluating Trigonometric Functions*

Find the values of the six trigonometric functions of θ, as shown in Figure 6.17.

Solution

By the Pythagorean Theorem, $(\text{hyp})^2 = (\text{opp})^2 + (\text{adj})^2$, it follows that

$$\text{hyp} = \sqrt{4^2 + 3^2} = \sqrt{25} = 5.$$

Thus, the six trigonometric functions of θ are

$$\sin \theta = \frac{\text{opp}}{\text{hyp}} = \frac{4}{5} \qquad \csc \theta = \frac{\text{hyp}}{\text{opp}} = \frac{5}{4}$$

$$\cos \theta = \frac{\text{adj}}{\text{hyp}} = \frac{3}{5} \qquad \sec \theta = \frac{\text{hyp}}{\text{adj}} = \frac{5}{3}$$

$$\tan \theta = \frac{\text{opp}}{\text{adj}} = \frac{4}{3} \qquad \cot \theta = \frac{\text{adj}}{\text{opp}} = \frac{3}{4}.$$

In Example 1, you were given the lengths of two sides of the right triangle, but not the angle θ. It is more common in trigonometry to be asked to find the trigonometric functions of a *given* acute angle θ. To do this, you can construct a right triangle having θ as one of its angles.

EXAMPLE 2 *Evaluating Trigonometric Functions of 45°*

Find the values of $\sin 45°$, $\cos 45°$, and $\tan 45°$.

Solution

Construct a right triangle having 45° as one of its acute angles, as shown in Figure 6.18. Choose the length of the adjacent side to be 1. From geometry, you know that the other acute angle is also 45°. Hence, the triangle is isosceles and the length of the opposite side is also 1. Using the Pythagorean Theorem, you find the length of the hypotenuse to be $\sqrt{2}$.

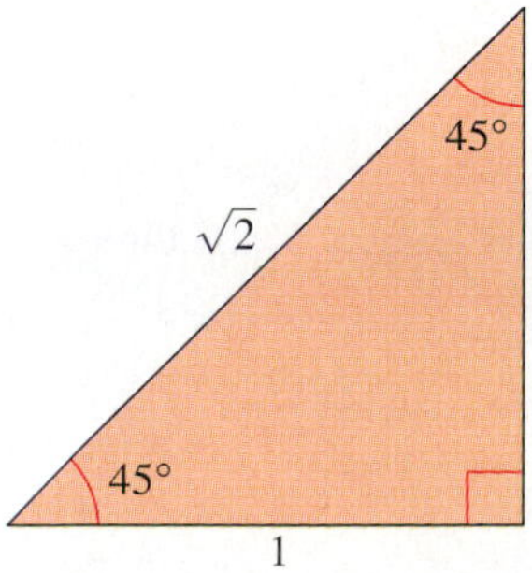

$$\sin 45° = \frac{\text{opp}}{\text{hyp}} = \frac{1}{\sqrt{2}} = \frac{\sqrt{2}}{2}$$

$$\cos 45° = \frac{\text{adj}}{\text{hyp}} = \frac{1}{\sqrt{2}} = \frac{\sqrt{2}}{2}$$

$$\tan 45° = \frac{\text{opp}}{\text{adj}} = \frac{1}{1} = 1$$

FIGURE **6.18**

EXAMPLE 3 *Evaluating Trigonometric Functions of 30° and 60°*

Use the equilateral triangle shown in Figure 6.19 to find the values of sin 60°, cos 60°, sin 30°, and cos 30°.

Solution

Try using the Pythagorean Theorem and the equilateral triangle in Figure 6.19 to verify the lengths of the sides given in Figure 6.19. For $\theta = 60°$, you have adj $= 1$, opp $= \sqrt{3}$, and hyp $= 2$. Therefore,

$$\sin 60° = \frac{\text{opp}}{\text{hyp}} = \frac{\sqrt{3}}{2} \quad \text{and} \quad \cos 60° = \frac{\text{adj}}{\text{hyp}} = \frac{1}{2}.$$

For $\theta = 30°$, adj $= \sqrt{3}$, opp $= 1$, and hyp $= 2$. Thus,

$$\sin 30° = \frac{\text{opp}}{\text{hyp}} = \frac{1}{2} \quad \text{and} \quad \cos 30° = \frac{\text{adj}}{\text{hyp}} = \frac{\sqrt{3}}{2}.$$

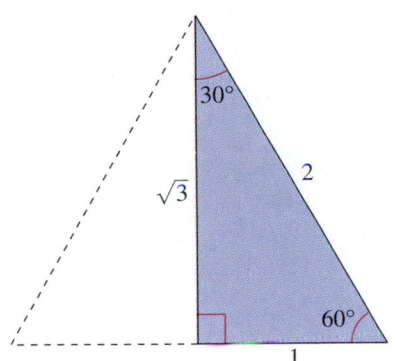

FIGURE 6.19

Because the angles 30°, 45°, and 60° ($\pi/6$, $\pi/4$, and $\pi/3$) occur frequently in trigonometry, we suggest that you learn to construct the triangles shown in Figures 6.18 and 6.19.

SINES, COSINES, AND TANGENTS OF SPECIAL ANGLES

$$\sin 30° = \sin \frac{\pi}{6} = \frac{1}{2} \qquad \cos 30° = \cos \frac{\pi}{6} = \frac{\sqrt{3}}{2} \qquad \tan 30° = \tan \frac{\pi}{6} = \frac{\sqrt{3}}{3}$$

$$\sin 45° = \sin \frac{\pi}{4} = \frac{\sqrt{2}}{2} \qquad \cos 45° = \cos \frac{\pi}{4} = \frac{\sqrt{2}}{2} \qquad \tan 45° = \tan \frac{\pi}{4} = 1$$

$$\sin 60° = \sin \frac{\pi}{3} = \frac{\sqrt{3}}{2} \qquad \cos 60° = \cos \frac{\pi}{3} = \frac{1}{2} \qquad \tan 60° = \tan \frac{\pi}{3} = \sqrt{3}$$

In the box, note that $\sin 30° = \frac{1}{2} = \cos 60°$. This occurs because 30° and 60° are complementary angles, and, in general, it can be shown from the right triangle definitions that *cofunctions of complementary angles are equal.* That is, if θ is an acute angle, the following relationships are true.

$$\sin(90° - \theta) = \cos \theta \qquad \cos(90° - \theta) = \sin \theta$$
$$\tan(90° - \theta) = \cot \theta \qquad \cot(90° - \theta) = \tan \theta$$
$$\sec(90° - \theta) = \csc \theta \qquad \csc(90° - \theta) = \sec \theta$$

Trigonometric Identities

In trigonometry, a great deal of time is spent studying relationships between trigonometric functions (identities).

FUNDAMENTAL TRIGONOMETRIC IDENTITIES

Reciprocal Identities

$$\sin\theta = \frac{1}{\csc\theta} \qquad \cos\theta = \frac{1}{\sec\theta} \qquad \tan\theta = \frac{1}{\cot\theta}$$

$$\csc\theta = \frac{1}{\sin\theta} \qquad \sec\theta = \frac{1}{\cos\theta} \qquad \cot\theta = \frac{1}{\tan\theta}$$

Quotient Identities

$$\tan\theta = \frac{\sin\theta}{\cos\theta} \qquad \cot\theta = \frac{\cos\theta}{\sin\theta}$$

Pythagorean Identities

$$\sin^2\theta + \cos^2\theta = 1 \qquad 1 + \tan^2\theta = \sec^2\theta$$

$$1 + \cot^2\theta = \csc^2\theta$$

NOTE   Note that $\sin^2\theta$ represents $(\sin\theta)^2$, $\cos^2\theta$ represents $(\cos\theta)^2$, and so on. ▪▪

EXAMPLE 4   *Applying Trigonometric Identities*

Let θ be an acute angle such that $\sin\theta = 0.6$. Find the values of (a) $\cos\theta$ and (b) $\tan\theta$ using trigonometric identities.

Solution

a. To find the value of $\cos\theta$, use the Pythagorean identity

$$\sin^2\theta + \cos^2\theta = 1.$$

Thus, you have

$$(0.6)^2 + \cos^2\theta = 1$$
$$\cos^2\theta = 1 - (0.6)^2 = 0.64$$
$$\cos\theta = \sqrt{0.64} = 0.8.$$

b. Now, knowing the sine and cosine of θ, you can find the tangent of θ to be

$$\tan\theta = \frac{\sin\theta}{\cos\theta} = \frac{0.6}{0.8} = 0.75.$$

Try using the definitions of $\cos\theta$ and $\tan\theta$, and the triangle shown in Figure 6.20, to check these results.

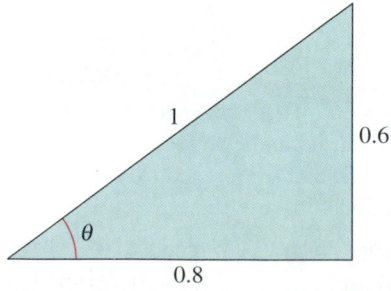

FIGURE 6.20

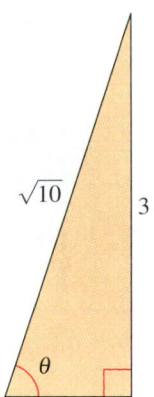

FIGURE 6.21

EXAMPLE 5   *Applying Trigonometric Identities*

Let θ be an acute angle such that $\tan \theta = 3$. Find the values of (a) $\cot \theta$ and (b) $\sec \theta$ using trigonometric identities.

Solution

a. $\cot \theta = \dfrac{1}{\tan \theta}$ Reciprocal identity

$\cot \theta = \dfrac{1}{3}$

b. $\sec^2 \theta = 1 + \tan^2 \theta$ Pythagorean identity

$\sec^2 \theta = 1 + 3^2 = 10$

$\sec \theta = \sqrt{10}$

Try using the definitions of $\cot \theta$ and $\sec \theta$, and the triangle shown in Figure 6.21, to check these results.

Evaluating Trigonometric Functions with a Calculator

When evaluating a trigonometric function with a calculator, you need to set the calculator to the desired *mode* of measurement (degrees or radians).

Most calculators do not have keys for the cosecant, secant, and cotangent functions. To evaluate these functions, you can use the $\boxed{x^{-1}}$ key with their respective reciprocal functions sine, cosine, and tangent. For example, to evaluate $\csc(\pi/8)$, use the fact that

$$\csc \frac{\pi}{8} = \frac{1}{\sin(\pi/8)}$$

and enter the following keystroke sequence in radian mode.

$\boxed{(}$ $\boxed{\text{SIN}}$ $\boxed{(}$ $\boxed{\pi}$ $\boxed{\div}$ 8 $\boxed{)}$ $\boxed{)}$ $\boxed{x^{-1}}$ $\boxed{\text{ENTER}}$ Display
2.6131259

EXAMPLE 6   *Using a Calculator*

Function	Mode	Graphing Calculator Keystrokes	Display
a. $\sin 76.4°$	Degree	$\boxed{\text{SIN}}$ 76.4 $\boxed{\text{ENTER}}$	0.9719610
b. $\cot 1.5$	Radian	$\boxed{(}$ $\boxed{\text{TAN}}$ 1.5 $\boxed{)}$ $\boxed{x^{-1}}$ $\boxed{\text{ENTER}}$	0.0709148

Applications Involving Right Triangles

Many applications of trigonometry involve a process called **solving right triangles.** In this type of application, you are usually given one side of a right triangle and one of the acute angles and asked to find one of the other sides, *or* you are given two sides and asked to find one of the acute angles.

EXAMPLE 7   *Using Trigonometry to Solve a Right Triangle*   *Real Life*

A surveyor is standing 50 feet from the base of a large tree, as shown in Figure 6.22. The surveyor measures the angle of elevation to the top of the tree as 71.5°. How tall is the tree?

Solution

From Figure 6.22, you see that

$$\tan 71.5° = \frac{\text{opp}}{\text{adj}} = \frac{y}{x}$$

where $x = 50$ and y is the height of the tree. Thus, the height of the tree is

$$y = x \tan 71.5° \approx 50(2.98868) \approx 149.4 \text{ feet.}$$

FIGURE 6.22

EXAMPLE 8   *Using Trigonometry to Solve a Right Triangle*   *Real Life*

A person is 200 yards from a river. Rather than walking directly to the river, the person walks 400 yards along a straight path to the river's edge. Find the acute angle θ between this path and the river's edge, as illustrated in Figure 6.23.

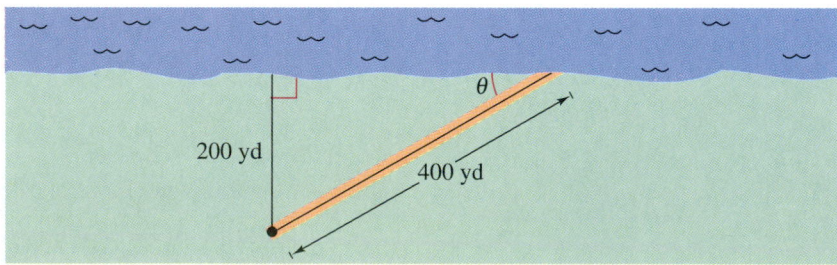

FIGURE 6.23

Solution

From Figure 6.23, you can see that the sine of the angle θ is

$$\sin \theta = \frac{\text{opp}}{\text{hyp}} = \frac{200}{400} = \frac{1}{2}.$$

Therefore, $\theta = 30°$.

Land surveyors use fixed boundaries to find areas of plots of ground. The transit, a small telescope mounted on a tripod, can measure horizontal and vertical angles within small fractions of degrees. *(Photo: "Images © 1995 PhotoDisc, Inc.")*

The *Interactive* CD-ROM offers graphing utility emulators of the *TI-82* and *TI-83*, which can be used with the Examples, Explorations, Technology notes, and Exercises.

In Example 8, you were able to recognize that the acute angle that satisfies the equation $\sin \theta = \frac{1}{2}$ is $\theta = 30°$. Suppose, however, that you were given the equation $\sin \theta = 0.6$ and asked to find the acute angle θ. Because

$$\sin 30° = \frac{1}{2} = 0.5000 \qquad \text{and} \qquad \sin 45° = \frac{1}{\sqrt{2}} \approx 0.7071$$

you might guess that θ lies somewhere between 30° and 45°. A more precise value of θ can be found using the *inverse* key on a calculator. To do this, you can use the following keystroke sequence in degree mode.

$\boxed{\text{SIN}^{-1}}$ .6 $\boxed{\text{ENTER}}$ Display 36.8699

Thus, you can conclude that if $\sin \theta = 0.6$, then $\theta \approx 36.87°$.

EXAMPLE 9   *Using Trigonometry to Solve a Right Triangle*   Real Life

A 12-meter flagpole casts a 9-meter shadow, as shown in Figure 6.24. Find θ, the angle of elevation of the sun.

Solution

Figure 6.24 shows that the *opposite* and *adjacent* sides are known. Thus,

$$\tan \theta = \frac{\text{opp}}{\text{adj}} = \frac{12}{9}.$$

With a calculator in degree mode you use the keystrokes

$\boxed{\text{TAN}^{-1}}$ $\boxed{(}$ 12 $\boxed{\div}$ 9 $\boxed{)}$ $\boxed{\text{ENTER}}$

to obtain $\theta \approx 53.13°$.

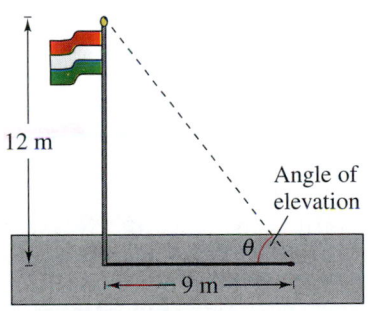

FIGURE 6.24

12 m

9 m

Angle of elevation

θ

GROUP ACTIVITY

ERROR ANALYSIS

Suppose you are tutoring a student in trigonometry. Your student is asked to evaluate the cosine of 30° and, using a calculator, obtains the following.

Keystrokes	*Display*
$\boxed{\text{COS}}$ 30 $\boxed{\text{ENTER}}$	0.1542514

Because you know that $\cos 30° = \sqrt{3}/2 \approx 0.866$, you realize that the answer is incorrect. What has your student done wrong?

Find the distance between the points.

1. $(3, 8), (1, 4)$

2. $(5, 2), (2, -7)$

3. $(-4, 0), (2, 8)$

4. $(-3, -3), (0, 0)$

Perform the operations. (Round your answer to two decimal places.)

5. 0.300×4.125

6. 7.30×43.50

7. $\dfrac{151.5}{2.40}$

8. $\dfrac{3740}{28.0}$

9. $\dfrac{19,500}{0.007}$

10. $\dfrac{(10.5)(3401)}{1240}$

6.2 Exercises

In Exercises 1–4, find the exact values of the six trigonometric functions of the angle θ given in the figure. (Use the Pythagorean Theorem to find the third side of the triangle.)

1.

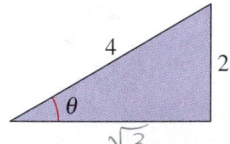

2.

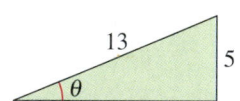

3.

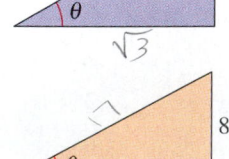

4.

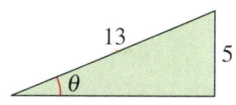

5.

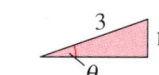

6.

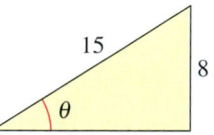

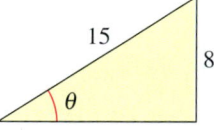

7.

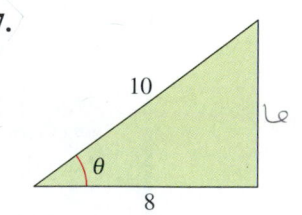

8.

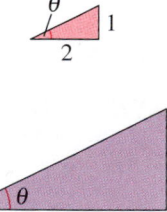

In Exercises 5–8, find the exact values of the six trigonometric functions of the angle θ for each of the triangles. Explain why the function values are the same.

In Exercises 9–16, sketch a right triangle corresponding to the trigonometric function of the acute angle θ. Use the Pythagorean Theorem to determine the third side and then find the other five trigonometric functions of θ.

9. $\sin \theta = \frac{2}{3}$ **10.** $\cot \theta = 5$

11. $\sec \theta = 2$ **12.** $\cos \theta = \frac{5}{7}$

13. $\tan \theta = 3$ **14.** $\csc \theta = \frac{17}{4}$

15. $\cot \theta = \frac{3}{2}$ **16.** $\sin \theta = \frac{3}{8}$

In Exercises 17–22, use the given function value(s), and trigonometric identities (including the relationship between a trigonometric function and its cofunction of a complementary angle) to find the indicated trigonometric functions.

17. $\sin 60° = \dfrac{\sqrt{3}}{2}$, $\cos 60° = \dfrac{1}{2}$

   (a) $\tan 60°$ (b) $\sin 30°$

   (c) $\cos 30°$ (d) $\cot 60°$

18. $\sin 30° = \dfrac{1}{2}$, $\tan 30° = \dfrac{\sqrt{3}}{3}$

   (a) $\csc 30°$ (b) $\cot 60°$

   (c) $\cos 30°$ (d) $\cot 30°$

19. $\csc \theta = 3$, $\sec \theta = \dfrac{3\sqrt{2}}{4}$

   (a) $\sin \theta$ (b) $\cos \theta$

   (c) $\tan \theta$ (d) $\sec(90° - \theta)$

20. $\sec \theta = 5$, $\tan \theta = 2\sqrt{6}$

   (a) $\cos \theta$ (b) $\cot \theta$

   (c) $\cot(90° - \theta)$ (d) $\sin \theta$

21. $\cos \alpha = \frac{1}{4}$

   (a) $\sec \alpha$ (b) $\sin \alpha$

   (c) $\cot \alpha$ (d) $\sin(90° - \alpha)$

22. $\tan \beta = 5$

   (a) $\cot \beta$ (b) $\cos \beta$

   (c) $\tan(90° - \beta)$ (d) $\csc \beta$

In Exercises 23–32, use trigonometric identities to transform one side of the equation into the other.

23. $\tan \theta \cot \theta = 1$ **24.** $\cos \theta \sec \theta = 1$

25. $\tan \alpha \cos \alpha = \sin \alpha$ **26.** $\cot \alpha \sin \alpha = \cos \alpha$

27. $(1 + \cos \theta)(1 - \cos \theta) = \sin^2 \theta$

28. $(1 + \sin \theta)(1 - \sin \theta) = \cos^2 \theta$

29. $(\sec \theta + \tan \theta)(\sec \theta - \tan \theta) = 1$

30. $\sin^2 \theta - \cos^2 \theta = 2\sin^2 \theta - 1$

31. $\dfrac{\sin \theta}{\cos \theta} + \dfrac{\cos \theta}{\sin \theta} = \csc \theta \sec \theta$

32. $\dfrac{\tan \beta + \cot \beta}{\tan \beta} = \csc^2 \beta$

In Exercises 33–36, evaluate the trigonometric function by memory or by constructing an appropriate triangle for the given special angle.

33. (a) $\cos 60°$ (b) $\tan \dfrac{\pi}{6}$

34. (a) $\csc 30°$ (b) $\sin \dfrac{\pi}{4}$

35. (a) $\cot 45°$ (b) $\cos 45°$

36. (a) $\sin \dfrac{\pi}{3}$ (b) $\csc 45°$

In Exercises 37–46, use a calculator to evaluate each function. Round your answers to four decimal places. (Be sure the calculator is in the correct mode.)

37. (a) $\sin 10°$ (b) $\cos 80°$

38. (a) $\tan 23.5°$ (b) $\cot 66.5°$

39. (a) $\sin 16.35°$ (b) $\csc 16.35°$

40. (a) $\cos 16° \, 18'$ (b) $\sin 73° \, 56'$

41. (a) $\sec 42° \, 12'$ (b) $\csc 48° \, 7'$

42. (a) $\cos 4° \, 50' \, 15''$ (b) $\sec 4° \, 50' \, 15''$

43. (a) $\cot (\pi/16)$ (b) $\tan (\pi/16)$

44. (a) $\sec 0.75$ (b) $\cos 0.75$

45. (a) $\csc 1$ (b) $\tan \frac{1}{2}$

46. (a) $\sec(\pi/2 - 1)$ (b) $\cot(\pi/2 - 1/2)$

In Exercises 47–52, find the values of θ in degrees $(0° < \theta < 90°)$ and radians $(0 < \theta < \pi/2)$ without the aid of a calculator.

47. (a) $\sin \theta = \dfrac{1}{2}$ (b) $\csc \theta = 2$

48. (a) $\cos \theta = \dfrac{\sqrt{2}}{2}$ (b) $\tan \theta = 1$

49. (a) $\sec \theta = 2$ (b) $\cot \theta = 1$

50. (a) $\tan \theta = \sqrt{3}$ (b) $\cos \theta = \dfrac{1}{2}$

51. (a) $\csc \theta = \dfrac{2\sqrt{3}}{3}$ (b) $\sin \theta = \dfrac{\sqrt{2}}{2}$

52. (a) $\cot \theta = \dfrac{\sqrt{3}}{3}$ (b) $\sec \theta = \sqrt{2}$

In Exercises 53–56, find the values of θ in degrees $(0° < \theta < 90°)$ and radians $(0 < \theta < \pi/2)$ by using a calculator.

53. (a) $\sin \theta = 0.8191$ (b) $\cos \theta = 0.0175$
54. (a) $\cos \theta = 0.9848$ (b) $\cos \theta = 0.8746$
55. (a) $\tan \theta = 1.1920$ (b) $\tan \theta = 0.4663$
56. (a) $\sin \theta = 0.3746$ (b) $\cos \theta = 0.3746$

In Exercises 57–64, solve for x, y, or r, as indicated.

57. Solve for y.

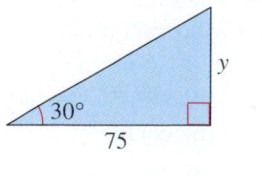

58. Solve for x.

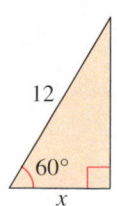

59. Solve for x.

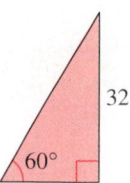

60. Solve for r.

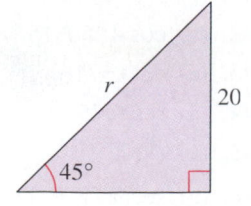

61. Solve for r.

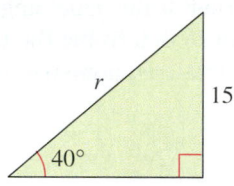

62. Solve for x.

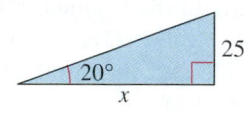

63. Solve for y.

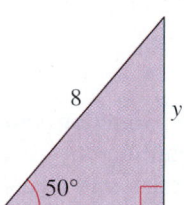

64. Solve for r.

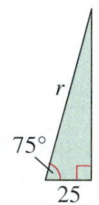

65. *Height*   A 6-foot person standing 15 feet from a streetlight casts an 8-foot shadow (see figure). What is the height of the streetlight?

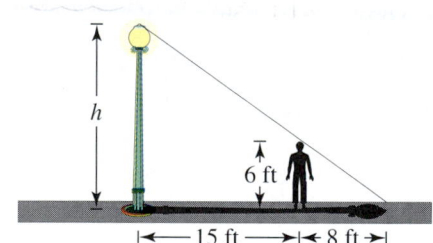

66. *Height*   A 6-foot person walked from the base of a broadcasting tower directly toward the tip of the shadow cast by the tower. When the person was 132 feet from the tower and 3 feet from the tip of the shadow, the person's shadow started to appear beyond the tower's shadow.

(a) Draw a right triangle that gives a visual representation of the problem. Show the known quantities of the triangle and use a variable to indicate the height of the tower.

(b) Write an equation involving the unknown.

(c) What is the height of the tower?

67. **Length** A 30-meter line is used to tether a helium-filled balloon. Because of a breeze, the line makes an angle of approximately 75° with the ground.

    (a) Draw a right triangle that gives a visual representation of the problem. Show the known quantities on the triangle and use a variable to indicate the height of the balloon.

    (b) Use a trigonometric function to write an equation involving the unknown.

    (c) What is the height of the balloon?

68. **Width of a River** A biologist wants to know the width w of a river in order to properly set instruments for studying the pollutants in the water. From point A, the biologist walks downstream 100 feet and sights to point C (see figure). From this sighting, it is determined that $\theta = 54°$. How wide is the river?

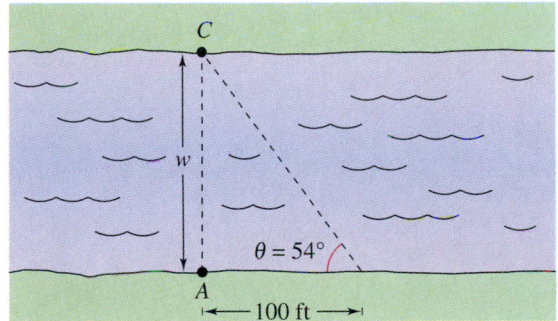

69. **Distance** From a 60-foot observation tower on the coast, a Coast Guard officer sights a boat in difficulty. The angle of depression of the boat is 3° (see figure). How far is the boat from the shoreline?

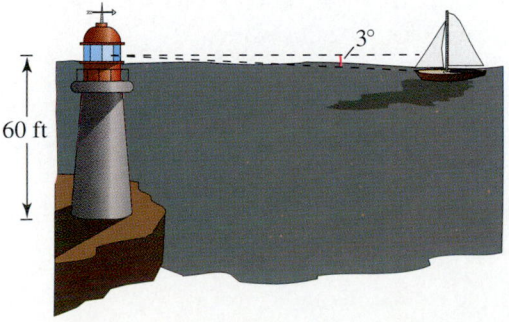

70. **Angle of Elevation** A ramp 20 feet in length rises to a loading platform that is $3\frac{1}{3}$ feet off the ground.

    (a) Draw a right triangle that gives a visual representation of the problem. Show the known quantities on the triangle and use a variable to indicate the angle of elevation of the ramp.

    (b) Use a trigonometric function to write an equation involving the unknown.

    (c) What is the angle of elevation of the ramp?

71. **Machine Shop Calculations** A steel plate has the form of one-fourth of a circle with a radius of 60 centimeters. Two 2-centimeter holes are to be drilled in the plate positioned as shown in the figure. Find the coordinates of the center of each hole.

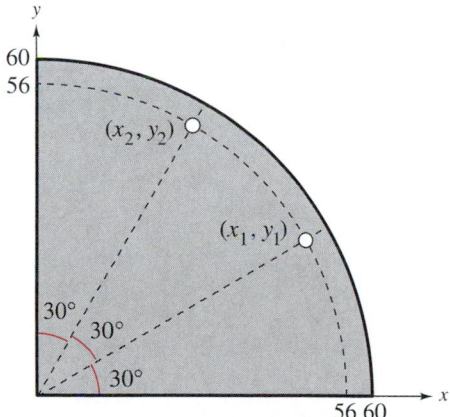

72. **Machine Shop Calculations** A tapered shaft has a diameter of 5 centimeters at the small end and is 15 centimeters long (see figure). If the taper is 3°, find the diameter d of the large end of the shaft.

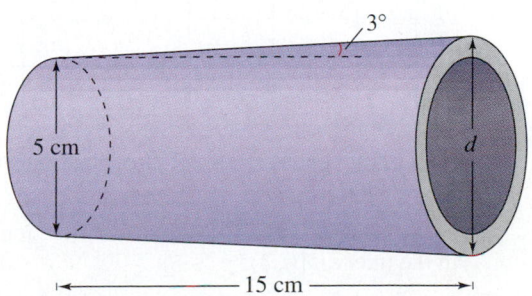

73. *Geometry* Use a compass to sketch a quarter of a circle of radius 10 centimeters. Using a protractor, construct an angle of 20° in standard position (see figure). Drop a perpendicular from the point of intersection of the terminal side of the angle and the arc of the circle. By actual measurement, calculate the coordinates (x, y) of the point of intersection and use these measurements to approximate the six trigonometric functions of a 20° angle.

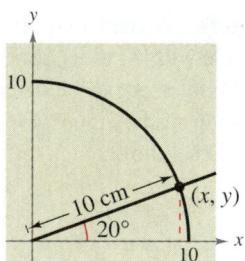

74. *Geometry* Repeat Exercise 73 using a 75° angle.

75. *Exploration*

(a) Complete the table below.

θ	0	0.1	0.2	0.3	0.4	0.5
$\sin \theta$						

(b) Is θ or $\sin \theta$ greater for θ in the interval $[0, 0.5]$?

(c) As θ approaches 0, how do θ and $\sin \theta$ compare? Explain.

76. *Exploration*

(a) Complete the table below.

θ	0	0.3	0.6	0.9	1.2	1.5
$\sin \theta$						
$\cos \theta$						

(b) Discuss the behavior of the sine function for θ in the interval $[0, 1.5]$.

(c) Discuss the behavior of the cosine function for θ in the interval $[0, 1.5]$.

(d) Use the definitions of the sine and cosine functions to explain the results of parts (b) and (c).

In Exercises 77–82, determine whether the statement is true or false, and give a reason for your answer.

77. $\sin 60° \csc 60° = 1$

78. $\sec 30° = \csc 60°$

79. $\sin 45° + \cos 45° = 1$

80. $\cot^2 10° - \csc^2 10° = -1$

81. $\dfrac{\sin 60°}{\sin 30°} = \sin 2°$

82. $\tan[(0.8)^2] = \tan^2(0.8)$

Review Solve Exercises 83–86 as a review of the skills and problem-solving techniques you learned in previous sections. Perform the operations and simplify.

83. $\dfrac{x^2 - 6x}{x^2 + 4x - 12} \cdot \dfrac{x^2 + 12x + 36}{x^2 - 36}$

84. $\dfrac{2t^2 + 5t - 12}{9 - 4t^2} \div \dfrac{t^2 - 16}{4t^2 + 12t + 9}$

85. $\dfrac{3}{x + 2} - \dfrac{2}{x - 2} + \dfrac{x}{x^2 + 4x + 4}$

86. $\dfrac{\left(\dfrac{3}{x} - \dfrac{1}{4}\right)}{\left(\dfrac{12}{x} - 1\right)}$

6.3   *Trigonometric Functions of Any Angle*

See Exercise 95 on page 491 for an example of how a trigonometric function can be used to model the average daily temperature in a city.

Introduction ◻ *Reference Angles* ◻ *Trigonometric Functions of Real Numbers*

Introduction

In Section 6.2, the definitions of trigonometric functions were restricted to acute angles. In this section, the definitions are extended to cover *any* angle.

DEFINITIONS OF TRIGONOMETRIC FUNCTIONS OF ANY ANGLE

Let θ be an angle in standard position with (x, y) a point on the terminal side of θ and $r = \sqrt{x^2 + y^2} \neq 0$, as shown in Figure 6.25.

$$\sin \theta = \frac{y}{r} \qquad\qquad \cos \theta = \frac{x}{r}$$

$$\tan \theta = \frac{y}{x}, \quad x \neq 0 \qquad\qquad \cot \theta = \frac{x}{y}, \quad y \neq 0$$

$$\sec \theta = \frac{r}{x}, \quad x \neq 0 \qquad\qquad \csc \theta = \frac{r}{y}, \quad y \neq 0$$

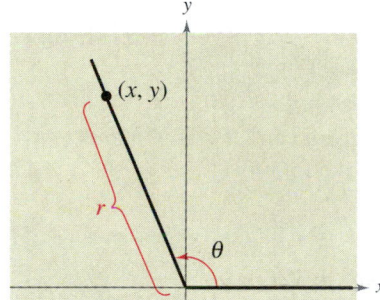

FIGURE 6.25

NOTE   If θ is an *acute* angle, these definitions coincide with those given in the previous section. ■■

Because $r = \sqrt{x^2 + y^2}$ *cannot* be zero, it follows that the sine and cosine functions are defined for any real value of θ. However, if $x = 0$, the tangent and secant of θ are undefined. For example, the tangent of 90° is undefined. Similarly, if $y = 0$, the cotangent and cosecant of θ are undefined.

EXAMPLE 1   *Evaluating Trigonometric Functions*

Let $(-3, 4)$ be a point on the terminal side of θ. Find the sine, cosine, and tangent of θ.

Solution

Referring to Figure 6.26, you see that $x = -3$, $y = 4$, and

$$r = \sqrt{x^2 + y^2} = \sqrt{(-3)^2 + 4^2} = \sqrt{25} = 5.$$

Thus, you have the following.

$$\sin \theta = \frac{y}{r} = \frac{4}{5}, \qquad \cos \theta = \frac{x}{r} = -\frac{3}{5}, \qquad \tan \theta = \frac{y}{x} = -\frac{4}{3}$$

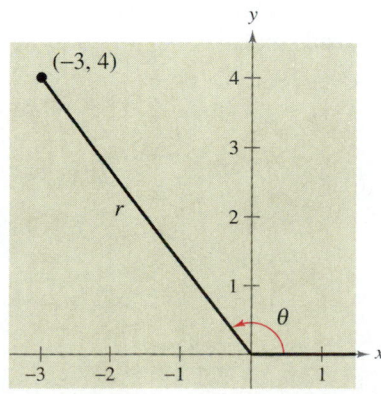

FIGURE 6.26

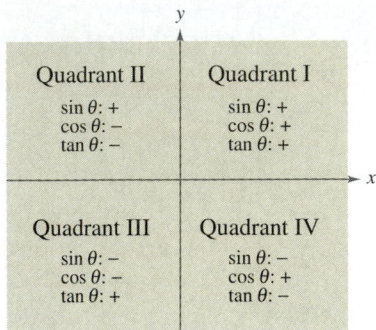

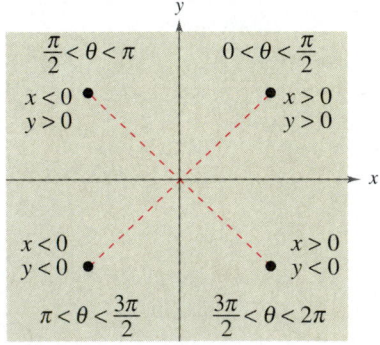

FIGURE 6.27

The *signs* of the trigonometric functions in the four quadrants can be determined easily from the definitions of the functions. For instance, because $\cos \theta = x/r$, it follows that $\cos \theta$ is positive wherever $x > 0$, which is in Quadrants I and IV. (Remember, r is always positive.) In a similar manner you can verify the results shown in Figure 6.27.

■

EXAMPLE 2 *Evaluating Trigonometric Functions*

Given $\tan \theta = -\frac{5}{4}$ and $\cos \theta > 0$, find $\sin \theta$ and $\sec \theta$.

Solution

Note that θ lies in Quadrant IV because that is the only quadrant in which the tangent is negative and the cosine is positive. Moreover, using

$$\tan \theta = \frac{y}{x} = -\frac{5}{4}$$

and the fact that y is negative in Quadrant IV, you can let $y = -5$ and $x = 4$. Hence, $r = \sqrt{16 + 25} = \sqrt{41}$ and you have the following.

$$\sin \theta = \frac{y}{r} = \frac{-5}{\sqrt{41}} \approx -0.7809$$

$$\sec \theta = \frac{r}{x} = \frac{\sqrt{41}}{4} \approx 1.6008$$

■

EXAMPLE 3 *Trigonometric Functions of Quadrant Angles*

Evaluate the sine function at the four quadrant angles 0, $\dfrac{\pi}{2}$, π, and $\dfrac{3\pi}{2}$.

Solution

To begin, choose a point on the terminal side of each angle, as shown in Figure 6.28. For each of the four given points, $r = 1$, and you have

$$\sin 0 = \frac{y}{r} = \frac{0}{1} = 0 \qquad \textcolor{red}{(x, y) = (1, 0)}$$

$$\sin \frac{\pi}{2} = \frac{y}{r} = \frac{1}{1} = 1 \qquad \textcolor{red}{(x, y) = (0, 1)}$$

$$\sin \pi = \frac{y}{r} = \frac{0}{1} = 0 \qquad \textcolor{red}{(x, y) = (-1, 0)}$$

$$\sin \frac{3\pi}{2} = \frac{y}{r} = \frac{-1}{1} = -1. \qquad \textcolor{red}{(x, y) = (0, -1)}$$

Try using Figure 6.28 to evaluate some of the other trigonometric functions at the four quadrant angles.

■

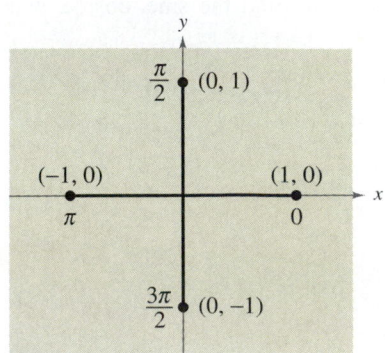

FIGURE 6.28

Reference Angles

The values of the trigonometric functions of angles greater than 90° (or less than 0°) can be determined from their values at corresponding acute angles called **reference angles.**

> ### DEFINITION OF REFERENCE ANGLE
>
> Let θ be an angle in standard position. Its **reference angle** is the acute angle θ' formed by the terminal side of θ and the horizontal axis.

Figure 6.29 shows the reference angles for θ in Quadrants II, III, and IV.

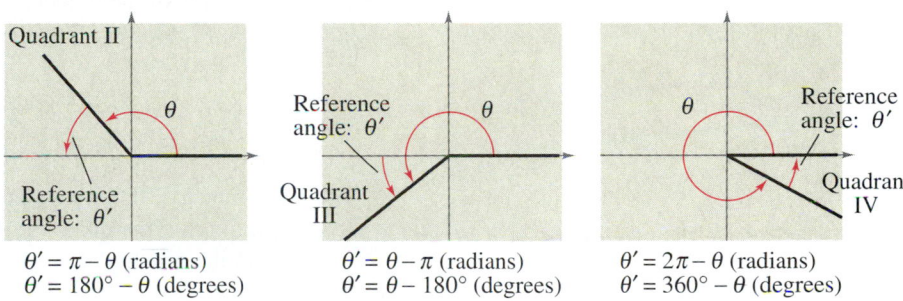

$\theta' = \pi - \theta$ (radians)
$\theta' = 180° - \theta$ (degrees)

$\theta' = \theta - \pi$ (radians)
$\theta' = \theta - 180°$ (degrees)

$\theta' = 2\pi - \theta$ (radians)
$\theta' = 360° - \theta$ (degrees)

FIGURE 6.29

EXAMPLE 4 Finding Reference Angles

Find the reference angle θ'.

a. $\theta = 300°$ **b.** $\theta = 2.3$ **c.** $\theta = -135°$

Solution

a. Because 300° lies in Quadrant IV, the angle it makes with the x-axis is

$$\theta' = 360° - 300° = 60°. \qquad \text{Degrees}$$

b. Because 2.3 lies between $\pi/2 \approx 1.5708$ and $\pi \approx 3.1416$, it follows that it is in Quadrant II and its reference angle is

$$\theta' = \pi - 2.3 \approx 0.8416. \qquad \text{Radians}$$

c. First, determine that $-135°$ is coterminal with 225°, which lies in Quadrant III. Hence, the reference angle is

$$\theta' = 225° - 180° = 45°. \qquad \text{Degrees}$$

Figure 6.30 shows each angle θ and its reference angle θ'.

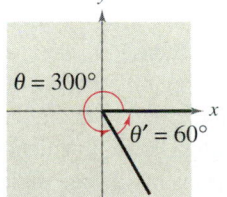

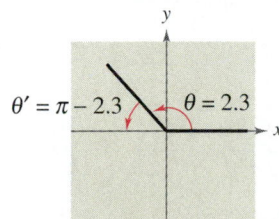

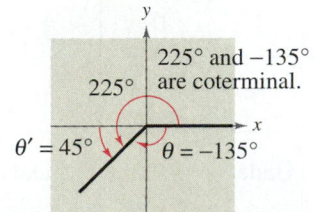

FIGURE 6.30

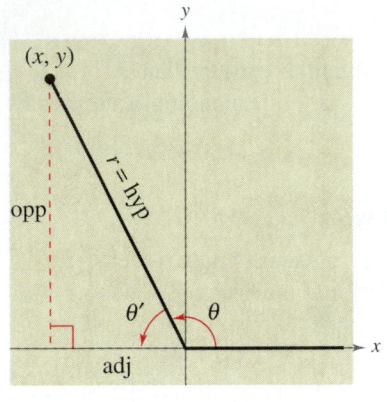

$$\text{opp} = |y|, \text{adj} = |x|$$

FIGURE 6.31

To see how a reference angle is used to evaluate a trigonometric function, consider the point (x, y) on the terminal side of θ, as shown in Figure 6.31. By definition, you know that

$$\sin \theta = \frac{y}{r} \quad \text{and} \quad \tan \theta = \frac{y}{x}.$$

For the right triangle with acute angle θ' and sides of lengths $|x|$ and $|y|$, you have

$$\sin \theta' = \frac{\text{opp}}{\text{hyp}} = \frac{|y|}{r} \quad \text{and} \quad \tan \theta' = \frac{\text{opp}}{\text{adj}} = \frac{|y|}{|x|}.$$

Thus, it follows that $\sin \theta$ and $\sin \theta'$ are equal, *except possibly in sign.* The same is true for $\tan \theta$ and $\tan \theta'$ *and* for the other four trigonometric functions. In all cases, the sign of the function value can be determined by the quadrant in which θ lies.

 EVALUATING TRIGONOMETRIC FUNCTIONS OF ANY ANGLE

To find the value of a trigonometric function of any angle θ,

1. determine the function value for the associated reference angle θ';
2. depending on the quadrant in which θ lies, prefix the appropriate sign to the function value.

By using reference angles and the special angles discussed in the previous section, you can greatly extend the scope of *exact* trigonometric values. For instance, knowing the function values of 30° means that you know the function values of all angles for which 30° is a reference angle. For convenience, the table gives the exact value of the trigonometric functions of special angles and quadrant angles.

Trigonometric Values of Common Angles

θ (degrees)	0°	30°	45°	60°	90°	180°	270°
θ (radians)	0	$\frac{\pi}{6}$	$\frac{\pi}{4}$	$\frac{\pi}{3}$	$\frac{\pi}{2}$	π	$\frac{3\pi}{2}$
$\sin \theta$	0	$\frac{1}{2}$	$\frac{\sqrt{2}}{2}$	$\frac{\sqrt{3}}{2}$	1	0	-1
$\cos \theta$	1	$\frac{\sqrt{3}}{2}$	$\frac{\sqrt{2}}{2}$	$\frac{1}{2}$	0	-1	0
$\tan \theta$	0	$\frac{\sqrt{3}}{3}$	1	$\sqrt{3}$	Undef.	0	Undef.

EXAMPLE 5 **Trigonometric Functions of Nonacute Angles**

Evaluate the following.

a. $\cos \dfrac{4\pi}{3}$ **b.** $\tan(-210°)$ **c.** $\csc \dfrac{11\pi}{4}$

Solution

a. Because $\theta = 4\pi/3$ lies in Quadrant III, the reference angle is $\theta' = (4\pi/3) - \pi = \pi/3$, as shown in Figure 6.32(a). Moreover, the cosine is negative in Quadrant III, so that

$$\cos \frac{4\pi}{3} = (-) \cos \frac{\pi}{3} = -\frac{1}{2}.$$

b. Because $-210° + 360° = 150°$, it follows that $-210°$ is coterminal with the second-quadrant angle $150°$. Therefore, the reference angle is $\theta' = 180° - 150° = 30°$, as shown in Figure 6.32(b). Finally, because the tangent is negative in Quadrant II, you have

$$\tan(-210°) = (-) \tan 30° = -\frac{\sqrt{3}}{3}.$$

c. Because $(11\pi/4) - 2\pi = 3\pi/4$, it follows that $11\pi/4$ is coterminal with the second-quadrant angle $3\pi/4$. Therefore, the reference angle is $\theta' = \pi - (3\pi/4) = \pi/4$, as shown in Figure 6.32(c). Because the cosecant is positive in Quadrant II, you have

$$\csc \frac{11\pi}{4} = (+) \csc \frac{\pi}{4} = \frac{1}{\sin(\pi/4)} = \sqrt{2}.$$

(a) 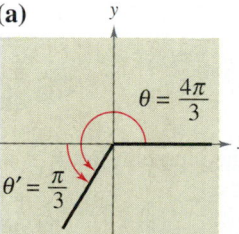   **(b)** 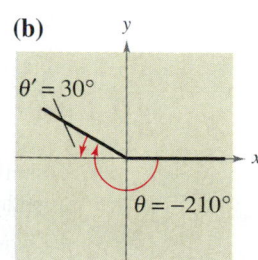   **(c)**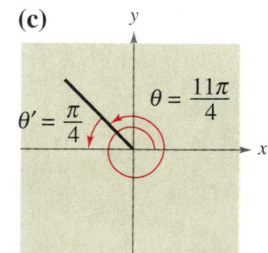

FIGURE 6.32

The fundamental trigonometric identities listed in the previous section (for an acute angle θ) are also valid when θ is any angle in the domain of the function.

EXAMPLE 6   ***Using Trigonometric Identities***

Let θ be an angle in Quadrant II such that $\sin \theta = \frac{1}{3}$. Find (a) $\cos \theta$ and (b) $\tan \theta$ by using trigonometric identities.

Solution

a. Using the Pythagorean identity $\sin^2 \theta + \cos^2 \theta = 1$, you obtain

$$\left(\frac{1}{3}\right)^2 + \cos^2 \theta = 1$$

$$\cos^2 \theta = 1 - \frac{1}{9} = \frac{8}{9}.$$

Because $\cos \theta < 0$ in quadrant II, you can use the negative root to obtain

$$\cos \theta = -\frac{\sqrt{8}}{\sqrt{9}} = -\frac{2\sqrt{2}}{3}.$$

b. Using the trigonometric identity $\tan \theta = \sin \theta / \cos \theta$, you obtain

$$\tan \theta = \frac{1/3}{-2\sqrt{2}/3} = -\frac{1}{2\sqrt{2}} = -\frac{\sqrt{2}}{4}.$$

EXAMPLE 7   ***Using a Calculator***

a. Use a calculator to evaluate $\cot 410°$ and $\sin(-7)$.

b. Use a calculator to solve $\tan \theta = 4.812$, $0 \le \theta < 2\pi$.

Solution

Function	*Mode*	*Graphing Calculator Keystrokes*	*Display*
a. $\cot 410°$	Degree	(TAN 410) x^{-1} ENTER	0.8390996
$\sin(-7)$	Radian	SIN ((-) 7) ENTER	−0.6569866

b. To solve the equation $\tan \theta = 4.812$, you can use the inverse tangent key, as follows.

Equation	*Mode*	*Graphing Calculator Keystrokes*	*Display*
$\tan \theta = 4.812$	Radian	TAN^{-1} 4.812 ENTER	1.365898912

The angle $\theta \approx 1.366$ lies in Quadrant I. A second value of θ lies in Quadrant III (tangent is positive) and is

$$\theta = \pi + 1.366 \approx 4.507.$$

Trigonometric Functions of Real Numbers

To define a trigonometric function of a real number (rather than an angle), let t represent any real number. Then imagine that the real number line is wrapped around a *unit circle* as shown in Figure 6.33. Note that positive numbers correspond to a counterclockwise wrapping and negative numbers correspond to a clockwise wrapping.

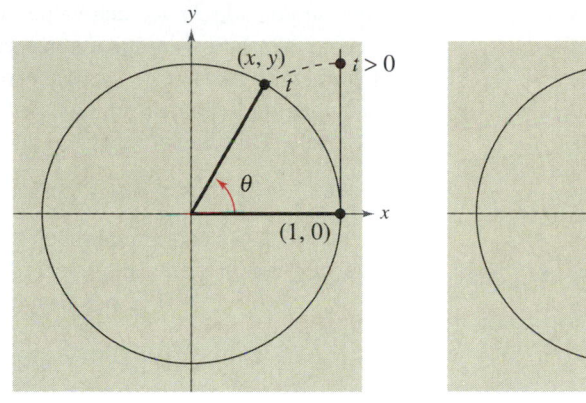

FIGURE 6.33

As the real line is wrapped around the unit circle, each real number t will correspond with a central angle θ. Moreover, because the circle has a radius of 1, the arc intercepted by the angle θ will have a length of t. The point is that *if θ is measured in radians, then $t = \theta$.* Thus, you can define $\sin t$ as

$$\sin t = \sin(t \text{ radians}).$$

Similarly, $\cos t = \cos(t \text{ radians})$, $\tan t = \tan(t \text{ radians})$, and so on.

EXAMPLE 8   *Evaluating Trigonometric Functions*

Evaluate $f(t) = \sin t$ for (a) $t = 1$, (b) $t = 3\pi/2$, and (c) $t = 2\pi$.

Solution

a. $f(1) = \sin 1 \approx 0.841471$ Radian mode

b. $f\left(\dfrac{3\pi}{2}\right) = \sin \dfrac{3\pi}{2} = -1$ Common angle

c. $f(2\pi) = \sin 2\pi = 0$ Reference angle: $\theta' = 0$

The *domain* of the sine and cosine functions is the set of all real numbers. To determine the *range* of these two functions, consider the unit circle shown in Figure 6.34. Because $r = 1$, it follows that $\sin t = y$ and $\cos t = x$. Moreover, because (x, y) is on the unit circle, you know that $-1 \le y \le 1$ and $-1 \le x \le 1$. Thus, the values of sine and cosine also range between -1 and 1.

$$-1 \le \ y \ \le 1 \qquad \text{and} \qquad -1 \le \ x \ \le 1$$
$$-1 \le \sin t \le 1 \qquad\qquad\qquad -1 \le \cos t \le 1$$

Suppose you add 2π to each value of t in the interval $[0, 2\pi]$, thus completing a second revolution around the unit circle, as shown in Figure 6.35. The values of $\sin(t + 2\pi)$ and $\cos(t + 2\pi)$ correspond to those of $\sin t$ and $\cos t$. Similar results can be obtained for repeated revolutions (positive or negative) on the unit circle. This leads to the general result

$$\sin(t + 2\pi n) = \sin t \qquad \text{and} \qquad \cos(t + 2\pi n) = \cos t$$

for any integer n and real number t. Functions that behave in such a repetitive (or cyclic) manner are called **periodic.**

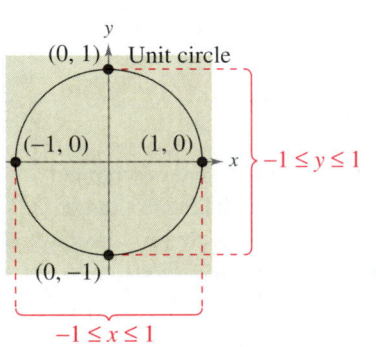

FIGURE 6.34

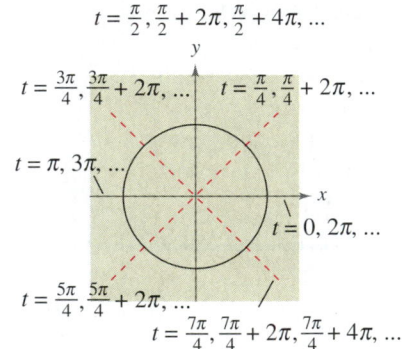

FIGURE 6.35

DEFINITION OF A PERIODIC FUNCTION

A function f is **periodic** if there exists a positive real number c such that

$$f(t + c) = f(t)$$

for all t in the domain of f. The smallest number c for which f is periodic is called the **period** of f.

From this definition it follows that the sine and cosine functions are periodic and have a period of 2π. The other four trigonometric functions are also periodic, and we will say more about that in Section 6.5.

Recall from Section 2.3 that a function f is *even* if $f(-t) = f(t)$, and is *odd* if $f(-t) = -f(t)$.

EVEN AND ODD TRIGONOMETRIC FUNCTIONS

The cosine and secant functions are *even*.

$$\cos(-t) = \cos t \qquad \sec(-t) = \sec t$$

The sine, cosecant, tangent, and cotangent functions are *odd*.

$$\sin(-t) = -\sin t \qquad \csc(-t) = -\csc t$$
$$\tan(-t) = -\tan t \qquad \cot(-t) = -\cot t$$

At this point, you have completed your introduction to basic trigonometry. You have measured angles in both degrees and radians. You have defined the six trigonometric functions from a right triangle perspective and as functions of real numbers. In your remaining work with trigonometry you should continue to rely on both perspectives. For instance, in the next two sections on graphing techniques, it helps to think of the trigonometric functions as functions of real numbers. Later, in Section 6.7, you will look at applications involving angles and triangles.

For your convenience, we have included on the inside back cover of this text a summary of basic trigonometry.

GROUP ACTIVITY

PATTERNS IN TRIGONOMETRIC FUNCTIONS

Complete the following table. Then identify and discuss any inherent patterns in the trigonometric functions. What can you conclude?

Function	Domain	Range	Even/Odd	Period	Zeros
Sine					
Cosine					
Tangent					
Cosecant					
Secant					
Cotangent					

6.3 Exercises

In Exercises 1–4, determine the exact values of the six trigonometric functions of the angle θ.

1. (a)
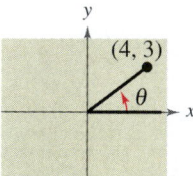
(b)
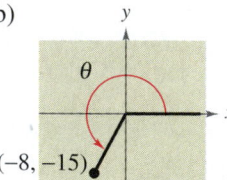

2. (a)
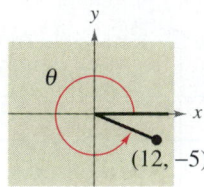
(b)
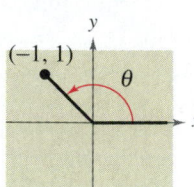

3. (a)
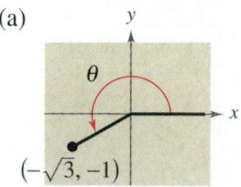
(b)
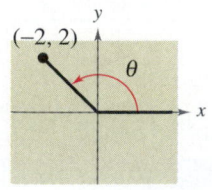

4. (a)
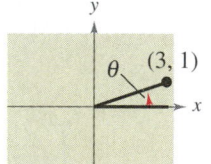
(b)
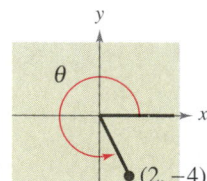

In Exercises 5–8, the point is on the terminal side of an angle in standard position. Determine the exact values of the six trigonometric functions of the angle.

5. $(7, 24)$ **6.** $(8, 15)$

7. $(-4, 10)$ **8.** $(-5, -2)$

In Exercises 9–12, state the quadrant in which θ lies.

9. $\sin \theta < 0$ and $\cos \theta < 0$

10. $\sin \theta > 0$ and $\cos \theta > 0$

11. $\sin \theta > 0$ and $\tan \theta < 0$

12. $\sec \theta > 0$ and $\cot \theta < 0$

In Exercises 13–22, find the values of the six trigono-metric functions of θ.

Function Value	Constraint
13. $\sin \theta = \frac{3}{5}$	θ lies in Quadrant II.
14. $\cos \theta = -\frac{4}{5}$	θ lies in Quadrant III.
15. $\tan \theta = -\frac{15}{8}$	$\sin \theta < 0$
16. $\cos \theta = \frac{8}{17}$	$\tan \theta < 0$
17. $\cot \theta = -3$	$\cos \theta > 0$
18. $\csc \theta = 4$	$\cot \theta < 0$
19. $\sec \theta = -2$	$\sin \theta > 0$
20. $\cot \theta$ is undefined.	$\frac{\pi}{2} \le \theta \le \frac{3\pi}{2}$
21. $\sin \theta = 0$	$\sec \theta = -1$
22. $\tan \theta$ is undefined.	$\pi \le \theta \le 2\pi$

In Exercises 23–26, the terminal side of θ lies on the given line in the specified quadrant. Find the values of the six trigonometric functions of θ.

Line	Quadrant
23. $y = -x$	Quadrant II
24. $y = \frac{1}{3}x$	Quadrant III
25. $y = 2x$	Quadrant III
26. $4x + 3y = 0$	Quadrant IV

In Exercises 27–34, evaluate the trigonometric function of the quadrant angle.

27. $\cos \pi$

28. $\cos \frac{3\pi}{2}$

29. $\sec \pi$

30. $\sec \frac{3\pi}{2}$

31. $\tan \frac{\pi}{2}$

32. $\tan \pi$

33. $\cot \frac{\pi}{2}$

34. $\csc \pi$

In Exercises 35–42, find the reference angle θ′, and sketch θ and θ′ in standard position.

35. $\theta = 203°$

36. $\theta = 309°$

37. $\theta = -245°$

38. $\theta = -145°$

39. $\theta = \frac{2\pi}{3}$

40. $\theta = \frac{7\pi}{4}$

41. $\theta = 3.5$

42. $\theta = \frac{11\pi}{3}$

In Exercises 43–52, evaluate the sine, cosine, and tangent of the angle without using a calculator.

43. $225°$

44. $300°$

45. $750°$

46. $-405°$

47. $\frac{4\pi}{3}$

48. $\frac{\pi}{4}$

49. $-\frac{\pi}{6}$

50. $-\frac{\pi}{2}$

51. $\frac{11\pi}{4}$

52. $\frac{10\pi}{3}$

In Exercises 53–62, use a calculator to evaluate the trigonometric function to four decimal places. (Be sure the calculator is set in the correct mode.)

53. $\sin 10°$

54. $\sec 225°$

55. $\cos(-110°)$

56. $\csc 330°$

57. $\tan 240°$

58. $\cot 1.35$

59. $\tan \frac{\pi}{9}$

60. $\tan\left(-\frac{\pi}{9}\right)$

61. $\sin 0.65$

62. $\sin(-0.65)$

In Exercises 63–68, find two solutions of the equation. Give your answers in degrees ($0° \le \theta < 360°$) and radians ($0 \le \theta < 2\pi$). Do not use a calculator.

63. (a) $\sin \theta = \frac{1}{2}$ (b) $\sin \theta = -\frac{1}{2}$

64. (a) $\cos \theta = \frac{\sqrt{2}}{2}$ (b) $\cos \theta = -\frac{\sqrt{2}}{2}$

65. (a) $\csc \theta = \frac{2\sqrt{3}}{3}$ (b) $\cot \theta = -1$

66. (a) $\sec \theta = 2$ (b) $\sec \theta = -2$

67. (a) $\tan \theta = 1$ (b) $\cot \theta = -\sqrt{3}$

68. (a) $\sin \theta = \frac{\sqrt{3}}{2}$ (b) $\sin \theta = -\frac{\sqrt{3}}{2}$

In Exercises 69 and 70, use a calculator to approximate two values of θ ($0° \leq \theta < 360°$) that satisfy the equation. Round to two decimal places.

69. $\sin \theta = 0.8191$ **70.** $\cos \theta = 0.8746$

In Exercises 71–74, use a calculator to approximate two values of θ ($0 \leq \theta < 2\pi$) that satisfy the equation. Round to three decimal places.

71. $\cos \theta = 0.9848$ **72.** $\sin \theta = 0.0175$

73. $\tan \theta = 1.192$ **74.** $\cot \theta = 5.671$

In Exercises 75–80, find the indicated trigonometric value in the specified quadrant.

	Function	*Quadrant*	*Trigonometric Value*
75.	$\sin \theta = -\frac{3}{5}$	IV	$\cos \theta$
76.	$\cot \theta = -3$	II	$\sin \theta$
77.	$\tan \theta = \frac{3}{2}$	III	$\sec \theta$
78.	$\csc \theta = -2$	IV	$\cot \theta$
79.	$\cos \theta = \frac{5}{8}$	I	$\sec \theta$
80.	$\sec \theta = -\frac{9}{4}$	III	$\tan \theta$

In Exercises 81–88, find the point (x, y) on the unit circle that corresponds to the real number t. Use the result to evaluate $\sin t$, $\cos t$, and $\tan t$.

81. $t = \dfrac{\pi}{4}$ **82.** $t = \dfrac{\pi}{3}$

83. $t = \dfrac{5\pi}{6}$ **84.** $t = \dfrac{5\pi}{4}$

85. $t = \dfrac{4\pi}{3}$ **86.** $t = \dfrac{11\pi}{6}$

87. $t = \dfrac{3\pi}{2}$ **88.** $t = \pi$

Estimation **In Exercises 89 and 90, use the figure and a straightedge to approximate the value of the trigonometric function.**

89. (a) $\sin 5$ (b) $\cos 2$

90. (a) $\sin 0.75$ (b) $\cos 2.5$

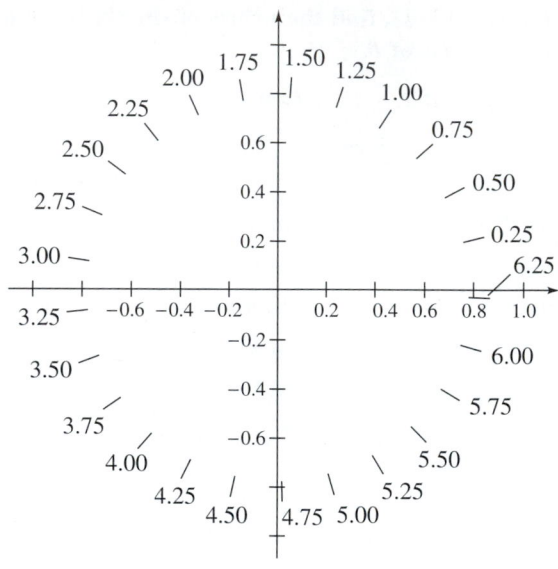

FIGURE FOR 89–92

Estimation **In Exercises 91 and 92, use the figure to approximate the solution of the equation where $0 \leq t < 2\pi$.**

91. (a) $\sin t = 0.25$ (b) $\cos t = -0.25$

92. (a) $\sin t = -0.75$ (b) $\cos t = 0.75$

93. *Think About It* Because $f(t) = \sin t$ is an odd function and $g(t) = \cos t$ is an even function, what can be said about the function $h(t) = f(t)g(t)$?

94. *Essay* Consider an angle in standard position with $r = 12$ centimeters, as shown in the figure. Write a short paragraph describing the change in the magnitudes of x, y, $\sin \theta$, $\cos \theta$, and $\tan \theta$ as θ increases continually from $0°$ to $90°$.

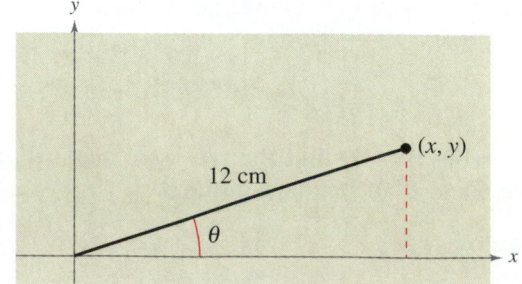

95. *Average Temperature*   The average daily temperature T (in degrees Fahrenheit) for a certain city is

$$T = 45 - 23 \cos\left[\frac{2\pi}{365}(t - 32)\right]$$

where t is the time in days, with $t = 1$ corresponding
to January 1. Find the average daily temperatures on
the following days.

(a) January 1

(b) July 4 ($t = 185$)

(c) October 18 ($t = 291$)

96. *Sales*   A company that produces a seasonal product
forecasts monthly sales over the next 2 years to be

$$S = 23.1 + 0.442t + 4.3 \sin\frac{\pi t}{6}$$

where S is measured in thousands of units and t is the
time in months, with $t = 1$ representing January
1996. Predict sales for the following months.

(a) February 1996   (b) February 1997

(c) September 1996   (d) September 1997

97. *Harmonic Motion*   The displacement from equilibrium of an oscillating weight suspended by a spring is
given by

$$y(t) = 2 \cos 6t$$

where y is the displacement in centimeters and t is the
time in seconds (see figure). Find the displacement
when (a) $t = 0$, (b) $t = 1/4$, and (c) $t = 1/2$.

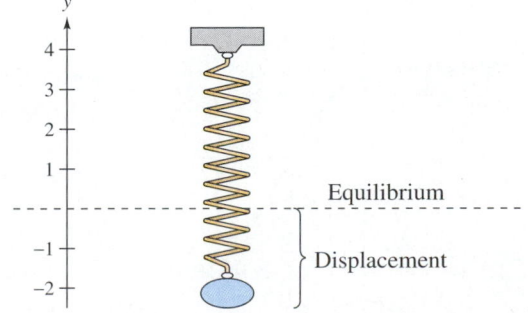

98. *Harmonic Motion*   The displacement from equilibrium of an oscillating weight suspended by a
spring and subject to the damping effect of friction
is given by

$$y(t) = 2e^{-t} \cos 6t$$

where y is the displacement in centimeters and t is
the time in seconds. Find the displacement when
(a) $t = 0$, (b) $t = 1/4$, and (c) $t = 1/2$.

99. *Electric Circuits*   The initial current and charge in
an electric circuit are zero. The current when 100
volts is applied to the circuit is given by

$$I = 5e^{-2t} \sin t$$

where the resistance, inductance, and capacitance
are 80 ohms, 20 henrys, and 0.01 farads, respectively. Approximate the current when $t = 0.7$ seconds
after the voltage is applied.

100. *Distance*   An airplane, flying at an altitude of 6
miles, is on a flight path that passes directly over an
observer (see figure). If θ is the angle of elevation
from the observer to the plane, find the distance d
from the observer to the plane when (a) $\theta = 30°$, (b)
$\theta = 90°$, and (c) $\theta = 120°$.

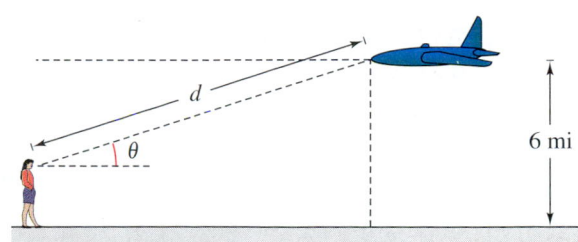

6 mi

Review   **Solve Exercises 101–104 as a review of the
skills and problem-solving techniques you learned in
previous sections. Graph the function.**

101. $y = 2^{x-1}$

102. $y = 3^{-x/2}$

103. $y = \ln(x - 1)$

104. $y = \ln x^4$

6.4 *Graphs of Sine and Cosine Functions*

See Exercise 89 on page 503 for an example of how a sine function can be used to model the daily high temperatures in Honolulu and Chicago.

Basic Sine and Cosine Curves ▫ *Amplitude and Period* ▫
Translations of Sine and Cosine Curves ▫ *Mathematical Modeling*

Basic Sine and Cosine Curves

In this section you will study techniques for sketching the graphs of the sine and cosine functions. The graph of the sine function is a **sine curve.** In Figure 6.36, the black portion of the graph represents one period of the function and is called **one cycle** of the sine curve. The gray portion of the graph indicates that the basic sine wave repeats indefinitely to the right and left. The graph of the cosine function is shown in Figure 6.37.

Recall from Section 6.3 that the domain of the sine and cosine functions is the set of all real numbers. Moreover, the range of each function is the interval $[-1, 1]$, and each function has a period of 2π. Do you see how this information is consistent with the basic graphs given in Figures 6.36 and 6.37?

NOTE Note in Figures 6.36 and 6.37 that the sine curve is symmetric with respect to the *origin*, whereas the cosine curve is symmetric with respect to the *y-axis*. These properties of symmetry follow from the fact that the sine function is odd whereas the cosine function is even. ▪▪

x	$\sin x$	$\cos x$
0	0	1
$\dfrac{\pi}{6}$	$\dfrac{1}{2}$	$\dfrac{\sqrt{3}}{2}$
$\dfrac{\pi}{4}$	$\dfrac{\sqrt{2}}{2}$	$\dfrac{\sqrt{2}}{2}$
$\dfrac{\pi}{3}$	$\dfrac{\sqrt{3}}{2}$	$\dfrac{1}{2}$
$\dfrac{\pi}{2}$	1	0
$\dfrac{3\pi}{4}$	$\dfrac{\sqrt{2}}{2}$	$-\dfrac{\sqrt{2}}{2}$
π	0	-1
$\dfrac{3\pi}{2}$	-1	0
2π	0	1

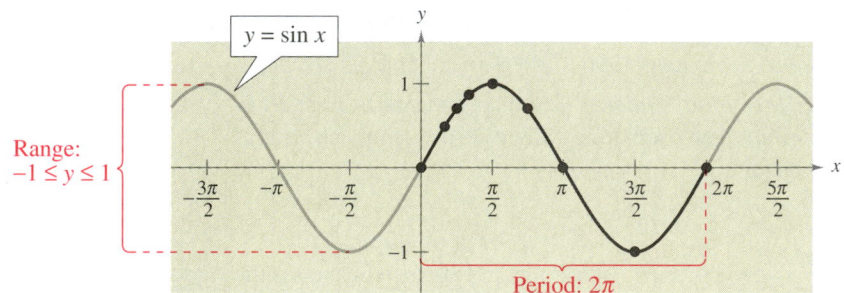

FIGURE 6.36

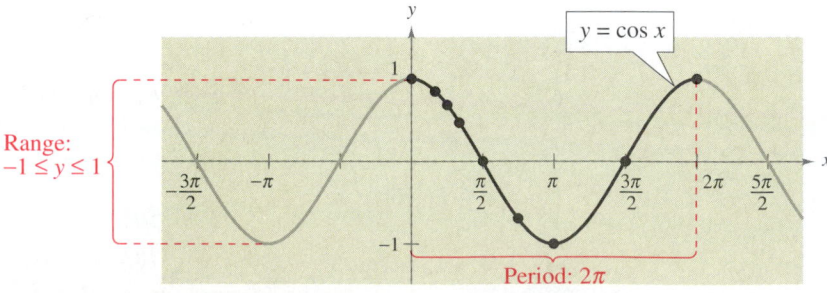

FIGURE 6.37

To sketch the graphs of the basic sine and cosine functions by hand, it helps to note five **key points** in one period of each graph: the *intercepts, maximum points,* and *minimum points.* See Figure 6.38.

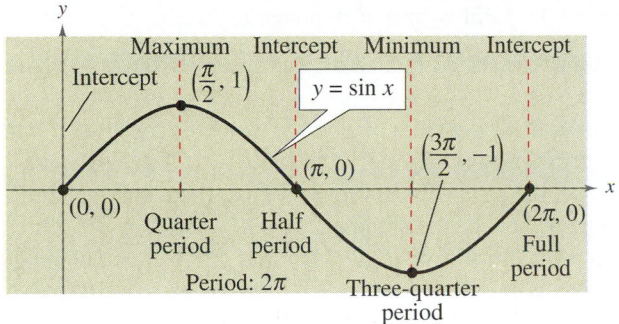 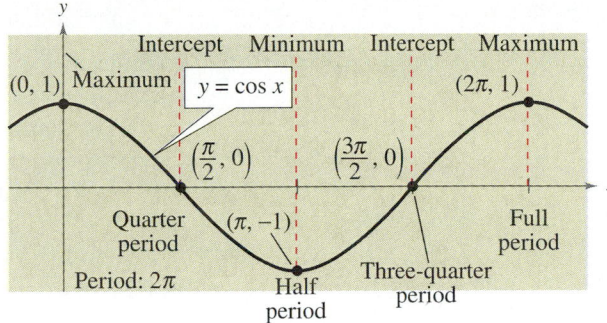

FIGURE 6.38

EXAMPLE 1   Using Key Points to Sketch a Sine Curve

Sketch the graph of $y = 2 \sin x$ on the interval $[-\pi, 4\pi]$.

Solution

Note that $y = 2 \sin x = 2(\sin x)$ indicates that the y-values for the key points will have twice the magnitude of the graph of $y = \sin x$. Divide the period 2π into four equal parts to get the following key points for $y = 2 \sin x$.

$$(0, 0), \quad \left(\frac{\pi}{2}, 2\right), \quad (\pi, 0), \quad \left(\frac{3\pi}{2}, -2\right), \quad \text{and} \quad (2\pi, 0)$$

By connecting these key points with a smooth curve and extending the curve in both directions over the interval $[-\pi, 4\pi]$, you obtain the graph shown in Figure 6.39.

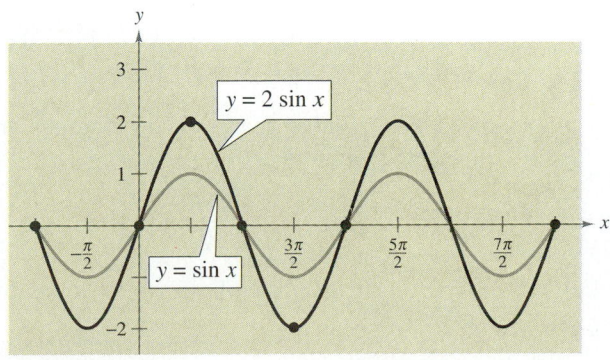

FIGURE 6.39

Amplitude and Period

In the rest of this section you will study the graphic effect of each of the constants *a, b, c,* and *d* in equations of the forms

$$y = d + a \sin(bx - c) \quad \text{and} \quad y = d + a \cos(bx - c).$$

A quick review of the transformations studied in Section 2.4 should help in this investigation.

The constant factor *a* in $y = a \sin x$ acts as a *scaling factor*—a *vertical stretch* or *vertical shrink* of the basic sine curve. If $|a| > 1$, the basic sine curve is stretched, and if $|a| < 1$, the basic sine curve is shrunk. The result is that the graph of $y = a \sin x$ ranges between $-a$ and a instead of between -1 and 1. The absolute value of *a* is the **amplitude** of the function $y = a \sin x$. The range of the function $y = a \sin x$ is $-a \leq y \leq a$.

DEFINITION OF AMPLITUDE OF SINE AND COSINE CURVES

The **amplitude** of $y = a \sin x$ and $y = a \cos x$ is the largest value of *y* and is given by

$$\text{Amplitude} = |a|.$$

EXAMPLE 2   *Scaling: Vertical Shrinking and Stretching*

On the same coordinate axes, sketch the graphs of

$$y = \frac{1}{2} \cos x \quad \text{and} \quad y = 3 \cos x.$$

Solution

Because the amplitude of $y = \frac{1}{2} \cos x$ is $\frac{1}{2}$, the maximum value is $\frac{1}{2}$ and the minimum value is $-\frac{1}{2}$. Divide one cycle, $0 \leq x \leq 2\pi$, into four equal parts to get the key points

$$\left(0, \frac{1}{2}\right), \quad \left(\frac{\pi}{2}, 0\right), \quad \left(\pi, -\frac{1}{2}\right), \quad \left(\frac{3\pi}{2}, 0\right), \quad \text{and} \quad \left(2\pi, \frac{1}{2}\right).$$

A similar analysis shows that the amplitude of $y = 3 \cos x$ is 3, and the key points are

$$(0, 3), \quad \left(\frac{\pi}{2}, 0\right), \quad (\pi, -3), \quad \left(\frac{3\pi}{2}, 0\right), \quad \text{and} \quad (2\pi, 3).$$

The graphs of these two functions are shown in Figure 6.40.

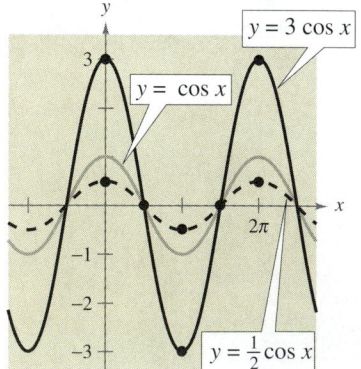

FIGURE 6.40

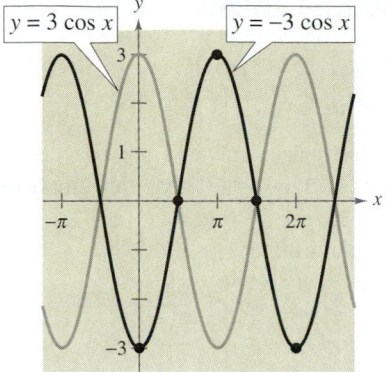

FIGURE 6.41

You know from Section 2.4 that the graph of $y = -f(x)$ is a **reflection** in the x-axis of the graph of $y = f(x)$. For instance, the graph of $y = -3 \cos x$ is a reflection of the graph of $y = 3 \cos x$, as shown in Figure 6.41.

Because $y = a \sin x$ completes one cycle from $x = 0$ to $x = 2\pi$, it follows that $y = a \sin bx$ completes one cycle from $x = 0$ to $x = 2\pi/b$.

> ### *PERIOD OF SINE AND COSINE FUNCTIONS*
>
> Let b be a positive real number. The **period** of $y = a \sin bx$ and $y = a \cos bx$ is $2\pi/b$.

Note that if $0 < b < 1$, the period of $y = a \sin bx$ is greater than 2π and represents a *horizontal stretching* of the graph of $y = a \sin x$. Similarly, if $b > 1$, the period of $y = a \sin bx$ is less than 2π and represents a *horizontal shrinking* of the graph of $y = a \sin x$. If b is negative, we use the identities $\sin(-x) = -\sin x$ and $\cos(-x) = \cos x$ to rewrite the function.

EXAMPLE 3 *Scaling: Horizontal Stretching*

Sketch the graph of $y = \sin \dfrac{x}{2}$.

Solution

The amplitude is 1. Moreover, because $b = \frac{1}{2}$, the period is

$$\frac{2\pi}{b} = \frac{2\pi}{\frac{1}{2}} = 4\pi.$$

Now, divide the period-interval $[0, 4\pi]$ into four equal parts with the values π, 2π, and 3π to obtain the following key points on the graph.

$$(0, 0), \quad (\pi, 1), \quad (2\pi, 0), \quad (3\pi, -1), \quad \text{and} \quad (4\pi, 0)$$

The graph is shown in Figure 6.42.

Study Tip

In general, to divide a period-interval into four equal parts, successively add "period/4," starting with the left endpoint of the interval. For instance, for the period-interval $[-\pi/6, \pi/2]$ of length $2\pi/3$, you would successively add

$$\frac{2\pi/3}{4} = \frac{\pi}{6}$$

to get $-\pi/6, 0, \pi/6, \pi/3,$ and $\pi/2$.

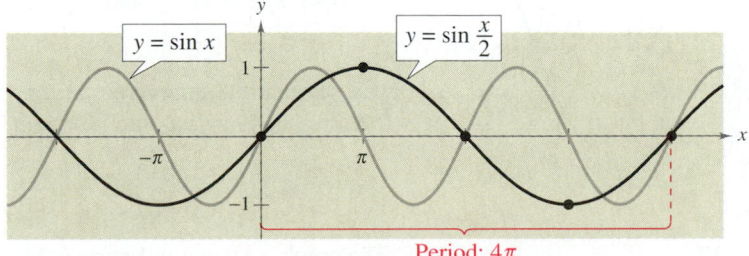

FIGURE 6.42

Translations of Sine and Cosine Curves

The constant c in the general equations

$$y = a \sin(bx - c) \quad \text{and} \quad y = a \cos(bx - c)$$

creates a *horizontal translation* (shift) of the basic sine and cosine curves. Comparing $y = a \sin bx$ with $y = a \sin(bx - c)$, we find that the graph of $y = a \sin(bx - c)$ completes one cycle from $bx - c = 0$ to $bx - c = 2\pi$. By solving for x, we find the interval for one cycle to be

Left endpoint   Right endpoint

$$\frac{c}{b} \le x \le \underbrace{\frac{c}{b} + \frac{2\pi}{b}}.$$

Period

This implies that the period of $y = a \sin(bx - c)$ is $2\pi/b$, and the graph of $y = a \sin bx$ is shifted by an amount c/b. The number c/b is the **phase shift.**

Exploration

Sketch the graph of $y = \sin(x + c)$, where $c = -\pi/4$, 0, and $\pi/4$. How does the value of c affect the graph?

> ### GRAPHS OF SINE AND COSINE FUNCTIONS
>
> The graphs of $y = a \sin(bx - c)$ and $y = a \cos(bx - c)$ have the following characteristics. (Assume $b > 0$.)
>
> $$\textbf{Amplitude} = |a| \qquad \textbf{Period} = 2\pi/b$$
>
> The left and right endpoints of a one-cycle interval can be determined by solving the equations $bx - c = 0$ and $bx - c = 2\pi$.

EXAMPLE 4   *Horizontal Translation*

Sketch the graph of $y = (1/2) \sin(x - \pi/3)$.

Solution

The amplitude is $\frac{1}{2}$ and the period is 2π. By solving the equations

$$x - \pi/3 = 0 \quad \text{and} \quad x - \pi/3 = 2\pi$$
$$x = \pi/3 \qquad\qquad\qquad x = 7\pi/3$$

you see that the interval $[\pi/3, 7\pi/3]$ corresponds to one cycle of the graph. Dividing this interval into four equal parts produces the following key points.

$$\left(\frac{\pi}{3}, 0\right), \quad \left(\frac{5\pi}{6}, \frac{1}{2}\right), \quad \left(\frac{4\pi}{3}, 0\right), \quad \left(\frac{11\pi}{6}, -\frac{1}{2}\right), \quad \text{and} \quad \left(\frac{7\pi}{3}, 0\right)$$

The graph is shown in Figure 6.43.

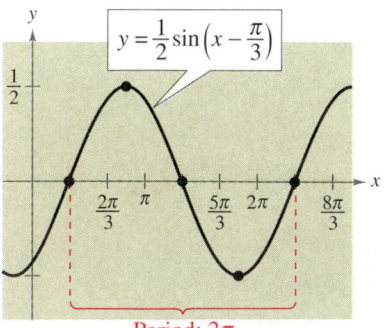

FIGURE 6.43

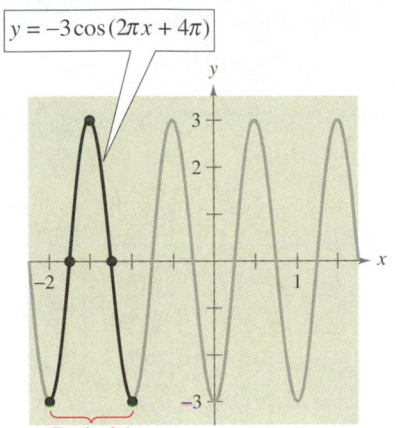

$y = -3\cos(2\pi x + 4\pi)$

Period 1

FIGURE 6.44

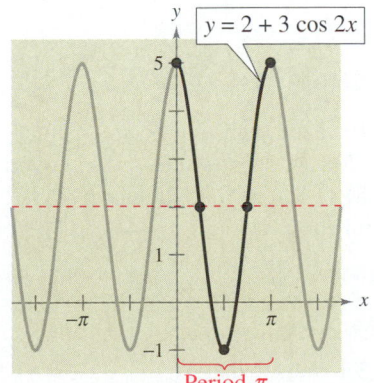

$y = 2 + 3\cos 2x$

Period π

FIGURE 6.45

EXAMPLE 5   Horizontal Translation

Sketch the graph of

$$y = -3\cos(2\pi x + 4\pi).$$

Solution

The amplitude is 3 and the period is $2\pi/2\pi = 1$. By solving the equations

$$2\pi x + 4\pi = 0 \qquad \text{and} \qquad 2\pi x + 4\pi = 2\pi$$
$$2\pi x = -4\pi \qquad\qquad\qquad 2\pi x = -2\pi$$
$$x = -2 \qquad\qquad\qquad x = -1$$

you see that the interval $[-2, -1]$ corresponds to one cycle of the graph. Dividing this interval into four equal parts produces the following key points.

$$(-2, -3), \quad \left(-\frac{7}{4}, 0\right), \quad \left(-\frac{3}{2}, 3\right), \quad \left(-\frac{5}{4}, 0\right), \quad \text{and} \quad (-1, -3)$$

The graph is shown in Figure 6.44.

The final type of transformation is the *vertical translation* caused by the constant d in the equations

$$y = d + a\sin(bx - c) \qquad \text{and} \qquad y = d + a\cos(bx - c).$$

The shift is d units upward for $d > 0$ and downward for $d < 0$. In other words, the graph oscillates about the horizontal line $y = d$ instead of the x-axis.

EXAMPLE 6   Vertical Translation

Sketch the graph of

$$y = 2 + 3\cos 2x.$$

Solution

The amplitude is 3 and the period is π. The key points over the interval $[0, \pi]$ are

$$(0, 5), \quad \left(\frac{\pi}{4}, 2\right), \quad \left(\frac{\pi}{2}, -1\right), \quad \left(\frac{3\pi}{4}, 2\right), \quad \text{and} \quad (\pi, 5).$$

The graph is shown in Figure 6.45.

Mathematical Modeling

Sine and cosine functions can be used to model many real-life situations, including electric currents, musical tones, radio waves, tides, sunrises, and weather patterns.

EXAMPLE 7   *Finding a Trigonometric Model*

Real Life

Throughout the day, the depth of water at the end of a dock varies with the tides. The table shows the depth (in meters) at various times during the morning.

t (time)	Midnight	2 A.M.	4 A.M.	6 A.M.	8 A.M.	10 A.M.	Noon
y (depth)	2.55	3.80	4.40	3.80	2.55	1.80	2.27

a. Use a trigonometric function to model this data.

b. Find the depth at 9 A.M. and 3 P.M.

c. A boat needs at least 3 meters of water to moor at the dock. During what times in the afternoon can it safely dock?

Solution

a. Begin by graphing the data, as shown in Figure 6.46. You can use either a sine or cosine model. Suppose you use a cosine model of the form

$$y = a \cos(bt - c) + d.$$

The amplitude is given by

$$a = \tfrac{1}{2}[(\text{high}) - (\text{low})] = \tfrac{1}{2}(4.4 - 1.8) = 1.3.$$

The period is

$$p = 2[(\text{low time}) - (\text{high time})] = 2(10 - 4) = 12$$

which implies that $b = 2\pi/p \approx 0.524$. Because high tide occurs 4 hours after midnight, you can conclude that $c/b = 4$, so $c \approx 2.094$. Moreover, because the average depth is $\tfrac{1}{2}(4.4 + 1.8) = 3.1$, it follows that $d = 3.1$. Thus, you can model the depth with the function

$$y = 1.3 \cos(0.542t - 2.094) + 3.1.$$

b. At 9 A.M. and 3 P.M. the depth is as follows.

$$y = 1.3 \cos(0.524 \cdot 9 - 2.094) + 3.1 \approx 1.97 \text{ meters} \qquad \text{9 A.M.}$$
$$y = 1.3 \cos(0.524 \cdot 15 - 2.094) + 3.1 \approx 4.23 \text{ meters} \qquad \text{3 P.M.}$$

c. Using a graphing utility, you can graph the model with the line $y = 3$, as shown in Figure 6.47. From the graph, it follows that the depth is at least 3 meters between 12:54 P.M. ($t \approx 12.9$) and 7:06 P.M. ($t \approx 19.1$).

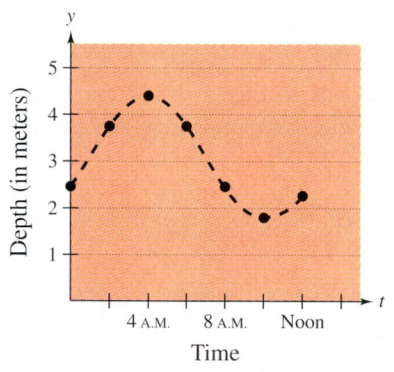

FIGURE 6.46

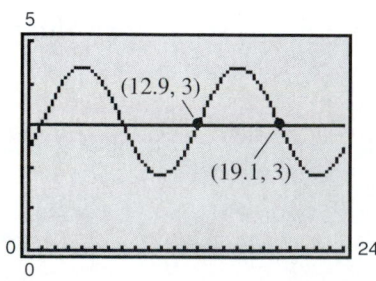

FIGURE 6.47

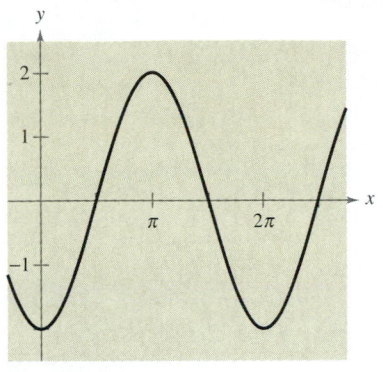

FIGURE 6.48

EXAMPLE 8   *Finding an Equation for a Graph*

Find the amplitude, period, and phase shift for the sine function whose graph is shown in Figure 6.48. Write an equation for this graph.

Solution

The amplitude for this sine curve is 2, the period is 2π, and there is a right phase shift of $\pi/2$. Thus, you can write

$$y = 2\sin\left(x - \frac{\pi}{2}\right).$$

Try finding a cosine function with this same graph.

GROUP ACTIVITY

A SINE SHOW

Enter the following program in a *TI-83* or a *TI-82* graphing calculator. (Program steps for other graphing calculators are given in the appendix.) This program simultaneously draws the unit circle and the corresponding points on the sine curve. After the circle and sine curve are drawn, you can connect the points on the unit circle with their corresponding points on the sine curve by pressing ENTER . Discuss the relationship that is illustrated.

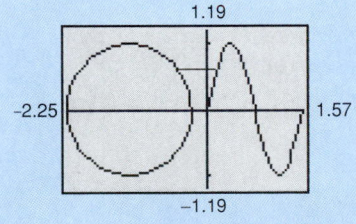

```
PROGRAM:SINESHOW
:Radian
:ClrDraw:FnOff
:Param:Simul
:-2.25→Xmin
:π/2→Xmax
:3→Xscl
:-1.19→Ymin
:1.19→Ymax
:1→Yscl
:0→Tmin
:6.3→Tmax
:.15→Tstep
:"-1.25+cos T"→X₁ₜ
```

```
:"sin T"→Y₁ₜ
:"T/4"→X₂ₜ
:"sin T"→Y₂ₜ
:DispGraph
:For(N,1,12)
:N*π/6.5→T
:-1.25+cos T→A
:sin T→B
:T/4→C
:Line(A,B,C,B)
:Pause
:End
:Pause:Func
:Sequential:Disp
```

6.4 Exercises

In Exercises 1–14, find the period and amplitude.

5. $y = \dfrac{2}{3} \sin \pi x$ **6.** $y = \dfrac{3}{2} \cos \dfrac{\pi x}{2}$

1. $y = 3 \sin 2x$ **2.** $y = 2 \cos 3x$

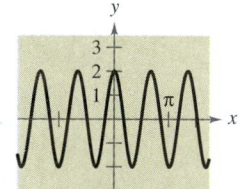   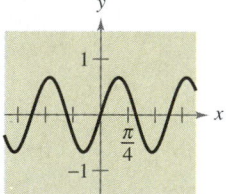   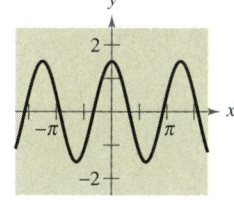

3. $y = \dfrac{5}{2} \cos \dfrac{x}{2}$ **4.** $y = -3 \sin \dfrac{x}{3}$

7. $y = -2 \sin x$ **8.** $y = -\cos \dfrac{2x}{3}$

9. $y = 3 \sin 10x$ **10.** $y = \dfrac{1}{3} \sin 8x$

11. $y = \dfrac{1}{2} \cos \dfrac{2x}{3}$ **12.** $y = \dfrac{5}{2} \cos \dfrac{x}{4}$

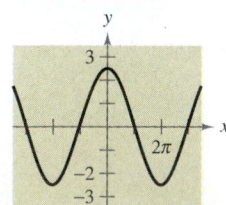   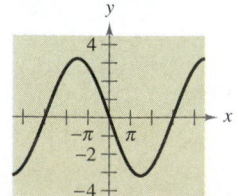

13. $y = 3 \sin 4\pi x$ **14.** $y = \dfrac{2}{3} \cos \dfrac{\pi x}{10}$

In Exercises 15–22, describe the relationship between the graphs of f and g.

15. $f(x) = \sin x$
$g(x) = \sin(x - \pi)$

16. $f(x) = \cos x$
$g(x) = \cos(x + \pi)$

17. $f(x) = \cos 2x$
$g(x) = -\cos 2x$

18. $f(x) = \sin 3x$
$g(x) = \sin(-3x)$

19. $f(x) = \cos x$
$g(x) = \cos 2x$

20. $f(x) = \sin x$
$g(x) = \sin 3x$

21. $f(x) = \sin x$
$g(x) = 2 + \sin x$

22. $f(x) = \cos 4x$
$g(x) = -2 + \cos 4x$

In Exercises 23–26, describe the relationship between the graphs of f and g.

23.

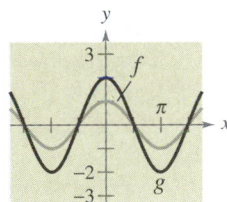

24.

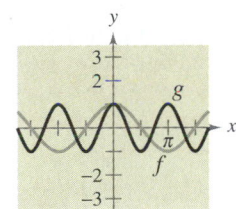

25.

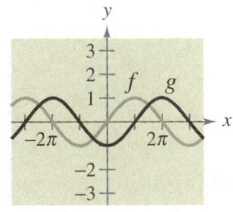

26.

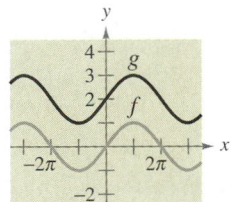

27. *Essay* Use a graphing utility to graph the function $y = a \sin x$ for $a = \frac{1}{2}$, $a = \frac{3}{2}$, and $a = -3$. Write a paragraph describing the changes in the graph corresponding to the specified changes in a.

28. *Essay* Use a graphing utility to graph the function $y = d + \sin x$ for $d = 2$, $d = 3.5$, and $d = -2$. Write a paragraph describing the changes in the graph corresponding to the specified changes in d.

29. *Essay* Use a graphing utility to graph the function $y = \sin bx$ for $b = \frac{1}{2}$, $b = \frac{3}{2}$, and $b = 4$. Write a paragraph describing the changes in the graph corresponding to the specified changes in b.

30. *Essay* Use a graphing utility to graph the function $y = \sin(x - c)$ for $c = 1$, $c = 3$, and $c = -2$. Write a paragraph describing the changes in the graph corresponding to the specified changes in c.

In Exercises 31–38, graph f and g on the same set of coordinate axes. (Include two full periods.)

31. $f(x) = -2 \sin x$
$g(x) = 4 \sin x$

32. $f(x) = \sin x$
$g(x) = \sin \dfrac{x}{3}$

33. $f(x) = \cos x$
$g(x) = 1 + \cos x$

34. $f(x) = 2 \cos 2x$
$g(x) = -\cos 4x$

35. $f(x) = -\dfrac{1}{2} \sin \dfrac{x}{2}$
$g(x) = 3 - \dfrac{1}{2} \sin \dfrac{x}{2}$

36. $f(x) = 4 \sin \pi x$
$g(x) = 4 \sin \pi x - 3$

37. $f(x) = 2 \cos x$
$g(x) = 2 \cos(x + \pi)$

38. $f(x) = -\cos x$
$g(x) = -\cos(x - \pi)$

Conjecture **In Exercises 39–42, graph f and g on the same set of coordinate axes. (Include two full periods.) Make a conjecture about the functions.**

39. $f(x) = \sin x, \quad g(x) = \cos\left(x - \dfrac{\pi}{2}\right)$

40. $f(x) = \sin x, \quad g(x) = -\cos\left(x + \dfrac{\pi}{2}\right)$

41. $f(x) = \cos x, \quad g(x) = -\sin\left(x - \dfrac{\pi}{2}\right)$

42. $f(x) = \cos x, \quad g(x) = -\cos(x - \pi)$

In Exercises 43–60, sketch the graph of the function. (Include two full periods.)

43. $y = -2 \sin 6x$

44. $y = -3 \cos 4x$

45. $y = \cos 2\pi x$

46. $y = \dfrac{3}{2} \sin \dfrac{\pi x}{4}$

47. $y = -\sin \dfrac{2\pi x}{3}$

48. $y = 10 \cos \dfrac{\pi x}{6}$

49. $y = \sin\left(x - \dfrac{\pi}{4}\right)$ **50.** $y = \dfrac{1}{2}\sin(x - \pi)$

51. $y = 3\cos(x + \pi)$ **52.** $y = 4\cos\left(x + \dfrac{\pi}{4}\right)$

53. $y = \dfrac{1}{10}\cos 60\pi x$ **54.** $y = -3 + 5\cos\dfrac{\pi t}{12}$

55. $y = 2 - \sin\dfrac{2\pi x}{3}$ **56.** $y = 2\cos x - 3$

57. $y = 3\cos(x + \pi) - 3$ **58.** $y = 4\cos\left(x + \dfrac{\pi}{4}\right) + 4$

59. $y = \dfrac{2}{3}\cos\left(\dfrac{x}{2} - \dfrac{\pi}{4}\right)$ **60.** $y = -3\cos(6x + \pi)$

 In Exercises 61–68, use a graphing utility to graph the function. (Include two full periods.)

61. $y = -2\sin(4x + \pi)$

62. $y = -4\sin\left(\dfrac{2}{3}x - \dfrac{\pi}{3}\right)$

63. $y = \cos\left(2\pi x - \dfrac{\pi}{2}\right) + 1$

64. $y = 3\cos\left(\dfrac{\pi x}{2} + \dfrac{\pi}{2}\right) - 2$

65. $y = -0.1\sin\left(\dfrac{\pi x}{10} + \pi\right)$

66. $y = 5\sin(\pi - 2x) + 10$

67. $y = 5\cos(\pi - 2x) + 2$

68. $y = \dfrac{1}{100}\sin 120\pi t$

Graphical Reasoning   **In Exercises 69–72, find *a* and *d* for the function $f(x) = a\cos x + d$ so that the graph of *f* matches the figure.**

69.

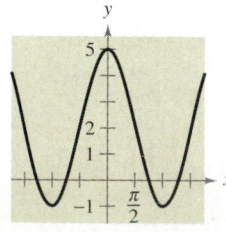

70.

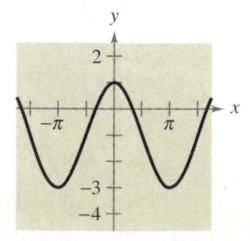

71.

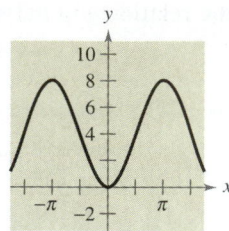

72.

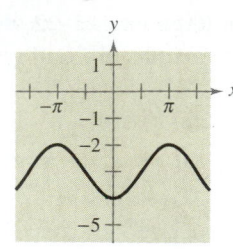

Graphical Reasoning   **In Exercises 73–76, find *a*, *b*, and *c* for the function $y = a\sin(bx - c)$ so that the graph of *f* matches the figure.**

73.

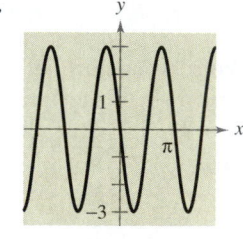

74.

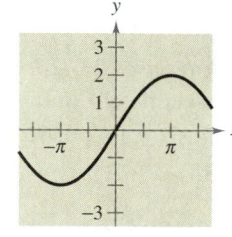

75.

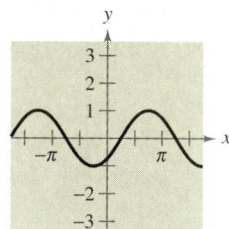

76.

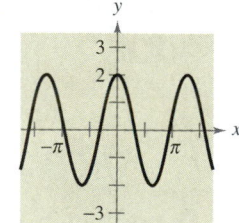

In Exercises 77–80, use a graphing utility to graph y_1 and y_2 in the interval $[-2\pi, 2\pi]$. Use the graphs to find real numbers x such that $y_1 = y_2$.

77. $y_1 = \sin x$
  $y_2 = -\dfrac{1}{2}$

78. $y_1 = \cos x$
  $y_2 = -1$

79. $y_1 = \cos x$
  $y_2 = \dfrac{\sqrt{2}}{2}$

80. $y_1 = \sin x$
  $y_2 = \dfrac{\sqrt{3}}{2}$

81. **Exploration** Use a graphing utility to graph h, and use the graph to decide whether h is even, odd, or neither.

   (a) $h(x) = \cos^2 x$ (b) $h(x) = \sin^2 x$

82. **Conjecture** If f is an even function and g is an odd function, use the results of Exercise 81 to make a conjecture about h where

   (a) $h(x) = [f(x)]^2$ (b) $h(x) = [g(x)]^2$.

83. **Respiratory Cycle** For a person at rest, the velocity v (in liters per second) of air flow during a respiratory cycle is

   $$v = 0.85 \sin \frac{\pi t}{3}$$

   where t is the time in seconds. (Inhalation occurs when $v > 0$, and exhalation occurs when $v < 0$.)

   (a) Find the time for one full respiratory cycle.

   (b) Find the number of cycles per minute.

   (c) Sketch the graph of the velocity function.

84. **Respiratory Cycle** After exercising for a few minutes, a person has a respiratory cycle for which the velocity of air flow is approximated by

   $$v = 1.75 \sin \frac{\pi t}{2}.$$

   Use this model to repeat Exercise 83.

85. **Piano Tuning** When tuning a piano, a technician strikes a tuning fork for the A above middle C and sets up wave motion that can be approximated by

   $$y = 0.001 \sin 880\pi t$$

   where t is the time in seconds.

   (a) What is the period of the function?

   (b) The frequency f is given by $f = 1/p$. What is the frequency of the note?

86. **Blood Pressure** The function

   $$P = 100 - 20 \cos \frac{5\pi t}{3}$$

   approximates the blood pressure P in millimeters of mercury at time t in seconds for a person at rest.

   (a) Find the period of the function.

   (b) Find the number of heartbeats per minute.

Sales In Exercises 87 and 88, use a graphing utility to graph the sales function over 1 year where S is the sales in thousands of units and t is the time in months, with $t = 1$ corresponding to January.

87. $S = 22.3 - 3.4 \cos \dfrac{\pi t}{6}$

88. $S = 74.50 + 43.75 \sin \dfrac{\pi t}{6}$

89. **Data Analysis** The table gives the normal daily high temperatures for Honolulu H and Chicago C (in degrees Fahrenheit) for month t with $t = 1$ corresponding to January. (Source: National Oceanic and Atmospheric Association)

t	1	2	3	4	5	6
H	80.1	80.5	81.6	82.8	84.7	86.5
C	29.0	33.5	45.8	58.6	70.1	79.6

t	7	8	9	10	11	12
H	87.5	88.7	88.5	86.9	84.1	81.2
C	83.7	81.8	74.8	63.3	48.4	34.0

   (a) A model for Honolulu is given by

   $$H(t) = 84.40 + 4.28 \sin\left(\frac{\pi t}{6} + 3.86\right).$$

   Find a trigonometric model for Chicago.

   (b) Use a graphing utility to graph the data points and the model for the temperatures in Honolulu. How well does the model fit?

   (c) Use a graphing utility to graph the data points and the model for the temperatures in Chicago. How well does the model fit?

   (d) Use the models to estimate the average annual temperature in each city. Which term of the models did you use? Explain.

   (e) What is the period of each model? Are they what you expected? Explain.

   (f) Which city has the greater variability in temperature throughout the year? Which factor of the models determines this variability? Explain.

90. Fuel Consumption The daily consumption C (in gallons) of diesel fuel on a farm is modeled by

$$C = 30.3 + 21.6 \sin\left(\frac{2\pi t}{365} + 10.9\right)$$

where t is the time in days, with $t = 1$ corresponding to January 1.

(a) What is the period of the model? Is it what you expected? Explain.

(b) What is the average daily fuel consumption? Which term of the model did you use? Explain.

(c) Use a graphing utility to graph the model. Use the graph to approximate the time of the year when consumption exceeds 40 gallons per day.

91. Exploration Using calculus, it can be shown that the sine and cosine functions can be approximated by the polynomials

$$\sin x \approx x - \frac{x^3}{3!} + \frac{x^5}{5!} \text{ and } \cos x \approx 1 - \frac{x^2}{2!} + \frac{x^4}{4!}$$

where x is in radians.

(a) Use a graphing utility to graph the sine function and its polynomial approximation on the same viewing rectangle. How do the graphs compare?

(b) Use a graphing utility to graph the cosine function and its polynomial approximation on the same viewing rectangle. How do the graphs compare?

(c) Study the patterns in the polynomial approximations of the sine and cosine functions and guess the next term in each. Then repeat parts (a) and (b). How did the accuracy of the approximations change when additional terms were added?

92. Exploration Use the polynomial approximations for the sine and cosine functions given in Exercise 91 to approximate the following functional values. Compare the results with those given by a calculator. Is the error in the approximation the same in each case? Explain.

(a) $\sin \dfrac{1}{2}$ (b) $\sin 1$ (c) $\sin \dfrac{\pi}{6}$

(d) $\cos(-0.5)$ (e) $\cos 1$ (f) $\cos \dfrac{\pi}{4}$

93. Data Analysis The percent y of the moon's face that is illuminated on day x of the year 1995, where $x = 300$ represents October 27, is given in the table. (Source: American Museum of Natural History)

x	303	311	319	326	333	341
y	0.5	1.0	0.5	0	0.5	1.0

(a) Create a scatter plot of the data.

(b) Find a trigonometric model that fits the data.

(c) Add the graph of your model in part (b) to the scatter plot. How well does the model fit the data?

(d) Find the moon's percent illumination for December 22, 1995.

Review   **Solve Exercises 94–97 as a review of the skills and problem-solving techniques you learned in previous sections. Use the properties of logarithms to write the expression as a sum, difference, and/or constant multiple of a logarithm.**

94. $\log_{10}\sqrt{x - 2}$

95. $\log_2[x^2(x - 3)]$

96. $\ln \dfrac{t^3}{t - 1}$

97. $\ln\sqrt{\dfrac{z}{z^2 + 1}}$

See Exercise 68 on page 514 for an example of how a trigonometric function can be used to analyze the distance between a television camera and a parade unit.

6.5 *Graphs of Other Trigonometric Functions*

Graph of the Tangent Function ❏ *Graph of the Cotangent Function* ❏
Graphs of the Reciprocal Functions ❏ *Damped Trigonometric Graphs*

Graph of the Tangent Function

Recall from Section 6.3 that the tangent function is odd. That is,

$$\tan(-x) = -\tan x.$$

Consequently, the graph of

$$y = \tan x$$

is symmetric with respect to the origin. You also know from the identity $\tan x = \sin x / \cos x$ that the tangent is undefined when $\cos x = 0$. Two such values are $x = \pm \pi/2 \approx \pm 1.5708$.

x	$-\dfrac{\pi}{2}$	-1.57	-1.5	-1	0	1	1.5	1.57	$\dfrac{\pi}{2}$
$\tan x$	Undef.	-1255.8	-14.1	-1.56	0	1.56	14.1	1255.8	Undef.

$\tan x$ approaches $-\infty$ as x approaches $-\pi/2$ from the right

$\tan x$ approaches ∞ as x approaches $\pi/2$ from the left

As indicated in the table, $\tan x$ increases without bound as x approaches $\pi/2$ from the left, and decreases without bound as x approaches $-\pi/2$ from the right. Thus, the graph of $y = \tan x$ has *vertical asymptotes* at $x = \pi/2$ and $x = -\pi/2$, as shown in Figure 6.49. Moreover, because the period of the tangent function is π, vertical asymptotes also occur when $x = \pi/2 + n\pi$, where n is an integer. The domain of the tangent function is the set of all real numbers other than $x = \pi/2 + n\pi$, and the range is the set of all real numbers.

Sketching the graph of a function of the form $y = a \tan(bx - c)$ is similar to sketching the graph of $y = a \sin(bx - c)$ in that you locate key points that identify the intercepts and asymptotes. Two consecutive asymptotes can be found by solving the equations

$$bx - c = -\frac{\pi}{2} \qquad \text{and} \qquad bx - c = \frac{\pi}{2}.$$

The midpoint between two consecutive asymptotes is an x-intercept of the graph. After plotting the asymptotes and the x-intercept, plot a few additional points between the two asymptotes and sketch one cycle. Finally sketch one or two additional cycles to the left and right.

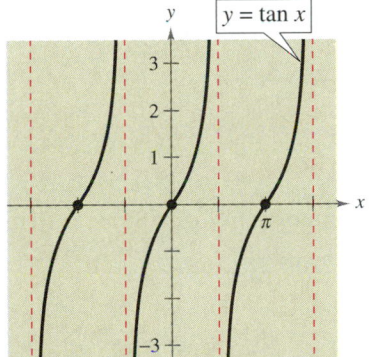

$y = \tan x$

Period: π
Domain: all $x \neq \frac{\pi}{2} + n\pi$
Range: $(-\infty, \infty)$
Vertical asymptotes: $x = \frac{\pi}{2} + n\pi$

FIGURE 6.49

NOTE The period of the function $y = a \tan(bx - c)$ is the distance between two consecutive asymptotes. The amplitude of a tangent function is not defined. ■■

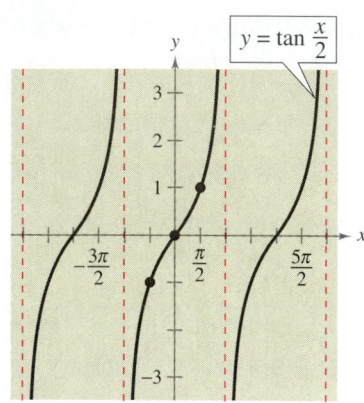

FIGURE 6.50

■

***EXAMPLE* 1**   *Sketching the Graph of a Tangent Function*

Sketch the graph of $y = \tan \dfrac{x}{2}$.

Solution

By solving the equations

$$\frac{x}{2} = -\frac{\pi}{2} \quad \text{and} \quad \frac{x}{2} = \frac{\pi}{2}$$

$$x = -\pi \qquad\qquad x = \pi$$

you can see that two consecutive asymptotes occur at $x = -\pi$ and $x = \pi$. Between these two asymptotes, plot a few points, including the x-intercept, as shown in the table. Three cycles of the graph are shown in Figure 6.50.

x	$-\dfrac{\pi}{2}$	0	$\dfrac{\pi}{2}$
$\tan\dfrac{x}{2}$	-1	0	1

■

***EXAMPLE* 2**   *Sketching the Graph of a Tangent Function*

Sketch the graph of $y = -3\tan 2x$.

Solution

By solving the equations

$$2x = -\frac{\pi}{2} \quad \text{and} \quad 2x = \frac{\pi}{2}$$

$$x = -\frac{\pi}{4} \qquad\qquad x = \frac{\pi}{4}$$

you can see that two consecutive asymptotes occur at $x = -\pi/4$ and $x = \pi/4$. Between these two asymptotes, plot a few points, including the x-intercept, as shown in the table. Four cycles of the graph are shown in Figure 6.51.

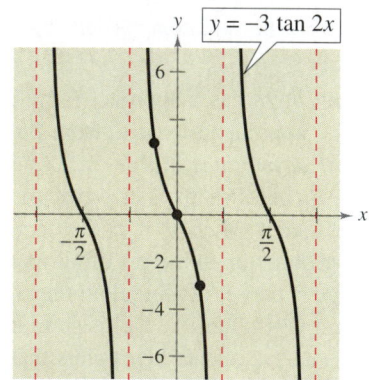

FIGURE 6.51

x	$-\dfrac{\pi}{8}$	0	$\dfrac{\pi}{8}$
$-3\tan 2x$	3	0	-3

By comparing the graphs in Examples 1 and 2, you can see that the graph of $y = a \tan(bx - c)$ is increasing between consecutive vertical asymptotes if $a > 0$, and decreasing between consecutive vertical asymptotes if $a < 0$. In other words, the graph for $a < 0$ is a reflection in the x-axis of the graph for $a > 0$.

Graph of the Cotangent Function

The graph of the cotangent function is similar to the graph of the tangent function. It also has a period of π. However, from the identity

$$y = \cot x = \frac{\cos x}{\sin x}$$

you can see that the cotangent function has vertical asymptotes at $x = n\pi$ where n is an integer, because $\sin x$ is zero at these x-values. The graph of the cotangent function is shown in Figure 6.52.

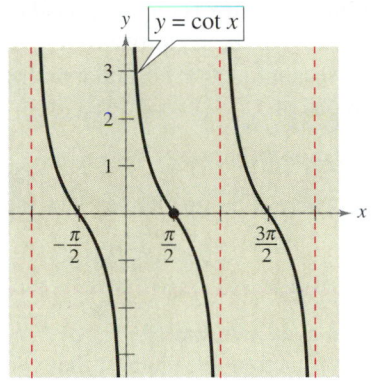

Period: π
Domain: all $x \neq n\pi$
Range: $(-\infty, \infty)$
Vertical asymptotes: $x = n\pi$

FIGURE 6.52

EXAMPLE 3   *Sketching the Graph of a Cotangent Function*

Sketch the graph of $y = 2 \cot \dfrac{x}{3}$.

Solution

To locate two consecutive vertical asymptotes of the graph, solve the equations $x/3 = 0$ and $x/3 = \pi$, as follows.

$$\frac{x}{3} = 0 \qquad \text{and} \qquad \frac{x}{3} = \pi$$

$$x = 0 \qquad\qquad x = 3\pi$$

Then, between these two asymptotes, plot a few points, including the x-intercept, as shown in the table. Three cycles of the graph are shown in Figure 6.53. (Note that the period is 3π, the distance between consecutive asymptotes.)

x	$\dfrac{3\pi}{4}$	$\dfrac{3\pi}{2}$	$\dfrac{9\pi}{4}$
$2 \cot \dfrac{x}{3}$	2	0	-2

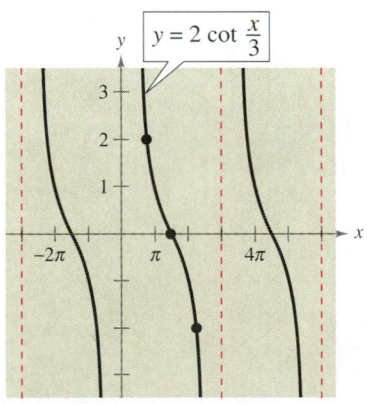

FIGURE 6.53

Graphs of the Reciprocal Functions

The graphs of the two remaining trigonometric functions can be obtained from the graphs of the sine and cosine functions using the reciprocal identities

$$\csc x = 1/\sin x \qquad \text{and} \qquad \sec x = 1/\cos x.$$

For instance, at a given value of x, the y-coordinate of $\sec x$ is the reciprocal of the y-coordinate of $\cos x$. Of course, when $\cos x = 0$, the reciprocal does not exist. Near such values of x, the behavior of the secant function is similar to that of the tangent function. In other words, the graphs of

$$\tan x = \sin x/\cos x \qquad \text{and} \qquad \sec x = 1/\cos x$$

have vertical asymptotes at $x = \pi/2 + n\pi$, where n is an integer and the cosine is zero at these x-values. Similarly,

$$\cot x = \cos x/\sin x \qquad \text{and} \qquad \csc x = 1/\sin x$$

have vertical asymptotes where $\sin x = 0$—that is, at $x = n\pi$.

To sketch the graph of a secant or cosecant function, we suggest that you first make a sketch of its reciprocal function. For instance, the sketch the graph of $y = \csc x$, first sketch the graph of $y = \sin x$. Then take reciprocals of the y-coordinates to obtain points on the graph of $y = \csc x$. We use this procedure to obtain the graphs shown in Figure 6.54.

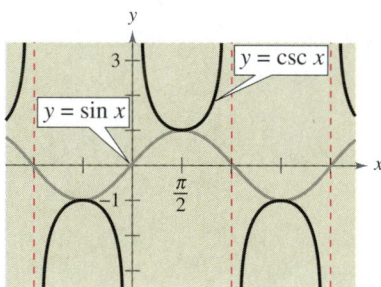

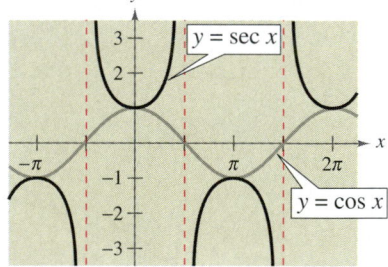

Period: 2π
Domain: all $x \neq n\pi$
Range: $(-\infty, -1]$ and $[1, \infty)$
Vertical asymptotes: $x = n\pi$
Symmetry: origin

Period: 2π
Domain: all $x \neq \frac{\pi}{2} + n\pi$
Range: $(-\infty, -1]$ and $[1, \infty)$
Vertical asymptotes: $x = \frac{\pi}{2} + n\pi$
Symmetry: y-axis

FIGURE 6.54

In comparing the graphs of the secant and cosecant functions with those of the sine and cosine functions, note that the "hills" and "valleys" are interchanged. For example, a hill (or maximum point) on the sine curve corresponds to a valley (a local minimum) on the cosecant curve. Similarly, a valley (or minimum point) on the sine curve corresponds to a hill (a local maximum) on the cosecant curve, as shown in Figure 6.55.

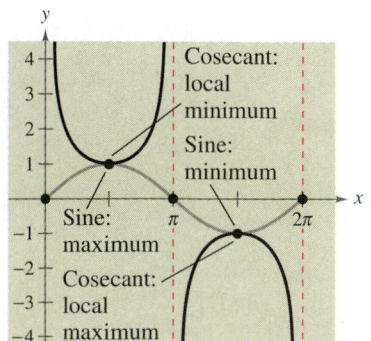

FIGURE 6.55

EXAMPLE 4   *Sketching the Graph of a Cosecant Function*

Sketch the graph of $y = 2 \csc\left(x + \dfrac{\pi}{4}\right)$.

Solution

Begin by sketching the graph of

$$y = 2 \sin\left(x + \frac{\pi}{4}\right).$$

For this function, the amplitude is 2 and the period is 2π. By solving the equations

$$x + \frac{\pi}{4} = 0 \qquad \text{and} \qquad x + \frac{\pi}{4} = 2\pi$$

$$x = -\frac{\pi}{4} \qquad\qquad\qquad x = \frac{7\pi}{4}$$

you can see that one cycle of the sine function corresponds to the interval from $x = -\pi/4$ to $x = 7\pi/4$. The graph of this sine function is represented by the gray curve in Figure 6.56. Because the sine function is zero at the endpoints of this interval, the corresponding cosecant function

$$y = 2 \csc\left(x + \frac{\pi}{4}\right) = 2\left(\frac{1}{\sin[x + (\pi/4)]}\right)$$

has vertical asymptotes at $x = -\pi/4, x = 3\pi/4, x = 7\pi/4$, etc. The graph of the cosecant function is represented by the black curve in Figure 6.56.

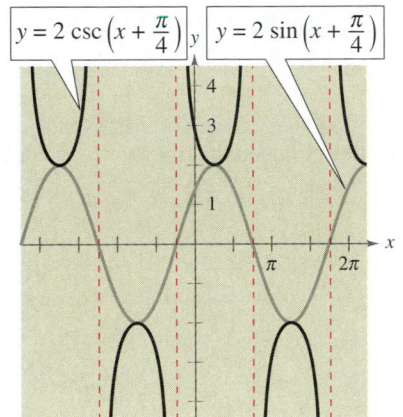

FIGURE 6.56

EXAMPLE 5   *Sketching the Graph of a Secant Function*

Sketch the graph of $y = \sec 2x$.

Solution

Begin by sketching the graph of $y = \cos 2x$, as indicated by the gray curve in Figure 6.57. Then, form the graph of $y = \sec 2x$ as the black curve in the figure. Note that the x-intercepts of $y = \cos 2x$

$$\left(\frac{\pi}{4}, 0\right), \quad \left(\frac{3\pi}{4}, 0\right), \quad \left(\frac{5\pi}{4}, 0\right), \dots$$

correspond to the vertical asymptotes

$$x = \frac{\pi}{4}, \quad x = \frac{3\pi}{4}, \quad x = \frac{5\pi}{4}, \dots$$

of the graph of $y = \sec 2x$.

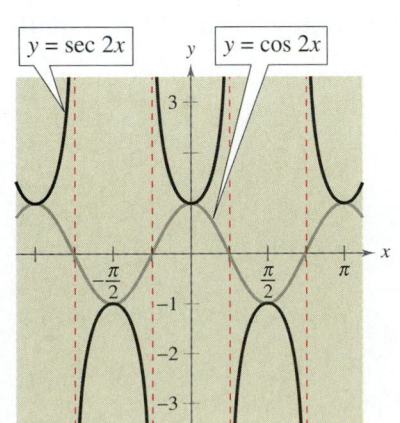

FIGURE 6.57

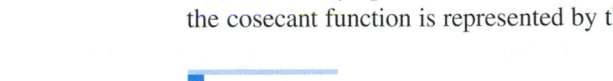

Damped Trigonometric Graphs

A *product* of two functions can be graphed using properties of the individual functions. For instance, consider the function

$$f(x) = x \sin x$$

as the product of the functions $y = x$ and $y = \sin x$. Using properties of absolute value and the fact that $|\sin x| \leq 1$, we have $0 \leq |x||\sin x| \leq |x|$. Consequently,

$$-|x| \leq x \sin x \leq |x|$$

which means that the graph of $f(x) = x \sin x$ lies between the lines $y = -x$ and $y = x$. Furthermore, because

$$f(x) = x \sin x = \pm x \qquad \text{at} \qquad x = \frac{\pi}{2} + n\pi$$

and

$$f(x) = x \sin x = 0 \qquad \text{at} \qquad x = n\pi$$

the graph of f touches the line $y = -x$ or the line $y = x$ at $x = \pi/2 + n\pi$ and has x-intercepts at $x = n\pi$. A sketch of f is shown in Figure 6.58. In the function $f(x) = x \sin x$, the factor x is called the **damping factor.**

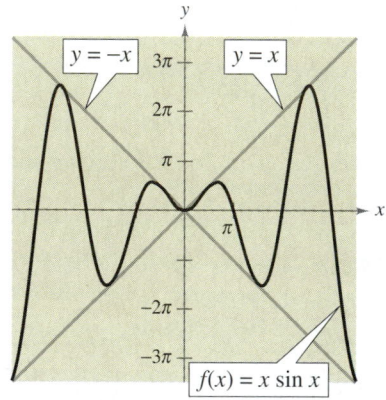

FIGURE 6.58

EXAMPLE 6   *Damped Sine Wave*

Sketch the graph of $f(x) = e^{-x} \sin 3x$.

Solution

Consider $f(x)$ as the product of the two functions

$$y = e^{-x} \qquad \text{and} \qquad y = \sin 3x$$

each of which has the set of real numbers as its domain. For any real number x, you know that $e^{-x} \geq 0$ and $|\sin 3x| \leq 1$. Therefore, $e^{-x}|\sin 3x| \leq e^{-x}$, which means that

$$-e^{-x} \leq e^{-x} \sin 3x \leq e^{-x}.$$

Furthermore, because

$$f(x) = e^{-x} \sin 3x = \pm e^{-x} \qquad \text{at} \qquad x = \frac{\pi}{6} + \frac{n\pi}{3}$$

and

$$f(x) = e^{-x} \sin 3x = 0 \qquad \text{at} \qquad x = \frac{n\pi}{3}$$

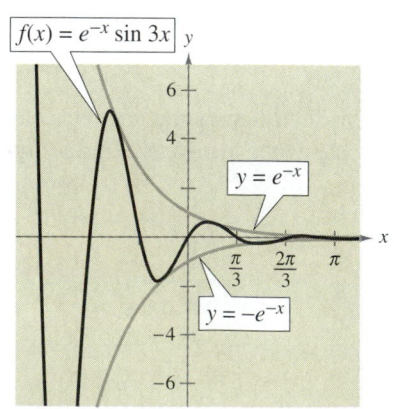

FIGURE 6.59

the graph of f touches the curve $y = -e^{-x}$ and $y = e^{-x}$ at $x = \pi/6 + n\pi/3$ and has intercepts at $x = n\pi/3$. A sketch is shown in Figure 6.59.

Figure 6.60 summarizes the six basic trigonometric functions.

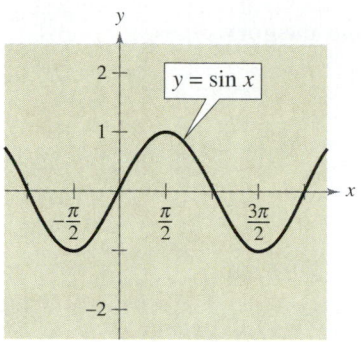

Domain: all reals
Range: $[-1, 1]$
Period: 2π

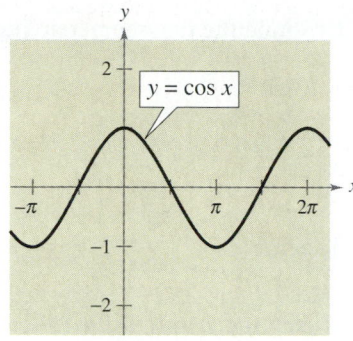

Domain: all reals
Range: $[-1, 1]$
Period: 2π

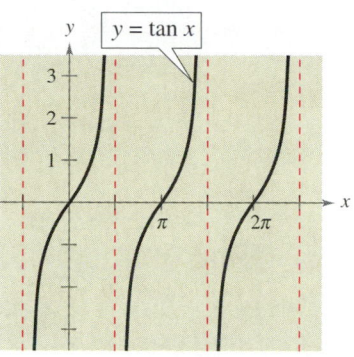

Domain: all $x \neq \frac{\pi}{2} + n\pi$
Range: $(-\infty, \infty)$
Period: π

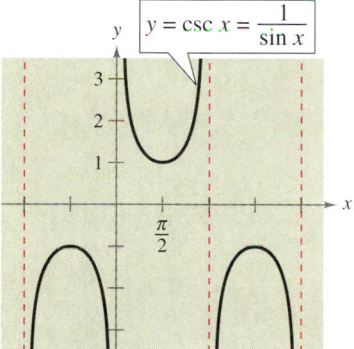

Domain: all $x \neq n\pi$
Range: $(-\infty, -1]$ and $[1, \infty)$
Period: 2π

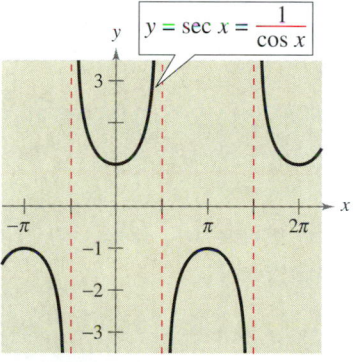

Domain: all $x \neq \frac{\pi}{2} + n\pi$
Range: $(-\infty, -1]$ and $[1, \infty)$
Period: 2π

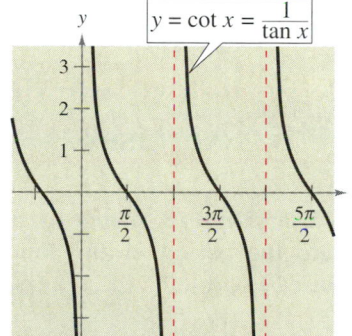

Domain: all $x \neq n\pi$
Range: $(-\infty, \infty)$
Period: π

FIGURE 6.60

GROUP ACTIVITY

COMBINING TRIGONOMETRIC FUNCTIONS

Recall from Section 2.4 that functions can be combined arithmetically. This also applies to trigonometric functions. For each of the functions $h(x) = x + \sin x$ and $h(x) = \cos x - \sin 3x$, (a) identify two simpler functions f and g that comprise the combination, (b) use a table to show how to obtain the numerical values of $h(x)$ from the numerical values of $f(x)$ and $g(x)$, and (c) use a graph of f and g to show how h may be formed.

Can you find functions $f(x) = d + a \sin(bx + c)$ and $g(x) = d + a \cos(bx + c)$ such that $f(x) + g(x) = 0$ for all x?

Evaluate the trigonometric function from memory.

1. $\tan 0$

2. $\cos \dfrac{\pi}{4}$

3. $\tan \dfrac{\pi}{4}$

4. $\cot \dfrac{\pi}{2}$

5. $\sin \pi$

6. $\cos \dfrac{\pi}{2}$

Sketch the graph of the function. (Include two full periods.)

7. $y = -2 \cos 2x$

8. $y = 3 \sin \dfrac{x}{4}$

9. $y = \dfrac{3}{2} \sin 2\pi x$

10. $y = -2 \cos \dfrac{\pi x}{2}$

6.5 Exercises

In Exercises 1–8, match the function with its graph. State the period of the function. [The graphs are labeled (a), (b), (c), (d), (e), (f), (g), and (h).]

(a)
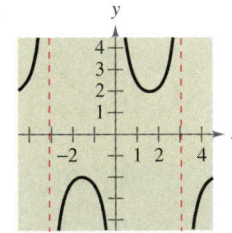

(b)
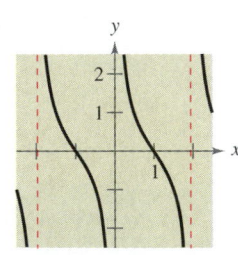

(c)
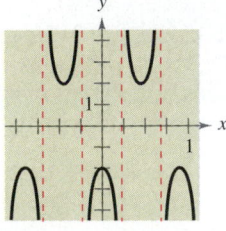

(d)
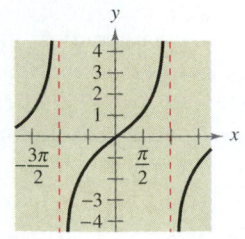

(e)
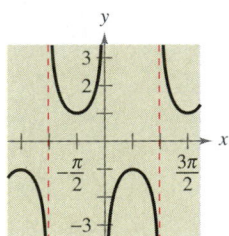

(f)
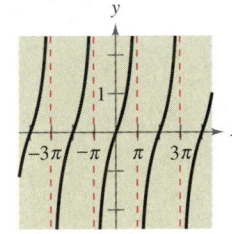

(g)
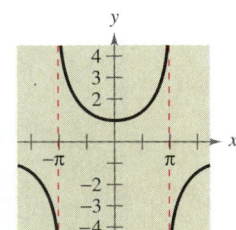

(h)
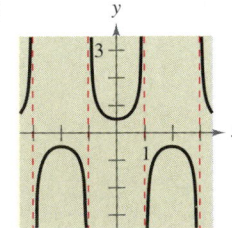

1. $y = \sec \dfrac{x}{2}$

2. $y = \tan \dfrac{x}{2}$

3. $y = \tan 2x$

4. $y = 2 \csc x$

5. $y = \cot \dfrac{\pi x}{2}$

6. $y = \dfrac{1}{2} \sec \dfrac{\pi x}{2}$

7. $y = -\csc x$

8. $y = -2 \sec 2\pi x$

In Exercises 9–30, sketch the graph of the function. (Include two full periods.)

9. $y = \frac{1}{3} \tan x$

10. $y = \frac{1}{4} \tan x$

11. $y = \tan 2x$

12. $y = -3 \tan \pi x$

13. $y = -\frac{1}{2} \sec x$

14. $y = \frac{1}{4} \sec x$

15. $y = \sec \pi x$

16. $y = 2 \sec 4x$

17. $y = \sec \pi x - 1$

18. $y = -2 \sec 4x + 2$

19. $y = \csc \frac{x}{2}$

20. $y = \csc \frac{x}{3}$

21. $y = \cot \frac{x}{2}$

22. $y = 3 \cot \frac{\pi x}{2}$

23. $y = \frac{1}{2} \sec 2x$

24. $y = -\frac{1}{2} \tan x$

25. $y = \tan \frac{\pi x}{4}$

26. $y = \sec(x + \pi)$

27. $y = \csc(\pi - x)$

28. $y = \sec(\pi - x)$

29. $y = \frac{1}{4} \csc\left(x + \frac{\pi}{4}\right)$

30. $y = 2 \cot\left(x + \frac{\pi}{2}\right)$

In Exercises 31–40, use a graphing utility to graph the function. (Include two full periods.)

31. $y = \tan \frac{x}{3}$

32. $y = -\tan 2x$

33. $y = -2 \sec 4x$

34. $y = \sec \pi x$

35. $y = \tan\left(x - \frac{\pi}{4}\right)$

36. $y = -\csc(4x - \pi)$

37. $y = \frac{1}{4} \cot\left(x - \frac{\pi}{2}\right)$

38. $y = 0.1 \tan\left(\frac{\pi x}{4} + \frac{\pi}{4}\right)$

39. $y = 2 \sec(2x - \pi)$

40. $y = \frac{1}{3} \sec\left(\frac{\pi x}{2} + \frac{\pi}{2}\right)$

In Exercises 41–44, use a graph to solve the equation on the interval $[-2\pi, 2\pi]$.

41. $\tan x = 1$

42. $\cot x = -\sqrt{3}$

43. $\sec x = -2$

44. $\csc x = \sqrt{2}$

In Exercises 45 and 46, use the graph of the function to determine whether the function is even, odd, or neither.

45. $f(x) = \sec x$

46. $f(x) = \tan x$

47. *Essay* Describe the behavior of $f(x) = \tan x$ as x approaches $\pi/2$ from the left and from the right.

48. *Essay* Describe the behavior of $f(x) = \csc x$ as x approaches π from the left and from the right.

49. *Graphical Reasoning* Consider the functions

$$f(x) = 2 \sin x \quad \text{and} \quad g(x) = \frac{1}{2} \csc x$$

on the interval $(0, \pi)$.

(a) Graph f and g in the same coordinate plane.

(b) Approximate the interval where $f > g$.

(c) Describe the behavior of each of the functions as x approaches π. How is the behavior of g related to the behavior of f as x approaches π?

50. *Graphical Reasoning* Consider the functions

$$f(x) = \tan \frac{\pi x}{2} \quad \text{and} \quad g(x) = \frac{1}{2} \sec \frac{\pi x}{2}$$

on the interval $(-1, 1)$.

(a) Use a graphing utility to graph f and g on the same viewing rectangle.

(b) Approximate the interval where $f < g$.

(c) Approximate the interval where $2f < 2g$. How does the result compare with that of part (b)? Explain.

In Exercises 51–54, use a graphing utility to graph the two equations on the same viewing rectangle. Use the graphs to lend evidence that the expressions are equivalent. Verify the results analytically.

51. $y_1 = \sin x \csc x, \quad y_2 = 1$

52. $y_1 = \sin x \sec x, \quad y_2 = \tan x$

53. $y_1 = \frac{\cos x}{\sin x}, \quad y_2 = \cot x$

54. $y_1 = \sec^2 x - 1, \quad y_2 = \tan^2 x$

In Exercises 55–58, match the function with its graph. Describe the behavior of the function as x approaches zero. [The graphs are labeled (a), (b), (c), and (d).]

(a)

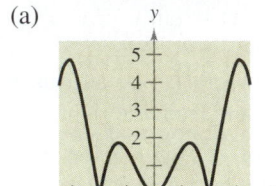

(b)

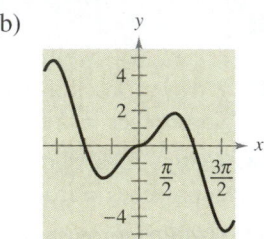

(c)

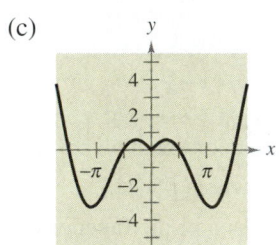

(d)

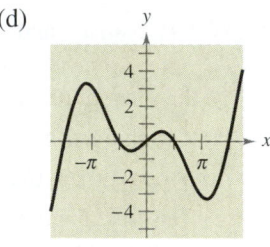

55. $f(x) = x \cos x$

56. $f(x) = |x \sin x|$

57. $g(x) = |x| \sin x$

58. $g(x) = |x| \cos x$

Conjecture **In Exercises 59–62, graph the functions of f and g. Use the graphs to make a conjecture about the relationship between the functions.**

59. $f(x) = \sin x + \cos\left(x + \dfrac{\pi}{2}\right), \quad g(x) = 0$

60. $f(x) = \sin x - \cos\left(x + \dfrac{\pi}{2}\right), \quad g(x) = 2 \sin x$

61. $f(x) = \sin^2 x, \quad g(x) = \dfrac{1}{2}(1 - \cos 2x)$

62. $f(x) = \cos^2 \dfrac{\pi x}{2}, \quad g(x) = \dfrac{1}{2}(1 + \cos \pi x)$

In Exercises 63–66, use a graphing utility to graph the function and the damping factor of the function on the same viewing rectangle. Describe the behavior of the function as x increases without bound.

63. $f(x) = 2^{-x/4} \cos \pi x$

64. $f(x) = e^{-x} \cos x$

65. $g(x) = e^{-x^2/2} \sin x$

66. $h(x) = 2^{-x^2/4} \sin x$

67. *Distance* A plane flying at an altitude of 5 miles over level ground will pass directly over a radar antenna (see figure). Let d be the ground distance from the antenna to the point directly under the plane and let x be the angle of elevation to the plane from the antenna. Write d as a function of x and graph the function over the interval $0 < x < \pi$.

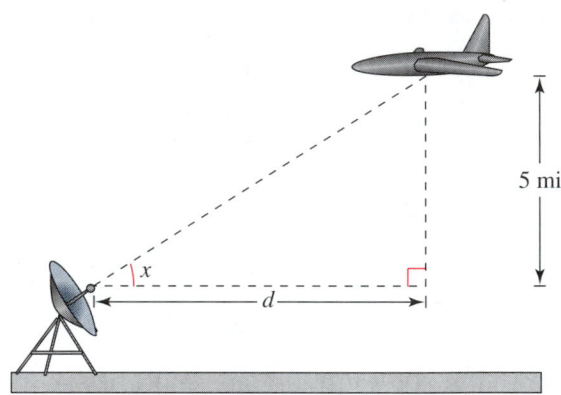

68. *Television Coverage* A television camera is on a reviewing platform 36 meters from the street on which a parade will be passing from left to right (see figure). Express the distance d from the camera to a particular unit in the parade as a function of the angle x, and graph the function over the interval $-\pi/2 < x < \pi/2$. (Consider x as negative when a unit in the parade approaches from the left.)

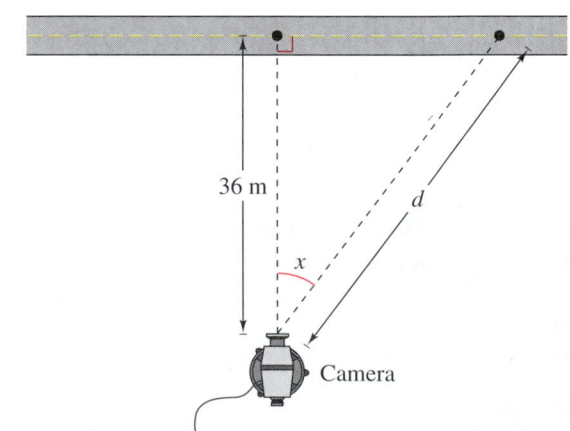

69. Predator-Prey Model Suppose the population of a certain predator at time t (in months) in a given region is estimated to be

$$P = 10,000 + 3000 \sin \frac{2\pi t}{24}.$$

and the population of its primary food source (its prey) is estimated to be

$$p = 15,000 + 5000 \cos \frac{2\pi t}{24}.$$

Use the graphs of the models to explain the oscillations in the size of each population.

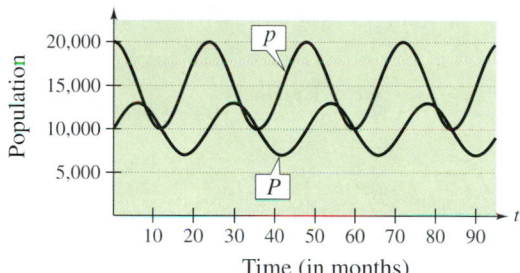

70. Normal Temperatures The normal monthly high temperatures in degrees Fahrenheit for Erie, Pennsylvania are approximated by

$$H(t) = 54.33 - 20.38 \cos \frac{\pi t}{6} - 15.69 \sin \frac{\pi t}{6}$$

and the normal monthly low temperatures are approximated by

$$L(t) = 39.36 - 15.70 \cos \frac{\pi t}{6} - 14.16 \sin \frac{\pi t}{6}$$

where t is the time in months, with $t = 1$ corresponding to January (see figure). (Source: National Oceanic and Atmospheric Association)

(a) What is the period of each function?

(b) During what part of the year is the difference between the normal high and low temperatures greatest? When is it smallest?

(c) The sun is northernmost in the sky around June 21, but the graph shows the warmest temperatures at a later date. Approximate the lag time of the temperatures relative to the position of the sun.

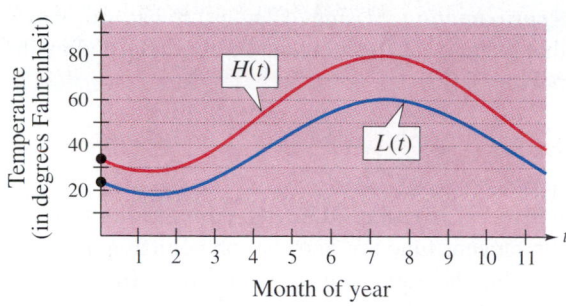

FIGURE FOR 70

71. Harmonic Motion An object weighing W pounds is suspended from the ceiling by a steel spring (see figure). The weight is pulled downward (positive direction) from its equilibrium position and released. The resulting motion of the weight is described by the function

$$y = \frac{1}{2}e^{-t/4} \cos 4t, \qquad t > 0$$

where y is the distance in feet and t is the time in seconds.

(a) Use a graphing utility to graph the function.

(b) Describe the behavior of the displacement function for increasing values of time t.

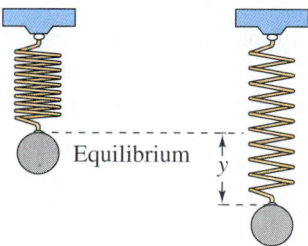

72. Exploration Consider the function

$$f(x) = x - \cos x.$$

(a) Use a graphing utility to graph the function and verify that there exists a zero between 0 and 1. Use the graph to approximate the zero.

(b) Starting with $x_0 = 1$, generate a sequence $x_1, x_2, x_3, \ldots$ where $x_n = \cos(x_{n-1})$. Verify that the sequence approaches the zero of f.

73. *Approximation* Using calculus, it can be shown that the tangent function can be approximated by the polynomial

$$\tan x \approx x + \frac{2x^3}{3!} + \frac{16x^5}{5!}$$

where x is in radians. Use a graphing utility to graph the tangent function and its polynomial approximation on the same viewing rectangle. How do the graphs compare?

74. *Approximation* Using calculus, it can be shown that the secant function can be approximated by the polynomial

$$\sec x \approx 1 + \frac{x^2}{2!} + \frac{5x^4}{4!}$$

where x is in radians. Use a graphing utility to graph the secant function and its polynomial approximation on the same viewing rectangle. How do the graphs compare?

75. *Pattern Recognition*

(a) Use a graphing utility to graph each function.

$$y_1 = \frac{4}{\pi}\left(\sin \pi x + \frac{1}{3}\sin 3\pi x\right)$$

$$y_2 = \frac{4}{\pi}\left(\sin \pi x + \frac{1}{3}\sin 3\pi x + \frac{1}{5}\sin 5\pi x\right)$$

(b) Identify the pattern started in part (a) and find a function y_3 that continues the pattern one more term. Use a graphing utility to graph y_3.

(c) The graphs of parts (a) and (b) approximate the periodic function in the figure. Find a function y_4 that is a better approximation.

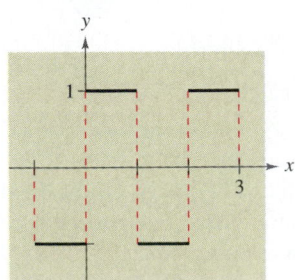

76. *Sales* The projected monthly sales S (in thousands of units) of a seasonal product are modeled by

$$S = 74 + 3x + 40 \sin \frac{\pi t}{6}$$

where t is the time in months, with $t = 1$ corresponding to January. Graph the sales function over 1 year.

Exploration In Exercises 77–82, use a graphing utility to graph the function. Describe the behavior of the function as x approaches zero.

77. $y = \dfrac{6}{x} + \cos x, \quad x > 0$

78. $y = \dfrac{4}{x} + \sin 2x, \quad x > 0$

79. $g(x) = \dfrac{\sin x}{x}$

80. $f(x) = \dfrac{1 - \cos x}{x}$

81. $f(x) = \sin \dfrac{1}{x}$

82. $h(x) = x \sin \dfrac{1}{x}$

Review Solve Exercises 83–86 as a review of the skills and problem-solving techniques you learned in previous sections. Solve the equation. (Round your solution to three decimal places.)

83. $e^{2x} = 54$

84. $\dfrac{300}{1 + e^{-x}} = 100$

85. $\ln(x^2 + 1) = 3.2$

86. $\log_8 x + \log_8(x - 1) = \frac{1}{3}$

6.6 *Inverse Trigonometric Functions*

See Exercise 85 on page 526 for an example of how an inverse trigonometric function can be used to analyze a photography setup.

Inverse Sine Function ❑ *Other Inverse Trigonometric Functions* ❑
Compositions of Functions

Inverse Sine Function

Recall from Section 2.5 that, for a function to have an inverse, it must pass the Horizontal Line Test. From Figure 6.61 it is obvious that $y = \sin x$ does not pass the test because different values of x yield the same y-value. However, if you restrict the domain to the interval $-\pi/2 \le x \le \pi/2$ (corresponding to the black portion of the graph in Figure 6.61), the following properties hold.

1. On the interval $[-\pi/2, \pi/2]$, the function $y = \sin x$ is increasing.
2. On the interval $[-\pi/2, \pi/2]$, $y = \sin x$ takes on its full range of values, $-1 \le \sin x \le 1$.
3. On the interval $[-\pi/2, \pi/2]$, $y = \sin x$ passes the Horizontal Line Test.

Thus, on the restricted domain $-\pi/2 \le x \le \pi/2$, $y = \sin x$ has a unique inverse called the **inverse sine function.** It is denoted by

$$y = \arcsin x \qquad \text{or} \qquad y = \sin^{-1} x.$$

The notation $\sin^{-1} x$ is consistent with the inverse function notation $f^{-1}(x)$. The arcsin x notation (read as "the arcsine of x") comes from the association of a central angle with its subtended *arc length* on a unit circle. Thus, arcsin x means the angle (or arc) whose sine is x. Both notations, arcsin x and $\sin^{-1}x$, are commonly used in mathematics, so remember that $\sin^{-1}x$ denotes the *inverse* sine function rather than $1/\sin x$. The values of arcsin x lie in the interval

$$-\frac{\pi}{2} \le \arcsin x \le \frac{\pi}{2}.$$

The graph of $y = \arcsin x$ is shown in Example 2.

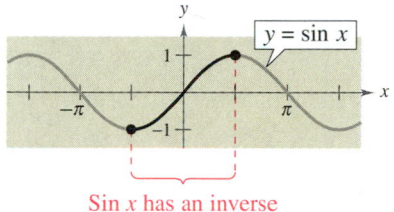

Sin x has an inverse on this interval.

FIGURE 6.61

DEFINITION OF INVERSE SINE FUNCTION

The **inverse sine function** is defined by

$$y = \arcsin x \qquad \text{if and only if} \qquad \sin y = x$$

where $-1 \le x \le 1$ and $-\pi/2 \le y \le \pi/2$. The domain of $y = \arcsin x$ is $[-1, 1]$, and the range is $[-\pi/2, \pi/2]$.

NOTE When evaluating the inverse sine function, it helps to remember the phrase "the arcsine of x is the angle (or number) whose sine is x." ∎∎

Study Tip

As with the trigonometric functions, much of the work with the inverse trigonometric functions can be done by *exact* calculations rather than by calculator approximations. Exact calculations help to increase your understanding of the inverse functions by relating them to the triangle definitions of the trigonometric functions.

EXAMPLE 1   *Evaluating the Inverse Sine Function*

If possible, find the exact value.

a. $\arcsin\left(-\dfrac{1}{2}\right)$ **b.** $\sin^{-1}\dfrac{\sqrt{3}}{2}$ **c.** $\sin^{-1}2$

Solution

a. Because $\sin(-\pi/6) = -\frac{1}{2}$ for $-\pi/2 \le y \le \pi/2$, it follows that

$$\arcsin\left(-\frac{1}{2}\right) = -\frac{\pi}{6}.$$

b. Because $\sin(\pi/3) = \sqrt{3}/2$ for $-\pi/2 \le y \le \pi/2$, it follows that

$$\sin^{-1}\frac{\sqrt{3}}{2} = \frac{\pi}{3}.$$

c. It is not possible to evaluate $y = \sin^{-1}x$ when $x = 2$ because there is no angle whose sine is 2. Remember that the domain of the inverse sine function is $[-1, 1]$.

EXAMPLE 2   *Graphing the Arcsine Function*

Sketch a graph of $y = \arcsin x$.

Solution

By definition, the equations

$$y = \arcsin x \qquad \text{and} \qquad \sin y = x$$

are equivalent for $-\pi/2 \le y \le \pi/2$. Hence, their graphs are the same. From the interval $[-\pi/2, \pi/2]$, you can assign values to y in the second equation to make a table of values.

y	$-\dfrac{\pi}{2}$	$-\dfrac{\pi}{4}$	$-\dfrac{\pi}{6}$	0	$\dfrac{\pi}{6}$	$\dfrac{\pi}{4}$	$\dfrac{\pi}{2}$
$x = \sin y$	-1	$-\dfrac{\sqrt{2}}{2}$	$-\dfrac{1}{2}$	0	$\dfrac{1}{2}$	$\dfrac{\sqrt{2}}{2}$	1

The resulting graph for $y = \arcsin x$ is shown in Figure 6.62. Note that it is the reflection (in line $y = x$) of the black portion of Figure 6.61. Be sure you see that Figure 6.62 shows the *entire* graph of the inverse sine function. Remember that the range of $y = \arcsin x$ is the closed interval $[-\pi/2, \pi/2]$.

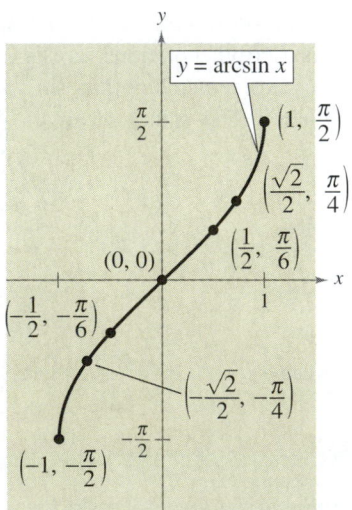

FIGURE 6.62

Other Inverse Trigonometric Functions

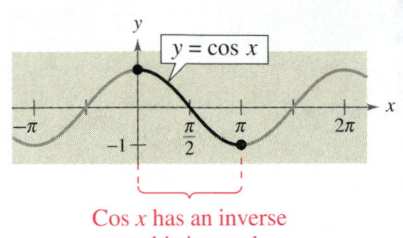

Cos x has an inverse
on this interval.

FIGURE 6.63

The cosine function is decreasing on the interval $0 \le x \le \pi$, as shown in Figure 6.63. Consequently, on this interval the cosine function has an inverse function—the **inverse cosine function**—denoted by

$$y = \arccos x \qquad \text{or} \qquad y = \cos^{-1} x.$$

Similarly, you can define an **inverse tangent function** by restricting the domain of $y = \tan x$ to the interval $(-\pi/2, \pi/2)$. The following list summarizes the definitions of the three most common inverse trigonometric functions. The remaining three are discussed in the exercise set. (The graphs, domains, and ranges of *all six* inverse trigonometric functions are summarized in the appendix.)

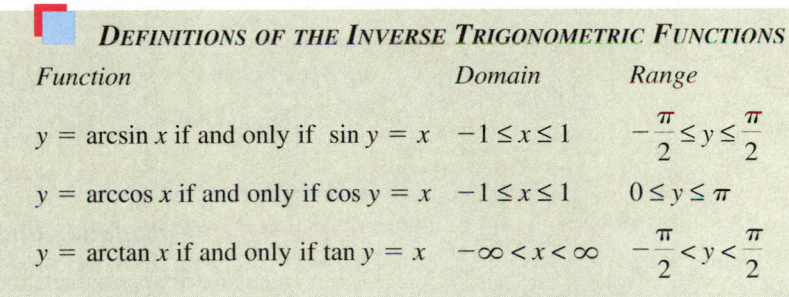

DEFINITIONS OF THE INVERSE TRIGONOMETRIC FUNCTIONS

Function	Domain	Range
$y = \arcsin x$ if and only if $\sin y = x$	$-1 \le x \le 1$	$-\dfrac{\pi}{2} \le y \le \dfrac{\pi}{2}$
$y = \arccos x$ if and only if $\cos y = x$	$-1 \le x \le 1$	$0 \le y \le \pi$
$y = \arctan x$ if and only if $\tan y = x$	$-\infty < x < \infty$	$-\dfrac{\pi}{2} < y < \dfrac{\pi}{2}$

The graphs of these three inverse trigonometric functions are shown in Figure 6.64.

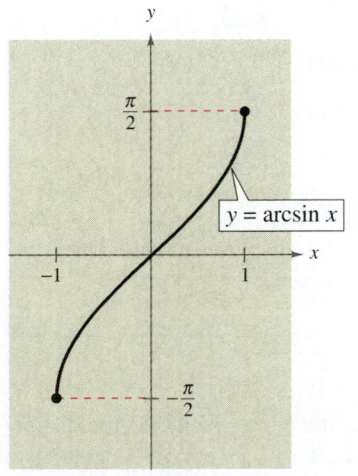

Domain: $[-1, 1]$
Range: $\left[-\frac{\pi}{2}, \frac{\pi}{2}\right]$

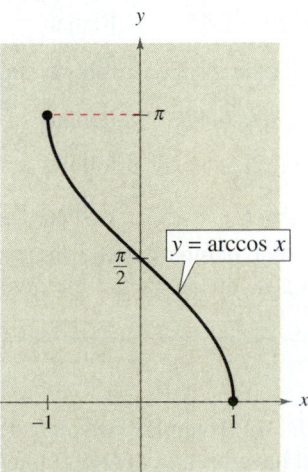

Domain: $[-1, 1]$
Range: $[0, \pi]$

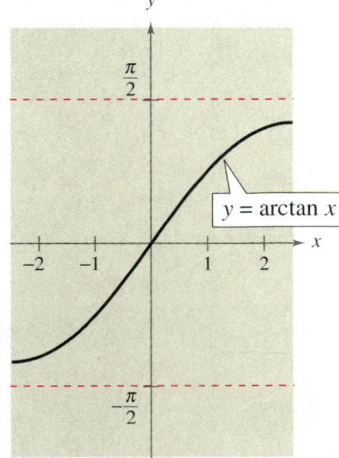

Domain: $(-\infty, \infty)$
Range: $\left(-\frac{\pi}{2}, \frac{\pi}{2}\right)$

FIGURE 6.64

---

*E*XAMPLE 3   *Evaluating Inverse Trigonometric Functions*

Find the exact value.

a. $\arccos \dfrac{\sqrt{2}}{2}$ **b.** $\arccos(-1)$ **c.** $\arctan 0$

Solution

a. Because $\cos(\pi/4) = \sqrt{2}/2$, and $\pi/4$ lies in $[0, \pi]$, it follows that

$$\arccos \frac{\sqrt{2}}{2} = \frac{\pi}{4}.$$

b. Because $\cos \pi = -1$, and π lies in $[0, \pi]$, it follows that

$$\arccos(-1) = \pi.$$

c. Because $\tan 0 = 0$, and 0 lies in $(-\pi/2, \pi/2)$, it follows that

$$\arctan 0 = 0.$$

---

*E*XAMPLE 4   *Calculators and Inverse Trigonometric Functions*

Use a calculator to approximate the value (if possible).

a. $\arctan(-8.45)$ **b.** $\arcsin 0.2447$ **c.** $\arccos 2$

Solution

Function	*Mode*	*Graphing Calculator Keystrokes*
a. $\arctan(-8.45)$	Radian	TAN⁻¹ ((-) 8.45) ENTER

From the display, it follows that $\arctan(-8.45) \approx -1.453001$.

b. $\arcsin 0.2447$	Radian	SIN⁻¹ 0.2447 ENTER

From the display, it follows that $\arcsin 0.2447 \approx 0.2472103$.

c. $\arccos 2$	Radian	COS⁻¹ 2 ENTER

In real number mode, the calculator should display an *error message* because the domain of the inverse cosine function is $[-1, 1]$.

NOTE   In Example 4, if you had set the calculator to degree mode, the display would have been in degrees rather than radians. This convention is peculiar to calculators. By definition, the values of inverse trigonometric functions are always *in radians*.

Compositions of Functions

Recall from Section 2.5 that inverse functions possess the properties

$$f(f^{-1}(x)) = x \qquad \text{and} \qquad f^{-1}(f(x)) = x.$$

The inverse trigonometric versions of these properties are given below.

INVERSE PROPERTIES

If $-1 \le x \le 1$ and $-\pi/2 \le y \le \pi/2$, then

$$\sin(\arcsin x) = x \qquad \text{and} \qquad \arcsin(\sin y) = y.$$

If $-1 \le x \le 1$ and $0 \le y \le \pi$, then

$$\cos(\arccos x) = x \qquad \text{and} \qquad \arccos(\cos y) = y.$$

If $-\pi/2 < y < \pi/2$, then

$$\tan(\arctan x) = x \qquad \text{and} \qquad \arctan(\tan y) = y.$$

NOTE Keep in mind that these inverse properties do not apply for arbitrary values of x and y. For instance,

$$\arcsin\left(\sin \frac{3\pi}{2}\right) = \arcsin(-1)$$

$$= -\frac{\pi}{2}$$

$$\neq \frac{3\pi}{2}.$$

In other words, the property

$$\arcsin(\sin y) = y$$

is not valid for values of y outside the interval $[-\pi/2, \pi/2]$. ■■

EXAMPLE 5 *Using Inverse Properties*

If possible, find the exact value.

a. $\tan[\arctan(-5)]$ **b.** $\arcsin\left(\sin \dfrac{5\pi}{3}\right)$ **c.** $\cos(\cos^{-1}\pi)$

Solution

a. Because -5 lies in the domain of the arctan x, the inverse property applies, and you have

$$\tan[\arctan(-5)] = -5.$$

b. In this case, $5\pi/3$ does not lie within the range of the arcsine function, $-\pi/2 \le y \le \pi/2$. However, $5\pi/3$ is coterminal with

$$\frac{5\pi}{3} - 2\pi = -\frac{\pi}{3}$$

which does lie in the range of the arcsine function, and you have

$$\arcsin\left(\sin \frac{5\pi}{3}\right) = \arcsin\left[\sin\left(-\frac{\pi}{3}\right)\right] = -\frac{\pi}{3}.$$

c. The expression $\cos(\cos^{-1}\pi)$ is not defined because $\cos^{-1}\pi$ is not defined. Remember that the domain of the inverse cosine function is $[-1, 1]$.

Example 6 shows how to use right triangles to find exact values of functions of inverse functions. Then, Example 7 shows how to use triangles to convert a trigonometric expression into an algebraic expression. This conversion technique is used frequently in calculus.

EXAMPLE 6   *Evaluating Compositions of Functions*

Find the exact value.

a. $\tan\left(\arccos\dfrac{2}{3}\right)$ **b.** $\cos\left[\arcsin\left(-\dfrac{3}{5}\right)\right]$

Solution

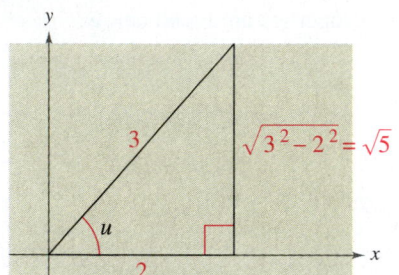

FIGURE 6.65

a. If you let $u = \arccos\frac{2}{3}$, then $\cos u = \frac{2}{3}$. Because $\cos u$ is positive, u is a *first*-quadrant angle. You can sketch and label angle u as shown in Figure 6.65. Consequently,

$$\tan\left(\arccos\frac{2}{3}\right) = \tan u = \frac{\text{opp}}{\text{adj}} = \frac{\sqrt{5}}{2}.$$

b. If you let $u = \arcsin -\frac{3}{5}$, then $\sin u = -\frac{3}{5}$. Because $\sin u$ is negative, u is a *fourth*-quadrant angle. You can sketch and label angle u as shown in Figure 6.66. Consequently,

$$\cos\left[\arcsin\left(-\frac{3}{5}\right)\right] = \cos u = \frac{\text{adj}}{\text{hyp}} = \frac{4}{5}.$$

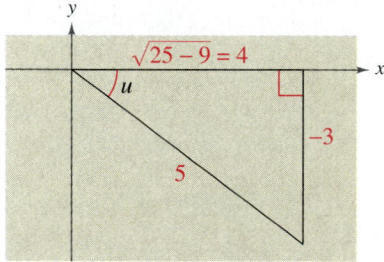

FIGURE 6.66

EXAMPLE 7   *Some Problems from Calculus*

Write each of the following as an algebraic expression in x.

a. $\sin(\arccos 3x), \quad 0 \le x \le \dfrac{1}{3}$ **b.** $\cot(\arccos 3x), \quad 0 \le x \le \dfrac{1}{3}$

Solution

If you let $u = \arccos 3x$, then $\cos u = 3x$. Because

$$\cos u = \frac{3x}{1} = \frac{\text{adj}}{\text{hyp}}$$

you can sketch a right triangle with acute angle u, as shown in Figure 6.67. From this triangle, you can easily convert each expression to algebraic form.

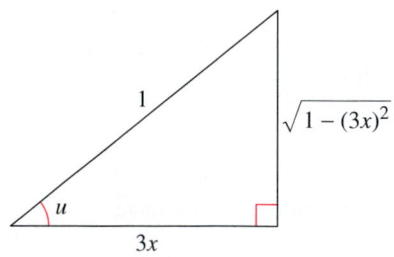

FIGURE 6.67

a. $\sin(\arccos 3x) = \sin u = \dfrac{\text{opp}}{\text{hyp}} = \sqrt{1 - 9x^2}, \quad 0 \le x \le \dfrac{1}{3}$

b. $\cot(\arccos 3x) = \cot u = \dfrac{\text{adj}}{\text{opp}} = \dfrac{3x}{\sqrt{1 - 9x^2}}, \quad 0 \le x \le \dfrac{1}{3}$

NOTE   In Example 7, a similar argument can be made for x-values lying in the interval $\left[-\frac{1}{3}, 0\right]$.

GROUP ACTIVITY

INVERSE FUNCTIONS

We have discussed inverse functions for several types of functions. Match the function in the left column with its inverse function in the right column.

1. $f(x) = x$
2. $f(x) = x^2, \quad 0 \le x$
3. $f(x) = x^3$
4. $f(x) = e^x$
5. $f(x) = \ln x$
6. $f(x) = \sin x, \quad -\dfrac{\pi}{2} \le x \le \dfrac{\pi}{2}$
7. $f(x) = \cos x, \quad 0 \le x \le \pi$
8. $f(x) = \tan x, \quad -\dfrac{\pi}{2} < x < \dfrac{\pi}{2}$

a. $f^{-1}(x) = \arcsin x$
b. $f^{-1}(x) = \ln x$
c. $f^{-1}(x) = \sqrt{x}$
d. $f^{-1}(x) = \arctan x$
e. $f^{-1}(x) = \arccos x$
f. $f^{-1}(x) = \sqrt[3]{x}$
g. $f^{-1}(x) = e^x$
h. $f^{-1}(x) = x$

Discuss reasons for your answers. Verify each pair of inverses algebraically, graphically, and numerically.

WARM UP

Evaluate the trigonometric function from memory.

1. $\sin\left(-\dfrac{\pi}{2}\right)$ 2. $\cos \pi$ 3. $\tan\left(-\dfrac{\pi}{4}\right)$ 4. $\sin\dfrac{\pi}{4}$

Find a real number x in the interval $[-\pi/2, \pi/2]$ that has the same sine value as the given value.

5. $\sin 2\pi$ 6. $\sin\dfrac{5\pi}{6}$

Find the real number x in the interval $[-\pi/2, \pi/2]$ that has the same cosine value as the given value.

7. $\cos 3\pi$ 8. $\cos\left(-\dfrac{\pi}{4}\right)$

Find a real number x in the interval $[-\pi/2, \pi/2]$ that has the same tangent value as the given value.

9. $\tan 4\pi$ 10. $\tan\dfrac{3\pi}{4}$

6.6 Exercises

1. *True or False?* Explain your reasoning.

$$\sin \frac{5\pi}{6} = \frac{1}{2} \quad\Longrightarrow\quad \arcsin \frac{1}{2} = \frac{5\pi}{6}$$

2. *True or False?* Explain your reasoning.

$$\tan \frac{5\pi}{4} = 1 \quad\Longrightarrow\quad \arctan 1 = \frac{5\pi}{4}$$

In Exercises 3–18, evaluate the expression without the aid of a calculator.

3. $\arcsin \frac{1}{2}$

4. $\arcsin 0$

5. $\arccos \frac{1}{2}$

6. $\arccos 0$

7. $\arctan \dfrac{\sqrt{3}}{3}$

8. $\arctan(-1)$

9. $\arccos\left(-\dfrac{\sqrt{3}}{2}\right)$

10. $\arcsin\left(-\dfrac{\sqrt{2}}{2}\right)$

11. $\arctan\left(-\sqrt{3}\right)$

12. $\arctan\sqrt{3}$

13. $\arccos\left(-\dfrac{1}{2}\right)$

14. $\arcsin\dfrac{\sqrt{2}}{2}$

15. $\arcsin \dfrac{\sqrt{3}}{2}$

16. $\arctan\left(-\dfrac{\sqrt{3}}{3}\right)$

17. $\arctan 0$

18. $\arccos 1$

In Exercises 19–30, use a calculator to approximate the expression. (Round your result to two decimal places.)

19. $\arccos 0.28$

20. $\arcsin 0.45$

21. $\arcsin(-0.75)$

22. $\arccos(-0.7)$

23. $\arctan(-3)$

24. $\arctan 15$

25. $\arcsin 0.31$

26. $\arccos 0.26$

27. $\arccos(-0.41)$

28. $\arcsin(-0.125)$

29. $\arctan 0.92$

30. $\arctan 2.8$

In Exercises 31 and 32, determine the missing coordinates of the points on the graph of the function.

31.
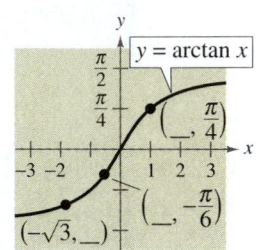

32.
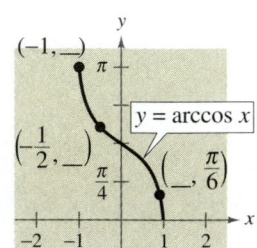

In Exercises 33 and 34, use a graphing utility to graph f, g, and $y = x$ on the same viewing rectangle to verify geometrically that g is the inverse of f. (Be sure to restrict the domain of f properly.)

33. $f(x) = \tan x$, $g(x) = \arctan x$

34. $f(x) = \sin x$, $g(x) = \arcsin x$

In Exercises 35–38, use an inverse trigonometric function to write θ as a function of x.

35.
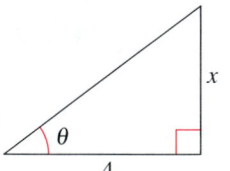

36.
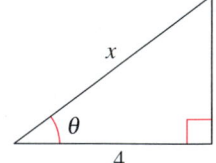

37.
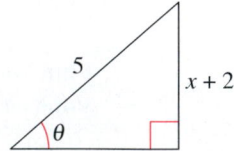

38.
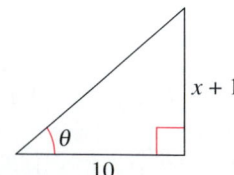

In Exercises 39–44, use the properties of inverse functions to evaluate the expression.

39. $\sin(\arcsin 0.3)$

40. $\tan(\arctan 25)$

41. $\cos[\arccos(-0.1)]$

42. $\sin[\arcsin(-0.2)]$

43. $\arcsin(\sin 3\pi)$

44. $\arccos\left(\cos \dfrac{7\pi}{2}\right)$

In Exercises 45–54, find the exact value of the expression. (*Hint:* Make a sketch of a right triangle.)

45. $\sin\left(\arctan \frac{3}{4}\right)$

46. $\sec\left(\arcsin \frac{4}{5}\right)$

47. $\cos(\arctan 2)$

48. $\sin\left(\arccos \dfrac{\sqrt{5}}{5}\right)$

49. $\cos\left(\arcsin \frac{5}{13}\right)$

50. $\csc\left[\arctan\left(-\frac{5}{12}\right)\right]$

51. $\sec\left[\arctan\left(-\frac{3}{5}\right)\right]$

52. $\tan\left[\arcsin\left(-\frac{3}{4}\right)\right]$

53. $\sin\left[\arccos\left(-\frac{2}{3}\right)\right]$

54. $\cot\left(\arctan \frac{5}{8}\right)$

In Exercises 55–64, write an algebraic expression that is equivalent to the expression. (*Hint:* Sketch a right triangle, as demonstrated in Example 7.)

55. $\cot(\arctan x)$

56. $\sin(\arctan x)$

57. $\cos(\arcsin 2x)$

58. $\sec(\arctan 3x)$

59. $\sin(\arccos x)$

60. $\sec[\arcsin(x - 1)]$

61. $\tan\left(\arccos \dfrac{x}{3}\right)$

62. $\cot\left(\arctan \dfrac{1}{x}\right)$

63. $\csc\left(\arctan \dfrac{x}{\sqrt{2}}\right)$

64. $\cos\left(\arcsin \dfrac{x - h}{r}\right)$

In Exercises 65 and 66, use a graphing utility to graph f and g on the same viewing rectangle to verify that the two are equal. Explain why they are equal. Identify any asymptotes of the graphs.

65. $f(x) = \sin(\arctan 2x),\quad g(x) = \dfrac{2x}{\sqrt{1 + 4x^2}}$

66. $f(x) = \tan\left(\arccos \dfrac{x}{2}\right),\quad g(x) = \dfrac{\sqrt{4 - x^2}}{x}$

In Exercises 67–70, fill in the blank.

67. $\arctan \dfrac{9}{x} = \arcsin(\quad),\quad x \neq 0$

68. $\arcsin \dfrac{\sqrt{36 - x^2}}{x} = \arccos(\quad),\quad 0 \leq x \leq 6$

69. $\arccos \dfrac{3}{\sqrt{x^2 - 2x + 10}} = \arcsin(\quad)$

70. $\arccos \dfrac{x - 2}{2} = \arctan(\quad),\quad |x - 2| \leq 2$

In Exercises 71–78, sketch a graph of the function.

71. $y = 2 \arccos x$

72. $y = \arcsin \dfrac{x}{2}$

73. $f(x) = \arcsin(x - 1)$

74. $g(t) = \arccos(t + 2)$

75. $f(x) = \arctan 2x$

76. $f(x) = \dfrac{\pi}{2} + \arctan x$

77. $h(v) = \tan(\arccos v)$

78. $f(x) = \arccos \dfrac{x}{4}$

In Exercises 79 and 80, write the given functions in terms of the sine function by using the identity

$$A \cos \omega t + B \sin \omega t = \sqrt{A^2 + B^2} \sin\left(\omega t + \arctan \dfrac{B}{A}\right).$$

Use a graphing utility to graph both forms of the function. What does the graph imply?

79. $f(t) = 3 \cos 2t + 3 \sin 2t$

80. $f(t) = 4 \cos \pi t + 3 \sin \pi t$

81. *Think About It* Consider the functions

$$f(x) = \sin x \quad \text{and} \quad f^{-1}(x) = \arcsin x.$$

(a) Use a graphing utility to graph the composite functions $f \circ f^{-1}$ and $f^{-1} \circ f$.

(b) Explain why the graphs of part (a) are not the graph of the line $y = x$. Why do the graphs of $f \circ f^{-1}$ and $f^{-1} \circ f$ differ?

82. *Think About It* Use a graphing utility to graph the functions $f(x) = \sqrt{x}$ and $g(x) = 6 \arctan x$. For $x > 0$ it appears that $g > f$. Explain why you know that there exists a positive real number a such that $g < f$ for $x > a$. Approximate the number a.

83. *Docking a Boat* A boat is pulled in by means of a winch located on a dock 10 feet above the deck of the boat (see figure). Let θ be the angle of elevation from the boat to the winch and let s be the length of the rope from the winch to the boat.

(a) Write θ as a function of s.

(b) Find θ when $s = 48$ feet and $s = 24$ feet.

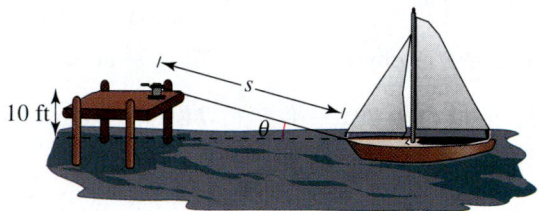

84. *Photography* A television camera at ground level is filming the lift-off of a space shuttle at a point 750 meters from the launch pad (see figure). Let θ be the angle of elevation to the shuttle and let s be the height of the shuttle.

(a) Write θ as a function of s.

(b) Find θ when $s = 300$ meters and $s = 1200$ meters.

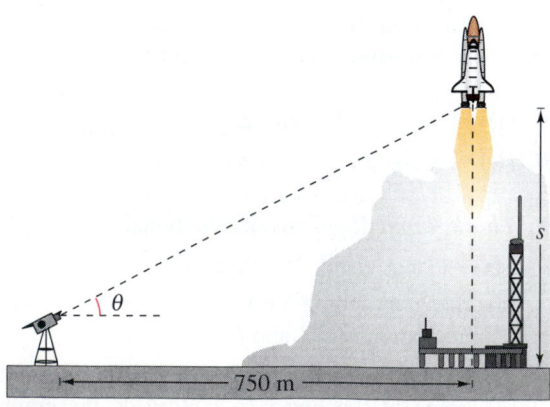

85. *Photography* A photographer is taking a picture of a 3-foot painting hung in an art gallery. The camera lens is 1 foot below the lower edge of the painting (see figure). The angle β subtended by the camera lens x feet from the painting is given by

$$\beta = \arctan \frac{3x}{x^2 + 4}, \qquad x > 0.$$

(a) Use a graphing utility to graph β as a function of x.

(b) Move the cursor along the graph to approximate the distance from the picture when β is maximum.

(c) Identify the asymptote of the graph and discuss its meaning in the context of the problem.

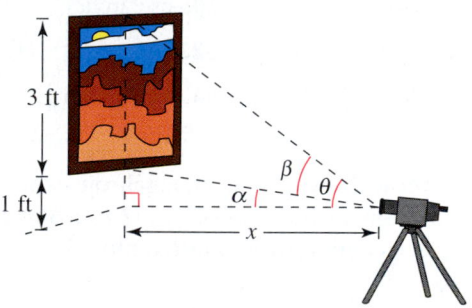

86. *Area* In calculus, it is shown that the area of the region bounded by the graphs of $y = 0$, $y = 1/(x^2 + 1)$, $x = a$, and $x = b$ is given by

Area $= \arctan b - \arctan a$

(see figure). Find the area for the following values of a and b.

(a) $a = 0, b = 1$ (b) $a = -1, b = 1$

(c) $a = 0, b = 3$ (d) $a = -1, b = 3$

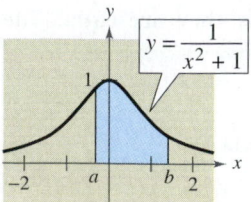

87. *Angle of Elevation*   An airplane flies at an altitude of 5 miles toward a point directly over an observer. Consider θ and x as shown in the figure.

(a) Write θ as a function of x.

(b) Find θ when $x = 10$ miles and $x = 3$ miles.

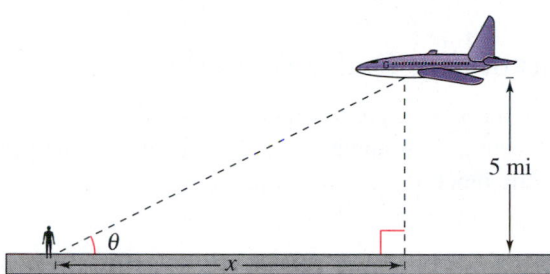

5 mi

88. *Security Patrol*   A security car with its spotlight on is parked 20 meters from a long warehouse. Consider θ and x as shown in the figure.

(a) Write θ as a function of x.

(b) Find θ when $x = 5$ meters and $x = 12$ meters.

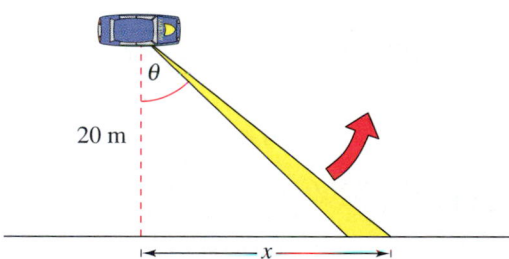

20 m

89. Define the inverse cotangent function by restricting the domain of the cotangent function to the interval $(0, \pi)$, and sketch its graph.

90. Define the inverse secant function by restricting the domain of the secant function to the intervals $[0, \pi/2)$ and $(\pi/2, \pi]$, and sketch its graph.

91. Define the inverse cosecant function by restricting the domain of the cosecant function to the intervals $[-\pi/2, 0)$ and $(0, \pi/2]$, and sketch its graph.

92. Use the results of Exercises 89–91 to evaluate the following without using a calculator.

(a) $\operatorname{arcsec} \sqrt{2}$ (b) $\operatorname{arcsec} 1$

(c) $\operatorname{arccot}(-\sqrt{3})$ (d) $\operatorname{arccsc} 2$

In Exercises 93–98, prove the identity.

93. $\arcsin(-x) = -\arcsin x$

94. $\arctan(-x) = -\arctan x$

95. $\arccos(-x) = \pi - \arccos x$

96. $\arctan x + \arctan \dfrac{1}{x} = \dfrac{\pi}{2}, \quad x > 0$

97. $\arcsin x + \arccos x = \dfrac{\pi}{2}$

98. $\arcsin x = \arctan \dfrac{x}{\sqrt{1 - x^2}}$

Review   **Solve Exercises 99–102 as a review of the skills and problem-solving techniques you learned in previous sections.**

99. *Buy Now or Wait?*   A sales representative indicates that if a customer waits another month before purchasing a new car that currently costs $23,500, the price will increase by 4%. However, the customer will pay an interest penalty of $725 for the early withdrawal of a certificate of deposit if the car is purchased now. Determine whether the customer should buy now or wait another month.

100. *Insurance Premium*   The annual insurance premium for a policyholder is normally $739. However, after having an automobile accident, the policyholder is charged an additional 30%. What is the new annual premium?

101. *Partnership Costs*   A group of people agree to share equally in the cost of a $250,000 endowment to a college. If they could find two more people to join the group, each person's share of the cost would decrease by $6250. How many people are presently in the group?

102. *Speed*   A boat travels at a speed of 18 miles per hour in still water. It travels 35 miles upstream and then returns to the starting point in a total of 4 hours. Find the speed of the current.

6.7 *Applications and Models*

See Exercise 58 on page 538 for an example of how a trigonometric function can be used to model harmonic motion.

Applications Involving Right Triangles ❑ *Trigonometry and Bearings* ❑
Harmonic Motion

Applications Involving Right Triangles

NOTE In this section the three angles of a right triangle are denoted by the letters A, B, and C (where C is the right angle), and the lengths of the sides opposite these angles by the letters a, b, and c (where c is the hypotenuse). ■■

In keeping with our twofold perspective of trigonometry, this section includes both right triangle applications and applications that emphasize the periodic nature of the trigonometric functions.

EXAMPLE 1 *Solving a Right Triangle*

Solve the right triangle shown in Figure 6.68.

Solution

Because $C = 90°$, it follows that $A + B = 90°$ and $B = 90° - 34.2° = 55.8°$. To solve for a, use the fact that

$$\tan A = \frac{\text{opp}}{\text{adj}} = \frac{a}{b} \qquad \Longrightarrow \qquad a = b \tan A.$$

Thus, $a = 19.4 \tan 34.2° \approx 13.18$. Similarly, to solve for c, use the fact that

$$\cos A = \frac{\text{adj}}{\text{hyp}} = \frac{b}{c} \qquad \Longrightarrow \qquad c = \frac{b}{\cos A}.$$

Thus, $c = \dfrac{19.4}{\cos 34.2°} \approx 23.46$.

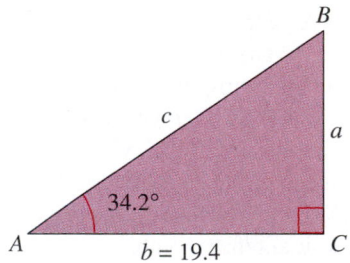

FIGURE 6.68

EXAMPLE 2 *Finding a Side of a Right Triangle*

Real Life

A safety regulation states that the maximum angle of elevation for a rescue ladder is 72°. If a fire department's longest ladder is 110 feet, what is the maximum safe rescue height?

Solution

A sketch is shown in Figure 6.69. From the equation $\sin A = a/c$, it follows that

$$a = c \sin A = 110 \sin 72° \approx 104.6.$$

Thus, the maximum safe rescue height is about 104.6 feet above the height of the fire truck.

FIGURE 6.69

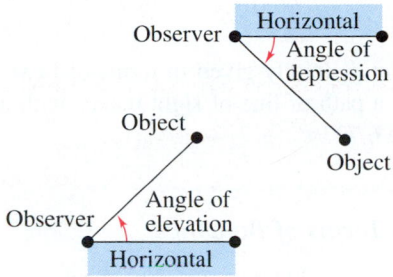

FIGURE 6.70

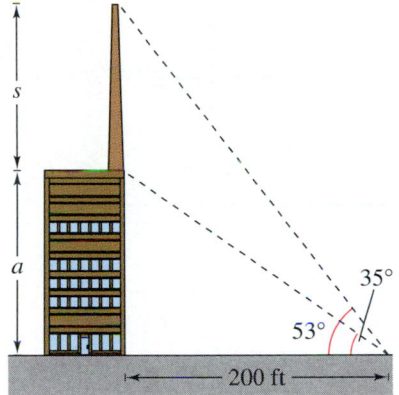

FIGURE 6.71

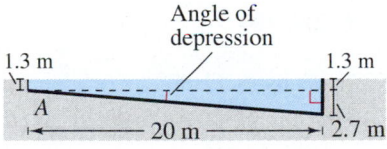

FIGURE 6.72

In Example 2, the term **angle of elevation** represents the angle from the horizontal upward to an object. For objects that lie below the horizontal, it is common to use the term **angle of depression,** as shown in Figure 6.70.

Real Life

EXAMPLE 3 *Finding a Side of a Right Triangle*

At a point 200 feet from the base of a building, the angle of elevation to the *bottom* of a smokestack is 35°, whereas the angle of elevation to the *top* is 53°, as shown in Figure 6.71. Find the height s of the smokestack alone.

Solution

Note from Figure 6.71 that this problem involves two right triangles. In the smaller right triangle, use the fact that $\tan 35° = a/200$ to conclude that the height of the building is

$$a = 200 \tan 35°.$$

In the larger right triangle, use the equation

$$\tan 53° = \frac{a + s}{200}$$

to conclude that $a + s = 200 \tan 53°$. Hence, the height of the smokestack is

$$s = 200 \tan 53° - a$$
$$= 200 \tan 53° - 200 \tan 35°$$
$$= 125.4 \text{ feet.}$$

Real Life

EXAMPLE 4 *Finding an Acute Angle of a Right Triangle*

A swimming pool is 20 meters long and 12 meters wide. The bottom of the pool is slanted so that the water depth is 1.3 meters at the shallow end and 4 meters at the deep end, as shown in Figure 6.72. Find the angle of depression of the bottom of the pool.

Solution

Using the tangent function, you see that

$$\tan A = \frac{\text{opp}}{\text{adj}} = \frac{2.7}{20} = 0.135.$$

Thus, the angle of depression is given by

$$A = \arctan 0.135$$
$$\approx 0.13419 \text{ radians}$$
$$\approx 7.69°.$$

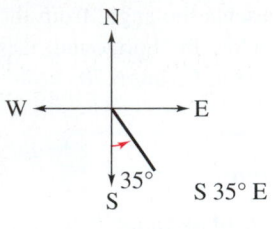

S 35° E

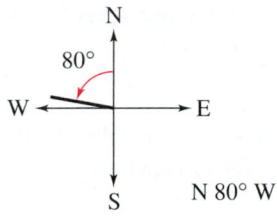

N 80° W

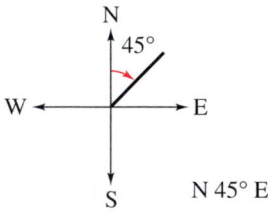

N 45° E

NOTE   The bearing of S 35° E in Figure 6.73 means 35 degrees east of south. ▪▪

Trigonometry and Bearings

In surveying and navigation, directions are generally given in terms of **bearings.** A bearing measures the acute angle a path or line of sight makes with a fixed north-south line, as shown in Figure 6.73.

*E*XAMPLE 5   *Finding Directions in Terms of Bearings*

A ship leaves port at noon and heads due west at 20 knots, or 20 nautical miles (nm) per hour. At 2 P.M. the ship changes course to N 54° W, as shown in Figure 6.74. Find the ship's bearing and distance from the port of departure at 3 P.M..

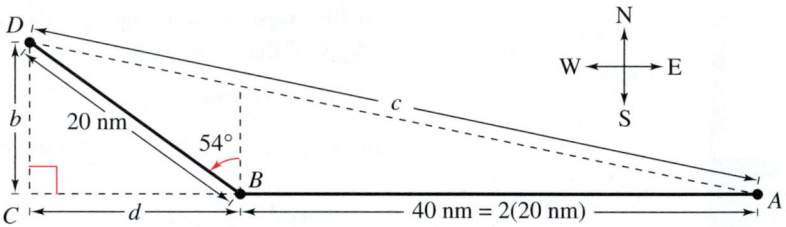

FIGURE 6.74

Solution

In triangle *BCD*, you have $B = 90° - 54° = 36°$. The two sides of this triangle can be determined to be

$$b = 20 \sin 36° \qquad \text{and} \qquad d = 20 \cos 36°.$$

In triangle *ACD*, you find angle *A* as follows.

$$\tan A = \frac{b}{d + 40} = \frac{20 \sin 36°}{20 \cos 36° + 40} \approx 0.2092494$$

$$A \approx \arctan 0.2092494 \approx 0.2062732 \text{ radians} \approx 11.82°$$

The angle with the north-south line is $90° - 11.82° = 78.18°$. Therefore, the bearing of the ship is

N 78.18° W.   Bearing

Finally, from triangle *ACD*, you have $\sin A = b/c$, which yields

$$c = \frac{b}{\sin A} = \frac{20 \sin 36°}{\sin 11.82°}$$

$$\approx 57.4 \text{ nautical miles.}$$ Distance from port

Harmonic Motion

The periodic nature of the trigonometric functions is useful for describing the motion of a point on an object that vibrates, oscillates, rotates, or is moved by wave motion.

For example, consider a ball that is bobbing up and down on the end of a spring, as shown in Figure 6.75. Suppose that 10 centimeters is the maximum distance the ball moves vertically upward or downward from its equilibrium (at rest) position. Suppose further that the time it takes for the ball to move from its maximum displacement above zero to its maximum displacement below zero and back again is $t = 4$ seconds. Assuming the ideal conditions of perfect elasticity and no friction or air resistance, the ball would continue to move up and down in a uniform and regular manner.

From this spring you can conclude that the period (time for one complete cycle) of the motion is

Period = 4 seconds

and that its amplitude (maximum displacement from equilibrium) is

Amplitude = 10 centimeters.

Motion of this nature can be described by a sine or cosine function, and is called **simple harmonic motion.**

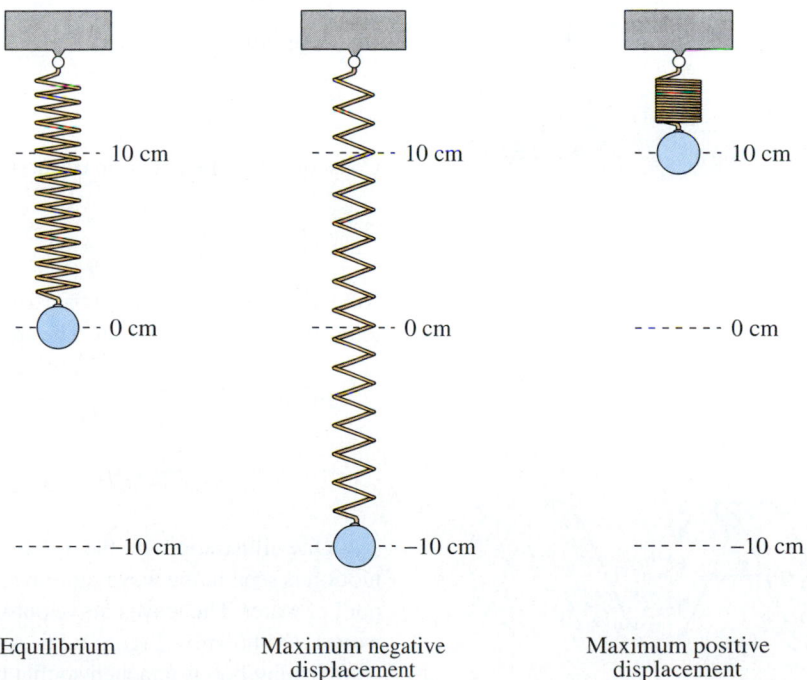

Equilibrium   Maximum negative   Maximum positive
                      displacement   displacement

Figure 6.75

> ### DEFINITION OF SIMPLE HARMONIC MOTION
>
> A point that moves on a coordinate line is said to be in **simple harmonic motion** if its distance d from the origin at time t is given by either
>
> $$d = a \sin \omega t \qquad \text{or} \qquad d = a \cos \omega t$$
>
> where a and ω are real numbers such that $\omega > 0$. The motion has **amplitude** $|a|$, **period** $2\pi/\omega$, and **frequency** $\omega/2\pi$.

Real Life

EXAMPLE 6   Simple Harmonic Motion

Write the equation for the simple harmonic motion of the ball described in Figure 6.75, where the period is 4 seconds. What is the frequency of this harmonic motion?

Solution

Because the spring is at equilibrium $(d = 0)$ when $t = 0$, you use the equation

$$d = a \sin \omega t.$$

Moreover, because the maximum displacement from zero is 10 and the period is 4, you have

$$\text{Amplitude} = |a| = 10$$

$$\text{Period} = \frac{2\pi}{\omega} = 4 \qquad \Longrightarrow \qquad \omega = \frac{\pi}{2}.$$

Consequently, the equation of motion is

$$d = 10 \sin \frac{\pi}{2} t.$$

Note that the choice of $a = 10$ or $a = -10$ depends on whether the ball initially moves up or down. The frequency is given by

$$\text{Frequency} = \frac{\omega}{2\pi} = \frac{\pi/2}{2\pi} = \frac{1}{4} \text{ cycle per second.}$$

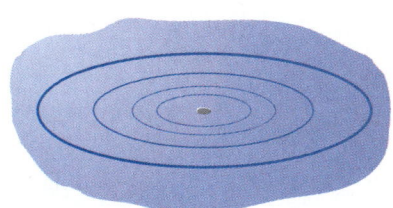

FIGURE 6.76

One illustration of the relationship between sine waves and harmonic motion is seen in the wave motion resulting when a stone is dropped into a calm pool of water. The waves move outward in roughly the shape of sine (or cosine) waves, as shown in Figure 6.76. As an example, suppose you are fishing and your fishing bob is attached so that it does not move horizontally. As the waves move outward from the dropped stone, your fishing bob will move up and down in simple harmonic motion, as shown in Figure 6.77.

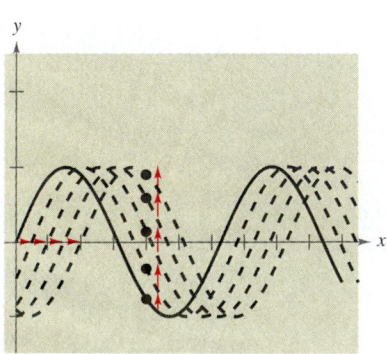

FIGURE 6.77

TECHNOLOGY

■■

Use the zero or root feature of a graphing utility to verify Example 7 (d). For an example of how to use the zero feature of a *TI-83* or the root feature of a *TI-82* graphing calculator, see page 275.

■

EXAMPLE 7   *Simple Harmonic Motion*

Given the equation for simple harmonic motion

$$d = 6 \cos \frac{3\pi}{4} t$$

find (a) the maximum displacement, (b) the frequency, (c) the value of d when $t = 4$, and (d) the least positive value of t for which $d = 0$.

Solution

The given equation has the form $d = a \cos \omega t$, with $a = 6$ and $\omega = 3\pi/4$.

a. The maximum displacement (from the point of equilibrium) is given by the amplitude. Thus, the maximum displacement is 6.

b. Frequency $= \dfrac{\omega}{2\pi} = \dfrac{3\pi/4}{2\pi} = \dfrac{3}{8}$ cycle per unit of time

c. $d = 6 \cos\left[\dfrac{3\pi}{4}(4)\right] = 6 \cos 3\pi = 6(-1) = -6$

d. To find the least positive value of t for which $d = 0$, solve the equation

$$d = 6 \cos \frac{3\pi}{4} t = 0$$

to obtain

$$\frac{3\pi}{4} t = \frac{\pi}{2}, \frac{3\pi}{2}, \frac{5\pi}{2}, \ldots \qquad \Longrightarrow \qquad t = \frac{2}{3}, 2, \frac{10}{3}, \ldots .$$

Thus, the least positive value of t is $t = \frac{2}{3}$.

■

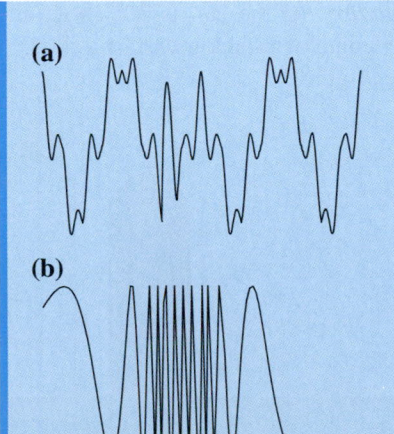

(a)

(b)

GROUP ACTIVITY

RADIO WAVES

Many different physical phenomena can be characterized by wave motion. These include electromagnetic waves such as radio waves, television waves, and microwaves. Radio waves transmit sound in two different ways. For an AM station, the *amplitude* of the wave is modified to carry sound. The letters AM stand for **amplitude modulation.** An FM radio signal has its *frequency* modified in order to carry sound, hence the term **frequency modulation.** Of the two graphs at the left, one shows an AM wave and the other shows an FM wave. Which is which? Explain your reasoning.

WARM UP

Evaluate the expression and round to two decimal places.

1. $20 \sin 25°$

2. $42 \tan 62°$

3. $\arcsin 0.8723$

4. $\arctan 2.8703$

Solve for x and round to two decimal places.

5. $\cos 22° = \dfrac{x + 13 \sin 22°}{13 \sin 54°}$

6. $\tan 36° = \dfrac{x + 85 \tan 18°}{85}$

Find the amplitude and period of the function.

7. $f(x) = -4 \sin 2x$

8. $f(x) = \frac{1}{2} \sin \pi x$

9. $g(x) = 3 \cos 3\pi x$

10. $g(x) = 0.2 \cot \dfrac{x}{4}$

6.7 Exercises

In Exercises 1–10, solve the right triangle shown in the figure. (Round to two decimal places.)

1. $A = 20°, \quad b = 10$

2. $B = 54°, \quad c = 15$

3. $B = 71°, \quad b = 24$

4. $A = 8.4°, \quad a = 40.5$

5. $a = 6, \quad b = 10$

6. $a = 25, \quad c = 35$

7. $b = 16, \quad c = 52$

8. $b = 1.32, \quad c = 9.45$

9. $A = 12°15', \quad c = 430.5$

10. $B = 65°12', \quad a = 14.2$

In Exercises 11 and 12, find the altitude of the isosceles triangle shown in the figure. (Round to two decimal places.)

11. $\theta = 52°, \quad b = 4$ inches

12. $\theta = 18°, \quad b = 10$ meters

13. *Length of a Shadow* If the sun is $30°$ above the horizon, find the length of a shadow cast by a silo that is 60 feet tall (see figure).

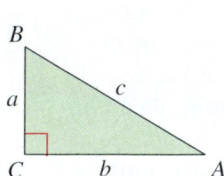

FIGURE FOR 1–10

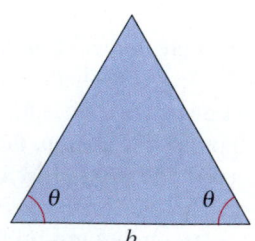

FIGURE FOR 11 AND 12

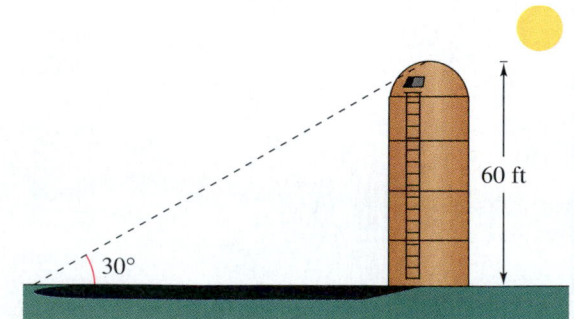

14. *Length of a Shadow*   If the sun is 20° above the horizon, find the length of a shadow cast by a building that is 600 feet tall.

15. *Height*   A ladder 16 feet long leans against the side of a house. Find the height h from the top of the ladder to the ground if the angle of elevation of the ladder is 74°.

16. *Height*   The length of a shadow of a tree is 125 feet when the angle of elevation of the sun is 33°. Approximate the height h of the tree.

17. *Height*   From a point 50 feet in front of a church, the angles of elevation to the base of the steeple and the top of the steeple are 35° and 47° 40′, respectively.

(a) Draw right triangles that represent the problem. Label the known and unknown quantities.

(b) Use a trigonometric function to write an equation involving the unknown.

(c) Find the height of the steeple.

18. *Height*   From a point 100 feet in front of the public library, the angles of elevation to the base of the flagpole and the top of the pole are 28° and 39° 45′, respectively. The flagpole is mounted on the front of the library's roof. Find the height of the pole.

19. *Depth of a Submarine*   The sonar of a navy cruiser detects a submarine that is 4000 feet from the cruiser. The angle between the water line and the submarine is 34° (see figure). How deep is the submarine?

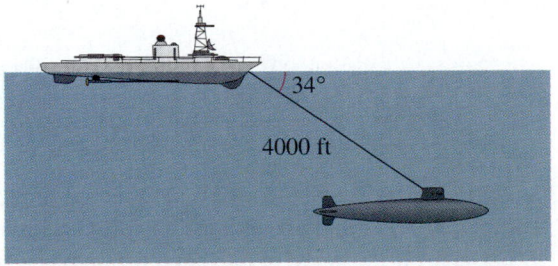

20. *Height of a Kite*   A 100-foot line is attached to a kite. When the kite has pulled the line taut, the angle of elevation to the kite is approximately 50°. Approximate the height of the kite.

21. *Angle of Elevation*   An amateur radio operator erects a 75-foot vertical tower for an antenna. Find the angle of elevation to the top of the tower at a point on level ground 50 feet from its base.

22. *Angle of Elevation*   The height of an outdoor basketball backboard is $12\frac{1}{2}$ feet, and the backboard casts a shadow $17\frac{1}{3}$ feet long.

(a) Draw a right triangle that represents the problem. Label the known and unknown quantities.

(b) Use a trigonometric function to write an equation involving the unknown.

(c) Find the angle of elevation of the sun.

23. *Angle of Depression*   A spacecraft is traveling in a circular orbit 150 miles above the surface of the earth (see figure). Find the angle of depression from the spacecraft to the horizon. Assume the radius of the earth is 4000 miles.

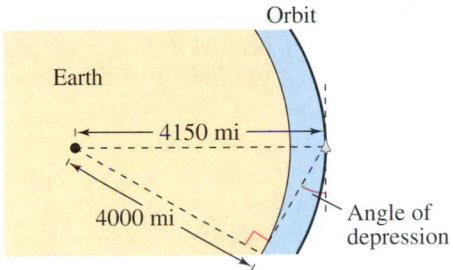

24. *Angle of Depression*   Find the angle of depression from the top of a lighthouse 250 feet above water level to the water line of a ship 2 miles offshore.

25. *Airplane Ascent*   When an airplane leaves the runway, its angle of climb is 18° and its speed is 275 feet per second. Find the plane's altitude after one minute.

26. *Airplane Ascent*   How long will it take the plane in Exercise 25 to climb to an altitude of 10,000 feet?

27. *Mountain Descent*   A sign on the roadway at the top of a mountain indicates that for the next 4 miles the grade is 10.5° (see figure). Find the change in elevation for a car descending the mountain.

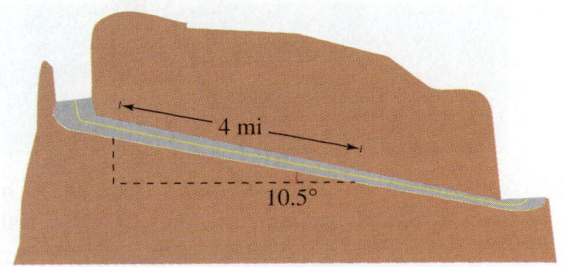

28. Mountain Descent A sign on the roadway at the top of a mountain indicates that for the next 4 miles the grade is 12%. Find the angle of the grade and the change in elevation for a car descending the mountain.

29. Navigation An airplane flying at 550 miles per hour has a bearing of N 52° E. After flying 1.5 hours, how far north and how far east will the plane have traveled from its point of departure?

30. Navigation A ship leaves port at noon and has a bearing of S 27° W. If its speed is 20 knots, how many nautical miles south and how many nautical miles west will the ship have traveled by 6:00 P.M.?

31. Surveying A surveyor wishes to find the distance across a swamp (see figure). The bearing from *A* to *B* is N 32° W. The surveyor walks 50 meters from *A*, and at the point *C* the bearing to *B* is N 68° W. Find (a) the bearing from *A* to *C* and (b) the distance from *A* to *B*.

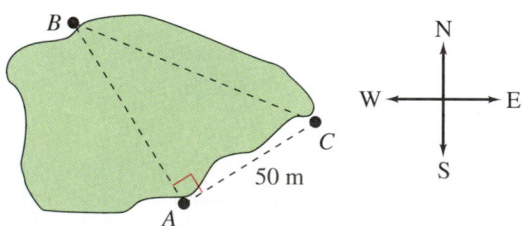

32. Location of a Fire Two fire towers are 30 kilometers apart, tower *A* being due west of tower *B*. A fire is spotted from the towers, and the bearings from *A* and *B* are E 14° N and W 34° N, respectively (see figure). Find the distance *d* of the fire from the line segment *AB*.

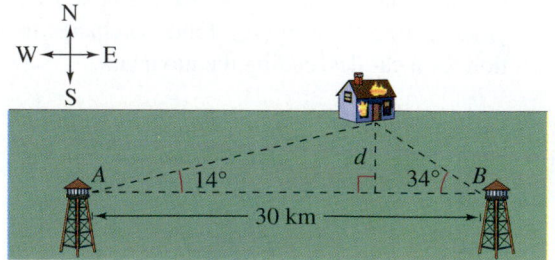

33. Navigation A ship is 45 miles east and 30 miles south of port. If the captain wants to sail directly to port, what bearing should be taken?

34. Navigation A plane is 120 miles north and 85 miles east of an airport. If the pilot wants to fly directly to the airport, what bearing should be taken?

35. Distance Between Ships An observer in a lighthouse 350 feet above sea level observes two ships directly offshore. The angles of depression to the ships are 4° and 6.5° (see figure). How far apart are the ships?

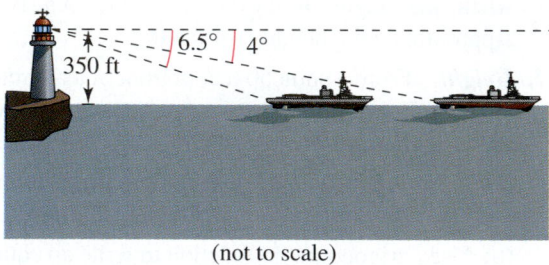

(not to scale)

36. Distance Between Towns A passenger in an airplane at an altitude of 10 kilometers sees two towns directly to the left of the plane. The angles of depression to the towns are 28° and 55° (see figure). How far apart are the towns?

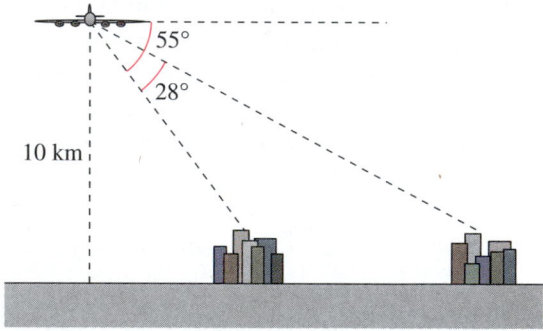

37. Altitude of a Plane A plane is observed approaching your home and you assume its speed is 550 miles per hour. If the angle of elevation of the plane is 16° at one time and 57° 1 minute later, approximate the altitude of the plane.

38. Height of a Mountain While traveling across flat land, you notice a mountain directly in front of you. The angle of elevation to the peak is 3.5°. After you drive 13 miles closer to the mountain, the angle of elevation is 9°. Approximate the height of the mountain.

Geometry In Exercises 39 and 40, find the angle α between two nonvertical lines L_1 and L_2. The angle α satisfies the equation

$$\tan \alpha = \left| \frac{m_2 - m_1}{1 + m_2 m_1} \right|$$

where m_1 and m_2 are the slopes of L_1 and L_2, respectively. (Assume $m_1 m_2 \neq -1$.)

39. L_1: $3x - 2y = 5$ **40.** L_1: $2x + y = 8$
    L_2: $x + y = 1$ L_2: $x - 5y = -4$

41. *Geometry* Determine the angle between the diagonal of the cube and the diagonal of its base, as shown in the figure.

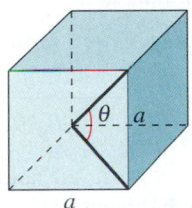

42. *Geometry* Determine the angle between the diagonal of the cube and its edge, as shown in the figure.

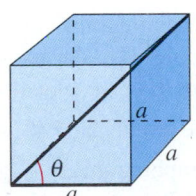

43. *Wrench Size* Express the distance y across the flat sides of the hexagonal nut as a function of r, as shown in the figure.

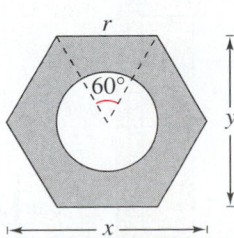

44. *Bolt Circle* The figure shows a circular sheet of diameter 40 centimeters, containing 12 equally spaced bolt holes. Determine the straight-line distance between the centers of consecutive bolt holes.

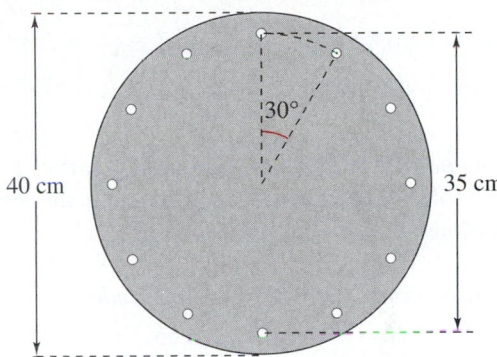

45. *Geometry* A regular pentagon is inscribed in a circle of radius 25 inches. Find the length of the sides of the pentagon.

46. *Geometry* A regular hexagon is inscribed in a circle of radius 25 inches. Find the length of the sides of the hexagon.

Trusses In Exercises 47 and 48, find the lengths of all the unknown members of the truss.

47.

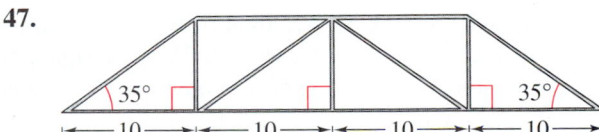

48.

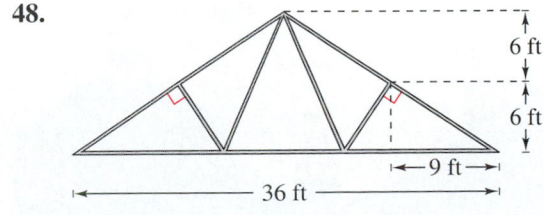

Harmonic Motion   **In Exercises 49–52, for the simple harmonic motion described by the trigonometric function, find (a) the maximum displacement, (b) the frequency, and (c) the least positive value of t for which $d = 0$.**

49. $d = 4 \cos 8\pi t$ **50.** $d = \frac{1}{2} \cos 20\pi t$

51. $d = \frac{1}{16} \sin 120\pi t$ **52.** $d = \frac{1}{64} \sin 792\pi t$

Harmonic Motion   **In Exercises 53–56, find a model for simple harmonic motion satisfying the specified conditions.**

	Displacement $(t = 0)$	Amplitude	Period
53.	0	4 cm	2 sec
54.	0	3 m	6 sec
55.	3 in.	3 in.	1.5 sec
56.	2 ft	2 ft	10 sec

57. *Tuning Fork*   A point on the end of a tuning fork moves in simple harmonic motion described by $d = a \sin \omega t$. Find ω given that the tuning fork for middle C has a frequency of 264 vibrations per second.

58. *Wave Motion*   A buoy oscillates in simple harmonic motion as waves go past. At a given time it is noted that the buoy moves a total of 3.5 feet from its low point to its high point (see figure), and that it returns to its high point every 10 seconds. Write an equation that describes the motion of the buoy if, at $t = 0$, it is at its high point.

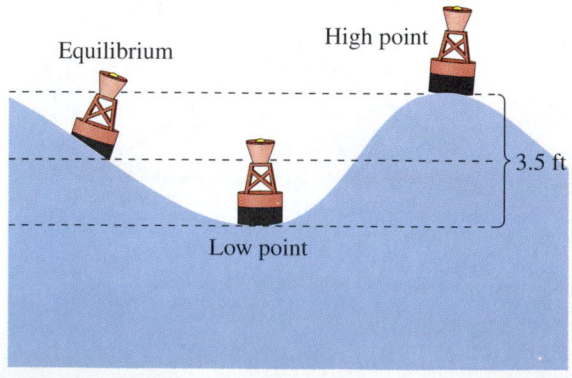

59. *Springs*   A weight stretches a spring 1.5 inches. The weight is pushed 3 inches above its equilibrium position and released. Its motion is modeled by

$$y = \frac{1}{4} \cos 16t, \qquad t > 0.$$

(a) Graph the function.

(b) What is the period of the oscillations?

(c) Determine the first time the weight passes the point of equilibrium $(y = 0)$.

60. *Numerical and Graphical Analysis*   A 2-meter-high fence is 3 meters from the side of a grain storage bin. A grain elevator must reach from ground level outside the fence to the storage bin (see figure). The objective is to determine the shortest elevator meeting the constraints.

(a) Complete four rows of the table.

θ	L_1	L_2	$L_1 + L_2$
0.1	$\dfrac{2}{\sin 0.1}$	$\dfrac{3}{\cos 0.1}$	23.0
0.2	$\dfrac{2}{\sin 0.2}$	$\dfrac{3}{\cos 0.2}$	13.1

(b) Use a graphing utility to generate additional rows of the table. Use the table to estimate the minimum length of the elevator.

(c) Write the length $L_1 + L_2$ as a function of θ.

(d) Use a graphing utility to graph the function. Use the graph to estimate the minimum length. How does your estimate compare with that of part (b)?

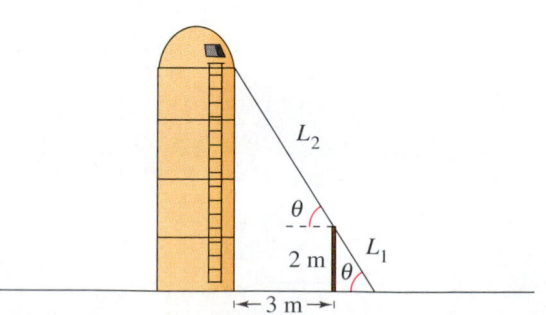

61. *Numerical and Graphical Analysis*   The cross-sections of an irrigation canal are isosceles trapezoids where the length of three of the sides is 8 feet (see figure). The objective is to find the angle θ that maximizes the area of the cross sections. [*Hint:* The area of a trapezoid is $(h/2)(b_1 + b_2)$.]

(a) Complete six rows of the table.

Base 1	Base 2	Altitude	Area
8	$8 + 16 \cos 10°$	$8 \sin 10°$	22.1
8	$8 + 16 \cos 20°$	$8 \sin 20°$	42.5

(b) Use a graphing utility to generate additional rows of the table. Use the table to estimate the maximum cross-sectional area.

(c) Write the area A as a function of θ.

(d) Use a graphing utility to graph the function. Use the graph to estimate the maximum cross-sectional area. How does your estimate compare with that of part (b)?

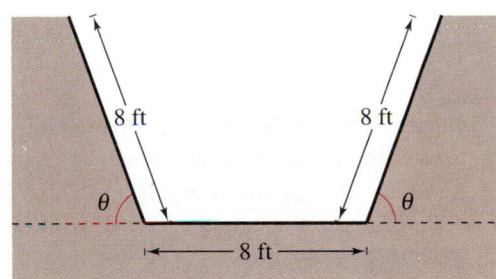

62. *Data Analysis*   The times S of sunset (Greenwich Mean Time) at 40° north latitude on the 15th of each month are: 1(16:59), 2(17:35), 3(18:06), 4(18:38), 5(19:08), 6(19:30), 7(19:28), 8(18:57), 9(18:09), 10(17:21), 11(16:44), 12(16:36). The month is represented by t, with $t = 1$ corresponding to January. A model (where minutes have been converted to the decimal part of an hour) for this data is

$$S(t) = 18.09 + 1.41 \sin\left(\frac{\pi t}{6} + 4.60\right).$$

(a) Use a graphing utility to graph the data points and the model on the same viewing rectangle.

(b) What is the period of the model? Is it what you expected? Explain.

(c) What is the amplitude of the function? What does it represent in the model? Explain.

63. *Data Analysis*   The table gives the average sales S (in millions) of an outerwear manufacturer for each month t, where $t = 1$ represents January.

t	1	2	3	4	5	6
S	13.46	11.15	8.00	4.85	2.54	1.70

t	7	8	9	10	11	12
S	2.54	4.85	8.00	11.15	13.46	14.30

(a) Create a scatter plot of the data.

(b) Find a trigonometric model that fits the data. Graph the model on your scatter plot. How well does the model fit?

(c) What is the period of the model? Do you think it is reasonable given the context? Explain your reasoning.

(d) Interpret the meaning of the model's amplitude in the context of the problem.

Review   Solve Exercises 64–67 as a review of the skills and problem-solving techniques you learned in previous sections. Graph the equation.

64. $3x - 2y = 4$

65. $(y - 2)^2 = 8(x + 2)$

66. $\dfrac{x^2}{4} + y^2 = 1$

67. $\dfrac{x^2}{4} - y^2 = 1$

FOCUS ON CONCEPTS

In this chapter, you studied trigonometry. Use the following questions to check your understanding of several of the basic concepts presented. The answers to these questions are given in the back of the book.

1. In your own words, explain the meaning of (a) an angle in standard position, (b) a negative angle, (c) a coterminal angle, and (d) an obtuse angle.

2. A fan motor turns at a given angular speed. How does the speed of the tips of the blades change if a fan of greater diameter is installed on the motor? Explain.

3. *True or False?* $y = \sin \theta$ is not a function because $\sin 30° = \sin 150°$. Explain.

4. In right triangle trigonometry, $\sin 30° = \frac{1}{2}$ regardless of the size of the triangle. Explain.

5. Describe the behavior of $f(\theta) = \sec \theta$ at the zeros of $g(\theta) = \cos \theta$. Explain.

6. Explain how reference angles are used to find the trigonometric functions of obtuse angles.

In Exercises 7–10, match the function $y = a \sin bx$ with its graph. Base your selection solely on your interpretation of the constants a and b. Explain your reasoning. [The graphs are labeled (a), (b), (c), and (d).]

7. $y = 3 \sin x$

8. $y = -3 \sin x$

9. $y = 2 \sin \pi x$

10. $y = 2 \sin \dfrac{x}{2}$

(a)
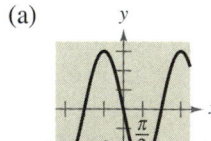

(b)
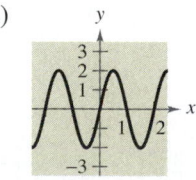

(c)
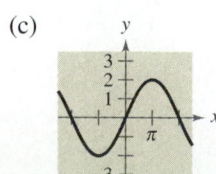

(d)
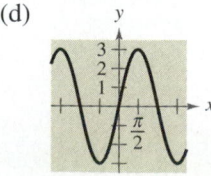

11. The function f is periodic, with period c. Therefore, $f(t + c) = f(t)$. Are the following equal? Explain.

(a) $f(t - 2c) \stackrel{?}{=} f(t)$

(b) $f\left(t + \frac{1}{2}c\right) \stackrel{?}{=} f\left(\frac{1}{2}t\right)$

(c) $f\left(\frac{1}{2}[t + c]\right) \stackrel{?}{=} f\left(\frac{1}{2}t\right)$

12. When graphing the sine and cosine functions, determining the amplitude is part of the analysis. Why is this not true for the other four trigonometric functions?

13. A weight is suspended from a ceiling by a steel spring. The weight is lifted (positive direction) from the equilibrium position and released. The resulting motion of the weight is modeled by

$$y = Ae^{-kt} \cos bt = \frac{1}{5}e^{-t/10} \cos 6t, \qquad t \geq 0$$

where y is the distance in feet from equilibrium and t is the time in seconds. The graph of the function is given in the figure. For each of the following, describe the change in the system without graphing the resulting function.

(a) A is changed from $\frac{1}{5}$ to $\frac{1}{3}$.

(b) k is changed from $\frac{1}{10}$ to $\frac{1}{3}$.

(c) b is changed from 6 to 9.

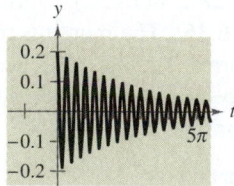

14. *True or False?* Because $\tan 3\pi/4 = -1$, arctan $(-1) = 3\pi/4$. Explain.

Review Exercises

In Exercises 1–4, sketch the angle in standard position. List one positive and one negative coterminal angle.

1. $\dfrac{11\pi}{4}$

2. $\dfrac{2\pi}{9}$

3. $-110°$

4. $-405°$

In Exercises 5–8, convert the measure to decimal degree form. Round to two decimal places.

5. $135°\,16'\,45''$

6. $-234°\,50''$

7. $5°\,22'\,53''$

8. $280°\,8'\,50''$

In Exercises 9–12, convert the measure to D° M′ S″ form.

9. $135.27°$

10. $25.1°$

11. $-85.15°$

12. $-327.85°$

In Exercises 13–16, convert the measure from radians to degrees. Round to two decimal places.

13. $\dfrac{5\pi}{7}$

14. $-\dfrac{3\pi}{5}$

15. -3.5

16. 1.75

In Exercises 17–20, convert the measure from degrees to radians. Round to four decimal places.

17. $480°$

18. $-16.5°$

19. $-33°\,45'$

20. $84°\,15'$

In Exercises 21–24, find the reference angle for the angle.

21. $252°$

22. $640°$

23. $-\dfrac{6\pi}{5}$

24. $\dfrac{17\pi}{3}$

In Exercises 25–30, find the six trigonometric functions of the angle θ (in standard position) whose terminal side passes through the given point.

25. $(12, 16)$

26. $(x, 4x), \quad x > 0$

27. $(-7, 2)$

28. $(4, -8)$

29. $(-4, -6)$

30. $\left(\dfrac{2}{3}, \dfrac{5}{2}\right)$

In Exercises 31–34, find the remaining five trigonometric functions of θ satisfying the given conditions.

31. $\sec\theta = \dfrac{6}{5}, \quad \tan\theta < 0$

32. $\tan\theta = -\dfrac{12}{5}, \quad \sin\theta > 0$

33. $\sin\theta = \dfrac{3}{8}, \quad \cos\theta < 0$

34. $\cos\theta = -\dfrac{2}{5}, \quad \sin\theta > 0$

In Exercises 35–40, evaluate the trigonometric function without using a calculator.

35. $\tan\dfrac{\pi}{3}$

36. $\sec\dfrac{\pi}{4}$

37. $\sin\dfrac{5\pi}{3}$

38. $\cot\left(-\dfrac{5\pi}{6}\right)$

39. $\cos 495°$

40. $\csc 270°$

In Exercises 41–44, use a calculator to evaluate the trigonometric function. Round to two decimal places.

41. $\tan 33°$

42. $\csc 105°$

43. $\sec\dfrac{12\pi}{5}$

44. $\sin\left(-\dfrac{\pi}{9}\right)$

In Exercises 45–48, find two values of θ in degrees $(0° \le \theta < 360°)$ and in radians $(0 \le \theta < 2\pi)$.

45. $\cos\theta = -\dfrac{\sqrt{2}}{2}$

46. $\sec\theta$ is undefined.

47. $\csc\theta = -2$

48. $\tan\theta = \dfrac{\sqrt{3}}{3}$

In Exercises 49–52, use a calculator to find two values of θ. Express both values in degrees ($0° \le \theta < 360°$) and in radians ($0 \le \theta < 2\pi$).

49. $\sin \theta = 0.8387$

50. $\cot \theta = -1.5399$

51. $\sec \theta = -1.0353$

52. $\csc \theta = 11.4737$

In Exercises 53–66, sketch a graph of the function.

53. $y = 3 \cos 2\pi x$

54. $y = -2 \sin \pi x$

55. $f(x) = 5 \sin \dfrac{2x}{5}$

56. $f(x) = 8 \cos\left(-\dfrac{x}{4}\right)$

57. $f(x) = -\dfrac{1}{4} \cos \dfrac{\pi x}{4}$

58. $f(x) = -\tan \dfrac{\pi x}{4}$

59. $g(t) = \dfrac{5}{2} \sin(t - \pi)$

60. $g(t) = 3 \cos(t + \pi)$

61. $h(t) = \tan\left(t - \dfrac{\pi}{4}\right)$

62. $h(t) = \sec\left(t - \dfrac{\pi}{4}\right)$

63. $f(t) = \csc\left(3t - \dfrac{\pi}{2}\right)$

64. $f(t) = 3 \csc\left(2t + \dfrac{\pi}{4}\right)$

65. $y = \arcsin \dfrac{x}{2}$

66. $y = 2 \arccos x$

In Exercises 67–76, use a graphing utility to graph the function. If the function is periodic, find the period.

67. $f(x) = \dfrac{x}{4} - \sin x$

68. $y = \dfrac{x}{3} + \cos \pi x$

69. $f(x) = \dfrac{\pi}{2} + \arctan x$

70. $y = 4 - \dfrac{x}{4} + \cos \pi x$

71. $h(\theta) = \theta \sin \pi\theta$

72. $f(\theta) = \cot \dfrac{\pi\theta}{8}$

73. $f(t) = 2.5e^{-t/4} \sin 2\pi t$

74. $f(x) = \arccos(x - \pi)$

75. $g(x) = 3 \sin\left(\dfrac{\pi x}{3} + 1\right)$

76. $E(t) = 110 \cos\left(120\pi t - \dfrac{\pi}{3}\right)$

In Exercises 77–80, use a graphing utility to graph the function. Use the graph to determine if the function is periodic. If the function is periodic, approximate any relative maximum or minimum points through one period.

77. $f(x) = e^{\sin x}$

78. $g(x) = \sin e^x$

79. $g(x) = 2 \sin x \cos^2 x$

80. $h(x) = 4 \sin^2 x \cos^2 x$

In Exercises 81–84, write an algebraic expression for the given expression.

81. $\sec[\arcsin(x - 1)]$

82. $\tan\left(\arccos \dfrac{x}{2}\right)$

83. $\sin\left(\arccos \dfrac{x^2}{4 - x^2}\right)$

84. $\csc(\arcsin 10x)$

85. *Altitude of a Triangle* Find the altitude of the triangle in the figure.

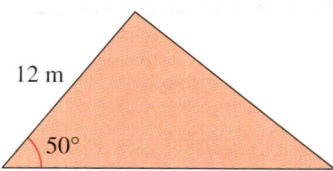

86. *Angle of Elevation* The height of a radio transmission tower is 70 meters, and it casts a shadow of length 30 meters (see figure). Find the angle of elevation of the sun.

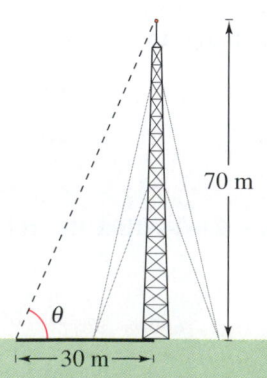

87. *Shuttle Height* An observer 2.5 miles from the launch pad of a space shuttle measures the angle of elevation to the base of the shuttle to be 25° soon after liftoff (see figure). How high is the shuttle at that instant? (Assume that the shuttle is still moving vertically.)

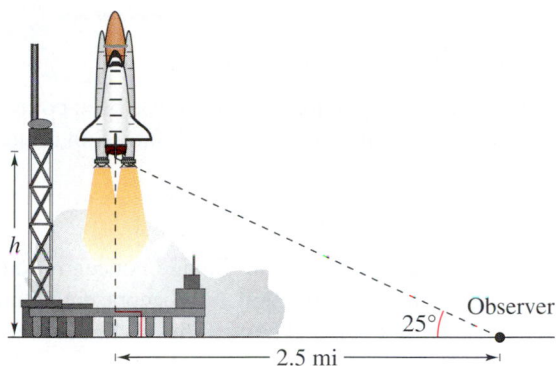

88. *Distance* From city *A* to city *B*, a plane flies 650 miles at a bearing of N 48° E. From city *B* to city *C* the plane flies 810 miles at a bearing of S 65° E. Find the distance from *A* to *C* and the bearing from *A* to *C*.

89. *Railroad Grade* A train travels 3.5 kilometers on a straight track with a grade of 1° 10′ (see figure). What is the vertical rise of the train in that distance?

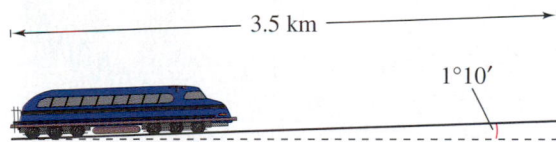

90. *Distance Between Towns* A passenger in an airplane at an altitude of 37,000 feet sees two towns directly to the left of the airplane. The angles of depression to the towns are 32° and 76° (see figure). How far apart are the towns?

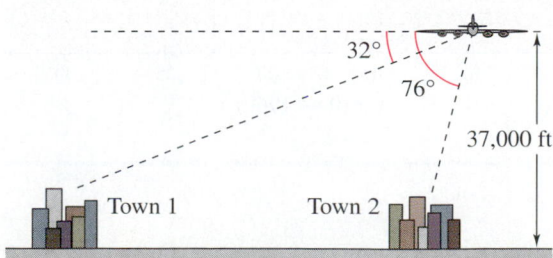

91. *Exploration* The base of the triangle in the figure is also the radius of a circular arc.

(a) Find the area *A* of the shaded region as a function of θ for $0 < \theta < \dfrac{\pi}{2}$.

(b) Use a graphing utility to graph the area function over the given domain. Interpret the graph in the context of the problem.

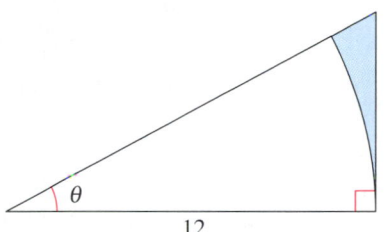

92. *Exploration* In calculus it can be shown that the arcsine and arctangent functions can be approximated by the polynomials

$$\arcsin x \approx x + \frac{x^3}{6} + \frac{3x^5}{40} + \frac{5x^7}{112}$$

$$\arctan x \approx x + \frac{x^3}{3} + \frac{x^5}{5} + \frac{x^7}{7}$$

where *x* is in radians.

(a) Use a graphing utility to graph the arcsine function and its polynomial approximation on the same viewing rectangle. How do the graphs compare?

(b) Use a graphing utility to graph the arctangent function and its polynomial approximation on the same viewing rectangle. How do the graphs compare?

(c) Study the pattern in the polynomial approximation of the arctangent function and guess the next term. Then repeat part (b). How did the accuracy of the approximation change when additional terms were added?

CHAPTER PROJECT: *Analyzing a Graph*

Graphs of functions that are combinations of algebraic functions and trigonometric functions can be difficult to sketch by hand. For such graphs, a graphing utility is helpful.

Real Life

EXAMPLE 1 *Sketching the Graph of a Function*

Since 1958, the Mauna Loa Climate Observatory in Hawaii has been collecting data on the carbon dioxide level of earth's atmosphere. A model that closely represents the data is

$$y = 316 + 0.654t + 0.0216t^2 + 2.5 \sin 2\pi t$$

where y represents the monthly average of carbon dioxide concentration (in parts per million) and $t = 0$ represents January 1960, $t = 1$ represents January 1961, etc. Sketch the graph of this function and explain the oscillations in the graph.

Solution

The graph of the function is shown below. From the graph, you can see that the carbon dioxide level fluctuates each year. The low level each year, which occurs toward the end of the summer in the northern hemisphere, is caused by the intake of carbon dioxide in growing plants.

Mauna Loa Climate Observatory
(Photo: NOAA/photo by Bernard G. Mendonca)

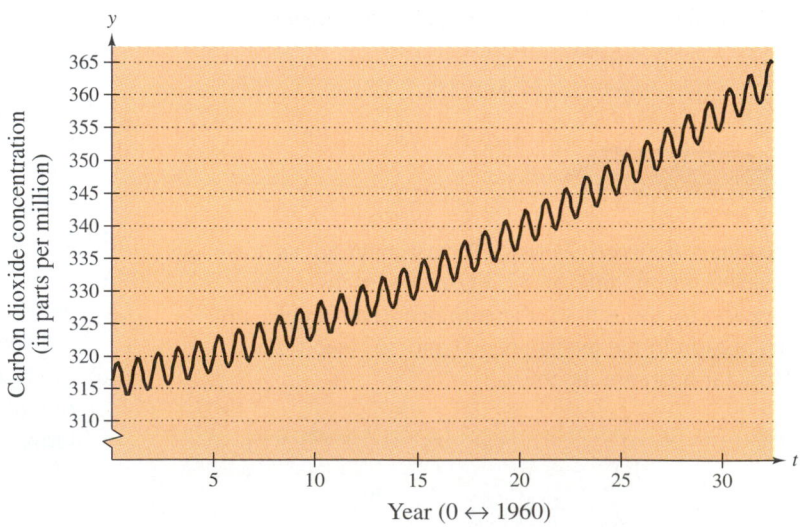

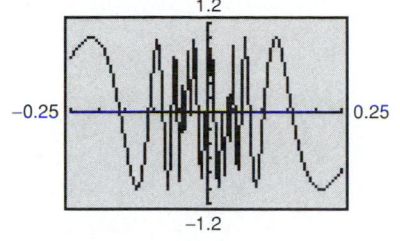

EXAMPLE 2 *Sketching the Graph of a Function*

Sketch the graph of $y = \sin \dfrac{1}{x}$. Describe the graph near the origin.

Solution

The graph is difficult to sketch, even with a graphing utility. The graph shown at the left was produced with a graphing utility. The low resolution of the utility gives a distorted image of the graph. To obtain a better image, we used high-resolution computer software and obtained the graph below. From the high-resolution graph, you can see that as x approaches the origin from the left or the right, the graph oscillates more and more quickly.

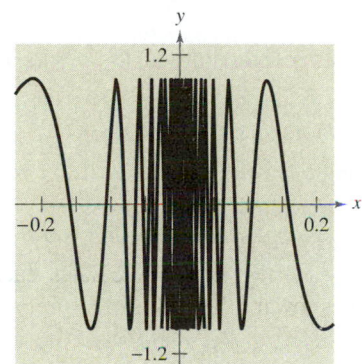

CHAPTER PROJECT INVESTIGATIONS

In Questions 1–6, use a graphing utility to graph the function. Choose a viewing rectangle that you think produces a good representation of the important features of the graph.

1. $y = x^2 + \sin x$

2. $y = x^2 \sin x$

3. $y = |\cos x|$

4. $y = 2 \sin x - \cos 2x$

5. $y = \sin^2 x + \sin x$

6. $y = \dfrac{\sin x}{x}$

7. *Carbon Dioxide Levels* Sketch the graph of the model given in Example 1 for $28 \le t \le 30$. Between January 1988 and January 1990, what were the highest and lowest levels of carbon dioxide? When did each occur?

8. *Throwing a Shot Put* The path of a shot put can be modeled by

$$y = -\frac{16}{v^2 \cos^2 \theta} x^2 + (\tan \theta)x + h$$

where y is the height of the shot put (in feet), x is the horizontal distance (in feet), v is the initial velocity (in feet per second), h is the initial height (in feet), and θ is the angle at which the shot put is thrown.

(a) Choose several values of v, h, and θ and sketch the corresponding graphs. Discuss your results.

(b) Of the graphs you sketched in part (a), which do you think best models a real-life shot-put event? Explain your reasoning.

Chapter Test

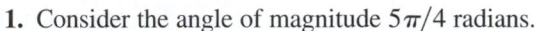

Take this test as you would take a test in class. After you are done, check your work against the answers given in the back of the book.

1. Consider the angle of magnitude $5\pi/4$ radians.
   (a) Sketch the angle in standard position.
   (b) Determine two coterminal angles (one positive and one negative).
   (c) Convert the angle to degree measure.

2. A truck is moving at a rate of 90 kilometers per hour, and the diameter of its wheels is 1 meter. Find the angular speed of the wheels in radians per minute.

3. Find the exact values of the six trigonometric functions of the angle θ shown in the figure.

4. Given that $\tan \theta = \frac{3}{2}$, find the other five trigonometric functions of θ.

5. Determine the reference angle θ' of the angle $\theta = 290°$ and sketch θ and θ' in standard position.

6. Determine the quadrant in which θ lies if $\sec \theta < 0$ and $\tan \theta > 0$.

7. Find two values of θ in degrees ($0 \le \theta < 360°$) if $\cos \theta = -\sqrt{3}/2$. (Do not use a calculator.)

8. Use a calculator to approximate two values of θ in radians ($0 \le \theta < 2\pi$) if $\csc \theta = 1.030$. Round the result to two decimal places.

In Exercises 9 and 10, graph the function through two full periods without the aid of a graphing utility.

9. $g(x) = -2 \sin\left(x - \dfrac{\pi}{4}\right)$ 10. $f(\alpha) = \dfrac{1}{2} \tan 2\alpha$

 In Exercises 11 and 12, use a graphing utility to graph the function. If the function is periodic, find its period.

11. $y = \sin 2\pi x + 2 \cos \pi x$ 12. $y = 6e^{-0.12t} \cos(0.25t), \quad 0 \le t \le 32$

13. Find a, b, and c for the function $f(x) = a \sin(bx + c)$ so that the graph of f matches the figure.

14. Find the exact value of $\tan\left(\arccos \frac{2}{3}\right)$ without the aid of a calculator.

15. Graph the function $f(x) = 2 \arcsin\left(\frac{1}{2}x\right)$.

16. A ship leaves port at noon and sails at a speed of 18 knots. Its bearing is N 16° W. If the port is positioned at the origin, determine the coordinates of the position of the ship at 3 P.M.

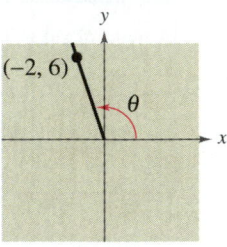

FIGURE FOR 3

The *Interactive* CD-ROM provides answers to the Chapter Tests and Cumulative Tests. It also offers Chapter Pre-Tests (which test key skills and concepts covered in previous chapters) and Chapter Post-Tests, both of which have randomly generated exercises with diagnostic capabilities.

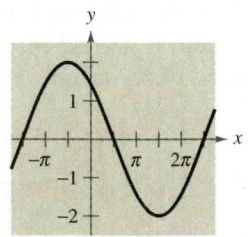

FIGURE FOR 13

The number of hours of daylight that occur at any location on earth depends on two things: the time of year and the latitude of the location.

Here are models for the number of hours of daylight in Seward, Alaska (60 degrees latitude) and New Orleans, Louisiana (30 degrees latitude).

Seward
$D = 12.2 - 6.4 \cos[\pi(t + 0.2)/6]$

New Orleans
$D = 12.2 - 1.9 \cos[\pi(t + 0.2)/6]$

In these models, D represents the number of hours of daylight and t represents the month, with $t = 0$ representing January 1.

To find the time of year that both cities receive the same amount of daylight, you can equate the two models and solve for t. When you do that, you obtain $t = 2.8$ (spring equinox) and $t = 8.8$ (fall equinox).

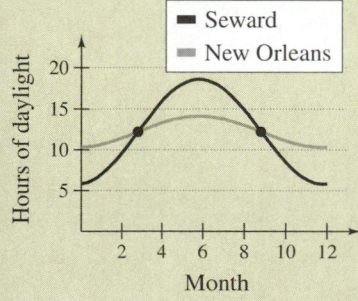

See Exercises 73 and 74 on page 563.

Photos: National Center for Atmospheric Research

7 *Analytic Trigonometry*

7.1 *Using Fundamental Identities*

7.2 *Verifying Trigonometric Identities*

7.3 *Solving Trigonometric Equations*

7.4 *Sum and Difference Formulas*

7.5 *Multiple-Angle and Product-to-Sum Formulas*

Since 1987, Dr. Warren Washington has directed the Climate Global Dynamics Division of NCAR (National Center for Atmospheric Research) in Boulder, Colorado. He has helped develop innovative computer models to predict long-term weather patterns.

7.1 *Using Fundamental Identities*

See Exercises 81 and 82 on page 555 for examples of how trigonometric identities can be used to simplify logarithmic expressions.

Introduction □ *Using the Fundamental Identities*

Introduction

In Chapter 6, you studied the basic definitions, properties, graphs, and applications of the individual trigonometric functions. In this chapter, you will learn how to use the fundamental identities to

1. evaluate trigonometric functions.
2. simplify trigonometric expressions.
3. develop additional trigonometric identities.
4. solve trigonometric equations.

NOTE   Pythagorean identities are sometimes used in radical form such as

$$\sin u = \pm\sqrt{1 - \cos^2 u}$$

or

$$\tan u = \pm\sqrt{\sec^2 u - 1}$$

where the sign depends on the choice of u. ∎∎

FUNDAMENTAL TRIGONOMETRIC IDENTITIES

Reciprocal Identities

$$\sin u = \frac{1}{\csc u} \qquad \cos u = \frac{1}{\sec u} \qquad \tan u = \frac{1}{\cot u}$$

$$\csc u = \frac{1}{\sin u} \qquad \sec u = \frac{1}{\cos u} \qquad \cot u = \frac{1}{\tan u}$$

Quotient Identities

$$\tan u = \frac{\sin u}{\cos u} \qquad \cot u = \frac{\cos u}{\sin u}$$

Pythagorean Identities

$$\sin^2 u + \cos^2 u = 1 \quad 1 + \tan^2 u = \sec^2 u \quad 1 + \cot^2 u = \csc^2 u$$

Cofunction Identities

$$\sin\left(\frac{\pi}{2} - u\right) = \cos u \qquad \cos\left(\frac{\pi}{2} - u\right) = \sin u$$

$$\tan\left(\frac{\pi}{2} - u\right) = \cot u \qquad \cot\left(\frac{\pi}{2} - u\right) = \tan u$$

$$\sec\left(\frac{\pi}{2} - u\right) = \csc u \qquad \csc\left(\frac{\pi}{2} - u\right) = \sec u$$

Even/Odd Identities

$$\sin(-u) = -\sin u, \qquad \cos(-u) = \cos u, \qquad \tan(-u) = -\tan u$$

$$\csc(-u) = -\csc u, \qquad \sec(-u) = \sec u, \qquad \cot(-u) = -\cot u$$

Using the Fundamental Identities

One common use of trigonometric identities is to use given values of trigonometric functions to evaluate other trigonometric functions.

The *Interactive* CD-ROM offers graphing utility emulators of the *TI-82* and *TI-83*, which can be used with the Examples, Explorations, Technology notes, and Exercises.

EXAMPLE 1   *Using Identities to Evaluate a Function*

Use the given values $\sec u = -\frac{3}{2}$ and $\tan u > 0$ to find the values of all six trigonometric functions.

Solution

Using a reciprocal identity, you have

$$\cos u = \frac{1}{\sec u} = \frac{1}{-3/2} = -\frac{2}{3}.$$

Using a Pythagorean identity, you have

$$\sin^2 u = 1 - \cos^2 u = 1 - \left(-\frac{2}{3}\right)^2 = 1 - \frac{4}{9} = \frac{5}{9}.$$

Because $\sec u < 0$ and $\tan u > 0$, it follows that u lies in Quadrant III. Moreover, because $\sin u$ is negative when u is in Quadrant III, you can choose the negative root and obtain $\sin u = -\sqrt{5}/3$. Now, knowing the values of the sine and cosine, you can find the values of all six trigonometric functions.

$$\sin u = -\frac{\sqrt{5}}{3} \qquad\qquad \csc u = \frac{1}{\sin u} = -\frac{3}{\sqrt{5}}$$

$$\cos u = -\frac{2}{3} \qquad\qquad \sec u = \frac{1}{\cos u} = -\frac{3}{2}$$

$$\tan u = \frac{\sin u}{\cos u} = \frac{-\sqrt{5}/3}{-2/3} = \frac{\sqrt{5}}{2} \qquad\qquad \cot u = \frac{1}{\tan u} = \frac{2}{\sqrt{5}}$$

TECHNOLOGY

You can use a graphing utility to check the result of Example 2. To do this, graph

$$y = \sin x \cos^2 x - \sin x$$

and

$$y = -\sin^3 x$$

on the same viewing rectangle, as shown below. Because Example 2 shows the equivalence algebraically and the two graphs appear to coincide, you can conclude that the expressions are equivalent.

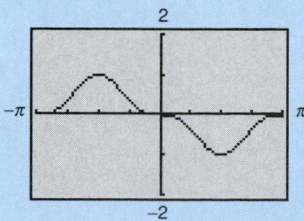

EXAMPLE 2   *Simplifying a Trigonometric Expression*

Simplify $\sin x \cos^2 x - \sin x$.

Solution

Factor out the common monomial factor and then use a fundamental identity.

$$
\begin{aligned}
\sin x \cos^2 x - \sin x &= \sin x (\cos^2 x - 1) &&\text{Monomial factor}\\
&= -\sin x (1 - \cos^2 x) &&\\
&= -\sin x (\sin^2 x) &&\text{Pythagorean identity}\\
&= -\sin^3 x &&\text{Multiply.}
\end{aligned}
$$

EXAMPLE 3   *Factoring Trigonometric Expressions*

Factor each expression.

a. $\sec^2 \theta - 1$ **b.** $4 \tan^2 \theta + \tan \theta - 3$

Solution

a. Here you have the difference of two squares, which factors as

$$\sec^2 \theta - 1 = (\sec \theta - 1)(\sec \theta + 1).$$

b. This expression has the polynomial form, $ax^2 + bx + c$, and it factors as

$$4 \tan^2 \theta + \tan \theta - 3 = (4 \tan \theta - 3)(\tan \theta + 1).$$

Study Tip

On occasion, factoring or simplifying can best be done by first rewriting the expression in terms of just *one* trigonometric function or in terms of *sine and cosine alone*. These strategies are illustrated in Examples 4 and 5, respectively.

EXAMPLE 4   *Factoring a Trigonometric Expression*

Factor $\csc^2 x - \cot x - 3$.

Solution

You can use the identity $\csc^2 x = 1 + \cot^2 x$ to rewrite the expression in terms of the cotangent alone.

$$\csc^2 x - \cot x - 3 = (1 + \cot^2 x) - \cot x - 3 \qquad \text{Pythagorean identity}$$
$$= \cot^2 x - \cot x - 2 \qquad \text{Combine like terms.}$$
$$= (\cot x - 2)(\cot x + 1) \qquad \text{Factor.}$$

EXAMPLE 5   *Simplifying a Trigonometric Expression*

Simplify $\sin t + \cot t \cos t$.

Solution

Begin by rewriting the expression in terms of sine and cosine.

$$\sin t + \cot t \cos t = \sin t + \left(\frac{\cos t}{\sin t}\right)\cos t \qquad \text{Quotient identity}$$
$$= \frac{\sin^2 t + \cos^2 t}{\sin t} \qquad \text{Add fractions.}$$
$$= \frac{1}{\sin t} \qquad \text{Pythagorean identity}$$
$$= \csc t \qquad \text{Reciprocal identity}$$

EXAMPLE 6 *Verifying a Trigonometric Identity*

Verify the identity $\dfrac{\sin \theta}{1 + \cos \theta} + \dfrac{\cos \theta}{\sin \theta} = \csc \theta.$

Solution

$$\frac{\sin \theta}{1 + \cos \theta} + \frac{\cos \theta}{\sin \theta} = \frac{(\sin \theta)(\sin \theta) + (\cos \theta)(1 + \cos \theta)}{(1 + \cos \theta)(\sin \theta)}$$

$$= \frac{\sin^2 \theta + \cos^2 \theta + \cos \theta}{(1 + \cos \theta)(\sin \theta)} \qquad \text{Multiply.}$$

$$= \frac{\cancel{1 + \cos \theta}}{\cancel{(1 + \cos \theta)}(\sin \theta)} \qquad \text{Pythagorean identity}$$

$$= \frac{1}{\sin \theta} \qquad \text{Cancel common factor.}$$

$$= \csc \theta \qquad \text{Reciprocal identity}$$

The last two examples in this section involve techniques for rewriting expressions into forms that are useful in calculus.

EXAMPLE 7 *Rewriting a Trigonometric Expression*

Rewrite $\dfrac{1}{1 + \sin x}$ so that it is *not* in fractional form.

Solution

From the Pythagorean identity $\cos^2 x = 1 - \sin^2 x = (1 - \sin x)(1 + \sin x)$, you can see that by multiplying both the numerator and the denominator by $(1 - \sin x)$ you produce a monomial denominator.

$$\frac{1}{1 + \sin x} = \frac{1}{1 + \sin x} \cdot \frac{1 - \sin x}{1 - \sin x} \qquad \begin{array}{l} \text{Multiply numerator and} \\ \text{denominator by } (1 - \sin x). \end{array}$$

$$= \frac{1 - \sin x}{1 - \sin^2 x} \qquad \text{Multiply.}$$

$$= \frac{1 - \sin x}{\cos^2 x} \qquad \text{Pythagorean identity}$$

$$= \frac{1}{\cos^2 x} - \frac{\sin x}{\cos^2 x} \qquad \text{Separate fractions.}$$

$$= \frac{1}{\cos^2 x} - \frac{\sin x}{\cos x} \cdot \frac{1}{\cos x} \qquad \text{Separate fractions.}$$

$$= \sec^2 x - \tan x \sec x \qquad \text{Identities}$$

EXAMPLE 8   *Trigonometric Substitution*

Use the substitution $x = 2 \tan \theta$, $0 < \theta < \pi/2$, to express

$$\sqrt{4 + x^2}$$

as a trigonometric function of θ.

Solution

Begin by letting $x = 2 \tan \theta$. Then, you can obtain the following.

$$\sqrt{4 + x^2} = \sqrt{4 + (2 \tan \theta)^2}$$ Substitute $2 \tan \theta$ for x.

$$= \sqrt{4(1 + \tan^2 \theta)}$$

$$= \sqrt{4 \sec^2 \theta}$$ Pythagorean identity

$$= 2 \sec \theta$$ $\sec \theta > 0$ for $0 < \theta < \frac{\pi}{2}$

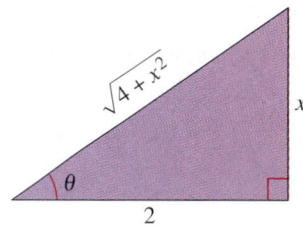

FIGURE 7.1

Figure 7.1 shows the right angle illustration of the trigonometric substitution in Example 8. For $0 < \theta < \pi/2$, you have

$$\text{opp} = x, \quad \text{adj} = 2, \quad \text{and} \quad \text{hyp} = \sqrt{4 + x^2}.$$

With these expressions, you can write the following.

$$\sec \theta = \frac{\sqrt{4 + x^2}}{2}$$

$$2 \sec \theta = \sqrt{4 + x^2}$$

GROUP ACTIVITY

REMEMBERING TRIGONOMETRIC IDENTITIES

Most people find the Pythagorean identity involving sine and cosine to be fairly easy to remember: $\sin^2 u + \cos^2 u = 1$. The one involving tangent and secant, however, tends to give some people trouble. They can't remember if the identity is

$$1 + \tan^2 u \stackrel{?}{=} \sec^2 u \qquad \text{or} \qquad 1 + \sec^2 u \stackrel{?}{=} \tan^2 u.$$

Which of these two is the correct Pythagorean identity involving tangent and secant? Discuss how to remember (or derive) this identity. Can you think of easy ways to remember other fundamental trigonometric identities?

WARM UP

Use a right triangle to evaluate the other five trigonometric functions of the acute angle θ.

1. $\tan \theta = \frac{3}{2}$

2. $\sec \theta = 3$

The point is on the terminal side of an angle θ in standard position. Find the exact values of the six trigonometric functions of θ.

3. $(7, -3)$

4. $(-10, 5)$

Simplify the expression.

5. $\sqrt{1 - \left(\frac{\sqrt{3}}{2}\right)^2}$

6. $\sqrt{\left(\frac{3}{4}\right)^2 + 1}$

7. $\sqrt{1 + \left(\frac{3}{8}\right)^2}$

8. $\sqrt{1 - \left(\frac{\sqrt{5}}{3}\right)^2}$

Perform the operations and simplify.

9. $\dfrac{4}{1 + x} + \dfrac{x}{4}$

10. $\dfrac{3}{1 - x} - \dfrac{5}{1 + x}$

7.1 Exercises

In Exercises 1–14, use the given values to evaluate (if possible) the other four trigonometric functions.

1. $\sin x = \frac{1}{2}, \quad \cos x = \frac{\sqrt{3}}{2}$

2. $\tan x = \frac{\sqrt{3}}{3}, \quad \cos x = -\frac{\sqrt{3}}{2}$

3. $\sec \theta = \sqrt{2}, \quad \sin \theta = -\frac{\sqrt{2}}{2}$

4. $\csc \theta = \frac{5}{3}, \quad \tan \theta = \frac{3}{4}$

5. $\tan x = \frac{5}{12}, \quad \sec x = -\frac{13}{12}$

6. $\cot \phi = -3, \quad \sin \phi = \frac{\sqrt{10}}{10}$

7. $\sec \phi = -1, \quad \sin \phi = 0$

8. $\cos\left(\frac{\pi}{2} - x\right) = \frac{3}{5}, \quad \cos x = \frac{4}{5}$

9. $\sin(-x) = -\frac{2}{3}, \quad \tan x = -\frac{2\sqrt{2}}{5}$

10. $\csc x = 5, \quad \cos x > 0$

11. $\tan \theta = 2, \quad \sin \theta < 0$

12. $\sec \theta = -3, \quad \tan \theta < 0$

13. $\sin \theta = -1, \quad \cot \theta = 0$

14. $\tan \theta$ is undefined, $\quad \sin \theta > 0$

In Exercises 15–18, fill in the blanks. (*Note:* The notation $x \to c^+$ indicates that x approaches c from the right and $x \to c^-$ indicates that x approaches c from the left.)

15. As $x \to \dfrac{\pi}{2}^-$, $\sin x \to$ ▢ and $\csc x \to$ ▢ .

16. As $x \to 0^+$, $\cos x \to$ ▢ and $\sec x \to$ ▢ .

17. As $x \to \dfrac{\pi}{2}^-$, $\tan x \to$ ▢ and $\cot x \to$ ▢ .

18. As $x \to \pi^+$, $\sin x \to$ ▢ and $\csc x \to$ ▢ .

In Exercises 19–24, match the trigonometric expression with one of the following.

(a) -1 (b) $\cos x$ (c) $\cot x$

(d) 1 (e) $-\tan x$ (f) $\sin x$

19. $\sec x \cos x$

20. $\cot x \sin x$

21. $\tan^2 x - \sec^2 x$

22. $(1 - \cos^2 x)(\csc x)$

23. $\dfrac{\sin(-x)}{\cos(-x)}$

24. $\dfrac{\sin[(\pi/2) - x]}{\cos[(\pi/2) - x]}$

In Exercises 25–30, match the trigonometric expression with one of the following.

(a) $\csc x$ (b) $\tan x$ (c) $\sin^2 x$

(d) $\sin x \tan x$ (e) $\sec^2 x$ (f) $\sec^2 x + \tan^2 x$

25. $\sin x \sec x$

26. $\cos^2 x(\sec^2 x - 1)$

27. $\sec^4 x - \tan^4 x$

28. $\cot x \sec x$

29. $\dfrac{\sec^2 x - 1}{\sin^2 x}$

30. $\dfrac{\cos^2[(\pi/2) - x]}{\cos x}$

In Exercises 31–44, use the fundamental identities to simplify the expression.

31. $\tan \phi \csc \phi$

32. $\sin \phi(\csc \phi - \sin \phi)$

33. $\cos \beta \tan \beta$

34. $\sec^2 x(1 - \sin^2 x)$

35. $\dfrac{\cot x}{\csc x}$

36. $\dfrac{\csc \theta}{\sec \theta}$

37. $\sec \alpha \cdot \dfrac{\sin \alpha}{\tan \alpha}$

38. $\dfrac{1}{\tan^2 x + 1}$

39. $\dfrac{\sin(-x)}{\cos x}$

40. $\dfrac{\tan^2 \theta}{\sec^2 \theta}$

41. $\cos\!\left(\dfrac{\pi}{2} - x\right)\sec x$

42. $\cot\!\left(\dfrac{\pi}{2} - x\right)\cos x$

43. $\dfrac{\cos^2 y}{1 - \sin y}$

44. $\cos t(1 + \tan^2 t)$

In Exercises 45–52, factor the expression and use the fundamental identities to simplify.

45. $\tan^2 x - \tan^2 x \sin^2 x$

46. $\sec^2 x \tan^2 x + \sec^2 x$

47. $\sin^2 x \sec^2 x - \sin^2 x$

48. $\dfrac{\sec^2 x - 1}{\sec x - 1}$

49. $\tan^4 x + 2 \tan^2 x + 1$

50. $1 - 2 \cos^2 x + \cos^4 x$

51. $\sin^4 x - \cos^4 x$

52. $\csc^3 x - \csc^2 x - \csc x + 1$

In Exercises 53–56, perform the multiplication and use the fundamental identities to simplify.

53. $(\sin x + \cos x)^2$

54. $(\cot x + \csc x)(\cot x - \csc x)$

55. $(\sec x + 1)(\sec x - 1)$

56. $(3 - 3 \sin x)(3 + 3 \sin x)$

In Exercises 57–60, perform the addition or subtraction and use the fundamental identities to simplify.

57. $\dfrac{1}{1 + \cos x} + \dfrac{1}{1 - \cos x}$

58. $\dfrac{1}{\sec x + 1} - \dfrac{1}{\sec x - 1}$

59. $\dfrac{\cos x}{1 + \sin x} + \dfrac{1 + \sin x}{\cos x}$

60. $\tan x - \dfrac{\sec^2 x}{\tan x}$

In Exercises 61–64, rewrite the expression so that it is *not* in fractional form.

61. $\dfrac{\sin^2 y}{1 - \cos y}$

62. $\dfrac{5}{\tan x + \sec x}$

63. $\dfrac{3}{\sec x - \tan x}$

64. $\dfrac{\tan^2 x}{\csc x + 1}$

Numerical and Graphical Analysis In Exercises 65–68, use a graphing utility to complete the table and graph the functions. Make a conjecture about y_1 and y_2.

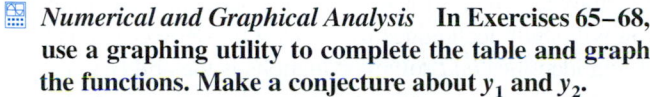

x	0.2	0.4	0.6	0.8	1.0	1.2	1.4
y_1							
y_2							

65. $y_1 = \cos\left(\dfrac{\pi}{2} - x\right)$, $\quad y_2 = \sin x$

66. $y_1 = \cos x + \sin x \tan x$, $\quad y_2 = \sec x$

67. $y_1 = \dfrac{\cos x}{1 - \sin x}$, $\quad y_2 = \dfrac{1 + \sin x}{\cos x}$

68. $y_1 = \sec^4 x - \sec^2 x$, $\quad y_2 = \tan^2 x + \tan^4 x$

In Exercises 69 and 70, use a graphing utility to determine which of the six trigonometric functions is equal to the expression.

69. $\cos x \cot x + \sin x$

70. $\dfrac{1}{2}\left(\dfrac{1 + \sin \theta}{\cos \theta} + \dfrac{\cos \theta}{1 + \sin \theta}\right)$

In Exercises 71–76, use the trigonometric substitution to write the algebraic expression as a trigonometric function of θ, where $0 < \theta < \pi/2$.

71. $\sqrt{25 - x^2}$, $\quad x = 5 \sin \theta$

72. $\sqrt{16 - 4x^2}$, $\quad x = 2 \sin \theta$

73. $\sqrt{x^2 - 9}$, $\quad x = 3 \sec \theta$

74. $\sqrt{x^2 - 4}$, $\quad x = 2 \sec \theta$

75. $\sqrt{x^2 + 25}$, $\quad x = 5 \tan \theta$

76. $\sqrt{x^2 + 100}$, $\quad x = 10 \tan \theta$

In Exercises 77–80, use a graphing utility to solve the equation for θ, where $0 \le \theta < 2\pi$.

77. $\sin \theta = \sqrt{1 - \cos^2 \theta}$

78. $\cos \theta = -\sqrt{1 - \sin^2 \theta}$

79. $\sec \theta = \sqrt{1 + \tan^2 \theta}$

80. $\tan \theta = \sqrt{\sec^2 \theta - 1}$

In Exercises 81 and 82, rewrite the expression as a single logarithm and simplify the result.

81. $\ln|\cos \theta| - \ln|\sin \theta|$

82. $\ln|\cot t| + \ln(1 + \tan^2 t)$

In Exercises 83–86, determine whether or not the equation is an identity, and give a reason for your answer.

83. $(\sin k\theta)/(\cos k\theta) = \tan \theta$, $\quad k$ is a constant.

84. $1/(5 \cos \theta) = 5 \sec \theta$

85. $\sin \theta \csc \theta = 1$

86. $\sin \theta \csc \phi = 1$

In Exercises 87–90, use a calculator to demonstrate the identity for the given values of θ.

87. $\csc^2 \theta - \cot^2 \theta = 1$, (a) $\theta = 132°$, (b) $\theta = \dfrac{2\pi}{7}$

88. $\tan^2 \theta + 1 = \sec^2 \theta$, (a) $\theta = 346°$, (b) $\theta = 3.1$

89. $\cos\left(\dfrac{\pi}{2} - \theta\right) = \sin \theta$, (a) $\theta = 80°$, (b) $\theta = 0.8$

90. $\sin(-\theta) = -\sin \theta$, (a) $\theta = 250°$, (b) $\theta = \dfrac{1}{2}$

91. Express each of the other trigonometric functions of θ in terms of $\sin \theta$.

92. Express each of the other trigonometric functions of θ in terms of $\cos \theta$.

Review Solve Exercises 93–96 as a review of the skills and problem-solving techniques you learned in previous sections. Perform the operations and simplify.

93. $\left(\sqrt{x} + 5\right)\left(\sqrt{x} - 5\right)$

94. $\sqrt{v}\left(\sqrt{20} - \sqrt{5}\right)$

95. $\left(2\sqrt{z} + 3\right)^2$

96. $50x/\left(\sqrt{30} - 5\right)$

7.2 *Verifying Trigonometric Identities*

See Exercise 72 on page 563 for an example of how trigonometric identities can be used to solve a problem dealing with the coefficient of friction for an object on an inclined plane.

Introduction □ *Verifying Trigonometric Identities*

Introduction

In this section, you will study techniques for verifying trigonometric identities. In the next section, you will study techniques for solving trigonometric equations. The key to verifying identities *and* solving equations is the ability to use the fundamental identities and the rules of algebra to rewrite trigonometric expressions.

Remember that a *conditional equation* is an equation that is true for only some of the values in its domain. For example, the conditional equation

$$\sin x = 0 \qquad\qquad \text{Conditional equation}$$

is true only for $x = n\pi$. When you find these values, you are *solving* the equation. On the other hand, an equation that is true for all real values in the domain of the variable is an *identity*. For example, the familiar equation

$$\sin^2 x = 1 - \cos^2 x \qquad\qquad \text{Identity}$$

is true for all real numbers x. Hence, it is an identity.

Although there are similarities, proving that a trigonometric equation is an identity is quite different from solving an equation. There is no well-defined set of rules to follow in verifying trigonometric identities, and the process is best learned by practice.

GUIDELINES FOR VERIFYING TRIGONOMETRIC IDENTITIES

1. Work with one side of the equation at a time. It is often better to work with the more complicated side first.
2. Look for opportunities to factor an expression, add fractions, square a binomial, or create a monomial denominator.
3. Look for opportunities to use the fundamental identities. Note which functions are in the final expression you want. Sines and cosines pair up well, as do secants and tangents, and cosecants and cotangents.
4. If the preceding guidelines do not help, try converting all terms to sines and cosines.
5. Do not just sit and stare at the problem. Try something! Even paths that lead to dead ends give you insights.

Verifying Trigonometric Identities

■
EXAMPLE 1   *Verifying a Trigonometric Identity*

Verify the identity $\dfrac{\sec^2 \theta - 1}{\sec^2 \theta} = \sin^2 \theta.$

Solution

Because the left side is more complicated, start with it.

$$\frac{\sec^2 \theta - 1}{\sec^2 \theta} = \frac{(\tan^2 \theta + 1) - 1}{\sec^2 \theta} \qquad \textcolor{red}{\text{Pythagorean identity}}$$

$$= \frac{\tan^2 \theta}{\sec^2 \theta} \qquad \textcolor{red}{\text{Simplify.}}$$

$$= \tan^2 \theta (\cos^2 \theta) \qquad \textcolor{red}{\text{Reciprocal identity}}$$

$$= \frac{\sin^2 \theta}{\cos^2 \theta} (\cos^2 \theta) \qquad \textcolor{red}{\text{Quotient identity}}$$

$$= \sin^2 \theta \qquad \textcolor{red}{\text{Simplify.}}$$

NOTE   Here is another way to verify the identity in Example 1.

$$\frac{\sec^2 \theta - 1}{\sec^2 \theta} = \frac{\sec^2 \theta}{\sec^2 \theta} - \frac{1}{\sec^2 \theta}$$

$$= 1 - \cos^2 \theta$$

$$= \sin^2 \theta \quad \blacksquare\blacksquare$$

As you can see from the note at the left, there can be more than one way to verify an identity. Your method may differ from that used by your instructor or fellow students. Here is a good chance to be creative and establish your own style, but try to be as efficient as possible.

■
EXAMPLE 2   *Combining Fractions Before Using Identities*

Verify the identity

$$\frac{1}{1 - \sin \alpha} + \frac{1}{1 + \sin \alpha} = 2\sec^2 \alpha.$$

Solution

$$\frac{1}{1 - \sin \alpha} + \frac{1}{1 + \sin \alpha} = \frac{1 + \sin \alpha + 1 - \sin \alpha}{(1 - \sin \alpha)(1 + \sin \alpha)} \qquad \textcolor{red}{\text{Add fractions.}}$$

$$= \frac{2}{1 - \sin^2 \alpha} \qquad \textcolor{red}{\text{Simplify.}}$$

$$= \frac{2}{\cos^2 \alpha} \qquad \textcolor{red}{\text{Pythagorean identity}}$$

$$= 2\sec^2 \alpha \qquad \textcolor{red}{\text{Reciprocal identity}}$$

EXAMPLE 3 *Verifying a Trigonometric Identity*

Verify the identity

$$(\tan^2 x + 1)(\cos^2 x - 1) = -\tan^2 x.$$

Solution

By applying identities before multiplying, you obtain the following.

$$(\tan^2 x + 1)(\cos^2 x - 1) = (\sec^2 x)(-\sin^2 x) \qquad \text{Pythagorean identities}$$

$$= -\frac{\sin^2 x}{\cos^2 x} \qquad \text{Reciprocal identity}$$

$$= -\left(\frac{\sin x}{\cos x}\right)^2 \qquad \text{Rule of exponents}$$

$$= -\tan^2 x \qquad \text{Quotient identity}$$

EXAMPLE 4 *Converting to Sines and Cosines*

Verify the identity

$$\tan x + \cot x = \sec x \csc x.$$

Solution

In this case there appear to be no fractions to add, no products to find, and no opportunity to use one of the Pythagorean identities. Hence, try converting the left side into sines and cosines to see what happens.

$$\tan x + \cot x = \frac{\sin x}{\cos x} + \frac{\cos x}{\sin x} \qquad \text{Quotient identities}$$

$$= \frac{\sin^2 x + \cos^2 x}{\cos x \sin x} \qquad \text{Add fractions.}$$

$$= \frac{1}{\cos x \sin x} \qquad \text{Pythagorean identity}$$

$$= \frac{1}{\cos x} \cdot \frac{1}{\sin x} \qquad \text{Product of fractions}$$

$$= \sec x \csc x \qquad \text{Reciprocal identities}$$

Recall from algebra that *rationalizing the denominator* is, on occasion, a powerful simplification technique. A related form of this technique works for simplifying trigonometric expressions as well.

EXAMPLE 5 *Verifying Trigonometric Identities*

Verify the identity $\sec y + \tan y = \dfrac{\cos y}{1 - \sin y}$.

Solution

Write with the *right* side. Note that you can create a monomial denominator by multiplying the numerator and denominator by $(1 + \sin y)$.

$$\frac{\cos y}{1 - \sin y} = \frac{\cos y}{1 - \sin y}\left(\frac{1 + \sin y}{1 + \sin y}\right) \qquad \text{Multiply numerator and denominator by } (1 + \sin y).$$

$$= \frac{\cos y + \cos y \sin y}{1 - \sin^2 y} \qquad \text{Multiply.}$$

$$= \frac{\cos y + \cos y \sin y}{\cos^2 y} \qquad \text{Pythagorean identity}$$

$$= \frac{\cos y}{\cos^2 y} + \frac{\cos y \sin y}{\cos^2 y} \qquad \text{Separate fractions.}$$

$$= \frac{1}{\cos y} + \frac{\sin y}{\cos y} \qquad \text{Simplify.}$$

$$= \sec y + \tan y \qquad \text{Identities}$$

EXAMPLE 6 *Working with Each Side Separately*

Verify the identity $\dfrac{\cot^2 \theta}{1 + \csc \theta} = \dfrac{1 - \sin \theta}{\sin \theta}$.

Solution

Working with the left side, you have

$$\frac{\cot^2 \theta}{1 + \csc \theta} = \frac{\csc^2 \theta - 1}{1 + \csc \theta} \qquad \text{Pythagorean identity}$$

$$= \frac{(\csc \theta - 1)(\csc \theta + 1)}{1 + \csc \theta} \qquad \text{Factor.}$$

$$= \csc \theta - 1. \qquad \text{Simplify.}$$

Now, simplifying the right side, you have

$$\frac{1 - \sin \theta}{\sin \theta} = \frac{1}{\sin \theta} - \frac{\sin \theta}{\sin \theta} \qquad \text{Separate fractions.}$$

$$= \csc \theta - 1. \qquad \text{Reciprocal identity}$$

The identity is verified because both sides are equal to $\csc \theta - 1$.

Study Tip

In Examples 1 through 5, you have been verifying trigonometric identities by working with one side of the equation and converting to the form given on the other side. On occasion it is practical to work with each side *separately,* to obtain one common form equivalent to both sides. This is illustrated in Example 6.

In Example 7, powers of trigonometric functions are rewritten as more complicated sums of products of trigonometric functions. This is a common procedure used in calculus.

EXAMPLE 7   Two Examples from Calculus

Verify each identity.

a. $\tan^4 x = \tan^2 x \sec^2 x - \tan^2 x$

b. $\sin^3 x \cos^4 x = (\cos^4 x - \cos^6 x) \sin x$

Solution

a.
$$\tan^4 x = (\tan^2 x)(\tan^2 x) \qquad \text{Separate factors.}$$
$$= \tan^2 x(\sec^2 x - 1) \qquad \text{Pythagorean identity}$$
$$= \tan^2 x \sec^2 x - \tan^2 x \qquad \text{Multiply.}$$

b.
$$\sin^3 x \cos^4 x = \sin^2 x \cos^4 x \sin x \qquad \text{Separate factors.}$$
$$= (1 - \cos^2 x)\cos^4 x \sin x \qquad \text{Pythagorean identity}$$
$$= (\cos^4 x - \cos^6 x)\sin x \qquad \text{Multiply.}$$

GROUP ACTIVITY

ERROR ANALYSIS

Suppose you are tutoring a student in trigonometry. One of the homework problems your student encounters asks whether the following statement is an identity.

$$\tan^2 x \sin^2 x \stackrel{?}{=} \frac{5}{6} \tan^2 x$$

Your student does not attempt to verify the equivalence algebraically, but mistakenly uses only a graphical approach. Using range settings of Xmin = -3π, Xmax = 3π, Xscl = $\pi/2$, Ymin = -20, Ymax = 20, and Yscl = 1, your student graphs both sides of the expression on a graphing utility and concludes that the statement is an identity.

What is wrong with your student's reasoning? Explain.

Factor each expression and, if possible, simplify the result.

1. (a) $x^2 - x^2 y^2$

   (b) $\sin^2 x - \sin^2 x \cos^2 x$

2. (a) $x^2 + x^2 y^2$

   (b) $\cos^2 x + \cos^2 x \tan^2 x$

3. (a) $x^4 - 1$

   (b) $\tan^4 x - 1$

4. (a) $z^3 + 1$

   (b) $\tan^3 x + 1$

5. (a) $x^3 - x^2 + x - 1$

   (b) $\cot^3 x - \cot^2 x + \cot x - 1$

6. (a) $x^4 - 2x^2 + 1$

   (b) $\sin^4 x - 2 \sin^2 x + 1$

Perform the operations and, if possible, simplify the result.

7. (a) $\dfrac{y^2}{x} - x$

   (b) $\dfrac{\csc^2 x}{\cot x} - \cot x$

8. (a) $1 - \dfrac{1}{x^2}$

   (b) $1 - \dfrac{1}{\sec^2 x}$

9. (a) $\dfrac{y}{1+z} + \dfrac{1+z}{y}$

   (b) $\dfrac{\sin x}{1 + \cos x} + \dfrac{1 + \cos x}{\sin x}$

10. (a) $\dfrac{y}{z} - \dfrac{z}{1+y}$

   (b) $\dfrac{\tan x}{\sec x} - \dfrac{\sec x}{1 + \tan x}$

7.2 Exercises

In Exercises 1–44, verify the identity.

1. $\sin t \csc t = 1$

2. $\tan y \cot y = 1$

3. $(1 + \sin \alpha)(1 - \sin \alpha) = \cos^2 \alpha$

4. $\cot^2 y(\sec^2 y - 1) = 1$

5. $\cos^2 \beta - \sin^2 \beta = 1 - 2 \sin^2 \beta$

6. $\cos^2 \beta - \sin^2 \beta = 2 \cos^2 \beta - 1$

7. $\tan^2 \theta + 4 = \sec^2 \theta + 3$

8. $2 - \sec^2 z = 1 - \tan^2 z$

9. $\sin^2 \alpha - \sin^4 \alpha = \cos^2 \alpha - \cos^4 \alpha$

10. $\cos x + \sin x \tan x = \sec x$

11. $\dfrac{\sec^2 x}{\tan x} = \sec x \csc x$

12. $\dfrac{\cot^3 t}{\csc t} = \cos t (\csc^2 t - 1)$

13. $\dfrac{\cot^2 t}{\csc t} = \csc t - \sin t$

14. $\dfrac{1}{\sin x} - \sin x = \dfrac{\cos^2 x}{\sin x}$

15. $\sin^{1/2} x \cos x - \sin^{5/2} x \cos x = \cos^3 x \sqrt{\sin x}$

16. $\sec^6 x(\sec x \tan x) - \sec^4 x(\sec x \tan x) = \sec^5 x \tan^3 x$

17. $\dfrac{1}{\sec x \tan x} = \csc x - \sin x$

18. $\dfrac{\sec\theta - 1}{1 - \cos\theta} = \sec\theta$

19. $\csc x - \sin x = \cos x \cot x$

20. $\sec x - \cos x = \sin x \tan x$

21. $\cos x + \sin x \tan x = \sec x$

22. $\dfrac{\sec x + \tan x}{\sec x - \tan x} = (\sec x + \tan x)^2$

23. $\dfrac{1}{\tan x} + \dfrac{1}{\cot x} = \tan x + \cot x$

24. $\dfrac{1}{\sin x} - \dfrac{1}{\csc x} = \csc x - \sin x$

25. $\dfrac{\cos\theta \cot\theta}{1 - \sin\theta} - 1 = \csc\theta$

26. $\dfrac{1 + \sin\theta}{\cos\theta} + \dfrac{\cos\theta}{1 + \sin\theta} = 2\sec\theta$

27. $\dfrac{1}{\cot x + 1} + \dfrac{1}{\tan x + 1} = 1$

28. $\cos x - \dfrac{\cos x}{1 - \tan x} = \dfrac{\sin x \cos x}{\sin x - \cos x}$

29. $\cos\!\left(\dfrac{\pi}{2} - x\right)\csc x = 1$

30. $\dfrac{\cos[(\pi/2) - x]}{\sin[(\pi/2) - x]} = \tan x$

31. $\dfrac{\csc(-x)}{\sec(-x)} = -\cot x$

32. $(1 + \sin y)[1 + \sin(-y)] = \cos^2 y$

33. $\dfrac{\cos(-\theta)}{1 + \sin(-\theta)} = \sec\theta + \tan\theta$

34. $\dfrac{1 + \sec(-\theta)}{\sin(-\theta) + \tan(-\theta)} = -\csc\theta$

35. $\dfrac{\sin x \cos y + \cos x \sin y}{\cos x \cos y - \sin x \sin y} = \dfrac{\tan x + \tan y}{1 - \tan x \tan y}$

36. $\dfrac{\tan x + \tan y}{1 - \tan x \tan y} = \dfrac{\cot x + \cot y}{\cot x \cot y - 1}$

37. $\dfrac{\tan x + \cot y}{\tan x \cot y} = \tan y + \cot x$

38. $\dfrac{\cos x - \cos y}{\sin x + \sin y} + \dfrac{\sin x - \sin y}{\cos x + \cos y} = 0$

39. $\sqrt{\dfrac{1 + \sin\theta}{1 - \sin\theta}} = \dfrac{1 + \sin\theta}{|\cos\theta|}$

40. $\sqrt{\dfrac{1 - \cos\theta}{1 + \cos\theta}} = \dfrac{1 - \cos\theta}{|\sin\theta|}$

41. $\sin^2 x + \sin^2\!\left(\dfrac{\pi}{2} - x\right) = 1$

42. $\sec^2 y - \cot^2\!\left(\dfrac{\pi}{2} - y\right) = 1$

43. $\csc x \cos\!\left(\dfrac{\pi}{2} - x\right) = 1$

44. $\sec^2\!\left(\dfrac{\pi}{2} - x\right) - 1 = \cot^2 x$

In Exercises 45–56, verify the identity algebraically, and use a graphing utility to confirm it graphically.

45. $2\sec^2 x - 2\sec^2 x \sin^2 x - \sin^2 x - \cos^2 x = 1$

46. $\csc x(\csc x - \sin x) + \dfrac{\sin x - \cos x}{\sin x} + \cot x = \csc^2 x$

47. $2 + \cos^2 x - 3\cos^4 x = \sin^2 x(2 + 3\cos^2 x)$

48. $4\tan^4 x + \tan^2 x - 3 = \sec^2 x(4\tan^2 x - 3)$

49. $\csc^4 x - 2\csc^2 x + 1 = \cot^4 x$

50. $\sin x(1 - 2\cos^2 x + \cos^4 x) = \sin^5 x$

51. $\sec^4\theta - \tan^4\theta = 1 + 2\tan^2\theta$

52. $\csc^4\theta - \cot^4\theta = 2\csc^2\theta - 1$

53. $\dfrac{\sin\beta}{1 - \cos\beta} = \dfrac{1 + \cos\beta}{\sin\beta}$

54. $\dfrac{\cot\alpha}{\cot\alpha - 1} = \dfrac{\csc\alpha + 1}{\cot\alpha}$

55. $\dfrac{\tan^3\alpha - 1}{\tan\alpha - 1} = \tan^2\alpha + \tan\alpha + 1$

56. $\dfrac{\sin^3\beta + \cos^3\beta}{\sin\beta + \cos\beta} = 1 - \sin\beta\cos\beta$

In Exercises 57–60, use the properties of logarithms and trigonometric identities to verify the identity.

57. $\ln|\tan \theta| = \ln|\sin \theta| - \ln|\cos \theta|$

58. $\ln|\sec \theta| = -\ln|\cos \theta|$

59. $-\ln(1 + \cos \theta) = \ln(1 - \cos \theta) - 2\ln|\sin \theta|$

60. $-\ln|\sec \theta + \tan \theta| = \ln|\sec \theta - \tan \theta|$

Think About It   In Exercises 61–64, explain why the equation is *not* an identity and find one value of the variable for which the equation is not true.

61. $\sin \theta = \sqrt{1 - \cos^2 \theta}$

62. $\tan \theta = \sqrt{\sec^2 \theta - 1}$

63. $\sqrt{\tan^2 x} = \tan x$

64. $\sqrt{\sin^2 x + \cos^2 x} = \sin x + \cos x$

In Exercises 65–68, use the cofunction identities to evaluate the expression without the aid of a calculator.

65. $\sin^2 25° + \sin^2 65°$

66. $\cos^2 18° + \cos^2 72°$

67. $\cos^2 20° + \cos^2 52° + \cos^2 38° + \cos^2 70°$

68. $\sin^2 12° + \sin^2 40° + \sin^2 50° + \sin^2 78°$

69. Verify that for all integers n

$$\cos\left[\frac{(2n + 1)\pi}{2}\right] = 0.$$

70. Verify that for all integers n

$$\sin\left[\frac{(12n + 1)\pi}{6}\right] = \frac{1}{2}.$$

71. *Rate of Change*   The rate of change of the function

$$f(x) = \sin x + \csc x$$

with respect to change in the variable x is given by the expression

$$\cos x - \csc x \cot x.$$

Show that the expression for the rate of change can also be given by

$$-\cos x \cot^2 x.$$

72. *Friction*   The forces acting on an object weighing W units on an inclined plane positioned at an angle of θ with the horizontal (see figure) is modeled by

$$\mu W \cos \theta = W \sin \theta$$

where μ is the coefficient of friction. Solve the equation for μ and simplify the result.

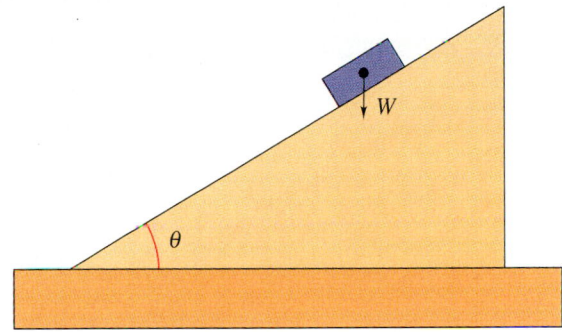

73. *Chapter Opener*   Which city has the greater variation in the number of daylight hours? Which constant in each model would you use to determine the difference between the greatest and least number of hours of daylight?

74. *Chapter Opener*   Determine the period of each model.

Review   Solve Exercises 75–78 as a review of the skills and problem-solving techniques you learned in previous sections. Perform the operations and simplify.

75. $(2 + 3i) - \sqrt{-26}$

76. $(2 - 5i)^2$

77. $\sqrt{-16}\left(1 + \sqrt{-4}\right)$

78. $(3 + 2i)^3$

See Exercise 69 on page 574 for an example of how solving a trigonometric equation can help answer questions about the unemployment rate in the United States.

7.3 Solving Trigonometric Equations

Introduction ◻ *Equations of Quadratic Type* ◻ *Functions Involving Multiple Angles* ◻ *Using Inverse Functions*

Introduction

To solve a trigonometric equation, use standard algebraic techniques such as collecting like terms and factoring. Your preliminary goal is to isolate the trigonometric function involved in the equation.

EXAMPLE 1 Solving a Trigonometric Equation

$$2 \sin x - 1 = 0 \qquad \text{Original equation}$$
$$2 \sin x = 1 \qquad \text{Add 1 to both sides.}$$
$$\sin x = \tfrac{1}{2} \qquad \text{Divide both sides by 2.}$$

To solve for x, note in Figure 7.2 that the equation $\sin x = \tfrac{1}{2}$ has solutions $x = \pi/6$ and $x = 5\pi/6$ in the interval $[0, 2\pi)$. Moreover, because $\sin x$ has a period of 2π, there are infinitely many other solutions, which can be written as

$$x = \pi/6 + 2n\pi \qquad \text{and} \qquad x = 5\pi/6 + 2n\pi \qquad \text{General solution}$$

where n is an integer, as shown in Figure 7.2.

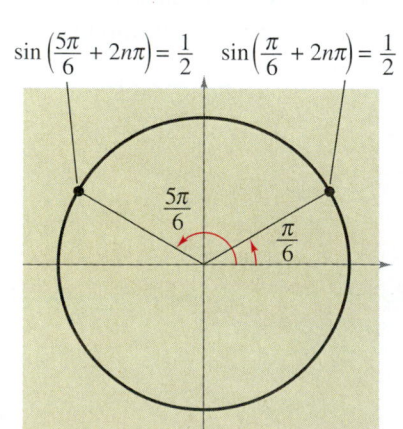

$$\sin\left(\frac{5\pi}{6} + 2n\pi\right) = \frac{1}{2} \qquad \sin\left(\frac{\pi}{6} + 2n\pi\right) = \frac{1}{2}$$

FIGURE 7.3

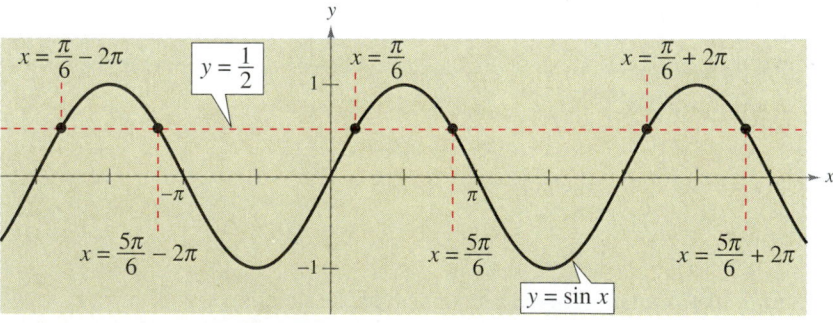

$$x = \frac{\pi}{6} - 2\pi \qquad y = \frac{1}{2} \qquad x = \frac{\pi}{6} \qquad x = \frac{\pi}{6} + 2\pi$$

$$x = \frac{5\pi}{6} - 2\pi \qquad x = \frac{5\pi}{6} \qquad x = \frac{5\pi}{6} + 2\pi$$

$$y = \sin x$$

FIGURE 7.2

Another way to see that the equation $\sin x = \tfrac{1}{2}$ has infinitely many solutions is indicated in Figure 7.3. For $0 \le x < 2\pi$, the solutions are $x = \pi/6$ and $x = 5\pi/6$. Any angles that are coterminal with $\pi/6$ or $5\pi/6$ will also be solutions of the equation.

EXAMPLE 2   *Collecting Like Terms*

Solve $\sin x + \sqrt{2} = -\sin x$.

Solution

Your goal is to rewrite the equation so that $\sin x$ is isolated on one side of the equation.

$$\sin x + \sqrt{2} = -\sin x \qquad \text{Original equation}$$
$$\sin x + \sin x = -\sqrt{2} \qquad \text{Add } \sin x \text{ to, and subtract } \sqrt{2} \text{ from, both sides.}$$
$$2 \sin x = -\sqrt{2} \qquad \text{Combine like terms.}$$
$$\sin x = -\frac{\sqrt{2}}{2} \qquad \text{Divide both sides by 2.}$$

Because $\sin x$ has a period of 2π, first find all solutions in the interval $[0, 2\pi)$. These are $x = 5\pi/4$ and $x = 7\pi/4$. Finally, add $2n\pi$ to each of these solutions to get the general form

$$x = \frac{5\pi}{4} + 2n\pi \qquad \text{and} \qquad x = \frac{7\pi}{4} + 2n\pi \qquad \text{General solution}$$

where n is an integer.

EXAMPLE 3   *Extracting Square Roots*

Solve $3 \tan^2 x - 1 = 0$.

Solution

Your goal is to rewrite the equation so that $\tan x$ is isolated on one side of the equation.

$$3 \tan^2 x - 1 = 0 \qquad \text{Original equation}$$
$$3 \tan^2 x = 1 \qquad \text{Add 1 to both sides.}$$
$$\tan^2 x = \frac{1}{3} \qquad \text{Divide both sides by 3.}$$
$$\tan x = \pm \frac{1}{\sqrt{3}} \qquad \text{Extract square roots.}$$

Because $\tan x$ has a period of π, first find all solutions in the interval $[0, \pi)$. These are $x = \pi/6$ and $x = 5\pi/6$. Finally, add $n\pi$ to each of these solutions to get the general form

$$x = \frac{\pi}{6} + n\pi \qquad \text{and} \qquad x = \frac{5\pi}{6} + n\pi \qquad \text{General solution}$$

where n is an integer.

TECHNOLOGY

The solutions in Examples 2 and 3 are obtained analytically. You can use a graphing utility to confirm the solutions graphically. For instance, to confirm the solutions found in Example 3, sketch the graph of

$$y = 3 \tan^2 x - 1$$

as shown below.

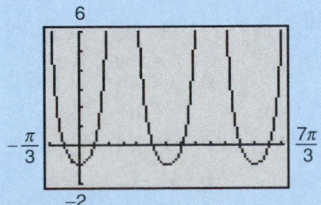

The equations in Examples 1, 2, and 3 involved only one trigonometric function. When two or more functions occur in the same equation, collect all terms on one side and try to separate the functions by factoring or by using appropriate identities. This may produce factors that yield no solutions, as illustrated in Example 4.

EXAMPLE 4   *Factoring*

Solve $\cot x \cos^2 x = 2 \cot x$.

Solution

$$\cot x \cos^2 x = 2 \cot x \qquad \text{Original equation}$$

$$\cot x \cos^2 x - 2 \cot x = 0 \qquad \text{Subtract } 2 \cot x \text{ from both sides.}$$

$$\cot x(\cos^2 x - 2) = 0 \qquad \text{Factor.}$$

By setting each of these factors equal to zero, you obtain the following.

$$\cot x = 0 \qquad \text{and} \qquad \cos^2 x - 2 = 0$$

$$x = \frac{\pi}{2} \qquad\qquad\qquad \cos^2 x = 2$$

$$\cos x = \pm\sqrt{2}$$

No solution is obtained from $\cos x = \pm\sqrt{2}$ because $\pm\sqrt{2}$ are outside the range of the cosine function. Therefore, the general form of the solution is obtained by adding multiples of π to $x = \pi/2$, to get

$$x = \frac{\pi}{2} + n\pi \qquad\qquad \text{General solution}$$

where n is an integer. You can confirm this graphically by sketching the graph of $y = \cot x \cos^2 x - 2 \cot x$, as shown in Figure 7.4.

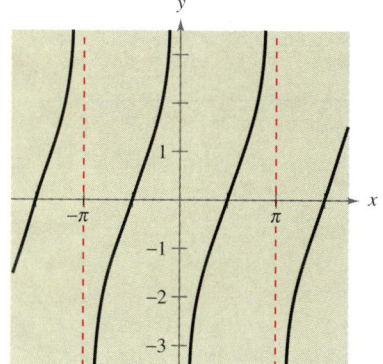

FIGURE 7.4

NOTE   In Example 4, don't make the mistake of dividing both sides of the equation by $\cot x$. If you do this, you lose the solutions. Can you see why? ∎

Equations of Quadratic Type

Many trigonometric equations are of quadratic type. Here are a couple of examples.

Quadratic in sin x	*Quadratic in sec x*
$2 \sin^2 x - \sin x - 1 = 0$	$\sec^2 x - 3 \sec x - 2 = 0$
$2(\sin x)^2 - (\sin x) - 1 = 0$	$(\sec x)^2 - 3(\sec x) - 2 = 0$

To solve equations of this type, factor the quadratic or, if this is not possible, use the Quadratic Formula.

EXAMPLE 5 *Factoring an Equation of Quadratic Type*

Find all solutions of $2 \sin^2 x - \sin x - 1 = 0$ in the interval $[0, 2\pi)$.

Solution

Begin by treating the equation as a quadratic in $\sin x$ and factoring.

$$2 \sin^2 x - \sin x - 1 = 0 \qquad \text{Original equation}$$
$$(2 \sin x + 1)(\sin x - 1) = 0 \qquad \text{Factor.}$$

Setting each factor equal to zero, you obtain the following solutions.

$$2 \sin x + 1 = 0 \qquad \text{and} \qquad \sin x - 1 = 0$$
$$\sin x = -\frac{1}{2} \qquad\qquad\qquad \sin x = 1$$
$$x = \frac{7\pi}{6}, \frac{11\pi}{6} \qquad\qquad\qquad x = \frac{\pi}{2}$$

NOTE In Example 5, the general solution would be

$$x = \frac{7\pi}{6} + 2n\pi$$
$$x = \frac{11\pi}{6} + 2n\pi$$
$$x = \frac{\pi}{2} + 2n\pi$$

where n is an integer. ■■

When working with an equation of quadratic type, be sure that the equation involves a *single* trigonometric function, as shown in the next example.

EXAMPLE 6 *Rewriting with a Single Trigonometric Function*

Solve $2 \sin^2 x + 3 \cos x - 3 = 0$.

Solution

This equation contains both sine and cosine functions. You can rewrite the equation so that it has only cosine functions by using the identity $\sin^2 x = 1 - \cos^2 x$.

$$2 \sin^2 x + 3 \cos x - 3 = 0 \qquad \text{Original equation}$$
$$2(1 - \cos^2 x) + 3 \cos x - 3 = 0 \qquad \text{Pythagorean identity}$$
$$2 \cos^2 x - 3 \cos x + 1 = 0 \qquad \text{Multiply both sides by } -1.$$
$$(2 \cos x - 1)(\cos x - 1) = 0 \qquad \text{Factor.}$$

By setting each factor equal to zero, you can find the solutions in the interval $[0, 2\pi)$ to be $x = 0$, $x = \pi/3$, and $x = 5\pi/3$. The general solution is therefore

$$x = 2n\pi, \qquad x = \frac{\pi}{3} + 2n\pi, \qquad x = \frac{5\pi}{3} + 2n\pi \qquad \text{General solution}$$

where n is an integer.

Sometimes you must square both sides of an equation to obtain a quadratic, as demonstrated in the next example. Because this procedure can introduce extraneous solutions, you should check any solutions in the original equation to see if they are valid or extraneous.

EXAMPLE 7   *Squaring and Converting to Quadratic Type*

Find all solutions of $\cos x + 1 = \sin x$ in the interval $[0, 2\pi)$.

Solution

It is not clear how to rewrite this equation in terms of a single trigonometric function. See what happens when you square both sides of the equation.

$\cos x + 1 = \sin x$	Original equation
$\cos^2 x + 2 \cos x + 1 = \sin^2 x$	Square both sides.
$\cos^2 x + 2 \cos x + 1 = 1 - \cos^2 x$	Pythagorean identity
$2 \cos^2 x + 2 \cos x = 0$	Combine like terms.
$2 \cos x(\cos x + 1) = 0$	Factor.

Setting each factor equal to zero produces the following.

$$2 \cos x = 0 \qquad \text{and} \qquad \cos x + 1 = 0$$

$$\cos x = 0 \qquad\qquad\qquad \cos x = -1$$

$$x = \frac{\pi}{2}, \frac{3\pi}{2} \qquad\qquad\qquad x = \pi$$

NOTE In Example 7, the general solution would be

$$x = \frac{\pi}{2} + 2n\pi$$

$$x = \pi + 2n\pi$$

where n is an integer. ∎

Because you squared the original equation, check for extraneous solutions. Of the three possible solutions, $x = 3\pi/2$ is extraneous. (Try checking this.) Thus, in the interval $[0, 2\pi)$, the only two solutions are $x = \pi/2$ and $x = \pi$.

▶ *Exploration*

Use a graphing utility to confirm the solutions found in Example 7 in two different ways. Do both methods produce the same x-values? Which method do you prefer? Why?

1. Graph both sides of the equation and find the x-coordinates of the points at which the graphs intersect.

   Left side: $y = \cos x + 1$ *Right side:* $y = \sin x$

2. Graph the equation $y = \cos x + 1 - \sin x$ and find the x-intercepts of the graph.

Functions Involving Multiple Angles

EXAMPLE 8   *Functions of Multiple Angles*

Find all solutions of $2 \cos 3t - 1 = 0$.

Solution

$$2 \cos 3t - 1 = 0 \qquad \text{Original equation}$$

$$2 \cos 3t = 1 \qquad \text{Add 1 to both sides.}$$

$$\cos 3t = \frac{1}{2} \qquad \text{Divide both sides by 2.}$$

In the interval $[0, 2\pi)$, you know that $3t = \pi/3$ and $3t = 5\pi/3$ are the only solutions so that, in general, you have

$$3t = \frac{\pi}{3} + 2n\pi \qquad \text{and} \qquad 3t = \frac{5\pi}{3} + 2n\pi.$$

Dividing this result by 3, you obtain the general solution

$$t = \frac{\pi}{9} + \frac{2n\pi}{3} \qquad \text{and} \qquad t = \frac{5\pi}{9} + \frac{2n\pi}{3} \qquad \text{General solution}$$

where n is an integer.

EXAMPLE 9   *Functions of Multiple Angles*

Find all solutions of $3 \tan(x/2) + 3 = 0$.

Solution

$$3 \tan \frac{x}{2} + 3 = 0 \qquad \text{Original equation}$$

$$3 \tan \frac{x}{2} = -3 \qquad \text{Subtract 3 from both sides.}$$

$$\tan \frac{x}{2} = -1 \qquad \text{Divide both sides by 3.}$$

In the interval $[0, \pi)$, you know that $x/2 = 3\pi/4$ is the only solution so that, in general, you have

$$\frac{x}{2} = \frac{3\pi}{4} + n\pi.$$

Multiplying this result by 2, you obtain the general solution

$$x = \frac{3\pi}{2} + 2n\pi \qquad \text{General solution}$$

where n is an integer.

Using Inverse Functions

EXAMPLE 10   *Using Inverse Functions*

Find all solutions of $\sec^2 x - 2\tan x = 4$.

Solution

$$\sec^2 x - 2\tan x = 4 \qquad \text{Original equation}$$

$$1 + \tan^2 x - 2\tan x - 4 = 0 \qquad \text{Pythagorean identity}$$

$$\tan^2 x - 2\tan x - 3 = 0 \qquad \text{Combine like terms.}$$

$$(\tan x - 3)(\tan x + 1) = 0 \qquad \text{Factor.}$$

Setting each factor equal to zero, you obtain two solutions in the interval $(-\pi/2, \pi/2)$. [Recall that the range of the inverse tangent function is $(-\pi/2, \pi/2)$.]

$$\tan x = 3, \qquad\qquad \tan x = -1$$

$$x = \arctan 3 \qquad\qquad x = -\frac{\pi}{4}$$

Finally, by adding multiples of π, you obtain the general solution

$$x = \arctan 3 + n\pi \qquad \text{and} \qquad x = -\frac{\pi}{4} + n\pi \qquad\qquad \text{General solution}$$

where n is an integer.

From 1985 through 1994, the unemployment rate varied between 5.3% and 7.4%. To see the cyclical nature of the unemployment rate during these years, see Exercise 69 on page 574.

GROUP ACTIVITY

EQUATIONS WITH NO SOLUTIONS

One of the following equations has solutions and the other two don't. Which two equations do not have solutions?

a. $\sin^2 x - 5\sin x + 6 = 0$

b. $\sin^2 x - 4\sin x + 6 = 0$

c. $\sin^2 x - 5\sin x - 6 = 0$

Can you find conditions involving the constants b and c that will guarantee that the equation

$$\sin^2 x + b\sin x + c = 0$$

has at least one solution on some interval of length 2π?

WARM UP

Solve for θ in the interval $0 \leq \theta < 2\pi$.

1. $\cos \theta = -\dfrac{1}{2}$

2. $\sin \theta = \dfrac{\sqrt{3}}{2}$

3. $\cos \theta = \dfrac{\sqrt{2}}{2}$

4. $\sin \theta = -\dfrac{\sqrt{2}}{2}$

5. $\tan \theta = \sqrt{3}$

6. $\tan \theta = -1$

Solve for x.

7. $\dfrac{x}{3} + \dfrac{x}{5} = 1$

8. $2x(x + 3) - 5(x + 3) = 0$

9. $2x^2 - 4x - 5 = 0$

10. $\dfrac{1}{x} = \dfrac{x}{2x + 3}$

7.3 Exercises

In Exercises 1–4, find the x-intercepts of the graph.

1. $y = \sin \dfrac{\pi x}{2} + 1$

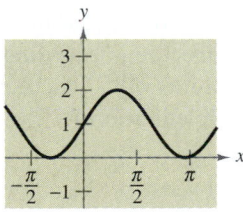

2. $y = \sin \pi x + \cos \pi x$

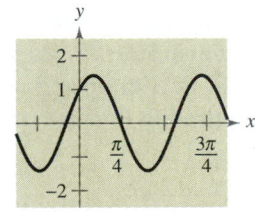

3. $y = \tan^2\left(\dfrac{\pi x}{6}\right) - 3$

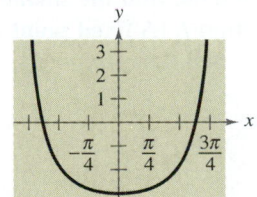

4. $y = \sec^4\left(\dfrac{\pi x}{8}\right) - 4$

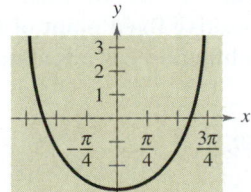

In Exercises 5–10, verify that the x-values are solutions.

5. $2 \cos x - 1 = 0$

(a) $x = \dfrac{\pi}{3}$ (b) $x = \dfrac{5\pi}{3}$

6. $\csc x - 2 = 0$

(a) $x = \dfrac{\pi}{6}$ (b) $x = \dfrac{5\pi}{6}$

7. $3 \tan^2 2x - 1 = 0$

(a) $x = \dfrac{\pi}{12}$ (b) $x = \dfrac{5\pi}{12}$

8. $2 \cos^2 4x - 1 = 0$

(a) $x = \dfrac{\pi}{16}$ (b) $x = \dfrac{3\pi}{16}$

9. $2 \sin^2 x - \sin x - 1 = 0$

(a) $x = \dfrac{\pi}{2}$ (b) $x = \dfrac{7\pi}{6}$

10. $\sec^4 x - 4 \sec^2 x = 0$

(a) $x = \dfrac{2\pi}{3}$ (b) $x = \dfrac{5\pi}{3}$

In Exercises 11–24, solve the equation.

11. $2 \cos x + 1 = 0$ **12.** $2 \sin x - 1 = 0$

13. $\sqrt{3} \csc x - 2 = 0$ **14.** $\tan x + 1 = 0$

15. $3 \sec^2 x - 4 = 0$ **16.** $\csc^2 x - 2 = 0$

17. $2 \sin^2 2x = 1$ **18.** $\tan^2 3x = 3$

19. $4 \sin^2 x - 3 = 0$

20. $\sin x(\sin x + 1) = 0$

21. $\sin^2 x = 3 \cos^2 x$

22. $\tan 3x(\tan x - 1) = 0$

23. $(3 \tan^2 x - 1)(\tan^2 x - 3) = 0$

24. $\cos 2x(2 \cos x + 1) = 0$

In Exercises 25–40, find all solutions of the equation in the interval $[0, 2\pi)$.

25. $\cos^3 x = \cos x$ **26.** $\tan^2 x - 1 = 0$

27. $3 \tan^3 x = \tan x$ **28.** $2 \sin^2 x = 2 + \cos x$

29. $\sec^2 x - \sec x = 2$ **30.** $\sec x \csc x = 2 \csc x$

31. $2 \sin x + \csc x = 0$ **32.** $\sin 2x = -\dfrac{\sqrt{3}}{2}$

33. $\csc x + \cot x = 1$ **34.** $\tan 3x = 1$

35. $\cos \dfrac{x}{2} = \dfrac{\sqrt{2}}{2}$ **36.** $\sec 4x = 2$

37. $\dfrac{1 + \cos x}{1 - \cos x} = 0$

38. $2 \sin^2 x + 3 \sin x + 1 = 0$

39. $2 \sec^2 x + \tan^2 x - 3 = 0$

40. $\cos x + \sin x \tan x = 2$

In Exercises 41 and 42, solve both equations. How do the solutions of the algebraic equation compare to the solutions of the trigonometric equation?

41. $6y^2 - 13y + 6 = 0$

    $6 \cos^2 x - 13 \cos x + 6 = 0$

42. $y^2 + y - 20 = 0$

    $\sin^2 x + \sin x - 20 = 0$

In Exercises 43–56, use a graphing utility to approximate the solutions of the equation in the interval $[0, 2\pi)$.

43. $2 \cos x - \sin x = 0$

44. $4 \sin^3 x + 2 \sin^2 x - 2 \sin x - 1 = 0$

45. $\dfrac{1 + \sin x}{\cos x} + \dfrac{\cos x}{1 + \sin x} = 4$

46. $\dfrac{\cos x \cot x}{1 - \sin x} = 3$

47. $2 \sin x - x = 0$

48. $x \cos x - 1 = 0$

49. $\sec^2 x + 0.5 \tan x - 1 = 0$

50. $\csc^2 x + 0.5 \cot x - 5 = 0$

51. $2 \tan^2 x + 7 \tan x - 15 = 0$

52. $12 \cos^2 x + 5 \cos x - 3 = 0$

53. $12 \sin^2 x - 13 \sin x + 3 = 0$

54. $3 \tan^2 x + 4 \tan x - 4 = 0$

55. $\sin^2 x + 2 \sin x - 1 = 0$

56. $4 \cos^2 x - 4 \cos x - 1 = 0$

In Exercises 57 and 58, (a) use a graphing utility to graph the function and approximate the maximum and minimum points on the graph in the interval $[0, 2\pi)$, and (b) solve the trigonometric equation and demonstrate that its solutions are the x-coordinates of the maximum and minimum points of f. (Calculus is required to find the trigonometric equation.)

Function	*Trigonometric Equation*
57. $f(x) = \sin x + \cos x$	$\cos x - \sin x = 0$
58. $f(x) = 2 \sin x + \cos 2x$	$2 \cos x - 4 \sin x \cos x = 0$

Fixed Point **In Exercises 59 and 60, find the smallest positive fixed point of the function f. [A fixed point of a function f is a real number c such that $f(c) = c$.]**

59. $f(x) = \tan \dfrac{\pi x}{4}$

60. $f(x) = \cos x$

61. Graphical Reasoning Consider the function

$$f(x) = \cos \frac{1}{x}$$

and its graph shown in the figure.

(a) What is the domain of the function?

(b) Identify any symmetry or asymptotes of the graph.

(c) Describe the behavior of the function as $x \to 0$.

(d) How many solutions does the equation

$$\cos \frac{1}{x} = 0$$

have in the interval $[-1, 1]$?

(e) Does the equation $\cos(1/x) = 0$ have a greatest solution? If so, approximate the solution. If not, explain.

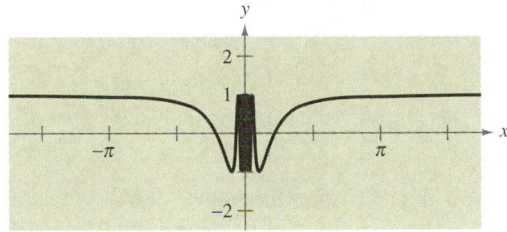

62. Graphical Reasoning Consider the function

$$f(x) = \frac{\sin x}{x}$$

and its graph shown in the figure.

(a) What is the domain of the function?

(b) Identify any symmetry or asymptotes of the graph.

(c) Describe the behavior of the function as $x \to 0$.

(d) How many solutions does the equation

$$\frac{\sin x}{x} = 0$$

have in the interval $[-8, 8]$? Find the solutions.

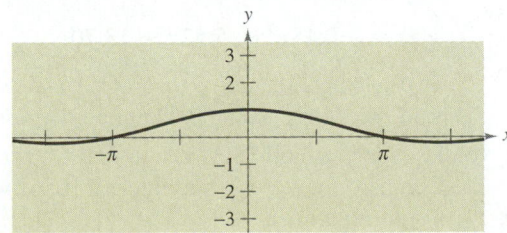

63. Harmonic Motion A weight is oscillating on the end of a spring (see figure). The position of the weight relative to the point of equilibrium is given by

$$y = \frac{1}{12}(\cos 8t - 3 \sin 8t)$$

where y is the displacement in meters and t is the time in seconds. Find the times when the weight is at the point of equilibrium $(y = 0)$ for $0 \le t \le 1$.

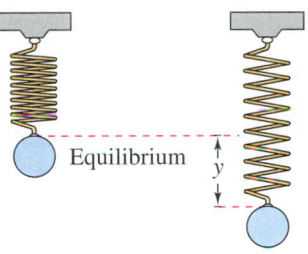

Equilibrium

64. Sales The monthly sales (in thousands of units) of a seasonal product are approximated by

$$S = 74.50 + 43.75 \sin \frac{\pi t}{6}$$

where t is the time in months, with $t = 1$ corresponding to January. Determine the months when sales exceed 100,000 units.

65. Projectile Motion A batted baseball leaves the bat at an angle of θ with the horizontal and an initial velocity of $v_0 = 100$ feet per second. The ball is caught by an outfielder 300 feet from home plate (see figure). Find θ if the range r of a projectile is given by

$$f = \frac{1}{32}v_0^2 \sin 2\theta.$$

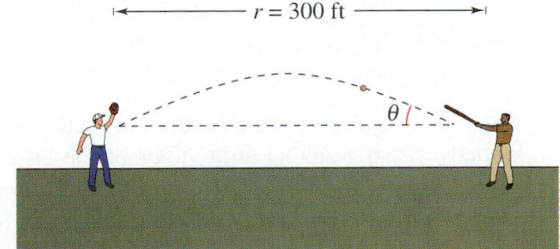

$r = 300$ ft

66. Projectile Motion   A sharpshooter intends to hit a target at a distance of 1000 yards with a gun that has a muzzle velocity of 1200 feet per second (see figure). Neglecting air resistance, determine the minimum angle of elevation of the gun if the range is given by

$$f = \frac{1}{32} v_0^2 \sin 2\theta.$$

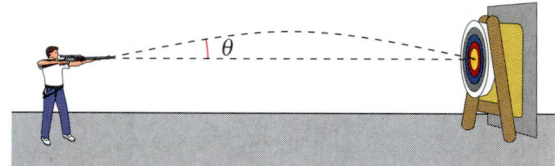

r = 1000 yd

67. Area   The area of a rectangle (see figure) inscribed in one arch of the graph of $y = \cos x$ is given by

$$A = 2x \cos x, \qquad -\frac{\pi}{2} < x < \frac{\pi}{2}.$$

(a) Use a graphing utility to graph the area function, and approximate the area of the largest inscribed rectangle.

(b) Determine the values of x for which $A \geq 1$.

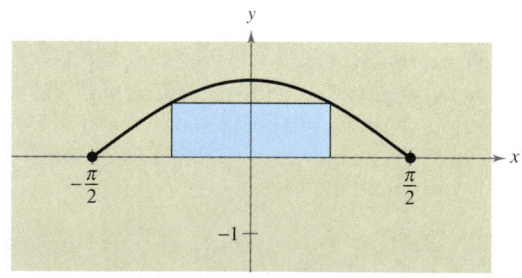

68. Damped Harmonic Motion   The displacement from equilibrium of a weight oscillating on the end of a spring is given by

$$y = 1.56 e^{-0.22t} \cos 4.9t,$$

where y is the displacement in feet and t is the time in seconds. Use a graphing utility to graph the displacement function for $0 \leq t \leq 10$. Find the time beyond which the displacement does not exceed 1 inch from equilibrium.

69. Data Analysis   The table gives the unemployment rate r for the years 1985 through 1994 in the United States. The time t is measured in years, with $t = 0$ corresponding to 1990. (Source: U.S. Bureau of Labor Statistics)

t	-5	-4	-3	-2	-1
r	7.2	7.0	6.2	5.5	5.3

t	0	1	2	3	4
r	5.5	6.7	7.4	6.8	6.1

(a) Create a scatter plot of the data.

(b) Which of the following models best represents the data? Explain your reasoning.

  (1) $r = 1.5 \cos(t + 3.9) + 6.37$

  (2) $r = 1.03 \sin(0.9t + 0.44) + 6.19$

  (3) $r = 1.05 \sin[0.95(t + 6.32)] + 6.20$

  (4) $r = 1.5 \sin[0.5(t + 2.8)] + 6.25$

(c) What term in the model gives the average unemployment rate? What is the rate?

(d) Economists study the lengths of business cycles such as unemployment rates. Based on this short span of time, use the model to give the length of this cycle.

(e) Use the model to estimate the next time the unemployment rate will be 6% or less.

70. Quadratic Approximation   Consider the function

$$f(x) = 3 \sin(0.6x - 2).$$

(a) Approximate the zero of the function in the interval $[0, 6]$.

(b) A quadratic approximation agreeing with f at $x = 5$ is given by

$$g(x) = -0.45x^2 + 5.52x - 13.70.$$

Use a graphing utility to graph f and g on the same viewing rectangle. Describe the result.

(c) Use the Quadratic Formula to find the zeros of g. Compare the zero in the interval $[0, 6]$ with the result of part (a).

7.4 Sum and Difference Formulas

See Exercise 79 on page 582 for an example of how a sum formula can be used to help analyze the harmonic motion of a spring.

Introduction ◻ *Using Sum and Difference Formulas*

Introduction

In this and the following section, you will study the derivations and uses of several trigonometric identities and formulas.

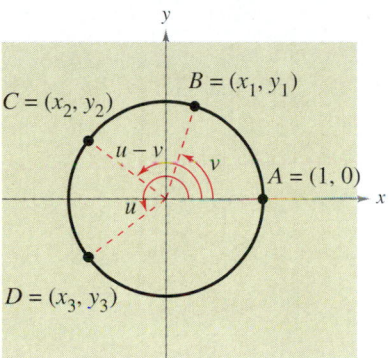

FIGURE 7.5

> ### SUM AND DIFFERENCE FORMULAS
>
> $\sin(u + v) = \sin u \cos v + \cos u \sin v$ $\qquad \tan(u + v) = \dfrac{\tan u + \tan v}{1 - \tan u \tan v}$
>
> $\sin(u - v) = \sin u \cos v - \cos u \sin v$
>
> $\cos(u + v) = \cos u \cos v - \sin u \sin v$
>
> $\cos(u - v) = \cos u \cos v + \sin u \sin v$ $\qquad \tan(u - v) = \dfrac{\tan u - \tan v}{1 + \tan u \tan v}$

PROOF ▪▪

Here are proofs for the formulas for $\cos(u \pm v)$. In Figure 7.5, let A be the point $(1, 0)$ and then use u and v to locate the points $B = (x_1, y_1)$, $C = (x_2, y_2)$, and $D = (x_3, y_3)$ on the unit circle. Thus, $x_i^2 + y_i^2 = 1$ for $i = 1, 2, 3$. For convenience, assume that $0 < v < u < 2\pi$. In Figure 7.6, note that arcs AC and BD have the same length. Hence, *line segments AC and BD are also equal in length*, which implies that

$$\sqrt{(x_2 - 1)^2 + (y_2 - 0)^2} = \sqrt{(x_3 - x_1)^2 + (y_3 - y_1)^2}$$
$$x_2^2 - 2x_2 + 1 + y_2^2 = x_3^2 - 2x_1x_3 + x_1^2 + y_3^2 - 2y_1y_3 + y_1^2$$
$$(x_2^2 + y_2^2) + 1 - 2x_2 = (x_3^2 + y_3^2) + (x_1^2 + y_1^2) - 2x_1x_3 - 2y_1y_3$$
$$1 + 1 - 2x_2 = 1 + 1 - 2x_1x_3 - 2y_1y_3$$
$$x_2 = x_3x_1 + y_3y_1.$$

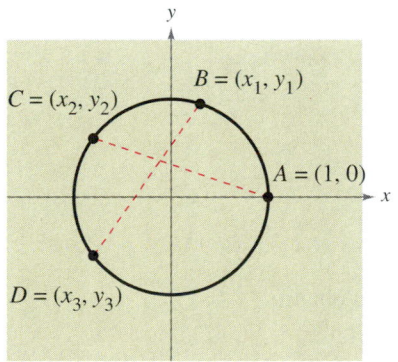

FIGURE 7.6

NOTE Note that $\sin(u + v) \neq \sin u + \sin v$. Similar statements can be made for $\cos(u + v)$ and $\tan(u + v)$. ▪▪

Finally, by substituting the values $x_2 = \cos(u - v)$, $x_3 = \cos u$, $x_1 = \cos v$, $y_3 = \sin u$, and $y_1 = \sin v$, you obtain

$$\cos(u - v) = \cos u \cos v + \sin u \sin v.$$

The formula for $\cos(u + v)$ can be established by considering $u + v = u - (-v)$ and using the formula just derived to obtain

$$\cos(u + v) = \cos[u - (-v)]$$
$$= \cos u \cos(-v) + \sin u \sin(-v)$$
$$= \cos u \cos v - \sin u \sin v. \qquad ▪▪$$

Using Sum and Difference Formulas

In the remainder of this section, you will study a variety of uses of sum and difference formulas. For instance, Examples 1 and 2 show how sum and difference formulas can be used to find exact values of trigonometric functions involving sums or differences of special angles.

EXAMPLE 1   *Evaluating a Trigonometric Function*

Find the exact value of cos 75°.

Solution

To find the *exact* value of cos 75°, use the fact that $75° = 30° + 45°$. Consequently, the formula for $\cos(u + v)$ yields

$$\cos 75° = \cos(30° + 45°)$$
$$= \cos 30° \cos 45° - \sin 30° \sin 45°$$
$$= \frac{\sqrt{3}}{2}\left(\frac{\sqrt{2}}{2}\right) - \frac{1}{2}\left(\frac{\sqrt{2}}{2}\right)$$
$$= \frac{\sqrt{6} - \sqrt{2}}{4}.$$

Hipparchus, considered the most eminent of Greek astronomers, was born about 160 B.C. in Nicaea. He was credited with the invention of trigonometry. He also derived the sum and difference formulas for $\sin(A \pm B)$ and $\cos(A \pm B)$. *(Illustration: The Granger Collection, New York)*

NOTE   Try checking the result obtained in Example 1 on your calculator. You will find that $\cos 75° \approx 0.259$. ■■

EXAMPLE 2   *Evaluating a Trigonometric Function*

Find the exact value of $\sin \dfrac{\pi}{12}$.

Solution

Using the fact that

$$\frac{\pi}{12} = \frac{\pi}{3} - \frac{\pi}{4}$$

together with the formula for $\sin(u - v)$, you obtain

$$\sin \frac{\pi}{12} = \sin\left(\frac{\pi}{3} - \frac{\pi}{4}\right)$$
$$= \sin \frac{\pi}{3} \cos \frac{\pi}{4} - \cos \frac{\pi}{3} \sin \frac{\pi}{4}$$
$$= \frac{\sqrt{3}}{2}\left(\frac{\sqrt{2}}{2}\right) - \frac{1}{2}\left(\frac{\sqrt{2}}{2}\right)$$
$$= \frac{\sqrt{6} - \sqrt{2}}{4}.$$

Exploration

Graph $y = \cos(x + 2)$ and $y = \cos x + \cos 2$ on the same coordinate plane. What can you conclude about the graphs? Is it true that $\cos(x + 2) = \cos x + \cos 2$?

Graph $y = \sin(x + 4)$ and $y = \sin x + \sin 4$ on the same coordinate plane. What can you conclude about the graphs? Is it true that $\sin(x + 4) = \sin x + \sin 4$?

EXAMPLE 3 *Evaluating a Trigonometric Expression*

Find the exact value of $\sin 42° \cos 12° - \cos 42° \sin 12°$.

Solution

Recognizing that this expression fits the formula for $\sin(u - v)$, you can write

$$\sin 42° \cos 12° - \cos 42° \sin 12° = \sin(42° - 12°)$$

$$= \sin 30° = \frac{1}{2}.$$

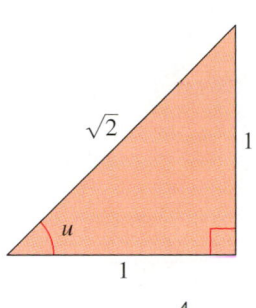

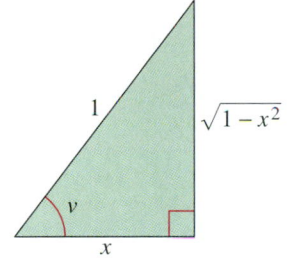

FIGURE 7.7

EXAMPLE 4 *An Application of a Sum Formula*

Evaluate $\cos(\arctan 1 + \arccos x)$.

Solution

This expression fits the formula for $\cos(u + v)$. Angles $u = \arctan 1$ and $v = \arccos x$ are shown in Figure 7.7. Then

$$\cos(u + v) = \cos(\arctan 1)\cos(\arccos x) - \sin(\arctan 1)\sin(\arccos x)$$

$$= \frac{1}{\sqrt{2}} \cdot x - \frac{1}{\sqrt{2}} \cdot \sqrt{1 - x^2}$$

$$= \frac{x - \sqrt{1 - x^2}}{\sqrt{2}}.$$

EXAMPLE 5 *Proving a Cofunction Identity*

Prove the cofunction identity $\cos\left(\frac{\pi}{2} - x\right) = \sin x$.

Solution

Using the formula for $\cos(u - v)$, you have

$$\cos\left(\frac{\pi}{2} - x\right) = \cos\frac{\pi}{2}\cos x + \sin\frac{\pi}{2}\sin x$$

$$= (0)(\cos x) + (1)(\sin x)$$

$$= \sin x.$$

Sum and difference formulas can be used to derive **reduction formulas** involving expressions such as

$$\sin\left(\theta + \frac{n\pi}{2}\right) \quad \text{and} \quad \cos\left(\theta + \frac{n\pi}{2}\right), \quad \text{where } n \text{ is an integer.}$$

EXAMPLE 6   *Deriving Reduction Formulas*

Simplify each expression.

a. $\cos\left(\theta - \dfrac{3\pi}{2}\right)$ **b.** $\tan(\theta + 3\pi)$

Solution

a. Using the formula for $\cos(u - v)$, you have

$$\cos\left(\theta - \frac{3\pi}{2}\right) = \cos\theta\cos\frac{3\pi}{2} + \sin\theta\sin\frac{3\pi}{2}$$

$$= (\cos\theta)(0) + (\sin\theta)(-1)$$

$$= -\sin\theta.$$

b. Using the formula for $\tan(u + v)$, you have

$$\tan(\theta + 3\pi) = \frac{\tan\theta + \tan 3\pi}{1 - \tan\theta\tan 3\pi}$$

$$= \frac{\tan\theta + 0}{1 - (\tan\theta)(0)}$$

$$= \tan\theta.$$

The next example was taken from calculus. It is used to derive the derivative of the sine function.

EXAMPLE 7   *An Application from Calculus*

Verify that

$$\frac{\sin(x + h) - \sin x}{h} = (\cos x)\left(\frac{\sin h}{h}\right) - (\sin x)\left(\frac{1 - \cos h}{h}\right)$$

where $h \neq 0$.

Solution

Using the formula for $\sin(u + v)$, you have

$$\frac{\sin(x + h) - \sin x}{h} = \frac{\sin x\cos h + \cos x\sin h - \sin x}{h}$$

$$= \frac{\cos x\sin h - \sin x(1 - \cos h)}{h}$$

$$= (\cos x)\left(\frac{\sin h}{h}\right) - (\sin x)\left(\frac{1 - \cos h}{h}\right).$$

EXAMPLE 8   *Solving a Trigonometric Equation*

Find all solutions of

$$\sin\left(x + \frac{\pi}{4}\right) + \sin\left(x - \frac{\pi}{4}\right) = -1$$

in the interval $[0, 2\pi)$.

Solution

Using sum and difference formulas, rewrite the given equation as

$$\sin x \cos \frac{\pi}{4} + \cos x \sin \frac{\pi}{4} + \sin x \cos \frac{\pi}{4} - \cos x \sin \frac{\pi}{4} = -1$$

$$2 \sin x \cos \frac{\pi}{4} = -1$$

$$2(\sin x)\left(\frac{\sqrt{2}}{2}\right) = -1$$

$$\sin x = -\frac{1}{\sqrt{2}}$$

$$\sin x = -\frac{\sqrt{2}}{2}$$

Therefore, the only solutions in the interval $[0, 2\pi)$ are

$$x = \frac{5\pi}{4} \qquad \text{and} \qquad x = \frac{7\pi}{4}.$$

These solutions are checked graphically in Figure 7.8.

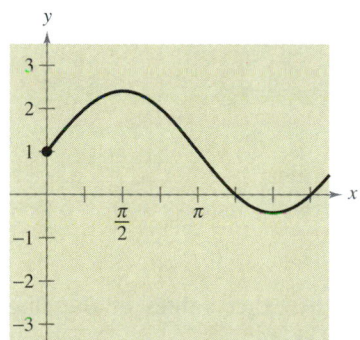

FIGURE 7.8

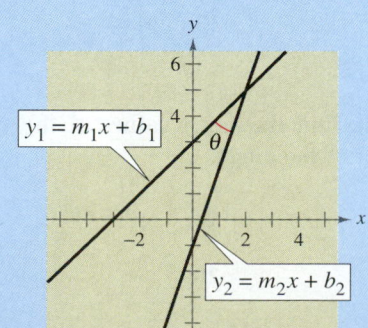

GROUP ACTIVITY

THE ANGLE BETWEEN TWO LINES

The figure at the left shows two lines whose equations are

$$y_1 = m_1 x + b_1 \qquad \text{and} \qquad y_2 = m_2 x + b_2.$$

Assume that both lines have positive slopes, as shown in the figure. With others in your group, derive a formula for the angle between the two lines. Then use your formula to find the angle between the following pairs of lines.

a. $y = x$ and $y = \sqrt{3}x$ **b.** $y = x$ and $y = \dfrac{1}{\sqrt{3}}x$

WARM UP

Use the given information to find $\sin \theta$.

1. $\tan \theta = \frac{1}{3}$, θ in Quadrant I
2. $\cot \theta = \frac{3}{5}$, θ in Quadrant III
3. $\cos \theta = \frac{3}{4}$, θ in Quadrant IV
4. $\sec \theta = -3$, θ in Quadrant II

Find all solutions in the interval $[0, 2\pi)$.

5. $\sin x = \dfrac{\sqrt{2}}{2}$
6. $\cos x = 0$

Simplify the expression.

7. $\tan x \sec^2 x - \tan x$
8. $\dfrac{\cos x \csc x}{\tan x}$
9. $\dfrac{\cos x}{1 - \sin x} - \tan x$
10. $\dfrac{\cos^4 x - \sin^4 x}{\cos^2 x}$

7.4 Exercises

In Exercises 1–4, find the exact value of each expression.

1. (a) $\cos\left(\dfrac{\pi}{4} + \dfrac{\pi}{3}\right)$ (b) $\cos\dfrac{\pi}{4} + \cos\dfrac{\pi}{3}$

2. (a) $\sin\left(\dfrac{3\pi}{4} + \dfrac{5\pi}{6}\right)$ (b) $\sin\dfrac{3\pi}{4} + \sin\dfrac{5\pi}{6}$

3. (a) $\sin\left(\dfrac{7\pi}{6} - \dfrac{\pi}{3}\right)$ (b) $\sin\dfrac{7\pi}{6} - \sin\dfrac{\pi}{3}$

4. (a) $\cos\left(\dfrac{2\pi}{3} - \dfrac{\pi}{6}\right)$ (b) $\cos\dfrac{2\pi}{3} + \cos\dfrac{\pi}{6}$

5. **Think About It**   Use the results of Exercises 1–4 to determine if the following are true or false. Explain.

(a) $\sin(u \pm v) = \sin u \pm \sin v$

(b) $\cos(u \pm v) = \cos u \pm \cos v$

6. **True or False?** $\cos\left(x - \dfrac{\pi}{2}\right) = -\sin x$

In Exercises 7–16, find the exact values of the sine, cosine, and tangent of the angle.

7. $75° = 30° + 45°$
8. $15° = 45° - 30°$
9. $105° = 60° + 45°$
10. $165° = 135° + 30°$
11. $195° = 225° - 30°$
12. $255° = 300° - 45°$
13. $\dfrac{11\pi}{12} = \dfrac{3\pi}{4} + \dfrac{\pi}{6}$
14. $\dfrac{7\pi}{12} = \dfrac{\pi}{3} + \dfrac{\pi}{4}$
15. $\dfrac{17\pi}{12} = \dfrac{9\pi}{4} - \dfrac{5\pi}{6}$
16. $-\dfrac{\pi}{12} = \dfrac{\pi}{6} - \dfrac{\pi}{4}$

In Exercises 17–20, find the exact values of the sine, cosine, and tangent of the angle.

17. $285°$
18. $-105°$
19. $-\dfrac{13\pi}{12}$
20. $\dfrac{5\pi}{12}$

In Exercises 21–30, write the expression as the sine, cosine, or tangent of an angle.

21. $\cos 25° \cos 15° - \sin 25° \sin 15°$

22. $\sin 140° \cos 50° + \cos 140° \sin 50°$

23. $\sin 230° \cos 30° - \cos 230° \sin 30°$

24. $\cos 20° \cos 30° + \sin 20° \sin 30°$

25. $\dfrac{\tan 325° - \tan 86°}{1 + \tan 325° \tan 86°}$ **26.** $\dfrac{\tan 140° - \tan 60°}{1 + \tan 140° \tan 60°}$

27. $\sin 3 \cos 1.2 - \cos 3 \sin 1.2$

28. $\cos \dfrac{\pi}{7} \cos \dfrac{\pi}{5} - \sin \dfrac{\pi}{7} \sin \dfrac{\pi}{5}$

29. $\dfrac{\tan 2x + \tan x}{1 - \tan 2x \tan x}$

30. $\cos 3x \cos 2y + \sin 3x \sin 2y$

In Exercises 31–38, find the exact value of the trigonometric function given that $\sin u = \frac{5}{13}$ and $\cos v = -\frac{3}{5}$. (Both u and v are in Quadrant II.)

31. $\sin(u + v)$ **32.** $\cos(v - u)$

33. $\cos(u + v)$ **34.** $\sin(u - v)$

35. $\sec(u + v)$ **36.** $\csc(u - v)$

37. $\tan(u - v)$ **38.** $\cot(u + v)$

In Exercises 39–44, find the exact value of the trigonometric function given that $\sin u = -\frac{7}{25}$ and $\cos v = -\frac{4}{5}$. (Both u and v are in Quadrant III.)

39. $\cos(u + v)$ **40.** $\sin(u + v)$

41. $\sin(v - u)$ **42.** $\cos(u - v)$

43. $\csc(u + v)$ **44.** $\sec(v - u)$

In Exercises 45–58, verify the identity.

45. $\sin(3\pi - x) = \sin x$ **46.** $\sin\left(\dfrac{\pi}{2} + x\right) = \cos x$

47. $\sin\left(\dfrac{\pi}{6} + x\right) = \dfrac{1}{2}(\cos x + \sqrt{3} \sin x)$

48. $\cos\left(\dfrac{5\pi}{4} - x\right) = -\dfrac{\sqrt{2}}{2}(\cos x + \sin x)$

49. $\cos(\pi - \theta) + \sin\left(\dfrac{\pi}{2} + \theta\right) = 0$

50. $\tan\left(\dfrac{\pi}{4} - \theta\right) = \dfrac{1 - \tan \theta}{1 + \tan \theta}$

51. $\cos(x + y) \cos(x - y) = \cos^2 x - \sin^2 y$

52. $\sin(x + y) \sin(x - y) = \sin^2 x - \cos^2 y$

53. $\sin(x + y) + \sin(x - y) = 2 \sin x \cos y$

54. $\cos(x + y) + \cos(x - y) = 2 \cos x \cos y$

55. $\cos(n\pi + \theta) = (-1)^n \cos \theta$, n is an integer.

56. $\sin(n\pi + \theta) = (-1)^n \sin \theta$, n is an integer.

57. $a \sin B\theta + b \cos B\theta = \sqrt{a^2 + b^2} \sin(B\theta + C)$, where $C = \arctan(b/a)$, $a > 0$

58. $a \sin B\theta + b \cos B\theta = \sqrt{a^2 + b^2} \cos(B\theta - C)$, where $C = \arctan(a/b)$, $b > 0$

In Exercises 59–62, verify the identity algebraically and use a graphing utility to confirm it graphically.

59. $\cos\left(\dfrac{3\pi}{2} - x\right) = -\sin x$

60. $\cos(\pi + x) = -\cos x$

61. $\sin\left(\dfrac{3\pi}{2} + \theta\right) + \sin(\pi - \theta) = \sin \theta - \cos \theta$

62. $\tan(\pi + \theta) = \tan \theta$

In Exercises 63–66, use the formulas given in Exercises 57 and 58 to write the trigonometric expression in the following forms.

(a) $\sqrt{a^2 + b^2} \sin(B\theta + C)$

(b) $\sqrt{a^2 + b^2} \cos(B\theta - C)$

63. $\sin \theta + \cos \theta$ **64.** $3 \sin 2\theta + 4 \cos 2\theta$

65. $12 \sin 3\theta + 5 \cos 3\theta$ **66.** $\sin 2\theta - \cos 2\theta$

In Exercises 67 and 68, use the formulas given in Exercises 57 and 58 to write the trigonometric expression in the form $a \sin B\theta + b \cos B\theta$.

67. $2 \sin\left(\theta + \dfrac{\pi}{2}\right)$ **68.** $5 \cos\left(\theta + \dfrac{3\pi}{4}\right)$

In Exercises 69 and 70, write the trigonometric expression as an algebraic expression.

69. $\sin(\arcsin x + \arccos x)$

70. $\sin(\arctan 2x - \arccos x)$

In Exercises 71–74, find all solutions of the equation in the interval $[0, 2\pi)$.

71. $\sin\left(x + \dfrac{\pi}{3}\right) + \sin\left(x - \dfrac{\pi}{3}\right) = 1$

72. $\sin\left(x + \dfrac{\pi}{6}\right) - \sin\left(x - \dfrac{\pi}{6}\right) = \dfrac{1}{2}$

73. $\cos\left(x + \dfrac{\pi}{4}\right) - \cos\left(x - \dfrac{\pi}{4}\right) = 1$

74. $\tan(x + \pi) + 2\sin(x + \pi) = 0$

In Exercises 75 and 76, use a graphing utility to approximate the solutions in the interval $[0, 2\pi)$.

75. $\cos\left(x + \dfrac{\pi}{4}\right) + \cos\left(x - \dfrac{\pi}{4}\right) = 1$

76. $\tan(x + \pi) - \cos\left(x + \dfrac{\pi}{2}\right) = 0$

77. *Conjecture* Consider the function

$$f(\theta) = \sin^2\left(\theta + \dfrac{\pi}{4}\right) + \sin^2\left(\theta - \dfrac{\pi}{4}\right).$$

Graph the function and use the graph to create an identity. Prove your conjecture.

78. *Conjecture* Three squares of side s are placed side by side (see figure). Make a conjecture about the relationship between the sum $u + v$ and w. Prove your conjecture by using the identity for the tangent of the sum of two angles.

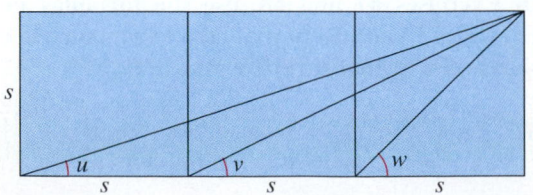

79. *Harmonic Motion* A weight is attached to a spring suspended vertically from a ceiling. When a driving force is applied to the system, the weight moves vertically from its equilibrium position, and this motion is modeled by

$$y = \dfrac{1}{3}\sin 2t + \dfrac{1}{4}\cos 2t$$

where y is the distance from equilibrium measured in feet and t is the time in seconds.

(a) Write the model in the form

$$y = \sqrt{a^2 + b^2}\,\sin(Bt + C).$$

(See Exercise 57.)

(b) Find the amplitude of the oscillations of the weight.

(c) Find the frequency of the oscillations of the weight.

80. *Standing Waves* The equation of a standing wave is obtained by adding the displacements of two waves traveling in opposite directions (see figure). Assume that each of the waves has amplitude A, period T, and wavelength λ. If the models for these waves are

$$y_1 = A\cos 2\pi\left(\dfrac{t}{T} - \dfrac{x}{\lambda}\right) \quad \text{and}$$

$$y_2 = A\cos 2\pi\left(\dfrac{t}{T} + \dfrac{x}{\lambda}\right)$$

show that

$$y_1 + y_2 = 2A\cos\dfrac{2\pi t}{T}\cos\dfrac{2\pi x}{\lambda}.$$

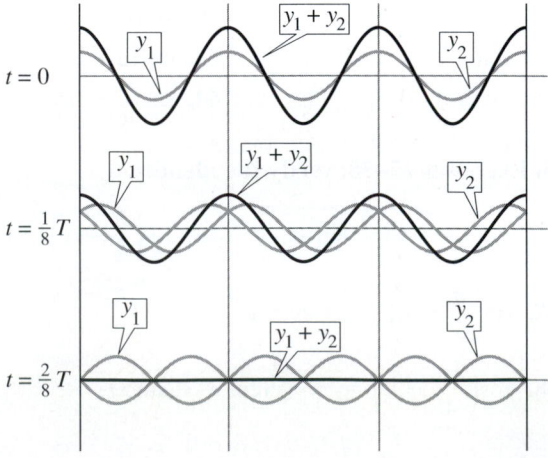

7.5 Multiple-Angle and Product-to-Sum Formulas

See Exercise 116 on page 594 for an example of how a double-angle formula can help analyze the motion of a projectile.

Multiple-Angle Formulas □ *Power-Reducing Formulas* □
Half-Angle Formulas □ *Product-to-Sum Formulas*

Multiple-Angle Formulas

In this section you will study four other categories of trigonometric identities.

1. The first category involves functions of multiple angles such as $\sin ku$ or $\cos ku$.
2. The second category involves squares of trigonometric functions such as $\sin^2 u$.
3. The third category involves functions of half-angles such as $\sin(u/2)$.
4. The fourth category involves products of trigonometric functions such as $\sin u \cos v$.

The most commonly used multiple-angle formulas are the **double-angle formulas.** They are used often, so you should learn them.

DOUBLE-ANGLE FORMULAS

$$\sin 2u = 2 \sin u \cos u \qquad \tan 2u = \frac{2 \tan u}{1 - \tan^2 u}$$

$$\cos 2u = \cos^2 u - \sin^2 u$$
$$= 2 \cos^2 u - 1$$
$$= 1 - 2 \sin^2 u$$

PROOF ■■

To prove the first formula, let $v = u$ in the formula for $\sin(u + v)$.

$$\sin 2u = \sin(u + u)$$
$$= \sin u \cos u + \cos u \sin u$$
$$= 2 \sin u \cos u$$

NOTE Note that $\sin 2u \neq 2 \sin u$. Similar statements can be made for $\cos 2u$ and $\tan 2u$. ■■

To prove the second formula, let $v = u$ in the formula for $\cos(u + v)$.

$$\cos 2u = \cos(u + u)$$
$$= \cos u \cos u - \sin u \sin u$$
$$= \cos^2 u - \sin^2 u$$

The tangent double-angle formula can be proven in a similar way. ■■

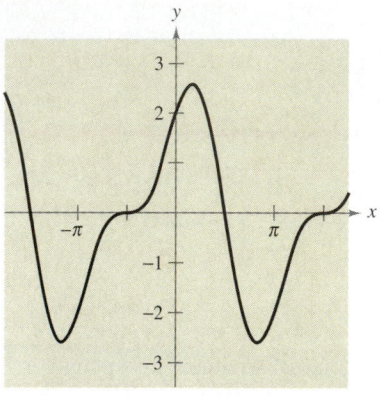

FIGURE 7.9

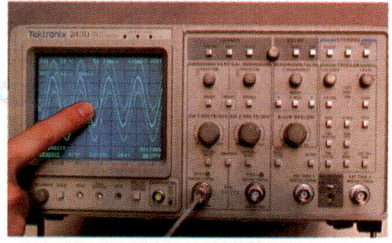

An oscilloscope is an electronic instrument that displays changes in electrical or sound waves on a fluorescent screen. See Exercise 80 on page 582. *(Photo: Peter Aprahamian/ Science Photo Library/Photo Researchers)*

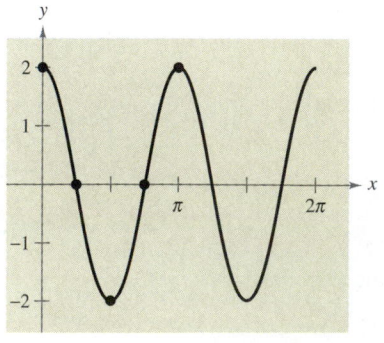

FIGURE 7.10

■
EXAMPLE 1   *Solving a Trigonometric Equation*

Find all solutions of $2 \cos x + \sin 2x = 0$.

Solution

Begin by rewriting the equation so that it involves functions of x (rather than $2x$). Then factor and solve as usual.

$2 \cos x + \sin 2x = 0$	Original equation
$2 \cos x + 2 \sin x \cos x = 0$	Double-angle formula
$2 \cos x(1 + \sin x) = 0$	Factor.
$\cos x = 0, \quad 1 + \sin x = 0$	Set factors equal to zero.
$x = \dfrac{\pi}{2}, \dfrac{3\pi}{2} \qquad\qquad x = \dfrac{3\pi}{2}$	Solutions in $[0, 2\pi)$

Therefore, the general solution is

$$x = \frac{\pi}{2} + 2n\pi \qquad \text{and} \qquad x = \frac{3\pi}{2} + 2n\pi$$

where n is an integer. The graph of $y = 2 \cos x + \sin 2x$, as shown in Figure 7.9, allows you to verify these solutions graphically.

■
EXAMPLE 2   *Using Double-Angle Formulas in Sketching Graphs*

Sketch the graph of $y = 4 \cos^2 x - 2$ over the interval $[0, 2\pi]$.

Solution

Using a double-angle formula, you can rewrite the given function as

$$\begin{aligned} y &= 4 \cos^2 x - 2 \\ &= 2(2 \cos^2 x - 1) \\ &= 2 \cos 2x. \end{aligned}$$

Using the techniques discussed in Section 6.4, you can recognize that the graph of this function has an amplitude of 2 and a period of π. The key points in the interval $[0, \pi]$ are as follows.

Maximum	Intercept	Minimum	Intercept	Maximum
$(0, 2)$	$\left(\dfrac{\pi}{4}, 0\right)$	$\left(\dfrac{\pi}{2}, -2\right)$	$\left(\dfrac{3\pi}{4}, 0\right)$	$(\pi, 2)$

Two cycles of the graph are shown in Figure 7.10.

■

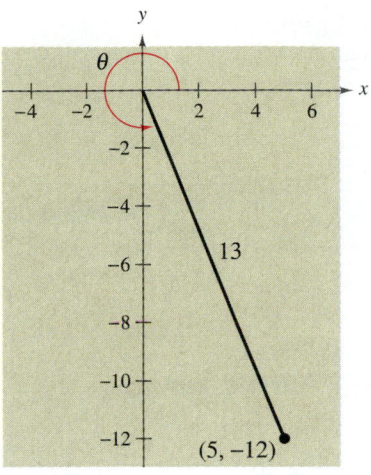

(5, −12)

FIGURE 7.11

EXAMPLE 3   *Evaluating Functions Involving Double Angles*

Use the following to find $\sin 2\theta$, $\cos 2\theta$, and $\tan 2\theta$.

$$\cos \theta = \frac{5}{13}, \qquad \frac{3\pi}{2} < \theta < 2\pi$$

Solution

From Figure 7.11, you can see that $\sin \theta = y/r = -12/13$. Consequently, you can write the following.

$$\sin 2\theta = 2 \sin \theta \cos \theta = 2\left(\frac{-12}{13}\right)\left(\frac{5}{13}\right) = -\frac{120}{169}$$

$$\cos 2\theta = 2 \cos^2 \theta - 1 = 2\left(\frac{25}{169}\right) - 1 = -\frac{119}{169}$$

$$\tan 2\theta = \frac{\sin 2\theta}{\cos 2\theta} = \frac{120}{119}$$

The double-angle formulas are not restricted to angles 2θ and θ. Other *double* combinations, such as 4θ and 2θ or 6θ and 3θ, are also valid. Here are two examples.

$$\sin 4\theta = 2 \sin 2\theta \cos 2\theta \qquad \text{and} \qquad \cos 6\theta = \cos^2 3\theta - \sin^2 3\theta$$

By using double-angle formulas together with the sum formulas derived in the preceding section, you can form other multiple-angle formulas.

EXAMPLE 4   *Deriving a Triple-Angle Formula*

Express $\sin 3x$ in terms of $\sin x$.

Solution

$$\begin{aligned}
\sin 3x &= \sin(2x + x) \\
&= \sin 2x \cos x + \cos 2x \sin x \\
&= 2 \sin x \cos x \cos x + (1 - 2 \sin^2 x)\sin x \\
&= 2 \sin x \cos^2 x + \sin x - 2 \sin^3 x \\
&= 2\sin x(1 - \sin^2 x) + \sin x - 2 \sin^3 x \\
&= 2 \sin x - 2 \sin^3 x + \sin x - 2 \sin^3 x \\
&= 3\sin x - 4 \sin^3 x
\end{aligned}$$

Power-Reducing Formulas

The double-angle formulas can be used to obtain the following **power-reducing formulas.**

POWER-REDUCING FORMULAS

$$\sin^2 u = \frac{1 - \cos 2u}{2} \qquad \cos^2 u = \frac{1 + \cos 2u}{2} \qquad \tan^2 u = \frac{1 - \cos 2u}{1 + \cos 2u}$$

PROOF ▪▪

The first two formulas can be verified by solving for $\sin^2 u$ and $\cos^2 u$, respectively, in the double-angle formulas

$$\cos 2u = 1 - 2\sin^2 u \qquad \text{and} \qquad \cos 2u = 2\cos^2 u - 1.$$

The third formula can be verified using the fact that

$$\tan^2 u = \frac{\sin^2 u}{\cos^2 u}. \qquad ▪▪$$

Example 5 shows a typical power reduction that is used in calculus.

EXAMPLE 5 *Reducing the Power of a Trigonometric Function*

Rewrite $\sin^4 x$ as a sum of first powers of the cosines of multiple angles.

Solution

Note the repeated use of power-reducing formulas.

$$\sin^4 x = (\sin^2 x)^2 = \left(\frac{1 - \cos 2x}{2}\right)^2$$

$$= \frac{1}{4}(1 - 2\cos 2x + \cos^2 2x)$$

$$= \frac{1}{4}\left(1 - 2\cos 2x + \frac{1 + \cos 4x}{2}\right)$$

$$= \frac{1}{4} - \frac{1}{2}\cos 2x + \frac{1}{8} + \frac{1}{8}\cos 4x$$

$$= \frac{3}{8} - \frac{1}{2}\cos 2x + \frac{1}{8}\cos 4x$$

$$= \frac{1}{8}(3 - 4\cos 2x + \cos 4x)$$

Half-Angle Formulas

You can derive some useful alternative forms of the power-reducing formulas by replacing u with $u/2$. The results are called **half-angle formulas.**

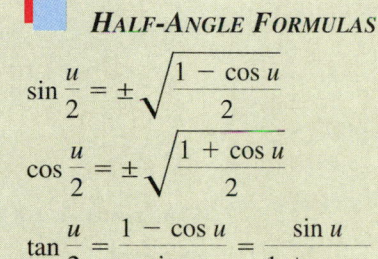

HALF-ANGLE FORMULAS

$$\sin \frac{u}{2} = \pm \sqrt{\frac{1 - \cos u}{2}}$$

$$\cos \frac{u}{2} = \pm \sqrt{\frac{1 + \cos u}{2}}$$

$$\tan \frac{u}{2} = \frac{1 - \cos u}{\sin u} = \frac{\sin u}{1 + \cos u}$$

The signs of $\sin(u/2)$ and $\cos(u/2)$ depend on the quadrant in which $u/2$ lies.

EXAMPLE 6   *Using a Half-Angle Formula*

Find the exact value of $\sin 105°$.

Solution

Begin by noting that $105°$ is half of $210°$. Then, using the half-angle formula for $\sin(u/2)$ and the fact that $105°$ lies in Quadrant II, you have

$$\sin 105° = \sqrt{\frac{1 - \cos 210°}{2}}$$

$$= \sqrt{\frac{1 - (-\cos 30°)}{2}}$$

$$= \sqrt{\frac{1 + \left(\sqrt{3}/2\right)}{2}}$$

$$= \frac{\sqrt{2 + \sqrt{3}}}{2}.$$

The positive square root is chosen because $\sin \theta$ is positive in Quadrant II.

NOTE   Use your calculator to verify the result obtained in Example 6. That is, evaluate $\sin 105°$ and $\left(\sqrt{2 + \sqrt{3}}\right)/2$ and you will see that both values are approximately 0.9659258. ■■

EXAMPLE 7   *Solving a Trigonometric Equation*

Find all solutions of $2 - \sin^2 x = 2 \cos^2 \dfrac{x}{2}$ in the interval $[0, 2\pi]$.

Solution

$$2 - \sin^2 x = 2 \cos^2 \frac{x}{2} \qquad \text{Original equation}$$

$$2 - \sin^2 x = 2\left(\frac{1 + \cos x}{2}\right) \qquad \text{Half-angle formula}$$

$$2 - \sin^2 x = 1 + \cos x \qquad \text{Simplify.}$$

$$2 - (1 - \cos^2 x) = 1 + \cos x \qquad \text{Pythagorean identity}$$

$$\cos^2 x - \cos x = 0 \qquad \text{Simplify.}$$

$$\cos x (\cos x - 1) = 0 \qquad \text{Factor.}$$

By setting the factors $\cos x$ and $(\cos x - 1)$ equal to zero, you find that the solutions in the interval $[0, 2\pi]$ are $x = \pi/2$, $x = 3\pi/2$, and $x = 0$.

Product-to-Sum Formulas

Each of the following **product-to-sum formulas** is easily verified using the sum and difference formulas discussed in the preceding section.

> **PRODUCT-TO-SUM FORMULAS**
>
> $$\sin u \sin v = \frac{1}{2}\left[\cos(u - v) - \cos(u + v)\right]$$
>
> $$\cos u \cos v = \frac{1}{2}\left[\cos(u - v) + \cos(u + v)\right]$$
>
> $$\sin u \cos v = \frac{1}{2}\left[\sin(u + v) + \sin(u - v)\right]$$
>
> $$\cos u \sin v = \frac{1}{2}\left[\sin(u + v) - \sin(u - v)\right]$$

EXAMPLE 8   *Writing Products as Sums*

Rewrite $\cos 5x \sin 4x$ as a sum or difference.

Solution

$$\cos 5x \sin 4x = \tfrac{1}{2}\left[\sin(5x + 4x) - \sin(5x - 4x)\right] = \tfrac{1}{2}\sin 9x - \tfrac{1}{2}\sin x$$

Occasionally, it is useful to reverse the procedure and write a sum of trigonometric functions as a product. This can be accomplished with the following **sum-to-product formulas.**

SUM-TO-PRODUCT FORMULAS

$$\sin x + \sin y = 2 \sin\left(\frac{x + y}{2}\right) \cos\left(\frac{x - y}{2}\right)$$

$$\sin x - \sin y = 2 \cos\left(\frac{x + y}{2}\right) \sin\left(\frac{x - y}{2}\right)$$

$$\cos x + \cos y = 2 \cos\left(\frac{x + y}{2}\right) \cos\left(\frac{x - y}{2}\right)$$

$$\cos x - \cos y = -2 \sin\left(\frac{x + y}{2}\right) \sin\left(\frac{x - y}{2}\right)$$

PROOF ∎

To prove the first formula, let $x = u + v$ and $y = u - v$. Then substitute $u = (x + y)/2$ and $v = (x - y)/2$ in the product-to-sum formula.

$$\sin u \cos v = \frac{1}{2}[\sin(u + v) + \sin(u - v)]$$

$$\sin\left(\frac{x + y}{2}\right) \cos\left(\frac{x - y}{2}\right) = \frac{1}{2}(\sin x + \sin y)$$

$$2 \sin\left(\frac{x + y}{2}\right) \cos\left(\frac{x - y}{2}\right) = \sin x + \sin y \quad ∎$$

EXAMPLE 9   Using a Sum-to-Product Formula

Find the exact value of $\cos 195° + \cos 105°$.

Solution

Using the appropriate sum-to-product formula, you obtain

$$\cos 195° + \cos 105° = 2 \cos\left(\frac{195° + 105°}{2}\right) \cos\left(\frac{195° - 105°}{2}\right)$$

$$= 2 \cos 150° \cos 45°$$

$$= 2\left(-\frac{\sqrt{3}}{2}\right)\left(\frac{\sqrt{2}}{2}\right)$$

$$= -\frac{\sqrt{6}}{2}.$$

EXAMPLE 10 *Solving a Trigonometric Equation*

Find all solutions of $\sin 5x + \sin 3x = 0$.

Solution

$$\sin 5x + \sin 3x = 0 \qquad \text{Original equation}$$

$$2 \sin\left(\frac{5x + 3x}{2}\right) \cos\left(\frac{5x - 3x}{2}\right) = 0 \qquad \text{Sum-to-product formula}$$

$$2 \sin 4x \cos x = 0 \qquad \text{Simplify.}$$

By setting the factor $\sin 4x$ equal to zero, you can find that the solutions in the interval $[0, 2\pi)$ are

$$x = 0, \frac{\pi}{4}, \frac{\pi}{2}, \frac{3\pi}{4}, \pi, \frac{5\pi}{4}, \frac{3\pi}{2}, \frac{7\pi}{4}.$$

The equation $\cos x = 0$ yields no additional solutions, and you can conclude that the solutions are of the form

$$x = \frac{n\pi}{4}$$

where n is an integer. These solutions are verified graphically in Figure 7.12.

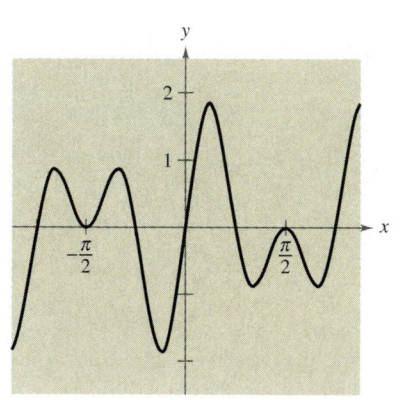

FIGURE 7.12

EXAMPLE 11 *Verifying a Trigonometric Identity*

Verify the identity

$$\frac{\sin t + \sin 3t}{\cos t + \cos 3t} = \tan 2t.$$

Solution

Using appropriate sum-to-product formulas, you have

$$\frac{\sin t + \sin 3t}{\cos t + \cos 3t} = \frac{2 \sin 2t \cos(-t)}{2 \cos 2t \cos(-t)} = \frac{\sin 2t}{\cos 2t} = \tan 2t.$$

GROUP ACTIVITY

DERIVING AN AREA FORMULA

With others in your group, discuss how you can use a double-angle formula or a half-angle formula to derive a formula for the area of an isosceles triangle. Use a labeled sketch to illustrate your derivation. Then write two examples that show how your formula can be used.

Factor the trigonometric expression.

1. $2 \sin x + \sin x \cos x$

2. $\cos^2 x - \cos x - 2$

Find all solutions of the equation in the interval $[0, 2\pi)$.

3. $\sin 2x = 0$

4. $\cos 2x = 0$

5. $\cos \dfrac{x}{2} = 0$

6. $\sin \dfrac{x}{2} = 0$

Simplify the expression.

7. $\dfrac{1 - \cos(\pi/4)}{2}$

8. $\dfrac{1 + \cos(\pi/3)}{2}$

9. $\dfrac{2 \sin 3x \cos x}{2 \cos 3x \cos x}$

10. $\dfrac{(1 - 2 \sin^2 x) \cos x}{2 \sin^2 x \cos x}$

7.5 Exercises

In Exercises 1–8, use the figure to find the exact value of the trigonometric function.

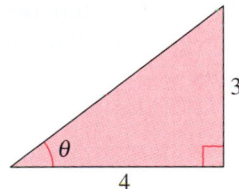

1. $\sin \theta$ **2.** $\tan \theta$

3. $\cos 2\theta$ **4.** $\sin 2\theta$

5. $\tan 2\theta$ **6.** $\sec 2\theta$

7. $\csc 2\theta$ **8.** $\cot 2\theta$

In Exercises 9–18, find the exact solutions of the equation in the interval $[0, 2\pi)$.

9. $\sin 2x - \sin x = 0$ **10.** $\sin 2x + \cos x = 0$

11. $4 \sin x \cos x = 1$ **12.** $\sin 2x \sin x = \cos x$

13. $\cos 2x - \cos x = 0$ **14.** $\cos 2x + \sin x = 0$

15. $\tan 2x - \cot x = 0$ **16.** $\tan 2x - 2 \cos x = 0$

17. $\sin 4x = -2 \sin 2x$ **18.** $(\sin 2x + \cos 2x)^2 = 1$

In Exercises 19–22, use a double-angle formula to rewrite the expression.

19. $6 \sin x \cos x$

20. $4 \sin x \cos x + 2$

21. $4 - 8 \sin^2 x$

22. $(\cos x + \sin x)(\cos x - \sin x)$

In Exercises 23–28, find the exact values of $\sin 2u$, $\cos 2u$, and $\tan 2u$ using the double-angle formulas.

23. $\sin u = \dfrac{3}{5}, \quad 0 < u < \dfrac{\pi}{2}$

24. $\cos u = -\dfrac{2}{3}, \quad \dfrac{\pi}{2} < u < \pi$

25. $\tan u = \dfrac{1}{2}, \quad \pi < u < \dfrac{3\pi}{2}$

26. $\cot u = -4, \quad \dfrac{3\pi}{2} < u < 2\pi$

27. $\sec u = -\dfrac{5}{2}, \quad \dfrac{\pi}{2} < u < \pi$

28. $\csc u = 3, \quad \dfrac{\pi}{2} < u < \pi$

In Exercises 29–34, use the power-reducing formulas to rewrite the expression in terms of the first power of the cosine.

29. $\cos^4 x$ **30.** $\sin^4 x$

31. $\sin^2 x \cos^2 x$ **32.** $\cos^2 x$

33. $\sin^2 x \cos^4 x$ **34.** $\sin^4 x \cos^2 x$

In Exercises 35–40, use the figure to find the exact value of the trigonometric function.

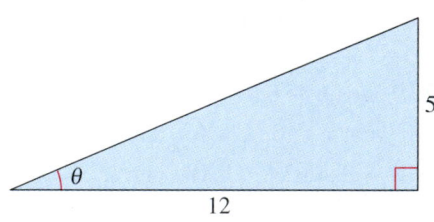

35. $\cos \dfrac{\theta}{2}$ **36.** $\sin \dfrac{\theta}{2}$

37. $\tan \dfrac{\theta}{2}$ **38.** $2 \sin \dfrac{\theta}{2} \cos \dfrac{\theta}{2}$

39. $\csc \dfrac{\theta}{2}$ **40.** $\cot \dfrac{\theta}{2}$

In Exercises 41–46, use the half-angle formulas to determine the exact values of the sine, cosine, and tangent of the angle.

41. $105°$ **42.** $165°$

43. $112° \; 30'$ **44.** $67° \; 30'$

45. $\dfrac{\pi}{8}$ **46.** $\dfrac{\pi}{12}$

In Exercises 47–52, find the exact values of $\sin(u/2)$, $\cos(u/2)$, and $\tan(u/2)$ using the half-angle formulas.

47. $\sin u = \dfrac{5}{13}, \quad \dfrac{\pi}{2} < u < \pi$

48. $\cos u = \dfrac{3}{5}, \quad 0 < u < \dfrac{\pi}{2}$

49. $\tan u = -\dfrac{5}{8}, \quad \dfrac{3\pi}{2} < u < 2\pi$

50. $\cot u = 3, \quad \pi < u < \dfrac{3\pi}{2}$

51. $\csc u = -\dfrac{5}{3}, \quad \pi < u < \dfrac{3\pi}{2}$

52. $\sec u = -\dfrac{7}{2}, \quad \dfrac{\pi}{2} < u < \pi$

In Exercises 53–56, use the half-angle formulas to simplify the expression.

53. $\sqrt{\dfrac{1 - \cos 6x}{2}}$ **54.** $\sqrt{\dfrac{1 + \cos 4x}{2}}$

55. $-\sqrt{\dfrac{1 - \cos 8x}{1 + \cos 8x}}$ **56.** $-\sqrt{\dfrac{1 - \cos(x - 1)}{2}}$

In Exercises 57–60, find the exact zeros of the function in the interval $[0, 2\pi)$. Use the graphing utility to graph the function and verify the zeros.

57. $f(x) = \sin \dfrac{x}{2} + \cos x$

58. $h(x) = \sin \dfrac{x}{2} + \cos x - 1$

59. $h(x) = \cos \dfrac{x}{2} - \sin x$ **60.** $g(x) = \tan \dfrac{x}{2} - \sin x$

In Exercises 61–70, use the product-to-sum formulas to write the product as a sum or difference.

61. $6 \sin \dfrac{\pi}{4} \cos \dfrac{\pi}{4}$ **62.** $4 \sin \dfrac{\pi}{3} \cos \dfrac{5\pi}{6}$

63. $\sin 5\theta \cos 3\theta$ **64.** $3 \sin 2\alpha \sin 3\alpha$

65. $5 \cos(-5\beta) \cos 3\beta$ **66.** $\cos 2\theta \cos 4\theta$

67. $\sin(x + y)\sin(x - y)$

68. $\sin(x + y)\cos(x - y)$

69. $\sin(\theta + \pi)\cos(\theta - \pi)$

70. $10\cos 75° \cos 15°$

In Exercises 71–80, use the sum-to-product formulas to write the sum or difference as a product.

71. $\sin 60° + \sin 30°$

72. $\cos 120° + \cos 30°$

73. $\cos \dfrac{3\pi}{4} - \cos \dfrac{\pi}{4}$

74. $\sin 5\theta - \sin 3\theta$

75. $\cos 6x + \cos 2x$

76. $\sin x + \sin 5x$

77. $\sin(\alpha + \beta) - \sin(\alpha - \beta)$

78. $\cos\left(\theta + \dfrac{\pi}{2}\right) - \cos\left(\theta - \dfrac{\pi}{2}\right)$

79. $\cos(\phi + 2\pi) + \cos \phi$

80. $\sin\left(x + \dfrac{\pi}{2}\right) + \sin\left(x - \dfrac{\pi}{2}\right)$

In Exercises 81–84, find the exact zeros of the function in the interval $[0, 2\pi)$. Use a graphing utility to graph the function and verify the zeros.

81. $g(x) = \sin 6x + \sin 2x$

82. $h(x) = \cos 2x - \cos 6x$

83. $f(x) = \dfrac{\cos 2x}{\sin 3x - \sin x} - 1$

84. $f(x) = \sin^2 3x - \sin^2 x$

In Exercises 85–88, use the figure to find the exact value of the trigonometric function in two ways.

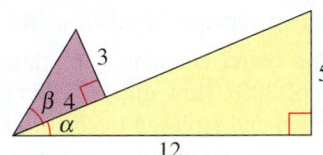

85. $\sin^2 \alpha$

86. $\cos^2 \alpha$

87. $\sin \alpha \cos \beta$

88. $\cos \alpha \sin \beta$

In Exercises 89–102, verify the identity.

89. $\csc 2\theta = \dfrac{\csc \theta}{2\cos \theta}$

90. $\sec 2\theta = \dfrac{\sec^2 \theta}{2 - \sec^2 \theta}$

91. $\cos^2 2\alpha - \sin^2 2\alpha = \cos 4\alpha$

92. $\cos^4 x - \sin^4 x = \cos 2x$

93. $(\sin x + \cos x)^2 = 1 + \sin 2x$

94. $\sin \dfrac{\alpha}{3} \cos \dfrac{\alpha}{3} = \dfrac{1}{2}\sin \dfrac{2\alpha}{3}$

95. $1 + \cos 10y = 2\cos^2 5y$

96. $\dfrac{\cos 3\beta}{\cos \beta} = 1 - 4\sin^2 \beta$

97. $\sec \dfrac{u}{2} = \pm\sqrt{\dfrac{2\tan u}{\tan u + \sin u}}$

98. $\tan \dfrac{u}{2} = \csc u - \cot u$

99. $\dfrac{\cos 4x + \cos 2x}{\sin 4x + \sin 2x} = \cot 3x$

100. $\dfrac{\sin x \pm \sin y}{\sin x + \cos y} = \tan \dfrac{x \pm y}{2}$

101. $\dfrac{\cos t + \cos 3t}{\sin 3t - \sin t} = \cot t$

102. $\sin\left(\dfrac{\pi}{6} + x\right) + \sin\left(\dfrac{\pi}{6} - x\right) = \cos x$

In Exercises 103–106, verify the identity algebraically. Use a graphing utility to confirm the identity.

103. $\cos 3\beta = \cos^3 \beta - 3\sin^2 \beta \cos \beta$

104. $\sin 4\beta = 4\sin \beta \cos \beta(1 - \sin^2 \beta)$

105. $(\cos 4x - \cos 2x)/(2\sin 3x) = -\sin x$

106. $(\cos 3x - \cos x)/(\sin 3x - \sin x) = -\tan x$

In Exercises 107 and 108, graph the function by using the power-reducing formulas.

107. $f(x) = \sin^2 x$

108. $f(x) = \cos^2 x$

In Exercises 109 and 110, (a) use a graphing utility to graph the function and approximate the maximum and minimum points on the graph in the interval $[0, 2\pi)$. (b) Solve the trigonometric equation and verify that its solutions are the x-coordinates of the maximum and minimum points of f. (Calculus is required to find the trigonometric equation.)

Function	Trigonometric Equation

109. $f(x) = 4 \sin \dfrac{x}{2} + \cos x$ $2 \cos \dfrac{x}{2} - \sin x = 0$

110. $f(x) = \cos 2x - 2 \sin x$ $-2 \cos x(2 \sin x + 1) = 0$

111. *Conjecture* Consider the function

$$f(x) = 2 \sin x[2 \cos^2(x/2) - 1].$$

(a) Use a graphing utility to graph the function.

(b) Make a conjecture about the function that is an identity with f.

(c) Verify your conjecture analytically.

112. *Exploration* Consider the function

$$f(x) = \sin^4 x + \cos^4 x.$$

(a) Use the power-reducing formulas to write the function in terms of cosine to the first power.

(b) Determine another way of rewriting the function. Use a graphing utility to rule out incorrectly rewritten functions.

(c) Add a trigonometric term to the function so that it becomes a perfect square trinomial. Rewrite the function as a perfect square trinomial minus the term that you added. Use a graphing utility to rule out incorrectly rewritten functions.

(d) Rewrite the result of part (c) in terms of the sine of a double angle. Use a graphing utility to rule out incorrectly rewritten functions.

(e) When you rewrite a trigonometric expression, the result may not be the same as a friend's. Does this mean that one of you is wrong? Explain.

In Exercises 113 and 114, write the trigonometric expression as an algebraic expression.

113. $\sin(2 \arcsin x)$ **114.** $\cos(2 \arccos x)$

115. *Area* The length of each of the two equal sides of an isosceles triangle is 10 meters (see figure). The angle between the two sides is θ.

(a) Express the area of the triangle as a function of $\theta/2$.

(b) Express the area as a function of θ. Determine the value of θ so that the area is a maximum.

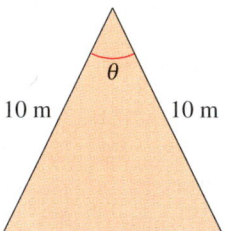

116. *Projectile Motion* The range of a projectile fired at an angle θ with the horizontal and with an initial velocity of v_0 feet per second is given by

$$f = \frac{1}{32} v_0 \sin 2\theta$$

where r is measured in feet. Determine the expression for the range in terms of θ.

Review **Solve Exercises 117–120 as a review of the skills and problem-solving techniques you learned in previous sections.**

117. The total profit for a company in October was 16% higher than it was in September. The total profit for the two months was $507,600. Find the profit for each month.

118. Two cars start at a given point and travel in the same direction at average speeds of 48 miles per hour and 56 miles per hour. How much time must elapse before the two cars are 12 miles apart?

119. A 55-gallon barrel contains a mixture with a concentration of 30%. How much of this mixture must be withdrawn and replaced by 100% concentrate to bring the mixture up to 50% concentration?

120. *Baseball Diamond* A baseball diamond has the shape of a square where the distance between each of the consecutive bases is 90 feet. Approximate the distance from home plate to second base.

FOCUS ON CONCEPTS

In this chapter, you studied the fundamental identities of trigonometry. Use the following questions to check your understanding of several of these basic concepts presented. The answers to these questions are given in the back of the book.

1. In your own words, describe the difference between an identity and a conditional equation.

2. Describe the difference between verifying an identity and solving an equation.

3. List the reciprocal identities, quotient identities, and Pythagorean identities from memory.

4. Is $\cos \theta = \sqrt{1 - \sin^2 \theta}$ an identity? Explain.

5. *True or False?*   Usually there is only one correct set of steps to verify an identity. Explain.

6. By observation, determine which of the following is an identity. Explain.

(a) $\tan(\theta + \pi) \overset{?}{=} \tan \theta$

(b) $\cos(\theta + \pi) \overset{?}{=} \cos \theta$

(c) $\sec \theta \csc \theta \overset{?}{=} 1$

(d) $\tan \theta \cot \theta \overset{?}{=} 1$

(e) $\sin(\theta - \pi) \overset{?}{=} -\sin(\pi - \theta)$

In Exercises 7 and 8, use the graph of y_1 and y_2 to determine how to change one function to form the identity $y_1 = y_2$.

7. $y_1 = \sec^2\left(\dfrac{\pi}{2} - x\right)$

$y_2 = \cot^2 x$

8. $y_1 = \dfrac{\cos 3x}{\cos x}$

$y_2 = (2 \sin x)^2$

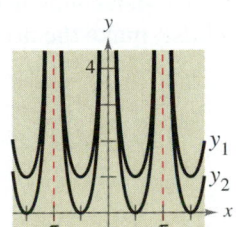

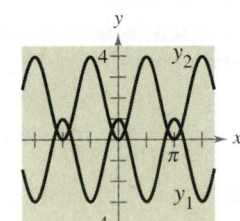

In Exercises 9 and 10, use the graph to determine the number of points of intersection of the graphs of y_1 and y_2.

9. $y_1 = 2 \sin x$

$y_2 = 3x + 1$

10. $y_1 = 2 \sin x$

$y_2 = \frac{1}{2}x + 1$

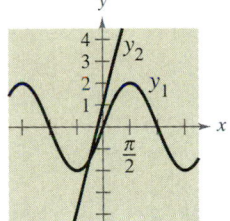

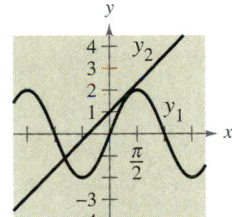

In Exercises 11 and 12, use the graph to determine the number of zeros of the function.

11. $y = \sqrt{x + 3} + 4 \cos x$

12. $y = 2 - \dfrac{1}{2}x^2 + 3 \sin \dfrac{\pi x}{2}$

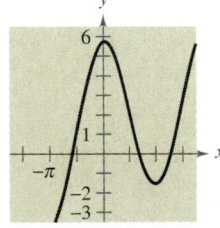

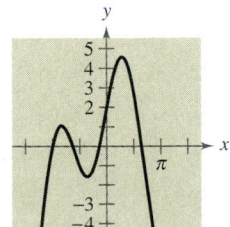

13. Sales of a product are seasonal and can be modeled by the function $y = a + bt + c \sin(dt + e)$, where t is the time in years. What is the value of d ?

Review Exercises

In Exercises 1–8, simplify the trigonometric expression.

1. $\dfrac{1}{\cot^2 x + 1}$

2. $\dfrac{\sin 2\alpha}{\cos^2 \alpha - \sin^2 \alpha}$

3. $\dfrac{\sin^2 \alpha - \cos^2 \alpha}{\sin^2 \alpha - \sin \alpha \cos \alpha}$

4. $\dfrac{\sin^3 \beta + \cos^3 \beta}{\sin \beta + \cos \beta}$

5. $\tan^2 \theta(\csc^2 \theta - 1)$

6. $1 - 4 \sin^2 x \cos^2 x$

7. $\dfrac{2 \tan(x + 1)}{1 - \tan^2(x + 1)}$

8. $\sqrt{\dfrac{1 - \cos^2 x}{1 + \cos x}}$

In Exercises 9–26, verify the identity.

9. $\tan x(1 - \sin^2 x) = \frac{1}{2} \sin 2x$

10. $\cos x(\tan^2 x + 1) = \sec x$

11. $\sec^2 x \cot x - \cot x = \tan x$

12. $\sin^3 \theta + \sin \theta \cos^2 \theta = \sin \theta$

13. $\sin^5 x \cos^2 x = (\cos^2 x - 2 \cos^4 x + \cos^6 x)\sin x$

14. $\cos^3 x \sin^2 x = (\sin^2 x - \sin^4 x)\cos x$

15. $\sin 3\theta \sin \theta = \frac{1}{2}(\cos 2\theta - \cos 4\theta)$

16. $\sin 3x \cos 2x = \frac{1}{2}(\sin 5x + \sin x)$

17. $\sqrt{\dfrac{1 - \sin \theta}{1 + \sin \theta}} = \dfrac{1 - \sin\theta}{|\cos\theta|}$

18. $\sqrt{1 - \cos x} = \dfrac{|\sin x|}{\sqrt{1 + \cos x}}$

19. $\cos 3x = 4 \cos^3 x - 3 \cos x$

20. $\cos(x + \pi/2) = -\sin x$

21. $\cot(\pi/2 - x) = \tan x$

22. $\sin(\pi - x) = \sin x$

23. $\dfrac{\sec x - 1}{\tan x} = \tan \dfrac{x}{2}$

24. $\dfrac{2 \cos 3x}{\sin 4x - \sin 2x} = \csc x$

25. $2 \sin y \cos y \sec 2y = \tan 2y$

26. $\dfrac{\sin(\alpha + \beta)}{\cos \alpha \cos \beta} = \tan \alpha + \tan \beta$

In Exercises 27–30, verify the identity algebraically and use a graphing utility to confirm it graphically.

27. $\sin\left(x - \dfrac{3\pi}{2}\right) = \cos x$

28. $\sin 4x = 8 \cos^3 x \sin x - 4 \cos x \sin x$

29. $\tan^2 x = \dfrac{1 - \cos 2x}{1 + \cos 2x}$

30. $\cos^2 5x - \cos^2 x = -\sin 4x \sin 6x$

In Exercises 31–34, find the exact value of the trigonometric function by using the sum, difference, or half-angle formulas.

31. $\sin \dfrac{5\pi}{12} = \sin\left(\dfrac{2\pi}{3} - \dfrac{\pi}{4}\right)$

32. $\cos 285° = \cos(225° + 60°)$

33. $\cos(157° \, 30') = \cos \dfrac{315°}{2}$

34. $\sin \dfrac{3\pi}{8} = \sin\left[\dfrac{1}{2}\left(\dfrac{3\pi}{4}\right)\right]$

In Exercises 35–40, find the exact value of the trigonometric function given that $\sin u = \frac{3}{4}$, $\cos v = -\frac{5}{13}$, and u and v are in Quadrant II.

35. $\sin(u + v)$

36. $\tan(u + v)$

37. $\cos(u - v)$

38. $\sin 2v$

39. $\cos \dfrac{u}{2}$

40. $\tan 2v$

True or False? **In Exercises 41–44, determine if the statement is true or false. If it is false, make the necessary correction.**

41. If $\dfrac{\pi}{2} < \theta < \pi$, then $\cos \dfrac{\theta}{2} < 0$.

42. $\sin(x + y) = \sin x + \sin y$

43. $4 \sin(-x) \cos(-x) = -2 \sin 2x$

44. $4 \sin 45° \cos 15° = 1 + \sqrt{3}$

In Exercises 45–50, find all solutions of the equation in the interval $[0, 2\pi)$.

45. $\sin x - \tan x = 0$ **46.** $\csc x - 2 \cot x = 0$

47. $\sin 2x + \sqrt{2} \sin x = 0$ **48.** $\cos 4x - 7 \cos 2x = 8$

49. $\cos^2 x + \sin x = 1$ **50.** $\sin 4x - \sin 2x = 0$

In Exercises 51–54, use a graphing utility to graph the function and approximate its zeros in the interval $[0, 2\pi)$. If possible, find the exact values of the zeros algebraically.

51. $y = \dfrac{1 + \sin x}{\cos x} + \dfrac{\cos x}{1 + \sin x} - 4$

52. $y = \cos x - \cos \dfrac{x}{2}$

53. $y = \tan^3 x - \tan^2 x + 3 \tan x - 3$

54. $h(s) = \sin s + \sin 3s + \sin 5s$

55. *Think About It*   If a trigonometric equation has an infinite number of solutions, is it true that the equation is an identity? Explain.

56. *Think About It*   Explain why you know from observation that the equation $a \sin x - b = 0$ has no solution if $|a| < |b|$.

In Exercises 57 and 58, write the trigonometric expression as a product.

57. $\cos 3\theta + \cos 2\theta$

58. $\sin(x + \pi/4) - \sin(x - \pi/4)$

In Exercises 59 and 60, write the trigonometric expression as a sum or difference.

59. $\sin 3\alpha \sin 2\alpha$ **60.** $\cos \dfrac{x}{2} \cos \dfrac{x}{4}$

In Exercises 61 and 62, write the trigonometric expression as an algebraic expression.

61. $\cos(2 \arccos 2x)$ **62.** $\sin(2 \arctan x)$

63. *Rate of Change*   The rate of change of the function $f(x) = 2\sqrt{\sin x}$ is given by the expression $\sin^{-1/2} x \cos x$. Show that this expression can also be written as $\cot x \sqrt{\sin x}$.

64. *Projectile Motion*   A baseball leaves the hand of the person at first base at an angle of θ with the horizontal and an initial velocity of $v_0 = 80$ feet per second. The ball is caught by the person at second base 100 feet away. Find θ if the range r of a projectile is given by

$$r = \frac{1}{32} v_0^2 \sin 2\theta.$$

65. *Harmonic Motion*   A weight is attached to a spring suspended vertically from a ceiling. When a driving force is applied to the system, the weight moves vertically from its equilibrium position, and this motion is described by the model

$$y = 1.5 \sin 8t - 0.5 \cos 8t$$

where y is the distance from equilibrium measured in feet and t is the time in seconds.

(a) Write the model in the form

$$y = \sqrt{a^2 + b^2} \sin(Bt + C).$$

(b) Find the amplitude of the oscillations of the weight.

(c) Find the frequency of the oscillations of the weight.

66. *Volume*   A trough for feeding cattle is 4 meters long and its cross sections are isosceles triangles with the two equal sides being $\frac{1}{2}$ meter (see figure). The angle between the two sides is θ.

(a) Express the trough's volume as a function of $\theta/2$.

(b) Express the volume of the trough as a function of θ and determine the value of θ so that the volume is maximum.

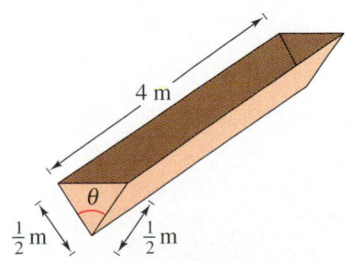

CHAPTER PROJECT: *Solving Equations Graphically*

Equations that involve both algebraic and trigonometric expressions are often difficult to solve analytically. For instance, how would you solve the equation

$$x = \cos x?$$

None of the standard techniques, such as factoring or using a trigonometric identity, can be used to solve this equation. In such cases, a graphing utility can be used to approximate the solution.

EXAMPLE 1   *Approximating Solutions of an Equation*

Approximate all solutions of $x = \cos x$.

Solution

There are several ways that a graphing utility can be used to solve this equation. One way is to graph

$$y = x \quad \text{and} \quad y = \cos x \qquad \text{First method}$$

on the same viewing rectangle and use the trace feature of the graphing utility to approximate the x-coordinate of the point of intersection. From the graph shown below on the left, you can see that the two graphs intersect when x is approximately 0.74. More accuracy can be obtained by using the zoom feature of the graphing utility. Another way to solve the equation is to collect all nonzero terms on one side of the equation, $x - \cos x = 0$. Then, use a graphing utility to graph

$$y = x - \cos x \qquad \text{Second method}$$

and approximate the zeros of the function. By using a viewing rectangle in which $0.735 \leq x \leq 0.745$ and $-0.01 \leq y \leq 0.01$, you can obtain the graph shown on the right. From this graph, you can see that the solution $x \approx 0.739$, which is accurate to three decimal places.

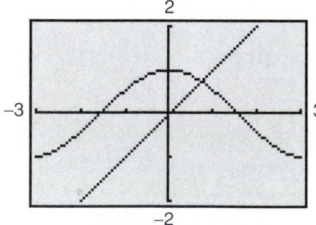
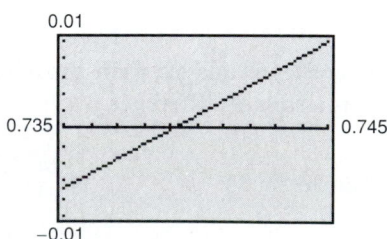

EXAMPLE 2   *Approximating Solutions of an Equation*

Approximate all solutions of $0.5x + 1.13 = \cos x$.

Solution

When the graphs of $y = 0.5x + 1.13$ and $y = \cos x$ are sketched on the same viewing rectangle, it is clear that one solution occurs when $x \approx -3.8$. From the screen shown below on the left, however, it is unclear whether the two graphs have other points of intersection near $x = -0.5$. To determine whether there are other points of intersection, use the zoom feature. After doing this, you can see that, for x-values near -0.5, the line lies above the cosine curve. Thus, there is only one point of intersection, which occurs when $x \approx -3.819$.

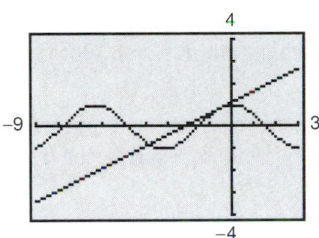

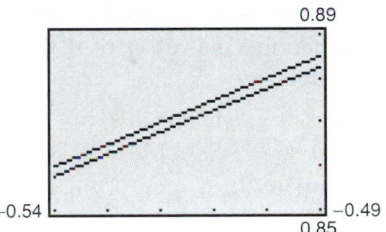

CHAPTER PROJECT INVESTIGATIONS

In Questions 1–4, sketch the graphs of all three functions on the same coordinate plane. Which two graphs are the same? Identify the trigonometric identity that is supported by your discovery.

1. (a) $y = \sin^2 x$

    (b) $y = \frac{1}{2}(1 - \cos 2x)$

    (c) $y = \frac{1}{2}(1 + \cos 2x)$

2. (a) $y = 2\cos^2 x$

    (b) $y = 1 - \cos 2x$

    (c) $y = 1 + \cos 2x$

3. (a) $y = 2\sin x \cos 2x$

    (b) $y = \sin 3x + \sin x$

    (c) $y = \sin 3x - \sin x$

4. (a) $y = \sin 2x$

    (b) $y = 2\sin x$

    (c) $y = 2\sin x \cos x$

In Exercises 5 and 6, without finding the solutions, decide whether the equation has infinitely many solutions or a finite number of solutions. Explain.

5. $\sin 2x = x$

6. $\tan 2x = x$

In Questions 7–12, use a graphing utility to approximate all solutions of the equation. List the solutions correct to three decimal places.

7. $x + \sin x = 1$

8. $x^2 + \cos x = 2$

9. $5 \cos \dfrac{x}{x^2 + 1} = 3$

10. $|x| + \sec \dfrac{1}{x^2 + 1} = 3$

11. $x + 1.25 = \arctan x$

12. $(x^2 + 1)\cos x = 1$

13. *Seasonal Sales*   During 1994, the national monthly sales S (in thousands of units) of a lawn furniture company can be modeled by

$$S = 74.50 + 43.75 \sin \frac{\pi t}{6}$$

where t represents the time in months, with $t = 0$ corresponding to January 1. During which months did sales exceed 100,000 units?

Chapter Test

Take this test as you would take a test in class. After you are done, check your work against the answers given in the back of the book.

The *Interactive* CD-ROM provides answers to the Chapter Tests and Cumulative Tests. It also offers Chapter Pre-Tests (which test key skills and concepts covered in previous chapters) and Chapter Post-Tests, both of which have randomly generated exercises with diagnostic capabilities.

1. If $\tan \theta = \frac{3}{2}$ and $\cos \theta < 0$, use the fundamental identities to evaluate the other five trigonometric functions of θ.

2. Use the fundamental identities to simplify $\csc^2 \beta (1 - \cos^2 \beta)$.

3. Factor and simplify $\dfrac{\sec^4 x - \tan^4 x}{\sec^2 x + \tan^2 x}$.

4. Add and simplify $\dfrac{\cos \theta}{\sin \theta} + \dfrac{\sin \theta}{\cos \theta}$.

5. Determine the values of θ, $0 \le \theta < 2\pi$, for which $\tan \theta = -\sqrt{\sec^2 \theta - 1}$ is true.

6. Use a graphing utility to graph the functions $y_1 = \cos x + \sin x \tan x$ and $y_2 = \sec x$. Make a conjecture about y_1 and y_2. Verify the result analytically.

In Exercises 7–12, verify the identity.

7. $\sin \theta \sec \theta = \tan \theta$

8. $\sec^2 x \tan^2 x + \sec^2 x = \sec^4 x$

9. $\dfrac{\csc \alpha + \sec \alpha}{\sin \alpha + \cos \alpha} = \cot \alpha + \tan \alpha$

10. $\cos\left(x + \dfrac{\pi}{2}\right) = -\sin x$

11. $\sin(nx + \theta) = (-1)^2 \sin \theta$, n is an integer.

12. $(\sin x + \cos x)^2 = 1 + \sin 2x$

In Exercises 13–16, find all solutions of the equation in the interval $[0, 2\pi)$.

13. $\tan^2 x + \tan x = 0$

14. $\sin 2\alpha - \cos \alpha = 0$

15. $4 \cos^2 x - 3 = 0$

16. $\csc^2 x - \csc x - 2 = 0$

17. Use a graphing utility to approximate the solutions of the equation $3 \cos x - x = 0$ accurate to three decimal places.

18. Explain why the equation $\cos^2 x + \cos x - 6 = 0$ has no solution.

19. Find the exact value of $\cos 105°$ using the fact that $105° = 135° - 30°$.

20. Use the figure to find the exact values of $\sin 2u$ and $\tan 2u$.

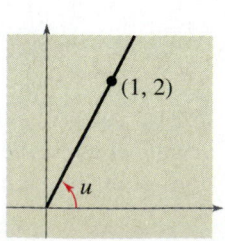

FIGURE FOR 20

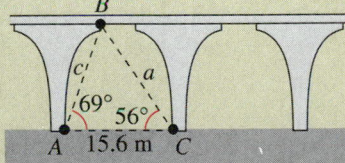

Photos: Courtesy of Keystone Aerial Survey; BET Consultants (inset)

Julio Esquivel is a civil engineer who specializes in surveying. He often uses aerial photographs such as this one to create topographic maps.

8.1 Law of Sines

See Exercise 33 on page 610 for an example of how the Law of Sines can be used to help locate a forest fire.

Introduction □ *The Ambiguous Case (SSA)* □ *Area of an Oblique Triangle* □ *Application*

Introduction

In Chapter 6 you looked at techniques for solving right triangles. In this section and the next, you will solve **oblique triangles**—triangles that have no right angles. As standard notation, the angles of a triangle are labeled as *A, B,* and *C,* and their opposite sides as *a, b,* and *c,* as shown in Figure 8.1.

To solve an oblique triangle, you need to know the measure of at least one side and any two other parts of the triangle—either two sides, two angles, or one angle and one side. This breaks down into the following four cases.

1. Two angles and any side (AAS or ASA)
2. Two sides and an angle opposite one of them (SSA)
3. Three sides (SSS)
4. Two sides and their included angle (SAS)

The first two cases can be solved using the **Law of Sines,** whereas the last two cases require the **Law of Cosines** (Section 8.2).

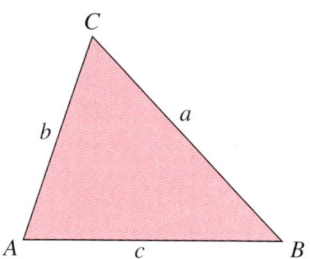

FIGURE 8.1

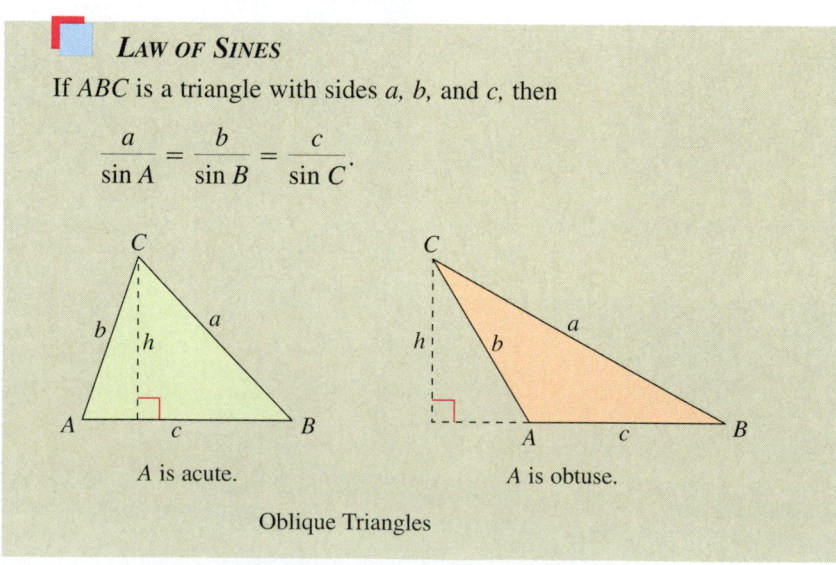

> ◼ **LAW OF SINES**
>
> If *ABC* is a triangle with sides *a, b,* and *c,* then
>
> $$\frac{a}{\sin A} = \frac{b}{\sin B} = \frac{c}{\sin C}.$$
>
> *A* is acute.   *A* is obtuse.
>
> Oblique Triangles

THINK ABOUT THE PROOF

To prove the Law of Sines, let *h* be the altitude of either triangle shown in the figure at the right. Then you have

$$\sin A = \frac{h}{b} \text{ or } h = b \sin A$$

$$\sin B = \frac{h}{a} \text{ or } h = a \sin B.$$

By equating the two values of *h,* you can establish part of the Law of Sines. Can you see how to establish the other part? The details of the proof are given in the appendix.

NOTE The Law of Sines can also be written in the reciprocal form

$$\frac{\sin A}{a} = \frac{\sin B}{b} = \frac{\sin C}{c}. \quad ▪▪$$

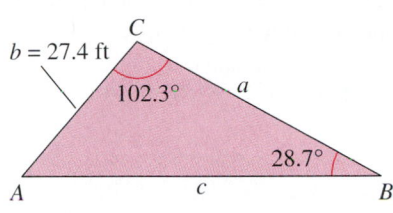

$b = 27.4$ ft

C

$102.3°$

a

$28.7°$

c

A B

FIGURE 8.2

Study Tip

When solving triangles, a careful sketch is useful as a quick test for the feasibility of an answer. Remember that the longest side lies opposite the largest angle, and the shortest side lies opposite the smallest angle.

EXAMPLE 1 *Given Two Angles and One Side—AAS*

For the triangle in Figure 8.2, $C = 102.3°$, $B = 28.7°$, and $b = 27.4$ feet. Find the remaining angle and sides.

Solution

The third angle of the triangle is

$$A = 180° - B - C = 180° - 28.7° - 102.3° = 49.0°.$$

By the Law of Sines, you have

$$\frac{a}{\sin 49°} = \frac{b}{\sin 28.7°} = \frac{c}{\sin 102.3°}.$$

Using $b = 27.4$ produces

$$a = \frac{27.4}{\sin 28.7°}(\sin 49°) \approx 43.06 \text{ feet}$$

and

$$c = \frac{27.4}{\sin 28.7°}(\sin 102.3°) \approx 55.75 \text{ feet}.$$

EXAMPLE 2 *Given Two Angles and One Side—ASA*

Real Life

A pole tilts *toward* the sun at an 8° angle from the vertical, and it casts a 22-foot shadow. The angle of elevation from the tip of the shadow to the top of the pole is 43°. How tall is the pole?

Solution

From Figure 8.3, note that $A = 43°$ and $B = 90° + 8° = 98°$. Thus, the third angle is

$$C = 180° - A - B = 180° - 43° - 98° = 39°.$$

By the Law of Sines, you have

$$\frac{a}{\sin 43°} = \frac{c}{\sin 39°}.$$

Because $c = 22$ feet, the length of the pole is

$$a = \frac{22}{\sin 39°}(\sin 43°) \approx 23.84 \text{ feet}.$$

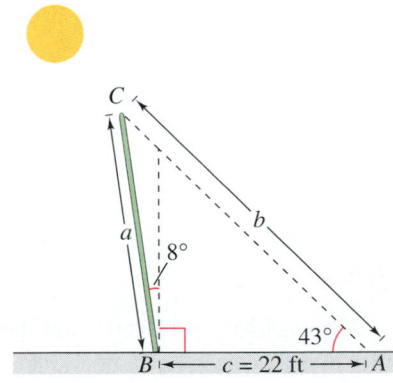

C

a

$8°$

b

$43°$

B ←— $c = 22$ ft —→ A

FIGURE 8.3

NOTE For practice, try reworking Example 2 for a pole that tilts *away from* the sun under the same conditions. ■■

The Ambiguous Case (SSA)

In Examples 1 and 2 you saw that two angles and one side determine a unique triangle. However, if two sides and one opposite angle are given, three possible situations can occur: (1) no such triangle exists, (2) one such triangle exists, or (3) two distinct triangles may satisfy the conditions.

THE AMBIGUOUS CASE (SSA)

Consider a triangle in which you are given a, b, and A. ($h = b \sin A$)

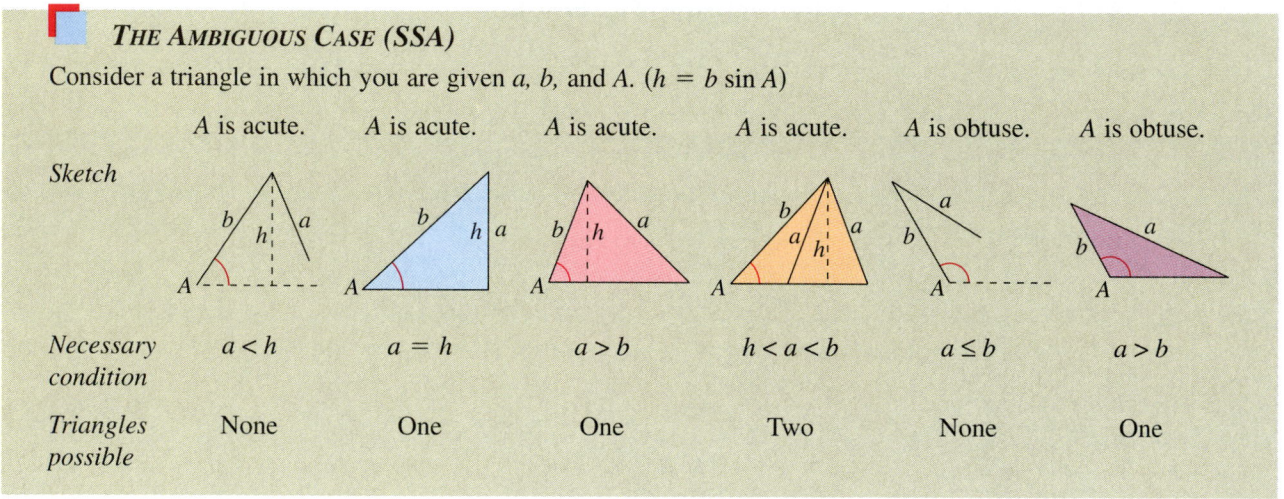

	A is acute.	A is acute.	A is acute.	A is acute.	A is obtuse.	A is obtuse.
Necessary condition	$a < h$	$a = h$	$a > b$	$h < a < b$	$a \le b$	$a > b$
Triangles possible	None	One	One	Two	None	One

EXAMPLE 3 *Single-Solution Case—SSA*

For the triangle in Figure 8.4, $a = 22$ inches, $b = 12$ inches, and $A = 42°$. Find the remaining side and angles.

Solution

By the Law of Sines, you have

$$\frac{22}{\sin 42°} = \frac{12}{\sin B}$$

$$\sin B = 12\left(\frac{\sin 42°}{22}\right) \approx 0.3649803$$

$$B \approx 21.41° \qquad\qquad \text{\textcolor{red}{B is acute.}}$$

Now, you can determine that $C \approx 180° - 42° - 21.41° = 116.59°$, and the remaining side is given by

$$\frac{c}{\sin 116.59°} = \frac{22}{\sin 42°}$$

$$c = \sin 116.59°\left(\frac{22}{\sin 42°}\right) \approx 29.40 \text{ inches.}$$

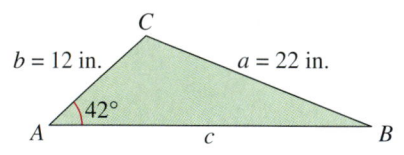

$b = 12$ in. C $a = 22$ in.

A $42°$ c B

One solution: $a > b$

FIGURE 8.4

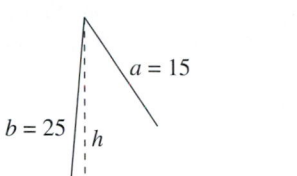

No solution: $a < h$

FIGURE 8.5

■
EXAMPLE 4   *No-Solution Case—SSA*

Show that there is no triangle for which $a = 15$, $b = 25$, and $A = 85°$.

Solution

Begin by making the sketch shown in Figure 8.5. From this figure it appears that no triangle is formed. You can verify this using the Law of Sines.

$$\frac{a}{\sin A} = \frac{b}{\sin B}$$

$$\frac{15}{\sin 85°} = \frac{25}{\sin B}$$

$$\sin B = 25\left(\frac{\sin 85°}{15}\right) \approx 1.660 > 1$$

This contradicts the fact that $|\sin B| \leq 1$. Hence, no triangle can be formed having sides $a = 15$ and $b = 25$ and an angle of $A = 85°$.

■
EXAMPLE 5   *Two-Solution Case—SSA*

Find two triangles for which $a = 12$ meters, $b = 31$ meters, and $A = 20.5°$.

Solution

By the Law of Sines, you have

$$\frac{a}{\sin A} = \frac{b}{\sin B}$$

$$\sin B = b\left(\frac{\sin A}{a}\right) = 31\left(\frac{\sin 20.5°}{12}\right) \approx 0.9047.$$

There are two angles $B_1 \approx 64.8°$ and $B_2 \approx 115.2°$ between $0°$ and $180°$ whose sine is 0.9047. For $B_1 \approx 64.8°$, you obtain

$$C \approx 180° - 20.5° - 64.8° = 94.7°$$

$$c = \frac{a}{\sin A}(\sin C) = \frac{12}{\sin 20.5°}(\sin 94.7°) \approx 34.15 \text{ meters.}$$

For $B_2 \approx 115.2°$, you obtain

$$C \approx 180° - 20.5° - 115.2° = 44.3°$$

$$c = \frac{a}{\sin A}(\sin C) = \frac{12}{\sin 20.5°}(\sin 44.3°) \approx 23.93 \text{ meters.}$$

The resulting triangles are shown in Figure 8.6.

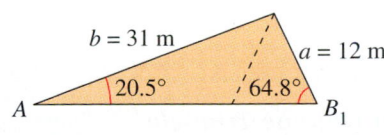

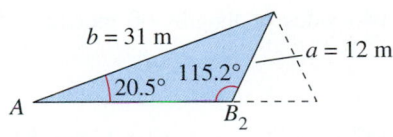

FIGURE 8.6

Area of an Oblique Triangle

The procedure used to prove the Law of Sines leads to a simple formula for the area of an oblique triangle. Referring to Figure 8.7, note that each triangle has a height of $h = b \sin A$. Consequently, the area of each triangle is given by

$$\text{Area} = \frac{1}{2}(\text{base})(\text{height}) = \frac{1}{2}(c)(b \sin A) = \frac{1}{2}bc \sin A.$$

By similar arguments, you can develop the formulas

$$\text{Area} = \frac{1}{2}ab \sin C = \frac{1}{2}ac \sin B.$$

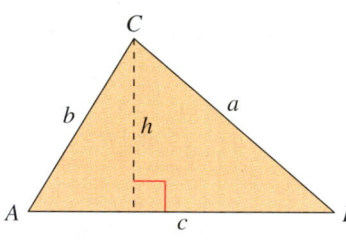 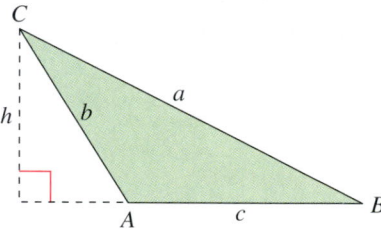

A is acute.   A is obtuse.

FIGURE 8.7

NOTE   Note that if angle A is $90°$, the formula gives the area for a right triangle:

$$\text{Area} = \frac{1}{2}bc = \frac{1}{2}(\text{base})(\text{height}).$$

Similar results are obtained for angles C and B equal to $90°$. ■■

> ◼   **AREA OF AN OBLIQUE TRIANGLE**
>
> The area of any triangle is given by one-half the product of the lengths of two sides times the sine of their included angle. That is,
>
> $$\text{Area} = \frac{1}{2}bc \sin A = \frac{1}{2}ab \sin C = \frac{1}{2}ac \sin B.$$

Real Life

EXAMPLE 6   *Finding the Area of an Oblique Triangle*

Find the area of a triangular lot having two sides of lengths 90 meters and 52 meters and an included angle of $102°$.

Solution

Consider $a = 90$ m, $b = 52$ m, and angle $C = 102°$, as shown in Figure 8.8. Then the area of the triangle is

$$\text{Area} = \frac{1}{2}ab \sin C = \frac{1}{2}(90)(52)(\sin 102°) \approx 2289 \text{ square meters.}$$

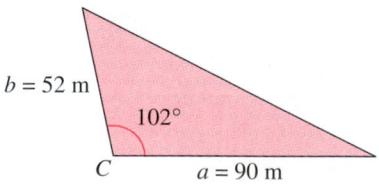

FIGURE 8.8

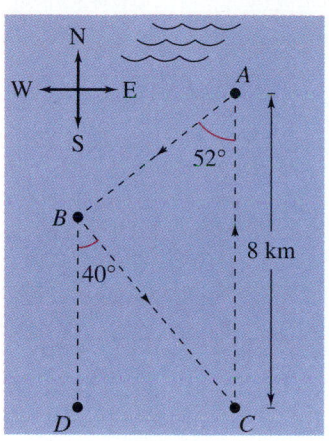

FIGURE 8.9

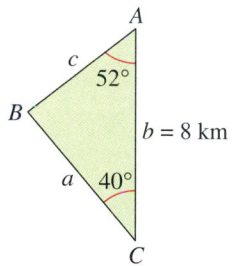

FIGURE 8.10

Application

*E*XAMPLE 7   *An Application of the Law of Sines*

The course for a boat race starts at point A and proceeds in the direction S 52° W to point B, then in the direction S 40° E to point C, and finally back to A, as shown in Figure 8.9. The point C lies 8 kilometers directly south of point A. Approximate the total distance of the race course.

Solution

Because lines BD and AC are parallel, it follows that $\angle BCA \cong \angle DBC$. Consequently, triangle ABC has the measures shown in Figure 8.10. For angle B, you have $B = 180° - 52° - 40° = 88°$. Using the Law of Sines

$$\frac{a}{\sin 52°} = \frac{b}{\sin 88°} = \frac{c}{\sin 40°}$$

you can let $b = 8$ and obtain the following.

$$a = \frac{8}{\sin 88°}(\sin 52°) \approx 6.308 \qquad c = \frac{8}{\sin 88°}(\sin 40°) \approx 5.145$$

The total length of the course is approximately

$$\text{Length} \approx 8 + 6.308 + 5.145 = 19.453 \text{ kilometers.}$$

GROUP ACTIVITY

USING THE LAW OF SINES

In this section, you have been using the Law of Sines to solve *oblique* triangles. Can the Law of Sines also be used to solve a right triangle? If so, write a short paragraph explaining how to use the Law of Sines to solve the following two triangles. Is there an easier way to solve these triangles?

a. (AAS)

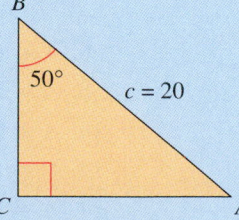

b. (ASA)

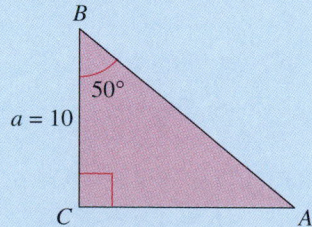

The *Interactive* CD-ROM provides additional help with Warm-Up exercises by providing a hypertext link to the section in which the concept was introduced.

WARM UP

Solve the right triangle. (c is the hypotenuse.)

1. $a = 3$, $c = 6$ **2.** $a = 5$, $b = 5$

3. $b = 15$, $c = 17$ **4.** $A = 42°$, $a = 7.5$

5. $B = 10°$, $b = 4$ **6.** $B = 72°15'$, $c = 150$

Find the altitude of the triangle.

7. 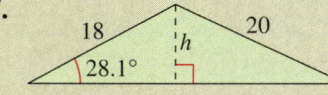   **8.**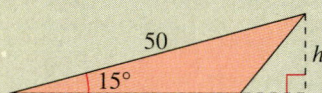

The *Interactive* CD-ROM contains step-by-step solutions to all odd-numbered Section and Review Exercises. It also provides Tutorial Exercises, which link to Guided Examples for additional help.

Solve the equation for x.

9. $\dfrac{2}{\sin 30°} = \dfrac{9}{x}$ **10.** $\dfrac{100}{\sin 72°} = \dfrac{x}{\sin 60°}$

8.1 Exercises

In Exercises 1–16, use the given information to solve the triangle.

1.

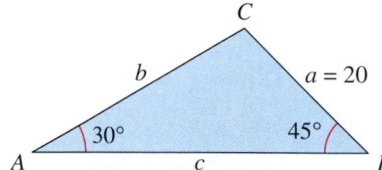

2.

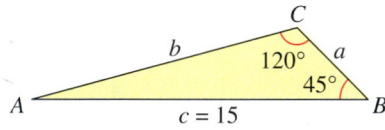

3.

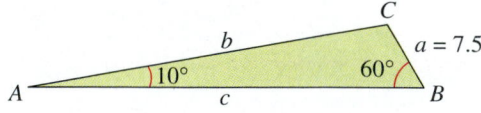

4.

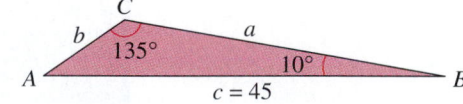

5. $A = 36°$, $a = 8$, $b = 5$ **6.** $A = 60°$, $a = 9$, $c = 10$

7. $A = 150°$, $C = 20°$, $a = 200$ **8.** $A = 24.3°$, $C = 54.6°$, $c = 2.68$

9. $A = 83°20'$, $C = 54.6°$, $c = 18.1$ **10.** $A = 5°40'$, $B = 8°15'$, $b = 4.8$

11. $B = 15°30'$, $a = 4.5$, $b = 6.8$ **12.** $C = 85°20'$, $a = 35$, $c = 50$

13. $C = 145°$, $b = 4$, $c = 14$ **14.** $A = 100°$, $a = 125$, $c = 10$

15. $A = 110°15'$, $a = 48$, $b = 16$ **16.** $B = 2°45'$, $b = 6.2$, $c = 5.8$

In Exercises 17–22, use the given information to solve (if possible) the triangle. If two solutions exist, find both.

17. $A = 58°$, $a = 4.5$, $b = 12.8$
18. $A = 58°$, $a = 11.4$, $b = 12.8$
19. $A = 58°$, $a = 4.5$, $b = 5$
20. $A = 58°$, $a = 42.4$, $b = 50$
21. $A = 110°$, $a = 125$, $b = 200$
22. $A = 110°$, $a = 125$, $b = 100$

In Exercises 23 and 24, find a value for b such that the triangle has (a) one solution, (b) two solutions, and (c) no solution.

23. $A = 36°$, $a = 5$ 24. $A = 60°$, $a = 10$

25. *Height* A flagpole is located on a slope that makes an angle of 12° with the horizontal. The pole's shadow is 16 meters long and points directly up the slope. The angle of elevation of the sun is 20°.

(a) Draw a triangle that represents the problem. Show the known quantities on the triangle and use a variable to indicate the height of the flagpole.

(b) Write an equation involving the unknown.

(c) Find the height of the flagpole.

26. *Height* Because of prevailing winds, a tree grew so that it was leaning 6° from the vertical. At a point 30 meters from the tree, the angle of elevation to the top of the tree is 22° 50′ (see figure). Find the height h of the tree.

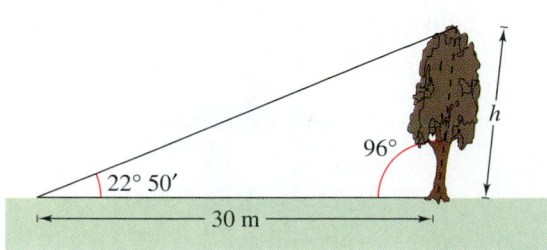

27. *Angle of Elevation* A 10-meter telephone pole casts a 17-meter shadow directly down a slope when the angle of elevation of the sun is 42° (see figure). Find, $θ$, the angle of elevation of the ground.

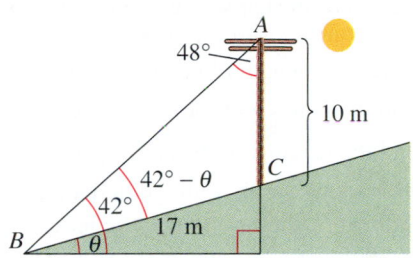

28. *Flight Path* A plane flies 500 kilometers with a bearing of N 44° W from B to C (see figure). The plane then flies 720 kilometers from C to A. Find the bearing of the flight from C to A.

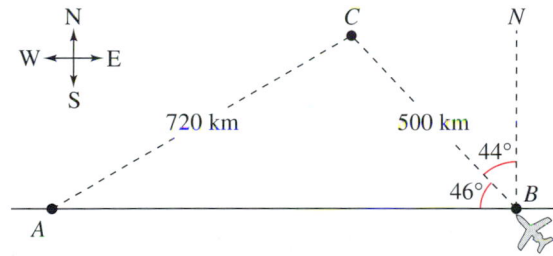

29. *Bridge Design* A bridge is to be built across a small lake from B to C (see figure). The bearing from B to C is S 41° W. From a point A, 100 meters from B, the bearings to B and C are S 74° E and S 28° E, respectively. Find the distance from B to C.

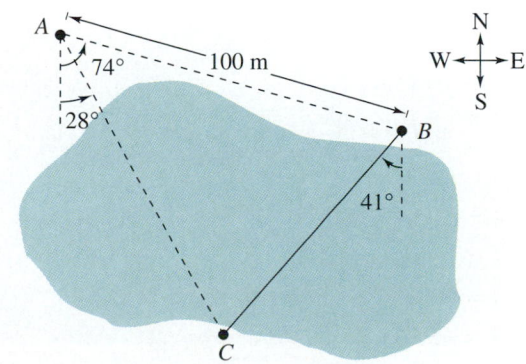

30. **Railroad Track Design** The circular arc of a railroad curve has a chord of length 3000 feet and a central angle of 40°.

    (a) Draw a figure that visually represents the problem. Show the known quantities on the figure and use variables r and s to represent the radius of the arc and the length of the arc, respectively.

    (b) Find the radius r of the circular arc.

    (c) Find the length s of the circular arc.

31. **Glide Path** A pilot has just started on the glide path for landing at an airport where the length of the runway is 9000 feet. The angles of depression from the plane to the ends of the runway are 17.5° and 18.8°.

    (a) Draw a figure that visually represents the problem.

    (b) Find the air distance the plane must travel until touching down on the near end of the runway.

    (c) Find the ground distance the plane must travel until touching down.

    (d) Find the altitude of the plane when the pilot begins the descent.

32. **Altitude** The angles of elevation to an airplane from two points A and B on level ground are 51° and 68°, respectively. The points A and B are 2.5 miles apart, and the airplane is east of both points in the same vertical plane. Find the altitude of the plane.

33. **Locating a Fire** Two fire towers A and B are 30 kilometers apart. The bearing from A to B is N 65° E. A fire is spotted by a ranger in each tower, and its bearings from A and B are N 28° E and N 16.5° W, respectively (see figure). Find the distance of the fire from each tower.

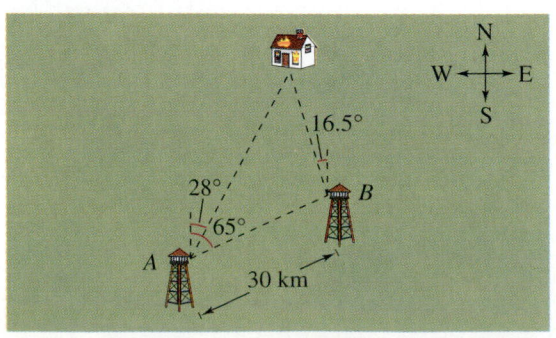

34. **Distance** A boat is sailing due east parallel to the shoreline at a speed of 10 miles per hour. At a given time the bearing to the lighthouse is S 70° E, and 15 minutes later the bearing is S 63° E (see figure). Find the distance from the boat to the shoreline if the lighthouse is at the shoreline.

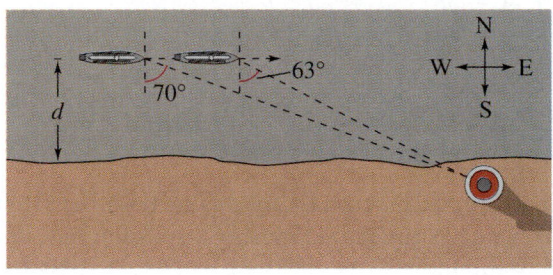

35. **Distance** A family is traveling due west on a road that passes a famous landmark. At a given time the bearing to the landmark is N 62° W, and after the family travels 5 miles farther the bearing is N 38° W. What is the closest the family will come to the landmark while on the road?

36. **Engine Design** The connecting rod in an engine is 6 inches in length and the radius of the crankshaft is $1\frac{1}{2}$ inches (see figure). Let d be the distance the piston is from the top of its stroke for the angle θ.

    (a) Complete the table.

θ	0°	45°	90°	135°	180°
d					

    (b) The spark plug fires at $\theta = 5°$ before top dead center. How far is the piston from the top of its stroke at this time?

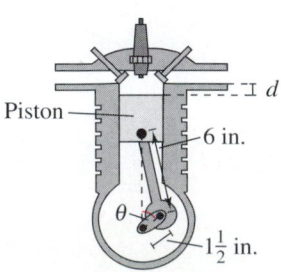

37. *Graphical and Numerical Analysis*   In the figure, α and β are positive angles.

(a) Write α as a function of β.

(b) Use a graphing utility to graph the function. Determine its domain and range.

(c) Use the result of part (a) to write c as a function of β.

(d) Use a graphing utility to graph the function in part (c). Determine its domain and range.

(e) Complete the table. What can you infer?

β	0	0.4	0.8	1.2	1.6	2.0	2.4	2.8
α								
c								

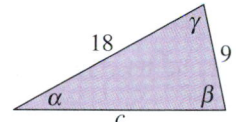

38. *Distance*   The angles of elevation θ and ϕ to an airplane are being continuously monitored at two observation points A and B that are 2 miles apart (see figure). Write an equation giving the distance d between the plane and point B in terms of θ and ϕ.

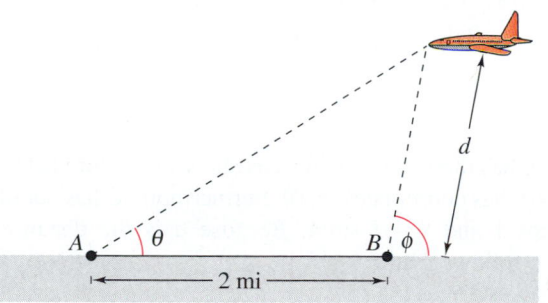

In Exercises 39–44, find the area of the triangle having the indicated sides and angle.

39. $C = 120°, \quad a = 4, \quad b = 6$

40. $B = 72° \, 30', \quad a = 105, \quad c = 64$

41. $A = 43° \, 45', \quad b = 57, \quad c = 85$

42. $A = 5° \, 15', \quad b = 4.5, \quad c = 22$

43. $B = 130, \quad a = 62, \quad c = 20$

44. $C = 84° \, 30', \quad a = 16, \quad b = 20$

45. *Graphical Analysis*

(a) Write the area A of the shaded region in the figure as a function of θ.

(b) Use a graphing utility to graph the area function.

(c) Determine the domain of the area function. Explain how the area of the region and the domain of the function would change if the 8-centimeter line segment were decreased in length.

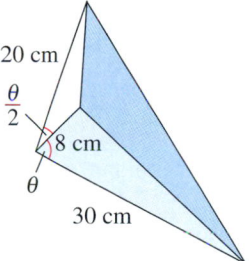

8.2 *Law of Cosines*

See Exercise 36 on page 620 for an example of how the Law of Cosines can be used to analyze the design of a paper manufacturing machine.

Introduction ◻ *Heron's Formula*

Introduction

Two cases remain in the list of conditions needed to solve an oblique triangle—SSS and SAS. The Law of Sines does not work in either of these cases. To see why, consider the three ratios given in the Law of Sines.

$$\frac{a}{\sin A} = \frac{b}{\sin B} = \frac{c}{\sin C}$$

To use the Law of Sines, you must know at least one side and its opposite angle. If you are given three sides (SSS), or two sides and their included angle (SAS), none of the above ratios would be complete. In such cases you can use the **Law of Cosines.**

▪ LAW OF COSINES

Standard Form   *Alternative Form*

$$a^2 = b^2 + c^2 - 2bc \cos A \qquad \cos A = \frac{b^2 + c^2 - a^2}{2bc}$$

$$b^2 = a^2 + c^2 - 2ac \cos B \qquad \cos B = \frac{a^2 + c^2 - b^2}{2ac}$$

$$c^2 = a^2 + b^2 - 2ab \cos C \qquad \cos C = \frac{a^2 + b^2 - c^2}{2ab}$$

PROOF ▪▪

Consider a triangle that has three acute angles, as shown in Figure 8.11. In the figure, note that vertex B has coordinates $(c, 0)$. Furthermore, C has coordinates (x, y), where $x = b \cos A$ and $y = b \sin A$. Because a is the distance from vertex C to vertex B, it follows that

$$a = \sqrt{(x - c)^2 + (y - 0)^2}$$
$$a^2 = (b \cos A - c)^2 + (b \sin A)^2$$
$$a^2 = b^2 \cos^2 A - 2bc \cos A + c^2 + b^2 \sin^2 A$$
$$a^2 = b^2(\sin^2 A + \cos^2 A) + c^2 - 2ab \cos A$$
$$a^2 = b^2 + c^2 - 2bc \cos A. \qquad \qquad \text{\textcolor{red}{$\sin^2 A + \cos^2 A = 1$}}$$

Similar arguments can be used to establish the other two equations. ▪▪

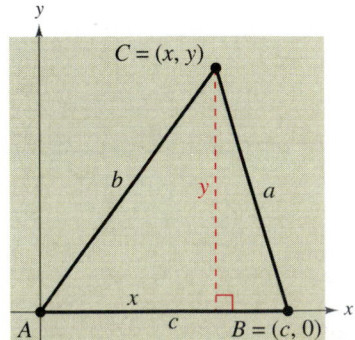

FIGURE 8.11

Note that if $A = 90°$ in Figure 8.11, then $\cos A = 0$ and the first form of the Law of Cosines becomes the Pythagorean Theorem.

$$a^2 = b^2 + c^2$$

Thus, the Pythagorean Theorem is actually just a special case of the more general Law of Cosines.

EXAMPLE 1   Three Sides of a Triangle—SSS

Find the three angles of the triangle whose sides have lengths $a = 8$ feet, $b = 19$ feet, and $c = 14$ feet.

Solution

It is a good idea first to find the angle opposite the longest—side b in this case (see Figure 8.12). Using the Law of Cosines, you find that

$$\cos B = \frac{a^2 + c^2 - b^2}{2ac} = \frac{8^2 + 14^2 - 19^2}{2(8)(14)} \approx -0.45089.$$

Because $\cos B$ is negative, you know that B is an *obtuse* angle given by $B \approx 116.80°$. At this point you could use the Law of Cosines to find $\cos A$ and $\cos C$. However, knowing that $B \approx 116.80°$, it is simpler to use the Law of Sines to obtain the following.

$$\frac{b}{\sin B} = \frac{a}{\sin A}$$

$$\sin A = a\left(\frac{\sin B}{b}\right) \approx 8\left(\frac{\sin 116.80°}{19}\right) \approx 0.37582$$

Because B is obtuse, you know that A must be acute, because a triangle can have, at most, one obtuse angle. Thus, $A \approx 22.08°$ and

$$C \approx 180° - 22.08° - 116.80°$$
$$= 41.12°.$$

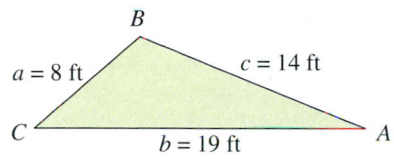

B

$a = 8$ ft $c = 14$ ft

C $b = 19$ ft A

FIGURE 8.12

The *Interactive* CD-ROM shows every example with its solution; clicking on the *Try It!* button brings up similar problems. Guided Examples and Integrated Examples show step-by-step solutions to additional examples. Integrated Examples are related to several concepts in the section.

Do you see why it was wise to find the largest angle *first* in Example 1? Knowing the cosine of an angle, you can determine whether the angle is acute or obtuse. That is,

$\cos\theta > 0$ for $0° < \theta < 90°$ Acute

$\cos\theta < 0$ for $90° < \theta < 180°$.   Obtuse

So, in Example 1, once you found that angle B was obtuse, you knew that angles A and C were both acute. If the largest angle is acute, the remaining two angles are acute also.

EXAMPLE 2   *Two Sides and the Included Angle—SAS*

The pitcher's mound on a softball field is 46 feet from home plate and the distance between the bases in 60 feet, as shown in Figure 8.13. (The pitcher's mound is not halfway between home plate and second base.) How far is the pitcher's mound from first base?

Solution

In triangle *HPF*, $H = 45°$ (line *HP* bisects the right angle at *H*), $f = 46$, and $p = 60$. Using the Law of Cosines for this SAS case, you have

$$h^2 = f^2 + p^2 - 2fp \cos H$$
$$= 46^2 + 60^2 - 2(46)(60) \cos 45°$$
$$\approx 1812.8.$$

Therefore, the approximate distance from the pitcher's mound to first base is

$$h \approx \sqrt{1812.8} \approx 42.58 \text{ feet.}$$

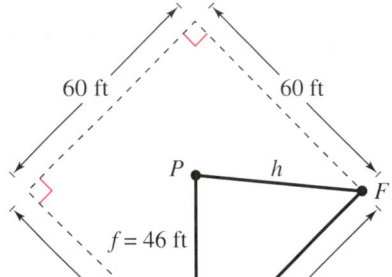

FIGURE 8.13

EXAMPLE 3   *Two Sides and the Included Angle—SAS*

A ship travels 60 miles due east, then adjusts its course 15° northward, as shown in Figure 8.14. After traveling 80 miles in that direction, how far is the ship from its point of departure?

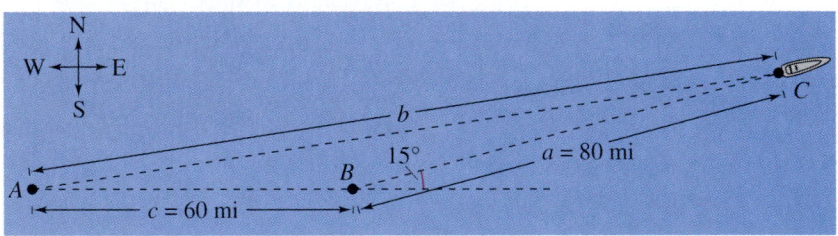

FIGURE 8.14

Solution

You have $c = 60$, $B = 180° - 15° = 165°$, and $a = 80$. Consequently, by the Law of Cosines, it follows that

$$b^2 = a^2 + c^2 - 2ac \cos B$$
$$= 80^2 + 60^2 - 2(80)(60) \cos 165°$$
$$\approx 19,273.$$

Therefore, the distance *b* is

$$b \approx \sqrt{19,273} \approx 138.8 \text{ miles.}$$

The world's largest ship is the Norwegian oil tanker, Jahre Viking, which is 1503 feet (458 meters) long. It takes 5 minutes to walk the ship's length.

Heron's Formula

The Law of Cosines can be used to establish the following formula for the area of a triangle. This formula is credited to the Greek mathematician Heron (c. 100 B.C.).

HERON'S AREA FORMULA

Given any triangle with sides of lengths a, b, and c, the area of the triangle is

$$\text{Area} = \sqrt{s(s-a)(s-b)(s-c)}$$

where $s = (a+b+c)/2$.

PROOF ■■

From the previous section, you know that

$$\text{Area} = \frac{1}{2}bc \sin A$$

$$= \sqrt{\frac{1}{4}b^2c^2 \sin^2 A}$$

$$= \sqrt{\frac{1}{4}b^2c^2(1 - \cos^2 A)}$$

$$= \sqrt{\left[\frac{1}{2}bc(1 + \cos A)\right]\left[\frac{1}{2}bc(1 - \cos A)\right]}.$$

Using the Law of Cosines, you can show that

$$\frac{1}{2}bc(1 + \cos A) = \frac{a+b+c}{2} \cdot \frac{-a+b+c}{2}$$

and

$$\frac{1}{2}bc(1 - \cos A) = \frac{a-b+c}{2} \cdot \frac{a+b-c}{2}.$$

(See Exercises 49 and 50.) Letting $s = (a+b+c)/2$, these two equations can be rewritten as

$$\frac{1}{2}bc(1 + \cos A) = s(s-a)$$

and

$$\frac{1}{2}bc(1 - \cos A) = (s-b)(s-c).$$

Thus, you can conclude that

$$\text{Area} = \sqrt{s(s-a)(s-b)(s-c)}. \quad ■■$$

EXAMPLE 4   *Using Heron's Area Formula*

Find the area of the triangular region having sides of lengths $a = 43$ meters, $b = 53$ meters, and $c = 72$ meters.

Solution

Because

$$s = \frac{1}{2}(a + b + c) = \frac{168}{2} = 84,$$

Heron's Formula yields

$$
\begin{aligned}
\text{Area} &= \sqrt{s(s - a)(s - b)(s - c)} \\
&= \sqrt{84(41)(31)(12)} \\
&\approx 1131.89 \text{ square meters.}
\end{aligned}
$$

GROUP ACTIVITY

THE AREA OF A TRIANGLE

You have now discussed three different formulas for the area of a triangle.

Standard Formula   Area $= \frac{1}{2}bh$

Oblique Triangle   Area $= \frac{1}{2}bc \sin A = \frac{1}{2}ab \sin C = \frac{1}{2}ac \sin B$

Heron's Formula   Area $= \sqrt{s(s - a)(s - b)(s - c)}$

Use the most appropriate formula to find the area of each triangle. Show your work and give your reasons for choosing each formula.

a.
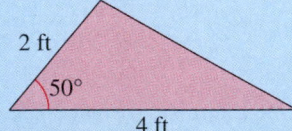

b.
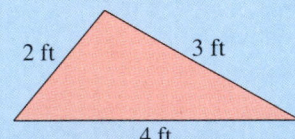

c.
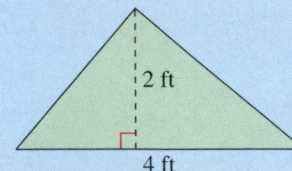

d.
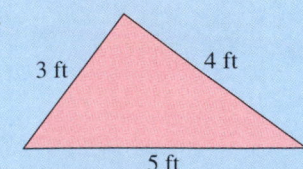

Simplify the expression.

1. $\sqrt{(7-3)^2 + [1-(-5)]^2}$ **2.** $\sqrt{[-2-(-5)]^2 + (12-6)^2}$

Find the distance between the two points.

3. $(4, -2), (8, 10)$ **4.** $(1, 3), (7, 12)$

Find the area of the triangle.

5.

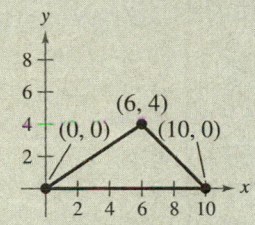

6.

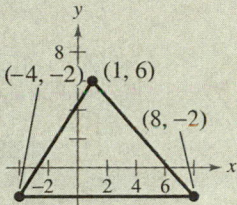

Find (if possible) the remaining sides and angles of the triangle.

7. $A = 10°, C = 100°, b = 25$ **8.** $A = 20°, C = 90°, c = 100$

9. $B = 30°, b = 6.5, c = 15$ **10.** $A = 30°, b = 6.5, a = 10$

8.2 Exercises

In Exercises 1–14, use the Law of Cosines to solve the triangle.

1.

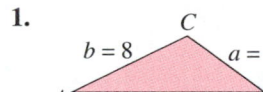

2.

3.

4.

5. $a = 9$, $b = 12$, $c = 15$

6. $a = 55$, $b = 25$, $c = 72$

7. $a = 75.4$, $b = 52$, $c = 52$

8. $a = 1.42$, $b = 0.75$, $c = 1.25$

9. $A = 120°$, $b = 3$, $c = 10$

10. $A = 55°$, $b = 3$, $c = 10$

11. $B = 8° 45'$, $a = 25$, $c = 15$

12. $B = 75° 20'$, $a = 6.2$, $c = 9.5$

13. $B = 125° 40'$, $a = 32$, $c = 32$

14. $C = 15°$, $a = 6.25$, $b = 2.15$

In Exercises 15–20, complete the table by solving the parallelogram shown in the figure. (The lengths of the diagonals are given by *c* and *d*.)

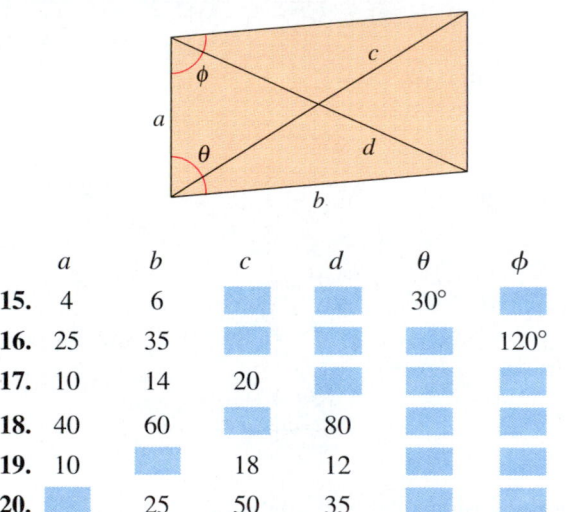

	a	*b*	*c*	*d*	*θ*	*φ*
15.	4	6			30°	
16.	25	35				120°
17.	10	14	20			
18.	40	60		80		
19.	10		18	12		
20.		25	50	35		

21. **Navigation** A boat race runs along a triangular course marked by buoys *A*, *B*, and *C*. The race starts with the boats headed west for 2500 meters. The other two sides of the course lie to the north of the first side, and their lengths are 1100 meters and 2000 meters. Draw a figure that gives a visual representation of the problem, and find the bearings for the last two legs of the race.

22. **Navigation** A plane flies 810 miles from *A* to *B* with a bearing of N 75° E. Then it flies 648 miles from *B* to *C* with a bearing of N 32° E. Draw a figure that visually represents the problem, and find the straight-line distance and bearing from *C* to *A*.

23. **Surveying** To approximate the length of a marsh, a surveyor walks 300 meters from point *A* to point *B*, then turns 80° and walks 250 meters to point *C* (see figure). Approximate the length *AC* of the marsh.

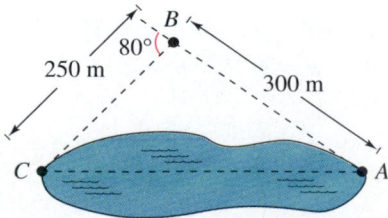

24. **Surveying** A triangular parcel of land has 115 meters of frontage, and the other boundaries have lengths of 76 meters and 92 meters. What angles does the frontage make with the two other boundaries?

25. **Surveying** A triangular parcel of ground has sides of length 725 feet, 650 feet, and 575 feet. Find the measure of the largest angle.

26. **Streetlight Design** Determine the angle *θ* in the design of the streetlight shown in the figure.

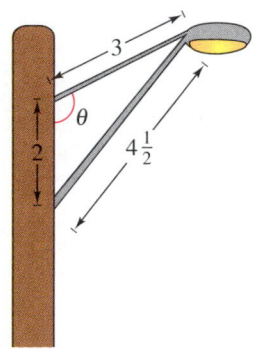

27. **Distance** Two ships leave a port at 9 A.M. One travels at a bearing of N 53° W at 12 miles per hour and the other travels at a bearing of S 67° W at 16 miles per hour. Approximate how far apart they are at noon that day.

28. **Distance** A 100-foot vertical tower is to be erected on the side of a hill that makes a 6° angle with the horizontal (see figure). Find the length of each of the two guy wires that will be anchored 75 feet uphill and downhill from the base of the tower.

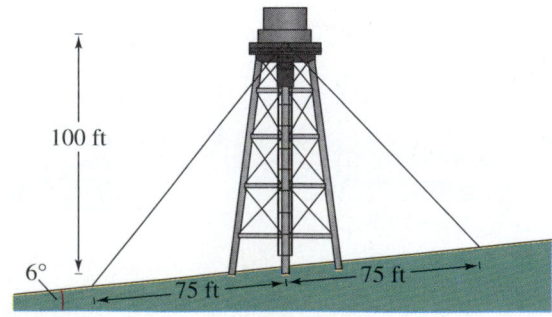

29. *Navigation* On a map, Orlando is 178 millimeters due south of Niagara Falls, Denver is 273 millimeters from Orlando, and Denver is 235 millimeters from Niagara Falls (see figure).

(a) Find the bearing of Denver from Orlando.

(b) Find the bearing of Denver from Niagara Falls.

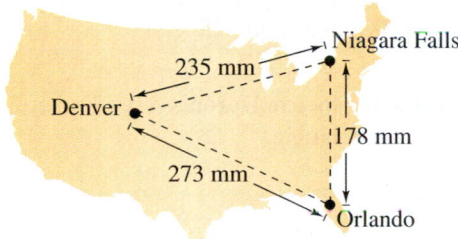

30. *Navigation* On a map, Minneapolis is 165 millimeters due west of Albany, Phoenix is 216 millimeters from Minneapolis, and Phoenix is 368 millimeters from Albany.

(a) Find the bearing of Minneapolis from Phoenix.

(b) Find the bearing of Albany from Phoenix.

31. *Baseball* On a baseball diamond with 90-foot sides, the pitcher's mound is 60.5 feet from home plate. How far is it from the pitcher's mound to third base?

32. *Baseball* The baseball player in center field is playing approximately 330 feet from the television camera that is behind home plate. A batter hits a fly ball that goes to the wall 420 feet from the camera (see figure). Approximate the number of feet that the center fielder has to run to make the catch if the camera turns 8° to follow the play.

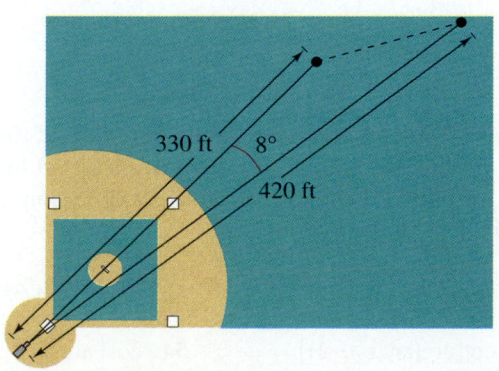

33. *Engineering* If Q is the midpoint of the line segment $\overline{PR}$, find the lengths of the line segments $\overline{PQ}$, $\overline{QS}$, and $\overline{RS}$ on the truss rafter shown in the figure.

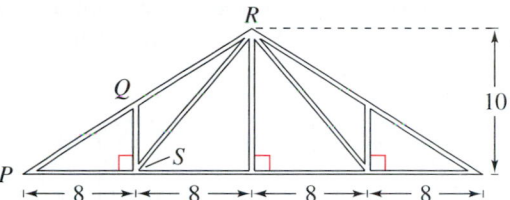

34. *Aircraft Tracking* To determine the distance between two aircraft, a tracking station continuously determines the distance to each aircraft and the angle A between them. Determine the distance a between the planes when $A = 42°$, $b = 35$ miles, and $c = 20$ miles.

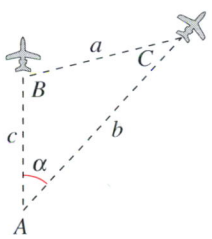

35. *Engine Design* An engine has a 7-inch connecting rod fastened to a crank (see figure).

(a) Use the Law of Cosines to write an equation giving the relationship between x and θ.

(b) Write x as a function of θ. (Select the sign that yields positive values of x.)

(c) Use a graphing utility to graph the function in part (b).

(d) Use the graph in part (c) to determine the maximum distance the piston moves in one cycle.

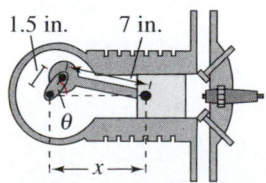

36. *Paper Manufacturing* In a certain process with continuous paper, the paper passes across three rollers of radii 3 inches, 4 inches, and 6 inches (see figure). The centers of the 3-inch and 6-inch rollers are d inches apart, and the length of the arc in contact with the paper on the 4-inch roller is s inches. Complete the following table.

d (inches)	9	10	12	13	14	15	16
θ (degrees)							
s (inches)							

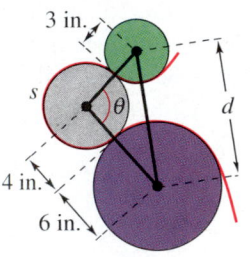

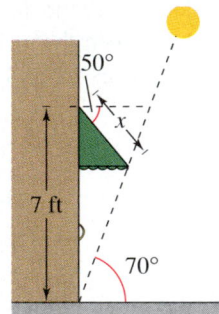

FIGURE FOR 36   **FIGURE FOR 37**

37. *Awning Design* A retractable awning lowers at an angle of 50° from the top of a patio door that is 7 feet tall (see figure). Find the length x of the awning if no direct sunlight is to enter the door when the angle of elevation of the sun is greater than 70°.

38. *Circumscribed and Inscribed Circles* Let R and r be the radii of the circumscribed and inscribed circles of a triangle ABC, respectively, and let $s = (a + b + c)/2$. Prove the following.

(a) $2R = \dfrac{a}{\sin A} = \dfrac{b}{\sin B} = \dfrac{c}{\sin C}$

(b) $r = \sqrt{\dfrac{(s - a)(s - b)(s - c)}{s}}$

Circumscribed and Inscribed Circles **In Exercises 39 and 40, use the results of Exercise 38.**

39. Given the triangle with $a = 25$, $b = 55$, and $c = 72$, find the area of (a) the triangle, (b) the circumscribed circle, and (c) the inscribed circle.

40. Find the length of the largest circular track that can be built on a triangular piece of property whose sides are 200 feet, 250 feet, and 325 feet.

In Exercises 41–46, use Heron's Area Formula to find the area of the triangle.

41. $a = 5$, $b = 7$, $c = 10$

42. $a = 2.5$, $b = 10.2$, $c = 9$

43. $a = 12$, $b = 15$, $c = 9$

44. $a = 75.4$, $b = 52$, $c = 52$

45. $a = 20$, $b = 20$, $c = 10$

46. $a = 4.25$, $b = 1.55$, $c = 3.00$

47. *Area* The lengths of the sides of a triangular parcel of land are approximately 200 feet, 500 feet, and 600 feet. Approximate the area of the parcel.

48. *Area* The lengths of two adjacent sides of a parallelogram are 4 meters and 6 meters. Find the area of the parallelogram if the angle between the two sides is 30°.

49. Use the Law of Cosines to prove that

$$\frac{1}{2} bc(1 + \cos A) = \frac{a + b + c}{2} \cdot \frac{-a + b + c}{2}.$$

50. Use the Law of Cosines to prove that

$$\frac{1}{2} bc(1 - \cos A) = \frac{a - b + c}{2} \cdot \frac{a + b - c}{2}.$$

Review **Solve Exercises 51–54 as a review of the skills and problem-solving techniques you learned in previous sections. Write an algebraic expression that is equivalent to the expression.**

51. $\sec(\arcsin 2x)$

52. $\tan(\arccos 3x)$

53. $\cot[\arctan(x - 2)]$

54. $\cos\left(\arcsin \dfrac{x - 1}{2}\right)$

8.3 *Vectors in the Plane*

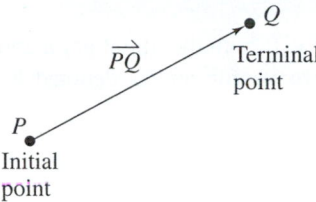

See Exercises 77 and 78 on page 633 for examples of how vectors can be used to analyze the direction and speed of an airplane.

Introduction □ *Component Form of a Vector* □ *Vector Operations* □
Unit Vectors □ *Direction Angles* □ *Applications of Vectors*

Introduction

Many quantities in geometry and physics, such as area, time, and temperature, can be represented by a single real number. Other quantities, such as force and velocity, involve both *magnitude* and *direction* and cannot be completely characterized by a single real number. To represent such a quantity, you can use a **directed line segment,** as shown in Figure 8.15. The directed line segment $\overrightarrow{PQ}$ has **initial point** P and **terminal point** Q. Its **length** is denoted by $\|\overrightarrow{PQ}\|$.

Two directed line segments that have the same length (or magnitude) and direction are called **equivalent.** For example, the directed line segments in Figure 8.16 are all equivalent. The set of all directed line segments that are equivalent to given directed line segment $\overrightarrow{PQ}$ is a **vector v in the plane,** written $\mathbf{v} = \overrightarrow{PQ}$. Vectors are denoted by lowercase, boldface letters such as $\mathbf{u}$, $\mathbf{v}$, and $\mathbf{w}$.

Be sure you see that a vector in the plane can be represented by many different directed line segments.

FIGURE 8.15

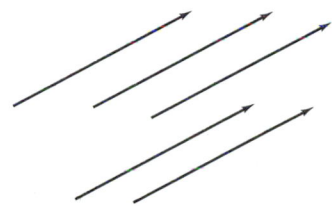

FIGURE 8.16

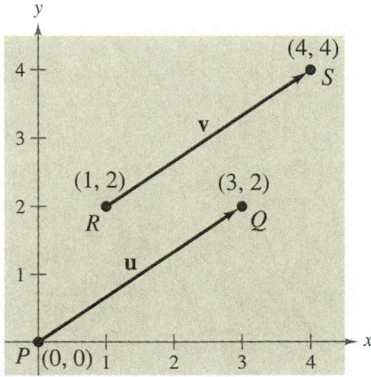

FIGURE 8.17

EXAMPLE 1 *Vector Representation by Directed Line Segments*

Let $\mathbf{u}$ be represented by the directed line segment from $P = (0, 0)$ to $Q = (3, 2)$, and let $\mathbf{v}$ be represented by the directed line segment from $R = (1, 2)$ to $S = (4, 4)$, as shown in Figure 8.17. Show that $\mathbf{u} = \mathbf{v}$.

Solution

From the distance formula, it follows that $\overrightarrow{PQ}$ and $\overrightarrow{RS}$ have the *same length.*

$$\|\overrightarrow{PQ}\| = \sqrt{(3 - 0)^2 + (2 - 0)^2} = \sqrt{13}$$
$$\|\overrightarrow{RS}\| = \sqrt{(4 - 1)^2 + (4 - 2)^2} = \sqrt{13}$$

Moreover, both line segments have the *same direction* because they are both directed toward the upper right on lines having a slope of $\frac{2}{3}$. Thus, $\overrightarrow{PQ}$ and $\overrightarrow{RS}$ have the same length and direction, and it follows that $\mathbf{u} = \mathbf{v}$.

Component Form of a Vector

The directed line segment whose initial point is the origin is often the most convenient representative of a set of equivalent directed line segments. This representative of the vector **v** is in **standard position.**

A vector whose initial point is at the origin $(0, 0)$ can be uniquely represented by the coordinates of its terminal point (v_1, v_2). This is the **component form of a vector v,** written

$$\mathbf{v} = \langle v_1, v_2 \rangle.$$

The coordinates v_1 and v_2 are the **components** of **v**. If both the initial point and the terminal point lie at the origin, **v** is the **zero vector** and is denoted by $\mathbf{0} = \langle 0, 0 \rangle$.

NOTE   Two vectors $\mathbf{u} = \langle u_1, u_2 \rangle$ and $\mathbf{v} = \langle v_1, v_2 \rangle$ are **equal** if and only if $u_1 = v_1$ and $u_2 = v_2$. For instance, in Example 1, the vector **u** from $P = (0, 0)$ to $Q = (3, 2)$ is

$$\mathbf{u} = \overrightarrow{PQ} = \langle 3 - 0, 2 - 0 \rangle = \langle 3, 2 \rangle$$

and the vector **v** from $R = (1, 2)$ to $S = (4, 4)$ is $\mathbf{v} = \overrightarrow{RS} = \langle 4 - 1, 4 - 2 \rangle = \langle 3, 2 \rangle.$ ■ ■

 COMPONENT FORM OF A VECTOR

The component form of the vector with initial point $P = (p_1, p_2)$ and terminal point $Q = (q_1, q_2)$ is

$$\overrightarrow{PQ} = \langle q_1 - p_1, q_2 - p_2 \rangle = \langle v_1, v_2 \rangle = \mathbf{v}.$$

The **length** (or magnitude) of **v** is given by

$$\|\mathbf{v}\| = \sqrt{(q_1 - p_1)^2 + (q_2 - p_2)^2} = \sqrt{v_1^2 + v_2^2}.$$

If $\|\mathbf{v}\| = 1$, **v** is a **unit vector.** Moreover, $\|\mathbf{v}\| = 0$ if and only if **v** is the zero vector **0**.

EXAMPLE 2   *Finding the Component Form of a Vector*

Find the component form and length of the vector **v** that has initial point $(4, -7)$ and terminal point $(-1, 5)$.

Solution

Let $P = (4, -7) = (p_1, p_2)$ and $Q = (-1, 5) = (q_1, q_2)$. Then, the components of $\mathbf{v} = \langle v_1, v_2 \rangle$ are given by

$$v_1 = q_1 - p_1 = -1 - 4 = -5$$
$$v_2 = q_2 - p_2 = 5 - (-7) = 12.$$

Thus, $\mathbf{v} = \langle -5, 12 \rangle$ and the length of **v** is

$$\|\mathbf{v}\| = \sqrt{(-5)^2 + 12^2} = \sqrt{169} = 13,$$

is shown in Figure 8.18.

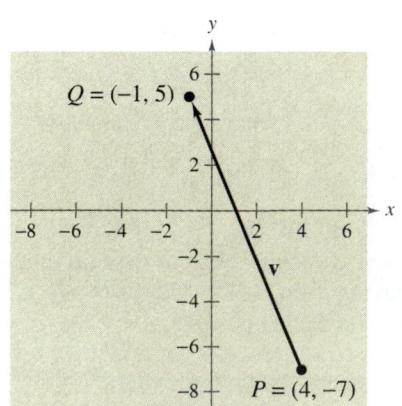

FIGURE 8.18

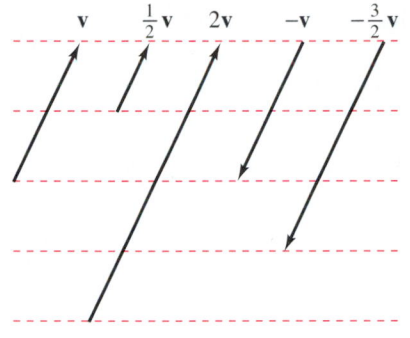

FIGURE 8.19

Vector Operations

The two basic vector operations are **scalar multiplication** and **vector addition.** Geometrically, the product of a vector **v** and a scalar k is the vector that is $|k|$ times as long as **v**. If k is positive, $k\mathbf{v}$ has the same direction as **v**, and if k is negative, $k\mathbf{v}$ has the direction opposite that of **v**, as shown in Figure 8.19.

To add two vectors geometrically, position them (without changing length or direction) so that the initial point of one coincides with the terminal point of the other. The sum $\mathbf{u} + \mathbf{v}$ is formed by joining the initial point of the second vector **v** with the terminal point of the first vector **u**, as shown in Figure 8.20. This technique is called the **parallelogram law** for vector addition because the vector $\mathbf{u} + \mathbf{v}$, often called the **resultant** of vector addition, is the diagonal of a parallelogram having **u** and **v** as its adjacent sides.

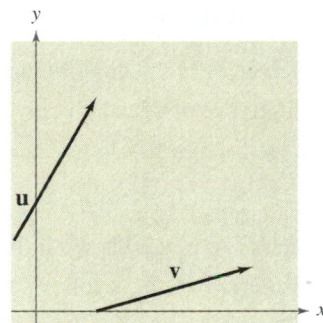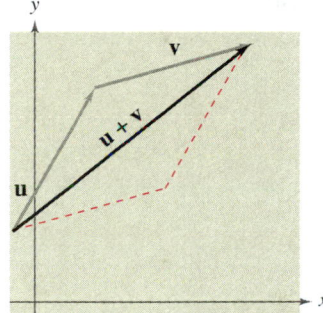

FIGURE 8.20

> ### DEFINITIONS OF VECTOR ADDITION AND SCALAR MULTIPLICATION
>
> Let $\mathbf{u} = \langle u_1, u_2 \rangle$ and $\mathbf{v} = \langle v_1, v_2 \rangle$ be vectors and let k be a scalar (a real number). Then the **sum** of **u** and **v** is the vector
>
> $$\mathbf{u} + \mathbf{v} = \langle u_1 + v_1, u_2 + v_2 \rangle \qquad \text{Sum}$$
>
> and the **scalar multiple** of k times **u** is the vector
>
> $$k\mathbf{u} = k\langle u_1, u_2 \rangle = \langle ku_1, ku_2 \rangle. \qquad \text{Scalar multiple}$$

The **negative** of $\mathbf{v} = \langle v_1, v_2 \rangle$ is

$$-\mathbf{v} = (-1)\mathbf{v} = \langle -v_1, -v_2 \rangle \qquad \text{Negative}$$

and the **difference** of **u** and **v** is

$$\mathbf{u} - \mathbf{v} = \mathbf{u} + (-\mathbf{v}) = \langle u_1 - v_1, u_2 - v_2 \rangle. \qquad \text{Difference}$$

To represent $\mathbf{u} - \mathbf{v}$ graphically, you can use directed line segments with the *same* initial point. The difference $\mathbf{u} - \mathbf{v}$ is the vector from the terminal point of **v** to the terminal point of **u**, as shown in Figure 8.21.

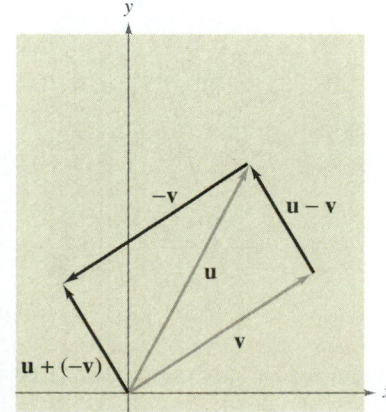

FIGURE 8.21

The component definitions of vector addition and scalar multiplication are illustrated in Example 3. In this example, notice that each of the vector operations can be interpreted geometrically.

EXAMPLE 3   *Vector Operations*

Let $\mathbf{v} = \langle -2, 5 \rangle$ and $\mathbf{w} = \langle 3, 4 \rangle$, and find each of the following vectors.

a. $2\mathbf{v}$ **b.** $\mathbf{w} - \mathbf{v}$ **c.** $\mathbf{v} + 2\mathbf{w}$

Solution

a. Because $\mathbf{v} = \langle -2, 5 \rangle$, you have

$$2\mathbf{v} = \langle 2(-2), 2(5) \rangle$$
$$= \langle -4, 10 \rangle.$$

A sketch of $2\mathbf{v}$ is shown in Figure 8.22(a).

b. The difference of $\mathbf{w}$ and $\mathbf{v}$ is given by

$$\mathbf{w} - \mathbf{v} = \langle 3 - (-2), 4 - 5 \rangle$$
$$= \langle 5, -1 \rangle.$$

A sketch of $\mathbf{w} - \mathbf{v}$ is shown in Figure 8.22(b).

c. Because $2\mathbf{w} = \langle 6, 8 \rangle$, it follows that

$$\mathbf{v} + 2\mathbf{w} = \langle -2, 5 \rangle + \langle 6, 8 \rangle$$
$$= \langle -2 + 6, 5 + 8 \rangle$$
$$= \langle 4, 13 \rangle.$$

A sketch of $\mathbf{v} + 2\mathbf{w}$ is shown in Figure 8.22(c).

(a)
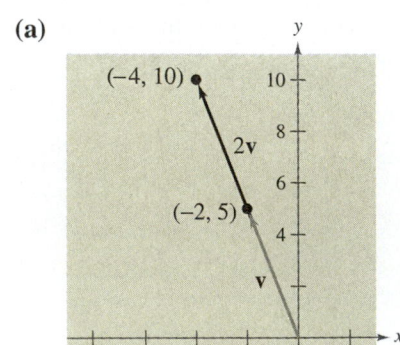

(b)
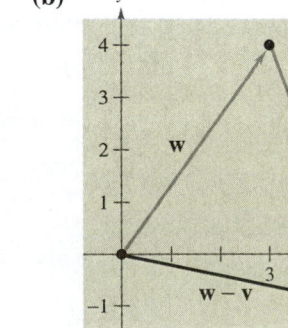

(c)
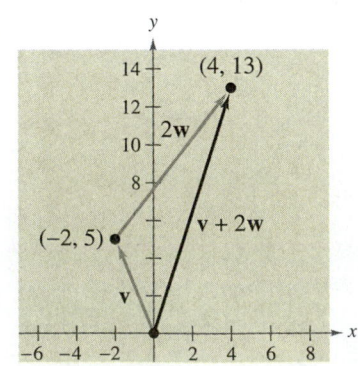

FIGURE 8.22

Vector addition and scalar multiplication share many of the properties of ordinary arithmetic.

> **PROPERTIES OF VECTOR ADDITION AND SCALAR MULTIPLICATION**
>
> Let **u**, **v**, and **w** be vectors and let c and d be scalars. Then the following properties are true.
>
> 1. $\mathbf{u} + \mathbf{v} = \mathbf{v} + \mathbf{u}$
> 2. $(\mathbf{u} + \mathbf{v}) + \mathbf{w} = \mathbf{u} + (\mathbf{v} + \mathbf{w})$
> 3. $\mathbf{u} + \mathbf{0} = \mathbf{u}$
> 4. $\mathbf{u} + (-\mathbf{u}) = \mathbf{0}$
> 5. $c(d\mathbf{u}) = (cd)\mathbf{u}$
> 6. $(c + d)\mathbf{u} = c\mathbf{u} + d\mathbf{u}$
> 7. $c(\mathbf{u} + \mathbf{v}) = c\mathbf{u} + c\mathbf{v}$
> 8. $1(\mathbf{u}) = \mathbf{u},\ 0(\mathbf{u}) = \mathbf{0}$
> 9. $\|c\mathbf{v}\| = |c|\,\|\mathbf{v}\|$

NOTE Property 9 can be stated as follows: the length of the vector $c\mathbf{v}$ is the absolute value of c times the length of **v**. ■■

Unit Vectors

In many applications of vectors it is useful to find a unit vector that has the same direction as a given nonzero vector **v**. To do this, you can divide **v** by its length to obtain

$$\mathbf{u} = \text{unit vector} = \frac{\mathbf{v}}{\|\mathbf{v}\|} = \left(\frac{1}{\|\mathbf{v}\|}\right)\mathbf{v}.$$

Note that **u** is a scalar multiple of **v**. The vector **u** has length 1 and the same direction as **v**. The vector **u** is called a **unit vector in the direction of v.**

Some of the earliest work with vectors was done by the Irish mathematician William Rowan Hamilton (1805–1865). Hamilton spent many years developing a system of vector-like quantities called quaternions. Although Hamilton was convinced of the benefits of quaternions, the operations he defined did not produce good models for physical phenomena. It wasn't until the latter half of the nineteenth century that the Scottish physicist James Maxwell (1831–1879) restructured Hamilton's quaternions in a form useful for representing physical quantities such as force, velocity, and acceleration.

EXAMPLE 4   *Finding a Unit Vector*

Find a unit vector in the direction of $\mathbf{v} = \langle -2, 5 \rangle$ and verify that the result has length 1.

Solution

The unit vector in the direction of **v** is

$$\frac{\mathbf{v}}{\|\mathbf{v}\|} = \frac{\langle -2, 5 \rangle}{\sqrt{(-2)^2 + (5)^2}}$$

$$= \frac{1}{\sqrt{29}}\langle -2, 5 \rangle = \left\langle \frac{-2}{\sqrt{29}}, \frac{5}{\sqrt{29}} \right\rangle.$$

This vector has length 1 because

$$\sqrt{\left(\frac{-2}{\sqrt{29}}\right)^2 + \left(\frac{5}{\sqrt{29}}\right)^2} = \sqrt{\frac{4}{29} + \frac{25}{29}} = \sqrt{\frac{29}{29}} = 1.$$

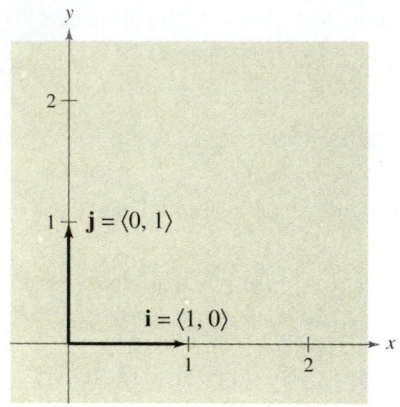

FIGURE 8.23

The unit vectors $\langle 1, 0 \rangle$ and $\langle 0, 1 \rangle$ are called the **standard unit vectors** and are denoted by

$$\mathbf{i} = \langle 1, 0 \rangle \quad \text{and} \quad \mathbf{j} = \langle 0, 1 \rangle,$$

as shown in Figure 8.23. (Note that the lowercase letter $\mathbf{i}$ is written in boldface to distinguish it from the imaginary number $i = \sqrt{-1}$.) These vectors can be used to represent any vector $\mathbf{v} = \langle v_1, v_2 \rangle$ as follows.

$$\begin{aligned}
\mathbf{v} &= \langle v_1, v_2 \rangle \\
&= v_1 \langle 1, 0 \rangle + v_2 \langle 0, 1 \rangle \\
&= v_1 \mathbf{i} + v_2 \mathbf{j}
\end{aligned}$$

The scalars v_1 and v_2 are called the **horizontal** and **vertical components of v,** respectively. The vector sum $v_1 \mathbf{i} + v_2 \mathbf{j}$ is called a **linear combination** of the vectors $\mathbf{i}$ and $\mathbf{j}$. Any vector in the plane can be expressed as a linear combination of the standard unit vectors $\mathbf{i}$ and $\mathbf{j}$.

EXAMPLE 5   *Writing a Linear Combination of Unit Vectors*

Let $\mathbf{u}$ be the vector with initial point $(2, -5)$ and terminal point $(-1, 3)$. Write $\mathbf{u}$ as a linear combination of the standard unit vectors $\mathbf{i}$ and $\mathbf{j}$.

Solution

Begin by writing the component form of the vector $\mathbf{u}$.

$$\begin{aligned}
\mathbf{u} &= \langle -1 - 2, 3 + 5 \rangle \\
&= \langle -3, 8 \rangle \\
&= -3\mathbf{i} + 8\mathbf{j}
\end{aligned}$$

This result is shown graphically in Figure 8.24.

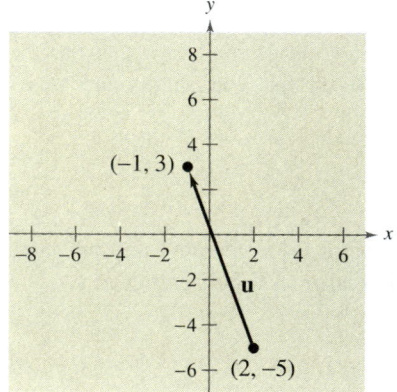

FIGURE 8.24

EXAMPLE 6   *Vector Operations*

Let $\mathbf{u} = -3\mathbf{i} + 8\mathbf{j}$ and $\mathbf{v} = 2\mathbf{i} - \mathbf{j}$. Find $2\mathbf{u} - 3\mathbf{v}$.

Solution

You could solve this problem by converting $\mathbf{u}$ and $\mathbf{v}$ to component form. This, however, is not necessary. It is just as easy to perform the operations in unit vector form.

$$\begin{aligned}
2\mathbf{u} - 3\mathbf{v} &= 2(-3\mathbf{i} + 8\mathbf{j}) - 3(2\mathbf{i} - \mathbf{j}) \\
&= -6\mathbf{i} + 16\mathbf{j} - 6\mathbf{i} + 3\mathbf{j} \\
&= -12\mathbf{i} + 19\mathbf{j}
\end{aligned}$$

Direction Angles

If **u** is a *unit vector* such that θ is the angle (measured counterclockwise) from the positive x-axis to **u**, the terminal point of **u** lies on the unit circle and you have

$$\mathbf{u} = \langle \cos\theta, \sin\theta \rangle = (\cos\theta)\mathbf{i} + (\sin\theta)\mathbf{j}$$

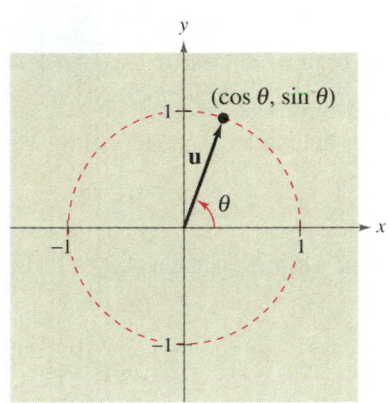

FIGURE 8.25

as shown in Figure 8.25. The angle θ is the **direction angle** of the vector **u**.

Suppose that **u** is a unit vector with direction angle θ. If **v** is any vector that makes an angle θ with the positive x-axis, it has the same direction as **u** and you can write

$$\mathbf{v} = \|\mathbf{v}\| \langle \cos\theta, \sin\theta \rangle$$
$$= \|\mathbf{v}\| (\cos\theta)\mathbf{i} + \|\mathbf{v}\| (\sin\theta)\mathbf{j}.$$

Because $\mathbf{v} = a\mathbf{i} + b\mathbf{j} = \|\mathbf{v}\| (\cos\theta)\mathbf{i} + \|\mathbf{v}\| (\sin\theta)\mathbf{j}$, it follows that the direction angle θ for **v** is determined from

$$\tan\theta = \frac{\sin\theta}{\cos\theta} = \frac{\|\mathbf{v}\| \sin\theta}{\|\mathbf{v}\| \cos\theta} = \frac{b}{a}.$$

EXAMPLE 7   *Finding Direction Angles of Vectors*

Find the direction angle of each vector.

a. u $= 3\mathbf{i} + 3\mathbf{j}$ **b. v** $= 3\mathbf{i} - 4\mathbf{j}$

Solution

a. The direction angle is given by

$$\tan\theta = \frac{b}{a} = \frac{3}{3} = 1.$$

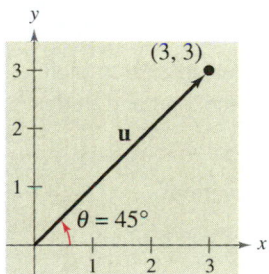

FIGURE 8.26

Therefore, $\theta = 45°$, as shown in Figure 8.26.

b. The direction angle is given by

$$\tan\theta = \frac{b}{a} = \frac{-4}{3}.$$

Moreover, because $\mathbf{v} = 3\mathbf{i} - 4\mathbf{j}$ lies in Quadrant IV, θ lies in Quadrant IV and its reference angle is

$$\theta = \left| \arctan\left(-\frac{4}{3}\right) \right| \approx |-53.13°| = 53.13°.$$

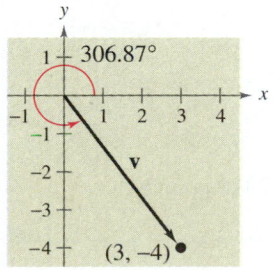

FIGURE 8.27

Therefore, it follows that $\theta \approx 360° - 53.13° = 306.87°$, as shown in Figure 8.27.

Applications of Vectors

EXAMPLE 8   *Finding the Component Form of a Vector*

Find the component form of the vector that represents the velocity of an airplane descending at a speed of 100 miles per hour at an angle 30° below the horizontal, as shown in Figure 8.28.

Solution

The velocity vector **v** has a magnitude of 100 and a direction angle of $\theta = 210°$.

$$\mathbf{v} = \|\mathbf{v}\| (\cos \theta)\mathbf{i} + \|\mathbf{v}\| (\sin \theta)\mathbf{j}$$
$$= 100(\cos 210°)\mathbf{i} + 100(\sin 210°)\mathbf{j}$$
$$= 100\left(\frac{-\sqrt{3}}{2}\right)\mathbf{i} + 100\left(\frac{-1}{2}\right)\mathbf{j}$$
$$= -50\sqrt{3}\,\mathbf{i} - 50\mathbf{j}$$
$$= \langle -50\sqrt{3}, -50 \rangle$$

You should check to see that $\|\mathbf{v}\| = 100$.

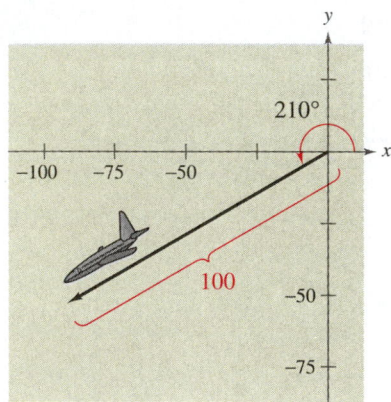

FIGURE 8.28

EXAMPLE 9   *An Application*

A force of 600 pounds is required to pull a boat and trailer up a ramp inclined at 15° from the horizontal. Find the combined weight of the boat and trailer.

Solution

Based on Figure 8.29, you can make the following observations.

$\|\overrightarrow{BA}\|$ = force of gravity = combined weight of boat and trailer

$\|\overrightarrow{BC}\|$ = force against ramp

$\|\overrightarrow{AC}\|$ = force required to move boat up ramp   = 600 pounds

By construction, triangles *BWD* and *ABC* are similar. Hence, angle *ABC* is 15°. Therefore, in triangle *ABC* you have

$$\sin 15° = \frac{\|\overrightarrow{AC}\|}{\|\overrightarrow{BA}\|} = \frac{600}{\|\overrightarrow{BA}\|}$$

$$\|BA\| = \frac{600}{\sin 15°} \approx 2318.$$

Consequently, the combined weight is approximately 2318 pounds.

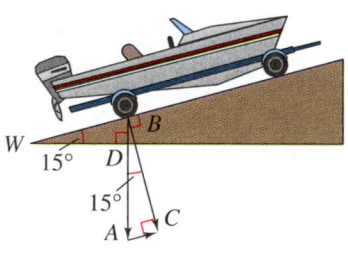

FIGURE 8.29

NOTE   In Figure 8.29, note that $\overrightarrow{AC}$ is parallel to the ramp. ∎∎

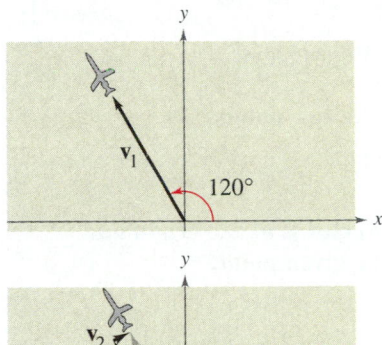

FIGURE 8.30

Real
Life

EXAMPLE 10   An Application

An airplane is traveling at a fixed altitude with a negligible wind factor. The airplane is headed N 30° W at a speed of 500 miles per hour, as shown in Figure 8.30. As the airplane reaches a certain point, it encounters a wind with a velocity of 70 miles per hour in the direction N 45° E. What are the resultant speed and direction of the airplane?

Solution

Using Figure 8.30, the velocity of the airplane (alone) is given by

$$\mathbf{v}_1 = 500\langle\cos 120°, \sin 120°\rangle = \langle-250, 250\sqrt{3}\rangle$$

and the velocity of the wind is given by

$$\mathbf{v}_2 = 70\langle\cos 45°, \sin 45°\rangle = \langle 35\sqrt{2}, 35\sqrt{2}\rangle.$$

Thus, the velocity of the airplane (in the wind) is given by

$$\mathbf{v} = \mathbf{v}_1 + \mathbf{v}_2 = \langle-250 + 35\sqrt{2}, 250\sqrt{3} + 35\sqrt{2}\rangle \approx \langle-200.5, 482.5\rangle$$

and the speed of the airplane is

$$\|\mathbf{v}\| = \sqrt{(-200.5)^2 + (482.5)^2} \approx 522.5 \text{ miles per hour.}$$

Finally, if θ if the direction angle of the flight path, you have

$$\tan \theta = \frac{482.5}{-200.5} \approx -2.4065$$

which implies that

$$\theta \approx 180° + \arctan(-2.4065) \approx 180° - 67.4° = 112.6°.$$

GROUP ACTIVITY

VERIFYING THE ASSOCIATIVITY OF VECTOR ADDITION

On page 625, you learned that vector addition is associative—that is, for vectors **u**, **v**, and **w**, $(\mathbf{u} + \mathbf{v}) + \mathbf{w} = \mathbf{u} + (\mathbf{v} + \mathbf{w})$. Use graph paper and the information in the table to demonstrate geometrically that the resultant vector is the same regardless of whether you add **u** and **v** or **v** and **w** first.

	u	**v**	**w**
Initial Point	$(-3, 2)$	$(1, 6)$	$(2, -2)$
Terminal Point	$(5, -3)$	$(4, -3)$	$(-4, 3)$

WARM UP

Find the distance between the points.

1. $(-2, 6), (5, -15)$ **2.** $(0, 0), (-3, -7)$

Find an equation of the line through the two points.

3. $(3, 1), (-2, 4)$ **4.** $(-2, -3), (4, 5)$

Find an angle θ ($0 \leq \theta \leq 360°$) whose vertex is at the origin and whose terminal side passes through the given point.

5. $(-2, 5)$ **6.** $(4, -3)$

Find the sine and cosine of the angle θ.

7. $\theta = 30°$ **8.** $\theta = 120°$

9. $\theta = 300°$ **10.** $\theta = 210°$

8.3 Exercises

In Exercises 1–10, find the component form and the magnitude of the vector v.

1.
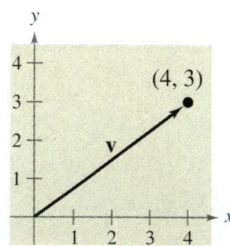

2.
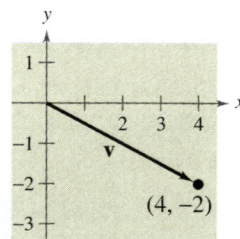

5.
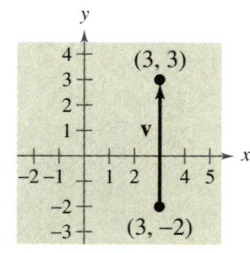

6.
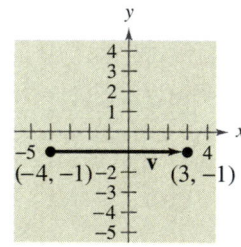

3.
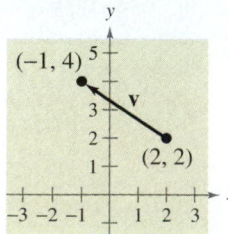

4.
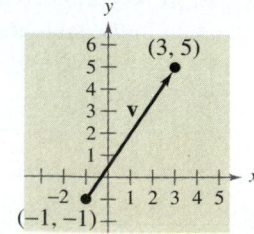

	Initial Point	Terminal Point
7.	$(-1, 5)$	$(15, 12)$
8.	$(1, 11)$	$(9, 3)$
9.	$(-3, -5)$	$(5, 1)$
10.	$(-3, 11)$	$(9, 40)$

In Exercises 11–16, use the figure to sketch a graph of the specified vector.

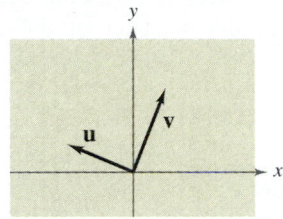

11. $-\mathbf{v}$

12. $3\mathbf{v}$

13. $\mathbf{u} + \mathbf{v}$

14. $\mathbf{u} + 2\mathbf{v}$

15. $\mathbf{u} - \mathbf{v}$

16. $\mathbf{v} - \frac{1}{2}\mathbf{u}$

In Exercises 17–24, find (a) $\mathbf{u} + \mathbf{v}$, (b) $\mathbf{u} - \mathbf{v}$, and (c) $2\mathbf{u} - 3\mathbf{v}$.

17. $\mathbf{u} = \langle 1, 2 \rangle$, $\mathbf{v} = \langle 3, 1 \rangle$

18. $\mathbf{u} = \langle 2, 3 \rangle$, $\mathbf{v} = \langle 4, 0 \rangle$

19. $\mathbf{u} = \langle 4, -2 \rangle$, $\mathbf{v} = \langle 0, 0 \rangle$

20. $\mathbf{u} = \langle 0, 0 \rangle$, $\mathbf{v} = \langle 2, 1 \rangle$

21. $\mathbf{u} = \mathbf{i} + \mathbf{j}$, $\mathbf{v} = 2\mathbf{i} - 3\mathbf{j}$

22. $\mathbf{u} = 2\mathbf{i} - \mathbf{j}$, $\mathbf{v} = -\mathbf{i} + \mathbf{j}$

23. $\mathbf{u} = 2\mathbf{i}$, $\mathbf{v} = \mathbf{j}$

24. $\mathbf{u} = 3\mathbf{j}$, $\mathbf{v} = 2\mathbf{i}$

In Exercises 25–32, find a unit vector in the direction of the given vector.

25. $\mathbf{u} = \langle 5, 0 \rangle$

26. $\mathbf{u} = \langle 0, -3 \rangle$

27. $\mathbf{v} = \langle -2, 2 \rangle$

28. $\mathbf{v} = \langle 5, -12 \rangle$

29. $\mathbf{v} = 4\mathbf{i} - 3\mathbf{j}$

30. $\mathbf{v} = \mathbf{i} + \mathbf{j}$

31. $\mathbf{w} = 2\mathbf{j}$

32. $\mathbf{w} = \mathbf{i} - 2\mathbf{j}$

In Exercises 33–36, find the vector $\mathbf{v}$ with the given magnitude and the same direction as $\mathbf{u}$.

	Magnitude	Direction
33.	$\|\mathbf{v}\| = 5$	$\mathbf{u} = \langle 3, 3 \rangle$
34.	$\|\mathbf{v}\| = 3$	$\mathbf{u} = \langle 4, -4 \rangle$
35.	$\|\mathbf{v}\| = 7$	$\mathbf{u} = \langle -3, 4 \rangle$
36.	$\|\mathbf{v}\| = 10$	$\mathbf{u} = \langle -10, 0 \rangle$

In Exercises 37–42, find the component form of $\mathbf{v}$ and sketch the specified vector operations geometrically, where $\mathbf{u} = 2\mathbf{i} - \mathbf{j}$ and $\mathbf{w} = \mathbf{i} + 2\mathbf{j}$.

37. $\mathbf{v} = \frac{3}{2}\mathbf{u}$

38. $\mathbf{v} = \mathbf{u} + \mathbf{w}$

39. $\mathbf{v} = \mathbf{u} + 2\mathbf{w}$

40. $\mathbf{v} = -\mathbf{u} + \mathbf{w}$

41. $\mathbf{v} = \frac{1}{2}(3\mathbf{u} + \mathbf{w})$

42. $\mathbf{v} = \mathbf{u} - 2\mathbf{w}$

In Exercises 43–46, find the magnitude and direction angle of the vector $\mathbf{v}$.

43. $\mathbf{v} = 5(\cos 30°\mathbf{i} + \sin 30°\mathbf{j})$

44. $\mathbf{v} = 8(\cos 135°\mathbf{i} + \sin 135°\mathbf{j})$

45. $\mathbf{v} = 6\mathbf{i} - 6\mathbf{j}$

46. $\mathbf{v} = -2\mathbf{i} + 5\mathbf{j}$

In Exercises 47–54, find the component form of $\mathbf{v}$ given its magnitude and the angle it makes with the positive x-axis. Sketch $\mathbf{v}$.

	Magnitude	Angle
47.	$\|\mathbf{v}\| = 3$	$\theta = 0°$
48.	$\|\mathbf{v}\| = 1$	$\theta = 45°$
49.	$\|\mathbf{v}\| = 1$	$\theta = 150°$
50.	$\|\mathbf{v}\| = \frac{5}{2}$	$\theta = 45°$
51.	$\|\mathbf{v}\| = 3\sqrt{2}$	$\theta = 150°$
52.	$\|\mathbf{v}\| = 9$	$\theta = 90°$
53.	$\|\mathbf{v}\| = 2$	$\mathbf{v}$ in the direction $\mathbf{i} + 3\mathbf{j}$
54.	$\|\mathbf{v}\| = 3$	$\mathbf{v}$ in the direction $3\mathbf{i} + 4\mathbf{j}$

In Exercises 55–58, find the component form of the sum of $\mathbf{u}$ and $\mathbf{v}$ with direction angles $\theta_{\mathbf{u}}$ and $\theta_{\mathbf{v}}$.

	Magnitude	Angle
55.	$\|\mathbf{u}\| = 5$	$\theta_{\mathbf{u}} = 0°$
	$\|\mathbf{v}\| = 5$	$\theta_{\mathbf{v}} = 90°$
56.	$\|\mathbf{u}\| = 2$	$\theta_{\mathbf{u}} = 30°$
	$\|\mathbf{v}\| = 2$	$\theta_{\mathbf{v}} = 90°$
57.	$\|\mathbf{u}\| = 20$	$\theta_{\mathbf{u}} = 45°$
	$\|\mathbf{v}\| = 50$	$\theta_{\mathbf{v}} = 180°$
58.	$\|\mathbf{u}\| = 35$	$\theta_{\mathbf{u}} = 25°$
	$\|\mathbf{v}\| = 50$	$\theta_{\mathbf{v}} = 120°$

In Exercises 59–62, use the Law of Cosines to find the angle α between the given vectors. (Assume $0° \le \alpha \le 180°$.)

59. $v = i + j$, $w = 2(i - j)$

60. $v = 3i + j$, $w = 2i - j$

61. $v = i + j$, $w = 3i - j$

62. $v = i + 2j$, $w = 2i - j$

In Exercises 63 and 64, find the angle between the forces given the magnitude of their resultant. (*Hint:* Write force one as a vector in the direction of the positive x-axis and force two as a vector at an angle θ with the positive x-axis.)

	Force One	*Force Two*	*Resultant Force*
63.	45 pounds	60 pounds	90 pounds
64.	3000 pounds	1000 pounds	3750 pounds

65. *Think About It*   Consider two forces of equal magnitude acting on a point.

(a) If the magnitude of the resultant is the sum of the magnitudes of the two forces, make a conjecture about the angle between the forces.

(b) If the resultant of the forces is **0,** make a conjecture about the angle between the forces.

(c) Can the magnitude of the resultant be greater than the sum of the magnitudes of the two forces? Explain.

66. *Graphical Reasoning*   Consider two forces

$\mathbf{F}_1 = \langle 10, 0 \rangle$ and $\mathbf{F}_2 = 5\langle \cos \theta, \sin \theta \rangle$.

(a) Find $\|\mathbf{F}_1 + \mathbf{F}_2\|$.

(b) Determine the magnitude of the resultant as a function of θ. Use a graphing utility to graph the function for $0 \le \theta < 2\pi$.

(c) Use the graph in part (b) to determine the range of the function. What is its maximum, and for what value of θ does it occur? What is its minimum, and for what value of θ does it occur?

(d) Explain why the magnitude of the resultant is never 0.

67. *Resultant Force*   Forces with magnitudes of 150 newtons and 220 newtons act on a hook (see figure). The angle between the two forces is 30°. Find the direction and magnitude of the resultant of these forces.

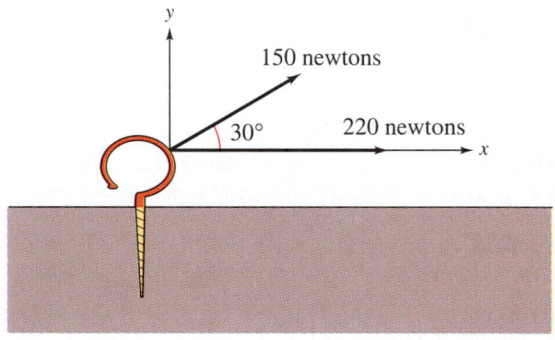

68. *Resultant Force*   Forces with magnitudes of 2000 newtons and 900 newtons act on a machine part at angles of 30° and −45°, respectively, with the x-axis (see figure). Find the direction and magnitude of the resultant of these forces.

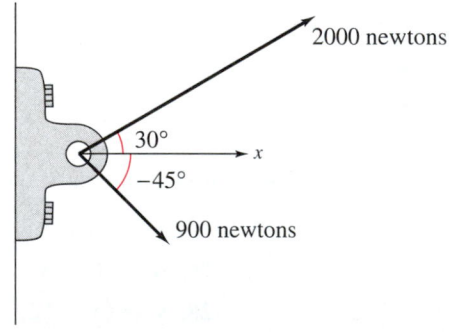

69. *Resultant Force*   Three forces with magnitudes of 75 pounds, 100 pounds, and 125 pounds act on an object at angles of 30°, 45°, 120°, respectively, with the positive x-axis. Find the direction and magnitude of the resultant of these forces.

70. *Resultant Force*   Three forces with magnitudes of 70 pounds, 40 pounds, and 60 pounds act on an object at angles of −30°, 45°, and 135°, respectively, with the positive x-axis. Find the direction and magnitude of the resultant of these forces.

71. *Horizontal and Vertical Components of Velocity* A ball is thrown with an initial velocity of 80 feet per second, at an angle of 40° with the horizontal (see figure). Find the vertical and horizontal components of the velocity.

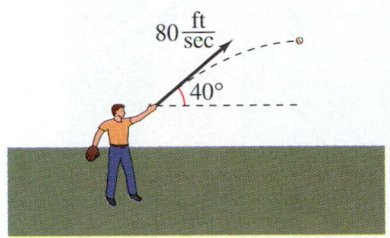

72. *Horizontal and Vertical Components of Velocity* A gun with a muzzle velocity of 1200 feet per second is fired at an angle of 6° with the horizontal. Find the vertical and horizontal components of the velocity.

Cable Tension **In Exercises 73 and 74, use the figure to determine the tension in each cable supporting the given load.**

73.

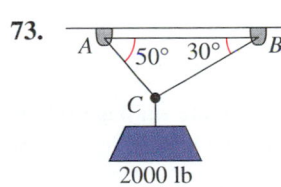

74.

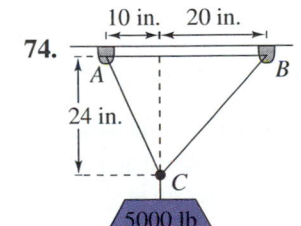

75. *Barge Towing* A loaded barge is being towed by two tugboats, and the magnitude of the resultant is 6000 pounds directed along the axis of the barge (see figure). Find the tension in the tow lines if they each make an 18° angle with the axis of the barge.

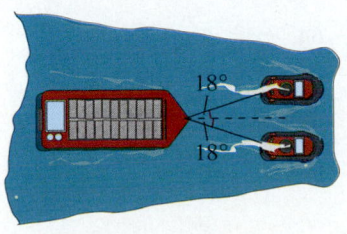

76. *Shared Load* To carry a 100-pound cylindrical weight, two people lift on the ends of short ropes that are tied to an eyelet on the top center of the cylinder. Each rope makes a 20° angle with the vertical. Draw a figure that gives a visual representation of the problem, and find the tension in the ropes.

77. *Navigation* An airplane is flying in the direction S 32° E, with an airspeed of 875 kilometers per hour. Because of the wind, its ground speed and direction are 800 kilometers per hour and S 40° E, respectively (see figure). Find the direction and speed of the wind.

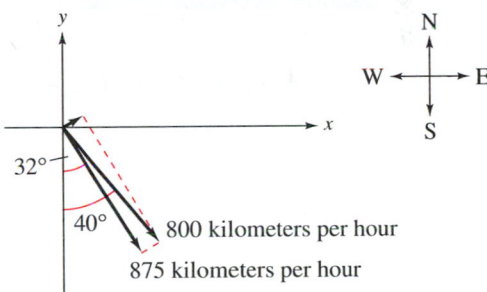

78. *Navigation* An airplane's velocity with respect to the air is 580 miles per hour, and it is heading N 60° W. The wind, at the altitude of the plane, is from the southwest and has a velocity of 60 miles per hour. Draw a figure that gives a visual representation of the problem. What is the true direction of the plane, and what is its speed with respect to the ground?

79. *Work* A heavy implement is pulled 20 feet across the floor, using a force of 85 pounds. Find the work done if the direction of the force is 60° above the horizontal (see figure). (Use the formula for work, $W = FD$, where F is the component of the force in the direction of motion and D is the distance.)

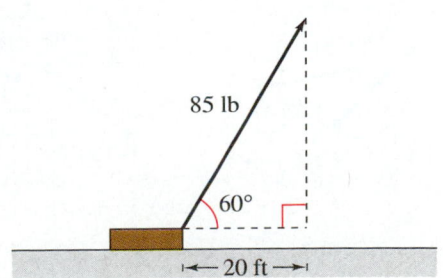

80. *Tether Ball* A tether ball weighing 1 pound is pulled outward from the pole by a horizontal force **u** until the rope makes a 45° angle with the pole (see figure). Determine the resulting tension in the rope and the magnitude of **u**.

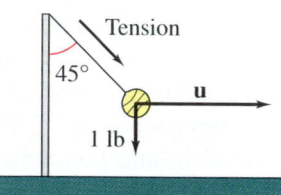

True or False? **In Exercises 81–84, decide whether the statement is true or false. If it is false, explain why or give an example that shows it is false.**

81. If **u** and **v** have the same magnitude and direction, then **u** = **v**.

82. If **u** is a unit vector in the direction of **v**, then **v** = $\|\mathbf{v}\|\,\mathbf{u}$.

83. If $\mathbf{v} = a\mathbf{i} + b\mathbf{j} = \mathbf{0}$, then $a = -b$.

84. If $\mathbf{u} = a\mathbf{i} + b\mathbf{j}$ is a unit vector, then $a^2 + b^2 = 1$.

85. Prove that $(\cos\theta)\mathbf{i} + (\sin\theta)\mathbf{j}$ is a unit vector for any value of θ.

86. *Technology* Write a program for your graphing utility that graphs two vectors and their difference given the vectors in component form.

In Exercises 87 and 88, use the result of Exercise 86 to find the difference of the vectors shown in the figure.

87.

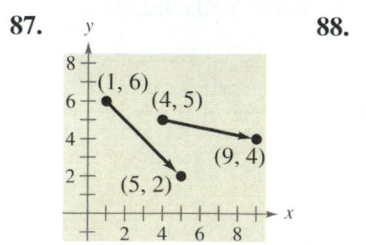

88.

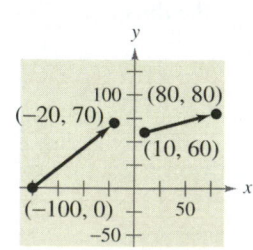

Geodetic surveying is a method of determining the position of points on the earth's surface and the dimension of areas so large that the curvature of the earth must be taken into account. *(Photo: Zigy Kaluzny/ Tony Stone Images)*

89. *Chapter Opener* Turn to the figure on page 601 to approximate the distance from the bridge deck to the water level.

90. *Chapter Opener* Approximate a in the figure on page 601 if the distance between A and C is 27 meters.

Review **Solve Exercises 91–94 as a review of the skills and problem-solving techniques you learned in previous sections. Use the specified trigonometric substitution to write the algebraic expression as a trigonometric function of θ, where $0 < \theta < \pi/2$.**

91. $\sqrt{x^2 - 64}, \quad x = 8\sec\theta$

92. $\sqrt{64 - x^2}, \quad x = 8\sin\theta$

93. $\sqrt{x^2 + 36}, \quad x = 6\tan\theta$

94. $\sqrt{(x^2 - 25)^3}, \quad x = 5\sec\theta$

8.4 *Vectors and Dot Products*

See Exercise 43 on page 644 for an example of how the dot product can be used to find the force necessary to keep a truck from rolling down a hill.

The Dot Product of Two Vectors □ *The Angle Between Two Vectors* □
Finding Vector Components □ *Work*

The Dot Product of Two Vectors

So far you have studied two vector operations—vector addition and multiplication by a scalar—each of which yields another vector. In this section you will study a third vector operation, the **dot product.** This product yields a scalar, rather than a vector.

DEFINITION OF DOT PRODUCT

The **dot product** of $\mathbf{u} = \langle u_1, u_2 \rangle$ and $\mathbf{v} = \langle v_1, v_2 \rangle$ is

$$\mathbf{u} \cdot \mathbf{v} = u_1 v_1 + u_2 v_2.$$

PROPERTIES OF THE DOT PRODUCT

Let $\mathbf{u}$, $\mathbf{v}$, and $\mathbf{w}$ be vectors in the plane or in space and let c be a scalar.

1. $\mathbf{u} \cdot \mathbf{v} = \mathbf{v} \cdot \mathbf{u}$
2. $\mathbf{0} \cdot \mathbf{v} = 0$
3. $\mathbf{u} \cdot (\mathbf{v} + \mathbf{w}) = \mathbf{u} \cdot \mathbf{v} + \mathbf{u} \cdot \mathbf{w}$
4. $\mathbf{v} \cdot \mathbf{v} = \|\mathbf{v}\|^2$
5. $c(\mathbf{u} \cdot \mathbf{v}) = c\mathbf{u} \cdot \mathbf{v} = \mathbf{u} \cdot c\mathbf{v}$

THINK ABOUT THE PROOF

To prove the second, third, and fifth properties of the dot product, consider the component forms of vectors $\mathbf{u}$, $\mathbf{v}$, and $\mathbf{w}$. The details of the proof are given in the appendix.

PROOF ▪▪

To prove the first property, let $\mathbf{u} = \langle u_1, u_2 \rangle$ and $\mathbf{v} = \langle v_1\ v_2 \rangle$. Then

$$\mathbf{u} \cdot \mathbf{v} = u_1 v_1 + u_2 v_2$$
$$= v_1 u_1 + v_2 u_2$$
$$= \mathbf{v} \cdot \mathbf{u}.$$

For the fourth property, let $\mathbf{v} = \langle v_1, v_2 \rangle$. Then

$$\mathbf{v} \cdot \mathbf{v} = v_1^2 + v_2^2$$
$$= \left(\sqrt{v_1^2 + v_2^2} \right)^2$$
$$= \|\mathbf{v}\|^2. \quad ▪▪$$

NOTE In Example 1, be sure you see that the dot product of two vectors is a scalar (a real number), not a vector. Moreover, notice that the dot product can be positive, zero, or negative. ■■

TECHNOLOGY
■■

A graphing utility can be used to find the angle between two vectors. The following program for the *TI-83* or the *TI-82* sketches two vectors $\mathbf{u} = \langle a, b \rangle$ and $\mathbf{v} = \langle c, d \rangle$ in standard position and finds the measure of the angle between them. Use the program to verify Example 4. (Before running the program, set an appropriate viewing rectangle.)

```
:VECANGL
:ClrHome
:Disp "ENTER(A,B)"
:Input "ENTER A",A
:Input "ENTER B",B
:ClrHome
:Disp "ENTER(C,D)"
:Input "ENTER C",C
:Input "ENTER D",D
:Line(0,0,A,B)
:Line(0,0,C,D)
:Pause
:AC+BD→E
:√(A²+B²)→U
:√(C²+D²)→V
:cos⁻¹(E/(UV))→θ
:ClrDraw:ClrHome
:Disp "θ=",θ
:Stop
```

EXAMPLE 1   *Finding Dot Products*

Find each dot product.

a. $\langle 4, 5 \rangle \cdot \langle 2, 3 \rangle$

b. $\langle 2, -1 \rangle \cdot \langle 1, 2 \rangle$

c. $\langle 0, 3 \rangle \cdot \langle 4, -2 \rangle$

Solution

a. $\langle 4, 5 \rangle \cdot \langle 2, 3 \rangle = 4(2) + 5(3) = 8 + 15 = 23$

b. $\langle 2, -1 \rangle \cdot \langle 1, 2 \rangle = 2(1) + (-1)(2) = 2 - 2 = 0$

c. $\langle 0, 3 \rangle \cdot \langle 4, -2 \rangle = 0(4) + 3(-2) = 0 - 6 = -6$

EXAMPLE 2   *Using Properties of Dot Products*

Let $\mathbf{u} = \langle -1, 3 \rangle$, $\mathbf{v} = \langle 2, -4 \rangle$, and $\mathbf{w} = \langle 1, -2 \rangle$. Find each dot product.

a. $(\mathbf{u} \cdot \mathbf{v})\mathbf{w}$ **b.** $\mathbf{u} \cdot 2\mathbf{v}$

Solution

Begin by finding the dot product of $\mathbf{u}$ and $\mathbf{v}$.

$$\mathbf{u} \cdot \mathbf{v} = \langle -1, 3 \rangle \cdot \langle 2, -4 \rangle$$
$$= (-1)(2) + 3(-4)$$
$$= -14$$

a. $(\mathbf{u} \cdot \mathbf{v})\mathbf{w} = -14\langle 1, -2 \rangle = \langle -14, 28 \rangle$

b. $\mathbf{u} \cdot 2\mathbf{v} = 2(\mathbf{u} \cdot \mathbf{v}) = 2(-14) = -28$

Notice that the first product is a vector, whereas the second is a scalar. Can you see why?

EXAMPLE 3   *Dot Product and Length*

The dot product of $\mathbf{u}$ with itself is 5. What is the length of $\mathbf{u}$?

Solution

Because $\|\mathbf{u}\|^2 = \mathbf{u} \cdot \mathbf{u} = 5$, it follows that

$$\|\mathbf{u}\| = \sqrt{\mathbf{u} \cdot \mathbf{u}}$$
$$= \sqrt{5}.$$

The Angle Between Two Vectors

The **angle between two nonzero vectors** is the angle θ, $0 \leq \theta \leq \pi$, between their respective standard position vectors, as shown in Figure 8.31. This angle can be found using the dot product. (Note that the angle between the zero vector and another vector is not defined.)

> ### ANGLE BETWEEN TWO VECTORS
>
> If θ is the angle between two nonzero vectors $\mathbf{u}$ and $\mathbf{v}$, then
> $$\cos \theta = \frac{\mathbf{u} \cdot \mathbf{v}}{\|\mathbf{u}\| \, \|\mathbf{v}\|}.$$

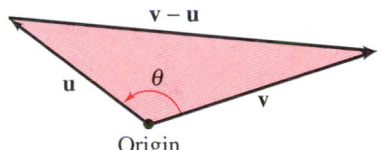

FIGURE 8.31

PROOF ▪▪

Consider the triangle determined by vectors $\mathbf{u}$, $\mathbf{v}$, and $\mathbf{v} - \mathbf{u}$, as shown in Figure 8.31. By the Law of Cosines, you can write

$$\|\mathbf{v} - \mathbf{u}\|^2 = \|\mathbf{u}\|^2 + \|\mathbf{v}\|^2 - 2\|\mathbf{u}\| \, \|\mathbf{v}\| \cos \theta$$

$$(\mathbf{v} - \mathbf{u}) \cdot (\mathbf{v} - \mathbf{u}) = \|\mathbf{u}\|^2 + \|\mathbf{v}\|^2 - 2\|\mathbf{u}\| \, \|\mathbf{v}\| \cos \theta$$

$$(\mathbf{v} - \mathbf{u}) \cdot \mathbf{v} - (\mathbf{v} - \mathbf{u}) \cdot \mathbf{u} = \|\mathbf{u}\|^2 + \|\mathbf{v}\|^2 - 2\|\mathbf{u}\| \, \|\mathbf{v}\| \cos \theta$$

$$\mathbf{v} \cdot \mathbf{v} - \mathbf{u} \cdot \mathbf{v} - \mathbf{v} \cdot \mathbf{u} + \mathbf{u} \cdot \mathbf{u} = \|\mathbf{u}\|^2 + \|\mathbf{v}\|^2 - 2\|\mathbf{u}\| \, \|\mathbf{v}\| \cos \theta$$

$$\|\mathbf{v}\|^2 - 2\mathbf{u} \cdot \mathbf{v} + \|\mathbf{u}\|^2 = \|\mathbf{u}\|^2 + \|\mathbf{v}\|^2 - 2\|\mathbf{u}\| \, \|\mathbf{v}\| \cos \theta$$

$$-2\mathbf{u} \cdot \mathbf{v} = -2\|\mathbf{u}\| \, \|\mathbf{v}\| \cos \theta$$

$$\cos \theta = \frac{\mathbf{u} \cdot \mathbf{v}}{\|\mathbf{u}\| \, \|\mathbf{v}\|}. \qquad ▪▪$$

EXAMPLE 4   *Finding the Angle Between Two Vectors*

Find the angle between $\mathbf{u} = \langle 4, 3 \rangle$ and $\mathbf{v} = \langle 3, 5 \rangle$.

Solution

$$\cos \theta = \frac{\mathbf{u} \cdot \mathbf{v}}{\|\mathbf{u}\| \, \|\mathbf{v}\|} = \frac{\langle 4, 3 \rangle \cdot \langle 3, 5 \rangle}{\|\langle 4, 3 \rangle\| \, \|\langle 3, 5 \rangle\|} = \frac{27}{5\sqrt{34}}$$

This implies that the angle between the two vectors is

$$\theta = \arccos \frac{27}{5\sqrt{34}} \approx 22.2°$$

as shown in Figure 8.32.

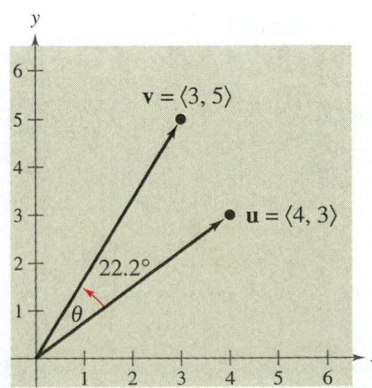

FIGURE 8.32

Rewriting the expression for the angle between two vectors in the form

$$\mathbf{u} \cdot \mathbf{v} = \|\mathbf{u}\| \, \|\mathbf{v}\| \cos \theta \qquad \text{\color{red}Alternative form of dot product}$$

produces an alternative way to calculate the dot product. From this form, you can see that because $\|\mathbf{u}\|$ and $\|\mathbf{v}\|$ are always positive, $\mathbf{u} \cdot \mathbf{v}$ and $\cos \theta$ will always have the same sign. Figure 8.33 shows the five possible orientations of two vectors.

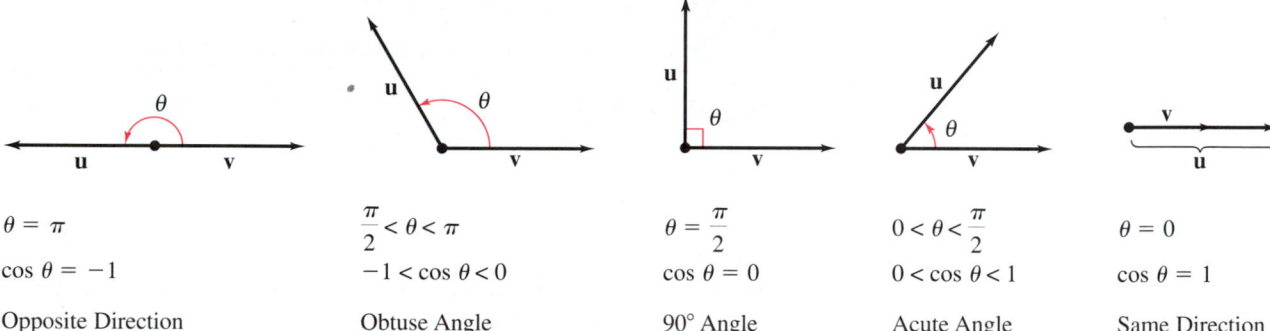

$\theta = \pi$	$\dfrac{\pi}{2} < \theta < \pi$	$\theta = \dfrac{\pi}{2}$	$0 < \theta < \dfrac{\pi}{2}$	$\theta = 0$
$\cos \theta = -1$	$-1 < \cos \theta < 0$	$\cos \theta = 0$	$0 < \cos \theta < 1$	$\cos \theta = 1$
Opposite Direction	Obtuse Angle	90° Angle	Acute Angle	Same Direction

FIGURE 8.33

> ### DEFINITION OF ORTHOGONAL VECTORS
> The vectors $\mathbf{u}$ and $\mathbf{v}$ are **orthogonal** if $\mathbf{u} \cdot \mathbf{v} = 0$.

The terms "orthogonal" and "perpendicular" mean essentially the same thing—meeting at right angles. By definition, however, the zero vector is orthogonal to every vector $\mathbf{u}$, because $\mathbf{0} \cdot \mathbf{u} = 0$.

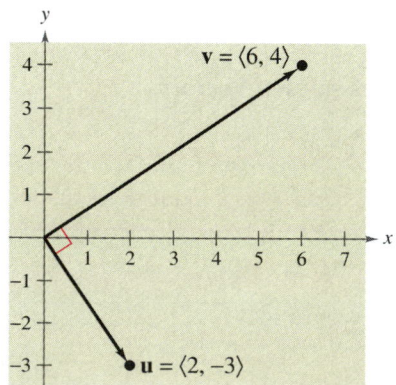

FIGURE 8.34

EXAMPLE 5   Determining Orthogonal Vectors

Are the vectors $\mathbf{u} = \langle 2, -3 \rangle$ and $\mathbf{v} = \langle 6, 4 \rangle$ orthogonal?

Solution

Begin by finding the dot product of the two vectors.

$$\begin{aligned}
\mathbf{u} \cdot \mathbf{v} &= \langle 2, -3 \rangle \cdot \langle 6, 4 \rangle \\
&= 2(6) + (-3)(4) \\
&= 0
\end{aligned}$$

Because the dot product is 0, the two vectors are orthogonal, as shown in Figure 8.34.

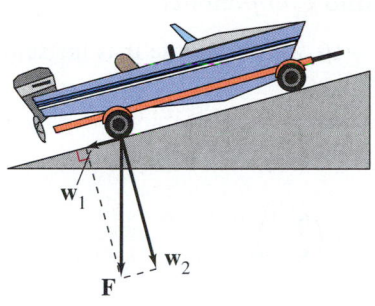

FIGURE 8.35

Finding Vector Components

You have already seen applications in which two vectors are added to produce a resultant vector. Many applications in physics and engineering pose the reverse problem—decomposing a given vector into the sum of two **vector components.**

Consider a boat on an inclined ramp, as shown in Figure 8.35. The force **F** due to gravity pulls the boat *down* the ramp and *against* the ramp. These two orthogonal forces, $\mathbf{w}_1$ and $\mathbf{w}_2$, are vector components of **F**. That is,

$$\mathbf{F} = \mathbf{w}_1 + \mathbf{w}_2.\qquad \text{Vector components of } \mathbf{F}$$

The negative of component $\mathbf{w}_1$ represents the force needed to keep the boat from rolling down the ramp, whereas $\mathbf{w}_2$ represents the force that the tires must withstand against the ramp. A procedure for finding $\mathbf{w}_1$ and $\mathbf{w}_2$ is shown below.

DEFINITION OF VECTOR COMPONENTS

Let **u** and **v** be nonzero vectors such that

$$\mathbf{u} = \mathbf{w}_1 + \mathbf{w}_2$$

where $\mathbf{w}_1$ and $\mathbf{w}_2$ are orthogonal and $\mathbf{w}_1$ is parallel to **v**, as shown in Figure 8.36. The vectors $\mathbf{w}_1$ and $\mathbf{w}_2$ are called **vector components** of **u**. The vector $\mathbf{w}_1$ is the **projection** of **u** onto **v** and is denoted by

$$\mathbf{w}_1 = \text{proj}_{\mathbf{v}}\mathbf{u}.$$

The vector $\mathbf{w}_2$ is given by $\mathbf{w}_2 = \mathbf{u} - \mathbf{w}_1$.

Study Tip

From the definition of vector components, you can see that it is easy to find the component $\mathbf{w}_2$ once you have found the projection of **u** onto **v**. To find the projection, you can use the dot product.

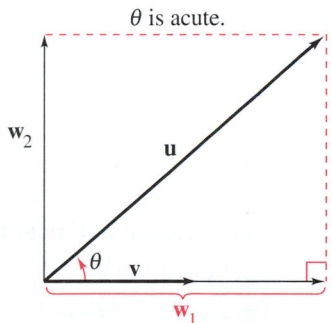

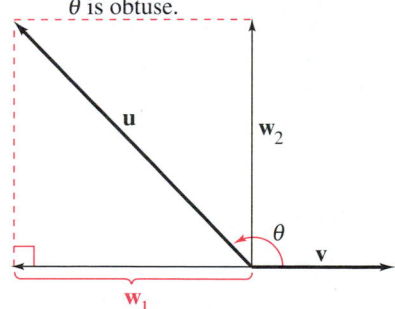

FIGURE 8.36

PROJECTION OF U ONTO V

Let **u** and **v** be nonzero vectors. The projection of **u** onto **v** is

$$\text{proj}_{\mathbf{v}}\mathbf{u} = \left(\frac{\mathbf{u} \cdot \mathbf{v}}{\|\mathbf{v}\|^2}\right)\mathbf{v}.$$

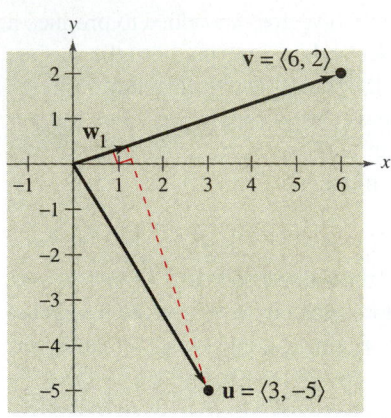

FIGURE 8.37

EXAMPLE 6   *Decomposing a Vector into Components*

Find the projection of $\mathbf{u} = \langle 3, -5 \rangle$ onto $\mathbf{v} = \langle 6, 2 \rangle$. Then write $\mathbf{u}$ as the sum of two orthogonal vectors, one of which is $\text{proj}_\mathbf{v}\mathbf{u}$.

Solution

The projection of $\mathbf{u}$ onto $\mathbf{v}$ is

$$\mathbf{w}_1 = \text{proj}_\mathbf{v}\mathbf{u} = \left(\frac{\mathbf{u} \cdot \mathbf{v}}{\|\mathbf{v}\|^2}\right)\mathbf{v} = \left(\frac{8}{40}\right)\langle 6, 2 \rangle = \left\langle \frac{6}{5}, \frac{2}{5} \right\rangle,$$

as shown in Figure 8.37. The other component, $\mathbf{w}_2$, is

$$\mathbf{w}_2 = \mathbf{u} - \mathbf{w}_1 = \langle 3, -5 \rangle - \left\langle \frac{6}{5}, \frac{2}{5} \right\rangle = \left\langle \frac{9}{5}, -\frac{27}{5} \right\rangle.$$

Thus, $\mathbf{u} = \mathbf{w}_1 + \mathbf{w}_2 = \left\langle \frac{6}{5}, \frac{2}{5} \right\rangle + \left\langle \frac{9}{5}, -\frac{27}{5} \right\rangle = \langle 3, -5 \rangle.$

EXAMPLE 7   *Finding a Force*

Real Life

A 600-pound boat sits on a ramp inclined at 30°, as shown in Figure 8.38. What force is required to keep the boat from rolling down the ramp?

Solution

Because the force due to gravity is vertical and downward, you can represent the gravitational force by the vector

$$\mathbf{F} = -600\mathbf{j}. \qquad \text{\color{red}Force due to gravity}$$

To find the force required to keep the boat from rolling down the ramp, project $\mathbf{F}$ onto a unit vector $\mathbf{v}$ in the direction of the ramp, as follows.

$$\mathbf{v} = (\cos 30°)\mathbf{i} + (\sin 30°)\mathbf{j} = \frac{\sqrt{3}}{2}\mathbf{i} + \frac{1}{2}\mathbf{j} \qquad \text{\color{red}Unit vector along ramp}$$

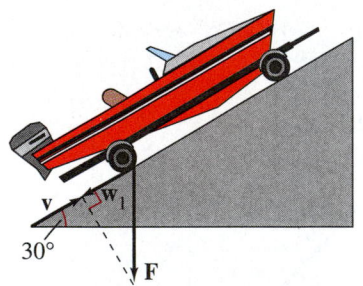

FIGURE 8.38

Therefore, the projection of $\mathbf{F}$ onto $\mathbf{v}$ is given by

$$\mathbf{w}_1 = \text{proj}_\mathbf{v}\mathbf{F}$$
$$= \left(\frac{\mathbf{F} \cdot \mathbf{v}}{\|\mathbf{v}\|^2}\right)\mathbf{v}$$
$$= (\mathbf{F} \cdot \mathbf{v})\mathbf{v} = (-600)\left(\frac{1}{2}\right)\mathbf{v} = -300\left(\frac{\sqrt{3}}{2}\mathbf{i} + \frac{1}{2}\mathbf{j}\right).$$

The magnitude of this force is 300, and therefore a force of 300 pounds is required to keep the boat from rolling down the ramp.

Work

(a)

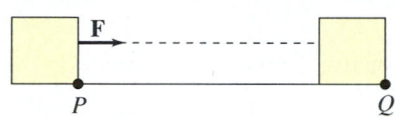

(b)

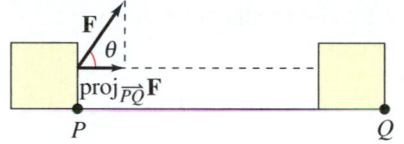

FIGURE 8.39

The work W done by a constant force $\mathbf{F}$ acting along the line of motion of an object is given by

$$W = (\text{magnitude of force})(\text{distance})$$
$$= \|\mathbf{F}\|\,\|\overrightarrow{PQ}\|$$

as shown in Figure 8.39(a). If the constant force $\mathbf{F}$ is not directed along the line of motion, as shown in Figure 8.39(b), the work W done by the force is

$$W = \|\text{proj}_{\overrightarrow{PQ}}\,\mathbf{F}\|\,\|\overrightarrow{PQ}\| = (\cos\theta)\|\mathbf{F}\|\,\|\overrightarrow{PQ}\| = \mathbf{F}\cdot PQ.$$

This notion of work is summarized in the following definition.

 DEFINITION OF WORK

The **work** W done by a constant force $\mathbf{F}$ as its point of application moves along the vector $\overrightarrow{PQ}$ is given by either of the following.

1. $W = \|\text{proj}_{\overrightarrow{PQ}}\,\mathbf{F}\|\,\|\overrightarrow{PQ}\|$ Projection form

2. $W = \mathbf{F}\cdot\overrightarrow{PQ}$ Dot product form

EXAMPLE 8   Finding Work

To close a sliding door, a person pulls on a rope with a constant force of 50 pounds at a constant angle of $60°$, as shown in Figure 8.40. Find the work done in moving the door 12 feet to its closed position.

Solution

Using a projection, you can calculate the work as follows.

$$W = \|\text{proj}_{\overrightarrow{PQ}}\,\mathbf{F}\|\,\|\overrightarrow{PQ}\|$$
$$= (\cos 60°)\|\mathbf{F}\|\,\|\overrightarrow{PQ}\|$$
$$= \frac{1}{2}(50)(12)$$
$$= 300 \text{ foot–pounds}$$

Thus, the work done is 300 foot-pounds.

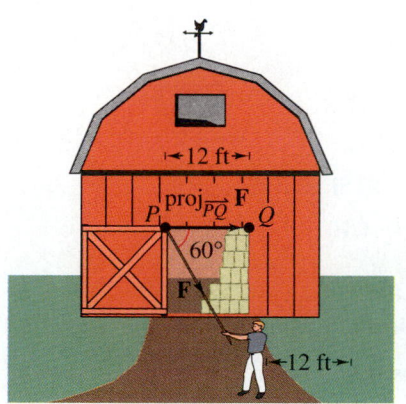

FIGURE 8.40

GROUP ACTIVITY

THE SIGN OF THE DOT PRODUCT

On page 638, you were given the alternative form of the dot product of two vectors.

$$\mathbf{u} \cdot \mathbf{v} = \|\mathbf{u}\| \, \|\mathbf{v}\| \cos \theta \qquad \text{Alternative form of dot product}$$

Use this form to determine the sign of the dot product of **u** and **v** for the vectors shown below. Explain your reasoning.

a. 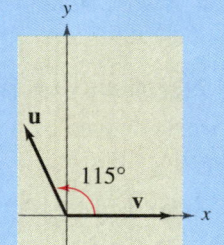   **b.** 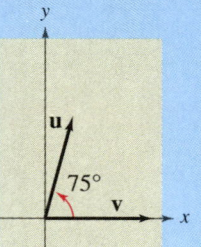   **c.**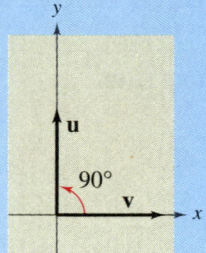

WARM UP

Find (a) u + 2v and (b) ‖u‖.

1. $\mathbf{u} = \langle 6, -3 \rangle$
   $\mathbf{v} = \langle -10, -1 \rangle$

2. $\mathbf{u} = \langle \frac{3}{8}, \frac{4}{5} \rangle$
   $\mathbf{v} = \langle \frac{5}{2}, -\frac{1}{10} \rangle$

3. $\mathbf{u} = 4\mathbf{i} - 16\mathbf{j}$
   $\mathbf{v} = -5\mathbf{i} + 10\mathbf{j}$

4. $\mathbf{u} = 0.5\mathbf{i} + 1.4\mathbf{j}$
   $\mathbf{v} = 4.1\mathbf{i} - 1.8\mathbf{j}$

Find the values of θ in the interval $0 \le \theta \le 2\pi$ that satisfy the equation. Round the result to two decimal places.

5. $\cos \theta = -\frac{1}{2}$

6. $\cos \theta = 0$

7. $\cos \theta = 0.5403$

8. $\cos \theta = -0.9689$

Find a unit vector (a) in the direction of u and (b) in the direction opposite that of u.

9. $\mathbf{u} = \langle 120, -50 \rangle$

10. $\mathbf{u} = \langle \frac{4}{5}, \frac{1}{3} \rangle$

8.4 Exercises

In Exercises 1–4, find the dot product of u and v.

1. $\mathbf{u} = \langle 3, 4 \rangle$
   $\mathbf{v} = \langle 2, -3 \rangle$

2. $\mathbf{u} = \langle 5, 12 \rangle$
   $\mathbf{v} = \langle -3, 2 \rangle$

3. $\mathbf{u} = 4\mathbf{i} - 2\mathbf{j}$
   $\mathbf{v} = \mathbf{i} - \mathbf{j}$

4. $\mathbf{u} = 2\mathbf{i} + 5\mathbf{j}$
   $\mathbf{v} = 9\mathbf{i} - 3\mathbf{j}$

In Exercises 5–8, use the vectors $\mathbf{u} = \langle 2, 2 \rangle$ and $\mathbf{v} = \langle -3, 4 \rangle$ to find the indicated quantity. State whether the result is a vector or a scalar.

5. $\mathbf{u} \cdot \mathbf{u}$

6. $\|\mathbf{u}\| - 2$

7. $(\mathbf{u} \cdot \mathbf{v})\mathbf{v}$

8. $\mathbf{u} \cdot 2\mathbf{v}$

In Exercises 9–12, use the dot product to find the length of u.

9. $\mathbf{u} = \langle -5, 12 \rangle$

10. $\mathbf{u} = \langle 2, -4 \rangle$

11. $\mathbf{u} = 20\mathbf{i} + 25\mathbf{j}$

12. $\mathbf{u} = 6\mathbf{j}$

13. *Revenue* The vector $\mathbf{u} = \langle 1245, 2600 \rangle$ gives the numbers of units of two products produced by a company. The vector $\mathbf{v} = \langle 12.20, 8.50 \rangle$ gives the price (in dollars) of each unit, respectively. Find the dot product, $\mathbf{u} \cdot \mathbf{v}$, and explain what information it gives.

14. *Revenue* Repeat Exercise 13 after increasing the prices by 5%. Identify the vector operation used to increase the prices by 5%.

In Exercises 15–20, find the angle θ between the vectors.

15. $\mathbf{u} = \langle 1, 0 \rangle$
   $\mathbf{v} = \langle 0, -2 \rangle$

16. $\mathbf{u} = \langle 4, 4 \rangle$
   $\mathbf{v} = \langle 2, 0 \rangle$

17. $\mathbf{u} = 3\mathbf{i} + 4\mathbf{j}$
   $\mathbf{v} = -2\mathbf{j}$

18. $\mathbf{u} = 2\mathbf{i} - 3\mathbf{j}$
   $\mathbf{v} = \mathbf{i} - 2\mathbf{j}$

19. $\mathbf{u} = \cos\left(\dfrac{\pi}{3}\right)\mathbf{i} + \sin\left(\dfrac{\pi}{3}\right)\mathbf{j}$

   $\mathbf{v} = \cos\left(\dfrac{3\pi}{4}\right)\mathbf{i} + \sin\left(\dfrac{3\pi}{4}\right)\mathbf{j}$

20. $\mathbf{u} = \cos\left(\dfrac{\pi}{4}\right)\mathbf{i} + \sin\left(\dfrac{\pi}{4}\right)\mathbf{j}$

   $\mathbf{v} = \cos\left(\dfrac{\pi}{2}\right)\mathbf{i} + \sin\left(\dfrac{\pi}{2}\right)\mathbf{j}$

In Exercises 21–24, use a graphing utility to sketch the vectors and find the degree measure of the angle between the vectors.

21. $\mathbf{u} = 3\mathbf{i} + 4\mathbf{j}$
   $\mathbf{v} = -7\mathbf{i} + 5\mathbf{j}$

22. $\mathbf{u} = -6\mathbf{i} - 3\mathbf{j}$
   $\mathbf{v} = -8\mathbf{i} + 4\mathbf{j}$

23. $\mathbf{u} = 5\mathbf{i} + 5\mathbf{j}$
   $\mathbf{v} = -6\mathbf{i} + 6\mathbf{j}$

24. $\mathbf{u} = 2\mathbf{i} - 3\mathbf{j}$
   $\mathbf{v} = 4\mathbf{i} + 3\mathbf{j}$

In Exercises 25 and 26, use vectors to find the interior angles of the triangle with the given vertices.

25. $(1, 2), (3, 4), (2, 5)$

26. $(-3, 0), (2, 2), (0, 6)$

In Exercises 27 and 28, find $\mathbf{u} \cdot \mathbf{v}$, where θ is the angle between u and v.

27. $\|\mathbf{u}\| = 4, \|\mathbf{v}\| = 10, \theta = \dfrac{2\pi}{3}$

28. $\|\mathbf{u}\| = 100, \|\mathbf{v}\| = 250, \theta = \dfrac{\pi}{6}$

In Exercises 29–34, determine whether u and v are orthogonal, parallel, or neither.

29. $\mathbf{u} = \langle -12, 30 \rangle$
   $\mathbf{v} = \langle \frac{1}{2}, -\frac{5}{4} \rangle$

30. $\mathbf{u} = \langle 15, 45 \rangle$
   $\mathbf{v} = \langle -5, 12 \rangle$

31. $\mathbf{u} = \frac{1}{4}(3\mathbf{i} - \mathbf{j})$
   $\mathbf{v} = 5\mathbf{i} + 6\mathbf{j}$

32. $\mathbf{u} = \mathbf{j}$
   $\mathbf{v} = \mathbf{i} - 2\mathbf{j}$

33. $\mathbf{u} = 2\mathbf{i} - 2\mathbf{j}$
   $\mathbf{v} = -\mathbf{i} - \mathbf{j}$

34. $\mathbf{u} = \langle \cos\theta, \sin\theta \rangle$
   $\mathbf{v} = \langle \sin\theta, -\cos\theta \rangle$

In Exercises 35–38, find the projection of u onto v, and the vector component of u orthogonal to v.

35. $\mathbf{u} = \langle 3, 4 \rangle$
    $\mathbf{v} = \langle 8, 2 \rangle$

36. $\mathbf{u} = \langle 4, 2 \rangle$
    $\mathbf{v} = \langle 1, -2 \rangle$

37. $\mathbf{u} = \langle 0, 3 \rangle$
    $\mathbf{v} = \langle 2, 15 \rangle$

38. $\mathbf{u} = \langle -5, -1 \rangle$
    $\mathbf{v} = \langle -1, 1 \rangle$

In Exercises 39–42, find two vectors in opposite directions that are orthogonal to the vector u. (The answers are not unique.)

39. $\mathbf{u} = \langle 3, 5 \rangle$

40. $\mathbf{u} = \langle -8, 3 \rangle$

41. $\mathbf{u} = \frac{1}{2}\mathbf{i} - \frac{2}{3}\mathbf{j}$

42. $\mathbf{u} = -\frac{5}{2}\mathbf{i} - 3\mathbf{j}$

43. *Braking Load* A truck with a gross weight of 36,000 pounds is parked on a 10° slope (see figure). Assume that the only force to overcome is the force of gravity.

(a) Find the force required to keep the truck from rolling down the hill.

(b) Find the force perpendicular to the hill.

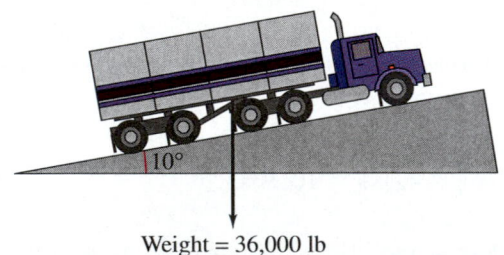

Weight = 36,000 lb

44. *Braking Load* Rework Exercise 43 for a truck that is parked on a 12° slope.

45. *Think About It* What is known about θ, the angle between two nonzero vectors $\mathbf{u}$ and $\mathbf{v}$, under the following conditions?

(a) $\mathbf{u} \cdot \mathbf{v} = 0$ (b) $\mathbf{u} \cdot \mathbf{v} > 0$ (c) $\mathbf{u} \cdot \mathbf{v} < 0$

46. *Think About It* What can be said about the vectors $\mathbf{u}$ and $\mathbf{v}$ under the following conditions?

(a) The projection of $\mathbf{u}$ onto $\mathbf{v}$ equals $\mathbf{u}$.

(b) The projection of $\mathbf{u}$ onto $\mathbf{v}$ equals $\mathbf{0}$.

47. *Work* A 25-kilogram (245-newton) bag of sugar is lifted 3 meters. Determine the work done.

48. *Work* Determine the work done by a crane lifting a 2400-pound car 5 feet.

49. *Work* A force of 45 pounds in the direction of 30° above the horizontal is required to slide an implement across a floor (see figure). Find the work done if the implement is dragged 20 feet.

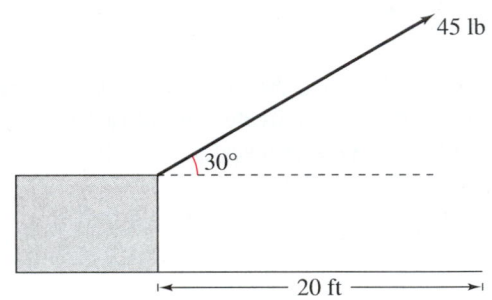

45 lb

30°

20 ft

50. *Work* A tractor pulls a log 800 meters and the tension in the cable connecting the tractor and log is approximately 1600 kilograms (15,691 newtons). Approximate the work done if the direction of the force is 35° above the horizontal.

Work **In Exercises 51 and 52, find the work done in moving a particle from P to Q if the magnitude and direction of the force are given by v.**

51. $P = (0, 0)$, $Q = (4, 7)$, $\mathbf{v} = \langle 1, 4 \rangle$

52. $P = (1, 3)$, $Q = (-3, 5)$, $\mathbf{v} = -2\mathbf{i} + 3\mathbf{j}$

53. Use vectors to prove that the diagonals of a rhombus are perpendicular.

54. Prove the following.

$$\|\mathbf{u} - \mathbf{v}\| = \|\mathbf{u}\| + \|\mathbf{v}\| - 2\mathbf{u} \cdot \mathbf{v}$$

55. Prove the following properties of the dot product.

(a) $\mathbf{0} \cdot \mathbf{v} = 0$

(b) $\mathbf{u} \cdot (\mathbf{v} + \mathbf{w}) = \mathbf{u} \cdot \mathbf{v} + \mathbf{u} \cdot \mathbf{w}$

(c) $c(\mathbf{u} \cdot \mathbf{v}) = c\mathbf{u} \cdot \mathbf{v} = \mathbf{u} \cdot c\mathbf{v}$

56. Prove that if $\mathbf{u}$ is orthogonal to $\mathbf{v}$ and $\mathbf{w}$, then $\mathbf{u}$ is orthogonal to $c\mathbf{v} + d\mathbf{w}$ for any scalars c and d.

8.5   *DeMoivre's Theorem*

See Exercises 105–112
on page 655 to see how
Demoivre's Theorem can be used
to solve a polynomial equation.

*The Complex Plane   ▫   Trigonometric Form of a Complex Number   ▫
Multiplication and Division of Complex Numbers   ▫   Powers of
Complex Numbers   ▫   Roots of Complex Numbers*

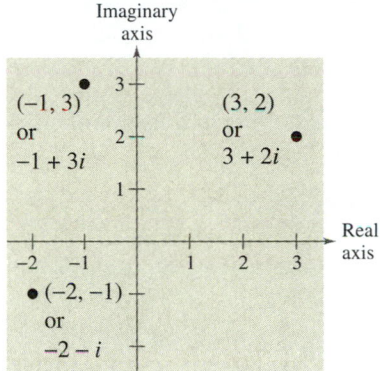

FIGURE 8.41

NOTE   If the complex number
$a + bi$ is a real number (that is, if
$b = 0$), then this definition agrees
with that given for the absolute
value of a real number

$$|a + 0i| = \sqrt{a^2 + 0^2} = |a|. \quad ∎∎$$

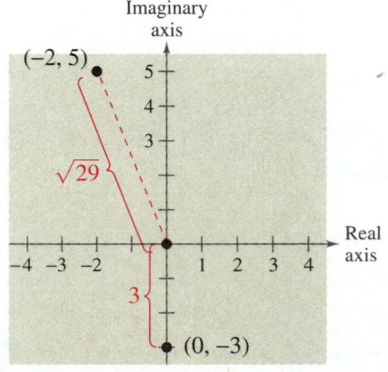

FIGURE 8.42

The Complex Plane

Just as real numbers can be represented by points on the real number line, you
can represent a complex number

$$z = a + bi$$

as the point (a, b) in a coordinate plane (the **complex plane**). The horizontal
axis is called the **real axis** and the vertical axis is called the **imaginary axis,** as
shown in Figure 8.41.

The **absolute value** of the complex number $a + bi$ is defined as the distance between the origin $(0, 0)$ and the point (a, b).

> **DEFINITION OF THE ABSOLUTE VALUE**
> **OF A COMPLEX NUMBER**
>
> The **absolute value** of the complex number $z = a + bi$ is given by
>
> $$|a + bi| = \sqrt{a^2 + b^2}.$$

EXAMPLE 1   *Finding the Absolute Value of a Complex Number*

Plot each complex number and find its absolute value.

a. $z = -3i$ **b.** $z = -2 + 5i$

Solution

The points are shown in Figure 8.42.

a. The complex number $z = 0 + (-3)i$ has an absolute value of

$$|z| = \sqrt{0^2 + (-3)^2}$$
$$= 3.$$

b. The complex number $z = -2 + 5i$ has an absolute value of

$$|z| = \sqrt{(-2)^2 + 5^2}$$
$$= \sqrt{29}.$$

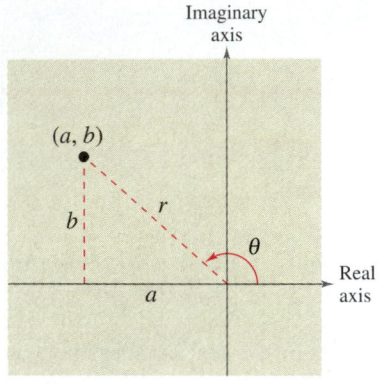

FIGURE 8.43

NOTE The trigonometric form of a complex number is also called the **polar form.** Because there are infinitely many choices for θ, the trigonometric form of a complex number is not unique. Normally, θ is restricted to the interval $0 \leq \theta < 2\pi$, although on occasion it is convenient to use $\theta < 0$. ∎∎

Trigonometric Form of a Complex Number

In Section 1.5 you learned how to add, subtract, multiply, and divide complex numbers. To work effectively with *powers* and *roots* of complex numbers, it is helpful to write complex numbers in **trigonometric form.** In Figure 8.43, consider the nonzero complex number $a + bi$. By letting θ be the angle from the positive x-axis (measured counterclockwise) to the line segment connecting the origin and the point (a, b), you can write

$$a = r \cos \theta \qquad \text{and} \qquad b = r \sin \theta$$

where $r = \sqrt{a^2 + b^2}$. Consequently, you have

$$a + bi = (r \cos \theta) + (r \sin \theta)i$$

from which you can obtain the **trigonometric form of a complex number.**

  TRIGONOMETRIC FORM OF A COMPLEX NUMBER

The **trigonometric form** of the complex number $z = a + bi$ is

$$z = r(\cos \theta + i \sin \theta)$$

where $a = r \cos \theta, b = r \sin \theta, r = \sqrt{a^2 + b^2},$ and $\tan \theta = b/a.$ The number r is the **modulus** of z, and θ is called an **argument** of z.

EXAMPLE 2 *Writing a Complex Number in Trigonometric Form*

Write the complex number $z = -2 - 2\sqrt{3}i$ in trigonometric form.

Solution

The absolute value of z is

$$r = \left| -2 - 2\sqrt{3}i \right| = \sqrt{(-2)^2 + \left(-2\sqrt{3}\right)^2} = \sqrt{16} = 4$$

and the angle θ is given by

$$\tan \theta = \frac{b}{a} = \frac{-2\sqrt{3}}{-2} = \sqrt{3}.$$

Because $\tan(\pi/3) = \sqrt{3}$ and $z = -2 - 2\sqrt{3}i$ lies in Quadrant III, you choose θ to be $\theta = \pi + \pi/3 = 4\pi/3$. Thus the trigonometric form is

$$z = r(\cos \theta + i \sin \theta) = 4\left(\cos \frac{4\pi}{3} + i \sin \frac{4\pi}{3} \right).$$

See Figure 8.44.

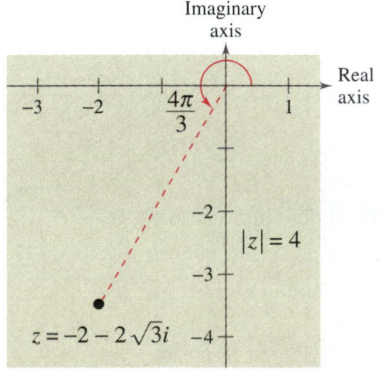

FIGURE 8.44

TECHNOLOGY

A graphing utility can be used to convert a complex number in polar form to rectangular form and vice versa. For instance, to illustrate Example 3 on a *TI-83* or a *TI-82*, use the following steps.

1. Press ANGLE and choose P▷Rx(.
2. Enter the values for r and θ as (r, θ). Press ENTER to obtain the x-coordinate.
3. Press ANGLE and choose P▷Ry(.
4. Enter the values for r and θ as (r, θ). Press ENTER to obtain the y-coordinate.

Write the result in rectangular form. Convert the complex number $-1 + \sqrt{3}i$ to trigonometric form using R▷Pr(and R▷Pθ(.

EXAMPLE 3   *Writing a Complex Number in Standard Form*

Write the complex number in standard form $a + bi$.

$$z = \sqrt{8}\left[\cos\left(-\frac{\pi}{3}\right) + i\sin\left(-\frac{\pi}{3}\right)\right]$$

Solution

Because $\cos(-\pi/3) = 1/2$ and $\sin(-\pi/3) = -\sqrt{3}/2$, you can write

$$z = \sqrt{8}\left[\cos\left(-\frac{\pi}{3}\right) + i\sin\left(-\frac{\pi}{3}\right)\right]$$

$$= 2\sqrt{2}\left[\frac{1}{2} - \frac{\sqrt{3}}{2}i\right]$$

$$= \sqrt{2} - \sqrt{6}i.$$

Multiplication and Division of Complex Numbers

The trigonometric form adapts nicely to multiplication and division of complex numbers. Suppose you are given two complex numbers

$$z_1 = r_1(\cos\theta_1 + i\sin\theta_1) \qquad \text{and} \qquad z_2 = r_2(\cos\theta_2 + i\sin\theta_2).$$

The product of z_1 and z_2 is

$$z_1 z_2 = r_1 r_2(\cos\theta_1 + i\sin\theta_1)(\cos\theta_2 + i\sin\theta_2)$$

$$= r_1 r_2[(\cos\theta_1\cos\theta_2 - \sin\theta_1\sin\theta_2) + i(\sin\theta_1\cos\theta_2 + \cos\theta_1\sin\theta_2)].$$

Using the sum and difference formulas for cosine and sine, you can rewrite this equation as

$$z_1 z_2 = r_1 r_2[\cos(\theta_1 + \theta_2) + i\sin(\theta_1 + \theta_2)].$$

This establishes the first part of the following rule. The second part is left to you (see Exercise 63).

PRODUCT AND QUOTIENT OF TWO COMPLEX NUMBERS

Let $z_1 = r_1(\cos\theta_1 + i\sin\theta_1)$ and $z_2 = r_2(\cos\theta_2 + i\sin\theta_2)$ be complex numbers.

$$z_1 z_2 = r_1 r_2[\cos(\theta_1 + \theta_2) + i\sin(\theta_1 + \theta_2)] \qquad \text{Product}$$

$$\frac{z_1}{z_2} = \frac{r_1}{r_2}[\cos(\theta_1 - \theta_2) + i\sin(\theta_1 - \theta_2)], \quad z_2 \neq 0 \qquad \text{Quotient}$$

Note that this rule says that to multiply two complex numbers you multiply moduli and add arguments, whereas to divide two complex numbers you divide moduli and subtract arguments.

EXAMPLE 4 **Multiplying Complex Numbers in Trigonometric Form**

Find the product of the complex numbers.

$$z_1 = 2\left(\cos\frac{2\pi}{3} + i\sin\frac{2\pi}{3}\right) \qquad z_2 = 8\left(\cos\frac{11\pi}{6} + i\sin\frac{11\pi}{6}\right)$$

Solution

$$z_1 z_2 = 2\left(\cos\frac{2\pi}{3} + i\sin\frac{2\pi}{3}\right) \cdot 8\left(\cos\frac{11\pi}{6} + i\sin\frac{11\pi}{6}\right)$$

$$= 16\left[\cos\left(\frac{2\pi}{3} + \frac{11\pi}{6}\right) + i\sin\left(\frac{2\pi}{3} + \frac{11\pi}{6}\right)\right]$$

$$= 16\left[\cos\frac{5\pi}{2} + i\sin\frac{5\pi}{2}\right]$$

$$= 16\left[\cos\frac{\pi}{2} + i\sin\frac{\pi}{2}\right]$$

$$= 16[0 + i(1)] = 16i$$

Check this result by first converting to the standard forms $z_1 = -1 + \sqrt{3}i$ and $z_2 = 4\sqrt{3} - 4i$ and then multiplying algebraically, as in Section 1.5.

EXAMPLE 5 **Dividing Complex Numbers in Trigonometric Form**

Find the quotient, z_1/z_2, of the complex numbers.

$$z_1 = 24(\cos 300° + i\sin 300°) \qquad z_2 = 8(\cos 75° + i\sin 75°)$$

Solution

$$\frac{z_1}{z_2} = \frac{24(\cos 300° + i\sin 300°)}{8(\cos 75° + i\sin 75°)}$$

$$= \frac{24}{8}[\cos(300° - 75°) + i\sin(300° - 75°)]$$

$$= 3[\cos 225° + i\sin 225°]$$

$$= 3\left[\left(-\frac{\sqrt{2}}{2}\right) + i\left(-\frac{\sqrt{2}}{2}\right)\right]$$

$$= -\frac{3\sqrt{2}}{2} - \frac{3\sqrt{2}}{2}i$$

Powers of Complex Numbers

To raise a complex number to a power, consider repeated use of the multiplication rule.

$$z = r(\cos \theta + i \sin \theta)$$

$$z^2 = r(\cos \theta + i \sin \theta)r(\cos \theta + i \sin \theta) = r^2(\cos 2\theta + i \sin 2\theta)$$

$$z^3 = r^2(\cos 2\theta + i \sin 2\theta)r(\cos \theta + i \sin \theta) = r^3(\cos 3\theta + i \sin 3\theta)$$

$$z^4 = r^4(\cos 4\theta + i \sin 4\theta)$$

$$z^5 = r^5(\cos 5\theta + i \sin 5\theta)$$

$$\vdots$$

This pattern leads to the following important theorem, which is named after the French mathematician Abraham DeMoivre (1667–1754).

DEMOIVRE'S THEOREM

If $z = r(\cos \theta + i \sin \theta)$ is a complex number and n is a positive integer, then

$$z^n = [r(\cos \theta + i \sin \theta)]^n = r^n(\cos n\theta + i \sin n\theta).$$

EXAMPLE 6   *Finding Powers of a Complex Number*

Use DeMoivre's Theorem to find $\left(-1 + \sqrt{3}i\right)^{12}$.

Solution

First convert to trigonometric form.

$$-1 + \sqrt{3}i = 2\left(\cos \frac{2\pi}{3} + i \sin \frac{2\pi}{3}\right)$$

Then, by DeMoivre's Theorem, you have

$$(-1 + \sqrt{3}i)^{12} = \left[2\left(\cos \frac{2\pi}{3} + i \sin \frac{2\pi}{3}\right)\right]^{12}$$

$$= 2^{12}\left[\cos(12) \frac{2\pi}{3} + i \sin(12) \frac{2\pi}{3}\right]$$

$$= 4096(\cos 8\pi + i \sin 8\pi)$$

$$= 4096(1 + 0)$$

$$= 4096.$$

Are you surprised to see a real number as the answer?

Roots of Complex Numbers

Recall that a consequence of the Fundamental Theorem of Algebra is that a polynomial equation of degree n has n solutions in the complex number system. Hence, the equation $x^6 = 1$ has six solutions, and in this particular case you can find the six solutions by factoring and using the Quadratic Formula.

$$x^6 - 1 = (x^3 - 1)(x^3 + 1)$$
$$= (x - 1)(x^2 + x + 1)(x + 1)(x^2 - x + 1) = 0$$

Consequently, the solutions are

$$x = \pm 1, \qquad x = \frac{-1 \pm \sqrt{3}i}{2}, \qquad \text{and} \qquad x = \frac{1 \pm \sqrt{3}i}{2}.$$

Each of these numbers is a sixth root of 1. In general, the **nth root** of a complex number is defined as follows.

> **DEFINITION OF NTH ROOT OF A COMPLEX NUMBER**
>
> The complex number $u = a + bi$ is an **nth root** of the complex number z if
>
> $$z = u^n = (a + bi)^n.$$

To find a formula for an nth root of a complex number, let u be an nth root of z, where

$$u = s(\cos \beta + i \sin \beta) \qquad \text{and} \qquad z = r(\cos \theta + i \sin \theta).$$

By DeMoivre's Theorem and the fact that $u^n = z$, you have

$$s^n (\cos n\beta + i \sin n\beta) = r(\cos \theta + i \sin \theta).$$

Taking the absolute values of both sides of this equation, it follows that $s^n = r$. Substituting back into the previous equation and dividing by r, you get

$$\cos n\beta + i \sin n\beta = \cos \theta + i \sin \theta.$$

Thus, it follows that

$$\cos n\beta = \cos \theta \qquad \text{and} \qquad \sin n\beta = \sin \theta.$$

Because both sine and cosine have a period of 2π, these last two equations have solutions if and only if the angles differ by a multiple of 2π. Consequently, there must exist an integer k such that

$$n\beta = \theta + 2\pi k$$
$$\beta = \frac{\theta + 2\pi k}{n}.$$

By substituting this value for β into the trigonometric form of u, you get the result stated on the following page.

Exploration

The nth roots of a complex number are useful for solving some polynomial equations. For instance, explain how you can use DeMoivre's Theorem to solve the polynomial equation

$$x^4 + 16 = 0.$$

[*Hint*: Write -16 as $16(\cos \pi + i \sin \pi)$.]

NOTE When k exceeds $n - 1$, the roots begin to repeat. For instance, if $k = n$, the angle

$$\frac{\theta + 2\pi n}{n} = \frac{\theta}{n} + 2\pi$$

is coterminal with θ/n, which is also obtained when $k = 0$. ■■

> **NTH ROOTS OF A COMPLEX NUMBER**
>
> For a positive integer n, the complex number $z = r(\cos\theta + i\sin\theta)$ has exactly n distinct nth roots given by
>
> $$\sqrt[n]{r}\left(\cos\frac{\theta + 2\pi k}{n} + i\sin\frac{\theta + 2\pi k}{n}\right)$$
>
> where $k = 0, 1, 2, \ldots, n - 1$.

This formula for the nth roots of a complex number z has a nice geometrical interpretation, as shown in Figure 8.45. Note that because the nth roots of z all have the same magnitude $\sqrt[n]{r}$, they all lie on a circle of radius $\sqrt[n]{r}$ with center at the origin. Furthermore, because successive nth roots have arguments that differ by $2\pi/n$, the n roots are equally spaced along the circle.

You have already found the sixth roots of 1 by factoring and by using the Quadratic Formula. Example 7 shows how you can solve the same problem with the formula for nth roots.

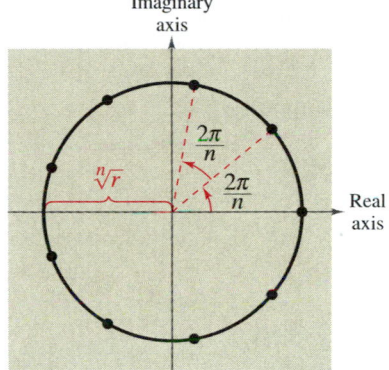

FIGURE 8.45

EXAMPLE 7   Finding nth Roots of a Real Number

Find all the sixth roots of 1.

Solution

First write 1 in the trigonometric form $1 = 1(\cos 0 + i\sin 0)$. Then, by the nth root formula, with $n = 6$ and $r = 1$, the roots have the form

$$\sqrt[6]{1}\left(\cos\frac{0 + 2\pi k}{6} + i\sin\frac{0 + 2\pi k}{6}\right)$$

or simply $\cos(\pi k/3) + i\sin(\pi k/3)$. Thus, for $k = 0, 1, 2, 3, 4$, and 5, the sixth roots are as follows. (See Figure 8.46.)

$$\cos 0 + i\sin 0 = 1$$

$$\cos\frac{\pi}{3} + i\sin\frac{\pi}{3} = \frac{1}{2} + \frac{\sqrt{3}}{2}i$$

$$\cos\frac{2\pi}{3} + i\sin\frac{2\pi}{3} = -\frac{1}{2} + \frac{\sqrt{3}}{2}i$$

$$\cos\pi + i\sin\pi = -1$$

$$\cos\frac{4\pi}{3} + i\sin\frac{4\pi}{3} = -\frac{1}{2} - \frac{\sqrt{3}}{2}i$$

$$\cos\frac{5\pi}{3} + i\sin\frac{5\pi}{3} = \frac{1}{2} - \frac{\sqrt{3}}{2}i$$

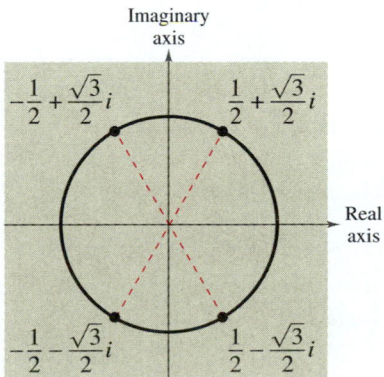

FIGURE 8.46

Use a graphing utility, set in parametric and radian modes, to display the graph given by

$$X_{1T} = \cos T \text{ and}$$

$$Y_{1T} = \sin T.$$

Set the viewing rectangle so that $-1.5 \le X \le 1.5$ and $-1 \le Y \le 1$. Then, using $0 \le T \le 2\pi$, set the "Tstep" to $2\pi/n$ for various values of n. Explain how the graphing utility can be used to obtain the nth roots of unity.

In Figure 8.46, notice that the roots obtained in Example 7 all have a magnitude of 1 and are equally spaced around the unit circle. Also notice that the complex roots occur in conjugate pairs, as discussed in Section 3.5. The n distinct nth roots of 1 are called the **nth roots of unity.**

EXAMPLE 8   Finding the nth Roots of a Complex Number

Find the three cube roots of $z = -2 + 2i$.

Solution

Because z lies in Quadrant II, the trigonometric form for z is

$$z = -2 + 2i = \sqrt{8}\,(\cos 135° + i \sin 135°).$$

By the formula for nth roots, the cube roots have the form

$$\sqrt[6]{8}\left(\cos \frac{135° + 360°k}{3} + i \sin \frac{135° + 360°k}{3}\right).$$

Finally, for $k = 0, 1,$ and 2, you obtain the roots

$$\sqrt{2}(\cos 45° + i \sin 45°) = 1 + i$$
$$\sqrt{2}(\cos 165° + i \sin 165°) \approx -1.3660 + 0.3660i$$
$$\sqrt{2}(\cos 285° + i \sin 285°) \approx 0.3660 - 1.3660i.$$

GROUP ACTIVITY

A FAMOUS MATHEMATICAL FORMULA

The famous formula

$$e^{a + bi} = e^a(\cos b + i \sin b)$$

is called Euler's Formula, after the German mathematician Leonhard Euler (1707–1783). Although the interpretation of this formula is beyond the scope of this text, we decided to include it because it gives rise to one of the most wonderful equations in mathematics.

$$e^{\pi i} + 1 = 0$$

This elegant equation relates the five most famous numbers in mathematics—0, 1, π, e, and i—in a single equation. Show how Euler's Formula can be used to derive this equation.

WARM UP

Write the complex number in standard form.

1. $-5 - \sqrt{-100}$

2. $7 + \sqrt{-54}$

3. $-4i + i^2$

4. $3i^3$

Perform the operations and write the answer in standard form.

5. $(3 - 10i) - (-3 + 4i)$

6. $(2 + \sqrt{-50}) + (4 - \sqrt{2}i)$

7. $(4 - 2i)(-6 + i)$

8. $(3 - 2i)(3 + 2i)$

9. $\dfrac{1 + 4i}{1 - i}$

10. $\dfrac{3 - 5i}{2i}$

8.5 Exercises

In Exercises 1–6, plot the complex number and find its absolute value.

1. $-5i$

2. -5

3. $-4 + 4i$

4. $5 - 12i$

5. $6 - 7i$

6. $-8 + 3i$

In Exercises 7–10, write in trigonometric form.

7.

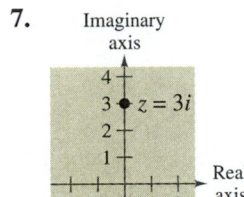

8.

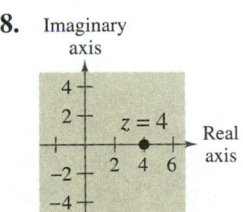

9.

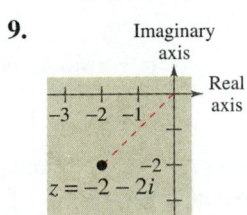

10.

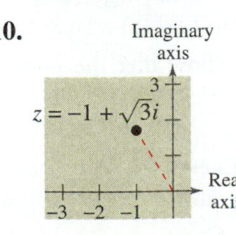

In Exercises 11–26, represent the complex number graphically, and find the trigonometric form of the number.

11. $3 - 3i$

12. $2 + 2i$

13. $\sqrt{3} + i$

14. $-1 + \sqrt{3}i$

15. $-2(1 + \sqrt{3}i)$

16. $\frac{5}{2}(\sqrt{3} - i)$

17. $6i$

18. 4

19. $-7 + 4i$

20. $3 - i$

21. 7

22. $-2i$

23. $1 + 6i$

24. $2\sqrt{2} - i$

25. $-3 - i$

26. $1 + 3i$

In Exercises 27–30, use a graphing utility to represent the complex number in trigonometric form.

27. $5 + 2i$

28. $-3 + i$

29. $3\sqrt{2} - 7i$

30. $-8 - 5\sqrt{3}i$

In Exercises 31–40, represent the complex number graphically, and find the standard form of the number.

31. $2(\cos 150° + i \sin 150°)$

32. $5(\cos 135° + i \sin 135°)$

33. $\frac{3}{2}(\cos 300° + i \sin 300°)$

34. $\frac{3}{4}(\cos 315° + i \sin 315°)$

35. $3.75\left(\cos \frac{3\pi}{4} + i \sin \frac{3\pi}{4}\right)$

36. $8\left(\cos \frac{\pi}{12} + i \sin \frac{\pi}{12}\right)$

37. $4\left(\cos \frac{3\pi}{2} + i \sin \frac{3\pi}{2}\right)$

38. $7(\cos 0 + i \sin 0)$

39. $3[\cos(18° \; 45') + i \sin(18° \; 45')]$

40. $6[\cos(230° \; 30') + i \sin(230° \; 30')]$

In Exercises 41–44, use a graphing utility to represent the complex number in standard form.

41. $5\left(\cos \frac{\pi}{9} + i \sin \frac{\pi}{9}\right)$

42. $12\left(\cos \frac{3\pi}{5} + i \sin \frac{3\pi}{5}\right)$

43. $4(\cos 216.5° + i \sin 216.5°)$

44. $9(\cos 58° + i \sin 58°)$

In Exercises 45–56, perform the operation and leave the result in trigonometric form.

45. $\left[3\left(\cos \frac{\pi}{3} + i \sin \frac{\pi}{3}\right)\right]\left[4\left(\cos \frac{\pi}{6} + i \sin \frac{\pi}{6}\right)\right]$

46. $\left[\frac{3}{2}\left(\cos \frac{\pi}{2} + i \sin \frac{\pi}{2}\right)\right]\left[6\left(\cos \frac{\pi}{4} + i \sin \frac{\pi}{4}\right)\right]$

47. $\left[\frac{5}{3}(\cos 140° + i \sin 140°)\right]\left[\frac{2}{3}(\cos 60° + i \sin 60°)\right]$

48. $[0.5(\cos 100° + i \sin 100°)]$
$[0.8(\cos 300° + i \sin 300°)]$

49. $[0.45(\cos 310° + i \sin 310°)]$
$[0.60(\cos 200° + i \sin 200°)]$

50. $(\cos 5° + i \sin 5°)(\cos 20° + i \sin 20°)$

51. $\dfrac{\cos 40° + i \sin 40°}{\cos 10° + i \sin 10°}$

52. $\dfrac{2(\cos 120° + i \sin 120°)}{4(\cos 40° + i \sin 40°)}$

53. $\dfrac{\cos(5\pi/3) + i \sin(5\pi/3)}{\cos \pi + i \sin \pi}$

54. $\dfrac{5(\cos 4.3 + i \sin 4.3)}{4(\cos 2.1 + i \sin 2.1)}$

55. $\dfrac{12(\cos 52° + i \sin 52°)}{3(\cos 110° + i \sin 110°)}$

56. $\dfrac{9(\cos 20° + i \sin 20°)}{5(\cos 75° + i \sin 75°)}$

In Exercises 57–62, (a) give the trigonometric form of the complex number, (b) perform the indicated operation using the trigonometric form, and (c) perform the indicated operation using the standard form, and check your result with that of part (b).

57. $(2 + 2i)(1 - i)$

58. $(\sqrt{3} + i)(1 + i)$

59. $-2i(1 + i)$

60. $\dfrac{3 + 4i}{1 - \sqrt{3}i}$

61. $\dfrac{5}{2 + 3i}$

62. $\dfrac{4i}{-4 + 2i}$

63. Given two complex numbers $z_1 = r_1(\cos \theta_1 + i \sin \theta_1)$ and $z_2 = r_2(\cos \theta_2 + i \sin \theta_2)$, $z_2 \neq 0$, prove that
$$\frac{z_1}{z_2} = \frac{r_1}{r_2}[\cos(\theta_1 - \theta_2) + i \sin(\theta_1 - \theta_2)].$$

64. Show that $\bar{z} = r[\cos(-\theta) + i \sin(-\theta)]$ is the complex conjugate of $z = r(\cos \theta + i \sin \theta)$.

65. Use the trigonometric forms of z and $\bar{z}$ in Exercise 64 to find (a) $z\bar{z}$ and (b) $z/\bar{z}$, $\bar{z} \neq 0$.

66. Show that the negative of $z = r(\cos \theta + i \sin \theta)$ is $-z = r[\cos(\theta + \pi) + i \sin(\theta + \pi)]$.

In Exercises 67 and 68, sketch the graphs of all complex numbers z satisfying the given condition.

67. $|z| = 2$

68. $\theta = \pi/6$

In Exercises 69–80, use DeMoivre's Theorem to find the indicated power of the complex number. Express the result in standard form.

69. $(1 + i)^5$ **70.** $(2 + 2i)^6$

71. $(-1 + i)^{10}$ **72.** $(1 - i)^{12}$

73. $2(\sqrt{3} + i)^7$ **74.** $4(1 - \sqrt{3}i)^3$

75. $[5(\cos 20° + i \sin 20°)]^3$

76. $[3(\cos 150° + i \sin 150°)]^4$

77. $\left(\cos \dfrac{5\pi}{4} + i \sin \dfrac{5\pi}{4}\right)^{10}$

78. $\left[2\left(\cos \dfrac{\pi}{2} + i \sin \dfrac{\pi}{2}\right)\right]^8$

79. $[5(\cos 3.2 + i \sin 3.2)]^4$

80. $(\cos 0 + i \sin 0)^{20}$

In Exercises 81–84, use a graphing utility and DeMoivre's Theorem to find the indicated power of the complex number. Express the result in standard form.

81. $(3 - 2i)^5$

82. $(\sqrt{5} - 4i)^3$

83. $[3(\cos 15° + i \sin 15°)]^4$

84. $\left[2\left(\cos \dfrac{\pi}{10} + i \sin \dfrac{\pi}{10}\right)\right]^5$

85. Show that $-\frac{1}{2}(1 + \sqrt{3}i)$ is a sixth root of 1.

86. Show that $2^{-1/4}(1 - i)$ is a fourth root of -2.

Graphical Reasoning In Exercises 87 and 88, use the graph of the roots of a complex number. (a) Write each of the roots in trigonometric form. (b) Identify the complex number whose roots are given. (c) Use a graphing utility to verify the results of part (b).

87.
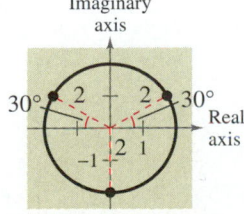

88.
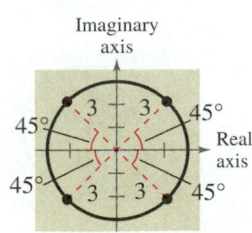

In Exercises 89–100, (a) use the theorem on page 651 to find the indicated roots of the complex number, (b) represent each of the roots graphically, and (c) express each of the roots in standard form.

89. Square roots of $5(\cos 120° + i \sin 120°)$

90. Square roots of $16(\cos 60° + i \sin 60°)$

91. Fourth roots of $16\left(\cos \dfrac{4\pi}{3} + i \sin \dfrac{4\pi}{3}\right)$

92. Fifth roots of $32\left(\cos \dfrac{5\pi}{6} + i \sin \dfrac{5\pi}{6}\right)$

93. Square roots of $-25i$

94. Fourth roots of $625i$

95. Cube roots of $-\dfrac{125}{2}(1 + \sqrt{3}i)$

96. Cube roots of $-4\sqrt{2}(1 - i)$

97. Cube roots of 8

98. Fourth roots of i

99. Fifth roots of 1

100. Cube roots of 1000

In Exercises 101–104, (a) use the theorem on page 651 and a graphing utility to find the indicated roots of the complex number, (b) represent each of the roots graphically, and (c) express each of the roots in standard form.

101. Cube roots of -125

102. Fourth roots of -4

103. Fifth roots of $128(-1 + i)$

104. Sixth roots of $64i$

In Exercises 105–112, use the theorem on page 651 to find all the solutions of the equation and represent the solutions graphically.

105. $x^4 - i = 0$ **106.** $x^3 + 1 = 0$

107. $x^5 + 243 = 0$ **108.** $x^4 - 81 = 0$

109. $x^3 + 64i = 0$ **110.** $x^6 - 64i = 0$

111. $x^3 - (1 - i) = 0$ **112.** $x^4 + (1 + i) = 0$

FOCUS ON CONCEPTS

In this chapter, you studied the methods for solving oblique triangles and vectors in the plane. Use the following questions to check your understanding of several of the basic concepts presented. The answers to these questions are given in the back of the book.

1. State the Law of Sines from memory.

2. State the Law of Cosines from memory.

3. *True or False*   The Law of Sines is true if one of the angles in the triangle is a right angle.

4. If one of the angles in the triangle is a right angle, the Law of Cosines simplifies to what famous theorem?

5. *True or False*   When the Law of Sines is used, the solution is always unique. Explain.

6. What characterizes a vector in the plane?

7. Which vectors in the figure appear to be equivalent?

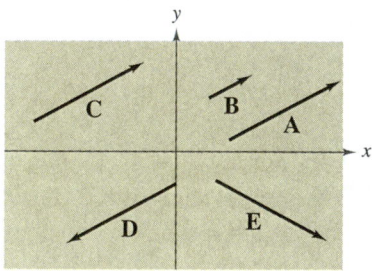

8. The vectors **u** and **v** have the same magnitudes in the two figures. In which figure will the magnitude of the resultant be greater? Give a reason for your answer.

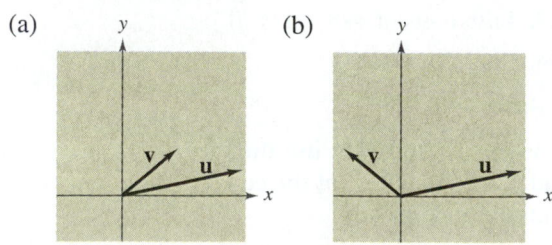

9. Give a geometric description of the scalar multiple $k\mathbf{u}$ of the vector **u**.

10. Give a geometric description of the sum of the vectors **u** and **v**.

11. Which of the two figures shows the difference $\mathbf{u} - \mathbf{v}$? Give a geometric description of the difference and state how you determine its direction.

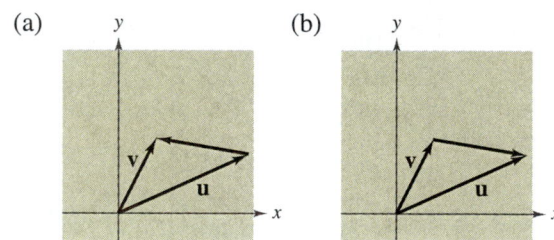

12. The figure shows z_1 and z_2. Describe $z_1 \cdot z_2$ and z_1/z_2.

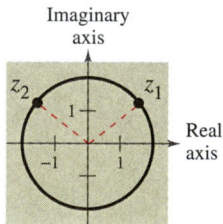

13. One of the fourth roots of a complex number z is shown in the figure.

(a) How many roots are not shown?

(b) Describe the other roots.

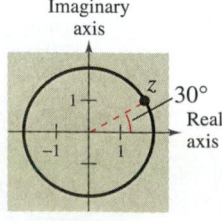

Review Exercises

In Exercises 1–16, use the given information to solve the triangle (if possible). If two solutions exist, list both.

1. $a = 5$, $b = 8$,
$c = 10$

2. $a = 6$, $b = 9$,
$C = 45°$

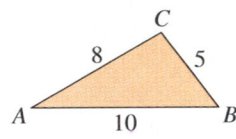

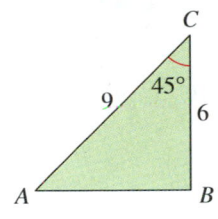

3. $A = 12°$, $B = 58°$, $a = 5$

4. $B = 110°$, $C = 30°$, $c = 10.5$

5. $B = 110°$, $a = 4$, $c = 4$

6. $a = 80$, $b = 60$, $c = 100$

7. $A = 75°$, $a = 2.5$, $b = 16.5$

8. $A = 130°$, $a = 50$, $b = 30$

9. $B = 115°$, $a = 7$, $b = 14.5$

10. $C = 50°$, $a = 25$, $c = 22$

11. $A = 15°$, $a = 5$, $b = 10$

12. $B = 150°$, $a = 64$, $b = 10$

13. $B = 150°$, $a = 10$, $c = 20$

14. $a = 2.5$, $b = 15.0$, $c = 4.5$

15. $B = 25°$, $a = 6.2$, $b = 4$

16. $B = 90°$, $a = 5$, $c = 12$

In Exercises 17–20, find the area of the triangle.

17. $a = 4$, $b = 5$, $c = 7$

18. $a = 15$, $b = 8$, $c = 10$

19. $A = 27°$, $b = 5$, $c = 8$

20. $B = 80°$, $a = 4$, $c = 8$

21. *Height* From a certain distance, the angle of elevation to the top of a building is $17°$. At a point 50 meters closer to the building, the angle of elevation is $31°$. Approximate the height of the building.

22. *Geometry* The lengths of the diagonals of a parallelogram are 10 feet and 16 feet. Find the lengths of the sides of the parallelogram if the diagonals intersect at an angle of $28°$.

23. *Height of a Tree* Find the height of a tree that stands on a hillside of slope $28°$ (from the horizontal) if from a point 75 feet down the hill the angle of elevation to the top of the tree is $45°$ (see figure).

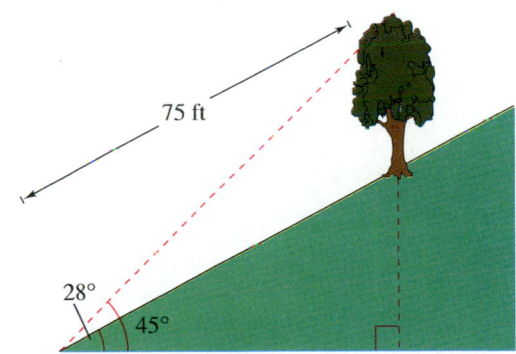

24. *Surveying* To approximate the length of a marsh, a surveyor walks 425 meters from point A to point B. Then the surveyor turns $65°$ and walks 300 meters to point C. Approximate the length AC of the marsh (see figure).

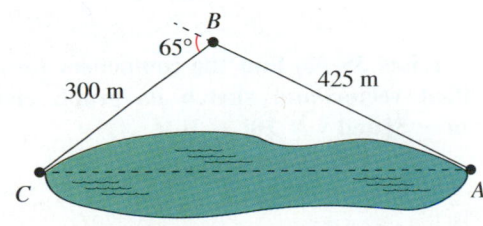

25. *Navigation*   Two planes leave an airport at approximately the same time. One is flying 425 miles per hour at a bearing of N 5° W, and the other is flying 530 miles per hour at a bearing of N 67° E. Draw a figure that gives a visual representation of the problem and determine the distance between the planes after they have flown for 2 hours.

26. *River Width*   Determine the width of a river that flows due east, if a tree on the opposite bank has a bearing of N 22° 30′ E and if, after walking 400 feet downstream, a surveyor finds that the tree has a bearing of N 15° W.

In Exercises 27–32, find the component form of the vector v satisfying the given conditions.

27.

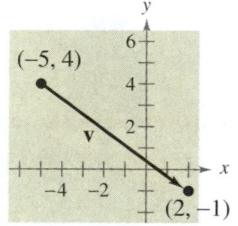

28.

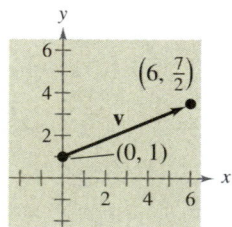

29. Initial point: (0, 10), Terminal point: (7, 3)

30. Initial point: (1, 5), Terminal point: (15, 9)

31. $\|\mathbf{v}\| = 8$, $\theta = 120°$

32. $\|\mathbf{v}\| = \frac{1}{2}$, $\theta = 225°$

In Exercises 33 and 34, write the vector v in the form $\|\mathbf{v}\|(\mathbf{i} \sin\theta + \mathbf{j} \cos\theta)$.

33. $\mathbf{v} = -10\mathbf{i} + 10\mathbf{j}$ **34.** $\mathbf{v} = 4\mathbf{i} - \mathbf{j}$

In Exercises 35–38, find the component form of the specified vector and sketch its graph given that $\mathbf{u} = 6\mathbf{i} - 5\mathbf{j}$ and $\mathbf{v} = 10\mathbf{i} + 3\mathbf{j}$.

35. $\dfrac{1}{\|\mathbf{u}\|}\mathbf{u}$ **36.** $3\mathbf{v}$

37. $4\mathbf{u} - 5\mathbf{v}$ **38.** $\frac{1}{2}\mathbf{v}$

In Exercises 39 and 40, use a graphing utility to graph the vectors and the resultant of the vectors. Find the magnitude and direction of the resultant.

39.

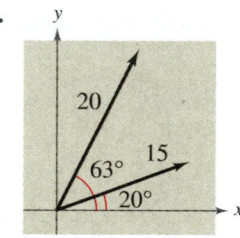

40.

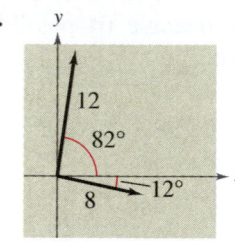

41. *Resultant Force*   Find the direction and magnitude of the resultant of the three forces shown in the figure.

FIGURE FOR 41   FIGURE FOR 43

$\tan\beta = \frac{3}{4}$ $\tan\alpha = \frac{12}{5}$

42. *Resultant Force*   Forces of 85 pounds and 50 pounds act on a single point. The angle between the forces is 15°. Describe the resultant force.

43. *Rope Tension*   A 180-pound weight is supported by two ropes, as shown in the figure. Find the tension exerted on each rope.

44. *Braking Force*   A 500-pound motorcycle is headed up a hill inclined at 12°. What force is required to keep the motorcycle from rolling back down the hill when stopped at a red light?

45. *Navigation*   An airplane has an airspeed of 724 kilometers per hour at a bearing of N 30° E. If the wind velocity is 32 kilometers per hour from the west, find the ground speed and the direction of the plane.

46. *Angle Between Forces*   Forces of 60 pounds and 100 pounds have a resultant force of 125 pounds. Find the angle between the two forces.

In Exercises 47 and 48, find a unit vector in the direction of $\overrightarrow{PQ}$.

47. $P(7, -4), Q(-3, 2)$ **48.** $P(0, 3), Q(5, -8)$

In Exercises 49 and 50, decide whether the vectors are orthogonal, parallel, or neither.

49. $\mathbf{u} = \langle 39, -12 \rangle$ **50.** $\mathbf{u} = \langle 8, 5 \rangle$
$\quad\ \ \mathbf{v} = \langle -26, 8 \rangle$ $\qquad\ \mathbf{v} = \langle -2, 4 \rangle$

In Exercises 51–54, find the angle between **u** and **v**.

51. $\mathbf{u} = \cos \dfrac{7\pi}{4} \mathbf{i} + \sin \dfrac{7\pi}{4} \mathbf{j}, \quad \mathbf{v} = \cos \dfrac{5\pi}{6} \mathbf{i} + \sin \dfrac{5\pi}{6} \mathbf{j}$

52. $\mathbf{u} = \langle -6, -3 \rangle, \quad \mathbf{v} = \langle 4, 2 \rangle$

53. $\mathbf{u} = \langle 2\sqrt{2}, -4 \rangle, \quad \mathbf{v} = \langle -\sqrt{2}, 1 \rangle$

54. $\mathbf{u} = \langle 3, 1 \rangle, \quad \mathbf{v} = \langle 4, 5 \rangle$

In Exercises 55–58, find $\text{proj}_{\mathbf{v}}\mathbf{u}$.

55. $\mathbf{u} = \langle -4, 3 \rangle, \quad \mathbf{v} = \langle -8, -2 \rangle$

56. $\mathbf{u} = \langle 5, 6 \rangle, \quad \mathbf{v} = \langle 10, 0 \rangle$

57. $\mathbf{u} = \langle 2, 7 \rangle, \quad \mathbf{v} = \langle 1, -1 \rangle$

58. $\mathbf{u} = \langle -3, 5 \rangle, \quad \mathbf{v} = \langle -5, 2 \rangle$

In Exercises 59–62, find the trigonometric form of the complex number.

59. $5 - 5i$ **60.** $-3\sqrt{3} + 3i$

61. $5 + 12i$ **62.** -7

In Exercises 63–66, write the complex number in standard form.

63. $100(\cos 240° + i \sin 240°)$

64. $24(\cos 330° + i \sin 330°)$

65. $13(\cos 0 + i \sin 0)$

66. $8\left(\cos \dfrac{5\pi}{6} + i \sin \dfrac{5\pi}{6}\right)$

In Exercises 67 and 68, (a) express the two complex numbers in trigonometric form, and (b) use the trigonometric form to find $z_1 z_2$ and z_1/z_2.

67. $z_1 = 2\sqrt{3} - 2i, \quad z_2 = -10i$

68. $z_1 = -3(1 + i), \quad z_2 = 2(\sqrt{3} + i)$

In Exercises 69–72, use DeMoivre's Theorem to find the indicated power of the complex number. Express the result in standard form.

69. $\left[5\left(\cos \dfrac{\pi}{12} + i \sin \dfrac{\pi}{12}\right)\right]^4$

70. $\left[2\left(\cos \dfrac{4\pi}{15} + i \sin \dfrac{4\pi}{15}\right)\right]^5$

71. $(2 + 3i)^6$

72. $(1 - i)^8$

⊞ *Graphical Reasoning*   In Exercises 73 and 74, use the graph of the roots of a complex number. (a) Write each of the roots in trigonometric form. (b) Identify the complex number whose roots are given. (c) Use a graphing utility to verify the results of part (b).

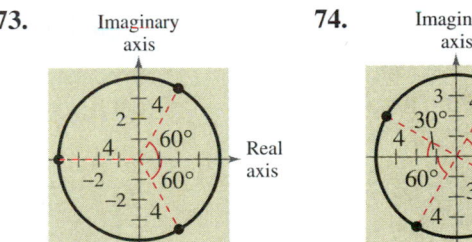

73.   **74.**

In Exercises 75 and 76, use the theorem on page 651 to find the roots of the complex number.

75. Sixth roots of $-729i$ **76.** Fourth roots of 256

In Exercises 77–80, find all solutions of the equation and represent the solutions graphically.

77. $x^4 + 81 = 0$ **78.** $x^5 - 32 = 0$

79. $x^3 + 8i = 0$ **80.** $(x^3 - 1)(x^2 + 1) = 0$

CHAPTER PROJECT: *Adding Vectors Graphically*

The following program is written for a *TI–83* or a *TI–82* graphing calculator. The program sketches two vectors $\mathbf{u} = a\mathbf{i} + b\mathbf{j}$ and $\mathbf{v} = c\mathbf{i} + d\mathbf{j}$ in standard position. Then, using the parallelogram law for vector addition, the program also sketches the vector sum $\mathbf{u} + \mathbf{v}$. *Before* running the program, you should set values that produce an appropriate viewing rectangle.

TI–83 or TI–82 Program

```
PROGRAM:ADDVECT        :Line(0,0,A,B)       :Line(A,B,E,F)
:Input "ENTER A",A      :Line(0,0,C,D)       :Line(C,D,E,F)
:Input "ENTER B",B      :A+C→E               :Pause
:Input "ENTER C",C      :B+D→F               :ClrDraw
:Input "ENTER D",D      :Line(0,0,E,F)       :Stop
```

EXAMPLE 1   Sketching a Vector Sum

Use the program listed above to sketch the sum of the vectors $\mathbf{u} = 5\mathbf{i} + 2\mathbf{j}$ and $\mathbf{v} = -4\mathbf{i} + 3\mathbf{j}$.

Solution

To show both vectors and their sum, you can use the viewing rectangle $-6 \le x \le 6$, $-2 \le y \le 6$. Note that this is a "square" setting. That is, the spacing on the horizontal and vertical axes is the same. After running the program and entering $A = 5$, $B = 2$, $C = -4$, and $D = 3$, you should obtain the screen shown below. Note that the vector sum

$$\mathbf{u} + \mathbf{v} = \mathbf{i} + 5\mathbf{j}$$

appears as the diagonal of the parallelogram. The vectors $\mathbf{u}$ and $\mathbf{v}$ appear as two of the sides of the parallelogram.

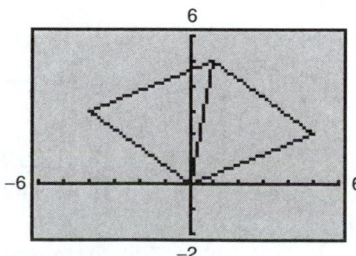

EXAMPLE 2 *Finding an Airplane's Speed and Direction*   Real Life

An airplane is headed N 60° W at a speed of 400 miles per hour. The airplane encounters wind of velocity 75 miles per hour in the direction N 40° E. What are the resultant speed and direction of the airplane? (See Example 10 on page 629.)

Solution

The velocity of the airplane can be represented by the vector

$$\mathbf{v}_1 = 400\langle\cos 150°, \sin 150°\rangle$$

and the velocity of the wind by the vector

$$\mathbf{v}_2 = 75\langle\cos 50°, \sin 50°\rangle.$$

Thus, the resultant velocity of the airplane can be represented by $\mathbf{v}_1 + \mathbf{v}_2$. With the program listed on page 660, you do not need to evaluate the numerical values of the vector coordinates. Simply enter the following.

$$A = 400 \cos 150° \qquad B = 400 \sin 150°$$
$$C = 75 \cos 50° \qquad D = 75 \sin 50°$$

Using $-400 \le x \le 200$ and $-100 \le y \le 300$, you should obtain the screen shown at the left.

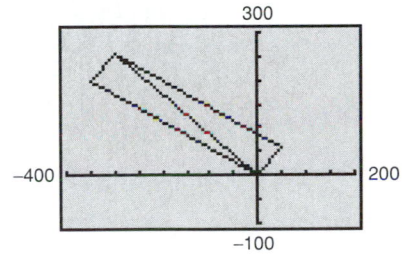

CHAPTER PROJECT INVESTIGATIONS

In Questions 1–4, use the program on page 660 (or a comparable program on some other graphing utility) to sketch the sum of the vectors. Use the result to estimate graphically the components of the sum. Then check your result analytically. (Use $-9 \le x \le 9$ and $-6 \le y \le 6$.)

1. $\mathbf{u} = 3\mathbf{i} + 4\mathbf{j}$, $\mathbf{v} = -5\mathbf{i} + \mathbf{j}$
2. $\mathbf{u} = 5\mathbf{i} - 4\mathbf{j}$, $\mathbf{v} = 3\mathbf{i} + 2\mathbf{j}$
3. $\mathbf{u} = -4\mathbf{i} + 4\mathbf{j}$, $\mathbf{v} = -2\mathbf{i} - 6\mathbf{j}$
4. $\mathbf{u} = 7\mathbf{i} + 3\mathbf{j}$, $\mathbf{v} = -2\mathbf{i} - 6\mathbf{j}$

5. *Airplane Speed* After encountering the wind, is the airplane in Example 2 traveling at a higher speed or a lower speed? Explain.

6. *Airplane Speed* Consider the airplane described in Example 2, headed N 60° W at a speed of 400 miles per hour. What wind velocity, in the direction of N 40° E, will produce a resultant direction of N 50° W? Explain how to use the program on page 660 to obtain the answer *experimentally*. Then explain how to obtain the answer analytically.

7. *Airplane Speed* Consider the airplane described in Example 2, headed N 60° W at a speed of 400 miles per hour. What wind direction, at a speed of 75 miles per hour, will produce a resultant direction of N 50° W? Explain how to use the program on page 660 to obtain the answer *experimentally*. Then explain how to obtain the answer analytically.

Cumulative Test for Chapters 6–8

Take this test as you would take a test in class. After you are done, check your work against the answers given in the back of the book.

The *Interactive* CD-ROM provides answers to the Chapter Tests and Cumulative Tests. It also offers Chapter Pre-Tests (which test key skills and concepts covered in previous chapters) and Chapter Post-Tests, both of which have randomly generated exercises with diagnostic capabilities.

1. Consider the angle $\theta = -120°$.

   (a) Sketch the angle in standard position.

   (b) Determine a coterminal angle in the interval $[0°, 360°)$.

   (c) Convert the angle to radian measure.

   (d) Find the reference angle θ'.

   (e) Find the exact values of the six trigonometric functions of θ.

2. Convert the angle of magnitude 2.35 radians to degrees. Round the answer to one decimal place.

3. Find $\cos \theta$ if $\tan \theta = -\frac{4}{3}$ and $\sin \theta < 0$.

4. Sketch the graphs of (a) $f(x) = 3 - 2 \sin \pi x$ and (b) $g(x) = \frac{1}{2} \tan\left(x - \frac{\pi}{2}\right)$.

5. Find a, b, and c so that the graph of the function $h(x) = a \cos(bx + c)$ matches the graph in the figure.

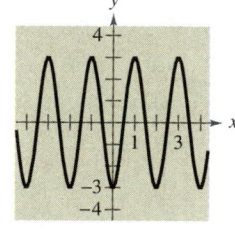

FIGURE FOR 5

6. Write an algebraic expression equivalent to $\sin(\arccos 2x)$.

7. Subtract and simplify: $\dfrac{\sin \theta - 1}{\cos \theta} - \dfrac{\cos \theta}{\sin \theta - 1}$.

8. Prove the identities.

   (a) $\cot^2 \alpha(\sec^2 \alpha - 1) = 1$ (b) $\sin(x + y) \sin(x - y) = \sin^2 x - \sin^2 y$

   (c) $\sin^2 x \cos^2 x = \frac{1}{8}(1 - \cos 4x)$

9. Find all solutions of the equations in the interval $[0, 2\pi)$.

   (a) $2 \cos^2 \beta - \cos \beta = 0$ (b) $3 \tan \theta - \cot \theta = 0$

10. Find the remaining sides and angles of the triangle shown in the figure.

   (a) $A = 30°$, $a = 9$, $b = 8$ (b) $A = 30°$, $b = 8$, $c = 10$

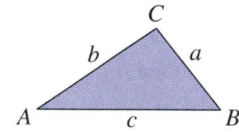

FIGURE FOR 10

11. Find the trigonometric form of the complex number $-2 + 2i$.

12. Find the product of $[4(\cos 30° + i \sin 30°)][6(\cos 120° + i \sin 120°)]$. Write the answer in standard form.

13. Use DeMoivre's Theorem to find the three cube roots of 1.

14. From a point 200 feet from a flagpole, the angles of elevation to the bottom and top of the flag are $16° \, 45'$ and $18°$, respectively. Approximate the height of the flagpole to the nearest foot.

15. An airplane's velocity with respect to the air is 500 kilometers per hour, with a bearing of N 30° E. The wind at the altitude of the plane has a velocity of 50 kilometers per hour with a bearing of N 60° E. What is the true direction of the plane, and what is its speed relative to the ground?

In 1975, pickups, sport utilities, and vans accounted for only 21% of all light vehicles sold in the United States. By 1995, they accounted for almost 40%.

Each of these three types of "trucks" comes in full-sized models and compact models. From 1990 to 1994, the sales (in thousands) of compact pickups P and compact sport utilities S were as follows.

Year	P	S
1990	1102.9	756.3
1991	980.7	807.9
1992	887.8	998.4
1993	954.1	1203.2
1994	1080.8	1359.1

These sales can be modeled by the following quadratic and linear models.

$$P = 47t^2 - 195t + 1109$$
$$S = 160t + 705$$

The graphs of these models show that compact sport utilities began outselling compact pickups in 1992.

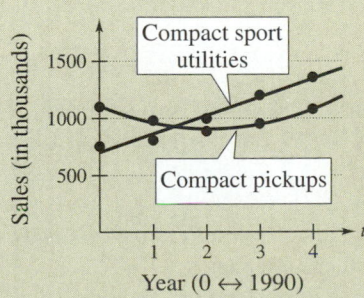

See Exercises 77 and 78 on page 699.

Photo: Alan Levenson / Tony Stone Images

Systems of Equations and Inequalities

9

9.1 *Solving Systems of Equations*
9.2 *Two-Variable Linear Systems*
9.3 *Multivariable Linear Systems*
9.4 *Systems of Inequalities*
9.5 *Linear Programming*

Dave Smith, John Sodana, and Tim Anness (from left to right) use a CAD-CAM system to design vehicles for Chrysler. Among other models, Chrysler manufactures Dodge Ram pickups and Jeep sport utilities.

9.1 Solving Systems of Equations

See Exercise 75 on page 674 for an example of how a system of equations can be used to compare two models for newsprint production in the United States.

The Method of Substitution ▫ Graphical Approach to Finding Solutions ▫ Applications

The Method of Substitution

Up to this point in the book, most problems have involved either a function of one variable or a single equation in two variables. However, many problems in science, business, and engineering involve two or more equations in two or more variables. To solve such problems, you need to find solutions of a **system of equations.** Here is an example of a system of two equations in x and y.

$$2x + y = 5 \qquad \text{Equation 1}$$
$$3x - 2y = 4 \qquad \text{Equation 2}$$

A **solution** of this system is an ordered pair that satisfies each equation in the system. Finding the set of all solutions is called **solving the system of equations.** For instance, the ordered pair $(2, 1)$ is a solution of this system. To check this, you can substitute 2 for x and 1 for y in *each* equation.

$$2x + y = 5 \qquad \text{Equation 1}$$
$$2(2) + 1 \stackrel{?}{=} 5 \qquad \text{Substitute 2 for } x \text{ and 1 for } y.$$
$$4 + 1 = 5 \qquad \text{Solution checks in Equation 1. } ✔$$
$$3x - 2y = 4 \qquad \text{Equation 2}$$
$$3(2) - 2(1) \stackrel{?}{=} 4 \qquad \text{Substitute 2 for } x \text{ and 1 for } y.$$
$$6 - 2 = 4 \qquad \text{Solution checks in Equation 2. } ✔$$

In this chapter you will study three ways to solve equations, beginning with the **method of substitution.**

METHOD OF SUBSTITUTION

1. *Solve* one of the equations for one variable in terms of the other.
2. *Substitute* the expression found in Step 1 into the other equation to obtain an equation in one variable.
3. *Solve* the equation obtained in Step 2.
4. *Back-substitute* the solution in Step 3 into the expression obtained in Step 1 to find the value of the other variable.
5. *Check* that the solution satisfies *each* of the original equations.

The *Interactive* CD-ROM offers graphing utility emulators of the *TI-82* and *TI-83*, which can be used with the Examples, Explorations, Technology notes, and Exercises.

Study Tip

Because many steps are required to solve a system of equations, it is very easy to make errors in arithmetic. Thus, we *strongly* suggest that you always *check your solution by substituting it into each equation in the original system.*

EXAMPLE 1 *Solving a System of Equations*

Solve the system of equations.

$x + y = 4$	Equation 1
$x - y = 2$	Equation 2

Solution

Begin by solving for y in Equation 1.

$y = 4 - x$	Solve for y in Equation 1.

Next, substitute this expression for y into Equation 2 and solve the resulting single-variable equation for x.

$x - y = 2$	Equation 2
$x - (4 - x) = 2$	Substitute $4 - x$ for y.
$x - 4 + x = 2$	Simplify.
$2x = 6$	Combine like terms.
$x = 3$	Divide both sides by 2.

Finally, you can solve for y by *back-substituting* $x = 3$ into the equation $y = 4 - x$, to obtain

$y = 4 - x$	Revised Equation 1
$y = 4 - 3$	Substitute 3 for x.
$y = 1.$	Solve for y.

The solution is the ordered pair $(3, 1)$. You can check this as follows.

Check

$x + y = 4$	Equation 1
$3 + 1 \overset{?}{=} 4$	Substitute for x and y.
$4 = 4$	Solution checks in Equation 1. ✔
$x - y = 2$	Equation 2
$3 - 1 \overset{?}{=} 2$	Substitute for x and y.
$2 = 2$	Solution checks in Equation 2. ✔

NOTE The term *back-substitution* implies that you work *backwards*. First you solve for one of the variables, and then you substitute that value *back* into one of the equations in the system to find the value of the other variable. ▪▪

EXAMPLE 2   *Solving a System by Substitution*

A total of $12,000 is invested in two funds paying 9% and 11% simple interest. The yearly interest is $1180. How much is invested at each rate?

Solution

Verbal Model:

9% fund	+	11% fund	=	Total investment

9% interest	+	11% interest	=	Total interest

Labels: Amount in 9% fund $= x$ *(dollars)*
Interest for 9% fund $= 0.09x$ *(dollars)*
Amount in 11% fund $= y$ *(dollars)*
Interest for 11% fund $= 0.11y$ *(dollars)*
Total investment $= \$12,000$ *(dollars)*
Total interest $= \$1180$ *(dollars)*

System: $x + \quad y = 12,000$ Equation 1
$0.09x + 0.11y = 1180$ Equation 2

To begin, it is convenient to multiply both sides of Equation 2 by 100 to obtain $9x + 11y = 118,000$. This eliminates the need to work with decimals.

$9x + 11y = 118,000$ Revised Equation 2

To solve this system, you can solve for x in Equation 1.

$x = 12,000 - y$ Revised Equation 1

Then, substitute this expression for x into revised Equation 2 and solve the resulting equation for y.

$$9x + 11y = 118,000 \qquad \text{Revised Equation 2}$$
$$9(12,000 - y) + 11y = 118,000 \qquad \text{Substitute } 12,000 - y \text{ for } x.$$
$$108,000 - 9y + 11y = 118,000 \qquad \text{Distributive Property}$$
$$2y = 10,000 \qquad \text{Combine like terms.}$$
$$y = 5000 \qquad \text{Divide both sides by 2.}$$

Next, back-substitute the value $y = 5000$ to solve for x.

$$x = 12,000 - y \qquad \text{Revised Equation 1}$$
$$x = 12,000 - 5000 \qquad \text{Substitute 5000 for } y.$$
$$x = 7000 \qquad \text{Simplify.}$$

The solution is (7000, 5000). Check this in the original problem.

The *Interactive* CD-ROM shows every example with its solution; clicking on the *Try It!* button brings up similar problems. Guided Examples and Integrated Examples show step-by-step solutions to additional examples. Integrated Examples are related to several concepts in the section.

The equations in Examples 1 and 2 are linear. Substitution can also be used to solve systems in which one or both of the equations are nonlinear.

EXAMPLE 3   *Substitution: Two-Solution Case*

Solve the system of equations.

$$x^2 - x - y = 1 \qquad \text{Equation 1}$$
$$-x + y = -1 \qquad \text{Equation 2}$$

Solution

Begin by solving for y in Equation 2 to obtain $y = x - 1$. Next, substitute this expression for y into Equation 1 and solve for x.

$$x^2 - x - y = 1 \qquad \text{Equation 1}$$
$$x^2 - x - (x - 1) = 1 \qquad \text{Substitute } x - 1 \text{ for } y.$$
$$x^2 - 2x + 1 = 1 \qquad \text{Simplify.}$$
$$x^2 - 2x = 0 \qquad \text{Standard form}$$
$$x(x - 2) = 0 \qquad \text{Factor.}$$
$$x = 0, 2 \qquad \text{Solve for } x.$$

Back-substituting these values of x to solve for the corresponding values of y produces the solutions $(0, -1)$ and $(2, 1)$. Check these in the original system.

EXAMPLE 4   *Substitution: No-Real-Solution Case*

Solve the system of equations.

$$-x + y = 4 \qquad \text{Equation 1}$$
$$x^2 + y = 3 \qquad \text{Equation 2}$$

Solution

Begin by solving for y in Equation 1 to obtain $y = x + 4$. Next, substitute this expression for y into Equation 2 and solve for x.

$$x^2 + y = 3 \qquad \text{Equation 2}$$
$$x^2 + (x + 4) = 3 \qquad \text{Substitute } x + 4 \text{ for } y.$$
$$x^2 + x + 1 = 0 \qquad \text{Simplify.}$$
$$x = \frac{-1 \pm \sqrt{1^2 - 4(1)(1)}}{2} \qquad \text{Quadratic Formula}$$

Because the discriminant is negative, the equation $x^2 + x + 1 = 0$ has no (real) solution. Hence, this system has no (real) solution.

NOTE Example 5 shows the value of a graphical approach to solving systems of equations in two variables. Notice what would happen if you tried only the substitution method in Example 5. It would be difficult to solve this equation for x using standard algebraic techniques. ▪▪

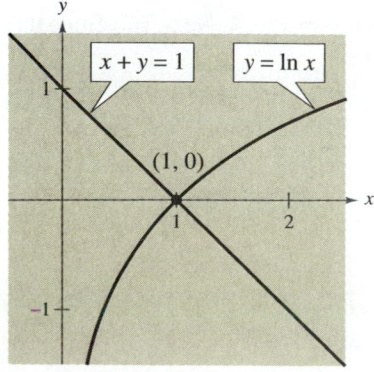

FIGURE 9.2

Graphical Approach to Finding Solutions

From Examples 2, 3, and 4, you can see that a system of two equations in two unknowns can have exactly one solution, more than one solution, or no solution. In practice, you can gain insight about the location(s) and number of solution(s) of a system of equations by graphing each of the equations in the same coordinate plane. The solutions of the system correspond to the **points of intersection** of the graphs. For instance, the two equations in Figure 9.1(a) graph as two lines with a *single point* of intersection; the two equations in Example 3 graph as a parabola and a line with *two points* of intersection, as shown in Figure 9.1(b); and the two equations in Example 4 graph as a line and a parabola that happen to have no points of intersection, as shown in Figure 9.1(c).

(a) One intersection point   **(b)** Two intersection points   **(c)** No intersection points

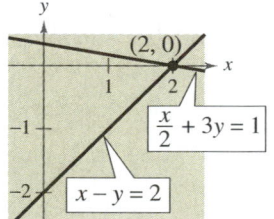 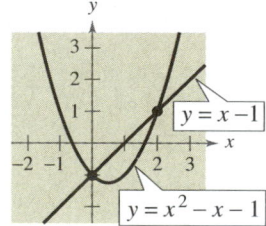 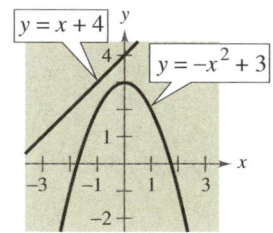

FIGURE 9.1

EXAMPLE 5   *Solving a System of Equations*

Solve the system of equations.

$$y = \ln x \qquad\qquad \text{Equation 1}$$
$$x + y = 1 \qquad\qquad \text{Equation 2}$$

Solution

The graphs of the two equations are shown in Figure 9.2. From the graphs, it is clear that there is only one point of intersection and that $(1, 0)$ is the solution point. You can confirm this by substituting in *both* equations.

Check

$$0 = \ln 1 \qquad\qquad \text{Equation 1 checks.} ✔$$
$$1 + 0 = 1 \qquad\qquad \text{Equation 2 checks.} ✔$$

Applications

The total cost C of producing x units of a product typically has two components—the initial cost and the cost per unit. When enough units have been sold so that the total revenue R equals the total cost, the sales are said to have reached the **break-even point.** You will find that the break-even point corresponds to the point of intersection of the cost and revenue curves.

EXAMPLE 6   *An Application: Break-Even Analysis*

A small business invests $10,000 in equipment to produce a product. Each unit of the product costs $0.65 to produce and is sold for $1.20. How many items must be sold before the business breaks even?

Solution

The total cost of producing x units is

$$\boxed{\text{Total cost}} = \boxed{\text{Cost per unit}} \cdot \boxed{\begin{array}{c}\text{Number} \\ \text{of units}\end{array}} + \boxed{\begin{array}{c}\text{Initial} \\ \text{cost}\end{array}}$$

$$C = 0.65x + 10,000. \qquad \text{Equation 1}$$

The revenue obtained by selling x units is

$$\boxed{\begin{array}{c}\text{Total} \\ \text{revenue}\end{array}} = \boxed{\begin{array}{c}\text{Price per} \\ \text{unit}\end{array}} \cdot \boxed{\begin{array}{c}\text{Number} \\ \text{of units}\end{array}}$$

$$R = 1.2x. \qquad \text{Equation 2}$$

Because the break-even point occurs when $R = C$, you have

$1.2x = 0.65x + 10,000$	Equate R and C.
$0.55x = 10,000$	Subtract $0.65x$ from both sides.
$x = \dfrac{10,000}{0.55}$	Divide both sides by 0.55.
$x \approx 18,182$ units.	Use a calculator.

Note in Figure 9.3 that sales less than the break-even point correspond to an overall loss, whereas sales greater than the break-even point correspond to a profit.

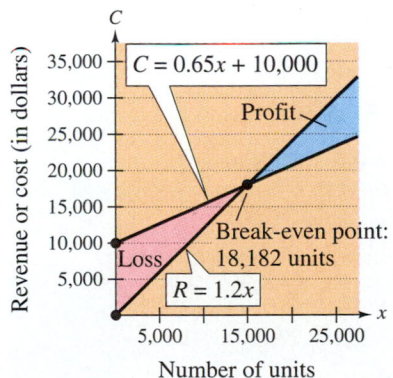

FIGURE 9.3

NOTE   Another way to view the solution in Example 6 is to consider the profit function $P = R - C$. The break-even point occurs when the profit is 0, which is the same as saying that $R = C$. ∎

EXAMPLE 7 *State Population*

From 1985 to 1993, the population of Arizona was increasing at a faster rate than the population of South Carolina. Models that approximate the two populations P are

$$P = 2785.8 + 88.8t \qquad \text{Arizona}$$
$$P = 3079.3 + 42.9t \qquad \text{South Carolina}$$

where $t = 5$ represents 1985 (see Figure 9.4). According to these two models, when would you expect the population of Arizona to have exceeded the population of South Carolina?   (Source: U.S. Bureau of the Census)

Solution

Because the first equation has already been solved for P in terms of t, substitute this value into the second equation and solve for t, as follows.

$$2785.8 + 88.8t = 3079.3 + 42.9t$$
$$88.8t - 42.9t = 3079.3 - 2785.8$$
$$45.9t = 293.5$$
$$t \approx 6.4$$

Thus, from the given models, you would expect that the population of Arizona exceeded the population of South Carolina sometime during 1986.

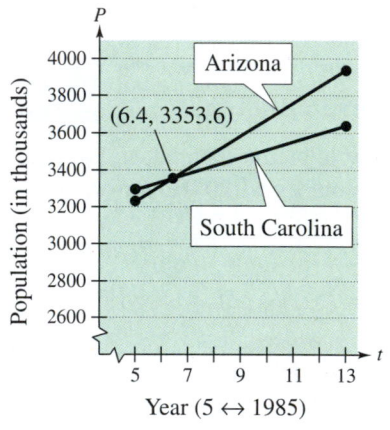

FIGURE 9.4

GROUP ACTIVITY

INTERPRETING POINTS OF INTERSECTION

You plan to rent a 14-foot truck for a 2-day local move. At truck rental agency A, you can rent a truck for $29.95 per day plus $0.49 per mile. At agency B, you can rent a truck for $50 per day plus $0.25 per mile. The total cost y (in dollars) for the truck from agency A is

$$y = (\$29.95 \text{ per day})(2 \text{ days}) + 0.49x = 59.90 + 0.49x$$

where x is the total number of miles the truck is driven. Write a total cost equation in terms of x and y for the total cost of the truck from agency B. Use a graphing utility to graph the two equations and find the point of intersection. Interpret the meaning of the point of intersection in the context of the problem. Which agency should you choose if you plan to travel a total of 100 miles over the 2-day move? Why? How does the situation change if you plan to drive 200 miles over the 2-day move?

The *Interactive* CD-ROM provides additional help with Warm-Up exercises by providing a hypertext link to the section in which the concept was introduced.

WARM UP

Sketch the graph of the equation.

1. $y = -\frac{1}{3}x + 6$ **2.** $y = 2(x - 3)$

3. $x^2 + y^2 = 4$ **4.** $y = 5 - (x - 3)^2$

Perform the operations and simplify.

5. $(3x + 2y) - 2(x + y)$ **6.** $(-10u + 3v) + 5(2u - 8v)$

7. $x^2 + (x - 3)^2 + 6x$ **8.** $y^2 - (y + 1)^2 + 2y$

Solve the equation.

9. $3x + (x - 5) = 15 + 4$ **10.** $y^2 + (y - 2)^2 = 2$

 The *Interactive* CD-ROM contains step-by-step solutions to all odd-numbered Section and Review Exercises. It also provides Tutorial Exercises, which link to Guided Examples for additional help.

9.1 Exercises

In Exercises 1–10, solve the system by substitution. Check your solution graphically.

1. $2x + y = 6$
    $-x + y = 0$

2. $x - y = -4$
    $x + 2y = 5$

5. $x - 3y = 15$
    $x^2 + y^2 = 25$

6. $x + y = 0$
    $x^3 - 5x - y = 0$

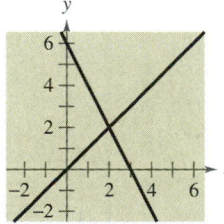

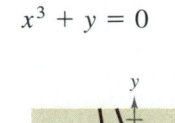

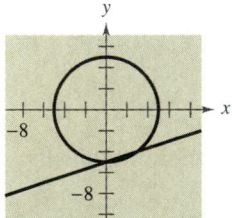

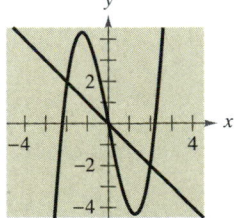

3. $x - y = -4$
    $x^2 - y = -2$

4. $3x + y = 2$
    $x^3 + y = 0$

7. $x^2 + y = 0$
    $x^2 - 4x - y = 0$

8. $y = -2x^2 + 2$
    $y = 2(x^4 - 2x^2 + 1)$

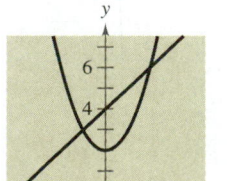

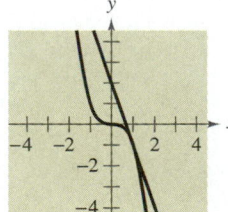

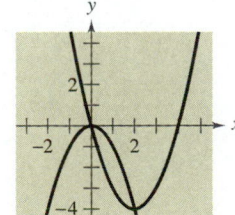

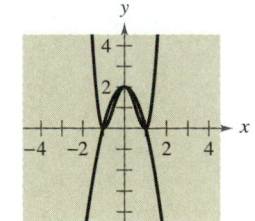

9. $x - 6y = -8$
$x^2 - 4y^3 = 0$

10. $y = x^3 - 3x^2 + 4$
$y = -2x + 4$

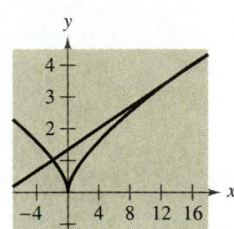

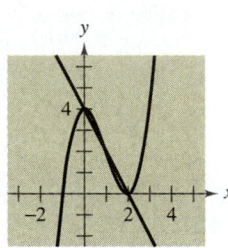

In Exercises 11–22, solve the system by the method of substitution.

11. $x - y = 0$
$5x - 3y = 10$

12. $x + 2y = 1$
$5x - 4y = -23$

13. $2x - y + 2 = 0$
$4x + y - 5 = 0$

14. $6x - 3y - 4 = 0$
$x + 2y - 4 = 0$

15. $30x - 40y - 33 = 0$
$10x + 20y - 21 = 0$

16. $1.5x + 0.8y = 2.3$
$0.3x - 0.2y = 0.1$

17. $\frac{1}{5}x + \frac{1}{2}y = 8$
$x + y = 20$

18. $\frac{1}{2}x + \frac{3}{4}y = 10$
$\frac{3}{4}x - y = 4$

19. $2x - y = 4$
$-4x + 2y = -12$

20. $-\frac{2}{3}x + y = -2$
$2x - 3y = 6$

21. $x - y = 0$
$2x + y = 0$

22. $x - 2y = 0$
$3x - y = 0$

In Exercises 23–28, solve the system graphically.

23. $-x + 2y = 2$
$3x + y = 15$

24. $x + y = 0$
$3x - 2y = 10$

25. $x - 3y = -2$
$5x + 3y = 17$

26. $-x + 2y = 1$
$x - y = 2$

27. $x + y = 4$
$x^2 + y^2 - 4x = 0$

28. $x - y + 3 = 0$
$x^2 - 4x + 7 = y$

In Exercises 29–34, solve the system graphically. Use a graphing utility to verify your results.

29. $7x + 8y = 24$
$x - 8y = 8$

30. $x - y = 0$
$5x - 2y = 6$

31. $3x - 2y = 0$
$x^2 - y^2 = 4$

32. $2x - y + 3 = 0$
$x^2 + y^2 - 4x = 0$

33. $x^2 + y^2 = 8$
$y = x^2$

34. $x^2 + y^2 = 25$
$(x - 8)^2 + y^2 = 41$

In Exercises 35–40, use a graphing utility to approximate all points of intersection of the graphs.

35. $y = e^x$
$x - y + 1 = 0$

36. $x + 2y = 8$
$y = \log_2 x$

37. $y = \sqrt{x}$
$y = x$

38. $x - y = 3$
$x - y^2 = 1$

39. $x^2 + y^2 = 169$
$x^2 - 8y = 104$

40. $x^2 + y^2 = 4$
$2x^2 - y = 2$

In Exercises 41–52, solve the system graphically or algebraically. Explain your choice of method.

41. $y = 2x$
$y = x^2 + 1$

42. $x + y = 4$
$x^2 + y = 2$

43. $3x - 7y + 6 = 0$
$x^2 - y^2 = 4$

44. $x^2 + y^2 = 25$
$2x + y = 10$

45. $x - 2y = 4$
$x^2 - y = 0$

46. $y = (x + 1)^3$
$y = \sqrt{x - 1}$

47. $y - e^{-x} = 1$
$y - \ln x = 3$

48. $y = x^3 - 2x^2 + x - 1$
$y = -x^2 + 3x - 1$

49. $y = x^4 - 2x^2 + 1$
$y = 1 - x^2$

50. $x^2 + y = 4$
$e^x - y = 0$

51. $xy - 1 = 0$
$2x - 4y + 7 = 0$

52. $x - 2y = 1$
$y = \sqrt{x - 1}$

53. *Think About It*   When solving a system of equations by substitution, how do you recognize that the system has no solution?

54. *Essay*   Write a brief paragraph describing any advantages of substitution over the graphical method of solving a system of equations.

***Break-Even Analysis*   In Exercises 55–58, find the sales necessary to break even ($R = C$) for the given cost C of x units and the given revenue R obtained by selling x units. (Round to the nearest whole unit.)**

55. $C = 8650x + 250,000, \quad R = 9950x$

56. $C = 5.5\sqrt{x} + 10,000, \quad R = 3.29x$

57. $C = 2.65x + 350,000, \quad R = 4.15x$

58. $C = 0.08x + 50,000, \quad R = 0.25x$

59. *Break-Even Point*   A small business invests $16,000 to produce an item that will sell for $5.95. Each unit can be produced for $3.45.

(a) How many units must be sold to break even?

(b) How many units must be sold to make a profit of $6000?

60. *Break-Even Point*   A small business has an initial investment of $5000. The unit cost of the product is $21.60, and the selling price is $34.10. How many units must be sold to break even?

61. *Investment Portfolio*   A total of $25,000 is invested in two funds paying 6% and 8.5% simple interest. The 6% investment has a lower risk. The investor wants a yearly interest check of $2000 from the investment.

(a) Write a system of equations in which one equation represents the total amount invested and the other equation represents the $2000 required in interest. Let x and y represent the amounts invested at 6% and 8.5%, respectively.

(b) Use a graphing utility to graph the two equations. As the amount invested at 6% increases, how does the amount invested at 8.5% change and how does the amount of interest change? Explain.

(c) What is the most that can be invested at 6% to meet the requirement of $2000 per year in interest?

62. *Investment Portfolio*   A total of $20,000 is invested in two funds paying 6.5% and 8.5% simple interest. The 6.5% investment has a lower risk. The investor wants a yearly interest check of $1600 from the investment. What is the most that can be invested at 6.5% to meet this requirement?

63. *Choice of Two Jobs*   You are offered two jobs selling dental supplies. One company offers a straight commission of 6% of sales. The other company offers a salary of $250 per week plus 3% of sales. How much would you have to sell in a week in order to make the straight commission offer better?

64. *Choice of Two Jobs*   You are offered two different jobs selling college textbooks. One company offers an annual salary of $20,000 plus a year-end bonus of 1% of your total sales. The other company offers an annual salary of $15,000 plus a year-end bonus of 2% of your total sales. Determine the annual sales required to make the second offer better.

65. *Log Volume*   You are offered two different rules for estimating the number of board feet in a 16-foot log. One is the *Doyle Log Rule* and is modeled by

$$V = (D - 4)^2, \qquad 5 \le D \le 40.$$

The other is the *Scribner Log Rule* and is modeled by

$$V = 0.79D^2 - 2D - 4, \qquad 5 \le D \le 40$$

where D is the diameter of the log and V is its volume in board feet.

(a) Use a graphing utility to graph the two log rules on the same viewing rectangle.

(b) For what diameter do the two scales agree?

(c) If you were selling large logs, which scale would you use?

66. *Market Equilibrium*   The supply and demand curves for a business dealing with wheat are given by

Supply: $p = 1.45 + 0.00014x^2$
Demand: $p = (2.388 - 0.007x)^2$

where p is the price in dollars per bushel and x is the quantity in bushels per day. Use a graphing utility to graph the supply and demand equations and find the market equilibrium. (The *market equilibrium* is the point of intersection of the graphs for $x > 0$.)

Geometry **In Exercises 67–70, find the dimensions of the rectangle meeting the specified conditions.**

67. The perimeter is 30 meters and the length is 3 meters greater than the width.

68. The perimeter is 280 centimeters and the width is 20 centimeters less than the length.

69. The perimeter is 42 inches and the width is three-fourths the length.

70. The perimeter is 210 feet and the length is $1\frac{1}{2}$ times the width.

71. *Geometry* What are the dimensions of a rectangular tract of land if its perimeter is 40 kilometers and its area is 96 square kilometers?

72. *Geometry* What are the dimensions of an isosceles right triangle with a 2-inch hypotenuse and an area of 1 square inch?

73. *Hyperbolic Mirror* In a hyperbolic mirror, light rays directed to one focus are reflected to the other focus. The mirror in the figure has the equation

$$\frac{x^2}{25} - \frac{y^2}{36} = 1.$$

At which point on the mirror will light from the point $(0, 10)$ reflect to the focus?

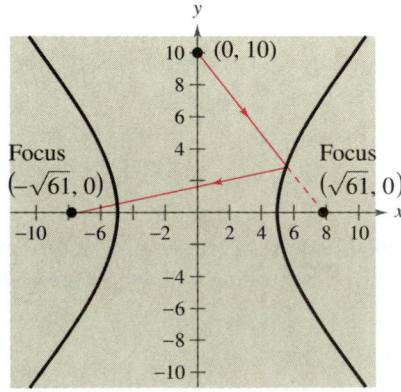

74. *Exploration* Find an equation of a line whose graph intersects the graph of the parabola $y = x^2$ at (a) two points, (b) one point, and (c) no points. (There is more than one correct answer.)

75. *Data Analysis* The table gives the amount y, in millions of short tons, of newsprint produced in the years 1990 through 1993 in the United States. (Source: American Paper Institute)

Year	1990	1991	1992	1993
y	6.6	6.8	7.1	7.1

(a) Use a graphing utility to find a linear model and a quadratic model that represent the data in the interval from 1990 through 1993. (Let $t = 0$ represent 1990.)

(b) Use a graphing utility to graph the data and the two models on the same viewing rectangle.

(c) Approximate the points of intersection of the graphs of the models.

(d) Use the models to estimate newsprint production in 1994. Which model do you think gives the more accurate estimate? Explain.

76. *Conjecture* In parts (a) and (b), use a graphing utility to make a conjecture about exponential and polynomial equations.

(a) Use a graphing utility to graph the system of equations

$$y = b^x$$
$$y = x^b$$

for $b = 2$ and $b = 4$.

(b) For a fixed value of $b > 1$, make a conjecture about the number of points of intersection of the graphs in part (a).

Review **Solve Exercises 77–82 as a review of the skills and problem-solving techniques you learned in previous sections. Find the general form of the equation of the line through the two points.**

77. $(-2, 7), (5, 5)$

78. $(3.5, 4), (10, 6)$

79. $(6, 3), (10, 3)$

80. $(4, -2), (4, 5)$

81. $\left(\frac{3}{5}, 0\right), (4, 6)$

82. $\left(-\frac{7}{3}, 8\right), \left(\frac{5}{2}, \frac{1}{2}\right)$

See Exercise 74 on page 686 for an example of how a system of linear equations can be used to determine the relationship between wheat yield and the amount of fertilizer applied to a field.

9.2 Two-Variable Linear Systems

The Method of Elimination □ *Graphical Interpretation of Solutions* □ *Applications*

The Method of Elimination

In Section 9.1, you studied two methods for solving a system of equations: substitution and graphing. Now you will study the **method of elimination.** The key step in this method is to obtain, for one of the variables, coefficients that differ only in sign so that *adding* the equations eliminates the variable.

$$
\begin{aligned}
3x + 5y &= 7 \qquad &\text{Equation 1}\\
-3x - 2y &= -1 \qquad &\text{Equation 2}\\
\hline
3y &= 6 \qquad &\text{Add equations.}
\end{aligned}
$$

Note that by adding the two equations, you eliminate the variable x and obtain a single equation in y. Solving this equation for y produces $y = 2$, which you can then back-substitute into one of the original equations to solve for x.

EXAMPLE 1 The Method of Elimination

Solve the system of linear equations.

$$
\begin{aligned}
3x + 2y &= 4 \qquad &\text{Equation 1}\\
5x - 2y &= 8 \qquad &\text{Equation 2}
\end{aligned}
$$

Solution

Because the coefficients for y differ only in sign, you can eliminate y by adding the two equations.

$$
\begin{aligned}
3x + 2y &= 4 \qquad &\text{Equation 1}\\
5x - 2y &= 8 \qquad &\text{Equation 2}\\
\hline
8x \phantom{{}+2y} &= 12 \qquad &\text{Add equations.}
\end{aligned}
$$

Therefore, $x = \frac{3}{2}$. By back-substituting this value into Equation 1, you can solve for y, as follows.

$$
\begin{aligned}
3x + 2y &= 4 \qquad &\text{Equation 1}\\
3\left(\tfrac{3}{2}\right) + 2y &= 4 \qquad &\text{Substitute } \tfrac{3}{2} \text{ for } x.\\
y &= -\tfrac{1}{4} \qquad &\text{Solve for } y.
\end{aligned}
$$

The solution is $\left(\frac{3}{2}, -\frac{1}{4}\right)$. Check this in the original system.

NOTE Try using the method of substitution to solve the system given in Example 1. Which method do you think is easier? Many people find that the method of elimination is more efficient. ■■

Study Tip

To obtain coefficients (for one of the variables) that differ only in sign, you often need to multiply one or both of the equations by suitably chosen constants.

EXAMPLE 2 *The Method of Elimination*

Solve the system of linear equations.

$$2x - 3y = -7 \qquad \text{Equation 1}$$
$$3x + y = -5 \qquad \text{Equation 2}$$

Solution

For this system, you can obtain coefficients that differ only in sign by multiplying Equation 2 by 3.

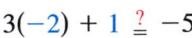

$$2x - 3y = -7 \qquad\qquad 2x - 3y = \ -7 \qquad \text{Equation 1}$$
$$3x + y = -5 \qquad\qquad \underline{9x + 3y = -15} \qquad \text{Multiply Equation 2 by 3.}$$
$$11x \qquad\ \ = -22 \qquad \text{Add equations.}$$

Thus, you can see that $x = -2$. By back-substituting this value of x into Equation 1, you can solve for y, as follows.

$$2x - 3y = -7 \qquad \text{Equation 1}$$
$$2(-2) - 3y = -7 \qquad \text{Substitute } -2 \text{ for } x.$$
$$-3y = -3 \qquad \text{Collect like terms.}$$
$$y = 1 \qquad \text{Solve for } y.$$

The solution is $(-2, 1)$. Check this in the original system, as follows.

Check

$$2(-2) - 3(1) \stackrel{?}{=} -7 \qquad \text{Substitute into Equation 1.}$$
$$-4 - 3 = -7 \qquad \text{Equation 1 checks.} \ \checkmark$$

$$3(-2) + 1 \stackrel{?}{=} -5 \qquad \text{Substitute into Equation 2.}$$
$$-6 + 1 = -5 \qquad \text{Equation 2 checks.} \ \checkmark$$

In Example 2, the two systems of linear equations

$$2x - 3y = -7 \qquad \text{and} \qquad 2x - 3y = \ -7$$
$$3x + y = -5 \qquad\qquad\qquad 9x + 3y = -15$$

are called **equivalent** because they have precisely the same solution set. The operations that can be performed on a system of linear equations to produce an equivalent system are (1) interchanging any two equations, (2) multiplying an equation by a nonzero constant, and (3) adding a multiple of one equation to any other equation in the system.

▶ *Exploration*

Sketch the graph of each of the following systems of equations.

a. $y = 5x + 1$

$y - x = -5$

b. $3y = 4x - 1$

$-8x + 2 = -6y$

c. $2y = -x + 3$

$-4 = y + \frac{1}{2}x$

Determine the number of solutions each system has. Explain your reasoning.

■ **THE METHOD OF ELIMINATION**

To use the **method of elimination** to solve a system of two linear equations in x and y, use the following steps.

1. Obtain coefficients for x (or y) that differ only in sign by multiplying all terms of one or both equations by suitably chosen constants.

2. Add the equations to eliminate one variable and solve the resulting equation.

3. Back-substitute the value obtained in Step 2 into either of the original equations and solve for the other variable.

4. Check your solution in both of the original equations.

EXAMPLE 3   *The Method of Elimination*

Solve the system of linear equations.

$$5x + 3y = 9 \qquad\qquad \text{Equation 1}$$

$$2x - 4y = 14 \qquad\qquad \text{Equation 2}$$

Solution

You can obtain coefficients that differ only in sign by multiplying Equation 1 by 4 and multiplying Equation 2 by 3.

$5x + 3y = 9$ ⟹ $20x + 12y = 36$ Multiply Equation 1 by 4.

$2x - 4y = 14$ ⟹ $\underline{6x - 12y = 42}$ Multiply Equation 2 by 3.

$26x \qquad\quad = 78$ Add equations.

From this equation, you can see that $x = 3$. By back-substituting this value of x into Equation 2, you can solve for y, as follows.

$$2x - 4y = 14 \qquad\qquad \text{Equation 2}$$

$$2(3) - 4y = 14 \qquad\qquad \text{Substitute 3 for } x.$$

$$-4y = 8 \qquad\qquad \text{Collect like terms.}$$

$$y = -2 \qquad\qquad \text{Solve for } y.$$

The solution is $(3, -2)$. Check this in the original system.

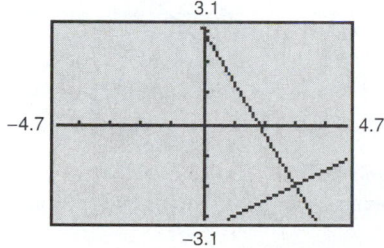

FIGURE 9.5

NOTE   Remember that you can check the solution of a system of equations graphically. For instance, to check the solution found in Example 3, sketch the graphs of both equations on the same viewing rectangle, as shown in Figure 9.5. Notice that the two lines intersect at $(3, -2)$. ■■

Graphical Interpretation of Solutions

It is possible for a *general* system of equations to have exactly one solution, two or more solutions, or no solution. If a system of *linear* equations has two different solutions, it must have an *infinite* number of solutions. To see why this is true, consider the following graphical interpretations of a system of two linear equations in two variables. (Remember that the graph of a linear equation in two variables is a straight line.)

> ### GRAPHICAL INTERPRETATIONS OF SOLUTIONS
>
> For a system of two linear equations in two variables, the number of solutions is given by one of the following.
>
> | *Number of Solutions* | *Graphical Interpretation* |
> |---|---|
> | **1.** Exactly one solution | The two lines intersect at one point. |
> | **2.** Infinitely many solutions | The two lines are identical. |
> | **3.** No solution | The two lines are parallel. |

The graphical interpretations described above are further illustrated in Figure 9.6. In the second graph, note that the two lines coincide. This case is illustrated in Example 5. In the third graph, note that the two lines are parallel. This case is illustrated in Example 4.

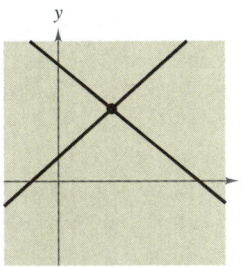 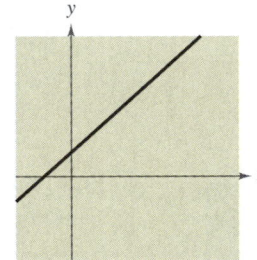 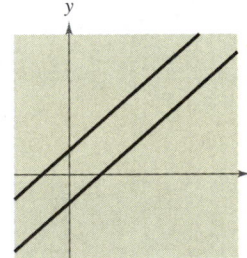

Consistent
Two lines that intersect
One point of intersection

Consistent
Two lines that coincide
Infinitely many points

Inconsistent
Two parallel lines
No point of intersection

FIGURE 9.6

A system of linear equations is called **consistent** if it has at least one solution, and it is called **inconsistent** if it has no solution. In Examples 4 and 5, note how you can use the method of elimination to determine that a system of linear equations has no solution or infinitely many solutions.

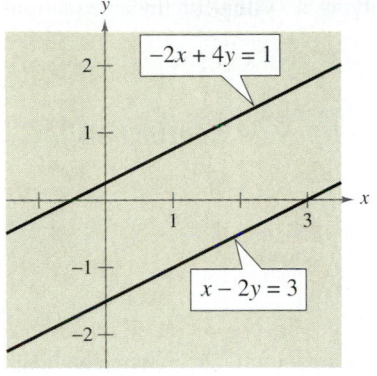

$-2x + 4y = 1$

$x - 2y = 3$

FIGURE 9.7

EXAMPLE 4   The Method of Elimination: No-Solution Case

Solve the system of linear equations.

$$x - 2y = 3 \qquad \text{Equation 1}$$
$$-2x + 4y = 1 \qquad \text{Equation 2}$$

Solution

To obtain coefficients that differ only in sign, multiply Equation 1 by 2.

$$x - 2y = 3 \quad\Longrightarrow\quad 2x - 4y = 6 \qquad \text{Multiply Equation 1 by 2.}$$
$$-2x + 4y = 1 \quad\Longrightarrow\quad -2x + 4y = 1 \qquad \text{Equation 2}$$
$$\overline{ \qquad\qquad 0 = 7} \qquad \text{False statement.}$$

Because there are no values of x and y for which $0 = 7$, you can conclude that the system is inconsistent and has no solution. The lines corresponding to the two equations in this system are shown in Figure 9.7. Note that the two lines are parallel, and therefore have no point of intersection.

In Example 4, note that the occurrence of a false statement, such as $0 = 7$, indicates that the system has no solution. In the next example, note that the occurrence of a statement that is true for all values of the variables, such as $0 = 0$, indicates that the system has infinitely many solutions.

EXAMPLE 5   The Method of Elimination: Many-Solutions Case

Solve the system of linear equations.

$$2x - y = 1 \qquad \text{Equation 1}$$
$$4x - 2y = 2 \qquad \text{Equation 2}$$

Solution

To obtain coefficients that differ only in sign, multiply Equation 2 by $-\frac{1}{2}$.

$$2x - y = 1 \quad\Longrightarrow\quad 2x - y = 1 \qquad \text{Equation 1}$$
$$4x - 2y = 2 \quad\Longrightarrow\quad -2x + y = -1 \qquad \text{Multiply Equation 2 by } -\frac{1}{2}.$$
$$\overline{ \qquad\qquad 0 = 0} \qquad \text{Add equations.}$$

Because the two equations turn out to be equivalent (have the same solution set), you can conclude that the system has infinitely many solutions. The solution set consists of all points (x, y) lying on the line

$$2x - y = 1. \qquad \text{See Figure 9.8.}$$

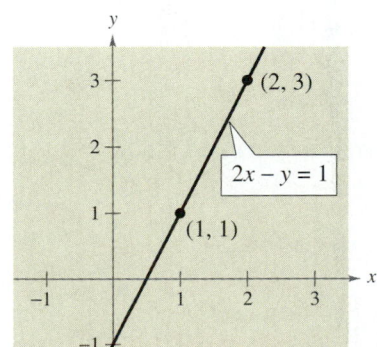

$(2, 3)$

$2x - y = 1$

$(1, 1)$

FIGURE 9.8

Example 6 illustrates a strategy for solving a system of linear equations that has decimal coefficients.

EXAMPLE 6   *A Linear System Having Decimal Coefficients*

Solve the system of linear equations.

$$0.02x - 0.05y = -0.38 \qquad \text{Equation 1}$$
$$0.03x + 0.04y = 1.04 \qquad \text{Equation 2}$$

Solution

Because the coefficients in this system have two decimal places, you can begin by multiplying each equation by 100. (This produces a system in which the coefficients are all integers.)

$$2x - 5y = -38 \qquad \text{Revised Equation 1}$$
$$3x + 4y = 104 \qquad \text{Revised Equation 2}$$

Now, to obtain coefficients that differ only in sign, multiply Equation 1 by 3 and multiply Equation 2 by -2.

$2x - 5y = -38$	$6x - 15y = -114$	Multiply Equation 1 by 3.
$3x + 4y = 104$	$-6x - 8y = -208$	Multiply Equation 2 by -2.
	$-23y = -322$	Add equations.

Thus, you can conclude that

$$y = \frac{-322}{-23} = 14.$$

Back-substituting this value into Equation 2 produces the following.

$$3x + 4y = 104 \qquad \text{Revised Equation 2}$$
$$3x + 4(14) = 104 \qquad \text{Substitute 14 for } y.$$
$$3x = 48 \qquad \text{Collect like terms.}$$
$$x = 16 \qquad \text{Solve for } x.$$

The solution is $(16, 14)$. Check this in the original system, as follows.

Check

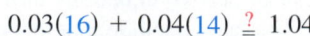

$$0.02(16) - 0.05(14) \stackrel{?}{=} -0.38 \qquad \text{Substitute into Equation 1.}$$
$$0.32 - 0.70 = -0.38 \qquad \text{Equation 1 checks.} \checkmark$$

$$0.03(16) + 0.04(14) \stackrel{?}{=} 1.04 \qquad \text{Substitute into Equation 2.}$$
$$0.48 + 0.56 = 1.04 \qquad \text{Equation 2 checks.}$$

Applications

At this point, you may be asking the question "How can I tell which application problems can be solved using a system of linear equations?" The answer comes from the following considerations.

1. Does the problem involve more than one unknown quantity?
2. Are there two (or more) equations or conditions to be satisfied?

If one or both of these conditions occur, the appropriate mathematical model for the problem may be a system of linear equations. Example 7 shows how to construct such a model.

Real Life

EXAMPLE 7 An Application of a Linear System

An airplane flying into a headwind travels the 2000-mile flying distance between two cities in 4 hours and 24 minutes. On the return flight, the same distance is traveled in 4 hours. Find the air speed of the plane and the speed of the wind, assuming that both remain constant.

Solution

The two unknown quantities are the speeds of the wind and the plane. If r_1 is the speed of the plane and r_2 is the speed of the wind, then

$$r_1 - r_2 = \text{speed of the plane } against \text{ the wind}$$
$$r_1 + r_2 = \text{speed of the plane } with \text{ the wind}$$

as shown in Figure 9.9. Using the formula

$$\text{Distance} = (\text{rate})(\text{time})$$

for these two speeds, you obtain the following equations.

$$2000 = (r_1 - r_2)\left(4 + \frac{24}{60}\right)$$
$$2000 = (r_1 + r_2)(4)$$

These two equations simplify as follows.

$$5000 = 11r_1 - 11r_2 \qquad \text{Equation 1}$$
$$500 = r_1 + r_2 \qquad \text{Equation 2}$$

By elimination, the solution is

$$r_1 = \frac{5250}{11} \approx 477.27 \text{ miles per hour} \qquad \text{Speed of plane}$$

$$r_2 = \frac{250}{11} \approx 22.73 \text{ miles per hour.} \qquad \text{Speed of wind}$$

The supersonic BAC/Aérospatiale Concorde, the fastest jet airliner, can travel at Mach 2. The New York to London record is 2 hours, 54 minutes, 30 seconds, set on April 14, 1990. *(Photo: Tony Stone Images)*

Original flight

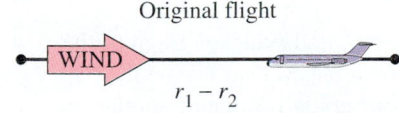

$r_1 - r_2$

Return flight

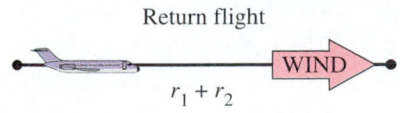

$r_1 + r_2$

FIGURE 9.9

NOTE   In a free market, the demand for many products is related to the price of the product. As the price decreases, the demand by consumers increases and the amount that producers are able or willing to supply decreases. ■■

EXAMPLE 8   *Finding the Point of Equilibrium*

The demand and supply functions for a certain type of calculator are given by

$$p = 150 - 0.00001x \qquad \text{Demand equation}$$
$$p = 60 + 0.00002x \qquad \text{Supply equation}$$

where p is the price in dollars and x represents the number of units. Find the point of equilibrium for this market. The point of equilibrium is the price p and number of units x that satisfy both the demand and supply equations.

Solution

Begin by substituting the value of p given in the supply equation into the demand equation.

$$p = 150 - 0.00001x \qquad \text{Demand equation}$$
$$60 + 0.00002x = 150 - 0.00001x \qquad \text{Substitute } 60 + 0.00002x \text{ for } p.$$
$$0.00003x = 90 \qquad \text{Collect like terms.}$$
$$x = 3,000,000 \qquad \text{Solve for } x.$$

Thus, the point of equilibrium occurs when the demand and supply are each 3 million units. (See Figure 9.10.) The price that corresponds to this x-value is obtained by back-substituting $x = 3,000,000$ into either of the original equations. For instance, back-substituting into the demand equation produces

$$p = 150 - 0.00001(3,000,000) = 150 - 30 = \$120.$$

Try back-substituting $x = 3,000,000$ into the supply equation to see that you obtain the same price.

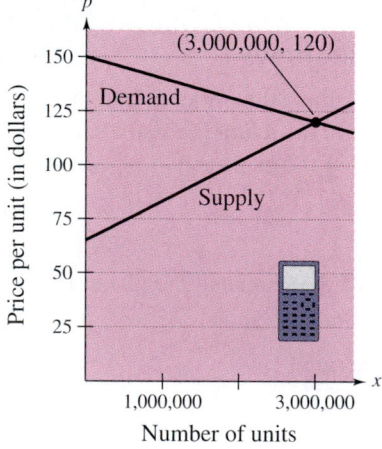

FIGURE 9.10

GROUP ACTIVITY

GRAPHING SYSTEMS OF EQUATIONS

Make up a system of two linear equations, and sketch a graph of the system as accurately as possible on graph paper. Don't write the equations on your graph. Exchange your graph for that of another student. Try to reconstruct the equations of the system that are represented by the graph that you received. Algebraically solve the system, and verify your solution with the graph. Compare results with the person who used your graph.

Sketch the graph of the equation.

1. $2x + y = 4$ 　　　　　　**2.** $5x - 2y = 3$

3. $x - y = 3$ 　　　　　　**4.** $x + y = 3$

5. $3x + 6y = 4$ 　　　　　**6.** $7x - 4y = 10$

Decide whether the lines are parallel, perpendicular, or neither.

7. $2x - 3y = -10$ 　　　　**8.**　$4x - 12y = 5$

　　$3x + 2y = -11$ 　　　　　　$-2x + 16y = 3$

9. $5x + 2y = 2$ 　　　　　**10.**　$x - 3y = 2$

　　$3x + 2y = 1$ 　　　　　　$6x + 2y = 4$

9.2 Exercises

In Exercises 1–10, solve by elimination. Label each line with its equation.

1. $2x + y = 5$
　$x - y = 1$

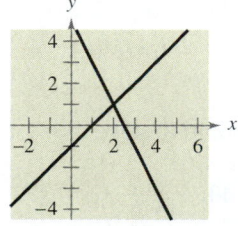

2. 　$x + 3y = 1$
　$-x + 2y = 4$

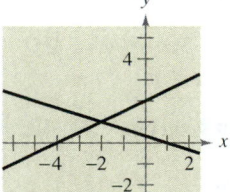

5. 　$x - y = 2$
　$-2x + 2y = 5$

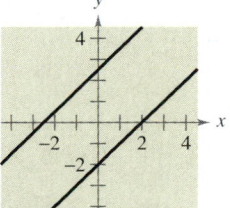

6. $3x + 2y = 3$
　$6x + 4y = 14$

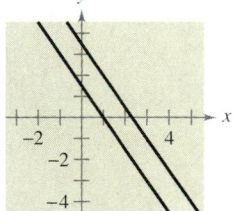

3. 　$x + y = 0$
　$3x + 2y = 1$

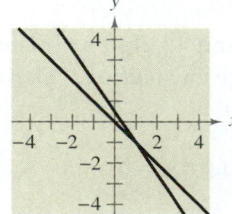

4. $2x - y = 3$
　$4x + 3y = 21$

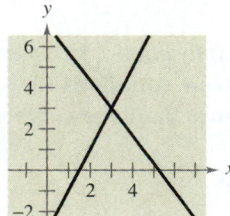

7. 　$3x - 2y = 5$
　$-6x + 4y = -10$

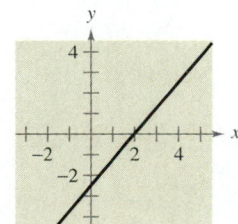

8. 　$x - 2y = 4$
　$6x + 2y = 10$

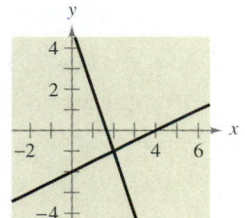

9. $9x + 3y = 1$
$3x - 6y = 5$

10. $5x + 3y = -18$
$2x - 6y = 1$

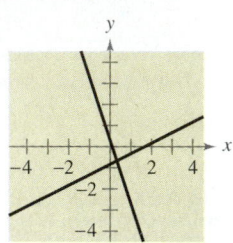

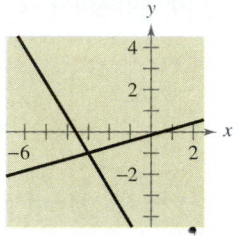

In Exercises 11–30, solve the system by elimination and check any solution algebraically.

11. $x + 2y = 4$
$x - 2y = 1$

12. $3x - 5y = 2$
$2x + 5y = 13$

13. $2x + 3y = 18$
$5x - y = 11$

14. $x + 7y = 12$
$3x - 5y = 10$

15. $3x + 2y = 10$
$2x + 5y = 3$

16. $8r + 16s = 20$
$16r + 50s = 55$

17. $2u + v = 120$
$u + 2v = 120$

18. $5u + 6v = 24$
$3u + 5v = 18$

19. $6r - 5s = 3$
$10s - 12r = 5$

20. $1.8x + 1.2y = 4$
$9x + 6y = 3$

21. $\dfrac{x}{4} + \dfrac{y}{6} = 1$
$x - y = 3$

22. $\dfrac{2}{3}x + \dfrac{1}{6}y = \dfrac{2}{3}$
$4x + y = 4$

23. $\dfrac{x + 3}{4} + \dfrac{y - 1}{3} = 1$
$2x - y = 12$

24. $\dfrac{x - 1}{2} + \dfrac{y + 2}{3} = 4$
$x - 2y = 5$

25. $2.5x - 3y = 1.5$
$10x - 12y = 6$

26. $0.02x - 0.05y = -0.19$
$0.03x + 0.04y = 0.52$

27. $0.05x - 0.03y = 0.21$
$0.07x + 0.02y = 0.16$

28. $0.2x - 0.5y = -27.8$
$0.3x + 0.4y = 68.7$

29. $4b + 3m = 3$
$3b + 11m = 13$

30. $3b + 3m = 7$
$3b + 5m = 3$

In Exercises 31–34, use a graphing utility to graph the lines in the system. Use the graphs to determine if the system is consistent or inconsistent. If the system is consistent, determine the number of solutions.

31. $\dfrac{1}{5}x - \dfrac{1}{3}y = 1$
$-3x + 5y = 9$

32. $2x + y = 5$
$x - 2y = -1$

33. $2x - 5y = 0$
$x - y = 3$

34. $4x - 6y = 7$
$2x - 3y = 3.5$

In Exercises 35–38, use a graphing utility to graph the two equations. Use the graphs to approximate the solution of the system.

35. $8x + 9y = 42$
$6x - y = 16$

36. $\dfrac{3}{2}x - \dfrac{1}{5}y = 8$
$-2x + 3y = 3$

37. $4y = -8$
$7x - 2y = 25$

38. $0.5x + 2.2y = 9$
$6x + 0.4y = -22$

In Exercises 39–42, use any method to solve the system.

39. $3x - 5y = 7$
$2x + y = 9$

40. $-x + 3y = 17$
$4x + 3y = 7$

41. $y = 2x - 5$
$y = 5x - 11$

42. $7x + 3y = 16$
$y = x + 2$

Exploration **In Exercises 43 and 44, find a system of linear equations that has the given solution. (There is more than one correct answer.)**

43. $\left(3, \dfrac{5}{2}\right)$

44. $(8, -2)$

Think About It   In Exercises 45 and 46, the graphs of the two equations appear to be parallel. Yet, when the system is solved algebraically, you find that the system does have a solution. Find the solution and explain why it does not appear on the portion of the graph that is shown.

45. $100y - x = 200$
$99y - x = -198$

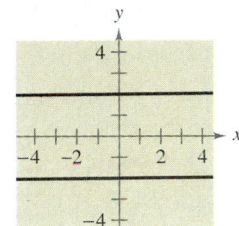

46. $21x - 20y = 0$
$13x - 12y = 120$

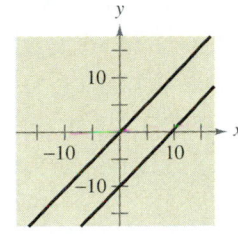

47. *Essay*   Briefly explain whether or not it is possible for a consistent system of linear equations to have exactly two solutions.

48. *Think About It*   Give examples of (a) a system of linear equations that has no solution and (b) a system that has an infinite number of solutions.

Exploration   In Exercises 49 and 50, find the value of k such that the system of linear equations is inconsistent.

49. $4x - 8y = -3$
$2x + ky = 16$

50. $15x + 3y = 6$
$-10x + ky = 9$

51. *Airplane Speed*   An airplane flying into a headwind travels the 1800-mile flying distance between two cities in 3 hours and 36 minutes. On the return flight, the distance is traveled in 3 hours. Find the air speed of the plane and the speed of the wind, assuming that both remain constant.

52. *Airplane Speed*   Two planes start from the same airport and fly in opposite directions. The second plane starts one-half hour after the first plane, but its speed is 80 kilometers per hour faster. Find the air speed of each plane if 2 hours after the first plane departs the planes are 3200 kilometers apart.

53. *Acid Mixture*   Ten liters of a 30% acid solution is obtained by mixing a 20% solution with a 50% solution.

(a) Write a system of equations in which one equation represents the amount of final mixture required and the other represents the amount of acid in the final mixture. Let x and y represent the amounts of 20% and 50% solutions, respectively.

(b) Use a graphing utility to graph the two equations in part (a). As the amount of the 20% solution increases, how does the amount of the 50% solution change?

(c) How much of each solution is required to obtain the specified concentration of the final mixture?

54. *Fuel Mixture*   Five hundred gallons of 89 octane gasoline is obtained by mixing 87 octane gasoline with 92 octane gasoline. How much of each must be used?

55. *Investment Portfolio*   A total of $12,000 is invested in two corporate bonds that pay 10.5% and 12% simple interest. The investor wants an annual interest income of $1350 from the investments. What is the most that can be invested in the 10.5% bond?

56. *Investment Portfolio*   A total of $32,000 is invested in two municipal bonds that pay 5.75% and 6.25% simple interest. The investor wants an annual interest income of $1900 from the investments. What is the most that can be invested in the 5.75% bond?

57. *Ticket Sales*   Five hundred tickets were sold for a certain performance of a play. The tickets for adults and children sold for $7.50 and $4.00, respectively, and the total receipts for the performance were $3312.50. How many of each kind of ticket were sold?

58. *Shoe Sales*   On Saturday night the manager of a shoe store evaluates the receipts of the previous week's sales. Two hundred and forty pairs of two different styles of tennis shoes were sold. One style sold for $66.95 and the other sold for $84.95. The total receipts were $17,652. The cash register that was supposed to record the number of each type of shoe sold malfunctioned. Can you recover the information? If so, how many shoes of each type were sold?

59. *Driving Distances*   On a trip of 300 kilometers, two people drive. One person drives three times as far as the other. Find the distance that each person drives.

60. *Truck Scheduling*   A contractor hires two trucking companies to haul 1600 tons of crushed stone for a highway construction project. The contracts state that one company is to haul four times as much as the other. Find the amount hauled by each.

Supply and Demand   **In Exercises 61–64, find the point of equilibrium of the demand and supply equations.**

Demand	*Supply*
61. $p = 50 - 0.5x$	$p = 0.125x$
62. $p = 100 - 0.05x$	$p = 25 + 0.1x$
63. $p = 140 - 0.00002x$	$p = 80 + 0.00001x$
64. $p = 400 - 0.0002x$	$p = 225 + 0.0005x$

Fitting a Line to Data   **In Exercises 65–72, find the least squares regression line $y = ax + b$ for the points**

$$(x_1, y_1), (x_2, y_2), \ldots, (x_n, y_n).$$

To find the line, solve the system for a and b.

$$nb + \left(\sum_{i=1}^{n} x_i\right)a = \sum_{i=1}^{n} y_i$$

$$\left(\sum_{i=1}^{n} x_i\right)b + \left(\sum_{i=1}^{n} x_i^2\right)a = \sum_{i=1}^{n} x_i y_i$$

Then use the linear regression capabilities of a graphing utility to confirm the result.

65. $5b + 10a = 20.2$ **66.** $5b + 10a = 11.7$
     $10b + 30a = 50.1$ $10b + 30a = 25.6$

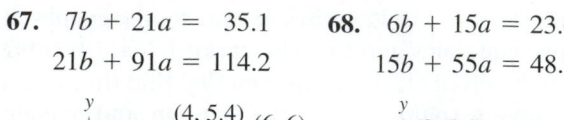

67. $7b + 21a = 35.1$ **68.** $6b + 15a = 23.6$
     $21b + 91a = 114.2$ $15b + 55a = 48.8$

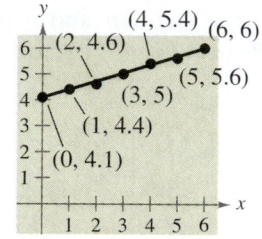
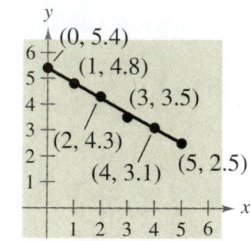

69. $(-2, 0), (0, 1), (2, 3)$

70. $(-3, 0), (-1, 1), (1, 1), (3, 2)$

71. $(0, 4), (1, 3), (1, 1), (2, 0)$

72. $(1, 0), (2, 0), (3, 0), (3, 1), (4, 1), (4, 2), (5, 2), (6, 2)$

73. *Data Analysis*   A store manager wants to know the demand for a certain product as a function of the price. The daily sales for the different prices of the product are given in the table.

Price (x)	\$1.00	\$1.25	\$1.50
Demand (y)	450	375	330

Use the technique demonstrated in Exercises 65–72 to find the line that best fits the data. Then use the line to predict the demand when the price is \$1.40.

74. *Data Analysis*   A farmer used four test plots to determine the relationship between wheat yield in bushels per acre and the amount of fertilizer in hundreds of pounds per acre. The results are given in the table.

Fertilizer (x)	1.0	1.5	2.0	2.5
Yield (y)	32	41	48	53

Use the technique demonstrated in Exercises 65–72 to find the line that best fits the data. Then use the line to estimate the yield for a fertilizer application of 160 pounds per acre.

9.3   *Multivariable Linear Systems*

See Exercise 71 on page 699 for an example of how a system of linear equations in three variables can be used to analyze an automobile's braking system.

Row-Echelon Form and Back-Substitution □ *Gaussian Elimination*
Nonsquare Systems □ *Applications*

Row-Echelon Form and Back-Substitution

The method of elimination can be applied to a system of linear equations in more than two variables. In fact, this method easily adapts to computer use for solving linear systems with dozens of variables.

When elimination is used to solve a system of linear equations, the goal is to rewrite the system in a form to which back-substitution can be applied. To see how this works, consider the following two systems of linear equations.

$$x - 2y + 3z = 9 \qquad\qquad x - 2y + 3z = 9$$
$$-x + 3y \quad\quad = -4 \qquad\qquad y + 3z = 5$$
$$2x - 5y + 5z = 17 \qquad\qquad z = 2$$

The system on the right is said to be in **row-echelon form,** which means that it has a "stair-step" pattern with leading coefficients of 1. After comparing the two systems, it should be clear that it is easier to solve the system on the right.

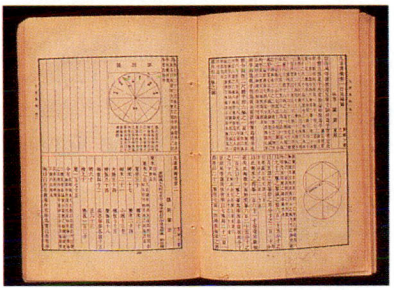

One of the most influential Chinese mathematics books was the Chui-chang suan-shu or Nine Chapters on the Mathematical Art (written in approximately 250 B.C.). Chapter Eight of the Nine Chapters contained solutions of systems of linear equations using positive and negative numbers. One such system was

$$3x + 2y + z = 39$$
$$2x + 3y + z = 34$$
$$x + 2y + 3z = 26.$$

This system was solved using column operations on a matrix. Matrices (plural for matrix) will be discussed in the next chapter. *(Photo: Christopher Lui/China Stock)*

EXAMPLE 1   *Using Back-Substitution*

Solve the system of linear equations.

$$x - 2y + 3z = 9 \qquad \text{Equation 1}$$
$$y + 3z = 5 \qquad \text{Equation 2}$$
$$z = 2 \qquad \text{Equation 3}$$

Solution

From Equation 3, you know the value of z. To solve for y, substitute $z = 2$ into Equation 2 to obtain

$$y + 3(2) = 5 \qquad \text{Substitute 2 for } z.$$
$$y = -1. \qquad \text{Solve for } y.$$

Finally, substitute $y = -1$ and $z = 2$ into Equation 1 to obtain

$$x - 2(-1) + 3(2) = 9 \qquad \text{Substitute } -1 \text{ for } y \text{ and 2 for } z.$$
$$x = 1. \qquad \text{Solve for } x.$$

The solution is $x = 1$, $y = -1$, and $z = 2$, which can be written as the **ordered triple** $(1, -1, 2)$. Check this in the original system of equations.

Gaussian Elimination

Two systems of equations are **equivalent** if they have the same solution set. To solve a system that is not in row-echelon form, first convert it to an *equivalent* system that is in row-echelon form. To see how this is done, let's take another look at the method of elimination, as applied to a system of two linear equations.

EXAMPLE 2   *The Method of Elimination*

Solve the system of linear equations.

$$\begin{aligned} 3x - 2y &= -1 \\ x - y &= 0 \end{aligned}$$

Solution

There are two strategies that seem reasonable: eliminate the variable x or eliminate the variable y. The following steps show how to use the first strategy.

$$\begin{aligned} x - y &= 0 \\ 3x - 2y &= -1 \end{aligned}$$ You can interchange two equations in the system.

$$-3x + 3y = 0$$ Multiply the first equation by -3.

$$\begin{aligned} -3x + 3y &= 0 \\ \underline{3x - 2y = -1} \\ y &= -1 \end{aligned}$$ You can add the multiple of the first equation to the second equation to obtain a new equation.

$$\begin{aligned} x - y &= 0 \\ y &= -1 \end{aligned}$$ New system in row-echelon form

Now, using back-substitution, you can determine that the solution is $y = -1$ and $x = -1$, which can be written as the ordered pair $(-1, -1)$. Check this in the original system of equations.

NOTE   As shown in Example 2, rewriting a system of linear equations in row-echelon form usually involves a *chain* of equivalent systems, each of which is obtained by using one of the three basic row operations. This process is called **Gaussian elimination,** after the German mathematician Carl Friedrich Gauss (1777–1855).

OPERATIONS THAT PRODUCE EQUIVALENT SYSTEMS

Each of the following **row operations** on a system of linear equations produces an *equivalent* system of linear equations.

1. Interchange two equations.
2. Multiply one of the equations by a nonzero constant.
3. Add a multiple of one of the equations to another equation to replace the latter equation.

EXAMPLE 3   *Using Elimination to Solve a System*

Solve the system of linear equations.

$$x - 2y + 3z = 9 \qquad \text{Equation 1}$$
$$-x + 3y \qquad = -4 \qquad \text{Equation 2}$$
$$x - 5y + 5z = 17 \qquad \text{Equation 3}$$

Solution

There are many ways to begin, but we suggest working from the upper left corner, saving the x in the upper left position and eliminating the other x's from the first column.

$$x - 2y + 3z = 9$$
$$y + 3z = 5$$
$$2x - 5y + 5z = 17$$

> Adding the first equation to the second equation produces a new second equation.

$$x - 2y + 3z = 9$$
$$y + 3z = 5$$
$$-y - z = -1$$

> Adding -2 times the first equation to the third equation produces a new third equation.

Now that all but the first x have been eliminated from the first column, go to work on the second column. (You need to eliminate y from the third equation.)

$$x - 2y + 3z = 9$$
$$y + 3z = 5$$
$$2z = 4$$

> Adding the second equation to the third equation produces a new third equation.

Finally, you need a coefficient of 1 for z in the third equation.

$$x - 2y + 3z = 9$$
$$y + 3z = 5$$
$$z = 2$$

> Multiplying the third equation by $\frac{1}{2}$ produces a new third equation.

This is the same system that was solved in Example 1, and, as in that example, you can conclude that the solution is

$$x = 1, \qquad y = -1, \qquad \text{and} \qquad z = 2.$$

In Example 3, you can check the solution by substituting $x = 1$, $y = -1$, and $z = 2$ into each original equation, as follows.

Equation 1: $1 - 2(-1) + 3(2) = 9$ ✓

Equation 2: $-1 + 3(-1) \qquad = -4$ ✓

Equation 3: $2(1) - 5(-1) + 5(2) = 17$ ✓

The next example involves an inconsistent system—one that has no solution. The key to recognizing an inconsistent system is that at some stage in the elimination process, you obtain a false statement such as $0 = -2$.

■

EXAMPLE 4   *An Inconsistent System*

Solve the system of linear equations.

$$x - 3y + z = 1 \qquad \text{Equation 1}$$
$$2x - y - 2z = 2 \qquad \text{Equation 2}$$
$$x + 2y - 3z = -1 \qquad \text{Equation 3}$$

Solution

$$x - 3y + z = 1$$
$$5y - 4z = 0$$
$$x + 2y - 3z = -1$$

> Adding -2 times the first equation to the second equation produces a new second equation.

$$x - 3y + z = 1$$
$$5y - 4z = 0$$
$$5y - 4z = -2$$

> Adding -1 times the first equation to the third equation produces a new third equation.

$$x - 3y + z = 1$$
$$5y - 4z = 0$$
$$0 = -2$$

> Adding -1 times the second equation to the third equation produces a new third equation.

Because the third "equation" is impossible, you can conclude that this system is inconsistent and therefore has no solution. Moreover, because this system is equivalent to the original system, you can conclude that the original system also has no solution.

■

As with a system of linear equations in two variables, the solution(s) of a system of linear equations in more than two variables must fall into one of three categories.

THE NUMBER OF SOLUTIONS OF A LINEAR SYSTEM

For a system of linear equations, exactly one of the following is true.

1. There is exactly one solution.
2. There are infinitely many solutions.
3. There is no solution.

EXAMPLE 5   A System with Infinitely Many Solutions

Solve the system of linear equations.

$$
\begin{aligned}
x + y - 3z &= -1 & &\text{Equation 1} \\
y - z &= 0 & &\text{Equation 2} \\
-x + 2y &= 1 & &\text{Equation 3}
\end{aligned}
$$

Solution

$$
\begin{aligned}
x + y - 3z &= -1 \\
y - z &= 0 \\
3y - 3z &= 0
\end{aligned}
$$

> Adding the first equation to the third equation produces a new third equation.

$$
\begin{aligned}
x + y - 3z &= -1 \\
y - z &= 0 \\
0 &= 0
\end{aligned}
$$

> Adding -3 times the second equation to the third equation produces a new third equation.

This means that Equation 3 depends on Equations 1 and 2 in the sense that it gives us no additional information about the variables. Thus, the original system is equivalent to the system

$$
\begin{aligned}
x + y - 3z &= -1 \\
y - z &= 0.
\end{aligned}
$$

In this last equation, solve for y in terms of z to obtain $y = z$. Back-substituting for y into the previous equation produces $x = 2z - 1$. Finally, letting $z = a$, you can see that the solutions to the given system are all of the form

$$
x = 2a - 1, \qquad y = a \qquad \text{and} \qquad z = a
$$

where a is a real number. Thus, every ordered triple of the form

$$
(2a - 1, a, a), \qquad a \text{ is a real number}
$$

is a solution of the system.

In Example 5, there are other ways to write the same infinite set of solutions. For instance, the solutions could have been written as

$$
\left(b, \tfrac{1}{2}(b + 1), \tfrac{1}{2}(b + 1)\right), \qquad b \text{ is a real number.}
$$

Try convincing yourself of this by substituting $a = 0$, $a = 1$, $a = 2$, and $a = 3$ into the solution listed in Example 5. Then substitute $b = -1$, $b = 1$, $b = 3$, and $b = 5$ into the solution listed above. In both cases, you should obtain the same ordered triples. Thus, when comparing descriptions of an infinite solution set, keep in mind that there is more than one way to describe the set.

Nonsquare Systems

So far, each system of linear equations has been **square,** which means that the number of equations is equal to the number of variables. In a **nonsquare** system, the number of equations differs from the number of variables. A system of linear equations cannot have a unique solution unless there are at least as many equations as there are variables in the system.

EXAMPLE 6   *A System with Fewer Equations Than Variables*

Solve the system of linear equations.

$$x - 2y + z = 2 \qquad \text{Equation 1}$$
$$2x - y - z = 1 \qquad \text{Equation 2}$$

Solution

Begin by rewriting the system in row-echelon form, as follows.

$$x - 2y + z = 2$$
$$3y - 3z = -3$$

> Adding -2 times the first equation to the second equation produces a new second equation.

$$x - 2y + z = 2$$
$$y - z = -1$$

> Multiplying the second equation by $\frac{1}{3}$ produces a new second equation.

Solving for y in terms of z, you get $y = z - 1$, and back-substitution into Equation 1 yields

$$x - 2(z - 1) + z = 2$$
$$x - 2z + 2 + z = 2$$
$$x = z.$$

Finally, by letting $z = a$, you have the solution

$$x = a, \qquad y = a - 1, \qquad \text{and} \qquad z = a$$

where a is a real number. Thus, every ordered triple of the form

$$(a, a - 1, a), \qquad a \text{ is a real number}$$

is a solution of the system.

NOTE   In Example 6, try choosing some values of a to obtain different solutions of the system, such as $(1, 0, 1)$, $(2, 1, 2)$, and $(3, 2, 3)$. Then check each of the solutions in the original system. ■■

When you use a system of linear equations to solve an application, it is wise to interpret your solution in the context of the problem to see if it makes sense. For instance, in Example 7 the solution results in a position equation of $s = -16t^2 + 48t + 20$ and implies that the object was thrown upward at a velocity of 48 feet per second from a height of 20 feet. The object underwent a constant downward acceleration of 32 feet per second squared. (Physics will tell you that this is the value of the acceleration due to gravity.)

Applications

EXAMPLE 7 *Vertical Motion*

Real Life

The height at time t of an object that is moving in a (vertical) line with constant acceleration a is given by the **position equation**

$$s = \tfrac{1}{2}at^2 + v_0 t + s_0.$$

The height s is measured in feet, t is measured in seconds, v_0 is the initial velocity (at $t = 0$), and s_0 is the initial height. Find the values of a, v_0, and s_0 if $s = 52$ at $t = 1$, $s = 52$ at $t = 2$, and $s = 20$ at $t = 3$.

Solution

You can obtain three linear equations in a, v_0, and s_0 as follows.

When $t = 1$: $\tfrac{1}{2}a(1)^2 + v_0(1) + s_0 = 52$ ⟹ $a + 2v_0 + 2s_0 = 104$

When $t = 2$: $\tfrac{1}{2}a(2)^2 + v_0(2) + s_0 = 52$ ⟹ $2a + 2v_0 + s_0 = 52$

When $t = 3$: $\tfrac{1}{2}a(3)^2 + v_0(3) + s_0 = 20$ ⟹ $9a + 6v_0 + 2s_0 = 40$

Solving this system yields $a = -32$, $v_0 = 48$, and $s_0 = 20$.

EXAMPLE 8 *Partial Fractions*

Write the partial fraction decomposition for $\dfrac{3x + 4}{x^3 - 2x - 4}$.

Solution

Because $x^3 - 2x - 4 = (x - 2)(x^2 + 2x + 2)$, you can write

$$\frac{3x + 4}{x^3 - 2x - 4} = \frac{A}{x - 2} + \frac{Bx + C}{x^2 + 2x + 2}$$

$$3x + 4 = A(x^2 + 2x + 2) + (Bx + C)(x - 2)$$

$$3x + 4 = (A + B)x^2 + (2A - 2B + C)x + (2A - 2C).$$

NOTE Be sure you see that these coefficients give the partial fraction decomposition for

$$\frac{3x + 4}{x^3 - 2x - 4}$$

as

$$\frac{1}{x - 2} + \frac{-x - 1}{x^2 + 2x + 2}$$

which equals

$$\frac{1}{x - 2} - \frac{x + 1}{x^2 + 2x + 2}.$$

By equating coefficients of like powers on both sides of the expanded equation, you obtain the following system in A, B, and C.

$$A + B = 0$$
$$2A - 2B + C = 3$$
$$2A - 2C = 4$$

You can solve this system to find that $A = 1$, $B = -1$, and $C = -1$.

EXAMPLE 9 Data Analysis: Curve-Fitting

Find a quadratic equation, $y = ax^2 + bx + c$, whose graph passes through the points $(-1, 3)$, $(1, 1)$, and $(2, 6)$.

Solution

Because the graph of $y = ax^2 + bx + c$ passes through the points $(-1, 3)$, $(1, 1)$, and $(2, 6)$, you can write the following.

When $x = -1$, $y = 3$: $a(-1)^2 + b(-1) + c = 3$
When $x = 1$, $y = 1$: $a(1)^2 + b(1) + c = 1$
When $x = 2$, $y = 6$: $a(2)^2 + b(2) + c = 6$

This produces the following system of linear equations.

$$a - b + c = 3 \qquad \text{Equation 1}$$
$$a + b + c = 1 \qquad \text{Equation 2}$$
$$4a + 2b + c = 6 \qquad \text{Equation 3}$$

The solution of this system is $a = 2$, $b = -1$, and $c = 0$. Thus, the equation of the parabola is $y = 2x^2 - x$, as shown in Figure 9.11.

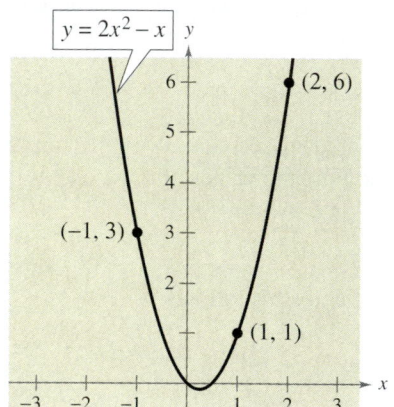

$y = 2x^2 - x$

(2, 6)

(−1, 3)

(1, 1)

FIGURE 9.11

GROUP ACTIVITY

MATHEMATICAL MODELING

x	1	3	5
y	59	49	59

Suppose you work for an outerwear manufacturer, and the marketing department is concerned about sales trends in Georgia. Your manager has asked you to investigate climatic data in hopes of explaining sales patterns and gives you the table to the left, which represents the average monthly temperature y in degrees Fahrenheit for Savannah, Georgia for month x, where $x = 1$ corresponds to November.

Construct a scatter plot of the data. Decide what type of mathematical model might be appropriate for the data and use the methods you have learned thus far to fit an appropriate model. Your manager would like to know the average monthly temperatures for December and February. Explain to your manager how you found your model, what it represents, and how it may be used to find the December and February average temperatures. Investigate the usefulness of this model for the rest of the year. Would you recommend using the model to predict average monthly temperatures for the whole year or just part of the year? Explain your reasoning. (Source: National Climatic Data Center)

WARM UP

Solve the system of linear equations.

1. $x + y = 25$
   $y = 10$

2. $2x - 3y = 4$
   $6x = -12$

3. $x + y = 32$
   $x - y = 24$

4. $2r - s = 5$
   $r + 2s = 10$

Decide whether the ordered triple is a solution of the equation.

5. $5x - 3y + 4z = 2$
   $(-1, -2, 1)$

6. $x - 2y + 12z = 9$
   $(6, 3, 2)$

7. $2x - 5y + 3z = -9$
   $(a - 2, a + 1, a)$

8. $-5x + y + z = 21$
   $(a - 4, 4a + 1, a)$

Solve for x in terms of a.

9. $x + 2y - 3z = 4$
   $y = 1 - a, z = a$

10. $x - 3y + 5z = 4$
   $y = 2a + 3, z = a$

9.3 Exercises

In Exercises 1–6, use back-substitution to solve the system of linear equations.

1. $2x - y + 5z = 24$
   $y + 2z = 6$
   $z = 4$

2. $4x - 3y - 2z = 21$
   $6y - 5z = -8$
   $z = -2$

3. $2x + y - 3z = 10$
   $y = 2$
   $y - z = 4$

4. $x = 8$
   $2x + 3y = 10$
   $x - y + 2z = 22$

5. $4x - 2y + z = 8$
   $2z = 4$
   $- y + z = 4$

6. $5x - 8z = 22$
   $3y - 5z = 10$
   $z = -4$

In Exercises 7 and 8, perform the row operation and write the equivalent system.

7. Add Equation 1 to Equation 2.

   $x - 2y + 3z = 5$ Equation 1
   $-x + 3y - 5z = 4$ Equation 2
   $2x - 3z = 0$ Equation 3

   What did this operation accomplish?

8. Add -2 times Equation 1 to Equation 3.

   $x - 2y + 3z = 5$ Equation 1
   $-x + 3y - 5z = 4$ Equation 2
   $2x - 3z = 0$ Equation 3

   What did this operation accomplish?

In Exercises 9–34, solve the system of linear equations and check any solution algebraically.

9.
$$x + y + z = 6$$
$$2x - y + z = 3$$
$$3x \quad - z = 0$$

10.
$$x + y + z = 2$$
$$-x + 3y + 2z = 8$$
$$4x + y \quad = 4$$

11.
$$2x \quad + 2z = 2$$
$$5x + 3y \quad = 4$$
$$3y - 4z = 4$$

12.
$$4x + y - 3z = 11$$
$$2x - 3y + 2z = 9$$
$$x + y - z = -3$$

13.
$$6y + 4z = -12$$
$$3x + 3y \quad = 9$$
$$2x \quad - 3z = 10$$

14.
$$2x + 4y + z = -4$$
$$2x - 4y + 6z = 13$$
$$4x - 2y + z = 6$$

15.
$$3x - 2y + 4z = 1$$
$$x + y - 2z = 3$$
$$2x - 3y + 6z = 8$$

16.
$$5x - 3y + 2z = 3$$
$$2x + 4y - z = 7$$
$$x - 11y + 4z = 3$$

17.
$$3x + 3y + 5z = 1$$
$$3x + 5y + 9z = 0$$
$$5x + 9y + 17z = 0$$

18.
$$2x + y + 3z = 1$$
$$2x + 6y + 8z = 3$$
$$6x + 8y + 18z = 5$$

19.
$$x + 2y - 7z = -4$$
$$2x + y + z = 13$$
$$3x + 9y - 36z = -33$$

20.
$$2x + y - 3z = 4$$
$$4x \quad + 2z = 10$$
$$-2x + 3y - 13z = -8$$

21.
$$3x - 3y + 6z = 6$$
$$x + 2y - z = 5$$
$$5x - 8y + 13z = 7$$

22.
$$x \quad + 4z = 13$$
$$4x - 2y + z = 7$$
$$2x - 2y - 7z = -19$$

23.
$$x - 2y + 5z = 2$$
$$4x \quad - z = 0$$

24.
$$x - 3y + 2z = 18$$
$$5x - 13y + 12z = 80$$

25.
$$2x - 3y + z = -2$$
$$-4x + 9y \quad = 7$$

26.
$$2x + 3y + 3z = 7$$
$$4x + 18y + 15z = 44$$

27.
$$x \quad + 3w = 4$$
$$2y - z - w = 0$$
$$3y \quad - 2w = 1$$
$$2x - y + 4z \quad = 5$$

28.
$$x + y + z + w = 6$$
$$2x + 3y \quad - w = 0$$
$$-3x + 4y + z + 2w = 4$$
$$x + 2y - z + w = 0$$

29.
$$x \quad + 4z = 1$$
$$x + y + 10z = 10$$
$$2x - y + 2z = -5$$

30.
$$3x - 2y - 6z = -4$$
$$-3x + 2y + 6z = 1$$
$$x - y - 5z = -3$$

31.
$$2x + 3y \quad = 0$$
$$4x + 3y - z = 0$$
$$8x + 3y + 3z = 0$$

32.
$$4x + 3y + 17z = 0$$
$$5x + 4y + 22z = 0$$
$$4x + 2y + 19z = 0$$

33.
$$12x + 5y + z = 0$$
$$23x + 4y - z = 0$$

34.
$$5x + 5y - z = 0$$
$$10x + 5y + 2z = 0$$
$$5x + 15y - 9z = 0$$

35. *Think About It* Are the two systems of equations equivalent? Give reasons for your answer.

$$x + 3y - z = 6 \qquad x + 3y - z = 6$$
$$2x - y + 2z = 1 \qquad -7y + 4z = 1$$
$$3x + 2y - z = 2 \qquad -7y - 4z = -16$$

36. *Think About It* When using Gaussian elimination to solve a system of linear equations, how can you recognize that the system has no solution? Give an example that illustrates your answer.

Exploration **In Exercises 37 and 38, find a system of linear equations that has the given solution. (The answer is not unique.)**

37. $(4, -1, 2)$

38. $\left(-\frac{3}{2}, 4, -7\right)$

In Exercises 39–42, find the equation of the parabola

$$y = ax^2 + bx + c$$

that passes through the given points. To verify your result, use a graphing utility to plot the points and graph the parabola.

39. $(0, 0), (2, -2), (4, 0)$

40. $(0, 3), (1, 4), (2, 3)$

41. $(2, 0), (3, -1), (4, 0)$

42. $(1, 3), (2, 2), (3, -3)$

In Exercises 43–46, find the equation of the circle

$$x^2 + y^2 + Dx + Ey + F = 0$$

that passes through the given points. To verify your result, use a graphing utility to plot the points and graph the circle.

43. $(0, 0), (2, 2), (4, 0)$

44. $(0, 0), (0, 6), (3, 3)$

45. $(-3, -1), (2, 4), (-6, 8)$

46. $(0, 0), (0, -2), (3, 0)$

Vertical Motion **In Exercises 47–50, find the position equation** $s = \frac{1}{2}at^2 + v_0 t + s_0$ **for an object at the given heights moving vertically at the specified times.**

47. At $t = 1$ second, $s = 128$ feet
At $t = 2$ seconds, $s = 80$ feet
At $t = 3$ seconds, $s = 0$ feet

48. At $t = 1$ second, $s = 48$ feet
At $t = 2$ seconds, $s = 64$ feet
At $t = 3$ seconds, $s = 48$ feet

49. At $t = 1$ second, $s = 452$ feet
At $t = 2$ seconds, $s = 372$ feet
At $t = 3$ seconds, $s = 260$ feet

50. At $t = 1$ second, $s = 132$ feet
At $t = 2$ seconds, $s = 100$ feet
At $t = 3$ seconds, $s = 36$ feet

51. *Investments* An inheritance of $16,000 was divided among three investments yielding a total of $990 in interest per year. The interest rates were 5%, 6%, and 7%. Find the amount in each investment if the 5% and 6% investments were $3000 and $2000 less than the 7% investment, respectively.

52. *Investments* A total of $1520 a year is received in interest from three investments. The interest rates for the three investments are 5%, 7%, and 8%. The 5% investment is half of the 7% investment, and the 7% investment is $1500 less than the 8% investment. Find the amount in each investment.

53. *Borrowing* A small corporation borrowed $775,000 to expand its product line. Some of the money was borrowed at 8%, some at 9%, and some at 10%. How much was borrowed at each rate if the annual interest was $67,500 and the amount borrowed at 8% was four times the amount borrowed at 10%?

54. *Borrowing* A small corporation borrowed $800,000 to expand its product line. Some of the money was borrowed at 8%, some at 9%, and some at 10%. How much was borrowed at each rate if the annual interest was $67,000 and the amount borrowed at 8% was five times the amount borrowed at 10%?

Investment Portfolio **In Exercises 55 and 56, consider an investor with a portfolio totaling $500,000 that is invested in certificates of deposit, municipal bonds, blue-chip stocks, and growth or speculative stocks. How much is put in each type of investment?**

55. The certificates of deposit pay 10% annually, and the municipal bonds pay 8% annually. Over a 5-year period, the investor expects the blue-chip stocks to return 12% annually and the growth stocks to return 13% annually. The investor wants a combined annual return of 10% and also wants to have only one-fourth of the portfolio invested in stocks.

56. The certificates of deposit pay 9% annually, and the municipal bonds pay 5% annually. Over a 5-year period, the investor expects the blue-chip stocks to return 12% annually and the growth stocks to return 14% annually. The investor wants a combined annual return of 10% and also wants to have only one-fourth of the portfolio invested in stocks.

57. Crop Spraying   A mixture of 12 liters of chemical A, 16 liters of chemical B, and 26 liters of chemical C is required to kill a certain destructive crop insect. Commercial spray X contains 1, 2, and 2 parts, respectively, of these chemicals. Commercial spray Y contains only chemical C. Commercial spray Z contains only chemicals A and B in equal amounts. How much of each type of commercial spray is needed to get the desired mixture?

58. Chemistry   A chemist needs 10 liters of a 25% acid solution. The solution is to be mixed from three solutions whose concentrations are 10%, 20%, and 50%. How many liters of each solution should the chemist use to satisfy the following?

(a) Use as little as possible of the 50% solution.

(b) Use as much as possible of the 50% solution.

(c) Use 2 liters of the 50% solution.

59. Truck Scheduling   A small company that manufactures products A and B has an order for 15 units of product A and 16 units of product B. The company has trucks of three different sizes that can haul the products, as shown in the table.

Truck	Large	Medium	Small
Product A	6	4	0
Product B	3	4	3

How many trucks of each size are needed to deliver the order? Give *two* possible solutions.

60. Electrical Network   Applying Kirchhoff's Laws to the electrical network in the figure, the currents I_1, I_2, and I_3 are the solution of the system

$$I_1 - I_2 + I_3 = 0$$
$$3I_1 + 2I_2 \quad\quad = 7$$
$$\quad\quad 2I_2 + 4I_3 = 8.$$

Find the currents.

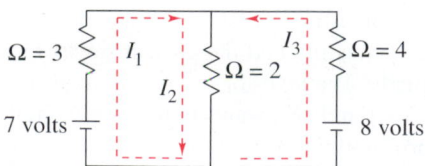

61. Pulley System   A system of pulleys is leaded with 128-pound and 32-pound weights (see figure). The tensions t_1 and t_2 in the ropes and the acceleration a of the 32-pound weight are found by solving the system

$$t_1 - 2t_2 \quad\quad = \quad 0$$
$$t_1 \quad\quad - 2a = 128$$
$$\quad\quad t_2 + \quad a = \quad 32$$

where t_1 and t_2 are measured in pounds and a is in feet per second squared. Solve the system.

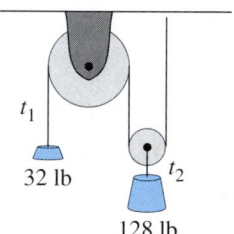

62. Pulley System   If the 32-pound weight is replaced by a 64-pound weight in the pulley system of Exercise 61, it is modeled by the following system of equations.

$$t_1 - 2t_2 \quad\quad = \quad 0$$
$$t_1 \quad\quad - 2a = 128$$
$$\quad\quad t_2 + 2a = \quad 64$$

Solve the system and use your answer for the acceleration to describe what (if anything) is happening in the system.

Partial Fraction   In Exercises 63–66, write the partial fraction decomposition for the rational fraction.

63. $\dfrac{1}{x^3 - x} = \dfrac{A}{x} + \dfrac{B}{x - 1} + \dfrac{C}{x + 1}$

64. $\dfrac{3}{x^2 + x - 2} = \dfrac{A}{x - 1} + \dfrac{B}{x + 2}$

65. $\dfrac{x^2 - 3x - 3}{x(x - 2)(x + 3)} = \dfrac{A}{x} + \dfrac{B}{x - 2} + \dfrac{C}{x + 3}$

66. $\dfrac{12}{x(x - 2)(x + 3)} = \dfrac{A}{x} + \dfrac{B}{x - 2} + \dfrac{C}{x + 3}$

Fitting a Parabola   In Exercises 67–70, find the least squares regression parabola $y = ax^2 + bx + c$ for the points $(x_1, y_1), (x_2, y_2), \ldots, (x_n, y_n)$. To find the parabola, solve the following system of linear equations for a, b, and c. Then use the least squares regression capabilities of a graphing utility to confirm the result.

$$nc + \left(\sum_{i=1}^{n} x_i\right)b + \left(\sum_{i=1}^{n} x_i^2\right)a = \sum_{i=1}^{n} y_i$$

$$\left(\sum_{i=1}^{n} x_i\right)c + \left(\sum_{i=1}^{n} x_i^2\right)b + \left(\sum_{i=1}^{n} x_i^3\right)a = \sum_{i=1}^{n} x_i y_i$$

$$\left(\sum_{i=1}^{n} x_i^2\right)c + \left(\sum_{i=1}^{n} x_i^3\right)b + \left(\sum_{i=1}^{n} x_i^4\right)a = \sum_{i=1}^{n} x_i^2 y_i$$

67.

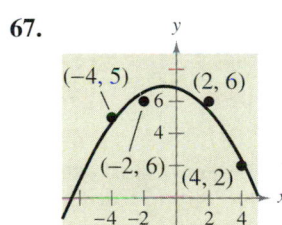

68.

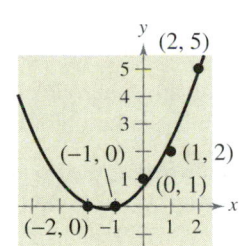

69.

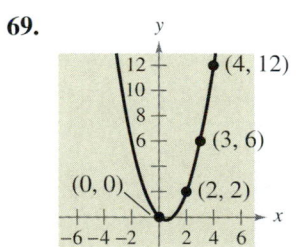

70.

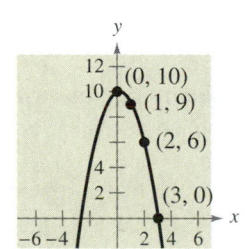

71. *Data Analysis*   In testing a new braking system on an automobile, the speed in miles per hour and the stopping distance in feet were recorded in the table.

Speed (x)	20	30	40	50	60
Stopping Distance (y)	25	55	105	188	300

(a) Find the least squares regression parabola for the data.

(b) Graph the parabola and the data on the same set of axes.

(c) Use the model to estimate the stopping distance if the speed is 70 miles per hour.

72. *Data Analysis*   A wildlife management team studied the reproduction rates of deer in 5 tracts of a wildlife preserve. Each tract contained 5 acres. In each tract the number of females and the percent of females that had offspring the following year were recorded. The results are given in the table.

Number (x)	80	100	120	140	160
Percent (y)	80	75	68	55	30

(a) Find the least squares regression parabola for the data.

(b) Use a graphing utility to graph the parabola and the data on the same viewing rectangle.

(c) Use the model to estimate the percent of females that had offspring if $x = 170$.

Advanced Applications   In Exercises 73–76, find x, y, and λ satisfying the system. These systems arise in certain optimization problems in calculus, and λ is called a *Lagrange multiplier*.

73.
$$y + \lambda = 0$$
$$x + \lambda = 0$$
$$x + y - 10 = 0$$

74.
$$2x + \lambda = 0$$
$$2y + \lambda = 0$$
$$x + y - 4 = 0$$

75.
$$2x - 2x\lambda = 0$$
$$-2y + \lambda = 0$$
$$y - x^2 = 0$$

76.
$$2 + 2y + 2\lambda = 0$$
$$2x + 1 + \lambda = 0$$
$$2x + y - 100 = 0$$

77. *Chapter Opener*   Interpret the slope in the model for sales of compact sport utilities on page 455.

78. *Chapter Opener*   If the models on page 455 are assumed to be accurate in forecasting future sales, is there a time in the future when compact pickup sales will again exceed compact utility sales? If so, when?

Review   Solve Exercises 79–82 as a review of the skills and problem-solving techniques you learned in previous sections.

79. What is $7\frac{1}{2}\%$ of 85?

80. 225 is what percent of 150?

81. 0.5% of what number is 400?

82. 48% of what number is 132?

9.4   *Systems of Inequalities*

See Exercises 55 and 56 on page 709 for examples of how systems of linear inequalities can be used to analyze the compositions of dietary supplements.

The Graph of an Inequality ▫ *Systems of Inequalities* ▫ *Applications*

The Graph of an Inequality

The following statements are inequalities in two variables:

$$3x - 2y < 6 \quad \text{and} \quad 2x^2 + 3y^2 \geq 6.$$

An ordered pair (a, b) is a **solution of an inequality** in x and y if the inequality is true when a and b are substituted for x and y, respectively. The **graph of an inequality** is the collection of all solutions of the inequality. To sketch the graph of an inequality, begin by sketching the graph of the *corresponding equation*. The graph of the equation will normally separate the plane into two or more regions. In each such region, one of the following must be true.

1. *All* points in the region are solutions of the inequality.
2. *No* point in the region is a solution of the inequality.

Thus, you can determine whether the points in an entire region satisfy the inequality by simply testing *one* point in the region.

SKETCHING THE GRAPH OF AN INEQUALITY IN TWO VARIABLES

1. Replace the inequality sign by an equal sign, and sketch the graph of the resulting equation. (Use a dashed line for < or > and a solid line for ≤ or ≥.)
2. Test one point in each of the regions formed by the graph in Step 1. If the point satisfies the inequality, shade the entire region to denote that every point in the region satisfies the inequality.

EXAMPLE 1   *Sketching the Graph of an Inequality*

Sketch the graph of $y \geq x^2 - 1$.

Solution

The graph of the corresponding *equation* $y = x^2 - 1$ is a parabola, as shown in Figure 9.12. By testing a point *above* the parabola $(0, 0)$ and a point *below* the parabola $(0, -2)$, you can see that the points that satisfy the inequality are those lying above (or on) the parabola.

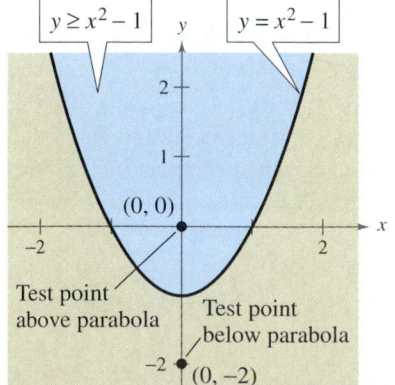

FIGURE 9.12

The inequality given in Example 1 is a nonlinear inequality in two variables. Most of the following examples involve **linear inequalities** such as $ax + by < c$. The graph of a linear inequality is a half-plane lying on one side of the line $ax + by = c$.

EXAMPLE 2 *Sketching the Graph of a Linear Inequality*

Sketch the graph of each linear inequality.

a. $x > -2$ **b.** $y \leq 3$

Solution

a. The graph of the corresponding equation $x = -2$ is a vertical line. The points that satisfy the inequality $x > -2$ are those lying to the right of this line, as shown in Figure 9.13.

b. The graph of the corresponding equation $y = 3$ is a horizontal line. The points that satisfy the inequality $y \leq 3$ are those lying below (or on) this line, as shown in Figure 9.14.

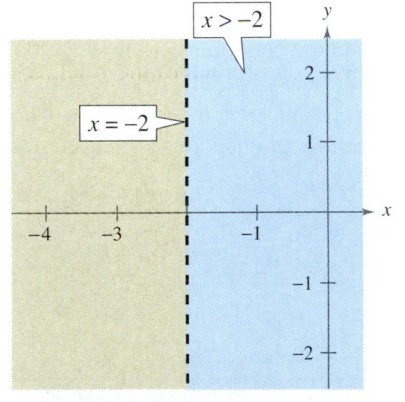

FIGURE 9.13

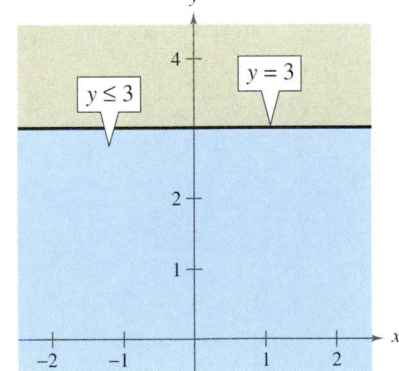

FIGURE 9.14

EXAMPLE 3 *Sketching the Graph of a Linear Inequality*

Sketch the graph of $x - y < 2$.

Solution

The graph of the corresponding equation $x - y = 2$ is a line, as shown in Figure 9.15. Because the origin $(0, 0)$ satisfies the inequality, the graph consists of the half-plane lying above the line. (Try checking a point below the line. Regardless of which point you choose, you will see that it does not satisfy the inequality.)

Study Tip

To graph a linear inequality, it can help to write the inequality in slope-intercept form. For instance, by writing $x - y < 2$ in the form

$$y > x - 2$$

you can see that the solution points lie *above* the line $x - y = 2$ (or $y = x - 2$), as shown in Figure 9.15.

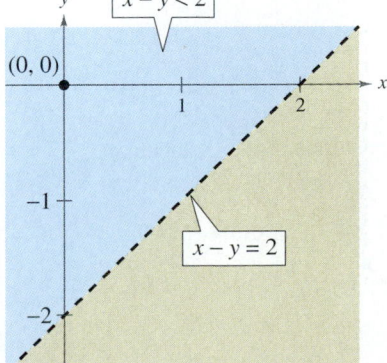

FIGURE 9.15

Systems of Inequalities

Many practical problems in business, science, and engineering involve systems of linear inequalities. A **solution** of a system of inequalities in x and y is a point (x, y) that satisfies each inequality in the system.

To sketch the graph of a system of inequalities in two variables, first sketch the graph of each individual inequality (on the same coordinate system) and then find the region that is *common* to every graph in the system. For systems of *linear* inequalities, it is helpful to find the vertices of the solution region.

EXAMPLE 4   *Solving a System of Inequalities*

Sketch the graph (and label the vertices) of the solution set of the system.

$$x - y < 2$$
$$x > -2$$
$$y \leq 3$$

Solution

The graphs of these inequalities are shown in Figures 9.13 to 9.15. The triangular region common to all three graphs can be found by superimposing the graphs on the same coordinate system, as shown in Figure 9.16. To find the vertices of the region, solve the three systems of corresponding equations obtained by taking *pairs* of equations representing the boundaries of the individual regions.

Vertex A: $(-2, -4)$ *Vertex B:* $(5, 3)$ *Vertex C:* $(-2, 3)$

$$\begin{aligned} x - y &= 2 \\ x &= -2 \end{aligned} \qquad \begin{aligned} x - y &= 2 \\ y &= 3 \end{aligned} \qquad \begin{aligned} x &= -2 \\ y &= 3 \end{aligned}$$

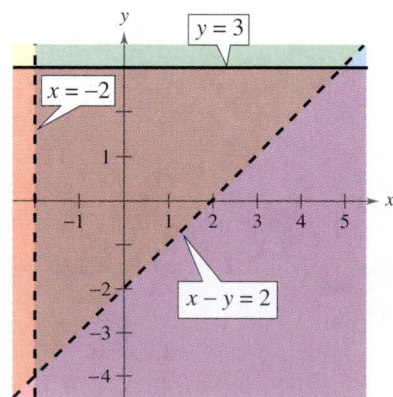
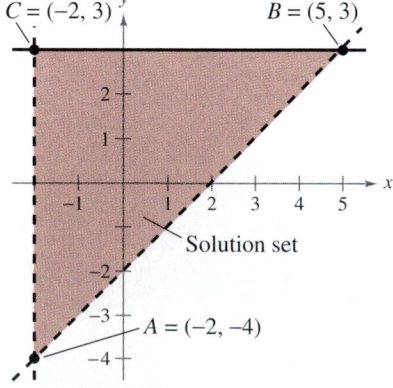

FIGURE 9.16

For the triangular region shown in Figure 9.16, each point of intersection of a pair of boundary lines corresponds to a vertex. With more complicated regions, two border lines can sometimes intersect at a point that is not a vertex of the region, as shown in Figure 9.17. To keep track of which points of intersection are actually vertices of the region, we suggest that you sketch the region and refer to your sketch as you find each point of intersection.

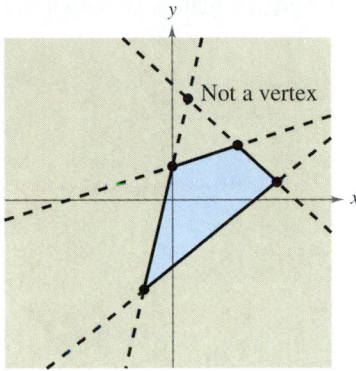

FIGURE 9.17

EXAMPLE 5   *Solving a System of Inequalities*

Sketch the region containing all points that satisfy the system.

$$x^2 - y \leq 1$$
$$-x + y \leq 1$$

Solution

As shown in Figure 9.18, the points that satisfy the inequality $x^2 - y \leq 1$ are the points lying above (or on) the parabola given by

$$y = x^2 - 1. \qquad \text{Parabola}$$

The points satisfying the inequality $-x + y \leq 1$ are the points lying below (or on) the line given by

$$y = x + 1. \qquad \text{Line}$$

To find the points of intersection of the parabola and the line, solve the system of corresponding equations.

$$x^2 - y = 1$$
$$-x + y = 1$$

Using the method of substitution, you can find the solutions to be $(-1, 0)$ and $(2, 3)$, as shown in Figure 9.18.

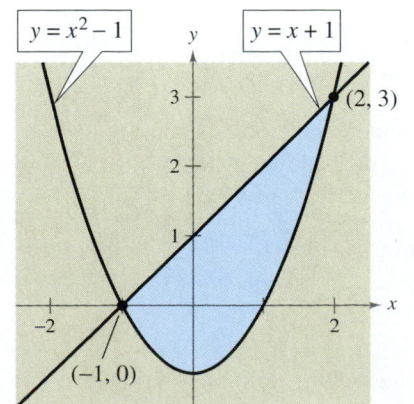

FIGURE 9.18

When solving a system of inequalities, you should be aware that the system might have no solution. For instance, the system

$$x + y > 3$$
$$x + y < -1$$

has no solution points, because the quantity $(x + y)$ cannot be both less than -1 and greater than 3, as shown in Figure 9.19.

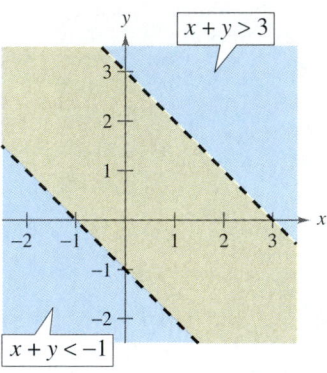

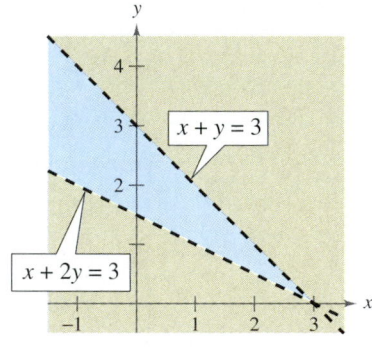

FIGURE 9.19   **FIGURE 9.20**

Another possibility is that the solution set of a system of inequalities can be unbounded. For instance, the solution set of

$$x + y < 3$$
$$x + 2y > 3$$

forms an *infinite wedge,* as shown in Figure 9.20.

TECHNOLOGY

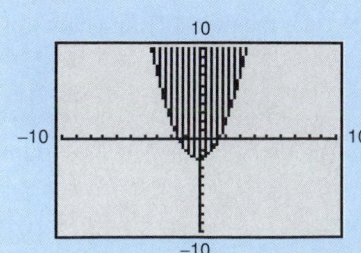

A graphing utility can be used to graph an inequality. For instance, to graph $y \geq x^2 - 2$ on the *TI-83*, you can use the following steps.

1. Press $\boxed{Y=}$ and enter $x^2 - 2$ for Y_1.

2. Move the cursor to the icon to the left of Y_1.

3. Press $\boxed{\text{ENTER}}$ until the ◥ icon appears.

4. Press $\boxed{\text{GRAPH}}$.

The graph is shown at the left. Try using a graphing utility to graph the following inequalities.

    a. $y \leq 2x + 2$ **b.** $y \geq \frac{1}{2}x^2 - 4$

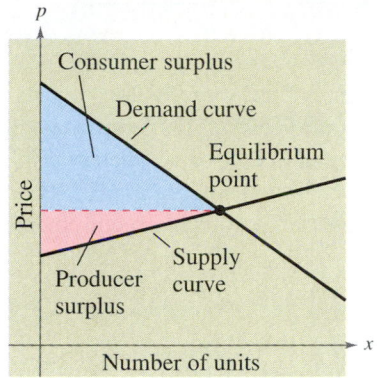

FIGURE 9.21

Applications

Example 8 in Section 9.2 discussed the *point of equilibrium* for a system of demand and supply functions. The next example discusses two related concepts that economists call **consumer surplus** and **producer surplus.** As shown in Figure 9.21, the consumer surplus is defined as the area of the region that lies *below* the demand curve, *above* the horizontal line passing through the equilibrium point, and to the right of the *p*-axis. Similarly, the producer surplus is defined as the area of the region that lies *above* the supply curve, *below* the horizontal line passing through the equilibrium point, and to the right of the *p*-axis. The consumer surplus is a measure of the amount that consumers would have been willing to pay *above what they actually paid,* whereas the producer surplus is a measure of the amount that producers would have been willing to receive *below what they actually received.*

EXAMPLE 6   **Consumer Surplus and Producer Surplus**

The demand and supply functions for a certain type of calculator are given by

$$p = 150 - 0.00001x \qquad \text{Demand equation}$$
$$p = 60 + 0.00002x \qquad \text{Supply equation}$$

where p is the price in dollars and x represents the number of units. Find the consumer surplus and producer surplus for these two equations.

Solution

Begin by finding the point of equilibrium by solving the equation

$$60 + 0.00002x = 150 - 0.00001x.$$

In Example 8 in Section 9.2, you saw that the solution is $x = 3,000,000$, which corresponds to an equilibrium price of $p = \$120$. Thus, the consumer surplus and producer surplus are the areas of the following triangular regions.

Consumer Surplus	*Producer Surplus*
$p \le 150 - 0.00001x$	$p \ge 60 + 0.00002x$
$p \ge 120$	$p \le 120$
$x \ge 0$	$x \ge 0$

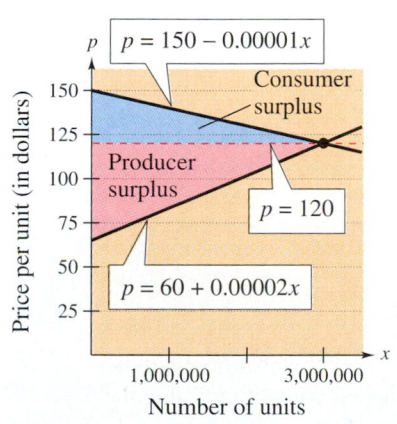

FIGURE 9.22

In Figure 9.22, you can see that the consumer and producer surpluses are

$$\text{Consumer surplus} = \tfrac{1}{2}(\text{base})(\text{height}) = \tfrac{1}{2}(30)(3,000,000) = \$45,000,000$$

$$\text{Producer surplus} = \tfrac{1}{2}(\text{base})(\text{height}) = \tfrac{1}{2}(60)(3,000,000) = \$90,000,000.$$

■
EXAMPLE 7   *Nutrition*

The liquid portion of a diet is to provide at least 300 calories, 36 units of vitamin A, and 90 units of vitamin C daily. A cup of dietary drink X provides 60 calories, 12 units of vitamin A, and 10 units of vitamin C. A cup of dietary drink Y provides 60 calories, 6 units of vitamin A, and 30 units of vitamin C. Set up a system of linear inequalities that describes the minimum daily requirements for calories and vitamins.

Solution

Begin by letting x and y represent the following.

x = number of cups of dietary drink X

y = number of cups of dietary drink Y

To meet the minimum daily requirements, the following inequalities must be satisfied.

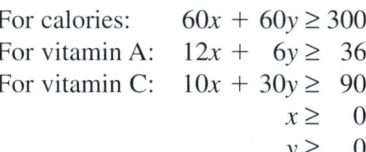

For calories: $60x + 60y \geq 300$
For vitamin A: $12x + 6y \geq 36$
For vitamin C: $10x + 30y \geq 90$
$x \geq 0$
$y \geq 0$

The last two inequalities are included because x and y cannot be negative. The graph of this system of inequalities is shown in Figure 9.23. (More is said about this application in Example 7 in Section 9.5.)

■

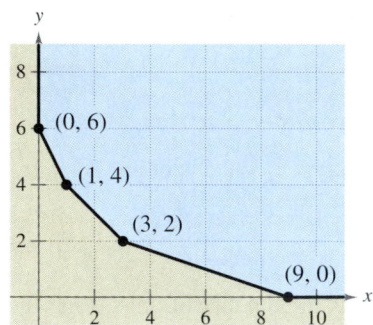

FIGURE 9.23

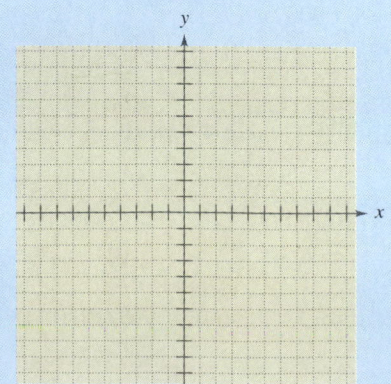

GROUP ACTIVITY
USING SYSTEMS OF INEQUALITIES

Try the following activity. One person picks a point with whole-number coordinates from a grid such as the one at the left without revealing the coordinates. A second person writes a system of two linear equations both of which pass through the region. The first person graphs the system on the grid and rewrites the equations as a system of inequalities to indicate in which region on the graph the secret point lies. The first person continues writing and graphing systems until the second person is able to guess the coordinates of the secret point. Switch roles and try the activity again.

Identify the graph as a line, parabola, circle, or ellipse.

1. $x + y = 3$ **2.** $4x - y = 8$

3. $y = x^2 - 4$ **4.** $y = -x^2 + 1$

5. $x^2 + y^2 = 9$ **6.** $\dfrac{x^2}{4} + \dfrac{y^2}{9} = 1$

Solve the system of equations.

7. $\begin{aligned} x + 2y &= 3 \\ 4x - 7y &= -3 \end{aligned}$ **8.** $\begin{aligned} 2x - 3y &= 4 \\ x + 5y &= 2 \end{aligned}$

9. $\begin{aligned} x^2 + y &= 5 \\ 2x - 4y &= 0 \end{aligned}$ **10.** $\begin{aligned} x^2 + y^2 &= 13 \\ x + y &= 5 \end{aligned}$

9.4 Exercises

In Exercises 1–12, sketch the graph of the inequality.

1. $x \geq 2$ **2.** $x \leq 4$

3. $y \geq -1$ **4.** $y \leq 3$

5. $y < 2 - x$ **6.** $y > 2x - 4$

7. $2y - x \geq 4$ **8.** $5x + 3y \geq -15$

9. $(x + 1)^2 + (y - 2)^2 < 9$

10. $y^2 - x < 0$

11. $y \leq \dfrac{1}{1 + x^2}$ **12.** $y < \ln x$

In Exercises 13–16, use a graphing utility to graph the inequality. Shade the region representing the solution.

13. $y \geq \frac{2}{3}x - 1$ **14.** $y \leq 6 - \frac{3}{2}x$

15. $x^2 + 5y - 10 \leq 0$ **16.** $2x^2 - y - 3 > 0$

In Exercises 17–20, write an inequality for the shaded region shown in the figure.

17.
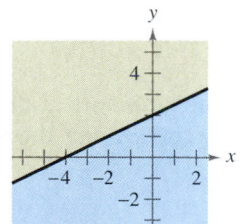

18.
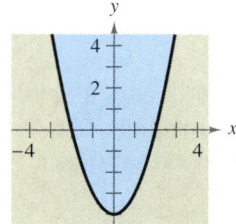

19.
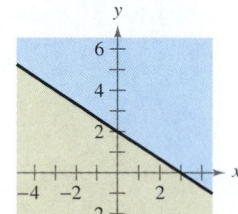

20.
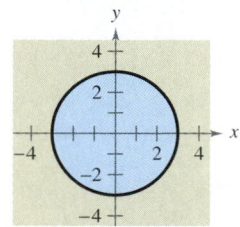

In Exercises 21–34, sketch the graph of the solution of the system of inequalities.

21. $x + y \le 1$
$-x + y \le 1$
$y \ge 0$

22. $3x + 2y < 6$
$x > 0$
$y > 0$

23. $x + y \le 5$
$x \ge 2$
$y \ge 0$

24. $2x^2 + y \ge 2$
$x \le 2$
$y \le 1$

25. $-3x + 2y < 6$
$x - 4y > -2$
$2x + y < 3$

26. $x - 7y > -36$
$5x + 2y > 5$
$6x - 5y > 6$

27. $2x + y > 2$
$6x + 3y < 2$

28. $x - 2y < -6$
$5x - 3y > -9$

29. $x \ge 1$
$x - 2y \le 3$
$3x + 2y \ge 9$
$x + y \le 6$

30. $x - y^2 > 0$
$x - y < 2$

31. $x^2 + y^2 \le 9$
$x^2 + y^2 \ge 1$

32. $x^2 + y^2 \le 25$
$4x - 3y \le 0$

33. $x > y^2$
$x < y + 2$

34. $x < 2y - y^2$
$0 < x + y$

In Exercises 35–40, use a graphing utility to graph the inequalities. Shade the region representing the solution of the system.

35. $y \le \sqrt{3x} + 1$
$y \ge x^2 + 1$

36. $y < -x^2 + 2x + 3$
$y > x^2 - 4x + 3$

37. $y < x^3 - 2x + 1$
$y > -2x$
$x \le 1$

38. $y \ge x^4 - 2x^2 + 1$
$y \le 1 - x^2$

39. $x^2 y \ge 1$
$0 < x \le 4$
$y \le 4$

40. $y \le e^{-x^2/2}$
$y \ge 0$
$-2 \le x \le 2$

In Exercises 41–50, derive a set of inequalities to describe the region.

41.

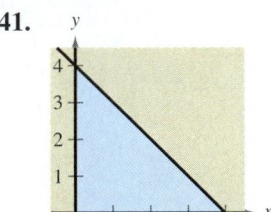

42.

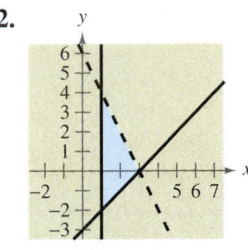

43.

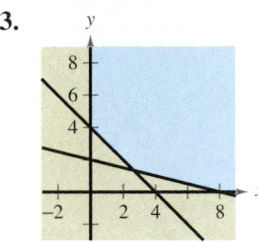

44.

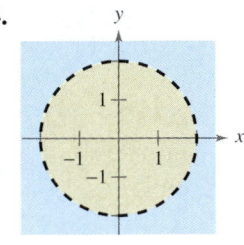

45.

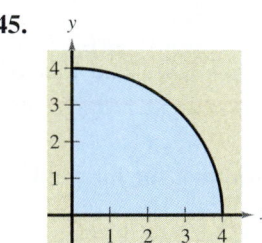

46.

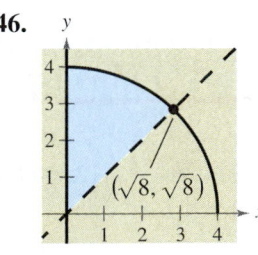

47. Rectangle: Vertices at $(2, 1)$, $(5, 1)$, $(5, 7)$, $(2, 7)$

48. Parallelogram: Vertices at $(0, 0)$, $(4, 0)$, $(1, 4)$, $(5, 4)$

49. Triangle: Vertices at $(0, 0)$, $(5, 0)$, $(2, 3)$

50. Triangle: Vertices at $(-1, 0)$, $(1, 0)$, $(0, 1)$

51. *Furniture Production* A furniture company can sell all the tables and chairs it produces. Each table requires 1 hour in the assembly center and $1\frac{1}{3}$ hours in the finishing center. Each chair requires $1\frac{1}{2}$ hours in the assembly center and $1\frac{1}{2}$ hours in the finishing center. The company's assembly center is available 12 hours per day, and its finishing center is available 15 hours per day. Find and graph a system of inequalities describing all possible production levels.

52. *Computer Inventory* A store sells two models of computers. Because of the demand, the store stocks twice as many units of model *A* as of model *B*. The cost to the store for the two models is $800 and $1200, respectively. The management does not want more than $20,000 in computer inventory at any one time, and it wants at least four model *A* computers and two model *B* computers in inventory at all times. Devise a system of inequalities describing all possible inventory levels, and graph the system.

53. *Investment* A person plans to invest $20,000 in two different interest-bearing accounts. Each account is to contain at least $5000. Moreover, the amount in one account should be at least twice the amount in the other account. Find a system of inequalities to describe the various amounts that can be deposited in each account, and graph the system.

54. *Concert Ticket Sales* One type of concert ticket costs $15 per ticket and another costs $25 per ticket. The promoter of the concert must sell at least 15,000 tickets, including 8000 of the $15 tickets and 4000 of the $25 tickets. Moreover, the gross receipts must total at least $275,000. Find a system of inequalities describing the different numbers of tickets that can be sold, and graph the system.

55. *Diet Supplement* A dietitian is asked to design a special diet supplement using two different foods. Each ounce of food *X* contains 20 units of calcium, 15 units of iron, and 10 units of vitamin B. Each ounce of food *Y* contains 10 units of calcium, 10 units of iron, and 20 units of vitamin B. The minimum daily requirements of the diet are 280 units of calcium, 160 units of iron, and 180 units of vitamin B. Find and graph a system of inequalities describing the different amounts of food *X* and food *Y* that can be used.

56. *Diet Supplement* A dietitian is asked to design a special diet supplement using two different foods. Each ounce of food *X* contains 20 units of calcium, 15 units of iron, and 10 units of vitamin B. Each ounce of food *Y* contains 10 units of calcium, 10 units of iron, and 20 units of vitamin B. The minimum daily requirements of the diet are 300 units of calcium, 150 units of iron, and 200 units of vitamin B. Find and graph a system of inequalities describing the different amounts of food *X* and food *Y* that can be used.

57. *Physical Fitness Facility* An indoor running track is to be constructed with a space for body-building equipment inside the track (see figure). The track must be at least 125 meters long, and the body-building space must have an area of at least 500 square meters. Find a system of inequalities describing the requirements of the facility. Graph the system.

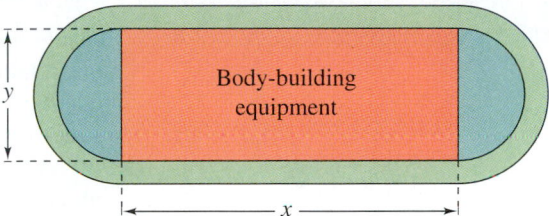

58. *Graphical Reasoning* Two concentric circles have radii x and y, where $y > x$. The area between the circles must be at least 10 square units.

(a) Find an inequality describing the constraints on the circles.

(b) Use a graphing utility to graph the inequality in part (a). Graph the line $y = x$ on the same viewing rectangle.

(c) Identify the graph of the line in relation to the boundary of the inequality. Explain its meaning in the context of the problem.

Consumer Surplus and Producer Surplus **In Exercises 59–62, find the consumer surplus and producer surplus for the supply and demand equations.**

	Demand	*Supply*
59.	$p = 50 - 0.5x$	$p = 0.125x$
60.	$p = 100 - 0.05x$	$p = 25 + 0.1x$
61.	$p = 140 - 0.00002x$	$p = 80 + 0.00001x$
62.	$p = 400 - 0.0002x$	$p = 225 + 0.0005x$

63. *Think About It* After graphing the boundary of an inequality in x and y, how do you decide on which side of the boundary the solution set of the inequality lies?

64. *Essay* Explain the difference between the graph of the inequality $x \le 4$ on the real number line and the rectangular coordinate system.

9.5 Linear Programming

See Exercise 39 on page 719 for an example of how linear programming can be used to analyze the profitability of two models of compact disc players.

Linear Programming: A Graphical Approach ▫ *Applications*

Linear Programming: A Graphical Approach

Many applications in business and economics involve a process called **optimization,** in which you are asked to find the minimum or maximum of a quantity. In this section you will study an optimization strategy called **linear programming.**

A two-dimensional linear programming problem consists of a linear **objective function** and a system of linear inequalities called **constraints.** The objective function gives the quantity that is to be maximized (or minimized), and the constraints determine the set of **feasible solutions.** For example, suppose you are asked to maximize the value of

$$z = ax + by \qquad \text{Objective function}$$

subject to a set of constraints that determines the region in Figure 9.24. Because every point in the region satisfies each constraint, it is not clear how you should go about finding the point that yields a maximum value of z. Fortunately, it can be shown that if there is an optimal solution, it must occur at one of the vertices. This means that *you can find the maximum value by testing z at each of the vertices.*

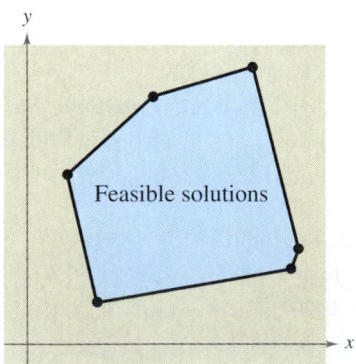

Feasible solutions

FIGURE 9.24

> **OPTIMAL SOLUTION OF A LINEAR PROGRAMMING PROBLEM**
>
> If a linear programming problem has a solution, it must occur at a vertex of the set of feasible solutions. If there is more than one solution, at least one of them must occur at a such a vertex. In either case, the value of the objective function is unique.

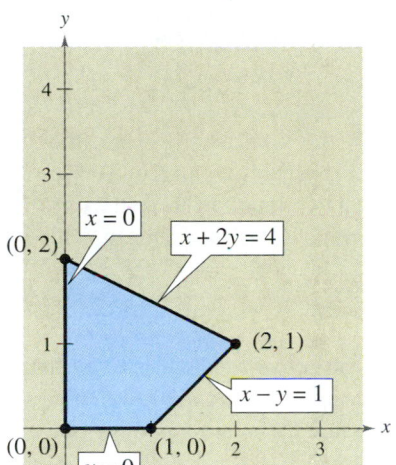

FIGURE 9.25

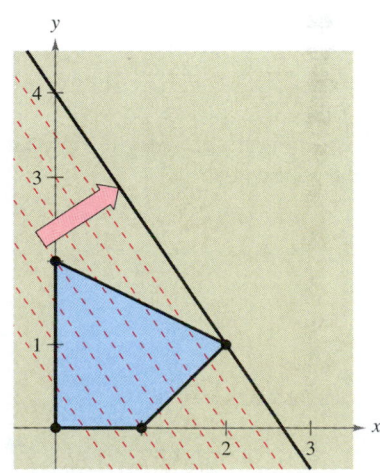

FIGURE 9.26

EXAMPLE 1   *Solving a Linear Programming Problem*

Find the maximum value of

$$z = 3x + 2y \qquad \text{Objective function}$$

subject to the following constraints.

$$\left.\begin{array}{r} x \geq 0 \\ y \geq 0 \\ x + 2y \leq 4 \\ x - y \leq 1 \end{array}\right\} \quad \text{Constraints}$$

Solution

The constraints form the region shown in Figure 9.25. At the four vertices of this region, the objective function has the following values.

At $(0, 0)$: $z = 3(0) + 2(0) = 0$
At $(1, 0)$: $z = 3(1) + 2(0) = 3$
At $(2, 1)$: $z = 3(2) + 2(1) = 8$ Maximum value of z
At $(0, 2)$: $z = 3(0) + 2(2) = 4$

Thus, the maximum value of z is 8, and this occurs when $x = 2$ and $y = 1$.

NOTE In Example 1, try testing some of the *interior* points in the region. You will see that the corresponding values of z are less than 8. Here are some examples.

At $(1, 1)$: $z = 3(1) + 2(1) = 5$
At $\left(1, \frac{1}{2}\right)$: $z = 3(1) + 2\left(\frac{1}{2}\right) = 4$
At $\left(\frac{1}{2}, \frac{3}{2}\right)$: $z = 3\left(\frac{1}{2}\right) + 2\left(\frac{3}{2}\right) = \frac{9}{2}$ ■■

To see why the maximum value of the objective function in Example 1 must occur at a vertex, consider writing the objective function in the form

$$y = -\frac{3}{2}x + \frac{z}{2} \qquad \text{Family of lines}$$

where $z/2$ is the y-intercept of the objective function. This equation represents a family of lines, each of slope $-\frac{3}{2}$. Of these infinitely many lines, you want the one that has the largest z-value while still intersecting the region determined by the constraints. In other words, of all the lines whose slope is $-\frac{3}{2}$, you want the one that has the largest y-intercept *and* intersects the given region, as shown in Figure 9.26. It should be clear that such a line will pass through one (or more) of the vertices of the region.

SOLVING A LINEAR PROGRAMMING PROBLEM

To solve a linear programming problem involving two variables by the graphical method, use the following steps.

1. Sketch the region corresponding to the system of constraints. (The points inside or on the boundary of the region are called *feasible solutions*.)
2. Find the vertices of the region.
3. Test the objective function at each of the vertices and select the values of the variables that optimize the objective function. For a bounded region, both a minimum and a maximum value will exist. (For an unbounded region, *if* an optimal solution exists, it will occur at a vertex.)

Study Tip

Remember that a vertex of a region can be found using a system of linear equations. The system will consist of the equations of the lines passing through the vertex.

These guidelines will work whether the objective function is to be maximized or minimized. For instance, the same test used in Example 1 to find the maximum value of z can be used to conclude that the minimum value of z is 0 and that this value occurs at the vertex (0, 0).

EXAMPLE 2   *Solving a Linear Programming Problem*

Find the maximum value of

$$z = 4x + 6y \qquad \text{Objective function}$$

where $x \geq 0$ and $y \geq 0$, subject to the following constraints.

$$\left. \begin{array}{r} -x + y \leq 11 \\ x + y \leq 27 \\ 2x + 5y \leq 90 \end{array} \right\} \quad \text{Constraints}$$

Solution

The region bounded by the constraints is shown in Figure 9.27. By testing the objective function at each vertex, you obtain the following.

At $(0, 0)$: $z = 4(0) + 6(0) = 0$
At $(0, 11)$: $z = 4(0) + 6(11) = 66$
At $(5, 16)$: $z = 4(5) + 6(16) = 116$
At $(15, 12)$: $z = 4(15) + 6(12) = 132$ Maximum value of z
At $(27, 0)$: $z = 4(27) + 6(0) = 108$

FIGURE 9.27

Thus, the maximum value of z is 132, and this occurs when $x = 15$ and $y = 12$.

The next example shows that the same basic procedure can be used to solve a problem in which the objective function is to be *minimized*.

EXAMPLE 3   *Minimizing an Objective Function*

Find the minimum value of

$$z = 5x + 7y \qquad \text{Objective function}$$

where $x \geq 0$ and $y \geq 0$, subject to the following constraints.

$$\left. \begin{aligned} 2x + 3y &\geq 6 \\ 3x - y &\leq 15 \\ -x + y &\leq 4 \\ 2x + 5y &\leq 27 \end{aligned} \right\} \quad \text{Constraints}$$

Solution

The region bounded by the constraints is shown in Figure 9.28. By testing the objective function at each vertex, you obtain the following.

At $(0, 2)$: $z = 5(0) + 7(2) = 14$ Minimum value of z
At $(0, 4)$: $z = 5(0) + 7(4) = 28$
At $(1, 5)$: $z = 5(1) + 7(5) = 40$
At $(6, 3)$: $z = 5(6) + 7(3) = 51$
At $(5, 0)$: $z = 5(5) + 7(0) = 25$
At $(3, 0)$: $z = 5(3) + 7(0) = 15$

Thus, the minimum value of z is 14, and this occurs when $x = 0$ and $y = 2$.

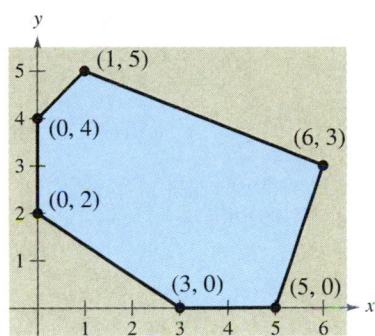

FIGURE 9.28

EXAMPLE 4   *Maximizing an Objective Function*

Find the maximum value of

$$z = 5x + 7y \qquad \text{Objective function}$$

where $x \geq 0$ and $y \geq 0$, subject to the following constraints.

$$\left. \begin{aligned} 2x + 3y &\geq 6 \\ 3x - y &\leq 15 \\ -x + y &\leq 4 \\ 2x + 5y &\leq 27 \end{aligned} \right\} \quad \text{Constraints}$$

Solution

This linear programming problem is identical to that given in Example 3 above, *except* that the objective function is maximized instead of minimized. Using the values of z at the vertices shown above, you can conclude that the maximum value of z is 51, and that this value occurs when $x = 6$ and $y = 3$.

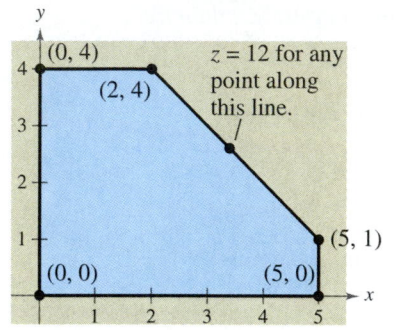

FIGURE 9.29

It is possible for the maximum (or minimum) value in a linear programming problem to occur at *two* different vertices. For instance, at the vertices of the region shown in Figure 9.29, the objective function

$$z = 2x + 2y \qquad \text{Objective function}$$

has the following values.

At $(0, 0)$: $z = 2(0) + 2(0) = 0$
At $(0, 4)$: $z = 2(0) + 2(4) = 8$
At $(2, 4)$: $z = 2(2) + 2(4) = 12$ Maximum value of z
At $(5, 1)$: $z = 2(5) + 2(1) = 12$ Maximum value of z
At $(5, 0)$: $z = 2(5) + 2(0) = 10$

In this case, you can conclude that the objective function has a maximum value not only at the vertices $(2, 4)$ and $(5, 1)$; it also has a maximum value (of 12) at *any point on the line segment connecting these two vertices*. Note that the objective function

$$y = -x + \tfrac{1}{2}z$$

has the same slope as the line through the vertices $(2, 4)$ and $(5, 1)$.

Some linear programming problems have no optimal solutions. This can occur if the region determined by the constraints is *unbounded*. Example 5 illustrates such a problem.

EXAMPLE 5   *An Unbounded Region*

Find the maximum value of

$$z = 4x + 2y \qquad \text{Objective function}$$

where $x \geq 0$ and $y \geq 0$, subject to the following constraints.

$$\left. \begin{array}{r} x + 2y \geq 4 \\ 3x + y \geq 7 \\ -x + 2y \leq 7 \end{array} \right\} \quad \text{Constraints}$$

Solution

The region determined by the constraints is shown in Figure 9.30. For this unbounded region, there is no maximum value of z. To see this, note that the point $(x, 0)$ lies in the region for all values of $x \geq 4$. By choosing x to be large, you can obtain values of

$$z = 4(x) + 2(0) = 4x$$

that are as large as you want. Thus, there is no maximum value of z. For this problem, there *is* a minimum value of $z = 10$, which occurs at the vertex $(2, 1)$.

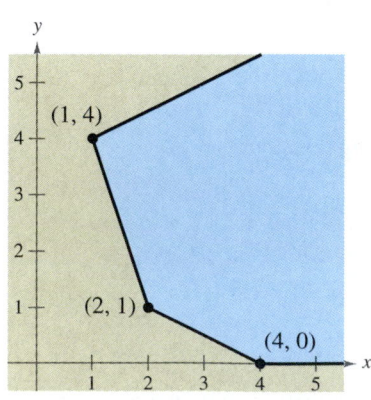

FIGURE 9.30

Applications

Example 6 shows how linear programming can be used to find the maximum profit in a business application.

EXAMPLE 6 *Maximum Profit*

A manufacturer wants to maximize the profit for two products. Product I yields a profit of $1.50 per unit, and product II yields a profit of $2.00 per unit. Market tests and available resources have indicated the following constraints.

1. The combined production level should not exceed 1200 units per month.
2. The demand for product II is no more than half the demand for product I.
3. The production level of product I is less than or equal to 600 units plus three times the production level of product II.

Solution

If you let x be the number of units of product I and y be the number of units of product II, the objective function (for the combined profit) is given by

$$P = 1.5x + 2y. \qquad \text{Objective function}$$

The three constraints translate into the following linear inequalities.

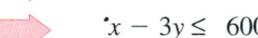

1. $x + y \le 1200$ $x + y \le 1200$
2. $y \le \frac{1}{2}x$ $-x + 2y \le 0$
3. $x \le 3y + 600$ $x - 3y \le 600$

Because neither x nor y can be negative, you also have the two additional constraints of $x \ge 0$ and $y \ge 0$. Figure 9.31 shows the region determined by the constraints. To find the maximum profit, test the values of P at the vertices of the region.

$$
\begin{aligned}
\text{At } (0, 0): \quad & P = 1.5(0) &+ 2(0) &= 0 \\
\text{At } (800, 400): \quad & P = 1.5(800) &+ 2(400) &= 2000 \qquad \text{Maximum profit} \\
\text{At } (1050, 150): \quad & P = 1.5(1050) &+ 2(150) &= 1875 \\
\text{At } (600, 0): \quad & P = 1.5(600) &+ 2(0) &= 900
\end{aligned}
$$

Thus, the maximum profit is $2000, and it occurs when the monthly production consists of 800 units of product I and 400 units of product II.

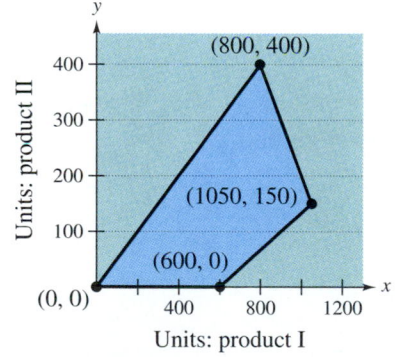

FIGURE 9.31

NOTE In Example 6, suppose the manufacturer improved the production of product I so that it yielded a profit of $2.50 per unit. How would this affect the number of units the manufacturer should sell to obtain a maximum profit? ■■

The cherry-size acerola, fruit of the Barbados cherry tree, is the richest known source of vitamin C. Commercial canneries use acerola in fruit juice mixes and preserves to increase the natural vitamin C content. *(Illustration: From THE WORLD BOOK ENCYCLOPEDIA. © 1996, World Book, Inc. by permission of the publisher)*

Real Life

EXAMPLE 7   *Minimum Cost*

The liquid portion of a diet is to provide at least 300 calories, 36 units of vitamin A, and 90 units of vitamin C daily. A cup of dietary drink X costs $0.12 and provides 60 calories, 12 units of vitamin A, and 10 units of vitamin C. A cup of dietary drink Y costs $0.15 and provides 60 calories, 6 units of vitamin A, and 30 units of vitamin C. How many cups of each drink should be consumed each day to minimize the cost and still meet the daily requirements?

Solution

As in Example 7 on page 706, let x be the number of cups of dietary drink X and let y be the number of cups of dietary drink Y.

For calories: $60x + 60y \geq 300$
For vitamin A: $12x + \ \ 6y \geq \ \ 36$
For vitamin C: $10x + 30y \geq \ \ 90$ } Constraints
$\qquad\qquad\qquad\qquad x \geq \ \ \ 0$
$\qquad\qquad\qquad\qquad y \geq \ \ \ 0$

The cost C is given by $C = 0.12x + 0.15y$.   Objective function

The graph of the region corresponding to the constraints is shown in Figure 9.32. To determine the minimum cost, test C at each vertex of the region.

At $(0, 6)$: $C = 0.12(0) + 0.15(6) = 0.90$
At $(1, 4)$: $C = 0.12(1) + 0.15(4) = 0.72$
At $(3, 2)$: $C = 0.12(3) + 0.15(2) = 0.66$ Minimum value of C
At $(9, 0)$: $C = 0.12(9) + 0.15(0) = 1.08$

Thus, the minimum cost is $0.66 per day, and this occurs when three cups of drink X and two cups of drink Y are consumed each day.

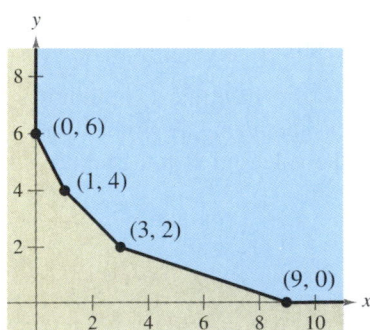

FIGURE 9.32

GROUP ACTIVITY

CREATING A LINEAR PROGRAMMING PROBLEM

Sketch the region determined by the constraints: $x \geq 0$, $y \geq 0$, $x + 2y \leq 8$, and $x + y \leq 5$. Find, if possible, an objective function of the form $z = ax + by$ that has a maximum at the indicated vertex of the region.

a. Maximum at $(0, 4)$ **b.** Maximum at $(2, 3)$

c. Maximum at $(5, 0)$ **d.** Maximum at $(0, 0)$

Sketch the graph of the linear equation.

1. $y + x = 3$ **2.** $y - x = 12$

3. $x = 0$ **4.** $y = 4$

Solve the system of equations.

5. $x + y = 4$ **6.** $x + 2y = 12$
       $x \quad\;\; = 0$ $y = 0$

7. $x + \;\; y = 4$ **8.** $x + 2y = 12$
       $2x + 3y = 9$ $2x + \;\; y = 9$

Sketch the graph of the inequality.

9. $2x + 3y \geq 18$ **10.** $4x + 3y \geq 12$

9.5 Exercises

In Exercises 1–12, find the minimum and maximum values of the objective function, subject to the indicated constraints. (For each exercise, the graph of the region determined by the constraints is provided.)

1. Objective function:

  $z = 4x + 5y$

  Constraints:

  $x \geq 0$
  $y \geq 0$
  $x + y \leq 6$

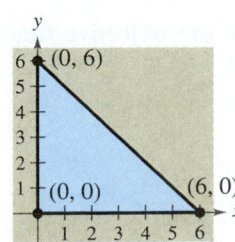

2. Objective function:

  $z = 2x + 8y$

  Constraints:

  $x \geq 0$
  $y \geq 0$
  $2x + y \leq 4$

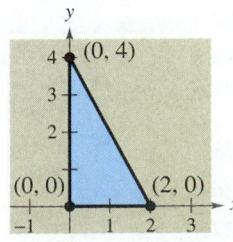

3. Objective function:

  $z = 10x + 6y$

  Constraints:
  (See Exercise 1.)

4. Objective function:

  $z = 7x + 3y$

  Constraints:
  (See Exercise 2.)

5. Objective function:

  $z = 3x + 2y$

  Constraints:

  $x \geq 0$
  $y \geq 0$
  $x + 3y \leq 15$
  $4x + \;\; y \leq 16$

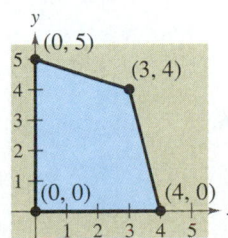

6. Objective function:

  $z = 4x + 3y$

  Constraints:

  $x \geq 0$
  $2x + 3y \geq 6$
  $3x - 2y \leq 9$
  $x + 5y \leq 20$

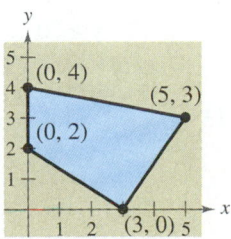

7. Objective function:

$z = 5x + 0.5y$

Constraints:
(See Exercise 5.)

9. Objective function:

$z = 10x + 7y$

Constraints:

$0 \le x \le 60$

$0 \le y \le 45$

$5x + 6y \le 420$

8. Objective function:

$z = x + 6y$

Constraints:
(See Exercise 6.)

10. Objective function:

$z = 50x + 35y$

Constraints:

$x \ge 0$

$y \ge 0$

$8x + 9y \le 7200$

$8x + 9y \ge 5400$

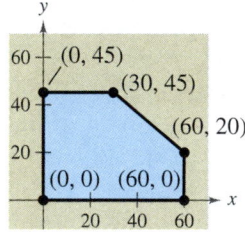

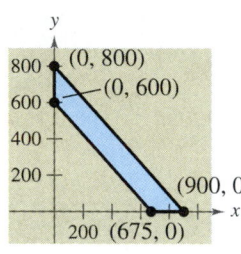

11. Objective function:

$z = 25x + 30y$

Constraints:
(See Exercise 9.)

12. Objective function:

$z = 16x + 18y$

Constraints:
(See Exercise 10.)

In Exercises 13–20, sketch the constraint region. Then find the minimum and maximum values of the objective function, subject to the constraints.

13. Objective function:

$z = 6x + 10y$

Constraints:

$x \ge 0$

$y \ge 0$

$2x + 5y \le 10$

14. Objective function:

$z = 7x + 8y$

Constraints:

$x \ge 0$

$y \ge 0$

$x + \frac{1}{2}y \le 4$

15. Objective function:

$z = 9x + 24y$

Constraints:
(See Exercise 13.)

16. Objective function:

$z = 7x + 2y$

Constraints:
(See Exercise 14.)

17. Objective function:

$z = 4x + 5y$

Constraints:

$x \ge 0$

$y \ge 0$

$x + y \ge 8$

$3x + 5y \ge 30$

19. Objective function:

$z = 2x + 7y$

Constraints:
(See Exercise 17.)

18. Objective function:

$z = 4x + 5y$

Constraints:

$x \ge 0$

$y \ge 0$

$2x + 2y \le 10$

$x + 2y \le 6$

20. Objective function:

$z = 2x - y$

Constraints:
(See Exercise 18.)

In Exercises 21–24, use a graphing utility to sketch the region determined by the constraints. Then find the minimum and maximum values of the objective function, subject to the constraints.

21. Objective function:

$z = 4x + y$

Constraints:

$x \ge 0$

$y \ge 0$

$x + 2y \le 40$

$2x + 3y \ge 72$

22. Objective function:

$z = x$

Constraints:

$x \ge 0$

$y \ge 0$

$2x + 3y \le 60$

$2x + y \le 28$

$4x + y \le 48$

23. Objective function:

$z = x + 4y$

Constraints:
(See Exercise 21.)

24. Objective function:

$z = y$

Constraints:
(See Exercise 22.)

In Exercises 25–28, maximize the objective function subject to the constraints $x \ge 0$, $y \ge 0$, $3x + y \le 15$, and $4x + 3y \le 30$.

25. $z = 2x + y$

26. $z = 5x + y$

27. $z = x + y$

28. $z = 3x + y$

In Exercises 29–32, maximize the objective function subject to the constraints $x \geq 0$, $y \geq 0$, $x + 4y \leq 20$, $x + y \leq 18$, and $2x + 2y \leq 21$.

29. $z = x + 5y$

30. $z = 2x + 4y$

31. $z = 4x + 5y$

32. $z = 4x + y$

Think About It In Exercises 33–36, find an objective function that has a maximum or minimum value at the indicated vertex of the constraint region shown below. (There are many correct answers.)

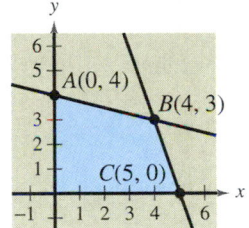

33. The maximum occurs at vertex A.

34. The maximum occurs at vertex B.

35. The maximum occurs at vertex C.

36. The minimum occurs at vertex C.

37. *Maximum Profit* A manufacturer produces two models of bicycles. The times (in hours) required for assembling, painting, and packaging each model are as follows.

Process	Model A	Model B
Assembling	2	2.5
Painting	4	1
Packaging	1	0.75

The total times available for assembling, painting, and packaging are 4000 hours, 4800 hours, and 1500 hours, respectively. The profits per unit are $45 for model A and $50 for model B. How many of each type should be produced to maximize profit?

38. *Maximum Profit* A manufacturer produces two models of bicycles. The times (in hours) required for assembling, painting, and packaging each model are as follows.

Process	Model A	Model B
Assembling	2.5	3
Painting	2	1
Packaging	0.75	1.25

The total times available for assembling, painting, and packaging are 4000 hours, 2500 hours, and 1500 hours, respectively. The profits per unit are $50 for model A and $52 for model B. How many of each type should be produced to maximize profit?

39. *Maximum Profit* A merchant plans to sell two models of compact disc players at costs of $250 and $400. The $250 model yields a profit of $45 and the $400 model yields a profit of $50. The merchant estimates that the total monthly demand will not exceed 250 units. The merchant does not want to invest more than $70,000 in inventory for these products. Find the number of units of each model that should be stocked in order to maximize profit.

40. *Maximum Profit* A fruit grower has 150 acres of land available to raise two crops, A and B. It takes 1 day to trim an acre of crop A, 2 days to trim an acre of crop B, and there are 240 days per year available for trimming. It takes 0.3 day to pick an acre of crop A, 0.1 day to pick an acre of crop B, and there are 30 days available for picking. The profit is $140 per acre for crop A and $235 per acre for crop B. Find the number of acres of each fruit that should be planted to maximize profit.

41. *Minimum Cost* A farming cooperative mixes two brands of cattle feed. Brand X costs $25 per bag and contains two units of nutritional element A, two units of element B, and two units of element C. Brand Y costs $20 per bag and contains one unit of nutritional element A, nine units of element B, and three units of element C. Find the number of bags of each brand that should be mixed to produce a mixture having a minimum cost per bag. The minimum requirements of nutrients A, B, and C are 12 units, 36 units, and 24 units, respectively.

42. *Minimum Cost*   Two gasolines, type A and type B, have octane ratings of 80 and 92, respectively. Type A costs $1.13 per gallon and type B costs $1.28 per gallon. Determine the blend of minimum cost with an octane rating of at least 90. (*Hint:* Let x be the fraction of each gallon that is type A and let y be the fraction that is type B.)

43. *Maximum Revenue*   An accounting firm has 900 hours of staff time and 100 hours of reviewing time available each week. The firm charges $2000 for an audit and $300 for a tax return. Each audit requires 100 hours of staff time and 10 hours of review time. Each tax return requires 12.5 hours of staff time and 2.5 hours of review time. What numbers of audits and tax returns will yield the maximum revenue?

44. *Maximum Revenue*   The accounting firm in Exercise 43 lowers its charge for an audit to $1000. What numbers of audits and tax returns will yield the maximum revenue?

In Exercises 45–50, the linear programming problem has an unusual characteristic. Sketch a graph of the solution region for the problem and describe the unusual characteristic. The objective function is to be maximized in each case.

45. Objective function:

$z = 2.5x + y$

Constraints:

$x \geq 0$

$y \geq 0$

$3x + 5y \leq 15$

$5x + 2y \leq 10$

46. Objective function:

$z = x + y$

Constraints:

$x \geq 0$

$y \geq 0$

$-x + y \leq 1$

$-x + 2y \leq 4$

47. Objective function:

$z = -x + 2y$

Constraints:

$x \geq 0$

$y \geq 0$

$x \leq 10$

$x + y \leq 7$

48. Objective function:

$z = x + y$

Constraints:

$x \geq 0$

$y \geq 0$

$-x + y \leq 0$

$-3x + y \geq 3$

49. Objective function:

$z = 3x + 4y$

Constraints:

$x \geq 0$

$y \geq 0$

$x + y \leq 1$

$2x + y \leq 4$

50. Objective function:

$z = x + 2y$

Constraints:

$x \geq 0$

$y \geq 0$

$x + 2y \leq 4$

$2x + y \leq 4$

In Exercises 51 and 52, determine values of t such that the objective function has a maximum value at the indicated vertex.

51. Objective function:

$z = 3x + ty$

Constraints:

$x \geq 0$

$y \geq 0$

$x + 3y \leq 15$

$4x + y \leq 16$

(a) $(0, 5)$

(b) $(3, 4)$

52. Objective function:

$z = 3x + ty$

Constraints:

$x \geq 0$

$y \geq 0$

$x + 2y \geq 4$

$x - y \leq 1$

(a) $(2, 1)$

(b) $(0, 2)$

Review   Solve Exercises 53–56 as a review of the skills and problem-solving techniques you learned in previous sections. Simplify the compound fraction.

53. $\dfrac{\dfrac{9}{x}}{\left(\dfrac{6}{x} + 2\right)}$

54. $\dfrac{\left(1 + \dfrac{2}{x}\right)}{\left(x - \dfrac{4}{x}\right)}$

55. $\dfrac{\left(\dfrac{4}{x^2 - 9} + \dfrac{2}{x - 2}\right)}{\left(\dfrac{1}{x + 3} + \dfrac{1}{x - 3}\right)}$

56. $\dfrac{\left(\dfrac{1}{x + 1} + \dfrac{1}{2}\right)}{\left(\dfrac{3}{2x^2 + 4x + 2}\right)}$

FOCUS ON CONCEPTS

In this chapter, you studied several concepts that are required for solving systems of equations and inequalities. Answer the following questions to check your understanding of several of these basic concepts. The answers to these questions are given in the back of the book.

1. What is meant by a solution of a system of equations in two variables?

2. When solving a system of equations by substitution, how do you recognize that the system has no solution?

3. When solving a system of equations by elimination, how do you recognize that the system has no solution?

4. Describe any advantages of the algebraic method over the graphical method of solving a system of equations.

5. A system of two equations in two unknowns is solved and has a finite number of solutions. Determine the maximum number of solutions of the system satisfying each of the following.

  (a) Both equations are linear.

  (b) One equation is linear and the other is quadratic.

  (c) Both equations are quadratic.

6. Explain what is meant by an inconsistent system of linear equations.

7. How can you tell graphically that a system of linear equations in two variables has no solution? Give an example.

8. Describe the operations on a system of linear equations that produce an equivalent system of equations.

9. Are the following two systems of equations equivalent? Give reasons for your answer.

$$x + 3y - z = 6 \qquad x + 3y - z = 6$$
$$2x - y + 2z = 1 \qquad -7y + 4z = 1$$
$$3x + 2y - z = 2 \qquad -7y - 4z = -16$$

10. One of the following systems is inconsistent and the other has one solution. How can you identify each by observation?

$$3x - 5y = 3 \qquad 3x - 5y = 3$$
$$-12x + 20y = 8 \qquad 9x - 20y = 6$$

In Exercises 11–14, match the system of inequalities with the graph of its solution. [The graphs are labeled (a), (b), (c), and (d).]

(a)
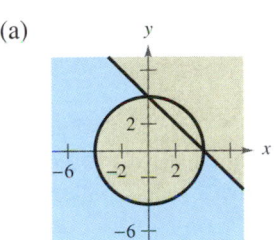

(b)
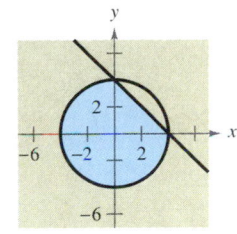

(c)
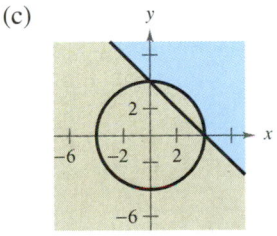

(d)
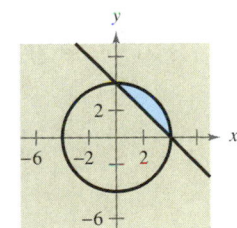

11. $x^2 + y^2 \le 16$
   $x + y \ge 4$

12. $x^2 + y^2 \le 16$
   $x + y \le 4$

13. $x^2 + y^2 \ge 16$
   $x + y \ge 4$

14. $x^2 + y^2 \ge 16$
   $x + y \le 4$

15. The graph of the solution of the inequality $x + 2y < 6$ is given in the figure. Describe how the solution set would change for each of the following.

  (a) $x + 2y \le 6$
  (b) $x + 2y > 6$

Review Exercises

In Exercises 1–6, solve the system by the method of substitution.

1. $x + y = 2$
$x - y = 0$

2. $2x = 3(y - 1)$
$y = x$

3. $x^2 - y^2 = 9$
$x - y = 1$

4. $x^2 + y^2 = 169$
$3x + 2y = 39$

5. $y = 2x^2$
$y = x^4 - 2x^2$

6. $x = y + 3$
$x = y^2 + 1$

In Exercises 7–10, use a graphing utility to solve the system of equations. If you cannot identify the exact solution, find the solution accurate to two decimal places.

7. $y^2 - 2y + x = 0$
$x + y = 0$

8. $y = 2x^2 - 4x + 1$
$y = x^2 - 4x + 3$

9. $y = 2(6 - x)$
$y = 2^{x-2}$

10. $y = \ln(x - 1) - 3$
$y = 4 - \frac{1}{2}x$

In Exercises 11–18, solve the system by elimination.

11. $2x - y = 2$
$6x + 8y = 39$

12. $40x + 30y = 24$
$20x - 50y = -14$

13. $0.2x + 0.3y = 0.14$
$0.4x + 0.5y = 0.20$

14. $12x + 42y = -17$
$30x - 18y = 19$

15. $3x - 2y = 0$
$3x + 2(y + 5) = 10$

16. $7x + 12y = 63$
$2x + 3y = 15$

17. $1.25x - 2y = 3.5$
$5x - 8y = 14$

18. $1.5x + 2.5y = 8.5$
$6x + 10y = 24$

In Exercises 19 and 20, find a system of linear equations having the given solution. (There is more than one correct answer.)

19. $\left(\frac{4}{3}, 3\right)$

20. $(-6, 8)$

21. *Break-Even Point*   You set up a business and make an initial investment of \$10,000. The unit cost of the product is \$2.85 and the selling price is \$4.95. How many units must you sell to break even?

22. *Choice of Two Jobs*   You are offered two sales jobs. One company offers an annual salary of \$22,500 plus a year-end bonus of 1.5% of your total sales. The other company offers an annual salary of \$20,000 plus a year-end bonus of 2% of your total sales. What sales will make the second offer better? Explain.

23. *Acid Mixture*   One hundred liters of a 60% acid solution is obtained by mixing a 75% solution with a 50% solution. How many liters of each must be used to obtain the desired mixture?

24. *Cassette Tape Sales*   Suppose you are the manager of a music store. At the end of one week you are going over receipts for the previous week's sales. Six hundred and fifty cassette tapes were sold. One type of cassette sold for \$9.95 and another sold for \$14.95. The total cassette receipts were \$7717.50. The cash register that was supposed to record the number of each type of cassette sold malfunctioned. Can you recover the information? If so, how many of each type of cassette were sold?

25. *Flying Speeds*   Two planes leave Pittsburgh and Philadelphia at the same time, each going to the other city. One plane flies 25 miles per hour faster than the other. Find the air speed of each plane if the cities are 275 miles apart and the planes pass one another after 40 minutes of flying time.

26. *Dimensions of a Rectangle*   The perimeter of a rectangle is 480 meters and its length is 150% of its width. Find the dimensions of the rectangle.

Supply and Demand   In Exercises 27 and 28, find the point of equilibrium.

Demand Function	Supply Function
27. $p = 37 - 0.0002x$	$p = 22 + 0.00001x$
28. $p = 120 - 0.0001x$	$p = 45 + 0.0002x$

In Exercises 29–34, solve the system of equations.

29.
$$x + 2y + 6z = 4$$
$$-3x + 2y - z = -4$$
$$4x + 2z = 16$$

30.
$$x + 3y - z = 13$$
$$2x - 5z = 23$$
$$4x - y - 2z = 14$$

31.
$$x - 2y + z = -6$$
$$2x - 3y = -7$$
$$-x + 3y - 3z = 11$$

32.
$$2x + 6z = -9$$
$$3x - 2y + 11z = -16$$
$$3x - y + 7z = -11$$

33.
$$2x + 5y - 19z = 34$$
$$3x + 8y - 31z = 54$$

34.
$$2x + y + z + 2w = -1$$
$$5x - 2y + z - 3w = 0$$
$$-x + 3y + 2z + 2w = 1$$
$$3x + 2y + 3z - 5w = 12$$

Exploration   **In Exercises 35 and 36, find a system of linear equations having the given solution. (There is more than one correct answer.)**

35. $(4, -1, 3)$

36. $\left(5, \frac{3}{2}, 2\right)$

In Exercises 37 and 38, find the equation of the parabola

$$y = ax^2 + bx + c$$

that passes through the given points. Use a graphing utility to verify your result.

37.

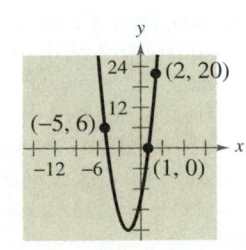

38.

In Exercises 39 and 40, find the equation of the circle

$$x^2 + y^2 + Dx + Ey + F = 0$$

that passes through the given points. Use a graphing utility to verify your result.

39.

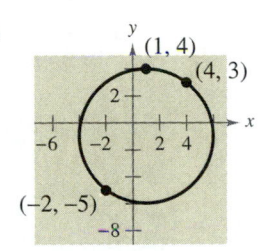

40.

41. *Crop Spraying*   A mixture of 6 gallons of chemical A, 8 gallons of chemical B, and 13 gallons of chemical C is required to kill a certain destructive crop insect. Commercial spray X contains 1, 2, and 2 parts, respectively, of these chemicals. Commercial spray Y contains only chemical C. Commercial spray Z contains chemicals A, B, and C in equal amounts. How much of each type of commercial spray is needed to get the desired mixture?

42. *Investments*   An inheritance of $20,000 was divided among three investments yielding $1818 in interest per year. The interest rates for the three investments were 7%, 9%, and 11%. Find the amount placed in each investment if the second and third were $3000 and $1000 less than the first, respectively.

43. *Fitting a Line to Data*   Solve the system of equations to find the least squares regression line $y = ax + b$ for the points in the figure.

$$5b + 10a = 17.8$$
$$10b + 30a = 45.7$$

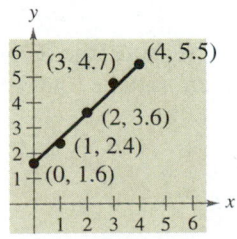

44. *Fitting a Parabola to Data* Solve the system of equations to find the least squares regression parabola $y = ax^2 + bx + c$ for the points in the figure.

$$5c \quad + 10a = \quad 9.1$$
$$10b \qquad = \quad 8.0$$
$$10c \quad + 34a = \quad 19.8$$

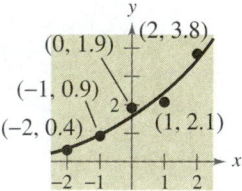

45. *Data Analysis* Let x and y represent the median ages at first marriage for women and men, respectively. In a sample of 7 years since 1970, these median ages are given by the following ordered pairs. (Source: U.S. Center for Health Statistics)

(20.6, 22.5),   (20.8, 22.7),   (21.8, 23.6),

(23.0, 24.8),   (23.3, 25.1),   (23.6, 25.3),

(23.7, 25.5)

(a) Find the least squares regression line $y = ax + b$ for the data by solving the following system of linear equations.

$$7b + \quad 156.8a = \quad 169.5$$
$$156.8b + 3522.78a = 3806.8$$

(b) Use a graphing utility to plot the data and graph the regression line on the same viewing rectangle.

(c) Use the graph to determine whether the line is a good model for the data. Explain.

(d) What information is given by the slope of the regression line? Explain.

46. *Exploration* Find k_1 and k_2 such that the following system of equations has an infinite number of solutions.

$$3x - \quad 5y = 8$$
$$2x + k_1y = k_2$$

In Exercises 47–54, sketch a graph of the solution set of the system of inequalities.

47.
$$x + 2y \le 160$$
$$3x + \quad y \le 180$$
$$x \ge \quad 0$$
$$y \ge \quad 0$$

48.
$$2x + 3y \le 24$$
$$2x + \quad y \le 16$$
$$x \ge \quad 0$$
$$y \ge \quad 0$$

49.
$$3x + 2y \ge 24$$
$$x + 2y \ge 12$$
$$2 \le \quad x \le 15$$
$$y \le 15$$

50.
$$2x + \quad y \ge 16$$
$$x + 3y \ge 18$$
$$0 \le \quad x \le 25$$
$$0 \le \quad y \le 25$$

51.
$$y < x + 1$$
$$y > x^2 - 1$$

52.
$$y \le 6 - 2x - x^2$$
$$y \ge x + 6$$

53.
$$2x - 3y \ge 0$$
$$2x - \quad y \le 8$$
$$y \ge 0$$

54.
$$x^2 + y^2 \le 9$$
$$(x - 3)^2 + y^2 \le 9$$

In Exercises 55 and 56, derive a set of inequalities to describe the region.

55. *Parallelogram:* Vertices at $(1, 5)$, $(3, 1)$, $(6, 10)$, $(8, 6)$

56. *Triangle:* Vertices at $(1, 2)$, $(6, 7)$, $(8, 1)$

In Exercises 57 and 58, determine a system of inequalities that models the description. Use a graphing utility to graph and shade the solution of the system.

57. *Fruit Distribution* A Pennsylvania fruit grower has 1500 bushels of apples that are to be divided between markets in Harrisburg and Philadelphia. These two markets need at least 400 bushels and 600 bushels, respectively.

58. *Inventory Costs* A warehouse operator has 24,000 square feet of floor space in which to store two products. Each unit of product I requires 20 square feet of floor space and costs $12 per day to store. Each unit of product II requires 30 square feet of floor space and costs $8 per day to store. The total storage cost per day cannot exceed $12,400.

In Exercises 59 and 60, find the consumer surplus and producer surplus for the supply and demand equations.

Demand	Supply
59. $p = 160 - 0.0001x$	$p = 70 + 0.0002x$
60. $p = 130 - 0.0002x$	$p = 30 + 0.0003x$

In Exercises 61–64, find the required optimum value of the objective function subject to the indicated constraints.

61. Maximize:

$z = 3x + 4y$

Constraints:

$x \geq 0$

$y \geq 0$

$2x + 5y \leq 50$

$4x + y \leq 28$

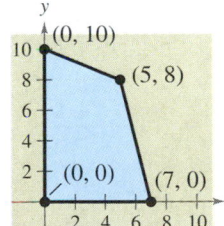

62. Minimize:

$z = 10x + 7y$

Constraints:

$x \geq 0$

$y \geq 0$

$2x + y \geq 100$

$x + y \geq 75$

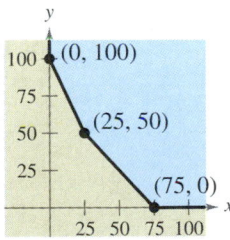

63. Minimize:

$z = 1.75x + 2.25y$

Constraints:

$x \geq 0$

$y \geq 0$

$2x + y \geq 25$

$3x + 2y \geq 45$

64. Maximize:

$z = 50x + 70y$

Constraints:

$x \geq 0$

$y \geq 0$

$x + 2y \leq 1500$

$5x + 2y \leq 3500$

65. *Maximum Revenue*   A student is working part time as a cosmetologist to pay college expenses. The student may work no more than 24 hours per week. Haircuts cost $17 and require an average of 20 minutes, and permanents cost $60 and require an average of 1 hour and 10 minutes. What combination of haircuts and/or permanents will yield a maximum revenue?

66. *Maximum Profit*   A manufacturer produces products A and B yielding profits of $18 and $24, respectively. Each product must go through three processes with the required times per unit shown in the table.

Process	Hours for Product A	Hours for Product B	Hours Available per Day
I	4	2	24
II	1	2	9
III	1	1	8

Find the daily production level for each unit to maximize the profit.

67. *Minimum Cost*   A pet supply company mixes two brands of dry dog food. Brand X costs $15 per bag and contains eight units of nutritional element A, one unit of nutritional element B, and two units of nutritional element C. Brand Y costs $30 per bag and contains two units of nutritional element A, one unit of nutritional element B, and seven units of nutritional element C. Each bag of mixed dog food must contain at least 16 units, 5 units, and 20 units of nutritional elements A, B, and C, respectively. Find the number of bags of brands X and Y that should be mixed to produce a mixture meeting the minimum nutritional requirements and having a minimum cost per bag.

68. *Minimum Cost*   Two gasolines, type A and type B, have octane ratings of 80 and 92, respectively. Type A costs $1.25 per gallon and type B costs $1.55 per gallon. Determine the blend of minimum cost with an octane rating of at least 88. (*Hint:* Let x be the fraction of each gallon that is type A and let y be the fraction that is type B.)

CHAPTER PROJECT: *Fitting Models to Data*

Many of the models in this book were created with a statistical method called *least squares regression analysis.* This procedure is tedious to perform by hand, but can be performed easily with a computer or graphing calculator.

EXAMPLE 1 *Fitting a Line to Data*

The numbers (in millions) of morning and evening newspapers sold each day in the United States from 1978 through 1992 are shown in the table. Use the data to project the numbers of morning and evening newspapers that will be sold each day in 1998. In the table, $t = 0$ represents 1980. (Source: Editor and Publisher Company)

Year, t	−2	−1	0	1	2	3	4	5
Morning	27.7	28.6	29.4	30.6	33.2	33.8	35.4	36.4
Evening	34.3	33.6	32.8	30.9	29.3	28.8	27.7	26.4

Year, t	6	7	8	9	10	11	12
Morning	37.4	39.1	40.4	40.7	41.3	41.5	42.4
Evening	25.1	23.7	22.2	21.8	21.0	19.2	17.8

Solution

Begin by finding a computer or graphing calculator that will perform linear regression analysis. After entering the data and running the program, you should obtain the following models. (Both models have correlation coefficients whose absolute values are greater than 0.98, which means that the models are very good fits for the data.)

$y = 30.246 + 1.123t$ Morning newspaper circulation
$y = 32.225 - 1.184t$ Evening newspaper circulation

With these models, you can project the 1998 newspaper sales. If the sales through 1998 continue to follow the pattern from 1978 through 1992, the 1998 sales of newspapers should be about

$y = 30.246 + 1.123(18) \approx 50.5$ million   Morning newspaper prediction
$y = 32.225 + 1.184(18) \approx 10.9$ million.   Evening newspaper prediction

The graphs of the data and models are shown at the left.

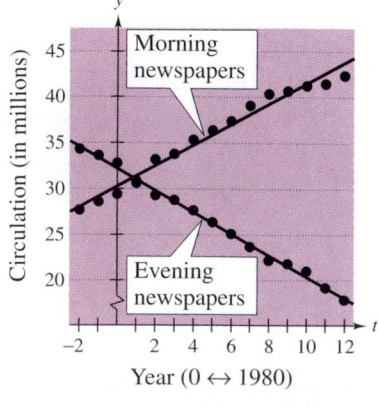

Year (0 ↔ 1980)

CHAPTER PROJECT INVESTIGATIONS

1. *Sunday Paper Circulation*   The numbers (in millions) of Sunday newspapers sold each week in the United States from 1981 through 1992 are shown in the table below. Find a linear model that represents the data. Use your model to project the number of Sunday newspapers to be sold each week in 1998.

2. *Newspaper Companies*   The numbers of newspaper companies in the United States from 1981 through 1992 are shown in the table below. Find a linear model for each of these sets of data.

3. *Average Sales*   From 1981 through 1992, the circulation of morning newspapers increased. However, because the number of morning newspaper companies also increased, the competition for morning newspaper readers became keener. Did the average circulation per morning newspaper company increase or decrease? Explain.

4. *Average Sales*   From 1981 through 1992, the circulation of evening newspapers decreased. However, because the number of evening newspaper companies also decreased, the competition for evening newspaper readers became less keen. Did the average circulation per evening newspaper company increase or decrease? Explain.

5. *Average Sales*   From 1981 through 1992, the circulation of Sunday newspapers increased. However, because the number of Sunday newspaper companies also increased, the competition for Sunday newspaper readers became keener. Did the average circulation per Sunday newspaper company increase or decrease? Explain.

6. *Which Would You Choose?*   If you had the opportunity to invest in a company that published only one type of newspaper (morning, evening, or Sunday), which would you choose? Explain your reasoning.

7. *Households and Population*   Figures for population and number of households in the United States (both in millions) from 1981 through 1992 are given in the table below. From this information would you say that the percent of Americans who read newspapers was increasing or decreasing from 1981 through 1992? Explain your reasoning.

Year, t	1	2	3	4	5	6	7	8	9	10	11	12
Sunday	55.2	56.3	56.7	57.5	58.8	58.9	60.1	61.5	62.0	62.6	62.1	62.2

Table for Exercise 1 ($t = 0$ represents 1980.)

Year, t	1	2	3	4	5	6	7	8	9	10	11	12
Morning	408	434	446	458	482	499	511	529	530	559	571	596
Evening	1352	1310	1284	1257	1220	1188	1166	1141	1125	1084	1042	996
Sunday	755	768	772	783	798	802	820	840	847	863	875	891

Table for Exercise 2 ($t = 0$ represents 1980.)

Year, t	1	2	3	4	5	6	7	8	9	10	11	12
Households	81.6	83.0	83.9	85.4	86.8	88.5	89.5	91.1	92.8	93.3	95.7	96.4
Population	230.0	232.2	234.3	236.3	238.5	240.7	242.8	245.0	247.3	249.9	252.6	255.5

Table for Exercise 7 ($t = 0$ represents 1980.)

Chapter Test

Take this test as you would take a test in class. After you are done, check your work against the answers given in the back of the book.

 The *Interactive* CD-ROM provides answers to the Chapter Tests and Cumulative Tests. It also offers Chapter Pre-Tests (which test key skills and concepts covered in previous chapters) and Chapter Post-Tests, both of which have randomly generated exercises with diagnostic capabilities.

In Exercises 1–3, solve the system by the method of substitution.

1. $x - y = 4$
$3x + 2y = 2$

2. $y = x - 1$
$y = (x - 1)^3$

3. $2x - y^2 = 0$
$x - y = 4$

In Exercises 4–6, solve the system graphically.

4. $2x - 3y = 0$
$2x + 3y = 12$

5. $y = 9 - x^2$
$y = x + 3$

6. $y = \log_3 x$
$y = -\frac{1}{3}x + 2$

In Exercises 7 and 8, solve the linear system by elimination.

7. $2x + 3y = 17$
$5x - 4y = -15$

8. $x - 2y + 3z = 11$
$2x \quad - z = 3$
$3y + z = -8$

9. Find a system of linear equations that has the solution $\left(\frac{4}{3}, -5\right)$.

10. Find the equation of the parabola $y = ax^2 + bx + c$ passing through the points $(0, 6)$, $(-2, 2)$, and $\left(3, \frac{9}{2}\right)$.

In Exercises 11 and 12, use a graphing utility to graph the inequalities and shade the region representing the solution.

11. $2x + y \leq 4$
$2x - y \geq 0$
$x \geq 0$

12. $y < -x^4 + x^2 + 4$
$y > 4x$

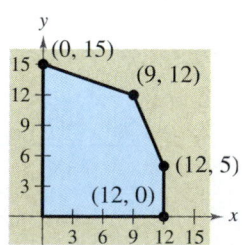

FIGURE FOR 13

13. Derive a set of inequalities to describe the region in the figure.

14. Find the maximum value of the objective function $z = 20x + 12y$ subject to the constraints $x \geq 0$, $y \geq 0$, $x + 4y \leq 32$, and $3x + 2y \leq 36$.

15. A merchant plans to sell two models of compact disc players at costs of $275 and $400. The $275 model yields a profit of $55 and the $400 model yields a profit of $75. The merchant estimates that the total monthly demand will not exceed 300 units. The merchant does not want to invest more than $100,000 in inventory for these products. Find the number of units of each model that should be stocked in order to maximize profit.

The times (in minutes) for the winning men's and women's 1000-meter speed skating events at the winter Olympics are shown below. (In 1994, the winter Olympics was held only 2 years after the previous winter Olympics.)

Year	Men	Women
1976	1.322	1.474
1980	1.253	1.402
1984	1.263	1.360
1988	1.217	1.294
1992	1.248	1.358
1994	1.207	1.312

The best-fitting linear models for these two data sets are

$s = 1.279 - 0.0049t$ Men

$s = 1.411 - 0.0078t$ Women

where s is the time (in minutes) and t is the year, with $t = 0$ representing 1980.

According to these two models, the women's times are decreasing a little more rapidly than the men's times.

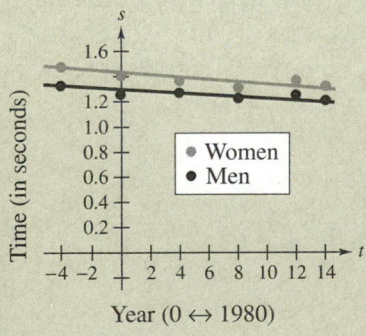

See Exercises 67 and 68 on page 768.

Photo: Bob Martin/Allsport

10.1 *Matrices and Systems of Equations*
10.2 *Operations with Matrices*
10.3 *The Inverse of a Square Matrix*
10.4 *The Determinant of a Square Matrix*
10.5 *Applications of Matrices and Determinants*

Bonnie Blair won the 1000-meter women's speed skating event in the 1992 and 1994 winter Olympics. These were the first times this event was ever won by an American.

729

10.1 | *Matrices and Systems of Equations*

See Exercise 71 on page 744 for an example of how matrices can be used to help find a model for the parabolic path of a baseball.

Matrices □ *Elementary Row Operations* □ *Gaussian Elimination with Back-Substitution* □ *Gauss-Jordan Elimination*

Matrices

In this section you will study a streamlined technique for solving systems of linear equations. This technique involves the use of a rectangular array of real numbers called a **matrix.**

> ### DEFINITION OF MATRIX
>
> If m and n are positive integers, an $m \times n$ **matrix** (read "m by n") is a rectangular array
>
> $$\begin{bmatrix} a_{11} & a_{12} & a_{13} & \cdots & a_{1n} \\ a_{21} & a_{22} & a_{23} & \cdots & a_{2n} \\ a_{31} & a_{32} & a_{33} & \cdots & a_{3n} \\ \vdots & \vdots & \vdots & & \vdots \\ a_{m1} & a_{m2} & a_{m3} & \cdots & a_{mn} \end{bmatrix} \quad m \text{ rows}$$
>
> n columns
>
> in which each **entry,** a_{ij}, of the matrix is a number. An $m \times n$ matrix has m **rows** (horizontal lines) and n **columns** (vertical lines).

NOTE The plural of matrix is *matrices.* ■■

The entry in the ith row and jth column is denoted by the *double subscript* notation a_{ij}. A matrix having m rows and n columns is said to be of **order** $m \times n$. If $m = n$, the matrix is **square** of order n. For a square matrix, the entries $a_{11}, a_{22}, a_{33}, \ldots$ are the **main diagonal** entries.

NOTE A matrix that has only one row is called a **row matrix,** and a matrix that has only one column is called a **column matrix.** ■■

EXAMPLE 1   *Examples of Matrices*

a. *Order:* 1×1

$[2]$

b. *Order:* 1×4

$\begin{bmatrix} 1 & -3 & 0 & \frac{1}{2} \end{bmatrix}$

c. *Order:* 2×2

$\begin{bmatrix} 0 & 0 \\ 0 & 0 \end{bmatrix}$

d. *Order:* 3×2

$\begin{bmatrix} 5 & 0 \\ 2 & -2 \\ -7 & 4 \end{bmatrix}$

A matrix derived from a system of linear equations (each written in standard form with the constant term on the right) is the **augmented matrix** of the system. Moreover, the matrix derived from the coefficients of the system (but not including the constant terms) is the **coefficient matrix** of the system.

$$
\begin{array}{ccc}
\textit{System} & \textit{Augmented Matrix} & \textit{Coefficient Matrix} \\[6pt]
\begin{aligned}
x - 4y + 3z &= 5 \\
-x + 3y - z &= -3 \\
2x \qquad - 4z &= 6
\end{aligned}
&
\left[
\begin{array}{ccc:c}
1 & -4 & 3 & 5 \\
-1 & 3 & -1 & -3 \\
2 & 0 & -4 & 6
\end{array}
\right]
&
\left[
\begin{array}{ccc}
1 & -4 & 3 \\
-1 & 3 & -1 \\
2 & 0 & -4
\end{array}
\right]
\end{array}
$$

NOTE   Note the use of 0 for the missing y-variable in the third equation, and also note the fourth column of constant terms in the augmented matrix. ■■

When forming either the coefficient matrix or the augmented matrix of a system, you should begin by vertically aligning the variables in the equations and using zeros for the missing variables.

$$
\begin{array}{ccc}
\textit{Given System} & \textit{Line Up Variables} & \textit{Form Augmented Matrix} \\[6pt]
\begin{aligned}
x + 3y &= 9 \\
-y + 4z &= -2 \\
x - 5z &= 0
\end{aligned}
&
\begin{aligned}
x + \quad 3y \quad &= 9 \\
-y + 4z &= -2 \\
x \qquad - 5z &= 0
\end{aligned}
&
\left[
\begin{array}{ccc:c}
1 & 3 & 0 & 9 \\
0 & -1 & 4 & -2 \\
1 & 0 & -5 & 0
\end{array}
\right]
\end{array}
$$

Elementary Row Operations

In Section 9.3, you studied three operations that can be used on a system of linear equations to produce an equivalent system.

1. Interchange two equations.
2. Multiply an equation by a nonzero constant.
3. Add a multiple of an equation to another equation.

In matrix terminology these three operations correspond to **elementary row operations.** An elementary row operation on an augmented matrix of a given system of linear equations produces a new augmented matrix corresponding to a new (but equivalent) system of linear equations. Two matrices are **row-equivalent** if one can be obtained from the other by a sequence of elementary row operations.

ELEMENTARY ROW OPERATIONS

1. Interchange two rows.
2. Multiply a row by a nonzero constant.
3. Add a multiple of a row to another row.

Although elementary row operations are simple to perform, they involve a lot of arithmetic. Because it is easy to make a mistake, we suggest that you get in the habit of noting the elementary row operations performed in each step so that you can go back and check your work.

EXAMPLE 2   *Elementary Row Operations*

a. Interchange the first and second rows.

Original Matrix

$$\begin{bmatrix} 0 & 1 & 3 & 4 \\ -1 & 2 & 0 & 3 \\ 2 & -3 & 4 & 1 \end{bmatrix}$$

New Row-Equivalent Matrix

$$\begin{matrix} R_2 \\ R_1 \end{matrix} \begin{bmatrix} -1 & 2 & 0 & 3 \\ 0 & 1 & 3 & 4 \\ 2 & -3 & 4 & 1 \end{bmatrix}$$

b. Multiply the first row by $\frac{1}{2}$.

Original Matrix

$$\begin{bmatrix} 2 & -4 & 6 & -2 \\ 1 & 3 & -3 & 0 \\ 5 & -2 & 1 & 2 \end{bmatrix}$$

New Row-Equivalent Matrix

$$\frac{1}{2}R_1 \rightarrow \begin{bmatrix} 1 & -2 & 3 & -1 \\ 1 & 3 & -3 & 0 \\ 5 & -2 & 1 & 2 \end{bmatrix}$$

c. Add -2 times the first row to the third row.

Original Matrix

$$\begin{bmatrix} 1 & 2 & -4 & 3 \\ 0 & 3 & -2 & -1 \\ 2 & 1 & 5 & -2 \end{bmatrix}$$

New Row-Equivalent Matrix

$$\begin{matrix} \\ \\ -2R_1 + R_3 \rightarrow \end{matrix} \begin{bmatrix} 1 & 2 & -4 & 3 \\ 0 & 3 & -2 & -1 \\ 0 & -3 & 13 & -8 \end{bmatrix}$$

Note that the elementary row operation is written beside the row that is *changed.*

TECHNOLOGY

Most graphing utilities can perform elementary row operations on matrices. For instance, on a *TI-83* or a *TI-82* you can perform the elementary row operation shown in Example 2(c) as follows.

1. Use the matrix edit feature to enter the matrix as [A].
2. Choose the "* row + (" feature in the matrix math menu.

   * row + (−2, [A], 1, 3) $\boxed{\text{ENTER}}$

The new row-equivalent matrix will be displayed. To do a sequence of row operations, use $\boxed{\text{ANS}}$ in place of [A] in each operation. If you want to save this new matrix, you must do this with separate steps.

In Example 3 of Section 9.3, you used Gaussian elimination with back-substitution to solve a system of linear equations. The next example demonstrates the matrix version of Gaussian elimination. The two methods are essentially the same. The basic difference is that with matrices you do not need to keep writing the variables.

EXAMPLE 3   Using Elementary Row Operations

Linear System	Associated Augmented Matrix

$$x - 2y + 3z = 9$$
$$-x + 3y \quad\quad = -4$$
$$2x - 5y + 5z = 17$$

$$\begin{bmatrix} 1 & -2 & 3 & \vdots & 9 \\ -1 & 3 & 0 & \vdots & -4 \\ 2 & -5 & 5 & \vdots & 17 \end{bmatrix}$$

Add the first equation to the second equation.

Add the first row to the second row.

$$x - 2y + 3z = 9$$
$$y + 3z = 5$$
$$2x - 5y + 5z = 17$$

$$R_1 + R_2 \rightarrow \begin{bmatrix} 1 & -2 & 3 & \vdots & 9 \\ 0 & 1 & 3 & \vdots & 5 \\ 2 & -5 & 5 & \vdots & 17 \end{bmatrix}$$

Add -2 times the first equation to the third equation.

Add -2 times the first row to the third row.

$$x - 2y + 3z = 9$$
$$y + 3z = 5$$
$$-y - z = -1$$

$$-2R_1 + R_3 \rightarrow \begin{bmatrix} 1 & -2 & 3 & \vdots & 9 \\ 0 & 1 & 3 & \vdots & 5 \\ 0 & -1 & -1 & \vdots & -1 \end{bmatrix}$$

Add the second equation to the third equation.

Add the second row to the third row.

$$x - 2y + 3z = 9$$
$$y + 3z = 5$$
$$2z = 4$$

$$R_2 + R_3 \rightarrow \begin{bmatrix} 1 & -2 & 3 & \vdots & 9 \\ 0 & 1 & 3 & \vdots & 5 \\ 0 & 0 & 2 & \vdots & 4 \end{bmatrix}$$

Multiply the third equation by $\frac{1}{2}$.

Multiply the third row by $\frac{1}{2}$.

$$x - 2y + 3z = 9$$
$$y + 3z = 5$$
$$z = 2$$

$$\tfrac{1}{2}R_3 \rightarrow \begin{bmatrix} 1 & -2 & 3 & \vdots & 9 \\ 0 & 1 & 3 & \vdots & 5 \\ 0 & 0 & 1 & \vdots & 2 \end{bmatrix}$$

At this point, you can use back-substitution to find that the solution is $x = 1$, $y = -1$, and $z = 2$, as was done in Example 3 in Section 9.3.

NOTE   Remember that you can check a solution by substituting the values of x, y, and z into each equation in the original system. ■■

The last matrix in Example 3 is said to be in **row-echelon form.** The term *echelon* refers to the stair-step pattern formed by the nonzero elements of the matrix. To be in this form, a matrix must have the following properties.

ROW-ECHELON FORM AND REDUCED ROW-ECHELON FORM

A matrix in **row-echelon form** has the following properties.

1. All rows consisting entirely of zeros occur at the bottom of the matrix.
2. For each row that does not consist entirely of zeros, the first nonzero entry is 1 (called a **leading 1**).
3. For two successive (nonzero) rows, the leading 1 in the higher row is farther to the left than the leading 1 in the lower row.

A matrix in *row-echelon form* is in **reduced row-echelon form** if every column that has a leading 1 has zeros in every position above and below its leading 1.

EXAMPLE 4   *Row-Echelon Form*

The following matrices are in row-echelon form.

a. $\begin{bmatrix} 1 & 2 & -1 & 4 \\ 0 & 1 & 0 & 3 \\ 0 & 0 & 1 & -2 \end{bmatrix}$ **b.** $\begin{bmatrix} 0 & 1 & 0 & 5 \\ 0 & 0 & 1 & 3 \\ 0 & 0 & 0 & 0 \end{bmatrix}$

c. $\begin{bmatrix} 1 & -5 & 2 & -1 & 3 \\ 0 & 0 & 1 & 3 & -2 \\ 0 & 0 & 0 & 1 & 4 \\ 0 & 0 & 0 & 0 & 1 \end{bmatrix}$ **d.** $\begin{bmatrix} 1 & 0 & 0 & -1 \\ 0 & 1 & 0 & 2 \\ 0 & 0 & 1 & 3 \\ 0 & 0 & 0 & 0 \end{bmatrix}$

The matrices in (b) and (d) also happen to be in *reduced* row-echelon form. The following matrices are not in row-echelon form.

e. $\begin{bmatrix} 1 & 2 & -3 & 4 \\ 0 & 2 & 1 & -1 \\ 0 & 0 & 1 & -3 \end{bmatrix}$ **f.** $\begin{bmatrix} 1 & 2 & -1 & 2 \\ 0 & 0 & 0 & 0 \\ 0 & 1 & 2 & -4 \end{bmatrix}$

Every matrix is row-equivalent to a matrix in row-echelon form. For instance, in Example 4, you can change the matrix in part (e) to row-echelon form by multiplying its second row by $\frac{1}{2}$. What elementary row operation could you perform on the matrix in part (f) so that it would be in row-echelon form?

Gaussian Elimination with Back-Substitution

EXAMPLE 5 *Gaussian Elimination with Back-Substitution*

Solve the system.

$$
\begin{aligned}
y + z - 2w &= -3 \\
x + 2y - z &= 2 \\
2x + 4y + z - 3w &= -2 \\
x - 4y - 7z - w &= -19
\end{aligned}
$$

Solution

$\begin{array}{c} R_2 \\ R_1 \end{array}$
$\left[\begin{array}{cccc:c}
1 & 2 & -1 & 0 & 2 \\
0 & 1 & 1 & -2 & -3 \\
2 & 4 & 1 & -3 & -2 \\
1 & -4 & -7 & -1 & -19
\end{array}\right]$
First column has leading 1 in upper left corner.

$\begin{array}{c} \\ \\ -2R_1 + R_3 \rightarrow \\ -R_1 + R_4 \rightarrow \end{array}$
$\left[\begin{array}{cccc:c}
1 & 2 & -1 & 0 & 2 \\
0 & 1 & 1 & -2 & -3 \\
0 & 0 & 3 & -3 & -6 \\
0 & -6 & -6 & -1 & -21
\end{array}\right]$
First column has zeros below its leading 1.

$\begin{array}{c} \\ \\ \\ 6R_2 + R_4 \rightarrow \end{array}$
$\left[\begin{array}{cccc:c}
1 & 2 & -1 & 0 & 2 \\
0 & 1 & 1 & -2 & -3 \\
0 & 0 & 3 & -3 & -6 \\
0 & 0 & 0 & -13 & -39
\end{array}\right]$
Second column has zeros below its leading 1.

$\begin{array}{c} \\ \\ \frac{1}{3}R_3 \rightarrow \\ \end{array}$
$\left[\begin{array}{cccc:c}
1 & 2 & -1 & 0 & 2 \\
0 & 1 & 1 & -2 & -3 \\
0 & 0 & 1 & -1 & -2 \\
0 & 0 & 0 & -13 & -39
\end{array}\right]$
Third column has zeros below its leading 1.

$\begin{array}{c} \\ \\ \\ -\frac{1}{13}R_4 \rightarrow \end{array}$
$\left[\begin{array}{cccc:c}
1 & 2 & -1 & 0 & 2 \\
0 & 1 & 1 & -2 & -3 \\
0 & 0 & 1 & -1 & -2 \\
0 & 0 & 0 & 1 & 3
\end{array}\right]$
Fourth column has a leading 1.

The matrix is now in row-echelon form, and the corresponding system is

$$
\begin{aligned}
x + 2y - z &= 2 \\
y + z - 2w &= -3 \\
z - w &= -2 \\
w &= 3.
\end{aligned}
$$

Using back-substitution, you can determine that the solution is $x = -1$, $y = 2$, $z = 1$, and $w = 3$. Check this in the original system of equations.

Study Tip

Gaussian elimination with back-substitution works well for solving systems of linear equations by hand or with a computer. For this algorithm, the order in which the elementary row operations are performed is important. We suggest operating from *left to right by columns,* using elementary row operations to obtain zeros in all entries directly below the leading 1's.

GAUSSIAN ELIMINATION WITH BACK-SUBSTITUTION

1. Write the augmented matrix of the system of linear equations.
2. Use elementary row operations to rewrite the augmented matrix in row-echelon form.
3. Write the system of linear equations corresponding to the matrix in row-echelon form, and use back-substitution to find the solution.

When solving a system of linear equations, remember that it is possible for the system to have no solution. If, in the elimination process, you obtain a row with zeros except for the last entry, it is unnecessary to continue the elimination process. You can simply conclude that the system is inconsistent.

EXAMPLE 6   *A System with No Solution*

Solve the system.

$$
\begin{aligned}
x - \ y + 2z &= 4 \\
x \quad \ + \ z &= 6 \\
2x - 3y + 5z &= 4 \\
3x + 2y - \ z &= 1
\end{aligned}
$$

Solution

$$
\begin{bmatrix}
1 & -1 & 2 & \vdots & 4 \\
1 & 0 & 1 & \vdots & 6 \\
2 & -3 & 5 & \vdots & 4 \\
3 & 2 & -1 & \vdots & 1
\end{bmatrix}
\begin{matrix}
\\
-R_1 + R_2 \to \\
-2R_1 + R_3 \to \\
-3R_1 + R_4 \to
\end{matrix}
\begin{bmatrix}
1 & -1 & 2 & \vdots & 4 \\
0 & 1 & -1 & \vdots & 2 \\
0 & -1 & 1 & \vdots & -4 \\
0 & 5 & -7 & \vdots & -11
\end{bmatrix}
$$

$$
\begin{matrix}
\\
\\
R_2 + R_3 \to \\
\\
\end{matrix}
\begin{bmatrix}
1 & -1 & 2 & \vdots & 4 \\
0 & 1 & -1 & \vdots & 2 \\
0 & 0 & 0 & \vdots & -2 \\
0 & 5 & -7 & \vdots & -11
\end{bmatrix}
$$

Note that the third row of this matrix consists of zeros except for the last entry. This means that the original system of linear equation is *inconsistent*. You can see why this is true by converting back to a system of linear equations.

$$
\begin{aligned}
x - \ y + 2z &= \quad 4 \\
y - \ z &= \quad 2 \\
0 &= \ -2 \\
5y - 7z &= -11
\end{aligned}
$$

Because the third equation is not possible, the system has no solution.

Gauss-Jordan Elimination

With Gaussian elimination, elementary row operations are applied to a matrix to obtain a (row-equivalent) row-echelon form. A second method of elimination, called **Gauss-Jordan elimination,** after Carl Friedrich Gauss and Wilhelm Jordan (1842–1899), continues the reduction process until a *reduced* row-echelon form is obtained. This procedure is demonstrated in the following example.

EXAMPLE 7 *Gauss-Jordan Elimination*

Use Gauss-Jordan elimination to solve the system.

$$\begin{aligned} x - 2y + 3z &= 9 \\ -x + 3y &= -4 \\ 2x - 5y + 5z &= 17 \end{aligned}$$

Solution

In Example 3, Gaussian elimination was used to obtain the row-echelon form

$$\begin{bmatrix} 1 & -2 & 3 & \vdots & 9 \\ 0 & 1 & 3 & \vdots & 5 \\ 0 & 0 & 1 & \vdots & 2 \end{bmatrix}.$$

Now, rather than using back-substitution, apply elementary row operations until you obtain a matrix in reduced row-echelon form. To do this, you must produce zeros above each of the leading 1's, as follows.

$$2R_2 + R_1 \rightarrow \begin{bmatrix} 1 & 0 & 9 & \vdots & 19 \\ 0 & 1 & 3 & \vdots & 5 \\ 0 & 0 & 1 & \vdots & 2 \end{bmatrix}$$

Second column has zeros above its leading 1.

$$\begin{matrix} -9R_3 + R_1 \rightarrow \\ -3R_3 + R_2 \rightarrow \end{matrix} \begin{bmatrix} 1 & 0 & 0 & \vdots & 1 \\ 0 & 1 & 0 & \vdots & -1 \\ 0 & 0 & 1 & \vdots & 2 \end{bmatrix}$$

Third column has zeros above its leading 1.

Now, converting back to a system of linear equations, you have

$$\begin{aligned} x &= 1 \\ y &= -1 \\ z &= 2. \end{aligned}$$

The beauty of Gauss-Jordan elimination is that, from the reduced row-echelon form, you can simply read the solution.

NOTE Which technique do you prefer: Gaussian elimination or Gauss-Jordan elimination? ▪▪

The elimination procedures described in this section employ an algorithmic approach that is easily adapted to computer use. However, the procedure makes no effort to avoid fractional coefficients. For instance, if the system given in Example 7 had been listed as

$$2x - 5y + 5z = 17$$
$$x - 2y + 3z = 9$$
$$-x + 3y \quad = -4$$

the procedure would have required multiplication of the first row by $\frac{1}{2}$, which would have introduced fractions in the first row. For hand computations, fractions can sometimes be avoided by judiciously choosing the order in which the elementary row operations are applied.

EXAMPLE 8   *A System with an Infinite Number of Solutions*

Solve the system.

$$2x + 4y - 2z = 0$$
$$3x + 5y \quad = 1$$

Solution

$$\begin{bmatrix} 2 & 4 & -2 & \vdots & 0 \\ 3 & 5 & 0 & \vdots & 1 \end{bmatrix} \qquad \frac{1}{2}R_1 \rightarrow \begin{bmatrix} 1 & 2 & -1 & \vdots & 0 \\ 3 & 5 & 0 & \vdots & 1 \end{bmatrix}$$

$$-3R_1 + R_2 \rightarrow \begin{bmatrix} 1 & 2 & -1 & \vdots & 0 \\ 0 & -1 & 3 & \vdots & 1 \end{bmatrix}$$

$$-R_2 \rightarrow \begin{bmatrix} 1 & 2 & -1 & \vdots & 0 \\ 0 & 1 & -3 & \vdots & -1 \end{bmatrix}$$

$$-2R_2 + R_1 \rightarrow \begin{bmatrix} 1 & 0 & 5 & \vdots & 2 \\ 0 & 1 & -3 & \vdots & -1 \end{bmatrix}$$

The corresponding system of equations is

$$x \quad + 5z = 2$$
$$y - 3z = -1.$$

Solving for x and y in terms of z, you have $x = -5z + 2$ and $y = 3z - 1$. Then, letting $z = a$, the solution set has the form

$$(-5a + 2, 3a - 1, a)$$

where a is a real number. Try substituting values for a to obtain a few solutions. Then check each solution in the original system of equations.

NOTE You have seen that the row-echelon form of a given matrix *is not* unique; however, the *reduced* row-echelon form of a given matrix *is* unique. Try applying Gauss-Jordan elimination to the row-echelon matrix given at the right to see that you obtain the same reduced row-echelon form as in Example 7. ∎∎

It is worth noting that the row-echelon form of a matrix is not unique. That is, two different sequences of elementary row operations may yield different row-echelon forms. For instance, the following sequence of elementary row operations on the matrix in Example 3 produces a slightly different row-echelon form.

$$
\begin{bmatrix}
1 & -2 & 3 & \vdots & 9 \\
-1 & 3 & 0 & \vdots & -4 \\
2 & -5 & 5 & \vdots & 17
\end{bmatrix}
\quad
\begin{matrix} R_2 \\ R_1 \end{matrix}
\begin{bmatrix}
-1 & 3 & 0 & \vdots & -4 \\
1 & -2 & 3 & \vdots & 9 \\
2 & -5 & 5 & \vdots & 17
\end{bmatrix}
$$

$$
-R_1 \rightarrow
\begin{bmatrix}
1 & -3 & 0 & \vdots & 4 \\
1 & -2 & 3 & \vdots & 9 \\
2 & -5 & 5 & \vdots & 17
\end{bmatrix}
$$

$$
\begin{matrix} \\ -R_1 + R_2 \rightarrow \\ -2R_1 + R_3 \rightarrow \end{matrix}
\begin{bmatrix}
1 & -3 & 0 & \vdots & 4 \\
0 & 1 & 3 & \vdots & 5 \\
0 & 1 & 5 & \vdots & 9
\end{bmatrix}
$$

$$
\begin{matrix} \\ \\ -R_2 + R_3 \rightarrow \end{matrix}
\begin{bmatrix}
1 & -3 & 0 & \vdots & 4 \\
0 & 1 & 3 & \vdots & 5 \\
0 & 0 & 2 & \vdots & 4
\end{bmatrix}
$$

$$
\begin{matrix} \\ \\ \tfrac{1}{2}R_3 \rightarrow \end{matrix}
\begin{bmatrix}
1 & -3 & 0 & \vdots & 4 \\
0 & 1 & 3 & \vdots & 5 \\
0 & 0 & 1 & \vdots & 2
\end{bmatrix}
$$

The corresponding system of linear equations is

$$
\begin{aligned}
x - 3y \quad &= 4 \\
y + 3z &= 5 \\
z &= 2.
\end{aligned}
$$

Try using back-substitution on this system to see that you obtain the same solution that was obtained in Example 3.

GROUP ACTIVITY

ERROR ANALYSIS

One of your students has handed in the following steps for solving a system by Gauss-Jordan elimination. Find the error(s) in the solution and discuss how to explain the error(s) to your student.

$$
\begin{bmatrix}
1 & 1 & \vdots & 4 \\
2 & 3 & \vdots & 5
\end{bmatrix}
\; -2R_1 + R_2 \rightarrow
\begin{bmatrix}
1 & 1 & \vdots & 4 \\
0 & 1 & \vdots & 5
\end{bmatrix}
$$

$$
-1R_2 + R_1 \rightarrow
\begin{bmatrix}
1 & 0 & \vdots & 4 \\
0 & 1 & \vdots & 5
\end{bmatrix}
$$

The *Interactive* CD-ROM provides additional help with Warm-Up exercises by providing a hypertext link to the section in which the concept was introduced.

The *Interactive* CD-ROM contains step-by-step solutions to all odd-numbered Section and Review Exercises. It also provides Tutorial Exercises, which link to Guided Examples for additional help.

WARM UP

Evaluate the expression.

1. $2(-1) - 3(5) + 7(2)$

2. $-4(-3) + 6(7) + 8(-3)$

3. $11\left(\frac{1}{2}\right) - 7\left(-\frac{3}{2}\right) - 5(2)$

4. $\frac{2}{3}\left(\frac{1}{2}\right) + \frac{4}{3}\left(-\frac{1}{3}\right)$

Decide whether $(1, 3, -1)$ is a solution of the system.

5.
$$4x - 2y + 3z = -5$$
$$x + 3y - z = 11$$
$$-x + 2y = 5$$

6.
$$-x + 2y + z = 4$$
$$2x - 3z = 5$$
$$3x + 5y - 2z = 21$$

Use back-substitution to solve the system of linear equations.

7.
$$2x - 3y = 4$$
$$y = 2$$

8.
$$5x + 4y = 0$$
$$y = -3$$

9.
$$x - 3y + z = 0$$
$$y - 3z = 8$$
$$z = 2$$

10.
$$2x - 5y + 3z = -2$$
$$y - 4z = 0$$
$$z = 1$$

10.1 Exercises

In Exercises 1–6, determine the order of the matrix.

1. $\begin{bmatrix} 4 & -2 \\ 7 & 0 \\ 0 & 8 \end{bmatrix}$

2. $[5 \quad -3 \quad 8 \quad 7]$

3. $\begin{bmatrix} 2 \\ 36 \\ 3 \end{bmatrix}$

4. $\begin{bmatrix} -3 & 7 & 15 & 0 \\ 0 & 0 & 3 & 3 \\ 1 & 1 & 6 & 7 \end{bmatrix}$

5. $\begin{bmatrix} 33 & 45 \\ -9 & 20 \end{bmatrix}$

6. $[4]$

In Exercises 7–10, form the augmented matrix for the system of linear equations.

7.
$$4x - 3y = -5$$
$$-x + 3y = 12$$

8.
$$7x + 4y = 22$$
$$5x - 9y = 15$$

9.
$$x + 10y - 2z = 2$$
$$5x - 3y + 4z = 0$$
$$2x + y = 6$$

10.
$$7x - 5y + z = 13$$
$$19x - 8z = 10$$

In Exercises 11–14, write the system of linear equations represented by the augmented matrix. (Use variables x, y, z, and w.)

11. $\left[\begin{array}{cc:c} 1 & 2 & 7 \\ 2 & -3 & 4 \end{array}\right]$

12. $\left[\begin{array}{cc:c} 7 & -5 & 0 \\ 8 & 3 & -2 \end{array}\right]$

13. $\left[\begin{array}{ccc:c} 2 & 0 & 5 & -12 \\ 0 & 1 & -2 & 7 \\ 6 & 3 & 0 & 2 \end{array}\right]$

14. $\begin{bmatrix} 9 & 12 & 3 & 0 & \vdots & 0 \\ -2 & 18 & 5 & 2 & \vdots & 10 \\ 1 & 7 & -8 & 0 & \vdots & -4 \end{bmatrix}$

In Exercises 15–18, determine whether the matrix is in row-echelon form. If it is, determine if it is also in reduced row-echelon form.

15. $\begin{bmatrix} 1 & 0 & 0 & 0 \\ 0 & 1 & 1 & 5 \\ 0 & 0 & 0 & 0 \end{bmatrix}$ **16.** $\begin{bmatrix} 1 & 3 & 0 & 0 \\ 0 & 0 & 1 & 8 \\ 0 & 0 & 0 & 0 \end{bmatrix}$

17. $\begin{bmatrix} 2 & 0 & 4 & 0 \\ 0 & -1 & 3 & 6 \\ 0 & 0 & 1 & 5 \end{bmatrix}$ **18.** $\begin{bmatrix} 1 & 0 & 2 & 1 \\ 0 & 1 & -3 & 10 \\ 0 & 0 & 1 & 0 \end{bmatrix}$

In Exercises 19–22, fill in the blank(s) using elementary row operations to form a row-equivalent matrix.

19. $\begin{bmatrix} 1 & 4 & 3 \\ 2 & 10 & 5 \end{bmatrix}$ **20.** $\begin{bmatrix} 3 & 6 & 8 \\ 4 & -3 & 6 \end{bmatrix}$

$\begin{bmatrix} 1 & 4 & 3 \\ 0 & \boxed{} & -1 \end{bmatrix}$ $\begin{bmatrix} 1 & \boxed{} & \frac{8}{3} \\ 4 & -3 & 6 \end{bmatrix}$

21. $\begin{bmatrix} 1 & 1 & 4 & -1 \\ 3 & 8 & 10 & 3 \\ -2 & 1 & 12 & 6 \end{bmatrix}$

$\begin{bmatrix} 1 & 1 & 4 & -1 \\ 0 & 5 & \boxed{} & \boxed{} \\ 0 & 3 & \boxed{} & \boxed{} \end{bmatrix}$

$\begin{bmatrix} 1 & 1 & 4 & -1 \\ 0 & 1 & -\frac{2}{5} & \frac{6}{5} \\ 0 & 3 & \boxed{} & \boxed{} \end{bmatrix}$

22. $\begin{bmatrix} 2 & 4 & 8 & 3 \\ 1 & -1 & -3 & 2 \\ 2 & 6 & 4 & 9 \end{bmatrix}$

$\begin{bmatrix} 1 & \boxed{} & \boxed{} & \boxed{} \\ 5 & -1 & -3 & 2 \\ 2 & 6 & 4 & 9 \end{bmatrix}$

$\begin{bmatrix} 1 & 2 & 4 & \frac{3}{2} \\ 0 & \boxed{} & -7 & \frac{1}{2} \\ 0 & 2 & \boxed{} & \boxed{} \end{bmatrix}$

23. Perform the *sequence* of row operations on the matrix. What did the operations accomplish?

$\begin{bmatrix} 1 & 2 & 3 \\ 2 & -1 & -4 \\ 3 & 1 & -1 \end{bmatrix}$

(a) Add (-2) times Row 1 to Row 2.

(b) Add (-3) times Row 1 to Row 3.

(c) Add (-1) times Row 2 to Row 3.

(d) Multiply Row 2 by $\left(-\frac{1}{5}\right)$.

(e) Add (-2) times Row 2 to Row 1.

24. Perform the *sequence* of row operations on the matrix. What did the operations accomplish?

$\begin{bmatrix} 7 & 1 \\ 0 & 2 \\ -3 & 4 \\ 4 & 1 \end{bmatrix}$

(a) Add Row 3 to Row 4.

(b) Interchange Rows 1 and 4.

(c) Add (3) times Row 1 to Row 3.

(d) Add (-7) times Row 1 to Row 4.

(e) Multiply Row 2 by $\frac{1}{2}$.

(f) Add the appropriate multiple of Row 2 to Rows 1, 3, and 4.

In Exercises 25–28, write the matrix in row-echelon form. Remember that the row-echelon form for a matrix is not unique.

25. $\begin{bmatrix} 1 & 1 & 0 & 5 \\ -2 & -1 & 2 & -10 \\ 3 & 6 & 7 & 14 \end{bmatrix}$

26. $\begin{bmatrix} 1 & 2 & -1 & 3 \\ 3 & 7 & -5 & 14 \\ -2 & -1 & -3 & 8 \end{bmatrix}$

27. $\begin{bmatrix} 1 & -1 & -1 & 1 \\ 5 & -4 & 1 & 8 \\ -6 & 8 & 18 & 0 \end{bmatrix}$

28. $\begin{bmatrix} 1 & -3 & 0 & -7 \\ -3 & 10 & 1 & 23 \\ 4 & -10 & 2 & -24 \end{bmatrix}$

In Exercises 29–32, use the matrix capabilities of a graphing utility to write the matrix in *reduced* row-echelon form.

$$29. \begin{bmatrix} 3 & 3 & 3 \\ -1 & 0 & -4 \\ 2 & 4 & -2 \end{bmatrix} \qquad 30. \begin{bmatrix} 1 & 3 & 2 \\ 5 & 15 & 9 \\ 2 & 6 & 10 \end{bmatrix}$$

$$31. \begin{bmatrix} 1 & 2 & 3 & -5 \\ 1 & 2 & 4 & -9 \\ -2 & -4 & -4 & 3 \\ 4 & 8 & 11 & -14 \end{bmatrix} \qquad 32. \begin{bmatrix} 1 & -3 \\ -1 & 8 \\ 0 & 4 \\ -2 & 10 \end{bmatrix}$$

In Exercises 33–36, write the system of linear equations represented by the augmented matrix. Then use back-substitution to solve. (Use variables x, y, and z.)

$$33. \begin{bmatrix} 1 & -2 & \vdots & 4 \\ 0 & 1 & \vdots & -3 \end{bmatrix}$$

$$34. \begin{bmatrix} 1 & 5 & \vdots & 0 \\ 0 & 1 & \vdots & -1 \end{bmatrix}$$

$$35. \begin{bmatrix} 1 & -1 & 2 & \vdots & 4 \\ 0 & 1 & -1 & \vdots & 2 \\ 0 & 0 & 1 & \vdots & -2 \end{bmatrix}$$

$$36. \begin{bmatrix} 1 & 2 & -2 & \vdots & -1 \\ 0 & 1 & 1 & \vdots & 9 \\ 0 & 0 & 1 & \vdots & -3 \end{bmatrix}$$

In Exercises 37–40, an augmented matrix that represents a system of linear equations (in variables x, y, and z) has been reduced using Gauss-Jordan elimination. Write the solution represented by the augmented matrix.

$$37. \begin{bmatrix} 1 & 0 & \vdots & 7 \\ 0 & 1 & \vdots & -5 \end{bmatrix}$$

$$38. \begin{bmatrix} 1 & 0 & \vdots & -2 \\ 0 & 1 & \vdots & 4 \end{bmatrix}$$

$$39. \begin{bmatrix} 1 & 0 & 0 & \vdots & -4 \\ 0 & 1 & 0 & \vdots & -8 \\ 0 & 0 & 1 & \vdots & 2 \end{bmatrix}$$

$$40. \begin{bmatrix} 1 & 0 & 0 & \vdots & 3 \\ 0 & 1 & 0 & \vdots & -1 \\ 0 & 0 & 1 & \vdots & 0 \end{bmatrix}$$

In Exercises 41–56, solve the system of equations. Use Gaussian elimination with back-substitution or Gauss-Jordan elimination.

41. $x + 2y = 7$
     $2x + y = 8$

42. $2x + 6y = 16$
     $2x + 3y = 7$

43. $-3x + 5y = -22$
     $3x + 4y = 4$
     $4x - 8y = 32$

44. $x + 2y = 0$
     $x + y = 6$
     $3x - 2y = 8$

45. $8x - 4y = 7$
     $5x + 2y = 1$

46. $2x - y = -0.1$
     $3x + 2y = 1.6$

47. $-x + 2y = 1.5$
     $2x - 4y = 3$

48. $x - 3y = 5$
     $-2x + 6y = -10$

49. $x - 3z = -2$
     $3x + y - 2z = 5$
     $2x + 2y + z = 4$

50. $2x - y + 3z = 24$
     $2y - z = 14$
     $7x - 5y = 6$

51. $x + y - 5z = 3$
     $x - 2z = 1$
     $2x - y - z = 0$

52. $2x + 3z = 3$
     $4x - 3y + 7z = 5$
     $8x - 9y + 15z = 9$

53. $x + 2y + z = 8$
     $3x + 7y + 6z = 26$

54. $4x + 12y - 7z - 20w = 22$
     $3x + 9y - 5z - 28w = 30$

55. $x + 2y = 0$
     $-x - y = 0$

56. $x + 2y = 0$
     $2x + 4y = 0$

In Exercises 57–62, use the matrix capabilities of a graphing utility to reduce the augmented matrix corresponding to the system of equations, and solve the system.

57.
$$3x + 3y + 12z = 6$$
$$x + y + 4z = 2$$
$$2x + 5y + 20z = 10$$
$$-x + 2y + 8z = 4$$

58.
$$2x + 10y + 2z = 6$$
$$x + 5y + 2z = 6$$
$$x - 5y + z = 3$$
$$-3x - 15y - 3z = -9$$

59.
$$2x + y - z + 2w = -6$$
$$3x + 4y + w = 1$$
$$x + 5y + 2z + 6w = -3$$
$$5x + 2y - z - w = 3$$

60.
$$x + 2y + 2z + 4w = 11$$
$$3x + 6y + 5z + 12w = 30$$

61.
$$x + y + z = 0$$
$$2x + 3y + z = 0$$
$$3x + 5y + z = 0$$

62.
$$x + 2y + z + 3w = 0$$
$$x - y + w = 0$$
$$y - z + 2w = 0$$

63. **Think About It** The augmented matrix represents a system of linear equations (in variables x, y, and z) that has been reduced using Gauss-Jordan elimination. Write a system of equations with nonzero coefficients that is represented by the reduced matrix. (The answer is not unique.)

$$\begin{bmatrix} 1 & 0 & 3 & \vdots & -2 \\ 0 & 1 & 4 & \vdots & 1 \\ 0 & 0 & 0 & \vdots & 0 \end{bmatrix}$$

64. **Think About It**

(a) Describe the row-echelon form of an augmented matrix that corresponds to a system of linear equations that is inconsistent.

(b) Describe the row echelon form of an augmented matrix that corresponds to a system of linear equations that has an infinite number of solutions.

65. **Borrowing Money** A small corporation borrowed $1,500,000 to expand its product line. Some of the money was borrowed at 8%, some at 9%, and some at 12%. How much was borrowed at each rate if the annual interest was $133,000 and the amount borrowed at 8% was 4 times the amount borrowed at 12%?

66. **Borrowing Money** A small corporation borrowed $500,000 to expand its product line. Some of the money was borrowed at 9%, some at 10%, and some at 12%. How much was borrowed at each rate if the annual interest was $52,000 and the amount borrowed at 10% was $2\frac{1}{2}$ times the amount borrowed at 9%?

67. **Partial Fractions** Write the partial fraction decomposition for the rational expression

$$\frac{4x^2}{(x+1)^2(x-1)} = \frac{A}{x-1} + \frac{B}{x+1} + \frac{C}{(x+1)^2}.$$

68. **Electrical Network** The currents in an electrical network are given by the solution of the system

$$I_1 - I_2 + I_3 = 0$$
$$2I_1 + 2I_2 = 7$$
$$2I_2 + 4I_3 = 8$$

where I_1, I_2, and I_3 are measured in amperes. Solve the system of equations.

In Exercises 69 and 70, find the specified equation that passes through the points. Use a graphing utility to verify your results.

69. Parabola:

$$y = ax^2 + bx + c$$

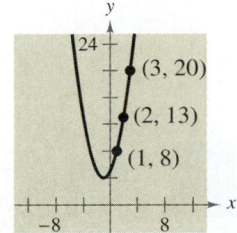

70. Parabola:

$$y = ax^2 + bx + c$$

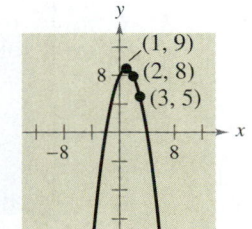

71. *Mathematical Modeling* A videotape of the path of a ball thrown by a baseball player was analyzed with a grid covering the TV screen (see figure). The tape was paused three times, and the position of the ball was measured each time. The coordinates were approximately $(0, 5.0)$, $(15, 9.6)$, and $(30, 12.4)$. (The x-coordinate measures the horizontal distance from the player in feet, and the y-coordinate is the height of the ball in feet.)

(a) Find the equation of the parabola $y = ax^2 + bx + c$ that passes through the three points.

(b) Use a graphing utility to graph the parabola. Approximate the maximum height of the ball and the point at which the ball struck the ground.

(c) Find analytically the maximum height of the ball and the point at which it struck the ground.

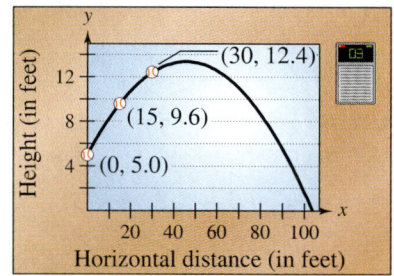

72. *Reading a Graph* The bar graph gives the value y, in millions of dollars, for new orders of civil jet transport aircraft built by U.S. companies for the years 1990 through 1992. (Source: Aerospace Industries Association of America)

(a) Find the equation of the parabola that passes through the points. Let $t = 0$ represent 1990.

(b) Use a graphing utility to graph the parabola.

(c) Use the equation in part (a) to estimate y in 1993.

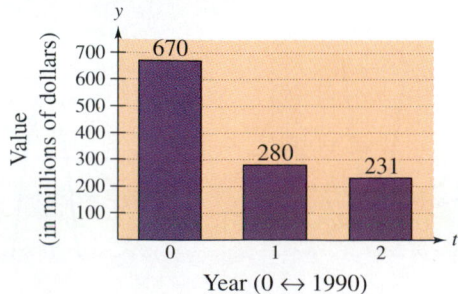

Network Analysis In Exercises 73 and 74, answer the questions about the specified network. (In a network it is assumed that the total flow into each junction is equal to the total flow out of each junction.)

73. Water flowing through a network of pipes (in thousands of cubic meters per hour) is shown in the figure.

(a) Solve this system for the water flow represented by x_i, $i = 1, 2, \ldots, 7$.

(b) Find the network flow pattern when $x_6 = x_7 = 0$.

(c) Find the network flow pattern when $x_5 = 1000$ and $x_6 = 0$.

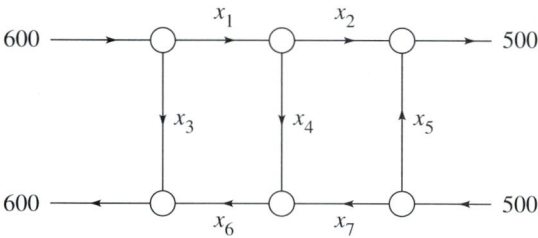

74. The flow of traffic (in vehicles per hour) through a network of streets is shown in the figure.

(a) Solve this system for the traffic flow represented by x_i, $i = 1, 2, \ldots, 5$.

(b) Find the traffic flow when $x_2 = 200$ and $x_3 = 50$.

(c) Find the traffic flow when $x_2 = 150$ and $x_3 = 0$.

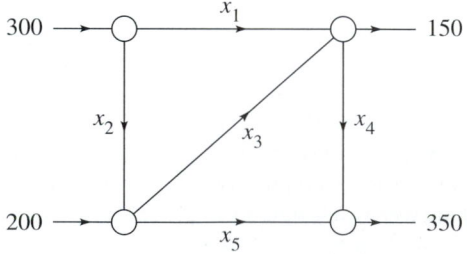

Review Solve Exercises 75–78 as a review of the skills and problem-solving techniques you learned in previous sections. Graph the function.

75. $f(x) = 2^{x-1}$

76. $g(x) = 3^{-x/2}$

77. $h(x) = \log_2(x - 1)$

78. $f(x) = 3 + \ln x$

10.2 *Operations with Matrices*

See Exercise 61 on page 758 for an example of how matrix multiplication can be used to help analyze the labor and wage requirements for a boat manufacturer.

Equality of Matrices □ *Matrix Addition and Scalar Multiplication* □
Matrix Multiplication □ *Applications*

Equality of Matrices

In Section 10.1, you used matrices to solve systems of linear equations. Matrices, however, can do much more than this. There is a rich mathematical theory of matrices, and its applications are numerous. This section and the next two introduce some fundamentals of matrix theory. It is standard mathematical convention to represent matrices in any of the following three ways.

1. A matrix can be denoted by an uppercase letter such as *A*, *B*, or *C*.
2. A matrix can be denoted by a representative element enclosed in brackets, such as $[a_{ij}]$, $[b_{ij}]$, or $[c_{ij}]$.
3. A matrix can be denoted by a rectangular array of numbers such as

$$A = [a_{ij}] = \begin{bmatrix} a_{11} & a_{12} & a_{13} & \cdots & a_{1n} \\ a_{21} & a_{22} & a_{23} & \cdots & a_{2n} \\ a_{31} & a_{32} & a_{33} & \cdots & a_{3n} \\ \vdots & \vdots & \vdots & & \vdots \\ a_{m1} & a_{m2} & a_{m3} & \cdots & a_{mn} \end{bmatrix}.$$

Two matrices $A = [a_{ij}]$ and $B = [b_{ij}]$ are **equal** if they have the same order $(m \times n)$ and $a_{ij} = b_{ij}$ for $1 \le i \le m$ and $1 \le j \le n$. In other words, two matrices are equal if their corresponding entries are equal.

A British mathematician, Arthur Cayley, invented matrices around 1858. Cayley was a Cambridge University graduate and a lawyer by profession. His ground-breaking work on matrices was begun as he studied the theory of transformations. Cayley also was instrumental in the development of determinants. Cayley and two American mathematicians, Benjamin Peirce (1809–1880) and his son Charles S. Peirce (1839–1914), are credited with developing "matrix algebra."

EXAMPLE 1   *Equality of Matrices*

Solve for a_{11}, a_{12}, a_{21}, and a_{22} in the following matrix equation.

$$\begin{bmatrix} a_{11} & a_{12} \\ a_{21} & a_{22} \end{bmatrix} = \begin{bmatrix} 2 & -1 \\ -3 & 0 \end{bmatrix}$$

Solution

Because two matrices are equal only if their corresponding entries are equal, you can conclude that

$$a_{11} = 2, \quad a_{12} = -1, \quad a_{21} = -3, \quad \text{and} \quad a_{22} = 0.$$

Matrix Addition and Scalar Multiplication

You can **add** two matrices (of the same order) by adding their corresponding entries.

> **DEFINITION OF MATRIX ADDITION**
>
> If $A = [a_{ij}]$ and $B = [b_{ij}]$ are matrices of order $m \times n$, their **sum** is the $m \times n$ matrix given by
>
> $$A + B = [a_{ij} + b_{ij}].$$
>
> The sum of two matrices of different orders is undefined.

EXAMPLE 2   Addition of Matrices

a. $\begin{bmatrix} -1 & 2 \\ 0 & 1 \end{bmatrix} + \begin{bmatrix} 1 & 3 \\ -1 & 2 \end{bmatrix} = \begin{bmatrix} -1+1 & 2+3 \\ 0-1 & 1+2 \end{bmatrix} = \begin{bmatrix} 0 & 5 \\ -1 & 3 \end{bmatrix}$

b. $\begin{bmatrix} 0 & 1 & -2 \\ 1 & 2 & 3 \end{bmatrix} + \begin{bmatrix} 0 & 0 & 0 \\ 0 & 0 & 0 \end{bmatrix} = \begin{bmatrix} 0 & 1 & -2 \\ 1 & 2 & 3 \end{bmatrix}$

c. $\begin{bmatrix} 1 \\ -3 \\ -2 \end{bmatrix} + \begin{bmatrix} -1 \\ 3 \\ 2 \end{bmatrix} = \begin{bmatrix} 0 \\ 0 \\ 0 \end{bmatrix}$

d. The sum of

$$A = \begin{bmatrix} 2 & 1 & 0 \\ 4 & 0 & -1 \\ 3 & -2 & 2 \end{bmatrix} \quad \text{and} \quad B = \begin{bmatrix} 0 & 1 \\ -1 & 3 \\ 2 & 4 \end{bmatrix}$$

is undefined.

In operations with matrices, numbers are usually referred to as **scalars.** In this text, scalars will always be real numbers. You can multiply a matrix A by a scalar c by multiplying each entry in A by c.

> **DEFINITION OF SCALAR MULTIPLICATION**
>
> If $A = [a_{ij}]$ is an $m \times n$ matrix and c is a scalar, the **scalar multiple** of A by c is the $m \times n$ matrix given by
>
> $$cA = [ca_{ij}].$$

The symbol $-A$ represents the scalar product $(-1)A$. Moreover, if A and B are of the same order, then $A - B$ represents the sum of A and $(-1)B$. That is,

$$A - B = A + (-1)B.$$ Subtraction of matrices

EXAMPLE 3 Scalar Multiplication and Matrix Subtraction

For the following matrices, find (a) $3A$, (b) $-B$, and (c) $3A - B$.

$$A = \begin{bmatrix} 2 & 2 & 4 \\ -3 & 0 & -1 \\ 2 & 1 & 2 \end{bmatrix} \quad \text{and} \quad B = \begin{bmatrix} 2 & 0 & 0 \\ 1 & -4 & 3 \\ -1 & 3 & 2 \end{bmatrix}$$

Solution

a. $3A = 3 \begin{bmatrix} 2 & 2 & 4 \\ -3 & 0 & -1 \\ 2 & 1 & 2 \end{bmatrix}$ Scalar multiplication

$= \begin{bmatrix} 3(2) & 3(2) & 3(4) \\ 3(-3) & 3(0) & 3(-1) \\ 3(2) & 3(1) & 3(2) \end{bmatrix}$ Multiply each entry by 3.

$= \begin{bmatrix} 6 & 6 & 12 \\ -9 & 0 & -3 \\ 6 & 3 & 6 \end{bmatrix}$ Simplify.

b. $-B = (-1) \begin{bmatrix} 2 & 0 & 0 \\ 1 & -4 & 3 \\ -1 & 3 & 2 \end{bmatrix}$ Definition of negation

$= \begin{bmatrix} -2 & 0 & 0 \\ -1 & 4 & -3 \\ 1 & -3 & -2 \end{bmatrix}$ Multiply each entry by -1.

c. $3A - B = \begin{bmatrix} 6 & 6 & 12 \\ -9 & 0 & -3 \\ 6 & 3 & 6 \end{bmatrix} - \begin{bmatrix} 2 & 0 & 0 \\ 1 & -4 & 3 \\ -1 & 3 & 2 \end{bmatrix}$ Matrix subtraction

$= \begin{bmatrix} 4 & 6 & 12 \\ -10 & 4 & -6 \\ 7 & 0 & 4 \end{bmatrix}$ Subtract corresponding entries.

It is often convenient to rewrite the scalar multiple cA by factoring c out of every entry in the matrix. For instance, in the following example, the scalar $\frac{1}{2}$ has been factored out of the matrix.

$$\begin{bmatrix} \frac{1}{2} & -\frac{3}{2} \\ \frac{5}{2} & \frac{1}{2} \end{bmatrix} = \frac{1}{2} \begin{bmatrix} 1 & -3 \\ 5 & 1 \end{bmatrix}$$

The properties of matrix addition and scalar multiplication are similar to those of addition and multiplication of real numbers.

PROPERTIES OF MATRIX ADDITION AND SCALAR MULTIPLICATION

Let A, B, and C be $m \times n$ matrices and let c and d be scalars.

1. $A + B = B + A$	Commutative Property of Matrix Addition
2. $A + (B + C) = (A + B) + C$	Associative Property of Matrix Addition
3. $(cd)A = c(dA)$	Associative Property of Scalar Multiplication
4. $1A = A$	Scalar Identity
5. $c(A + B) = cA + cB$	Distributive Property
6. $(c + d)A = cA + dA$	Distributive Property

Note that the Associative Property of Matrix Addition allows you to write expressions such as $A + B + C$ without ambiguity because the same sum occurs no matter how the matrices are grouped. In other words, you obtain the same sum whether you group $A + B + C$ as $(A + B) + C$ or as $A + (B + C)$. This same reasoning applies to sums of four or more matrices.

EXAMPLE 4   *Addition of More Than Two Matrices*

By adding corresponding entries, you obtain the following sum of four matrices.

$$\begin{bmatrix} 1 \\ 2 \\ -3 \end{bmatrix} + \begin{bmatrix} -1 \\ -1 \\ 2 \end{bmatrix} + \begin{bmatrix} 0 \\ 1 \\ 4 \end{bmatrix} + \begin{bmatrix} 2 \\ -3 \\ -2 \end{bmatrix} = \begin{bmatrix} 2 \\ -1 \\ 1 \end{bmatrix}$$

TECHNOLOGY

Most graphing utilities can add and subtract matrices and multiply matrices by scalars. For instance, on a *TI-83* or a *TI-82*, you can find the sum of the matrices

$$A = \begin{bmatrix} 2 & -3 \\ -1 & 0 \end{bmatrix} \quad \text{and} \quad B = \begin{bmatrix} -1 & 4 \\ 2 & -5 \end{bmatrix}$$

by entering the matrices and then using the following keystrokes.

$[A]$ $\boxed{+}$ $[B]$ $\boxed{\text{ENTER}}$

One important property of addition of real numbers is that the number 0 is the additive identity. That is, $c + 0 = c$ for any real number c. For matrices, a similar property holds. That is, if A is an $m \times n$ matrix and O is the $m \times n$ **zero matrix** consisting entirely of zeros, then $A + O = A$.

In other words, O is the **additive identity** for the set of all $m \times n$ matrices. For example, the following matrices are the additive identities for the set of all 2×3 and 2×2 matrices.

$$O = \begin{bmatrix} 0 & 0 & 0 \\ 0 & 0 & 0 \end{bmatrix} \quad \text{and} \quad O = \begin{bmatrix} 0 & 0 \\ 0 & 0 \end{bmatrix}$$

Zero 2×3 matrix $\qquad$ Zero 2×2 matrix

NOTE The algebra of real numbers and the algebra of matrices also have important differences, which will be discussed later. ■■

The algebra of real numbers and the algebra of matrices have many similarities. For example, compare the following solutions.

Real Numbers
(Solve for x.)

$x + a = b$
$x + a + (-a) = b + (-a)$
$x + 0 = b - a$
$x = b - a$

m × n Matrices
(Solve for X.)

$X + A = B$
$X + A + (-A) = B + (-A)$
$X + O = B - A$
$X = B - A$

EXAMPLE 5 *Solving a Matrix Equation*

Solve for X in the equation $3X + A = B$, where

$$A = \begin{bmatrix} 1 & -2 \\ 0 & 3 \end{bmatrix} \quad \text{and} \quad B = \begin{bmatrix} -3 & 4 \\ 2 & 1 \end{bmatrix}.$$

Solution

Begin by solving the equation for X to obtain

$$3X = B - A \quad \Longrightarrow \quad X = \frac{1}{3}(B - A).$$

Now, using the matrices A and B, you have

$$X = \frac{1}{3}\left(\begin{bmatrix} -3 & 4 \\ 2 & 1 \end{bmatrix} - \begin{bmatrix} 1 & -2 \\ 0 & 3 \end{bmatrix}\right)$$
$$= \frac{1}{3}\begin{bmatrix} -4 & 6 \\ 2 & -2 \end{bmatrix}$$
$$= \begin{bmatrix} -\frac{4}{3} & 2 \\ \frac{2}{3} & -\frac{2}{3} \end{bmatrix}.$$

Matrix Multiplication

The third basic matrix operation is **matrix multiplication.** At first glance the definition may seem unusual. You will see later, however, that this definition of the product of two matrices has many practical applications.

NOTE The definition of matrix multiplication indicates a *row-by-column* multiplication, where the entry in the ith row and jth column of the product AB is obtained by multiplying the entries in the ith row of A by the corresponding entries in the jth column of B and then adding the results. Example 6 illustrates this process. ■■

> **DEFINITION OF MATRIX MULTIPLICATION**
>
> If $A = [a_{ij}]$ is an $m \times n$ matrix and $B = [b_{ij}]$ is an $n \times p$ matrix, the **product** AB is an $m \times p$ matrix
>
> $$AB = [c_{ij}]$$
>
> where $c_{ij} = a_{i1}b_{1j} + a_{i2}b_{2j} + a_{i3}b_{3j} + \cdots + a_{in}b_{nj}$.

TECHNOLOGY ■■

Some graphing utilities, such as the *TI-83* or the *TI-82*, are able to add, subtract, and multiply matrices. If you have such a graphing utility, enter the matrices

$$A = \begin{bmatrix} 1 & 2 & 3 \\ 2 & -5 & 1 \end{bmatrix} \text{ and}$$

$$B = \begin{bmatrix} -3 & 2 & 1 \\ 4 & -2 & 0 \\ 1 & 2 & 3 \end{bmatrix}$$

and use the following keystrokes to find the product of the matrices.

$$[A] \boxed{\times} [B] \boxed{\text{ENTER}}$$

You should get:

$$\begin{bmatrix} 8 & 4 & 10 \\ -25 & 16 & 5 \end{bmatrix}.$$

EXAMPLE 6   *Finding the Product of Two Matrices*

Find the product AB where

$$A = \begin{bmatrix} -1 & 3 \\ 4 & -2 \\ 5 & 0 \end{bmatrix} \text{ and } B = \begin{bmatrix} -3 & 2 \\ -4 & 1 \end{bmatrix}.$$

Solution

First, note that the product AB is defined because the number of columns of A is equal to the number of rows of B. Moreover, the product AB has order 3×2, and is of the form

$$\begin{bmatrix} -1 & 3 \\ 4 & -2 \\ 5 & 0 \end{bmatrix}\begin{bmatrix} -3 & 2 \\ -4 & 1 \end{bmatrix} = \begin{bmatrix} c_{11} & c_{12} \\ c_{21} & c_{22} \\ c_{31} & c_{32} \end{bmatrix}.$$

To find the entries of the product, multiply each row of A by each column of B, as follows. Use a graphing utility to check this result.

$$AB = \begin{bmatrix} -1 & 3 \\ 4 & -2 \\ 5 & 0 \end{bmatrix}\begin{bmatrix} -3 & 2 \\ -4 & 1 \end{bmatrix}$$

$$= \begin{bmatrix} (-1)(-3) + (3)(-4) & (-1)(2) + (3)(1) \\ (4)(-3) + (-2)(-4) & (4)(2) + (-2)(1) \\ (5)(-3) + (0)(-4) & (5)(2) + (0)(1) \end{bmatrix}$$

$$= \begin{bmatrix} -9 & 1 \\ -4 & 6 \\ -15 & 10 \end{bmatrix}$$

Be sure you understand that for the product of two matrices to be defined, the number of columns of the first matrix must equal the number of rows of the second matrix. That is, the middle two indices must be the same and the outside two indices give the order of the product, as shown below.

$$A \qquad B \qquad = \qquad AB$$
$$m \times n \qquad n \times p \qquad \qquad m \times p$$

Equal
Order of AB

EXAMPLE 7 Matrix Multiplication

a. $\begin{bmatrix} 1 & 0 & 3 \\ 2 & -1 & -2 \end{bmatrix} \begin{bmatrix} -2 & 4 & 2 \\ 1 & 0 & 0 \\ -1 & 1 & -1 \end{bmatrix} = \begin{bmatrix} -5 & 7 & -1 \\ -3 & 6 & 6 \end{bmatrix}$

$\qquad 2 \times 3 \qquad\qquad 3 \times 3 \qquad\qquad 2 \times 3$

b. $\begin{bmatrix} 3 & 4 \\ -2 & 5 \end{bmatrix} \begin{bmatrix} 1 & 0 \\ 0 & 1 \end{bmatrix} = \begin{bmatrix} 3 & 4 \\ -2 & 5 \end{bmatrix}$

$\qquad 2 \times 2 \quad 2 \times 2 \qquad 2 \times 2$

c. $\begin{bmatrix} 1 & 2 \\ 1 & 1 \end{bmatrix} \begin{bmatrix} -1 & 2 \\ 1 & -1 \end{bmatrix} = \begin{bmatrix} 1 & 0 \\ 0 & 1 \end{bmatrix}$

$\qquad 2 \times 2 \quad 2 \times 2 \qquad 2 \times 2$

d. $\begin{bmatrix} 1 & -2 & -3 \end{bmatrix} \begin{bmatrix} 2 \\ -1 \\ 1 \end{bmatrix} = \begin{bmatrix} 1 \end{bmatrix}$

$\qquad 1 \times 3 \qquad 3 \times 1 \quad 1 \times 1$

e. $\begin{bmatrix} 2 \\ -1 \\ 1 \end{bmatrix} \begin{bmatrix} 1 & -2 & -3 \end{bmatrix} = \begin{bmatrix} 2 & -4 & -6 \\ -1 & 2 & 3 \\ 1 & -2 & -3 \end{bmatrix}$

$\quad 3 \times 1 \qquad 1 \times 3 \qquad\qquad 3 \times 3$

f. The product AB for the following matrices is not defined.

$$A = \begin{bmatrix} -2 & 1 \\ 1 & -3 \\ 1 & 4 \end{bmatrix} \text{ and } B = \begin{bmatrix} -2 & 3 & 1 & 4 \\ 0 & 1 & -1 & 2 \\ 2 & -1 & 0 & 1 \end{bmatrix}$$

$\qquad\qquad 3 \times 2 \qquad\qquad\qquad\qquad 3 \times 4$

NOTE In parts (d) and (e) of Example 7, note that the two products are different. Matrix multiplication is not, in general, commutative. That is, for most matrices, $AB \neq BA$. ■■

The general pattern for matrix multiplication is as follows. To obtain the entry in the ith row and the jth column of the product AB, use the ith row of A and the jth column of B.

$$\begin{bmatrix} a_{11} & a_{12} & a_{13} & \cdots & a_{1n} \\ a_{21} & a_{22} & a_{23} & \cdots & a_{2n} \\ a_{31} & a_{32} & a_{33} & \cdots & a_{3n} \\ \vdots & \vdots & \vdots & & \vdots \\ a_{i1} & a_{i2} & a_{i3} & \cdots & a_{in} \\ \vdots & \vdots & \vdots & & \vdots \\ a_{m1} & a_{m2} & a_{m3} & \cdots & a_{mn} \end{bmatrix} \begin{bmatrix} b_{11} & b_{12} & \cdots & b_{1j} & \cdots & b_{1p} \\ b_{21} & b_{22} & \cdots & b_{2j} & \cdots & b_{2p} \\ b_{31} & b_{32} & \cdots & b_{3j} & \cdots & b_{3p} \\ \vdots & \vdots & & \vdots & & \vdots \\ b_{n1} & b_{n2} & \cdots & b_{nj} & \cdots & b_{np} \end{bmatrix} = \begin{bmatrix} c_{11} & c_{12} & \cdots & c_{1j} & \cdots & c_{1p} \\ c_{21} & c_{22} & \cdots & c_{2j} & \cdots & c_{2p} \\ \vdots & \vdots & & \vdots & & \vdots \\ c_{i1} & c_{i2} & \cdots & c_{ij} & \cdots & c_{ip} \\ \vdots & \vdots & & \vdots & & \vdots \\ c_{m1} & c_{m2} & \cdots & c_{mj} & \cdots & c_{mp} \end{bmatrix}$$

$$a_{i1}b_{1j} + a_{i2}b_{2j} + a_{i3}b_{3j} + \cdots + a_{in}b_{nj} = c_{ij}$$

PROPERTIES OF MATRIX MULTIPLICATION

Let A, B, and C be matrices and let c be a scalar.

1. $A(BC) = (AB)C$ — Associative Property of Multiplication
2. $A(B + C) = AB + AC$ — Distributive Property
3. $(A + B)C = AC + BC$ — Distributive Property
4. $c(AB) = (cA)B = A(cB)$

The $n \times n$ matrix that consists of 1's on its main diagonal and 0's elsewhere is called the **identity matrix of order n** and is denoted by

$$I_n = \begin{bmatrix} 1 & 0 & 0 & \cdots & 0 \\ 0 & 1 & 0 & \cdots & 0 \\ 0 & 0 & 1 & \cdots & 0 \\ \vdots & \vdots & \vdots & & \vdots \\ 0 & 0 & 0 & \cdots & 1 \end{bmatrix}.$$ Identity matrix

Note that an identity matrix must be *square*. When the order is understood to be n, you can denote I_n simply by I. If A is an $n \times n$ matrix, the identity matrix has the property that $AI_n = A$ and $I_nA = A$. For example,

$$\begin{bmatrix} 3 & -2 & 5 \\ 1 & 0 & 4 \\ -1 & 2 & -3 \end{bmatrix} \begin{bmatrix} 1 & 0 & 0 \\ 0 & 1 & 0 \\ 0 & 0 & 1 \end{bmatrix} = \begin{bmatrix} 3 & -2 & 5 \\ 1 & 0 & 4 \\ -1 & 2 & -3 \end{bmatrix}$$

and

$$\begin{bmatrix} 1 & 0 & 0 \\ 0 & 1 & 0 \\ 0 & 0 & 1 \end{bmatrix} \begin{bmatrix} 3 & -2 & 5 \\ 1 & 0 & 4 \\ -1 & 2 & -3 \end{bmatrix} = \begin{bmatrix} 3 & -2 & 5 \\ 1 & 0 & 4 \\ -1 & 2 & -3 \end{bmatrix}.$$

Applications

One application of matrix multiplication is representation of a system of linear equations. Note how the system

$$a_{11}x_1 + a_{12}x_2 + a_{13}x_3 = b_1$$
$$a_{21}x_1 + a_{22}x_2 + a_{23}x_3 = b_2$$
$$a_{31}x_1 + a_{32}x_2 + a_{33}x_3 = b_3$$

can be written as the matrix equation $AX = B$, where A is the *coefficient matrix* of the system, and X and B are column matrices.

$$\begin{bmatrix} a_{11} & a_{12} & a_{13} \\ a_{21} & a_{22} & a_{23} \\ a_{31} & a_{32} & a_{33} \end{bmatrix} \begin{bmatrix} x_1 \\ x_2 \\ x_3 \end{bmatrix} = \begin{bmatrix} b_1 \\ b_2 \\ b_3 \end{bmatrix}$$

$$A \quad\quad \times \quad X \;=\; B$$

EXAMPLE 8 *Solving a System of Linear Equations*

Solve the matrix equation $AX = B$ for X, where

Coefficient matrix   Constant matrix

$$A = \begin{bmatrix} 1 & -2 & 1 \\ 0 & 1 & 2 \\ 2 & 3 & -2 \end{bmatrix} \quad \text{and} \quad B = \begin{bmatrix} -4 \\ 4 \\ 2 \end{bmatrix}.$$

Solution

As a system of linear equations, $AX = B$ is as follows.

$$x_1 - 2x_2 + x_3 = -4$$
$$x_2 + 2x_3 = 4$$
$$2x_1 + 3x_2 - 2x_3 = 2$$

Using Gauss-Jordan elimination on the augmented matrix of this system, you obtain the following reduced row-echelon matrix.

$$\begin{bmatrix} 1 & 0 & 0 & \vdots & -1 \\ 0 & 1 & 0 & \vdots & 2 \\ 0 & 0 & 1 & \vdots & 1 \end{bmatrix}$$

Thus, the solution of the system of linear equations is $x_1 = -1$, $x_2 = 2$, and $x_3 = 1$, and the solution of the matrix equation is

$$X = \begin{bmatrix} x_1 \\ x_2 \\ x_3 \end{bmatrix} = \begin{bmatrix} -1 \\ 2 \\ 1 \end{bmatrix}.$$

Check this solution in the original system of equations.

*E*XAMPLE 9   *Softball Team Expenses*

Two softball teams submit equipment lists to their sponsors.

	Women's Team	Men's Team
Bats	12	15
Balls	45	38
Gloves	15	17

Each bat costs $48, each ball costs $4, and each glove costs $42. Use matrices to find the total cost of equipment for each team.

The U.S.A. Softball Women's National Team won the International Softball Federation World Championship for the third consecutive time in 1994. Their 1995 Pan American Games win pushed the team's international win streak to 105–0. *(Photo: USA Softball Photo)*

Solution

The equipment lists and the costs per item can be written in matrix form as

$$E = \begin{bmatrix} 12 & 15 \\ 45 & 38 \\ 15 & 17 \end{bmatrix} \quad \text{and} \quad C = \begin{bmatrix} 48 & 4 & 42 \end{bmatrix}.$$

The total cost of equipment for each team is given by the product

$$CE = \begin{bmatrix} 48 & 4 & 42 \end{bmatrix} \begin{bmatrix} 12 & 15 \\ 45 & 38 \\ 15 & 17 \end{bmatrix} = \begin{bmatrix} 1386 & 1586 \end{bmatrix}.$$

Thus, the total cost of equipment for the women's team is $1386, and the total cost of equipment for the men's team is $1586.

GROUP ACTIVITY

PROBLEM POSING

Write a matrix multiplication application problem that uses the matrix

$$A = \begin{bmatrix} 20 & 42 & 33 \\ 17 & 30 & 50 \end{bmatrix}.$$

Exchange problems with another student in your class. Form the matrices that represent the problem, and solve the problem. Interpret your solution in the context of the problem. Check with the creator of the problem to see if you are correct. Discuss other ways to represent and/or approach the problem.

Evaluate the expression.

1. $-3\left(-\frac{5}{6}\right) + 10\left(-\frac{3}{4}\right)4$

2. $-22\left(\frac{5}{2}\right) + 6(8)$

Decide whether the matrix is in *reduced* row-echelon form.

3. $\begin{bmatrix} 0 & 1 & 0 & -5 \\ 1 & 0 & 3 & 2 \\ 0 & 0 & 1 & 0 \end{bmatrix}$

4. $\begin{bmatrix} 1 & 0 & 0 & 2 & 3 \\ 0 & 0 & 0 & 0 & 0 \\ 0 & 1 & 1 & 3 & 10 \end{bmatrix}$

Write the augmented matrix for the system of linear equations.

5. $-5x + 10y = 12$
$7x - 3y = 0$
$-x + 7y = 25$

6. $10x + 15y - 9z = 42$
$6x - 5y = 0$

Solve the system of linear equations represented by the matrix.

7. $\begin{bmatrix} 1 & 0 & \vdots & 0 \\ 0 & 1 & \vdots & 2 \end{bmatrix}$

8. $\begin{bmatrix} 1 & 0 & -1 & \vdots & 2 \\ 0 & 1 & 1 & \vdots & 3 \end{bmatrix}$

9. $\begin{bmatrix} 1 & 2 & 1 & \vdots & 0 \\ 0 & 0 & 1 & \vdots & -1 \\ 0 & 0 & 0 & \vdots & 0 \end{bmatrix}$

10. $\begin{bmatrix} 1 & -1 & 0 & \vdots & 3 \\ 0 & 1 & -2 & \vdots & 1 \\ 0 & 0 & 1 & \vdots & -1 \end{bmatrix}$

10.2 Exercises

In Exercises 1–4, find x and y.

1. $\begin{bmatrix} x & -2 \\ 7 & y \end{bmatrix} = \begin{bmatrix} -4 & -2 \\ 7 & 22 \end{bmatrix}$

2. $\begin{bmatrix} -5 & x \\ y & 8 \end{bmatrix} = \begin{bmatrix} -5 & 13 \\ 12 & 8 \end{bmatrix}$

3. $\begin{bmatrix} 16 & 4 & 5 & 4 \\ -3 & 13 & 15 & 6 \\ 0 & 2 & 4 & 0 \end{bmatrix} = \begin{bmatrix} 16 & 4 & 2x+1 & 4 \\ -3 & 13 & 15 & 3x \\ 0 & 2 & 3y-5 & 0 \end{bmatrix}$

4. $\begin{bmatrix} x+2 & 8 & -3 \\ 1 & 2y & 2x \\ 7 & -2 & y+2 \end{bmatrix} = \begin{bmatrix} 2x+6 & 8 & -3 \\ 1 & 18 & 8 \\ 7 & -2 & 11 \end{bmatrix}$

In Exercises 5–10, find (a) $A + B$, (b) $A - B$, (c) $3A$, and (d) $3A - 2B$.

5. $A = \begin{bmatrix} 1 & -1 \\ 2 & -1 \end{bmatrix}$, $B = \begin{bmatrix} 2 & -1 \\ -1 & 8 \end{bmatrix}$

6. $A = \begin{bmatrix} 1 & 2 \\ 2 & 1 \end{bmatrix}$, $B = \begin{bmatrix} -3 & -2 \\ 4 & 2 \end{bmatrix}$

7. $A = \begin{bmatrix} 6 & -1 \\ 2 & 4 \\ -3 & 5 \end{bmatrix}$, $B = \begin{bmatrix} 1 & 4 \\ -1 & 5 \\ 1 & 10 \end{bmatrix}$

8. $A = \begin{bmatrix} 2 & 1 & 1 \\ -1 & -1 & 4 \end{bmatrix}$, $B = \begin{bmatrix} 2 & -3 & 4 \\ -3 & 1 & -2 \end{bmatrix}$

9. $A = \begin{bmatrix} 2 & 2 & -1 & 0 & 1 \\ 1 & 1 & -2 & 0 & -1 \end{bmatrix}$,

$B = \begin{bmatrix} 1 & 1 & -1 & 1 & 0 \\ -3 & 4 & 9 & -6 & -7 \end{bmatrix}$

10. $A = \begin{bmatrix} 3 \\ 2 \\ -1 \end{bmatrix}$, $B = \begin{bmatrix} -4 \\ 6 \\ 2 \end{bmatrix}$

In Exercises 11–14, solve for X given

$$A = \begin{bmatrix} -2 & -1 \\ 1 & 0 \\ 3 & -4 \end{bmatrix} \text{ and } B = \begin{bmatrix} 0 & 3 \\ 2 & 0 \\ -4 & -1 \end{bmatrix}.$$

11. $X = 3A - 2B$ **12.** $2X = 2A - B$

13. $2X + 3A = B$ **14.** $2A + 4B = -2X$

In Exercises 15–20, find (a) AB, (b) BA, and, if possible, (c) A^2. (*Note:* $A^2 = AA$.)

15. $A = \begin{bmatrix} 1 & 2 \\ 4 & 2 \end{bmatrix}$, $B = \begin{bmatrix} 2 & -1 \\ -1 & 8 \end{bmatrix}$

16. $A = \begin{bmatrix} 2 & -1 \\ 1 & 4 \end{bmatrix}$, $B = \begin{bmatrix} 0 & 0 \\ 3 & -3 \end{bmatrix}$

17. $A = \begin{bmatrix} 3 & -1 \\ 1 & 3 \end{bmatrix}$, $B = \begin{bmatrix} 1 & -3 \\ 3 & 1 \end{bmatrix}$

18. $A = \begin{bmatrix} 1 & -1 \\ 1 & 1 \end{bmatrix}$, $B = \begin{bmatrix} 1 & 3 \\ -3 & 1 \end{bmatrix}$

19. $A = \begin{bmatrix} 1 & -1 & 7 \\ 2 & -1 & 8 \\ 3 & 1 & -1 \end{bmatrix}$, $B = \begin{bmatrix} 1 & 1 & 2 \\ 2 & 1 & 1 \\ 1 & -3 & 2 \end{bmatrix}$

20. $A = \begin{bmatrix} 3 & 2 & 1 \end{bmatrix}$, $B = \begin{bmatrix} 2 \\ 3 \\ 0 \end{bmatrix}$

In Exercises 21–28, find AB, if possible.

21. $A = \begin{bmatrix} 2 & 1 \\ -3 & 4 \\ 1 & 6 \end{bmatrix}$, $B = \begin{bmatrix} 0 & -1 & 0 \\ 4 & 0 & 2 \\ 8 & -1 & 7 \end{bmatrix}$

22. $A = \begin{bmatrix} 0 & -1 & 0 \\ 4 & 0 & 2 \\ 8 & -1 & 7 \end{bmatrix}$, $B = \begin{bmatrix} 2 & 1 \\ -3 & 4 \\ 1 & 6 \end{bmatrix}$

23. $A = \begin{bmatrix} -1 & 3 \\ 4 & -5 \\ 0 & 2 \end{bmatrix}$, $B = \begin{bmatrix} 1 & 2 \\ 0 & 7 \end{bmatrix}$

24. $A = \begin{bmatrix} 1 & 0 & 0 \\ 0 & 4 & 0 \\ 0 & 0 & -2 \end{bmatrix}$, $B = \begin{bmatrix} 3 & 0 & 0 \\ 0 & -1 & 0 \\ 0 & 0 & 5 \end{bmatrix}$

25. $A = \begin{bmatrix} 5 & 0 & 0 \\ 0 & -8 & 0 \\ 0 & 0 & 7 \end{bmatrix}$, $B = \begin{bmatrix} \frac{1}{5} & 0 & 0 \\ 0 & -\frac{1}{8} & 0 \\ 0 & 0 & \frac{1}{2} \end{bmatrix}$

26. $A = \begin{bmatrix} 0 & 0 & 5 \\ 0 & 0 & -3 \\ 0 & 0 & 4 \end{bmatrix}$, $B = \begin{bmatrix} 6 & -11 & 4 \\ 8 & 16 & 4 \\ 0 & 0 & 0 \end{bmatrix}$

27. $A = \begin{bmatrix} 10 \\ 12 \end{bmatrix}$, $B = \begin{bmatrix} 6 & -2 & 1 & 6 \end{bmatrix}$

28. $A = \begin{bmatrix} 1 & 0 & 3 & -2 \\ 6 & 13 & 8 & -17 \end{bmatrix}$, $B = \begin{bmatrix} 1 & 6 \\ 4 & 2 \end{bmatrix}$

In Exercises 29–34, use the matrix capabilities of a graphing utility to find AB.

29. $A = \begin{bmatrix} 5 & 6 & -3 \\ -2 & 5 & 1 \\ 10 & -5 & 5 \end{bmatrix}$, $B = \begin{bmatrix} 1 & -1 & 2 \\ 8 & 1 & 4 \\ 4 & -2 & 9 \end{bmatrix}$

30. $A = \begin{bmatrix} 11 & -12 & 4 \\ 14 & 10 & 12 \\ 6 & -2 & 9 \end{bmatrix}$, $B = \begin{bmatrix} 12 & 10 \\ -5 & 12 \\ 15 & 16 \end{bmatrix}$

31. $A = \begin{bmatrix} -3 & 8 & -6 & 8 \\ -12 & 15 & 9 & 6 \\ 5 & -1 & 1 & 5 \end{bmatrix}$, $B = \begin{bmatrix} 3 & 1 & 6 \\ 24 & 15 & 14 \\ 16 & 10 & 21 \\ 8 & -4 & 10 \end{bmatrix}$

32. $A = \begin{bmatrix} -2 & 4 & 8 \\ 21 & 5 & 6 \\ 13 & 2 & 6 \end{bmatrix}$, $B = \begin{bmatrix} 2 & 0 \\ -7 & 15 \\ 32 & 14 \\ 0.5 & 1.6 \end{bmatrix}$

33. $A = \begin{bmatrix} 9 & 10 & -38 & 18 \\ 100 & -50 & 250 & 75 \end{bmatrix}$,

$B = \begin{bmatrix} 52 & -85 & 27 & 45 \\ 40 & -35 & 60 & 82 \end{bmatrix}$

34. $A = \begin{bmatrix} 15 & -18 \\ -4 & 12 \\ -8 & 22 \end{bmatrix}$, $B = \begin{bmatrix} -7 & 22 & 1 \\ 8 & 16 & 24 \end{bmatrix}$

In Exercises 35–38, find matrices A, X, and B such that the system of linear equations can be written as the matrix equation $AX = B$. Solve the system of equations.

35. $-x + y = 4$
$-2x + y = 0$

36. $x - 2y + 3z = 9$
$-x + 3y - z = -6$
$2x - 5y + 5z = 17$

37. $2x + 3y = 5$
$x + 4y = 10$

38. $x + y - 3z = -1$
$-x + 2y = 1$
$-y + z = 0$

⊞ In Exercises 39–42, use the matrix capabilities of a graphing utility to find

$$f(A) = a_0 I_n + a_1 A + a_2 A^2 + \cdots + a_n A^n.$$

39. $f(x) = x^2 - 5x + 2$, $A = \begin{bmatrix} 2 & 0 \\ 4 & 5 \end{bmatrix}$

40. $f(x) = x^2 - 7x + 6$, $A = \begin{bmatrix} 5 & 4 \\ 1 & 2 \end{bmatrix}$

41. $f(x) = x^3 - 10x^2 + 31x - 30$, $A = \begin{bmatrix} 3 & 1 & 4 \\ 0 & 2 & 6 \\ 0 & 0 & 5 \end{bmatrix}$

42. $f(x) = x^2 - 10x + 24$, $A = \begin{bmatrix} 8 & -4 \\ 2 & 2 \end{bmatrix}$

43. *Think About It* If a, b, and c are real numbers such that $c \neq 0$ and $ac = bc$, then $a = b$. However, if A, B, and C are nonzero matrices such that $AC = BC$, then A is *not* necessarily equal to B. Illustrate this using the following matrices.

$$A = \begin{bmatrix} 0 & 1 \\ 0 & 1 \end{bmatrix}, \quad B = \begin{bmatrix} 1 & 0 \\ 1 & 0 \end{bmatrix}, \quad C = \begin{bmatrix} 2 & 3 \\ 2 & 3 \end{bmatrix}$$

44. *Think About It* If a and b are real numbers such that $ab = 0$, then $a = 0$ or $b = 0$. However, if A and B are matrices such that $AB = O$, it is *not* necessarily true that $A = O$ or $B = O$. Illustrate this using the following matrices.

$$A = \begin{bmatrix} 3 & 3 \\ 4 & 4 \end{bmatrix}, \quad B = \begin{bmatrix} 1 & -1 \\ -1 & 1 \end{bmatrix}$$

Think About It In Exercises 45–54, let matrices A, B, C, and D be of order 2×3, 2×3, 3×2, and 2×2, respectively. Determine whether the matrices are of proper order to perform the operation(s). If so, give the order of the answer.

45. $A + 2C$

46. $B - 3C$

47. AB

48. BC

49. $BC - D$

50. $CB - D$

51. $(CA)D$

52. $(BC)D$

53. $D(A - 3B)$

54. $(BC - D)A$

55. *Factory Production* A certain corporation has three factories, each of which manufactures two products. The number of units of product i produced at factory j in one day is represented by a_{ij} in the matrix

$$A = \begin{bmatrix} 60 & 40 & 20 \\ 30 & 90 & 60 \end{bmatrix}.$$

Find the production levels if production is increased by 20%. (*Hint*: Because an increase of 20% corresponds to 100% + 20%, multiply the given matrix by 1.2.)

56. *Factory Production* A certain corporation has four factories, each of which manufactures two products. The number of units of product i produced at factory j in one day is represented by a_{ij} in the matrix

$$A = \begin{bmatrix} 100 & 90 & 70 & 30 \\ 40 & 20 & 60 & 60 \end{bmatrix}.$$

Find the production levels if production is increased by 10%.

57. *Crop Production* A fruit grower raises two crops, which are shipped to three outlets. The number of units of product i that are shipped to outlet j is represented by a_{ij} in the matrix

$$A = \begin{bmatrix} 100 & 75 & 75 \\ 125 & 150 & 100 \end{bmatrix}.$$

The profit per unit is represented by the matrix

$$B = [\$3.75 \quad \$7.00].$$

Find the product BA, and state what each entry of the product represents.

58. *Revenue* A manufacturer produces three models of a product, which are shipped to two warehouses. The number of units of model i that are shipped to warehouse j is represented by a_{ij} in the matrix

$$A = \begin{bmatrix} 5,000 & 4,000 \\ 6,000 & 10,000 \\ 8,000 & 5,000 \end{bmatrix}.$$

The price per unit is represented by the matrix

$$B = [\$20.50 \quad \$26.50 \quad \$29.50].$$

Compute BA and interpret the result.

59. *Inventory Levels* A company sells five models of computers through three retail outlets. The inventories are given by S.

Model

$$\begin{array}{cccccc} & A & B & C & D & E \\ S = & \begin{bmatrix} 3 & 2 & 2 & 3 & 0 \\ 0 & 2 & 3 & 4 & 3 \\ 4 & 2 & 1 & 3 & 2 \end{bmatrix} & \begin{matrix} 1 \\ 2 \\ 3 \end{matrix} \end{array}$$ Outlet

The wholesale and retail prices are given by T.

Price

$$\begin{array}{ccc} & \text{Wholesale} & \text{Retail} \\ T = & \begin{bmatrix} \$840 & \$1100 \\ \$1200 & \$1350 \\ \$1450 & \$1650 \\ \$2650 & \$3000 \\ \$3050 & \$3200 \end{bmatrix} & \begin{matrix} A \\ B \\ C \\ D \\ E \end{matrix} \end{array}$$ Model

Compute ST and interpret the result.

60. *Voting Preferences* The matrix

From

$$\begin{array}{cccc} & R & D & I \\ P = & \begin{bmatrix} 0.6 & 0.1 & 0.1 \\ 0.2 & 0.7 & 0.1 \\ 0.2 & 0.2 & 0.8 \end{bmatrix} & \begin{matrix} R \\ D \\ I \end{matrix} \end{array}$$ To

is called a stochastic matrix. Each entry $p_{ij}(i \neq j)$ represents the proportion of the voting population that changes from party i to party j, and p_{ii} represents the proportion that remains loyal to the party from one election to the next. Compute and interpret P^2.

61. *Labor/Wage Requirements* A company that manufactures boats has the following labor-hour and wage requirements.

Labor per boat

Department

$$\begin{array}{cccc} & \text{Cutting} & \text{Assembly} & \text{Packaging} \\ S = & \begin{bmatrix} 1.0 \text{ hr} & 0.5 \text{ hr} & 0.2 \text{ hr} \\ 1.6 \text{ hr} & 1.0 \text{ hr} & 0.2 \text{ hr} \\ 2.5 \text{ hr} & 2.0 \text{ hr} & 0.4 \text{ hr} \end{bmatrix} & \begin{matrix} \text{Small} \\ \text{Medium} \\ \text{Large} \end{matrix} \end{array}$$ Boat size

Wages per hour

Plant

$$\begin{array}{ccc} & A & B \\ T = & \begin{bmatrix} \$12 & \$10 \\ \$9 & \$8 \\ \$6 & \$5 \end{bmatrix} & \begin{matrix} \text{Cutting} \\ \text{Assembly} \\ \text{Packaging} \end{matrix} \end{array}$$ Department

Compute ST and interpret the result.

62. *Voting Preference* Use a graphing utility to find P^3, P^4, P^5, P^6, P^7, and P^8 for the matrix given in Exercise 60. Can you detect a pattern as P is raised to higher powers?

Exploration **In Exercises 63 and 64, let $i = \sqrt{-1}$.**

63. Consider the matrix

$$A = \begin{bmatrix} i & 0 \\ 0 & i \end{bmatrix}.$$

Find A^2, A^3, and A^4. Identify any similarities with i^2, i^3, and i^4.

64. Consider the matrix

$$A = \begin{bmatrix} 0 & -i \\ i & 0 \end{bmatrix}.$$

Find and identify A^2.

65. *Exploration* Let A and B be unequal diagonal matrices of the same order. (A *diagonal matrix* is a square matrix in which each entry not on the main diagonal is zero.) Determine the products AB for several pairs of such matrices. Make a conjecture about a quick rule for such products.

See Exercises 63 and 64 on page 768 for examples of how an inverse matrix can be used to help analyze an electrical circuit.

▶ **10.3** *The Inverse of a Square Matrix*

The Inverse of a Matrix ▫ *Finding Inverse Matrices* ▫
The Inverse of a 2 × 2 Matrix ▫ *Systems of Linear Equations*

The Inverse of a Matrix

This section further develops the algebra of matrices. To begin, consider the real number equation $ax = b$. To solve this equation for x, multiply both sides of the equation by a^{-1} (provided that $a \neq 0$).

$$ax = b$$
$$(a^{-1}a)x = a^{-1}b$$
$$(1)x = a^{-1}b$$
$$x = a^{-1}b$$

The number a^{-1} is called the *multiplicative inverse of a* because $a^{-1}a = 1$. The definition of the multiplicative inverse of a matrix is similar.

NOTE The symbol A^{-1} is read "*A* inverse." ∎∎

DEFINITION OF THE INVERSE OF A SQUARE MATRIX

Let A be an $n \times n$ matrix. If there exists matrix A^{-1} such that

$$AA^{-1} = I_n = A^{-1}A$$

A^{-1} is called the **inverse** of A.

EXAMPLE 1 *The Inverse of a Matrix*

Show that B is the inverse of A, where

$$A = \begin{bmatrix} -1 & 2 \\ -1 & 1 \end{bmatrix} \quad \text{and} \quad B = \begin{bmatrix} 1 & -2 \\ 1 & -1 \end{bmatrix}.$$

Solution

To show that B is the inverse of A, show that $AB = I = BA$, as follows.

$$AB = \begin{bmatrix} -1 & 2 \\ -1 & 1 \end{bmatrix}\begin{bmatrix} 1 & -2 \\ 1 & -1 \end{bmatrix} = \begin{bmatrix} -1 + 2 & 2 - 2 \\ -1 + 1 & 2 - 1 \end{bmatrix} = \begin{bmatrix} 1 & 0 \\ 0 & 1 \end{bmatrix}$$

$$BA = \begin{bmatrix} 1 & -2 \\ 1 & -1 \end{bmatrix}\begin{bmatrix} -1 & 2 \\ -1 & 1 \end{bmatrix} = \begin{bmatrix} -1 + 2 & 2 - 2 \\ -1 + 1 & 2 - 1 \end{bmatrix} = \begin{bmatrix} 1 & 0 \\ 0 & 1 \end{bmatrix}$$

NOTE Recall that it is not always true that $AB = BA$, even if both products are defined. However, if A and B are both square matrices and $AB = I_n$, it can be shown that $BA = I_n$. Hence, in Example 1, you need only to check that $AB = I_2$. ∎∎

If a matrix A has an inverse, A is called **invertible** (or **nonsingular**); otherwise, A is called **singular.** A nonsquare matrix cannot have an inverse. To see this, note that if A is of order $m \times n$ and B is of order $n \times m$ (where $m \neq n$), the products AB and BA are of different orders and therefore cannot be equal to each other. Not all square matrices have inverses (see the matrix at the bottom of page 762). If, however, a matrix does have an inverse, that inverse is unique. The following example shows how to use a system of equations to find the inverse of a matrix.

EXAMPLE 2   *Finding the Inverse of a Matrix*

Find the inverse of

$$A = \begin{bmatrix} 1 & 4 \\ -1 & -3 \end{bmatrix}.$$

Solution

To find the inverse of A, try to solve the matrix equation $AX = I$ for X.

$$\overset{A}{\begin{bmatrix} 1 & 4 \\ -1 & -3 \end{bmatrix}} \overset{X}{\begin{bmatrix} x_{11} & x_{12} \\ x_{21} & x_{22} \end{bmatrix}} = \overset{I}{\begin{bmatrix} 1 & 0 \\ 0 & 1 \end{bmatrix}}$$

$$\begin{bmatrix} x_{11} + 4x_{21} & x_{12} + 4x_{22} \\ -x_{11} - 3x_{21} & -x_{12} - 3x_{22} \end{bmatrix} = \begin{bmatrix} 1 & 0 \\ 0 & 1 \end{bmatrix}$$

Equating corresponding entries, you obtain the following two systems of linear equations.

$$\begin{aligned} x_{11} + 4x_{21} &= 1 & x_{12} + 4x_{22} &= 0 \\ -x_{11} - 3x_{21} &= 0 & -x_{12} - 3x_{22} &= 1 \end{aligned}$$

From the first system you can determine that $x_{11} = -3$ and $x_{21} = 1$, and from the second system you can determine that $x_{12} = -4$ and $x_{22} = 1$. Therefore, the inverse of A is

$$X = A^{-1} = \begin{bmatrix} -3 & -4 \\ 1 & 1 \end{bmatrix}.$$

You can use matrix multiplication to check this result.

Check

$$AA^{-1} = \begin{bmatrix} 1 & 4 \\ -1 & -3 \end{bmatrix} \begin{bmatrix} -3 & -4 \\ 1 & 1 \end{bmatrix} = \begin{bmatrix} 1 & 0 \\ 0 & 1 \end{bmatrix} \checkmark$$

$$A^{-1}A = \begin{bmatrix} -3 & -4 \\ 1 & 1 \end{bmatrix} \begin{bmatrix} 1 & 4 \\ -1 & -3 \end{bmatrix} = \begin{bmatrix} 1 & 0 \\ 0 & 1 \end{bmatrix} \checkmark$$

Finding Inverse Matrices

In Example 2, note that the two systems of linear equations have the *same coefficient matrix A*. Rather than solve the two systems represented by

$$\begin{bmatrix} 1 & 4 & \vdots & 1 \\ -1 & -3 & \vdots & 0 \end{bmatrix} \quad \text{and} \quad \begin{bmatrix} 1 & 4 & \vdots & 0 \\ -1 & -3 & \vdots & 1 \end{bmatrix}$$

separately, you can solve them *simultaneously* by **adjoining** the identity matrix to the coefficient matrix to obtain

$$\overset{A}{}\qquad\overset{I}{}$$

$$\begin{bmatrix} 1 & 4 & \vdots & 1 & 0 \\ -1 & -3 & \vdots & 0 & 1 \end{bmatrix}.$$

Then, applying Gauss-Jordan elimination to this matrix, you can solve *both* systems with a single elimination process, as follows.

$$\begin{bmatrix} 1 & 4 & \vdots & 1 & 0 \\ -1 & -3 & \vdots & 0 & 1 \end{bmatrix}$$

$$R_1 + R_2 \rightarrow \begin{bmatrix} 1 & 4 & \vdots & 1 & 0 \\ 0 & 1 & \vdots & 1 & 1 \end{bmatrix}$$

$$-4R_2 + R_1 \rightarrow \begin{bmatrix} 1 & 0 & \vdots & -3 & -4 \\ 0 & 1 & \vdots & 1 & 1 \end{bmatrix}$$

Thus, from the "doubly augmented" matrix $[A : I]$, you obtained the matrix $[I : A^{-1}]$.

$$\overset{A}{}\qquad\overset{I}{}\qquad\qquad\overset{I}{}\qquad\overset{A^{-1}}{}$$

$$\begin{bmatrix} 1 & 4 & \vdots & 1 & 0 \\ -1 & -3 & \vdots & 0 & 1 \end{bmatrix} \Longrightarrow \begin{bmatrix} 1 & 0 & \vdots & -3 & -4 \\ 0 & 1 & \vdots & 1 & 1 \end{bmatrix}$$

This procedure (or algorithm) works for an arbitrary square matrix that has an inverse.

FINDING AN INVERSE MATRIX

Let A be a square matrix of order n.

1. Write the $n \times 2n$ matrix that consists of the given matrix A on the left and the $n \times n$ identity matrix I on the right to obtain $[A : I]$. Note that we separate the matrices A and I by a dotted line. We call this process **adjoining** the matrices A and I.
2. If possible, row reduce A to I using elementary row operations on the *entire* matrix $[A : I]$. The result will be the matrix $[I : A^{-1}]$. If this is not possible, A is not invertible.
3. Check your work by multiplying to see that $AA^{-1} = I = A^{-1}A$.

■

EXAMPLE 3   Finding the Inverse of a Matrix

Find the inverse of

$$A = \begin{bmatrix} 1 & -1 & 0 \\ 1 & 0 & -1 \\ 6 & -2 & -3 \end{bmatrix}.$$

Solution

Begin by adjoining the identity matrix A to form the matrix

$$[A \;\vdots\; I] = \begin{bmatrix} 1 & -1 & 0 & \vdots & 1 & 0 & 0 \\ 1 & 0 & -1 & \vdots & 0 & 1 & 0 \\ 6 & -2 & -3 & \vdots & 0 & 0 & 1 \end{bmatrix}.$$

Use elementary row operations to obtain the form $[I \;\vdots\; A^{-1}]$, as follows.

$$\begin{matrix} \\ -R_1 + R_2 \rightarrow \\ -6R_1 + R_3 \rightarrow \end{matrix} \begin{bmatrix} 1 & -1 & 0 & \vdots & 1 & 0 & 0 \\ 0 & 1 & -1 & \vdots & -1 & 1 & 0 \\ 0 & 4 & -3 & \vdots & -6 & 0 & 1 \end{bmatrix}$$

$$\begin{matrix} R_2 + R_1 \rightarrow \\ \\ -4R_2 + R_3 \rightarrow \end{matrix} \begin{bmatrix} 1 & 0 & -1 & \vdots & 0 & 1 & 0 \\ 0 & 1 & -1 & \vdots & -1 & 1 & 0 \\ 0 & 0 & 1 & \vdots & -2 & -4 & 1 \end{bmatrix}$$

$$\begin{matrix} R_3 + R_1 \rightarrow \\ R_3 + R_2 \rightarrow \\ \\ \end{matrix} \begin{bmatrix} 1 & 0 & 0 & \vdots & -2 & -3 & 1 \\ 0 & 1 & 0 & \vdots & -3 & -3 & 1 \\ 0 & 0 & 1 & \vdots & -2 & -4 & 1 \end{bmatrix}$$

Therefore, the matrix A is invertible and its inverse is

$$A^{-1} = \begin{bmatrix} -2 & -3 & 1 \\ -3 & -3 & 1 \\ -2 & -4 & 1 \end{bmatrix}.$$

Try confirming this result by multiplying A and A^{-1} to obtain I.

■

The process shown in Example 3 applies to any $n \times n$ matrix A. If A has an inverse, this process will find it. On the other hand, if A does not have an inverse (if A is *singular*), the process will tell us so. For instance, the following matrix has no inverse.

$$A = \begin{bmatrix} 1 & 2 & 0 \\ 3 & -1 & 2 \\ -2 & 3 & -2 \end{bmatrix}$$

Explain how the elimination process shows that this matrix is singular.

▶ *Exploration*

Use a graphing utility with matrix operations to find the inverse of the matrix

$$A = \begin{bmatrix} 1 & -3 \\ -2 & 6 \end{bmatrix}.$$

What message appears on the screen? Why does the graphing utility display this message?

The Inverse of a 2 × 2 Matrix

Using Gauss-Jordan elimination to find the inverse of a matrix works well (even as a computer technique) for matrices of order 3 × 3 or greater. For 2 × 2 matrices, however, many people prefer to use a formula for the inverse rather than Gauss-Jordan elimination. This simple formula, which works *only* for 2 × 2 matrices, is explained as follows. If A is a 2 × 2 matrix given by

$$A = \begin{bmatrix} a & b \\ c & d \end{bmatrix}$$

then A is invertible if and only if $ad - bc \neq 0$. Moreover, if $ad - bc \neq 0$, the inverse is given by

$$A^{-1} = \frac{1}{ad - bc}\begin{bmatrix} d & -b \\ -c & a \end{bmatrix}.$$

Try verifying this inverse by multiplication.

NOTE   The denominator $ad - bc$ is called the **determinant** of the 2 × 2 matrix A. You will study determinants in the next section. ■■

---

*E*XAMPLE 4   *Finding the Inverse of a 2 × 2 Matrix*

If possible, find the inverse of the matrix.

a. $A = \begin{bmatrix} 3 & -1 \\ -2 & 2 \end{bmatrix}$ **b.** $B = \begin{bmatrix} 3 & -1 \\ -6 & 2 \end{bmatrix}$

Solution

a. For the matrix A, apply the formula for the inverse of a 2 × 2 matrix to obtain

$$ad - bc = (3)(2) - (-1)(-2) = 4.$$

Because this quantity is not zero, the inverse is formed by interchanging the entries on the main diagonal, changing the signs of the other two entries, and multiplying by the scalar $\frac{1}{4}$, as follows.

$$A^{-1} = \frac{1}{4}\begin{bmatrix} 2 & 1 \\ 2 & 3 \end{bmatrix} = \begin{bmatrix} \frac{1}{2} & \frac{1}{4} \\ \frac{1}{2} & \frac{3}{4} \end{bmatrix}$$

b. For the matrix B, you have

$$ad - bc = (3)(2) - (-1)(-6) = 0$$

which means that B is not invertible.

Systems of Linear Equations

You know that a system of linear equations can have exactly one solution, infinitely many solutions, or no solution. If the coefficient matrix A of a *square* system (a system that has the same number of equations as variables) is invertible, the system has a unique solution, which is defined as follows.

> **A SYSTEM OF EQUATIONS WITH A UNIQUE SOLUTION**
>
> If A is an invertible matrix, the system of linear equations represented by $AX = B$ has a unique solution given by
>
> $$X = A^{-1}B.$$

EXAMPLE 5   *Solving a System of Equations Using an Inverse*

Use an inverse matrix to solve the system.

$$2x + 3y + z = -1$$
$$3x + 3y + z = 1$$
$$2x + 4y + z = -2$$

Solution

NOTE   Use Gauss-Jordan elimination or a graphing utility to verify A^{-1} for the system of equations in Example 5. ■■

$$X = A^{-1}B = \begin{bmatrix} -1 & 1 & 0 \\ -1 & 0 & 1 \\ 6 & -2 & -3 \end{bmatrix} \begin{bmatrix} -1 \\ 1 \\ -2 \end{bmatrix} = \begin{bmatrix} 2 \\ -1 \\ -2 \end{bmatrix}$$

Thus, the solution is $x = 2$, $y = -1$, and $z = -2$.

GROUP ACTIVITY

FINDING AN INVERSE MATRIX

Explain how to use a graphing utility to find the inverse matrix for each of the following.

a. $A = \begin{bmatrix} -3 & 2 \\ 7 & 4 \end{bmatrix}$ **b.** $B = \begin{bmatrix} 1 & -4 & 2 \\ 2 & -9 & 5 \\ 1 & -5 & 4 \end{bmatrix}$ **c.** $C = \begin{bmatrix} 3 & 1 & 0 \\ 1 & 1 & 1 \\ 1 & -1 & 2 \end{bmatrix}$

Perform the matrix operations.

1. $4\begin{bmatrix} 1 & 6 \\ 0 & -4 \\ 12 & 2 \end{bmatrix}$

2. $\dfrac{1}{2}\begin{bmatrix} 11 & 10 & 48 \\ 1 & 0 & 16 \\ 0 & 2 & 8 \end{bmatrix}$

3. $\begin{bmatrix} 1 & -10 \\ 4 & 1 \end{bmatrix} - 2\begin{bmatrix} 3 & -4 \\ 0 & 7 \end{bmatrix}$

4. $\begin{bmatrix} 5 & 20 \\ -7 & 15 \end{bmatrix} - 3\begin{bmatrix} 6 & 3 \\ 4 & -2 \end{bmatrix}$

5. $\begin{bmatrix} 1 & -2 \\ -1 & 3 \end{bmatrix}\begin{bmatrix} 3 & 2 \\ 1 & 1 \end{bmatrix}$

6. $\begin{bmatrix} 1 & 0 \\ 0 & 1 \end{bmatrix}\begin{bmatrix} 6 & 5 \\ 3 & -2 \end{bmatrix}$

7. $\begin{bmatrix} 1 & 1 & 0 \\ 1 & 0 & 1 \\ 6 & 2 & 3 \end{bmatrix}\begin{bmatrix} 2 & 3 & 1 \\ 3 & 3 & 1 \\ 2 & 4 & 1 \end{bmatrix}$

8. $\begin{bmatrix} 2 & 0 & 0 \\ 0 & -1 & 0 \\ 0 & 0 & 3 \end{bmatrix}\begin{bmatrix} \frac{1}{2} & 0 & 0 \\ 0 & -1 & 0 \\ 0 & 0 & \frac{1}{3} \end{bmatrix}$

Rewrite the matrix in reduced row-echelon form.

9. $\begin{bmatrix} 3 & -2 & 1 & 0 \\ 4 & -3 & 0 & 1 \end{bmatrix}$

10. $\begin{bmatrix} 1 & 1 & 2 & 1 & 0 & 0 \\ -1 & 0 & 3 & 0 & 1 & 0 \\ 1 & 2 & 8 & 0 & 0 & 1 \end{bmatrix}$

10.3 Exercises

In Exercises 1–8, show that B is the inverse of A.

1. $A = \begin{bmatrix} 2 & 1 \\ 5 & 3 \end{bmatrix}$, $B = \begin{bmatrix} 3 & -1 \\ -5 & 2 \end{bmatrix}$

2. $A = \begin{bmatrix} 1 & -1 \\ -1 & 2 \end{bmatrix}$, $B = \begin{bmatrix} 2 & 1 \\ 1 & 1 \end{bmatrix}$

3. $A = \begin{bmatrix} 1 & 2 \\ 3 & 4 \end{bmatrix}$, $B = \begin{bmatrix} -2 & 1 \\ \frac{3}{2} & -\frac{1}{2} \end{bmatrix}$

4. $A = \begin{bmatrix} 1 & -1 \\ 2 & 3 \end{bmatrix}$, $B = \begin{bmatrix} \frac{3}{5} & \frac{1}{5} \\ -\frac{2}{5} & \frac{1}{5} \end{bmatrix}$

5. $A = \begin{bmatrix} -2 & 2 & 3 \\ 1 & -1 & 0 \\ 0 & 1 & 4 \end{bmatrix}$, $B = \dfrac{1}{3}\begin{bmatrix} -4 & -5 & 3 \\ -4 & -8 & 3 \\ 1 & 2 & 0 \end{bmatrix}$

6. $A = \begin{bmatrix} 2 & -17 & 11 \\ -1 & 11 & -7 \\ 0 & 3 & -2 \end{bmatrix}$, $B = \begin{bmatrix} 1 & 1 & 2 \\ 2 & 4 & -3 \\ 3 & 6 & -5 \end{bmatrix}$

7. $A = \begin{bmatrix} 2 & 0 & 1 & 1 \\ 3 & 0 & 0 & 1 \\ -1 & 1 & -2 & 1 \\ 4 & -1 & 1 & 0 \end{bmatrix}$,

$B = \begin{bmatrix} -1 & 2 & -1 & -1 \\ -4 & 9 & -5 & -6 \\ 0 & 1 & -1 & -1 \\ 3 & -5 & 3 & 3 \end{bmatrix}$

8. $A = \begin{bmatrix} -1 & 1 & 0 & -1 \\ 1 & -1 & 1 & 0 \\ -1 & 1 & 2 & 0 \\ 0 & -1 & 1 & 1 \end{bmatrix}$,

$B = \dfrac{1}{3}\begin{bmatrix} -3 & 1 & 1 & -3 \\ -3 & -1 & 2 & -3 \\ 0 & 1 & 1 & 0 \\ -3 & -2 & 1 & 0 \end{bmatrix}$

In Exercises 9–24, find the inverse of the matrix (if it exists).

9. $\begin{bmatrix} 2 & 0 \\ 0 & 3 \end{bmatrix}$

10. $\begin{bmatrix} 1 & 2 \\ 3 & 7 \end{bmatrix}$

11. $\begin{bmatrix} 1 & -2 \\ 2 & -3 \end{bmatrix}$

12. $\begin{bmatrix} -7 & 33 \\ 4 & -19 \end{bmatrix}$

13. $\begin{bmatrix} -1 & 1 \\ -2 & 1 \end{bmatrix}$

14. $\begin{bmatrix} 11 & 1 \\ -1 & 0 \end{bmatrix}$

15. $\begin{bmatrix} 2 & 4 \\ 4 & 8 \end{bmatrix}$

16. $\begin{bmatrix} 2 & 3 \\ 1 & 4 \end{bmatrix}$

17. $\begin{bmatrix} 2 & 7 & 1 \\ -3 & -9 & 2 \end{bmatrix}$

18. $\begin{bmatrix} -2 & 5 \\ 6 & -15 \\ 0 & 1 \end{bmatrix}$

19. $\begin{bmatrix} 1 & 1 & 1 \\ 3 & 5 & 4 \\ 3 & 6 & 5 \end{bmatrix}$

20. $\begin{bmatrix} 1 & 2 & 2 \\ 3 & 7 & 9 \\ -1 & -4 & -7 \end{bmatrix}$

21. $\begin{bmatrix} 1 & 0 & 0 \\ 3 & 4 & 0 \\ 2 & 5 & 5 \end{bmatrix}$

22. $\begin{bmatrix} 1 & 0 & 0 \\ 3 & 0 & 0 \\ 2 & 5 & 5 \end{bmatrix}$

23. $\begin{bmatrix} -8 & 0 & 0 & 0 \\ 0 & 1 & 0 & 0 \\ 0 & 0 & 4 & 0 \\ 0 & 0 & 0 & -5 \end{bmatrix}$

24. $\begin{bmatrix} 1 & 3 & -2 & 0 \\ 0 & 2 & 4 & 6 \\ 0 & 0 & -2 & 1 \\ 0 & 0 & 0 & 5 \end{bmatrix}$

In Exercises 25–34, use the matrix capabilities of a graphing utility to find the inverse of the matrix (if it exists).

25. $\begin{bmatrix} 1 & 2 & -1 \\ 3 & 7 & -10 \\ -5 & -7 & -15 \end{bmatrix}$

26. $\begin{bmatrix} 10 & 5 & -7 \\ -5 & 1 & 4 \\ 3 & 2 & -2 \end{bmatrix}$

27. $\begin{bmatrix} 1 & 1 & 2 \\ 3 & 1 & 0 \\ -2 & 0 & 3 \end{bmatrix}$

28. $\begin{bmatrix} 3 & 2 & 2 \\ 2 & 2 & 2 \\ -4 & 4 & 3 \end{bmatrix}$

29. $\begin{bmatrix} 0.1 & 0.2 & 0.3 \\ -0.3 & 0.2 & 0.2 \\ 0.5 & 0.4 & 0.4 \end{bmatrix}$

30. $\begin{bmatrix} 2 & 0 & 0 \\ 0 & 3 & 0 \\ 0 & 0 & 5 \end{bmatrix}$

31. $\begin{bmatrix} 1 & 0 & 3 & 0 \\ 0 & 2 & 0 & 4 \\ 1 & 0 & 3 & 0 \\ 0 & 2 & 0 & 4 \end{bmatrix}$

32. $\begin{bmatrix} -1 & 0 & 1 & 0 \\ 0 & 2 & 0 & -1 \\ 2 & 0 & -1 & 0 \\ 0 & -1 & 0 & 1 \end{bmatrix}$

33. $\begin{bmatrix} 1 & -2 & -1 & -2 \\ 3 & -5 & -2 & -3 \\ 2 & -5 & -2 & -5 \\ -1 & 4 & 4 & 11 \end{bmatrix}$

34. $\begin{bmatrix} 4 & 8 & -7 & 14 \\ 2 & 5 & -4 & 6 \\ 0 & 2 & 1 & -7 \\ 3 & 6 & -5 & 10 \end{bmatrix}$

35. If A is a 2×2 matrix given by

$$A = \begin{bmatrix} a & b \\ c & d \end{bmatrix}$$

then A is invertible if and only if $ad - bc \neq 0$. If $ad - bc \neq 0$, verify that the inverse is given by

$$A^{-1} = \frac{1}{ad - bc} \begin{bmatrix} d & -b \\ -c & a \end{bmatrix}.$$

36. Use the result of Exercise 35 to find the inverse of each matrix.

(a) $\begin{bmatrix} 5 & -2 \\ 2 & 3 \end{bmatrix}$

(b) $\begin{bmatrix} 7 & 12 \\ -8 & -5 \end{bmatrix}$

In Exercises 37–40, use an inverse matrix to solve the system of linear equations. (Use the inverse matrix found in Exercise 11.)

37. $x - 2y = 5$
    $2x - 3y = 10$

38. $x - 2y = 0$
    $2x - 3y = 3$

39. $x - 2y = 4$
    $2x - 3y = 2$

40. $x - 2y = 1$
    $2x - 3y = -2$

In Exercises 41 and 42, use an inverse matrix to solve the system of linear equations. (Use the inverse matrix found in Exercise 19.)

41. $x + y + z = 0$
$3x + 5y + 4z = 5$
$3x + 6y + 5z = 2$

42. $x + y + z = -1$
$3x + 5y + 4z = 2$
$3x + 6y + 5z = 0$

In Exercises 43 and 44, use an inverse matrix to solve the system of linear equations. (Use the inverse matrix found in Exercise 33.)

43. $x_1 - 2x_2 - x_3 - 2x_4 = 0$
$3x_1 - 5x_2 - 2x_3 - 3x_4 = 1$
$2x_1 - 5x_2 - 2x_3 - 5x_4 = -1$
$-x_1 + 4x_2 + 4x_3 + 11x_4 = 2$

44. $x_1 - 2x_2 - x_3 - 2x_4 = 1$
$3x_1 - 5x_2 - 2x_3 - 3x_4 = -2$
$2x_1 - 5x_2 - 2x_3 - 5x_4 = 0$
$-x_1 + 4x_2 + 4x_3 + 11x_4 = -3$

In Exercises 45–52, use an inverse matrix to solve (if possible) the system of linear equations.

45. $3x + 4y = -2$
$5x + 3y = 4$

46. $18x + 12y = 13$
$30x + 24y = 23$

47. $-0.4x + 0.8y = 1.6$
$2x - 4y = 5$

48. $13x - 6y = 17$
$26x - 12y = 8$

49. $3x + 6y = 6$
$6x + 14y = 11$

50. $3x + 2y = 1$
$2x + 10y = 6$

51. $4x - y + z = -5$
$2x + 2y + 3z = 10$
$5x - 2y + 6z = 1$

52. $4x - 2y + 3z = -2$
$2x + 2y + 5z = 16$
$8x - 5y - 2z = 4$

In Exercises 53–56, use the matrix capabilities of a graphing utility to solve (if possible) the system of linear equations.

53. $5x - 3y + 2z = 2$
$2x + 2y - 3z = 3$
$x - 7y + 8z = -4$

54. $2x + 3y + 5z = 4$
$3x + 5y + 9z = 7$
$5x + 9y + 17z = 13$

55. $7x - 3y + 2w = 41$
$-2x + y - w = -13$
$4x + z - 2w = 12$
$-x + y - w = -8$

56. $2x + 5y + w = 11$
$x + 4y + 2z - 2w = -7$
$2x - 2y + 5z + w = 3$
$x - 3w = -1$

Bond Investments In Exercises 57–60, consider a person who invests in AAA-rated bonds, A-rated bonds, and B-rated bonds. The average yields are 6.5% on AAA bonds, 7% on A bonds, and 9% on B bonds. The person invests twice as much in B bonds as in A bonds. Let x, y, and z represent the amounts invested in AAA, A, and B bonds, respectively.

$$x + y + z = \text{(total investment)}$$
$$0.065x + 0.07y + 0.09z = \text{(annual return)}$$
$$2y - z = 0$$

Use the inverse of the coefficient matrix of this system to find the amount invested in each type of bond.

57. Total investment = $25,000
Annual return = $1900

58. Total investment = $45,000
Annual return = $3750

59. Total investment = $12,000
Annual return = $835

60. Total investment = $500,000
Annual return = $38,000

61. Essay Write a brief paragraph explaining the advantage of using inverse matrices to solve the systems of linear equations in Exercises 37–44.

62. True or False? Multiplication of an invertible matrix and its inverse is commutative. Give an example that supports your answer.

Circuit Analysis **In Exercises 63 and 64, consider the circuit in the figure. The currents I_1, I_2, and I_3, in amperes, are given by the solution of the system of linear equations**

$$2I_1 \qquad + 4I_3 = E_1$$
$$I_2 + 4I_3 = E_2$$
$$I_1 + I_2 - \ \ I_3 = 0$$

where E_1 and E_2 are voltages. Use the inverse of the coefficient matrix of this system to find the unknown currents for the given voltages.

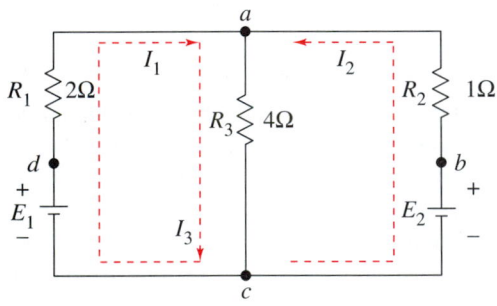

63. $E_1 = 14$ volts, $E_2 = 28$ volts

64. $E_1 = 10$ volts, $E_2 = 10$ volts

65. Exploration Consider the matrices of the form

$$A = \begin{bmatrix} a_{11} & 0 & 0 & 0 & \cdots & 0 \\ 0 & a_{22} & 0 & 0 & \cdots & 0 \\ 0 & 0 & a_{33} & 0 & \cdots & 0 \\ \vdots & \vdots & \vdots & \vdots & \cdots & \vdots \\ 0 & 0 & 0 & 0 & \cdots & a_{nn} \end{bmatrix}.$$

(a) Write a 2 × 2 matrix and a 3 × 3 matrix in the form of A. Find the inverse of each.

(b) Use the result of part (a) to make a conjecture about the inverses of matrices of the form of A.

66. Trigonometry Consider the point (x_1, y_1) on a line passing through the origin and the matrix product given by

$$\begin{bmatrix} x_1' \\ y_1' \end{bmatrix} = \begin{bmatrix} \cos\theta & -\sin\theta \\ \sin\theta & \cos\theta \end{bmatrix} \cdot \begin{bmatrix} x_1 \\ y_1 \end{bmatrix}.$$

(a) Perform the matrix product for $(x_1, y_1) = (3, 1)$ and $\theta = 30°$.

(b) Draw the two lines that pass through the origin and the points (x_1, y_1) and (x_1', y_1'), respectively.

(c) Repeat part (b) for several different points and angles. Make a conjecture about the effect of the matrix product.

(d) Using your conjecture from part (c), make a conjecture about the application of the inverse of the trigonometric matrix.

(e) Find the inverse of the trigonometric matrix using the techniques of this section.

(f) Using the conjecture from part (d), determine another method for finding the inverse. Try the method and determine if the result is equivalent to the result of part (b).

67. Chapter Opener Use the models on page 729 to estimate the men's and women's winning times in the 1000-meter speed skating events in the year 2002.

68. Chapter Opener If the models on page 729 continue to represent the winning times in the 1000-meter speed skating events, in which winter Olympics will the women's time be less than the men's time?

Review **Solve Exercises 69–72 as a review of the skills and problem-solving techniques you learned in previous sections. Solve the equation.**

69. $3^{x/2} = 315$

70. $2000e^{-x/5} = 400$

71. $\log_2 x - 2 = 4.5$

72. $\ln x + \ln(x - 1) = 0$

10.4   *The Determinant of a Square Matrix*

See Exercises 41–48 on page 776 for examples of how determinants can be evaluated using the matrix capabilities of a graphing utility.

The Determinant of a 2 × 2 Matrix   □   *Minors and Cofactors*   □
The Determinant of a Square Matrix   □   *Triangular Matrices*

The Determinant of a 2 × 2 Matrix

Every *square* matrix can be associated with a real number called its **determinant.** Determinants have many uses, and several will be discussed in this and the next section. Historically, the use of determinants arose from special number patterns that occur when systems of linear equations are solved. For instance, the system

$$a_1 x + b_1 y = c_1$$
$$a_2 x + b_2 y = c_2$$

has a solution given by

$$x = \frac{c_1 b_2 - c_2 b_1}{a_1 b_2 - a_2 b_1} \quad \text{and} \quad y = \frac{a_1 c_2 - a_2 c_1}{a_1 b_2 - a_2 b_1}$$

provided that $a_1 b_2 - a_2 b_1 \neq 0$. Note that the denominators of the two fractions are the same. This denominator is called the **determinant** of the coefficient matrix of the system.

Coefficient Matrix   *Determinant*

$$A = \begin{bmatrix} a_1 & b_1 \\ a_2 & b_2 \end{bmatrix} \qquad \det(A) = a_1 b_2 - a_2 b_1$$

The determinant of the matrix A can also be denoted by vertical bars on both sides of the matrix, as indicated in the following definition.

DEFINITION OF THE DETERMINANT OF A 2 × 2 MATRIX

The **determinant** of the matrix

$$A = \begin{bmatrix} a_1 & b_1 \\ a_2 & b_2 \end{bmatrix}$$

is given by

$$\det(A) = |A| = \begin{vmatrix} a_1 & b_1 \\ a_2 & b_2 \end{vmatrix} = a_1 b_2 - a_2 b_1.$$

NOTE   In this book, $\det(A)$ and $|A|$ are used interchangeably to represent the determinant of A. Although vertical bars are also used to denote the absolute value of a real number, the context will show which use is intended. ■■

A convenient method for remembering the formula for the determinant of a 2×2 matrix is shown in the following diagram.

$$\det(A) = \begin{vmatrix} a_1 & b_1 \\ a_2 & b_2 \end{vmatrix} = a_1 b_2 - a_2 b_1$$

Note that the determinant is given by the difference of the products of the two diagonals of the matrix.

EXAMPLE 1   *The Determinant of a 2×2 Matrix*

Find the determinant of each matrix

a. $A = \begin{bmatrix} 2 & -3 \\ 1 & 2 \end{bmatrix}$ **b.** $B = \begin{bmatrix} 2 & 1 \\ 4 & 2 \end{bmatrix}$ **c.** $C = \begin{bmatrix} 0 & \frac{3}{2} \\ 2 & 4 \end{bmatrix}$

Solution

NOTE   Notice in Example 1 that the determinant of a matrix can be positive, zero, or negative. ■■

a. $\det(A) = \begin{vmatrix} 2 & -3 \\ 1 & 2 \end{vmatrix} = 2(2) - 1(-3) = 4 + 3 = 7$

b. $\det(B) = \begin{vmatrix} 2 & 1 \\ 4 & 2 \end{vmatrix} = 2(2) - 4(1) = 4 - 4 = 0$

c. $\det(C) = \begin{vmatrix} 0 & \frac{3}{2} \\ 2 & 4 \end{vmatrix} = 0(4) - 2\left(\frac{3}{2}\right) = 0 - 3 = -3$

The determinant of a matrix of order 1×1 is defined simply as the entry of the matrix. For instance, if $A = [-2]$, then $\det(A) = -2$.

TECHNOLOGY
■■

Most graphing utilities can evaluate the determinant of a matrix. For instance, on a *TI-83* or a *TI-82,* you can evaluate the determinant of

$$A = \begin{bmatrix} 2 & -3 \\ 1 & 2 \end{bmatrix}$$

by entering the matrix as $[A]$ and then choosing the "det" feature in the matrix math menu.

det $[A]$ ENTER

The result should be 7, as in Example 1(a). Try evaluating determinants of other matrices. What happens when you try to evaluate the determinant of a nonsquare matrix?

Minors and Cofactors

To define the determinant of a square matrix of order 3×3 or higher, it is convenient to introduce the concepts of **minors** and **cofactors.**

> ### MINORS AND COFACTORS OF A SQUARE MATRIX
>
> If A is a square matrix, the **minor** M_{ij} of the entry a_{ij} is the determinant of the matrix obtained by deleting the ith row and jth column of A. The **cofactor** C_{ij} of the entry a_{ij} is given by
>
> $$C_{ij} = (-1)^{i+j}M_{ij}.$$

Sign Pattern for Cofactors

$$\begin{bmatrix} + & - & + \\ - & + & - \\ + & - & + \end{bmatrix}$$

3×3 matrix

$$\begin{bmatrix} + & - & + & - \\ - & + & - & + \\ + & - & + & - \\ - & + & - & + \end{bmatrix}$$

4×4 matrix

$$\begin{bmatrix} + & - & + & - & + & \cdots \\ - & + & - & + & - & \cdots \\ + & - & + & - & + & \cdots \\ - & + & - & + & - & \cdots \\ + & - & + & - & + & \cdots \\ \vdots & \vdots & \vdots & \vdots & \vdots & \end{bmatrix}$$

$n \times n$ matrix

NOTE In the sign pattern for cofactors above, notice that *odd* positions (where $i + j$ is odd) have negative signs and *even* positions (where $i + j$ is even) have positive signs. ■■

EXAMPLE 2   *Finding the Minors and Cofactors of a Matrix*

Find all the minors and cofactors of

$$A = \begin{bmatrix} 0 & 2 & 1 \\ 3 & -1 & 2 \\ 4 & 0 & 1 \end{bmatrix}.$$

Solution

To find the minor M_{11}, delete the first row and first column of A and evaluate the determinant of the resulting matrix.

$$\begin{bmatrix} 0 & 2 & 1 \\ 3 & -1 & 2 \\ 4 & 0 & 1 \end{bmatrix}, \quad M_{11} = \begin{vmatrix} -1 & 2 \\ 0 & 1 \end{vmatrix} = -1(1) - 0(2) = -1$$

Similarly, to find M_{12}, delete the first row and second column.

$$\begin{bmatrix} 0 & 2 & 1 \\ 3 & -1 & 2 \\ 4 & 0 & 1 \end{bmatrix}, \quad M_{12} = \begin{vmatrix} 3 & 2 \\ 4 & 1 \end{vmatrix} = 3(1) - 4(2) = -5$$

Continuing this pattern, you obtain the following minors.

$$\begin{array}{lll} M_{11} = -1 & M_{12} = -5 & M_{13} = 4 \\ M_{21} = 2 & M_{22} = -4 & M_{23} = -8 \\ M_{31} = 5 & M_{32} = -3 & M_{33} = -6 \end{array}$$

Now, to find the cofactors, combine the checkerboard pattern of signs for a 3×3 matrix (at left) with these minors to obtain the following.

$$\begin{array}{lll} C_{11} = -1 & C_{12} = 5 & C_{13} = 4 \\ C_{21} = -2 & C_{22} = -4 & C_{23} = 8 \\ C_{31} = 5 & C_{32} = 3 & C_{33} = -6 \end{array}$$

The Determinant of a Square Matrix

The definition given below is called **inductive** because it uses determinants of matrices of order $n - 1$ to define the determinant of a matrix of order n.

NOTE   Try checking that for a

2×2 matrix $A = \begin{bmatrix} a_1 & b_1 \\ a_2 & b_2 \end{bmatrix}$ this

definition yields

$|A| = a_1b_2 - a_2b_1$

as previously defined. ■ ■

DETERMINANT OF A SQUARE MATRIX

If A is a square matrix (of order 2×2 or greater), the determinant of A is the sum of the entries in any row (or column) of A multiplied by their respective cofactors. For instance, expanding along the first row yields

$$|A| = a_{11}C_{11} + a_{12}C_{12} + \cdots + a_{1n}C_{1n}.$$

Applying this definition to find a determinant is called **expanding by cofactors.**

EXAMPLE 3   *The Determinant of a Matrix of Order 3×3*

Find the determinant of

$$A = \begin{bmatrix} 0 & 2 & 1 \\ 3 & -1 & 2 \\ 4 & 0 & 1 \end{bmatrix}.$$

Solution

Note that this is the same matrix that was given in Example 2. There you found the cofactors of the entries in the first row to be

$C_{11} = -1, \quad C_{12} = 5, \quad \text{and} \quad C_{13} = 4.$

Therefore, by the definition of a determinant, you have the following.

$$\begin{aligned} |A| &= a_{11}C_{11} + a_{12}C_{12} + a_{13}C_{13} &&\text{First-row expansion} \\ &= 0(-1) + 2(5) + 1(4) \\ &= 14 \end{aligned}$$

In Example 3 the determinant was found by expanding by the cofactors in the first row. You could have used any row or column. For instance, you could have expanded along the second row to obtain

$$\begin{aligned} |A| &= a_{21}C_{21} + a_{22}C_{22} + a_{23}C_{23} &&\text{Second-row expansion} \\ &= 3(-2) + (-1)(-4) + 2(8) \\ &= 14. \end{aligned}$$

When expanding by cofactors, you do not need to find cofactors of zero entries, because zero times its cofactor is zero.

$$a_{ij}C_{ij} = (0)C_{ij} = 0$$

Thus, the row (or column) containing the most zeros is usually the best choice for expansion by cofactors. This is demonstrated in the next example.

EXAMPLE 4   The Determinant of a Matrix of Order 4 × 4

Find the determinant of

$$A = \begin{bmatrix} 1 & -2 & 3 & 0 \\ -1 & 1 & 0 & 2 \\ 0 & 2 & 0 & 3 \\ 3 & 4 & 0 & 2 \end{bmatrix}.$$

Solution

After inspecting this matrix, you can see that three of the entries in the third column are zeros. Thus, you can eliminate some of the work in the expansion by using the third column.

$$|A| = 3(C_{13}) + 0(C_{23}) + 0(C_{33}) + 0(C_{43})$$

Because $C_{23}, C_{33},$ and C_{43} have zero coefficients, you need only find the cofactor C_{13}. To do this, delete the first row and third column of A and evaluate the determinant of the resulting matrix.

$$C_{13} = (-1)^{1+3} \begin{vmatrix} -1 & 1 & 2 \\ 0 & 2 & 3 \\ 3 & 4 & 2 \end{vmatrix} \qquad \text{\color{red}Delete 1st row and 3rd column.}$$

$$= \begin{vmatrix} -1 & 1 & 2 \\ 0 & 2 & 3 \\ 3 & 4 & 2 \end{vmatrix} \qquad \text{\color{red}Simplify.}$$

Expanding by cofactors in the second row yields the following.

$$C_{13} = 0(-1)^3 \begin{vmatrix} 1 & 2 \\ 4 & 2 \end{vmatrix} + 2(-1)^4 \begin{vmatrix} -1 & 2 \\ 3 & 2 \end{vmatrix} + 3(-1)^5 \begin{vmatrix} -1 & 1 \\ 3 & 4 \end{vmatrix}$$

$$= 0 + 2(1)(-8) + 3(-1)(-7)$$

$$= 5$$

Thus, you obtain

$$|A| = 3C_{13} = 3(5) = 15.$$

NOTE   Try using a graphing utility to confirm the result of Example 4.

Triangular Matrices

Evaluating determinants of matrices of order 4 or higher can be tedious. There is, however, an important exception: the determinant of a **triangular** matrix. A square matrix is **upper triangular** if it has all zero entries below its main diagonal and **lower triangular** if it has all zero entries above its main diagonal. A matrix that is both upper and lower triangular is called **diagonal.** That is, a diagonal matrix is one in which all entries above and below the main diagonal are zero.

Upper Triangular Matrix

$$\begin{bmatrix} a_{11} & a_{12} & a_{13} & \cdots & a_{1n} \\ 0 & a_{22} & a_{23} & \cdots & a_{2n} \\ 0 & 0 & a_{33} & \cdots & a_{3n} \\ \vdots & \vdots & \vdots & & \vdots \\ 0 & 0 & 0 & \cdots & a_{nm} \end{bmatrix}$$

Lower Triangular Matrix

$$\begin{bmatrix} a_{11} & 0 & 0 & \cdots & 0 \\ a_{21} & a_{22} & 0 & \cdots & 0 \\ a_{31} & a_{32} & a_{33} & \cdots & 0 \\ \vdots & \vdots & \vdots & & \vdots \\ a_{n1} & a_{n2} & a_{n3} & \cdots & a_{nm} \end{bmatrix}$$

To find the determinant of a triangular matrix of any order, simply form the product of the entries on the main diagonal.

EXAMPLE 5 The Determinant of a Triangular Matrix

a.
$$\begin{vmatrix} 2 & 0 & 0 & 0 \\ 4 & -2 & 0 & 0 \\ -5 & 6 & 1 & 0 \\ 1 & 5 & 3 & 3 \end{vmatrix} = (2)(-2)(1)(3) = -12$$

b.
$$\begin{vmatrix} -1 & 0 & 0 & 0 & 0 \\ 0 & 3 & 0 & 0 & 0 \\ 0 & 0 & 2 & 0 & 0 \\ 0 & 0 & 0 & 4 & 0 \\ 0 & 0 & 0 & 0 & -2 \end{vmatrix} = (-1)(3)(2)(4)(-2) = 48$$

GROUP ACTIVITY

THE DETERMINANT OF A TRIANGULAR MATRIX

Write an argument that explains why the determinant of a 3×3 triangular matrix is the product of its main-diagonal entries.

$$\begin{bmatrix} a_{11} & a_{12} & a_{13} \\ 0 & a_{22} & a_{23} \\ 0 & 0 & a_{33} \end{bmatrix} = a_{11} a_{22} a_{33}$$

WARM UP

Perform the arithmetic operations.

1. $[(1)(3) + (-3)(2)] - [(1)(4) + (3)(5)]$

2. $[(4)(4) + (-1)(-3)] - [(-1)(2) + (-2)(7)]$

3. $\dfrac{4(7) - 1(-2)}{(-5)(-2) - 3(4)}$ **4.** $\dfrac{3(6) - 2(7)}{6(-5) - 2(1)}$

5. $-5(-1)^2[6(-2) - 7(-3)]$ **6.** $4(-1)^3[3(6) - 2(7)]$

Write the matrix in row-echelon form.

7. $\begin{bmatrix} 2 & -6 \\ 5 & 20 \end{bmatrix}$ **8.** $\begin{bmatrix} -7 & 21 \\ 3 & 5 \end{bmatrix}$

9. $\begin{bmatrix} 1 & 3 & 4 \\ 0 & 1 & 1 \\ 2 & 4 & 6 \end{bmatrix}$ **10.** $\begin{bmatrix} 4 & 8 & 16 \\ 3 & -1 & 2 \\ -2 & 10 & 12 \end{bmatrix}$

10.4 Exercises

In Exercises 1–16, find the determinant of the matrix.

1. $[5]$

2. $[-8]$

3. $\begin{bmatrix} 2 & 1 \\ 3 & 4 \end{bmatrix}$

4. $\begin{bmatrix} -3 & 1 \\ 5 & 2 \end{bmatrix}$

5. $\begin{bmatrix} 5 & 2 \\ -6 & 3 \end{bmatrix}$

6. $\begin{bmatrix} 2 & -2 \\ 4 & 3 \end{bmatrix}$

7. $\begin{bmatrix} -7 & 6 \\ \frac{1}{2} & 3 \end{bmatrix}$

8. $\begin{bmatrix} 4 & -3 \\ 0 & 0 \end{bmatrix}$

9. $\begin{bmatrix} 2 & 6 \\ 0 & 3 \end{bmatrix}$

10. $\begin{bmatrix} 2 & -3 \\ -6 & 9 \end{bmatrix}$

11. $\begin{bmatrix} 2 & -1 & 0 \\ 4 & 2 & 1 \\ 4 & 2 & 1 \end{bmatrix}$

12. $\begin{bmatrix} -2 & 2 & 3 \\ 1 & -1 & 0 \\ 0 & 1 & 4 \end{bmatrix}$

13. $\begin{bmatrix} 6 & 3 & -7 \\ 0 & 0 & 0 \\ 4 & -6 & 3 \end{bmatrix}$

14. $\begin{bmatrix} 1 & 1 & 2 \\ 3 & 1 & 0 \\ -2 & 0 & 3 \end{bmatrix}$

15. $\begin{bmatrix} -1 & 2 & -5 \\ 0 & 3 & 4 \\ 0 & 0 & 3 \end{bmatrix}$

16. $\begin{bmatrix} 1 & 0 & 0 \\ -4 & -1 & 0 \\ 5 & 1 & 5 \end{bmatrix}$

In Exercises 17–20, use the matrix capabilities of a graphing utility to find the determinant of the matrix.

17. $\begin{bmatrix} 0.3 & 0.2 & 0.2 \\ 0.2 & 0.2 & 0.2 \\ -0.4 & 0.4 & 0.3 \end{bmatrix}$

18. $\begin{bmatrix} 0.1 & 0.2 & 0.3 \\ -0.3 & 0.2 & 0.2 \\ 0.5 & 0.4 & 0.4 \end{bmatrix}$

19. $\begin{bmatrix} 1 & 4 & -2 \\ 3 & 6 & -6 \\ -2 & 1 & 4 \end{bmatrix}$

20. $\begin{bmatrix} 2 & 3 & 1 \\ 0 & 5 & -2 \\ 0 & 0 & -2 \end{bmatrix}$

In Exercises 21–24, find all (a) minors and (b) cofactors of the matrix.

21. $\begin{bmatrix} 3 & 4 \\ 2 & -5 \end{bmatrix}$

22. $\begin{bmatrix} 11 & 0 \\ -3 & 2 \end{bmatrix}$

23. $\begin{bmatrix} 3 & -2 & 8 \\ 3 & 2 & -6 \\ -1 & 3 & 6 \end{bmatrix}$ **24.** $\begin{bmatrix} -2 & 9 & 4 \\ 7 & -6 & 0 \\ 6 & 7 & -6 \end{bmatrix}$

In Exercises 25–30, find the determinant of the matrix by the method of expansion by cofactors. Expand using the indicated row or column.

25. $\begin{bmatrix} -3 & 2 & 1 \\ 4 & 5 & 6 \\ 2 & -3 & 1 \end{bmatrix}$ **26.** $\begin{bmatrix} -3 & 4 & 2 \\ 6 & 3 & 1 \\ 4 & -7 & -8 \end{bmatrix}$

    (a) Row 1   (a) Row 2

    (b) Column 2   (b) Column 3

27. $\begin{bmatrix} 5 & 0 & -3 \\ 0 & 12 & 4 \\ 1 & 6 & 3 \end{bmatrix}$ **28.** $\begin{bmatrix} 10 & -5 & 5 \\ 30 & 0 & 10 \\ 0 & 10 & 1 \end{bmatrix}$

    (a) Row 2   (a) Row 3

    (b) Column 2   (b) Column 1

29. $\begin{bmatrix} 6 & 0 & -3 & 5 \\ 4 & 13 & 6 & -8 \\ -1 & 0 & 7 & 4 \\ 8 & 6 & 0 & 2 \end{bmatrix}$

    (a) Row 2

    (b) Column 2

30. $\begin{bmatrix} 10 & 8 & 3 & -7 \\ 4 & 0 & 5 & -6 \\ 0 & 3 & 2 & 7 \\ 1 & 0 & -3 & 2 \end{bmatrix}$

    (a) Row 3

    (b) Column 1

In Exercises 31–40, find the determinant of the matrix. Expand by cofactors on the row or column that appears to make the computations easiest.

31. $\begin{bmatrix} 1 & 4 & -2 \\ 3 & 2 & 0 \\ -1 & 4 & 3 \end{bmatrix}$ **32.** $\begin{bmatrix} 2 & -1 & 3 \\ 1 & 4 & 4 \\ 1 & 0 & 2 \end{bmatrix}$

33. $\begin{bmatrix} 2 & 4 & 6 \\ 0 & 3 & 1 \\ 0 & 0 & -5 \end{bmatrix}$ **34.** $\begin{bmatrix} -3 & 0 & 0 \\ 7 & 11 & 0 \\ 1 & 2 & 2 \end{bmatrix}$

35. $\begin{bmatrix} 2 & 6 & 6 & 2 \\ 2 & 7 & 3 & 6 \\ 1 & 5 & 0 & 1 \\ 3 & 7 & 0 & 7 \end{bmatrix}$ **36.** $\begin{bmatrix} 3 & 6 & -5 & 4 \\ -2 & 0 & 6 & 0 \\ 1 & 1 & 2 & 2 \\ 0 & 3 & -1 & -1 \end{bmatrix}$

37. $\begin{bmatrix} 5 & 3 & 0 & 6 \\ 4 & 6 & 4 & 12 \\ 0 & 2 & -3 & 4 \\ 0 & 1 & -2 & 2 \end{bmatrix}$ **38.** $\begin{bmatrix} 1 & 4 & 3 & 2 \\ -5 & 6 & 2 & 1 \\ 0 & 0 & 0 & 0 \\ 3 & -2 & 1 & 5 \end{bmatrix}$

39. $\begin{bmatrix} 3 & 2 & 4 & -1 & 5 \\ -2 & 0 & 1 & 3 & 2 \\ 1 & 0 & 0 & 4 & 0 \\ 6 & 0 & 2 & -1 & 0 \\ 3 & 0 & 5 & 1 & 0 \end{bmatrix}$

40. $\begin{bmatrix} 5 & 2 & 0 & 0 & -2 \\ 0 & 1 & 4 & 3 & 2 \\ 0 & 0 & 2 & 6 & 3 \\ 0 & 0 & 3 & 4 & 1 \\ 0 & 0 & 0 & 0 & 2 \end{bmatrix}$

In Exercises 41–48, use the matrix capabilities of a graphing utility to evaluate the determinant.

41. $\begin{vmatrix} 3 & 8 & -7 \\ 0 & -5 & 4 \\ 8 & 1 & 6 \end{vmatrix}$ **42.** $\begin{vmatrix} 5 & -8 & 0 \\ 9 & 7 & 4 \\ -8 & 7 & 1 \end{vmatrix}$

43. $\begin{vmatrix} 7 & 0 & -14 \\ -2 & 5 & 4 \\ -6 & 2 & 12 \end{vmatrix}$ **44.** $\begin{vmatrix} 3 & 0 & 0 \\ -2 & 5 & 0 \\ 12 & 5 & 7 \end{vmatrix}$

45. $\begin{vmatrix} 1 & -1 & 8 & 4 \\ 2 & 6 & 0 & -4 \\ 2 & 0 & 2 & 6 \\ 0 & 2 & 8 & 0 \end{vmatrix}$ **46.** $\begin{vmatrix} 0 & -3 & 8 & 2 \\ 8 & 1 & -1 & 6 \\ -4 & 6 & 0 & 9 \\ -7 & 0 & 0 & 14 \end{vmatrix}$

47. $\begin{vmatrix} 3 & -2 & 4 & 3 & 1 \\ -1 & 0 & 2 & 1 & 0 \\ 5 & -1 & 0 & 3 & 2 \\ 4 & 7 & -8 & 0 & 0 \\ 1 & 2 & 3 & 0 & 2 \end{vmatrix}$

48. $\begin{vmatrix} -2 & 0 & 0 & 0 & 0 \\ 0 & 3 & 0 & 0 & 0 \\ 0 & 0 & -1 & 0 & 0 \\ 0 & 0 & 0 & 2 & 0 \\ 0 & 0 & 0 & 0 & -4 \end{vmatrix}$

In Exercises 49–52, evaluate the determinants to verify the equation.

49. $\begin{vmatrix} w & x \\ y & z \end{vmatrix} = -\begin{vmatrix} y & z \\ w & x \end{vmatrix}$

50. $\begin{vmatrix} w & cx \\ y & cz \end{vmatrix} = c\begin{vmatrix} w & x \\ y & z \end{vmatrix}$

51. $\begin{vmatrix} w & x \\ y & z \end{vmatrix} = \begin{vmatrix} w & x + cw \\ y & z + cy \end{vmatrix}$

52. $\begin{vmatrix} w & x \\ cw & cx \end{vmatrix} = 0$

In Exercises 53 and 54, evaluate the determinant to verify the equation.

53. $\begin{vmatrix} 1 & x & x^2 \\ 1 & y & y^2 \\ 1 & z & z^2 \end{vmatrix} = (y - x)(z - x)(z - y)$

54. $\begin{vmatrix} a + b & a & a \\ a & a + b & a \\ a & a & a + b \end{vmatrix} = b^2(3a + b)$

In Exercises 55 and 56, solve for x.

55. $\begin{vmatrix} x - 1 & 2 \\ 3 & x - 2 \end{vmatrix} = 0$

56. $\begin{vmatrix} x - 2 & -1 \\ -3 & x \end{vmatrix} = 0$

In Exercises 57–62, evaluate the determinant where the entries are functions. Determinants of this type occur in calculus.

57. $\begin{vmatrix} 4u & -1 \\ -1 & 2v \end{vmatrix}$

58. $\begin{vmatrix} 3x^2 & -3y^2 \\ 1 & 1 \end{vmatrix}$

59. $\begin{vmatrix} e^{2x} & e^{3x} \\ 2e^{2x} & 3e^{3x} \end{vmatrix}$

60. $\begin{vmatrix} e^{-x} & xe^{-x} \\ -e^{-x} & (1 - x)e^{-x} \end{vmatrix}$

61. $\begin{vmatrix} x & \ln x \\ 1 & 1/x \end{vmatrix}$

62. $\begin{vmatrix} x & x \ln x \\ 1 & 1 + \ln x \end{vmatrix}$

In Exercises 63–66, find (a) $|A|$, (b) $|B|$, (c) AB, and (d) $|AB|$.

63. $A = \begin{bmatrix} -1 & 0 \\ 0 & 3 \end{bmatrix}$, $B = \begin{bmatrix} 2 & 0 \\ 0 & -1 \end{bmatrix}$

64. $A = \begin{bmatrix} -2 & 1 \\ 4 & -2 \end{bmatrix}$, $B = \begin{bmatrix} 1 & 2 \\ 0 & -1 \end{bmatrix}$

65. $A = \begin{bmatrix} -1 & 2 & 1 \\ 1 & 0 & 1 \\ 0 & 1 & 0 \end{bmatrix}$, $B = \begin{bmatrix} -1 & 0 & 0 \\ 0 & 2 & 0 \\ 0 & 0 & 3 \end{bmatrix}$

66. $A = \begin{bmatrix} 2 & 0 & 1 \\ 1 & -1 & 2 \\ 3 & 1 & 0 \end{bmatrix}$, $B = \begin{bmatrix} 2 & -1 & 4 \\ 0 & 1 & 3 \\ 3 & -2 & 1 \end{bmatrix}$

67. *Exploration* Find square matrices A and B to demonstrate that

$$|A + B| \neq |A| + |B|.$$

68. *Exploration* Consider square matrices in which the entries are consecutive integers. An example of such a matrix is

$$\begin{bmatrix} 4 & 5 & 6 \\ 7 & 8 & 9 \\ 10 & 11 & 12 \end{bmatrix}.$$

(a) Use a graphing utility to evaluate four determinants of matrices of this type. Make a conjecture based on the results.

(b) Verify your conjecture.

69. *Essay* Write a brief paragraph explaining the difference between a square matrix and its determinant.

70. *Think About It* If A is a matrix of order 3×3 such that $|A| = 5$, is it possible to find $|2A|$? Explain.

Review **Solve Exercises 71–74 as a review of the skills and problem-solving techniques you learned in previous sections. Find the equation of the conic satisfying the given conditions.**

71. Parabola: Vertex: $(0, 3)$; Focus: $(2, 3)$

72. Ellipse: Vertices: $(0, \pm 4)$; Foci: $(0, \pm 3)$

73. Ellipse: Vertices: $(\pm 8, 0)$; Foci: $(\pm 6, 0)$

74. Hyperbola: Vertices: $(\pm 5, 0)$; Foci: $(\pm 6, 0)$

See Exercises 35 and 36 on page 789 for examples of how matrices can be used in cryptography.

▶ *10.5*   ***Applications of Matrices and Determinants***

Cramer's Rule ◻ *Area of a Triangle* ◻ *Lines in the Plane* ◻ *Cryptography*

Cramer's Rule

So far, you have studied three methods for solving a system of linear equations: substitution, elimination with equations, and elimination with matrices. In this section, you will study one more method, **Cramer's Rule,** named after Gabriel Cramer (1704–1752). This rule uses determinants to write the solution of a system of linear equations. To see how Cramer's Rule works, take another look at the solution described at the beginning of Section 10.4. There, it was pointed out that the system

$$a_1x + b_1y = c_1$$
$$a_2x + b_2y = c_2$$

has a solution given by

$$x = \frac{c_1b_2 - c_2b_1}{a_1b_2 - a_2b_1} \quad \text{and} \quad y = \frac{a_1c_2 - a_2c_1}{a_1b_2 - a_2b_1}$$

provided that $a_1b_2 - a_2b_1 \neq 0$. Each numerator and denominator in this solution can be expressed as a determinant, as follows.

$$x = \frac{c_1b_2 - c_2b_1}{a_1b_2 - a_2b_1} = \frac{\begin{vmatrix} c_1 & b_1 \\ c_2 & b_2 \end{vmatrix}}{\begin{vmatrix} a_1 & b_1 \\ a_2 & b_2 \end{vmatrix}}, \quad y = \frac{a_1c_2 - a_2c_1}{a_1b_2 - a_2b_1} = \frac{\begin{vmatrix} a_1 & c_1 \\ a_2 & c_2 \end{vmatrix}}{\begin{vmatrix} a_1 & b_1 \\ a_2 & b_2 \end{vmatrix}}$$

Relative to the original system, the denominator for x and y is simply the determinant of the *coefficient* matrix of the system. This determinant is denoted by D. The numerators for x and y are denoted by D_x and D_y, respectively. They are formed by using the column of constants as replacements for the coefficients of x and y, as follows.

Coefficient Matrix	D	D_x	D_y
$\begin{bmatrix} a_1 & b_1 \\ a_2 & b_2 \end{bmatrix}$	$\begin{bmatrix} a_1 & b_1 \\ a_2 & b_2 \end{bmatrix}$	$\begin{bmatrix} c_1 & b_1 \\ c_2 & b_2 \end{bmatrix}$	$\begin{bmatrix} a_1 & c_1 \\ a_2 & c_2 \end{bmatrix}$

EXAMPLE 1   **Using Cramer's Rule for a 2 × 2 System**

Use Cramer's Rule to solve the following system of linear equations.

$$4x - 2y = 10$$
$$3x - 5y = 11$$

Solution

To begin, find the determinant of the coefficient matrix.

$$D = \begin{vmatrix} 4 & -2 \\ 3 & -5 \end{vmatrix} = -20 - (-6) = -14$$

Because this determinant is not zero, you can apply Cramer's Rule to find the solution, as follows.

$$x = \frac{D_x}{D} = \frac{\begin{vmatrix} 10 & -2 \\ 11 & -5 \end{vmatrix}}{-14} = \frac{(-50) - (-22)}{-14} = \frac{-28}{-14} = 2$$

$$y = \frac{D_y}{D} = \frac{\begin{vmatrix} 4 & 10 \\ 3 & 11 \end{vmatrix}}{-14} = \frac{44 - 30}{-14} = \frac{14}{-14} = -1$$

Therefore, the solution is $x = 2$ and $y = -1$. Check this in the original system.

Cramer's Rule generalizes easily to systems of n equations in n variables. The value of each variable is given as the quotient of two determinants. The denominator is the determinant of the coefficient matrix, and the numerator is the determinant of the matrix formed by replacing the column corresponding to the variable (being solved for) with the column representing the constants. For instance, the solution for x_3 in the system

$$a_{11}x_1 + a_{12}x_2 + a_{13}x_3 = b_1$$
$$a_{21}x_1 + a_{22}x_2 + a_{23}x_3 = b_2$$
$$a_{31}x_1 + a_{32}x_2 + a_{33}x_3 = b_3$$

is given by

$$x_3 = \frac{|A_3|}{|A|} = \frac{\begin{vmatrix} a_{11} & a_{12} & b_1 \\ a_{21} & a_{22} & b_2 \\ a_{31} & a_{32} & b_3 \end{vmatrix}}{\begin{vmatrix} a_{11} & a_{12} & a_{13} \\ a_{21} & a_{22} & a_{23} \\ a_{31} & a_{32} & a_{33} \end{vmatrix}}.$$

> ◼ **CRAMER'S RULE**
>
> If a system of n linear equations in n variables has a coefficient matrix A with a *nonzero* determinant $|A|$, the solution of the system is given by
>
> $$x_1 = \frac{|A_1|}{|A|}, \quad x_2 = \frac{|A_2|}{|A|}, \quad \dots, \quad x_n = \frac{|A_n|}{|A|}$$
>
> where the ith column of A_i is the column of constants in the system of equations. If the coefficient matrix is zero, the system has either no solution or infinitely many solutions.

EXAMPLE 2   *Using Cramer's Rule for a 3 × 3 System*

Use Cramer's Rule to solve the following system of linear equations.

$$\begin{aligned} -x + 2y - 3z &= 1 \\ 2x \quad\quad + z &= 0 \\ 3x - 4y + 4z &= 2 \end{aligned}$$

Solution

$$\begin{aligned} D &= 2(-1)^3 \begin{vmatrix} 2 & -3 \\ -4 & 4 \end{vmatrix} + 0(-1)^4 \begin{vmatrix} -1 & -3 \\ 3 & 4 \end{vmatrix} + 1(-1)^5 \begin{vmatrix} -1 & 2 \\ 3 & -4 \end{vmatrix} \\ &= 2(4) + 0 + (1)(2) \\ &= 10 \end{aligned}$$

NOTE   The coefficient matrix

$$\begin{vmatrix} -1 & 2 & -3 \\ 2 & 0 & 1 \\ 3 & -4 & 4 \end{vmatrix}$$

in Example 2 is expanded along the second row. ◼◼

Because this determinant is not zero, you can apply Cramer's Rule to find the solution, as follows.

$$x = \frac{D_x}{D} = \frac{\begin{vmatrix} 1 & 2 & -3 \\ 0 & 0 & 1 \\ 2 & -4 & 4 \end{vmatrix}}{10} = \frac{8}{10} = \frac{4}{5}$$

$$y = \frac{D_y}{D} = \frac{\begin{vmatrix} -1 & 1 & -3 \\ 2 & 0 & 1 \\ 3 & 2 & 4 \end{vmatrix}}{10} = \frac{-15}{10} = -\frac{3}{2}$$

NOTE   When using Cramer's Rule, remember that the method *does not* apply if the determinant of the coefficient matrix is zero. ◼◼

$$z = \frac{D_z}{D} = \frac{\begin{vmatrix} -1 & 2 & 1 \\ 2 & 0 & 0 \\ 3 & -4 & 2 \end{vmatrix}}{10} = \frac{-16}{10} = -\frac{8}{5}$$

The solution is $\left(\frac{4}{5}, -\frac{3}{2}, -\frac{8}{5}\right)$. Check this in the original system.

Area of a Triangle

Another application of matrices and determinants is finding the area of a triangle whose vertices are given as points on a coordinate plane.

> **AREA OF A TRIANGLE**
>
> The area of a triangle with vertices (x_1, y_1), (x_2, y_2), and (x_3, y_3) is given by
>
> $$\text{Area} = \pm \frac{1}{2} \begin{vmatrix} x_1 & y_1 & 1 \\ x_2 & y_2 & 1 \\ x_3 & y_3 & 1 \end{vmatrix}$$
>
> where the symbol $\pm$ indicates that the appropriate sign should be chosen to yield a positive area.

EXAMPLE 3 *Finding the Area of a Triangle*

Find the area of a triangle whose vertices are $(1, 0)$, $(2, 2)$, and $(4, 3)$, as shown in Figure 10.1.

Solution

Let $(x_1, y_1) = (1, 0)$, $(x_2, y_2) = (2, 2)$, and $(x_3, y_3) = (4, 3)$. Then, to find the area of a triangle, evaluate the determinant.

$$\begin{vmatrix} x_1 & y_1 & 1 \\ x_2 & y_2 & 1 \\ x_3 & y_3 & 1 \end{vmatrix} = \begin{vmatrix} 1 & 0 & 1 \\ 2 & 2 & 1 \\ 4 & 3 & 1 \end{vmatrix}$$

$$= 1(-1)^2 \begin{vmatrix} 2 & 1 \\ 3 & 1 \end{vmatrix} + 0(-1)^3 \begin{vmatrix} 2 & 1 \\ 4 & 1 \end{vmatrix} + 1(-1)^4 \begin{vmatrix} 2 & 2 \\ 4 & 3 \end{vmatrix}$$

$$= 1(-1) + 0 + 1(-2)$$

$$= -3.$$

Using this value, you can conclude that the area of the triangle is

$$\text{Area} = -\frac{1}{2} \begin{vmatrix} 1 & 0 & 1 \\ 2 & 2 & 1 \\ 4 & 3 & 1 \end{vmatrix} = -\frac{1}{2}(-3) = \frac{3}{2}.$$

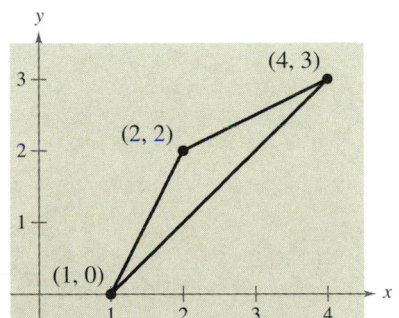

FIGURE 10.1

Lines in a Plane

Suppose the three points in Example 3 had been on the same line. What would have happened had the area formula been applied to three such points? The answer is that the determinant would have been zero. Consider, for instance, the three collinear points $(0, 1)$, $(2, 2)$, and $(4, 3)$, as shown in Figure 10.2. The area of the "triangle" that has these three points as vertices is

$$\frac{1}{2}\begin{vmatrix} 0 & 1 & 1 \\ 2 & 2 & 1 \\ 4 & 3 & 1 \end{vmatrix} = \frac{1}{2}\left[0(-1)^2 \begin{vmatrix} 2 & 1 \\ 3 & 1 \end{vmatrix} + 1(-1)^3 \begin{vmatrix} 2 & 1 \\ 4 & 1 \end{vmatrix} + 1(-1)^4 \begin{vmatrix} 2 & 2 \\ 4 & 3 \end{vmatrix} \right]$$

$$= \frac{1}{2}[0(-1) + 1(2) + 1(-2)] = 0.$$

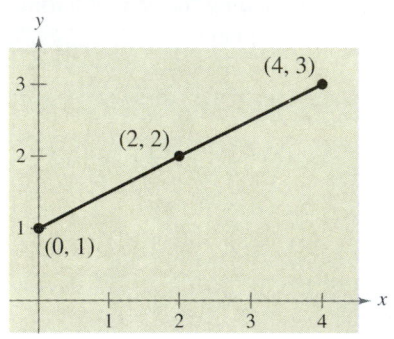

FIGURE 10.2

The result is generalized as follows.

TEST FOR COLLINEAR POINTS

Three points (x_1, y_1), (x_2, y_2), and (x_3, y_3) are collinear (lie on the same line) if and only if

$$\begin{vmatrix} x_1 & y_1 & 1 \\ x_2 & y_2 & 1 \\ x_3 & y_3 & 1 \end{vmatrix} = 0.$$

EXAMPLE 4 Testing for Collinear Points

Determine whether the points $(-2, -2)$, $(1, 1)$, and $(7, 5)$ lie on the same line. (See Figure 10.3.)

Solution

Letting $(x_1, y_1) = (-2, -2)$, $(x_2, y_2) = (1, 1)$, and $(x_3, y_3) = (7, 5)$, you have

$$\begin{vmatrix} x_1 & y_1 & 1 \\ x_2 & y_2 & 1 \\ x_3 & y_3 & 1 \end{vmatrix} = \begin{vmatrix} -2 & -2 & 1 \\ 1 & 1 & 1 \\ 7 & 5 & 1 \end{vmatrix}$$

$$= -2(-1)^2 \begin{vmatrix} 1 & 1 \\ 5 & 1 \end{vmatrix} + (-2)(-1)^3 \begin{vmatrix} 1 & 1 \\ 7 & 1 \end{vmatrix} + 1(-1)^4 \begin{vmatrix} 1 & 1 \\ 7 & 5 \end{vmatrix}$$

$$= -2(-4) + (-2)(6) + 1(-2)$$

$$= -6.$$

Because the value of this determinant is *not* zero, you can conclude that the three points do not lie on the same line.

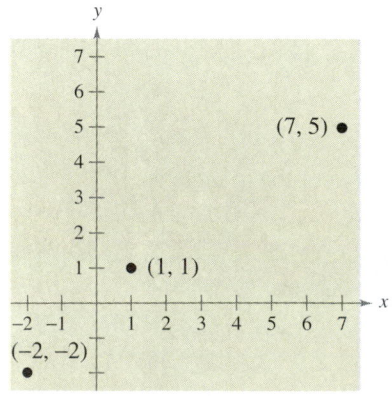

FIGURE 10.3

The test for collinear points can be adapted to another use. That is, if you are given two points on a rectangular coordinate system, you can find an equation of the line passing through the two points, as follows.

TWO-POINT FORM OF THE EQUATION OF A LINE

An equation of the line passing through the distinct points (x_1, y_1) and (x_2, y_2) is given by

$$\begin{vmatrix} x & y & 1 \\ x_1 & y_1 & 1 \\ x_2 & y_2 & 1 \end{vmatrix} = 0.$$

EXAMPLE 5 *Finding an Equation of a Line*

Find an equation of the line passing through the two points $(2, 4)$ and $(-1, 3)$, as shown in Figure 10.4.

Solution

Applying the determinant formula for the equation of a line produces

$$\begin{vmatrix} x & y & 1 \\ 2 & 4 & 1 \\ -1 & 3 & 1 \end{vmatrix} = 0.$$

To evaluate this determinant, you can expand by cofactors along the first row to obtain the following.

$$x(-1)^2\begin{vmatrix} 4 & 1 \\ 3 & 1 \end{vmatrix} + y(-1)^3\begin{vmatrix} 2 & 1 \\ -1 & 1 \end{vmatrix} + 1(-1)^4\begin{vmatrix} 2 & 4 \\ -1 & 3 \end{vmatrix} = x - 3y + 10 = 0$$

Therefore, an equation of the line is

$$x - 3y + 10 = 0.$$

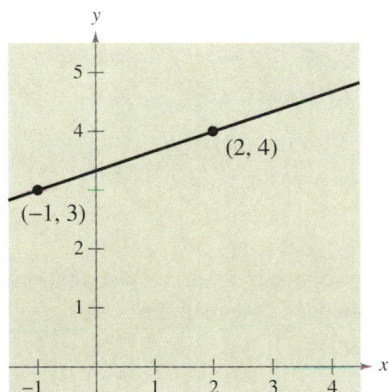

FIGURE 10.4

Note that this method of finding the equation of a line works for all lines, including horizontal and vertical lines. For instance, the equation of the vertical line through $(2, 0)$ and $(2, 2)$ is

$$\begin{vmatrix} x & y & 1 \\ 2 & 0 & 1 \\ 2 & 2 & 1 \end{vmatrix} = 0$$

$$4 - 2x = 0$$

$$x = 2.$$

Cryptography

A **cryptogram** is a message written according to secret code. (The Greek word *kryptos* means "hidden.") Matrix multiplication can be used to **encode** and **decode** messages. To begin, you need to assign a number to each letter in the alphabet (with 0 assigned to a blank space), as follows.

0 = _	9 = I	18 = R
1 = A	10 = J	19 = S
2 = B	11 = K	20 = T
3 = C	12 = L	21 = U
4 = D	13 = M	22 = V
5 = E	14 = N	23 = W
6 = F	15 = O	24 = X
7 = G	16 = P	25 = Y
8 = H	17 = Q	26 = Z

Then the message is converted to numbers and partitioned into **uncoded row matrices,** each having n entries, as demonstrated in Example 6.

The Enigma, a German cipher machine, was heavily used during World War II for intelligence work. Battery powered, reliable, portable, and well adapted to radio work, the Enigma was cryptanalytically secure if the keys were changed three times a day. *(Photo: Courtesy of Marshall Foundation, Lexington, VA)*

EXAMPLE 6   *Forming Uncoded Row Matrices*

Write the uncoded row matrices of order 1×3 for the message

    MEET ME MONDAY.

Solution

Partitioning the message (including blank spaces, but ignoring punctuation) into groups of three produces the following uncoded row matrices.

$$[13 \quad 5 \quad 5] \quad [20 \quad 0 \quad 13] \quad [5 \quad 0 \quad 13] \quad [15 \quad 14 \quad 4] \quad [1 \quad 25 \quad 0]$$
$$\text{M} \quad \text{E} \quad \text{E} \quad \text{T} \qquad \text{M} \quad \text{E} \qquad \text{M} \quad \text{O} \quad \text{N} \quad \text{D} \quad \text{A} \quad \text{Y}$$

Note that a blank space is used to fill out the last uncoded row matrix.

To **encode** a message, choose $n \times n$ invertible matrix A and multiply the uncoded row matrices by A (on the right) to obtain **coded row matrices.** Here is an example.

Uncoded Matrix Encoding Matrix A   Coded Matrix

$$[13 \quad 5 \quad 5] \begin{bmatrix} 1 & -2 & 2 \\ -1 & 1 & 3 \\ 1 & -1 & -4 \end{bmatrix} = [13 \quad -26 \quad 21]$$

This technique is further illustrated in Example 7.

EXAMPLE 7   Encoding a Message

Use the following matrix to encode the message MEET ME MONDAY.

$$A = \begin{bmatrix} 1 & -2 & 2 \\ -1 & 1 & 3 \\ 1 & -1 & -4 \end{bmatrix}$$

Solution

The coded row matrices are obtained by multiplying each of the uncoded row matrices found in Example 6 by the matrix A, as follows.

Uncoded Matrix   Encoding Matrix A   Coded Matrix

$$\begin{bmatrix} 13 & 5 & 5 \end{bmatrix} \begin{bmatrix} 1 & -2 & 2 \\ -1 & 1 & 3 \\ 1 & -1 & -4 \end{bmatrix} = \begin{bmatrix} 13 & -26 & 21 \end{bmatrix}$$

$$\begin{bmatrix} 20 & 0 & 13 \end{bmatrix} \begin{bmatrix} 1 & -2 & 2 \\ -1 & 1 & 3 \\ 1 & -1 & -4 \end{bmatrix} = \begin{bmatrix} 33 & -53 & -12 \end{bmatrix}$$

$$\begin{bmatrix} 5 & 0 & 13 \end{bmatrix} \begin{bmatrix} 1 & -2 & 2 \\ -1 & 1 & 3 \\ 1 & -1 & -4 \end{bmatrix} = \begin{bmatrix} 18 & -23 & -42 \end{bmatrix}$$

$$\begin{bmatrix} 15 & 14 & 4 \end{bmatrix} \begin{bmatrix} 1 & -2 & 2 \\ -1 & 1 & 3 \\ 1 & -1 & -4 \end{bmatrix} = \begin{bmatrix} 5 & -20 & 56 \end{bmatrix}$$

$$\begin{bmatrix} 1 & 25 & 0 \end{bmatrix} \begin{bmatrix} 1 & -2 & 2 \\ -1 & 1 & 3 \\ 1 & -1 & -4 \end{bmatrix} = \begin{bmatrix} -24 & 23 & 77 \end{bmatrix}$$

Thus, the sequence of coded row matrices is

$$\begin{bmatrix} 13 & -26 & 21 \end{bmatrix}\begin{bmatrix} 33 & -53 & -12 \end{bmatrix}\begin{bmatrix} 18 & -23 & -42 \end{bmatrix}\begin{bmatrix} 5 & -20 & 56 \end{bmatrix}\begin{bmatrix} -24 & 23 & 77 \end{bmatrix}.$$

Finally, removing the matrix notation produces the following cryptogram.

$$13 \ -26 \ 21 \ 33 \ -53 \ -12 \ 18 \ -23 \ -42 \ 5 \ -20 \ 56 \ -24 \ 23 \ 77$$

For those who do not know the matrix A, decoding the cryptogram found in Example 7 is difficult. But for an authorized receiver who knows the matrix A, decoding is simple. The receiver need only multiply the coded row matrices by A^{-1} (on the right) to retrieve the uncoded row matrices. Here is an example.

$$\underbrace{\begin{bmatrix} 13 & -26 & 21 \end{bmatrix}}_{\text{Coded}} A^{-1} = \underbrace{\begin{bmatrix} 13 & 5 & 5 \end{bmatrix}}_{\text{Uncoded}}$$

EXAMPLE 8 *Decoding a Message*

Use the inverse of the matrix $A = \begin{bmatrix} 1 & -2 & 2 \\ -1 & 1 & 3 \\ 1 & -1 & -4 \end{bmatrix}$ to decode the cryptogram

13 −26 21 33 −53 −12 18 −23 −42 5 −20 56 −24 23 77.

Solution

Partition the message into groups of three to form the coded row matrices. Then, multiply each coded row matrix by A^{-1} (on the right).

Coded Matrix   *Decoding Matrix* A^{-1} *Decoded Matrix*

$$[13 \quad -26 \quad 21] \begin{bmatrix} -1 & -10 & -8 \\ -1 & -6 & -5 \\ 0 & -1 & -1 \end{bmatrix} = [13 \quad 5 \quad 5]$$

$$[33 \quad -53 \quad -12] \begin{bmatrix} -1 & -10 & -8 \\ -1 & -6 & -5 \\ 0 & -1 & -1 \end{bmatrix} = [20 \quad 0 \quad 13]$$

$$[18 \quad -23 \quad -42] \begin{bmatrix} -1 & -10 & -8 \\ -1 & -6 & -5 \\ 0 & -1 & -1 \end{bmatrix} = [5 \quad 0 \quad 13]$$

$$[5 \quad -20 \quad 56] \begin{bmatrix} -1 & -10 & -8 \\ -1 & -6 & -5 \\ 0 & -1 & -1 \end{bmatrix} = [15 \quad 14 \quad 4]$$

$$[-24 \quad 23 \quad 77] \begin{bmatrix} -1 & -10 & -8 \\ -1 & -6 & -5 \\ 0 & -1 & -1 \end{bmatrix} = [1 \quad 25 \quad 0]$$

Thus, the message is as follows.

[13 5 5] [20 0 13] [5 0 13] [15 14 4] [1 25 0]

M E E T   M E   M O N D A Y

GROUP ACTIVITY

CRYPTOGRAPHY

Create your own numeric code for the alphabet (see page 784), and use it to convert a message into numbers. Create an invertible $n \times n$ matrix A to encode your message. Exchange your numeric code, encoded message, and matrix A with another group. Find the necessary decoding matrix and decode the message you received.

WARM UP

Evaluate the determinant.

1. $\begin{vmatrix} 4 & 3 \\ -3 & -2 \end{vmatrix}$

2. $\begin{vmatrix} 10 & -20 \\ -1 & 2 \end{vmatrix}$

3. $\begin{vmatrix} 4 & 0 \\ -3 & -2 \end{vmatrix}$

4. $\begin{vmatrix} x & x^2 \\ 1 & 2x \end{vmatrix}$

5. $\begin{vmatrix} 4 & 0 & -2 \\ 3 & 1 & 2 \\ -8 & 0 & 6 \end{vmatrix}$

6. $\begin{vmatrix} 3 & 2 & 5 \\ 0 & 0 & -4 \\ -6 & 1 & 1 \end{vmatrix}$

Find the inverse of the matrix.

7. $A = \begin{bmatrix} 1 & 3 \\ 2 & 7 \end{bmatrix}$

8. $A = \begin{bmatrix} 10 & 5 & -2 \\ -4 & -2 & 1 \\ 1 & 1 & 0 \end{bmatrix}$

Perform the matrix multiplication.

9. $\begin{bmatrix} 0.1 & 0.2 & 0.2 \\ 0.4 & 0.3 & 0.5 \\ 0.5 & 0.5 & 0.3 \end{bmatrix} \begin{bmatrix} 0.4 \\ 0.5 \\ 0.1 \end{bmatrix}$

10. $\begin{bmatrix} 2 & 5 & 8 \end{bmatrix} \begin{bmatrix} 1 & 2 & -1 \\ 1 & 2 & 2 \\ 2 & 5 & 0 \end{bmatrix}$

10.5 Exercises

In Exercises 1–4, use Cramer's Rule to solve (if possible) the system of equations.

1. $3x + 4y = -2$
   $5x + 3y = 4$

2. $-0.4x + 0.8y = 1.6$
   $0.2x + 0.3y = 2.2$

3. $4x - y + z = -5$
   $2x + 2y + 3z = 10$
   $5x - 2y + 6z = 1$

4. $4x - 2y + 3z = -2$
   $2x + 2y + 5z = 16$
   $8x - 5y - 2z = 4$

In Exercises 5 and 6, use a graphing utility and Cramer's Rule to solve (if possible) the system of equations.

5. $3x + 3y + 5z = 1$
   $3x + 5y + 9z = 2$
   $5x + 9y + 17z = 4$

6. $2x + 3y + 5z = 4$
   $3x + 5y + 9z = 7$
   $5x + 9y + 17z = 13$

In Exercises 7–16, use a determinant to find the area of the triangle with the given vertices.

7.

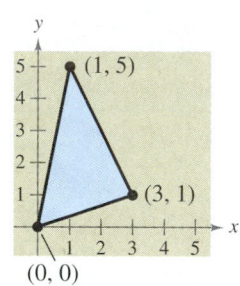

8.

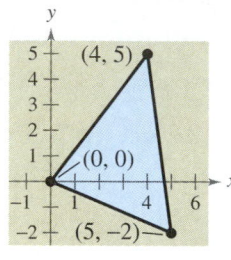

9.

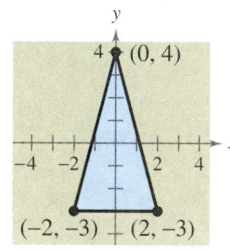

10.

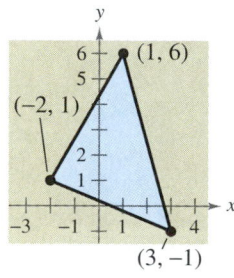

11.

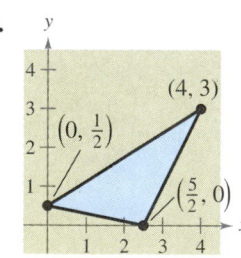

12.

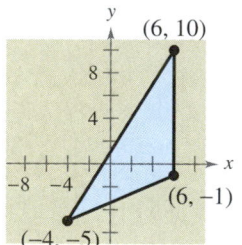

13. $(-2, 4), (2, 3), (-1, 5)$

14. $(0, -2), (-1, 4), (3, 5)$

15. $(-3, 5), (2, 6), (3, -5)$

16. $(-2, 4), (1, 5), (3, -2)$

In Exercises 17 and 18, find a value of x such that the triangle has an area of 4.

17. $(-5, 1), (0, 2), (-2, x)$

18. $(-4, 2), (-3, 5), (-1, x)$

19. *Area of a Region*   A large region of forest has been infected with gypsy moths. The region is roughly triangular, as shown in the figure. From the northernmost vertex A of the region, the distances to the other vertices are 25 miles south and 10 miles east (for vertex B), and 20 miles south and 28 miles east (for vertex C). Use a graphing utility to approximate the number of square miles in this region.

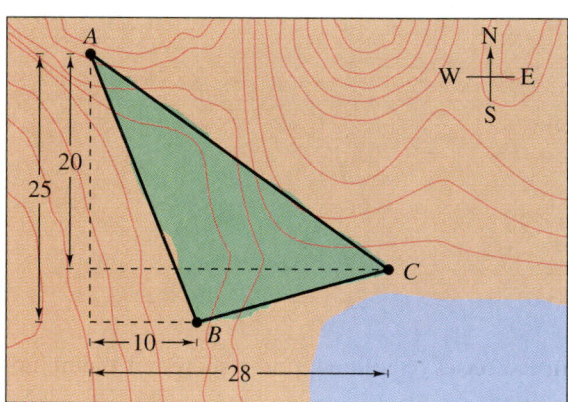

20. *Area of a Region*   You own a triangular tract of land, as shown in the figure. To estimate the number of square feet in the tract, you start at one vertex, walk 65 feet east and 50 feet north to the second vertex, and then walk 85 feet west and 30 feet north to the third vertex. Use a graphing utility to determine how many square feet there are in the tract of land.

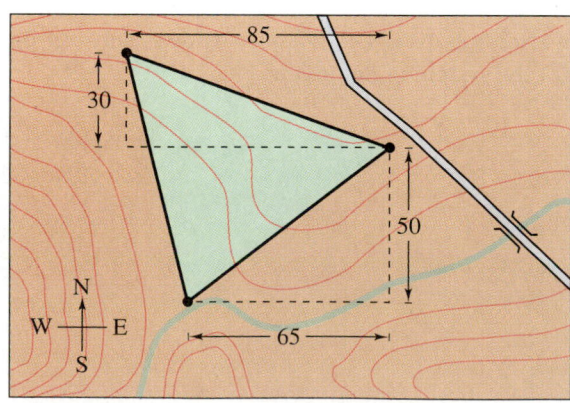

In Exercises 21–26, use a determinant to determine whether the points are collinear.

21. $(3, -1), (0, -3), (12, 5)$
22. $(-3, -5), (6, 1), (10, 2)$
23. $(2, -\frac{1}{2}), (-4, 4), (6, -3)$
24. $(0, 1), (4, -2), (-8, 7)$
25. $(0, 2), (1, 2.4), (-1, 1.6)$
26. $(2, 3), (3, 3.5), (-1, 2)$

In Exercises 27–32, use a determinant to find an equation of the line through the points.

27. $(0, 0), (5, 3)$
28. $(0, 0), (-2, 2)$
29. $(-4, 3), (2, 1)$
30. $(10, 7), (-2, -7)$
31. $(-\frac{1}{2}, 3), (\frac{5}{2}, 1)$
32. $(\frac{2}{3}, 4), (6, 12)$

In Exercises 33 and 34, find x such that the points are collinear.

33. $(2, -5), (4, x), (5, -2)$
34. $(-6, 2), (-5, x), (-3, 5)$

In Exercises 35 and 36, find the uncoded 1×3 row matrices for the message. Then encode the message using the matrix.

Message	Matrix
35. TROUBLE IN RIVER CITY	$\begin{bmatrix} 1 & -1 & 0 \\ 1 & 0 & -1 \\ -6 & 2 & 3 \end{bmatrix}$
36. PLEASE SEND MONEY	$\begin{bmatrix} 4 & 2 & 1 \\ -3 & -3 & -1 \\ 3 & 2 & 1 \end{bmatrix}$

In Exercises 37–40, write a cryptogram for the message using the matrix.

$$A = \begin{bmatrix} 1 & 2 & 2 \\ 3 & 7 & 9 \\ -1 & -4 & -7 \end{bmatrix}.$$

37. LANDING SUCCESSFUL

38. BEAM ME UP SCOTTY
39. HAPPY BIRTHDAY
40. OPERATION OVERLORD

In Exercises 41 and 42, use A^{-1} to decode the cryptogram.

41. $A = \begin{bmatrix} 1 & 2 \\ 3 & 5 \end{bmatrix}$

11, 21, 64, 112, 25, 50, 29, 53 23, 46, 40, 75, 55, 92

42. $A = \begin{bmatrix} 1 & -1 & 0 \\ 1 & 0 & -1 \\ -6 & 2 & 3 \end{bmatrix}$

9, −1, −9, 38, −19, −19, 28, −9, −19, −80, 25, 41, −64, 21, 31, −7, −4, 7

In Exercises 43 and 44, decode the cryptogram by using the inverse of the matrix

$$A = \begin{bmatrix} 1 & 2 & 2 \\ 3 & 7 & 9 \\ -1 & -4 & -7 \end{bmatrix}.$$

43. 20, 17, −15, −12, −56, −104, 1, −25, −65, 62, 143, 181
44. 13, −9, −59, 61, 112, 106, −17, −73, −131, 11, 24, 29, 65, 144, 172

45. The following cryptogram was encoded with a 2×2 matrix.

8, 21, −15, −10, −13, −13, 5, 10, 5, 25, 5, 19, −1, 6, 20, 40, −18, −18, 1, 16

The last word of the message is _RON. What is the message?

46. The following cryptogram was encoded with a 2×2 matrix.

5, 2, 25, 11, −2, −7, −15, −15, 32, 14, −8, −13, 38, 19, −19, −19, 37, 16

The last word of the message is _SUE. What is the message?

FOCUS ON CONCEPTS

In this chapter, you studied matrices and determinants and their applications. Answer the following questions to check your understanding of several of the basic concepts discussed in this chapter. The answers to these questions are given in the back of the book.

1. Describe the three elementary row operations that can be performed on an augmented matrix.

2. What is the relationship between the three elementary row operations on an augmented matrix and the operations that lead to equivalent systems of equations?

3. In your own words describe the difference between a matrix in row-echelon form and a matrix in reduced row-echelon form.

In Exercises 4–7, the row-echelon form of an augmented matrix that corresponds to a system of linear equations is given. Use the matrix to determine whether the system is consistent or inconsistent, and if consistent, determine the number of solutions.

4. $\begin{bmatrix} 1 & 2 & 3 & \vdots & 9 \\ 0 & 1 & -2 & \vdots & 2 \\ 0 & 0 & 0 & \vdots & 0 \end{bmatrix}$

5. $\begin{bmatrix} 1 & 2 & 3 & \vdots & 9 \\ 0 & 1 & -2 & \vdots & 2 \\ 0 & 0 & 0 & \vdots & 8 \end{bmatrix}$

6. $\begin{bmatrix} 1 & 2 & 3 & \vdots & 9 \\ 0 & 1 & -2 & \vdots & 2 \\ 0 & 0 & 1 & \vdots & -3 \end{bmatrix}$

7. $\begin{bmatrix} 1 & 2 & 3 & 10 & 6 & \vdots & 0 \\ 0 & 1 & -5 & -2 & 0 & \vdots & 5 \\ 0 & 0 & 1 & 12 & 0 & \vdots & -2 \\ 0 & 0 & 0 & 1 & 1 & \vdots & 0 \end{bmatrix}$

In Exercises 8–10, determine if the matrix operations (a) $A + 3B$ and (b) AB can be performed. If not, state why.

8. $A = \begin{bmatrix} 2 & -2 \\ 3 & 5 \end{bmatrix}$, $B = \begin{bmatrix} -3 & 10 \\ 12 & 8 \end{bmatrix}$

9. $A = \begin{bmatrix} 5 & 4 \\ -7 & 2 \\ 11 & 2 \end{bmatrix}$, $B = \begin{bmatrix} 4 & 12 \\ 20 & 40 \\ 15 & 30 \end{bmatrix}$

10. $A = \begin{bmatrix} 5 & 4 \\ -7 & 2 \\ 11 & 2 \end{bmatrix}$, $B = \begin{bmatrix} 4 & 12 \\ 20 & 40 \end{bmatrix}$

11. Under what conditions does a matrix have an inverse?

12. Explain the difference between a square matrix and its determinant.

13. Is it possible to find the determinant of a 4×5 matrix? Explain.

14. What is meant by the cofactor of an entry of a matrix? How is it used to find the determinant of the matrix?

15. Three people were asked to solve a system of equations using an augmented matrix. Each person reduced the matrix to row-echelon form. The reduced matrices were

$\begin{bmatrix} 1 & 2 & \vdots & 3 \\ 0 & 1 & \vdots & 1 \end{bmatrix}$

$\begin{bmatrix} 1 & 0 & \vdots & 1 \\ 0 & 1 & \vdots & 1 \end{bmatrix}$

and

$\begin{bmatrix} 1 & 2 & \vdots & 3 \\ 0 & 0 & \vdots & 0 \end{bmatrix}$.

Can all three be right? Explain.

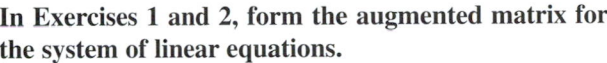

Review Exercises

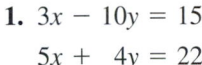

In Exercises 1 and 2, form the augmented matrix for the system of linear equations.

1. $3x - 10y = 15$
$\quad 5x + 4y = 22$

2. $8x - 7y + 4z = 12$
$\quad 3x - 5y + 2z = 20$
$\quad 5x + 3y - 3z = 26$

In Exercises 3 and 4, write the system of linear equations represented by the augmented matrix. (Use variables x, y, z, and w.)

3. $\begin{bmatrix} 5 & 1 & 7 & \vdots & -9 \\ 4 & 2 & 0 & \vdots & 10 \\ 9 & 4 & 2 & \vdots & 3 \end{bmatrix}$

4. $\begin{bmatrix} 13 & 16 & 7 & 3 & \vdots & 2 \\ 1 & 21 & 8 & 5 & \vdots & 12 \\ 4 & 10 & -4 & 3 & \vdots & -1 \end{bmatrix}$

In Exercises 5 and 6, write the matrix in *reduced* row-echelon form.

5. $\begin{bmatrix} 0 & 1 & 1 \\ 1 & 2 & 3 \\ 2 & 2 & 2 \end{bmatrix}$

6. $\begin{bmatrix} 1 & 1 & 1 & 0 \\ 1 & 1 & 0 & 1 \\ 1 & 0 & 1 & 1 \\ 0 & 1 & 1 & 1 \end{bmatrix}$

In Exercises 7–18, use matrices and elementary row operations to solve (if possible) the system of equations.

7. $5x + 4y = 2$
$\quad -x + y = -22$

8. $2x - 5y = 2$
$\quad 3x - 7y = 1$

9. $2x + y = 0.3$
$\quad 3x - y = -1.3$

10. $0.2x - 0.1y = 0.07$
$\quad 0.4x - 0.5y = -0.01$

11. $-x + y + 2z = 1$
$\quad 2x + 3y + z = -2$
$\quad 5x + 4y + 2z = 4$

12. $2x + 3y + z = 10$
$\quad 2x - 3y - 3z = 22$
$\quad 4x - 2y + 3z = -2$

13. $4x + 4y + 4z = 5$
$\quad 4x - 2y - 8z = 1$
$\quad 5x + 3y + 8z = 6$

14. $2x + 3y + 3z = 3$
$\quad 6x + 6y + 12z = 13$
$\quad 12x + 9y - z = 2$

15. $2x + y + 2z = 4$
$\quad 2x + 2y = 5$
$\quad 2x - y + 6z = 2$

16. $3x + 21y - 29z = -1$
$\quad 2x + 15y - 21z = 0$

17. $x + 2y + 6z = 1$
$\quad 2x + 5y + 15z = 4$
$\quad 3x + y + 3z = -6$

18. $x + 2y + w = 3$
$\quad -3y + 3z = 0$
$\quad 4x + 4y + z + 2w = 0$
$\quad 2x + z = 3$

19. *Think About It* Describe the row-echelon form of an augmented matrix that corresponds to a system of linear equations that has a unique solution.

20. *Partial Fractions* Write the partial fraction decomposition for the rational expression

$$\frac{x + 9}{(x + 1)(x + 2)^2} = \frac{A}{x + 1} + \frac{B}{x + 2} + \frac{C}{(x + 2)^2}.$$

In Exercises 21–28, perform the matrix operations. If it is not possible, explain why.

21. $\begin{bmatrix} 2 & 1 & 0 \\ 0 & 5 & -4 \end{bmatrix} - 3\begin{bmatrix} 5 & 3 & -6 \\ 0 & -2 & 5 \end{bmatrix}$

22. $-2\begin{bmatrix} 1 & 2 \\ 5 & -4 \\ 6 & 0 \end{bmatrix} + 8\begin{bmatrix} 7 & 1 \\ 1 & 2 \\ 1 & 4 \end{bmatrix}$

23. $\begin{bmatrix} 1 & 2 \\ 5 & -4 \\ 6 & 0 \end{bmatrix}\begin{bmatrix} 6 & -2 & 8 \\ 4 & 0 & 0 \end{bmatrix}$

24. $\begin{bmatrix} 1 & 5 & 6 \\ 2 & -4 & 0 \end{bmatrix}\begin{bmatrix} 6 & -2 & 8 \\ 4 & 0 & 0 \end{bmatrix}$

25. $\begin{bmatrix} 1 & 5 & 6 \\ 2 & -4 & 0 \end{bmatrix}\begin{bmatrix} 6 & 4 \\ -2 & 0 \\ 8 & 0 \end{bmatrix}$

26. $\begin{bmatrix} 4 \\ 6 \end{bmatrix}[6 \quad -2]$

27. $\begin{bmatrix} 1 & 3 & 2 \\ 0 & 2 & -4 \\ 0 & 0 & 3 \end{bmatrix}\begin{bmatrix} 4 & -3 & 2 \\ 0 & 3 & -1 \\ 0 & 0 & 2 \end{bmatrix}$

28. $\begin{bmatrix} 2 & 1 \\ 6 & 0 \end{bmatrix}\left(\begin{bmatrix} 4 & 2 \\ -3 & 1 \end{bmatrix} + \begin{bmatrix} -2 & 4 \\ 0 & 4 \end{bmatrix}\right)$

In Exercises 29–32, use a graphing utility to perform the matrix operations.

29. $3\begin{bmatrix} 8 & -2 & 5 \\ 1 & 3 & -1 \end{bmatrix} + 6\begin{bmatrix} 4 & -2 & -3 \\ 2 & 7 & 6 \end{bmatrix}$

30. $-5\begin{bmatrix} 2 & 0 \\ 7 & -2 \\ 8 & 2 \end{bmatrix} + 4\begin{bmatrix} 4 & -2 \\ 6 & 11 \\ -1 & 3 \end{bmatrix}$

31. $\begin{bmatrix} 4 & 1 \\ 11 & -7 \\ 12 & 3 \end{bmatrix}\begin{bmatrix} 3 & -5 & 6 \\ 2 & -2 & -2 \end{bmatrix}$

32. $\begin{bmatrix} -2 & 3 & 10 \\ 4 & -2 & 2 \end{bmatrix}\begin{bmatrix} 1 & 1 \\ -5 & 2 \\ 3 & 2 \end{bmatrix}$

In Exercises 33–36, solve for X given

$$A = \begin{bmatrix} -4 & 0 \\ 1 & -5 \\ -3 & 2 \end{bmatrix} \quad \text{and} \quad B = \begin{bmatrix} 1 & 2 \\ -2 & 1 \\ 4 & 4 \end{bmatrix}.$$

33. $X = 3A - 2B$

34. $6X = 4A + 3B$

35. $3X + 2A = B$

36. $2A - 5B = 3X$

37. Write the system of linear equations represented by the matrix equation

$$\begin{bmatrix} 5 & 4 \\ -1 & 1 \end{bmatrix}\begin{bmatrix} x \\ y \end{bmatrix} = \begin{bmatrix} 2 \\ -22 \end{bmatrix}.$$

38. Write the matrix equation $AX = B$ for the following system of linear equations.

$$2x + 3y + z = 10$$
$$2x - 3y - 3z = 22$$
$$4x - 2y + 3z = -2$$

In Exercises 39–42, use a graphing utility to find the inverse of the matrix (if it exists).

39. $\begin{bmatrix} 2 & 6 \\ 3 & -6 \end{bmatrix}$

40. $\begin{bmatrix} 3 & -10 \\ 4 & 2 \end{bmatrix}$

41. $\begin{bmatrix} 2 & 0 & 3 \\ -1 & 1 & 1 \\ 2 & -2 & 1 \end{bmatrix}$

42. $\begin{bmatrix} 1 & 4 & 6 \\ 2 & -3 & 1 \\ -1 & 18 & 16 \end{bmatrix}$

In Exercises 43–46, evaluate the determinant.

43. $\begin{vmatrix} 50 & -30 \\ 10 & 5 \end{vmatrix}$

44. $\begin{vmatrix} 8 & 5 \\ 2 & -4 \end{vmatrix}$

45. $\begin{vmatrix} 3 & 0 & -4 & 0 \\ 0 & 8 & 1 & 2 \\ 6 & 1 & 8 & 2 \\ 0 & 3 & -4 & 1 \end{vmatrix}$

46. $\begin{vmatrix} -5 & 6 & 0 & 0 \\ 0 & 1 & -1 & 2 \\ -3 & 4 & -5 & 1 \\ 1 & 6 & 0 & 3 \end{vmatrix}$

In Exercises 47–54, use a graphing utility to solve (if possible) the system of linear equations using the inverse of the coefficient matrix.

47. $x + 2y = -1$
$3x + 4y = -5$

48. $x + 3y = 23$
$-6x + 2y = -18$

49. $-3x - 3y - 4x = 2$
$y + z = -1$
$4x + 3y + 4z = -1$

50. $x - 3y - 2z = 8$
$-2x + 7y + 3z = -19$
$x - y - 3z = 3$

51. $x + 3y + 2z = 2$
$-2x - 5y - z = 10$
$2x + 4y = -12$

52.
$$2x + 4y \qquad = -12$$
$$3x + 4y - 2z = -14$$
$$-x + y + 2z = -6$$

53.
$$-x + y + z = 6$$
$$4x - 3y + z = 20$$
$$2x - y + 3z = 8$$

54.
$$2x + 3y - 4z = 1$$
$$x - y + 2z = -4$$
$$3x + 7y - 10z = 0$$

In Exercises 55–58, use Cramer's Rule to solve.

55. *Mixture Problem*   A florist wants to arrange a dozen flowers consisting of two varieties: carnations and roses. Carnations cost $0.75 each and roses cost $1.50 each. How many of each should the florist use so the arrangement will cost $12.00?

56. *Mixture Problem*   One hundred liters of a 60% acid solution is obtained by mixing a 75% solution with a 50% solution. How many liters of each must be used to obtain the desired mixture?

57. *Fitting a Parabola to Three Points*   Find an equation of the parabola $y = ax^2 + bx + c$ that passes through the points $(-1, 2)$, $(0, 3)$, and $(1, 6)$.

58. *Break-Even Point*   A small business invests $25,000 in equipment to produce a product. Each unit of the product costs $3.75 to produce and is sold for $5.25. How many items must be sold before the business breaks even?

59. *Data Analysis*   The median prices y (in thousands of dollars) of one-family houses sold in the United States for the years 1981 through 1993 are shown in the figure. The least squares regression line $y = a + bt$ for this data is found by solving the system

$$13a + 91b = 1107$$
$$91a + 819b = 8404.7$$

where $t = 1$ represents 1981. (Source: National Association of Realtors)

(a) Use a graphing utility to solve this system.

(b) Use a graphing utility to graph the regression line.

(c) Interpret the meaning of the slope of the regression line in the context of the problem.

(d) Use the regression line to estimate the median price of homes in 1995.

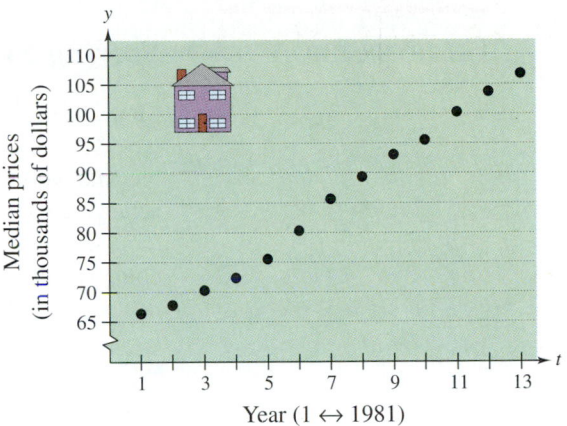

FIGURE FOR 59

60. Solve the equation $\begin{vmatrix} 2 - \lambda & 5 \\ 3 & -8 - \lambda \end{vmatrix} = 0$.

In Exercises 61–64, use a determinant to find the area of the triangle with the given vertices.

61. $(1, 0), (5, 0), (5, 8)$

62. $(-4, 0), (4, 0), (0, 6)$

63. $(1, 2), (4, -5), (3, 2)$

64. $\left(\frac{3}{2}, 1\right), \left(4, -\frac{1}{2}\right), (4, 2)$

In Exercises 65–68, use a determinant to find an equation of the line through the given points.

65. $(-4, 0), (4, 4)$

66. $(2, 5), (6, -1)$

67. $\left(-\frac{5}{2}, 3\right), \left(\frac{7}{2}, 1\right)$

68. $(-0.8, 0.2), (0.7, 3.2)$

69. Verify that

$$\begin{vmatrix} a_{11} & a_{12} & a_{13} \\ a_{21} & a_{22} & a_{23} \\ a_{31} + c_1 & a_{32} + c_2 & a_{33} + c_3 \end{vmatrix} =$$

$$\begin{vmatrix} a_{11} & a_{12} & a_{13} \\ a_{21} & a_{22} & a_{23} \\ a_{31} & a_{32} & a_{33} \end{vmatrix} + \begin{vmatrix} a_{11} & a_{12} & a_{13} \\ a_{21} & a_{22} & a_{23} \\ c_1 & c_2 & c_3 \end{vmatrix}.$$

CHAPTER PROJECT: *Solving Systems of Equations*

Many matrices used to solve problems in engineering, business, and science have several rows and columns. Because it is so easy to make a mistake, people who work with such matrices almost always use technology to perform the actual matrix calculations. Matrices have always been a powerful mathematical tool—especially for dealing with problems that involve a lot of data. With technology available to perform the calculations, matrices have also become a practical mathematical tool.

Real Life

EXAMPLE 1   Solving a Metallurgy Problem

Three iron alloys contain different percents of carbon, chromium, and iron. Alloy X is a type of wrought iron, alloy Y is a type of stainless steel, and alloy Z is a type of cast iron. How much of each of the three alloys can you make with 15 tons carbon, 39 tons of chromium, and 546 tons of iron?

	Alloy X	Alloy Y	Alloy Z
Carbon	1%	1%	4%
Chromium	0%	15%	3%
Iron	99%	84%	93%

Solution

Let x, y, and z represent the amounts of the three iron alloys. You can model the situation with the following linear system.

$$0.01x + 0.01y + 0.04z = 15 \qquad \text{Carbon}$$
$$0.15y + 0.03z = 39 \qquad \text{Chromium}$$
$$0.99x + 0.84y + 0.93z = 546 \qquad \text{Iron}$$

The matrix equation that represents this system is as follows.

$$AX = B$$

$$\begin{bmatrix} 0.01 & 0.01 & 0.04 \\ 0 & 0.15 & 0.03 \\ 0.99 & 0.84 & 0.93 \end{bmatrix} \begin{bmatrix} x \\ y \\ z \end{bmatrix} = \begin{bmatrix} 15 & 39 & 546 \end{bmatrix}$$

NOTE   In the solution at the right, the entries of A^{-1} are displayed. In practice, however, this is not necessary—simply instruct your calculator or computer to display $A^{-1}B$. ∎∎

With a graphing utility or computer, you can solve the equation as follows.

$$X = A^{-1}B = \begin{bmatrix} -25.4 & -5.4 & 1.267 \\ -6.6 & 6.733 & 0.067 \\ 33 & -0.333 & -0.333 \end{bmatrix} \begin{bmatrix} 15 \\ 39 \\ 546 \end{bmatrix} = \begin{bmatrix} 100 \\ 200 \\ 300 \end{bmatrix}$$

Thus, you can make 100 tons of alloy X, 200 tons of alloy Y, and 300 tons of alloy Z.

CHAPTER PROJECT INVESTIGATIONS

In Questions 1–4, use a graphing utility or a computer to solve the linear system.

1.
$$x - y + 10z = 2$$
$$3x + z = 4$$
$$7x + 2y + z = 0$$

2.
$$4x + 2z = 3$$
$$-x + 2y + 5z = -1$$
$$3x + y - 7z = 10$$

3.
$$2x + 6y - z = 4$$
$$3x + y + 2z = -4$$
$$6x - y + 3z = 1$$

4.
$$5x - 3y + 4z = 25$$
$$2x + 2y - z = -1$$
$$x - y + 2z = 9$$

5. *Shoe Purchases*   The bar graph shows the percents (by age group) of the total amounts spent on three types of shoes in 1992. For instance, 17% of the total amount spent on gym shoes was spent by people in the 14–17 age group, 5% was spent by people in the 18–24 age group, and 9% was spent by people in the 25–34 age group. The amounts (in millions of dollars) spent for *all three* types of shoes are shown in the matrix. How many dollars worth of gym shoes, jogging shoes, and walking shoes were sold in 1992? (Source: National Sporting Goods Association)

Amount (millions) Spent on Shoes

$$\text{Age Group} \begin{cases} 14\text{--}17 \\ 18\text{--}24 \\ 25\text{--}34 \end{cases} \begin{bmatrix} \$268.37 \\ \$160.04 \\ \$346.73 \end{bmatrix}$$

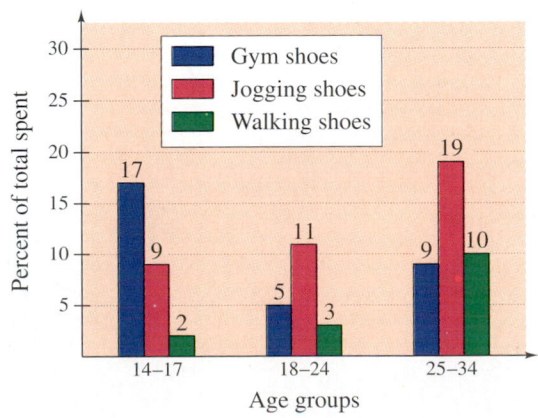

6. *Skiing*   The total number of people participating in downhill and cross-country skiing in 1992 was approximately 14.26 million. The total number of females participating in both types of skiing was approximately 6.02 million. Let x represent the number of people who participated in downhill skiing and let y represent the number of people who participated in cross-country skiing. From the bar graph, you can write the following system.

$$0.40x + 0.50y = 6.02$$
$$0.60x + 0.50y = 8.24$$

How many people participated in each type of skiing? (Source: National Sporting Goods Association)

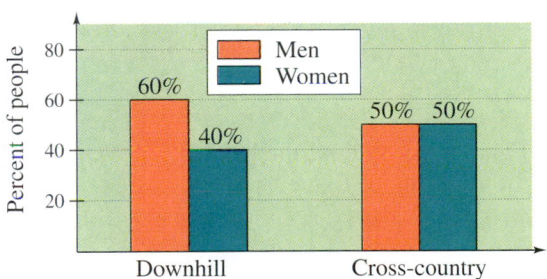

7. *Gold Alloys*   "Gold" jewelry is seldom made of pure gold because it is soft *and* expensive. Instead, gold is mixed with other metals to produce harder, less expensive gold alloys. The amount of gold (by weight) in the alloy is measured in karats. A 24-karat gold mixture is 100% gold. An 18-karat gold mixture is 75% gold, and so on. Three different gold alloys contain the percents of gold, copper, and silver shown in the matrix. You have 20,144 grams of gold, 766 grams of copper, and 1990 grams of silver. How much of each alloy can you make?

Percent by Weight

	Alloy X	Alloy Y	Alloy Z
Gold	94%	92%	80%
Copper	4%	2%	4%
Silver	2%	6%	16%

Chapter Test

Take this test as you would take a test in class. After you are done, check your work against the answers in the back of the book.

In Exercises 1 and 2, write the matrix in reduced row-echelon form. Use a graphing utility to verify your result.

1. $\begin{bmatrix} 1 & -1 & 5 \\ 6 & 2 & 3 \\ 5 & 3 & -3 \end{bmatrix}$

2. $\begin{bmatrix} 1 & 0 & -1 & 2 \\ -1 & 1 & 1 & -3 \\ 1 & 1 & -1 & 1 \\ 3 & 2 & -3 & 4 \end{bmatrix}$

3. Use the matrix capabilities of a graphing utility to reduce the augmented matrix corresponding to the system of equations, and solve the system.

$$4x + 3y - 2z = 14$$
$$-x - y + 2z = -5$$
$$3x + y - 4z = 8$$

4. Find the equation of the parabola $y = ax^2 + bx + c$ that passes through the points in the figure.

5. Find (a) $A - B$, (b) $3A$, (c) $3A - 2B$.

$$A = \begin{bmatrix} 5 & 4 & 4 \\ -4 & -4 & 0 \end{bmatrix}, \quad B = \begin{bmatrix} 4 & -1 & 6 \\ -4 & 0 & -3 \end{bmatrix}$$

6. Find AB, if possible.

$$A = \begin{bmatrix} 2 & -2 & 6 \\ 3 & -1 & 7 \\ 2 & 0 & -2 \end{bmatrix}, \quad B = \begin{bmatrix} 4 & 4 \\ 3 & 2 \\ 1 & -2 \end{bmatrix}$$

7. Find A^{-1} if

$$A = \begin{bmatrix} -6 & 4 \\ 10 & -5 \end{bmatrix}.$$

8. Use the result of Exercise 7 to solve the system.

$$-6x + 4y = 10$$
$$10x - 5y = 20$$

9. Evaluate the determinant of the matrix

$$\begin{bmatrix} 4 & 0 & 3 \\ 1 & -8 & 2 \\ 3 & 2 & 2 \end{bmatrix}.$$

10. Use a determinant to find the area of the triangle in the figure.

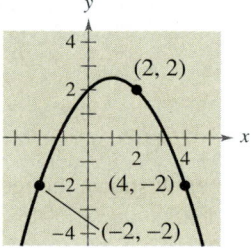

FIGURE FOR 4

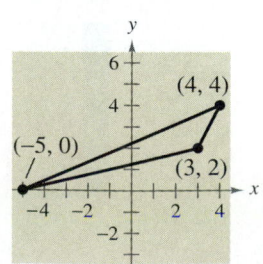

FIGURE FOR 10

Burney Le Boeuf, shown below, is a marine biologist at the University of California at Santa Cruz. He has spent many years studying elephant seals.

Of all mammals, reptiles and birds, elephant seals are the champions of deep sea dives. One bull elephant seal was measured at a depth of nearly one mile! "Our theories were blown clear out of the water," says Le Boeuf. (Source: Smithsonian Magazine, Sept., 1995, "Elephant Seals, the Champion Divers of the Deep")

The frequency distribution below shows the recorded depths of over 10,000 dives by female elephant seals. From this data, you can make predictions about the depth of a particular dive. For instance, because 54.6% of all dives were in depths greater than 400 meters, you can conclude that the probability that a particular dive will exceed 400 meters is 0.546.

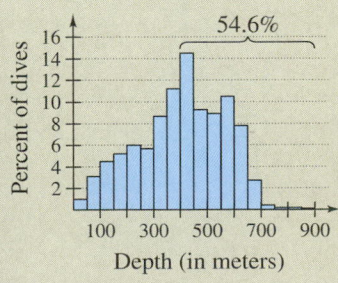

See Exercises 59 and 60 on page 871.

Photos: Frans Lanting—Minden Pictures

11 Sequences and Probability

11.1 *Sequences and Summation Notation*

11.2 *Arithmetic Sequences*

11.3 *Geometric Sequences*

11.4 *Mathematical Induction*

11.5 *The Binomial Theorem*

11.6 *Counting Principles*

11.7 *Probability*

Researchers from the University of California at Santa Cruz are trying to block a tranquilized bull elephant seal from entering the water. They will then attach instruments to the bull to measure the depths and durations of its dives.

▶ 11.1   *Sequences and Summation Notation*

See Exercise 95 on page 809 for an example of how a sequence can be used to model the net income of Wal-Mart from 1985 through 1994.

Sequences   □   *Factorial Notation*   □   *Summation Notation*   □
Application

Sequences

NOTE   On occasion it is convenient to begin subscripting a sequence with 0 instead of 1 so that the terms of the sequence become

$$a_0, a_1, a_2, a_3, \ldots$$

In mathematics, the word *sequence* is used in much the same way as in ordinary English. Saying that a collection is listed *in sequence* means that it is ordered so that it has a first member, a second member, a third member, and so on.

   Mathematically, you can think of a sequence as a *function* whose domain is the set of positive integers. Rather than using function notation, however, sequences are usually written using subscript notation, as indicated in the following definition.

> ### DEFINITION OF A SEQUENCE
>
> An **infinite sequence** is a function whose domain is the set of positive integers. The function values
>
> $$a_1, a_2, a_3, a_4, \ldots, a_n, \ldots$$
>
> are the **terms** of the sequence. If the domain of the function consists of the first n positive integers only, the sequence is a **finite sequence.**

TECHNOLOGY

To graph a sequence using a *TI-83* or a *TI-82*, set the mode to *seq* and *dot* and enter the sequence into Un (press [Y=]). The graph of the sequence in Example 1(a) is shown below.

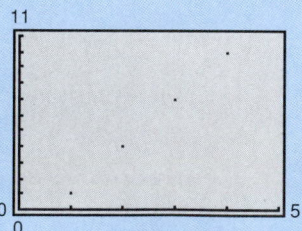

Graph the sequence in Example 1(b) and use the trace key to identify its terms.

EXAMPLE 1   *Finding Terms of a Sequence*

a. The first four terms of the sequence given by $a_n = 3n - 2$ are

$a_1 = 3(1) - 2 = 1$	1st term
$a_2 = 3(2) - 2 = 4$	2nd term
$a_3 = 3(3) - 2 = 7$	3rd term
$a_4 = 3(4) - 2 = 10.$	4th term

b. The first four terms of the sequence given by $a_n = 3 + (-1)^n$ are

$a_1 = 3 + (-1)^1 = 3 - 1 = 2$	1st term
$a_2 = 3 + (-1)^2 = 3 + 1 = 4$	2nd term
$a_3 = 3 + (-1)^3 = 3 - 1 = 2$	3rd term
$a_4 = 3 + (-1)^4 = 3 + 1 = 4.$	4th term

EXAMPLE 2   *Finding Terms of a Sequence*

The first four terms of a sequence given by $a_n = \dfrac{(-1)^n}{2n-1}$ are

$$a_1 = \frac{(-1)^1}{2(1)-1} = \frac{-1}{2-1} = -1 \qquad \text{1st term}$$

$$a_2 = \frac{(-1)^2}{2(2)-1} = \frac{1}{4-1} = \frac{1}{3} \qquad \text{2nd term}$$

$$a_3 = \frac{(-1)^3}{2(3)-1} = \frac{-1}{6-1} = -\frac{1}{5} \qquad \text{3rd term}$$

$$a_4 = \frac{(-1)^4}{2(4)-1} = \frac{1}{8-1} = \frac{1}{7}. \qquad \text{4th term}$$

Simply listing the first few terms is not sufficient to define a unique sequence—the *n*th term *must be given*. To see this, consider the following sequences, both of which have the same first three terms.

$$\frac{1}{2}, \frac{1}{4}, \frac{1}{8}, \frac{1}{16}, \cdots, \frac{1}{2^n}, \cdots$$

$$\frac{1}{2}, \frac{1}{4}, \frac{1}{8}, \frac{1}{15}, \cdots, \frac{6}{(n+1)(n^2-n+6)}, \cdots$$

EXAMPLE 3   *Finding the nth Term of a Sequence*

Write an expression for the apparent *n*th term (a_n) of each sequence.

a. 1, 3, 5, 7, . . .   **b.** 2, 5, 10, 17, . . .

Solution

a.   *n:* 1 2 3 4 . . . *n*

   Terms: 1 3 5 7 . . . a_n

   Apparent pattern: Each term is 1 less than twice *n*, which implies that

   $$a_n = 2n - 1.$$

b.   *n:* 1 2 3 4 . . . *n*

   Terms: 2 5 10 17 . . . a_n

   Apparent pattern: Each term is 1 more than the square of *n*, which implies that

   $$a_n = n^2 + 1.$$

NOTE   Some sequences are defined **recursively.** To define a sequence recursively, you need to be given one or more of the first few terms. All other terms of the sequence are then defined using previous terms. A well-known example is the Fibonacci Sequence shown in Example 4. ■■

EXAMPLE 4   *A Sequence That Is Defined Recursively*

The Fibonacci Sequence is defined recursively, as follows.

$$a_0 = 1, a_1 = 1, a_k = a_{k-2} + a_{k-1} \text{ where } k \geq 2$$

Write the first six terms of this sequence.

Solution

$a_0 = 1$	0th term is given.
$a_1 = 1$	1st term is given.
$a_2 = a_0 + a_1 = 1 + 1 = 2$	Use recursive formula.
$a_3 = a_1 + a_2 = 1 + 2 = 3$	Use recursive formula.
$a_4 = a_2 + a_3 = 2 + 3 = 5$	Use recursive formula.
$a_5 = a_3 + a_4 = 3 + 5 = 8$	Use recursive formula.

TECHNOLOGY

Some graphing utilities, such as the *TI-83* or the *TI-82*, have spread sheet capabilities that create tables. For instance, you can use the following steps to create a table that shows the terms of the sequence given by $a_n = 2^n + 1$.

1. Enter the sequence in  as $Y_1 = 2\text{^}X + 1$.
2. With the TblSet menu, set TblStart = 1 or TblMin = 1, and ΔTbl = 1.
3. View the terms of the sequence using the keystrokes ⌷TABLE⌷.

When viewing the table, you can use the cursor to look at additional terms.

Factorial Notation

Some very important sequences in mathematics involve terms that are defined with special types of products called **factorials.**

> **DEFINITION OF FACTORIAL**
>
> If n is a positive integer, *n* **factorial** is defined by
>
> $$n! = 1 \cdot 2 \cdot 3 \cdot 4 \cdots (n-1) \cdot n.$$
>
> As a special case, zero factorial is defined as $0! = 1$.

Here are some values of $n!$ for the first several nonnegative integers. Notice that $0!$ is 1 by definition.

$$0! = 1$$
$$1! = 1$$
$$2! = 1 \cdot 2 = 2$$
$$3! = 1 \cdot 2 \cdot 3 = 6$$
$$4! = 1 \cdot 2 \cdot 3 \cdot 4 = 24$$
$$5! = 1 \cdot 2 \cdot 3 \cdot 4 \cdot 5 = 120$$

The value of n does not have to be very large before the value of $n!$ becomes huge. For instance, $10! = 3{,}628{,}800$.

Factorials follow the same conventions for order of operations as do exponents. For instance,

$$2n! = 2(n!) = 2(1 \cdot 2 \cdot 3 \cdot 4 \cdots n)$$

whereas $(2n)! = 1 \cdot 2 \cdot 3 \cdot 4 \cdots 2n.$

EXAMPLE 5   *Finding Terms of a Sequence Involving Factorials*

List the first five terms of the sequence given by $a_n = 2^n /n!$. Begin with $n = 0$.

Solution

$$a_0 = \frac{2^0}{0!} = \frac{1}{1} = 1 \qquad\qquad \text{0th term}$$

$$a_1 = \frac{2^1}{1!} = \frac{2}{1} = 2 \qquad\qquad \text{1st term}$$

$$a_2 = \frac{2^2}{2!} = \frac{4}{2} = 2 \qquad\qquad \text{2nd term}$$

$$a_3 = \frac{2^3}{3!} = \frac{8}{6} = \frac{4}{3} \qquad\qquad \text{3rd term}$$

$$a_4 = \frac{2^4}{4!} = \frac{16}{24} = \frac{2}{3} \qquad\qquad \text{4th term}$$

When working with fractions involving factorials, you will often find that the fractions can be reduced.

EXAMPLE 6   *Evaluating Factorial Expressions*

Evaluate each factorial expression.

a. $\dfrac{8!}{2! \cdot 6!}$ **b.** $\dfrac{2! \cdot 6!}{3! \cdot 5!}$ **c.** $\dfrac{n!}{(n-1)!}$

Solution

a. $\dfrac{8!}{2! \cdot 6!} = \dfrac{1 \cdot 2 \cdot 3 \cdot 4 \cdot 5 \cdot 6 \cdot 7 \cdot 8}{1 \cdot 2 \cdot 1 \cdot 2 \cdot 3 \cdot 4 \cdot 5 \cdot 6} = \dfrac{7 \cdot 8}{2} = 28$

b. $\dfrac{2! \cdot 6!}{3! \cdot 5!} = \dfrac{1 \cdot 2 \cdot 1 \cdot 2 \cdot 3 \cdot 4 \cdot 5 \cdot 6}{1 \cdot 2 \cdot 3 \cdot 1 \cdot 2 \cdot 3 \cdot 4 \cdot 5} = \dfrac{6}{3} = 2$

c. $\dfrac{n!}{(n-1)!} = \dfrac{1 \cdot 2 \cdot 3 \cdots (n-1) \cdot n}{1 \cdot 2 \cdot 3 \cdots (n-1)} = n$

Summation Notation

There is a convenient notation for the sum of the terms of a finite sequence. It is called **summation notation** or **sigma notation** because it involves the use of the uppercase Greek letter sigma, written as Σ.

 DEFINITION OF SUMMATION NOTATION

The sum of the first n terms of a sequence is represented by

$$\sum_{i=1}^{n} a_i = a_1 + a_2 + a_3 + a_4 + \cdots + a_n$$

where i is called the **index of summation,** n is the **upper limit of summation,** and 1 is the **lower limit of summation.**

 The *Interactive* CD-ROM shows every example with its solution; clicking on the *Try It!* button brings up similar problems. Guided Examples and Integrated Examples show step-by-step solutions to additional examples. Integrated Examples are related to several concepts in the section.

EXAMPLE 7   **Summation Notation for Sums**

a. $\displaystyle\sum_{i=1}^{5} 3i = 3(1) + 3(2) + 3(3) + 3(4) + 3(5)$

$\qquad\qquad = 3(1 + 2 + 3 + 4 + 5)$

$\qquad\qquad = 3(15) = 45$

b. $\displaystyle\sum_{k=3}^{6} (1 + k^2) = (1 + 3^2) + (1 + 4^2) + (1 + 5^2) + (1 + 6^2)$

$\qquad\qquad\qquad = 10 + 17 + 26 + 37 = 90$

c. $\displaystyle\sum_{i=0}^{8} \frac{1}{i!} = \frac{1}{0!} + \frac{1}{1!} + \frac{1}{2!} + \frac{1}{3!} + \frac{1}{4!} + \frac{1}{5!} + \frac{1}{6!} + \frac{1}{7!} + \frac{1}{8!}$

$\qquad\qquad = 1 + 1 + \frac{1}{2} + \frac{1}{6} + \frac{1}{24} + \frac{1}{120} + \frac{1}{720} + \frac{1}{5040} + \frac{1}{40{,}320}$

$\qquad\qquad \approx 2.71828$

For this summation, note that the sum is very close to the irrational number $e \approx 2.718281828$. It can be shown that as more terms of the sequence whose nth term is $1/n!$ are added, the sum becomes closer and closer to e.

NOTE   In Example 7, note that the lower limit of a summation does not have to be 1. Also note that the index of summation does not have to be the letter i. For instance, in part (b), the letter k is the index of summation. ∎

THINK ABOUT THE PROOF

Use the Distributive Property of real numbers to prove Property 1 of the Properties of Sums. The details of the proof are given in the appendix.

PROPERTIES OF SUMS

1. $\displaystyle\sum_{i=1}^{n} ca_i = c \sum_{i=1}^{n} a_i,$ c is any constant.

2. $\displaystyle\sum_{i=1}^{n} (a_i + b_i) = \sum_{i=1}^{n} a_i + \sum_{i=1}^{n} b_i$

3. $\displaystyle\sum_{i=1}^{n} (a_i - b_i) = \sum_{i=1}^{n} a_i - \sum_{i=1}^{n} b_i$

Variations in the upper and lower limits of summation can produce quite different-looking summation notations for *the same sum.* For example, consider the following two sums.

$$\sum_{i=1}^{5} 3(2^i) = 3\sum_{i=1}^{5} 2^i = 3(2^1 + 2^2 + 2^3 + 2^4 + 2^5)$$

$$\sum_{i=0}^{4} 3(2^{i+1}) = 3\sum_{i=0}^{4} 2^{i+1} = 3(2^1 + 2^2 + 2^3 + 2^4 + 2^5)$$

Notice that both sums have identical terms.

The sum of the terms of any *finite* sequence must be a finite number. An important discovery in mathematics was that for some special types of *infinite* sequences, the sum of *all* the terms is a finite number. For instance, it can be shown that the sum of all of the terms of the sequence whose nth term is $1/2^n$ (with n beginning at 0) is

$$\sum_{n=0}^{\infty} \frac{1}{2^n} = 1 + \frac{1}{2} + \frac{1}{4} + \frac{1}{8} + \frac{1}{16} + \frac{1}{32} + \cdots$$

$$= 2.$$

NOTE   The summation of the terms of a sequence is a **series.** The summation

$$\sum_{i=1}^{\infty} a_i = a_1 + a_2 + a_3 + \cdots$$

is an **infinite series.** Infinite series have important uses in calculus. ▪▪

EXAMPLE 8   *Finding the Sum of an Infinite Sequence*

$$\sum_{n=1}^{\infty} \frac{3}{10^n} = \frac{3}{10^1} + \frac{3}{10^2} + \frac{3}{10^3} + \frac{3}{10^4} + \frac{3}{10^5} + \cdots$$

$$= 0.3 + 0.03 + 0.003 + 0.0003 + 0.00003 + \cdots$$

$$= 0.33333 \cdots$$

$$= \frac{1}{3}$$

Application

Sequences have many applications in business and science. One is illustrated in Example 9.

EXAMPLE 9 *Population of the United States*

For the years 1950 to 1993, the resident population of the United States can be approximated by the model

$$a_n = \sqrt{23{,}107 + 847.7n + 2.74n^2}, \qquad n = 0, 1, \ldots, 43$$

where a_n is the population in millions and n represents the calendar year, with $n = 0$ corresponding to 1950. Find the last five terms of this finite sequence. (Source: U.S. Bureau of Census)

Solution

The last five terms of this finite sequence are as follows.

$$a_{39} = \sqrt{23{,}107 + 874.7(39) + 2.74(39)^2} \approx 247.8 \qquad \text{1989 population}$$

$$a_{40} = \sqrt{23{,}107 + 874.7(40) + 2.74(40)^2} \approx 250.0 \qquad \text{1990 population}$$

$$a_{41} = \sqrt{23{,}107 + 874.7(41) + 2.74(41)^2} \approx 252.1 \qquad \text{1991 population}$$

$$a_{42} = \sqrt{23{,}107 + 874.7(42) + 2.74(42)^2} \approx 254.3 \qquad \text{1992 population}$$

$$a_{43} = \sqrt{23{,}107 + 874.7(43) + 2.74(43)^2} \approx 256.5 \qquad \text{1993 population}$$

The bar graph in Figure 11.1 graphically represents the population given by this sequence for the entire 44-year period from 1950 to 1993.

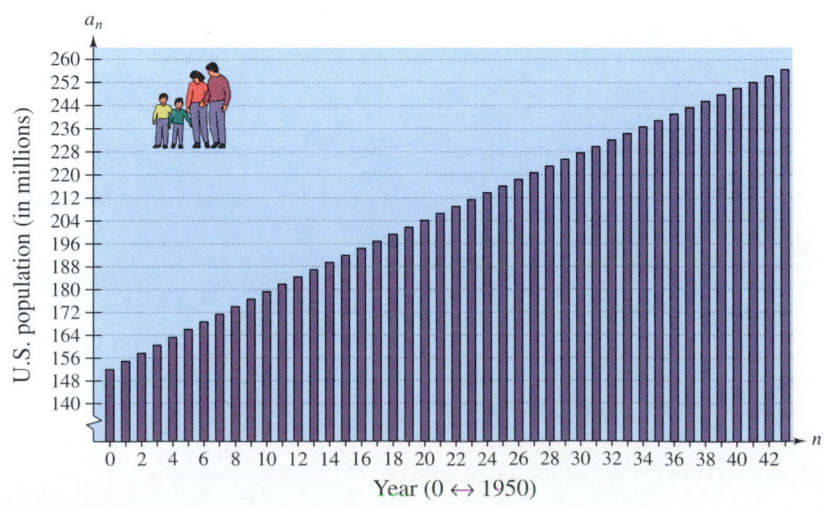

FIGURE 11.1

GROUP ACTIVITY

A SUMMATION PROGRAM

A graphing calculator may be programmed to calculate the sum of the first n terms of a sequence. For instance, the program below lists the program steps for the *TI-83* or the *TI-82*. See the appendix for programs for other graphing calculator models.

PROGRAM:SUM	Title of program
:Prompt M	M is the lower limit of summation.
:Prompt N	N is the upper limit of summation.
:0 → S	Initialize S for use as stored sum.
:For (X, M, N)	Loop from lower limit to upper limit.
:S + Y₁ → S	Evaluate function stored in Y₁ and add to sum.
:Disp S	Display partial sum.
:End	End loop.

To use this program to find the sum of a sequence, store the *n*th term of the sequence as Y₁ and run the program. For instance, to find the sum

$$\sum_{i=1}^{10} 2^i$$

you should press Y= , enter 2^X, and then use the keystroke sequence

PRGM , cursor to program SUM, ENTER ENTER

1 ENTER 10 ENTER .

The result should be 2046. (Note that the calculator displayed 10 numbers—the last number displayed is the final sum.) To pause the program after each partial sum, insert :Pause after :Disp S. Then press ENTER to display the next partial sum. Try using this program to find the following sums.

a. $\displaystyle\sum_{i=1}^{15} (i-2)^2$ **b.** $\displaystyle\sum_{i=1}^{20} 3^{(i-1)}$ **c.** $\displaystyle\sum_{i=1}^{10} \sqrt{e^i}$

Another way to calculate the sum of a sequence using the *TI-83* or the *TI-82* is to combine the **sum** and **seq(** commands under the LIST menu. However, this method does not display any of the partial sums leading up to the final sum. Try using this method to find the sums in (a), (b), and (c).

The *Interactive* CD-ROM provides additional help with Warm-Up exercises by providing a hypertext link to the section in which the concept was introduced.

Find the required value of the function.

1. $f(n) = \dfrac{2n}{n^2 + 1}, \quad f(2)$

2. $f(n) = \dfrac{4}{3(n + 1)}, \quad f(3)$

Factor the expression.

3. $4n^2 - 1$

4. $4n^2 - 8n + 3$

5. $n^2 - 3n + 2$

6. $n^2 + 3n + 2$

Perform the operations and/or simplify.

7. $\left(\dfrac{2}{3}\right)\left(\dfrac{3}{4}\right)\left(\dfrac{4}{5}\right)\left(\dfrac{5}{6}\right)$

8. $\dfrac{2 \cdot 4 \cdot 6 \cdot 8}{2^4}$

9. $\dfrac{1}{2 \cdot 2} + \dfrac{1}{2 \cdot 3} + \dfrac{1}{2 \cdot 4}$

10. $\dfrac{1}{1 \cdot 2} + \dfrac{1}{2 \cdot 3} + \dfrac{1}{3 \cdot 4}$

The *Interactive* CD-ROM contains step-by-step solutions to all odd-numbered Section and Review Exercises. It also provides Tutorial Exercises, which link to Guided Examples for additional help.

11.1 Exercises

In Exercises 1–22, write the first five terms of the sequence. (Assume that n begins with 1.)

1. $a_n = 2n + 1$

2. $a_n = 4n - 3$

3. $a_n = 2^n$

4. $a_n = \left(\dfrac{1}{2}\right)^n$

5. $a_n = (-2)^n$

6. $a_n = \left(-\dfrac{1}{2}\right)^n$

7. $a_n = \dfrac{n + 1}{n}$

8. $a_n = \dfrac{n}{n + 1}$

9. $a_n = \dfrac{6n}{3n^2 - 1}$

10. $a_n = \dfrac{3n^2 - n + 4}{2n^2 + 1}$

11. $a_n = \dfrac{1 + (-1)^n}{n}$

12. $a_n = 1 + (-1)^n$

13. $a_n = 3 - \dfrac{1}{2^n}$

14. $a_n = \dfrac{3^n}{4^n}$

15. $a_n = \dfrac{1}{n^{3/2}}$

16. $a_n = \dfrac{10}{n^{2/3}}$

17. $a_n = \dfrac{3^n}{n!}$

18. $a_n = \dfrac{n!}{n}$

19. $a_n = \dfrac{(-1)^n}{n^2}$

20. $a_n = (-1)^n \left(\dfrac{n}{n + 1}\right)$

21. $a_n = \dfrac{2}{3}$

22. $a_n = n(n - 1)(n - 2)$

In Exercises 23 and 24, find the indicated term of the sequence.

23. $a_n = (-1)^n(3n - 2)$

$a_{25} = $ ▮

24. $a_n = \dfrac{2^n}{n!}$

$a_{10} = $ ▮

In Exercises 25–28, write the first five terms of the sequence defined recursively.

25. $a_1 = 28, \quad a_{k+1} = a_k - 4$

26. $a_1 = 15, \quad a_{k+1} = a_k + 3$

27. $a_1 = 3, \quad a_{k+1} = 2(a_k - 1)$

28. $a_1 = 32, \quad a_{k+1} = \dfrac{1}{2}a_k$

In Exercises 29–34, use a graphing utility to graph the first ten terms of the sequence.

29. $a_n = \dfrac{2}{3}n$

30. $a_n = 2 - \dfrac{4}{n}$

31. $a_n = 16(-0.5)^{n-1}$

32. $a_n = 8(0.75)^{n-1}$

33. $a_n = \dfrac{2n}{n+1}$

34. $a_n = \dfrac{3n^2}{n^2+1}$

In Exercises 35–38, match the sequence with the graph of its first ten terms. [The graphs are labeled (a), (b), (c), and (d).]

(a)
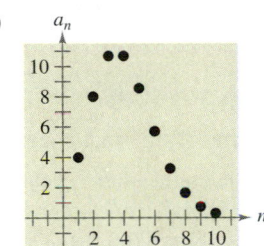

(b)
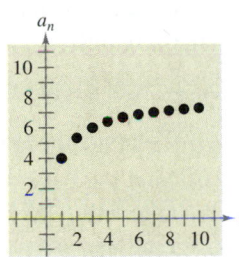

(c)
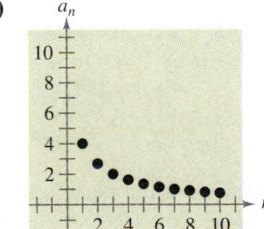

(d)
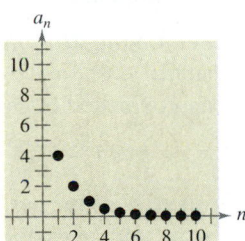

35. $a_n = \dfrac{8}{n+1}$

36. $a_n = \dfrac{8n}{n+1}$

37. $a_n = 4(0.5)^{n-1}$

38. $a_n = \dfrac{4^n}{n!}$

In Exercises 39–46, simplify the ratio of factorials.

39. $\dfrac{4!}{6!}$

40. $\dfrac{4!}{7!}$

41. $\dfrac{10!}{8!}$

42. $\dfrac{25!}{23!}$

43. $\dfrac{(n+1)!}{n!}$

44. $\dfrac{(n+2)!}{n!}$

45. $\dfrac{(2n-1)!}{(2n+1)!}$

46. $\dfrac{(2n+2)!}{(2n)!}$

In Exercises 47–60, write an expression for the *most apparent* nth term of the sequence. (Assume that n begins with 1.)

47. $1, 4, 7, 10, 13, \ldots$

48. $3, 7, 11, 15, 19, \ldots$

49. $0, 3, 8, 15, 24, \ldots$

50. $1, \dfrac{1}{4}, \dfrac{1}{9}, \dfrac{1}{16}, \dfrac{1}{25}, \ldots$

51. $\dfrac{2}{3}, \dfrac{3}{4}, \dfrac{4}{5}, \dfrac{5}{6}, \dfrac{6}{7}, \ldots$

52. $\dfrac{2}{1}, \dfrac{3}{3}, \dfrac{4}{5}, \dfrac{5}{7}, \dfrac{6}{9}, \ldots$

53. $\dfrac{1}{2}, \dfrac{-1}{4}, \dfrac{1}{8}, \dfrac{-1}{16}, \ldots$

54. $\dfrac{1}{3}, \dfrac{2}{9}, \dfrac{4}{27}, \dfrac{8}{81}, \ldots$

55. $1 + \dfrac{1}{1}, 1 + \dfrac{1}{2}, 1 + \dfrac{1}{3}, 1 + \dfrac{1}{4}, 1 + \dfrac{1}{5}, \ldots$

56. $1 + \dfrac{1}{2}, 1 + \dfrac{3}{4}, 1 + \dfrac{7}{8}, 1 + \dfrac{15}{16}, 1 + \dfrac{31}{32}, \ldots$

57. $1, \dfrac{1}{2}, \dfrac{1}{6}, \dfrac{1}{24}, \dfrac{1}{120}, \ldots$

58. $2, -4, 6, -8, 10, \ldots$

59. $1, -1, 1, -1, 1, \ldots$

60. $1, 2, \dfrac{2^2}{2}, \dfrac{2^3}{6}, \dfrac{2^4}{24}, \dfrac{2^5}{120}, \ldots$

In Exercises 61–64, write the first five terms of the sequence defined recursively. Use the pattern to write the nth term of the sequence as a function of n. (Assume that n begins with 1.)

61. $a_1 = 6, \quad a_{k+1} = a_k + 2$

62. $a_1 = 25, \quad a_{k+1} = a_k - 5$

63. $a_1 = 81, \quad a_{k+1} = \dfrac{1}{3}a_k$

64. $a_1 = 14, \quad a_{k+1} = (-2)a_k$

In Exercises 65–76, find the sum.

65. $\displaystyle\sum_{i=1}^{5}(2i+1)$

66. $\displaystyle\sum_{i=1}^{6}(3i-1)$

67. $\displaystyle\sum_{k=1}^{4}10$

68. $\displaystyle\sum_{k=1}^{5}6$

69. $\displaystyle\sum_{i=0}^{4}i^2$

70. $\displaystyle\sum_{i=0}^{5}3i^2$

71. $\displaystyle\sum_{k=0}^{3}\dfrac{1}{k^2+1}$

72. $\displaystyle\sum_{j=3}^{5}\dfrac{1}{j}$

73. $\displaystyle\sum_{i=1}^{4}[(i-1)^2 + (i+1)^3]$

74. $\displaystyle\sum_{k=2}^{5}(k+1)(k-3)$

75. $\displaystyle\sum_{i=1}^{4}2^i$

76. $\displaystyle\sum_{j=0}^{4}(-2)^j$

In Exercises 77–80, use a calculator to find the sum.

77. $\displaystyle\sum_{j=1}^{6}(24-3j)$ **78.** $\displaystyle\sum_{j=1}^{10}\frac{3}{j+1}$

79. $\displaystyle\sum_{k=0}^{4}\frac{(-1)^k}{k+1}$ **80.** $\displaystyle\sum_{k=0}^{4}\frac{(-1)^k}{k!}$

In Exercises 81–90, use sigma notation to write the sum.

81. $\dfrac{1}{3(1)} + \dfrac{1}{3(2)} + \dfrac{1}{3(3)} + \cdots + \dfrac{1}{3(9)}$

82. $\dfrac{5}{1+1} + \dfrac{5}{1+2} + \dfrac{5}{1+3} + \cdots + \dfrac{5}{1+15}$

83. $\left[2\left(\frac{1}{8}\right)+3\right] + \left[2\left(\frac{2}{8}\right)+3\right] + \cdots + \left[2\left(\frac{8}{8}\right)+3\right]$

84. $\left[1-\left(\frac{1}{6}\right)^2\right] + \left[1-\left(\frac{2}{6}\right)^2\right] + \cdots + \left[1-\left(\frac{6}{6}\right)^2\right]$

85. $3 - 9 + 27 - 81 + 243 - 729$

86. $1 - \dfrac{1}{2} + \dfrac{1}{4} - \dfrac{1}{8} + \cdots - \dfrac{1}{128}$

87. $\dfrac{1}{1^2} - \dfrac{1}{2^2} + \dfrac{1}{3^2} - \dfrac{1}{4^2} + \cdots + \dfrac{1}{20^2}$

88. $\dfrac{1}{1 \cdot 3} + \dfrac{1}{2 \cdot 4} + \dfrac{1}{3 \cdot 5} + \cdots + \dfrac{1}{10 \cdot 12}$

89. $\dfrac{1}{4} + \dfrac{3}{8} + \dfrac{7}{16} + \dfrac{15}{32} + \dfrac{31}{64}$

90. $\dfrac{1}{2} + \dfrac{2}{4} + \dfrac{6}{8} + \dfrac{24}{16} + \dfrac{120}{32} + \dfrac{720}{64}$

91. *Compound Interest*   A deposit of $5000 is made in an account that earns 8% interest compounded quarterly. The balance in the account after n quarters is given by

$$A_n = 5000\left(1 + \frac{0.08}{4}\right)^n, \qquad n = 1, 2, 3, \ldots .$$

  (a) Compute the first eight terms of this sequence.

  (b) Find the balance in this account after 10 years by computing the 40th term of the sequence.

92. *Compound Interest*   A deposit of $100 is made *each month* in an account that earns 12% interest compounded monthly. The balance in the account after n months is given by

$$A_n = 100(101)[(1.01)^n - 1], \qquad n = 1, 2, 3, \ldots .$$

  (a) Compute the first six terms of this sequence.

  (b) Find the balance in this account after 5 years by computing the 60th term of the sequence.

  (c) Find the balance in this account after 20 years by computing the 240th term of the sequence.

93. *Per Capita Hospital Care*   The average per capita annual cost for hospital care from 1981 to 1991 is approximated by the model

$$a_n = 510.13 + 16.37n + 3.23n^2, \quad n = 1, \ldots, 11$$

where a_n is the per capita cost in dollars and n is the year, with $n = 1$ corresponding to 1981. Find the terms of this finite sequence and use a graphing utility to construct a bar graph that represents the sequence. (Source: U.S. Health Care Financing Administration)

94. *Federal Debt*   It took more than 200 years for the United States to accumulate a $1 trillion debt. Then it took just 8 years to get to a $3 trillion debt. The federal debt during the decade of the 1980s is approximated by the model

$$a_n = 0.1\sqrt{82 + 9n^2}, \qquad n = 0, \ldots, 10$$

where a_n is the debt in trillions and n is the year, with $n = 0$ corresponding to 1980. Find the terms of this finite sequence and use a graphing utility to construct a bar graph that represents the sequence. (Source: Treasury Department)

95. Corporate Income The net incomes a_n (in millions of dollars) of Wal-Mart for the years 1985 through 1994 are shown in the figure. These incomes can be approximated by the model

$$a_n = 129.9 + 0.9n^3, \qquad n = 5, \ldots, 14$$

where $n = 5$ represents 1985. Use this model to approximate the total net income from 1985 through 1994. Compare this sum with the result of adding the incomes shown in the figure.   (Source: Wal-Mart)

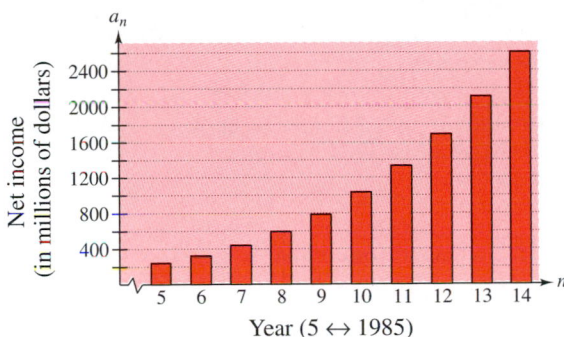

Year (5 ↔ 1985)

96. Corporate Dividends The dividends a_n declared per share of common stock of Procter & Gamble Company for the years 1985 through 1994 are shown in the figure. These dividends can be approximated by the model

$$a_n = 1.39 + 0.18n - 1.02 \ln n,$$

$$n = 5, \ldots, 14$$

where $n = 5$ represents 1985. Use this model to approximate the total dividends per share of common stock from 1985 through 1994. Compare this sum with the result of adding the dividends shown in the figure.   (Source: Procter & Gamble Company)

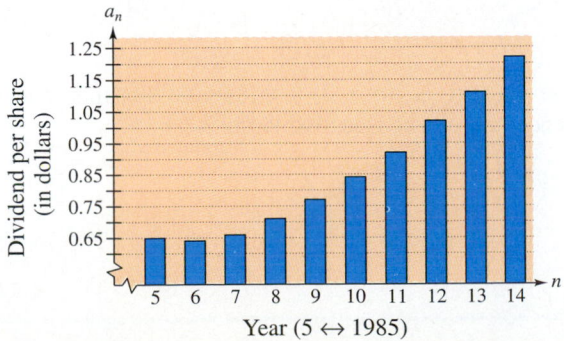

Year (5 ↔ 1985)

Fibonacci Sequence **In Exercises 97 and 98, use the Fibonacci Sequence. (See Example 4.)**

97. Write the first 12 terms of the Fibonacci sequence a_n and the first 10 terms of the sequence given by

$$b_n = \frac{a_{n+1}}{a_n}, \qquad n \geq 1.$$

98. Using the definition for b_n in Exercise 97, show that b_n can be defined recursively by

$$b_n = 1 + \frac{1}{b_{n-1}}.$$

Arithmetic Mean **In Exercises 99–102, use the following definition of the arithmetic mean $\bar{x}$ of a set of n measurements $x_1, x_2, x_3, \ldots, x_n$.**

$$\bar{x} = \frac{1}{n} \sum_{i=1}^{n} x_i$$

99. Find the arithmetic mean of the six checking account balances $327.15, $785.69, $433.04, $265.38, $604.12, and $590.30. Use statistical capabilities of a graphing utility to verify your result.

100. Find the arithmetic mean of the following prices per gallon for regular unleaded gasoline at five gasoline stations in a city: $1.279, $1.259, $1.289, $1.329, and $1.349. Use the statistical capabilities of a graphing utility to verify your result.

101. Prove that $\displaystyle\sum_{i=1}^{n} (x_i - \bar{x}) = 0$.

102. Prove that $\displaystyle\sum_{i=1}^{n} (x_i - \bar{x})^2 - \sum_{i=1}^{n} x_i^2 - \frac{1}{n}\left(\sum_{i=1}^{n} x_i\right)^2$.

Think About It **In Exercises 103 and 104, determine whether the statement is true. Explain your reasoning.**

103. $\displaystyle\sum_{i=1}^{4} (i^2 + 2i) = \sum_{i=1}^{4} i^2 + 2\sum_{i=1}^{4} i$

104. $\displaystyle\sum_{j=1}^{4} 2^j = \sum_{j=3}^{6} 2^{j-2}$

11.2   *Arithmetic Sequences*

See Exercise 83 on page 818 for an example of how an arithmetic sequence can be used to find the number of bricks needed to lay a brick patio.

Arithmetic Sequences   □   *The Sum of an Arithmetic Sequence*   □
Applications

Arithmetic Sequences

A sequence whose consecutive terms have a common difference is called an **arithmetic sequence.**

>   **DEFINITION OF AN ARITHMETIC SEQUENCE**
>
> A sequence is **arithmetic** if the differences between consecutive terms are the same. Thus, the sequence
>
> $$a_1, a_2, a_3, a_4, \ldots , a_n, \ldots$$
>
> is arithmetic if there is a number d such that
>
> $$a_2 - a_1 = d, \quad a_3 - a_2 = d, \quad a_4 - a_3 = d$$
>
> and so on. The number d is the **common difference** of the arithmetic sequence.

EXAMPLE 1   *Examples of Arithmetic Sequences*

a. The sequence whose nth term is $4n + 3$ is arithmetic. For this sequence, the common difference between consecutive terms is 4.

$$7, 11, 15, 19, \ldots , 4n + 3, \ldots$$

$$11 - 7 = 4$$

b. The sequence whose nth term is $7 - 5n$ is arithmetic. For this sequence, the common difference between consecutive terms is -5.

$$2, -3, -8, -13, \ldots , 7 - 5n, \ldots$$

$$-3 - 2 = -5$$

c. The sequence whose nth term is $\frac{1}{4}(n + 3)$ is arithmetic. For this sequence, the common difference between consecutive terms is $\frac{1}{4}$.

$$1, \frac{5}{4}, \frac{3}{2}, \frac{7}{4}, \ldots , \frac{n + 3}{4}, \ldots$$

$$\tfrac{5}{4} - 1 = \tfrac{1}{4}$$

In Example 1, notice that each of the arithmetic sequences has an nth term that is of the form $dn + c$, where the common difference of the sequence is d. This result is summarized as follows.

THE nTH TERM OF AN ARITHMETIC SEQUENCE

The nth term of an arithmetic sequence has the form

$$a_n = dn + c$$

where d is the common difference between consecutive terms of the sequence and $c = a_1 - d$.

EXAMPLE 2 *Finding the nth Term of an Arithmetic Sequence*

Find a formula for the nth term of the arithmetic sequence whose common difference is 3 and whose first term is 2.

Solution

Because the sequence is arithmetic, you know that the formula for the nth term is of the form $a_n = dn + c$. Moreover, because the common difference is $d = 3$, the formula must have the form

$$a_n = 3n + c.$$

Because $a_1 = 2$, it follows that

$$c = a_1 - d = 2 - 3 = -1.$$

Thus, the formula for the nth term is

$$a_n = 3n - 1.$$

The sequence therefore has the following form.

$$2, 5, 8, 11, 14, . . . , 3n - 1, . . .$$

Another way to find a formula for the nth term of the sequence in Example 2 is to begin by writing the terms of the sequence.

a_1	a_2	a_3	a_4	a_5	a_6	a_7	
2	$2 + 3$	$5 + 3$	$8 + 3$	$11 + 3$	$14 + 3$	$17 + 3$	. . .
2	5	8	11	14	17	20	. . .

From these terms, you can reason that the nth term is of the form

$$a_n = dn + c = 3n - 1.$$

EXAMPLE 3 *Finding the nth Term of an Arithmetic Sequence*

The fourth term of an arithmetic sequence is 20, and the 13th term is 65. Write the first several terms of this sequence.

Solution

The fourth and 13th terms of the sequence are related by

$$a_{13} = a_4 + 9d.$$

Using $a_4 = 20$ and $a_{13} = 65$, you can conclude that $d = 5$, which implies that the sequence is as follows.

a_1	a_2	a_3	a_4	a_5	a_6	a_7	a_8	a_9	a_{10}	a_{11}
5,	10,	15,	20,	25,	30,	35,	40,	45,	50,	55, . . .

Study Tip

If you substitute $a_1 - d$ for c in the formula $a_n = dn + c$, the nth term of an arithmetic sequence has the alternative recursive formula

$$a_n = a_1 + (n - 1)d.$$

Use this formula to solve Example 4 and Example 9.

If you know the nth term of an arithmetic sequence *and* you know the common difference of the sequence, you can find the $(n + 1)$th term by using the *recursive formula*

$$a_{n+1} = a_n + d.$$

With this formula, you can find any term of an arithmetic sequence, *provided* that you know the previous term. For instance, if you know the first term, you can find the second term. Then, knowing the second term, you can find the third term, and so on.

EXAMPLE 4 *Using a Recursive Formula*

Find the ninth term of the arithmetic sequence that begins with 2 and 9.

Solution

For this sequence, the common difference is $d = 9 - 2 = 7$. There are two ways to find the ninth term. One way is simply to write out the first nine terms (by repeatedly adding 7).

2, 9, 16, 23, 30, 37, 44, 51, 58

Another way to find the ninth term is first to find a formula for the nth term. Because the first term is 2, it follows that

$$c = a_1 - d = 2 - 7 = -5.$$

Therefore, a formula for the nth term is $a_n = 7n - 5$, which implies that the ninth term is $a_9 = 7(9) - 5 = 58$.

The Sum of an Arithmetic Sequence

There is a simple formula for the *sum* of a finite arithmetic sequence.

NOTE Be sure you see that this formula works only for *arithmetic* sequences. ■■

> **THE SUM OF A FINITE ARITHMETIC SEQUENCE**
>
> The sum of a finite arithmetic sequence with n terms is given by
>
> $$S_n = \frac{n}{2}(a_1 + a_n).$$

EXAMPLE 5   *Finding the Sum of an Arithmetic Sequence*

Find the sum: $1 + 3 + 5 + 7 + 9 + 11 + 13 + 15 + 17 + 19$.

Solution

To begin, notice that the sequence is arithmetic (with a common difference of 2). Moreover, the sequence has 10 terms. Thus, the sum of the sequence is

$$S_n = 1 + 3 + 5 + 7 + 9 + 11 + 13 + 15 + 17 + 19$$

$$= \frac{n}{2}(a_1 + a_n)$$

$$= \frac{10}{2}(1 + 19) \qquad n = 10$$

$$= 5(20)$$

$$= 100.$$

THINK ABOUT THE PROOF

To prove the formula for the sum of a finite arithmetic sequence, consider writing the sum in two different ways. One way is to write the sum as

$$S_n = a_1 + (a_1 + d) + (a_1 + 2d) + \cdots + [a_1 + (n - 1)d].$$

In the second way, you repeatedly subtract d from the nth term to obtain

$$S_n = a_n + (a_n - d) + (a_n - 2d) + \cdots + [a_n - (n - 1)d].$$

Can you discover a way to combine these two versions of S to obtain the following formula?

$$S_n = \frac{n}{2}(a_1 + a_n)$$

The details of this proof are shown in the appendix.

EXAMPLE 6   *Finding the Sum of an Arithmetic Sequence*

Find the sum of the integers from 1 to 100.

Solution

The integers from 1 to 100 form an arithmetic sequence that has 100 terms. Thus, you can use the formula for the sum of an arithmetic sequence, as follows.

$$S_n = 1 + 2 + 3 + 4 + 5 + 6 + \cdots + 99 + 100$$

$$= \frac{n}{2}(a_1 + a_n)$$

$$= \frac{100}{2}(1 + 100) \qquad n = 100$$

$$= 50(101)$$

$$= 5050$$

EXAMPLE 7   *Finding the Sum of an Arithmetic Sequence*

Find the sum of the first 150 terms of the arithmetic sequence

$$5, 16, 27, 38, 49, \ldots .$$

Solution

For this arithmetic sequence, $a_1 = 5$ and $d = 16 - 5 = 11$. Thus, $c = a_1 - d = 5 - 11 = -6$, and the nth term is

$$a_n = 11n - 6.$$

Therefore, $a_{150} = 11(150) - 6 = 1644$, and the sum of the first 150 terms is

$$S_n = \frac{n}{2}(a_1 + a_n)$$

$$= \frac{150}{2}(5 + 1644)$$

$$= 75(1649)$$

$$= 123{,}675.$$

NOTE   The sum of the first n terms of an infinite sequence is called the ***n*th partial sum.** ■■

Applications

EXAMPLE 8   *Seating Capacity*

Real
Life

An auditorium has 20 rows of seats. There are 20 seats in the first row, 21 seats in the second row, 22 seats in the third row, and so on (see Figure 11.2). How many seats are there in all 20 rows?

Solution

The numbers of seats in the 20 rows form an arithmetic sequence in which the common difference is $d = 1$. Because $c = a_1 - d = 20 - 1 = 19$, you can determine that the formula for the nth term of the sequence is $a_n = n + 19$. Therefore, the 20th term in the sequence is $a_{20} = 20 + 19 = 39$, and the total number of seats is

$$S_n = 20 + 21 + 22 + \cdots + 39$$

$$= \frac{n}{2}(a_1 + a_{20})$$

$$= \frac{20}{2}(20 + 39)$$

$$= 10(59)$$

$$= 590.$$

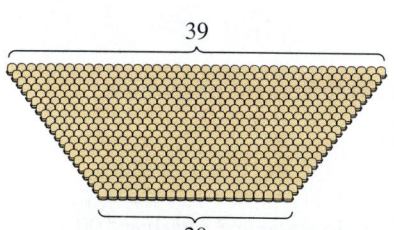

39

20

FIGURE 11.2

EXAMPLE 9   ***Total Sales***

A small business sells $10,000 worth of products during its first year. The owner of the business has set a goal of increasing annual sales by $7500 each year for 9 years. Assuming that this goal is met, find the total sales during the first 10 years this business is in operation.

Solution

The annual sales form an arithmetic sequence in which $a_1 = 10,000$ and $d = 7500$. Thus, $c = a_1 - d = 10,000 - 7500 = 2500$, and the *n*th term of the sequence is

$$a_n = 7500n + 2500.$$

This implies that the 10th term of the sequence is $a_{10} = 77,500$. The sum of the first 10 terms of the sequence is

$$S_n = \frac{n}{2}(a_1 + a_{10})$$

$$= \frac{10}{2}(10,000 + 77,500)$$

$$= 5(87,500)$$

$$= 437,500.$$

Thus, the total sales for the first 10 years is $437,500.

GROUP ACTIVITY

NUMERICAL RELATIONSHIPS

Decide whether it is possible to fill in the blanks in each of the following such that the resulting sequence is arithmetic. If so, find a recursive formula for the sequence.

a. −7, ⬜ , ⬜ , ⬜ , ⬜ , ⬜ , 11

b. 17, ⬜ , ⬜ , ⬜ , ⬜ , ⬜ , ⬜ , ⬜ , ⬜ , 71

c. 2, 6, ⬜ , ⬜ , 162

d. 4, 7.5, ⬜ , ⬜ , ⬜ , ⬜ , ⬜ , ⬜ , ⬜ , 39

e. 8, 12, ⬜ , ⬜ , ⬜ , 60.75

Find the sum.

1. $\displaystyle\sum_{i=1}^{6}(2i-1)$

2. $\displaystyle\sum_{i=1}^{10}(4i+2)$

Find the distance between the two real numbers.

3. $\frac{5}{2}, 8$

4. $\frac{4}{3}, \frac{14}{3}$

Find the indicated value of the function.

5. $f(n) = 10 + (n-1)4, \ f(3)$

6. $f(n) = 1 + (n-1)\frac{1}{3}, \ f(10)$

Evaluate the expression.

7. $\frac{11}{2}(1+25)$

8. $\frac{16}{2}(4+16)$

9. $\frac{20}{2}[2(5)+(12-1)3]$

10. $\frac{8}{2}[2(-3)+(15-1)5]$

11.2 Exercises

In Exercises 1–10, determine whether the sequence is arithmetic. If it is, find the common difference.

1. $10, 8, 6, 4, 2, \ldots$

2. $4, 7, 10, 13, 16, \ldots$

3. $1, 2, 4, 8, 16, \ldots$

4. $3, \frac{5}{2}, 2, \frac{3}{2}, 1, \ldots$

5. $\frac{9}{4}, 2, \frac{7}{4}, \frac{3}{2}, \frac{5}{4}, \ldots$

6. $\frac{1}{3}, \frac{2}{3}, \frac{4}{3}, \frac{8}{3}, \frac{16}{3}, \ldots$

7. $-12, -8, -4, 0, 4, \ldots$

8. $\ln 1, \ln 2, \ln 3, \ln 4, \ln 5, \ldots$

9. $5.3, 5.7, 6.1, 6.5, 6.9, \ldots$

10. $1^2, 2^2, 3^2, 4^2, 5^2, \ldots$

In Exercises 11–18, write the first five terms of the sequence. Determine whether the sequence is arithmetic, and if it is, find the common difference.

11. $a_n = 5 + 3n$

12. $a_n = (2^n)n$

13. $a_n = \dfrac{1}{n+1}$

14. $a_n = 1 + (n-1)4$

15. $a_n = 100 - 3n$

16. $a_n = 2^{n-1}$

17. $a_n = 3 + \dfrac{(-1)^n 2}{n}$

18. $a_n = (-1)^n$

In Exercises 19–24, write the first five terms of the arithmetic sequence. Find the common difference and write the nth term of the sequence as a function of n.

19. $a_1 = 15, \quad a_{k+1} = a_k + 4$

20. $a_1 = 2, \quad a_{k+1} = a_k + \frac{2}{3}$

21. $a_1 = 200, \quad a_{k+1} = a_k - 10$

22. $a_1 = 72, \quad a_{k+1} = a_k - 6$

23. $a_1 = \frac{3}{2}, \quad a_{k+1} = a_k - \frac{1}{4}$

24. $a_1 = 0.375, \quad a_{k+1} = a_k + 0.25$

In Exercises 25–32, write the first five terms of the arithmetic sequence.

25. $a_1 = 5, d = 6$

26. $a_1 = 5, d = -\frac{3}{4}$

27. $a_1 = -2.6, d = -0.4$

28. $a_1 = 16.5, d = 0.25$

29. $a_1 = 2, a_{12} = 46$

30. $a_4 = 16, a_{10} = 46$

31. $a_8 = 26, a_{12} = 42$

32. $a_3 = 19, a_{15} = -1.7$

In Exercises 33–44, find a formula for a_n for the arithmetic sequence.

33. $a_1 = 1, d = 3$

34. $a_1 = 15, d = 4$

35. $a_1 = 100, d = -8$

36. $a_1 = 0, d = -\frac{2}{3}$

37. $a_1 = x, d = 2x$

38. $a_1 = -y, d = 5y$

39. $4, \frac{3}{2}, -1, -\frac{7}{2}, \ldots$

40. $10, 5, 0, -5, -10, \ldots$

41. $a_1 = 5, a_4 = 15$

42. $a_1 = -4, a_5 = 16$

43. $a_3 = 94, a_6 = 85$

44. $a_5 = 190, a_{10} = 115$

In Exercises 45–48, match the sequence with its graph. [The graphs are labeled (a), (b), (c), and (d).]

(a)

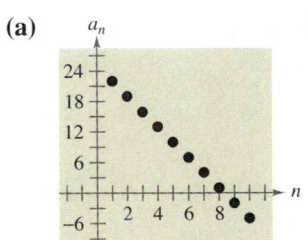

(b)

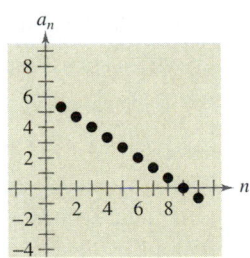

(c)

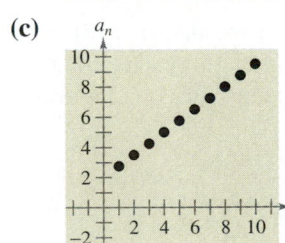

(d)

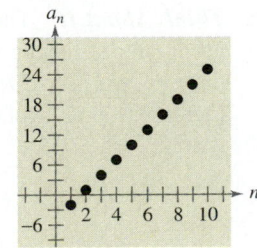

45. $a_n = -\frac{2}{3}n + 6$

46. $a_n = 3n - 5$

47. $a_n = 2 + \frac{3}{4}n$

48. $a_n = 25 - 3n$

In Exercises 49–52, use a graphing utility to graph the first 10 terms of the sequence.

49. $a_n = 15 - \frac{3}{2}n$

50. $a_n = -5 + 2n$

51. $a_n = 0.2n + 3$

52. $a_n = -0.3n + 8$

53. *Essay* Describe the geometric pattern of the graph of an arithmetic sequence. Explain.

54. *Essay* Explain how to use the first two terms of an arithmetic sequence to find the nth term.

In Exercises 55–62, find the sum of the first n terms of the arithmetic sequence.

55. $8, 20, 32, 44, \ldots,\quad n = 10$

56. $2, 8, 14, 20, \ldots,\quad n = 25$

57. $-6, -2, 2, 6, \ldots,\quad n = 50$

58. $0.5, 0.9, 1.3, 1.7, \ldots,\quad n = 10$

59. $40, 37, 34, 31, \ldots,\quad n = 10$

60. $1.50, 1.45, 1.40, 1.35, \ldots,\quad n = 20$

61. $a_1 = 100, a_{25} = 220,\quad n = 25$

62. $a_1 = 15, a_{100} = 307,\quad n = 100$

In Exercises 63–70, find the sum.

63. $\displaystyle\sum_{n=1}^{50} n$

64. $\displaystyle\sum_{n=1}^{100} 2n$

65. $\displaystyle\sum_{n=1}^{100} 5n$

66. $\displaystyle\sum_{n=51}^{100} 7n$

67. $\displaystyle\sum_{n=11}^{30} n - \sum_{n=1}^{10} n$

68. $\displaystyle\sum_{n=51}^{100} n - \sum_{n=1}^{50} n$

69. $\displaystyle\sum_{n=1}^{500} (n + 3)$

70. $\displaystyle\sum_{n=1}^{250} (1000 - n)$

In Exercises 71–76, use a calculator to find the sum.

71. $\displaystyle\sum_{n=1}^{20} (2n + 5)$

72. $\displaystyle\sum_{n=1}^{100} \frac{n + 4}{2}$

73. $\displaystyle\sum_{n=0}^{50}(1000-5n)$ **74.** $\displaystyle\sum_{n=0}^{100}\dfrac{8-3n}{16}$

75. $\displaystyle\sum_{i=1}^{60}\left(250-\tfrac{8}{3}i\right)$ **76.** $\displaystyle\sum_{j=1}^{200}(4.5+0.025j)$

77. Find the sum of the first 100 odd integers.

78. Find the sum of the integers from -10 to 50.

Job Offer   **In Exercises 79 and 80, consider a job offer with the given starting salary and the given annual raise.**

(a) Determine the salary during the sixth year of employment.

(b) Determine the total compensation from the company through six full years of employment.

	Starting Salary	*Annual Raise*
79.	$32,500	$1500
80.	$36,800	$1750

81. *Seating Capacity*   Determine the seating capacity of an auditorium with 30 rows of seats if there are 20 seats in the first row, 24 seats in the second row, 28 seats in the third row, and so on.

82. *Seating Capacity*   Determine the seating capacity of an auditorium with 36 rows of seats if there are 15 seats in the first row, 18 seats in the second row, 21 seats in the third row, and so on.

83. *Brick Pattern*   A brick patio has the approximate shape of a trapezoid (see figure). The patio has 18 rows of bricks. The first row has 14 bricks and the 18th row has 31 bricks. How many bricks are in the patio?

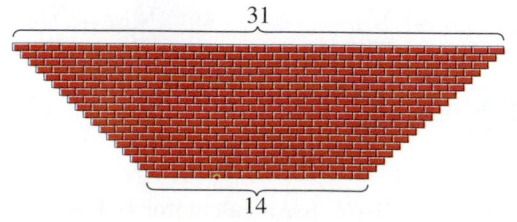

31

14

84. *Falling Object*   An object (with negligible air resistance) is dropped from a plane. During the first second of fall, the object falls 4.9 meters; during the second second, it falls 14.7 meters; during the third second, it falls 24.5 meters; during the fourth second, it falls 34.3 meters. If this arithmetic pattern continues, how many meters will the object fall in 10 seconds?

85. *Pattern Recognition*

(a) Compute the following sums of positive odd integers.

$1+3=$ ▭

$1+3+5=$ ▭

$1+3+5+7=$ ▭

$1+3+5+7+9=$ ▭

$1+3+5+7+9+11=$ ▭

(b) Use the sums in part (a) to make a conjecture about the sums of positive odd integers. Check your conjecture for the sum

$1+3+5+7+9+11+13=$ ▭ .

(c) Verify your conjecture analytically.

86. *Think About It*   The following operations are performed on each term of an arithmetic sequence. Determine if the resulting sequence is arithmetic, and if so, state the common difference.

(a) A constant C is added to each term.

(b) Each term is multiplied by a nonzero constant C.

(c) Each term is squared.

87. *Think About It*   The sum of the first 20 terms of an arithmetic sequence with a common difference of 3 is 650. Find the first term.

88. *Think About It*   The sum of the first n terms of an arithmetic sequence with first term a_1 and common difference d is S_n. Determine the sum if each term is increased by 5. Explain.

11.3   *Geometric Sequences*

See Exercise 75 on page 827 for an example of how a geometric sequence can be used to analyze a decreasing annuity.

Geometric Sequences   □   The Sum of a Geometric Sequence   □ | Application

Geometric Sequences

In Section 11.2, you learned that a sequence whose consecutive terms have a common *difference* is an arithmetic sequence. In this section, you will study another important type of sequence called a **geometric sequence.** Consecutive terms of a geometric sequence have a common *ratio.*

> **DEFINITION OF A GEOMETRIC SEQUENCE**
>
> A sequence is **geometric** if the ratios of consecutive terms are the same.
>
> $$\frac{a_2}{a_1} = r, \quad \frac{a_3}{a_2} = r, \quad \frac{a_4}{a_3} = r, \ldots, \qquad r \neq 0$$
>
> The number r is the **common ratio** of the sequence.

EXAMPLE 1   *Examples of Geometric Sequences*

a. The sequence whose nth term is 2^n is geometric. For this sequence, the common ratio of consecutive terms is 2.

$$2, 4, 8, 16, \ldots, 2^n, \ldots$$

$$\frac{4}{2} = 2$$

b. The sequence whose nth term is $4(3^n)$ is geometric. For this sequence, the common ratio of consecutive terms is 3.

$$12, 36, 108, 324, \ldots, 4(3^n), \ldots$$

$$\frac{36}{12} = 3$$

c. The sequence whose nth term is $\left(-\frac{1}{3}\right)^n$ is geometric. For this sequence, the common ratio of consecutive terms is $-\frac{1}{3}$.

$$-\frac{1}{3}, \frac{1}{9}, -\frac{1}{27}, \frac{1}{81}, \ldots, \left(-\frac{1}{3}\right)^n, \ldots$$

$$\frac{1/9}{-1/3} = -\frac{1}{3}$$

In Example 1, notice that each of the geometric sequences has an nth term that is of the form ar^n, where the common ratio of the sequence is r.

NOTE   If you know the nth term of a geometric sequence, you can find the $(n + 1)$th term by multiplying by r. That is, $a_{n+1} = ra_n$. ▪▪

THE NTH TERM OF A GEOMETRIC SEQUENCE

The nth term of a geometric sequence has the form

$$a_n = a_1 r^{n-1}$$

where r is the common ratio of consecutive terms of the sequence. Thus, every geometric sequence can be written in the following form.

$$a_1, \quad a_2, \quad a_3, \quad a_4, \quad a_5, \ldots \ldots, a_n, \ldots \ldots$$
$$\downarrow \quad \downarrow \quad \downarrow \quad \downarrow \quad \downarrow \cdots \cdots, \quad \downarrow \cdots \cdots$$
$$a_1, a_1 r, a_1 r^2, a_1 r^3, a_1 r^4, \ldots, a_1 r^{n-1}, \ldots$$

TECHNOLOGY

You can generate the geometric sequence in Example 2 with a graphing utility using the following steps.

3 | ENTER |

2 | × | | ANS |

Now press | ENTER | repeatedly to generate the terms of the sequence.

EXAMPLE 2   *Finding the Terms of a Geometric Sequence*

Write the first five terms of the geometric sequence whose first term is $a_1 = 3$ and whose common ratio is $r = 2$.

Solution

Starting with 3, repeatedly multiply by 2 to obtain the following.

$a_1 = 3$	1st term
$a_2 = 3(2^1) = 6$	2nd term
$a_3 = 3(2^2) = 12$	3rd term
$a_4 = 3(2^3) = 24$	4th term
$a_5 = 3(2^4) = 48$	5th term

EXAMPLE 3   *Finding a Term of a Geometric Sequence*

Find the 15th term of the geometric sequence whose first term is 20 and whose common ratio is 1.05.

Solution

$a_{15} = a_1 r^{n-1}$	Formula for geometric sequence
$= 20(1.05)^{15-1}$	Substitute for a_1, r, and n.
≈ 39.599	Use a calculator.

EXAMPLE 4 *Finding a Term of a Geometric Sequence*

Find the 12th term of the geometric sequence

$$5, 15, 45, \ldots \ldots$$

Solution

The common ratio of this sequence is $r = 15/5 = 3$. Because the first term is $a_1 = 5$, you can determine the 12th term ($n = 12$) to be

$$
\begin{aligned}
a_{12} &= a_1 r^{n-1} && \text{Formula for geometric sequence}\\
&= 5(3)^{12-1} && \text{Substitute for } a_1, r, \text{ and } n.\\
&= 5(177,147) && \text{Use a calculator.}\\
&= 885,735. && \text{Simplify.}
\end{aligned}
$$

If you know any two terms of a geometric sequence, you can use that information to find a formula for the *n*th term of the sequence.

EXAMPLE 5 *Finding a Term of a Geometric Sequence*

The fourth term of a geometric sequence is 125, and the 10th term is 125/64. Find the 14th term. (Assume that the terms of the sequence are positive.)

Solution

The 10th term is related to the fourth term by the equation

$$a_{10} = a_4 r^6 \qquad \text{Multiply 4th term by } r^{10-4}.$$

Because $a_{10} = 125/64$ and $a_4 = 125$, you can solve for r as follows.

$$\frac{125}{64} = 125 r^6$$

$$\frac{1}{64} = r^6$$

$$\frac{1}{2} = r$$

You can obtain the 14th term by multiplying the 10th term by r^4.

$$
\begin{aligned}
a_{14} &= a_{10} r^4\\
&= \frac{125}{64}\left(\frac{1}{2}\right)^4 = \frac{125}{1024}
\end{aligned}
$$

Study Tip

Remember that r is the common ratio of consecutive terms of a sequence. So in Example 5

$$
\begin{aligned}
a_{10} &= a_1 r^9\\
&= a_1 \cdot r \cdot r \cdot r \cdot r^6\\
&= a_1 \cdot \frac{a_2}{a_1} \cdot \frac{a_3}{a_2} \cdot \frac{a_4}{a_3} \cdot r^6\\
&= a_4 r^6.
\end{aligned}
$$

The Sum of a Geometric Sequence

The formula for the sum of a *finite* geometric sequence is as follows.

To prove the formula for the sum of a finite geometric sequence, consider the following two equations.

$$S_n = a_1 + a_1 r + a_1 r^2 +$$
$$\cdots + a_1 r^{n-2} +$$
$$a_1 r^{n-1}$$
$$rS_n = a_1 r + a_1 r^2 + a_1 r^3 +$$
$$\cdots + a_1 r^{n-1} + a_1 r^n$$

Can you discover a way to simplify the difference of these two equations to obtain the formula for the sum of a geometric sequence? The details of the proof are given in the appendix.

■ **THE SUM OF A FINITE GEOMETRIC SEQUENCE**

The sum of the geometric sequence

$$a_1, \ a_1 r, \ a_1 r^2, \ a_1 r^3, \ a_1 r^4, \ \ldots, a_1 r^{n-1}$$

with common ratio $r \neq 1$ is given by

$$S_n = a_1 \left(\frac{1 - r^n}{1 - r} \right).$$

EXAMPLE 6   *Finding the Sum of a Finite Geometric Sequence*

Find the sum $\sum_{n=1}^{12} 4(0.3)^n$.

Solution

By writing out a few terms, you have

$$\sum_{n=1}^{12} 4(0.3)^n = 4(0.3) + 4(0.3)^2 + 4(0.3)^3 + \cdots + 4(0.3)^{12}.$$

Now, because $a_1 = 4(0.3)$, $r = 0.3$, and $n = 12$, you can apply the formula for the sum of a finite geometric sequence to obtain

$$\sum_{n=1}^{12} 4(0.3)^n = a_1 \left(\frac{1 - r^n}{1 - r} \right)$$

$$= 4(0.3) \left[\frac{1 - (0.3)^{12}}{1 - 0.3} \right]$$

$$\approx 1.714.$$

When using the formula for the sum of a geometric sequence, be careful to check that the index begins at $i = 1$. If the index begins at $i = 0$, you must adjust the formula for the nth partial sum. For instance, if the index in Example 6 had begun with $n = 0$, the sum would have been

$$\sum_{n=0}^{12} 4(0.3)^n = 4 + \sum_{n=1}^{12} 4(0.3)^n \approx 4 + 1.714 = 5.714.$$

The formula for the sum of a *finite* geometric sequence can, depending on the value of r, be extended to produce a formula for the sum of an *infinite* geometric sequence. Specifically, if the common ratio r has the property that $|r| < 1$, It can be shown that r^n becomes arbitrarily close to zero as n increases without bound. Consequently,

$$a_1\left(\frac{1 - r^n}{1 - r}\right) \longrightarrow a_1\left(\frac{1 - 0}{1 - r}\right) \quad \text{as} \quad n \longrightarrow \infty.$$

This result is summarized as follows.

THE SUM OF AN INFINITE GEOMETRIC SEQUENCE

If $|r| < 1$, the infinite geometric sequence

$$a_1, a_1 r, a_1 r^2, a_1 r^3, \ldots a_1 r^{n-1}, \ldots$$

has the sum

$$S = \frac{a_1}{1 - r}.$$

EXAMPLE 7   *Finding the Sum of an Infinite Geometric Sequence*

Find the sum of each infinite geometric sequence.

a. $\displaystyle\sum_{n=1}^{\infty} 4(0.6)^{n-1}$ **b.** $\displaystyle\sum_{n=1}^{\infty} 3(0.1)^{n-1}$

Solution

a. $\displaystyle\sum_{n=1}^{\infty} 4(0.6)^{n-1} = 4 + 4(0.6) + 4(0.6)^2 + 4(0.6)^3 + \cdots + 4(0.6)^{n-1} + \cdots$

$$= \frac{4}{1 - (0.6)} \qquad \frac{a_1}{1 - r}$$

$$= 10$$

b. $\displaystyle\sum_{n=1}^{\infty} 3(0.1)^{n-1} = 3 + 3(0.1) + 3(0.1)^2 + 3(0.1)^3 + \cdots + 3(0.1)^{n-1} + \cdots$

$$= \frac{3}{1 - (0.1)} \qquad \frac{a_1}{1 - r}$$

$$= \frac{10}{3}$$

$$\approx 3.33$$

Application

NOTE   The type of investment
account in Example 8 is an *annuity*.
■■

EXAMPLE 8   *Compound Interest*

Real
Life

A deposit of $50 is made on the first day of each month in a savings account that pays 6% compounded monthly. What is the balance at the end of 2 years?

Solution

The first deposit will gain interest for 24 months, and its balance will be

$$A_{24} = 50\left(1 + \frac{0.06}{12}\right)^{24} = 50(1.005)^{24}.$$

The second deposit will gain interest for 23 months, and its balance will be

$$A_{23} = 50\left(1 + \frac{0.06}{12}\right)^{23} = 50(1.005)^{23}.$$

The last deposit will gain interest for only 1 month, and its balance will be

$$A_1 = 50\left(1 + \frac{0.06}{12}\right)^1 = 50(1.005).$$

The total balance in the account will be the sum of the balances of the 24 deposits. Using the formula for the sum of a finite geometric sequence, with $A_1 = 50(1.005)$ and $r = 1.005$, you have

$$S_n = 50(1.005)\left[\frac{1 - (1.005)^{24}}{1 - 1.005}\right] = \$1277.96.$$

GROUP ACTIVITY

AN EXPERIMENT

You will need a piece of string or yarn, a pair of scissors, and a tape measure. Measure out any length of string at least 5 feet long. Double over the string and cut it in half. Take one of the resulting halves, double it over, and cut it in half. Continue this process until you are no longer able to cut a length of string in half. How many cuts were you able to make? Construct a sequence of the resulting string lengths after each cut, starting with the original length of the string. Find a formula for the *n*th term of this sequence. How many cuts could you theoretically make? Discuss why you were not able to make that many cuts.

WARM UP

Evaluate the expression.

1. $\left(\frac{4}{5}\right)^3$

2. $\left(\frac{3}{4}\right)^2$

3. 2^{-4}

4. $\dfrac{5}{3^4}$

Simplify the expression.

5. $(2n)(3n^2)$

6. $n(3n)^3$

7. $\dfrac{4n^5}{n^2}$

8. $\dfrac{(2n)^3}{8n}$

9. $[2(3)^{-4}]^n$

10. $3(4^2)^{-n}$

11.3 Exercises

In Exercises 1–10, determine whether the sequence is geometric. If it is, find the common ratio.

1. $5, 15, 45, 135, \ldots$

2. $3, 12, 48, 192, \ldots$

3. $3, 12, 21, 30, \ldots$

4. $1, -2, 4, -8, \ldots$

5. $1, -\frac{1}{2}, \frac{1}{4}, -\frac{1}{8}, \ldots$

6. $5, 1, 0.2, 0.04, \ldots$

7. $\frac{1}{2}, \frac{2}{3}, \frac{3}{4}, \frac{4}{5}, \ldots$

8. $9, -6, 4, -\frac{8}{3}, \ldots$

9. $1, \frac{1}{2}, \frac{1}{3}, \frac{1}{4}, \ldots$

10. $\frac{1}{5}, \frac{2}{7}, \frac{3}{9}, \frac{4}{11}, \ldots$

In Exercises 11–20, write the first five terms of the geometric sequence.

11. $a_1 = 2, r = 3$

12. $a_1 = 6, r = 2$

13. $a_1 = 1, r = \frac{1}{2}$

14. $a_1 = 1, r = \frac{1}{3}$

15. $a_1 = 5, r = -\frac{1}{10}$

16. $a_1 = 6, r = -\frac{1}{4}$

17. $a_1 = 1, r = e$

18. $a_1 = 2, r = \sqrt{3}$

19. $a_1 = 3, r = \dfrac{x}{2}$

20. $a_1 = 5, r = 2x$

In Exercises 21–26, write the first five terms of the geometric sequence. Determine the common ratio and write the nth term of the sequence as a function of n.

21. $a_1 = 64, \quad a_{k+1} = \frac{1}{2}a_k$

22. $a_1 = 81, \quad a_{k+1} = \frac{1}{3}a_k$

23. $a_1 = 4, \quad a_{k+1} = 3a_k$

24. $a_1 = 5, \quad a_{k+1} = -2a_k$

25. $a_1 = 6, \quad a_{k+1} = -\frac{3}{2}a_k$

26. $a_1 = 36, \quad a_{k+1} = -\frac{2}{3}a_k$

In Exercises 27–38, find the nth term of the geometric sequence.

27. $a_1 = 4, r = \frac{1}{2}, n = 10$

28. $a_1 = 5, r = \frac{3}{2}, n = 8$

29. $a_1 = 6, r = -\frac{1}{3}, n = 12$

30. $a_1 = 8, r = \sqrt{5}, n = 9$

31. $a_1 = 100, r = e^x, n = 9$

32. $a_1 = 1, r = -\dfrac{x}{3}, n = 7$

33. $a_1 = 500, r = 1.02, n = 40$

34. $a_1 = 1000, r = 1.005, n = 60$

35. $a_1 = 16, a_4 = \dfrac{27}{4}, n = 3$

36. $a_2 = 3, a_5 = \dfrac{3}{64}, n = 1$

37. $a_2 = -18, a_5 = \dfrac{2}{3}, n = 6$

38. $a_3 = \dfrac{16}{3}, a_5 = \dfrac{64}{27}, n = 7$

In Exercises 39–42, match the sequence with its graph. [The graphs are labeled (a), (b), (c), and (d).]

(a)
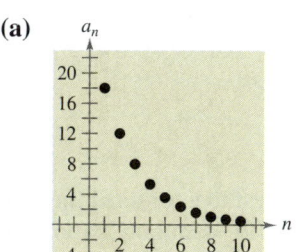

(b)
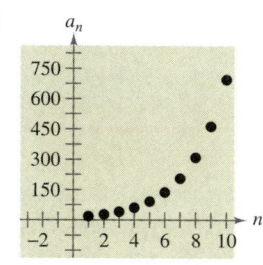

(c)
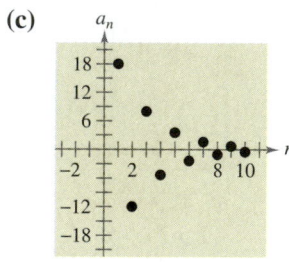

(d)
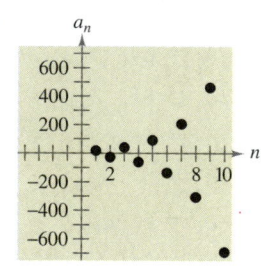

39. $a_n = 18\left(\dfrac{2}{3}\right)^{n-1}$

40. $a_n = 18\left(-\dfrac{2}{3}\right)^{n-1}$

41. $a_n = 18\left(\dfrac{3}{2}\right)^{n-1}$

42. $a_n = 18\left(-\dfrac{3}{2}\right)^{n-1}$

In Exercises 43–46, use a graphing utility to graph the first 10 terms of the sequence.

43. $a_n = 12(-0.75)^{n-1}$

44. $a_n = 12(-0.4)^{n-1}$

45. $a_n = 2(1.3)^{n-1}$

46. $a_n = 2(-1.4)^{n-1}$

47. *Essay* Write a brief paragraph explaining why the terms of a geometric sequence decrease in magnitude when $-1 < r < 1$.

48. *Essay* Write a brief paragraph explaining how to use the first two terms of a geometric sequence to find the nth term.

49. *Compound Interest* A principal of $1000 is invested at 10% interest. Find the amount after 10 years if the interest is compounded (a) annually, (b) semiannually, (c) quarterly, (d) monthly, and (e) daily.

50. *Compound Interest* A principal of $2500 is invested at 12% interest. Find the amount after 20 years if the interest is compounded (a) annually, (b) semiannually, (c) quarterly, (d) monthly, and (e) daily.

51. *Depreciation* A company buys a machine for $135,000 and it depreciates at a rate of 30% per year. (In other words, at the end of each year the depreciated value is 70% of what it was at the beginning of the year.) Find the depreciated value of the machine after five full years.

52. *Population Growth* A city of 250,000 people is growing at a rate of 1.3% per year. Estimate the population of the city 30 years from now.

In Exercises 53 and 54, find the first four terms of the sequence of partial sums of the geometric sequence. In a sequence of partial sums, the term S_n is the sum of the first n terms of the sequence. For instance, S_2 is the sum of the first two terms.

53. $8, -4, 2, -1, \dfrac{1}{2}, \ldots$

54. $8, 12, 18, 27, \dfrac{81}{2}, \ldots$

In Exercises 55–64, find the sum.

55. $\displaystyle\sum_{n=1}^{9} 2^{n-1}$

56. $\displaystyle\sum_{n=1}^{9} (-2)^{n-1}$

57. $\displaystyle\sum_{i=1}^{7} 64\left(-\dfrac{1}{2}\right)^{i-1}$

58. $\displaystyle\sum_{i=1}^{6} 32\left(\dfrac{1}{4}\right)^{i-1}$

59. $\displaystyle\sum_{n=0}^{20} 3\left(\dfrac{3}{2}\right)^{n}$

60. $\displaystyle\sum_{n=0}^{15} 2\left(\dfrac{4}{3}\right)^{n}$

61. $\displaystyle\sum_{i=1}^{10} 8\left(-\dfrac{1}{4}\right)^{i-1}$

62. $\displaystyle\sum_{i=1}^{10} 5\left(-\dfrac{1}{3}\right)^{i-1}$

63. $\displaystyle\sum_{n=0}^{5} 300(1.06)^{n}$

64. $\displaystyle\sum_{n=0}^{6} 500(1.04)^{n}$

In Exercises 65 and 66, use summation notation to express the sum.

65. $5 + 15 + 45 + \cdots + 3645$

66. $2 - \frac{1}{2} + \frac{1}{8} - \cdots + \frac{1}{2048}$

67. **Annuities** A deposit of $100 is made at the beginning of each month in an account that pays 10%, compounded monthly. The balance A in the account at the end of 5 years is given by

$$A = 100\left(1 + \frac{0.10}{12}\right)^1 + \cdots + 100\left(1 + \frac{0.10}{12}\right)^{60}.$$

Find A.

68. **Annuities** A deposit of $50 is made at the beginning of each month in an account that pays 12%, compounded monthly. The balance A in the account at the end of 5 years is given by

$$A = 50\left(1 + \frac{0.12}{12}\right)^1 + \cdots + 50\left(1 + \frac{0.10}{12}\right)^{60}.$$

Find A.

69. **Annuities** A deposit of P dollars is made at the beginning of each month in an account earning an annual interest rate r, compounded monthly. The balance A after t years is

$$A = P\left(1 + \frac{r}{12}\right) + P\left(1 + \frac{r}{12}\right)^2 + \cdots +$$
$$P\left(1 + \frac{r}{12}\right)^{12t}.$$

Show that the balance is given by

$$A = P\left[\left(1 + \frac{r}{12}\right)^{12t} - 1\right]\left(1 + \frac{12}{r}\right).$$

70. **Annuities** A deposit of P dollars is made at the beginning of each month in an account earning an annual interest rate r, compounded continuously. The balance A after t years is

$$A = Pe^{r/12} + Pe^{2r/12} + \cdots + Pe^{12tr/12}.$$

Show that the balance is given by

$$A = \frac{Pe^{r/12}(e^{rt} - 1)}{e^{r/12} - 1}.$$

Annuities In Exercises 71–74, consider making monthly deposits of P dollars in a savings account earning an annual interest rate r. Use the results of Exercises 69 and 70 to find the balance A after t years if the interest is compounded (a) monthly and (b) continuously.

71. $P = \$50$, $r = 7\%$, $t = 20$ years

72. $P = \$75$, $r = 9\%$, $t = 25$ years

73. $P = \$100$, $r = 10\%$, $t = 40$ years

74. $P = \$20$, $r = 6\%$, $t = 50$ years

75. **Annuities** Consider an initial deposit of P dollars in an account earning an annual interest rate r, compounded monthly. At the end of each month a withdrawal of W dollars will occur and the account will be depleted in t years. The amount of the initial deposit required is given by

$$P = W\left(1 + \frac{r}{12}\right)^{-1} + W\left(1 + \frac{r}{12}\right)^{-2} + \cdots +$$
$$W\left(1 + \frac{r}{12}\right)^{-12t}.$$

Show that the initial deposit is given by

$$P = W\left(\frac{12}{r}\right)\left[1 - \left(1 + \frac{r}{12}\right)^{-12t}\right].$$

76. **Annuities** Determine the amount required in an individual retirement account for an individual who retires at age 65 and wants an income of $2000 from the account each month for 20 years. Use the result of Exercise 75 and assume that the account earns 9% compounded monthly.

77. **Geometry** The sides of a square are 16 inches in length. A new square is formed by connecting the midpoints of the sides of the original square, and two of the triangles are shaded (see figure). If this process is repeated five more times, determine the total area of the shaded region.

 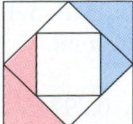  

78. *Corporate Revenue*   The annual revenues a_n (in billions of dollars) for the Coca-Cola Company for 1985 through 1994 can be approximated by the model

$$a_n = 3.49e^{0.108n}, \quad n = 5, 6, 7, \ldots, 14$$

where $n = 5$ represents 1985. Use this model and the formula for the sum of a geometric sequence to approximate the total revenue earned during this 10-year period.   (Source: The Coca-Cola Company)

79. *Think About It*   Suppose you work for a company that pays \$0.01 the first day, \$0.02 the second day, \$0.04 the third day, and so on. If the daily wage keeps doubling, what would your total income be for working (a) 29 days, (b) 30 days, and (c) 31 days?

80. *Salary*   A company has a job opening with a salary of \$30,000 for the first year. Suppose that during the next 39 years, there is a 5% raise each year. Find the total compensation over the 40-year period.

In Exercises 81–88, find the sum.

81. $\displaystyle\sum_{n=0}^{\infty} \left(\tfrac{1}{2}\right)^n$

82. $\displaystyle\sum_{n=0}^{\infty} 2\left(\tfrac{2}{3}\right)^n$

83. $\displaystyle\sum_{n=0}^{\infty} \left(-\tfrac{1}{2}\right)^n$

84. $\displaystyle\sum_{n=0}^{\infty} 2\left(-\tfrac{2}{3}\right)^n$

85. $\displaystyle\sum_{n=0}^{\infty} 4\left(\tfrac{1}{4}\right)^n$

86. $\displaystyle\sum_{n=0}^{\infty} \left(\tfrac{1}{10}\right)^n$

87. $8 + 6 + \tfrac{9}{2} + \tfrac{27}{8} + \cdots$

88. $3 - 1 + \tfrac{1}{3} - \tfrac{1}{9} + \cdots$

In Exercises 89–92, find the rational number representation of the repeating decimal.

89. $0.\overline{36}$

90. $0.\overline{297}$

91. $0.3\overline{18}$

92. $1.3\overline{8}$

Graphical Reasoning   **In Exercises 93 and 94, use a graphing utility to graph the function. Identify the horizontal asymptote of the graph and determine its relationship to the sum.**

93. $f(x) = 6\left[\dfrac{1 - (0.5)^x}{1 - (0.5)}\right], \qquad \displaystyle\sum_{n=0}^{\infty} 6\left(\tfrac{1}{2}\right)^n$

94. $f(x) = 2\left[\dfrac{1 - (0.8)^x}{1 - (0.8)}\right], \qquad \displaystyle\sum_{n=0}^{\infty} 2\left(\tfrac{4}{5}\right)^n$

95. *Distance*   A ball is dropped from a height of 16 feet. Each time it drops h feet, it rebounds $0.81h$ feet.

(a) Find the total distance traveled by the ball.

(b) The ball takes the following time for each fall.

$$s_1 = -16t^2 + 16, \qquad s_1 = 0 \text{ if } t = 1$$
$$s_2 = -16t^2 + 16(0.81), \qquad s_2 = 0 \text{ if } t = 0.9$$
$$s_3 = -16t^2 + 16(0.81)^2, \qquad s_3 = 0 \text{ if } t = (0.9)^2$$
$$s_4 = -16t^2 + 16(0.81)^3, \qquad s_4 = 0 \text{ if } t = (0.9)^3$$
$$\vdots \qquad\qquad\qquad \vdots$$
$$s_n = -16t^2 + 16(0.81)^{n-1}, \quad s_n = 0 \text{ if } t = (0.9)^{n-1}$$

Beginning with s_2, the ball takes the same amount of time to bounce up as it does to fall, and thus the total time elapsed before it comes to rest is

$$t = 1 + 2\sum_{n=1}^{\infty} (0.9)^n.$$

Find this total.

▶ 11.4   *Mathematical Induction*

See Exercises 39–48 on page 839 for examples of how mathematical induction can be used to prove properties of real numbers.

Introduction   ▫   *Sums of Powers of Integers*   ▫   *Pattern Recognition*   ▫   *Finite Differences*

Introduction

In this section you will study a form of mathematical proof called **mathematical induction.** It is important that you clearly see the logical need for it, so let's take a closer look at a problem discussed on page 605.

$$S_1 = 1 = 1^2$$
$$S_2 = 1 + 3 = 2^2$$
$$S_3 = 1 + 3 + 5 = 3^2$$
$$S_4 = 1 + 3 + 5 + 7 = 4^2$$
$$S_5 = 1 + 3 + 5 + 7 + 9 = 5^2$$

Judging from the pattern formed by these first five sums, it appears that the sum of the first n integers is

$$S_n = 1 + 3 + 5 + 7 + 9 + \cdots + (2n - 1) = n^2.$$

Although this particular formula *is* valid, it is important for you to see that recognizing a pattern and then simply *jumping to the conclusion* that the pattern must be true for all values of n is *not* a logically valid method of proof. There are many examples in which a pattern appears to be developing for small values of n and then at some point the pattern fails. One of the most famous cases of this was the conjecture by the French mathematician Pierre de Fermat (1601–1665), who speculated that all numbers of the form

$$F_n = 2^{2^n} + 1, \quad n = 0, 1, 2, \ldots$$

are prime. For $n = 0$, 1, 2, 3, and 4, the conjecture is true.

$$F_0 = 3, F_1 = 5, F_2 = 17, F_3 = 257, F_4 = 65{,}537$$

The size of the next Fermat number ($F_5 = 4{,}294{,}967{,}297$) is so great that it was difficult for Fermat to determine whether it was prime or not. However, another well-known mathematician, Leonhard Euler (1707–1783), later found the factorization

$$F_5 = 4{,}294{,}967{,}297 = 641(6{,}700{,}417)$$

which proved that F_5 is not prime and therefore Fermat's conjecture was false.

Just because a rule, pattern, or formula seems to work for several values of n, you cannot simply decide that it is valid for all values of n without going through a *legitimate proof*.

NOTE It is important to recognize that both parts of the Principle of Mathematical Induction are necessary. ■■

THE PRINCIPLE OF MATHEMATICAL INDUCTION

Let P_n be a statement involving the positive integer n. If

1. P_1 is true, and

2. the truth of P_k implies the truth of P_{k+1} for every positive k,

then P_n must be true for all positive integers n.

To apply the Principle of Mathematical Induction, you need to be able to determine the statement P_{k+1} for a given statement P_k.

EXAMPLE 1   *A Preliminary Example*

Find P_{k+1} for the following.

a. $P_k : S_k = \dfrac{k^2(k+1)^2}{4}$

b. $P_k : S_k = 1 + 5 + 9 + \cdots + [4(k-1) - 3] + (4k - 3)$

c. $P_k : 3^k \geq 2k + 1$

Solution

a. $P_{k+1} : S_{k+1} = \dfrac{(k+1)^2(k+1+1)^2}{4}$ Replace k by $k+1$.

$\qquad\qquad\quad = \dfrac{(k+1)^2(k+2)^2}{4}.$ Simplify.

b. $P_{k+1} : S_{k+1} = 1 + 5 + 9 + \cdots + \{4[(k+1) - 1] - 3\} + [4(k+1) - 3]$

$\qquad\qquad\quad = 1 + 5 + 9 + \cdots + (4k - 3) + (4k + 1).$

c. $P_{k+1} : 3^{k+1} \geq 2(k+1) + 1$

$\qquad\qquad 3^{k+1} \geq 2k + 3.$

A well-known illustration used to explain why the Principle of Mathematical Induction works is the unending line of dominoes shown in Figure 11.3. If the line actually contains infinitely many dominoes, it is clear that you could not knock the entire line down by knocking down only *one domino* at a time. However, suppose it were true that each domino would knock down the next one as it fell. Then you could knock them all down simply by pushing the first one and starting a chain reaction. Mathematical induction works in the same way. If the truth of P_k implies the truth of P_{k+1} and if P_1 is true, the chain reaction proceeds as follows: P_1 implies P_2, P_2 implies P_3, P_3 implies P_4, and so on.

FIGURE 11.3

◼

EXAMPLE 2   *Using Mathematical Induction*

Use mathematical induction to prove the following formula.

$$S_n = 1 + 3 + 5 + 7 + \cdots + (2n - 1)$$
$$= n^2$$

Solution

Mathematical induction consists of two distinct parts. First, you must show that the formula is true when $n = 1$.

1. When $n = 1$, the formula is valid, because

$$S_1 = 1 = 1^2.$$

The second part of mathematical induction has two steps. The first step is to assume that the formula is valid for *some* integer k. The second step is to use this assumption to prove that the formula is valid for the next integer, $k + 1$.

2. Assuming that the formula

$$S_k = 1 + 3 + 5 + 7 + \cdots + (2k - 1)$$
$$= k^2$$

is true, you must show that the formula $S_{k+1} = (k + 1)^2$ is true.

$$S_{k+1} = 1 + 3 + 5 + 7 + \cdots + (2k - 1) + [2(k + 1) - 1]$$
$$= [1 + 3 + 5 + 7 + \cdots + (2k - 1)] + (2k + 2 - 1)$$
$$= S_k + (2k + 1) \qquad \text{Group terms to form } S_k.$$
$$= k^2 + 2k + 1 \qquad \text{Replace } S_k \text{ by } k^2.$$
$$= (k + 1)^2$$

Combining the results of parts (1) and (2), you can conclude by mathematical induction that the formula is valid for *all* positive integer values of n.

◼

NOTE   When using mathematical induction to prove a *summation* formula (such as the one in Example 2), it is helpful to think of S_{k+1} as $S_{k+1} = S_k + a_{k+1}$, where a_{k+1} is the $(k + 1)$ term of the original sum. ◼◼

It occasionally happens that a statement involving natural numbers is not true for the first $k - 1$ positive integers but is true for all values of $n \geq k$. In these instances, you use a slight variation of the Principle of Mathematical Induction in which you verify P_k rather than P_1. This variation is called the **extended principle of mathematical induction.** To see the validity of this, note from Figure 11.3 that all but the first $k - 1$ dominoes can be knocked down by knocking over the kth domino. This suggests that you can prove a statement P_n to be true for $n \geq k$ by showing that P_k is true and that P_k implies P_{k+1}. In Exercises 35–38 of this section you are asked to apply this extension of mathematical induction.

■

EXAMPLE 3 *Using Mathematical Induction*

Use mathematical induction to prove the following formula.

$$S_n = 1^2 + 2^2 + 3^2 + 4^2 + \cdots + n^2$$
$$= \frac{n(n + 1)(2n + 1)}{6}$$

Solution

1. When $n = 1$, the formula is valid, because

$$S_1 = 1^2 = \frac{1(2)(3)}{6}.$$

2. Assuming that

$$S_k = 1^2 + 2^2 + 3^2 + 4^2 + \cdots + k^2$$
$$= \frac{k(k + 1)(2k + 1)}{6}$$

you must show that

$$S_{k+1} = \frac{(k + 1)(k + 2)(2k + 3)}{6}.$$

To do this, write the following.

$$S_{k+1} = S_k + a_{k+1}$$
$$= (1^2 + 2^2 + 3^2 + 4^2 + \cdots + k^2) + (k + 1)^2$$
$$= \frac{k(k + 1)(2k + 1)}{6} + (k + 1)^2$$
$$= \frac{k(k + 1)(2k + 1) + 6(k + 1)^2}{6}$$
$$= \frac{(k + 1)[k(2k + 1) + 6(k + 1)]}{6}$$
$$= \frac{(k + 1)(2k^2 + 7k + 6)}{6}$$
$$= \frac{(k + 1)(k + 2)(2k + 3)}{6}$$

Combining the results of parts (1) and (2), you can conclude by mathematical induction that the formula is valid for *all* $n \geq 1$.

■ ■

NOTE When proving a formula with mathematical induction, the only statement that you *need* to verify is P_1. As a check, however, it is good to try verifying other statements. For instance, in Example 3, try verifying S_2 and S_3.   ■ ■

Sums of Powers of Integers

The formula in Example 3 is one of a collection of useful summation formulas. We summarize this and other formulas dealing with the sums of various powers of the first n positive integers as follows.

SUMS OF POWERS OF INTEGERS

1. $1 + 2 + 3 + 4 + \cdots + n = \dfrac{n(n+1)}{2}$

2. $1^2 + 2^2 + 3^2 + 4^2 + \cdots + n^2 = \dfrac{n(n+1)(2n+1)}{6}$

3. $1^3 + 2^3 + 3^3 + 4^3 + \cdots + n^3 = \dfrac{n^2(n+1)^2}{4}$

4. $1^4 + 2^4 + 3^4 + 4^4 + \cdots + n^4 = \dfrac{n(n+1)(2n+1)(3n^2+3n-1)}{30}$

5. $1^5 + 2^5 + 3^5 + 4^5 + \cdots + n^5 = \dfrac{n^2(n+1)^2(2n^2+2n-1)}{12}$

NOTE Each of these formulas for sums can be proven by mathematical induction. (See Exercises 11–13, 15, and 16.) ■■

EXAMPLE 4 *Finding a Sum of Powers of Integers*

Find $\displaystyle\sum_{n=1}^{7} n^3 = 1^3 + 2^3 + 3^3 + 4^3 + 5^3 + 6^3 + 7^3$.

Solution

Using the formula for the sum of the cubes of the first n positive integers, you obtain the following.

$$\sum_{n=1}^{7} n^3 = 1^3 + 2^3 + 3^3 + 4^3 + 5^3 + 6^3 + 7^3$$

$$= \frac{7^2(7+1)^2}{4}$$

$$= \frac{49(64)}{4}$$

$$= 784$$

Check this sum by adding the numbers 1, 8, 27, 64, 125, 216, and 343.

EXAMPLE 5   *Proving an Inequality by Mathematical Induction*

Prove that $n < 2^n$ for all positive integers n.

Solution

1. For $n = 1$, the formula is true, because

   $1 < 2^1$.

2. Assuming that

   $k < 2^k$

   you need to show that $k + 1 < 2^{k+1}$. For $n = k$, you have

   $2^{k+1} = 2(2^k) > 2(k) = 2k.$ By assumption

   Because $2k = k + k > k + 1$ for all $k > 1$, it follows that

   $2^{k+1} > 2k > k + 1$

   or

   $k + 1 < 2^{k+1}$.

   Therefore, $n < 2^n$ for all integers $n \geq 1$.

Pattern Recognition

Although choosing a formula on the basis of a few observations does *not* guarantee the validity of the formula, pattern recognition *is* important. Once you have a pattern or formula that you think works, you can try using mathematical induction to prove your formula.

   FINDING A FORMULA FOR THE NTH TERM OF A SEQUENCE

To find a formula for the nth term of a sequence, consider the following guidelines.

1. Calculate the first several terms of the sequence. It is often a good idea to write the terms in both simplified and factored forms.
2. Try to find a recognizable pattern for the terms and write a formula for the nth term of the sequence. This is your *hypothesis* or *conjecture*. You might try computing one or two more terms in the sequence to test your hypothesis.
3. Use mathematical induction to prove your hypothesis.

EXAMPLE 6 *Finding a Formula for a Finite Sum*

Find a formula for the following finite sum.

$$\frac{1}{1 \cdot 2} + \frac{1}{2 \cdot 3} + \frac{1}{3 \cdot 4} + \frac{1}{4 \cdot 5} + \cdots + \frac{1}{n(n + 1)}$$

Solution

Begin by writing out the first few sums.

$$S_1 = \frac{1}{1 \cdot 2} = \frac{1}{2} = \frac{1}{1 + 1}$$

$$S_2 = \frac{1}{1 \cdot 2} + \frac{1}{2 \cdot 3} = \frac{4}{6} = \frac{2}{3} = \frac{2}{2 + 1}$$

$$S_3 = \frac{1}{1 \cdot 2} + \frac{1}{2 \cdot 3} + \frac{1}{3 \cdot 4} = \frac{9}{12} = \frac{3}{4} = \frac{3}{3 + 1}$$

$$S_4 = \frac{1}{1 \cdot 2} + \frac{1}{2 \cdot 3} + \frac{1}{3 \cdot 4} + \frac{1}{4 \cdot 5} = \frac{48}{60} = \frac{4}{5} = \frac{4}{4 + 1}$$

From this sequence, it appears that the formula for the kth sum is

$$S_k = \frac{1}{1 \cdot 2} + \frac{1}{2 \cdot 3} + \frac{1}{3 \cdot 4} + \frac{1}{4 \cdot 5} + \cdots + \frac{1}{k(k + 1)} = \frac{k}{k + 1}.$$

To prove the validity of this hypothesis, use mathematical induction, as follows. Note that you have already verified the formula for $n = 1$, so you can begin assuming that the formula is valid for $n = k$ and trying to show that it is valid for $n = k + 1$.

$$S_{k+1} = \left[\frac{1}{1 \cdot 2} + \frac{1}{2 \cdot 3} + \frac{1}{3 \cdot 4} + \frac{1}{4 \cdot 5} + \cdots + \frac{1}{k(k + 1)} \right] + \frac{1}{(k + 1)(k + 2)}$$

$$= \frac{k}{k + 1} + \frac{1}{(k + 1)(k + 2)}$$

$$= \frac{k(k + 2) + 1}{(k + 1)(k + 2)}$$

$$= \frac{k^2 + 2k + 1}{(k + 1)(k + 2)}$$

$$= \frac{(k + 1)^2}{(k + 1)(k + 2)}$$

$$= \frac{k + 1}{k + 2}$$

Thus, the hypothesis is valid.

Finite Differences

The **first differences** of a sequence are found by subtracting consecutive terms. The **second differences** are found by subtracting consecutive first differences. The first and second differences of the sequence 3, 5, 8, 12, 17, 23, . . . are as follows.

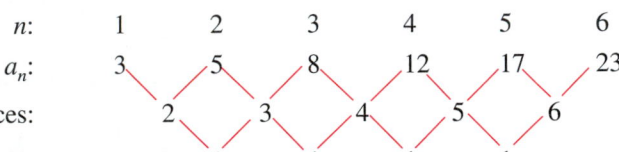

For this sequence, the second differences are all the same. When this happens, the sequence has a perfect quadratic model. If the first differences are all the same, the sequence has a linear model. That is, it is arithmetic.

EXAMPLE 7 *Finding a Quadratic Model*

Find the quadratic model for the sequence

$$3, 5, 8, 12, 17, 23, \ldots .$$

Solution

You know that the model has the form

$$a_n = an^2 + bn + c.$$

By substituting 1, 2, and 3 for n, you can obtain a system of three linear equations in three variables.

$a_1 = a(1)^2 + b(1) + c = 3$	Substitute 1 for n.
$a_2 = a(2)^2 + b(2) + c = 5$	Substitute 2 for n.
$a_3 = a(3)^2 + b(3) + c = 8$	Substitute 3 for n.

You now have a system of three equations in a, b, and c.

$a + b + c = 3$	Equation 1
$4a + 2b + c = 5$	Equation 2
$9a + 3b + c = 8$	Equation 3

Using the techniques discussed in Chapter 9, you can find the solution to be $a = \frac{1}{2}$, $b = \frac{1}{2}$, and $c = 2$. Thus, the quadratic model is

$$a_n = \frac{1}{2}n^2 + \frac{1}{2}n + 2.$$

Try checking the values of a_1, a_2, and a_3.

GROUP ACTIVITY

MATHEMATICAL MODELING

Use finite differences to determine whether each sequence in the table can be represented by either a linear or quadratic model.

n	1	2	3	4	5	6
a_n	12	14	22	36	56	82
b_n	-23.5	-20.0	-16.5	-13.0	-9.5	-6.0
c_n	7	13	20	26	33	39
d_n	0.8	4.2	9.2	15.8	24.0	33.8

If the sequence may be represented by a linear or quadratic model, find the appropriate model, and predict the value of the tenth term. Discuss how the finite differences technique complements other modeling techniques you have learned thus far.

WARM UP

Find the sum.

1. $\displaystyle\sum_{k=3}^{6} (2k - 3)$

2. $\displaystyle\sum_{j=1}^{5} (j^2 - j)$

3. $\displaystyle\sum_{k=2}^{5} \frac{1}{k}$

4. $\displaystyle\sum_{i=1}^{2} \left(1 + \frac{1}{i}\right)$

Simplify the expression.

5. $\dfrac{2(k + 1) + 3}{5}$

6. $\dfrac{3(k + 1) - 2}{6}$

7. $2 \cdot 2^{2(k+1)}$

8. $\dfrac{3^{2k}}{3^{2(k+1)}}$

9. $\dfrac{k + 1}{k^2 + k}$

10. $\dfrac{\sqrt{32}}{\sqrt{50}}$

11.4 Exercises

In Exercises 1– 4, find P_{k+1} for the given P_k.

1. $P_k = \dfrac{5}{k(k+1)}$

2. $P_k = \dfrac{1}{(k+1)(k+3)}$

3. $P_k = \dfrac{k^2(k+1)^2}{4}$

4. $P_k = \dfrac{k}{2}(3k-1)$

In Exercises 5–18, use mathematical induction to prove the formula for every positive integer n.

5. $2 + 4 + 6 + 8 + \cdots + 2n = n(n+1)$

6. $3 + 7 + 11 + 15 + \cdots + (4n-1) = n(2n+1)$

7. $2 + 7 + 12 + 17 + \cdots + (5n-3) = \dfrac{n}{2}(5n-1)$

8. $1 + 4 + 7 + 10 + \cdots + (3n-2) = \dfrac{n}{2}(3n-1)$

9. $1 + 2 + 2^2 + 2^3 + \cdots + 2^{n-1} = 2^n - 1$

10. $2(1 + 3 + 3^2 + 3^3 + \cdots + 3^{n-1}) = 3^n - 1$

11. $1 + 2 + 3 + 4 + \cdots + n = \dfrac{n(n+1)}{2}$

12. $1^2 + 2^2 + 3^2 + 4^2 + \cdots + n^2 = \dfrac{n(n+1)(2n+1)}{6}$

13. $1^3 + 2^3 + 3^3 + 4^3 + \cdots + n^3 = \dfrac{n^2(n+1)^2}{4}$

14. $\left(1 + \dfrac{1}{1}\right)\left(1 + \dfrac{1}{2}\right)\left(1 + \dfrac{1}{3}\right) \cdots \left(1 + \dfrac{1}{n}\right) = n + 1$

15. $\displaystyle\sum_{i=1}^{n} i^5 = \dfrac{n^2(n+1)^2(2n^2 + 2n - 1)}{12}$

16. $\displaystyle\sum_{i=1}^{n} i^4 = \dfrac{n(n+1)(2n+1)(3n^2 + 3n - 1)}{30}$

17. $\displaystyle\sum_{i=1}^{n} i(i+1) = \dfrac{n(n+1)(n+2)}{3}$

18. $\displaystyle\sum_{i=1}^{n} \dfrac{1}{(2i-1)(2i+1)} = \dfrac{n}{2n+1}$

In Exercises 19–28, find the sum using the formulas for the sums of powers of integers.

19. $\displaystyle\sum_{n=1}^{20} n$

20. $\displaystyle\sum_{n=1}^{50} n$

21. $\displaystyle\sum_{n=1}^{6} n^2$

22. $\displaystyle\sum_{n=1}^{10} n^3$

23. $\displaystyle\sum_{n=1}^{5} n^4$

24. $\displaystyle\sum_{n=1}^{8} n^5$

25. $\displaystyle\sum_{n=1}^{6} (n^2 - n)$

26. $\displaystyle\sum_{n=1}^{10} (n^3 - n^2)$

27. $\displaystyle\sum_{i=1}^{6} (6i - 8i^3)$

28. $\displaystyle\sum_{j=1}^{4} \left(2 + \tfrac{5}{2}j - \tfrac{3}{2}j^2\right)$

In Exercises 29–34, find a formula for the sum of the first n terms of the sequence.

29. $1, 5, 9, 13, \ldots$

30. $25, 22, 19, 16, \ldots$

31. $1, \dfrac{9}{10}, \dfrac{81}{100}, \dfrac{729}{1000}, \ldots$

32. $3, -\dfrac{9}{2}, \dfrac{27}{4}, -\dfrac{81}{8}, \ldots$

33. $\dfrac{1}{4}, \dfrac{1}{12}, \dfrac{1}{24}, \dfrac{1}{40}, \ldots, \dfrac{1}{2n(n-1)}, \ldots$

34. $\dfrac{1}{2 \cdot 3}, \dfrac{1}{3 \cdot 4}, \dfrac{1}{4 \cdot 5}, \dfrac{1}{5 \cdot 6}, \ldots, \dfrac{1}{(n+1)(n+2)}, \ldots$

In Exercises 35–38, prove the inequality for the indicated integer values of n.

35. $n! > 2^n, \quad n \geq 4$

36. $\left(\tfrac{4}{3}\right)^n > n, \quad n \geq 7$

37. $\dfrac{1}{\sqrt{1}} + \dfrac{1}{\sqrt{2}} + \dfrac{1}{\sqrt{3}} + \cdots + \dfrac{1}{\sqrt{n}} > \sqrt{n}, \quad n \geq 2$

38. $\left(\dfrac{x}{y}\right)^{n+1} < \left(\dfrac{x}{y}\right)^n, \quad$ if $n \geq 1$ and $0 < x < y$.

In Exercises 39–48, use mathematical induction to prove the given property for all positive integers n.

39. $(ab)^n = a^n b^n$

40. $\left(\dfrac{a}{b}\right)^n = \dfrac{a^n}{b^n}$

41. If $x_1 \neq 0,\ x_2 \neq 0, \ldots, x_n \neq 0$, then

$$(x_1 x_2 x_3 \cdots x_n)^{-1} = x_1^{-1} x_2^{-1} x_3^{-1} \cdots x_n^{-1}.$$

42. If $x_1 > 0,\ x_2 > 0, \ldots, x_n > 0$, then

$$\ln(x_1 x_2 x_3 \cdots x_n) = \ln x_1 + \ln x_2 + \ln x_3 + \cdots + \ln x_n.$$

43. Generalized Distributive Law:

$$x(y_1 + y_2 + \cdots + y_n) = xy_1 + xy_2 + \cdots + xy_n$$

44. $(a + bi)^n$ and $(a - bi)^n$ are complex conjugates for all $n \geq 1$.

45. *Trigonometry* $\quad \sin(x + n\pi) = (-1)^n \sin x$

46. *Trigonometry* $\quad \tan(x + n\pi) = \tan x$

47. A factor of $(n^3 + 3n^2 + 2n)$ is 3.

48. A factor of $(2^{2n-1} + 3^{2n-1})$ is 5.

49. *Essay* In your own words, explain what is meant by a proof by mathematical induction.

50. *Think About It* What conclusion can be drawn from the given information about the sequence of statements P_n?

(a) P_3 is true and P_k implies P_{k+1}.

(b) $P_1, P_2, P_3, \ldots, P_{50}$ are all true.

(c) $P_1, P_2,$ and P_3 are all true, but the truth of P_k does not imply that P_{k+1} is true.

(d) P_2 is true and P_{2k} implies P_{2k+2}.

In Exercises 51–54, write the first five terms of the sequence.

51. $a_0 = 1$
$a_n = a_{n-1} + 2$

52. $a_0 = 10$
$a_n = 4a_{n-1}$

53. $a_0 = 4$
$a_1 = 2$
$a_n = a_{n-1} - a_{n-2}$

54. $a_0 = 0$
$a_1 = 2$
$a_n = a_{n-1} + 2a_{n-2}$

In Exercises 55–64, write the first five terms of the sequence where $a_1 = f(1)$. Then calculate the first and second differences of the sequence. Does the sequence have a linear model, a quadratic model, or neither?

55. $f(1) = 0$
$a_n = a_{n-1} + 3$

56. $f(1) = 2$
$a_n = n - a_{n-1}$

57. $f(1) = 3$
$a_n = a_{n-1} - n$

58. $f(2) = -3$
$a_n = -2a_{n-1}$

59. $a_0 = 0$
$a_n = a_{n-1} + n$

60. $a_0 = 2$
$a_n = (a_{n-1})^2$

61. $f(1) = 2$
$a_n = a_{n-1} + 2$

62. $f(1) = 0$
$a_n = a_{n-1} + 2n$

63. $a_0 = 1$
$a_n = a_{n-1} + n^2$

64. $a_0 = 0$
$a_n = a_{n-1} - 1$

In Exercises 65–68, find a quadratic model for the sequence with the indicated terms.

65. $a_0 = 3,\ a_1 = 3, a_4 = 15$

66. $a_0 = 7,\ a_1 = 6,\ a_3 = 10$

67. $a_0 = -3,\ a_2 = 1,\ a_4 = 9$

68. $a_0 = 3,\ a_2 = 0,\ a_6 = 36$

Review Solve Exercises 69–72 as a review of the skills and problem-solving techniques you learned in previous sections. Solve the system of equations.

69.
$$y = x^2$$
$$-3x + 2y = 2$$

70.
$$x - y^3 = 0$$
$$x - 2y^2 = 0$$

71.
$$x - y \quad\quad = -1$$
$$x + 2y - 2z = 3$$
$$3x - y + 2z = 3$$

72.
$$2x + y - 2z = 1$$
$$x \quad\quad - z = 1$$
$$3x + 3y + z = 12$$

11.5   *The Binomial Theorem*

See Exercises 65–68 on page 846 for examples of how binomial coefficients can be used to find the probabilities of events.

Binomial Coefficients   ▫   *Pascal's Triangle*   ▫   *Binomial Expansions*

Binomial Coefficients

Recall that a **binomial** is a polynomial that has two terms. In this section, you will study a formula that gives a quick method of raising a binomial to a power. To begin, let's look at the expansion of $(x + y)^n$ for several values of *n*.

$$(x + y)^0 = 1$$
$$(x + y)^1 = x + y$$
$$(x + y)^2 = x^2 + 2xy + y^2$$
$$(x + y)^3 = x^3 + 3x^2y + 3xy^2 + y^3$$
$$(x + y)^4 = x^4 + 4x^3y + 6x^2y^2 + 4xy^3 + y^4$$
$$(x + y)^5 = x^5 + 5x^4y + 10x^3y^2 + 10x^2y^3 + 5xy^4 + y^5$$

There are several observations you can make about these expansions.

1. In each expansion, there are $n + 1$ terms.
2. In each expansion, *x* and *y* have symmetrical roles. The powers of *x* decrease by 1 in successive terms, whereas the powers of *y* increase by 1.
3. The sum of the powers of each term is *n*. For instance, in the expansion of $(x + y)^5$, the sum of the powers of each term is 5.

4 + 1 = 5   3 + 2 = 5

$$(x + y)^5 = x^5 + 5x^4y^1 + 10x^3y^2 + 10x^2y^3 + 5x^1y^4 + y^5$$

4. The coefficients increase and then decrease in a symmetric pattern.

The coefficients of a binomial expansion are called **binomial coefficients.** To find them, you can use the following theorem.

▪ **THINK ABOUT THE PROOF**

Use mathematical induction to prove the Binomial Theorem. The details of the proof are given in the appendix.

■ **THE BINOMIAL THEOREM**

In the expansion of $(x + y)^n$

$$(x + y)^n = x^n + nx^{n-1}y + \cdots + {_nC_r} \, x^{n-r} y^r + \cdots + nxy^{n-1} + y^n$$

the coefficient of $x^{n-r} y^r$ is given by

$$_nC_r = \frac{n!}{(n - r)!r!}.$$

NOTE   The symbol $\binom{n}{r}$ is often used in place of $_nC_r$ to denote binomial coefficients. ▪▪

EXAMPLE 1 *Finding Binomial Coefficients*

Find the binomial coefficients.

a. $_8C_2$ **b.** $_{10}C_3$ **c.** $_7C_0$ **d.** $_8C_8$

Solution

a. $_8C_2 = \dfrac{8!}{6! \cdot 2!} = \dfrac{(8 \cdot 7) \cdot \cancel{6!}}{\cancel{6!} \cdot 2!} = \dfrac{8 \cdot 7}{2 \cdot 1} = 28$

b. $_{10}C_3 = \dfrac{10!}{7! \cdot 3!} = \dfrac{(10 \cdot 9 \cdot 8) \cdot \cancel{7!}}{\cancel{7!} \cdot 3!} = \dfrac{10 \cdot 9 \cdot 8}{3 \cdot 2 \cdot 1} = 120$

c. $_7C_0 = \dfrac{\cancel{7!}}{\cancel{7!} \cdot 0!} = 1$

d. $_8C_8 = \dfrac{\cancel{8!}}{0! \cdot \cancel{8!}} = 1$

NOTE   When $r \neq 0$ and $r \neq n$, as in parts (a) and (b) above, there is a simple pattern for evaluating binomial coefficients.

$$_8C_2 = \underset{\text{2 factorial}}{\dfrac{\overset{\text{2 factors}}{8 \cdot 7}}{2 \cdot 1}} \quad \text{and} \quad _{10}C_3 = \underset{\text{3 factorial}}{\dfrac{\overset{\text{3 factors}}{10 \cdot 9 \cdot 8}}{3 \cdot 2 \cdot 1}}$$

EXAMPLE 2 *Finding Binomial Coefficients*

Find the binomial coefficients.

a. $_7C_3$ **b.** $_7C_4$ **c.** $_{12}C_1$ **d.** $_{12}C_{11}$

Solution

a. $_7C_3 = \dfrac{7 \cdot \cancel{6} \cdot 5}{\cancel{3} \cdot \cancel{2} \cdot 1} = 35$

b. $_7C_4 = \dfrac{7 \cdot \cancel{6} \cdot 5 \cdot \cancel{4}}{\cancel{4} \cdot \cancel{3} \cdot \cancel{2} \cdot 1} = 35$

c. $_{12}C_1 = \dfrac{12}{1} = 12$

d. $_{12}C_{11} = \dfrac{12!}{1! \cdot 11!} = \dfrac{(12) \cdot \cancel{11!}}{1! \cdot \cancel{11!}} = \dfrac{12}{1} = 12$

NOTE   It is not a coincidence that the results in parts (a) and (b) of Example 2 are the same and that the results in parts (c) and (d) are the same. In general, it is true that

$$_nC_r = _nC_{n-r}.$$

This shows the symmetric property of binomial coefficients that was identified earlier.

Pascal's Triangle

There is a convenient way to remember the pattern for binomial coefficients. By arranging the coefficients in a triangular pattern, you obtain the following array, which is called **Pascal's Triangle.** This triangle is named after the famous French mathematician Blaise Pascal (1623–1662).

$$
\begin{array}{ccccccccccccccc}
 & & & & & & & 1 & & & & & & & \\
 & & & & & & 1 & & 1 & & & & & & \\
 & & & & & 1 & & 2 & & 1 & & & & & \\
 & & & & 1 & & 3 & & 3 & & 1 & & & & \\
 & & & 1 & & 4 & & 6 & & 4 & & 1 & & & \\
 & & 1 & & 5 & & 10 & & 10 & & 5 & & 1 & & \\
 & 1 & & 6 & & 15 & & 20 & & 15 & & 6 & & 1 & \\
 1 & & 7 & & 21 & & 35 & & 35 & & 21 & & 7 & & 1
\end{array}
$$

NOTE The top row in Pascal's Triangle is called the *zero row* because it corresponds to the binomial expansion

$$(x + y)^0 = 1.$$

Similarly, the next row is called the *first row* because it corresponds to the binomial expansion

$$(x + y)^1 = 1(x) + 1(y).$$

In general, the *nth row* in Pascal's Triangle gives the coefficients of $(x + y)^n$. ∎∎

The first and last numbers in each row of Pascal's Triangle are 1. Every other number in each row is formed by adding the two numbers immediately above the number. Pascal noticed that numbers in this triangle are precisely the same numbers that are the coefficients of binomial expansions, as follows.

$$(x + y)^0 = 1$$
$$(x + y)^1 = 1x + 1y$$
$$(x + y)^2 = 1x^2 + 2xy + 1y^2$$
$$(x + y)^3 = 1x^3 + 3x^2y + 3xy^2 + 1y^3$$
$$(x + y)^4 = 1x^4 + 4x^3y + 6x^2y^2 + 4xy^3 + 1y^4$$
$$(x + y)^5 = 1x^5 + 5x^4y + 10x^3y^2 + 10x^2y^3 + 5xy^4 + 1y^5$$
$$(x + y)^6 = 1x^6 + 6x^5y + 15x^4y^2 + 20x^3y^3 + 15x^2y^4 + 6xy^5 + 1y^6$$
$$(x + y)^7 = 1x^7 + 7x^6y + 21x^5y^2 + 35x^4y^3 + 35x^3y^4 + 21x^2y^5 + 7xy^6 + 1y^7$$

EXAMPLE 3   *Using Pascal's Triangle*

Use the seventh row of Pascal's Triangle to find the binomial coefficients.

$$_8C_0,\ _8C_1,\ _8C_2,\ _8C_3,\ _8C_4,\ _8C_5,\ _8C_6,\ _8C_7,\ _8C_8$$

Solution

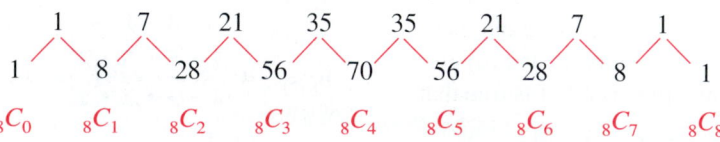

Binomial Expansions

As mentioned at the beginning of this section, when you write out the coefficients for a binomial that is raised to a power, you are **expanding a binomial.** The formulas for binomial coefficients give you an easy way to expand binomials, as demonstrated in the next three examples.

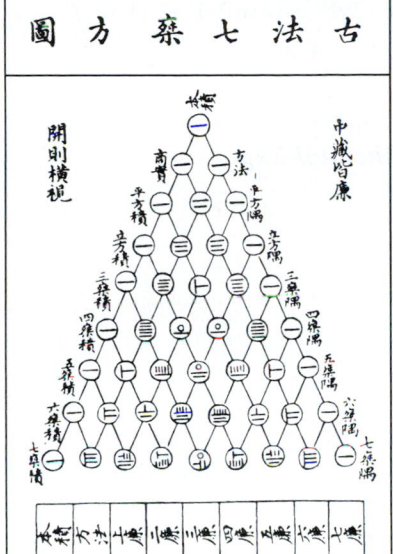

"Pascal's" Triangle and forms of the Binomial Theorem were known in Eastern cultures prior to the Western "discovery" of the theorem. A Chinese text entitled Precious Mirror contains a triangle of binomial expansions through the eighth power.

EXAMPLE 4   *Expanding a Binomial*

Write the expansion for the expression

$$(x + 1)^3.$$

Solution

The binomial coefficients from the third row of Pascal's Triangle are

$$1, 3, 3, 1.$$

Therefore, the expansion is as follows.

$$(x + 1)^3 = (1)x^3 + (3)x^2(1) + (3)x(1^2) + (1)(1^3)$$
$$= x^3 + 3x^2 + 3x + 1$$

To expand binomials representing *differences*, rather than sums, you alternate signs. Here are two examples.

$$(x - 1)^3 = x^3 - 3x^2 + 3x - 1$$
$$(x - 1)^4 = x^4 - 4x^3 + 6x^2 - 4x + 1$$

EXAMPLE 5   *Expanding a Binomial*

Write the expansion for the expression

$$(x - 3)^4.$$

Solution

The binomial coefficients from the fourth row of Pascal's Triangle are

$$1, 4, 6, 4, 1.$$

Therefore, the expansion is as follows.

$$(x - 3)^4 = (1)x^4 - (4)x^3(3) + (6)x^2(3^2) - (4)x(3^3) + (1)(3^4)$$
$$= x^4 - 12x^3 + 54x^2 - 108x + 81$$

EXAMPLE 6   *Expanding a Binomial*

Write the expansion for $(x - 2y)^4$.

Solution

Use the fourth row of Pascal's Triangle, as follows.

$$(x - 2y)^4 = (1)x^4 - (4)x^3(2y) + (6)x^2(2y)^2 - (4)x(2y)^3 + (1)(2y)^4$$
$$= x^4 - 8x^3y + 24x^2y^2 - 32xy^3 + 16y^4$$

EXAMPLE 7   *Finding a Term in a Binomial Expansion*

Find the sixth term of $(a + 2b)^8$.

Solution

The sixth term in the binomial expansion is

$$_8C_5 a^{8-5}(2b)^5 = 56 \cdot a^3 \cdot (2b)^5$$
$$= 56(2^5)a^3b^5$$
$$= 1792a^3b^5.$$

GROUP ACTIVITY

ERROR ANALYSIS

Suppose you are a math instructor and receive the following solutions from one of your students on a quiz. Find the error(s) in each solution. Discuss ways that your student could avoid the error(s) in the future.

a. Find the second term in the expansion of

$(2x - 3y)^5$.

$5(2x)^4(3y)^2 = 720x^4y^2$

b. Find the fourth term in the expansion of

$\left(\frac{1}{2}x + 7y\right)^6$.

$_6C_4\left(\frac{1}{2}x\right)^2(7y)^4 = 9003.75x^2y^4$

Perform the operations and/or simplify.

1. $5x^2(x^3 + 3)$

2. $(x + 5)(x^2 - 3)$

3. $(x + 4)^2$

4. $(2x - 3)^2$

5. $x^2y(3xy^{-2})$

6. $(-2z)^5$

Evaluate the expression.

7. $5!$

8. $\dfrac{8!}{5!}$

9. $\dfrac{10!}{7!}$

10. $\dfrac{6!}{3!3!}$

11.5 Exercises

In Exercises 1–10, evaluate $_nC_r$.

1. $_5C_3$

2. $_8C_6$

3. $_{12}C_0$

4. $_{20}C_{20}$

5. $_{20}C_{15}$

6. $_{12}C_5$

7. $_{100}C_{98}$

8. $_{10}C_4$

9. $_{100}C_2$

10. $_{10}C_6$

11. *Essay* In your own words, explain how to form the rows of Pascal's Triangle.

12. Form the first nine rows of Pascal's Triangle.

In Exercises 13–16, evaluate using Pascal's Triangle.

13. $_7C_4$ **14.** $_6C_3$ **15.** $_8C_5$ **16.** $_8C_7$

In Exercises 17–36, use the Binomial Theorem to expand and simplify the expression.

17. $(x + 1)^4$

18. $(x + 1)^6$

19. $(a + 2)^3$

20. $(a + 3)^4$

21. $(y - 2)^4$

22. $(y - 2)^5$

23. $(x + y)^5$

24. $(x + y)^6$

25. $(r + 3s)^6$

26. $(x + 2y)^4$

27. $(x - y)^5$

28. $(2x - y)^5$

29. $(1 - 2x)^3$

30. $(5 - 3y)^3$

31. $(x^2 + 5)^4$

32. $(x^2 + y^2)^6$

33. $\left(\dfrac{1}{x} + y\right)^5$

34. $\left(\dfrac{1}{x} + 2y\right)^6$

35. $2(x - 3)^4 + 5(x - 3)^2$

36. $3(x + 1)^5 - 4(x + 1)^3$

In Exercises 37–40, expand the binomial using Pascal's Triangle to determine the coefficients.

37. $(2t - s)^5$

38. $(x + 2y)^5$

39. $(3 - 2z)^4$

40. $(3y + 2)^5$

In Exercises 41–48, find the coefficient a of the given term in the expansion of the binomial.

Binomial	Term
41. $(x + 3)^{12}$	ax^5
42. $(x^2 + 3)^{12}$	ax^8
43. $(x - 2y)^{10}$	ax^8y^2
44. $(4x - y)^{10}$	ax^2y^8
45. $(3x - 2y)^9$	ax^4y^5
46. $(2x - 3y)^8$	ax^6y^2
47. $(x^2 + y)^{10}$	ax^8y^6
48. $(z^2 - 1)^{12}$	az^6

49. **Think About It**   How many terms are in the expansion of $(x + y)^n$?

50. **Think About It**   How do the expansions of $(x + y)^n$ and $(x - y)^n$ differ?

In Exercises 51–54, use the Binomial Theorem to expand and simplify the expression.

51. $\left(\sqrt{x} + 3\right)^4$

52. $\left(2\sqrt{t} - 1\right)^3$

53. $(x^{2/3} - y^{1/3})^3$

54. $(u^{3/5} + 2)^5$

In Exercises 55–58, expand the binomial in the difference quotient and simplify.

$$\frac{f(x + h) - f(x)}{h} \qquad \text{Difference quotient}$$

55. $f(x) = x^3$

56. $f(x) = x^4$

57. $f(x) = \sqrt{x}$

58. $f(x) = \dfrac{1}{x}$

In Exercises 59–64, use the Binomial Theorem to expand the complex number. Simplify your result.

59. $(1 + i)^4$

60. $(2 - i)^5$

61. $(2 - 3i)^6$

62. $\left(5 + \sqrt{-9}\right)^3$

63. $\left(-\dfrac{1}{2} + \dfrac{\sqrt{3}}{2}i\right)^3$

64. $\left(5 - \sqrt{3}i\right)^4$

Probability   In Exercises 65–68, consider n independent trials of an experiment in which each trial has two possible outcomes called success and failure. The probability of a success on each trial is p, and the probability of a failure is $q = 1 - p$. In this context, the term $_nC_k\, p^k q^{n-k}$ in the expansion of $(p + q)^n$ gives the probability of k successes in the n trials of the experiment.

65. A fair coin is tossed seven times. To find the probability of obtaining four heads, evaluate the term
$$_7C_4\left(\tfrac{1}{2}\right)^4\left(\tfrac{1}{2}\right)^3$$
in the expansion of $\left(\tfrac{1}{2} + \tfrac{1}{2}\right)^7$.

66. The probability of a baseball player getting a hit on any given time at bat is $\tfrac{1}{4}$. To find the probability that the player gets three hits during the next 10 times at bat, evaluate the term
$$_{10}C_3\left(\tfrac{1}{4}\right)^3\left(\tfrac{3}{4}\right)^7$$
in the expansion of $\left(\tfrac{1}{4} + \tfrac{3}{4}\right)^{10}$.

67. The probability of a sales representative making a sale with any one customer is $\tfrac{1}{3}$. The sales representative makes eight contacts a day. To find the probability of making four sales, evaluate the term
$$_8C_4\left(\tfrac{1}{3}\right)^4\left(\tfrac{2}{3}\right)^4$$
in the expansion of $\left(\tfrac{1}{3} + \tfrac{2}{3}\right)^8$.

68. To find the probability that the sales representative in Exercise 67 makes four sales if the probability of a sale with any one customer is $\tfrac{1}{2}$, evaluate the term
$$_8C_4\left(\tfrac{1}{2}\right)^4\left(\tfrac{1}{2}\right)^4$$
in the expansion of $\left(\tfrac{1}{2} + \tfrac{1}{2}\right)^8$.

Approximation   In Exercises 69–72, use the Binomial Theorem to approximate the given quantity accurate to three decimal places. For example, in Exercise 69, use the expansion

$$(1.02)^8 = (1 + 0.02)^8 = 1 + 8(0.02) + 28(0.02)^2 + \ldots.$$

69. $(1.02)^8$

70. $(2.005)^{10}$

71. $(2.99)^{12}$

72. $(1.98)^9$

Graphical Reasoning In Exercises 73 and 74, use a graphing utility to graph *f* and *g* on the same viewing rectangle. What is the relationship between the two graphs? Use the Binomial Theorem to write the polynomial function *g* in standard form.

73. $f(x) = x^3 - 4x, \quad g(x) = f(x + 4)$

74. $f(x) = -x^4 + 4x^2 - 1, \quad g(x) = f(x - 3)$

In Exercises 75–78, prove the given property for all integers *r* and *n* where $0 \leq r \leq n$.

75. $_nC_r = _nC_{n-r}$

76. $_nC_0 - _nC_1 + _nC_2 - \cdots \pm _nC_n = 0$

77. $_{n+1}C_r = _nC_r + _nC_{r-1}$

78. The sum of the numbers in the *n*th row of Pascal's Triangle is 2^n.

79. *Life Insurance* The average amount of life insurance per household $f(t)$ (in thousands of dollars) from 1970 through 1992 can be approximated by

$$f(t) = 0.1506t^2 + 0.7361t + 21.1374, \quad 0 \leq t \leq 22$$

where $t = 0$ represents 1970 (see figure). You want to adjust this model so that $t = 0$ corresponds to 1980 rather than 1970. To do this, you shift the graph of *f* 10 units *to the left* and obtain

$$g(t) = f(t + 10).$$

(Source: American Council of Life Insurance)

(a) Write $g(t)$ in standard form.

(b) Use a graphing utility to graph *f* and *g* on the same viewing rectangle.

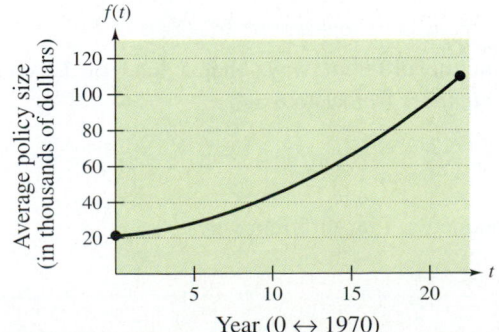

Year (0 ↔ 1970)

80. *Health Maintenance Organizations* The number of people $f(t)$ (in millions) enrolled in health maintenance organizations in the United States from 1976 through 1992 can be approximated by the model

$$f(t) = 0.1043t^2 + 0.7100t + 4.6852, \quad 0 \leq t \leq 16$$

where $t = 0$ represents 1976 (see figure). You want to adjust this model so that $t = 0$ corresponds to 1980 rather than 1976. To do this, you shift the graph of *f* 4 units *to the left* and obtain

$$g(t) = f(t + 4).$$

(Source: Group Health Insurance Association of America)

(a) Write $g(t)$ in standard form.

(b) Use a graphing utility to graph *f* and *g* on the same viewing rectangle.

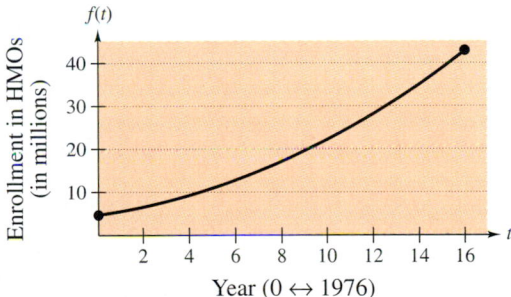

Year (0 ↔ 1976)

81. *Graphical Reasoning* Use a graphing utility to graph the functions in the given order and on the same viewing rectangle. Compare the graphs. Which two functions have identical graphs, and why?

(a) $f(x) = (1 - x)^3$

(b) $g(x) = 1 - 3x$

(c) $h(x) = 1 - 3x + 3x^2$

(d) $p(x) = 1 - 3x + 3x^2 - x^3$

Review Solve Exercises 82–85 as a review of the skills and problem-solving techniques you learned in previous sections. Describe the relationship between the graphs of *f* and *g*.

82. $g(x) = f(x) + 8$

83. $g(x) = f(x - 3)$

84. $g(x) = f(-x)$

85. $g(x) = -f(x)$

See Exercises 55 and 56 on page 857 for examples of how counting principles can be used to help analyze the probability of winning a state lottery.

11.6 *Counting Principles*

Simple Counting Problems □ *Counting Principles* □
Permutations □ *Combinations*

Simple Counting Problems

This section and Section 11.7 present a brief introduction to some of the basic counting principles and their application to probability. In Section 11.7 you will see that much of probability has to do with counting the number of ways an event can occur.

EXAMPLE 1 *Selecting Pairs of Numbers at Random*

Eight pieces of paper are numbered from 1 to 8 and placed in a box. One piece of paper is drawn from the box, its number is written down, and the piece of paper is replaced in the box. Then, a second piece of paper is drawn from the box, and its number is written down. Finally, the two numbers are added together. How many different ways can a total of 12 be obtained?

Solution

To solve this problem, count the different ways that a total of 12 can be obtained using two numbers from 1 to 8.

First number	4	5	6	7	8
Second number	8	7	6	5	4

From this list, you can see that a total of 12 can occur in five different ways.

EXAMPLE 2 *Selecting Pairs of Numbers at Random*

Eight pieces of paper are numbered from 1 to 8 and placed in a box. Two pieces of paper are drawn from the box, and the numbers on the papers are written down and totaled. How many different ways can a total of 12 be obtained?

Solution

To solve this problem, count the different ways that a total of 12 can be obtained *using two different numbers* from 1 to 8.

First number	4	5	7	8
Second number	8	7	5	4

Thus, a total of 12 can be obtained in four different ways.

NOTE The difference between the counting problems in Examples 1 and 2 can be expressed by saying that the random selection in Example 1 occurs **with replacement,** whereas the random selection in Example 2 occurs **without replacement,** which eliminates the possibility of choosing two 6's. ∎∎

Counting Principles

Examples 1 and 2 describe simple counting problems in which you can *list* each possible way that an event can occur. When it is possible, this is always the best way to solve a counting problem. However, some events can occur in so many different ways that it is not feasible to write out the entire list. In such cases, you must rely on formulas and counting principles. The most important of these is the **Fundamental Counting Principle.**

NOTE The Fundamental Counting Principle can be extended to three or more events. For instance, the number of ways that three events E_1, E_2, and E_3 can occur is $m_1 \cdot m_2 \cdot m_3$. ■■

> ### *FUNDAMENTAL COUNTING PRINCIPLE*
>
> Let E_1 and E_2 be two events. The first event E_1 can occur in m_1 different ways. After E_1 has occurred, E_2 can occur in m_2 different ways. The number of ways that the two events can occur is $m_1 \cdot m_2$.

Real Life

EXAMPLE 3   *Using the Fundamental Counting Principle*

How many different pairs of letters from the English alphabet are possible?

Solution

This experiment has two events. The first event is the choice of the first letter, and the second event is the choice of the second letter. Because the English alphabet contains 26 letters, it follows that the number of two-letter "words" is $26 \cdot 26 = 676$.

Real Life

EXAMPLE 4   *Using the Fundamental Counting Principle*

Telephone numbers in the United States have 10 digits. The first three are the *area code* and the next seven are the *local telephone number.* How many different telephone numbers are possible within each area code? (Note that a local telephone number cannot begin with 0 or 1.)

Solution

Because the first digit cannot be 0 or 1, there are only eight choices for the first digit. For each of the other six digits, there are 10 choices.

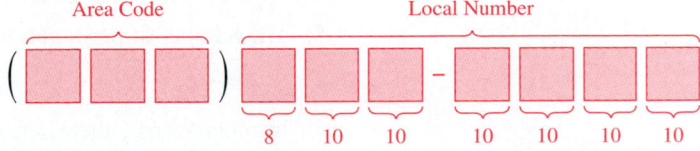

Thus, the number of local telephone numbers that are possible within each area code is $8 \cdot 10 \cdot 10 \cdot 10 \cdot 10 \cdot 10 \cdot 10 = 8{,}000{,}000$.

Permutations

One important application of the Fundamental Counting Principle is in determining the number of ways that n elements can be arranged (in order). An ordering of n elements is called a **permutation** of the elements.

> ### *DEFINITION OF PERMUTATION*
>
> A **permutation** of n different elements is an ordering of the elements such that one element is first, one is second, one is third, and so on.

EXAMPLE 5   *Finding the Number of Permutations of n Elements*

How many permutations are possible for the letters A, B, C, D, E, and F?

Solution

Consider the following reasoning.

> First position: Any of the *six* letters.
> Second position: Any of the remaining *five* letters.
> Third position: Any of the remaining *four* letters.
> Fourth position: Any of the remaining *three* letters.
> Fifth position: Any of the remaining *two* letters.
> Sixth position: The *one* remaining letter.

Thus, the numbers of choices for the six positions are as follows.

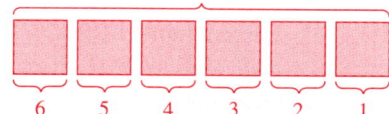

Permutations of six letters

The total number of permutations of the six letters is $6! = 720$.

> ### *NUMBER OF PERMUTATIONS OF N ELEMENTS*
>
> The number of permutations of n elements is given by
>
> $$n \cdot (n-1) \cdots 4 \cdot 3 \cdot 2 \cdot 1 = n!.$$
>
> In other words, there are $n!$ different ways that n elements can be ordered.

Occasionally, you are interested in ordering a *subset* of a collection of elements rather than the entire collection. For example, you might want to choose (and order) *r* elements out of a collection of *n* elements. Such an ordering is called a **permutation of *n* elements taken *r* at a time.**

Eleven thoroughbred racehorses hold the title of Triple Crown winner for winning in the Kentucky Derby, Preakness, and Belmont Stakes in the same year. Secretariat won both the Kentucky Derby and Belmont Stakes in 1973 with record-breaking times. These records still hold. *(Photo: Bob Coglianese Photos, Inc.)*

EXAMPLE 6   *Counting Horse Race Finishes*

Real Life

Eight horses are running in a race. In how many different ways can these horses come in first, second, and third? (Assume that there are no ties.)

Solution

Here are the different possibilities.

    Win (first position): *Eight* choices
    Place (second position): *Seven* choices
    Show (third position): *Six* choices

Using the Fundamental Counting Principle, multiply these three numbers together to obtain the following.

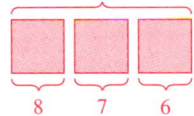

Thus, there are $8 \cdot 7 \cdot 6 = 336$ different orders.

> **PERMUTATIONS OF *N* ELEMENTS TAKEN *R* AT A TIME**
>
> The number of permutations of *n* elements taken *r* at a time is
>
> $$_nP_r = \frac{n!}{(n-r)!} = n(n-1)(n-2)\cdots(n-r+1).$$

Using this formula, you can rework Example 6 to find that the number of permutations of eight horses taken three at a time is

$$_8P_3 = \frac{8!}{5!}$$

$$= \frac{8 \cdot 7 \cdot 6 \cdot 5!}{5!}$$

$$= 336$$

which is the same answer obtained in the example.

Remember that for permutations, order is important. Thus, if you are looking at the possible permutations of the letters A, B, C, and D taken three at a time, the permutations (A, B, D) and (B, A, D) are different because the *order* of the elements is different.

Suppose, however, that you are asked to find the possible permutations of the letters A, A, B, and C. The total number of permutations of the four letters would be $_4P_4 = 4!$. However, not all of these arrangements would be *distinguishable* because there are two A's in the list. To find the number of distinguishable permutations, you can use the following formula.

> ### DISTINGUISHABLE PERMUTATIONS
>
> Suppose a set of n objects has n_1 of one kind of object, n_2 of a second kind, n_3 of a third kind, and so on, with $n = n_1 + n_2 + n_3 + \cdots + n_k$. Then the number of **distinguishable permutations** of the n objects is
>
> $$\frac{n!}{n_1! \cdot n_2! \cdot n_3! \cdots n_k!}.$$

EXAMPLE 7 *Distinguishable Permutations*

In how many distinguishable ways can the letters in BANANA be written?

Solution

This word has six letters, of which three are A's, two are N's, and one is a B. Thus, the number of distinguishable ways the letters can be written is

$$\frac{6!}{3! \cdot 2! \cdot 1!} = \frac{6 \cdot 5 \cdot 4 \cdot 3!}{3! \cdot 2!} = 60.$$

The 60 different "words" are as follows.

AAABNN	AAANBN	AAANNB	AABANN	AABNAN	AABNNA
AANABN	AANANB	AANBAN	AANBNA	AANNAB	AANNBA
ABAANN	ABANAN	ABANNA	ABNAAN	ABNANA	ABNNAA
ANAABN	ANAANB	ANABAN	ANABNA	ANANAB	ANANBA
ANBAAN	ANBANA	ANBNAA	ANNAAB	ANNABA	ANNBAA
BAAANN	BAANAN	BAANNA	BANAAN	BANANA	BANNAA
BNAAAN	BNAANA	BNANAA	BNNAAA	NAAABN	NAAANB
NAABAN	NAABNA	NAANAB	NAANBA	NABAAN	NABANA
NABNAA	NANAAB	NANABA	NANBAA	NBAAAN	NBAANA
NBANAA	NBNAAA	NNAAAB	NNAABA	NNABAA	NNBAAA

Combinations

When one counts the number of possible permutations of a set of elements, *order* is important. As a final topic in this section, we look at a method of selecting subsets of a larger set in which order is *not* important. Such subsets are called **combinations of *n* elements taken *r* at a time.** For instance, the combinations

$$\{A, B, C\} \quad \text{and} \quad \{B, A, C\}$$

are equivalent because both sets contain the same three elements, and the order in which the elements are listed is not important. Hence, you would count only one of the two sets. A common example of how a combination occurs is a card game in which the player is free to reorder the cards after they have been dealt.

EXAMPLE 8   *Combinations of n Elements Taken r at a time*

In how many different ways can three letters be chosen from the letters A, B, C, D, and E? (The order of the three letters is not important.)

Solution

The following subsets represent the different combinations of three letters that can be chosen from the five letters.

$$\{A, B, C\} \qquad \{A, B, D\}$$
$$\{A, B, E\} \qquad \{A, C, D\}$$
$$\{A, C, E\} \qquad \{A, D, E\}$$
$$\{B, C, D\} \qquad \{B, C, E\}$$
$$\{B, D, E\} \qquad \{C, D, E\}$$

From this list, you can conclude that there are 10 different ways that three letters can be chosen from five letters.

TECHNOLOGY

Most graphing utilities have keys that will evaluate the formulas for the numbers of permutations and combinations of *n* elements taken *r* at a time. For instance, on a *TI-83* or a *TI-82,* you can evaluate $_8C_5$ as follows.

8 $\boxed{\text{MATH}}$ (PRB) (3:*nCr*) 5 $\boxed{\text{ENTER}}$

The display should be 56. You can evaluate $_8P_5$ (the number of permutations of eight elements taken five at a time) in a similar way.

COMBINATIONS OF N ELEMENTS TAKEN R AT A TIME

The number of combinations of n elements taken r at a time is

$$_nC_r = \frac{n!}{(n-r)!r!}.$$

Note that the formula for $_nC_r$ is the same one given for binomial coefficients. To see how this formula is used, let's solve the counting problem in Example 8. In that problem, you are asked to find the number of combinations of five elements taken three at a time. Thus, $n = 5$, $r = 3$, and the number of combinations is

$$_5C_3 = \frac{5!}{2!3!} = \frac{5 \cdot \overset{2}{\cancel{4}} \cdot \cancel{3!}}{\cancel{2} \cdot 1 \cdot \cancel{3!}} = 10$$

which is the same answer obtained in the example.

EXAMPLE 9   *Counting Card Hands*

A standard poker hand consists of five cards dealt from a deck of 52. How many different poker hands are possible? (After the cards are dealt, the player may reorder them, and therefore order is not important.)

Solution

You can find the number of different poker hands by using the formula for the number of combinations of 52 elements taken five at a time, as follows.

$$_{52}C_5 = \frac{52!}{47!5!} = \frac{52 \cdot 51 \cdot 50 \cdot 49 \cdot 48 \cdot 47!}{5 \cdot 4 \cdot 3 \cdot 2 \cdot 1 \cdot 47!} = 2{,}598{,}960$$

GROUP ACTIVITY

PROBLEM POSING

According to NASA, each space shuttle astronaut consumes an average of 3000 calories per day. An evening meal normally consists of a main dish, a vegetable dish, and two different desserts. The space shuttle food list contains 10 items classified as main dishes, eight vegetable dishes, and 13 desserts. How many different evening meal menus are possible? Create two other problems that could be asked about the evening meal menus and solve them.   (Source: NASA)

Evaluate the expression.

1. $13 \cdot 8^2 \cdot 2^3$

2. $10^2 \cdot 9^3 \cdot 4$

3. $\dfrac{12!}{2!7!3!}$

4. $\dfrac{25!}{22!}$

Find the binomial coefficient.

5. $_{12}C_7$

6. $_{25}C_{22}$

Simplify the expression.

7. $\dfrac{n!}{(n-4)!}$

8. $\dfrac{(2n)!}{4(2n-3)!}$

9. $\dfrac{2 \cdot 4 \cdot 6 \cdot 8 \cdots (2n)}{2^n}$

10. $\dfrac{3 \cdot 6 \cdot 9 \cdot 12 \cdots (3n)}{3^n}$

11.6 Exercises

Random Selection In Exercises 1–8, determine the number of ways a computer can randomly generate one or more such integers from 1 through 12.

1. An odd integer

2. An even integer

3. A prime integer

4. An integer that is greater than 9

5. An integer that is divisible by 4

6. An integer that is divisible by 7

7. Two integers whose sum is 8

8. Two *distinct* integers whose sum is 8

9. *Entertainment Systems* A customer can choose one of two amplifiers, one of four compact disc players, and one of six speaker models for an entertainment system. Determine the number of possible system configurations.

10. *Computer Systems* A customer in a computer store can choose one of three monitors, one of two keyboards, and one of four computers. If all the choices are compatible, determine the number of possible system configurations.

11. *Job Applicants* A college needs two additional faculty members: a chemist and a statistician. In how many ways can these positions be filled if there are three applicants for the chemistry position and four applicants for the statistics position?

12. *Course Schedule* A college student is preparing a course schedule for the next semester. The student may select one of two mathematics courses, one of three science courses, and one of five courses from the social sciences and humanities. How many schedules are possible?

13. *True-False Exam* In how many ways can a six-question true-false exam be answered? (Assume that no questions are omitted.)

14. *True-False Exam* In how many ways can a 10-question true-false exam be answered? (Assume that no questions are omitted.)

15. *Toboggan Ride* Four people are lining up for a ride on a toboggan, but only two of the four are willing to take the first position. With that constraint, in how many ways can the four people be seated on the toboggan?

16. *Aircraft Boarding* Ten people are boarding an aircraft. Four have tickets for first class and board before those in the economy class. In how many ways can the 10 people board the aircraft?

17. *License Plate Numbers* In a certain state, each automobile license plate number consists of two letters followed by a four-digit number. How many distinct license plate numbers can be formed?

18. *License Plate Numbers* In a certain state, each automobile license plate number consists of two letters followed by a four-digit number. To avoid confusion between "O" and "zero" and "I" and "one," the letters "O" and "I" are not used. How many distinct license plate numbers can be formed?

19. *Three-Digit Numbers* How many three-digit numbers can be formed under the following conditions?

(a) The leading digit cannot be zero.

(b) The leading digit cannot be zero and no repetition of digits is allowed.

(c) The leading digit cannot be zero and the number must be a multiple of 5.

(d) The number is at least 400.

20. *Four-Digit Numbers* How many four-digit numbers can be formed under the following conditions?

(a) The leading digit cannot be zero.

(b) The leading digit cannot be zero and no repetition of digits is allowed.

(c) The leading digit cannot be zero and the number must be less than 5000.

(d) The leading digit cannot be zero and the number must be even.

21. *Combination Lock* A combination lock will open when the right choice of three numbers (from 1 to 40, inclusive) is selected. How many different lock combinations are possible?

22. *Combination Lock* A combination lock will open when the right choice of three numbers (from 1 to 50, inclusive) is selected. How many different lock combinations are possible?

23. *Concert Seats* Three couples have reserved seats in a given row for a concert. In how many different ways can they be seated if

(a) there are no seating restrictions?

(b) the two members of each couple wish to sit together?

24. *Single File* In how many orders can three girls and two boys walk through a doorway single file if

(a) there are no restrictions?

(b) the girls walk through before the boys?

In Exercises 25–30, evaluate $_nP_r$.

25. $_4P_4$ **26.** $_5P_5$

27. $_8P_3$ **28.** $_{20}P_2$

29. $_5P_4$ **30.** $_7P_4$

In Exercises 31 and 32, solve for n.

31. $14 \cdot {}_nP_3 = {}_{n+2}P_4$ **32.** $_nP_5 = 18 \cdot {}_{n-2}P_4$

In Exercises 33–38, evaluate using a calculator.

33. $_{20}P_5$ **34.** $_{100}P_5$

35. $_{100}P_3$ **36.** $_{10}P_8$

37. $_{20}C_5$ **38.** $_{10}C_7$

39. *Think About It* Can your calculator evaluate $_{100}P_{80}$? If not, explain why.

40. *Essay* Explain in words the meaning of $_nP_r$.

41. Write all permutations of the letters A, B, C, and D.

42. Write all the permutations of the letters A, B, C, and D if the letters B and C must remain between the letters A and D.

43. *Posing for a Photograph* In how many ways can five children line up in a row?

44. *Riding in a Car* In how many ways can six people sit in a six-passenger car?

45. *Choosing Officers*   From a pool of 12 candidates, the offices of president, vice-president, secretary, and treasurer will be filled. In how many different ways can the offices be filled?

46. *Assembly Line Production*   There are four processes involved in assembling a certain product, and these processes can be performed in any order. The management wants to test each order to determine which is the least time-consuming. How many different orders will have to be tested?

In Exercises 47–50, find the number of distinguishable permutations of the group of letters.

47. A, A, G, E, E, E, M

48. B, B, B, T, T, T, T, T

49. A, L, G, E, B, R, A

50. M, I, S, S, I, S, S, I, P, P, I

51. Write all the possible selections of two letters that can be formed from the letters A, B, C, D, E, and F. (The order of the two letters is not important.)

52. Write all the possible selections of three letters that can be formed from the letters A, B, C, D, E, and F. (The order of the three letters is not important.)

53. *Forming an Experimental Group*   In order to conduct a certain experiment, four students are randomly selected from a class of 20. How many different groups of four students are possible?

54. *Test Questions*   You can answer any 10 questions from a total of 12 questions on an exam. In how many different ways can you select the questions?

55. *Lottery Choices*   There are 40 numbers in a particular state lottery. In how many ways can a player select six of the numbers?

56. *Lottery Choices*   There are 50 numbers in a particular state lottery. In how many ways can a player select six of the numbers?

57. *Number of Subsets*   How many subsets of four elements can be formed from a set of 100 elements?

58. *Number of Subsets*   How many subsets of five elements can be formed from a set of 80 elements?

59. *Geometry*   Three points that are not on a line determine three lines. How many lines are determined by seven points, no three of which are on a line?

60. *Defective Units*   A shipment of 12 microwave ovens contains three defective units. In how many ways can a vending company purchase four of these units and receive (a) all good units, (b) two good units, and (c) at least two good units?

61. *Job Applicants*   An employer interviews eight people for four openings in the company. Three of the eight people are women. If all eight are qualified, in how many ways can the employer fill the four positions if (a) the selection is random and (b) exactly two are women?

62. *Poker Hand*   You are dealt five cards from an ordinary deck of 52 playing cards. In how many ways can you get a full house? (A full house consists of three of one kind and two of another. For example, A-A-A-5-5 and K-K-K-10-10 are full houses.)

63. *Forming a Committee*   Four people are to be selected at random from a group of four couples. In how many ways can this be done, given the following conditions?

(a) There are no restrictions.

(b) The group must have at least one couple.

(c) Each couple must be represented in the group.

64. *Interpersonal Relationships*   The complexity of the interpersonal relationships increases dramatically as the size of a group increases. Determine the number of different two-person relationships in a group of people of size (a) 3, (b) 8, (c) 12, and (d) 20.

In Exercises 65–68, find the number of diagonals of the polygon. (A line segment connecting any two non-adjacent vertices is called a *diagonal* of the polygon.)

65. Pentagon

66. Hexagon

67. Octagon

68. Decagon (10 sides)

In Exercises 69–72, prove the identity.

69. $_nP_{n-1} = \ _nP_n$

70. $_nC_n = \ _nC_0$

71. $_nC_{n-1} = \ _nC_1$

72. $_nC_r = \dfrac{_nP_r}{r!}$

11.7 *Probability*

See Exercise 29 on page 868 for an example of how probability can be used to help analyze the age distribution of employees who work for minimum wages.

The Probability of an Event ❑ *Mutually Exclusive Events* ❑ *Independent Events* ❑ *The Complement of an Event*

The Probability of an Event

Any happening the result of which is uncertain is called an **experiment.** The possible results of the experiment are **outcomes,** the set of all possible outcomes of the experiment is the **sample space** of the experiment, and any subcollection of a sample space is an **event.**

For instance, when a six-sided die is tossed, the sample space can be represented by the numbers 1 through 6. For this experiment, each of the outcomes is *equally likely.*

To describe sample spaces in such a way that each outcome is equally likely, you must sometimes distinguish between or among various outcomes in ways that appear artificial. Example 1 illustrates such a situation.

EXAMPLE 1 *Finding the Sample Space*

Real Life

Find the sample space for each of the following.

a. One coin is tossed.   **b.** Two coins are tossed.   **c.** Three coins are tossed.

Solution

a. Because the coin will land either heads up (denoted by H) or tails up (denoted by T), the sample space is $S = \{H, T\}$.

b. Because either coin can land heads up or tails up, the possible outcomes are as follows.

> HH = heads up on both coins
> HT = heads up on first coin and tails up on second coin
> TH = tails up on first coin and heads up on second coin
> TT = tails up on both coins

Thus, the sample space is $S = \{HH, HT, TH, TT\}$. Note that this list distinguishes between the two cases HT and TH, even though these two outcomes appear to be similar.

c. Following the notation of part (b), the sample space is

$$S = \{HHH, HHT, HTH, HTT, THH, THT, TTH, TTT\}.$$

To calculate the probability of an event, count the number of outcomes in the event and in the sample space. The *number of outcomes* in event E is denoted by $n(E)$, and the number of outcomes in the sample space S is denoted by $n(S)$. The probability that event E will occur is given by $n(E)/n(S)$.

THE PROBABILITY OF AN EVENT

If an event E has $n(E)$ equally likely outcomes and its sample space S has $n(S)$ equally likely outcomes, the **probability** of event E is

$$P(E) = \frac{n(E)}{n(S)}.$$

Because the number of outcomes in an event must be less than or equal to the number of outcomes in the sample space, the probability of an event must be a number between 0 and 1. That is,

$$0 \le P(E) \le 1.$$

If $P(E) = 0$, event E *cannot occur*, and E is called an **impossible event.** If $P(E) = 1$, event E *must occur*, and E is called a **certain event.**

Real
Life

EXAMPLE 2 **Finding the Probability of an Event**

a. Two coins are tossed. What is the probability that both land heads up?

b. A card is drawn from a standard deck of playing cards. What is the probability that it is an ace?

Solution

a. Following the procedure in Example 1(b), let

$$E = \{HH\}$$

and

$$S = \{HH, HT, TH, TT\}.$$

The probability of getting two heads is

$$P(E) = \frac{n(E)}{n(S)} = \frac{1}{4}.$$

b. Because there are 52 cards in a standard deck of playing cards and there are four aces (one in each suit), the probability of drawing an ace is

$$P(E) = \frac{n(E)}{n(S)} = \frac{4}{52} = \frac{1}{13}.$$

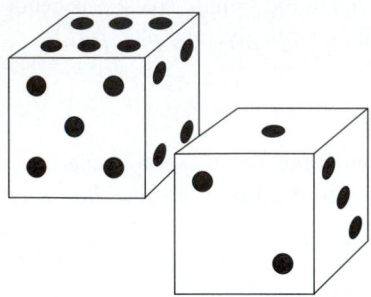

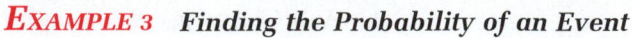

Real
Life

EXAMPLE 3   *Finding the Probability of an Event*

Two six-sided dice are tossed. What is the probability that the total of the two dice is 7? (See Figure 11.4.)

Solution

Because there are six possible outcomes on each die, you can use the Fundamental Counting Principle to conclude that there are 6 • 6 or 36 different outcomes when two dice are tossed. To find the probability of rolling a total of 7, you must first count the number of ways in which this can occur.

First Die	1	2	3	4	5	6
Second Die	6	5	4	3	2	1

Thus, a total of 7 can be rolled in six ways, which means that the probability of rolling a 7 is

$$P(E) = \frac{n(E)}{n(S)} = \frac{6}{36} = \frac{1}{6}.$$

FIGURE 11.4

You could have written out each sample space in Examples 2 and 3 and simply counted the outcomes in the desired events. For larger sample spaces, however, you should use the counting principles discussed in Section 11.6.

Real
Life

EXAMPLE 4   *Finding the Probability of an Event*

Twelve-sided dice, as shown in Figure 11.5, can be constructed (in the shape of regular dodecahedrons) so that each of the numbers from 1 to 6 appears twice on each die. Prove that these dice can be used in any game requiring ordinary six-sided dice without changing the probabilities of different outcomes.

Solution

For an ordinary six-sided die, each of the numbers 1, 2, 3, 4, 5, and 6 occurs only once, so the probability of any particular number coming up is

$$P(E) = \frac{n(E)}{n(S)} = \frac{1}{6}.$$

For one of the twelve-sided dice, each number occurs twice, so the probability of any particular number coming up is

$$P(E) = \frac{n(E)}{n(S)} = \frac{2}{12} = \frac{1}{6}.$$

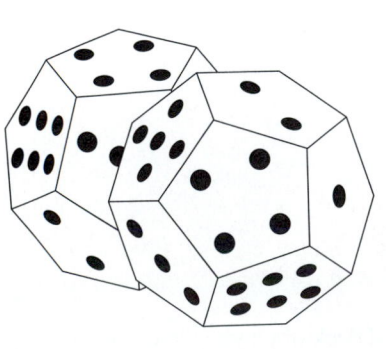

FIGURE 11.5

Although popular in the early 1800s, lotteries were banned in more and more states until, by 1894, no state allowed lotteries. In 1964, New Hampshire became the first state to reinstitute a state lottery. Today, lotteries are conducted by almost all states.
(Photo: Courtesy of Corel Professional Photos)

Real Life

EXAMPLE 5 *The Probability of Winning a Lottery*

In a state lottery, a player chooses six different numbers from 1 to 40. If these six numbers match the six numbers drawn by the lottery commission, the player wins (or shares) the top prize. What is the probability of winning?

Solution

To find the number of elements in the sample space, use the formula for the number of combinations of 40 elements taken six at a time.

$$n(S) = {}_{40}C_6 = \frac{40 \cdot 39 \cdot 38 \cdot 37 \cdot 36 \cdot 35}{6 \cdot 5 \cdot 4 \cdot 3 \cdot 2 \cdot 1} = 3{,}838{,}380$$

If a person buys only one ticket, the probability of winning is

$$P(E) = \frac{n(E)}{n(S)} = \frac{1}{3{,}838{,}380}.$$

Real Life

EXAMPLE 6 *Random Selection*

The numbers of colleges and universities in various regions of the United States in 1992 are shown in Figure 11.6. One institution is selected at random. What is the probability that the institution is in one of the three southern regions? (Source: U.S. National Center for Education Statistics)

Solution

From the figure, the total number of colleges and universities is 3628. Because there are $600 + 272 + 289 = 1161$ colleges and universities in the three southern regions, the probability that the institution is in one of these regions is

$$P(E) = \frac{n(E)}{n(S)} = \frac{1161}{3628} \approx 0.320.$$

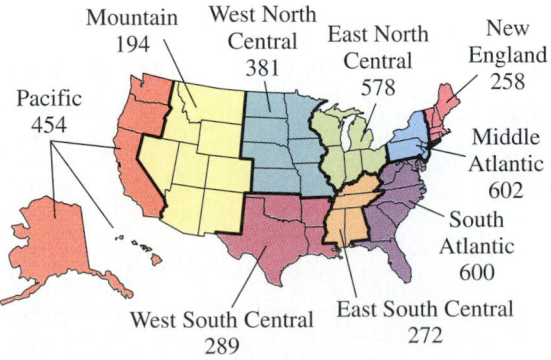

FIGURE 11.6

Mutually Exclusive Events

Two events A and B (from the same sample space) are **mutually exclusive** if A and B have no outcomes in common. In the terminology of sets, the **intersection of A and B** is the empty set, which is expressed as

$$P(A \cap B) = 0.$$

For instance, if two dice are tossed, the event A of rolling a total of 6 and the event B of rolling a total of 9 are mutually exclusive. To find the probability that one or the other of two mutually exclusive events will occur, you can *add* their individual probabilities.

PROBABILITY OF THE UNION OF TWO EVENTS

If A and B are events in the same sample space, the probability of A *or B* occurring is given by

$$P(A \cup B) = P(A) + P(B) - P(A \cap B).$$

If A and B are mutually exclusive, then

$$P(A \cup B) = P(A) + P(B).$$

EXAMPLE 7   *The Probability of a Union*

One card is selected from a standard deck of 52 playing cards. What is the probability that the card is either a heart or a face card?

Solution

Because the deck has 13 hearts, the probability of selecting a heart (event A) is $P(A) = \frac{13}{52}$. Similarly, because the deck has 12 face cards, the probability of selecting a face card (event B) is $P(B) = \frac{12}{52}$. Because three of the cards are hearts and face cards (see Figure 11.7), it follows that $P(A \cap B) = \frac{3}{52}$. Finally, applying the formula for the probability of the union of two events, you can conclude that the probability of selecting a heart or a face card is

$$P(A \cup B) = P(A) + P(B) - P(A \cap B)$$
$$= \frac{13}{52} + \frac{12}{52} - \frac{3}{52}$$
$$= \frac{22}{52}$$
$$\approx 0.423.$$

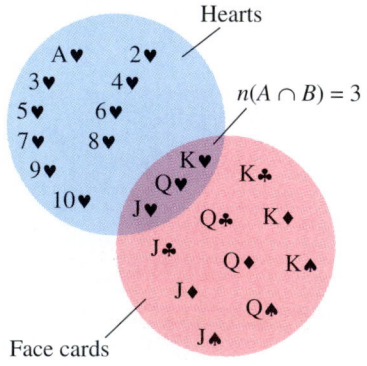

FIGURE 11.7

Real Life

EXAMPLE 8 *Probability of Mutually Exclusive Events*

The personnel department of a company has compiled data on the numbers of employees who have been with the company for various periods of time. The results are shown in the table.

Years of Service	Number of Employees
0–4	157
5–9	89
10–14	74
15–19	63
20–24	42
25–29	38
30–34	37
35–39	21
40–44	8

If an employee is chosen at random, what is the probability that the employee has had nine or fewer years of service?

Solution

To begin, add the number of employees and find that the total is 529. Next, let event A represent choosing an employee with 0 to 4 years of service and let event B represent choosing an employee with 5 to 9 years of service. Then

$$P(A) = \frac{157}{529} \quad \text{and} \quad P(B) = \frac{89}{529}.$$

Because A and B have no outcomes in common, you can conclude that these two events are mutually exclusive and that

$$P(A \cup B) = P(A) + P(B) = \frac{157}{529} + \frac{89}{529}$$
$$= \frac{246}{529}$$
$$\approx 0.465.$$

Thus, the probability of choosing an employee who has nine or fewer years of service is about 0.465.

Independent Events

Two events are **independent** if the occurrence of one has no effect on the occurrence of the other. For instance, rolling a total of 12 with two six-sided dice has no effect on the outcome of future rolls of the dice. To find the probability that two independent events will occur, *multiply* the probabilities of each.

> ### PROBABILITY OF INDEPENDENT EVENTS
>
> If A and B are independent events, the probability that both A and B will occur is
>
> $$P(A \text{ and } B) = P(A) \cdot P(B).$$

EXAMPLE 9 *Probability of Independent Events*

Real Life

A random number generator on a computer selects three integers from 1 to 20. What is the probability that all three numbers are less than or equal to 5?

Solution

The probability of selecting a number from 1 to 5 is

$$P(A) = \frac{5}{20} = \frac{1}{4}.$$

Thus, the probability that all three numbers are less than or equal to 5 is

$$P(A) \cdot P(A) \cdot P(A) = \left(\frac{1}{4}\right)\left(\frac{1}{4}\right)\left(\frac{1}{4}\right) = \frac{1}{64}.$$

EXAMPLE 10 *Probability of Independent Events*

Real Life

In 1992, 56% of the population of the United States was 30 years old or older. Suppose that in a survey, 10 people were chosen at random from the population. What is the probability that all 10 were 30 years old or older? (Source: U.S. Bureau of the Census)

Solution

Let A represent choosing a person who was 30 years old or older. Because the probability of choosing a person who was 30 years old or older was 0.56, you can conclude that the probability that all 10 people were 30 years old or older is

$$[P(A)]^{10} = (0.56)^{10} \approx 0.0030.$$

The Complement of an Event

The **complement of an event** A is the collection of all outcomes in the sample space that are *not* in A. The complement of event A is denoted by A'. Because $P(A \text{ or } A') = 1$ and because A and A' are mutually exclusive, it follows that $P(A) + P(A') = 1$. Therefore, the probability of A' is given by

$$P(A') = 1 - P(A).$$

For instance, if the probability of *winning* a certain game is

$$P(A) = \frac{1}{4}$$

the probability of *losing* the game is

$$P(A') = 1 - \frac{1}{4} = \frac{3}{4}.$$

> ### PROBABILITY OF A COMPLEMENT
>
> Let A be an event and let A' be its complement. If the probability of A is $P(A)$, the probability of the complement is
>
> $$P(A') = 1 - P(A).$$

EXAMPLE 11   *Finding the Probability of a Complement*   Real Life

A manufacturer has determined that a certain machine averages one faulty unit for every 1000 it produces. What is the probability that an order of 200 units will have one or more faulty units?

Solution

To solve this problem as stated, you would need to find the probabilities of having exactly one faulty unit, exactly two faulty units, exactly three faulty units, and so on. However, using complements, you can simply find the probability that all units are perfect and then subtract this value from 1. Because the probability that any given unit is perfect is 999/1000, the probability that all 200 units are perfect is

$$P(A) = \left(\frac{999}{1000}\right)^{200} \approx 0.8186.$$

Therefore, the probability that at least one unit is faulty is

$$P(A') = 1 - P(A) \approx 0.1814.$$

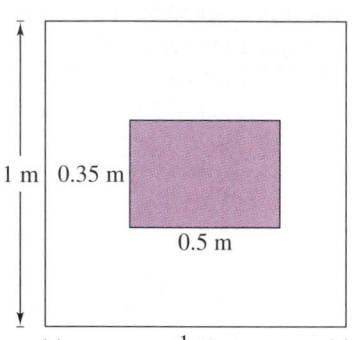

1 m | 0.35 m

0.5 m

|←———— 1 m ————→|

GROUP ACTIVITY

AN EXPERIMENT IN GEOMETRIC PROBABILITY

In this section you have been finding probabilities from a *theoretical* point of view. Another way to find probabilities is from an *experimental* point of view. For instance, suppose you wanted to find the probability of a thrown dart hitting the shaded portion of the rectangular target in the figure. (Assume that no dart lands outside the 1-meter square and that each dart is thrown randomly.)

(a) What is the theoretical probability that a thrown dart hits the shaded portion? Explain your reasoning.

(b) The following program for the *TI-83* or the *TI-82* can be used to simulate the outcome of a dart throw. (Programs for other calculator models may be found in the appendix.) Discuss how the program works. What does it do?

```
PROGRAM:DARTS
:0 → K
:0 → W                          :W+1 → W
:Input "HOW MANY THROWS?",N     :Lbl 2
:Lbl 1                          :If K=N
:K+1 → K                        :Then
:rand → X:rand → Y              :Disp "WINS=",W
:If X<.25 or X>.75              :Disp "LOSSES=",(N−W)
:Goto 2                         :Stop
:If Y<.3 or Y>.65               :End
:Goto 2                         :Goto 1
```

(c) Use the program to simulate 30 throws of a dart. What proportion of the throws landed inside the shaded area? This proportion is an experimental probability. How does it compare with the theoretical probability you calculated in part (a)?

(d) Combine your results with those of the rest of your class. What proportion of the pooled throws landed in the shaded region? How does this experimental probability compare with the theoretical probability and the experimental probability from part (c)?

(e) Can you make a conjecture about the relationship between the difference in the experimental and theoretical probabilities and the number of trials used to find the experimental probability? Explain how you might be able to get an experimental probability that more closely approximates the theoretical probability.

(f) Discuss the types of situations in which using simulations to estimate probabilities would be useful.

Evaluate the expression.

1. $\dfrac{1}{4} + \dfrac{5}{8} - \dfrac{5}{16}$

2. $\dfrac{4}{15} + \dfrac{3}{5} - \dfrac{1}{3}$

3. $\dfrac{5 \cdot 4}{5!}$

4. $\dfrac{5!22!}{27!}$

5. $\dfrac{4!}{8!12!}$

6. $\dfrac{9 \cdot 8 \cdot 7 \cdot 6 \cdot 5}{9!}$

7. $\dfrac{{}_5C_3}{{}_{10}C_3}$

8. $\dfrac{{}_{10}C_2 \cdot {}_{10}C_2}{{}_{20}C_4}$

Evaluate the expression. (Round your answer to three decimal places.)

9. $\left(\dfrac{99}{100}\right)^{100}$

10. $1 - \left(\dfrac{89}{100}\right)^{50}$

11.7 Exercises

In Exercises 1–6, determine the sample space for the given experiment.

1. A coin and a six-sided die are tossed.

2. A six-sided die is tossed twice and the sum of the points is recorded.

3. A taste tester has to rank three varieties of yogurt, A, B, and C, according to preference.

4. Two marbles are selected from a sack containing two red marbles, two blue marbles, and one black marble. The color of each marble is recorded.

5. Two county supervisors are selected from five supervisors, A, B, C, D, and E, to study a recycling plan.

6. A sales representative makes a presentation about a product in three homes per day. In each home there may be a sale (denote by S) or there may be no sale (denote by F).

Heads or Tails **In Exercises 7–10, find the probability in the experiment of tossing a coin three times. Use the sample space $S = \{HHH, HHT, HTH, HTT, THH, THT, TTH, TTT\}$.**

7. The probability of getting exactly one tail

8. The probability of getting a head on the first toss

9. The probability of getting at least one head

10. The probability of getting at least two heads

Drawing a Card **In Exercises 11–14, find the probability in the experiment of selecting one card from a standard deck of 52 playing cards.**

11. The card is a face card.

12. The card is not a face card.

13. The card is a red face card.

14. The card is a 6 or less.

Tossing a Die   **In Exercises 15–20, find the probability in the experiment of tossing a six-sided die twice.**

15. The sum is 4.

16. The sum is at least 7.

17. The sum is less than 11.

18. The sum is 2, 3, or 12.

19. The sum is odd and no more than 7.

20. The sum is odd or prime.

Drawing Marbles   **In Exercises 21–24, find the probability in the experiment of drawing two marbles (without replacement) from a bag containing one green, two yellow, and three red marbles.**

21. Both marbles are red.

22. Both marbles are yellow.

23. Neither marble is yellow.

24. The marbles are of different colors.

In Exercises 25 and 26, you are given the probability that an event *will* happen. Find the probability that the event *will not* happen.

25. $p = 0.7$ **26.** $p = 0.36$

In Exercises 27 and 28, you are given the probability that an event *will not* happen. Find the probability that the event *will* happen.

27. $p = 0.15$ **28.** $p = 0.84$

29. *Graphical Reasoning*   At the end of 1994 there were approximately 2.5 million minimum-wage workers in the United States. The figure gives the age profile of these workers. (Source: U.S. Bureau of Labor Statistics)

   (a) Estimate the number of minimum-wage workers in the age category 16–19.

   (b) What is the probability that a person selected at random from the population of minimum-wage workers is in the 25–34 age group?

   (c) What is the probability that a person selected at random from the population of minimum-wage workers is in the 35–54 age group?

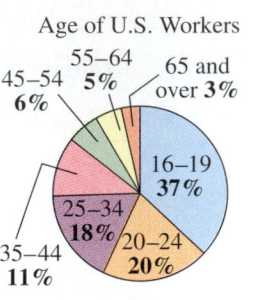

Age of U.S. Workers

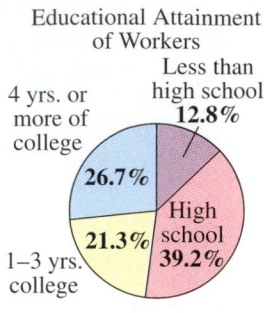

Educational Attainment of Workers

FIGURE FOR 29   **FIGURE FOR 30**

30. *Graphical Reasoning*   In 1991 there were approximately 101 million workers in the civilian labor force in the United States. The figure gives the educational attainment of these workers. (Source: U.S. Bureau of Labor Statistics)

   (a) Estimate the number of workers whose highest educational attainment was a high school education.

   (b) A person is selected at random from the civilian work force. What is the probability that the person has 1 to 3 years of college education?

   (c) A person is selected at random from the civilian work force. What is the probability that the person has more than a high school education?

31. *Data Analysis*   A study of the effectiveness of a flu vaccine was conducted with a sample of 500 people. Some in the study were given no vaccine, some were given one injection, and others were given two injections. The results of the study are given in the table.

	No Vaccine	One Injection	Two Injections	Total
Flu	7	2	13	22
No Flu	149	52	277	478
Total	156	54	290	500

A person is selected at random from the sample. Find the specified probability.

   (a) The person had two injections.

   (b) The person did not get the flu.

   (c) The person got the flu and had one injection.

32. *Data Analysis* One hundred college students were interviewed to determine their political party affiliations and whether they favored a balanced-budget amendment to the Constitution. The results of the study are given in the table.

	Favor	Not Favor	Unsure	Total
Democrat	23	25	7	55
Republican	32	9	4	45
Total	55	34	11	100

A person is selected at random from the sample. Find the probability that the described person is selected.

(a) A person who doesn't favor the amendment

(b) A Republican

(c) A Democrat who favors the amendment

33. *Alumni Association* A college sends a survey to selected members of the class of 1995. Of the 1254 people who graduated that year, 672 are women, of whom 124 went on to graduate school. Of the 582 male graduates, 198 went on to graduate school. If an alumni member is selected at random, what is the probability that the person is (a) female, (b) male, and (c) female and did not attend graduate school?

34. *Post–High School Education* In a high school graduating class of 72 students, 28 are on the honor roll. Of these, 18 are going on to college, and of the other 44 students, 12 are going on to college. If a student is selected at random from the class, what is the probability that the person chosen is (a) going to college, (b) not going to college, and (c) on the honor roll, but not going to college?

35. *Winning an Election* Taylor, Moore, and Jenkins are candidates for public office. It is estimated that Moore and Jenkins have about the same probability of winning, and Taylor is believed to be twice as likely to win as either of the others. Find the probability of each candidate winning the election.

36. *Winning an Election* Three people have been nominated for president of a class. From a poll, it is estimated that the first has a 37% chance of winning and the second has a 44% chance of winning. What is the probability that the third candidate will win?

In Exercises 37–48, the sample spaces are large and you should use the counting principles discussed in Section 11.6.

37. *Preparing for a Test* A class is given a list of 20 study problems from which 10 will be part of an upcoming exam. If a given student knows how to solve 15 of the problems, find the probability that the student will be able to answer (a) all 10 questions on the exam, (b) exactly eight questions on the exam, and (c) at least nine questions on the exam.

38. *Preparing for a Test* A class is given a list of eight study problems from which five will be part of an upcoming exam. If a given student knows how to solve six of the problems, find the probability that the student will be able to answer (a) all five questions on the exam, (b) exactly four questions on the exam, and (c) at least four questions on the exam.

39. *Letter Mix-Up* Four letters and envelopes are addressed to four different people. If the letters are randomly inserted into the envelopes, what is the probability that (a) exactly one will be inserted in the correct envelope and (b) at least one will be inserted in the correct envelope?

40. *Payroll Mix-Up* Five paychecks and envelopes are addressed to five different people. If the paychecks are randomly inserted into the envelopes, what is the probability that (a) exactly one will be inserted in the correct envelope and (b) at least one will be inserted in the correct envelope?

41. *Game Show* On a game show you are given five digits to arrange in the proper order to form the price of a car. If you are correct, you win the car. What is the probability of winning, given the following conditions?

(a) You guess the position of each digit.

(b) You know the first digit and guess the positions of the others.

42. *Game Show* On a game show you are given four digits to arrange in the proper order to form the price of a car. If you are correct, you win the car. What is the probability of winning, given the following conditions?

(a) You guess the position of each digit.

(b) You know the first digit and guess the others.

43. *Drawing Cards from a Deck*   Two cards are selected at random from an ordinary deck of 52 playing cards. Find the probability that two aces are selected, given the following conditions.

  (a) The cards are drawn in sequence, with the first card being replaced and the deck reshuffled prior to the second drawing.

  (b) The two cards are drawn consecutively, without replacement.

44. *Poker Hand*   Five cards are drawn from an ordinary deck of 52 playing cards. What is the probability that the hand drawn is a full house?

45. *Defective Units*   A shipment of 12 microwave ovens contains three defective units. A vending company has ordered four of these units, and because each is identically packaged, the selection will be random. What is the probability that (a) all four units are good, (b) exactly two units are good, and (c) at least two units are good?

46. *Defective Units*   A shipment of 20 compact disc players contains four defective units. A retail outlet has ordered five of these units. What is the probability that (a) all five units are good, (b) exactly four units are good, and (c) at least one unit is defective?

47. *Random Number Generator*   Two integers from 1 through 30 are chosen by a random number generator. What is the probability that (a) the numbers are both even, (b) one number is even and one is odd, (c) both numbers are less than 10, and (d) the same number is chosen twice?

48. *Random Number Generator*   Two integers from 1 through 40 are chosen by a random number generator. What is the probability that (a) the numbers are both even, (b) one number is even and one is odd, (c) both numbers are less than 30, and (d) the same number is chosen twice?

49. *Backup System*   A space vehicle has an independent backup system for one of its communication networks. The probability that either system will function satisfactorily during a flight is 0.985. What is the probability that during a given flight (a) both systems function satisfactorily, (b) at least one system functions satisfactorily, and (c) both systems fail?

50. *Backup Vehicle*   A fire company keeps two rescue vehicles. Because of the demand on the vehicles and the chance of mechanical failure, the probability that a specific vehicle is available when needed is 90%. If the availability of one vehicle is *independent* of the availability of the other, find the probability that (a) both vehicles are available at a given time, (b) neither vehicle is available at a given time, and (c) at least one vehicle is available at a given time.

51. *Making a Sale*   A sales representative makes a sale at approximately one-fourth of all calls. If, on a given day, the representative contacts five potential clients, what is the probability that a sale will be made with (a) each of the five contacts, (b) none of the contacts, and (c) at least one contact?

52. *A Boy or a Girl?*   Assume that the probability of the birth of a child of a particular sex is 50%. In a family with four children what is the probability that (a) all the children are boys, (b) all the children are the same sex, and (c) there is at least one boy?

53. *Will That Be Cash or Charge?*   Suppose that the methods used by shoppers to pay for merchandise are as shown in the circle graph. If two shoppers are chosen at random, what is the probability that both shoppers paid for their gifts only in cash?

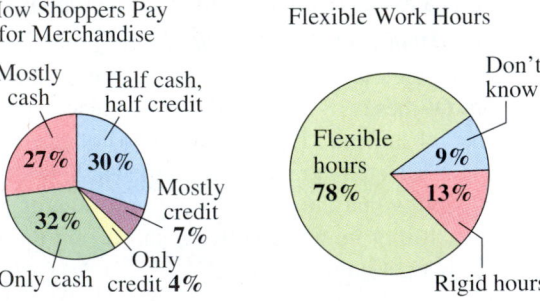

FIGURE FOR 53   FIGURE FOR 54

54. *Flexible Work Hours*   In a survey, people were asked if they would prefer to work flexible hours—even if it meant slower career advancement—so they could spend more time with their family. The results of the survey are shown in the figure above. Suppose that three people from the survey were chosen at random. What is the probability that all three people would prefer flexible work hours?

55. *Geometry*   You and a friend agree to meet at your favorite fast-food restaurant between 5:00 and 6:00 P.M. The one who arrives first will wait 15 minutes for the other, and then will leave (see figure). What is the probability that the two of you will actually meet, assuming that your arrival times are random within the hour?

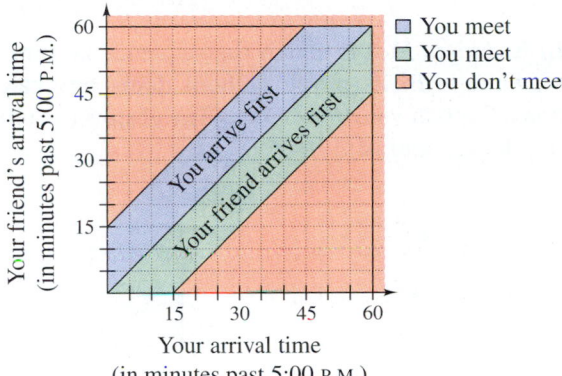

Your friend's arrival time (in minutes past 5:00 P.M.)

Your arrival time (in minutes past 5:00 P.M.)

☐ You meet
☐ You meet
☐ You don't meet

You arrive first

Your friend arrives first

56. *Estimating* π A coin of diameter d is dropped onto a paper that contains a grid of squares d units on a side (see figure).

(a) Find the probability that the coin covers a vertex of one of the squares on the grid.

(b) Perform the experiment 100 times and use the results to approximate π.

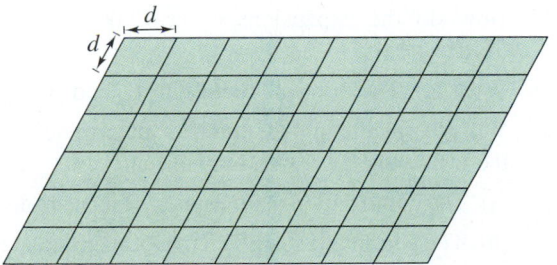

57. *Pattern Recognition and Exploration*   Consider a group of n people.

(a) Explain why the following pattern gives the probability that the n people have distinct birthdays.

$$n = 2: \quad \frac{365}{365} \cdot \frac{364}{365} = \frac{365 \cdot 364}{365^2}$$

$$n = 3: \quad \frac{365}{365} \cdot \frac{364}{365} \cdot \frac{363}{365} = \frac{365 \cdot 364 \cdot 363}{365^3}$$

(b) Use the pattern in part (a) to write an expression for the probability that $n = 4$ people have distinct birthdays.

(c) Let P_n be the probability that the n people have distinct birthdays. Verify that this probability can be obtained recursively by

$$P_1 = 1 \quad \text{and} \quad P_n = \frac{365 - (n - 1)}{365} P_{n-1}.$$

(d) Explain why $Q_n = 1 - P_n$ gives the probability that at least two people in a group of n people have the same birthday.

(e) Use the results of parts (c) and (d) to complete the table.

n	10	15	20	23	30	40	50
P_n							
Q_n							

(f) How many people must be in a group so that the probability of at least two of them having the same birthday is greater than $\frac{1}{2}$? Explain.

58. *Think About It*   A weather forecast indicates that the probability of rain is 40%. What does this mean?

59. *Chapter Opener*   If the probability that a dive of an elephant seal will exceed 400 meters is 0.546, determine the probability that a dive will be 400 meters or less.

60. *Chapter Opener*   Use the histogram on page 797 to estimate the probability that a dive will be between 300 and 500 meters.

FOCUS ON CONCEPTS

In this chapter, you studied sequences, counting principles, and probability. Answer the following questions to check your understanding of several of the basic concepts discussed in this chapter. The answers to these questions are given in the back of the book.

1. An infinite sequence is a function. What is the domain of the function?

2. How do the two sequences differ?

    (a) $a_n = \dfrac{(-1)^n}{n}$ (b) $a_n = \dfrac{(-1)^{n+1}}{n}$

True or False? **In Exercises 3–6, decide whether the statement is true or false. Explain your reasoning.**

3. $\dfrac{(n+2)!}{n!} = (n+2)(n+1)$

4. $\displaystyle\sum_{i=1}^{5}(i^3 + 2i) = \sum_{i=1}^{5}i^3 + \sum_{i=1}^{5}2i$

5. $\displaystyle\sum_{k=1}^{8}3k = 3\sum_{k=1}^{8}k$

6. $\displaystyle\sum_{j=1}^{6}2^j = \sum_{j=3}^{8}2^{j-2}$

7. In your own words, explain what makes a sequence (a) arithmetic and (b) geometric.

8. The graphs of two sequences are shown below. Identify each sequence as arithmetic or geometric. Explain your reasoning.

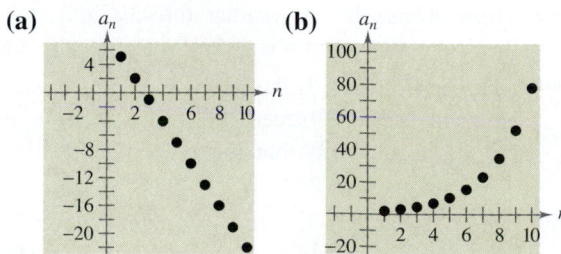

9. Explain what a recursion formula is.

10. Explain why the terms of a geometric sequence decrease when $0 < r < 1$.

In Exercises 11–14, match the sequence or sum of a sequence with its graph without doing any calculations. Explain your reasoning. [The graphs are labeled (a), (b), (c), and (d).]

(a) 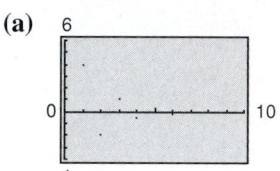   (b)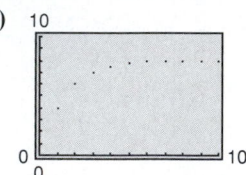

(c) 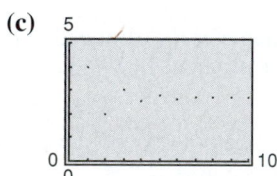   (d)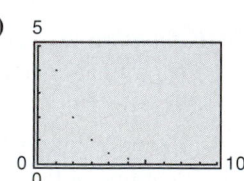

11. $a_n = 4\left(\tfrac{1}{2}\right)^{n-1}$ 12. $a_n = 4\left(-\tfrac{1}{2}\right)^{n-1}$

13. $a_n = \displaystyle\sum_{k=1}^{n}4\left(\tfrac{1}{2}\right)^{k-1}$ 14. $a_n = \displaystyle\sum_{k=1}^{n}4\left(-\tfrac{1}{2}\right)^{k-1}$

15. How do the expansions of $(x + y)^n$ and $(x - y)^n$ differ?

16. What is the relationship between $_nC_r$ and $_nC_{n-r}$?

17. Without calculating the numbers, determine which of the following is greater. Explain.

    (a) The combination of 10 elements taken six at a time

    (b) The permutation of 10 elements taken six at a time

18. The probability of an event must be a real number in what interval? Is the interval open or closed?

19. The probability of an event is $\tfrac{2}{3}$. What is the probability that the event does not occur? Explain.

20. A weather forecast indicates that the probability of rain is 60%. Explain what this means.

Review Exercises

In Exercises 1–4, write the first five terms of the sequence. (Assume that n begins with 1.)

1. $a_n = 2 + \dfrac{6}{n}$

2. $a_n = \dfrac{5n}{2n - 1}$

3. $a_n = \dfrac{72}{n!}$

4. $a_n = n(n - 1)$

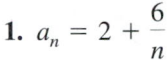 In Exercises 5–8, use a graphing utility to graph the first 10 terms of the sequence.

5. $a_n = \dfrac{3}{2}n$

6. $a_n = 4(0.4)^{n-1}$

7. $a_n = \dfrac{3n}{n + 2}$

8. $a_n = 5 - \dfrac{3}{n}$

In Exercises 9–12, use sigma notation to write the sum.

9. $\dfrac{1}{2(1)} + \dfrac{1}{2(2)} + \dfrac{1}{2(3)} + \cdots + \dfrac{1}{2(20)}$

10. $2(1^2) + 2(2^2) + 2(3^2) + \cdots + 2(9^2)$

11. $\dfrac{1}{2} + \dfrac{2}{3} + \dfrac{3}{4} + \cdots + \dfrac{9}{10}$

12. $1 - \dfrac{1}{3} + \dfrac{1}{9} - \dfrac{1}{27} + \cdots$

In Exercises 13–20, find the sum.

13. $\displaystyle\sum_{i=1}^{6} 5$

14. $\displaystyle\sum_{k=2}^{5} 4k$

15. $\displaystyle\sum_{j=1}^{4} \dfrac{6}{j^2}$

16. $\displaystyle\sum_{i=1}^{8} \dfrac{i}{i + 1}$

17. $\displaystyle\sum_{k=1}^{10} 2k^3$

18. $\displaystyle\sum_{j=0}^{4} (j^2 + 1)$

19. $\displaystyle\sum_{n=0}^{10} (n^2 + 3)$

20. $\displaystyle\sum_{n=1}^{100} \left(\dfrac{1}{n} - \dfrac{1}{n + 1}\right)$

In Exercises 21–24, write the first five terms of the arithmetic sequence.

21. $a_1 = 3, \ d = 4$

22. $a_1 = 8, \ d = -2$

23. $a_4 = 10, \ a_{10} = 28$

24. $a_2 = 14, \ a_6 = 22$

In Exercises 25–28, write the first five terms of the arithmetic sequence defined recursively. Determine the common difference and write the nth term of the sequence as a function of n.

25. $a_1 = 35 \qquad a_{k+1} = a_k - 3$

26. $a_1 = 15 \qquad a_{k+1} = a_k + \dfrac{5}{2}$

27. $a_1 = 9 \qquad a_{k+1} = a_k + 7$

28. $a_1 = 100 \qquad a_{k+1} = a_k - 5$

In Exercises 29 and 30, write an expression for the nth term of the arithmetic sequence and find the sum of the first 20 terms of the sequence.

29. $a_1 = 100, \ d = -3$

30. $a_1 = 10, \ a_3 = 28$

In Exercises 31–34, find the sum.

31. $\displaystyle\sum_{j=1}^{10} (2j - 3)$

32. $\displaystyle\sum_{j=1}^{8} (20 - 3j)$

33. $\displaystyle\sum_{k=1}^{11} \left(\dfrac{2}{3}k + 4\right)$

34. $\displaystyle\sum_{k=1}^{25} \left(\dfrac{3k + 1}{4}\right)$

35. Find the sum of the first 100 positive multiples of 5.

36. Find the sum of the integers from 20 to 80 (inclusive).

37. *Job Offer* The starting salary for a job is $34,000 with a guaranteed salary increase of $2250 per year. Determine (a) the salary during the fifth year and (b) the total compensation through five full years of employment.

38. *Baling Hay*   In the first two trips baling hay around a large field, a farmer obtains 123 bales and 112 bales, respectively. Because each round gets shorter, the farmer estimates that the same pattern will continue. Estimate the total number of bales made if there are another six trips around the field.

In Exercises 39–42, write the first five terms of the geometric sequence.

39. $a_1 = 4,\ r = -\frac{1}{4}$

40. $a_1 = 2,\ r = 2$

41. $a_1 = 9,\ a_3 = 4$

42. $a_1 = 2,\ a_3 = 12$

In Exercises 43–46, write the first five terms of the geometric sequence defined recursively. Determine the common ratio and write the nth term of the sequence as a function of n.

43. $a_1 = 120 \qquad a_{k+1} = \frac{1}{3}a_k$

44. $a_1 = 200 \qquad a_{k+1} = 0.1a_k$

45. $a_1 = 25 \qquad a_{k+1} = -\frac{3}{5}a_k$

46. $a_1 = 18 \qquad a_{k+1} = \frac{5}{3}a_k$

In Exercises 47 and 48, write an expression for the nth term of the geometric sequence and find the sum of the first 20 terms of the sequence.

47. $a_1 = 16,\ a_2 = -8$

48. $a_1 = 100,\ r = 1.05$

In Exercises 49–54, find the sum.

49. $\displaystyle\sum_{i=1}^{7} 2^{i-1}$

50. $\displaystyle\sum_{i=1}^{5} 3^{i-1}$

51. $\displaystyle\sum_{i=1}^{\infty} \left(\frac{7}{8}\right)^{i-1}$

52. $\displaystyle\sum_{i=1}^{\infty} \left(\frac{1}{3}\right)^{i-1}$

53. $\displaystyle\sum_{k=1}^{\infty} 4\left(\frac{2}{3}\right)^{k-1}$

54. $\displaystyle\sum_{k=1}^{\infty} 1.3\left(\frac{1}{10}\right)^{k-1}$

In Exercises 55 and 56, use a graphing utility to find the sum.

55. $\displaystyle\sum_{i=1}^{10} 10\left(\frac{3}{5}\right)^{i-1}$

56. $\displaystyle\sum_{i=1}^{25} 100(1.06)^{i-1}$

57. *Depreciation*   A company buys a machine for $120,000. During the next 5 years it will depreciate at a rate of 30% per year. (That is, at the end of each year the depreciated value will be 70% of what it was at the beginning of the year.)

   (a) Find the formula for the nth term of a geometric sequence that gives the value of the machine t full years after it was purchased.

   (b) Find the depreciated value of the machine at the end of five full years.

58. *Total Compensation*   A job pays a salary of $32,000 the first year. During the next 39 years, suppose there is a 5.5% raise each year. What would the total salary be over the 40-year period?

59. *Compound Interest*   A deposit of $200 is made at the beginning of each month for 2 years in an account that pays 6%, compounded monthly. What is the balance in the account at the end of 2 years?

60. *Compound Interest*   A deposit of $100 is made at the beginning of each month for 10 years in an account that pays 6.5%, compounded monthly. What is the balance in the account at the end of 10 years?

In Exercises 61–64, use mathematical induction to prove the formula for every positive integer n.

61. $1 + 4 + \cdots + (3n - 2) = \dfrac{n}{2}(3n - 1)$

62. $1 + \dfrac{3}{2} + 2 + \dfrac{5}{2} + \cdots + \dfrac{1}{2}(n + 1) = \dfrac{n}{4}(n + 3)$

63. $\displaystyle\sum_{i=0}^{n-1} ar^i = \dfrac{a(1 - r^n)}{1 - r}$

64. $\displaystyle\sum_{k=0}^{n-1} (a + kd) = \dfrac{n}{2}[2a + (n - 1)d]$

In Exercises 65–68, evaluate $_nC_r$ or $_nP_r$. Use the $_nC_r$ or the $_nP_r$ feature of a calculator to verify your answer.

65. $_6C_4$

66. $_{10}C_7$

67. $_8P_5$

68. $_{12}P_3$

In Exercises 69–74, use the Binomial Theorem to expand the binomial. Simplify your answer. (Remember that $i = \sqrt{-1}$.)

69. $\left(\dfrac{x}{2} + y\right)^4$

70. $(a - 3b)^5$

71. $\left(\dfrac{2}{x} - 3x\right)^6$

72. $(3x + y^2)^7$

73. $(5 + 2i)^4$

74. $(4 - 5i)^3$

75. *Amateur Radio*   A Novice Amateur Radio license consists of two letters, one digit, and then three more letters. How many different licenses can be issued if no restrictions are placed on the letters or digits?

76. *Morse Code*   In Morse code, all characters are transmitted using a sequence of dits and dahs. How many different characters can be formed by a sequence of three dits and dahs? (Repetition is allowed. For example, dit-dit-dit represents the letter "s.")

77. *Matching Socks*   A man has five pairs of socks of which no two pairs are the same color. If he randomly selects two socks from a drawer, what is the probability that he gets a matched pair?

78. *Bookshelf Order*   A child returns a five-volume set of books to a bookshelf. The child is not able to read, and hence cannot distinguish one volume from another. What is the probability that the books are shelved in the correct order?

79. *Roll of the Dice*   Are the chances of rolling a 3 with one die the same as the chances of rolling a total of 6 with two dice? If not, which has the higher probability?

80. *Roll of the Dice*   A six-sided die is rolled six times. What is the probability that each side appears exactly once?

81. *Tossing a Coin*   Find the probability of obtaining at least one tail when a coin is tossed five times.

82. *Parental Independence*   In a survey, senior citizens were asked if they would live with their children when they reached the point of not being able to live alone. The results are shown in the figure. Suppose three senior citizens who cannot live alone are randomly selected. What is the probability that all three are not living with their children?

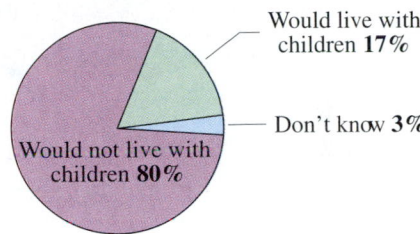

Would live with children **17%**

Don't know **3%**

Would not live with children **80%**

83. *Card Game*   Five cards are drawn from an ordinary deck of 52 playing cards. Find the probability of getting two pairs. (For example, the hand could be A-A-5-5-Q or 4-4-7-7-K.)

84. *Data Analysis*   A sample of college students, faculty, and administration were asked whether they favored a proposed increase in the annual activity fee to enhance student life on campus. The results of the study are given in the table.

	Students	Faculty	Admin.	Total
Favor	237	37	18	292
Oppose	163	38	7	208
Total	400	75	25	500

A person is selected at random from the sample. Find the specified probability.

(a) The person is not in favor of the proposal.

(b) The person is a student.

(c) The person is a faculty member and is in favor of the proposal.

CHAPTER PROJECT: *Recognizing Patterns in Data*

When creating a mathematical model for a real-life situation, you often do not expect the model to fit the data *exactly*. For instance, in the project in Chapter 9 (pages 726 and 727), you found linear models that approximately fit data for newspaper circulation. For that particular situation, there is no simple mathematical model that exactly fits the data.

Occasionally, however, real-life situations require models that do exactly fit the data. This often occurs in science and engineering, where you expect variables to be related by scientific or geometric principles.

EXAMPLE 1   *Finding an Exact Mathematical Model*

A polygon is *regular* if all its sides have the same length and all its angles have the same measure. The table gives the degree measures A_n of the interior angles of n-sided polygons $n = 3, 4, 5, 6, 7,$ and 8. Find a mathematical model for A_n in terms of n.

Number of Sides, n	3	4	5	6	7	8
Angle Measure, A_n	60°	90°	108°	120°	$128\frac{4}{7}°$	135°

Solution

From the table, it is not clear what pattern the angle measures are following. When hunting for a mathematical model, performing an operation on the data can help make the pattern more evident. For instance, you might take the natural logarithms of the angle measures, or square them, or take their square roots. For this data, multiplying A_n by n produces a recognizable pattern.

Number of Sides, n	3	4	5	6	7	8
Product nA_n	180°	360°	540°	720°	900°	1080°

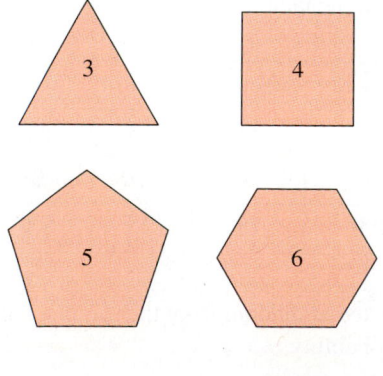

From this table, you can see that the products nA_n form an arithmetic sequence whose nth term is

$$nA_n = 180(n - 2)$$

which implies that

$$A_n = \frac{180(n - 2)}{n}.$$

This formula is your conjecture for the angle measure of *any* regular polygon. How could you prove that this conjecture is valid?

CHAPTER PROJECT INVESTIGATIONS

1. *Stars*   One way to form an *n*-pointed star is to begin with *n* equally spaced points on a circle and connect every second point. The angle measures of the tips of several stars are shown below. Find a mathematical model for these angle measures.

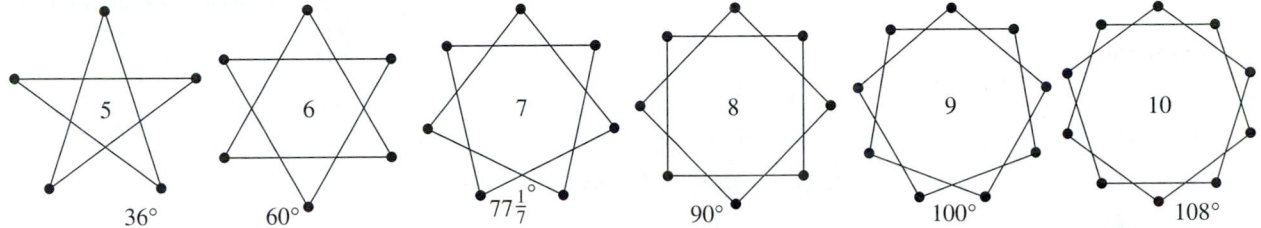

5	6	7	8	9	10
36°	60°	$77\frac{1}{7}^{\circ}$	90°	100°	108°

2. *More Stars*   Another way to form an *n*-pointed star is to begin with *n* equally spaced points on a circle and connect every third point. The angle measures of the tips of several stars are shown below. Find a mathematical model for these angle measures.

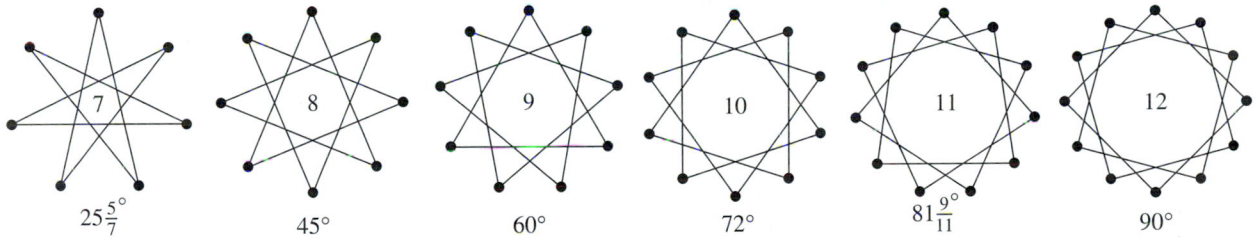

7	8	9	10	11	12
$25\frac{5}{7}^{\circ}$	45°	60°	72°	$81\frac{9}{11}^{\circ}$	90°

3. *Even More Stars*   A regular polygon can be considered to be a "star" formed by connecting adjacent points. Describe the pattern formed by the models in Example 1, Question 1, and Question 2.

Example 1	Form star by connecting adjacent points.
Question 1	Form star by connecting every second point.
Question 2	Form star by connecting every third point.

The stars below are formed by connecting every *fourth* point on a circle. Find a model for the angle measures of the star tips. Explain your reasoning.

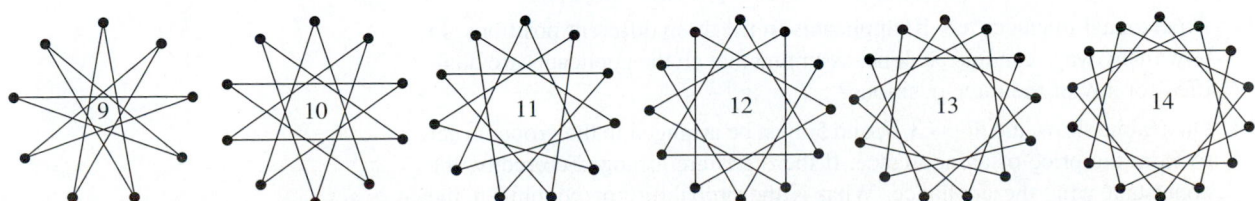

9	10	11	12	13	14

Cumulative Test for Chapters 9–11

Take this test as you would take a test in class. After you are done, check **your work against the answers given in the back of the book.**

The *Interactive* CD-ROM provides answers to the Chapter Tests and Cumulative Tests. It also offers Chapter Pre-Tests (which test key skills and concepts covered in previous chapters) and Chapter Post-Tests, both of which have randomly generated exercises with diagnostic capabilities.

In Exercises 1–4, solve the system by the specified method.

1. Substitution

$$y = 3 - x^2$$
$$2(y - 2) = x - 1$$

2. Elimination

$$x + 3y = -1$$
$$2x + 4y = 0$$

3. Elimination

$$-2x + 4y - z = 3$$
$$x - 2y + 2z = -6$$
$$x - 3y - z = 1$$

4. Gauss-Jordan Elimination

$$x + 3y - 2z = -7$$
$$-2x + y - z = -5$$
$$4x + y + z = 3$$

5. Sketch a graph of the solution of the constraints and maximize the objective function $z = 3x + 2y$ subject to the constraints.

$$x + 4y \le 20$$
$$2x + y \le 12$$
$$x \ge 0, \ y \ge 0$$

6. Find AB: $A = \begin{bmatrix} 4 & -3 \\ 2 & 1 \\ 5 & 0 \end{bmatrix}$, $B = \begin{bmatrix} 3 & -2 \\ 1 & -3 \end{bmatrix}$.

7. Find the inverse (if it exists): $\begin{bmatrix} 1 & 2 & -1 \\ 3 & 7 & -10 \\ -5 & -7 & -15 \end{bmatrix}$.

8. Sum the first 20 terms of the arithmetic sequence 8, 12, 16, 20, ...

9. Find the sum: $\displaystyle\sum_{i=0}^{\infty} 3\left(\tfrac{1}{2}\right)^i$.

10. Use mathematical induction to prove the formula

$$3 + 7 + 11 + 15 + \cdots + (4n - 1) = n(2n + 1).$$

11. Use the Binomial Theorem to expand and simplify $(z - 3)^4$.

12. A personnel manager has 10 applicants to fill three different positions. In how many ways can this be done, assuming that all the applicants are qualified for any of the three positions?

13. On a game show, the digits 3, 4, and 5 must be arranged in the proper order to form the price of an appliance. If the digits are arranged correctly, the contestant wins the appliance. What is the probability of winning if the contestant knows that the price is at least $400?

APPENDICES

Appendix A

Further Concepts in Statistics

Section A.1   Representing Data and Linear Modeling

Stem-and-Leaf Plots □ *Histograms and Frequency Distributions* □
Scatter Plots □ *Fitting a Line to Data*

Stem-and-Leaf Plots

Statistics is the branch of mathematics that studies techniques for collecting, organizing, and interpreting data. In this section, you will study several ways to organize and interpret data.

One type of plot that can be used to organize sets of numbers by hand is a **stem-and-leaf plot.** A set of test scores and the corresponding stem-and-leaf plot are shown below.

Test Scores

93, 70, 76, 58, 86, 93, 82, 78, 83, 86,
64, 78, 76, 66, 83, 83, 96, 74, 69, 76,
64, 74, 79, 76, 88, 76, 81, 82, 74, 70

Stems	Leaves
5	8
6	4 4 6 9
7	0 0 4 4 4 6 6 6 6 6 8 8 9
8	1 2 2 3 3 3 6 6 8
9	3 3 6

Note that the *leaves* represent the units digits of the numbers and the *stems* represent the tens digits. Stem-and-leaf plots can also be used to compare two sets of data, as shown in the following example.

■————

EXAMPLE 1   Comparing Two Sets of Data

Use a stem-and-leaf plot to compare the test scores given above with the following test scores. Which set of test scores is better?

90, 81, 70, 62, 64, 73, 81, 92, 73, 81, 92, 93, 83, 75, 76,
83, 94, 96, 86, 77, 77, 86, 96, 86, 77, 86, 87, 87, 79, 88

Solution

Begin by ordering the second set of scores.

62, 64, 70, 73, 73, 75, 76, 77, 77, 77, 79, 81, 81, 81, 83,
83, 86, 86, 86. 86, 87, 87, 88, 90, 92, 92, 93, 94, 96, 96

Now that the data has been ordered, you can construct a *double* stem-and-leaf plot by letting the leaves to the right of the stems represent the units digits for the first group of test scores and letting the leaves to the left of the stems represent the units digits for the second group of test scores.

Leaves (2nd Group)	Stems	Leaves (1st Group)
	5	8
4 2	6	4 4 6 9
9 7 7 7 6 5 3 3 0	7	0 0 4 4 4 6 6 6 6 8 8 9
8 7 7 6 6 6 6 3 3 1 1 1	8	1 2 2 3 3 3 6 6 8
6 6 4 3 2 2 0	9	3 3 6

By comparing the two sets of leaves, you can see that the second group of test scores is better than the first group.

EXAMPLE 2 *Using a Stem-and-Leaf Plot*

Table A.1 shows the percent of the population of each state and the District of Columbia that was at least 65 years old in 1989. Use a stem-and-leaf plot to organize the data.   (Source: U.S. Bureau of Census)

TABLE A.1

AK	4.1	AL	12.7	AR	14.8	AZ	13.1	CA	10.6
CO	9.8	CT	13.6	DC	12.5	DE	11.8	FL	18.0
GA	10.1	HI	10.7	IA	15.1	ID	11.9	IL	12.3
IN	12.4	KS	13.7	KY	12.7	LA	11.1	MA	13.8
MD	10.8	ME	13.4	MI	11.9	MN	12.6	MO	13.9
MS	12.4	MT	13.2	NC	12.1	ND	13.9	NE	13.9
NH	11.4	NJ	13.2	NM	10.5	NV	10.9	NY	13.0
OH	12.8	OK	13.3	OR	13.9	PA	15.1	RI	14.8
SC	11.1	SD	14.4	TN	12.6	TX	10.1	UT	8.6
VA	10.8	VT	11.9	WA	11.9	WI	13.4	WV	14.6
WY	9.8								

Solution

Begin by ordering the numbers, as shown below.

4.1, 8.6, 9.8, 9.8, 10.1, 10.1, 10.5, 10.6, 10.7, 10.8, 10.8,
10.9, 11.1, 11.1, 11.4, 11.8, 11.9, 11.9, 11.9, 11.9, 12.1, 12.3,
12.4, 12.4, 12.5, 12.6, 12.6, 12.7, 12.7, 12.8, 13.0, 13.1, 13.2,
13.2, 13.3, 13.4, 13.4, 13.6, 13.7, 13.8, 13.9, 13.9, 13.9, 13.9,
14.4, 14.6, 14.8, 14.8, 15.1, 15.1, 18.0

Next construct the stem-and-leaf plot using the leaves to represent the digits to the right of the decimal points.

Stems	Leaves	
4.	1	Alaska has the lowest percent.
5.		
6.		
7.		
8.	6	
9.	8 8	
10.	1 1 5 6 7 8 8 9	
11.	1 1 4 8 9 9 9 9	
12.	1 3 4 4 5 6 6 7 7 8	
13.	0 1 2 2 3 4 4 6 7 8 9 9 9 9	
14.	4 6 8 8	
15.	1 1	
16.		
17.		
18.	0	Florida has the highest percent.

TECHNOLOGY

Try using a computer or graphing calculator to create a histogram for the data at the right. How does the histogram change when the intervals change?

Histograms and Frequency Distributions

With data such as that given in Example 2, it is useful to group the numbers into intervals and plot the frequency of the data in each interval. For instance, the **frequency distribution** and **histogram** shown in Figure A.1 represent the data given in Example 2.

Frequency Distribution

Interval	Tally
[4, 6)	l
[6, 8)	
[8, 10)	lll
[10, 12)	HH HH HH l
[12, 14)	HH HH HH l HH llll
[14, 16)	HH l
[16, 8)	
[18, 20)	l

Histogram

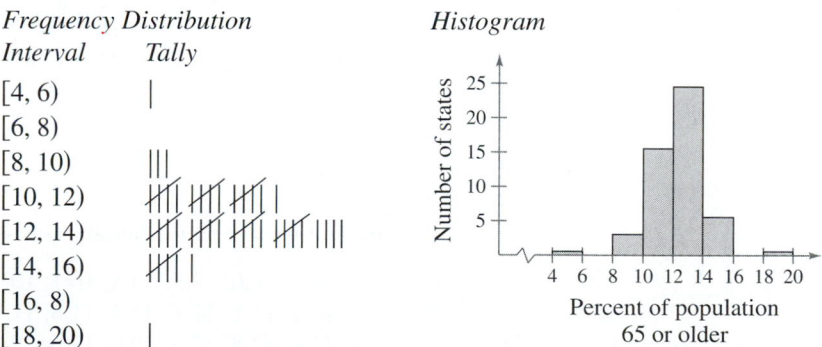

FIGURE A.1

A histogram has a portion of a real number line as its horizontal axis. A **bar graph** is similar to a histogram, except that the rectangles (bars) can be either horizontal or vertical and the labels of the bars are not necessarily numbers.

Another difference between a bar graph and a histogram is that the bars in a bar graph are usually separated by spaces, whereas the bars in a histogram are not separated by spaces.

EXAMPLE 3 Constructing a Bar Graph

The data below shows the average monthly precipitation (in inches) in Houston, Texas. Construct a bar graph for this data. What can you conclude? (Source: PC USA)

January	3.2	February	3.3	March	2.7
April	4.2	May	4.7	June	4.1
July	3.3	August	3.7	September	4.9
October	3.7	November	3.4	December	3.7

Solution

To create a bar graph, begin by drawing a vertical axis to represent the precipitation and a horizontal axis to represent the months. The bar graph is shown in Figure A.2. From the graph, you can see that Houston receives a fairly consistent amount of rain throughout the year, with the driest month tending to be March and the wettest month tending to be September.

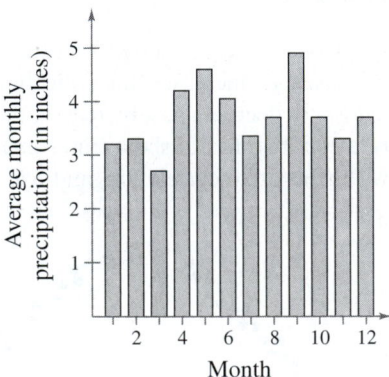

FIGURE A.2

Scatter Plots

Many real-life situations involve finding relationships between two variables, such as the year and the number of people in the labor force. In a typical situation, data is collected and written as a set of ordered pairs. The graph of such a set is called a **scatter plot.**

From the scatter plot in Figure A.3, it appears that the points describe a relationship that is nearly linear. (The relationship is not *exactly* linear because the labor force did not increase by precisely the same amount each year.) A mathematical equation that approximates the relationship between two variables is called a *mathematical model.* When developing a mathematical model, you strive for two (often conflicting) goals—accuracy and simplicity.

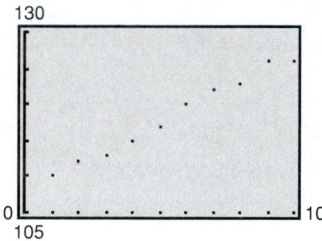

FIGURE A.3

Consider a collection of ordered pairs of the form (x, y). If y tends to increase as x increases, the collection is said to have a **positive correlation.** If y tends to decrease as x increases, the collection is said to have a **negative correlation.** Figure A.4 shows three examples: one with a positive correlation, one with a negative correlation, and one with no (discernible) correlation.

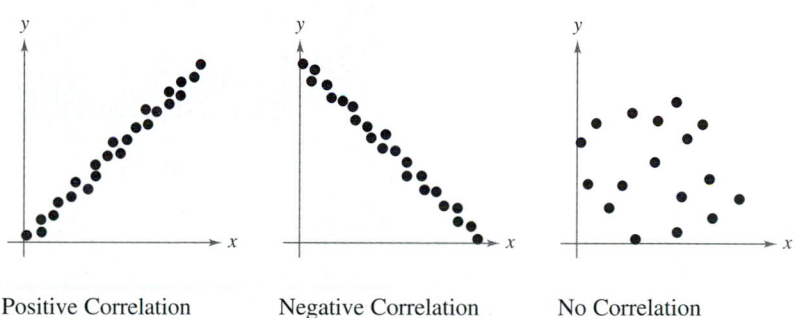

Positive Correlation   Negative Correlation   No Correlation

FIGURE A.4

Fitting a Line to Data

Finding a linear model that represents the relationship described by a scatter plot is called **fitting a line to data.** You can do this graphically by simply sketching the line that appears to fit the points, finding two points on the line, and then finding the equation of the line that passes through the two points.

■━━━━━━━

EXAMPLE 4 *Fitting a Line to Data*

Find a linear model that relates the year to the number of people P (in millions) who were part of the United States labor force from 1980 through 1990. In Table A.2, t represents the year, with $t = 0$ corresponding to 1980.   (Source: U.S. Bureau of Labor Statistics)

TABLE A.2

t	0	1	2	3	4	5	6	7	8	9	10
P	109	110	112	113	115	117	120	122	123	126	126

Solution

After plotting the data from Table A.2, draw the line that you think best represents the data, as shown in Figure A.5. Two points that lie on this line are (0, 109) and (9, 126). Using the point-slope form, you can find the equation of the line to be

$$P = \frac{17}{9}t + 109. \qquad \text{Linear model}$$

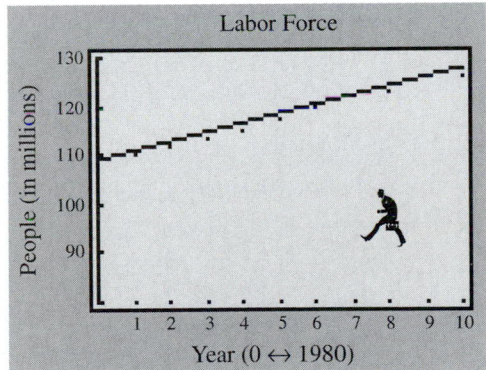

FIGURE A.5

━━━━━━━━━━━━━━━━━━━━━━━━━■

Once you have found a model, you can measure how well the model fits the data by comparing the actual values with the values given by the model, as shown in Table A.3.

TABLE A.3

	t	0	1	2	3	4	5	6	7	8	9	10
Actual →	P	109	110	112	113	115	117	120	122	123	126	126
Model →	P	109	110.9	112.8	114.7	116.6	118.4	120.3	122.2	124.1	126	127.9

The sum of the squares of the differences between the actual values and the model's values is the **sum of the squared differences.** The model that has the least sum is called the **least squares regression line** for the data. For the model in Example 4, the sum of the squared differences is 13.81. The least squares regression line for the data is

$$P = 1.864t + 108.2. \qquad \text{Best-fitting linear model}$$

Its sum of squared differences is 4.7.

LEAST SQUARES REGRESSION LINE

The least squares regression line, $y = ax + b$, for the points (x_1, y_1), (x_2, y_2), (x_3, y_3), . . . , (x_n, y_n) is given by

$$a = \frac{n\sum_{i=1}^{n} x_i y_i - \sum_{i=1}^{n} x_i \sum_{i=1}^{n} y_i}{n\sum_{i=1}^{n} x_i^2 - \left(\sum_{i=1}^{n} x_i\right)^2} \quad \text{and} \quad b = \frac{1}{n}\left(\sum_{i=1}^{n} y_i - a\sum_{i=1}^{n} x_i\right).$$

EXAMPLE 5   Finding a Least Squares Regression Line

Find the least squares regression line for the points $(-3, 0)$, $(-1, 1)$, $(0, 2)$, and $(2, 3)$.

Solution

Begin by constructing a table of values, as shown in Table A.4.

TABLE A.4

x	y	xy	x^2
-3	0	0	9
-1	1	-1	1
0	2	0	0
2	3	6	4
$\sum_{i=1}^{n} x_i = -2$	$\sum_{i=1}^{n} y_i = 6$	$\sum_{i=1}^{n} x_i y_i = 5$	$\sum_{i=1}^{n} x_i^2 = 14$

Applying the formulas for the least squares regression line with $n = 4$ produces

$$a = \frac{n\sum_{i=1}^{n} x_i y_i - \sum_{i=1}^{n} x_i \sum_{i=1}^{n} y_i}{n\sum_{i=1}^{n} x_i^2 - \left(\sum_{i=1}^{n} x_i\right)^2} = \frac{4(5) - (-2)(6)}{4(14) - (-2)^2} = \frac{32}{52} = \frac{8}{13} \quad \text{and}$$

$$b = \frac{1}{n}\left(\sum_{i=1}^{n} y_i - a\sum_{i=1}^{n} x_i\right) = \frac{1}{4}\left[6 - \frac{8}{13}(-2)\right] = \frac{47}{26}.$$

Thus, the least squares regression line is $y = \frac{8}{13}x + \frac{47}{26}$, as shown in Figure A.6.

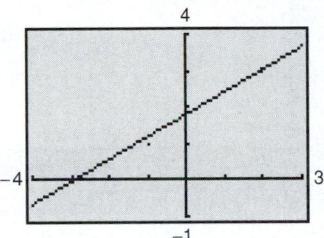

FIGURE A.6

Many calculators have "built-in" least squares regression programs. If your calculator has such a program, try using it to duplicate the results shown in the following example.

EXAMPLE 6   Finding a Least Squares Regression Line

The following ordered pairs (w, h) represent the shoe sizes w, and the heights h (in inches), of 25 men. Use a computer program or a statistical calculator to find the least squares regression line for this data.

(10.0, 70.0),	(10.5, 71.0),	(9.5, 70.0),	(11.0, 72.0),	(12.0, 74.0),
(8.5, 66.0),	(9.0, 68.5),	(13.0, 76.0),	(10.5, 71.5),	(10.5, 70.5),
(10.0, 72.0),	(9.5, 70.0),	(10.0, 71.0),	(10.5, 69.5),	(11.0, 71.5),
(12.0, 73.5),	(12.5, 74.0),	(11.0, 71.5),	(9.0, 67.5),	(10.0, 70.0),
(13.0, 73.5),	(10.5, 72.5),	(10.5, 71.0),	(11.0, 73.0),	(8.5, 68.0)

Solution

A scatter plot for the data is shown in Figure A.7. Note that the plot does not have 25 separate points because some of the ordered pairs graph as the same point. After entering the data into a statistical calculator, you can obtain

$$a = 1.67 \quad \text{and} \quad b = 53.57.$$

Thus, the least squares regression line for the data is

$$h = 1.67w + 53.57.$$

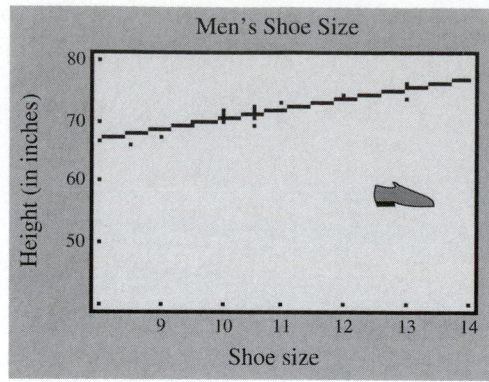

FIGURE A.7

If you use a statistical calculator or computer program to duplicate the results of Example 6, you will notice that the program also outputs a value of $r \approx 0.918$. This number is called the **correlation coefficient** of the data. Correlation coefficients vary between -1 and 1. Basically, the closer $|r|$ is to 1, the better the points can be described by a line. Three examples are shown in Figure A.8.

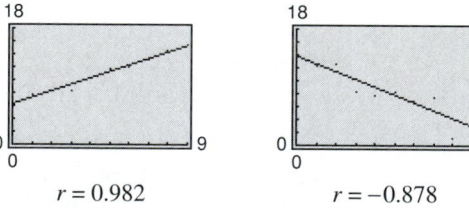
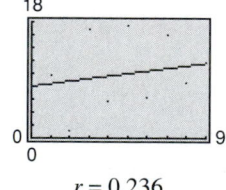

$r = 0.982$ $r = -0.878$ $r = 0.236$

FIGURE A.8

A.1 Exercises

Exam Scores   In Exercises 1 and 2, use the following scores from a math class of 30 students. The scores are given for two 100-point exams.

Exam #1:   77, 100, 77, 70, 83, 89, 87, 85, 81, 84, 81, 78, 89, 78, 88, 85, 90, 92, 75, 81, 85, 100, 98, 81, 78, 75, 85, 89, 82, 75

Exam #2:   76, 78, 73, 59, 70, 81, 71, 66, 66, 73, 68, 67, 63, 67, 77, 84, 87, 71, 78, 78, 90, 80, 77, 70, 80, 64, 74, 68, 68, 68

1. Construct a stem-and-leaf plot for Exam #1.

2. Construct a double stem-and-leaf plot to compare the scores for Exam #1 and Exam #2. Which set of test scores is higher?

3. *Educational Expenses*　The following table shows the per capita expenditures for public elementary and secondary education in the 50 states and the District of Columbia in 1991. Use a stem-and-leaf plot to organize the data. (Source: National Education Association)

AK	1626	AL	694	AR	668	AZ	892
CA	918	CO	841	CT	1151	DC	1010
DE	891	FL	862	GA	859	HI	784
IA	846	ID	725	IL	788	IN	925
KS	906	KY	725	LA	758	MA	866
MD	944	ME	1062	MI	926	MN	990
MO	742	MS	671	MT	983	NC	813
ND	719	NE	757	NH	881	NJ	1223
NM	915	NV	1004	NY	1186	OH	861
OK	776	OR	925	PA	889	RI	892
SC	835	SD	716	TN	618	TX	905
UT	828	VA	941	VT	992	WA	1095
WI	928	WV	883	WY	1178		

4. *Snowfall*　The data below shows the seasonal snowfall (in inches) at Erie, Pennsylvania for the years 1960 through 1989 (the amounts are listed in order by year). How would you organize this data? Explain your reasoning. (Source: National Oceanic and Atmospheric Administration)

69.6, 42.5, 75.9, 115.9, 92.9, 84.8, 68.6, 107.9, 79.7, 85.6, 120.0, 92.3, 53.7, 68.6, 66.7, 66.0, 111.5, 142.8, 76.5, 55.2, 89.4, 71.3, 41.2, 110.0, 106.3, 124.9, 68.2, 103.5, 76.5, 114.9

5. *Fruit Crops*　The data below shows the cash receipts (in millions of dollars) from fruit crops for farmers in 1990. Construct a bar graph for the data. (Source: U.S. Department of Agriculture)

Apples	1159	Peaches	365
Grapefruit	317	Pears	266
Grapes	1668	Plums and Prunes	293
Lemons	278	Strawberries	560
Oranges	1707		

6. *Travel to the United States*　The data below gives the places of origin and numbers of travelers (in millions) to the United States in 1991. Construct a horizontal bar graph for this data. (Source: U.S. Travel and Tourism Administration)

Canada	18.9	Mexico	7.0
Europe	7.4	Latin America	2.0
Other	6.8		

Crop Yield　**In Exercises 7–10, use the data in the table, where x is the number of units of fertilizer applied to sample plots and y is the yield (in bushels) of a crop.**

x	0	1	2	3	4	5	6	7	8
y	58	60	59	61	63	66	65	67	70

7. Sketch a scatter plot of the data.

8. Determine whether the points are positively correlated, are negatively correlated, or have no discernible correlation.

9. Sketch a linear model that you think best represents the data. Find an equation of the line you sketched. Use the line to predict the yield if 10 units of fertilizer are used.

10. Can the model found in Exercise 9 be used to predict yields for arbitrarily large values of x? Explain.

Speed of Sound　**In Exercises 11–14, use the data in the table, where h is the altitude in thousands of feet and v is the speed of sound in feet per second.**

h	0	5	10	15	20	25	30	35
v	1116	1097	1077	1057	1036	1015	995	973

11. Sketch a scatter plot of the data.

12. Determine whether the points are positively correlated, are negatively correlated, or have no discernible correlation.

13. Sketch a linear model that you think best represents the data. Find an equation of the line you sketched. Use the line to predict the speed of sound at an altitude of 27,000 feet.

14. The speed of sound at an altitude of 70,000 feet is approximately 971 feet per second. What does this suggest about the validity of using the model in Exercise 13 to extrapolate beyond the data given in the table?

In Exercises 15 and 16, (a) sketch a scatter plot of the points, (b) find an equation of the linear model you think best represents the data and find the sum of the squared differences, and (c) use the formulas in this section to find the least squares regression line and the sum of the squared differences.

15. $(-1, 0), (0, 1), (1, 3), (2, 3)$

16. $(0, 4), (1, 3), (2, 2), (4, 1)$

In Exercises 17–20, (a) sketch a scatter plot of the points, (b) use the formulas in this section to find the least squares regression line, and (c) sketch the graph of the line.

17. $(-2, 0), (-1, 1), (0, 1), (2, 2)$

18. $(-3, 1), (-1, 2), (0, 2), (1, 3), (3, 5)$

19. $(1, 5), (2, 8), (3, 13), (4, 16), (5, 22), (6, 26)$

20. $(1, 10), (2, 8), (3, 8), (4, 6), (5, 5), (6, 3)$

In Exercises 21–24, use a graphing utility to find the least squares regression line for the data. Sketch a scatter plot and the regression line.

21. $(0, 23), (1, 20), (2, 19), (3, 17), (4, 15), (5, 11),$
    $(6, 10)$

22. $(4, 52.8), (5, 54.7), (6, 55.7), (7, 57.8), (8, 60.2),$
    $(9, 63.1), (10, 66.5)$

23. $(-10, 5.1), (-5, 9.8), (0, 17.5), (2, 25.4), (4, 32.8),$
    $(6, 38.7), (8, 44.2), (10, 50.5)$

24. $(-10, 213.5), (-5, 174.9), (0, 141.7), (5, 119.7),$
    $(8, 102.4), (10, 87.6)$

25. *Advertising*   The management of a department store ran an experiment to determine if a relationship existed between sales S (in thousands of dollars) and the amount spent on advertising x (in thousands of dollars). The following data were collected.

x	1	2	3	4	5	6	7	8
S	405	423	455	466	492	510	525	559

(a) Use a graphing utility to find the least squares regression line. Use the equation to estimate sales if $4500 is spent on advertising.

(b) Make a scatter plot of the data and sketch the graph of the regression line.

(c) Use a computer or calculator to determine the correlation coefficient.

26. *School Enrollment*   The table gives the preprimary school enrollments y (in millions) for the years 1985 through 1991, where $t = 5$ corresponds to 1985. (Source: U.S. Bureau of the Census)

t	5	6	7	8	9	10	11
y	10.73	10.87	10.87	11.00	11.04	11.21	11.37

(a) Use a computer or calculator to find the least squares regression line. Use the equation to estimate enrollment in 1992.

(b) Make a scatter plot of the data and sketch the graph of the regression line.

(c) Use the computer or calculator to determine the correlation coefficient.

Section A.2   *Measures of Central Tendency and Dispersion*

Mean, Median, and Mode □ *Choosing a Measure of Central Tendency* □ *Variance and Standard Deviation*

Mean, Median, and Mode

In many real-life situations, it is helpful to describe data by a single number that is most representative of the entire collection of numbers. Such a number is called a **measure of central tendency.** The most commonly used measures are as follows.

1. The **mean,** or **average,** of n numbers is the sum of the numbers divided by n.
2. The **median** of n numbers is the middle number when the numbers are written in order. If n is even, the median is the average of the two middle numbers.
3. The **mode** of n numbers is the number that occurs most frequently. If two numbers tie for most frequent occurrence, the collection has two modes and is called **bimodal.**

EXAMPLE 1   *Comparing Measures of Central Tendency*

You are interviewing for a job. The interviewer tells you that the average income of the company's 25 employees is $60,849. The actual incomes of the 25 employees are shown below. What are the mean, median, and mode of the incomes? Was the person telling the truth?

$17,305,	$478,320,	$45,678,	$18,980,	$17,408,
$25,676,	$28,906,	$12,500,	$24,540,	$33,450,
$12,500,	$33,855,	$37,450,	$20,432,	$28,956,
$34,983,	$36,540,	$250,921,	$36,853,	$16,430,
$32,654,	$98,213,	$48,980,	$94,024,	$35,671

Solution

The mean of the incomes is

$$\text{Mean} = \frac{17{,}305 + 478{,}320 + 45{,}678 + 18{,}980 + \cdots + 35{,}671}{25}$$

$$= \frac{1{,}521{,}225}{25}$$

$$= \$60{,}849.$$

TECHNOLOGY

Statistical calculators have built-in programs for calculating the mean of a collection of numbers. Try using your calculator to find the mean given in Example 1. Then use your calculator to sort the incomes given in Example 1.

To find the median, order the incomes as follows.

$12,500,	$12,500,	$16,430,	$17,305,	$17,408,
$18,980,	$20,432,	$24,540,	$25,676,	$28,906,
$28,956,	$32,654,	$33,450,	$33,855,	$34,983,
$35,671,	$36,540,	$36,853,	$37,450,	$45,678,
$48,980,	$94,024,	$98,213,	$250,921,	$478,320

From this list, you can see that the median income (the middle number) is $33,450. From the same list, you can see that $12,500 is the only income that occurs more than once. Thus, the mode is $12,500. Technically, the person was telling the truth because the average is (generally) defined to be the mean. However, of the three measures of central tendency

Mean: $60,849   *Median:* $33,450   *Mode:* $12,500

it seems clear that the median is the most representative. The mean is inflated by the two highest salaries.

Choosing a Measure of Central Tendency

Which of the three measures of central tendency is the most representative? The answer is that it depends on the distribution of data *and* the way in which you plan to use the data.

For instance, in Example 1, the mean salary of $60,849 does not seem very representative to a potential employee. To a city income tax collector who wants to estimate 1% of the total income of the 25 employees, however, the mean is precisely the right measure.

EXAMPLE 2   Choosing a Measure of Central Tendency

Which measure of central tendency is the most representative of the data given in each of the following frequency distributions?

a. Number	Tally	b. Number	Tally	c. Number	Tally
1	7	1	9	1	6
2	20	2	8	2	1
3	15	3	7	3	2
4	11	4	6	4	3
5	8	5	5	5	5
6	3	6	6	6	5
7	2	7	7	7	4
8	0	8	8	8	3
9	15	9	9	9	0

Solution

a. For these data, the mean is 4.23, the median is 3, and the mode is 2. Of these, the mode is probably the most representative.

b. For these data, the mean and median are each 5 and the modes are 1 and 9 (the distribution is bimodal). Of these, the mean or median is the most representative.

c. For these data, the mean is 4.59, the median is 5, and the mode is 1. Of these, the mean or median is the most representative.

Variance and Standard Deviation

Very different sets of numbers can have the same mean. You will now study two **measures of dispersion,** which give you an idea of how much the numbers in the set differ from the mean of the set. These two measures are called the *variance* of the set and the *standard deviation* of the set.

> ### DEFINITIONS OF VARIANCE AND STANDARD DEVIATION
>
> Consider a set of numbers $\{x_1, x_2, \ldots, x_n\}$ with a mean of $\bar{x}$. The **variance** of the set is
>
> $$v = \frac{(x_1 - \bar{x})^2 + (x_2 - \bar{x})^2 + \cdots + (x_n - \bar{x})^2}{n}$$
>
> and the **standard deviation** of the set is
>
> $$\sigma = \sqrt{v}$$
>
> (σ is the lowercase Greek letter *sigma*).

The standard deviation of a set is a measure of how much a typical number in the set differs from the mean. The greater the standard deviation, the more the numbers in the set *vary* from the mean. For instance, each of the following sets has a mean of 5.

$$\{5, 5, 5, 5\}, \qquad \{4, 4, 6, 6\}, \qquad \text{and} \qquad \{3, 3, 7, 7\}$$

The standard deviations of the sets are 0, 1, and 2.

$$\sigma_1 = \sqrt{\frac{(5-5)^2 + (5-5)^2 + (5-5)^2 + (5-5)^2}{4}} = 0$$

$$\sigma_2 = \sqrt{\frac{(4-5)^2 + (4-5)^2 + (6-5)^2 + (6-5)^2}{4}} = 1$$

$$\sigma_3 = \sqrt{\frac{(3-5)^2 + (3-5)^2 + (7-5)^2 + (7-5)^2}{4}} = 2$$

■─────────────

EXAMPLE 3   Estimations of Standard Deviation

Consider the three sets of data represented by the following bar graphs. Which set has the smallest standard deviation? Which has the largest?

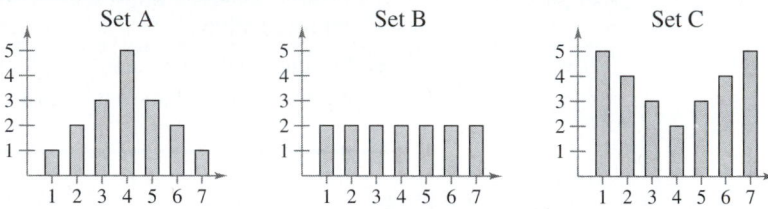

Solution

Of the three sets, the numbers in set A are grouped most closely to the center and the numbers in set C are the most dispersed. Thus, set A has the smallest standard deviation and set C has the largest standard deviation.

■─────────────

EXAMPLE 4   Finding Standard Deviation

Find the standard deviation of each set shown in Example 3.

Solution

Because of the symmetry of each bar graph, you can conclude that each has a mean of $\bar{x} = 4$. The standard deviation of set A is

$$\sigma = \sqrt{\frac{(-3)^2 + 2(-2)^2 + 3(-1)^2 + 5(0)^2 + 3(1)^2 + 2(2)^2 + (3)^2}{17}}$$

$$\approx 1.53.$$

The standard deviation of set B is

$$\sigma = \sqrt{\frac{2(-3)^2 + 2(-2)^2 + 2(-1)^2 + 2(0)^2 + 2(1)^2 + 2(2)^2 + 2(3)^2}{14}}$$

$$= 2.$$

The standard deviation of set C is

$$\sigma = \sqrt{\frac{5(-3)^2 + 4(-2)^2 + 3(-1)^2 + 2(0)^2 + 3(1)^2 + 4(2)^2 + 5(3)^2}{26}}$$

$$\approx 2.22.$$

These values confirm the results of Example 3. That is, set A has the smallest standard deviation and set C has the largest.

■─────────────

TECHNOLOGY
■■

If you have access to a computer or calculator with a standard deviation program, try using it to obtain the results given in Example 4. If you do this, the program will probably output two versions of the standard deviation. In one, the sum of the squared differences is divided by n, and in the other it is divided by $n - 1$. In this text, we always divide by n.

The following alternative formula provides a more efficient way to compute the standard deviation.

> **ALTERNATIVE FORMULA FOR STANDARD DEVIATION**
>
> The standard deviation of $\{x_1, x_2, \ldots , x_n\}$ is
>
> $$\sigma = \sqrt{\frac{x_1{}^2 + x_2{}^2 + \cdots + x_n{}^2}{n} - \bar{x}^2}.$$

Because of messy computations, this formula is difficult to verify. Conceptually, however, the process is straightforward. It consists of showing that the expressions

$$\sqrt{\frac{(x_1 - \bar{x})^2 + (x_2 - \bar{x})^2 + \cdots + (x_n - \bar{x})^2}{n}}$$

and

$$\sqrt{\frac{x_1{}^2 + x_2{}^2 + \ldots + x_n{}^2}{n} - \bar{x}^2}$$

are equivalent. Try verifying this equivalence for the set $\{x_1, x_2, x_3\}$ with $\bar{x} = (x_1 + x_2 + x_3)/3$.

EXAMPLE 5   *Using the Alternative Formula*

Use the alternative formula for standard deviation to find the standard deviation of the following set of numbers.

5, 6, 6, 7, 7, 8, 8, 8, 9, 10

Solution

Begin by finding the mean of the set, which is 7.4. Thus, the standard deviation is

$$\sigma = \sqrt{\frac{5^2 + 2(6)^2 + 2(7^2) + 3(8^2) + 9^2 + 10^2}{10} - (7.4)^2}$$

$$= \sqrt{\frac{568}{10} - 54.76}$$

$$= \sqrt{2.04}$$

$$\approx 1.43.$$

You can use the statistical features of a graphing utility to check this result.

A well-known theorem in statistics, called *Chebychev's Theorem*, states that at least $1 - (1/k^2)$ of the numbers in a distribution must lie within k standard deviations of the mean. Thus, 75% of the numbers in the collection must lie within two standard deviations of the mean, and at least 88.9% of the numbers must lie within three standard deviations of the mean. For most distributions, these percentages are low. For instance, in all three distributions shown in Example 3, 100% of the numbers lie within two standard deviations of the mean.

EXAMPLE 6   Describing a Distribution

Table A.5 shows the number of dentists (per 100,000 people) in each state and the District of Columbia. Find the mean and standard deviation of the numbers. What percent of the numbers lie within two standard deviations of the mean? (Source: American Dental Association)

TABLE A.5

AK	66	AL	40	AR	39	AZ	51	CA	62
CO	69	CT	80	DC	94	DE	44	FL	50
GA	46	HI	80	IA	55	ID	53	IL	61
IN	47	KS	51	KY	53	LA	45	MA	74
MD	68	ME	47	MI	62	MN	67	MO	53
MS	37	MT	62	NC	42	ND	47	NE	63
NH	59	NJ	77	NM	45	NV	49	NY	73
OH	55	OK	47	OR	70	PA	61	RI	56
SC	41	SD	49	TN	53	TX	47	UT	66
VA	54	VT	57	WA	68	WI	65	WV	43
WY	52								

Solution

Begin by entering the numbers into a computer or calculator that has a standard deviation program. After running the program, you should obtain

$$\bar{x} \approx 56.76 \quad \text{and} \quad \sigma \approx 12.14.$$

The interval that contains all numbers that lie with in two standard deviations of the mean is

$$[56.76 - 2(12.14), 56.76 + 2(12.14)] \quad \text{or} \quad [32.48, 81.04].$$

From the histogram in Figure A.9, you can see that all but one of the numbers (98%) lie in this interval—all but the number that corresponds to the number of dentists (per 100,000 people) in the District of Columbia.

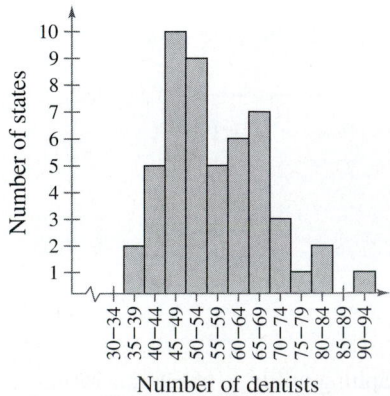

Number of dentists
(per 100,000 people)

FIGURE A.9

A.2 Exercises

In Exercises 1–6, find the mean, median, and mode of the set of measurements.

1. 5, 12, 7, 14, 8, 9, 7

2. 30, 37, 32, 39, 33, 34, 32

3. 5, 12, 7, 24, 8, 9, 7

4. 20, 37, 32, 39, 33, 34, 32

5. 5, 12, 7, 14, 9, 7

6. 30, 37, 32, 39, 34, 32

7. Compare your answers for Exercises 1 and 3 with those for Exercises 2 and 4. Which of the measures of central tendency is sensitive to extreme measurements? Explain your reasoning.

8. (a) Add 6 to each measurement in Exercise 1 and calculate the mean, median, and mode of the revised measurements. How are the measures of central tendency changed?

   (b) If a constant k is added to each measurement in a set of data, how will the measures of central tendency change?

9. **Electric Bills** A person had the following monthly bills for electricity. What are the mean and median of this collection of bills?

January	$67.92	February	$59.84
March	$52.00	April	$52.50
May	$57.99	June	$65.35
July	$81.76	August	$74.98
September	$87.82	October	$83.18
November	$65.35	December	$57.00

10. **Car Rental** A car rental company kept the following record of the number of miles driven by a car that was rented. What are the mean, median, and mode of this set of data?

Monday	410	Tuesday	260
Wednesday	320	Thursday	320
Friday	460	Saturday	150

11. **Six-Child Families** A study was done on families having six children. The table gives the number of families in the study with the indicated number of girls. Determine the mean, median, and mode of this set of data.

Number of Girls	0	1	2	3	4	5	6
Frequency	1	24	45	54	50	19	7

12. **Baseball** A baseball fan examined the records of a favorite baseball player's performance during his last 50 games. The number of games in which the player had 0, 1, 2, 3, and 4 hits are recorded in the table.

Number of Hits	0	1	2	3	4
Frequency	14	26	7	2	1

    (a) Determine the average number of hits per game.

    (b) Determine the player's batting average if he had 200 at bats during the 50-game series.

13. Construct a collection of numbers that has the following properties. If this is not possible, explain why it is not.

    Mean = 6,   Median = 4,   Mode = 4

14. Construct a collection of numbers that has the following properties. If this is not possible, explain why it is not.

    Mean = 6,   Median = 6,   Mode = 4

15. **Test Scores** A professor records the following scores for a 100-point exam.

    99, 64, 80, 77, 59, 72, 87, 79, 92, 88,
    90, 42, 20, 89, 42, 100, 98, 84, 78, 91

    Which measure of central tendency best describes these test scores?

16. **Shoe Sales** A salesman sold eight pairs of a certain style of men's shoes. The sizes of the eight pairs were as follows: $10\frac{1}{2}$, 8, 12, $10\frac{1}{2}$, 10, $9\frac{1}{2}$, 11, and $10\frac{1}{2}$. Which measure (or measures) of central tendency best describes the typical shoe size for this set of data?

In Exercises 17–24, find the mean, variance, and standard deviation of the numbers.

17. 4, 10, 8, 2

18. 3, 15, 6, 9, 2

19. 0, 1, 1, 2, 2, 2, 3, 3, 4

20. 2, 2, 2, 2, 2, 2

21. 1, 2, 3, 4, 5, 6, 7

22. 1, 1, 1, 5, 5, 5

23. 49, 62, 40, 29, 32, 70

24. 1.5, 0.4, 2.1, 0.7, 0.8

In Exercises 25–30, use the alternative formula to find the standard deviation of the numbers.

25. 2, 4, 6, 6, 13, 5

26. 10, 25, 50, 26, 15, 33, 29, 4

27. 246, 336, 473, 167, 219, 359

28. 6.0, 9.1, 4.4, 8.7, 10.4

29. 8.1, 6.9, 3.7, 4.2, 6.1

30. 9.0, 7.5, 3.3, 7.4, 6.0

31. Without calculating the standard deviation, explain why the set {4, 4, 20, 20} has a standard deviation of 8.

32. If the standard deviation of a set of numbers is 0, what does this imply about the set?

33. *Test Scores*   An instructor adds five points to each student's exam score. Will this change the mean or standard deviation of the exam scores? Explain.

34. Consider the four sets of data represented by the histograms. Order the sets from the smallest to the largest variance.

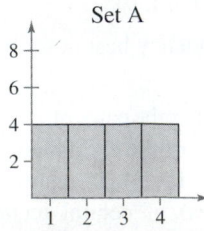
Set A

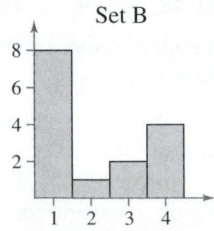
Set B

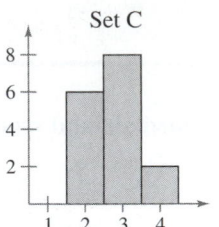
Set C

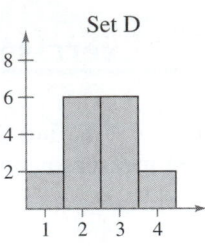
Set D

35. *Test Scores*   The scores on a mathematics exam given to 600 science and engineering students at a college had a mean of 235 and a standard deviation of 28. Use Chebychev's Theorem to determine the intervals containing at least $\frac{3}{4}$ and at least $\frac{8}{9}$ of the scores. How would the intervals change if the standard deviation were 16?

36. *Precipitation*   The following data represents the annual precipitation (in inches) at Erie, Pennsylvania, for the years 1960 through 1989. Use a computer or calculator to find the mean, variance, and standard deviation of the data. What percent of the data lies within two standard deviations of the mean? (Source: National Oceanic and Atmospheric Administration)

27.41,	36.50,	36.90,	28.11,	36.47,
38.41,	37.74,	37.78,	34.33,	36.58,
41.50,	34.06,	43.55,	38.04,	41.83,
43.03,	43.85,	61.70,	35.04,	55.31,
47.04,	41.97,	41.56,	46.25,	37.79,
45.87,	47.30,	44.86,	38.87,	41.88

Appendix B

Think About the Proof

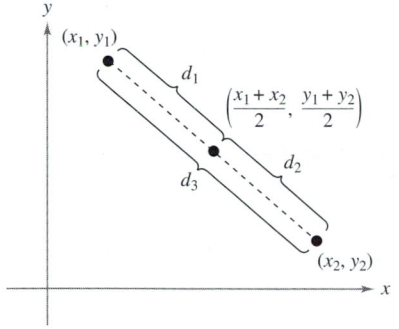

Midpoint Formula

SECTION P.7, PAGE 68

THE MIDPOINT FORMULA

The midpoint of the segment joining the points (x_1, y_1) and (x_2, y_2) is

$$\text{Midpoint} = \left(\frac{x_1 + x_2}{2}, \frac{y_1 + y_2}{2} \right).$$

PROOF

Using the figure, you must show that

$$d_1 = d_2 \quad \text{and} \quad d_1 + d_2 = d_3.$$

By the Distance Formula, you obtain

$$d_1 = \sqrt{\left(\frac{x_1 + x_2}{2} - x_1\right)^2 + \left(\frac{y_1 + y_2}{2} - y_1\right)^2} = \frac{1}{2}\sqrt{(x_2 - x_1)^2 + (y_2 - y_1)^2}$$

$$d_2 = \sqrt{\left(x_2 - \frac{x_1 + x_2}{2}\right)^2 + \left(y_2 - \frac{y_1 + y_2}{2}\right)^2} = \frac{1}{2}\sqrt{(x_2 - x_1)^2 + (y_2 - y_1)^2}$$

$$d_3 = \sqrt{(x_2 - x_1)^2 + (y_2 - y_1)^2}.$$

Thus, it follows that $d_1 = d_2$ and $d_1 + d_2 = d_3$. ■■

SECTION 3.3, PAGE 285

THE REMAINDER THEOREM

If a polynomial $f(x)$ is divided by $x - k$, the remainder is

$$r = f(k).$$

PROOF

From the Division Algorithm, you have

$$f(x) = (x - k)q(x) + r(x)$$

and because either $r(x) = 0$ or the degree of $r(x)$ is less than the degree of $x - k$, you know that $r(x)$ must be a constant. That is, $r(x) = r$. Now, by evaluating $f(x)$ at $x = k$, you have

$$f(k) = (k - k)q(k) + r = (0)q(k) + r = r. \quad ■■$$

SECTION 3.3, PAGE 285

THE FACTOR THEOREM

A polynomial $f(x)$ has a factor $(x - k)$ if and only if $f(k) = 0$.

PROOF

Using the Division Algorithm with the factor $(x - k)$, you have

$$f(x) = (x - k)q(x) + r(x).$$

By the Remainder Theorem, $r(x) = r = f(k)$, and you have

$$f(x) = (x - k)q(x) + f(k)$$

where $q(x)$ is a polynomial of lesser degree than $f(x)$. If $f(k) = 0$, then

$$f(x) = (x - k)q(x)$$

and you see that $(x - k)$ is a factor of $f(x)$. Conversely, if $(x - k)$ is a factor of $f(x)$, division of $f(x)$ by $(x - k)$ yields a remainder of 0. Hence, by the Remainder Theorem, you have $f(k) = 0$. ■■

SECTION 3.5, PAGE 302

LINEAR FACTORIZATION THEOREM

If $f(x)$ is a polynomial of degree n

$$f(x) = a_n x^n + a_{n-1}x^{n-1} + \cdots + a_1 x + a_0$$

where $n > 0$, then f has precisely n linear factors

$$f(x) = a_n(x - c_1)(x - c_2) \cdots (x - c_n)$$

where $c_1, c_2, \ldots, c_n$ are complex numbers and a_n is the leading coefficient of $f(x)$.

PROOF

Using the Fundamental Theorem, you know that f must have at least one zero, c_1. Consequently, $(x - c_1)$ is a factor of $f(x)$, and you have

$$f(x) = (x - c_1)f_1(x).$$

If the degree of $f_1(x)$ is greater than zero, you again apply the Fundamental Theorem to conclude that f_1 must have a zero c_2, which implies that

$$f(x) = (x - c_1)(x - c_2)f_2(x).$$

It is clear that the degree of $f_1(x)$ is $n - 1$, that the degree of $f_2(x)$ is $n - 2$, and that you can repeatedly apply the Fundamental Theorem n times until you obtain

$$f(x) = a_n(x - c_1)(x - c_2) \cdots (x - c_n)$$

where a_n is the leading coefficient of the polynomial $f(x)$. ■■

SECTION 3.5, PAGE 306

FACTORS OF A POLYNOMIAL

Every polynomial of degree $n > 0$ with real coefficients can be written as the product of linear and quadratic factors with real coefficients, where the quadratic factors have no real zeros.

PROOF

To begin, you use the Linear Factorization Theorem to conclude that $f(x)$ can be *completely* factored in the form

$$f(x) = d(x - c_1)(x - c_2)(x - c_3) \cdots (x - c_n).$$

If each c_i is real, there is nothing more to prove. If any c_i is complex ($c_i = a + bi$, $b \neq 0$), then, because the coefficients of $f(x)$ are real, you know that the conjugate $c_j = a - bi$ is also a zero. By multiplying the corresponding factors, you obtain

$$(x - c_i)(x - c_j) = [x - (a + bi)][x - (a - bi)]$$
$$= x^2 - 2ax + (a^2 + b^2)$$

where each coefficient is real. ■■

SECTION 4.4, PAGE 362

STANDARD EQUATION OF A PARABOLA (VERTEX AT ORIGIN)

The **standard form of the equation of a parabola** with vertex at $(0, 0)$ and directrix $y = -p$ is

$$x^2 = 4py, \qquad p \neq 0. \qquad\qquad \text{Vertical axis}$$

For directrix $x = -p$, the equation is

$$y^2 = 4px, \qquad p \neq 0. \qquad\qquad \text{Horizontal axis}$$

The focus is on the axis p units (directed distance) from the vertex.

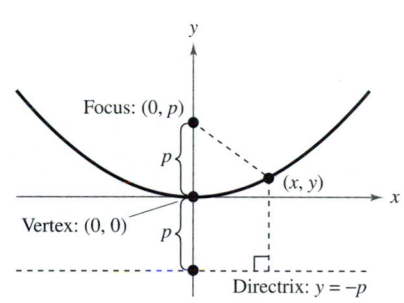

PROOF

Because the two cases are similar, a proof will be given for the first case only. Suppose the directrix ($y = -p$) is parallel to the x-axis. In the figure, you assume that $p > 0$, and because p is the directed distance from the vertex to the focus, the focus must lie above the vertex. Because the point (x, y) is equidistant from $(0, p)$ and $y = -p$, you can apply the Distance Formula to obtain

$$\sqrt{(x - 0)^2 + (y - p)^2} = y + p$$
$$x^2 + (y - p)^2 = (y + p)^2$$
$$x^2 + y^2 - 2py + p^2 = y^2 + 2py + p^2$$
$$x^2 = 4py. \qquad ■■$$

SECTION 5.3, PAGE 418

 PROPERTIES OF LOGARITHMS

Let a be a positive number such that $a \neq 1$, and let n be a real number. If u and v are positive real numbers, the following properties are true.

1. $\log_a(uv) = \log_a u + \log_a v$ **1.** $\ln(uv) = \ln u + \ln v$

2. $\log_a \dfrac{u}{v} = \log_a u - \log_a v$ **2.** $\ln \dfrac{u}{v} = \ln u - \ln v$

3. $\log_a u^n = n \log_a u$ **3.** $\ln u^n = n \ln u$

PROOF

To prove Property 1, let

$$x = \log_a u \quad \text{and} \quad y = \log_a v.$$

The corresponding exponential forms of these two equations are

$$a^x = u \quad \text{and} \quad a^y = v.$$

Multiplying u and v produces $uv = a^x a^y = a^{x+y}$. The corresponding logarithmic form of $uv = a^{x+y}$ is $\log_a(uv) = x + y$. Hence, $\log_a(uv) = \log_a u + \log_a v$. ∎

SECTION 8.1, PAGE 602

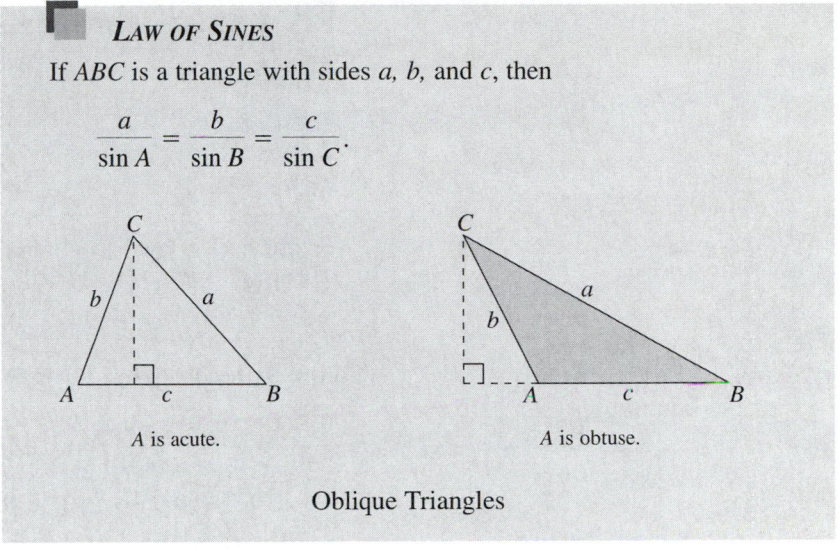

LAW OF SINES

If *ABC* is a triangle with sides *a*, *b*, and *c*, then

$$\frac{a}{\sin A} = \frac{b}{\sin B} = \frac{c}{\sin C}.$$

A is acute.   *A* is obtuse.

Oblique Triangles

Proof

Let *h* be the altitude of either triangle found in the figure showing oblique triangles, above. Then you have

$$\sin A = \frac{h}{b} \qquad \text{or} \qquad h = b \sin A$$

$$\sin B = \frac{h}{a} \qquad \text{or} \qquad h = a \sin B.$$

Equating these two values of *h*, you have

$$a \sin B = b \sin A \qquad \text{or} \qquad \frac{a}{\sin A} = \frac{b}{\sin B}.$$

Note that $\sin A \neq 0$ and $\sin B \neq 0$ because no angle of a triangle can have a measure of 0° or 180°. In a similar manner, by constructing an altitude from vertex *B* to side *AC* (extended), you can show that

$$\frac{a}{\sin A} = \frac{c}{\sin C}.$$

Hence, the Law of Sines is established.   ∎

SECTION 8.4, PAGE 635

> **PROPERTIES OF THE DOT PRODUCT**
>
> Let $\mathbf{u}$, $\mathbf{v}$, and $\mathbf{w}$ be vectors in the plane or in space and let c be a scalar.
>
> 1. $\mathbf{u} \cdot \mathbf{v} = \mathbf{v} \cdot \mathbf{u}$
> 2. $\mathbf{0} \cdot \mathbf{v} = 0$
> 3. $\mathbf{u} \cdot (\mathbf{v} + \mathbf{w}) = \mathbf{u} \cdot \mathbf{v} + \mathbf{u} \cdot \mathbf{w}$
>   4. $\mathbf{v} \cdot \mathbf{v} = \|\mathbf{v}\|^2$
> 5. $c(\mathbf{u} \cdot \mathbf{v}) = c\mathbf{u} \cdot \mathbf{v} = \mathbf{u} \cdot c\mathbf{v}$

Proof

To prove the second property, let $\mathbf{0} = \langle 0, 0 \rangle$ and $\mathbf{v} = \langle v_1, v_2 \rangle$. Then,

$$
\begin{aligned}
\mathbf{0} \cdot \mathbf{v} &= \langle 0, 0 \rangle \cdot \langle v_1, v_2 \rangle \\
&= 0 \cdot v_1 + 0 \cdot v_2 \\
&= 0 + 0 \\
&= 0.
\end{aligned}
$$

To prove the third property, let $\mathbf{u} = \langle u_1, u_2 \rangle$, $\mathbf{v} = \langle v_1, v_2 \rangle$, and $\mathbf{w} = \langle w_1, w_2 \rangle$. Then,

$$
\begin{aligned}
\mathbf{u} \cdot (\mathbf{v} + \mathbf{w}) &= \langle u_1, u_2 \rangle \cdot (\langle v_1, v_2 \rangle + \langle w_1, w_2 \rangle) \\
&= \langle u_1, u_2 \rangle \cdot \langle v_1 + w_1, v_2 + w_2 \rangle \\
&= u_1(v_1 + w_1) + u_2(v_2 + w_2) \\
&= u_1 v_1 + u_2 v_2 + u_1 w_1 + u_2 w_2 \\
&= \langle u_1, u_2 \rangle \cdot \langle v_1, v_2 \rangle + \langle u_1, u_2 \rangle \cdot \langle w_1, w_2 \rangle \\
&= \mathbf{u} \cdot \mathbf{v} + \mathbf{u} \cdot \mathbf{w}.
\end{aligned}
$$

To prove the fifth property, let $\mathbf{u} = \langle u_1, u_2 \rangle$ and $\mathbf{v} = \langle v_1, v_2 \rangle$ and let c be a scalar. Then,

$$
\begin{aligned}
c(\mathbf{u} \cdot \mathbf{v}) &= c(\langle u_1, u_2 \rangle \cdot \langle v_1, v_2 \rangle) \\
&= c(u_1 v_1 + u_2 v_2) \\
&= (cu_1)v_1 + (cu_2)v_2 \\
&= \langle cu_1, cu_2 \rangle \cdot \langle v_1, v_2 \rangle = c\mathbf{u} \cdot \mathbf{v} \\
&= u_1(cv_1) + u_2(cv_2) \\
&= \langle u_1, u_2 \rangle \cdot \langle cv_1, cv_2 \rangle = \mathbf{u} \cdot c\mathbf{v}. \quad \blacksquare\blacksquare
\end{aligned}
$$

SECTION 11.1, PAGE 803

 PROPERTIES OF SUMS

1. $\displaystyle\sum_{i=1}^{n} ca_i = c \sum_{i=1}^{n} a_i,$ c is any constant.

2. $\displaystyle\sum_{i=1}^{n} (a_i + b_i) = \sum_{i=1}^{n} a_i + \sum_{i=1}^{n} b_i$

3. $\displaystyle\sum_{i=1}^{n} (a_i - b_i) = \sum_{i=1}^{n} a_i - \sum_{i=1}^{n} b_i$

PROOF

Each of these properties follows directly from the Associative Property of Addition, the Commutative Property of Addition, and the Distributive Property of multiplication over addition. For example, note the use of the Distributive Property in the proof of Property 1.

$$\sum_{i=1}^{n} ca_i = ca_1 + ca_2 + ca_3 + \cdots + ca_n$$

$$= c(a_1 + a_2 + a_3 + \cdots + a_n) = c \sum_{i=1}^{n} a_i \quad \blacksquare\blacksquare$$

SECTION 11.2, PAGE 813

THE SUM OF A FINITE ARITHMETIC SEQUENCE

The sum of a finite arithmetic sequence with n terms is given by

$$S_n = \frac{n}{2}(a_1 + a_n).$$

PROOF

Begin by generating the terms of the arithmetic sequence in two ways. In the first way, repeatedly add d to the first term to obtain

$$S_n = a_1 + a_2 + a_3 + \cdots + a_{n-2} + a_{n-1} + a_n$$

$$= a_1 + [a_1 + d] + [a_1 + 2d] + \cdots + [a_1 + (n-1)d].$$

In the second way, repeatedly subtract d from the nth term to obtain

$$S_n = a_n + a_{n-1} + a_{n-2} + \cdots + a_3 + a_2 + a_1$$

$$= a_n + [a_n - d] + [a_n - 2d] + \cdots + [a_n - (n-1)d].$$

If you add these two versions of S_n, the multiples of d cancel and you obtain

$$\overbrace{2S_n = (a_1 + a_n) + (a_1 + a_n) + (a_1 + a_n) + \cdots + (a_1 + a_n)}^{n \text{ terms}}$$

$$= n(a_1 + a_n).$$

Thus, you have

$$S_n = \frac{n}{2}(a_1 + a_n). \quad \blacksquare\blacksquare$$

SECTION 11.3, PAGE 822

 THE SUM OF A FINITE GEOMETRIC SEQUENCE

The sum of the geometric sequence

$$a_1, \ a_1r, \ a_1r^2, \ a_1r^3, \ a_1r^4, \ \ldots, a_1r^{n-1}$$

with common ratio $r \neq 1$ is given by

$$S_n = a_1\left(\frac{1 - r^n}{1 - r}\right).$$

PROOF

Begin by writing out the nth partial sum.

$$S_n = a_1 + a_1r + a_1r^2 + \cdots + a_1r^{n-2} + a_1r^{n-1}$$

Multiplication by r yields

$$rS_n = a_1r + a_1r^2 + a_1r^3 + \cdots + a_1r^{n-1} + a_1r^n.$$

Subtracting the second equation from the first yields

$$S_n - rS_n = a_1 - a_1r^n.$$

Therefore, $S_n(1 - r) = a_1(1 - r^n)$, and, because $r \neq 1$, you have

$$S_n = a_1\left(\frac{1 - r^n}{1 - r}\right). \quad \blacksquare\blacksquare$$

SECTION 11.5, PAGE 840

THE BINOMIAL THEOREM

In the expansion of $(x + y)^n$

$$(x + y)^n = x^n + nx^{n-1}y + \cdots + {}_nC_r\, x^{n-r}y^r + \cdots + nxy^{n-1} + y^n$$

the coefficient of $x^{n-r}y^r$ is given by

$$_nC_r = \frac{n!}{(n-r)!r!}.$$

PROOF

The Binomial Theorem can be proved quite nicely using mathematical induction. The steps are straightforward but look a little messy, so we will present only an outline of the proof.

1. If $n = 1$, you have

$$(x + y)^1 = x^1 + y^1 = {}_1C_0\, x + {}_1C_1\, y$$

and the formula is valid.

2. Assuming that the formula is true for $n = k$, the coefficient of $x^{k-r}y^r$ is given by

$$_kC_r = \frac{k!}{(k-r)!r!} = \frac{k(k-1)(k-2)\cdots(k-r+1)}{r!}.$$

To show that the formula is true for $n = k + 1$, look at the coefficient of $x^{k+1-r}y^r$ in the expansion of

$$(x + y)^{k+1} = (x + y)^k(x + y).$$

From the right-hand side, you can determine that the term involving $x^{k+1-r}y^r$ is the sum of two products.

$$({}_kC_r x^{k-r}y^r)(x) + ({}_kC_{r-1} x^{k+1-r}y^{r-1})(y)$$

$$= \left[\frac{k!}{(k-r)!r!} + \frac{k!}{(k-r+1)!(r-1)!}\right]x^{k+1-r}y^r$$

$$= \left[\frac{(k+1-r)k!}{(k+1-r)!r!} + \frac{k!r}{(k+1-r)!r!}\right]x^{k+1-r}y^r$$

$$= \left[\frac{k!(k+1-r+r)}{(k+1-r)!r!}\right]x^{k+1-r}y^r$$

$$= \left[\frac{(k+1)!}{(k+1-r)!r!}\right]x^{k+1-r}y^r$$

$$= {}_{k+1}C_r\, x^{k+1-r}y^r$$

Thus, by mathematical induction, the Binomial Theorem is valid for all positive integers n. ■■

Appendix C

Programs

Evaluating an Algebraic Expression (Section P.3)

This program, shown in the Technology note on page 30, can be used to evaluate an algebraic expression in one variable at several values of the variable.

Note: On the *TI-83* and the *TI-82*, the "Lbl" and "Goto" commands may be entered through the "CTL" menu accessed by pressing the PRGM key. The "Disp" and "Input" commands may be entered through the "I/O" menu accessed by pressing the PRGM key. On the *TI-83*, the symbol "Y1" may be entered through the "Y-VARS" and "Function" menus accessed by pressing the VARS key. On the *TI-82,* the symbol "Y1" may be entered through the "Function" menu accessed by pressing the Y-VARS key. Keystroke sequences required for similar commands on other calculators will vary. Consult the user manual for your calculator.

TI-80
PROGRAM:EVALUAT
:LBL A
:INPUT "ENTER X",X
:DISP Y1
:GOTO A

To use this program, enter an expression in Y1. Expressions may also be evaluated directly on the *TI-80*'s home screen.

TI-81
Prgm1:EVALUATE
:Lbl 1
:Disp "ENTER X"
:Input X
:Disp Y1
:Goto 1

To use this program, enter an expression in Y1. Expressions may also be evaluated directly on the *TI-81*'s home screen.

TI-83
TI-82
PROGRAM:EVALUATE
:Lbl A
:Input "ENTER X",X
:Disp Y1
:Goto A

To use this program, enter an expression in Y1. Expressions may also be evaluated directly on the *TI-83* or *TI-82* home screen.

TI-85
PROGRAM:EVALUATE
:Lbl A
:Input "Enter x",x
:Disp y1
:Goto A

To use this program, enter an expression in y1.

TI-92

evaluate( )
Prgm
Lbl one
Input "enter x",x
Disp y1(x)
Goto one
EndPrgm

To use this program, enter an expression in y1. Expressions may also be evaluated on the *TI-92*'s home screen.

Casio fx-7700G

EVALUATE
Lbl 1
"X="?→X
"F(X)=":f1 ◢
Goto 1

To use this program, enter an expression in f1.

Casio fx-7700GE
Casio fx-9700GE
Casio CFX-9800G

EVALUATE ↵
Lbl 1 ↵
"X="?→X ↵
"F(X)=":f1 ◢
Goto 1

To use this program, enter an expression in f1.

Sharp EL 9200C
Sharp EL 9300C

evaluate
——————————REAL
Goto top
Label equation
Y=f(X)
Return
Label top
Input X
Gosub equation
Print Y
Goto top
End

To use this program, replace f(X) with your expression in X.

HP-38G

Use the Solve aplet to evaluate an expression.
1. Press ⎡LIB⎤. Highlight the Solve aplet. Press {{START}}.
2. Set your expression equal to *y*, enter the equation (*y = your expression*) in E1 and press {{OK}}. The equation should be checked.
3. Press ⎡NUM⎤.
4. Highlight the *x*-variable field. Enter a value for *x* and press {{OK}}.
5. Highlight the *y*-variable field and press {{SOLVE}}. The value of the expression will appear in the *y*-variable field.
6. Repeat steps 4 and 5 to evaluate the expression for other values of *x*.

Reflections and Shifts Program (Section 2.4)

This program, referenced in the Technology note on page 225, will sketch a graph of the function $y = R(x + H)^2 + V$, where $R = \pm 1$, H is an integer between -6 and 6, and V is an integer between -3 and 3. This program gives you practice working with reflections, horizontal shifts, and vertical shifts.

Note: On the *TI-83* and the *TI-82*, the "int" and "rand" commands may be entered through the "NUM" and "PRB" menus, respectively, accessed by pressing the MATH key. The "=" and "<" symbols may be entered through the "TEST" menu accessed by pressing the TEST key. Other commands, such as "If," "Then," "Else," and "End," may be entered through the "CTL" menu accessed by pressing the PRGM key. The commands "Xmin," "Xmax," "Xscl," "Ymin," "Ymax," and "Yscl" may be entered through the "Window" menu accessed by pressing the VARS key. The commands "DispGraph" and "Pause" may be entered through the "I/O" and "CTL" menus, respectively, accessed by pressing the PRGM key. For additional keystroke instructions, see previous programs in this appendix. Keystroke sequences for similar commands on other calculators will vary. Consult the user manual for your calculator.

TI-80

```
PROGRAM:PARABOL
:-6+INT (12RAND)→H
:-3+INT (6RAND)→V
:RAND→R
:IF R <.5
:THEN
:-1→R
:ELSE
:1→R
:END
:"R(X+H)²+V"→Y1
:-9→XMIN
:9→XMAX
:1→XSCL
:-6→YMIN
:6→YMAX
:1→YSCL
:DISPGRAPH
:PAUSE
:DISP "Y=R(X+H)²+V²"
:DISP "R=",R
:DISP "H=",H
:DISP "V=",V
:PAUSE
```

Press ENTER after the graph to display the values of the integers.

TI-81

```
Prgm2:PARABOLA
:Rand→H
:-6+Int (12H)→H
:Rand→V
:-3+Int (6V)→V
:Rand→R
:If R <.5
:-1→R
:If R >.49
:1→R
:"R(X+H)²+V"→Y1
:-9→Xmin
:9→Xmax
:1→Xscl
:-6→Ymin
:6→Ymax
:1→Yscl
:DispGraph
:Pause
:Disp "Y=R(X+H)²+V"
:Disp "R="
:Disp R
:Disp "H="
:Disp H
:Disp "V="
:Disp V
:End
```

Press ENTER after the graph to display the values of the integers.

TI-83
TI-82
PROGRAM:PARABOLA
:-6+int (12rand)→H
:-3+int (6rand)→V
:rand→R
:If R < .5
:Then
:-1→R
:Else
:1→R
:End
:"R(X+H)^2+V"→Y1
:-9→Xmin
:9→Xmax
:1→Xscl
:-6→Ymin
:6→Ymax
:1→Yscl
:DispGraph
:Pause
:Disp "Y=R(X+H)^2+V"
:Disp "R=",R
:Disp "H=",H
:Disp "V=",V
:Pause

Press ENTER after the graph to display the values of the integers.

TI-85
PROGRAM:PARABOLA
:rand→H
:-6+int (12H)→H
:rand→V
:-3+int (6V)→V
:rand→R
:If R < .5
:-1→R
:If R > .49
:1→R
:y1=R(x+H)^2+V
:-9→xMin
:9→xMax
:1→xScl
:-6→yMin
:6→yMax
:1→yScl
:DispG
:Pause
:Disp "Y=R(X+H)^2+V"
:Disp "R=",R
:Disp "H=",H
:Disp "V=",V
:Pause

Press ENTER after the graph to display the values of the integers.

TI-92

```
Parabola( )
Prgm
ClrHome
ClrIO
setMode("Split Screen",
   "Left-Right")
setMode("Split 1 App","Home")
setMode("Split 2 App","Graph")
-6+int (12rand( ))→h
-3+int (6rand( ))→v
rand( )→r
If r < .5 Then
   -1→r
      Else
   1→r
EndIf
r*(x+h)^2+v→y1(x)
-9→xmin
9→xmax
1→xscl
-6→ymin
6→ymax
1→yscl
DispG
Disp "y1(x)=r(x+h)^2+v"
Output 20,1, "r=":Output 20,11,r
Output 40,1, "h=":Output 40,11,h
Output 60,1, "v=":Output 60,11,v
Pause
setMode("Split Screen","Full")
EndPrgm
```

Casio fx-7700G

```
PARABOLA
-6+INT (12Ran#)→H
-3+INT (6Ran#)→V
-1→R:Ran#<0.5 ⇒1→R
Range -9,9,1,-6,6,1
Graph Y=R(X+H)²+V  ◢
"Y=R(X+H)²+V"
"R=":R  ◢
"H=":H  ◢
"V=":V
```

Press EXE after the graph to display the values of the integers.

Casio fx-7700GE
Casio fx-9700GE
Casio CFX-9800G

```
PARABOLA↵
-6+Int (12Ran#)→H↵
-3+Int (6Ran#)→V↵
Ran#→R↵
R< .5⇒-1→R↵
R≥ .5⇒1→R↵
Range -9,9,1,-6,6,1↵
Graph Y=R(X+H)²+V  ◢
"Y=R(X+H)²+V"↵
"R=":R  ◢
"H=":H  ◢
"V=":V
```

Press EXE after the graph to display the values of the integers.

Sharp EL-9200C
Sharp EL-9300C

```
parabola
──────────REAL
h=int (random*12) -6
v=int (random*6) -3
s=(random*2) -1
r=s/abs s
Range -9,9,1,-6,6,1
Graph r(X+h)²+v
Wait
Print "y=r(X+h)²+v
Print r
Print h
Print v
End
```

Press ENTER after the graph to display the values of the integers.

HP-38G

PARABOLA

PARABOLA PROGRAM
```
-6+INT(12RANDOM)▶H:
-3+INT(6RANDOM)▶V:
RANDOM ▶R:
IF R>.5
    THEN -1▶R:
    ELSE 1▶R:
END:
'R*(X+H)²+V'▶F1(X):
CHECK 1:
```

PARANS PROGRAM
```
ERASE:
DISP 2;"Y=R(X+H)²+V":
DISP 3;"R=":R:
DISP 4;"H=":H:
DISP 5;"V=":V:
FREEZE:
```

PARABOLA.SV PROGRAM
```
SETVIEWS "RUN
PARABOLA";PARABOLA;1;
"ANSWER";PARANS;1;
" ";PARABOLA.SV;0:
```

1. Press ⎡LIB⎤. Highlight the Function aplet. Press {{SAVE}}. Enter the name PARABOLA for the new aplet and press {{OK}}.
2. Press ■ [SETUP-PLOT] and set XRNG: from −12 to 12, YRNG: from −6 to 6, and XTICK: and YTICK: to 1.
3. Enter the 3 programs PARABOLA, PARANS, PARABOLA.SV.
4. Run the program PARABOLA.SV.
5. Enter the PARABOLA aplet.
6. Press ■ [VIEWS]. Highlight RUN PARABOLA and press {{OK}}.
7. After viewing the graph press ■ [VIEWS]. Highlight ANSWER and press {{OK}} to see the values of the integers.
8. Press {{OK}} to return to the graph.
9. Repeat steps 6, 7, and 8 for a new parabola.

Graph Reflection Program (Section 2.5)

This program, shown in the Technology note on page 241, will graph a function f and its reflection in the line $y = x$.

Note: On the *TI-83* and the *TI-82*, the "While" command may be entered through the "CTL" menu accessed by pressing the PRGM key. The "Pt-On(" command may be entered through the "POINTS" menu accessed by pressing the DRAW key. For additional keystroke instructions, see previous programs in this appendix. Keystroke sequences required for similar commands on other calculators will vary. Consult the user manual for your calculator.

TI-80

```
PROGRAM:REFLECT
:47XMIN/63→YMIN
:47XMAX/63→YMAX
:XSCL→YSCL
:"X"→Y2
:DISPGRAPH
:(XMAX−XMIN)/62→I
:XMIN→X
:LBL A
:PT-ON(Y1,X)
:X+I→X
:If X>XMAX
:STOP
:GOTO A
```

To use this program, enter the function in Y1 and set a viewing rectangle.

TI-81

```
Prgm3:REFLECT
:2Xmin/3→Ymin
:2Xmax/3→Ymax
:Xscl→Yscl
:"X"→Y2
:DispGraph
:(Xmax−Xmin)/95→I
:Xmin→X
:Lbl 1
:Pt-On(Y1,X)
:X+I→X
:If X>Xmax
:End
:Goto 1
```

To use this program, enter the function in Y1 and set a viewing rectangle.

TI-83
TI-82

```
PROGRAM:REFLECT
:63Xmin/95→Ymin
:63Xmax/95→Ymax
:Xscl→Yscl
:"X"→Y₂
:DispGraph
:(Xmax−Xmin)/94→I
:Xmin→X
:While X≤Xmax
:Pt-On(Y₁,X)
:X+I→X
:End
```

To use this program, enter the function in Y1 and set a viewing rectangle.

TI-85

```
PROGRAM:REFLECT
:63*xMin/127→yMin
:63*xMax/127→yMax
:xScl→yScl
:y2=x
:DispG
:(xMax−xMin)/126→I
:xMin→x
:Lbl A
:PtOn(y1,x)
:x+I→x
:If x>xMax
:Stop
:Goto A
```

To use this program, enter the function in y1 and set a viewing rectangle.

TI-92

```
Prgm
103xmin/239→ymin
103xmax/239→ymax
xscl→yscl
x→y2(x)
DispG
(xmax−xmin)/238→n
xmin→x
While x<xmax
    PtOn y1(x),x
    x+n→x
EndWhile
EndPrgm
```

To use this program, enter a function in y1 and set an appropriate viewing window.

Casio fx-7700G

```
REFLECTION
"GRAPH -A TO A"
"A="?→A
Range -A,A,1,-2A÷3,2A÷3,1
Graph Y=f₁
-A→B
Lbl 1
B→X
Plot f₁,B
B+A÷32→B
B≤A⇒Goto1 :Graph Y=X
```

To use this program, enter the function in f1.

Casio fx-7700GE

REFLECTION
"GRAPH -A TO A"↵
"A="?→A↵
Range -A,A,1,-2A÷3,2A÷3,1↵
Graph Y=f₁↵
-A→B↵
Lbl 1↵
B→X↵
Plot f₁,B↵
B+A÷32→B↵
B≤A⇒Goto1:Graph Y=X

To use this program, enter the function in f₁.

Casio fx-9700GE

REFLECTION↵
63Xmin÷127→A↵
63Xmax÷127→B↵
Xscl→C↵
Range , , , A, B, C↵
(Xmax−Xmin)÷126→I↵
Xmax→M↵
Xmin→D↵
Graph Y=f₁↵
Lbl 1↵
D→X↵
Plot f₁,D↵
D+I→D↵
D≤M⇒Goto 1:Graph Y=X

To use this program, enter a function in f₁ and set a viewing rectangle.

Casio CFX-9800G

REFLECTION↵
63Xmin÷95→A↵
63Xmax÷95→B↵
Xscl→C↵
Range , , , A, B, C↵
(Xmax−Xmin)÷94→I↵
Xmax→M↵
Xmin→D↵
Graph Y=f₁↵

Casio CFX-9800G

(Continued)

Lbl 1↵
D→X↵
Plot f₁,D↵
D+I→D↵
D≤M⇒Goto 1:Graph Y=X

To use this program, enter a function in f₁ and set a viewing rectangle.

Sharp EL 9200C
Sharp EL 9300C

reflection
————————REAL
Goto top
Label equation
Y=f(X)
Return
Label rng
xmin=-10
xmax=10
xstp=(xmax−xmin)/10
ymin=2xmin/3
ymax=2xmax/3
ystp=xstp
Range xmin,xmax,xstp,ymin,
    ymax,ystp
Return
Label top
Gosub rng
Graph X
step=(xmax−xmin)/(94*2)
X=xmin
Label 1
Gosub equation
Plot X,Y
Plot Y, X
X=X+step
If X<=xmax Goto 1
End

To use this program, replace f(X) with your expression in X.

Graphing a Sine Function (Section 6.4)

The program, shown in the Group Activity on page 499, will simultaneously draw a unit circle and the corresponding points on the sine curve. After the circle and sine curve are drawn, you can connect the points on the unit circle with their corresponding points on the sine curve by pressing ENTER or EXE .

TI-80

```
PROGRAM:SINESHO
:RADIAN
:CLRDRAW:FNOFF
:PARAM:SIMUL
:-2.25→XMIN
:π/2→XMAX
:3→XSCL
:-1.5→YMIN
:1.5→YMAX
:1→YSCL
:0→TMIN
:6.3→TMAX
:.15→TSTEP
:"-1.25+COS T"→X1T
:"SIN T"→Y1T
:"T/4"→X2T
:"SIN T"→Y2T
:DISPGRAPH
:FOR(N,1,12)
:Nπ/6.5→T
:"-1.25+COS T"→A
:SIN T→B
:T/4→C
:LINE(A,B,C,B)
:PAUSE
:END
:PAUSE:FUNC
:SEQUENTIAL:DISP
```

TI-81

```
PrgmA:SINESHOW
:Rad
:ClrDraw
:Param
:Simul
:-2.25→Xmin
:π/2→Xmax
:3→Xscl
:-1.19→Ymin
:1.19→Ymax
:1→Yscl
:0→Tmin
:6.3→Tmax
:.15→Tstep
:"-1.25+cos T"→X1T
:"sin T"→Y1T
:"T/4"→X2T
:"sin T"→Y2T
:DispGraph
:1→N
:Lbl 1
:IS>(N,12)
:Goto 2
:Pause
:Function
:Sequence
:Disp " "
:End
:Lbl 2
:Nπ/6.5→T
:-1.25+cos T→A
:sin T→B
:T/4→C
:Line(A,B,C,B)
:Pause
:Goto 1
```

TI-83
TI-82

```
PROGRAM:SINESHOW
:Radian
:ClrDraw:FnOff
:Param:Simul
:-2.25→Xmin
:π/2→Xmax
:3→Xscl
:-1.19→Ymin
:1.19→Ymax
:1→Yscl
:0→Tmin
:6.3→Tmax
:.15→Tstep
:"-1.25+cos (T)"→X1T
:"sin (T)"→Y1T
:"T/4"→X2T
:"sin (T)"→Y2T
:DispGraph
:For(N,1,12)
:Nπ/6.5→T
:-1.25+cos (T)→A
:sin(T)→B
:T/4→C
:Line(A,B,C,B)
:Pause
:End
:Pause :Func
:Sequential:Disp
```

TI-85

```
PROGRAM:SINESHOW
:Radian
:ClDrw:FnOff
:Param:SimulG
:-2.25→xMin
:π/2→xMax
:3→xScl
:-1.1→yMin
:1.1→yMax
:1→yScl
:0→tMin
:6.3→tMax
:.15→tStep
:xt1=-1.25+cos t
:yt1=sin t
:xt2=t/4
:yt2=sin t
:For(N,1,12)
:N*π/6.5→t
:-1.25+cos t→A
:sin t→B
:t/4→C
:Line(A,B,C,B)
:Pause
:End
:Pause :Func
:SeqG:Disp
```

TI-92

```
sineshow()
Prgm
Disp
ClrDraw:FnOff
setMode("Graph", "Parametric")
setGraph("Graph Order",
    "Simul")
-2.9→xmin
3π/4→xmax
3→xscl
-1.1→ymin
1.1→ymax
1→yscl
0→tmin
6.3→tmax
.15→tstep
-1.25+cos(t)→xt1(t)
sin(t)→yt1(t)
t/4→xt2(t)
sin(t)→yt2(t)
DispG
For N,1,12
N*π/6.5→t
-1.25+cos(t)→A
sin(t)→B
t/4→C
Line A,B,C,B
Pause
EndFor
Pause
setMode("Graph", "Function")
setGraph("Graph order",
    "Seq")
setMode("Split 1 App",
    "Home")
EndPrgm
```

Casio fx-7700G

```
SINESHOW
Rad
Range -2.25,π÷2,3,-1.19,1.19,
    10,6.3,.15
Graph(X,Y)=(-1.25+cos T,sinT)
Graph(X,Y)=(T÷4,sinT)
0→N
Lbl 1
N+1→N
Nπ÷6.5→T
-1.25+cos T→A
sin T→B
T÷4→C
Plot A,B
Plot C,B
Line◢
N<12⇒Goto 1
```

Press Mode Shift X to change to parametric mode when starting to write this program.

Casio fx-7700GE
Casio fx-9700GE
Casio CFX-9800G

SINESHOW↵
Rad↵
Range -2.25,$\pi\div 2$,3,-1.19,1.19,
    1,0,6.3,.15↵
Graph(X,Y)=(-1.25+cos T,sin T)↵
Graph(X,Y)=(T÷4,sin T)↵
0→N↵
Lbl 1↵
N+1→N↵
N$\pi\div$6.5→T↵
-1.25+cosT→A↵
sin T→B↵
T÷4→C↵
Plot A,B↵
Plot C,B↵
Line ◢
N<12⇒Goto 1↵
Cls

When starting to write this program press SHIFT SET UP and select PRM or PARM for the GRAPH TYPE to change to parametric mode.

Sharp EL-9200C
Sharp EL-9200C

sineshow
————————REAL
m=$\sin^{-1}$ 1/(π/2)
Range -2.25,π/2,3,-1.19,1.19,1
step=π/15
θ=0
xco=-.25
xso=0
yo=0
Label 1
θ=θ+step
xc=cos(mθ)−1.25
xs=θ/4
y=sin (mθ)
Line xco,yo,xc,y
Line xso,yo,xs,y
xco=xc
xso=xs
yo=y
If θ<(2π) Goto 1
step=π/6
θ=0
Label 2
θ=θ+step
xc=cos (mθ)−1.25
xs=θ/4
y=sin (mθ)
Line xc,y,xs,y
Wait
If θ<2π Goto 2
End

HP-38G Programs

SINESHOW PROGRAM
ASIN(1)/(π/2)▶M:
0▶T:
-.25▶A:
0▶B:
0▶C:
LINE -3;0;π/2;0:
LINE 0;-1.1;0;1.1:
FOR T=0 TO 31π/15
   STEP π/15;
   RUN "DRAW.SINE":
END:
0▶T:
FOR T=0 TO 2π
   STEP π/6;
   RUN "DRAW.LINE":
END

DRAW.SINE PROGRAM
COS(MT)$-$1.25▶D:
T/4▶E:
SIN(MT)▶F:
LINE A;C;D;F:
LINE B;C;E;F:
D▶A:
E▶B:
F▶C:

DRAW.LINE PROGRAM
COS(MT)$-$1.25▶D:
T/4▶E:
SIN(MT)▶F:
LINE D;F;E;F:
FREEZE

1. Enter the 3 programs
   SINESHOW, DRAW.SINE, and
   DRAW.LINE.
2. Set the plot range in the
   Function Aplet to $-3 \leq x \leq \pi/2$
   and $-1.1 \leq y \leq 1.1$. Set the angle
   measure to radians.
3. Run the SINESHOW program.

Finding the Angle Between Two Vectors (Section 8.4)

The program, shown in the Technology note on page 636, will sketch two vectors and calculate the measure of the angle between the vectors. Be sure to set an appropriate viewing rectangle.

TI-80

```
:PROGRAM:VECANGL
:CLRHOME
:DEGREE
:DISP "ENTER (A,B)"
:INPUT "ENTER A",A
:INPUT "ENTER B",B
:CLRHOME
:DISP "ENTER (C,D)"
:INPUT "ENTER C",C
:INPUT "ENTER D",D
:LINE(0,0,A,B)
:LINE(0,0,C,D)
:PAUSE
:AC+BD→E
:√(A²+B²)→U
:√(C²+D²)→V
:COS⁻¹(E/(UV))→θ
:DISP "θ=", θ
:CLRDRAW
```

TI-81

```
:PrgmB:VECANGL
:ClrHome
:Deg
:Disp "ENTER (A,B)"
:Disp "ENTER A"
:Input A
:Disp "ENTER B"
:Input B
:ClrHome
:Disp "ENTER (C,D)"
:Disp "ENTER C"
:Input C
:Disp "ENTER D"
:Input D
:Line(0,0,A,B)
:Line(0,0,C,D)
:Pause
:AC+BD→E
:√(A²+B²)→U
:√(C²+D²)→V
:cos⁻¹(E/(UV))→θ
:Disp "θ="
:Disp θ
:ClrDraw
:End
```

TI-83
TI-82

```
:PROGRAM:VECANGL
:ClrHome
:Degree
:Disp "ENTER (A,B)"
:Input "ENTER A",A
:Input "ENTER B",B
:ClrHome
:Disp "ENTER (C,D)"
:Input "ENTER C",C
:Input "ENTER D",D
:Line(0,0,A,B)
:Line(0,0,C,D)
:Pause
:AC+BD→E
:√ (A²+B²)→U
:√ (C²+D²)→V
:cos⁻¹(E/(UV))→θ
:ClrDraw:ClrHome
:Disp "θ=",θ
:Stop
```

TI-85

```
:PROGRAM:VECANGL
:ClLCD
:Radian
:Disp "enter (A,B)"
:Input "enter A",A
:Input "enter B",B
:ClLCD
:Disp "enter (C,D)"
:Input "enter C",C
:Input "enter D",D
:Line(0,0,A,B)
:Line(0,0,C,D)
:Pause
:A*C+B*D→E
:√ (A²+B²)→U
:√ (C²+D²)→V
:cos⁻¹(E/(U*V))→T
:T*180/π→T
:Disp "T=",T
:ClDrw
```

TI-92

vecangl( )
Prgm
FnOff
ClrHome:ClrDraw
SetMode("Split Screen",
    "Left-Right")
SetMode("Split 1 App", "Home")
SetMode("Split 2 App", "Graph")
SetMode("Exact/Approx",
    "Approximate")
ClrIO
Disp "ENTER (A,B)"
Input "ENTER A", A
Input "ENTER B", B
Line(0,0,A,B)
Pause
ClrIO
Disp "ENTER (C,D)"
Input "ENTER C",C
Input "ENTER D",D
Line(0,0,C,D)
Pause
ClrIO
A*C+B*D→E
$\sqrt{\ }$ ((A^2+B^2))→U
$\sqrt{\ }$ (C^2+D^2)→V
cos^{-1} (E/(U*V))→θ
Disp "θ=",θ
Pause
SetMode("Exact/Approx", "Auto")
SetMode("Split Screen", "Full")
SetMode("Split 1 App", "Home")
Stop
EndPrgm

Casio fx-7700G

VECANGL
Cls
Deg
"ENTER (A,B)"
"A="?→A
"B="?→B
"ENTER (C,D)"
"C="?→C
"D="?→D
Plot 0,0
Plot A,B
Line
Plot 0,0
Plot C,D
Line ◢
AC+BD→E
$\sqrt{\ }$ (A^2+B^2)→U
$\sqrt{\ }$ (C^2+D^2)→V
cos^{-1}(E÷UV)→θ
"θ="
θ

Casio fx-7700GE
Casio fx-9700GE
Casio CFX-9800G

VECANGL↵
Cls↵
Deg↵
"ENTER (A,B)"↵
"A="?→A↵
"B="?→B↵
"ENTER (C,D)"↵
"C="?→C↵
"D="?→D↵
Plot 0,0↵
Plot A,B↵
Line↵
Plot 0,0↵
Plot C,D↵
Line◢
AC+BD→E↵
$\sqrt{}$ (A^2+B^2)→U↵
$\sqrt{}$ (C^2+D^2)→V↵
cos^{-1}(E÷UV)→θ↵
"θ="↵
θ

Sharp EL-9200C
Sharp EL-9300C

vecangl
─────────REAL
ClrG
ClrT
Print"enter (a,b)"
Input a
Input b
ClrT
Print"enter (c,d)"
Input c
Input d
Line 0,0,a,b
Line 0,0,c,d
Wait
e=a*c+b*d
u=$\sqrt{}$ (a^2+b^2)
v=$\sqrt{}$ (c^2+d^2)
t=cos^{-1}(e/(u*v))
Print t
End

Set the calculator to degree mode before running the program.

HP-38G

VECANGL PROGRAM
INPUT A; "ENTER (A,B)";
    "ENTER A";;1:
INPUT B; "ENTER (A,B)";
    "ENTER B";;1:
INPUT C; "ENTER (C,D)";
    "ENTER C";;1:
INPUT D; "ENTER (C,D)";
    "ENTER D";;1:
ERASE:
LINE −10;0;10;0:
LINE 0;−10;0;10:
LINE 0;0;A;B:
LINE 0;0;C;D:
FREEZE:
AC+BD▶ E
$\sqrt{\;}$ (A^2+B^2)▶U:
$\sqrt{\;}$ (C^2+D^2)▶V:
ACOS(E/(UV))▶T:
ERASE:
DISP 3; "ANGLE= "T:
FREEZE

The Function Aplet should have a plot range of $-10 \le x \le 10$ and $-10 \le y \le 10$. Set the MODE to degrees before running the program.

Adding Vectors Graphically (Chapter 8 Project)

The program, shown in the Chapter 8 Project on page 660, will sketch two vectors in standard position. Using the parallelogram law for the vector addition, the program also sketches the vector sum. Be sure to set an appropriate viewing rectangle.

TI-80

```
:PROGRAM:ADDVECT
:CLRDRAW
:DISP "ENTER(A,B)"
:INPUT "ENTER A",A
:INPUT "ENTER B",B
:DISP "ENTER (C,D)"
:INPUT "ENTER C",C
:INPUT "ENTER D",D
:LINE(0,0,A,B)
:LINE(0,0,C,D)
:A+C→E
:B+D→F
:LINE(0,0,E,F)
:LINE(A,B,E,F)
:LINE(C,D,E,F)
:PAUSE
```

TI-81

```
:PrgmC:ADDVECT
:ClrDraw
:Disp "ENTER(A,B)"
:Disp "ENTER A"
:Input A
:Disp "ENTER B"
:Input B
:Disp "ENTER (C,D)"
:Disp "ENTER C"
:Input C
:Disp "ENTER D"
:Input D
:Line(0,0,A,B)
:Line(0,0,C,D)
:A+C→E
:B+D→F
:Line(0,0,E,F)
:Line(A,B,E,F)
:Line(C,D,E,F)
:Pause
:End
```

TI-83
TI-82

```
:PROGRAM:ADDVECT
:ClrDraw
:Input "ENTER A",A
:Input "ENTER B",B
:Input "ENTER C",C
:Input "ENTER D",D
:Line(0,0,A,B)
:Line(0,0,C,D)
:A+C→E
:B+D→F
:Line(0,0,E,F)
:Line(A,B,E,F)
:Line(C,D,E,F)
:Pause
:Stop
```

TI-85

```
:PROGRAM:ADDVECT
:ClrDraw
:Input "enter A",A
:Input "enter B",B
:Input "enter C",C
:Input "enter D",D
:Line(0,0,A,B)
:Line(0,0,C,D)
:A+C→E
:B+D→F
:Line(0,0,E,F)
:Line(A,B,E,F)
:Line(C,D,E,F)
:Pause
:Disp
```

TI-92

```
addvect( )
Prgm
ClrIO
Input "ENTER a ",a
Input "ENTER b ",b
Input "ENTER c ",c
Input "ENTER d ",d
ClrDraw
Line(0,0,a,b)
Line(0,0,c,d)
a+c→e
b+d→f
Line 0,0,e,f
Line a,b,e,f
Line c,d,e,f
Pause
setMode("Split 1 App","Home")
Stop
EndPrgm
```

Casio fx-7700G

```
ADDVECT
Cls
"A="?→A
"B="?→B
"C="?→C
"D="?→D
Plot 0,0
Plot A,B
Line
Plot 0,0
Plot C,D
Line ◢
A+C→E
B+D→F
Plot 0,0
Plot E,F
Line
Plot A,B
Plot E,F
Line
Plot C,D
Plot E,F
Line ◢
```

Casio fx-7700GE
Casio fx-9700GE
Casio CFX-9800G

```
ADDVECT↵
Cls↵
"A="?→A↵
"B="?→B↵
"C="?→C↵
"D="?→D↵
Plot 0,0↵
Plot A,B↵
Line↵
Plot 0,0↵
Plot C,D↵
Line◢
A+C→E↵
B+D→F↵
Plot 0,0↵
Plot E,F↵
Line↵
Plot A,B↵
Plot E,F↵
Line↵
Plot C,D↵
Plot E,F↵
Line◢
```

Sharp El-9200C
Sharp EL-9300C

```
addvect
——————REAL
ClrG
Input a
Input b
Input c
Input d
Line 0,0,a,b
Line 0,0,c,d
e=a+c
f=b+d
Line 0,0,e,f
Line a,b,e,f
Line c,d,e,f
Wait
End
```

HP-38G PROGRAMS

```
ADDVECT PROGRAM
INPUT A;; "ENTER A";;1:
INPUT B;; "ENTER B";;1:
INPUT C;; "ENTER C";;1:
INPUT D;; "ENTER D";;1:
ERASE:
LINE−10;0;10;0:
LINE 0;−10;0;10:
LINE 0;0;A;B:
LINE 0;0;C;D:
FREEZE:
A+C▶ E
B+D▶ F
LINE 0;0;E;F:
LINE A;B;E;F:
LINE C;D;E;F:
FREEZE
```

The Function Aplet should have a plot range of $-10 \le x \le 10$ and $-10 \le y \le 10$.

Systems of Linear Equations (Section 9.2)

This program, shown in the Technology note on page 680, will display the solution of a system of two linear equations in two variables of the form

$$ax + by = c$$
$$dx + ey = f$$

if a unique solution exists.

Note: For help with *TI-83* or *TI-82* keystrokes, see previous programs in this appendix.

TI-80

```
PROGRAM:SOLVE
:DISP "AX+BY=C"
:INPUT "ENTER A",A
:INPUT "ENTER B",B
:INPUT "ENTER C",C
:DISP "DX+EY=F"
:INPUT "ENTER D",D
:INPUT "ENTER E",E
:INPUT "ENTER F",F
:IF AE−DB=0
:THEN
:DISP "NO UNIQUE"
:DISP "SOLUTION"
:ELSE
:(CE−BF)/(AE−DB)→X
:(AF−CD)/(AE−DB)→Y
:DISP X
:DISP Y
:END
```

TI-81

```
Prgm4:SOLVE
:Disp "AX+BY=C"
:Input A
:Input B
:Input C
:Disp "DX+EY=F"
:Input D
:Input E
:Input F
:If AE−DB=0
:Goto 1
:(CE−BF)/(AE−DB)→X
:(AF−CD)/(AE−DB)→Y
:Disp X
:Disp Y
:End
:Lbl 1
:Disp "NO UNIQUE SOLUTION"
:End
```

TI-83
TI-82

```
PROGRAM:SOLVE
:Disp "AX+BY=C"
:Prompt A
:Prompt B
:Prompt C
:Disp "DX+EY=F"
:Prompt D
:Prompt E
:Prompt F
:If AE−DB=0
:Then
:Disp "NO UNIQUE"
:Disp "SOLUTION"
:Else
:(CE−BF)/(AE−DB)→X
:(AF−CD)/(AE−DB)→Y
:Disp X
:Disp Y
:End
```

TI-85

```
PROGRAM:SOLVE
:Disp "AX+BY=C"
:Input "ENTER A",A
:Input "ENTER B",B
:Input "ENTER C",C
:Disp "DX+EY=F"
:Input "ENTER D",D
:Input "ENTER E",E
:Input "ENTER F",F
:If A*E−D*B==0
:Goto A
:(C*E−B*F)/(A*E−D*B)→X
:(A*F−C*D)/(A*E−D*B)→Y
:Disp X
:Disp Y
:Stop
:Lbl A
:Disp "NO UNIQUE SOLUTION"
```

TI-92

```
Solve( )
Prgm
ClrIO
Disp "Ax+By=C"
Input "Enter A.",a
Input "Enter B.",b
Input "Enter C.",c
ClrIO
Disp "Dx+Ey=F"
Input "Enter D.",d
Input "Enter E.",e
Input "Enter F.",f
If a*e−d*b=0 Then
     Disp "No unique solution"
  Else
     (c*e−b*f)/(a*e−d*b)→x
     (a*f−c*d)/(a*e−d*b)→y
     Disp x
     Disp y
EndIf
EndPrgm
```

Casio fx-7700G

```
SOLVE
"AX+BY=C"
"A="?→A
"B="?→B
"C="?→C
"DX+EY=F"
"D="?→D
"E="?→E
"F="?→F
AE−DB=0⇒Goto 1
"X=":(CE−BF)÷(AE−DB) ◢
"Y=":(AF−CD)÷(AE−DB)
Goto 2
Lbl 1
"NO UNIQUE SOLUTION"
Lbl 2
```

Casio fx-7700GE
Casio fx-9700GE
Casio CFX-9800G

SOLVE↵
"AX+BY=C"↵
"A":?→A↵
"B":?→B↵
"C":?→C↵
"DX+EY=F"↵
"D":?→D↵
"E":?→E↵
"F":?→F↵
AE−DB=0⇒Goto 1↵
"X=":(CE−BF)÷(AE−DB)◢
"Y=":(AF−CD)÷(AE−DB)↵
Goto 2↵
Lbl 1↵
"NO UNIQUE SOLUTION"↵
Lbl 2

Solutions to systems of linear equations are also available directly from the Casio calculator's EQUATION MENU.

Sharp EL 9200C
Sharp EL 9300C

solve
————REAL
Print "AX+BY=C"
Input A
Input B
Input C
Print "DX+EY=F"
Input D
Input E
Input F
If A*E−D*B=0 Goto 1
X=(C*E−B*F)/(A*E−D*B)
Y=(A*F−C*D)/(A*E−D*B)
Print X
Print Y
End
Label 1
Print "no unique solution"
End

Equations must be entered in the form: AX + BY = C; DX + EY = F. Uppercase letters are used so that the values can be accessed in the calculation mode of the calculator.

HP-38G

SOLVE

SOLVE PROGRAM
INPUT A;"AX+BY=C";
    "ENTER A";" ";1:
INPUT B;"AX+BY=C";
    "ENTER B";" ";1:
INPUT C;"AX+BY=C";
    "ENTER C";" ";1:
INPUT D;"DX+EY=F";
    "ENTER D";" ";1:
INPUT E;"DX+EY=F";
    "ENTER E";" ";1:
INPUT F;"DX+EY=F";
    "ENTER F";" ";1:
ERASE:
IF AE−DB==0
THEN DISP 3; "NO UNIQUE
    SOLUTION":
ELSE RUN "SOLVE.SOLN":
END:
FREEZE:

SOLVE.SOLN PROGRAM
(CE−BF)/(AE−DB)▶X:
(AF−CD)/(AE−DB)▶Y:
DISP 3;"X="X:
DISP 5;"Y="Y:

1. Input the 2 programs SOLVE
   and SOLVE.SOLN.
2. Run the SOLVE program.

Visualizing Row Operations (Section 10.1)

This program, referenced in the Technology note on page 737, demonstrates how elementary matrix row operations used in Gauss-Jordan elimination may be interpreted graphically. It asks the user to enter a 2×3 matrix that corresponds to a system of two linear equations. (The matrix entries should not be equivalent to either vertical or horizontal lines. This demonstration is also most effective if the y-intercepts of the lines are between -10 and 10.)

While the demonstration is running, you should notice that each elementary row operation creates an equivalent system. This equivalence is reinforced graphically because while the equations of the lines change with each elementary row operation, the point of intersection remains the same. You may want to run this program a second time to notice the relationship between the row operations and the graphs of the lines of the system.

TI-81

```
Prgm6:ROWOPS
:Disp "ENTER A"
:Disp "2 BY 3 MATRIX"
:Disp "A B C"
:Disp "D E F"
:Input A
:Input B
:Input C
:Input D
:Input E
:Input F
:A→[A](1,1)
:B→[A](1,2)
:C→[A](1,3)
:D→[A](2,1)
:E→[A](2,2)
:F→[A](2,3)
:ClrHome
:Disp "ORIGINAL MATRIX"
:Disp [A]
:Pause
:"B⁻¹(C−AX)"→Y₂
:"E⁻¹(F−DX)"→Y₁
:-10→Xmin
:10→Xmax
:1→Xscl
:-10→Ymin
:10→Ymax
:1→Yscl
```

```
:DispGraph
:Pause
:ClrHome
:Disp "OBTAIN LEADING"
:Disp "1 IN ROW 1"
:*row(A⁻¹,[A],1)→[A]
:Disp [A]
:Pause
:ClrDraw
:"(A/B)(C/A−X)"→Y₂
:DispGraph
:Pause
:ClrHome
:Disp "OBTAIN 0 BELOW"
:Disp "LEADING 1 IN"
:Disp "COLUMN 1"
:*row+(-D,[A],1,2)→[A]
:Disp[A]
:Pause
:ClrDraw
:"(E−(BD/A))⁻¹(F−(DC/A))"→Y₁
:DispGraph
:Pause
:ClrHome
:[A](2,2)→G
:If G=0
:Goto 1
:*row(G⁻¹,[A],2)→[A]
:Disp "OBTAIN LEADING"
```

(Continued on next page)

```
:Disp "1 IN ROW 2"
:Disp [A]
:Pause
:ClrDraw
:DispGraph
:Pause
:ClrHome
:Disp "OBTAIN 0 ABOVE"
:Disp "LEADING 1 IN"
:Disp "COLUMN 2"
:[A](1,2)→H
:*row+(-H,[A],2,1)→[A]
:Disp [A]
:Pause
:ClrDraw
:Y2−Off
:Line([A](1,3),-10,[A](1,3),10)
:DispGraph
:Pause
:ClrHome
:Disp "THE POINT OF"
:Disp "INTERSECTION IS"
:Disp "X="
:Disp [A](1,3)
:Disp "Y="
:Disp [A](2,3)
:End
:Lbl 1
:If [A](2,3)=0
:Disp "INFINITELY MANY"
:Disp "SOLUTIONS"
:If [A](2,3) ≠ 0
:Disp "INCONSISTENT"
:Disp "SYSTEM"
:End
```

To use this program, dimension matrix [A] as a 2×3 matrix. Press ENTER after each screen display to continue the program.

TI-83
TI-82

PROGRAM: ROWOPS
:Disp "ENTER A"
:Disp "2 BY 3 MATRIX:"
:Disp "A B C"
:Disp "D E F"
:Prompt A,B,C
:Prompt D,E,F
:A→[A](1,1):B→[A](1,2)
:C→[A](1,3):D→[A](2,1)
:E→[A](2,2):F→[A](2,3)
:ClrHome
:Disp "ORIGINAL MATRIX:"
:Pause [A]
:"B^{-1}(C−AX)"→Y$_2$
:"E^{-1}(F−DX)"→Y$_1$
:ZStandard:Pause:ClrHome
:Disp "OBTAIN LEADING"
:Disp "1 IN ROW 1"
:*row(A^{-1},[A],1)→[A]
:Pause [A]:ClrDraw
:"(A/B)(C/A−X)"→Y$_2$
:DispGraph:Pause:ClrHome
:Disp "OBTAIN 0 BELOW"
:Disp "LEADING 1 IN"
:Disp "COLUMN 1"
:*row+(-D,[A],1,2)→[A]
:Pause [A]:ClrDraw
:"(E−(BD/A))$^{-1}$(F−(DC/A))"→Y$_1$
:DispGraph:Pause:ClrHome
:[A](2,2)→G
:If G=0
:Goto 1
:*row(G^{-1},[A],2)→[A]
:Disp "OBTAIN LEADING"
:Disp "1 IN ROW 2"
:Pause [A]:ClrDraw
:DispGraph:Pause:ClrHome
:Disp "OBTAIN 0 ABOVE"
:Disp "LEADING 1 IN"
:Disp "COLUMN 2"
:[A](1,2)→H
:*row+(-H,[A],2,1)→[A]
:Pause [A]:ClrDraw:FnOff 2

:Vertical -(B/A)(E−(BD/A))$^{-1}$
  (F−DC/A)+C/A
:DispGraph:Pause:ClrHome
:Disp "THE POINT OF"
:Disp "INTERSECTION IS"
:Disp "X=",[A](1,3),"Y=",[A](2,3)
:Stop
:Lbl 1
If [A](2,3)=0
:Then
:Disp "INFINITELY MANY"
:Disp "SOLUTIONS"
:Else
:Disp "INCONSISTENT"
:Disp "SYSTEM"
:End

To use this program, dimension matrix [A] as a 2×3 matrix. Press ENTER after each screen display to continue the program.

TI-85

PROGRAM:ROWOPS
:Disp "enter a"
:Disp "2 by 3 matrix:"
:Disp "A B C"
:Disp "D E F"
:Prompt A,B,C
:Prompt D,E,F
:A→TEMP(1,1):B→TEMP(1,2)
:C→TEMP(1,3):D→TEMP(2,1)
:E→TEMP(2,2):F→TEMP(2,3)
:CILCD
:Disp "original matrix:"
:Disp TEMP
:Pause
:y2=B^{-1}(C−A*x)
:y1=E^{-1}(F−D*x)
:ZStd:Pause:CILCD
:Disp "obtain leading"
:Disp "1 in row 1"
:multR(A^{-1},TEMP,1)→TEMP
:DispTEMP:Pause
:"(A/B)(C/A−X)"→y
:ClDrw:DispG:Pause:CILCD
:Disp "obtain 0 below"
:Disp "leading 1 in"
:Disp "column 1"
:mRAdd(-D,TEMP,1,2)→TEMP
:Disp TEMP:Pause
:If TEMP(2,2)==0
:Goto A
:y1=(E−(B*D/A))$^{-1}$(F−(D*C/A))
:ClDrw:DispG:Pause:CILCD
:TEMP(2,2)→G
:multR(G^{-1},TEMP,2)→TEMP
:Disp "obtain leading"
:Disp "1 in row 2"
:Disp TEMP:Pause
:ClDrw:DispG:Pause:CILCD
:Disp "obtain 0 above"
:Disp "leading 1 in"
:Disp "column 2"
:TEMP(1,2)→H
:mRAdd(-H,TEMP,2,1)→TEMP
:Disp TEMP

:Pause:FnOff 2:ClDrw
:Vert -(B/A)(E−(B*D/A))$^{-1}$
   (F−D*C/A)+C/A
:DispG:Pause:CILCD
:Disp "the point of"
:Disp "intersection is"
:Disp "X=",TEMP(1,3),
   "Y=",TEMP(2,3)
:Stop
:Lbl A
:If TEMP(2,3)==0
:Then
:Disp "infinitely many"
:Disp "solutions"
:Else
:Disp "inconsistent"
:Disp "system"
:End

To use this program, dimension matrix TEMP as a 2×3 matrix. Press ENTER after each screen display to continue the program.

TI-92

```
rowops( )
Prgm
ClrIO
ClrHome
setMode("Split Screen","Left-
Right")
setMode("Split 1 App","Home")
setMode("Split 2 App","Graph")
Disp "ENTER A"
Disp "2 BY 3 MATRIX:"
Disp "A B C"
Disp "D E F"
Prompt a,b,c
Prompt d,e,f
[[a,b,c][d,e,f]]→mat1
ClrIO
b^(-1)*(c-a*x)→y2(x)
e^(-1)*(f-d*x)→y1(x)
ZoomStd
Disp "ORIGINAL MATRIX:"
Pause mat1
ClrIO
a/b*(c/a-x)→y2(x)
Disp "OBTAIN LEADING"
Disp "1 IN ROW 1"
mRow(a^(-1),mat1,1)→mat1
Pause mat1
ClrIO
(e-b*d/a)^(-1)*(f-d*c/a)→y1(x)
DispG
Disp "OBTAIN 0 BELOW"
Disp "LEADING 1 IN"
Disp "COLUMN 1"
mRowAdd(-d,mat1,1,2)→mat1
Pause mat1
ClrIO
mat1[2,2]→g
If g=0
Goto a1
Disp "OBTAIN LEADING"
Disp "1 IN ROW 2"
mRow(g^(-1),mat1,2)→mat1
Pause mat1
ClrIO
mat1[1,2]→h
```

```
FnOff 2
LineVert -b/a*(e-b*d/a)^(-1)*
    (f-d*c/a)+c/a
Disp "OBTAIN 0 ABOVE"
Disp "LEADING 1 IN"
Disp "COLUMN 2"
mRowAdd(-h,mat1,2,1)→mat1
Pause mat1
ClrIO
Disp "THE POINT OF"
Disp "INTERSECTION IS"
Disp "X=",mat1[1,3],"Y=",
    mat1[2,3]
Goto A2
Lbl a1
If mat1[2,3]=0 Then
Disp "INFINITELY MANY"
Disp "SOLUTIONS"
Else
Disp "INCONSISTENT"
Disp "SYSTEM"
EndIf
Lbl A2
Pause
setMode("Split Screen","Full")
EndPrgm
```

Press ENTER after each screen display to continue the program.

Casio fx-7700GE
Casio fx-9700GE
Casio CFX-7800G

ROWOPS ↵
"ENTER A"↵
"2 BY 3 MATRIX:"↵
"A B C"↵
"D E F"↵
"A="?→A:"B="?→B:
"C="?→C:"D="?→D:
"E="?→E:"F="?→F:↵
[[A,B,C][D,E,F]]→Mat A↵
Cls↵
"ORIGINAL MATRIX:"▰
Mat A ▰
Range -10,10,1,-10,10,1↵
Graph Y=B^{-1}(C−AX)↵
Graph Y=E^{-1}(F−DX)▰
Cls↵
"OBTAIN LEADING"↵
"1 IN ROW 1"▰
∗Row A^{-1},A,1↵
Mat A ▰
Graph Y=(A÷B)(C÷A−X)↵
Graph Y=E^{-1}(F−DX) ▰
Cls↵
"OBTAIN 0 BELOW"↵
"LEADING 1 IN"↵
"COLUMN 1"▰
∗Row+ -D,A,1,2↵
Mat A ▰
Graph Y=(A÷B)(C÷A−X)↵
GraphY=(E−(BD÷A))$^{-1}$
   (F−(DC÷A))▰
Cls↵
Mat A[2,2]→G↵
G=0 ⇒ Goto 1↵
∗Row G^{-1},A,2↵
"OBTAIN LEADING"↵
"1 IN ROW 2"▰
Mat A ▰
Graph Y=(A÷B)(C÷A−X)↵
Graph Y=(E−(BD÷A))$^{-1}$
   (F−(DC÷A))▰

Cls↵
"OBTAIN 0 ABOVE"↵
"LEADING 1 IN"↵
"COLUMN 2"▰
Mat A[1,2]→H↵
∗Row+ -H,A,2,1↵
Mat A ▰
Mat A[1,3]→J↵
Mat A[2,3]→K↵
Graph Y=K↵
Plot J,-10:Plot J,10:Line
"THE POINT OF"↵
"INTERSECTION IS"↵
"X=":J ▰
"Y=":K ▰
Goto 3↵
Lbl 1↵
Mat A[2,3]=0 ⇒ Goto 2↵
"INCONSISTENT"↵
"SYSTEM"↵
Goto 3↵
Lbl 2↵
"INFINITELY MANY"↵
"SOLUTIONS"↵
Lbl 3

To use this program, dimension Mat A as a 2×3 matrix. Press $\boxed{\text{EXE}}$ after each screen display to continue the program.

Sum Program (Section 11.1, page 805)

TI-80

PROGRAM:SUM
:INPUT "ENTER M", M
:INPUT "ENTER N", N
:0→S
:FOR(X,M,N)
:S+Y1→S
:DISP S
:END

To use this program, first store the nth term of the sequence in Y1 (in terms of X).

TI-81

Prgm5:SUM
:Disp "ENTER M"
:Input M
:Disp "ENTER N"
:Input N
:0→S
:Lbl 1
:M→X
:S+Y1→S
:Disp S
:IS>(M,N)
:Goto 1
:End

To use this program, first store the nth term of the sequence in Y1 (in terms of X).

TI-83
TI-82

PROGRAM:SUM
:Prompt M
:Prompt N
:0→S
:For(X,M,N)
:S+Y1→S
:Disp S
:End

To use this program, first store the nth term of the sequence in Y1 (in terms of X).

TI-85

PROGRAM:SUMS
:Prompt M
:Prompt N
:0→S
:For(x,M,N)
:S+y1→S
:Disp S
:End

To use this program, first store the nth term of the sequence in y1 (in terms of x).

TI-92

Summatn( )
Prgm
Input "Enter lower limit.",m
Input "Enter upper limit.",n
Σ(y1(x),x,m,n)→s
Disp "The partial sum is",s
EndPrgm

Sharp EL-9200C
Sharp EL-9300C

sum
————————REAL
Goto top
Label sumit
s=s+(nth term)
Print s
m=m+1
Goto next
Label top
Input m
Input n
s=0
Label next
If m<=n Goto sumit
End

To use this program, first replace (nth term) with the nth term of the sequence in terms of m. For example, s=s+2^m, where 2^n is the nth term of the sequence.

Casio fx-7700G

SUM
"M="?→M
"N="?→N
0→S
Lbl 1
M→X
S+f₁→S
"S=":S◢
M+1→M
M≤N⇒1 Goto 1

To use this program, enter the nth term of the sequence into f₁ (in terms of X).

Casio fx-7700GE
Casio fx-9700GE
Casio CFX-9800G

SUM ↵
"M="?→M↵
"N="?→N↵
0→S↵
Lbl 1↵
M→X↵
S+f₁→S↵
"S=":S◢
M+1→M↵
M≤N⇒1 Goto 1

To use this program, enter the nth term of the sequence into f₁ (in terms of X).

HP-38

SUM
SUM PROGRAM
INPUT M;"LOWER BOUND";
   "ENTER M";"";1:
INPUT N;"UPPER BOUND";
   "ENTER N";"";1:
0▶S:
ERASE:
SELECT "Function":
FOR I=M TO N
STEP 1;
RUN "SUM.STEP":
END:

SUM.STEP PROGRAM
S+F1(I)▶S:
DISP 4;"   "S:
FREEZE:

1. Input the 2 programs SUM and SUM.STEP.
2. Store the nth term of the sequence in the F1 function (in terms of x) in the Function aplet. Be sure that F1 is checked.
3. Run the SUM program.

Darts Program (Section 11.7, page 866)

TI-80

```
PROGRAM:DARTS
:CLRHOME
:0→K
:0→W
:INPUT "HOW MANY
    THROWS?",N
:LBL 1
:K+1→K
:RAND→X:RAND→Y
:IF X<.25:GOTO 2
:IF X>.75:GOTO 2
:IF X<.3:GOTO 2
:IF X>.65:GOTO 2
:W+1→W
:LBL 2
:IF K=N
:THEN
:DISP "WINS=",W
:DISP "LOSSES=",(N−W)
:STOP
:END
:GOTO 1
```

TI-81

```
Prgm6:DARTS
:ClrHome
:0→K
:0→W
:Disp "HOW MANY
    THROWS?"
:Input N
:Lbl 1
:K+1→K
:Rand→X
:Rand→Y
:If X<.25
:Goto 2
:If X>.75
:Goto 2
:If Y<.3
:Goto 2
:If X>.65
:Goto 2
:W+1→W
:Lbl 2
:If K≠N
:Goto 1
:DISP "WINS="
:Disp W
:DISP "LOSSES="
:N−W→L
:Disp L
:End
```

TI-83
TI-82

```
PROGRAM;DARTS
:ClrHome
:0→K
:0→W
:Input "HOW MANY
    THROWS?",N
:Lbl 1
:K+1→K
:rand→X:rand→Y
:IF X<.25 or X>.75
:Goto 2
:IF Y<.3 or Y>.65
:Goto 2
:W+1→W
:Lbl 2
:If K=N
:Then
:Disp "WINS=",W
:Disp "LOSSES=",(N−W)
:Stop
:End
:Goto 1
```

TI-85

```
PROGRAM:DARTS
:CILCD
:0→K
:0→W
:Input "HOW MANY
    THROWS?",N
:Lbl A
:K+1→K
:rand→X:rand→Y
:If X<.25
:Goto B
:If X>.75
:Goto B
:If Y<.3
:Goto B
:If Y>.65
:Goto B
:W+1→W
:Lbl B
:If K≠N
:Goto A
:DISP "WINS=",W
:DISP "LOSSES=",(N−W)
:Stop
```

TI-92

```
darts( )
Prgm
ClrIO
0→k
0→w
Input "How many throws?",n
Lbl a
k+1→k
rand( )→x
rand( )→y
If x<.25 or x>.75
Goto b
If y<.3 or y>.65
Goto b
w+1→w
Lbl b
If k=n Then
    Disp "Wins=",w
    Disp "Losses=",(n−w)
    Stop
EndIf
Goto a
EndPrgm
```

Casio fx-7700G

```
DARTS
Cls
0→K
0→W
"HOW MANY THROWS?"
?→N
Lbl 1
K+1→K
Ran#→X
Ran#→Y
X<.25⇒Goto 2
X>.75⇒Goto 2
Y<.3⇒Goto 2
Y>.65⇒Goto 2
W+1→W
Lbl 2
K≠N⇒Goto 1
"WINS="
W◢
"LOSSES="
N−W
```

Casio fx-7700GE
Casio fx-9700GE
Casio CFX-9800G

```
DARTS↵
Cls↵
0→K↵
0→W↵
"HOW MANY THROWS?"↵
?→N↵
Lbl 1↵
K+1→K↵
Ran#→X↵
Ran#→Y↵
X<.25⇒Goto 2↵
X>.75⇒Goto 2↵
Y<.3⇒Goto 2↵
Y>.65⇒Goto 2↵
W+1→W↵
Lbl 2↵
K≠N⇒Goto 1↵
"WINS="↵
W◢
"LOSSES="↵
N−W↵
```

Sharp EL-9200C
Sharp EL-9300C

darts
─────────REAL
ClrT
k=0
w=0
Print "how many throws?"
Input n
Label a
k=k+1
x=random
y=random
If x<.25 Goto b
If x>.75 Goto b
If y<.3 Goto b
If y>.65 Goto b
w=w+1
If k≠n Goto a
Print "wins="
Print w
l=n−w
Print "losses="
Print 1

HP-38
DARTS

DARTS PROGRAM
0▶L:
INPUT N;"";"THROWS?";
"HOW MANY THROWS?";1:
FOR I=1 TO N
    STEP 1;
    RUN "DART.THROW":
END:
ERASE:
DISP 3;"WINS="N−L:
DISP 5;"LOSSES="L:
FREEZE:

DARTS.THROW PROGRAM
RANDOM▶X:
RANDOM▶Y:
CASE
IF X<.25 OR X>.75
THEN L+1▶L:
IF Y<.3 OR Y>.65
THEN L+1▶L:
END:

1. Input the 2 programs DARTS
   and DARTS.THROW.
2. Run the DARTS program.

Appendix D

Scientific Calculator Keystrokes

Chapter P

PAGE 14, EXAMPLE 3   CALCULATORS AND EXPONENTS

a. 13 $\boxed{y^x}$ 4 $\boxed{+}$ 5 $\boxed{=}$

b. 3 $\boxed{y^x}$ 2 $\boxed{+/-}$ $\boxed{+}$ 4 $\boxed{y^x}$ 1 $\boxed{+/-}$ $\boxed{=}$

c. $\boxed{(}$ 3 $\boxed{y^x}$ 5 $\boxed{+}$ 1 $\boxed{)}$ $\boxed{\div}$ $\boxed{(}$ 3 $\boxed{y^x}$ 5 $\boxed{-}$ 1 $\boxed{)}$ $\boxed{=}$

PAGE 15, EXAMPLE 5   USING SCIENTIFIC NOTATION WITH A CALCULATOR

6.5 $\boxed{EE}$ 4 $\boxed{\times}$ 3.4 $\boxed{EE}$ 9 $\boxed{=}$

PAGE 22, EXAMPLE 17   EVALUATING RADICALS WITH A CALCULATOR

1 $\boxed{\div}$ 3 $\boxed{=}$ $\boxed{STO}$ 56 $\boxed{y^x}$ $\boxed{RCL}$ $\boxed{=}$ Use memory key.

56 $\boxed{y^x}$ $\boxed{(}$ 1 $\boxed{\div}$ 3 $\boxed{)}$ $\boxed{=}$ Use parentheses.

56 $\boxed{y^x}$ 3 $\boxed{1/x}$ $\boxed{=}$ Use reciprocal key.

Chapter 5

PAGE 395, EXAMPLE 1   EVALUATING EXPONENTIAL EXPRESSIONS

a. 2 $\boxed{y^x}$ 3.1 $\boxed{+/-}$ $\boxed{=}$

b. 2 $\boxed{y^x}$ π $\boxed{+/-}$ $\boxed{=}$

PAGE 398, EXAMPLE 5   EVALUATING NATURAL EXPONENTIAL FUNCTIONS

a. 2 $\boxed{+/-}$ $\boxed{e^x}$ or   2 $\boxed{+/-}$ $\boxed{INV}$ $\boxed{\ln x}$

b. 1 $\boxed{+/-}$ $\boxed{e^x}$ or   1 $\boxed{+/-}$ $\boxed{INV}$ $\boxed{\ln x}$

c. 1 $\boxed{e^x}$ or   1 $\boxed{INV}$ $\boxed{\ln x}$

d. 2 $\boxed{e^x}$ or   2 $\boxed{INV}$ $\boxed{\ln x}$

PAGE 407, EXAMPLE 2   *EVALUATING LOGARITHMS ON A CALCULATOR*

a. 10 [log]

b. 2.5 [log] [×] 2 [=]

c. 2 [+/−] [log]

PAGE 411, EXAMPLE 8   *EVALUATING THE NATURAL LOGARITHMIC FUNCTION*

a. 2 [ln x]

b. .3 [ln x]

c. 2 [e^x] [ln x]

d. 1 [+/−] [ln x]

Chapter 6

PAGE 471,   *EVALUATING TRIGONOMETRIC FUNCTIONS WITH A CALCULATOR*

π [÷] 8 [=] [sin] [1/x]   In radian mode

PAGE 471, EXAMPLE 6   *USING A CALCULATOR*

a. 76.4 [sin]   **b.** 1.5 [tan] [1/x]

PAGE 473 (TOP OF PAGE)

.6 [INV] [sin]

PAGE 473, EXAMPLE 9   *USING TRIGONOMETRY TO SOLVE A RIGHT TRIANGLE*

12 [÷] 9 [=] [INV] [tan]

PAGE 473, GROUP ACTIVITY

30 [cos]

PAGE 484, EXAMPLE 7   USING A CALCULATOR

a. cot 410°   410 [tan] [1/x]

   sin(−7)   7 [+/−] [sin]

b. 4.812 [INV] [tan]

PAGE 520, EXAMPLE 4   CALCULATORS AND INVERSE TRIGONOMETRIC FUNCTIONS

a. 8.45 [+/−] [INV] [tan]

b. 0.2447 [INV] [sin]

c. 2 [INV] [cos]

ANSWERS TO WARM UPS, ODD-NUMBERED EXERCISES, FOCUS ON CONCEPTS, AND TESTS

CHAPTER P

Section P.1 *(page 10)*

1. (a) 5, 1   (b) $-9, 5, 0, 1$ (c) $-9, -\frac{7}{2}, 5, \frac{2}{3}, 0, 1$

(d) $\sqrt{2}$

3. (a) None   (b) -13 (c) $2.01, 0.666\ldots, -13$

(d) $0.010110111\ldots$

5. (a) $\frac{6}{3}$ (b) $\frac{6}{3}$ (c) $-\frac{1}{3}, \frac{6}{3}, -7.5$ (d) $-\pi, \frac{1}{2}\sqrt{2}$

7. 0.625   **9.** $0.\overline{123}$ **11.** $-1 < 2.5$

13. $\frac{3}{2} < 7$ **15.** $-4 > -8$

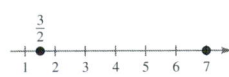   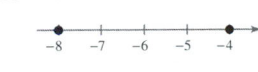

17. $\frac{5}{6} > \frac{2}{3}$

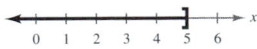

19. $x \le 5$ is the set of all real numbers less than or equal to 5. Unbounded

21. $x < 0$ is the set of all negative real numbers. Unbounded

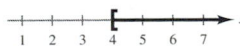

23. $x \ge 4$ is the set of all real numbers greater than or equal to 4. Unbounded

25. $-2 < x < 2$ is the set of all real numbers greater than -2 and less than 2. Bounded

27. $-1 \le x < 0$ is the set of all negative real numbers greater than or equal to -1. Bounded

29. $\frac{127}{90}, \frac{584}{413}, \frac{7071}{5000}, \sqrt{2}, \frac{47}{33}$ **31.** $x < 0$ **33.** $y \ge 0$

35. $A \ge 30$ **37.** 10   **39.** $\pi - 3 \approx 0.1416$

41. -1 **43.** -9 **45.** 3.75   **47.** $|-3| > -|-3|$

49. $-5 = -|5|$ **51.** $-|-2| = -|2|$ **53.** 4

55. $\frac{5}{2}$ **57.** 51   **59.** $\frac{128}{75}$ **61.** $|x - 5| \le 3$

63. $|7 - 18| = 11$ miles   **65.** $|y| \ge 6$

67. $|\$113{,}356 - \$112{,}700| = \$656 > \$500.00$

$0.05(\$112{,}700) = \$5635$

Because the actual expenses differ from the budget by more than $500.00, there is failure to meet the "budget variance test."

69. $|\$37{,}335 - \$37{,}640| = \$305 < \$500$

$0.05(\$37{,}640) = \$1882$

Because the difference between the actual expenses and the budget is less than $500 and less than 5% of the budgeted amount, there is compliance with the "budget variance test."

71. $|77.8 - 92.2| = 14.4$

There was a deficit of $14.4 billion.

73. $|1031.3 - 1252.7| = 221.4$

There was a deficit of $221.4 billion.

75. (a) No. If one is negative while the other is positive, they are unequal.

(b) $|u + v| \le |u| + |v|$

77. $7x, 4$ **79.** $4x^3, x, -5$

81. (a) -10 (b) -6 **83.** (a) 14   (b) 2

85. (a) Division by 0 is undefined.   (b) 0

87. Commutative Property of Addition

89. Multiplicative Inverse Property

91. Distributive Property

93. Multiplicative Identity Property

95. Associative and Commutative Properties of Multiplication

97. 0   **99.** Division by 0 is undefined.   **101.** 6   **103.** $\frac{1}{2}$

105. $\frac{3}{8}$ **107.** $\frac{3}{10}$ **109.** 48   **111.** -2.57 **113.** 1.56

115.

n	1	0.5	0.01	0.0001	0.000001
$5/n$	5	10	500	50,000	5,000,000

117.

n	1	10	100	10,000	100,000
$5/n$	5	0.5	0.05	0.0005	0.00005

Section P.2 *(page 23)*

Warm Up *(page 23)*

> **1.** <   **2.** >   **3.** >   **4.** <   **5.** 10   **6.** 12
>
> **7.** 14   **8.** -2 **9.** 21   **10.** $-\frac{13}{8}$

1. $-(0.4 \times 0.4 \times 0.4 \times 0.4 \times 0.4 \times 0.4)$

3. $(-10)^5$ **5.** (a) 48   (b) 81

7. (a) 729   (b) -9 **9.** (a) 243   (b) $-\frac{3}{4}$

11. -1600 **13.** 2.125   **15.** -24 **17.** 5

19. (a) $-125z^3$ (b) $5x^6$ **21.** (a) $24y^{10}$ (b) $3x^2$

23. (a) $\dfrac{7}{x}$ (b) $\dfrac{4}{3}(x+y)^2$ **25.** (a) 1   (b) $\dfrac{1}{4x^4}$

27. (a) $-2x^3$ (b) $\dfrac{10}{x}$ **29.** (a) $\dfrac{a^6}{64b^9}$ (b) $\dfrac{1}{625x^8y^8}$

31. (a) 3^{3n} (b) $\dfrac{b^5}{a^5}$ **33.** $9^{1/2} = 3$ **35.** $\sqrt[5]{32} = 2$

37. $\sqrt{196} = 14$ **39.** $(-216)^{1/3} = -6$ **41.** $\sqrt[3]{27^2} = 9$

43. $81^{3/4} = 27$ **45.** (a) 3   (b) 2   **47.** (a) 3   (b) $\frac{1}{2}$

49. (a) -125 (b) 3   **51.** (a) $\frac{1}{8}$ (b) $\frac{27}{8}$

53. (a) -4 (b) 2   **55.** (a) 7.550   (b) -7.225

57. (a) 14.499   (b) 0.528   **59.** (a) $2\sqrt{2}$ (b) $2\sqrt[3]{3}$

61. (a) $6x\sqrt{2x}$ (b) $\dfrac{18}{z\sqrt{z}}$

63. (a) $2x\sqrt[3]{2x^2}$ (b) $\dfrac{5|x|\sqrt{3}}{y^2}$ **65.** 625   **67.** $\dfrac{2}{x}$

69. $\dfrac{1}{x^3}$, $x > 0$ **71.** (a) $\dfrac{\sqrt{3}}{3}$ (b) $4\sqrt[3]{4}$

73. (a) $\dfrac{x(5+\sqrt{3})}{11}$ (b) $3(\sqrt{6}-\sqrt{5})$

75. (a) $\dfrac{2}{\sqrt{2}}$ (b) $\dfrac{3}{\sqrt[3]{75}}$

77. (a) $\dfrac{2}{3(\sqrt{5}-\sqrt{3})}$ (b) $-\dfrac{1}{2(\sqrt{7}+3)}$

79. (a) $3^{1/2} = \sqrt{3}$ (b) $(x+1)^{2/3} = \sqrt[3]{(x+1)^2}$

81. (a) $2\sqrt[4]{2}$ (b) $\sqrt[8]{2x}$ **83.** (a) $34\sqrt{2}$ (b) $22\sqrt{2}$

85. (a) $2\sqrt{x}$ (b) $4\sqrt{y}$ **87.** $\sqrt{5} + \sqrt{3} > \sqrt{5+3}$

89. $5 > \sqrt{3^2 + 2^2}$ **91.** 5.75×10^7 **93.** 8.99×10^{-5}

95. 524,000,000   **97.** 0.00000000048

99. (a) 954.448   (b) 3.077×10^{10}

101. (a) 67,082.039   (b) 39.791

103. When any positive integer is squared, the units digit is 0, 1, 4, 5, 6, or 9. Therefore, $\sqrt{5233}$ is not an integer.

105. $\dfrac{\pi}{2} \approx 1.57$ seconds   **107.** 0.280   **109.** $\frac{25}{3}$ minutes

Section P.3 *(page 32)*

Warm Up *(page 32)*

> **1.** $42x^3$ **2.** $-20z^2$ **3.** $-27x^6$ **4.** 1   **5.** $\frac{9}{4}z^3$
>
> **6.** $4\sqrt{3}$ **7.** $\dfrac{9}{4x^2}$ **8.** $\sqrt{2}$ **9.** 8   **10.** $-3x$

1. d   **3.** b   **5.** f

7. Degree: 2; Leading coefficient: 2

9. Degree: 5; Leading coefficient: 1

11. Degree: 5; Leading coefficient: -4

13. Polynomial: $-3x^3 + 2x + 8$

15. Not a polynomial because of the operation of division

17. Polynomial: $-y^4 + y^3 + y^2$

19. $-2x - 10$ **21.** $3x^3 - 2x + 2$

23. $8x^3 + 29x^2 + 11$ **25.** $12z + 8$

27. $3x^3 - 6x^2 + 3x$ **29.** $-15z^2 + 5z$

31. $-4x^4 + 4x$ **33.** $4x^3 - 2x^2 + 4$

35. $5x^2 - 4x + 11$ **37.** $-30x^3 + 57x^2 + 25x - 12$

39. $x^4 - x^3 + 5x^2 - 9x - 36$ **41.** $x^4 + x^2 + 1$

43. $x^2 + 7x + 12$ **45.** $6x^2 - 7x - 5$

47. $4x^2 + 12x + 9$ **49.** $4x^2 - 20xy + 25y^2$

51. $x^2 + 2xy + y^2 - 6x - 6y + 9$ **53.** $x^2 - 100$

55. $x^2 - 4y^2$ **57.** $m^2 - n^2 - 6m + 9$ **59.** $4r^4 - 25$

61. $x^3 + 3x^2 + 3x + 1$ **63.** $8x^3 - 12x^2y + 6xy^2 - y^3$

65. $16x^6 - 24x^3 + 9$ **67.** $2x^2 + 2x$ **69.** $u^4 - 16$

71. $x - y$ **73.** $x^3 + 4x^2 - 5x - 20$

75. No; $(x^2 + 1) + (-x^2 + 3) = 4$, which is not a second-degree polynomial.

77. (a) $x^2 - 1$ (b) $x^3 - 1$ (c) $x^4 - 1$

The result is $x^5 - 1$.

79. $(3 + 4)^2 = 49 \neq 25 = 3^2 + 4^2$

81. (a) $500r^2 + 1000r + 500$

(b)

r	$5\frac{1}{2}\%$	7%	8%
$500(1 + r)^2$	\$556.51	\$572.45	\$583.20

r	$8\frac{1}{2}\%$	9%
$500(1 + r)^2$	\$588.61	\$594.05

(c) Amount increases with increasing r.

83. $V = x(26 - 2x)(18 - 2x)$

$\quad = 4x(x - 13)(x - 9)$

x (cm)	1	2	3
V (cm³)	384	616	720

85. (a) $3x^2 + 8x$ (b) $30x^2$

87. (a) $T = 0.14x^2 - 3.33x + 58.40$

(b)

x (mi/hr)	30	40	55
T (ft)	84.50	149.20	298.75

(c) Stopping distance increases at an accelerating rate as speed increases.

89. $(x + 1)(x + 4) = x(x + 4) + 1(x + 4)$

Distributive Property

91. $m + n$ **93.** $(x - 3)^2 = x^2 - 6x + 9 \neq x^2 + 9$

Section P.4 *(page 41)*

Warm Up *(page 41)*

1. $15x^2 - 6x$ **2.** $-2y^2 - 2y$ **3.** $4x^2 + 12x + 9$

4. $9x^2 - 48x + 64$ **5.** $2x^2 + 13x - 24$

6. $4 - z - 5z^2$ **7.** $4y^2 - 1$ **8.** $x^2 - a^2$

9. $x^3 + 12x^2 + 48x + 64$ **10.** $8x^3 - 36x^2 + 54x - 27$

1. 30   **3.** $6x^2y$ **5.** $3(x + 2)$ **7.** $2x(x^2 - 3)$

9. $(x - 1)(x + 5)$ **11.** $(x + 6)(x - 6)$

13. $(4y + 3)(4y - 3)$ **15.** $(x + 1)(x - 3)$

17. $(x - 2)^2$ **19.** $(2t + 1)^2$ **21.** $(5y - 1)^2$

23. $-5(x^2 - 5)$ **25.** $-2(t^3 - 2t - 3)$

27. $(x + 2)(x - 1)$ **29.** $(s - 3)(s - 2)$

31. $-(5 + y)(y - 4)$ **33.** $(x - 20)(x - 10)$

35. $(3x - 2)(x - 1)$ **37.** $-(3z - 2)(3z + 1)$

39. $(5x + 1)(x + 5)$ **41.** $(x - 2)(x^2 + 2x + 4)$

43. $(y + 4)(y^2 - 4y + 16)$ **45.** $(2t - 1)(4t^2 + 2t + 1)$

47. $(x - 1)(x^2 + 2)$ **49.** $(2x - 1)(x^2 - 3)$

51. $(3 + x)(2 - x^3)$ **53.** $(x + 2)(3x + 4)$

55. $(2x - 1)(3x + 2)$ **57.** $(3x - 1)(5x - 2)$

59. $x(x + 3)(x - 3)$ **61.** $x^2(x - 4)$ **63.** $(x - 1)^2$

65. $(1 - 2x)^2$ **67.** $-2x(x + 1)(x - 2)$

69. $(9x + 1)(x + 1)$ **71.** $(3x + 1)(x^2 + 5)$

73. $x(x - 4)(x^2 + 1)$ **75.** $-z(z + 10)$

77. $(x + 1)^2(x - 1)^2$ **79.** $2(t - 2)(t^2 + 2t + 4)$

81. $(2x - 1)(6x - 1)$ **83.** $-(x + 1)(x - 3)(x + 9)$

85. $7(x^2 + 1)(3x^2 - 1)$ **87.** $-2x(x - 5)^3(x + 5)$

89. $-(x^2 + 1)^4\left(\dfrac{x^2}{2} + 1\right)$ **91.** b   **93.** a

95.

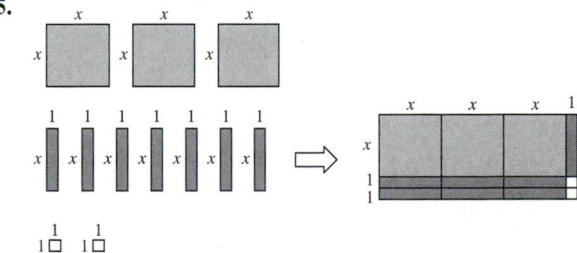

97.

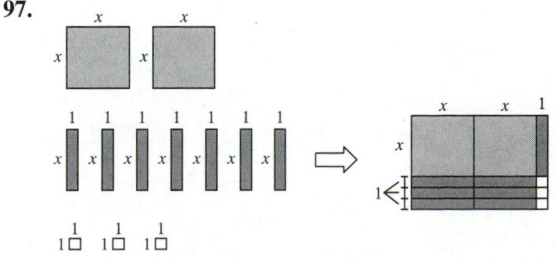

99. $4\pi(r + 1)$ **101.** $4(6 - x)(6 + x)$

103. $-14, 14, -2, 2$ **105.** $2, -12$

107. $9(x + 2)(x - 3)$

109. (a) $\pi h(R - r)(R + r)$ (b) $2\pi\left[\left(\dfrac{R + r}{2}\right)(R - r)\right]h$

Section P.5 *(page 52)*

Warm Up *(page 52)*

---
1. $5x^2(1 - 3x)$ **2.** $(4x + 3)(4x - 3)$ **3.** $(3x - 1)^2$

4. $(3 + 2y)^2$ **5.** $(z + 3)(z + 1)$ **6.** $(x - 10)(x - 5)$

7. $(1 + 3x)(3 - x)$ **8.** $(3x - 1)(x - 15)$

9. $(s + 2)(s - 2)(s + 1)$ **10.** $(y + 4)(y^2 - 4y + 16)$

---

1. All real numbers   **3.** All nonnegative real numbers

5. All real numbers x such that $x \neq 2$

7. All real numbers x such that $x \neq 0$ and $x \neq 4$

9. All real numbers x such that $x \geq -1$

11. $3x, \quad x \neq 0$ **13.** $x - 2, \quad x \neq 2$ **15.** $x, \quad x \neq 0$

17. $\dfrac{3x}{2}, \quad x \neq 0$ **19.** $\dfrac{3y}{y + 1}, \quad x \neq 0$ **21.** $-\dfrac{1}{2}, \quad x \neq 5$

23. $\dfrac{x(x + 3)}{x - 2}, \quad x \neq -2$ **25.** $\dfrac{y - 4}{y + 6}, \quad x \neq 3$

27. $-(x^2 + 1), \quad x \neq 2$ **29.** $z - 2$

31.

x	0	1	2	3	4	5	6
$\dfrac{x^2 - 2x - 3}{x - 3}$	1	2	3	undef.	5	6	7
$x + 1$	1	2	3	4	5	6	7

The expressions are equivalent except at $x = 3$.

33. The expression cannot be simplified.

35. $\dfrac{\pi}{4}, \quad r \neq 0$

37. $\dfrac{1}{5(x - 2)}, \quad x \neq 1$ **39.** $\dfrac{x - 3}{(x + 2)^2}, \quad x \neq -5$

41. $\dfrac{r + 1}{r}, \quad r \neq 1$ **43.** $\dfrac{t - 3}{(t + 3)(t - 2)}, \quad t \neq -2$

45. $\dfrac{x - y}{x(x + y)^2}, \quad x \neq -2y$ **47.** $\dfrac{3}{2}, \quad x \neq -y$

49. $x(x + 1), \quad x \neq -1, 0$ **51.** $\dfrac{x + 5}{x - 1}$ **53.** $\dfrac{6x + 13}{x + 3}$

55. $-\dfrac{2}{x - 2}$ **57.** $\dfrac{x - 4}{(x + 2)(x - 2)(x - 1)}$

59. $-\dfrac{x^2 + 3}{(x + 1)(x - 2)(x - 3)}$ **61.** $\dfrac{2 - x}{x^2 + 1}, \quad x \neq 0$

63. $\dfrac{-1}{(x^2 + 1)^5}$ **65.** $\dfrac{-2(x - 6)}{x + 2}$ **67.** $\dfrac{1}{2}, \quad x \neq 2$

69. $\dfrac{1}{x}, \quad x \neq -1$ **71.** $\dfrac{(x + 3)^3}{2x(x - 3)}, \quad x \neq -3$

73. $-\dfrac{2x + h}{x^2(x + h)^2}, \quad h \neq 0$ **75.** $\dfrac{2x - 1}{2x}, \quad x > 0$

77. $-\dfrac{1}{t^2\sqrt{t^2 + 1}}$ **79.** $-\dfrac{1}{x^2(x + 1)^{3/4}}$

81. $\dfrac{1}{\sqrt{x + 2} + \sqrt{x}}$

83. (a) $\dfrac{1}{16}$ minute   (b) $\dfrac{x}{16}$ minute(s)

(c) $\dfrac{60}{16} = \dfrac{15}{4}$ minutes

85. $\dfrac{11x}{30}$ **87.** (a) 12.65% (b) $\dfrac{288(MN - P)}{N(MN + 12P)}$

89. (a)

t	0	1	2	3	4	5
T	75	63.3	55.9	51.3	48.3	46.4

(b)

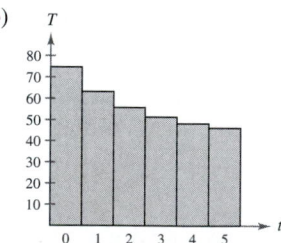

91. $\dfrac{x}{2(2x + 1)}$

Section P.6 *(page 61)*

Warm Up *(page 61)*

---

1. $a(a + 4)(a - 4)$ **2.** $(u + 5v)(u^2 - 5uv + 25v^2)$

3. $(1 + 4x)(2 - 3x)$ **4.** $(z + 3)(z + 2)(z - 2)$

5. $-\dfrac{2}{z^2}$ **6.** $\dfrac{(x + y)y}{(x - 4)(x - y)}$ **7.** 0 **8.** $\dfrac{9}{x - 2}$

9. -4 **10.** $\dfrac{-1}{2(2 + h)}$

---

1. $2x - (3y + 4) = 2x - 3y - 4$

3. $5z + 3(x - 2) = 5z + 3x - 6$

5. $-\dfrac{x - 3}{x - 1} = \dfrac{3 - x}{x - 1}$ **7.** $a\left(\dfrac{x}{y}\right) = \dfrac{ax}{y}$

9 $(4x)^2 = 16x^2$ **11.** $\sqrt{x + 9}$ cannot be simplified.

13. $\dfrac{6x + y}{6x - y}$ cannot be simplified.   **15.** $\dfrac{1}{x + y^{-1}} = \dfrac{y}{xy + 1}$

17. $x(2x - 1)^2 = x(4x^2 - 4x + 1)$

19. $\sqrt[3]{x^3 + 7x^2} = \sqrt[3]{x^2}\,\sqrt[3]{x + 7}$ **21.** $\dfrac{3}{x} + \dfrac{4}{y} = \dfrac{3y + 4x}{xy}$

23. $\dfrac{1}{2y} = \dfrac{1}{2} \cdot \dfrac{1}{y}$ **25.** $3x + 2$ **27.** $2x^2 + x + 15$

29. $\dfrac{1}{3}$ **31.** $\dfrac{1}{2}$ **33.** $-\dfrac{1}{4}$ **35.** 2 **37.** $\dfrac{1}{2}$ **39.** $\dfrac{1}{2x^2}$

41. $\dfrac{25}{9}, \dfrac{49}{16}$ **43.** $1, 2$ **45.** $1 + x$ **47.** $10x + 3$

49. -1 **51.** $3x - 1$ **53.** $\dfrac{16}{x} - 5 - x$

55. $4x^{8/3} - 7x^{5/3} + \dfrac{1}{x^{1/3}}$ **57.** $\dfrac{3}{\sqrt{x}} - 5x^{3/2} - x^{7/2}$

59. $\dfrac{-7x^2 - 4x + 9}{(x^2 - 3)^3(x + 1)^4}$ **61.** $\dfrac{27x^2 - 24x + 2}{(6x + 1)^4}$

63. $\dfrac{-1}{(x + 3)^{2/3}(x + 2)^{7/4}}$ **65.** $\dfrac{4x - 3}{(3x - 1)^{4/3}}$

67. (a) Answers will vary.

(b)

x	-2	-1	$-\frac{1}{2}$	0	1	2	$\frac{5}{2}$
y_1	-8.7	-2.9	-1.1	0	2.9	8.7	12.5
y_2	-8.7	-2.9	-1.1	0	2.9	8.7	12.5

69. Answers will vary. $y_2 = \dfrac{2x - 3x^3}{\sqrt{1 - x^2}}$

Section P.7 *(page 70)*

Warm Up *(page 70)*

---

1. 11.8 **2.** 13 **3.** 1 **4.** -2 **5.** 7.35

6. -4.3 **7.** 5 **8.** $3\sqrt{2}$ **9.** $3(\sqrt{2} + \sqrt{5})$

10. $2(\sqrt{3} + \sqrt{11})$

1.

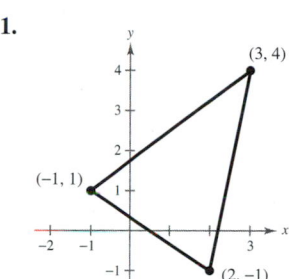

3.

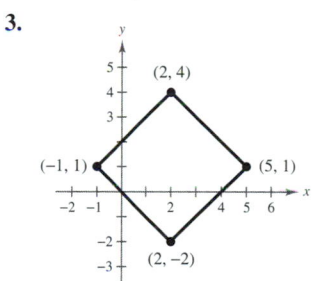

5. A: $(2, 6)$, B: $(-6, -2)$, C: $(4, -4)$, D: $(-3, 2)$

7. $(-3, 4)$ **9.** $(-5, -5)$

11. Point on x-axis: $y = 0$; Point on y-axis: $x = 0$

13. Quadrant IV   **15.** Quadrant II

17. Quadrant III or IV   **19.** Quadrant III

21. Quadrants I and III   **23.** $(0, 1)$, $(4, 2)$, $(1, 4)$

25.

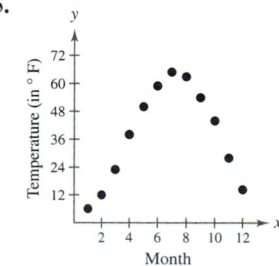

27.

x	-2	-1	$-\frac{1}{2}$	0	$\frac{1}{2}$	1	2
y	3	$\frac{5}{2}$	$\frac{9}{4}$	2	$\frac{7}{4}$	$\frac{3}{2}$	1

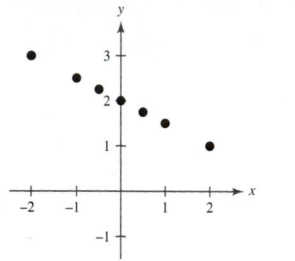

29. 1990: $13.70 per 100 pounds   **31.** 1900%

33. 1970s   **35.** 65   **37.** 8   **39.** 5

41. $4^2 + 3^2 = 5^2$ **43.** $10^2 + 3^2 = \left(\sqrt{109}\right)^2$

45. (a)

(b) 10

(c) (5, 4)

47. (a)

(b) 17

(c) $\left(0, \frac{5}{2}\right)$

49. (a)

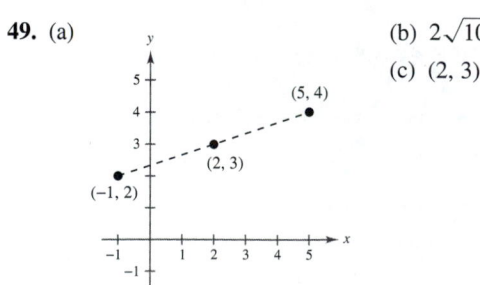

(b) $2\sqrt{10}$

(c) (2, 3)

51. (a)

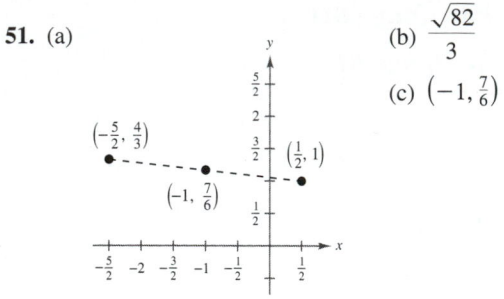

(b) $\dfrac{\sqrt{82}}{3}$

(c) $\left(-1, \frac{7}{6}\right)$

53. (a)

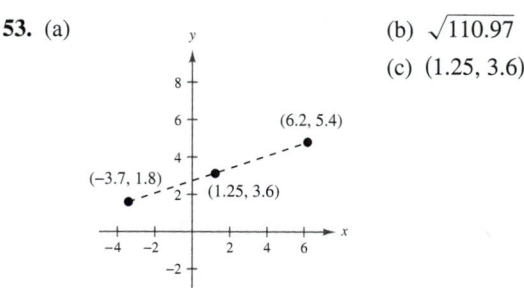

(b) $\sqrt{110.97}$

(c) (1.25, 3.6)

55. (a)

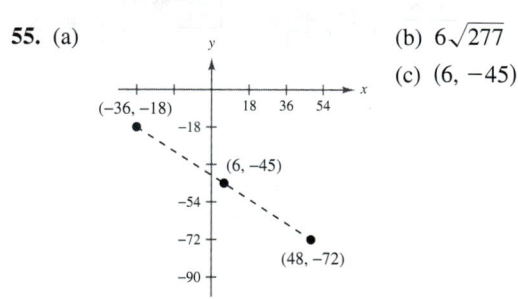

(b) $6\sqrt{277}$

(c) (6, -45)

57. $630,000

59. $\left(\sqrt{5}\right)^2 + \left(\sqrt{45}\right)^2 = \left(\sqrt{50}\right)^2$

61. All sides have a length of $\sqrt{5}$.

63. Opposite sides have equal lengths of $2\sqrt{5}$ and $\sqrt{85}$.

65. $(2x_m - x_1, 2y_m - y_1)$

67. $\left(\dfrac{3x_1 + x_2}{4}, \dfrac{3y_1 + y_2}{4}\right), \left(\dfrac{x_1 + x_2}{2}, \dfrac{y_1 + y_2}{2}\right),$

$\left(\dfrac{x_1 + 3x_2}{4}, \dfrac{y_1 + 3y_2}{4}\right)$

69. $5\sqrt{74} \approx 43$ yards

71.

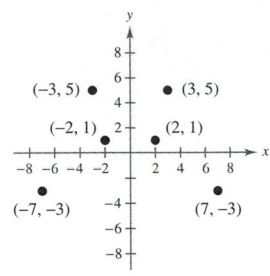

The point is reflected through the y-axis.

73. (a) Answers will vary.   (b) Answers will vary.

Focus on Concepts *(page 74)*

1.

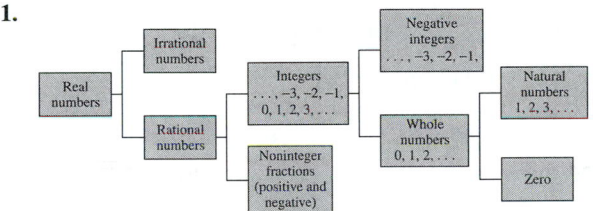

2. (a) Negative   (b) Positive   (c) Negative
(d) Positive

3. If $a < 0$, then $|a| = -a$. For example, $a = -7$ in $|-7|$.
Therefore, $|-7| = -(-7) = 7$.

4. (a) The base for the exponent -1 is $3x$. Therefore,
$$(3x)^{-1} = \frac{1}{3x}.$$

(b) When multiplying, add exponents. $y^3 \cdot y^2 = y^5$.

(c) Multiply the exponents to obtain $(a^2 b^3)^4 = a^8 b^{12}$

(d) The square of a binomial contains a cross-product term.
$(a + b)^2 = a^2 + 2ab + b^2$

(e) If $x < 0$, then $\sqrt{4x^2} > 0$ but $2x < 0$. $\sqrt{4x^2} = 2|x|$.

(f) Radicals cannot be combined unless the index and the radicand are the same.

5. No. A number written in scientific notation has the form $c \times 10^n$, where $1 \le c < 10$ and n is an integer. In true scientific notation, the number 52.7×10^5 is 5.27×10^6.

6. (a) Yes. $(x^4 + 2x - 2) + (x^3 + 2) = x^4 + 3x$

(b) No. When third- and fourth-degree polynomials are added, the fourth-degree term of the fourth-degree polynomial will be in the sum.

(c) No. The sum will be of fourth degree. The *product* of the two polynomials would be of seventh degree.

7. The polynomial is written as a product.

8. Factor the numerator and the denominator and cancel all common factors.

9. b   **10.** c   **11.** d   **12.** a

Review Exercises *(page 75)*

1. (a) 11   (b) 11, -14 (c) 11, -14, $-\frac{8}{9}$, $\frac{5}{2}$, 0.4
(d) $\sqrt{6}$

3. (a) $.8\overline{3}$ (b) 0.875

5. The set consists of all real numbers less than or equal to 7.

7. $|x - 7| \ge 4$ **9.** $|y + 30| < 5$ **11.** -11 **13.** $\frac{1}{12}$

15. -144 **17.** Associative Property of Addition

19. Multiplicative Inverse Property

21. (a) $\dfrac{3u^5}{v^4}$ (b) m^{-2} **23.** 3.0296×10^{10}

25. 483,300,000   **27.** (a) 11,414.125   (b) 18,380.160

29. $16^{1/2} = 4$ **31.** (a) $2x^2$ (b) $\dfrac{3|u|}{b} \sqrt{\dfrac{2}{b}}$

33. $2 + \sqrt{3}$ **35.** $2\sqrt{2}$ **37.** $192\sqrt{2}$ **39.** 280

41. $\frac{12}{7}$ **43.** -1 **45.** x^6 **47.** $x^4 - 3x^2$ **49.** 169

51. $\sqrt[6]{1372x^3}$ **53.** $-3x^2 - 7x + 1$

55. $4x^2 - 12x + 9$ **57.** $2x^5 + 3x^4 - x^3 - 9x^2 - 15x$

59. (a)   (b) $S = 2\pi r(r + h)$

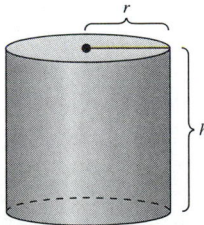

61. $x(x + 1)(x - 1)$ **63.** $(x + 10)(2x + 1)$

65. $(x - 1)(x^2 + 2)$ **67.** $16x^2 - 9x + 20$

69. $-5x^2 + 2x + 15$ **71.** $\dfrac{1}{x^2}$

73. $\dfrac{2x^3 - 4x^2 - 15x + 5}{(x - 4)(x + 2)}$ **75.** $\dfrac{3x}{(x - 1)(x^2 + x + 1)}$

77. $\dfrac{3ax^2}{(a^2 - x)(a - x)}$

79.

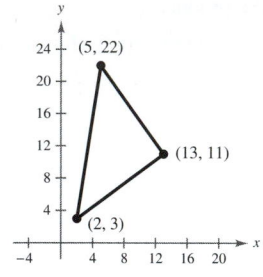

$$2\left(\sqrt{185}\right)^2 = \left(\sqrt{370}\right)^2$$

81. Quadrant IV   **83.** Quadrant IV

85. (a)

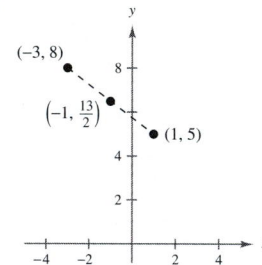

(b) 5

(c) $\left(-1, \frac{13}{2}\right)$

87. $(x + 3)(x + 5) = 5(x + 3) + x(x + 3)$

Distributive Property

89.

n	1	10	10^2	10^4	10^6	10^{10}
$\dfrac{5}{\sqrt{n}}$	5	1.5811	0.5	0.05	0.005	0.00005

$5/\sqrt{n}$ approaches 0 as n increases without bound.

Chapter Test *(page 80)*

1. $-\frac{10}{3} > -|-4|$ **2.** 9.15   **3.** (a) -18 (b) $\frac{4}{27}$

4. (a) $-\frac{27}{125}$ (b) $\frac{8}{729}$ **5.** (a) 25   (b) 6

6. (a) 1.8×10^5 (b) 2.7×10^{13}

7. (a) $12z^8$ (b) $(u - 2)^{-7}$ **8.** (a) $\dfrac{3x^2}{y^2}$ (b) $\dfrac{2}{v}\sqrt[3]{\dfrac{2}{v^2}}$

9. (a) $15z\sqrt{2z}$ (b) $-10\sqrt{y}$ **10.** $2x^2 - 3x - 5$

11. $x^2 - 5$ **12.** 8   **13.** $\dfrac{x - 1}{2x}$

14. (a) $x^2(2x + 1)(x - 2)$ (b) $(x - 2)(x + 2)^2$

15. (a) $4\sqrt[3]{4}$ (b) $-3\left(1 + \sqrt{3}\right)$

16.

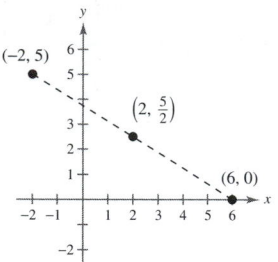

Midpoint: $\left(2, \frac{5}{2}\right)$; Distance: $\sqrt{89}$

17. $\frac{5}{6}\sqrt{3}\,x^2$

CHAPTER 1

Section 1.1 *(page 91)*

Warm Up *(page 91)*

1. $14x - 42$ **2.** $-17s$ **3.** $-24y^7$
4. $2(t + 1)(t - 1)(t + 2)$ **5.** $5x^2\sqrt{6}$
6. $2\sqrt{x + 3}$ **7.** $y = x^3 + 4x$ **8.** $x^2 + y^2 = 4$
9. $y = 4x^2 + 8$ **10.** $y^2 = -3x + 4$

1. (a) Yes   (b) Yes   **3.** (a) No   (b) Yes
5. (a) No   (b) Yes   **7.** (a) Yes   (b) Yes

9.

x	-1	0	1	$\frac{3}{2}$	2
y	5	3	1	0	-1

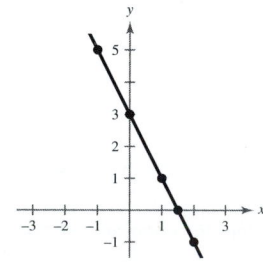

11.

x	-1	0	1	2	3
y	3	0	-1	0	3

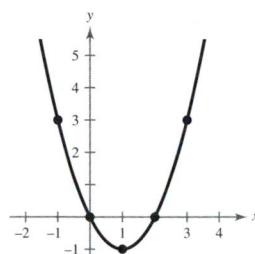

13.

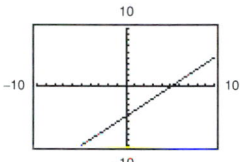

Intercepts: $(5, 0)$, $(0, -5)$

15.

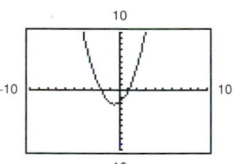

Intercepts: $(1, 0)$, $(-2, 0)$, $(0, -2)$

17.

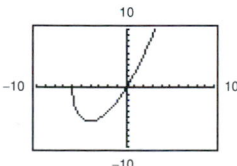

Intercepts: $(0, 0)$, $(-6, 0)$

19.

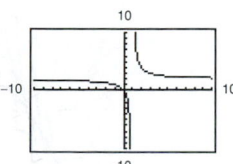

Intercept: $(0, 0)$

21. y-axis symmetry   **23.** x-axis symmetry

25. Origin symmetry   **27.** Origin symmetry

29.

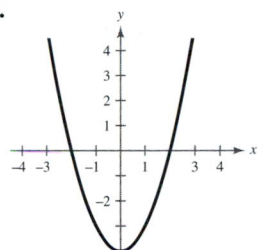

31.

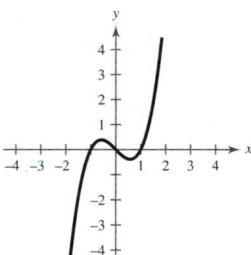

33. c   **35.** f   **37.** b

39.

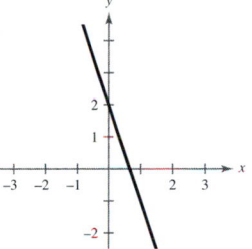

No symmetry

41.

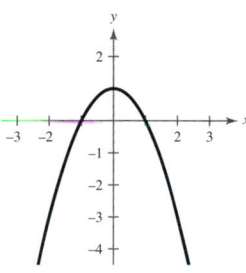

Symmetry: y-axis

43.

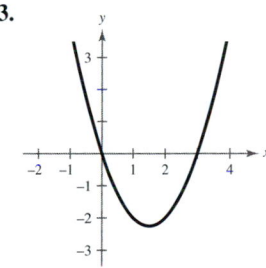

No symmetry

45.

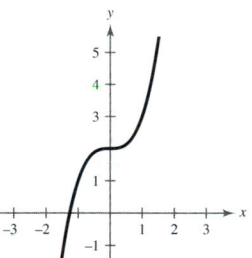

No symmetry

47.

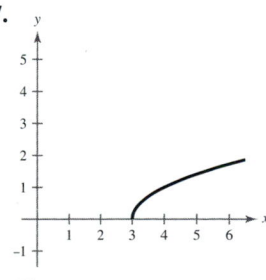

No symmetry

49.

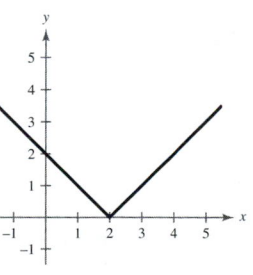

No symmetry

51.

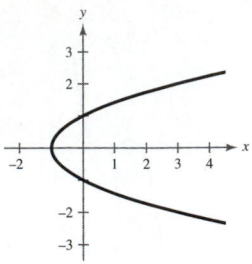

Symmetry: x-axis

53.

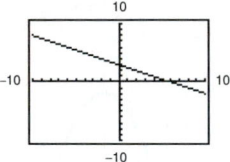

Intercepts: $(6, 0)$, $(0, 3)$

55.

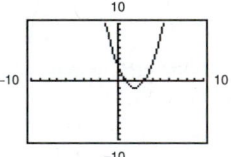

Intercepts: $(3, 0)$, $(1, 0)$, $(0, 3)$

57.

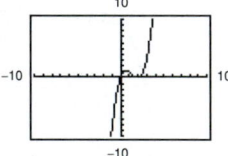

Intercepts: $(0, 0)$, $(2, 0)$

59.

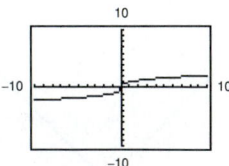

Intercept: $(0, 0)$

61.

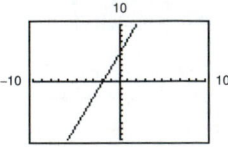 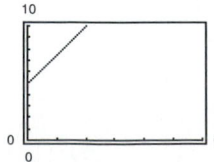

The standard setting gives a more complete graph.

63.

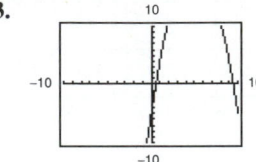 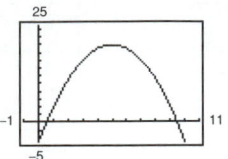

The specified setting gives a more complete graph.

65.
Xmin = -5
Xmax = 5
Xscl = 1
Ymin = -30
Ymax = 30
Yscl = 10

67.
Xmin = -10
Xmax = 20
Xscl = 5
Ymin = -5
Ymax = 30
Yscl = 5

69. $x^2 + y^2 = 9$

71. $(x - 2)^2 + (y + 1)^2 = 16$

73. $(x + 1)^2 + (y - 2)^2 = 5$

75. $(x - 3)^2 + (y - 4)^2 = 25$

77. $(x + 2)^2 + (y + 3)^2 = 4$

79. Center: $(0, 0)$; Radius: 2

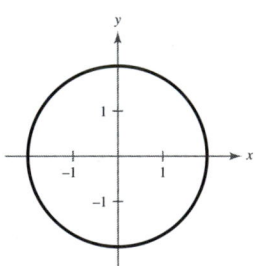

81. Center: $(1, -3)$; Radius: 2

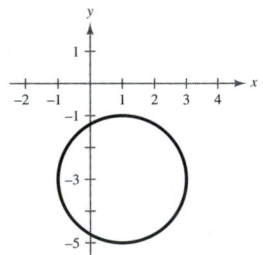

83. Center: $\left(\frac{1}{2}, \frac{1}{2}\right)$; Radius: $\frac{3}{2}$

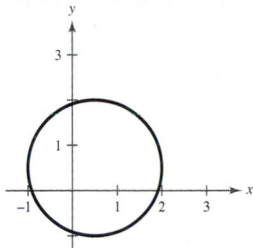

85.

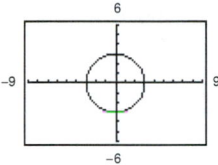

Circle

87. The graphs are identical. Distributive Property

89. The graphs are identical. Associative Property

91.

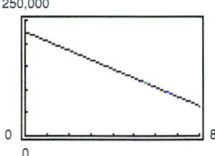

93. The viewing rectangle is incorrect. Change the viewing rectangle.

95. (a)

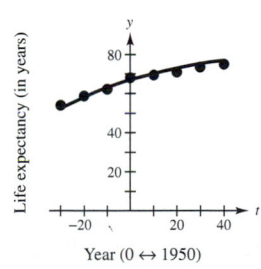

    (b) 77.7

    (c) 78.0

97.

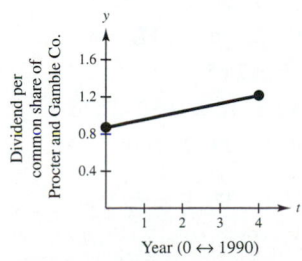

99. $9x^5$, $4x^3$, -7 **101.** False. $\dfrac{1}{3 \cdot 4^{-1}} = \dfrac{4}{3}$

103. $2\sqrt{2x}$ **105.** $\dfrac{10\sqrt{7x}}{x}$ **107.** $\sqrt[3]{|t|}$

Section 1.2 *(page 102)*

Warm Up *(page 102)*

1. $-3x - 10$ **2.** $5x - 12$ **3.** x **4.** $x + 26$

5. $\dfrac{8x}{15}$ **6.** $\dfrac{3x}{4}$ **7.** $-\dfrac{1}{x(x+1)}$ **8.** $\dfrac{5}{x}$

9. $\dfrac{7x - 8}{x(x - 2)}$ **10.** $-\dfrac{2}{x^2 - 1}$

1. (a) No (b) No (c) Yes (d) No

3. (a) Yes (b) Yes (c) No (d) No

5. (a) Yes (b) No (c) No (d) No

7. (a) No (b) No (c) No (d) Yes

9. Identity   **11.** Conditional equation   **13.** Identity

15. Identity   **17.** Conditional equation

19. The equations have the same solution, and one is derived from the other by steps for generating equivalent equations.

$2x = 5$, $2x + 3 = 8$

21. Original equation

Subtract 32 from both sides.

Simplify.

Divide both sides by 4.

Simplify.

23. 5   **25.** 6   **27.** $\frac{7}{3}$ **29.** 13   **31.** 5

33. -4 **35.** 3   **37.** 9   **39.** -26 **41.** -4

43. $-\frac{6}{5}$ **45.** 9

47.

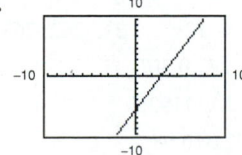

    $x = 3$

49.

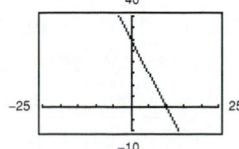

    $x = 10$

51. No solution   **53.** 10   **55.** 4   **57.** 3

59. 5   **61.** No solution   **63.** $\frac{11}{6}$ **65.** $\frac{5}{3}$

67. No solution   **69.** 0   **71.** All real numbers

73. $\dfrac{1}{3-a}$, $a \neq 3$ **75.** $\dfrac{5}{4+a}$, $a \neq -4$

77. 138.889   **79.** 62.372   **81.** 19.993

83. (a) 6.46   (b) $\dfrac{1.73}{0.27} \approx 6.41$

85. (a) 1.00   (b) $\dfrac{6.01}{5.98} \approx 1.01$

87. (a)

x	-1	0	1	2	3	4
$3.2x - 5.8$	-9	-5.8	-2.6	0.6	3.8	7

(b) $1 < x < 2$. The expression changes from negative to positive in this interval.

(c)

x	1.5	1.6	1.7	1.8	1.9	2
$3.2x - 5.8$	-1	-0.68	-0.36	0.04	0.28	0.6

(d) $1.8 < x < 1.9$. To improve accuracy, evaluate the expression in this interval and determine where the sign changes.

89. 61.2 inches   **91.** $T = 10,000 + \frac{1}{2}x$ **93.** $7600

95. $x = 10$ centimeters   **97.** 23,437.5 miles

Section 1.3 *(page 113)*

Warm Up *(page 113)*

1. 14   **2.** 4   **3.** -3 **4.** 4   **5.** -2
6. 1   **7.** $\frac{2}{5}$ **8.** $\frac{10}{3}$ **9.** 6   **10.** $-\frac{11}{5}$

1. A number increased by 4   **3.** A number divided by 5
5. A number decreased by 4 is divided by 5.
7. $n + (n + 1) = 2n + 1$
9. $(2n - 1)(2n + 1) = 4n^2 - 1$
11. $50t$ **13.** $0.20x$ **15.** $P = 2x + 2(2x) = 6x$
17. $25x + 1200$ **19.** $4x + 8x = 12x$ **21.** 262, 263
23. 37, 185   **25.** $-5, -4$ **27.** $0.30L$
29. $N = p(500)$ **31.** 13.5   **33.** 1192.5
35. 135%   **37.** 175   **39.** $22,316.98
41. Income taxes: $516 billion; Corporation taxes: $107 billion; Social Security taxes: $428 billion; Other: $96 billion
43. January: $71,590.00; February: $85,908.00
45. 72% increase   **47.** 43% increase

49. (a)

(b) $l = 1.5w$; $p = 5w$
(c) 7.5 meters $\times$ 5 meters

51. 97   **53.** 3 hours   **55.** $\frac{1}{3}$ hour
57. (a) 3.8 hours, 3.2 hours   (b) 1.1 hours
(c) 25.6 miles
59. $66\frac{2}{3}$ kilometers per hour   **61.** 1.29 seconds
63. 91.4 feet   **65.** 4.36 feet   **67.** $16,666.67
69. First three quarters: 11.5%; Last quarter: 10%
71. ≈ 32.1 gallons   **73.** 50 pounds of each kind
75. 8064 units   **77.** $x = 6$
79. $\dfrac{2A}{b}$ **81.** $\dfrac{S}{1+R}$ **83.** $\dfrac{A-P}{Pt}$
85. $\dfrac{2A - ah}{h}$ **87.** $\dfrac{3V + \pi h^3}{3\pi h^2}$ **89.** $\dfrac{L - L_0}{L_0 \Delta t}$
91. $\dfrac{Fr^2}{\alpha m_1}$ **93.** $\dfrac{(n-1)fR_2}{R_2 + (n-1)f}$ **95.** $\dfrac{L - a + d}{d}$
97. $\dfrac{S - a}{S - L}$ **99.** $\dfrac{x^2}{5}$ **101.** $\sqrt{10} + 2$

Section 1.4 *(page 128)*

Warm Up *(page 128)*

1. $x(3x + 7)$ **2.** $(2x + 5)(2x - 5)$
3. $-(x - 7)(x - 15)$ **4.** $(x - 2)(x + 9)$
5. $(5x - 1)(2x + 3)$ **6.** $(6x - 1)(x - 12)$
7. $3\sqrt{17}$ **8.** $2\sqrt{3}$ **9.** $4\sqrt{6}$ **10.** $3\sqrt{73}$

1. $2x^2 + 5x - 3 = 0$ **3.** $x^2 - 6x + 7 = 0$
5. $3x^2 - 60x - 10 = 0$ **7.** $0, -\frac{1}{2}$ **9.** $4, -2$
11. -5 **13.** $3, -\frac{1}{2}$ **15.** $2, -6$ **17.** $-a$
19. ± 4; ± 4.00 **21.** $\pm\sqrt{7}$; ± 2.65
23. $\pm 2\sqrt{3}$; ± 3.46 **25.** $12 \pm 3\sqrt{2}$; 16.24, 7.76
27. $-2 \pm 2\sqrt{3}$; 1.46, -5.46 **29.** 2; 2.00   **31.** 0, 2
33. $4, -8$ **35.** $-3 \pm \sqrt{7}$ **37.** $1 \pm \dfrac{\sqrt{6}}{3}$

39. $2 \pm 2\sqrt{3}$ **41.** $\dfrac{1}{(x+1)^2 + 4}$ **43.** $\dfrac{1}{\sqrt{9 - (x-3)^2}}$

45. 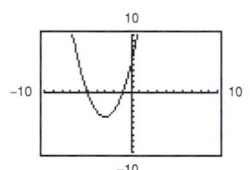   **47.**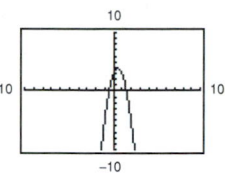

    $x = -1, -5$ $x = -\tfrac{1}{2}, \tfrac{3}{2}$

49. No real solution   **51.** Two real solutions

53. $\tfrac{1}{2}, -1$ **55.** $\tfrac{1}{4}, -\tfrac{3}{4}$ **57.** $1 \pm \sqrt{3}$

59. $-7 \pm \sqrt{5}$ **61.** $-4 \pm 2\sqrt{5}$

63. $\dfrac{2}{3} \pm \dfrac{\sqrt{7}}{3}$ **65.** $-\dfrac{3}{2} \pm \dfrac{\sqrt{13}}{2}$ **67.** $-\dfrac{1}{2} \pm \sqrt{2}$

69. $\dfrac{2}{7}$ **71.** $2 \pm \dfrac{\sqrt{6}}{2}$ **73.** $6 \pm \sqrt{11}$

75. $0.976, -0.643$ **77.** $1.687, -0.488$ **79.** $1 \pm \sqrt{2}$

81. $6, -12$ **83.** $\tfrac{1}{2} \pm \sqrt{3}$ **85.** $-\tfrac{1}{2}$ **87.** -11

89. False. The product must equal zero for the Zero-Factor Property to be used.

91. $x^2 - 2x - 24 = 0$

93. (a)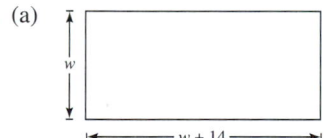

    (b) $w(w + 14) = 1632$

    (c) $w = 34$ feet

       $l = 48$ feet

95. 6 inches × 6 inches   **97.** 19.098 feet, 9.5 trips

99. (a) $s = -16t^2 + 1821$

    (b)

t	0	2	4	6	8	10
s	1821	1757	1565	1245	797	221

    (c) $t > 10$ seconds; $t = 10.67$ seconds

101. $\dfrac{5\sqrt{2}}{2} \approx 3.54$ centimeters

103. ≈ 550 miles per hour and 600 miles per hour

105. 50,000 units   **107.** 258 units

109. 653 units   **111.** 1990. Yes. No

113. (a)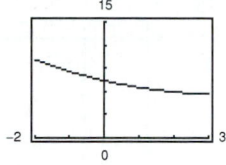

    (b) Yes; 1996

115. Associative Property of Multiplication

117. $\dfrac{5}{6u^2v^3}$ **119.** $x^2(x - 3)(x^2 + 3x + 9)$

121. $-(x + 10)$ **123.**

Xmin = -2
Xmax = 5
Xscl = 1
Ymin = -2
Ymax = 12
Yscl = 2

Section 1.5 *(page 138)*

Warm Up *(page 138)*

1. $2\sqrt{3}$ **2.** $10\sqrt{5}$ **3.** $\sqrt{5}$ **4.** $-6\sqrt{3}$

5. 12 **6.** 48 **7.** $\dfrac{\sqrt{3}}{3}$ **8.** $\sqrt{2}$

9. $\dfrac{1}{2} \pm \dfrac{\sqrt{5}}{2}$ **10.** $-1 \pm \sqrt{2}$

1. $a = -10, b = 6$ **3.** $a = 6, b = 5$

5. $4 + 3i$ **7.** $2 - 3\sqrt{3}i$ **9.** $5\sqrt{3}i$

11. $-1 - 6i$ **13.** 8 **15.** $0.3i$ **17.** $11 - i$

19. 4 **21.** $3 - 3\sqrt{2}i$ **23.** $-14 + 20i$

25. $\tfrac{1}{6} + \tfrac{7}{6}i$ **27.** $-2\sqrt{3}$ **29.** -10 **31.** $5 + i$

33. $12 + 30i$ **35.** 24 **37.** $-9 + 40i$ **39.** -10

41. $\sqrt{-6}\sqrt{-6} = \sqrt{6}i\sqrt{6}i = 6i^2 = -6$

43. $5 - 3i, 34$ **45.** $-2 + \sqrt{5}i, 9$

47. $-20i, 400$ **49.** $\sqrt{8}, 8$ **51.** $-6i$

53. $\tfrac{16}{41} + \tfrac{20}{41}i$ **55.** $\tfrac{3}{5} + \tfrac{4}{5}i$

57. $-7 - 6i$ **59.** $-\tfrac{9}{1681} + \tfrac{40}{1681}i$ **61.** $-\tfrac{1}{2} - \tfrac{5}{2}i$

63. $\tfrac{62}{949} + \tfrac{297}{949}i$ **65.** $1 \pm i$ **67.** $-2 \pm \tfrac{1}{2}i$

69. $-\dfrac{3}{2}, -\dfrac{5}{2}$ **71.** $\dfrac{1}{8} \pm \dfrac{\sqrt{11}}{8}i$

73.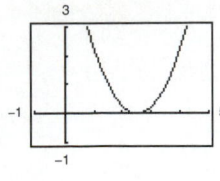

75.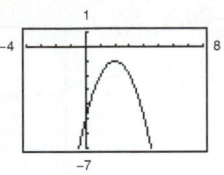

$x = \frac{5}{2}$ $x = 2 \pm i$

77. The number of x-intercepts of the graph corresponds to the number of real solutions of the equation. If there are no x-intercepts, the quadratic equation has two complex solutions.

79. $-1 + 6i$ **81.** $-5i$ **83.** $-375\sqrt{3}\,i$

85. i **87.** $8, 8, 8$ **89.–93.** Answers will vary.

95. $-x^2 - 3x + 12$ **97.** $4x^2 - 20x + 25$

99. $a = \frac{1}{2}\sqrt{\frac{3V}{\pi b}}$ **101.** 1 liter

Section 1.6 *(page 149)*

Warm Up *(page 149)*

1. 11 **2.** 20, -3 **3.** 5, -45 **4.** 0, $-\frac{1}{5}$	
5. $\frac{2}{3}, -2$ **6.** $\frac{11}{6}, -\frac{5}{2}$ **7.** 1, -5 **8.** $\frac{3}{2}, -\frac{5}{2}$	
9. $\frac{3 \pm \sqrt{5}}{2}$ **10.** $2 \pm \sqrt{2}$	

1. $0, \pm\frac{3\sqrt{2}}{2}$ **3.** $\pm 3, \pm 3i$ **5.** $-3, 0$

7. $3, 1, -1$ **9.** $\pm 1, \frac{1}{2} \pm \frac{\sqrt{3}}{2}i$ **11.** $\pm\sqrt{3}, \pm 1$

13. $\pm\frac{1}{2}, \pm 4$ **15.** $1, -2, 1 \pm \sqrt{3}i, -\frac{1}{2} \pm \frac{\sqrt{3}i}{2}$

17. $-\frac{1}{5}, -\frac{1}{3}$ **19.** $\frac{1}{4}$ **21.** $1, -\frac{125}{8}$

23.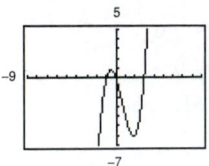

25.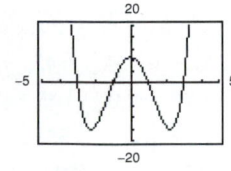

$x = 0, 3, -1$ $x = \pm 3, \pm 1$

27. 50   **29.** 26   **31.** -16 **33.** 2, -5

35. 0   **37.** 9   **39.** $\frac{101}{4}$ **41.** $-59, 69$

43. $-3 \pm 16\sqrt{2}$ **45.** $\pm\sqrt{69}, \pm\sqrt{59}\,i$ **47.** 1

49.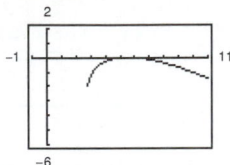

51.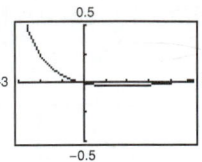

$x = 5, 6$ $x = 0, 4$

53. 4, -5 **55.** $\frac{-3 \pm \sqrt{21}}{6}$ **57.** $2, -\frac{3}{2}$ **59.** 1, -3

61. 3, -2 **63.** $\sqrt{3}, -3$ **65.** $3, \frac{-1 - \sqrt{17}}{2}$

67.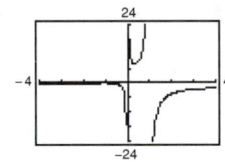

69.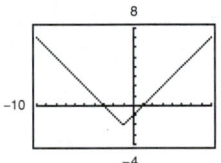

$x = -1$ $x = 1, -3$

71. ± 1.038 **73.** 16.756   **75.** $x^2 - 2x - 15 = 0$

77. $x^3 - 4x^2 - 2x + 8 = 0$ **79.** 34 students

81. 400 miles per hour   **83.** 7%   **85.** $x = 6, -4$

87. $y = \pm 15$ **89.** 26,250   **91.** 500 units

93. $x = 2$ miles or 0.382 mile

95. (a)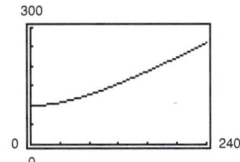

173

(b)

h	160	165	170	175	180	185
d	188.7	192.9	197.2	201.6	205.9	210.3

(c) 173.2

(d) Solving graphically or numerically yields an approximate solution. An exact solution is obtained algebraically.

97. $h = \frac{1}{\pi r}\sqrt{S^2 - \pi^2 r^4}$ **99.** $a = 4, b = 24$

101. $\frac{125y}{x}$ **103.** $\frac{25x(x-1)}{x+5}$ **105.** $\frac{25}{6t}$

107. $\frac{x-4}{(x-1)(x^2-4)}$

Section 1.7 *(page 160)*

Warm Up *(page 160)*

1. $-\frac{1}{2}$ **2.** $-\frac{1}{6}$ **3.** -3 **4.** $\frac{13}{2}$ **5.** $x \geq 0$
6. $-3 < z < 10$ **7.** $P \leq 2$ **8.** $W \geq 200$
9. $2, 7$ **10.** $0, 1$

1. A number is no more than 25.

3. $-1 \leq x \leq 3$. Bounded

5. $10 < x < \infty$. Unbounded

7. c   **9.** f   **11.** g   **13.** b

15. (a) Yes   (b) No   (c) Yes   (d) No

17. (a) Yes   (b) No   (c) No   (d) Yes

19. (a) Yes   (b) Yes   (c) Yes   (d) No

21. $x < 3$

23. $x > -4$

25. $x \geq 12$

27. $x < -2$

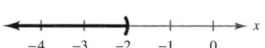

29. $x \geq \frac{1}{2}$

31. $x > \frac{1}{2}$

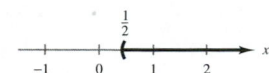

33. $-1 < x < 3$

35. $-\frac{9}{2} < x < \frac{15}{2}$

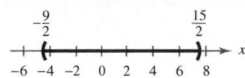

37. $-\frac{3}{4} < x < -\frac{1}{4}$

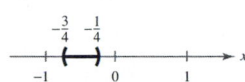

39. 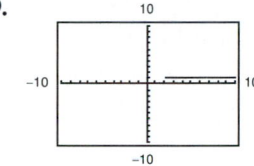   **41.**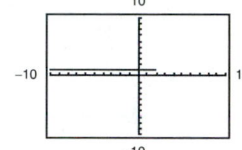

43.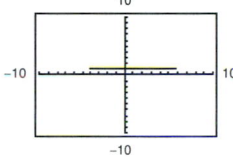

45.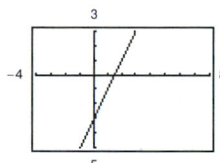

(a) $x \geq 2$

(b) $x \leq \frac{3}{2}$

47.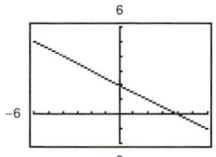

(a) $4 \geq x \geq -2$

(b) $x \leq 4$

49. $[5, \infty)$ **51.** $[-3, \infty)$ **53.** $\left(\infty, \frac{7}{2}\right]$

55. $-5 < x < 5$

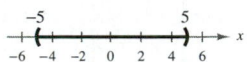

57. $x < -6, \ x > 6$

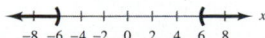

59. $16 \leq x \leq 24$

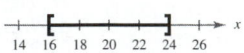

61. $x \leq 16$, $x \geq 24$

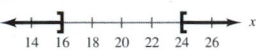

63. $x \leq -7$, $x \geq 13$

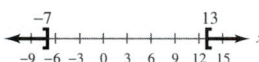

65. $4 < x < 5$

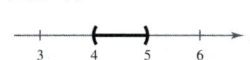

67. $x \leq -\frac{29}{2}$, $x \geq -\frac{11}{2}$

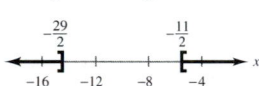

69. No solution

71.

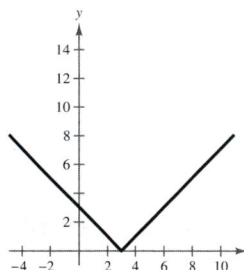

    (a) $1 \leq x \leq 5$

    (b) $x \leq -1$, $x \geq 7$

73. $|x| \leq 3$ **75.** $|x - 7| \geq 3$ **77.** $|x - 12| \leq 10$

79. $|x + 3| > 5$

81. All real numbers within eight units of 10

83. More than 400 miles   **85.** $r > 12.5\%$ **87.** $x \geq 36$

89. (a) and (b)

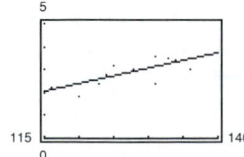

    (c) $x \geq 129$

    (d) IQ scores are not particularly good predictors of GPA.
Other factors such as study habits, class attendance, and
attitude influence college performance.

91. $106.864 \leq \text{area} \leq 109.464$

93. $65.8 \leq h \leq 71.2$

95. $a = 1$, $b = 5$, $c = 5$

97. ≈ 330 vibrations per second

99. ≈ 1.2 to ≈ 2.4 millimeters

101. $(-3, 10)$ **103.** $5\sqrt{5}$

Section 1.8 *(page 171)*

Warm Up *(page 171)*

1. $y < -6$	**2.** $z > -\frac{9}{2}$	**3.** $-3 \leq x < 1$
4. $x \leq -5$	**5.** $-3 < x$	**6.** $5 < x < 7$
7. $-\frac{7}{2} \leq x \leq \frac{7}{2}$	**8.** $x < 2$, $x > 4$	
9. $x < -6$, $x > -2$	**10.** $-2 \leq x \leq 6$	

1. (a) No   (b) Yes   (c) Yes   (d) No

3. (a) Yes   (b) No   (c) No   (d) Yes

5. $2, -\frac{3}{2}$ **7.** $\frac{7}{2}, 5$

9. $[-3, 3]$

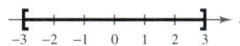

11. $(-\infty, -2)$, $(2, \infty)$

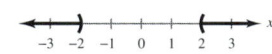

13. $(-7, 3)$

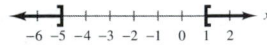

15. $(-\infty, -5]$, $[1, \infty)$

17. $(-3, 2)$

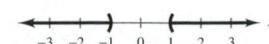

19. $(-\infty, -1)$, $(1, \infty)$

21. $(-3, 1)$

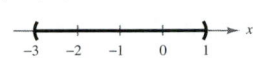

23. $(-\infty, 0), \left(0, \frac{3}{2}\right)$

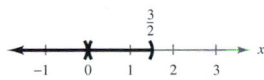

25. $[-2, 0], [2, \infty]$

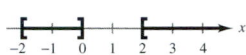

27. $[-2, \infty)$

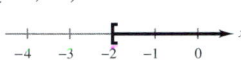

29.

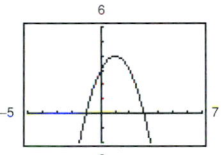

(a) $x \le -1, \ x \ge 3$

(b) $0 \le x \le 2$

31.

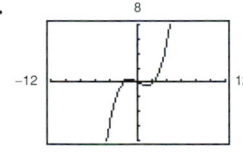

(a) $-2 \le x \le 0,$
   $2 \le x < \infty$

(b) $x \le 4$

33. $(-\infty, -1), \ (0, 1)$

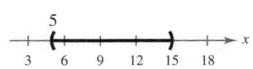

35. $(-\infty, -1), \ (4, \infty)$

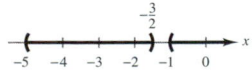

37. $(5, 15)$

39. $\left(-5, -\frac{3}{2}\right), \ (-1, \infty)$

41. $\left(-\frac{3}{4}, 3\right), \ [6, \infty)$

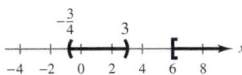

43. $(-3, -2], \ [0, 3)$

45. $(-\infty, -2), \ (-1, 1), \ (3, \infty)$

47.

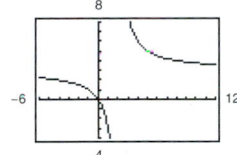

(a) $0 \le x < 2$

(b) $2 < x \le 4$

49.

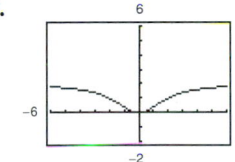

(a) $2 \le |x|$

(b) $-\infty < x < \infty$

51. $[-2, 2]$ **53.** $(-\infty, 3], \ [4, \infty)$ **55.** $[-4, 3]$

57. $(-3.51, 3.51)$ **59.** $(-0.13, 25.13)$ **61.** $(2.26, 2.39)$

63. $(-\infty, -4] \cup [4, \infty)$ **65.** $\left(-\infty, -2\sqrt{30}\,\right] \cup \left[2\sqrt{30}, \infty\right)$

67. If $a > 0$ and $c \le 0$, b can be any real number. If $a > 0$
   and $c > 0$, $b < -2\sqrt{ac}$ or $b > 2\sqrt{ac}$.

69. (a) $t = 10$ seconds   (b) 4 seconds $< t <$ 6 seconds

71. Between 13.8 and 36.2 meters   **73.** $R_1 \ge 2$ **75.** 1996

77. (a) $L = 6 - y + 2\sqrt{16 + y^2}$

(b) $0 \le y \le 6$

   $y = 0: \ L = 14$

   $y = 6: \ L = 4\sqrt{13} \approx 14.4$

   Decrease

(c)

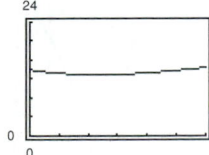

(d) $\frac{5}{3} < y < 3$

79. $(x + 7)(x - 1)$

81. $2x(x - 3)(x^2 + 3x + 9)$

Focus on Concepts *(page 174)*

1. a   **2.** c   **3.** d   **4.** b

5. Assuming that the graph does not go beyond the vertical limits of the display, you will see the graph for larger values of x.

6. Equivalent equations have the same solution(s). An equation can be transformed into an equivalent equation by the following steps.

  (a) Remove symbols of grouping, combine like terms, and/or reduce fractions.

  (b) Add (or subtract) the same quantity to (from) both sides of the equation.

  (c) Multiply (or divide) both sides of the equation by the same nonzero quantity.

  (d) Interchange the two sides of the equation.

7. (a) Negative   (b) Positive

8. (a) Neither   (b) Both   (c) Quadratic   (d) Neither

9. Dividing by x does not yield an equivalent equation. $x = 0$ is also a solution.

10. b

11. (a) $x = a, \quad x = b$

  (b)

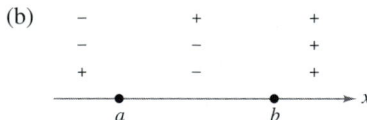

  (c) The real zeros of the polynomial.

Review Exercises *(page 175)*

1.

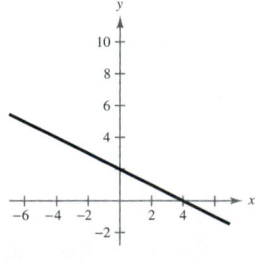

3.

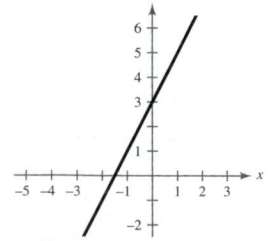

5.

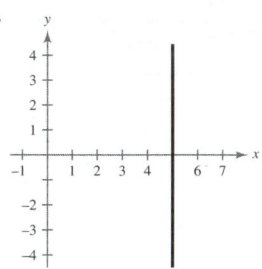

7.

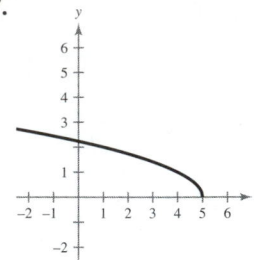

9.

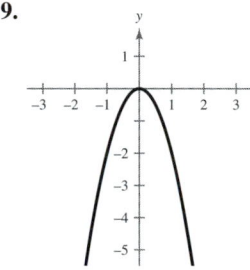

11.

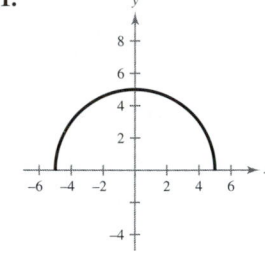

13.

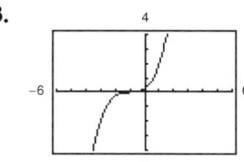

Intercepts: $(-1, 0), \left(0, \frac{1}{4}\right)$

15.

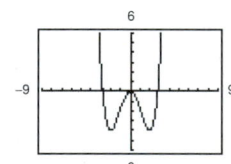

Intercepts: $(0, 0), \ (\pm 2\sqrt{2}, 0)$

17.

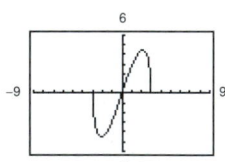

Intercepts: $(0, 0), \ (\pm 3, 0)$

19.

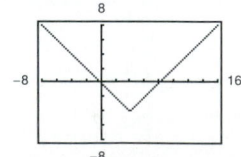

Intercepts: $(0, 0), \ (8, 0)$

21. $y = \pm\sqrt{25 - x^2}$

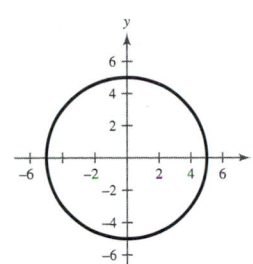

23.
```
Xmin = -2
Xmax = 3
Xscl = 1
Ymin = -20
Ymax = 15
Yscl = 5
```

105. (a)

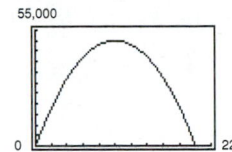

(b) $x = 0, 20$

(c) $x = 10$

(d) $x < 5.53,$

$x > 14.47$

107. $L \geq \dfrac{32}{\pi^2}$

25. Center: $(3, -1)$; Radius: 3

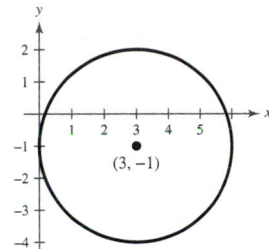

Chapter Test *(page 180)*

1.

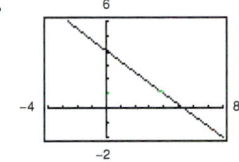

No symmetry

Intercepts: $(0, 4)$, $\left(\frac{16}{3}, 0\right)$

2.

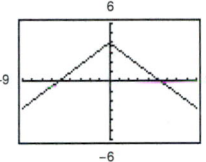

y-axis symmetry

Intercepts: $(0, 4)$, $\left(\pm\frac{16}{3}, 0\right)$

27. Identity   **29.** (a) No   (b) Yes   (c) Yes   (d) No

31. 20   **33.** $-\frac{1}{2}$ **35.** $\frac{1}{5}$ **37.** 0, 2   **39.** $-4 \pm 3\sqrt{2}$

41. $6 \pm \sqrt{6}$ **43.** $0, \frac{12}{5}$ **45.** 2, 6   **47.** 5   **49.** $\frac{25}{4}$

51. No solution   **53.** $-124, 126$ **55.** $-2 \pm \dfrac{\sqrt{95}}{5}, -4$

57. $-5, 15$ **59.** 1, 3

61.

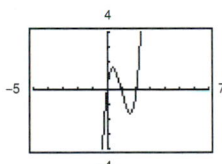

63.

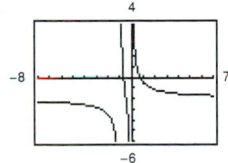

$x = 0, 1, 2$

$x = \pm\dfrac{\sqrt{2}}{2}$

65. $r = \sqrt{\dfrac{3V}{\pi h}}$ **67.** $p = \dfrac{k}{3\pi r^2 L}$ **69.** $C = 4$

71. $(-\infty, -1], [3, \infty)$ **73.** $(-\infty, 3), (5, \infty)$

75. $(1, 3)$ **77.** $(-\infty, 0], [3, \infty)$ **79.** $\left(-\infty, \frac{120}{7}\right)$

81. $(4, \infty)$ **83.** $[5, \infty)$ **85.** $3 + 7i$ **87.** $40 + 65i$

89. $-4 - 46i$ **91.** $1 - 6i$ **93.** $\frac{4}{3}i$ **95.** $\pm\sqrt{\frac{1}{3}}i$

97. September: \$325,000; October: \$364,000

99. $2\frac{6}{7}$ liters   **101.** 4   **103.** 56 miles per hour

3.

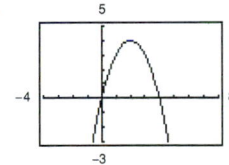

No symmetry

Intercepts: $(0, 0)$, $(4, 0)$

4.

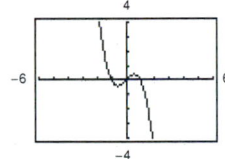

Origin symmetry

Intercepts: $(0, 0)$, $(1, 0)$, $(-1, 0)$

5.

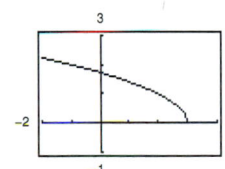

No symmetry

Intercepts: $\left(0, \sqrt{3}\right)$, $(3, 0)$

6.

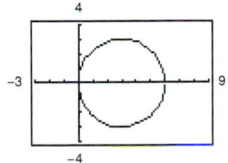

x-axis symmetry

Intercepts: $(0, 0)$, $(3, 0)$

7. $\frac{128}{11}$ **8.** $-4, 5$ **9.** No solution

10. $\pm\sqrt{2}, \pm\sqrt{3}i$ **11.** 0, 4   **12.** $-2, \frac{8}{3}$

13. $-\frac{11}{2} \leq x < 3$

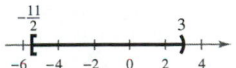

14. $x < -6$ or $0 < x < 4$

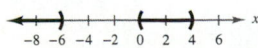

15. $93\frac{3}{4}$ kilometers per hour   **16.** $a = 80$, $b = 20$

17. (a) $-3 + 5i$ (b) 7   (c) $2 - i$

CHAPTER 2

Section 2.1 *(page 191)*

Warm Up *(page 191)*

1. $-\frac{9}{2}$ **2.** $-\frac{13}{3}$ **3.** $-\frac{5}{4}$ **4.** $\frac{1}{2}$

5. $y = \frac{2}{3}x - \frac{5}{3}$ **6.** $y = -2x$ **7.** $y = 3x - 1$

8. $y = \frac{2}{3}x + 5$ **9.** $y = -2x + 7$ **10.** $y = x + 3$

1. (a) L_2 (b) L_3 (c) L_1

3.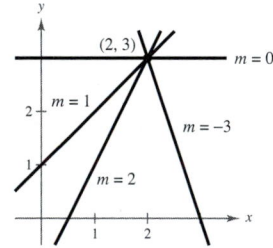

5. $\frac{8}{5}$ **7.** 0   **9.** -4

11. $m = 2$

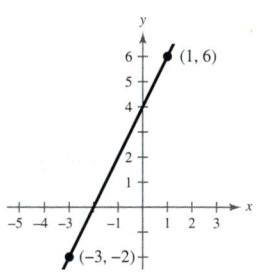

13. m is undefined.

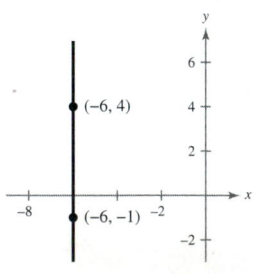

15. $m = \frac{4}{3}$

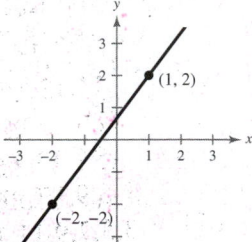

17. $(0, 1)$, $(3, 1)$, $(-1, 1)$ **19.** $(6, -5)$, $(7, -4)$, $(8, -3)$

21. $(-8, 0)$, $(-8, 2)$, $(-8, 3)$ **23.** Perpendicular

25. Parallel

27. Yes. The rate of change remains the same on a line.

29. (a) Sales increasing 135 units per year.

   (b) No change in sales.

   (c) Sales decreasing 40 units per year.

31. (a) 1989   (b) 1988

33. $16,666\frac{2}{3}$ feet ≈ 3.16 miles

35. $m = 5$; Intercept: $(0, 3)$

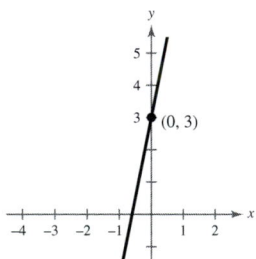

37. m is undefined. There is no y-intercept.

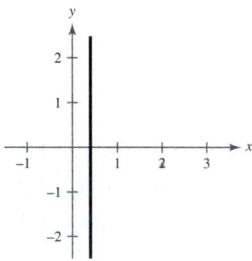

39. $m = -\frac{7}{6}$; Intercept: $(0, 5)$ **41.** $3x + 5y - 10 = 0$

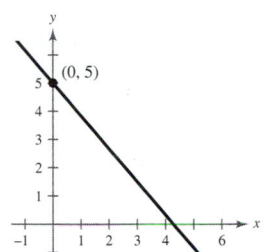

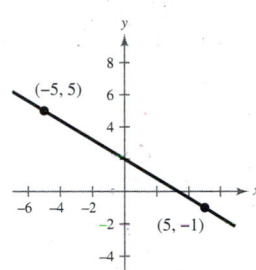

43. $x + 2y - 3 = 0$ **45.** $x + 8 = 0$

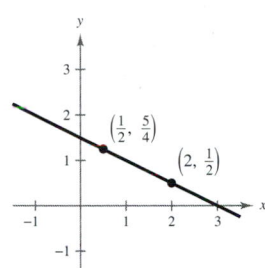

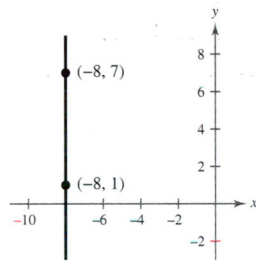

47. $2x - 5y + 1 = 0$ **49.** $3x - y - 2 = 0$

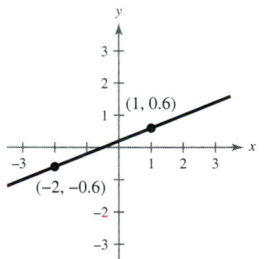

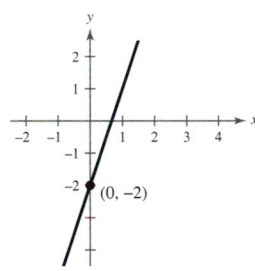

51. $2x + y = 0$ **53.** $x + 3y - 4 = 0$

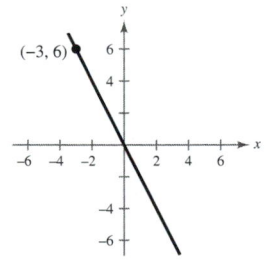

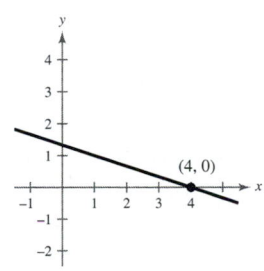

55. $x - 6 = 0$ **57.** $8x - 6y - 17 = 0$

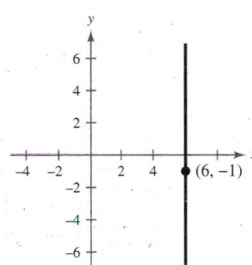

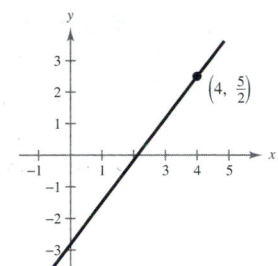

59. $3x + 2y - 6 = 0$ **61.** $12x + 3y + 2 = 0$

63. $x + y - 3 = 0$

65. (a) $2x - y - 3 = 0$ (b) $x + 2y - 4 = 0$

67. (a) $3x + 4y + 2 = 0$ (b) $4x - 3y + 36 = 0$

69. (a) $y = 0$ (b) $x + 1 = 0$

71. Parallel   **73.** Neither

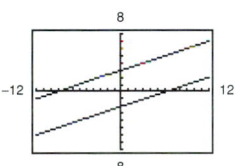

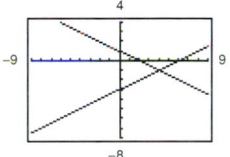

75. Perpendicular

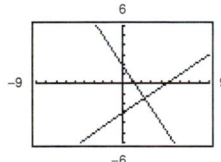

77.

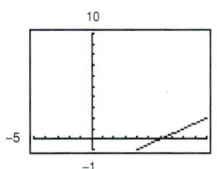

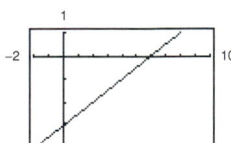

The second setting shows the x- and y-intercepts more clearly.

79.

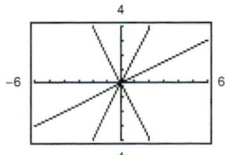

(b) is perpendicular to (c).

81.

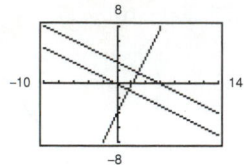

(a) is parallel to (b).

(c) is perpendicular to (a) and (b).

83. $V = 1790 + 125t$ **85.** b   **87.** a

89. $3x - 2y - 1 = 0$

91. $F = \frac{9}{5}C + 32$ **93.** $39,500$

95. $V = -175t + 875$ **97.** $S = 0.85L$

99. (a) $C = 16.75t + 36,500$ (b) $R = 27t$

(c) $P = 10.25t - 36,500$ (d) $t \approx 3561$ hours

101. (a)

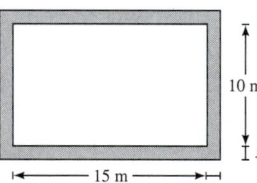

(b) $y = 8x + 50$

(c)

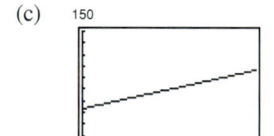

(d) 8 meters

103. $C = 120 + 0.26x$ **105.** $y = 91x + 164$

107. d   **109.** a

Section 2.2 *(page 205)*

Warm Up *(page 205)*

1. -73 **2.** 13 **3.** $2(x + 2)$ **4.** $-8(x - 2)$

5. $y = \frac{7}{5} - \frac{2}{5}x$ **6.** $y = \pm x$ **7.** $x \leq \frac{9}{2}$

8. $x \geq -\frac{2}{3}$ **9.** $-3 \leq x \leq 3$ **10.** $x \leq 1, x \geq 2$

1. Yes   **3.** No   **5.** Yes   **7.** No

9. (a) Function

(b) Not a function, because the element 1 in A corresponds to two elements, -2 and 1, in B.

(c) Function

(d) Not a function, because not every element in A is matched with an element in B.

11. Each is a function. For each year there corresponds one and only one circulation.

13. Not a function   **15.** Function   **17.** Function

19. Not a function   **21.** Function

23. (a) 4   (b) 0   (c) $4x$ (d) $(x + c)$

25. (a) -1 (b) -9 (c) $2x - 5$

27. (a) 0   (b) -0.75 (c) $x^2 + 2x$

29. (a) 1   (b) 2.5   (c) $3 - 2|x|$

31. (a) $-\frac{1}{9}$ (b) Undefined   (c) $\dfrac{1}{y^2 + 6y}$

33. (a) 1   (b) -1 (c) $\dfrac{|x - 1|}{x - 1}$

35. (a) -1 (b) 2   (c) 6

37.

x	-2	-1	0	1	2
$f(x)$	1	-2	-3	-2	1

39.

t	-5	-4	-3	-2	-1
$h(t)$	1	$\frac{1}{2}$	0	$\frac{1}{2}$	1

41.

x	-2	-1	0	1	2
$f(x)$	5	$\frac{9}{2}$	4	1	0

43. 5   **45.** ± 3 **47.** $2, -1$ **49.** $3, 0$

51. All real numbers x **53.** All real numbers $t \neq 0$

55. $y \geq 10$ **57.** $-1 \leq x \leq 1$

59. All real numbers $x \neq 0, -2$

61. $\{(-2, 4), (-1, 1), (0, 0), (1, 1), (2, 4)\}$

63. $\{(-2, 0), (-1, 1), (0, \sqrt{2}), (1, \sqrt{3}), (2, 2)\}$

65. The domain is the set of inputs of the function, and the range is the set of outputs.

67. $g(x) = -2x^2$ **69.** $r(x) = \dfrac{32}{x}$ **71.** $3 + h$

73. $3x^2 + 3xc + c^2$ **75.** 3   **77.** $A = \dfrac{C^2}{4\pi}$

79. (a)

Height, x	Length and Width	Volume, V
1	$24 - 2(1)$	$1[24 - 2(1)]^2 = 484$
2	$24 - 2(2)$	$2[24 - 2(2)]^2 = 800$
3	$24 - 2(3)$	$3[24 - 2(3)]^2 = 972$
4	$24 - 2(4)$	$4[24 - 2(4)]^2 = 1024$
5	$24 - 2(5)$	$5[24 - 2(5)]^2 = 980$
6	$24 - 2(6)$	$6[24 - 2(6)]^2 = 864$

Maximum when $x = 4$

(b)

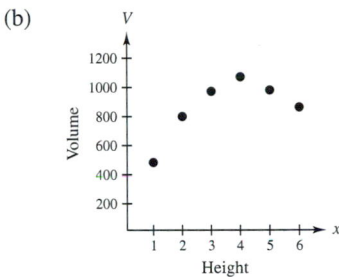

V is a function of x.

(c) $V = x(24 - 2x)^2, \quad 0 < x < 12$

81. $A = \dfrac{x^2}{x - 1}, \quad x > 1$

83. $V = x^2 y = x^2(108 - 4x) = 108x^2 - 4x^3,$
$0 < x < 27$

85. (a) $C = 12.30x + 98,000$

(b) $R = 17.98x$

(c) $P = 5.68x - 98,000$

87. (a) $R = \dfrac{240n - n^2}{20}$

(b)

n	90	100	110	120	130	140	150
R(n)	\$675	\$700	\$715	\$720	\$715	\$700	\$675

The revenue is maximum when 120 people take the trip.

89. (a)

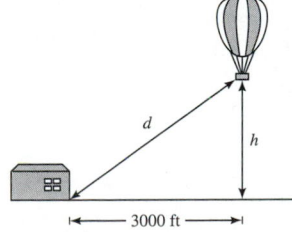

(b) $h = \sqrt{d^2 - 3000^2}, \quad [3000, \infty)$

91. $\dfrac{15}{8}$ **93.** $-\dfrac{1}{5}$

Section 2.3   *(page 218)*
Warm Up   *(page 218)*

1. -8 **2.** 0 **3.** $-\dfrac{3}{x}$ **4.** $x^2 + 3$ **5.** $0, \pm 4$

6. $\dfrac{1}{2}, 1$ **7.** All real numbers $x \neq 4$

8. All real numbers $x \neq 4, 5$ **9.** $t \leq \dfrac{5}{3}$

10. All real numbers

1. Domain: all real numbers
Range: $(-\infty, 1]$

3. Domain: $(-\infty, -1], [1, \infty)$
Range: $[0, \infty)$

5. Domain: $[-4, 4]$
Range: $[0, 4]$

7. Function   **9.** Not a function   **11.** Function

13. Yes. For each value of y there corresponds one and only one value of x.

15. Second setting   **17.** First setting

19. (a) Increasing on $(-\infty, \infty)$ (b) Odd function

21. (a) Increasing on $(-\infty, 0), (2, \infty)$
Decreasing on $(0, 2)$

(b) Neither even nor odd

23. (a)

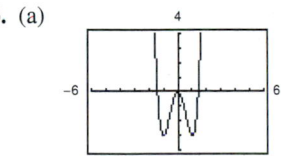

(b) Increasing on $(-1, 0), (1, \infty)$
Decreasing on $(-\infty, -1), (0, 1)$

(c) Even function

25. (a)

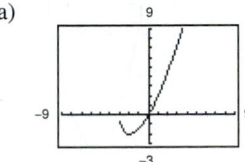

(b) Increasing on $(-2, \infty)$
Decreasing on $(-3, -2)$

(c) Neither even nor odd

27. Even   **29.** Odd   **31.** Neither even nor odd

33. (a) $\left(\frac{3}{2}, 4\right)$ (b) $\left(\frac{3}{2}, -4\right)$

35. Even

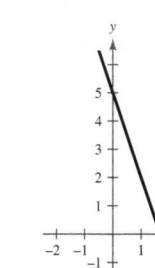

37. Neither even nor odd

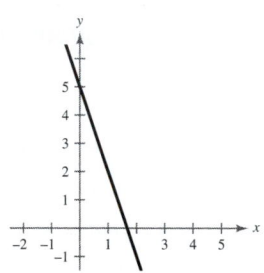

39. Even

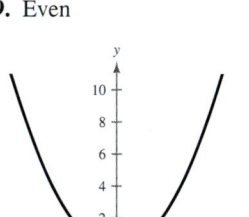

41. Neither even nor odd

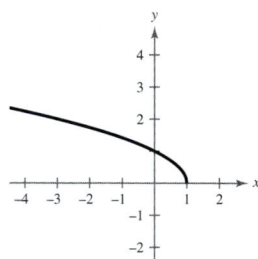

43. Neither even nor odd

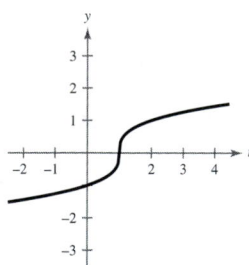

45. Neither even nor odd

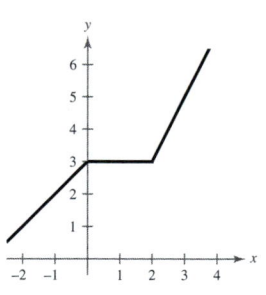

47. $(-\infty, 4]$

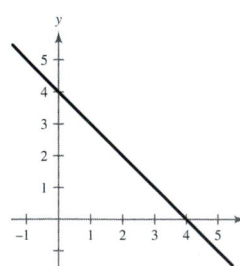

49. $(-\infty, -3], [3, \infty)$

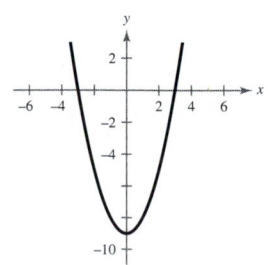

51. $[-1, 1]$

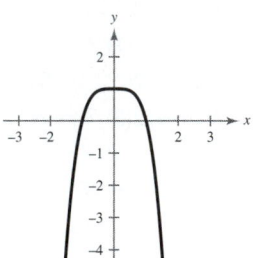

53. $(-\infty, \infty)$

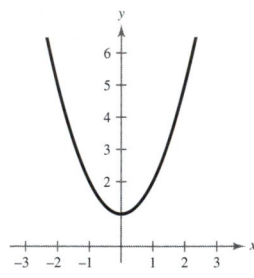

55. $f(x) < 0$ for all x

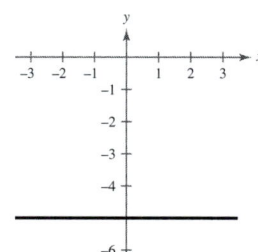

57.

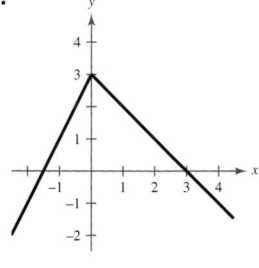

59.

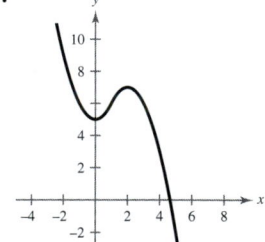

61.

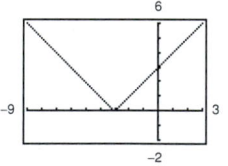

Domain: all real numbers
Range: $[0, \infty)$

63.

Domain: $(-\infty, \infty)$
Range: $[0, 2)$
Sawtooth pattern

65. (a)

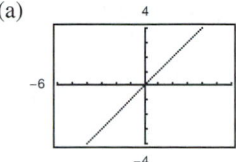

(b)

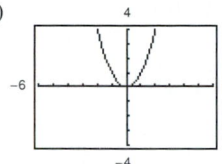

(c)

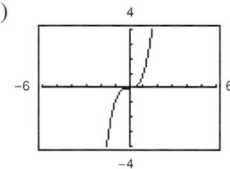

(d)

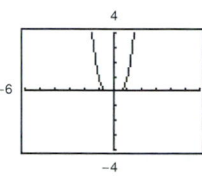

(e)

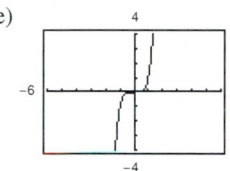

(f)

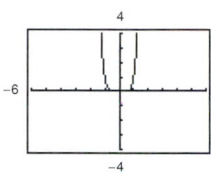

All the graphs pass through the origin. The graphs of the odd powers of x are symmetric with respect to the origin, and the graphs of the even powers are symmetric with respect to the y-axis. As the powers increase, the graphs become flatter in the interval $-1 < x < 1$.

67. (a) C_2 is the appropriate model, because the cost does not increase until after the next minute of conversation has started.

(b) $7.85

69. 350,000 units

71. $h = -x^2 + 4x - 3$　　**73.** $h = 2x - x^2$

75. $L = \frac{1}{2}y^2$　　**77.** $L = 4 - y^2$

79. (a) Domain: $-4 \le t \le 3$　　(c) 1992, 1991

(b)

(d) Because the balance would continue to decrease.

81. (a) $10,000　　(b) 50,000,000　　(c) 1%↓

83. Answers will vary.　　**85.** 0, 10　　**87.** 0, $\pm i$

Section 2.4 *(page 232)*

Warm Up *(page 232)*

1. $\dfrac{1}{x(1 - x)}$　　**2.** $-\dfrac{12}{(x + 3)(x - 3)}$　　**3.** $\dfrac{3x - 2}{x(x - 2)}$

4. $\dfrac{4x - 5}{3(x - 5)}$　　**5.** $\sqrt{\dfrac{x - 1}{x + 1}}$　　**6.** $\dfrac{x + 1}{x(x + 2)}$

7. $5(x - 2)$　　**8.** $\dfrac{x + 1}{(x - 2)(x + 3)}$　　**9.** $\dfrac{1 + 5x}{3x - 1}$

10. $\dfrac{x + 4}{4x}$

1. (a)

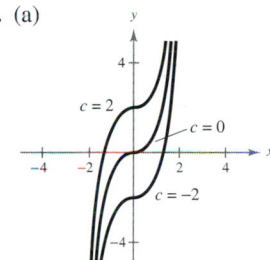

(b)

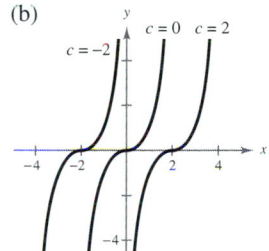

3. (a)

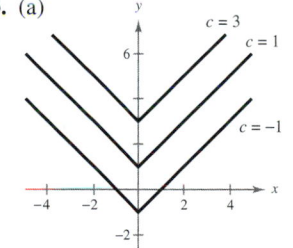

(b)

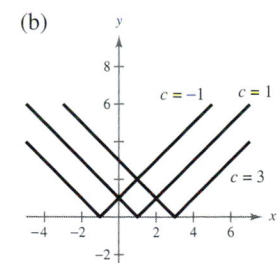

(c)

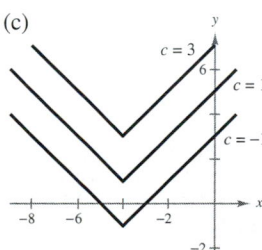

5. (a)

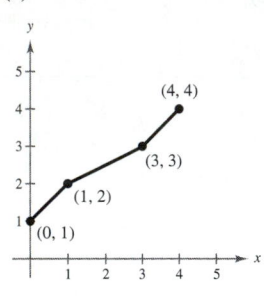

(b)

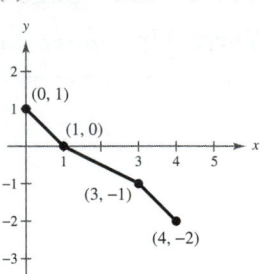

(c)

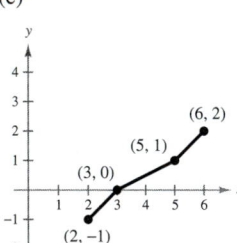

(d)

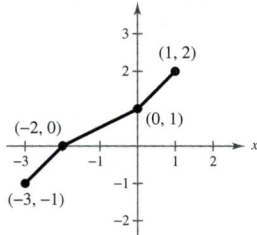

(e)

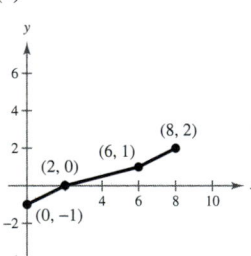

(f)

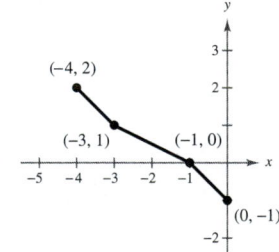

7. (a) $y = x^2 - 1$ (b) $y = 1 - (x + 1)^2$

9. Horizontal shift of $y = x^3$

$y = (x - 2)^3$

11. Reflection in the x-axis of $y = x^2$

$y = -x^2$

13. Reflection in the x-axis and a vertical shift of $y = \sqrt{x}$

$y = 1 - \sqrt{x}$

15.

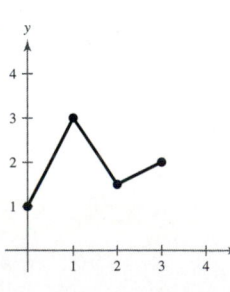

17.

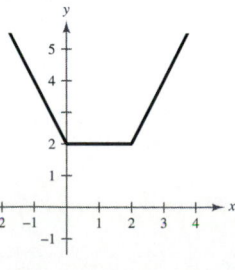

19. (a) $2x$ (b) 2 (c) $x^2 - 1$ (d) $\dfrac{x + 1}{x - 1}$, $x \neq 1$

21. (a) $x^2 - x + 1$ (b) $x^2 + x - 1$

(c) $x^2 - x^3$ (d) $\dfrac{x^2}{1 - x}$, $x \neq 1$

23. (a) $x^2 + 5 + \sqrt{1 - x}$ (b) $x^2 + 5 - \sqrt{1 - x}$

(c) $(x^2 + 5)\sqrt{1 - x}$ (d) $\dfrac{x^2 + 5}{\sqrt{1 - x}}$, $x < 1$

25. (a) $\dfrac{x + 1}{x^2}$ (b) $\dfrac{x - 1}{x^2}$ (c) $\dfrac{1}{x^3}$ (d) x, $x \neq 0$

27. 9   **29.** 5   **31.** $4t^2 - 2t + 5$ **33.** 0

35. 26   **37.** $\frac{3}{5}$

39.

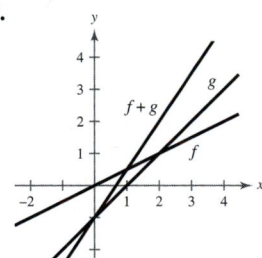

41.

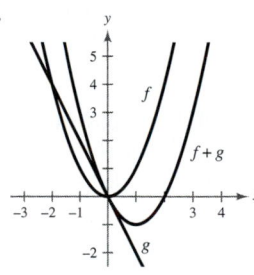

43.

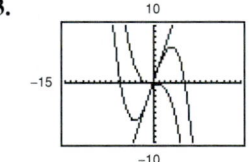

$f(x), g(x)$

45. $T = \frac{3}{4}x + \frac{1}{15}x^2$

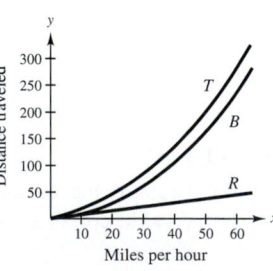

47.

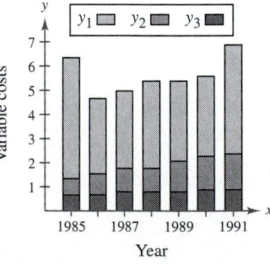

49. (a) For each time t there corresponds one and only one temperature T.

(b) $60°$, $72°$

(c) All the temperature changes would be 1 hour later.

(d) The temperature would be decreased by 1 degree.

51. (a) $(x - 1)^2$ (b) $x^2 - 1$ (c) x^4

53. (a) $20 - 3x$ (b) $-3x$ (c) $9x + 20$

55. (a) $\sqrt{x^2 + 4}$ (b) $x + 4$

57. (a) $x - \frac{8}{3}$ (b) $x - 8$ **59.** (a) $\sqrt[4]{x}$ (b) $\sqrt[4]{x}$

61. (a) $|x + 6|$ (b) $|x| + 6$

63. (a) 3 (b) 0 **65.** (a) 0 (b) 4

67. 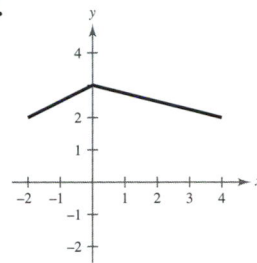   **69.**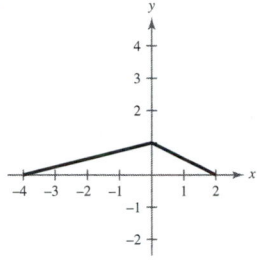

71. $f(x) = x^2$, $g(x) = 2x + 1$

73. $f(x) = \sqrt[3]{x}$, $g(x) = x^2 - 4$

75. $f(x) = \dfrac{1}{x}$, $g(x) = x + 2$

77. (a) $x \geq 0$ (b) All real numbers   (c) All real numbers

79. (a) All real numbers $x \neq \pm 1$ (b) All real numbers

(c) All real numbers $x \neq -2, 0$

81. 3 **83.** $\dfrac{-4}{x(x + h)}$

85. (a) $r(x) = \dfrac{x}{2}$ (b) $A(r) = \pi r^2$

(c) $(A \circ r)(x) = \pi \left(\dfrac{x}{2}\right)^2$;

$A \circ r$ represents the area of the circular base of the tank with radius $x/2$.

87. $(C \circ x)(t) = 3000t + 750$

$C \circ x$ represents the cost after t production hours.

89. (a) $R = p - 1200$ (b) $S = 0.92p$

(c) $(R \circ S)(p) = 0.92p - 1200$

$(S \circ R)(p) = 0.92(p - 1200)$

(d) $(R \circ S)(18,400) = 15,728$

$(S \circ R)(18,400) = 15,824$

91. Odd

Section 2.5 *(page 244)*

Warm Up *(page 244)*

1. All real numbers   **2.** $[-1, \infty)$

3. All real numbers $x \neq 0, 2$

4. All real numbers $x \neq -\frac{5}{3}$ **5.** x **6.** x

7. x **8.** x **9.** $x = \dfrac{3}{2}y + 3$ **10.** $x = \dfrac{y^3}{2} + 2$

1. c   **3.** a   **5.** $f^{-1}(x) = \frac{1}{8}x$

7. $f^{-1}(x) = x - 10$ **9.** $f^{-1}(x) = x^3$

11. (a) $f(g(x)) = f\left(\dfrac{x}{2}\right) = 2\left(\dfrac{x}{2}\right) = x$

$g(f(x)) = g(2x) = \dfrac{(2x)}{2} = x$

(b)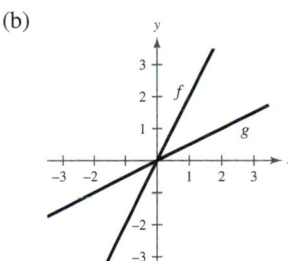

13. (a) $f(g(x)) = f\left(\dfrac{x - 1}{5}\right) = 5\left(\dfrac{x - 1}{5}\right) + 1 = x$

$g(f(x)) = g(5x + 1) = \dfrac{(5x + 1) - 1}{5} = x$

(b)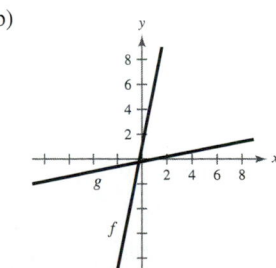

15. (a) $f(g(x)) = f\left(\sqrt[3]{x}\right) = \left(\sqrt[3]{x}\right)^3 = x$

$g(f(x)) = g(x^3) = \sqrt[3]{x^3} = x$

(b)

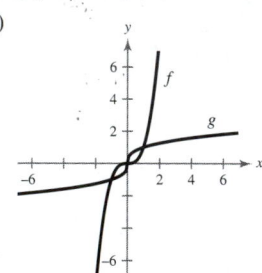

17. (a) $f(g(x)) = f(x^2 + 4), \quad x \geq 0$

$= \sqrt{(x^2 + 4) - 4} = x$

$g(f(x)) = g\left(\sqrt{x - 4}\right)$

$= \left(\sqrt{x - 4}\right)^2 + 4 = x$

(b)

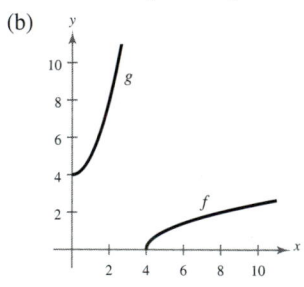

19. (a) $f(g(x)) = f\left(\sqrt{9 - x}\right), \quad x \leq 9$

$= 9 - \left(\sqrt{9 - x}\right)^2 = x$

$g(f(x)) = g(9 - x^2), \quad x \geq 0$

$= \sqrt{9 - (9 - x^2)} = x$

(b)

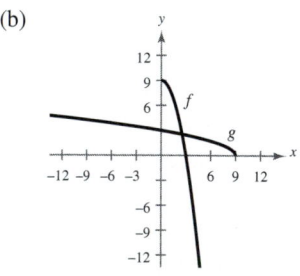

21. No   **23.** Yes   **25.** No

27.

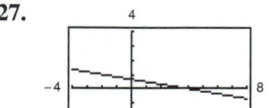

Yes

29.

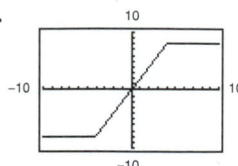

No

31.

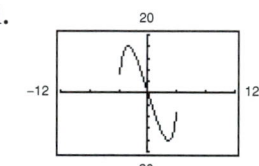

No

33. $f^{-1}(x) = \dfrac{x + 3}{2}$

35. $f^{-1}(x) = \sqrt[5]{x}$

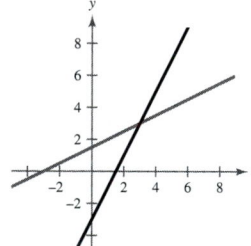

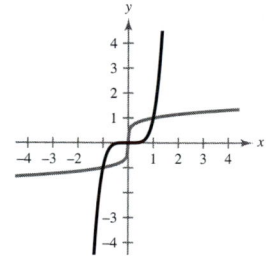

37. $f^{-1}(x) = x^2, \quad x \geq 0$

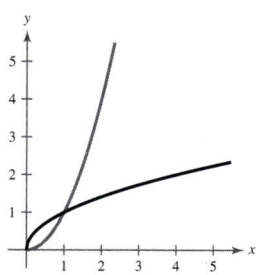

39. $f^{-1}(x) = \sqrt{4 - x^2}, \quad 0 \leq x \leq 2$

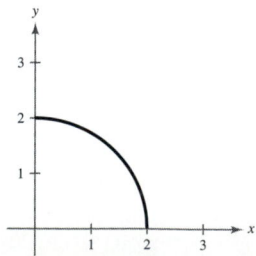

41. $f^{-1}(x) = x^3 + 1$

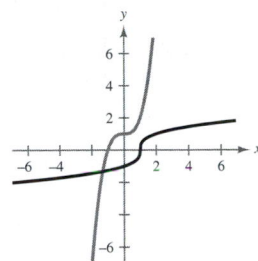

43. No inverse   **45.** $g^{-1}(x) = 8x$ **47.** No inverse

49. $f^{-1}(x) = \sqrt{x} - 3, \quad x \geq 0$ **51.** $h^{-1}(x) = \dfrac{1}{x}$

53. $f^{-1}(x) = \dfrac{x^2 - 3}{2}, \quad x \geq 0$ **55.** No inverse

57. $f^{-1}(x) = -\sqrt{25 - x}, \quad x \leq 25$

59. $y = \sqrt{x} + 2, \quad x \geq 0$ **61.** $y = x - 2, \quad x \geq 0$

63.

x	-4	-2	2	3
$f^{-1}(x)$	-2	-1	1	3

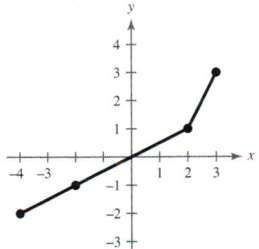

65. False. $f(x) = x^2$ **67.** True   **69.** 32

71. 600   **73.** $2\sqrt[3]{x + 3}$ **75.** $\dfrac{x + 1}{2}$ **77.** $\dfrac{x + 1}{2}$

79. (a) $y = \dfrac{x - 8}{0.75}$

   (b) y = number of units produced; x = hourly wage

   (c) 19

81. (a) $y = \sqrt{\dfrac{x - 254.50}{0.03}}$

   x = degrees Fahrenheit

   y = % load

(b)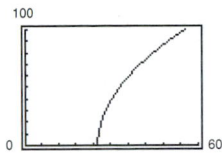
   100
   0   600
   0

(c) $0 < x < 90.46$

83. (a) Yes

   (b) f^{-1} yields the year for a given average fuel consumption.

   (c) 8

85. ± 8 **87.** $\frac{3}{2}$ **89.** $3 \pm \sqrt{5}$ **91.** $5, -\frac{10}{3}$

93. 16, 18   **95.** $b = h = 2\sqrt{5}$

Focus on Concepts *(page 248)*

1. No. The slope cannot be determined without knowing the scale on the y-axis. The slopes could be the same.

2. -4. The slope with the greatest magnitude corresponds to the steepest line.

3. V-intercept: Initial cost; Slope: Annual depreciation

4. No. The element 3 in the domain corresponds to two elements in the range.

5. (a)

Xmin = -15
Xmax = 6
Xscl = 3
Ymin = -18
Ymax = 6
Yscl = 3

(b)

Xmin = -24
Xmax = 36
Xscl = 6
Ymin = -54
Ymax = 12
Yscl = 6

6. (a) Even. The graph is a reflection in the x-axis.

   (b) Even. The graph is a reflection in the y-axis.

   (c) Even. The graph is a vertical translation of f.

   (d) Neither. The graph is a horizontal translation of f.

7. (a) $g(t) = \frac{3}{4} f(t)$ (b) $g(t) = f(t) + 10,000$

   (c) $g(t) = f(t - 2)$

Review Exercises *(page 249)*

1. (a) L_2 (b) L_3 (c) L_1

3.

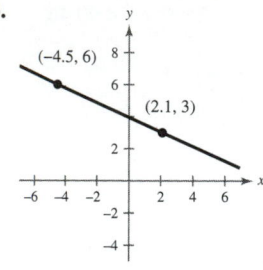

$$m = -\frac{5}{11}$$

5. $t = \frac{7}{3}$ **7.** $(6, 0)$, $(10, 1)$, $(-2, -2)$ **9.** $x = 0$

11.

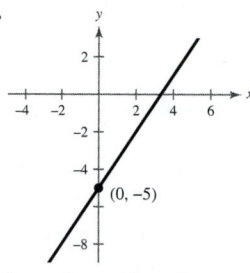

$$3x - 2y - 10 = 0$$

13. (a) $5x - 4y - 23 = 0$ (b) $4x + 5y - 2 = 0$

15. $y = 850x + 7400$ **17.** $x + y - 1 = 0$

19. $210,000

21. (a) is not a function, because 20 in the domain corresponds to two values in the range. (d) is not a function, because 30 is not matched with any element of B.

23. No   **25.** Yes

27. (a) 5   (b) 17   (c) $t^4 + 1$ (d) $-x^2 - 1$

29. $-5 \le x \le 5$ **31.** All real numbers $s \ne 3$

33. All real numbers $x \ne 3, 2$ **35.** Second setting

37. (a) Increasing: $(-2, 2)$ (b) Odd
        Constant: $(-\infty, -2], [2, \infty)$

39. (a) Increasing: $(-\infty, 3)$ (b) Neither
        Decreasing: $(3, \infty)$

41. (a) 16 feet per second   (b) 1.5 seconds
        (c) -16 feet per second

43. (a) $A = x(12 - x)$
        (b) $(0, 12)$

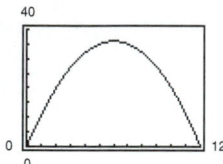

        (c) Maximum area: 36; 6×6; A square

45. (a) $R(n) = n[8 - 0.05(n - 80)]$, $n \ge 80$
        (b)

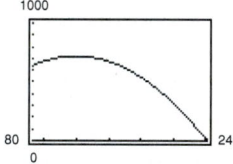

        120 passengers

47. (a) $f^{-1}(x) = 2x + 6$
        (b)

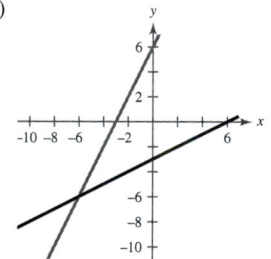

        (c) $f^{-1}(f(x)) = f^{-1}\left(\frac{1}{2}x - 3\right)$
            $= 2\left(\frac{1}{2}x - 3\right) + 6$
            $= x$
            $f(f^{-1}(x)) = f(2x + 6)$
            $= \frac{1}{2}(2x + 6) - 3$
            $= x$

49. (a) $f^{-1}(x) = x^2 - 1, \quad x \geq 0$

(b)

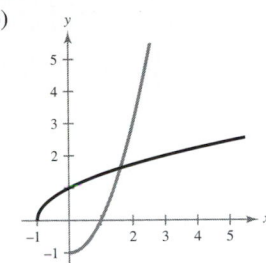

(c) $f^{-1}(f(x)) = f^{-1}\left(\sqrt{x+1}\right)$

$\qquad = (x+1) - 1$

$\qquad = x$

$f(f^{-1}(x)) = f(x^2 - 1), \quad x \geq 0$

$\qquad = \sqrt{x^2 - 1 + 1}$

$\qquad = x$

51. $x \geq 4, \quad f^{-1}(x) = \sqrt{\dfrac{x}{2} + 4}$ **53.** -7 **55.** 23

57. -55

Cumulative Test for Chapters P–2
(page 254)

1. $\dfrac{4x^3}{15y^5}$ **2.** $2x^2 y \sqrt{6y}$ **3.** $5x - 6$

4. $x^3 - x^2 - 5x + 6$ **5.** $\dfrac{s-1}{(s+1)(s+3)}$

6. $(x+3)(7-x)$ **7.** $x(x+1)(1-6x)$

8. $2(3-2x)(9+6x+4x^2)$

9.

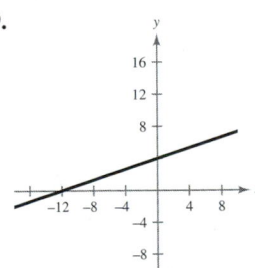

10.

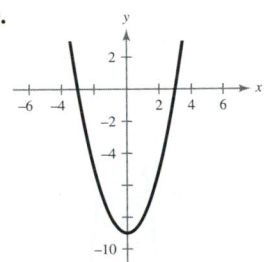

11.

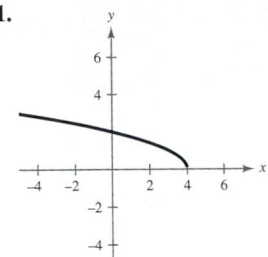

12. 7   **13.** $-1 \pm \frac{1}{3}\sqrt{3}$ **14.** 6

15. $-5 < x < 9$ **16.** $2x - y + 2 = 0$

17. For some value of x there correspond two values of y.

18. (a) $\frac{3}{2}$ (b) Division by 0 is undefined.

(c) $\dfrac{s+2}{s}$

19. (a) Vertical shrink by $\frac{1}{2}$

(b) Vertical shift of two units

(c) Horizontal shift of two units

20. $h^{-1}(x) = \frac{1}{5}(x+2)$ **21.** 9

CHAPTER 3

Section 3.1   *(page 263)*

Warm Up   *(page 263)*

1. $\frac{1}{2}, -6$ **2.** $-\frac{3}{5}, 3$ **3.** $\frac{3}{2}, -1$ **4.** -10

5. $3 \pm \sqrt{5}$ **6.** $-2 \pm \sqrt{3}$ **7.** $4 \pm \dfrac{\sqrt{14}}{2}$

8. $-5 \pm \dfrac{\sqrt{3}}{3}$ **9.** $-\dfrac{3}{2} \pm \dfrac{\sqrt{3}}{2} i$ **10** $-\dfrac{3}{2} \pm \dfrac{\sqrt{21}}{2}$

1. g   **3.** b   **5.** f   **7.** e

9.

(a)

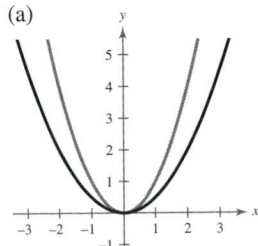

Vertical shrink

(b)

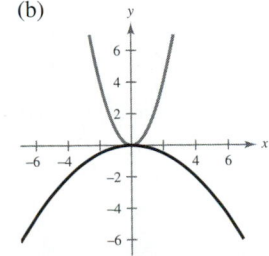

Vertical shrink and reflection in the x-axis

(c)

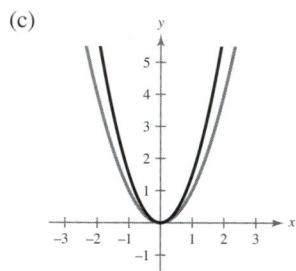

Vertical stretch

(d)

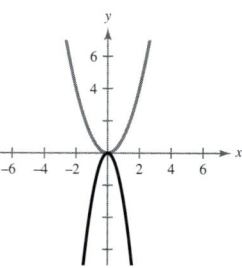

Vertical stretch and reflection in the x-axis

11. (a)

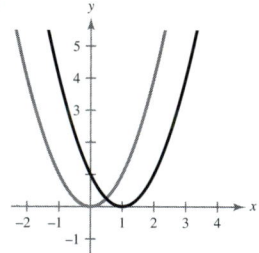

Horizontal translation

(b)

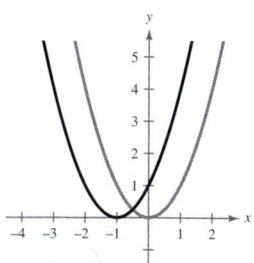

Horizontal translation

(c)

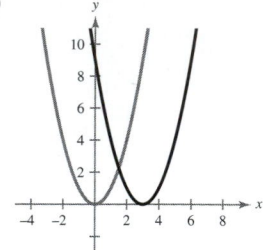

Horizontal translation

(d)

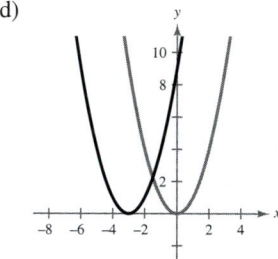

Horizontal translation

13. Vertex: $(0, -5)$

Intercepts: $(\pm\sqrt{5}, 0)$, $(0, -5)$

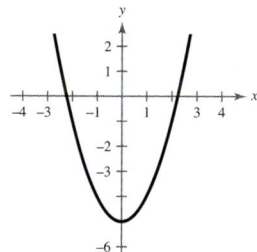

15. Vertex: $(0, 16)$

Intercepts: $(\pm 4, 0)$, $(0, 16)$

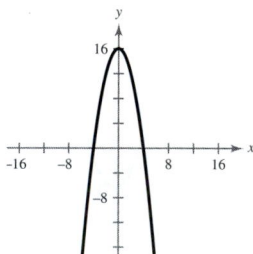

17. Vertex: $(-5, -6)$

Intercepts: $(-5 \pm \sqrt{6}, 0)$, $(0, 19)$

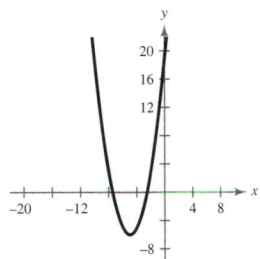

19. Vertex: $(4, 0)$

Intercepts: $(4, 0)$, $(0, 16)$

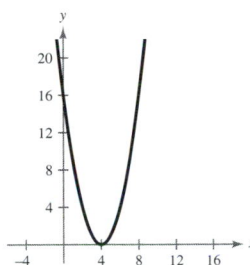

21. Vertex: $\left(\frac{1}{2}, 1\right)$

Intercept: $\left(0, \frac{5}{4}\right)$

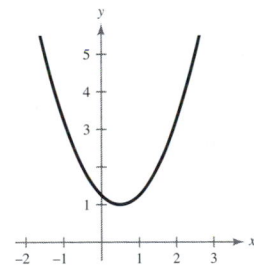

23. Vertex: $(1, 6)$

Intercepts: $\left(1 \pm \sqrt{6}, 0\right)$, $(0, 5)$

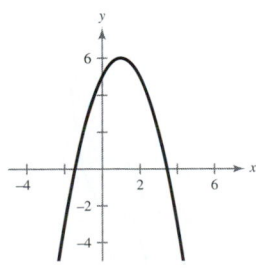

25. Vertex: $\left(\frac{1}{2}, 20\right)$

Intercept: $(0, 21)$

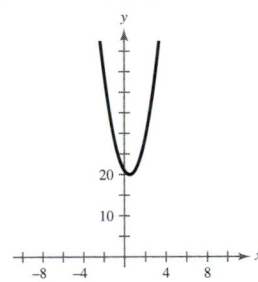

27. Vertex: $(-1, 4)$

Intercepts: $(1, 0)$, $(-3, 0)$, $(0, 3)$

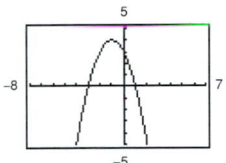

29. Vertex: $(4, -1)$

Intercepts: $\left(4 \pm \frac{1}{2}\sqrt{2}, 0\right)$, $(0, 31)$

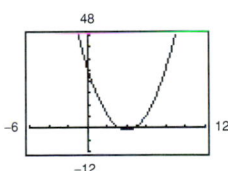

31. $y = (x - 1)^2$ **33.** $y = -(x + 1)^2 + 4$

35. $y = -2(x + 2)^2 + 2$ **37.** $f(x) = (x + 2)^2 + 5$

39. $f(x) = -\frac{1}{2}(x - 3)^2 + 4$ **41.** $f(x) = \frac{3}{4}(x - 5)^2 + 12$

43. $(\pm 4, 0)$ **45.** $(5, 0)$, $(-1, 0)$

47. 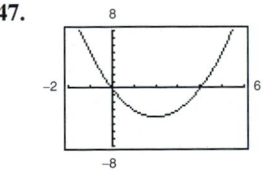   **49.**

$(0, 0)$, $(4, 0)$ $\left(-\frac{5}{2}, 0\right)$, $(6, 0)$

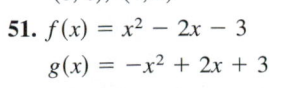

51. $f(x) = x^2 - 2x - 3$

$g(x) = -x^2 + 2x + 3$

(The answer is not unique.)

53. $f(x) = x^2 - 10x$

$g(x) = -x^2 + 10x$

(The answer is not unique.)

55. $f(x) = 2x^2 + 7x + 3$

$g(x) = -2x^2 - 7x - 3$

(The answer is not unique.)

57. 55, 55   **59.** 12, 6

61. (a) $A = x(50 - x), 0 < x < 50$

(b)

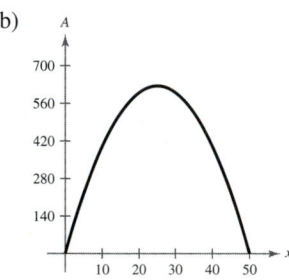

(c) 25 feet × 25 feet

63. (a)

x	y	Area
2	$\frac{1}{3}[200 - 4(2)]$	$(2)(2)\left(\frac{1}{3}\right)[200 - 4(2)] = 256$
4	$\frac{1}{3}[200 - 4(4)]$	$(2)(4)\left(\frac{1}{3}\right)[200 - 4(4)] \approx 491$
6	$\frac{1}{3}[200 - 4(6)]$	$(2)(6)\left(\frac{1}{3}\right)[200 - 4(6)] = 704$
8	$\frac{1}{3}[200 - 4(8)]$	$(2)(8)\left(\frac{1}{3}\right)[200 - 4(8)] = 896$
10	$\frac{1}{3}[200 - 4(10)]$	$(2)(10)\left(\frac{1}{3}\right)[200 - 4(10)] \approx 1067$
12	$\frac{1}{3}[200 - 4(12)]$	$(2)(12)\left(\frac{1}{3}\right)[200 - 4(12)] = 1216$

(b)

x	y	Area
20	$\frac{1}{3}[200 - 4(20)]$	$(2)(20)\left(\frac{1}{3}\right)[200 - 4(20)] = 1600$
22	$\frac{1}{3}[200 - 4(22)]$	$(2)(22)\left(\frac{1}{3}\right)[200 - 4(22)] \approx 1643$
24	$\frac{1}{3}[200 - 4(24)]$	$(2)(24)\left(\frac{1}{3}\right)[200 - 4(24)] = 1664$
26	$\frac{1}{3}[200 - 4(26)]$	$(2)(26)\left(\frac{1}{3}\right)[200 - 4(26)] = 1664$
28	$\frac{1}{3}[200 - 4(28)]$	$(2)(28)\left(\frac{1}{3}\right)[200 - 4(28)] \approx 1643$
30	$\frac{1}{3}[200 - 4(30)]$	$(2)(30)\left(\frac{1}{3}\right)[200 - 4(30)] = 1600$

(c) $A = \dfrac{8x(50 - x)}{3}$

(d)

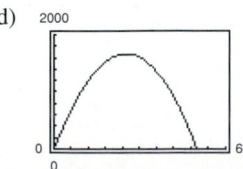

(e) $x = 25$ feet, $y = 33\frac{1}{3}$ feet

65. 4500 units   **67.** 20 fixtures   **69.** 350,000 units

71. (a) 4 feet   (b) 16 feet   (c) 25.86 feet

73. (a)

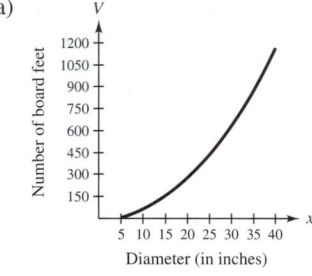

(b) 166.69 board feet

(c) 26.6 inches

75. (a)

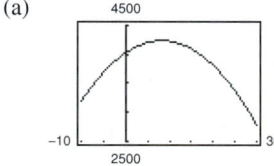

(b) 1968. Yes

(c) 4024.5 annually; 11 daily

77. Answers will vary.   **79.** $x + 3y - 5 = 0$

81. $y = \frac{5}{4}x + 3$

Section 3.2   *(page 277)*

Warm Up   *(page 277)*

1. $(3x - 2)(4x + 5)$ **2.** $x(5x - 6)^2$

3. $z^2(12z + 5)(z + 1)$ **4.** $(y + 5)(y^2 - 5y + 25)$

5. $(x + 3)(x + 2)(x - 2)$ **6.** $(x + 2)(x^2 + 3)$

7. No real solution   **8.** $3 \pm \sqrt{5}$ **9.** $-\frac{1}{2} \pm \sqrt{3}$

10. ± 3

1. c   **3.** h   **5.** a   **7.** d

9. (a)

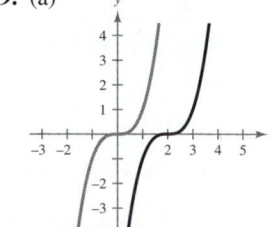

(b)

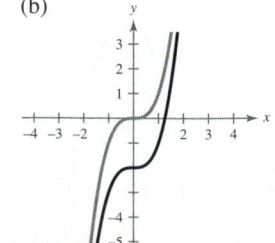

(c)

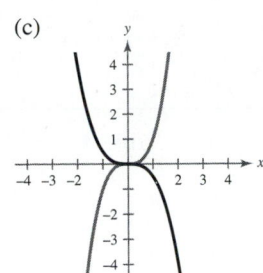

(d)

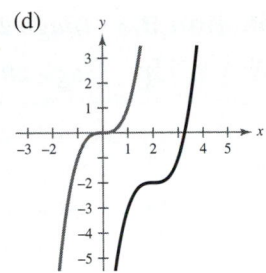

43.

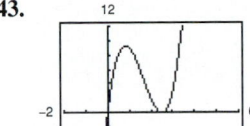

$(0, 0)$, $\left(\frac{5}{2}, 0\right)$

45.

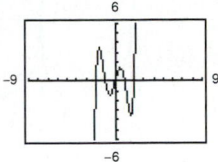

$(0, 0)$, $(\pm 1, 0)$, $(\pm 2, 0)$

47. $f(x) = x^2 - 10x$ **49.** $f(x) = x^2 + 4x - 12$

51. $f(x) = x^3 + 5x^2 + 6x$

53. $f(x) = x^4 - 4x^3 - 9x^2 + 36x$ **55.** $f(x) = x^2 - 2x - 2$

57. $(-1, 0)$, $(1, 2)$, $(2, 3)$ **59.** $(-2, -1)$, $(0, 1)$

11. (a)

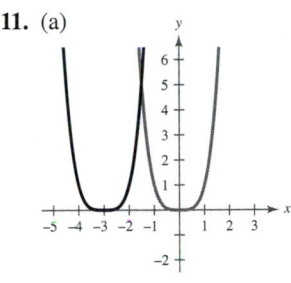

(b)

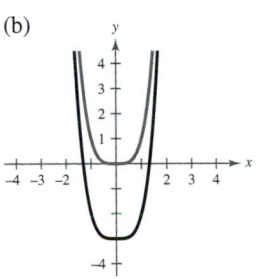

61.

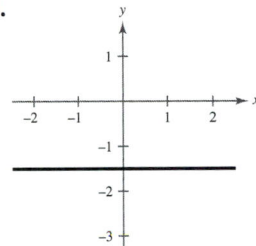

63.

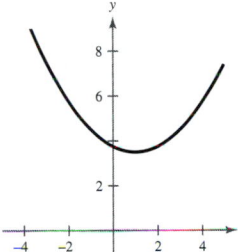

(c)

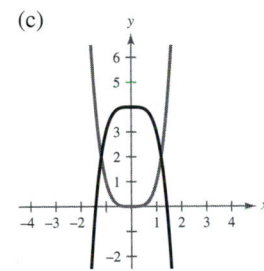

(d)

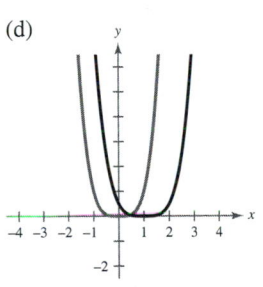

65.

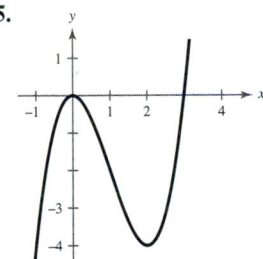

67.

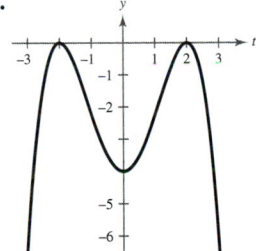

13. Falls to the left.
Rises to the right.

15. Falls to the left.
Falls to the right.

17. Rises to the left.
Falls to the right.

19. Rises to the left.
Falls to the right.

21. Falls to the left.
Falls to the right.

69.

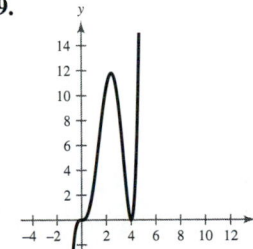

71.

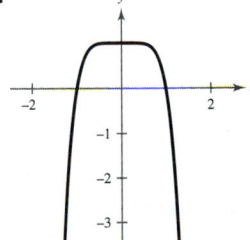

23.

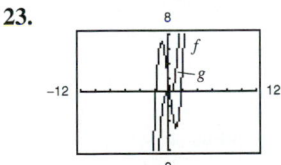

25.

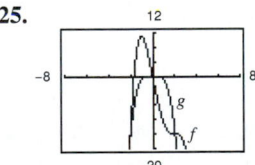

27. ± 5 **29.** 3 **31.** $1, -2$ **33.** $2 \pm \sqrt{3}$

35. $2, 0$ **37.** ± 1 **39.** $\pm \sqrt{5}$ **41.** No real zeros

73.

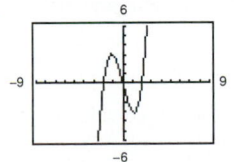

75.

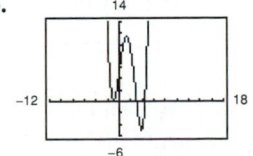

77. (a)

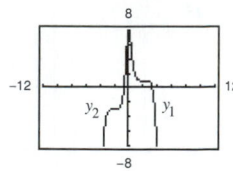

y_1 is decreasing. y_2 is increasing.

(b) Always increasing or decreasing, and it is determined by a.

(c)

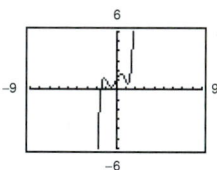

$$H \neq a(x - h)^5 + k$$

(d) Odd natural numbers

79.

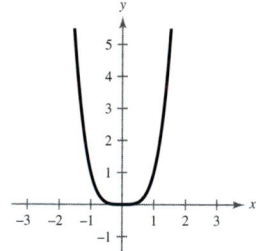

(a) Vertical shift of two units; Even

(b) Horizontal shift of two units;

    Neither even nor odd

(c) Reflection in the y-axis; Even

(d) Reflection in the x-axis; Even

(e) Vertical shrink; Even

(f) Vertical shrink; Even

(g) $g(x) = x^3$; Odd

(h) $g(x) = x^{16}$; Even

81. $x = 200$

83. (a) 0

(b) The graph does not pass the horizontal line test.

(c) 3

(d) 1

85. Yes. For each x there corresponds a unique value of y.

87. $x + 4$

Section 3.3 *(page 288)*

Warm Up *(page 288)*

---

1. $x^3 - x^2 + 2x + 3$ **2.** $2x^3 + 4x^2 - 6x - 4$

3. $x^4 - 2x^3 + 4x^2 - 2x - 7$

4. $2x^4 + 12x^3 - 3x^2 - 18x - 5$

5. $(x - 3)(x - 1)$ **6.** $2x(2x - 3)(x - 1)$

7. $x^3 - 7x^2 + 12x$ **8.** $x^2 + 5x - 6$

9. $x^3 + x^2 - 7x - 3$ **10.** $x^4 - 3x^3 - 5x^2 + 9x - 2$

---

1. Answers will vary.   **3.** Answers will vary.

5.

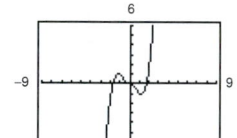

7. $2x + 4$ **9.** $x^2 - 3x + 1$ **11.** $x^3 + 3x^2 - 1$

13. $7 - \dfrac{11}{x + 2}$ **15.** $3x + 5 - \dfrac{2x - 3}{2x^2 + 1}$

17. $x^2 + 2x + 4 + \dfrac{2x - 11}{x^2 - 2x + 3}$

19. $x + 3 + \dfrac{6x^2 - 8x + 3}{(x - 1)^3}$ **21.** $x^{2n} + 6x^n + 9$

23. $3x^2 - 2x + 5$ **25.** $4x^2 - 9$ **27.** $-x^2 + 10x - 25$

29. $5x^2 + 14x + 56 + \dfrac{232}{x - 4}$

31. $10x^3 + 10x^2 + 60x + 360 + \dfrac{1360}{x - 6}$

33. $x^2 - 8x + 64$ **35.** $-3x^3 - 6x^2 - 12x - 24 - \dfrac{48}{x - 2}$

37. $-x^3 - 6x^2 - 36x - 36 - \dfrac{216}{x - 6}$

39. $4x^2 + 14x - 30$ **41.** The remainder is 0.

43. $c = -210$

45. $f(x) = (x - 4)(x^2 + 3x - 2) + 3$, $f(4) = 3$

47. $f(x) = (x - \sqrt{2})\left[x^2 + (3 + \sqrt{2})x + 3\sqrt{2}\right] - 8$,

    $f(\sqrt{2}) = -8$

49. (a) 1   (b) 4   (c) 4   (d) 1954

51. (a) 97   (b) $-\frac{5}{3}$ (c) 17   (d) -199

53. 0; $x + 3$ is a factor of f.

55. $(x - 2)(x + 3)(x - 1)$; zeros: 2, -3, 1

57. $(2x - 1)(x - 5)(x - 2)$; zeros: $\frac{1}{2}$, 5, 2

59. $\left(x + \sqrt{3}\right)\left(x - \sqrt{3}\right)(x + 2)$; zeros: $-\sqrt{3}, \sqrt{3}, -2$

61. $(x - 1)\left(x - 1 - \sqrt{3}\right)\left(x - 1 + \sqrt{3}\right)$;

zeros: 1, $1 + \sqrt{3}$, $1 - \sqrt{3}$

63. $f(x) = (x - 2)\left(x - \sqrt{5}\right)\left(x + \sqrt{5}\right)$

65. $h(t) = (t + 2)\left[t - \left(2 + \sqrt{3}\right)\right]\left[t - \left(2 - \sqrt{3}\right)\right]$

67. $2x^2 - x - 1$ **69.** $x^2 + 2x - 3$ **71.** $x^2 + 3x$

73. 3290 revolutions per minute

75. [Answers are not unique.]

(a) $f(x) = (x - 2)x^2 + 5 = x^3 - 2x^2 + 5$

(b) $f(x) = -(x + 3)x^2 + 1 = -x^3 - 3x^2 + 1$

77. $y = x^3 - 2$ **79.** $x \le -5, x \ge 5$

Section 3.4   *(page 298)*

Warm Up   *(page 298)*

1. $f(x) = 3x^3 - 8x^2 - 5x + 6$

2. $f(x) = 4x^4 - 3x^3 - 16x^2 + 12x$

3. $x^4 - 3x^3 + 5 + \dfrac{3}{x + 3}$ **4.** $3x^3 + 15x^2 - 9 - \dfrac{2}{x + \frac{2}{3}}$

5. $\frac{1}{2}, -3 \pm \sqrt{5}$ **6.** $10, -\frac{2}{3}, -\frac{3}{2}$ **7.** $-\frac{3}{4}, 2 \pm \sqrt{2}$

8. $\frac{2}{5}, -\frac{7}{2}, -2$ **9.** $\pm\sqrt{2}, \pm 1$ **10.** $\pm 2, \pm\sqrt{3}$

1. One negative zero

3. No positive or negative real zeros

5. No real zeros   **7.** One positive zero

9. One or three positive zeros   **11.** $\pm 1, \pm 3$

13. $\pm 1, \pm 3, \pm 5, \pm 9, \pm 15, \pm 45, \pm\frac{1}{2}, \pm\frac{3}{2}, \pm\frac{5}{2}, \pm\frac{9}{2}, \pm\frac{15}{2}, \pm\frac{45}{2}$

15. $\pm 1, \pm 2, \pm 4, \pm 8, \pm 16, \pm 32, \pm 5, \pm 10, \pm 20, \pm 40,$

$\pm 80, \pm 160, \pm 25, \pm 50, \pm 100, \pm 200, \pm 400, \pm 800,$

$\pm\frac{1}{2}, \pm\frac{5}{2}, \pm\frac{25}{2}, \pm\frac{1}{4}, \pm\frac{5}{4}, \pm\frac{25}{4}, \pm\frac{1}{5}, \pm\frac{2}{5}, \pm\frac{4}{5}, \pm\frac{8}{5}, \pm\frac{16}{5},$

$\pm\frac{32}{5}, \pm\frac{1}{10}, \pm\frac{1}{20}$

17. 1, 2, 3   **19.** 1, -1, 4   **21.** -1, -10 **23.** 1, 2

25. $\frac{1}{2}, -1$ **27.** $-2, 3, \pm\frac{2}{3}$ **29.** $-1, 2$ **31.** $-6, \frac{1}{2}, 1$

33. (a) $\pm 1, \pm 2, \pm 4$

(b)

(c) $-2, -1, 2$

35. (a) $\pm 1, \pm 3, \pm\frac{1}{2}, \pm\frac{3}{2}, \pm\frac{1}{4}, \pm\frac{3}{4}$

(b)

(c) $-\frac{1}{4}, 1, 3$

37. (a) $\pm 1, \pm 2, \pm 4, \pm 8, \pm\frac{1}{2}$

(b)

(c) $-\frac{1}{2}, 1, 2, 4$

39. (a) $\pm 1, \pm 3, \pm\frac{1}{2}, \pm\frac{3}{2}, \pm\frac{1}{4}, \pm\frac{3}{4}, \pm\frac{1}{8}, \pm\frac{3}{8}, \pm\frac{1}{16}, \pm\frac{3}{16},$

$\pm\frac{1}{32}, \pm\frac{3}{32}$

(b)

(c) $1, \frac{3}{4}, -\frac{1}{8}$

41. $\pm 1, \pm\sqrt{2}$

$f(x) = (x + 1)(x - 1)\left(x + \sqrt{2}\right)\left(x - \sqrt{2}\right)$

43. $x = 0, 3, 4, \pm\sqrt{2}$

$h(x) = x(x - 3)(x - 4)\left(x + \sqrt{2}\right)\left(x - \sqrt{2}\right)$

45. Answers will vary.   **47.** Answers will vary.

49. $1, -\frac{1}{2}$ **51.** $-\frac{3}{4}$ **53.** $\pm 2, \pm \frac{3}{2}$ **55.** $\pm 1, \frac{1}{4}$

57. d   **59.** b

61. (a)

(b) $V = x(9 - 2x)(15 - 2x)$

Domain: $0 < x < \frac{9}{2}$

(c)

Length of sides of
squares removed

$1.82 \times 5.36 \times 11.36$

(d) $\frac{1}{2}, \frac{7}{2}, 8$

8 is not in the domain of V.

63. r_1, r_2, r_3 **65.** $5 + r_1, 5 + r_2, 5 + r_3$

67. Cannot be determined.   **69.** $x \approx 38.4$

71. $x \approx 40$

73.

75.

77.

Section 3.5 (page 308)

Warm Up (page 308)

1. $4 - \sqrt{29}\,i, \; 4 + \sqrt{29}\,i$ **2.** $-5 - 12i, \; -5 + 12i$

3. $-1 + 4\sqrt{2}\,i, \; -1 - 4\sqrt{2}\,i$ **4.** $6 + \frac{1}{2}i, \; 6 - \frac{1}{2}i$

5. $-13 + 9i$ **6.** $12 + 16i$ **7.** $26 + 22i$

8. 29 **9.** i **10.** $-9 + 46i$

1. $0, 6, 6$ **3.** $3, 2, \pm 3i$ **5.** $(4, 0)$; Same

7. No intercepts; Same

9. $\pm 5i$; $(x + 5i)(x - 5i)$

11. $2 \pm \sqrt{3}$; $\left(x - 2 - \sqrt{3}\right)\left(x - 2 + \sqrt{3}\right)$

13. $\pm 3, \pm 3i$; $(x + 3)(x - 3)(x + 3i)(x - 3i)$

15. $1 \pm i$; $(z - 1 + i)(z - 1 - i)$

17. $2, 2 \pm i$; $(x - 2)(x - 2 + i)(x - 2 - i)$

19. $-5, 4 \pm 3i$; $(t + 5)(t - 4 + 3i)(t - 4 - 3i)$

21. $-2, 1 \pm \sqrt{2}\,i$; $(x + 2)\left(x - 1 + \sqrt{2}\,i\right)\left(x - 1 - \sqrt{2}\,i\right)$

23. $-\frac{1}{5}, 1 \pm \sqrt{5}\,i$; $(5x + 1)\left(x - 1 + \sqrt{5}\,i\right)\left(x - 1 - \sqrt{5}\,i\right)$

25. $2, \pm 2i$; $(x - 2)^2(x + 2i)(x - 2i)$

27. $\pm i, \pm 3i$; $(x + i)(x - i)(x + 3i)(x - 3i)$

29. $-10, -7 \pm 5i$ **31.** $-\frac{3}{4}, 1 \pm \frac{1}{2}i$ **33.** $-2, -\frac{1}{2}, \pm i$

35. $x^3 - x^2 + 25x - 25$ **37.** $x^3 + 4x^2 - 31x - 174$

39. $x^4 + 37x^2 + 36$ **41.** $3x^4 - 17x^3 + 25x^2 + 23x - 22$

43. $16x^4 + 36x^3 + 16x^2 + x - 30$

45. (a) $(x^2 + 9)(x^2 - 3)$

(b) $(x^2 + 9)\left(x + \sqrt{3}\right)\left(x - \sqrt{3}\right)$

(c) $(x + 3i)(x - 3i)\left(x + \sqrt{3}\right)\left(x - \sqrt{3}\right)$

47. (a) $(x^2 - 2x - 2)(x^2 - 2x + 3)$

(b) $\left(x - 1 + \sqrt{3}\right)\left(x - 1 - \sqrt{3}\right)(x^2 - 2x + 3)$

(c) $\left(x - 1 + \sqrt{3}\right)\left(x - 1 - \sqrt{3}\right)\left(x - 1 + \sqrt{2}\,i\right)$
$\left(x - 1 - \sqrt{2}\,i\right)$

49. $-\frac{3}{2}, \pm 5i$ **51.** $\pm 2i, 1, -\frac{1}{2}$ **53.** $-3 \pm i, \frac{1}{4}$

55. $2, -3 \pm \sqrt{2}\,i, 1$ **57.** $\frac{3}{4}, \frac{1}{2} \pm \frac{\sqrt{5}}{2}i$

59. (a) Answers will vary.

(b) f does not have real coefficients.

61. (a) $0 < k < 4$ (b) $k = 4$ (c) $k < 0$ (d) $k > 4$

63. No. Setting $h = 64$ and solving the resulting equation yields imaginary roots.

65. (a) $x^2 + b$ (b) $x^2 - 2ax + a^2 + b^2$

67.

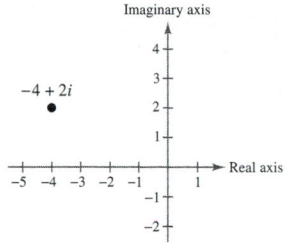

69. 630   **71.** 24 feet

Section 3.6 *(page 318)*

Warm Up *(page 318)*

1. $\frac{1}{3}$ **2.** $\frac{9}{16}$ **3.** $\frac{128}{3}$ **4.** 75   **5.** $\frac{275}{27}$
6. $\frac{105}{128}$ **7.** 27   **8.** $\frac{2}{9}$ **9.** $\frac{28}{13}$ **10.** 157.5

1.

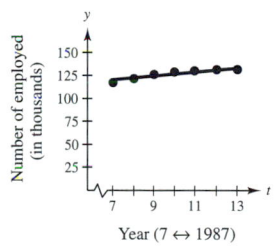

3. Inversely

5.

x	2	4	6	8	10
$y = kx^2$	4	16	36	64	100

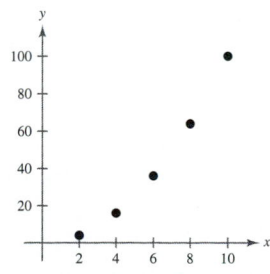

7.

x	2	4	6	8	10
$y = kx^2$	2	8	18	32	50

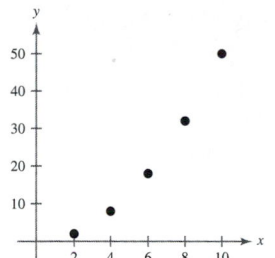

9.

x	2	4	6	8	10
$y = k/x^2$	$\frac{1}{2}$	$\frac{1}{8}$	$\frac{1}{18}$	$\frac{1}{32}$	$\frac{1}{50}$

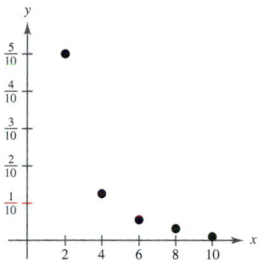

11.

x	2	4	6	8	10
$y = k/x^2$	$\frac{5}{2}$	$\frac{5}{8}$	$\frac{5}{18}$	$\frac{5}{32}$	$\frac{1}{10}$

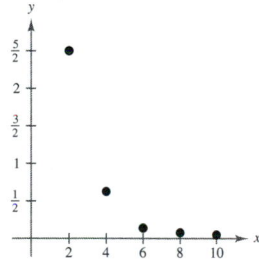

13. $y = \dfrac{5}{x}$ **15.** $y = -\dfrac{7}{10}x$ **17.** $y = \dfrac{12}{5}x$

19. $y = 205x$ **21.** $I = 0.075P$

23. $y = 0.0368x$ $3128

25.

Inches	5	10	20	25	30
Centimeters	12.7	25.4	50.8	63.5	76.2

27. (a) 0.05 meter   (b) $176\frac{2}{3}$ newtons   **29.** 39.47 pounds

31. $A = kr^2$ **33.** $y = \dfrac{k}{x^2}$ **35.** $z = k\sqrt[3]{u}$ **37.** $z = kuv$

39. $F = \dfrac{kg}{r^2}$ **41.** $P = \dfrac{k}{V}$ **43.** $F = \dfrac{km_1 m_2}{r^2}$

45. The area of a triangle is jointly proportional to the magnitude of the base and the height.

47. The volume of a sphere varies directly as the cube of its radius.

49. Average speed is directly proportional to the distance and inversely proportional to the time.

51. $A = \pi r^2$ **53.** $y = \dfrac{75}{x}$ **55.** $h = \dfrac{12}{t^3}$ **57.** $z = 2xy$

59. $F = 14rs^3$ **61.** $z = \dfrac{2x^2}{3y}$ **63.** $S = \dfrac{4L}{3(L - S)}$

65. 0.61 mile per hour   **67.** 506 feet   **69.** 400 feet

71. No. The 15-inch pizza is the best buy.

73. (a) The velocity is four-thirds the original.

(b) The velocity is decreased by one-fourth.

75. (a)   (b) Yes. $k \approx 44$

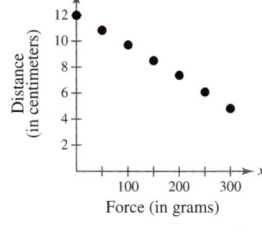

(c) 396 grams

77. (a)

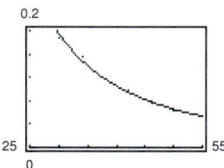

(b) 0.2857 microwatts per square centimeter

79. Good approximation   **81.** Poor approximation

83.

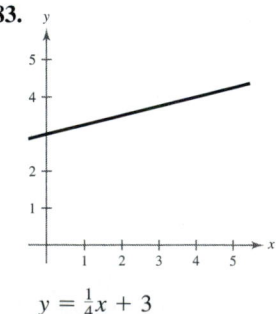

$y = \frac{1}{4}x + 3$

85.

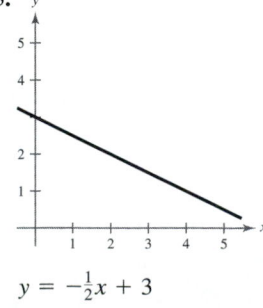

$y = -\frac{1}{2}x + 3$

87. (a) $y = 57.4t - 213.6$

(b)

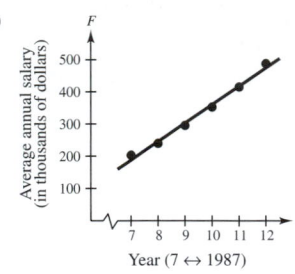

(c) 1994: 590.0   (d) Answers will vary.

1995: 647.4

1996: 704.8

89. (a) $y = 1.145t + 124.425$

(b)

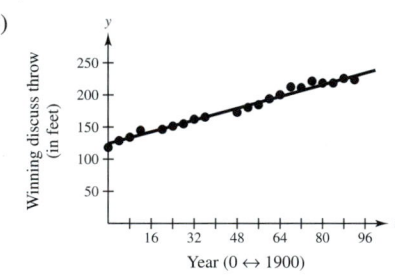

(c) 234.345

91. (a) $y = -0.74x + 106$

(b)

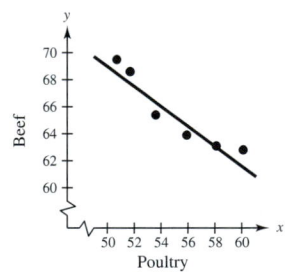

(c) 60.12 pounds

(d) For each 1 pound increase in per capita consumption of poultry, the per capita consumption of beef decreases by an average of 0.74 pounds.

Focus on Concepts *(page 325)*

1. Prefer the conditions (a) and (b) because profits would be increasing.

2. (a) Degree: 3; Leading coefficient: Positive

(b) Degree: 2; Leading coefficient: Positive

(c) Degree: 4; Leading coefficient: Positive

(d) Degree: 5; Leading coefficient: Positive

3. (a) No

(b) Yes   (c) Yes

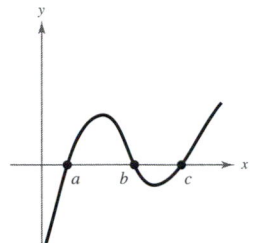   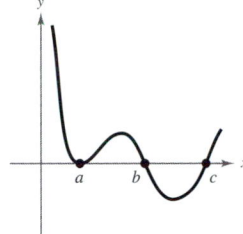

(d) Yes

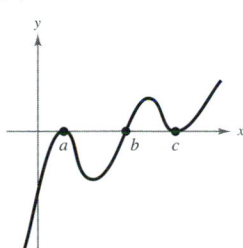

4. (a) First degree; Leading coefficient would be positive.

(b) Second degree; Leading coefficient would be positive.

(c) Second degree; Leading coefficient would be negative.

(d) First degree; Leading coefficient would be negative.

Review Exercises *(page 326)*

1.

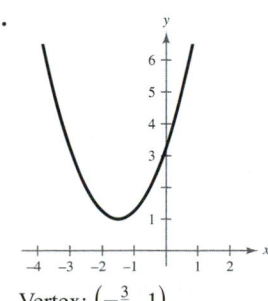

Vertex: $\left(-\frac{3}{2}, 1\right)$

Intercept: $\left(0, \frac{13}{4}\right)$

3.

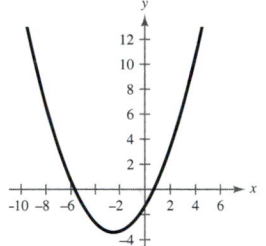

Vertex: $\left(-\frac{5}{2}, -\frac{41}{12}\right)$

Intercepts: $\left(0, -\frac{4}{3}\right), \left(\frac{-5 \pm \sqrt{41}}{2}, 0\right)$

5. $f(x) = -\frac{1}{2}(x - 4)^2 + 1$ **7.** $f(x) = (x - 1)^2 - 4$

9. (a)   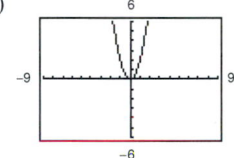   (b)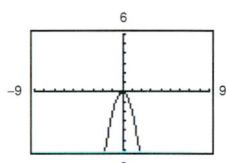

Vertical stretch   Vertical stretch and
                          reflection in the *x*-axis

(c)   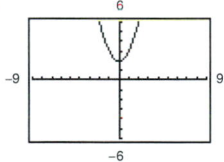   (d)

Vertical translation   Horizontal translation

11. Minimum: $(1, -1)$ **13.** Maximum: $(3, 9)$

15. Maximum: $(1, 3)$ **17.** Minimum: $\left(-\frac{5}{2}, -\frac{41}{4}\right)$

19. (a)

x	y	Area
1	$4 - \frac{1}{2}(1)$	$(1)[4 - \frac{1}{2}(1)] = \frac{7}{2}$
2	$4 - \frac{1}{2}(2)$	$(2)[4 - \frac{1}{2}(2)] = 6$
3	$4 - \frac{1}{2}(3)$	$(3)[4 - \frac{1}{2}(3)] = \frac{15}{2}$
4	$4 - \frac{1}{2}(4)$	$(4)[4 - \frac{1}{2}(4)] = 8$
5	$4 - \frac{1}{2}(5)$	$(5)[4 - \frac{1}{2}(5)] = \frac{15}{2}$
6	$4 - \frac{1}{2}(6)$	$(6)[4 - \frac{1}{2}(6)] = 6$

(b) $x = 4, \ y = 2$

(c) $A = x\left(\dfrac{8 - x}{2}\right), \quad 0 < x < 8$

(d)

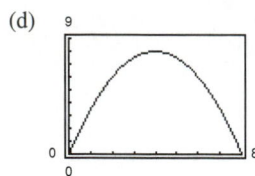

$x = 4, y = 2$

(e) $x = 4, y = 2$

21. $1020

23. Falls to the left.   **25.** Rises to the left.
Falls to the right.   Rises to the right.

27.

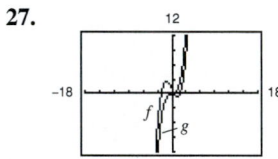

29.

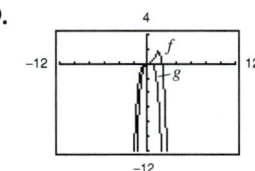

31.

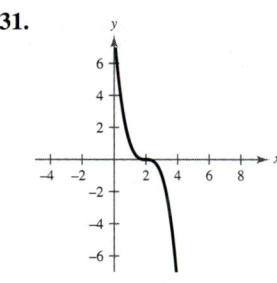

33.

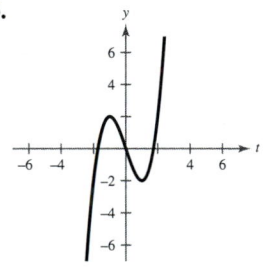

35. (a) $V = x^2(216 - 4x)$

(b)

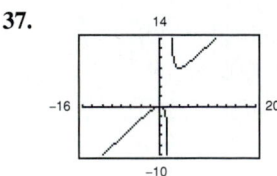

$x = 36$ centimeters, $y = 72$ centimeters

37.

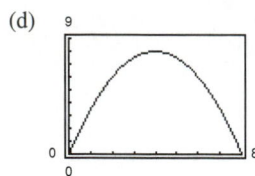

39. $8x + 5 + \dfrac{2}{3x - 2}$ **41.** $5x + 2$

43. $x^2 - 3x + 2 - \dfrac{1}{x^2 + 2}$ **45.** $6x^3 - 27x$

47. $2x^2 - (3 - 4i)x + (1 - 2i)$

49. (a) Yes   (b) Yes   (c) Yes   (d) No

51. (a) -421 (b) 96   **53.** $6x^4 + 13x^3 + 7x^2 - x - 1$

55. $3x^4 - 14x^3 + 17x^2 - 42x + 24$

57. Two or no positive zeros and one negative zero

59. $\pm 1, \pm 3, \pm 5, \pm 15, \pm\frac{1}{2}, \pm\frac{3}{2}, \pm\frac{5}{2}, \pm\frac{15}{2}, \pm\frac{1}{4}, \pm\frac{3}{4}, \pm\frac{5}{4}, \pm\frac{15}{4}$

61. $1, \frac{3}{4}$ **63.** $-1, \frac{3}{2}, 3, \frac{2}{3}$

65. (a)   **67.** (a)

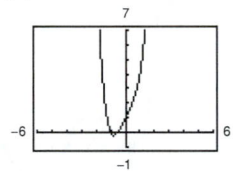   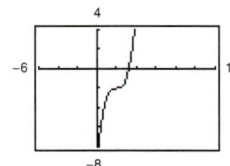

(b) two   (b) one

(c) $-1, -0.54$ (c) 3.26

69. (a)

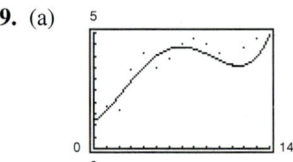

(b) Recession; Yes

(c) 0.47; More

(d) 6.72

71. $F = \frac{1}{3} x\sqrt{y}$ **73.** 2438.7 kilowatts

75.

Miles	2	5	10	12
Kilometers	3.2	8	16	19.2

Chapter Test *(page 332)*

1. (a) Reflection in the x-axis followed by a vertical translation

(b) Horizontal translation

2. Vertex: $(-2, -1)$

Intercepts: $(0, 3)$, $(-3, 0)$, $(-1, 0)$

3. $y = (x - 3)^2 - 6$

4. (a) 50 feet

(b) 5; Changing the constant term results in a vertical translation of the graph and, therefore, changes the maximum height.

5. Rises to the left.

Falls to the right.

6. $3x + \dfrac{x-1}{x^2+1}$ **7.** $2x^3 + 4x^2 + 3x + 6 + \dfrac{9}{x-2}$

8. $(4x-1)(x-\sqrt{3})(x+\sqrt{3})$

9. $\pm 1, \pm 2, \pm 3, \pm 4, \pm 6, \pm 8, \pm 12, \pm 24, \pm\frac{1}{2}, \pm\frac{3}{2}$

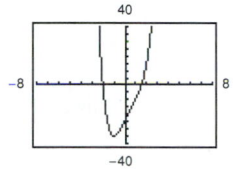

$-2, \frac{3}{2}$

10. $\pm 1, \pm 2, \pm\frac{1}{3}, \pm\frac{2}{3}$

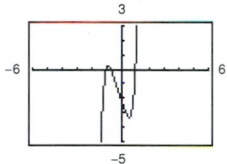

$\pm 1, -\frac{2}{3}$

11. $-0.819, 1.380$ **12.** $-1.414, 0.667, 1.414$

13. $f(x) = x^4 - 9x^3 + 28x^2 - 30x$

14. $f(x) = x^4 - 6x^3 + 16x^2 - 24x + 16$

CHAPTER 4

Section 4.1 *(page 339)*

Warm Up *(page 339)*

1. $(x-5)(x+2)$ **2.** $(x-5)(x-2)$

3. $x(x+1)(x+3)$ **4.** $(x^2-2)(x-4)$

5.
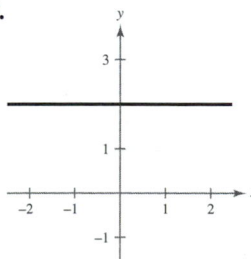

6.
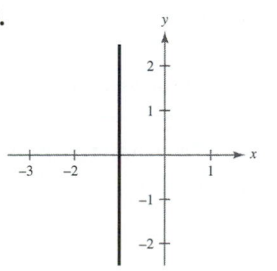

7.
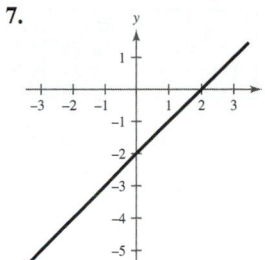

8.
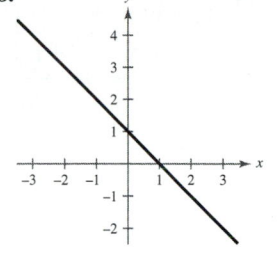

9. $x + 9 + \dfrac{42}{x-4}$ **10.** $x + 1 + \dfrac{2}{x+4}$

1. (a)

x	$f(x)$	x	$f(x)$	x	$f(x)$
0.5	-2	1.5	2	5	0.25
0.9	-10	1.1	10	10	$0.\overline{1}$
0.99	-100	1.01	100	100	$0.\overline{01}$
0.999	-1000	1.001	1000	1000	$0.\overline{001}$

(b) Vertical asymptote: $x = 1$

Horizontal asymptote: $y = 0$

(c) Domain: all real numbers $x \neq 1$

3. (a)

x	$f(x)$	x	$f(x)$	x	$f(x)$
0.5	3	1.5	9	5	3.75
0.9	27	1.1	33	10	$3.\overline{33}$
0.99	297	1.01	303	100	$3.\overline{03}$
0.999	2997	1.001	3003	1000	$3.\overline{003}$

(b) Vertical asymptote: $x = 1$

Horizontal asymptotes: $y = \pm 3$

(c) Domain: all real numbers $x \neq 1$

5. (a)

x	$f(x)$
0.5	-1
0.9	-12.79
0.99	-148.79
0.999	-1498

x	$f(x)$
1.5	5.4
1.1	17.29
1.01	152.3
1.001	1502.3

x	$f(x)$
5	3.125
10	$3.\overline{03}$
100	$3.\overline{0003}$
1000	3

   (b) Vertical asymptotes: $x = \pm 1$
   Horizontal asymptote: $y = 3$
   (c) Domain: all real numbers $x \neq \pm 1$

7. Domain: all real numbers $x \neq 0$
   Vertical asymptote: $x = 0$
   Horizontal asymptote: $y = 0$

9. Domain: all real numbers $x \neq 2$
   Vertical asymptote: $x = 2$
   Horizontal asymptote: $y = -1$

11. Domain: all real numbers $x \neq \pm 1$
   Vertical asymptotes: $x = \pm 1$

13. Domain: all real numbers
   Horizontal asymptote: $y = 3$

15. d   **17.** f   **19.** e

21. (a) Domain of f: all real numbers $x \neq -2$;
      Domain of g: all real numbers
   (b) Vertical asymptote: none
   (c)

x	-4	-3	-2.5	-2	-1.5	-1	0
$f(x)$	-6	-5	-4.5	Undef.	-3.5	-3	-2
$g(x)$	-6	-5	-4.5	-4	-3.5	-3	-2

Differ only where f is undefined.

23. (a) Domain of f: all real numbers $x \neq 0, 3$;
      Domain of g: all real numbers $x \neq 0$
   (b) Vertical asymptote: $x = 0$
   (c)

x	-1	-0.5	0	0.5	2	3	4
$f(x)$	-1	-2	Undef.	2	$\frac{1}{2}$	Undef.	$\frac{1}{4}$
$g(x)$	-1	-2	Undef.	2	$\frac{1}{2}$	$\frac{1}{3}$	$\frac{1}{4}$

Differ only where f is undefined and g is defined.

25. $f(x) = \dfrac{1}{x^2 + x - 2}$ **27.** $f(x) = \dfrac{2x^2}{1 + x^2}$

29. (a) 4   (b) Less than   (c) Greater than

31. (a) 2   (b) Greater than   (c) Less than

33. ± 2 **35.** 5

37.

M	200	400	600	800	1000
t	0.472	0.596	0.710	0.817	0.916

M	1200	1400	1600	1800	2000
t	1.009	1.096	1.168	1.178	1.328

The greater the mass, the more time required per oscillation.

39. (a) 28.33 million dollars
   (b) 170 million dollars
   (c) 765 million dollars
   (d) No. The function is undefined.

41. (a) 333 deer, 500 deer, 800 deer   (b) 1500

43. (a)

n	1	2	3	4	5	6
P	0.50	0.74	0.82	0.86	0.89	0.91

n	7	8	9	10
P	0.92	0.93	0.94	0.95

The percentage approaches 1 as n increases.
   (b) 100%

45. (a)

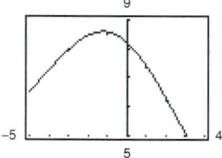

(b) 3.98%

(c) No. Model approaches 0.

Section 4.2 *(page 348)*

Warm Up *(page 348)*

1. $(0, 6)$, $(-3, 0)$ **2.** $\left(0, \frac{12}{5}\right)$, $(-3, 0)$

3. Odd   **4.** Even

5. Domain: all real numbers $x \neq 8$

Vertical asymptote: $x = 8$

Horizontal asymptote: $y = 0$

6. Domain: all real numbers $x \neq -\frac{1}{4}$

Vertical asymptote: $x = -\frac{1}{4}$

Horizontal asymptote: $y = \frac{3}{4}$

7. Domain: all real numbers $x \neq \pm 3$

Vertical asymptotes: $x = \pm 3$

Horizontal asymptote: $y = 2$

8. Domain: all real numbers $x \neq 0$

Vertical asymptote: $x = 0$

Horizontal asymptote: $y = 4$

9. $2x + \frac{7}{2} + \frac{23}{2(2x - 1)}$ **10.** $x + \frac{1}{x^2}$

1.

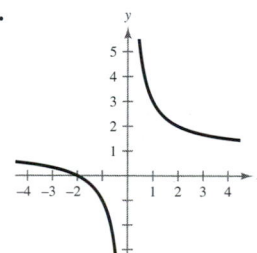

3.

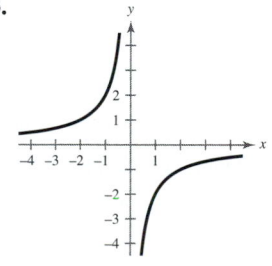

5.

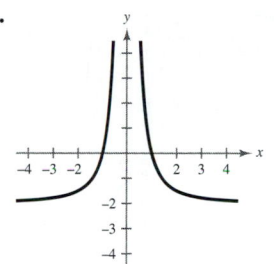

7.

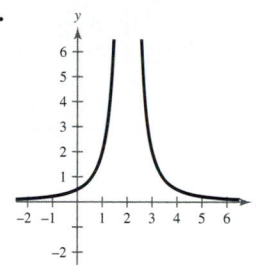

9.

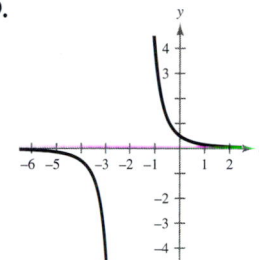

11.

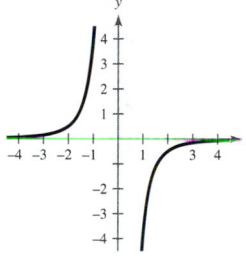

13.

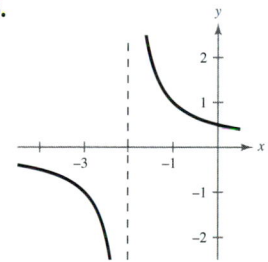

15.

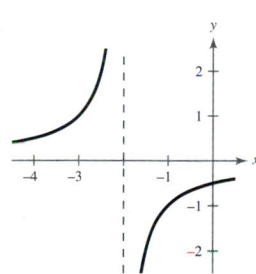

17.

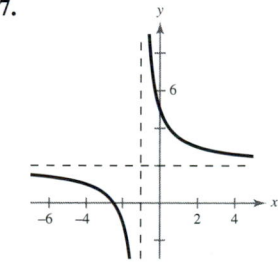

19.

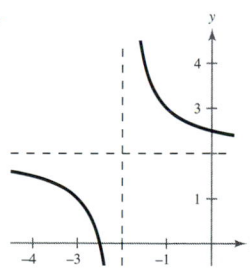

21.

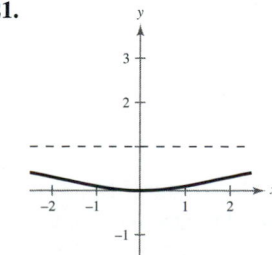

23.

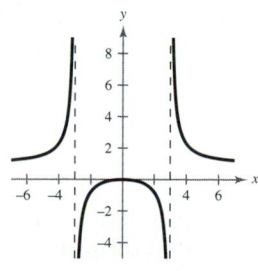

25.

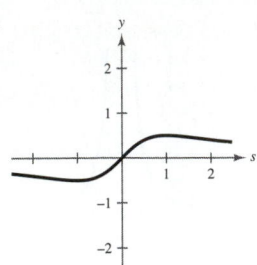

27.

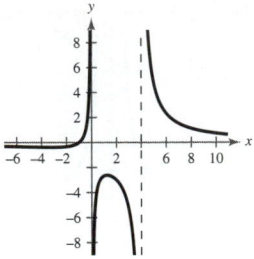

29.

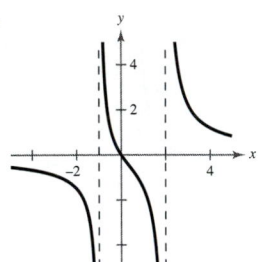

31.

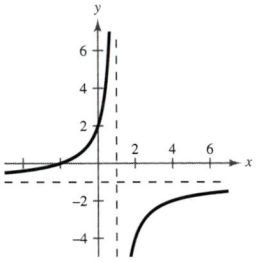

33.

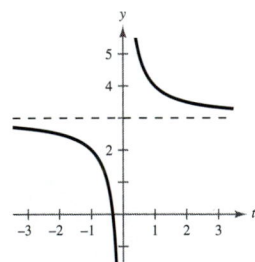

35. (a) Domain of f: all real numbers $x \neq -1$;

Domain of g: all real numbers

(b) Vertical asymptote: none

(c)

x	-3	-2	-1.5	-1	-0.5	0	1
$f(x)$	-4	-3	-2.5	Undef.	-1.5	-1	0
$g(x)$	-4	-3	-2.5	-2	-1.5	-1	0

(d)

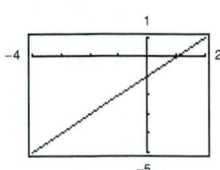

(e) Because there are only a finite number of pixels, the utility may not attempt to evaluate the function where it does not exist.

37. (a) Domain of f: all real numbers $x \neq 0, 2$;

Domain of g: all real numbers $x \neq 0$

(b) Vertical asymptote: $x = 0$

(c)

x	-0.5	0	0.5	1	1.5	2	3
$f(x)$	-2	Undef.	2	1	$\frac{2}{3}$	Undef.	$\frac{1}{3}$
$g(x)$	-2	Undef.	2	1	$\frac{2}{3}$	$\frac{1}{2}$	$\frac{1}{3}$

(d)

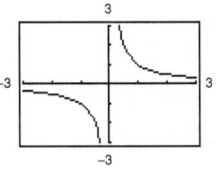

(e) Because there are only a finite number of pixels, the utility may not attempt to evaluate the function where it does not exist.

39.

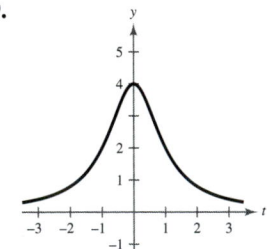

Domain: all real numbers

$y = 0$

41.

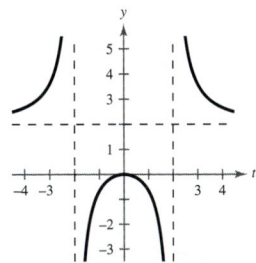

Domain: all real numbers $t \neq \pm 2$

$t = \pm 2, \ y = 2$

43.

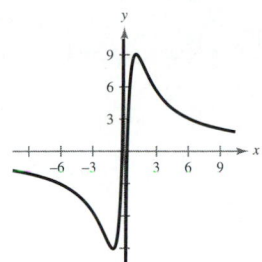

Domain: all real numbers $x \neq 0$

$x = 0, \ y = 0$

45.

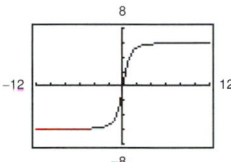

47.

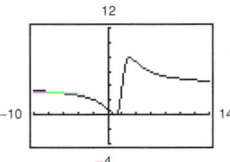

49.

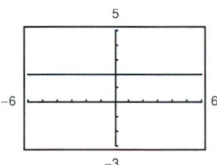

The fraction is not reduced.

51. No. There are two distinct branches of the graph.

53.

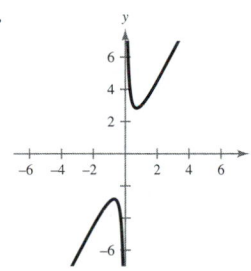

55.

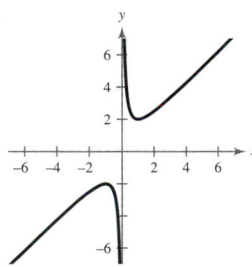

57.

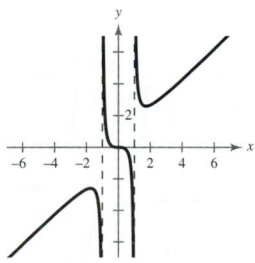

59.

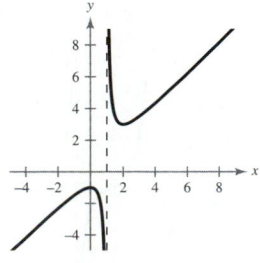

61.

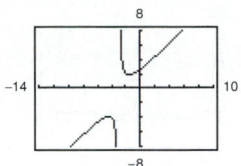

Domain: all real numbers $x \neq -3$

Vertical asymptote: $x = -3$

$y = x + 2$

63.

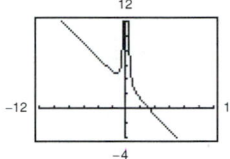

Domain: all real numbers $x \neq 0$

Vertical asymptote: $x = 0$

$y = -x + 3$

65. $(-1, 0)$ **67.** $(1, 0), \ (-1, 0)$ **69.** $(-4, 0)$

71. $(3, 0), \ (-2, 0)$

73. (a) Answers will vary.   (b) $[0, 950]$

(c)

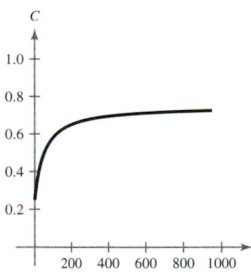

Increases more slowly; 0.75

75. (a) Answers will vary.

(b) $(4, \infty)$

(c)

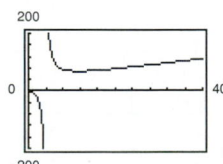

11.75×5.9 inches

77.

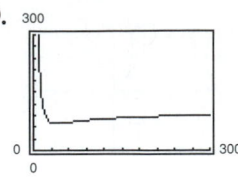

79.

Minimum: $(-2, -1)$ $x \approx 40$

Maximum: $(0, 3)$

81. (a) $C = 0$; The chemical will eventually dissipate.

(b)

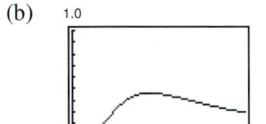

$t \approx 4.5$

83. $f(x) = \dfrac{x^2 - x - 6}{x - 2}$

85. $x \geq 3\frac{1}{3}$ **87.** $-3 < x < 7$

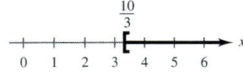

Section 4.3 *(page 359)*

Warm Up *(page 359)*

1. $\dfrac{5x + 2}{x(x + 1)}$ **2.** $\dfrac{2(4x + 3)}{x(x + 2)}$ **3.** $\dfrac{11x - 1}{(x - 2)(2x - 1)}$

4. $-\dfrac{3x + 1}{(x + 5)(x + 12)}$ **5.** $-\dfrac{5x + 6}{(x + 2)^2}$

6. $\dfrac{x^2 - 3x - 5}{(x - 3)^3}$ **7.** $-\dfrac{x + 9}{x(x^2 + 3)}$ **8.** $\dfrac{4x^2 + 5x + 31}{(x + 1)(x^2 + 5)}$

9. $\dfrac{x(3x + 1)}{(x^2 + 1)^2}$ **10.** $\dfrac{x^3 + x^2 + 1}{(x^2 + x + 1)^2}$

1. $\dfrac{A}{x} + \dfrac{B}{x - 14}$ **3.** $\dfrac{A}{x} + \dfrac{B}{x^2} + \dfrac{C}{x - 10}$

5. $\dfrac{A}{x} + \dfrac{Bx + C}{x^2 + 10}$ **7.** $\dfrac{1}{2}\left(\dfrac{1}{x - 1} - \dfrac{1}{x + 1}\right)$

9. $\dfrac{1}{x} - \dfrac{1}{x + 1}$ **11.** $\dfrac{1}{x} - \dfrac{2}{2x + 1}$ **13.** $\dfrac{1}{x - 1} - \dfrac{1}{x + 2}$

15. $-\dfrac{3}{x} - \dfrac{1}{x + 2} + \dfrac{5}{x - 2}$ **17.** $\dfrac{3}{x} - \dfrac{1}{x^2} + \dfrac{1}{x + 1}$

19. $\dfrac{3}{x - 3} + \dfrac{9}{(x - 3)^2}$ **21.** $-\dfrac{1}{x} + \dfrac{2x}{x^2 + 1}$

23. $\dfrac{1}{3(x^2 + 2)} - \dfrac{1}{6(x + 2)} + \dfrac{1}{6(x - 2)}$

25. $\dfrac{1}{8(2x + 1)} + \dfrac{1}{8(2x - 1)} - \dfrac{x}{2(4x^2 + 1)}$

27. $\dfrac{1}{x + 1} + \dfrac{2}{x^2 - 2x + 3}$

29. $x + 3 + \dfrac{6}{x - 1} + \dfrac{4}{(x - 1)^2} + \dfrac{1}{(x - 1)^3}$

31. $\dfrac{3}{2x - 1} - \dfrac{2}{x + 1}$ **33.** $\dfrac{2}{x} - \dfrac{1}{x^2} - \dfrac{2}{x + 1}$

35. $\dfrac{1}{x^2 + 2} + \dfrac{x}{(x^2 + 2)^2}$ **37.** $2x + \dfrac{1}{2}\left(\dfrac{3}{x - 4} - \dfrac{1}{x + 2}\right)$

39. $\dfrac{1}{2a}\left(\dfrac{1}{a + x} + \dfrac{1}{a - x}\right)$ **41.** $\dfrac{1}{a}\left(\dfrac{1}{y} + \dfrac{1}{a - y}\right)$

43. $\dfrac{3}{x} - \dfrac{2}{x - 4}$

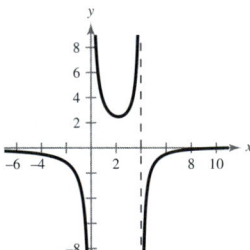
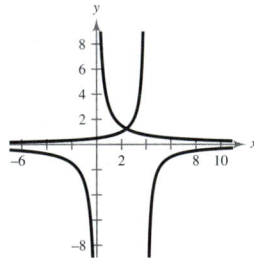

Vertical asymptotes are the same.

45. $\dfrac{3}{x - 3} + \dfrac{5}{x + 3}$

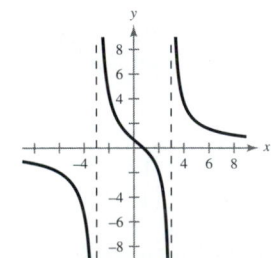
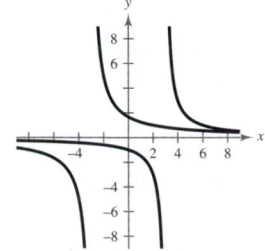

Vertical asymptotes are the same.

47. (a) $\dfrac{2000}{7 - 4x} - \dfrac{2000}{11 - 7x}$, $\quad 0 \le x \le 1$

(b)

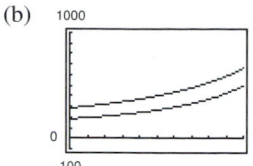

Section 4.4 *(page 370)*

Warm Up *(page 370)*

1. $9x^2 + 16y^2 = 144$ **2.** $x^2 + 4y^2 = 32$
3. $16x^2 - y^2 = 4$ **4.** $243x^2 + 4y^2 = 9$
5. $c = 2\sqrt{2}$ **6.** $c = \sqrt{13}$ **7.** $c = 2\sqrt{3}$
8. $c = \sqrt{5}$ **9.** 4   **10.** 2

1. Not shown   **3.** e   **5.** Not shown   **7.** f
9. Vertex: $(0, 0)$

    Focus: $\left(0, \frac{1}{2}\right)$

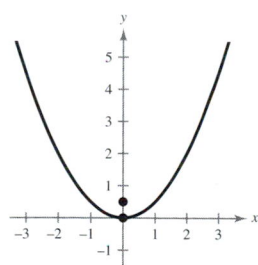

11. Vertex: $(0, 0)$ **13.** Vertex: $(0, 0)$

    Focus: $\left(-\frac{3}{2}, 0\right)$ Focus: $(0, -2)$

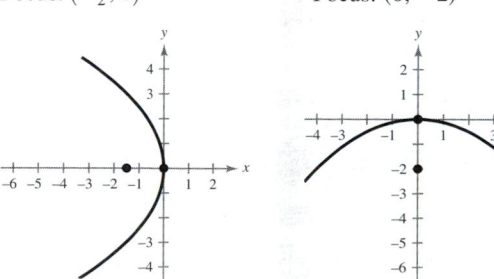

15.

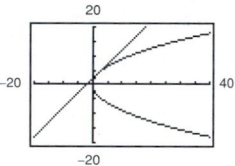

$(2, 4)$

17. $x^2 = -6y$ **19.** $y^2 = -8x$ **21.** $x^2 = 4y$
23. $x^2 = -8y$ **25.** $y^2 = 9x$
27. $y = \frac{2}{3}x^2$ Focus: $\left(0, \frac{3}{8}\right)$

29. $y = \dfrac{1}{14}x^2$ **31.** (a) $y = \dfrac{x^2}{12{,}288}$ (b) 22.6 feet

33. No. If the graph crossed the directrix, there would exist points nearer the directrix than the focus.

35. Center: $(0, 0)$ **37.** Center: $(0, 0)$

    Vertices: $(\pm 5, 0)$ Vertices: $(0, \pm 5)$

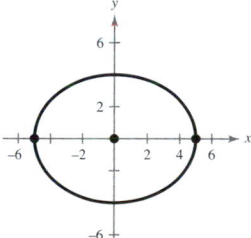 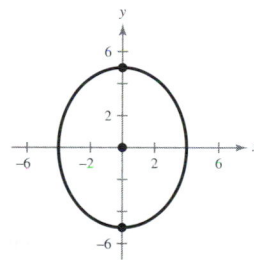

39. Center: $(0, 0)$ **41.**

    Vertices: $(\pm 3, 0)$

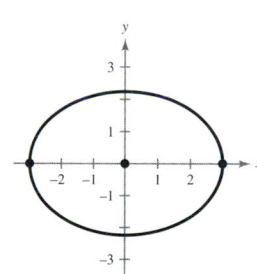 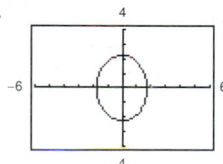

43. Left half   **45.** $\dfrac{x^2}{1} + \dfrac{y^2}{4} = 1$ **47.** $\dfrac{x^2}{25} + \dfrac{y^2}{21} = 1$

49. $\dfrac{x^2}{36} + \dfrac{y^2}{11} = 1$ **51.** $\dfrac{21x^2}{400} + \dfrac{y^2}{25} = 1$

53. (a) $2a$

(b) The sum of the distances from the two fixed points is constant.

55.

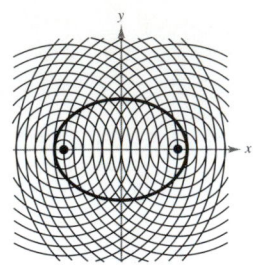

57. (a) $A = \pi a(20 - a)$

(b) $\dfrac{x^2}{196} + \dfrac{y^2}{36} = 1$

(c)

a	8	9	10	11	12	13
A	301.6	311.0	314.2	311.0	301.6	285.9

$a = 10$, circle

(d)

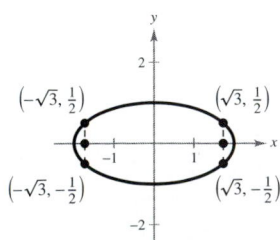

$a = 10$, circle

59. No. If it were an ellipse, the equation must be second degree.

61. The shape continuously changes from an ellipse with a vertical major axis of length 8 and minor axis of length 2 to a circle with a diameter of 8 and then to an ellipse with a horizontal major axis of length 16 and minor axis of length 8.

63.

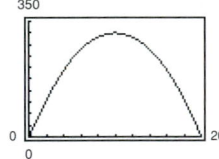

65.

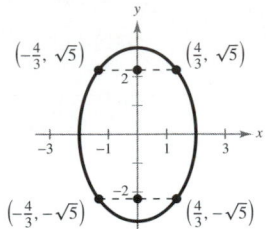

67. Center: $(0, 0)$
Vertices: $(\pm 1, 0)$

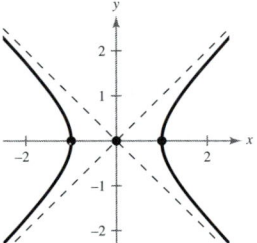

69. Center: $(0, 0)$
Vertices: $(0, \pm 1)$

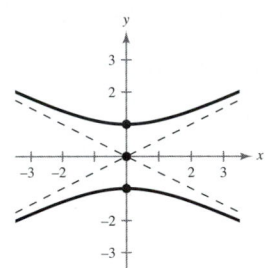

71. Center: $(0, 0)$
Vertices: $(0, \pm 5)$

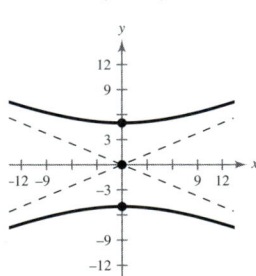

73.

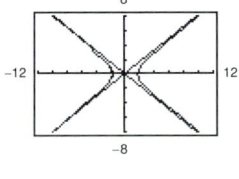

75. Bottom

77. $\dfrac{y^2}{4} - \dfrac{x^2}{12} = 1$

79. $\dfrac{x^2}{1} - \dfrac{y^2}{9} = 1$

81. $\dfrac{17y^2}{1024} - \dfrac{17x^2}{64} = 1$

83. $\dfrac{y^2}{9} - \dfrac{x^2}{9/4} = 1$

85. $\left(12\left(\sqrt{5} - 1\right), 0\right) \approx (14.83, 0)$

87. Answers will vary.

89. $x^3 - 7x^2 + 17x - 15$

91. $\pm 1, \pm 2, \pm 4, \pm 5, \pm 10, \pm 20, \pm\frac{1}{2}, \pm\frac{5}{2}, \pm\frac{1}{3}, \pm\frac{2}{3},$
$\pm\frac{4}{3}, \pm\frac{5}{3}, \pm\frac{10}{3}, \pm\frac{20}{3}, \pm\frac{1}{6}, \pm\frac{5}{6}$

Section 4.5 *(page 380)*

Warm Up *(page 380)*

1. Hyperbola	**2.** Ellipse	**3.** Parabola	**4.** Circle
5. Ellipse	**6.** Hyperbola	**7.** Parabola	
8. Parabola	**9.** Hyperbola	**10.** Ellipse	

1. Vertex: $(1, -2)$
Focus: $(1, -4)$
Directrix: $y = 0$

3. Vertex: $\left(5, -\frac{1}{2}\right)$
Focus: $\left(\frac{11}{2}, -\frac{1}{2}\right)$
Directrix: $x = \frac{9}{2}$

5. Vertex: $(1, 1)$
Focus: $(1, 2)$
Directrix: $y = 0$

7. Vertex: $(-2, -3)$
Focus: $(-4, -3)$
Directrix: $x = 0$

9. Vertex: $(-2, 1)$
Focus: $\left(-2, -\frac{1}{2}\right)$
Directrix: $y = \frac{5}{2}$

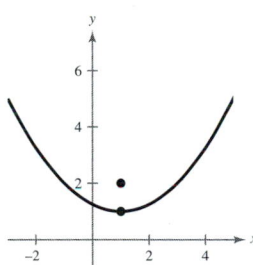

11. Vertex: $\left(\frac{1}{4}, -\frac{1}{2}\right)$
Focus: $\left(0, -\frac{1}{2}\right)$
Directrix: $x = \frac{1}{2}$

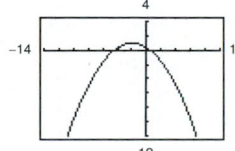

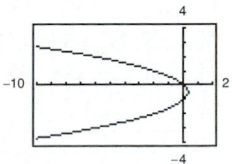

13. $(x - 3)^2 = -(y - 1)$ **15.** $y^2 = 2(x + 2)$

17. $(y - 2)^2 = -8(x - 3)$ **19.** $x^2 = 8(y - 4)$

21. $(y - 2)^2 = 8x$ **23.** $y = \sqrt{6(x + 1)} + 3$

25. (a) $17{,}500\sqrt{2}$ miles per hour
(b) $x^2 = -16{,}400(y - 4100)$

27. (a)

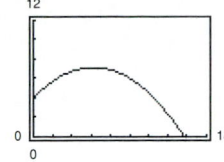

(b) $(6.25, 7.125)$, 15.69 feet

29. Center: $(1, 5)$
Vertices: $(1, 10)$, $(1, 0)$
Foci: $(1, 9)$, $(1, 1)$

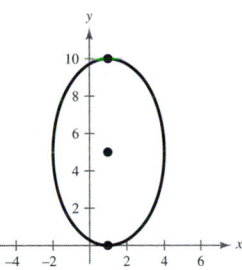

31. Center: $(-2, 3)$
Vertices: $(-2, 6)$, $(-2, 0)$
Foci: $\left(-2, 3 \pm \sqrt{5}\right)$

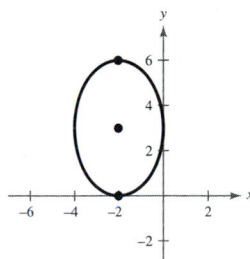

33. Center: $(1, -1)$
Vertices: $\left(\frac{9}{4}, -1\right)$, $\left(-\frac{1}{4}, -1\right)$
Foci: $\left(\frac{7}{4}, -1\right)$, $\left(\frac{1}{4}, -1\right)$

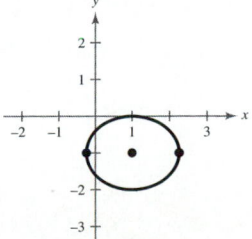

35. Center: $\left(\frac{1}{2}, -1\right)$
Vertices: $\left(\frac{1}{2} \pm \sqrt{5}, -1\right)$
Foci: $\left(\frac{1}{2} \pm \sqrt{2}, -1\right)$

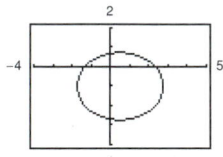

37. $\dfrac{(x-2)^2}{1} + \dfrac{(y-3)^2}{9} = 1$

39. $\dfrac{(x-2)^2}{9} + \dfrac{(y-2)^2}{4} = 1$

41. $\dfrac{(x-2)^2}{4} + \dfrac{(y-2)^2}{1} = 1$

43. $\dfrac{x^2}{48} + \dfrac{(y-4)^2}{64} = 1$

45. $\dfrac{(x-3)^2}{9} + \dfrac{(y-5)^2}{16} = 1$

47. $\dfrac{x^2}{16} + \dfrac{(y-4)^2}{12} = 1$

49. $x = \dfrac{3}{2}\left(2 + \sqrt{4 - y^2}\right)$

51. $\dfrac{x^2}{25} + \dfrac{y^2}{16} = 1$ **53.** 2,756,832,000; 4,575,168,000

55. $e = \dfrac{c}{a} \approx 0.052$

57. Center: $(1, -2)$
Vertices: $(3, -2), (-1, -2)$
Foci: $\left(1 \pm \sqrt{5}, -2\right)$

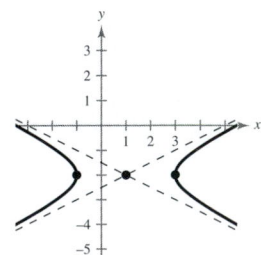

59. Center: $(2, -6)$
Vertices: $(2, -5), (2, -7)$
Foci: $\left(2, -6 \pm \sqrt{2}\right)$

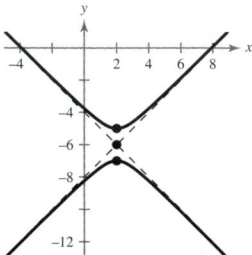

61. Center: $(2, -3)$
Vertices: $(3, -3), (1, -3)$
Foci: $\left(2 \pm \sqrt{10}, -3\right)$

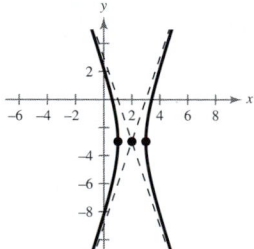

63. The graph of this equation is two lines intersecting at $(-1, -3)$.

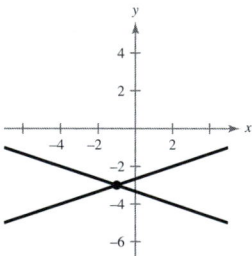

65. Center: $(1, -3)$
Vertices: $\left(1, -3 \pm \sqrt{2}\right)$
Foci: $\left(1, -3 \pm 2\sqrt{5}\right)$

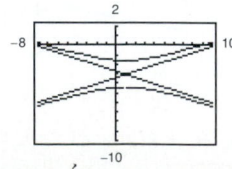

67. $(y - 1)^2 - x^2 = 1$ **69.** $\dfrac{(x - 3)^2}{4} - \dfrac{(y - 2)^2}{16/5} = 1$

71. $\dfrac{(x - 4)^2}{4} - \dfrac{y^2}{12} = 1$ **73.** $\dfrac{(y - 5)^2}{16} - \dfrac{(x - 4)^2}{9} = 1$

75. $\dfrac{y^2}{9} - \dfrac{4(x - 2)^2}{9} = 1$ **77.** $\dfrac{(x - 3)^2}{9} - \dfrac{(y - 2)^2}{4} = 1$

79. Left half   **81.** Circle   **83.** Hyperbola

85. Ellipse   **87.** Parabola

89.

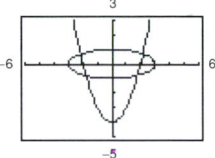

$(\pm 2.166, 0.692),\ (\pm 1.788, -0.803)$

91. $\dfrac{x^2}{328.15} + \dfrac{y^2}{19.39} = 1$

93. Additive Inverse Property   **95.** Distributive Property

Focus on Concepts *(page 385)*

1. $f(x) = \dfrac{9}{x^2 + 4}$, $g(x) = \dfrac{2}{x - 20}$

2. Line that the graph approaches as it moves farther and farther away from the origin.

3. No vertical asymptote: $f(x) = \dfrac{2}{x^2 + x + 1}$

No horizontal asymptote: $g(x) = \dfrac{x^2}{x + 2}$

4. (a) Solve $p(x) = 0$.   (b) Solve $q(x) = 0$.

   (c) $y = 0$

5. c

Vertical asymptote: $x = 2$

Horizontal asymptote: $y = 0$

No intercepts

6. b

Vertical asymptotes: $x = \pm\sqrt{2}$

Horizontal asymptote: $y = 0$

No intercepts

7. d

Vertical asymptote: $x = 2$

Horizontal asymptote: $y = 2$

Intercept: $(0, 0)$

8. a

Vertical asymptote: $x = 2$

Horizontal asymptote: $y = 2$

Intercept: $(3, 0)$

9. b   **10.** c   **11.** d   **12.** a

13. (a) Vertical translation

   (b) Horizontal translation

   (c) Reflection in the y-axis

   (d) Parabola opens more slowly.

14. (a) Major axis horizontal

   (b) Circle

   (c) Ellipse is flatter.

   (d) Horizontal translation

15. The extended diagonals of the central rectangle are asymptotes of the hyperbola.

16. 5. The ellipse becomes more circular and approaches a circle of radius 5.

Review Exercises *(page 386)*

1. Domain: all real numbers $x \neq -3$

Vertical asymptote: $x = -3$

Horizontal asymptote: $y = 0$

3. Domain: all real numbers $x \neq \pm 2$

Vertical asymptotes: $x = 2,\ x = -2$

Horizontal asymptote: $y = 1$

5.

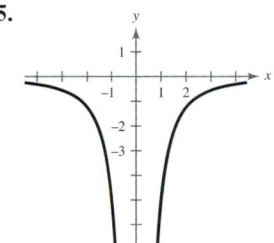

7.

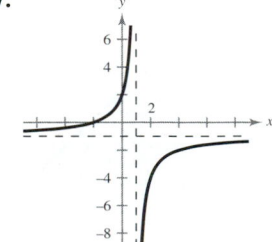

9.

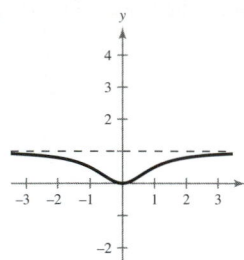

11.

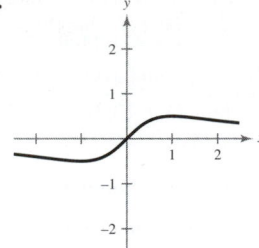

(c)

x	2.5	3	3.5	4	4.5
A	18.75	13.50	12.25	12	12.15

$x = 4$

(d)

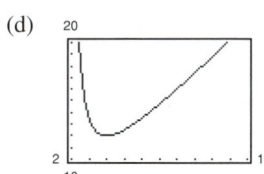

$x = 4$

(e) $y = \frac{3}{2}(x + 2)$. The area increases without bound as x increases.

13.

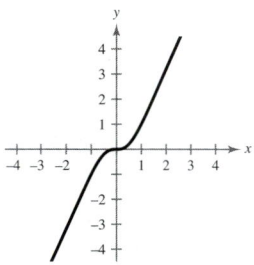

15.

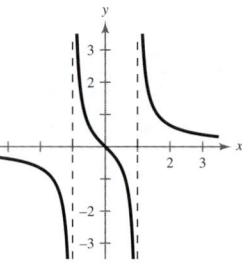

29.

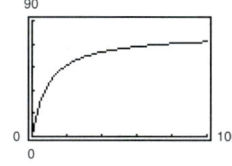

80.3 milligrams per square decimeter per hour

17.

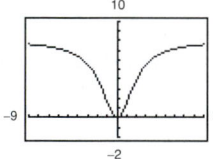

19.

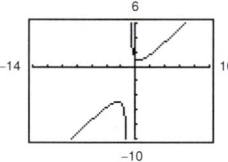

31. $\dfrac{3}{x + 2} - \dfrac{4}{x + 4}$ **33.** $1 - \dfrac{25}{8(x + 5)} + \dfrac{9}{8(x - 3)}$

35. $\dfrac{1}{2}\left(\dfrac{3}{x - 1} - \dfrac{x - 3}{x^2 + 1} \right)$ **37.** $\dfrac{3x}{x^2 + 1} + \dfrac{x}{(x^2 + 1)^2}$

21. $f(x) = \dfrac{2x^2}{x^2 - x - 12}$

23. As x increases, the cost approaches the horizontal asymptote, $\overline{C} = 0.5$.

25. (a) \$176 million

(b) \$528 million

(c) \$1584 million

(d) No

27. (a)

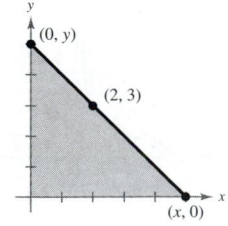

(b) Answers will vary.

39. Parabola   **41.** Parabola

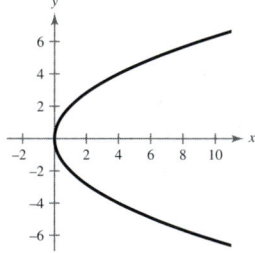

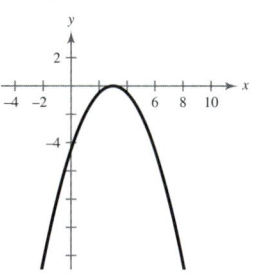

43. Degenerate circle (a point)

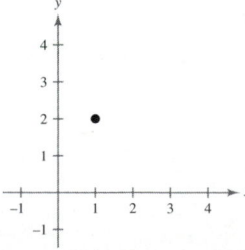

45. Ellipse

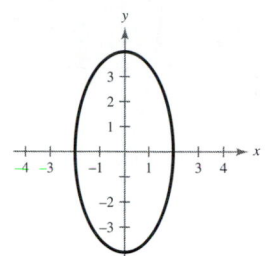

47. Ellipse

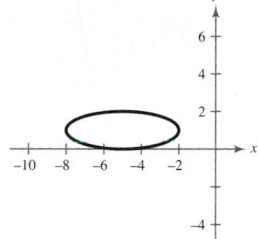

49. Hyperbola

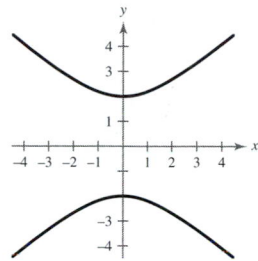

51. $y = 5x \pm \sqrt{24x^2 - 1}$

Hyperbola

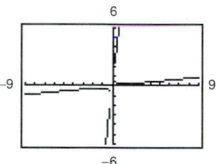

53. $(x + 6)^2 = -9(y - 4)$ **55.** $(x - 4)^2 = -8(y - 2)$

57. $(y - 2)^2 = -4x$ **59.** $\dfrac{(x - 5)^2}{25} + \dfrac{(y - 3)^2}{9} = 1$

61. $\dfrac{(x - 2)^2}{25} + \dfrac{y^2}{21} = 1$ **63.** $\dfrac{2x^2}{9} + \dfrac{y^2}{36} = 1$

65. $\dfrac{x^2}{1} - \dfrac{y^2}{4} = 1$ **67.** $\dfrac{y^2}{1} - \dfrac{x^2}{8} = 1$

69. $\dfrac{5(x - 4)^2}{16} - \dfrac{5y^2}{64} = 1$ **71.** $(0, 50)$

73. (a) $x^2 = -4(y - 4)$

   $x^2 + \left(y + 4\sqrt{3}\right)^2 = 64$

(b)

x	0	1	2	3	4
d	2.928	2.741	2.182	1.262	0

75. The foci should be placed 3 feet on either side of center and have the same height as the pillars.

Chapter Test *(page 392)*

1. Domain: all real numbers $x \neq 4$

   Vertical asymptote: $x = 4$

   Horizontal asymptote: $y = 0$

2. Domain: all real numbers

   Vertical asymptote: None

   Horizontal asymptote: $y = -1$

3. Domain: all real numbers $x \neq 2$

   Vertical asymptotes: $x = 2$

   Slant asymptote: $y = x + 4$

4.

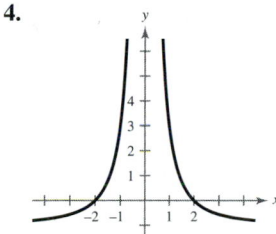

5.

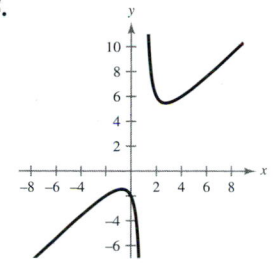

6. $f(x) = \dfrac{4x^2}{x^2 - 9}$

7. (a) Answers will vary.

(b) $A = \dfrac{x^2}{2(x - 2)}, \quad x > 2$

(c)

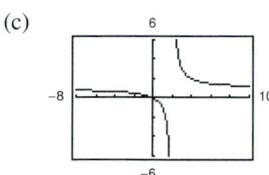

   $A = 4$

8. $\dfrac{3}{x - 2} - \dfrac{1}{x + 1}$ **9.** $\dfrac{2}{x^2} - \dfrac{3}{x - 2}$

10. $-\dfrac{1}{x} + \dfrac{1}{x - 1} + \dfrac{1}{x + 1}$ **11.** $-\dfrac{1}{x} + \dfrac{2x}{x^2 + 1}$

12.

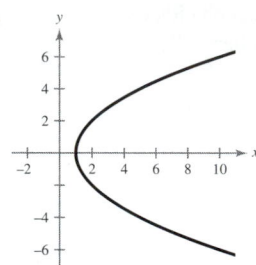

13.

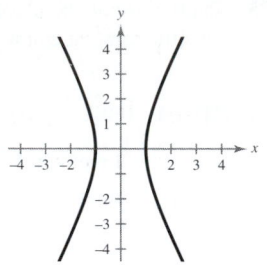

14.

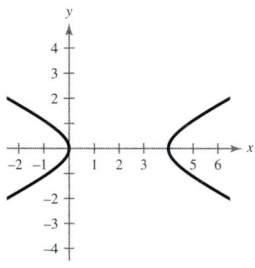

15. $\dfrac{(x-4)^2}{16} + \dfrac{(y-2)^2}{4} = 1$

16. $\dfrac{y^2}{9} - \dfrac{x^2}{4} = 1$

17.

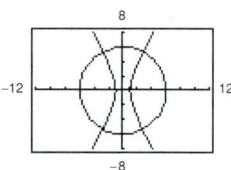

$\left(\pm 2\sqrt{2},\ \pm 2\sqrt{7}\right)$

CHAPTER 5

Section 5.1 *(page 402)*

Warm Up *(page 402)*

1. 5^x **2.** 3^{2x} **3.** 4^{3x} **4.** 10^x **5.** 4^{2x} **6.** 4^{10x}		
7. $\left(\dfrac{3}{2}\right)^x$ **8.** 4^{3x} **9.** 2^{-x} **10.** $16^{x/4}$		

1. 946.852   **3.** 7.352   **5.** 0.006   **7.** 673.639
9. 0.472   **11.** $f(x) = h(x)$ **13.** $f(x) = g(x) = h(x)$
15. d   **17.** a

19.

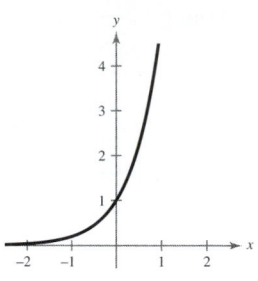

21.

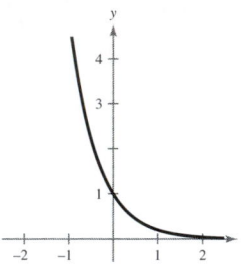

23.

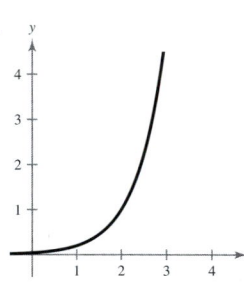

25.

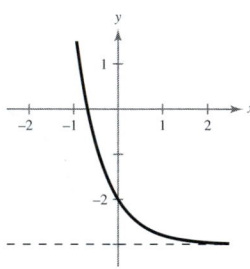

27.

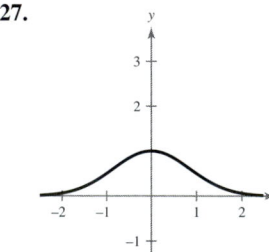

29.

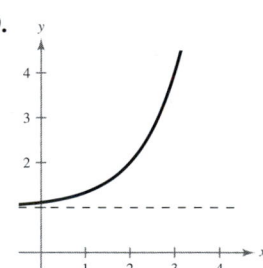

31.

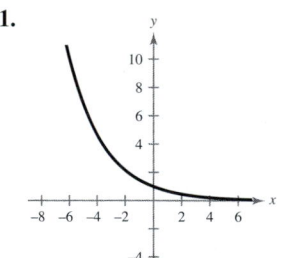

33.

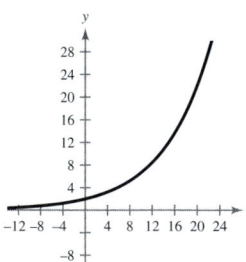

35.

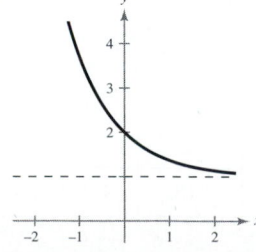

37. (a) $x < 0$ (b) $x > 0$

39. (a)

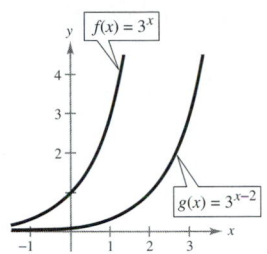

Horizontal shift two units to the right

(b)

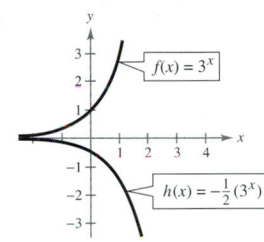

Vertical shrink and a reflection about the x-axis

(c)

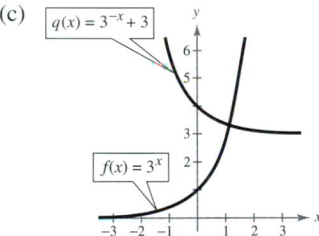

Reflection about the y-axis and a vertical translation

41. (a)

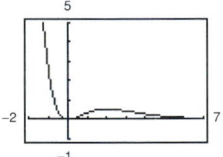

Decreasing: $(-\infty, 0),\ (2, \infty)$

Increasing: $(0, 2)$

Relative maximum: $(2, 4e^{-2})$

Relative minimum: $(0, 0)$

(b)

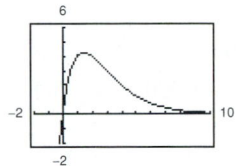

Decreasing: $(1.44, \infty)$

Increasing: $(-\infty, 1.44)$

Relative maximum: $(1.44, 4.25)$

43. The exponential function increases at a faster rate.

45.

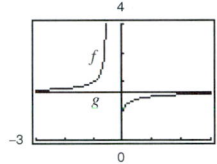

As $x \to \infty$, $f(x) \to g(x)$.

47.

n	1	2	4
A	\$7764.62	\$8017.84	\$8155.09

n	12	365	Continuous
A	\$8250.97	\$8298.66	\$8300.29

49.

n	1	2	4
A	\$24,115.73	\$25,714.29	\$26,602.23

n	12	365	Continuous
A	\$27,231.38	\$27,547.07	\$27,557.94

51.

t	1	10	20
P	\$91,393.12	\$40,656.97	\$16,529.89

t	30	40	50
P	\$6720.55	\$2732.37	\$1110.90

53. \$222,822.57

55. (a)

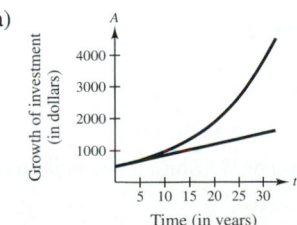

(b) $A = 500(1.07)^t$

$A = 500(0.07)t + 500$

57. \$35.45   **59.** (a) 100   (b) 300   (c) 900

61. (a) 25 units   (b) 16.30 units

(c)

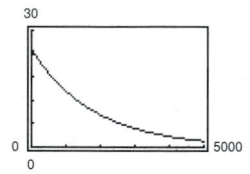

63. (a)

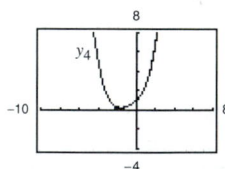

Altitude (in km)

(b)

h	0	5	10	15	20
P	10,958	5176	2445	1155	546

(c) 3300 kilograms per square meter

(d) 11.3 kilometers

65. False. e is an irrational number.

67. $1 < \sqrt{2} < 2$

$2^1 < 2^{\sqrt{2}} < 2^2$

69. $y_4 = 1 + \dfrac{x}{1!} + \dfrac{x^2}{2!} + \dfrac{x^3}{3!} + \dfrac{x^4}{4!}$

71. $y = \frac{1}{7}(2x + 14)$ **73.** $y = \pm\sqrt{25 - x^2}$

Section 5.2   *(page 413)*

Warm Up   *(page 413)*

1. 3   **2.** 0   **3.** −1   **4.** 1   **5.** 7.389

6. 0.368   **7.** Shift two units to the left

8. Reflection about the x-axis **9.** Shifted one unit downward

10. Reflection about the y-axis

1. $4^3 = 64$ **3.** $7^{-2} = \frac{1}{49}$ **5.** $32^{2/5} = 4$

7. $e^0 = 1$ **9.** $\log_5 125 = 3$ **11.** $\log_{81} 3 = \frac{1}{4}$

13. $\log_6 \frac{1}{36} = -2$ **15.** $\ln 20.0855 = 3$

17. $\ln 4 = x$ **19.** 4   **21.** $\frac{1}{2}$ **23.** 0

25. −2   **27.** $\frac{5}{3}$ **29.** 3   **31.** 2.538

33. 2.161   **35.** 2.913   **37.** 1.005   **39.** −1.139

41. 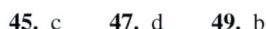 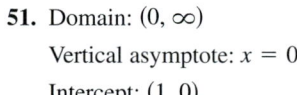   **43.**

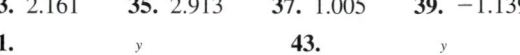

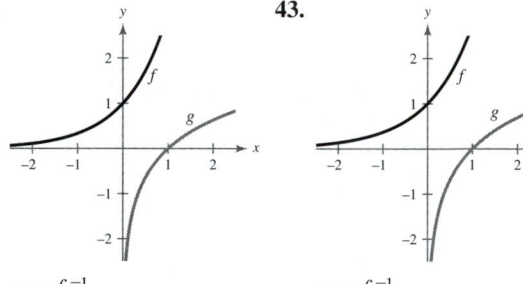

$g = f^{-1}$ $g = f^{-1}$

45. c   **47.** d   **49.** b

51. Domain: $(0, \infty)$

Vertical asymptote: $x = 0$

Intercept: $(1, 0)$

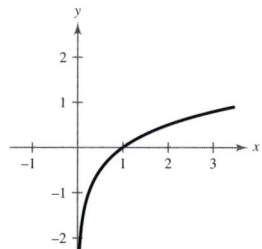

53. Domain: $(0, \infty)$

Vertical asymptote: $x = 0$

Intercept: $(9, 0)$

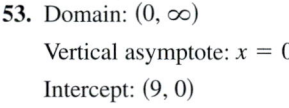 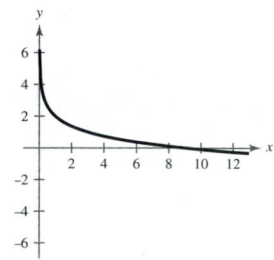

55. Domain: $(-2, \infty)$

Vertical asymptote: $x = -2$

Intercept: $(-1, 0)$

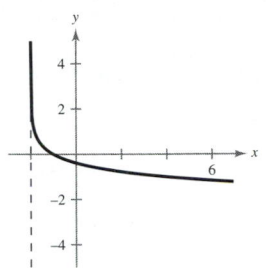

57. Domain: $(0, \infty)$

Vertical asymptote: $x = 0$

Intercept: $(5, 0)$

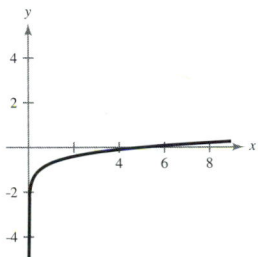

59. Domain: $(2, \infty)$

Vertical asymptote: $x = 2$

Intercept: $(3, 0)$

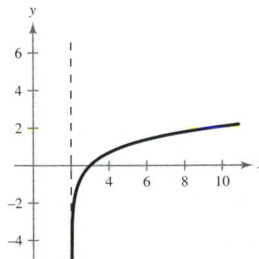

61. Domain: $(-\infty, 0)$

Vertical asymptote: $x = 0$

Intercept: $(-1, 0)$

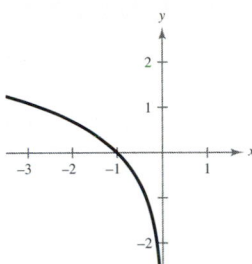

63. Decreasing: $(0, 1)$

Increasing: $(1, \infty)$

Relative minimum: $(1, 0)$

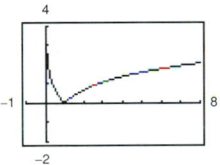

65. Decreasing: $(0, 2)$

Increasing: $(2, \infty)$

Relative minimum: $\left(2, 1 - \ln\frac{1}{2}\right)$

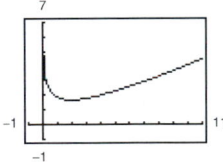

67. (a)

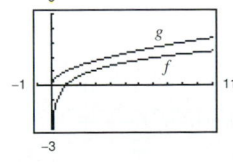

$g(x)$

(b)

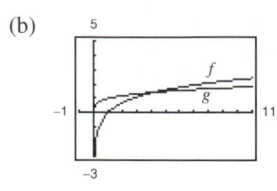

$g(x)$

As x increases without bound $\sqrt[n]{x}$ will eventually increase at a faster rate than $\ln x$.

69.

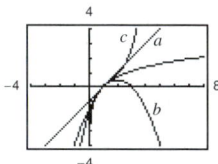

71. (a) 80   (b) 68.1   (c) 62.3

73.

r	0.005	0.010	0.015
t	138.6 yr	69.3 yr	46.2 yr

r	0.020	0.025	0.030
t	34.7 yr	27.7 yr	23.1 yr

75.

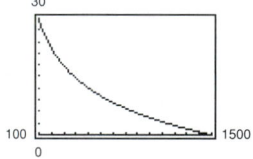

17.66 cubic feet per minute

77. 21,357 foot-pounds   **79.** 30 years

81. Total amount: $473,886

Interest: $323,886

83. (a)

x	1	5	10	10^2
$f(x)$	0	0.322	0.230	0.046

x	10^4	10^6
$f(x)$	0.00092	0.0000138

(b) 0

(c)

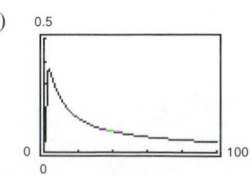

85. $8n - 3$ **87.** $83.95 + 37.50t$

Section 5.3 *(page 421)*

Warm Up *(page 421)*

1. 2	**2.** -5	**3.** -2	**4.** -3	**5.** e^5
6. $\dfrac{1}{e}$	**7.** e^6	**8.** 1	**9.** x^{-2}	**10.** $x^{1/2}$

1.

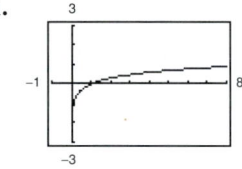

3. $\dfrac{\log_{10} 5}{\log_{10} 3}$ **5.** $\dfrac{\log_{10} x}{\log_{10} 2}$ **7.** $\dfrac{\ln 5}{\ln 3}$ **9.** $\dfrac{\ln x}{\ln 2}$ **11.** 1.771

13. -2.000 **15.** -0.417 **17.** 2.633

19. $\log_{10} 5 + \log_{10} x$ **21.** $\log_{10} 5 - \log_{10} x$ **23.** $4 \log_8 x$

25. $\frac{1}{2} \ln z$ **27.** $\ln x + \ln y + \ln z$ **29.** $\frac{1}{2} \ln(a - 1)$

31. $\ln z + 2 \ln(z - 1)$ **33.** $\frac{1}{3} \ln x - \frac{1}{3} \ln y$

35. $4 \ln x + \frac{1}{2} \ln y - 5 \ln z$

37. $2 \log_b x - 2 \log_b y - 3 \log_b z$

39.

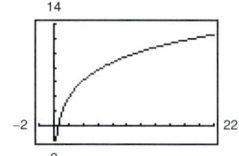

41. $\ln 2x$ **43.** $\log_4 \dfrac{z}{y}$ **45.** $\log_2 (x + 4)^2$

47. $\log_3 \sqrt[3]{5x}$ **49.** $\ln \dfrac{x}{(x + 1)^3}$ **51.** $\ln \dfrac{x - 2}{x + 2}$

53. $\ln \dfrac{x}{(x^2 - 4)^2}$ **55.** $\ln \sqrt[3]{\dfrac{x(x + 3)^2}{x^2 - 1}}$

57. $\ln \dfrac{\sqrt[3]{y}(y + 4)^2}{y - 1}$ **59.** $\ln \dfrac{9}{\sqrt{x^2 + 1}}$

61.

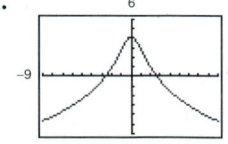

63.

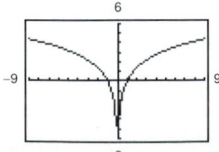

No. The domains differ.

65. $\log_2 \frac{32}{4} = \log_2 32 - \log_2 4$

67.

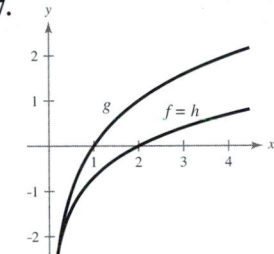

$f(x) = h(x)$

69. 2   **71.** 2.4   **73.** -9 is not in the domain of $\log_3 x$.

75. 2   **77.** -3 **79.** 0 is not in the domain of $\log_{10} x$.

81. 4.5   **83.** $\frac{3}{2}$ **85.** $-3 - \log_5 2$ **87.** $6 + \ln 5$

89. (a) 90   (b) 77   (c) 73   (d) 9 months

(e) $90 - \log_{10}(t + 1)^{15}$

(f)

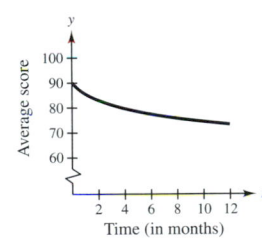

91. False. $\ln 1 = 0$ **93.** False. $f(x) - f(2) = \ln \frac{x}{2}$

95. False. $u = v^2$ **97.** Answers will vary.

99. $\frac{3x^4}{2y^3}$ **101.** 1

Section 5.4   *(page 431)*

Warm Up   *(page 431)*

1. $\frac{\ln 3}{\ln 2}$ **2.** $1 + \frac{2}{\ln 4}$ **3.** $\frac{e}{2}$ **4.** $2e$ **5.** $2 \pm i$

6. $\frac{1}{2}, 1$ **7.** $2x$ **8.** $3x$ **9.** $2x$ **10.** $-x^2$

1. (a) Yes   (b) No   **3.** (a) No   (b) Yes   (c) Yes

5. (a) No   (b) No   (c) Yes   **7.** (3, 8)   **9.** (9, 2)

11. 2   **13.** -2 **15.** 3   **17.** 64   **19.** $\frac{1}{10}$

21. x^2 **23.** $5x + 2$ **25.** x^2 **27.** $\ln 10 \approx 2.303$

29. 0   **31.** $\frac{\ln 12}{3} \approx 0.828$ **33.** $\ln \frac{5}{3} \approx 0.511$

35. $\ln 5 \approx 1.609$ **37.** $2 \ln 75 \approx 8.635$

39. $\log_{10} 42 \approx 1.623$ **41.** $\frac{\ln 80}{2 \ln 3} \approx 1.994$

43. 2   **45.** $\frac{\ln 8 - \ln 565}{\ln 2} \approx -6.142$

47.

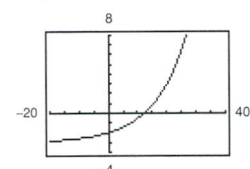

-0.427

49.

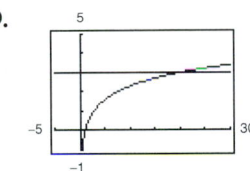

12.207

51. 0.059   **53.** 21.330   **55.** $e^{-3} \approx 0.050$

57. $\frac{e^{2.4}}{2} \approx 5.512$ **59.** $e^2 - 2 \approx 5.389$ **61.** 103

63. $1 + \sqrt{1 + e} \approx 2.928$ **65.** $\frac{-1 + \sqrt{17}}{2} \approx 1.562$

67. 4   **69.** No solution   **71.** 14.988

73. 33.115   **75.** 14.369

77.

2.807

79.

20.086

81. 8.2 years

83. Yes. Time to double: $t = \frac{\ln 2}{r}$;

Time to quadruple: $t = \frac{\ln 4}{r} = 2\left(\frac{\ln 4}{r}\right)$

85. 12.9 years   **87.** (a) 1426 units   (b) 1498 units

89. (a)

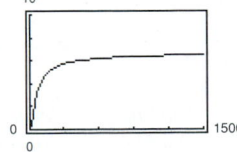

(b) $y = 6.7$. Yield will approach 6.7 million cubic feet per acre.

(c) 29.3 years

91. (a) $y = 100$ and $y = 0$

(b) Males: 69.71 inches   Females: 64.51 inches

93. (a) $y = 20$; Room temperature   (b) 0.81 hour

95. $4|x|y^2\sqrt{3y}$ **97.** $5\sqrt[3]{3}$

Section 5.5 *(page 442)*

Warm Up *(page 442)*

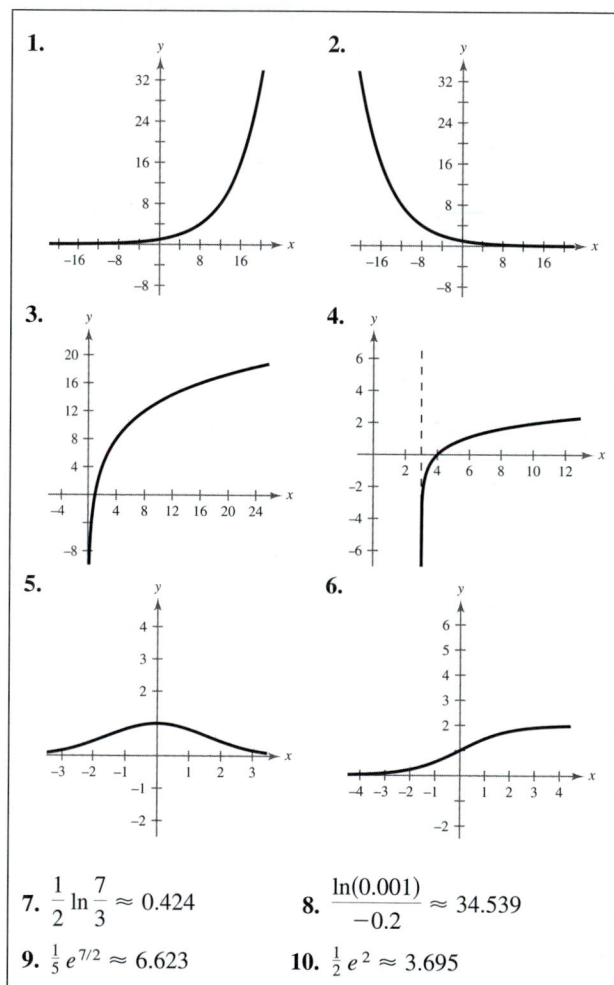

7. $\dfrac{1}{2}\ln\dfrac{7}{3} \approx 0.424$ **8.** $\dfrac{\ln(0.001)}{-0.2} \approx 34.539$

9. $\dfrac{1}{5}e^{7/2} \approx 6.623$ **10.** $\dfrac{1}{2}e^2 \approx 3.695$

1. c   **3.** a   **5.** d

	Initial Investment	Annual % Rate	Time to Double	Amount After 10 years
7.	$1000	12%	5.78 yr	$3,320.12
9.	$750	8.94%	7.75 yr	$1,833.67
11.	$500	9.5%	7.30 yr	$1,292.85
13.	$6376.28	4.5%	15.4 yr	$10,000.00
15.	$111,565.08			

17. (a) 6.642 years   (b) 6.330 years
    (c) 6.302 years   (d) 6.301 years

19.

r	2%	4%	6%	8%	10%	12%
t	54.93	27.47	18.31	13.73	10.99	9.16

21.

r	2%	4%	6%	8%	10%	12%
t	55.47	28.01	18.85	14.27	11.53	9.69

23.

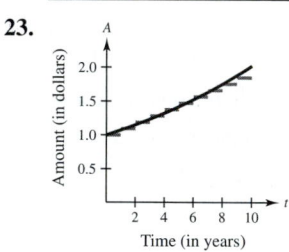

Continuous compounding

Isotope	Half-Life (Years)	Initial Quantity	Amount After 1000 years
25. Ra^{226}	1620	10 g	6.52 g
27. C^{14}	5730	2.26 g	2 g
29. Pu^{230}	24,360	2.16 g	2.1 g

31. $y = e^{0.7675x}$ **33.** $y = e^{-0.4621x}$

35. 2013   **37.** $k = 0.0137$, 3288

39. $y = 4.22e^{0.0430t}$, 9.97 million

41. $y = 3e^{-0.0091t}$, 2.50 million

43. The greater rate of growth, the greater the value of b.

45. 3.15 hours   **47.** 95.8%

49. (a) $V = -4500t + 22,000$ (b) $V = 22,000e^{-0.263t}$

    (c)

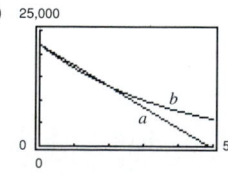

Exponential

    (d) 1 year. Straight-line: $17,500;
      Exponential: $16,912
      3 years. Straight-line: $8500;
      Exponential: $9995

    (e) Decreases $4500 per year.

51. (a) $S(t) = 100(1 - e^{-0.1625t})$

(b)

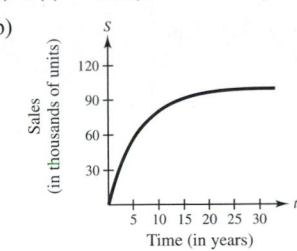

(c) 55,625

53. (a) $S = 10(1 - e^{-0.0575x})$ (b) 3314

55. (a) $N = 30(1 - e^{-0.050t})$ (b) 36 days

(c) No. It is not a linear function.

57. (a) 7.91   (b) 7.68

59. (a) 20   (b) 70   (c) 95   (d) 120   **61.** 95%

63. 4.64   **65.** 1.58×10^{-6} moles per liter   **67.** 10^7

69. (a)

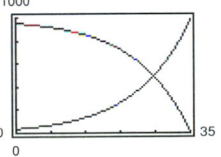

(b) Interest; $t \approx 28$ years

(c)

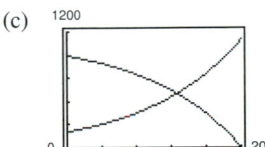

Interest; $t \approx 12.7$ years

71. (a) $t_3 = 0.2729s - 6.0143$

$t_4 = 1.5385e^{0.0291s}$

(b)

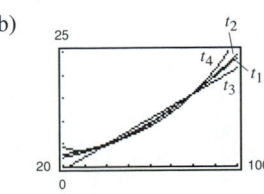

(c)

s	30	40	50	60	70	80	90
t_1	3.6	4.7	6.7	9.4	12.5	15.9	19.6
t_2	3.3	4.9	7.0	9.5	12.5	15.9	19.9
t_3	2.2	4.9	7.6	10.4	13.1	15.8	18.5
t_4	3.7	4.9	6.6	8.8	11.8	15.8	21.1

(d) Model: t_1; Sum = 1.9

Model: t_2; Sum = 1.1

Model: t_3; Sum = 5.6

Model: t_4; Sum = 2.6

Quadratic model fits best.

73. Answers will vary.   **75.** $4x^2 - 12x + 9$

77. $2x^2 + 3 + \dfrac{3}{x - 4}$

Focus on Concepts   *(page 448)*

1. $b < d < a < c$

b and d are negative.

2. (a) True. $\log_b uv = \log_b u + \log_b v$

(b) False.

$2.04 \approx \log_{10}(10 + 100) \neq (\log_{10}10)(\log_{10}100) = 2$

(c) False.

$1.95 \approx \log_{10}(100 - 10) \neq \log_{10}100 - \log_{10}10 = 1$

(d) True. $\log_b \dfrac{u}{v} = \log_b u - \log_b v$

3. Double the interest rate or time because it doubles the exponent in the exponential function.

4. (a) Logarithmic   (b) Logistic   (c) Exponential

(d) Linear   (e) None of the above   (f) Exponential

Review Exercises   *(page 449)*

1. e   **3.** a   **5.** d

7.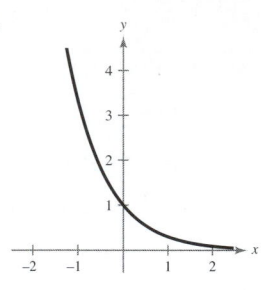

9.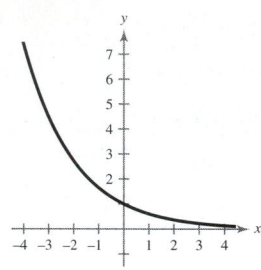

11.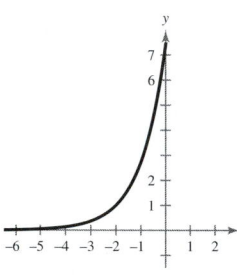

13.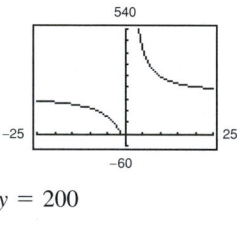

$y = 200$

15.

n	1	2	4	12
A	\$9499.28	\$9738.91	\$9867.22	\$9956.20

n	365	Continuous
A	\$10,000.27	\$10,001.78

17.

t	1	10	20
P	\$184,623.27	\$89,865.79	\$40,379.30

t	30	40	50
P	\$18,143.59	\$8152.44	\$3663.13

19. (a) 0.154   (b) 0.487   (c) 0.811
21. (a) \$1,069,047.14   (b) 7.9 years

23.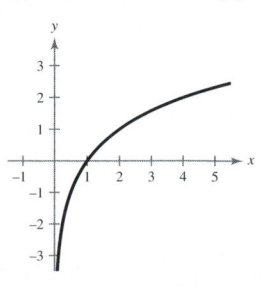

25.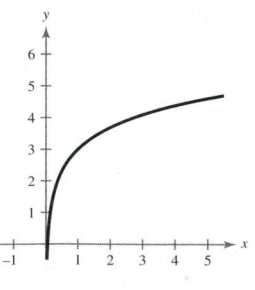

27.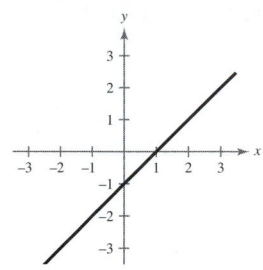

29.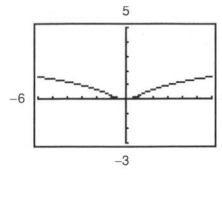

31. $\log_4 64 = 3$ **33.** 3   **35.** 7   **37.** 1.585
39. 2.132   **41.** $1 + 2\log_5 x$
43. $\log_{10} 5 + \frac{1}{2}\log_{10} y - 2\log_{10} x$ **45.** $\log_2 5x$
47. $\ln \dfrac{\sqrt{|2x - 1|}}{(x + 1)^2}$ **49.** True   **51.** False   **53.** True
55. 27.16 miles   **57.** $\ln 12 \approx 2.485$
59. $-\dfrac{\ln 44}{5} \approx -0.757$ **61.** $\ln 2 \approx 0.693$, $\ln 5 \approx 1.609$
63. $\frac{1}{3}e^{8.2} \approx 1213.650$ **65.** $3e^2 \approx 22.167$
67. No solution   **69.** 0.39, 7.48   **71.** 1.64
73. $y = 2e^{0.1014x}$ **75.** (a) 1151 units   (b) 1325 units
77. (a) 13.86%   (b) \$11,486.65
79. (a) $10^{8.4}$ (b) $10^{6.85}$ (c) $10^{9.1}$

Cumulative Test for Chapters 3–5
(page 454)

1.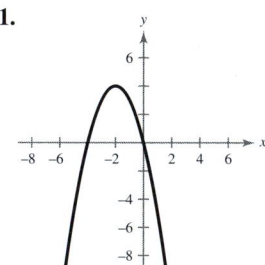

2.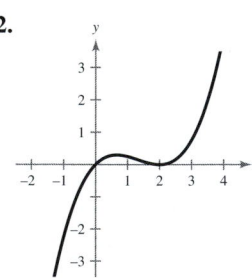

3.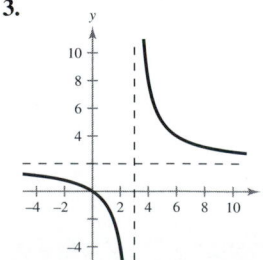

4.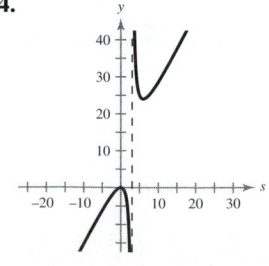

5.

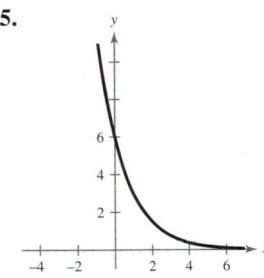

6.

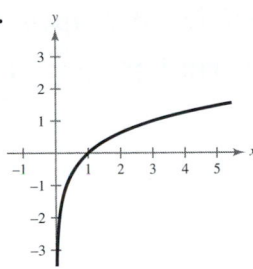

7. $3x - 2 - \dfrac{3x - 2}{2x^2 + 1}$ **8.** $-2, \pm 2i$ **9.** 1.20

10.

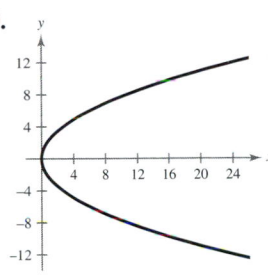

11.

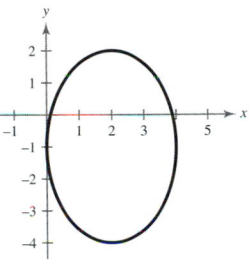

12. $2x^2 - 12x - 3y + 12 = 0$

13. $5x^2 - 26y^2 + 80y - 64 = 0$

14. $\ln \dfrac{x^2}{\sqrt{x + 5}}$

15.

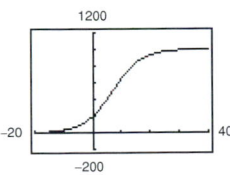

$y = 0, y = 1000$

16. $\frac{1}{2} \ln 12 \approx 1.2425$ **17.** $\frac{64}{5}$

18. \$2000   **19.** \$16,302.05

CHAPTER 6

Section 6.1   *(page 463)*

Warm Up   *(page 463)*

1. 45   **2.** 70   **3.** $\dfrac{\pi}{6}$ **4.** $\dfrac{\pi}{3}$ **5.** $\dfrac{\pi}{4}$ **6.** $\dfrac{4\pi}{3}$

7. $\dfrac{\pi}{9}$ **8.** $\dfrac{11\pi}{6}$ **9.** 45   **10.** 45

1. 210°   **3.** −45°

5. (a) Quadrant II   (b) Quadrant IV

7. (a) Quadrant III   (b) Quadrant I

9. (a)   (b)

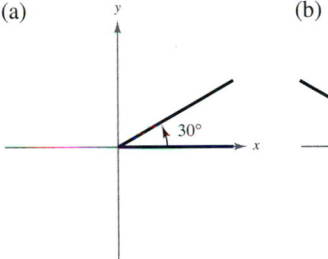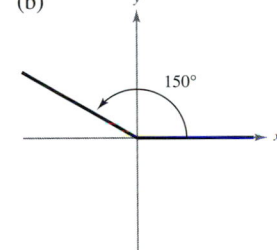

11. (a)   (b)

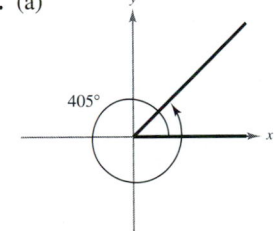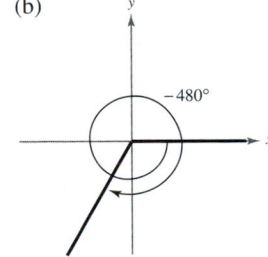

13. (a) 405°, −315°   (b) 324°, −396°

15. (a) 660°, −60°   (b) 20°, −340°

17. (a) 54.75°   (b) −128.5°

19. (a) 85.308°   (b) 330.007°

21. (a) 240° 36′   (b) −145° 48′

23. (a) 143° 14′ 22″   (b) −205° 7′ 8″

25. 2   **27.** −3   **29.** (a) Quadrant I   (b) Quadrant III

31. (a) Quadrant IV   (b) Quadrant II

33. (a) Quadrant III   (b) Quadrant II

35. (a)   (b)

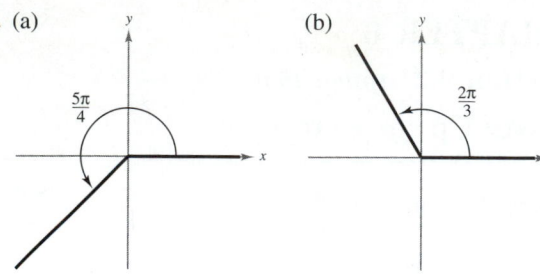

37. (a)   (b)

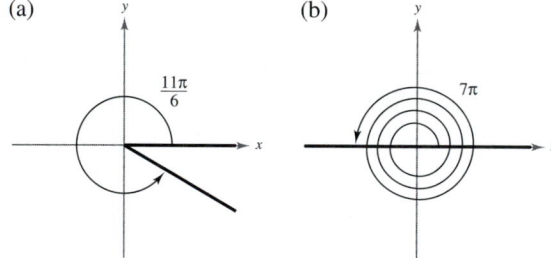

39. (a) $\dfrac{25\pi}{12},\ -\dfrac{23\pi}{12}$ (b) $\dfrac{8\pi}{3},\ -\dfrac{4\pi}{3}$

41. (a) $\dfrac{7\pi}{4},\ -\dfrac{\pi}{4}$ (b) $\dfrac{28\pi}{15},\ -\dfrac{32\pi}{15}$

43. (a) Complement: $72°$; Supplement: $162°$

   (b) Complement: None; Supplement: $65°$

45. (a) Complement: $\dfrac{\pi}{6}$; Supplement: $\dfrac{2\pi}{3}$

   (b) Complement: None; Supplement: $\dfrac{\pi}{4}$

47. (a) $\dfrac{\pi}{6}$ (b) $\dfrac{5\pi}{6}$ **49.** (a) $-\dfrac{\pi}{9}$ (b) $-\dfrac{4\pi}{3}$

51. (a) $270°$ (b) $210°$ **53.** (a) $420°$ (b) $-66°$

55. 2.007 **57.** -3.776 **59.** 9.285 **61.** -0.014

63. $25.714°$ **65.** $337.5°$ **67.** $-756°$

69. $-114.592°$ **71.** $\tfrac{6}{5}$ radians   **73.** $4\tfrac{4}{7}$ radians

75. $\tfrac{4}{15}$ radian   **77.** 1.724 radians

79. 15π inches ≈ 47.12 inches   **81.** 12 meters

83. 591.72 miles   **85.** 1141 miles

87. 0.094 radian $\approx 5.39°$ **89.** $\tfrac{5}{12}$ radian

91. (a) 560.2 revolutions per minute

   (b) 3520 radians per minute

93. Radian. 1 radian $\approx 57.3°$ **95.** 20.16π inches per second

97. Answers will vary.   **99.** ≈ 2.16 miles

Section 6.2 *(page 474)*

Warm Up *(page 474)*

1. $2\sqrt{5}$ **2.** $3\sqrt{10}$ **3.** 10 **4.** $3\sqrt{2}$ **5.** 1.24
6. 317.55 **7.** 63.13 **8.** 133.57
9. $2{,}785{,}714.29$ **10.** 28.80

1. $\sin\theta = \dfrac{1}{2}$

$\cos\theta = \dfrac{\sqrt{3}}{2}$

$\tan\theta = \dfrac{\sqrt{3}}{3}$

$\csc\theta = 2$

$\sec\theta = \dfrac{2\sqrt{3}}{3}$

$\cot\theta = \sqrt{3}$

3. $\sin\theta = \dfrac{8}{17}$

$\cos\theta = \dfrac{15}{17}$

$\tan\theta = \dfrac{8}{15}$

$\csc\theta = \dfrac{17}{8}$

$\sec\theta = \dfrac{17}{15}$

$\cot\theta = \dfrac{15}{8}$

5. $\sin\theta = \dfrac{1}{3}$

$\cos\theta = \dfrac{2\sqrt{2}}{3}$

$\tan\theta = \dfrac{1}{2\sqrt{2}}$

$\csc\theta = 3$

$\sec\theta = \dfrac{3}{2\sqrt{2}}$

$\cot\theta = 2\sqrt{2}$

7. $\sin\theta = \dfrac{3}{5}$

$\cos\theta = \dfrac{4}{5}$

$\tan\theta = \dfrac{3}{4}$

$\csc\theta = \dfrac{5}{3}$

$\sec\theta = \dfrac{5}{4}$

$\cot\theta = \dfrac{4}{3}$

The triangles are similar, and corresponding sides are proportional.

The triangles are similar, and corresponding sides are proportional.

9.

$\cos\theta = \dfrac{\sqrt{5}}{3}$

$\tan\theta = \dfrac{2\sqrt{5}}{5}$

$\csc\theta = \dfrac{3}{2}$

$\sec\theta = \dfrac{3\sqrt{5}}{5}$

$\cot\theta = \dfrac{\sqrt{5}}{2}$

11.

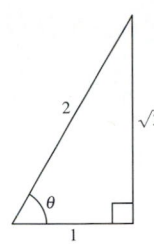

$\sin \theta = \dfrac{\sqrt{3}}{2}$

$\cos \theta = \dfrac{1}{2}$

$\tan \theta = \sqrt{3}$

$\csc \theta = \dfrac{2\sqrt{3}}{3}$

$\cot \theta = \dfrac{\sqrt{3}}{3}$

13.

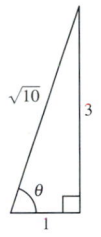

$\sin \theta = \dfrac{3\sqrt{10}}{10}$

$\cos \theta = \dfrac{\sqrt{10}}{10}$

$\csc \theta = \dfrac{\sqrt{10}}{3}$

$\sec \theta = \sqrt{10}$

$\cot \theta = \dfrac{1}{3}$

15.

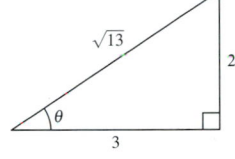

$\sin \theta = \dfrac{2\sqrt{13}}{13}$

$\cos \theta = \dfrac{3\sqrt{13}}{13}$

$\tan \theta = \dfrac{2}{3}$

$\csc \theta = \dfrac{\sqrt{13}}{2}$

$\sec \theta = \dfrac{\sqrt{13}}{3}$

17. (a) $\sqrt{3}$ (b) $\dfrac{1}{2}$ (c) $\dfrac{\sqrt{3}}{2}$ (d) $\dfrac{\sqrt{3}}{3}$

19. (a) $\dfrac{1}{3}$ (b) $\dfrac{2\sqrt{2}}{3}$ (c) $\dfrac{\sqrt{2}}{4}$ (d) 3

21. (a) 4   (b) $\dfrac{\sqrt{15}}{4}$ (c) $\dfrac{1}{\sqrt{15}}$ (d) $\dfrac{1}{4}$

23.–31. Answers will vary.

33. (a) $\dfrac{1}{2}$ (b) $\dfrac{\sqrt{3}}{3}$ **35.** (a) 1   (b) $\dfrac{\sqrt{2}}{2}$

37. (a) 0.1736   (b) 0.1736   **39.** (a) 0.2815   (b) 3.5523

41. (a) 1.3499   (b) 1.3432   **43.** (a) 5.0273   (b) 0.1989

45. (a) 1.1884   (b) 0.5463

47. (a) $30° = \dfrac{\pi}{6}$ (b) $30° = \dfrac{\pi}{6}$

49. (a) $60° = \dfrac{\pi}{3}$ (b) $45° = \dfrac{\pi}{4}$

51. (a) $60° = \dfrac{\pi}{3}$ (b) $45° = \dfrac{\pi}{4}$

53. (a) $55° \approx 0.96$ (b) $89° \approx 1.55$

55. (a) $50° \approx 0.873$ (b) $25° \approx 0.436$ **57.** $25\sqrt{3}$

59. $\dfrac{32\sqrt{3}}{3}$ **61.** 23.3   **63.** 6.1   **65.** $17\frac{1}{4}$ feet

67. (a)

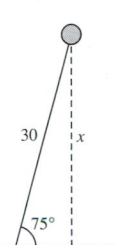

(b) $\sin 75° = \dfrac{x}{30}$

(c) 29.0 meters

69. 1144.9 feet   **71.** $(x_1, y_1) = \left(28\sqrt{3},\ 28\right)$

$(x_2, y_2) = \left(28,\ 28\sqrt{3}\right)$

73. $\sin 20° \approx 0.34$

$\cos 20° \approx 0.94$

$\tan 20° \approx 0.36$

$\csc 20° \approx 2.75$

$\sec 20° \approx 1.06$

$\cot 20° \approx 2.92$

75. (a)

θ	0	0.1	0.2	0.3	0.4	0.5
$\sin \theta$	0	0.0998	0.1987	0.2955	0.3894	0.4794

(b) θ

(c) $\sin \theta$ approaches θ as θ approaches 0.

77. True, $\csc x = \dfrac{1}{\sin x}$ **79.** False, $\dfrac{\sqrt{2}}{2} + \dfrac{\sqrt{2}}{2} \neq 1$

81. False, $1.7321 \neq \sin 2°$ **83.** $\dfrac{x}{x - 2}$

85. $\dfrac{2(x^2 - 5x - 10)}{(x - 2)(x + 2)^2}$

Section 6.3 *(page 488)*

Warm Up *(page 488)*

---

1. $\dfrac{1}{2}$ **2.** 1 **3.** $\dfrac{\sqrt{2}}{2}$ **4.** $\dfrac{\sqrt{3}}{3}$ **5.** $\dfrac{2\sqrt{3}}{3}$ **6.** $\sqrt{2}$

7. $\sin \theta = \dfrac{3\sqrt{13}}{13}$ **8.** $\sin \theta = \dfrac{\sqrt{5}}{3}$

   $\cos \theta = \dfrac{2\sqrt{13}}{13}$ $\tan \theta = \dfrac{\sqrt{5}}{2}$

   $\csc \theta = \dfrac{\sqrt{13}}{3}$ $\csc \theta = \dfrac{3\sqrt{5}}{5}$

   $\sec \theta = \dfrac{\sqrt{13}}{2}$ $\sec \theta = \dfrac{3}{2}$

   $\cot \theta = \dfrac{2}{3}$ $\cot \theta = \dfrac{2\sqrt{5}}{5}$

9. $\sin \theta = \dfrac{2\sqrt{6}}{5}$ **10.** $\sin \theta = \dfrac{2\sqrt{2}}{3}$

   $\cos \theta = \dfrac{\sqrt{6}}{12}$ $\cos \theta = \dfrac{1}{3}$

   $\csc \theta = 5$ $\tan \theta = 2\sqrt{2}$

   $\sec \theta = \dfrac{5\sqrt{6}}{12}$ $\csc \theta = \dfrac{3\sqrt{2}}{4}$

   $\cot \theta = 2\sqrt{6}$ $\cot \theta = \dfrac{\sqrt{2}}{4}$

---

1. (a) $\sin \theta = \dfrac{3}{5}$ (b) $\sin \theta = -\dfrac{15}{17}$

   $\cos \theta = \dfrac{4}{5}$ $\cos \theta = -\dfrac{8}{17}$

   $\tan \theta = \dfrac{3}{4}$ $\tan \theta = \dfrac{15}{8}$

   $\csc \theta = \dfrac{5}{3}$ $\csc \theta = -\dfrac{17}{15}$

   $\sec \theta = \dfrac{5}{4}$ $\sec \theta = -\dfrac{17}{8}$

   $\cot \theta = \dfrac{4}{3}$ $\cot \theta = \dfrac{8}{15}$

3. (a) $\sin \theta = -\dfrac{1}{2}$ (b) $\sin \theta = \dfrac{\sqrt{2}}{2}$

   $\cos \theta = -\dfrac{\sqrt{3}}{2}$ $\cos \theta = -\dfrac{\sqrt{2}}{2}$

   $\tan \theta = \dfrac{\sqrt{3}}{3}$ $\tan \theta = -1$

   $\csc \theta = -2$ $\csc \theta = \sqrt{2}$

   $\sec \theta = -\dfrac{2\sqrt{3}}{3}$ $\sec \theta = -\sqrt{2}$

   $\cot \theta = \sqrt{3}$ $\cot \theta = -1$

5. $\sin \theta = \dfrac{24}{25}$ **7.** $\sin \theta = \dfrac{5\sqrt{29}}{29}$

   $\cos \theta = \dfrac{7}{25}$ $\cos \theta = -\dfrac{2\sqrt{29}}{29}$

   $\tan \theta = \dfrac{24}{7}$ $\tan \theta = -\dfrac{5}{2}$

   $\csc \theta = \dfrac{25}{24}$ $\csc \theta = \dfrac{\sqrt{29}}{5}$

   $\sec \theta = \dfrac{25}{7}$ $\sec \theta = -\dfrac{\sqrt{29}}{2}$

   $\cot \theta = \dfrac{7}{24}$ $\cot \theta = -\dfrac{2}{5}$

9. Quadrant III   **11.** Quadrant II

13. $\sin \theta = \dfrac{3}{5}$ **15.** $\sin \theta = -\dfrac{15}{17}$

   $\cos \theta = -\dfrac{4}{5}$ $\cos \theta = \dfrac{8}{17}$

   $\tan \theta = -\dfrac{3}{4}$ $\tan \theta = -\dfrac{15}{8}$

   $\csc \theta = \dfrac{5}{3}$ $\csc \theta = -\dfrac{17}{15}$

   $\sec \theta = -\dfrac{5}{4}$ $\sec \theta = \dfrac{17}{8}$

   $\cot \theta = -\dfrac{4}{3}$ $\cot \theta = -\dfrac{8}{15}$

17. $\sin \theta = -\dfrac{\sqrt{10}}{10}$ **19.** $\sin \theta = \dfrac{\sqrt{3}}{2}$

   $\cos \theta = \dfrac{3\sqrt{10}}{10}$ $\cos \theta = -\dfrac{1}{2}$

   $\tan \theta = -\dfrac{1}{3}$ $\tan \theta = -\sqrt{3}$

   $\csc \theta = -\sqrt{10}$ $\csc \theta = \dfrac{2\sqrt{3}}{3}$

   $\sec \theta = \dfrac{\sqrt{10}}{3}$ $\sec \theta = -2$

   $\cot \theta = -3$ $\cot \theta = -\dfrac{\sqrt{3}}{3}$

21. $\sin \theta = 0$

$\cos \theta = -1$

$\tan \theta = 0$

$\csc \theta$ is undefined.

$\sec \theta = -1$

$\cot \theta$ is undefined.

23. $\sin \theta = \dfrac{\sqrt{2}}{2}$

$\cos \theta = -\dfrac{\sqrt{2}}{2}$

$\tan \theta = -1$

$\csc \theta = \sqrt{2}$

$\sec \theta = -\sqrt{2}$

$\cot \theta = -1$

25. $\sin \theta = -\dfrac{2\sqrt{5}}{5}$

$\cos \theta = -\dfrac{\sqrt{5}}{5}$

$\tan \theta = 2$

$\csc \theta = \dfrac{-\sqrt{5}}{2}$

$\sec \theta = -\sqrt{5}$

$\cot \theta = \dfrac{1}{2}$

27. -1 **29.** -1 **31.** Undefined   **33.** 0

35. $\theta' = 23°$

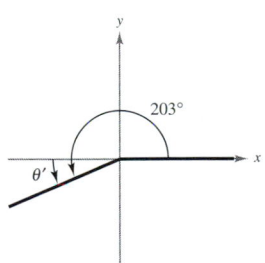

37. $\theta' = 65°$

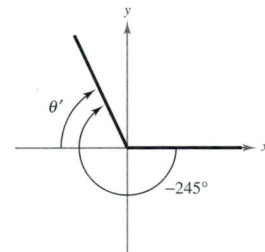

39. $\theta' = \dfrac{\pi}{3}$

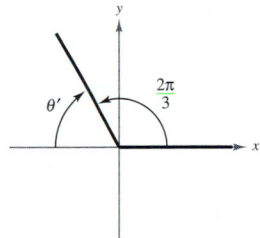

41. $\theta' = 3.5 - \pi$

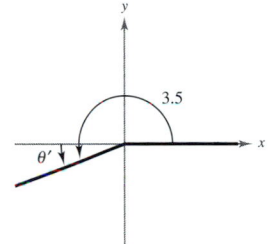

43. $\sin 225° = -\dfrac{\sqrt{2}}{2}$

$\cos 225° = -\dfrac{\sqrt{2}}{2}$

$\tan 225° = 1$

45. $\sin 750° = \dfrac{1}{2}$

$\cos 750° = \dfrac{\sqrt{3}}{2}$

$\tan 750° = \dfrac{\sqrt{3}}{3}$

47. $\sin \dfrac{4\pi}{3} = -\dfrac{\sqrt{3}}{2}$

$\cos \dfrac{4\pi}{3} = -\dfrac{1}{2}$

$\tan \dfrac{4\pi}{3} = \sqrt{3}$

49. $\sin\left(-\dfrac{\pi}{6}\right) = -\dfrac{1}{2}$

$\cos\left(-\dfrac{\pi}{6}\right) = \dfrac{\sqrt{3}}{2}$

$\tan\left(-\dfrac{\pi}{6}\right) = -\dfrac{\sqrt{3}}{3}$

51. $\sin \dfrac{11\pi}{4} = \dfrac{\sqrt{2}}{2}$

$\cos \dfrac{11\pi}{4} = -\dfrac{\sqrt{2}}{2}$

$\tan \dfrac{11\pi}{4} = -1$

53. 0.1736 **55.** -0.3420 **57.** 1.7321

59. 0.3640 **61.** 0.6052

63. (a) $30° = \dfrac{\pi}{6}$, $150° = \dfrac{5\pi}{6}$ (b) $210° = \dfrac{7\pi}{6}$, $330° = \dfrac{11\pi}{6}$

65. (a) $60° = \dfrac{\pi}{3}$, $120° = \dfrac{2\pi}{3}$ (b) $135° = \dfrac{3\pi}{4}$, $315° = \dfrac{7\pi}{4}$

67. (a) $45° = \dfrac{\pi}{4}$, $225° = \dfrac{5\pi}{4}$ (b) $150° = \dfrac{5\pi}{6}$, $330° = \dfrac{11\pi}{6}$

69. $54.99°$, $125.01°$ **71.** 0.175, 6.109

73. 0.873, 4.014 **75.** $\dfrac{4}{5}$ **77.** $-\dfrac{\sqrt{13}}{2}$ **79.** $\dfrac{8}{5}$

81. $\left(\dfrac{\sqrt{2}}{2}, \dfrac{\sqrt{2}}{2}\right)$ **83.** $\left(-\dfrac{\sqrt{3}}{2}, \dfrac{1}{2}\right)$

$\sin\dfrac{\pi}{4} = \dfrac{\sqrt{2}}{2}$ $\sin\dfrac{5\pi}{6} = \dfrac{1}{2}$

$\cos\dfrac{\pi}{4} = \dfrac{\sqrt{2}}{2}$ $\cos\dfrac{5\pi}{6} = -\dfrac{\sqrt{3}}{2}$

$\tan\dfrac{\pi}{4} = 1$ $\tan\dfrac{5\pi}{6} = -\dfrac{1}{\sqrt{3}}$

85. $\left(-\dfrac{1}{2}, -\dfrac{\sqrt{3}}{2}\right)$ **87.** $(0, -1)$

$\sin\dfrac{4\pi}{3} = -\dfrac{\sqrt{3}}{2}$ $\sin\dfrac{3\pi}{2} = -1$

$\cos\dfrac{4\pi}{3} = -\dfrac{1}{2}$ $\cos\dfrac{3\pi}{2} = 0$

$\tan\dfrac{4\pi}{3} = \sqrt{3}$ $\tan\dfrac{3\pi}{2}$ is undefined.

89. (a) -1 (b) -0.4 **91.** (a) 0.25 or 2.89

(b) 1.82 or 4.46

93. Odd **95.** (a) $25.2°F$ (b) $65.1°F$ (c) $50.8°F$

97. (a) 2 feet (b) 0.14 foot (c) -1.98 feet **99.** 0.79

101.

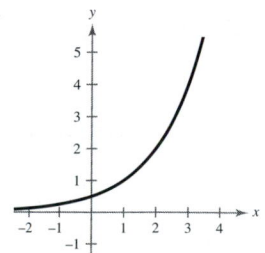

103.

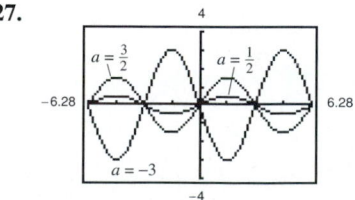

Section 6.4 *(page 500)*

Warm Up *(page 500)*

1. 6π **2.** $\dfrac{1}{2}$ **3.** $\dfrac{\pi}{6}$ **4.** $\dfrac{7\pi}{6}$ **5.** -2 **6.** $-\dfrac{4}{3}$

7. 1 **8.** 0 **9.** 1 **10.** 0

1. Period: π **3.** Period: 4π **5.** Period: 2

  Amplitude: 3 Amplitude: $\dfrac{5}{2}$ Amplitude: $\dfrac{2}{3}$

7. Period: 2π **9.** Period: $\dfrac{\pi}{5}$

  Amplitude: 2 Amplitude: 3

11. Period: 3π **13.** Period: $\dfrac{1}{2}$

  Amplitude: $\dfrac{1}{2}$ Amplitude: 3

15. g is a shift of f π units to the right.

17. g is a reflection of f about the x-axis.

19. The period of f is twice the period of g.

21. Shift the graph of f two units up to obtain the graph of g.

23. The graph of g has twice the amplitude of the graph of f.

25. The graph of g is a horizontal shift of the graph of f $\pi/2$ units to the right.

27.

Amplitude changes

29.

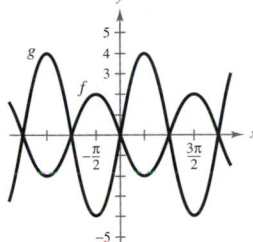

Period changes

31.

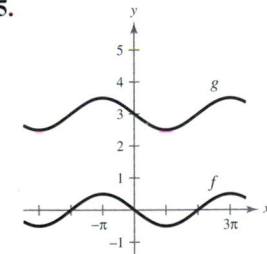

33.

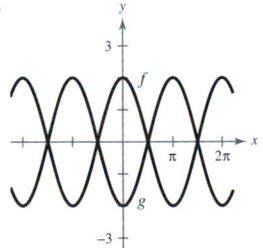

35.

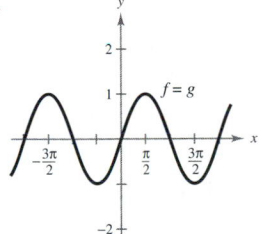

37.

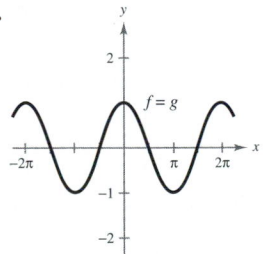

39.

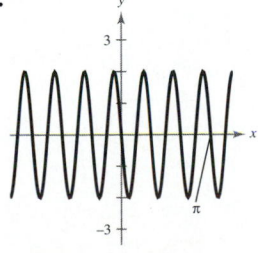

41.

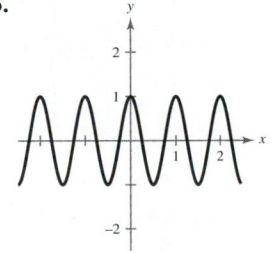

43.

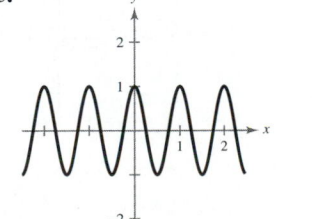

45.

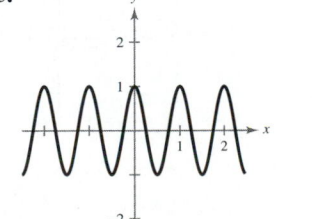

47.

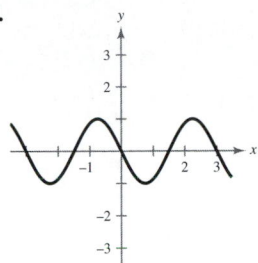

49.

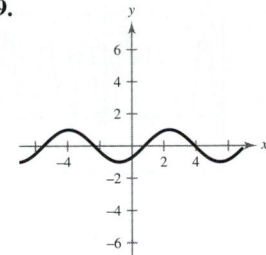

51.

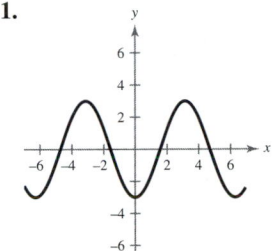

53.

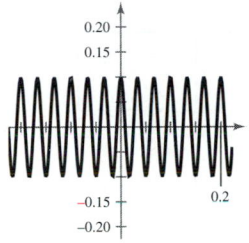

55.

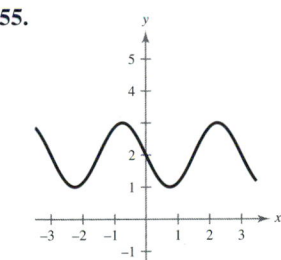

57.

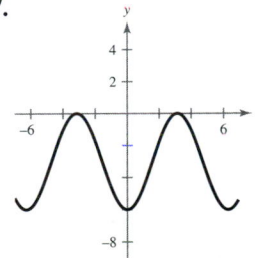

59.

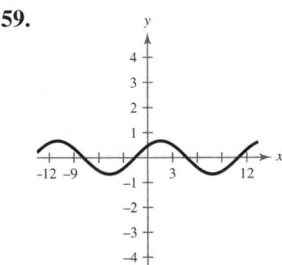

61.

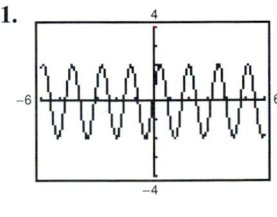

63.

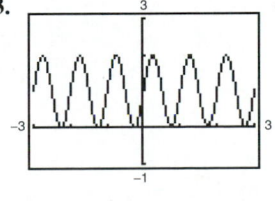

65.

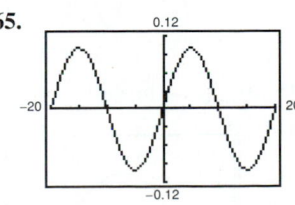

67.

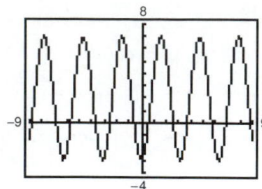

69. $y = 2 + 3 \cos x$ **71.** $y = 4 - 4 \cos x$

73. $y = -3 \sin(2x)$ **75.** $y = \sin\left(x - \dfrac{\pi}{4}\right)$

77.

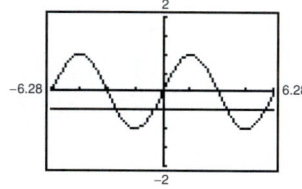

$$x = -\frac{\pi}{6}, \ -\frac{5\pi}{6}, \ \frac{7\pi}{6}, \ \frac{11\pi}{6}$$

79.

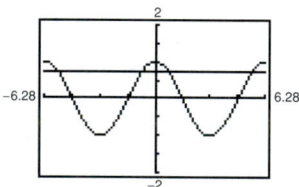

$$x = \pm\frac{\pi}{4}, \ \pm\frac{7\pi}{4}$$

81. (a) Even   (b) Even

83. (a) 6   (b) 10 cycles per minute

(c)

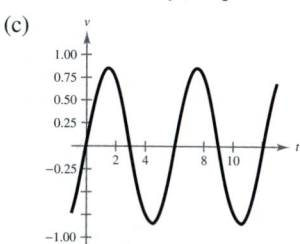

85 (a) $\frac{1}{440}$ (b) 440

87.

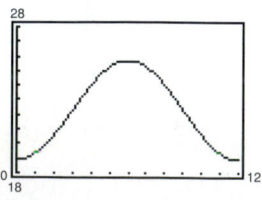

89. (a) $C(t) = 58.50 + 26.72 \sin\left(\dfrac{\pi t}{6} + 4.19\right)$

(b)

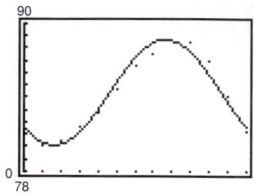

(c)

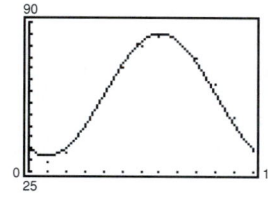

(d) Honolulu: 84.40; Chicago: 58.50

(e) 12. Yes. One full period is 1 year.

(f) Chicago, amplitude

91. (a)

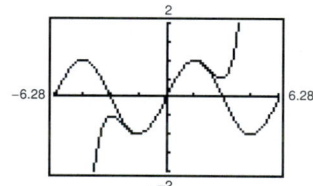

(b)

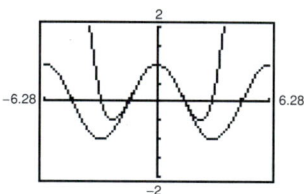

(c) $-\dfrac{x^7}{7!}$, $-\dfrac{x^6}{6!}$

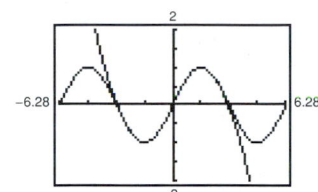

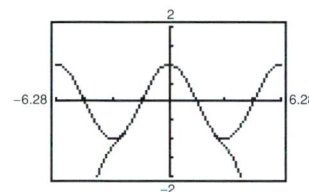

The accuracy increased.

93. (a)

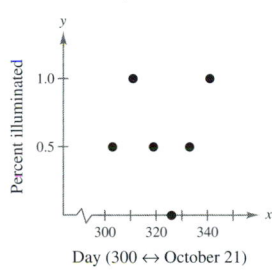

(b) $y = \dfrac{1}{2} + \dfrac{1}{2} \sin\left[\dfrac{\pi}{15}(t - 303)\right]$

(c)   (d) 0

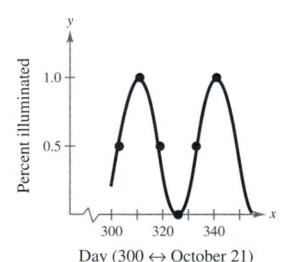

95. $2 \log_2 x + \log_2(x - 3)$ **97.** $\frac{1}{2} \ln z - \frac{1}{2} \ln(z^2 + 1)$

Section 6.5   *(page 512)*

Warm Up   *(page 512)*

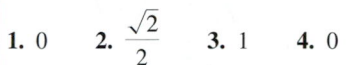

1. 0   **2.** $\dfrac{\sqrt{2}}{2}$ **3.** 1   **4.** 0   **5.** 0   **6.** 0

7.   **8.**

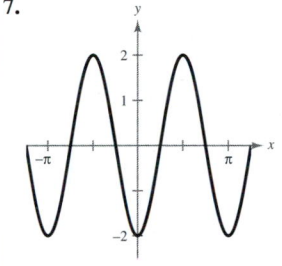   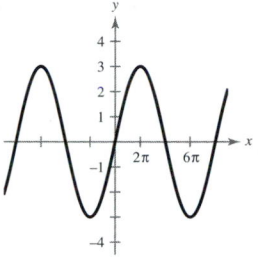

9.   **10.**

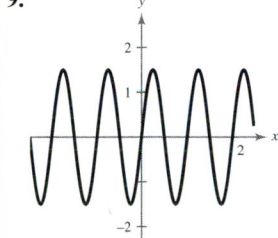   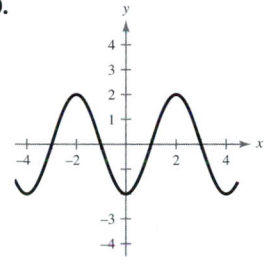

1. g, 4π **3.** f, $\dfrac{\pi}{2}$ **5.** b, 2   **7.** e, 2π

9.   **11.**

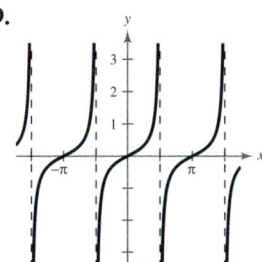   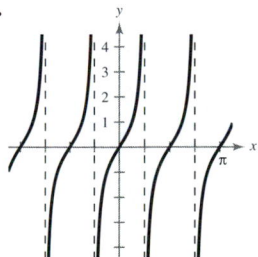

13.   **15.**

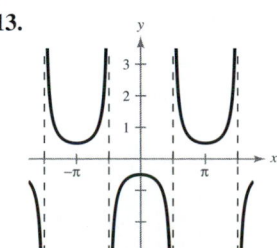   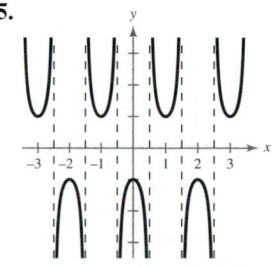

17.

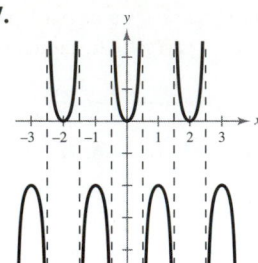

19.

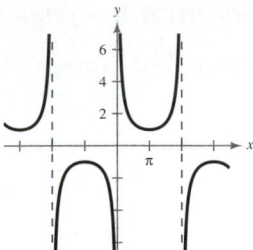

21.

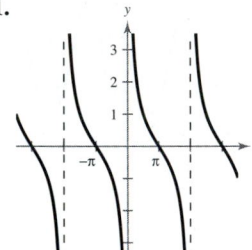

23.

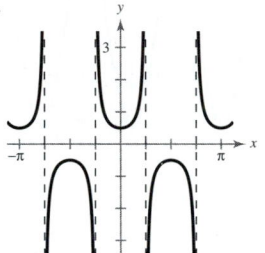

25.

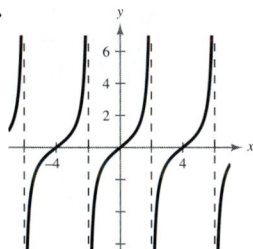

27.

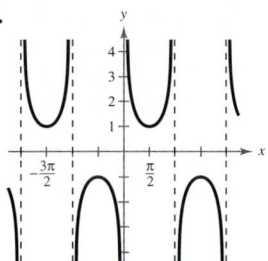

29.

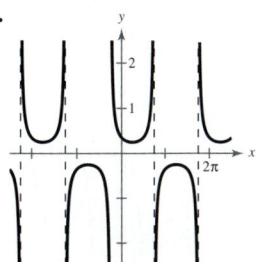

31.

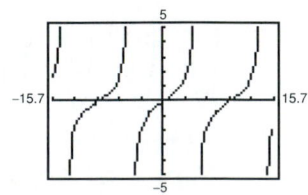

33.

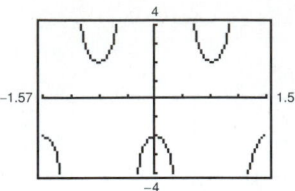

35.

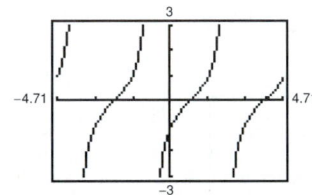

37.

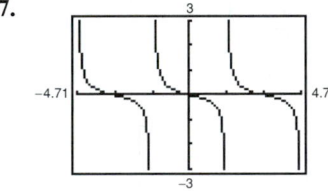

39.

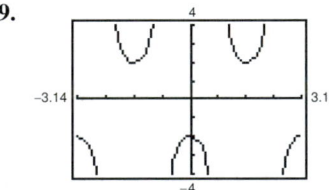

41. $-\dfrac{7\pi}{4}, -\dfrac{3\pi}{4}, \dfrac{\pi}{4}, \dfrac{5\pi}{4}$ **43.** $-\dfrac{4\pi}{3}, -\dfrac{2\pi}{3}, \dfrac{2\pi}{3}, \dfrac{4\pi}{3}$

45. Even   **47.** As x approaches $\pi/2$ from the left, f approaches ∞. As x approaches $\pi/2$ from the right, f approaches $-\infty$.

49. (a)

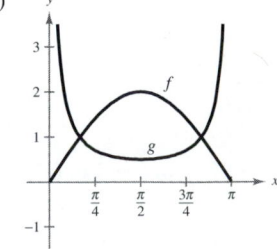

(b) $\dfrac{\pi}{6} < x < \dfrac{5\pi}{6}$

(c) Sine approaches 0 and cosecant approaches $\pm\infty$ because the cosecant is the reciprocal of the sine.

51.

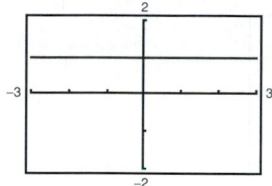

53.

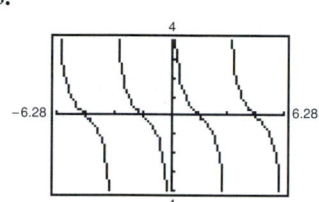

55. d   **57.** b

59.

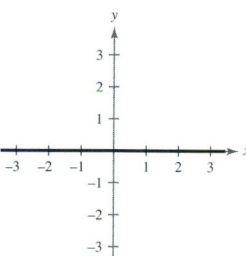

Equal

61.

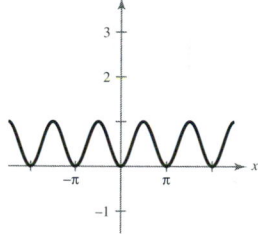

Equal

63.

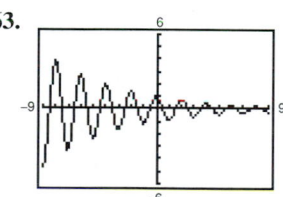

65.

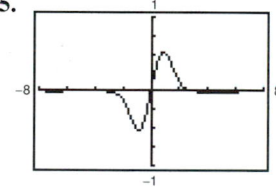

67. $\tan x = \dfrac{5}{d}$

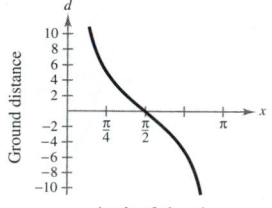

69. As the predator population increases, the number of prey decreases. When the number of prey is small, the number of predators decreases.

71. (a)

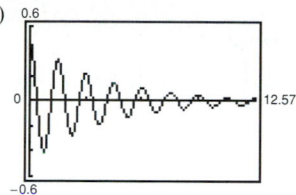

(b) Periodic but damped; goes to 0 as t increases.

73.

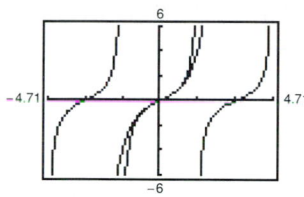

75. (a)

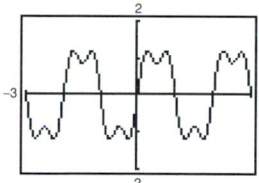

(b)
$$y_3 = \frac{4}{\pi}\left[\sin(\pi x) + \frac{1}{3}\sin(3\pi x) + \frac{1}{5}\sin(5\pi x) + \frac{1}{7}\sin(7\pi x)\right]$$

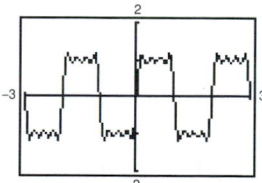

(c)
$$y_4 = \frac{4}{\pi}\left[\sin(\pi x) + \frac{1}{3}\sin(3\pi x) + \frac{1}{5}\sin(5\pi x) + \frac{1}{7}\sin(7\pi x)\right.$$
$$\left. + \frac{1}{9}\sin(9\pi x)\right]$$

77.

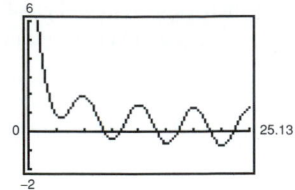

∞

79.

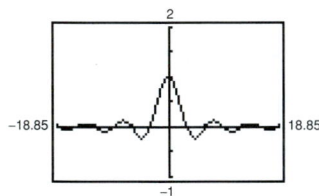

1

81.

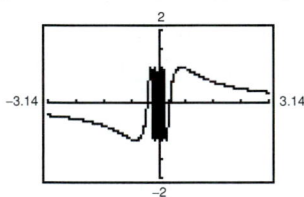

Oscillates

83. 1.994   **85.** ±4.851

Section 6.6 *(page 523)*

Warm Up *(page 523)*

1. −1   **2.** −1   **3.** −1   **4.** $\dfrac{\sqrt{2}}{2}$ **5.** 0   **6.** $\dfrac{\pi}{6}$

7. π **8.** $\dfrac{\pi}{4}$ **9.** 0   **10.** $-\dfrac{\pi}{4}$

1. False. $\dfrac{5\pi}{6}$ is not in the range of the arcsine.

3. $\dfrac{\pi}{6}$ **5.** $\dfrac{\pi}{3}$ **7.** $\dfrac{\pi}{6}$ **9.** $\dfrac{5\pi}{6}$ **11.** $-\dfrac{\pi}{3}$ **13.** $\dfrac{2\pi}{3}$

15. $\dfrac{\pi}{3}$ **17.** 0   **19.** 1.29   **21.** −0.85   **23.** −1.25

25. 0.32   **27.** 1.99   **29.** 0.74   **31.** $-\dfrac{\pi}{3}$, $-\dfrac{1}{\sqrt{3}}$, 1

33.

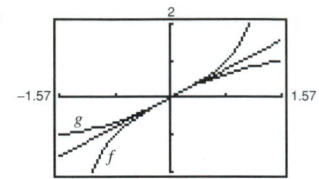

35. $\theta = \arctan \dfrac{x}{4}$ **37.** $\theta = \arcsin \dfrac{x+2}{5}$ **39.** 0.3

41. −0.1   **43.** 0   **45.** $\dfrac{3}{5}$ **47.** $\dfrac{\sqrt{5}}{5}$ **49.** $\dfrac{12}{13}$

51. $\dfrac{\sqrt{34}}{5}$ **53.** $\dfrac{\sqrt{5}}{3}$ **55.** $\dfrac{1}{x}$ **57.** $\sqrt{1-4x^2}$

59. $\sqrt{1-x^2}$ **61.** $\dfrac{\sqrt{9-x^2}}{x}$ **63.** $\dfrac{\sqrt{x^2+2}}{x}$

65.

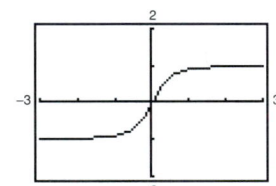

$y = \pm 1$

67. $\arcsin \dfrac{9}{\sqrt{x^2+81}}$ **69.** $\arcsin \dfrac{|x-1|}{\sqrt{x^2-2x+10}}$

71.

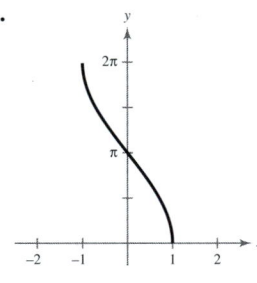

73.

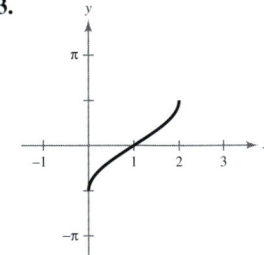

75.

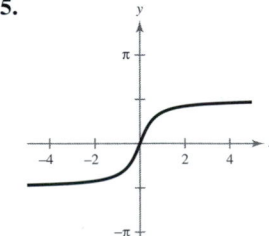

77.

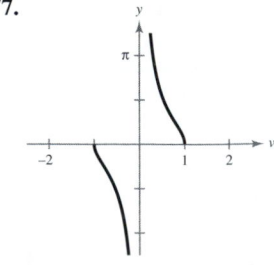

79. $3\sqrt{2} \sin (2t + \arctan 1)$

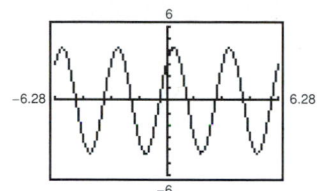

81. (a)

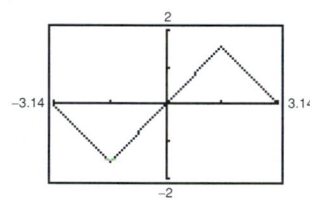

(b) The graphs of $f \circ f^{-1}$ and $f^{-1} \circ f$ differ because of the domains and ranges of f and f^{-1}.

83. (a) $\theta = \arcsin \dfrac{10}{s}$ (b) 0.21, 0.43

85. (a)

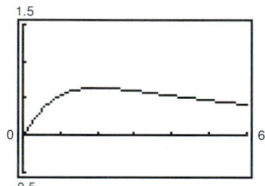

(b) 2 feet

(c) $\beta = 0$

87. (a) $\theta = \arctan \dfrac{5}{x}$ (b) 26.6°, 59.0°

89. Domain: $(-\infty, \infty)$

Range: $(0, \pi)$

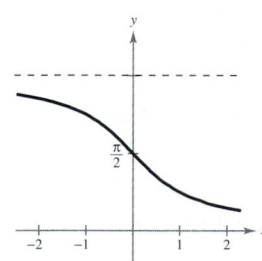

91. Domain: $(-\infty, -1] \cup [1, \infty)$

Range: $[-\pi/2, 0) \cup (0, \pi/2]$

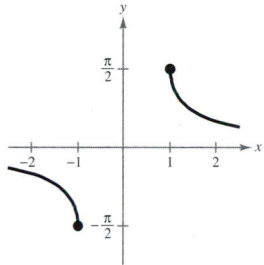

93.–97. Answers will vary.   **99.** Buy now.   **101.** 8

Section 6.7 *(page 534)*

Warm Up *(page 534)*

1. 8.45	**2.** 78.99	**3.** 1.06	**4.** 1.24	**5.** 4.88
6. 34.14	**7.** 4; π	**8.** $\frac{1}{2}$; 2	**9.** 3; $\frac{2}{3}$	**10.** 0.2; 8π

1. $a \approx 3.64$ **3.** $a \approx 8.26$ **5.** $c \approx 11.66$
$c \approx 10.64$ $c \approx 25.38$ $A \approx 30.96°$
$B = 70°$ $A = 19°$ $B \approx 59.04°$

7. $a \approx 49.48$ **9.** $a \approx 91.34$
$A \approx 72.08°$ $b \approx 420.70$
$B = 17.92°$ $B = 77°45'$

11. 2.56 inches   **13.** 103.9 feet   **15.** 15.4 feet

17. (a)

(b) $h = 50(\tan 47°40' - \tan 35°)$

(c) 19.9 feet

19. 2236.8 feet   **21.** 56.3°   **23.** 15.5°   **25.** 5099 feet

27. 0.73 mile   **29.** 508 miles north; 650 miles east

31. (a) N 58° E   (b) 68.82 meters   **33.** N 56.3° W

35. 1933.3 feet   **37.** $\approx$ 3.23 miles or $\approx$ 17,054 feet

39. 78.7°   **41.** 35.3°   **43.** $y = \sqrt{3}\,r$ **45.** 29.4 inches

47. $a \approx 7$, $c \approx 12.2$ **49.** (a) 4   (b) 4   (c) $\frac{1}{16}$

51. (a) $\frac{1}{16}$ (b) 60 (c) $\frac{1}{120}$ **53.** $y = 4 \sin (\pi t)$

55. $y = 3 \cos\left(\dfrac{4\pi t}{3}\right)$ **57.** $\omega = 528\pi$

59. (a)   (b) $\dfrac{\pi}{8}$ seconds

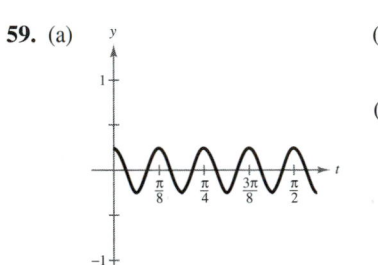

(c) $\dfrac{\pi}{32}$ seconds

61. (a) and (b)

Base 1	Base 2	Altitude	Area
8	$8 + 16 \cos 10°$	$8 \sin 10°$	22.1
8	$8 + 16 \cos 20°$	$8 \sin 20°$	42.5
8	$8 + 16 \cos 30°$	$8 \sin 30°$	59.7
8	$8 + 16 \cos 40°$	$8 \sin 40°$	72.7
8	$8 + 16 \cos 50°$	$8 \sin 50°$	80.5
8	$8 + 16 \cos 60°$	$8 \sin 60°$	83.1
8	$8 + 16 \cos 70°$	$8 \sin 70°$	80.7

83.1 (maximum cross-sectional area)

(c) $A = 64(1 + \cos \theta)(\sin \theta)$

(d)

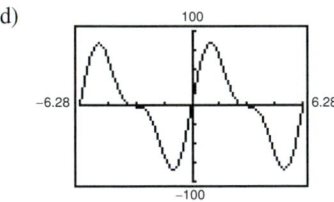

83.1

63. (a)

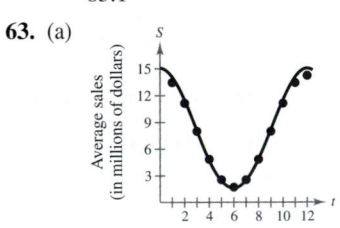

Month (1 ↔ January)

(b) $S = 8 + 6.3 \cos\left(\dfrac{\pi t}{6}\right)$

(c) 12. Yes, sales of outerwear are seasonal.

(d) Maximum displacement from average sales of 8.

65.   **67.**

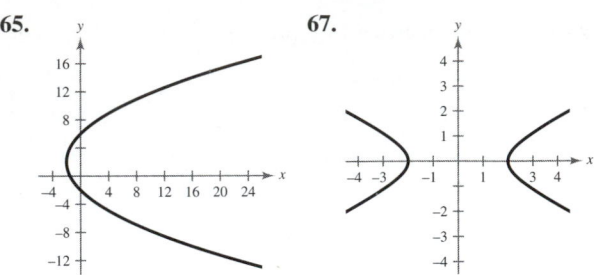

Focus on Concepts   *(page 540)*

1. (a) The vertex is at the origin and the initial side is on the positive *x*-axis.

  (b) Clockwise rotation of the terminal side.

  (c) Two angles in standard position where the terminal sides coincide.

  (d) The magnitude of the angle is between 90° and 180°.

2. Increases. The linear velocity is proportional to the radius.

3. False. For each θ there corresponds exactly one value of *y*.

4. Corresponding sides of similar triangles are proportional.

5. Undefined because $\sec \theta = 1/\cos \theta$.

6. Determine the trigonometric function of the reference angle and, depending on the quadrant in which the obtuse angle lies, prefix the appropriate sign.

7. d; the period is 2π and the amplitude is 3.

8. a; the period is 2π and, because $a < 0$, the graph is reflected about the *x*-axis.

9. b; the period is 2 and the amplitude is 2.

10. c; the period is 4π and the amplitude is 2.

11. (a) Equal; two-period shift

  (b) Not equal; $f\left(t + \frac{1}{2}c\right)$ is a horizontal translation and $f\left(\frac{1}{2}t\right)$ is a period change.

  (c) Equal; the period change is the same in each.

12. Their range is $(-\infty, \infty)$.

13. (a) The displacement is increased.

  (b) The friction damps the oscillations more quickly.

  (c) The frequency of the oscillations increases.

14. False. $3\pi/4$ is not in the range of the arctangent function.

Review Exercises *(page 541)*

1.

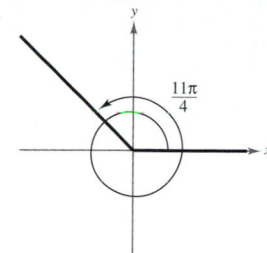

$$\frac{3\pi}{4}, \ -\frac{5\pi}{4}$$

3.

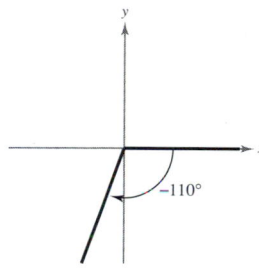

$250°, \ -470°$

5. $135.28°$ **7.** $5.38°$ **9.** $135°16'12''$ **11.** $-85°9'$

13. $128.57°$ **15.** $-200.54°$ **17.** 8.3776

19. -0.5890 **21.** $72°$ **23.** $\dfrac{\pi}{5}$

25. $\sin \theta = \dfrac{4}{5}$

$\cos \theta = \dfrac{3}{5}$

$\tan \theta = \dfrac{4}{3}$

$\csc \theta = \dfrac{5}{4}$

$\sec \theta = \dfrac{5}{3}$

$\cot \theta = \dfrac{3}{4}$

27. $\sin \theta = \dfrac{2\sqrt{53}}{53}$

$\cos \theta = -\dfrac{7\sqrt{53}}{53}$

$\tan \theta = -\dfrac{2}{7}$

$\csc \theta = \dfrac{\sqrt{53}}{2}$

$\sec \theta = -\dfrac{\sqrt{53}}{7}$

$\cot \theta = -\dfrac{7}{2}$

29. $\sin \theta = -\dfrac{3\sqrt{13}}{13}$

$\cos \theta = -\dfrac{2\sqrt{13}}{13}$

$\tan \theta = \dfrac{3}{2}$

$\csc \theta = -\dfrac{\sqrt{13}}{3}$

$\sec \theta = -\dfrac{\sqrt{13}}{2}$

$\cot \theta = \dfrac{2}{3}$

31. $\sin \theta = -\dfrac{\sqrt{11}}{6}$

$\cos \theta = \dfrac{5}{6}$

$\tan \theta = -\dfrac{\sqrt{11}}{5}$

$\csc \theta = -\dfrac{6\sqrt{11}}{11}$

$\cot \theta = -\dfrac{5\sqrt{11}}{11}$

33. $\cos \theta = -\dfrac{\sqrt{55}}{8}$

$\tan \theta = -\dfrac{3\sqrt{55}}{55}$

$\csc \theta = \dfrac{8}{3}$

$\sec \theta = -\dfrac{8\sqrt{55}}{55}$

$\cot \theta = -\dfrac{\sqrt{55}}{3}$

35. $\sqrt{3}$ **37.** $-\dfrac{\sqrt{3}}{2}$

39. $-\dfrac{\sqrt{2}}{2}$ **41.** 0.65 **43.** 3.24

45. $135° = \dfrac{3\pi}{4}, \ 225° = \dfrac{5\pi}{4}$

47. $210° = \dfrac{7\pi}{6}, \ 330° = \dfrac{11\pi}{6}$

49. $57° \approx 0.9948, \ 123° \approx 2.1468$

51. $165° \approx 2.8798, \ 195° \approx 3.4034$

53.

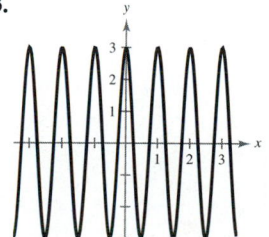

55.

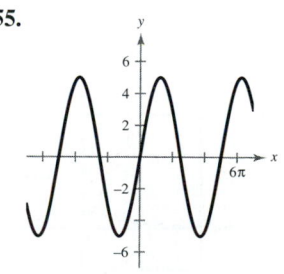

57.

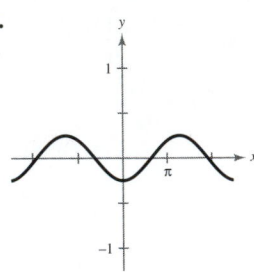

59.

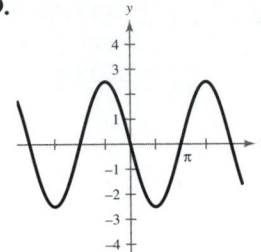

61.

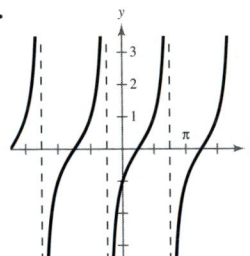

63.

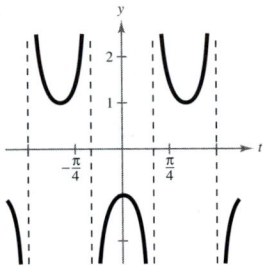

65.

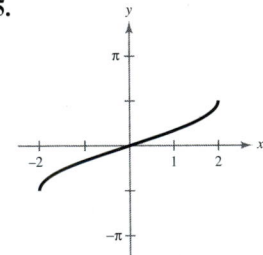

67.

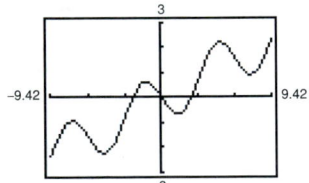

69.

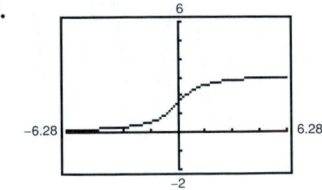

71.

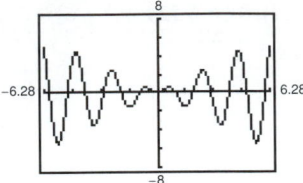

73.

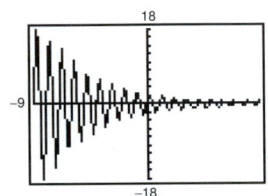

75.

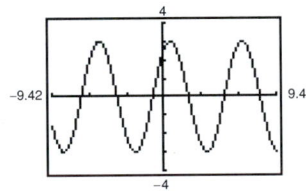

77.

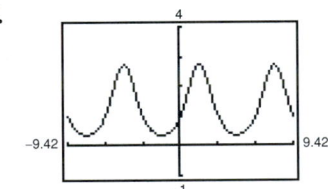

Periodic: $\left(\dfrac{\pi}{2}, e\right)$, $\left(\dfrac{3\pi}{2}, e^{-1}\right)$

79.

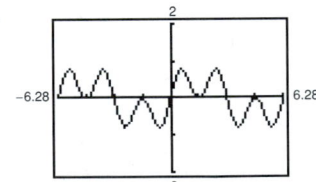

Periodic: $\left(\dfrac{\pi}{2}, 0\right)$, $\left(\dfrac{3\pi}{2}, 0\right)$, $(0.61, 0.77)$, $(2.53, 0.77)$, $(3.76, -0.77)$, $(5.67, -0.77)$

81. $\dfrac{\sqrt{-x^2 + 2x}}{-x^2 + 2x}$ **83.** $\dfrac{2\sqrt{4 - 2x^2}}{4 - x^2}$ **85.** 9.2 meters

87. 1.2 miles   **89.** 0.07 kilometer

91. (a) $A = 72(\tan \theta - \theta)$

(b)

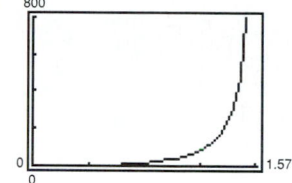

Area increases without bound as θ approaches $\pi/2$.

Chapter Test *(page 546)*

1. (a)

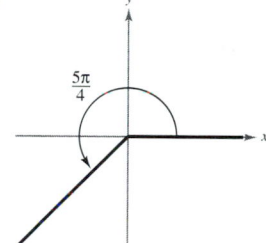

(b) $\dfrac{13\pi}{4}, \ -\dfrac{3\pi}{4}$

(c) $225°$

2. 1500 radians per minute

3. $\sin \theta = \dfrac{3}{\sqrt{10}}$

$\cos \theta = -\dfrac{1}{\sqrt{10}}$

$\tan \theta = -3$

$\csc \theta = \dfrac{\sqrt{10}}{3}$

$\sec \theta = -\sqrt{10}$

$\cot \theta = -\dfrac{1}{3}$

4. $\sin \theta = \dfrac{3}{\sqrt{13}}$

$\cos \theta = \dfrac{2}{\sqrt{13}}$

$\csc \theta = \dfrac{\sqrt{13}}{3}$

$\sec \theta = \dfrac{\sqrt{13}}{2}$

$\cot \theta = \dfrac{2}{3}$

5. $70°$

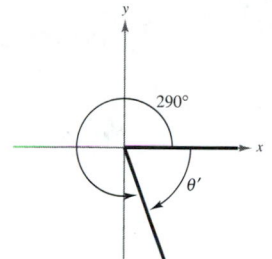

6. III   **7.** $150°, \ 210°$ **8.** $1.33, \ 1.81$

9.

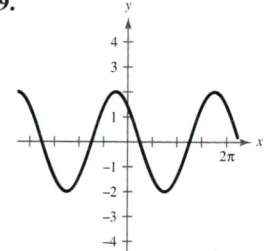

10.

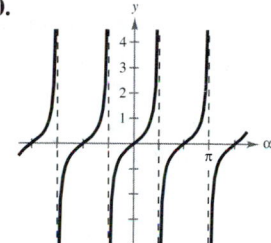

11.

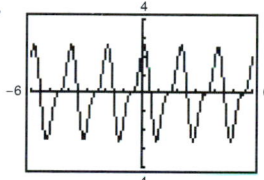

Period: 2

12.

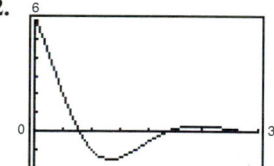

13. $y = -2 \sin\left(\dfrac{\pi}{2} - \dfrac{\pi}{4}\right)$

14. $\dfrac{\sqrt{5}}{2}$

15.

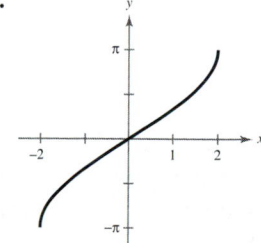

16. $(-14.88, \ 51.91)$

CHAPTER 7
Section 7.1 *(page 553)*
Warm Up *(page 553)*

1. $\sin \theta = \dfrac{3\sqrt{13}}{13}$

   $\cos \theta = \dfrac{2\sqrt{13}}{13}$

   $\tan \theta = \dfrac{3}{2}$

   $\csc \theta = \dfrac{\sqrt{13}}{3}$

   $\sec \theta = \dfrac{\sqrt{13}}{2}$

   $\cot \theta = \dfrac{2}{3}$

2. $\sin \theta = \dfrac{2\sqrt{2}}{3}$

   $\cos \theta = \dfrac{1}{3}$

   $\tan \theta = 2\sqrt{2}$

   $\csc \theta = \dfrac{3\sqrt{2}}{4}$

   $\sec \theta = 3$

   $\cot \theta = \dfrac{\sqrt{2}}{4}$

3. $\sin \theta = -\dfrac{3\sqrt{58}}{58}$

   $\cos \theta = \dfrac{7\sqrt{58}}{58}$

   $\tan \theta = -\dfrac{3}{7}$

   $\csc \theta = -\dfrac{\sqrt{58}}{3}$

   $\sec \theta = \dfrac{\sqrt{58}}{7}$

   $\cot \theta = -\dfrac{7}{3}$

4. $\sin \theta = \dfrac{\sqrt{5}}{5}$

   $\cos \theta = -\dfrac{2\sqrt{5}}{5}$

   $\tan \theta = -\dfrac{1}{2}$

   $\csc \theta = \sqrt{5}$

   $\sec \theta = -\dfrac{\sqrt{5}}{2}$

   $\cot \theta = -2$

5. $\dfrac{1}{2}$ 6. $\dfrac{5}{4}$ 7. $\dfrac{\sqrt{73}}{8}$ 8. $\dfrac{2}{3}$ 9. $\dfrac{x^2 + x + 16}{4(x + 1)}$

10. $\dfrac{8x - 2}{1 - x^2}$

1. $\sin x = \dfrac{1}{2}$

   $\cos x = \dfrac{\sqrt{3}}{2}$

   $\tan x = \dfrac{\sqrt{3}}{3}$

   $\csc x = 2$

   $\sec x = \dfrac{2\sqrt{3}}{3}$

   $\cot x = \sqrt{3}$

3. $\sin \theta = -\dfrac{\sqrt{2}}{2}$

   $\cos \theta = \dfrac{\sqrt{2}}{2}$

   $\tan \theta = -1$

   $\csc \theta = -\sqrt{2}$

   $\sec \theta = \sqrt{2}$

   $\cot \theta = -1$

5. $\sin x = -\dfrac{5}{13}$

   $\cos x = -\dfrac{12}{13}$

   $\tan x = \dfrac{5}{12}$

   $\csc x = -\dfrac{13}{5}$

   $\sec x = -\dfrac{13}{12}$

   $\cot x = \dfrac{12}{5}$

7. $\sin \phi = 0$

   $\cos \phi = -1$

   $\tan \phi = 0$

   $\csc \phi$ is undefined.

   $\sec \phi = -1$

   $\cot \phi$ is undefined.

9. $\sin x = \dfrac{2}{3}$

   $\cos x = -\dfrac{\sqrt{5}}{3}$

   $\tan x = -\dfrac{2\sqrt{5}}{5}$

   $\csc x = \dfrac{3}{2}$

   $\sec x = -\dfrac{3\sqrt{5}}{5}$

   $\cot x = -\dfrac{\sqrt{5}}{2}$

11. $\sin \theta = -\dfrac{2\sqrt{5}}{5}$

    $\cos \theta = -\dfrac{\sqrt{5}}{5}$

    $\tan \theta = 2$

    $\csc \theta = -\dfrac{\sqrt{5}}{2}$

    $\sec \theta = -\sqrt{5}$

    $\cot \theta = \dfrac{1}{2}$

13. $\sin \theta = -1$

    $\cos \theta = 0$

    $\tan \theta$ is undefined.

    $\csc \theta = -1$

    $\sec \theta$ is undefined.

    $\cot \theta = 0$

15. 1, 1   **17.** $\infty, 0$ **19.** d   **21.** a   **23.** e

25. b   **27.** f   **29.** e   **31.** $\sec \phi$ **33.** $\sin \beta$

35. $\cos x$ **37.** 1   **39.** $-\tan x$ **41.** $\tan x$

43. $1 + \sin y$ **45.** $\sin^2 x$ **47.** $\sin^2 x \tan^2 x$

49. $\sec^4 x$ **51.** $\sin^2 x - \cos^2 x$ **53.** $1 + 2 \sin x \cos x$

55. $\tan^2 x$ **57.** $2 \csc^2 x$ **59.** $2 \sec x$ **61.** $1 + \cos y$

63. $3(\sec x + \tan x)$

65.

x	0.2	0.4	0.6	0.8	1.0
y_1	0.1987	0.3894	0.5646	0.7174	0.8415
y_2	0.1987	0.3894	0.5646	0.7174	0.8415

x	1.2	1.4
y_1	0.9320	0.9854
y_2	0.9320	0.9854

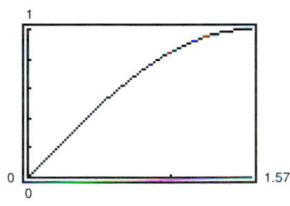

$y_1 = y_2$

67.

x	0.2	0.4	0.6	0.8	1.0
y_1	1.2230	1.5085	1.8958	2.4650	3.4082
y_2	1.2230	1.5085	1.8958	2.4650	3.4082

x	1.2	1.4
y_1	5.3319	11.6814
y_2	5.3319	11.6814

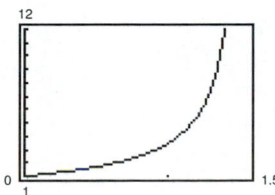

$y_1 = y_2$

69. $\csc x$ **71.** $5 \cos \theta$ **73.** $3 \tan \theta$ **75.** $5 \sec \theta$

77. $0 \le \theta \le \pi$ **79.** $0 \le \theta < \dfrac{\pi}{2}, \dfrac{3\pi}{2} < \theta < 2\pi$

81. $\ln |\cot \theta|$ **83.** Not an identity because $\dfrac{\sin k\theta}{\cos k\theta} = \tan k\theta$

85. Identity because $\sin \theta \cdot \dfrac{1}{\sin \theta} = 1$

87. (a) $\csc^2 132° - \cot^2 132° \approx 1.8107 - 0.8107 = 1$

(b) $\csc^2 \dfrac{2\pi}{7} - \cot^2 \dfrac{2\pi}{7} \approx 1.6360 - 0.6360 = 1$

89. (a) $\cos(90° - 80°) = \sin 80° \approx 0.9848$

(b) $\cos\left(\dfrac{\pi}{2} - 0.8\right) = \sin 0.8 \approx 0.7174$

91. $\cos \theta = \pm\sqrt{1 - \sin^2\theta}$

$\tan \theta = \pm\dfrac{\sin \theta}{\sqrt{1 - \sin^2\theta}}$

$\csc \theta = \dfrac{1}{\sin \theta}$

$\sec \theta = \pm\dfrac{1}{\sqrt{1 - \sin^2\theta}}$

$\cot \theta = \pm\dfrac{\sqrt{1 - \sin^2\theta}}{\sin \theta}$

93. $x - 25$ **95.** $4z + 12\sqrt{2} + 9$

Section 7.2   *(page 561)*

Warm Up   *(page 561)*

1. (a) $x^2(1 - y^2)$ (b) $\sin^4 x$

2. (a) $x^2(1 + y^2)$ (b) 1

3. (a) $(x^2 + 1)(x^2 - 1)$ (b) $\sec^2 x(\tan^2 x - 1)$

4. (a) $(z + 1)(z^2 - z + 1)$

(b) $(\tan x + 1)(\tan^2 x - \tan x + 1)$

5. (a) $(x - 1)(x^2 + 1)$ (b) $(\cot x - 1)\csc^2 x$

6. (a) $(x^2 - 1)^2$ (b) $\cos^4 x$

7. (a) $\dfrac{y^2 - x^2}{x}$ (b) $\tan x$ **8.** (a) $\dfrac{x^2 - 1}{x^2}$ (b) $\sin^2 x$

9. (a) $\dfrac{y^2 + (1 + z)^2}{y(1 + z)}$ (b) $2 \csc x$

10. (a) $\dfrac{y(1 + y) - z^2}{z(1 + y)}$ (b) $\dfrac{\tan x - 1}{\sec x(1 + \tan x)}$

1.–59. Answers will vary.

61. $\sin\theta = \pm\sqrt{1 - \cos^2\theta}$; $\dfrac{7\pi}{4}$

63. $\sqrt{\tan^2 x} = |\tan x|$; $\dfrac{3\pi}{4}$ **65.** 1   **67.** 2

69.–71. Answers will vary.

73. Seward; 6.4 and 1.9   **75.** $2 + i\left(3 - \sqrt{26}\right)$

77. $4i - 8$

Section 7.3 *(page 571)*

Warm Up *(page 571)*

1. $\dfrac{2\pi}{3}, \dfrac{4\pi}{3}$ **2.** $\dfrac{\pi}{3}, \dfrac{2\pi}{3}$ **3.** $\dfrac{\pi}{4}, \dfrac{7\pi}{4}$ **4.** $\dfrac{7\pi}{4}, \dfrac{5\pi}{4}$

5. $\dfrac{\pi}{3}, \dfrac{4\pi}{3}$ **6.** $\dfrac{3\pi}{4}, \dfrac{7\pi}{4}$ **7.** $\dfrac{15}{8}$ **8.** $-3, \dfrac{5}{2}$

9. $\dfrac{2 \pm \sqrt{14}}{2}$ **10.** $-1, 3$

1. $x = -1, 3$ **3.** $x = \pm 2$ **5.–9.** Answers will vary.

11. $\dfrac{2\pi}{3} + 2n\pi, \dfrac{4\pi}{3} + 2n\pi$ **13.** $\dfrac{\pi}{3} + 2n\pi, \dfrac{2\pi}{3} + 2n\pi$

15. $\dfrac{\pi}{6} + n\pi, \dfrac{5\pi}{6} + n\pi$

17. $\dfrac{\pi}{8} + n\pi, \dfrac{3\pi}{8} + n\pi, \dfrac{5\pi}{8} + n\pi, \dfrac{7\pi}{8} + n\pi$

19. $\dfrac{\pi}{3} + n\pi, \dfrac{2\pi}{3} + n\pi$ **21.** $\dfrac{\pi}{3} + n\pi, \dfrac{2\pi}{3} + n\pi$

23. $\dfrac{\pi}{6} + n\pi, \dfrac{5\pi}{6} + n\pi, \dfrac{\pi}{3} + n\pi, \dfrac{2\pi}{3} + n\pi$

25. $0, \dfrac{\pi}{2}, \pi, \dfrac{3\pi}{2}$ **27.** $0, \pi, \dfrac{\pi}{6}, \dfrac{5\pi}{6}, \dfrac{7\pi}{6}, \dfrac{11\pi}{6}$

29. $\dfrac{\pi}{3}, \dfrac{5\pi}{3}, \pi$ **31.** No solution   **33.** $\dfrac{\pi}{2}$ **35.** $\dfrac{\pi}{2}$

37. π **39.** $\dfrac{\pi}{6}, \dfrac{5\pi}{6}, \dfrac{7\pi}{6}, \dfrac{11\pi}{6}$

41. $\dfrac{2}{3}, \dfrac{3}{2}$; 0.8411, 5.4421

43. 1.1071, 4.2487   **45.** 1.0472, 5.2360   **47.** 0, 1.8955

49. 0, 2.6779, 3.1416, 5.8195

51. 0.9828, 1.7682, 4.1244, 4.9098

53. 0.3398, 0.8481, 2.2935, 2.8018   **55.** 0.4271, 2.7145

57.

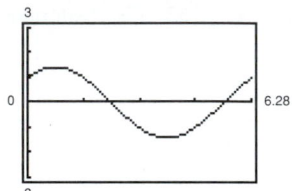

Maximum: $\left(\dfrac{\pi}{4}, \sqrt{2}\right)$

Minimum: $\left(\dfrac{5\pi}{4}, -\sqrt{2}\right)$

59. 0

61. (a) All real numbers except $x = 0$

(b) y-axis symmetry; horizontal asymptote: $y = 1$

(c) Oscillates

(d) Infinite solutions

(e) Yes, 0.6366

63. 0.04, 0.43, 0.83   **65.** 37°, 53°

67. (a)

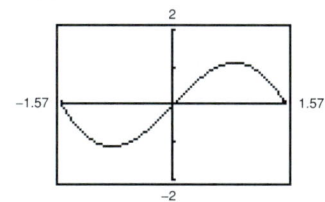

$x \approx 0.86$, $A \approx 1.12$

(b) $0.6 < x < 1.1$

69. (a)

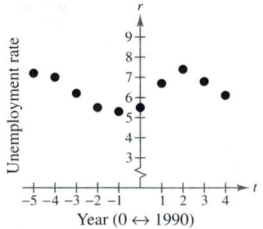

(b) (3)

(c) Constant: 6.20%

(d) 7 years

(e) 1996

Section 7.4 *(page 580)*

Warm Up *(page 580)*

---

1. $\dfrac{\sqrt{10}}{10}$ **2.** $\dfrac{-5\sqrt{34}}{34}$ **3.** $-\dfrac{\sqrt{7}}{4}$ **4.** $\dfrac{2\sqrt{2}}{3}$

5. $\dfrac{\pi}{4}, \dfrac{3\pi}{4}$ **6.** $\dfrac{\pi}{2}, \dfrac{3\pi}{2}$ **7.** $\tan^3 x$ **8.** $\cot^2 x$

9. $\sec x$ **10.** $1 - \tan^2 x$

---

1. (a) $\dfrac{\sqrt{2} - \sqrt{6}}{4}$ (b) $\dfrac{\sqrt{2} + 1}{2}$

3. (a) $\dfrac{1}{2}$ (b) $\dfrac{-\sqrt{3} - 1}{2}$

5. False. Parts (a) and (b) are unequal in Exercises 1–4.

7. $\sin 75° = \dfrac{\sqrt{2}}{4}\left(1 + \sqrt{3}\right)$

   $\cos 75° = \dfrac{\sqrt{2}}{4}\left(\sqrt{3} - 1\right)$

   $\tan 75° = \sqrt{3} + 2$

9. $\sin 105° = \dfrac{\sqrt{2}}{4}\left(\sqrt{3} + 1\right)$

   $\cos 105° = \dfrac{\sqrt{2}}{4}\left(1 - \sqrt{3}\right)$

   $\tan 105° = -2 - \sqrt{3}$

11. $\sin 195° = \dfrac{\sqrt{2}}{4}\left(1 - \sqrt{3}\right)$

   $\cos 195° = -\dfrac{\sqrt{2}}{4}\left(\sqrt{3} + 1\right)$

   $\tan 195° = 2 - \sqrt{3}$

13. $\sin \dfrac{11\pi}{12} = \dfrac{\sqrt{2}}{4}\left(\sqrt{3} - 1\right)$

   $\cos \dfrac{11\pi}{12} = -\dfrac{\sqrt{2}}{4}\left(\sqrt{3} + 1\right)$

   $\tan \dfrac{11\pi}{12} = -2 + \sqrt{3}$

15. $\sin \dfrac{17\pi}{12} = -\dfrac{\sqrt{2}}{4}\left(\sqrt{3} + 1\right)$

   $\cos \dfrac{17\pi}{12} = \dfrac{\sqrt{2}}{4}\left(1 - \sqrt{3}\right)$

   $\tan \dfrac{17\pi}{12} = 2 + \sqrt{3}$

17. $\sin 285° = -\dfrac{\sqrt{2}}{4}\left(\sqrt{3} + 1\right)$

   $\cos 285° = \dfrac{\sqrt{2}}{4}\left(\sqrt{3} - 1\right)$

   $\tan 285° = -\left(2 + \sqrt{3}\right)$

19. $\sin\left(-\dfrac{13\pi}{12}\right) = \dfrac{\sqrt{2}}{4}\left(\sqrt{3} - 1\right)$

   $\cos\left(-\dfrac{13\pi}{12}\right) = -\dfrac{\sqrt{2}}{4}\left(\sqrt{3} + 1\right)$

   $\tan\left(-\dfrac{13\pi}{12}\right) = -2 + \sqrt{3}$

21. $\cos 40°$ **23.** $\sin 200°$ **25.** $\tan 239°$ **27.** $\sin 1.8$

29. $\tan 3x$ **31.** $-\dfrac{63}{65}$ **33.** $\dfrac{16}{65}$ **35.** $\dfrac{65}{16}$

37. $\dfrac{33}{56}$ **39.** $\dfrac{3}{5}$ **41.** $\dfrac{44}{125}$ **43.** $\dfrac{5}{4}$

45.–61. Answers will vary.

63. (a) $\sqrt{2}\sin\left(\theta + \dfrac{\pi}{4}\right)$ (b) $\sqrt{2}\cos\left(\theta - \dfrac{\pi}{4}\right)$

65. (a) $13\sin(3\theta + 0.3948)$ (b) $13\cos(3\theta - 1.1760)$

67. $2\cos\theta$ **69.** 1 **71.** $\dfrac{\pi}{2}$ **73.** $\dfrac{5\pi}{4}, \dfrac{7\pi}{4}$

75. $\dfrac{\pi}{4}, \dfrac{7\pi}{4}$

77.

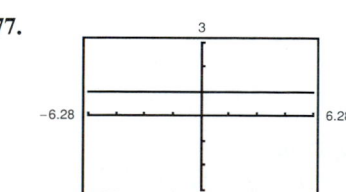

   $\sin^2\left(\theta + \dfrac{\pi}{4}\right) + \sin^2\left(\theta - \dfrac{\pi}{4}\right) = 1$

79. (a) $y = \dfrac{5}{12}\sin(2t + 0.6435)$ (b) $\dfrac{5}{12}$ (c) $\dfrac{1}{\pi}$

Section 7.5 (page 591)

Warm Up (page 591)

1. $\sin x(2 + \cos x)$ **2.** $(\cos x - 2)(\cos x + 1)$

3. $0, \dfrac{\pi}{2}, \pi, \dfrac{3\pi}{2}$ **4.** $\dfrac{\pi}{4}, \dfrac{3\pi}{4}, \dfrac{5\pi}{4}, \dfrac{7\pi}{4}$

5. π **6.** 0 **7.** $\dfrac{2 - \sqrt{2}}{4}$ **8.** $\dfrac{3}{4}$

9. $\tan 3x$ **10.** $\cos x(1 - 4\sin^2 x)$

1. $\dfrac{3}{5}$ **3.** $\dfrac{7}{25}$ **5.** $\dfrac{24}{7}$ **7.** $\dfrac{25}{24}$ **9.** $0, \dfrac{\pi}{3}, \pi, \dfrac{5\pi}{3}$

11. $\dfrac{\pi}{12}, \dfrac{5\pi}{12}, \dfrac{13\pi}{12}, \dfrac{17\pi}{12}$ **13.** $0, \dfrac{2\pi}{3}, \dfrac{4\pi}{3}$

15. $\dfrac{\pi}{2}, \dfrac{\pi}{6}, \dfrac{5\pi}{6}, \dfrac{7\pi}{6}, \dfrac{3\pi}{2}, \dfrac{11\pi}{6}$ **17.** $0, \dfrac{\pi}{2}, \pi, \dfrac{3\pi}{2}$

19. $f(x) = 3\sin 2x$ **21.** $g(x) = 4\cos 2x$

23. $\sin 2u = \dfrac{24}{25}$ **25.** $\sin 2u = \dfrac{4}{5}$
$\cos 2u = \dfrac{7}{25}$ $\cos 2u = \dfrac{3}{5}$
$\tan 2u = \dfrac{24}{7}$ $\tan 2u = \dfrac{4}{3}$

27. $\sin 2u = -\dfrac{4\sqrt{21}}{25}$

$\cos 2u = -\dfrac{17}{25}$

$\tan 2u = \dfrac{4\sqrt{21}}{17}$

29. $\dfrac{1}{8}(3 + 4\cos 2x + \cos 4x)$ **31.** $\dfrac{1}{8}(1 - \cos 4x)$

33. $\dfrac{1}{32}(2 + \cos 2x - 2\cos 4x - \cos 6x)$

35. $\dfrac{5}{\sqrt{26}}$ **37.** $\dfrac{1}{5}$ **39.** $\sqrt{26}$

41. $\sin 105° = \dfrac{1}{2}\sqrt{2 + \sqrt{3}}$

$\cos 105° = -\dfrac{1}{2}\sqrt{2 - \sqrt{3}}$

$\tan 105° = -2 - \sqrt{3}$

43. $\sin 112°\,30' = \dfrac{1}{2}\sqrt{2 + \sqrt{2}}$

$\cos 112°\,30' = -\dfrac{1}{2}\sqrt{2 - \sqrt{2}}$

$\tan 112°\,30' = -1 - \sqrt{2}$

45. $\sin \dfrac{\pi}{8} = \dfrac{1}{2}\sqrt{2 - \sqrt{2}}$ **47.** $\sin \dfrac{u}{2} = \dfrac{5\sqrt{26}}{26}$

$\cos \dfrac{\pi}{8} = \dfrac{1}{2}\sqrt{2 + \sqrt{2}}$ $\cos \dfrac{u}{2} = \dfrac{\sqrt{26}}{26}$

$\tan \dfrac{\pi}{8} = \sqrt{2} - 1$ $\tan \dfrac{u}{2} = 5$

49. $\sin \dfrac{u}{2} = \sqrt{\dfrac{89 - 8\sqrt{89}}{178}}$ **51.** $\sin \dfrac{u}{2} = \dfrac{3\sqrt{10}}{10}$

$\cos \dfrac{u}{2} = -\sqrt{\dfrac{89 + 8\sqrt{89}}{178}}$ $\cos \dfrac{u}{2} = -\dfrac{\sqrt{10}}{10}$

$\tan \dfrac{u}{2} = \dfrac{8 - \sqrt{89}}{5}$ $\tan \dfrac{u}{2} = -3$

53. $|\sin 3x|$ **55.** $-|\tan 4x|$ **57.** π

59. $\dfrac{\pi}{3}, \pi, \dfrac{5\pi}{3}$ **61.** $3\left(\sin \dfrac{\pi}{2} + \sin 0\right)$

63. $\dfrac{1}{2}(\sin 8\theta + \sin 2\theta)$ **65.** $\dfrac{5}{2}(\cos 8\beta + \cos 2\beta)$

67. $\dfrac{1}{2}(\cos 2y - \cos 2x)$ **69.** $\dfrac{1}{2}(\sin 2\theta + \sin 2\pi)$

71. $2\sin 45° \cos 15°$ **73.** $-2\sin \dfrac{\pi}{2}\sin \dfrac{\pi}{4}$

75. $2\cos 4x \cos 2x$ **77.** $2\cos \alpha \sin \beta$

79. $2\cos(\phi + \pi)\cos \pi$

81. $0, \dfrac{\pi}{4}, \dfrac{\pi}{2}, \dfrac{3\pi}{4}, \pi, \dfrac{5\pi}{4}, \dfrac{3\pi}{2}, \dfrac{7\pi}{4}$

83. $\dfrac{\pi}{6}, \dfrac{5\pi}{6}$ **85.** $\dfrac{25}{169}$ **87.** $\dfrac{4}{13}$

89.–101. Answers will vary.

103.

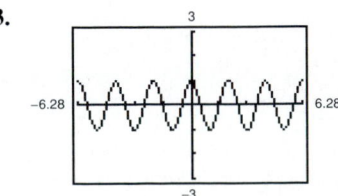

105.

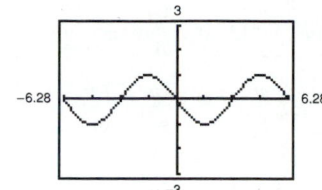

107.   **109.**

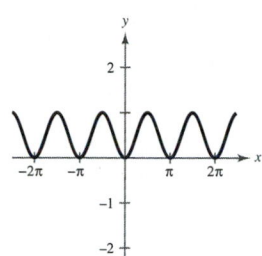   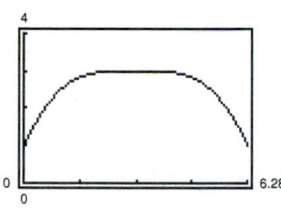

Maximum: $(\pi, 3)$

111. (a)

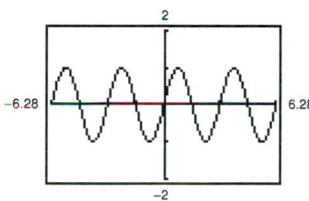

(b) $g(x) = \sin 2x$

(c) Answers will vary.

113. $2x\sqrt{1 - x^2}$

115. (a) $A = 100 \sin \dfrac{\theta}{2} \cos \dfrac{\theta}{2}$

(b) $A = 50 \sin \theta$

The area is maximum when $\theta = \dfrac{\pi}{2}$.

117. September: $235,000   **119.** 15.7 gallons
October: $272,000

Focus on Concepts *(page 595)*

1. An identity is true for all values of the variable and a conditional equation is true for some values of the variable.

2. When proving an identity you use the fundamental identities and rules of algebra to transform one expression into another. To solve a trigonometric equation, use standard algebraic techniques and identities to isolate a trigonometric function involved in the equation. Find the value of the variable by using the inverse of the trigonometric function.

3. Reciprocal identities: $\csc\theta = \dfrac{1}{\sin\theta}$, $\sec\theta = \dfrac{1}{\cos\theta}$,

$$\cot\theta = \dfrac{1}{\tan\theta}$$

Quotient identities: $\tan\theta = \dfrac{\sin\theta}{\cos\theta}$, $\cot\theta = \dfrac{\cos\theta}{\sin\theta}$

Pythagorean identities: $\sin^2\theta + \cos^2\theta = 1$,

$\tan^2\theta + 1 = \sec^2\theta$, $1 + \cot^2\theta = \csc^2\theta$

4. No. $\cos\theta = \pm\sqrt{1 - \sin^2\theta}$

5. False. The order in which algebraic operations and fundamental identities are done may vary.

6. (a) True. The period of tangent is π.

(b) False. The period of cosine is 2π.

(c) False. $\sec\theta\cos\theta = 1$

(d) True.

(e) True. $\sin(-\alpha) = -\sin\alpha$

7. $y_1 = y_2 + 1$ **8.** $y_1 = 1 - y_2$ **9.** 1   **10.** 3

11. 3   **12.** 4   **13.** $\dfrac{\pi}{6}$

Review Exercises *(page 596)*

1. $\sin^2 x$ **3.** $1 + \cot\alpha$ **5.** 1   **7.** $\tan(2x + 2)$

9.–25. Answers will vary.

27.

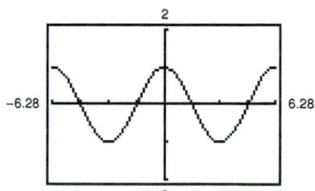

29.

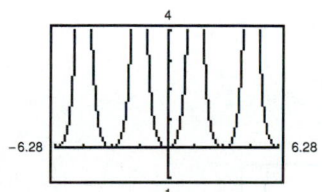

31. $\dfrac{\sqrt{2}}{4}(\sqrt{3} + 1)$ **33.** $-\dfrac{1}{2}\sqrt{2 + \sqrt{2}}$

35. $-\dfrac{3}{52}(5 + 4\sqrt{7})$ **37.** $\dfrac{1}{52}(36 + 5\sqrt{7})$

39. $\dfrac{1}{4}\sqrt{2\left(4-\sqrt{7}\right)}$

41. False. If $\dfrac{\pi}{2}<\theta<\pi$, then $\cos\dfrac{\theta}{2}>0$.

43. True   **45.** $0,\pi$ **47.** $0,\dfrac{3\pi}{4},\pi,\dfrac{5\pi}{4}$

49. $0,\dfrac{\pi}{2},\pi$ **51.** $\dfrac{\pi}{3},\dfrac{5\pi}{3}$ **53.** $\dfrac{\pi}{4},\dfrac{5\pi}{4}$

55. False. $\sin\theta=\frac{1}{2}$ has an infinite number of solutions but is not an identity.

57. $2\cos\dfrac{5\theta}{2}\cos\dfrac{\theta}{2}$ **59.** $\dfrac{1}{2}(\cos\alpha-\cos5\alpha)$

61. $8x^2-1$ **63.** Answers will vary.

65. (a) $y=\frac{1}{2}\sqrt{10}\sin\left(8t-\arctan\frac{1}{3}\right)$

     (b) $\frac{1}{2}\sqrt{10}$

     (c) $\dfrac{4}{\pi}$

Chapter Test *(page 600)*

1. $\sin\theta=-\dfrac{3}{\sqrt{13}}$ **2.** 1   **3.** 1

     $\cos\theta=-\dfrac{2}{\sqrt{13}}$

     $\csc\theta=-\dfrac{\sqrt{13}}{3}$

     $\sec\theta=-\dfrac{\sqrt{13}}{2}$

     $\cot\theta=\dfrac{2}{3}$

4. $\dfrac{1}{\sin\theta\cos\theta}$ **5.** $\dfrac{\pi}{2}<\theta<\pi$

                         $\dfrac{3\pi}{2}<\theta<2\pi$

6.

$y_1=y_2$

7.–12. Answers will vary.   **13.** $0,\dfrac{3\pi}{4},\pi,\dfrac{7\pi}{4}$

14. $\dfrac{\pi}{6},\dfrac{\pi}{2},\dfrac{5\pi}{6},\dfrac{3\pi}{2}$ **15.** $\dfrac{\pi}{6},\dfrac{5\pi}{6},\dfrac{7\pi}{6},\dfrac{11\pi}{6}$

16. $\dfrac{\pi}{6},\dfrac{5\pi}{6},\dfrac{3\pi}{2}$ **17.** $-2.938,-2.663,1.170$

18. $|\cos^2x+\cos x|\le2$ for all x

19. $\dfrac{\sqrt{2}-\sqrt{6}}{4}$ **20.** $\sin 2u=\dfrac{4}{5}$, $\tan 2u=-\dfrac{4}{3}$

CHAPTER 8

Section 8.1 *(page 608)*

Warm Up *(page 608)*

1. $b=3\sqrt{3}$, $A=30°$, $B=60°$

2. $c=5\sqrt{2}$, $A=45°$, $B=45°$

3. $a=8$, $A\approx28.07°$, $B\approx61.93°$

4. $b\approx8.33$, $c\approx11.21$, $B=48°$

5. $a\approx22.69$, $c\approx23.04$, $A=80°$

6. $a\approx45.73$, $b\approx142.86$, $A=17°45'$ **7.** 8.48

8. 12.94   **9.** 2.25   **10.** 91.06

1. $C=105°$, $b\approx28.28$, $c\approx38.64$

3. $C=110°$, $b\approx37.40$, $c\approx40.59$

5. $B\approx21.55°$, $C\approx122.45°$, $c\approx11.49$

7. $B=10°$, $b\approx69.46$, $c\approx136.81$

9. $B=42°4'$, $a\approx22.05$, $b\approx14.88$

11. $A\approx10°11'$, $C\approx154°19'$, $c\approx11.03$

13. $A\approx25.57°$, $B\approx9.43°$, $a\approx10.5$

15. $B\approx18°13'$, $C\approx51°32'$, $c\approx40.06$

17. No solution

19. Two solutions

     $B\approx70.4°$, $C\approx51.6°$, $c\approx4.16$

     $B\approx109.6°$, $C\approx12.4°$, $c\approx1.14$

21. No solution

23. (a) $b \leq 5$, $b = \dfrac{5}{\sin 36°}$

(b) $5 < b < \dfrac{5}{\sin 36°}$

(c) $b > \dfrac{5}{\sin 36°}$

25. (a)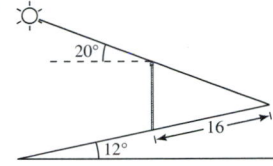

(b) $\dfrac{16}{\sin 70°} = \dfrac{h}{\sin 32°}$

(c) 9 meters

27. 16.1°   **29.** 77 meters

31. (a)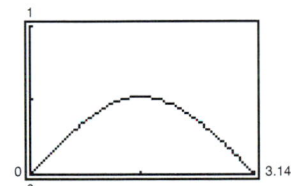

(b) 22.6 miles

(c) 21.4 miles

(d) 38,443 feet

33. 38.8 kilometers, 40.3 kilometers   **35.** 4.55 miles

37. (a) $\alpha = \arcsin(0.5 \sin \beta)$

(b)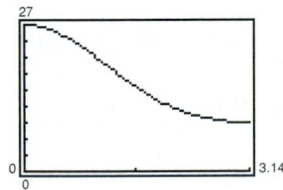

Domain: $0 \leq \beta \leq \pi$

Range: $0 \leq \alpha \leq \dfrac{\pi}{6}$

(c) $\dfrac{18 \sin[\pi - \beta - \arcsin(0.5 \sin \beta)]}{\sin \beta}$

(d)

Domain: $0 < \beta < \pi$

Range: $9 < c < 27$

(e)

β	0	0.4	0.8	1.2	1.6	2.0
α	0	0.1960	0.3669	0.4848	0.5234	0.4720
c	27	25.95	27.07	19.19	15.33	12.29

β	2.4	2.8
α	0.3445	0.1683
c	10.31	9.27

39. 10.4   **41.** 1675.2   **43.** 474.9

45. (a) $20\left[15 \sin \dfrac{3\theta}{2} - 4 \sin \dfrac{\theta}{2} - 6 \sin \theta\right]$

(b)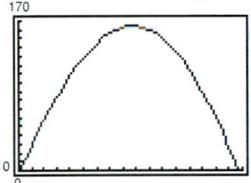

(c) Domain: $0 \leq \theta \leq 1.6690$

The domain would increase in length and the area would increase.

Section 8.2 *(page 617)*

Warm Up *(page 617)*

1. $2\sqrt{13}$ **2.** $3\sqrt{5}$ **3.** $4\sqrt{10}$ **4.** $3\sqrt{13}$

5. 20   **6.** 48   **7.** $a \approx 4.62$, $c \approx 26.20$, $B = 70°$

8. $a \approx 34.20$, $b \approx 93.97$, $B = 70°$

9. No solution   **10.** $a \approx 15.09$, $B \approx 18.97°$, $C \approx 131.03°$

1. $A \approx 26.4°$, $B \approx 36.3°$, $C \approx 117.3°$

3. $B \approx 23.8°$, $C \approx 126.2°$, $a \approx 18.6$

5. $A \approx 36.9°$, $B \approx 53.1°$, $C \approx 90°$

7. $A \approx 92.9°$, $B \approx 43.53°$, $C \approx 43.53°$

9. $a \approx 11.79$, $B \approx 12.7°$, $C \approx 47.3°$

11. $A \approx 158°37'$, $C \approx 12°38'$, $b \approx 10.4$

13. $A = 27°10'$, $B = 27°10'$, $c \approx 56.9$

	a	b	c	d	θ	ϕ
15.	4	6	9.67	3.23	30°	150°
17.	10	14	20	13.86	68.2°	111.8°
19.	10	11.57	18	12	67.1°	112.9°

21.

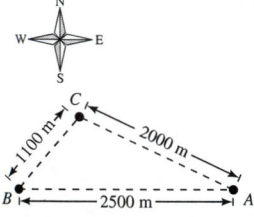

N 39° E, S 64.9° E

23. 422.5 meters   **25.** 72.3°   **27.** 43.3 miles

29. (a) N 58.4° W   (b) S 81.5° W   **31.** 63.7 feet

33. $\overline{PQ} \approx 9.4$, $\overline{QS} = 5$, $\overline{RS} \approx 12.8$

35. (a) $49 = 2.25 + x^2 - 3x \cos \theta$

(b) $x = \frac{1}{2}\left(3 \cos \theta + \sqrt{9 \cos^2\theta + 187}\right)$

(c)

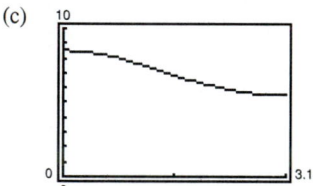

(d) 3 inches

37. 2.76 feet   **39.** (a) 570.60   (b) 5910.68   (c) 177.09

41. 16.25   **43.** 54   **45.** 96.82

47. 46,837.4 square feet   **49.** Answers will vary.

51. $\dfrac{1}{\sqrt{1 - 4x^2}}$ **53.** $\dfrac{1}{x - 2}$

Section 8.3 *(page 630)*

Warm Up *(page 630)*

1. $7\sqrt{10}$ **2.** $\sqrt{58}$ **3.** $3x + 5y - 14 = 0$

4. $4x - 3y - 1 = 0$ **5.** 111.8°   **6.** 323.1°

7. $\dfrac{1}{2}, \dfrac{\sqrt{3}}{2}$ **8.** $\dfrac{\sqrt{3}}{2}, -\dfrac{1}{2}$ **9.** $-\dfrac{\sqrt{3}}{2}, \dfrac{1}{2}$

10. $-\dfrac{1}{2}, -\dfrac{\sqrt{3}}{2}$

1. $\langle 4, 3 \rangle$, $\|\mathbf{v}\| = 5$ **3.** $\langle -3, 2 \rangle$, $\|\mathbf{v}\| = \sqrt{13}$

5. $\langle 0, 5 \rangle$, $\|\mathbf{v}\| = 5$ **7.** $\mathbf{v} = \langle 16, 7 \rangle$, $\|\mathbf{v}\| = \sqrt{305}$

9. $\mathbf{v} = \langle 8, 6 \rangle$, $\|\mathbf{v}\| = 10$

11.

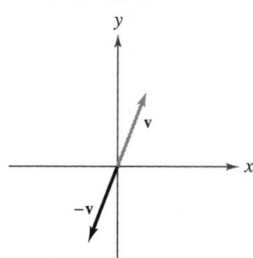

13.

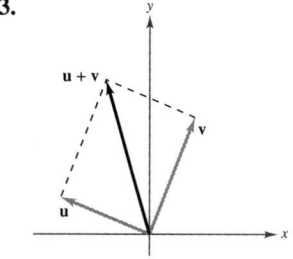

15.

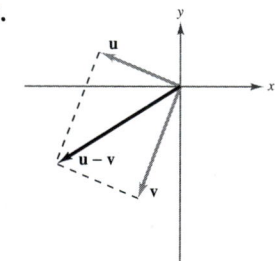

17. (a) $\langle 4, 3 \rangle$ (b) $\langle -2, 1 \rangle$ (c) $\langle -7, 1 \rangle$

19. (a) $\langle 4, -2 \rangle$ (b) $\langle 4, -2 \rangle$ (c) $\langle 8, -4 \rangle$

21. (a) $3\mathbf{i} - 2\mathbf{j}$ (b) $-\mathbf{i} + 4\mathbf{j}$ (c) $-4\mathbf{i} + 11\mathbf{j}$

23. (a) $2\mathbf{i} + \mathbf{j}$ (b) $2\mathbf{i} - \mathbf{j}$ (c) $4\mathbf{i} - 3\mathbf{j}$

25. $\langle 1, 0 \rangle$ **27.** $\left\langle -\dfrac{1}{\sqrt{2}}, \dfrac{1}{\sqrt{2}} \right\rangle$ **29.** $\dfrac{4}{5}\mathbf{i} - \dfrac{3}{5}\mathbf{j}$

31. $\mathbf{j}$ **33.** $\left\langle \dfrac{5}{\sqrt{2}}, \dfrac{5}{\sqrt{2}} \right\rangle$ **35.** $\left\langle -\dfrac{21}{5}, \dfrac{28}{5} \right\rangle$

37. $\mathbf{v} = \langle 3, -\frac{3}{2} \rangle$

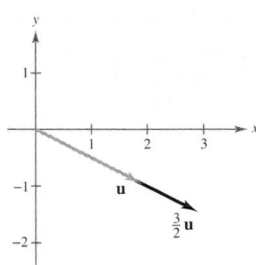

39. $\mathbf{v} = \langle 4, 3 \rangle$

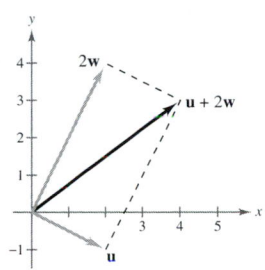

41. $\mathbf{v} = \langle \frac{7}{2}, -\frac{1}{2} \rangle$

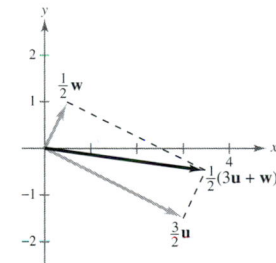

43. $\|\mathbf{v}\| = 5, \theta = 30°$ **45.** $\|\mathbf{v}\| = 6\sqrt{2}, \theta = 315°$
47. $\mathbf{v} = \langle 3, 0 \rangle$

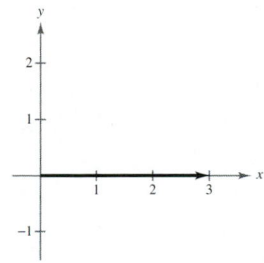

49. $\mathbf{v} = \left\langle -\dfrac{\sqrt{3}}{2}, \dfrac{1}{2} \right\rangle$

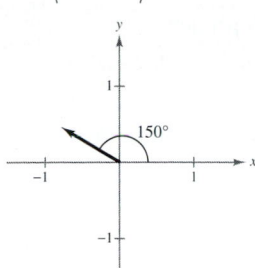

51. $\mathbf{v} = \left\langle -\dfrac{3\sqrt{6}}{2}, \dfrac{3\sqrt{2}}{2} \right\rangle$

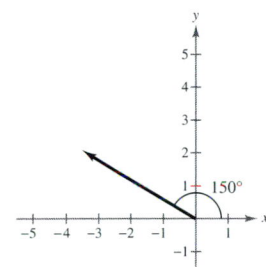

53. $\mathbf{v} = \left\langle \dfrac{\sqrt{10}}{5}, \dfrac{3\sqrt{10}}{5} \right\rangle$

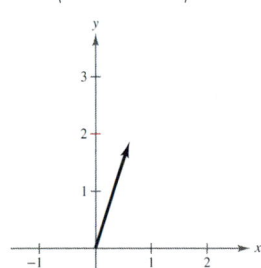

55. $\langle 5, 5 \rangle$ **57.** $\langle 10\sqrt{2} - 50, 10\sqrt{2} \rangle$
59. $90°$ **61.** $63.4°$ **63.** $62.7°$
65. (a) $0°$ (b) $180°$
   (c) No. Equal to the sum when the angle between the vectors is $0°$.
67. 357.85 newtons, $12.1°$ **69.** $71.3°$, 228.5 pounds
71. Horizontal component: $80 \cos 40° \approx 61.28$ feet per second
   Vertical component: $80 \sin 40° \approx 51.42$ feet per second

73. $T_{AC} \approx 1758.8$ pounds   **75.** 3154.4 pounds
$T_{BC} \approx 1305.4$ pounds

77. N 21.4° E, 138.7 kilometers per hour

79. 850 foot-pounds   **81.** True   **83.** False. $a = b = 0$

85. Answers will vary.   **87.** $\langle 1, 3 \rangle$ or $\langle -1, -3 \rangle$

89. 14.7 meters   **91.** 8 tan θ **93.** 6 sec θ

Section 8.4 *(page 642)*

Warm Up *(page 642)*

1. $\langle -14, -5 \rangle$ **2.** $\langle \frac{43}{8}, \frac{3}{5} \rangle$ **3.** $-6\mathbf{i} + 4\mathbf{j}$
$3\sqrt{5}$ $\dfrac{\sqrt{1249}}{40}$ $4\sqrt{17}$

4. $8.7\mathbf{i} - 2.2\mathbf{j}$ **5.** 2.09, 4.19   **6.** 1.57, 4.71
1.5

7. 1, 5.28   **8.** 2.89, 3.39

9. (a) $\langle \frac{12}{13}, -\frac{5}{13} \rangle$ (b) $\langle -\frac{12}{13}, \frac{5}{13} \rangle$

10. (a) $\langle \frac{12}{13}, \frac{5}{13} \rangle$ (b) $\langle -\frac{12}{13}, -\frac{5}{13} \rangle$

1. -6 **3.** 6   **5.** 8   **7.** $\langle -6, 8 \rangle$ **9.** 13

11. $5\sqrt{41}$ **13.** \$37,289 total revenue   **15.** 90°

17. 143.13°   **19.** $\dfrac{5\pi}{12}$

21.

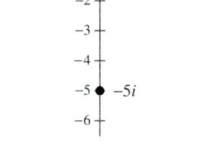

91.33°

23.

90°

25. 26.6°, 63.4°, 90°   **27.** -20 **29.** Parallel

31. Neither   **33.** Orthogonal   **35.** $\frac{16}{17}\langle 4, 1 \rangle, \frac{13}{17}\langle -1, 4 \rangle$

37. $\frac{45}{229}\langle 2, 15 \rangle, \frac{6}{229}\langle -15, 2 \rangle$ **39.** $\langle -5, 3 \rangle, \langle 5, -3 \rangle$

41. $\frac{2}{3}\mathbf{i} + \frac{1}{2}\mathbf{j}, -\frac{2}{3}\mathbf{i} - \frac{1}{2}\mathbf{j}$

43. (a) 6251.3 pounds   (b) 35,453.0 pounds

45. (a) $\theta = \dfrac{\pi}{2}$ (b) $0 \le \theta < \dfrac{\pi}{2}$ (c) $\dfrac{\pi}{2} < \theta \le \pi$

47. 735 newton-meters   **49.** 779.4 foot-pounds   **51.** 32

53.–55. Answers may vary.

Section 8.5 *(page 653)*

Warm Up *(page 653)*

1. $-5 - 10i$ **2.** $7 + 3\sqrt{6}i$ **3.** $-1 - 4i$ **4.** $-3i$

5. $6 - 14i$ **6.** $6 + 4\sqrt{2}i$ **7.** $-22 + 16i$ **8.** 13

9. $-\frac{3}{2} + \frac{5}{2}i$ **10.** $-\frac{5}{2} - \frac{3}{2}i$

1. 5

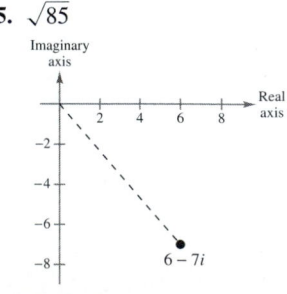

3. $4\sqrt{2}$

$-4 + 4i$

5. $\sqrt{85}$

$6 - 7i$

7. $3\left(\cos\dfrac{\pi}{2} + i\sin\dfrac{\pi}{2}\right)$ **9.** $2\sqrt{2}\left(\cos\dfrac{5\pi}{4} + i\sin\dfrac{5\pi}{4}\right)$ **17.** $6\left(\cos\dfrac{\pi}{2} + i\sin\dfrac{\pi}{2}\right)$

11. $3\sqrt{2}\left(\cos\dfrac{7\pi}{4} + i\sin\dfrac{7\pi}{4}\right)$

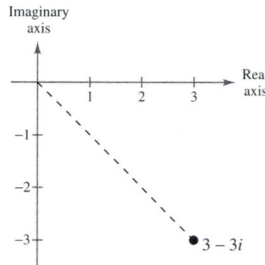

13. $2\left(\cos\dfrac{\pi}{6} + i\sin\dfrac{\pi}{6}\right)$

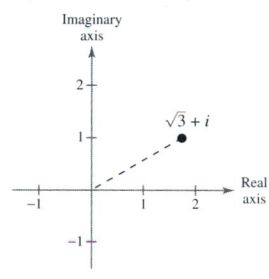

15. $4\left(\cos\dfrac{4\pi}{3} + i\sin\dfrac{4\pi}{3}\right)$

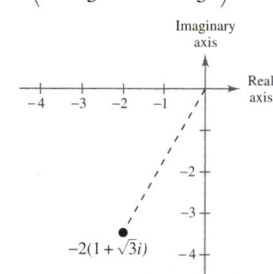

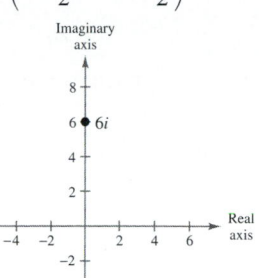

19. $\sqrt{65}\,(\cos 2.62 + i\sin 2.62)$

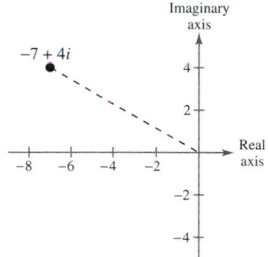

21. $7(\cos 0 + i\sin 0)$

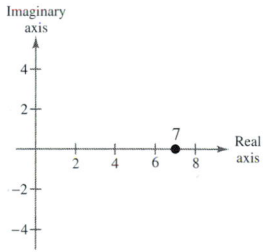

23. $\sqrt{37}\,(\cos 1.41 + i\sin 1.41)$

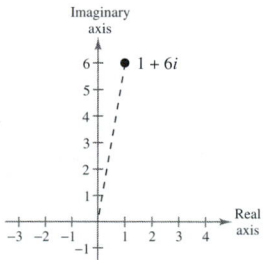

25. $\sqrt{10}(\cos 3.46 + i \sin 3.46)$

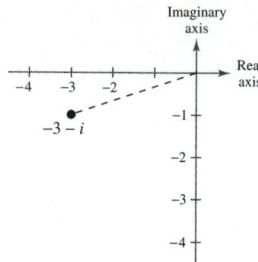

27. $5.39(\cos 0.38 + i \sin 0.38)$

29. $8.19(\cos 5.26 + i \sin 5.26)$

31. $-\sqrt{3} + i$

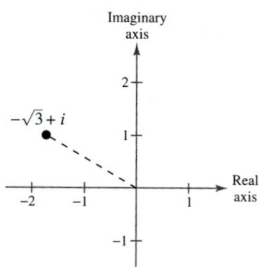

33. $\dfrac{3}{4} - \dfrac{3\sqrt{3}}{4}i$

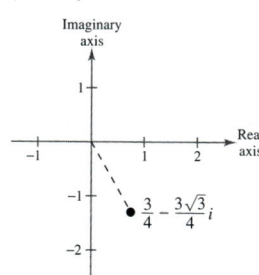

35. $-\dfrac{15\sqrt{2}}{8} + \dfrac{15\sqrt{2}}{8}i$

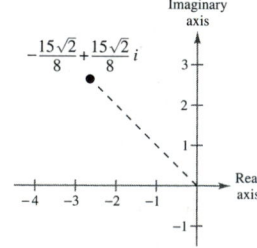

37. $-4i$

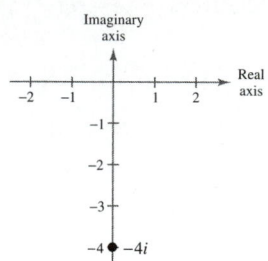

39. $2.8408 + 0.9643i$

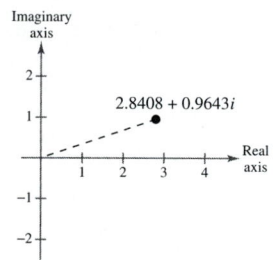

41. $4.70 + 1.71i$ **43.** $-3.22 - 2.38i$

45. $12\left(\cos \dfrac{\pi}{2} + i \sin \dfrac{\pi}{2}\right)$ **47.** $\dfrac{10}{9}(\cos 200° + i \sin 200°)$

49. $0.27(\cos 150° + i \sin 150°)$ **51.** $\cos 30° + i \sin 30°$

53. $\cos \dfrac{2\pi}{3} + i \sin \dfrac{2\pi}{3}$ **55.** $4[\cos(-58°) + i \sin (-58°)]$

57. (a) $2\sqrt{2}(\cos 45° + i \sin 45°)$;

   $\sqrt{2}[\cos(-45°) + i \sin (-45°)]$

   (b) $4(\cos 0° + i \sin 0°) = 4$

   (c) 4

59. (a) $2[\cos(-90°) + i \sin(-90°)]$;

   $\sqrt{2}(\cos 45° + i \sin 45°)$

   (b) $2\sqrt{2}[\cos(-45°) + i \sin(-45°)] = 2 - 2i$

   (c) $-2i - 2i^2 = -2i + 2 = 2 - 2i$

61. (a) $5(\cos 0° + i \sin 0°)$;

   $\sqrt{13}(\cos 56.31° + i \sin 56.31°)$

   (b) $\dfrac{5}{\sqrt{13}}[\cos(-56.31°) + i \sin (-56.31°)]$

   $\approx 0.7692 - 1.154i$

   (c) $\dfrac{10}{13} - \dfrac{15}{13}i \approx 0.7692 - 1.154i$

63.–65. Answers will vary.

67.

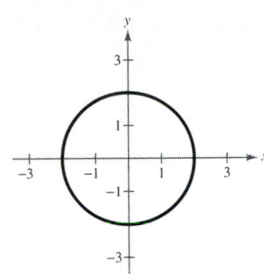

69. $-4 - 4i$ **71.** $-32i$ **73.** $-128\sqrt{3} - 128i$

75. $\dfrac{125}{2} + \dfrac{125\sqrt{3}}{2}i$ **77.** i **79.** $608.02 + 144.69i$

81. $-597 - 122i$ **83.** $\dfrac{81}{2} + \dfrac{81\sqrt{3}}{2}i$

85. Answers will vary.

87. (a) $2(\cos 30° + i \sin 30°)$

$2(\cos 150° + i \sin 150°)$

$2(\cos 270° + i \sin 270°)$

(b) $8i$

89. (a) $\sqrt{5}(\cos 60° + i \sin 60°)$

$\sqrt{5}(\cos 240° + i \sin 240°)$

(b)

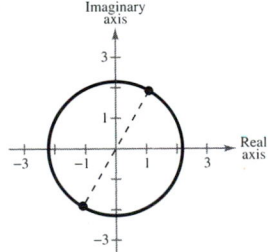

(c) $\dfrac{\sqrt{5}}{2} + \dfrac{\sqrt{15}}{2}i, \ -\dfrac{\sqrt{5}}{2} - \dfrac{\sqrt{15}}{2}i$

91. (a) $2\left(\cos \dfrac{\pi}{3} + i \sin \dfrac{\pi}{3}\right)$

$2\left(\cos \dfrac{5\pi}{6} + i \sin \dfrac{5\pi}{6}\right)$

$2\left(\cos \dfrac{4\pi}{3} + i \sin \dfrac{4\pi}{3}\right)$

$2\left(\cos \dfrac{11\pi}{6} + i \sin \dfrac{11\pi}{6}\right)$

(b)

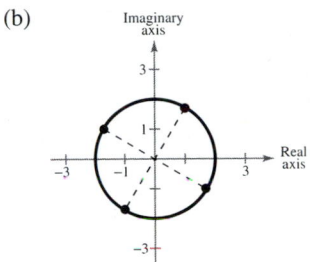

(c) $1 + \sqrt{3}i, \ -\sqrt{3} + i, \ -1 - \sqrt{3}i, \ \sqrt{3} - i$

93. (a) $5\left(\cos \dfrac{3\pi}{4} + i \sin \dfrac{3\pi}{4}\right)$

$5\left(\cos \dfrac{7\pi}{4} + i \sin \dfrac{7\pi}{4}\right)$

(b)

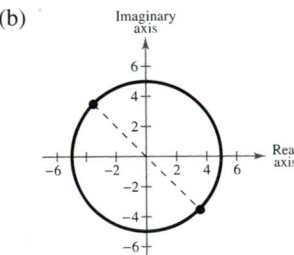

(c) $-\dfrac{5\sqrt{2}}{2} + \dfrac{5\sqrt{2}}{2}i, \ \dfrac{5\sqrt{2}}{2} - \dfrac{5\sqrt{2}}{2}i$

95. (a) $5\left(\cos\dfrac{4\pi}{9} + i\sin\dfrac{4\pi}{9}\right)$

$5\left(\cos\dfrac{10\pi}{9} + i\sin\dfrac{10\pi}{9}\right)$

$5\left(\cos\dfrac{16\pi}{9} + i\sin\dfrac{16\pi}{9}\right)$

(b)

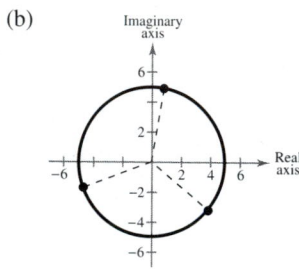

(c) $0.8682 + 4.924i,\ -4.698 - 1.710i,\ 3.830 - 3.214i$

97. (a) $2(\cos 0 + i\sin 0)$

$2\left(\cos\dfrac{2\pi}{3} + i\sin\dfrac{2\pi}{3}\right)$

$2\left(\cos\dfrac{4\pi}{3} + i\sin\dfrac{4\pi}{3}\right)$

(b)

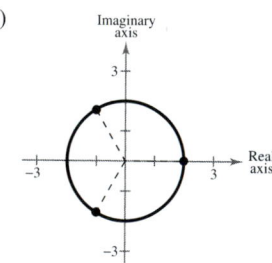

(c) $2,\ -1 + \sqrt{3}\,i,\ -1 - \sqrt{3}\,i$

99. (a) $\cos 0 + i\sin 0$

$\cos\dfrac{2\pi}{5} + i\sin\dfrac{2\pi}{5}$

$\cos\dfrac{4\pi}{5} + i\sin\dfrac{4\pi}{5}$

$\cos\dfrac{6\pi}{5} + i\sin\dfrac{6\pi}{5}$

$\cos\dfrac{8\pi}{5} + i\sin\dfrac{8\pi}{5}$

(b)

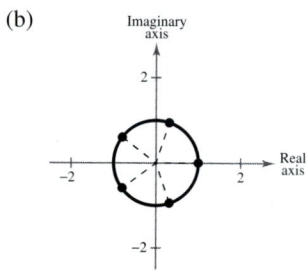

(c) $1,\ 0.3090 + 0.9511i,\ -0.8090 + 0.5878i,$
$-0.8090 - 0.5878i,\ 0.3090 - 0.9511i$

101. (a) $5(\cos 60° + i\sin 60°)$

$5(\cos 180° + i\sin 180°)$

$5(\cos 300° + i\sin 300°)$

(b)

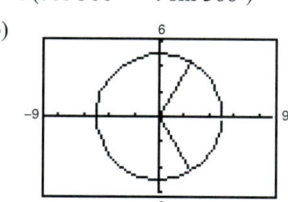

(c) $\dfrac{5}{2} + \dfrac{5\sqrt{3}}{2}\,i,\ -5,\ \dfrac{5}{2} - \dfrac{5\sqrt{3}}{2}\,i$

103. (a) $2\sqrt[5]{4\sqrt{2}}(\cos 27° + i \sin 27°)$

$2\sqrt[5]{4\sqrt{2}}(\cos 99° + i \sin 99°)$

$2\sqrt[5]{4\sqrt{2}}(\cos 171° + i \sin 171°)$

$2\sqrt[5]{4\sqrt{2}}(\cos 243° + i \sin 243°)$

$2\sqrt[5]{4\sqrt{2}}(\cos 315° + i \sin 315°)$

(b)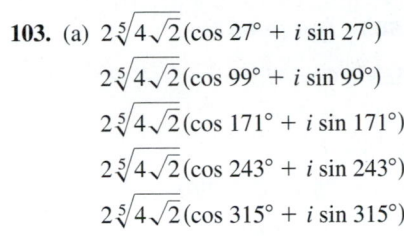

(c) $2.52 + 1.28i, \ -0.44 + 2.79i, \ -2.79 + 0.44i,$

$-1.28 - 2.52i, \ 2 - 2i$

105. $\cos\dfrac{\pi}{8} + i \sin\dfrac{\pi}{8}$

$\cos\dfrac{5\pi}{8} + i \sin\dfrac{5\pi}{8}$

$\cos\dfrac{9\pi}{8} + i \sin\dfrac{9\pi}{8}$

$\cos\dfrac{13\pi}{8} + i \sin\dfrac{13\pi}{8}$

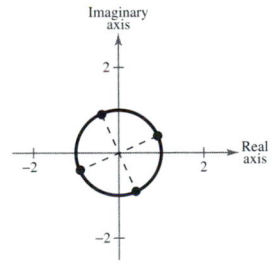

107. $3\left(\cos\dfrac{\pi}{5} + i \sin\dfrac{\pi}{5}\right)$

$3\left(\cos\dfrac{3\pi}{5} + i \sin\dfrac{3\pi}{5}\right)$

$3(\cos \pi + i \sin \pi)$

$3\left(\cos\dfrac{7\pi}{5} + i \sin\dfrac{7\pi}{5}\right)$

$3\left(\cos\dfrac{9\pi}{5} + i \sin\dfrac{9\pi}{5}\right)$

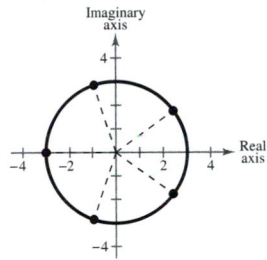

109. $4\left(\cos\dfrac{\pi}{2} + i \sin\dfrac{\pi}{2}\right)$

$4\left(\cos\dfrac{7\pi}{6} + i \sin\dfrac{7\pi}{6}\right)$

$4\left(\cos\dfrac{11\pi}{6} + i \sin\dfrac{11\pi}{6}\right)$

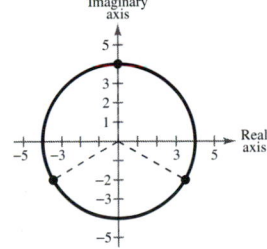

111. $\sqrt[6]{2}(\cos 105° + i \sin 105°)$
$\sqrt[6]{2}(\cos 225° + i \sin 225°)$
$\sqrt[6]{2}(\cos 345° + i \sin 345°)$

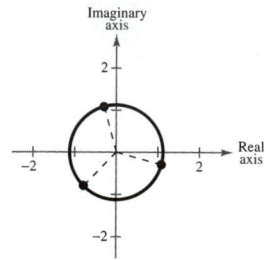

Focus on Concepts *(page 656)*

1. $\dfrac{a}{\sin A} = \dfrac{b}{\sin B} = \dfrac{c}{\sin C}$

2. $a^2 = b^2 + c^2 - 2bc \cos A$, $b^2 = a^2 + c^2 - 2ac \cos B$,
$c^2 = a^2 + b^2 - 2ab \cos C$

3. True   **4.** Pythagorean Theorem

5. False. There may be no solution, one solution, or two solutions.

6. Direction and magnitude   **7. A, C**

8. a. The angle between the vectors is acute.

9. If $k > 0$, the direction is the same and the magnitude is k times as great.

If $k < 0$, the result is a vector in the opposite direction and the magnitude is k times as great.

10. The diagonal of the parallelogram with **u** and **v** as its adjacent sides

11. b. Visualize the sum of **u** and $-$**v**.

12. $z_1 \cdot z_2 = -4$, $\dfrac{z_1}{z_2} = 1 - i$

13. (a) 3   (b) 120°, 210°, 300°

Review Exercises *(page 657)*

1. $A \approx 29.7°$, $B \approx 52.4°$, $C \approx 97.9°$

3. $C = 110°$, $b \approx 20.4$, $c \approx 22.6$

5. $A = 35°$, $C = 35°$, $b \approx 6.6$

7. No solution   **9.** $A \approx 25.9°$, $C \approx 39.1°$, $c \approx 10.1$

11. $B \approx 31.2°$, $C \approx 133.8°$, $c \approx 13.9$
$B \approx 148.8°$, $C \approx 16.2°$, $c \approx 5.39$

13. $A \approx 9.9°$, $C \approx 20.1°$, $b \approx 29.1$

15. $A \approx 40.9°$, $C \approx 114.1°$, $c \approx 8.6$
$A \approx 139.1°$, $C \approx 15.9°$, $c \approx 2.6$

17. 9.798   **19.** 9.08   **21.** 31.1 meters   **23.** 31.0 feet

25.

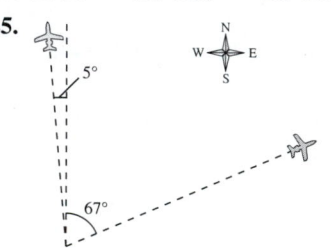

1135 miles

27. $\langle 7, -5 \rangle$ **29.** $\langle 7, -7 \rangle$ **31.** $\langle -4, 4\sqrt{3} \rangle$

33. $10\sqrt{2}(\mathbf{i} \sin 135° + \mathbf{j} \cos 135°)$

35. $\left\langle \dfrac{6}{\sqrt{61}}, -\dfrac{5}{\sqrt{61}} \right\rangle$

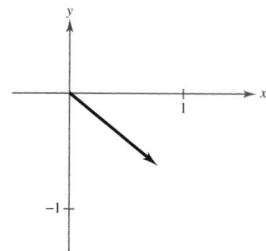

37. $\langle -26, -35 \rangle$

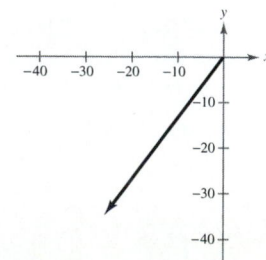

39.

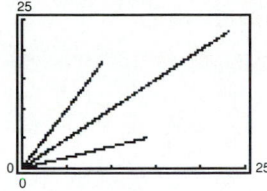

Magnitude: 32.62

Direction: 44.72°

41. 91.2 pounds, 79.9°   **43.** 180 pounds

45. 740.5 kilometers per hour, N 32.1° E

47. $\dfrac{1}{\sqrt{34}}\langle -5, 3\rangle$ **49.** Parallel   **51.** $\dfrac{11\pi}{12}$

53. 160.5°   **55.** $-\dfrac{13}{17}\langle 4, 1\rangle$ **57.** $-\dfrac{5}{2}\langle 1, -1\rangle$

59. $5\sqrt{2}\,(\cos 315° + i \sin 315°)$

61. $13(\cos 67.38° + i \sin 67.38°)$ **63.** $-50 - 50\sqrt{3}\,i$

65. 13

67. (a) $z_1 = 4(\cos 330° + i \sin 330°)$

$\quad z_2 = 10(\cos 270° + i \sin 270°)$

(b) $z_1 z_2 = 40(\cos 240° + i \sin 240°)$

$\quad \dfrac{z_1}{z_2} = \dfrac{2}{5}(\cos 60° + i \sin 60°)$

69. $\dfrac{625}{2} + \dfrac{625\sqrt{3}}{2}\,i$ **71.** $2035 - 828i$

73. (a) $4(\cos 60° + i \sin 60°)$

$\quad 4(\cos 180° + i \sin 180°)$

$\quad 4(\cos 300° + i \sin 300°)$

(b) -64

75. $3\left(\cos\dfrac{\pi}{4} + i \sin\dfrac{\pi}{4}\right)$

$\quad 3\left(\cos\dfrac{7\pi}{12} + i \sin\dfrac{7\pi}{12}\right)$

$\quad 3\left(\cos\dfrac{11\pi}{12} + i \sin\dfrac{11\pi}{12}\right)$

$\quad 3\left(\cos\dfrac{5\pi}{4} + i \sin\dfrac{5\pi}{4}\right)$

$\quad 3\left(\cos\dfrac{19\pi}{12} + i \sin\dfrac{19\pi}{12}\right)$

$\quad 3\left(\cos\dfrac{23\pi}{12} + i \sin\dfrac{23\pi}{12}\right)$

77. $3\left(\cos\dfrac{\pi}{4} + i \sin\dfrac{\pi}{4}\right) = \dfrac{3\sqrt{2}}{2} + \dfrac{3\sqrt{2}}{2}\,i$

$\quad 3\left(\cos\dfrac{3\pi}{4} + i \sin\dfrac{3\pi}{4}\right) = -\dfrac{3\sqrt{2}}{2} + \dfrac{3\sqrt{2}}{2}\,i$

$\quad 3\left(\cos\dfrac{5\pi}{4} + i \sin\dfrac{5\pi}{4}\right) = -\dfrac{3\sqrt{2}}{2} - \dfrac{3\sqrt{2}}{2}\,i$

$\quad 3\left(\cos\dfrac{7\pi}{4} + i \sin\dfrac{7\pi}{4}\right) = \dfrac{3\sqrt{2}}{2} - \dfrac{3\sqrt{2}}{2}\,i$

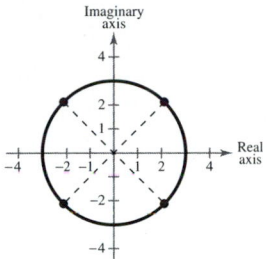

79. $2\left(\cos\dfrac{\pi}{2} + i \sin\dfrac{\pi}{2}\right) = 2i$

$\quad 2\left(\cos\dfrac{7\pi}{6} + i \sin\dfrac{7\pi}{6}\right) = -\sqrt{3} - i$

$\quad 2\left(\cos\dfrac{11\pi}{6} + i \sin\dfrac{11\pi}{6}\right) = \sqrt{3} - i$

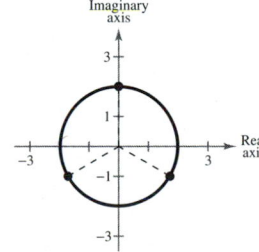

Cumulative Test for Chapters 6–8
(page 662)

1. (a)

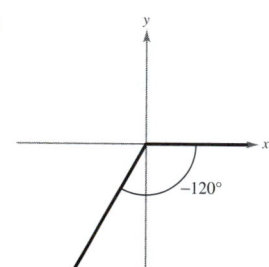

(b) $240°$

(c) $-\dfrac{2\pi}{3}$

(d) $60°$

(e) $\sin(-120°) = -\dfrac{\sqrt{3}}{2}$

$\cos(-120°) = -\dfrac{1}{2}$

$\tan(-120°) = \sqrt{3}$

$\csc(-120°) = -\dfrac{2\sqrt{3}}{3}$

$\sec(-120°) = -2$

$\cot(-120°) = \dfrac{\sqrt{3}}{3}$

2. $134.6°$ **3.** $\dfrac{3}{5}$

4. (a)   (b)

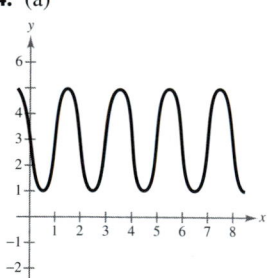

5. $h(x) = -3\cos(\pi x)$ **6.** $\sqrt{1 - 4x^2}$ **7.** $2\tan\theta$

8. Answers will vary.

9. (a) $\dfrac{\pi}{3}, \dfrac{\pi}{2}, \dfrac{3\pi}{2}, \dfrac{5\pi}{3}$ (b) $\dfrac{\pi}{6}, \dfrac{5\pi}{6}, \dfrac{7\pi}{6}, \dfrac{11\pi}{6}$

10. (a) $B \approx 26.4°,\ C \approx 123.6°,\ c \approx 15.0$

(b) $a \approx 5.0,\ B \approx 52.4°,\ C \approx 97.6°$

11. $2\sqrt{2}(\cos 135° + i\sin 135°)$

12. $24(\cos 150° + i\sin 150°)$

13. 1

$\cos 120° + i\sin 120° = -\dfrac{1}{2} + \dfrac{\sqrt{3}}{2}i$

$\cos 240° + i\sin 240° = -\dfrac{1}{2} - \dfrac{\sqrt{3}}{2}i$

14. 5 feet   **15.** N 32.6° E, 543.9 kilometers per hour

CHAPTER 9

Section 9.1   *(page 671)*

Warm Up   *(page 671)*

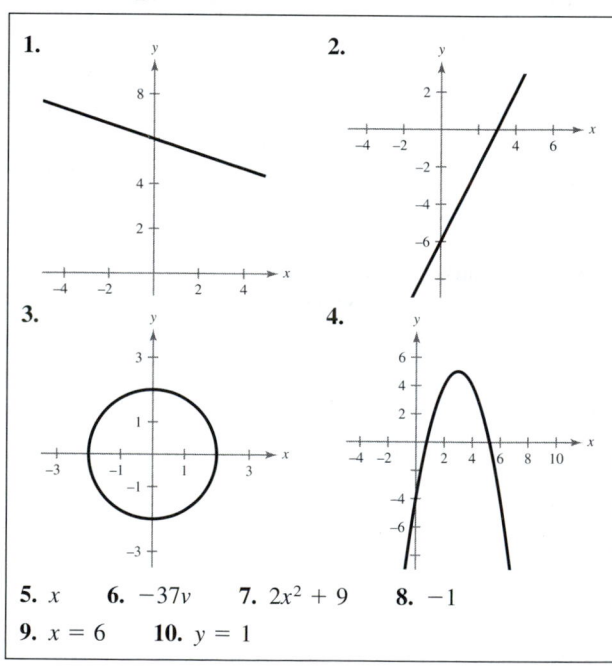

5. x **6.** $-37v$ **7.** $2x^2 + 9$ **8.** -1

9. $x = 6$ **10.** $y = 1$

1. $(2, 2)$ **3.** $(2, 6), (-1, 3)$ **5.** $(3, -4), (0, -5)$

7. $(0, 0), (2, -4)$ **9.** $(-2, 1), (16, 4)$ **11.** $(5, 5)$

13. $\left(\dfrac{1}{2}, 3\right)$ **15.** $\left(\dfrac{3}{2}, \dfrac{3}{10}\right)$ **17.** $\left(\dfrac{20}{3}, \dfrac{40}{3}\right)$ **19.** No solution

21. $(0, 0)$ **23.** $(4, 3)$ **25.** $\left(\dfrac{5}{2}, \dfrac{3}{2}\right)$ **27.** $(2, 2), (4, 0)$

29. $\left(4, -\dfrac{1}{2}\right)$ **31.** No points of intersection

33. $\left(\pm\sqrt{\dfrac{-1 + \sqrt{33}}{2}}, \dfrac{-1 + \sqrt{33}}{2}\right)$

35.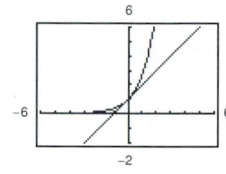

(0, 1)

37.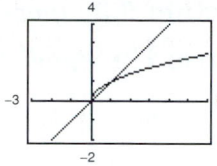

(0, 0), (1, 1)

39.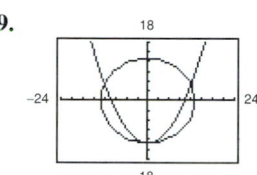

(0, −13), (±12, 5)

41. (1, 2)   **43.** (−2, 0), $\left(\frac{29}{10}, \frac{21}{10}\right)$ **45.** No solution

47. (0.287, 1.75)   **49.** (−1, 0), (0, 1), (1, 0)

51. $\left(\frac{1}{2}, 2\right)$, $\left(-4, -\frac{1}{4}\right)$

53. For a linear system the result will be a contradictory equation such as $0 = N$, where N is a nonzero real number. For a nonlinear system there may be an equation with imaginary roots.

55. 192 units   **57.** 233,333 units

59. (a) 6400 units   (b) 8800 units

61. (a) $x +$ $y = 25,000$
      $0.06x + 0.085y = 2,000$

(b)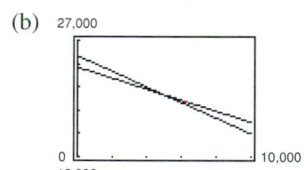

Decreases

(c) \$5000

63. \$8333.33

65. (a)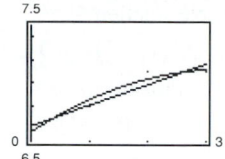

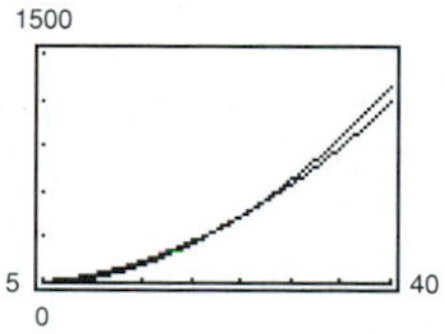

(b) 24.7 inches   (c) Doyle Log Rule

67. 6 × 9 meters   **69.** 9 × 12 inches

71. 8 × 12 kilometers   **73.** (5.55, 2.89)

75. (a) $f(t) = 0.18t + 6.63$
      $g(t) = -0.05t^2 + 0.33t + 6.58$

(b)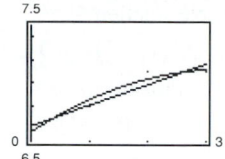

(c) (0.382, 6.70), (2.618, 7.10)

(d) $f(4) = 7.35$, $g(4) = 7.1$

$g(x)$ gives a more accurate prediction.

77. $2x + 7y - 45 = 0$ **79.** $y = 3$

81. $30x - 17y - 18 = 0$

Section 9.2   *(page 683)*

Warm Up   *(page 683)*

1.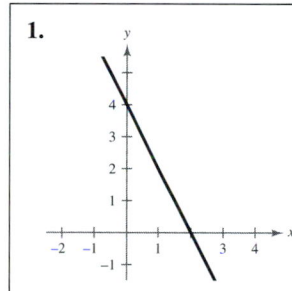

2.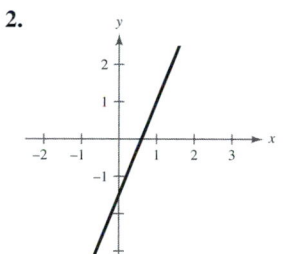

3.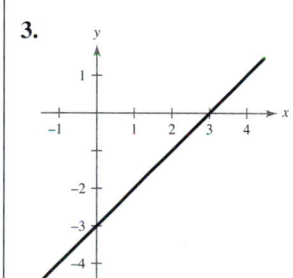

4.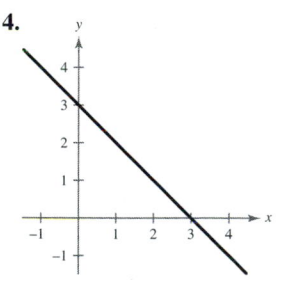

5.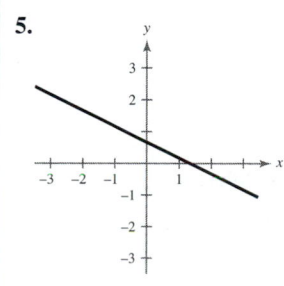

6.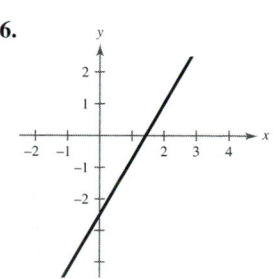

7. Perpendicular   **8.** Parallel   **9.** Neither

10. Perpendicular

1. $(2, 1)$ **3.** $(1, -1)$ **5.** No solution

7. Infinitely many solutions   **9.** $\left(\frac{1}{3}, -\frac{2}{3}\right)$ **11.** $\left(\frac{5}{2}, \frac{3}{4}\right)$

13. $(3, 4)$ **15.** $(4, -1)$ **17.** $(40, 40)$

19. Inconsistent   **21.** $\left(\frac{18}{5}, \frac{3}{5}\right)$ **23.** $(5, -2)$

25. $\left(a, \frac{5}{6}a - \frac{1}{2}\right)$ **27.** $\left(\frac{90}{31}, -\frac{67}{31}\right)$ **29.** $\left(-\frac{6}{35}, \frac{43}{35}\right)$

31. 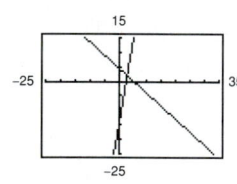   **33.**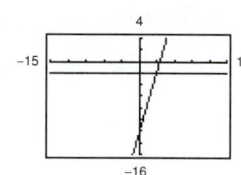

Inconsistent   Consistent, one solution

35.   **37.**

$(3, 2)$ $(3, -2)$

39. $(4, 1)$ **41.** $(2, -1)$ **43.** $2x + 2y = 11$

$x - 4y = -7$

45. $(39{,}600, 398)$. It is necessary to change the scale on the axes to see the point of intersection.

47. No. Two lines will intersect only once or coincide and the system will have infinitely many solutions.

49. $k = -4$ **51.** 550 miles per hour, 50 miles per hour

53. (a) $x + y = 10$

$0.2x + 0.5y = 3$

(b)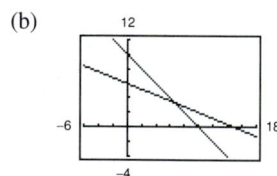

Decreases.

(c) 20% solution: $6\frac{2}{3}$ liters

50% solution: $3\frac{1}{3}$ liters

55. $6000   **57.** 375 adults, 125 children

59. 225 kilometers and 75 kilometers

61. $(80, 10)$ **63.** $(2{,}000{,}000, 100)$

65. $y = 0.97x + 2.10$ **67.** $y = 0.318x + 4.061$

69. $y = \frac{3}{4}x + \frac{4}{3}$ **71.** $y = -2x + 4$

73. $y = -240x + 685$; 349 units

Section 9.3 *(page 695)*

Warm Up *(page 695)*

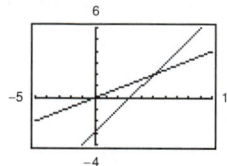

1. $(15, 10)$ **2.** $\left(-2, -\frac{8}{3}\right)$ **3.** $(28, 4)$ **4.** $(4, 3)$

5. Not a solution   **6.** Not a solution   **7.** Solution

8. Solution   **9.** $5a + 2$ **10.** $a + 13$

1. $(1, -2, 4)$ **3.** $(1, 2, -2)$ **5.** $\left(\frac{1}{2}, -2, 2\right)$

7. $x - 2y + 3z = 5$

$y - 2z = 9$

$2x \qquad - 3z = 0$

First step in putting the system in row-echelon form

9. $(1, 2, 3)$ **11.** $(-4, 8, 5)$ **13.** $(5, -2, 0)$

15. Inconsistent   **17.** $\left(1, -\frac{3}{2}, \frac{1}{2}\right)$

19. $(-3a + 10, 5a - 7, a)$ **21.** $(-a + 3, a + 1, a)$

23. $(2a, 21a - 1, 8a)$ **25.** $\left(\frac{1}{2} - \frac{3}{2}a, 1 - \frac{2}{3}a, a\right)$

27. $(1, 1, 1, 1)$ **29.** Inconsistent   **31.** $(0, 0, 0)$

33. $(9a, -35a, 67a)$

35. No. There are two arithmetic errors. They are the constant in the second equation and the coefficient of z in the third.

37. $3x + y - z = 9$

$x + 2y - z = 0$

$-x + y + 3z = 1$

39. $y = \frac{1}{2}x^2 - 2x$ **41.** $y = x^2 - 6x + 8$

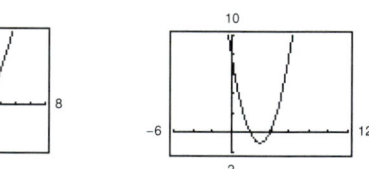

43. $x^2 + y^2 - 4x = 0$ **45.** $x^2 + y^2 + 6x - 8y = 0$

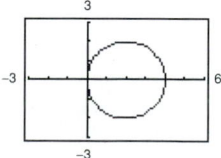   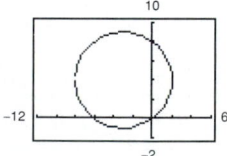

47. $s = -16t^2 + 144$ **49.** $s = -16t^2 - 32t + 500$

51. $4000 at 5%   **53.** $300,000 at 8%

$5000 at 6% $400,000 at 9%

$7000 at 7% $75,000 at 10%

55. $250,000 - \frac{1}{2}s$ in certificates of deposit,

$125,000 + \frac{1}{2}s$ in municipal bonds,

$125,000 - s$ in blue-chip stocks,

s in growth stocks

57. 20 liters of spray X,

18 liters of spray Y,

16 liters of spray Z

59. Use four medium trucks or use two large, one medium, and two small trucks

61. $t_1 = 96$ pounds

$t_2 = 48$ pounds

$a = -16$ feet per second squared

63. $\dfrac{1}{2}\left(-\dfrac{2}{x} + \dfrac{1}{x-1} + \dfrac{1}{x+1}\right)$

65. $\dfrac{1}{2}\left(\dfrac{1}{x} - \dfrac{1}{x-2} + \dfrac{2}{x+3}\right)$

67. $y = -\frac{5}{24}x^2 - \frac{3}{10}x + \frac{41}{6}$ **69.** $y = x^2 - x$

71. (a) $y = 0.14x^2 - 4.43x + 58.40$

(b)

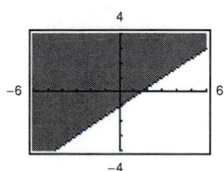

(c) 434.3

73. $x = 5$ **75.** $x = \pm\sqrt{2}/2$ or $x = 0$

$y = 5$ $y = \frac{1}{2}$ $y = 0$

$\lambda = -5$ $\lambda = 1$ $\lambda = 0$

77. The average increase in sales per year

79. 6.375   **81.** 80,000

Section 9.4   *(page 707)*

Warm Up   *(page 707)*

1. Line   **2.** Line   **3.** Parabola   **4.** Parabola

5. Circle   **6.** Ellipse   **7.** $(1, 1)$ **8.** $(2, 0)$

9. $(2, 1)$, $\left(-\frac{5}{2}, -\frac{5}{4}\right)$ **10.** $(2, 3)$, $(3, 2)$

1.

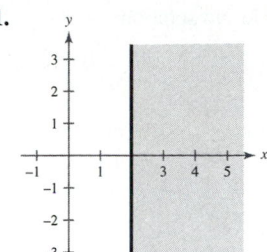

3.

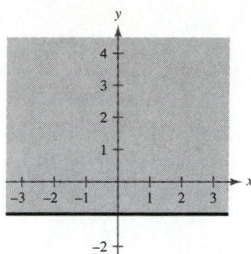

5.

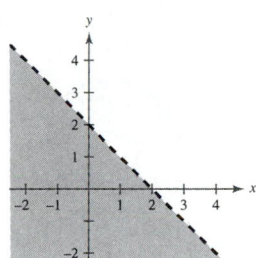

7.

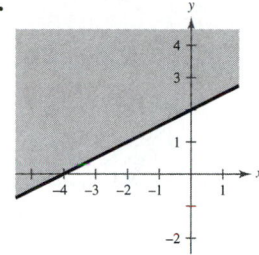

9.

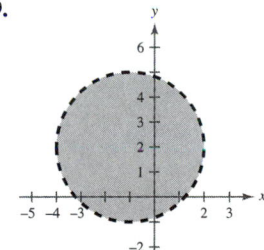

11.

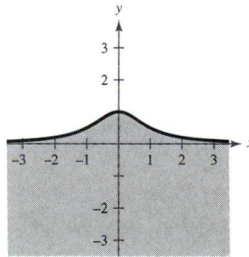

13.

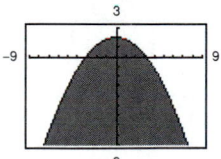

15.

17. $y \le \dfrac{1}{2}x + 2$ **19.** $\dfrac{x}{3} + \dfrac{y}{2} \ge 1$

21.

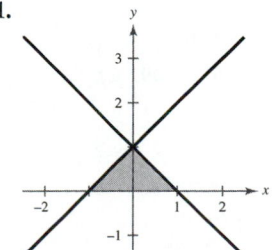

23.

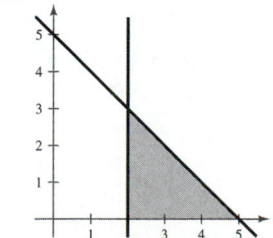

25.

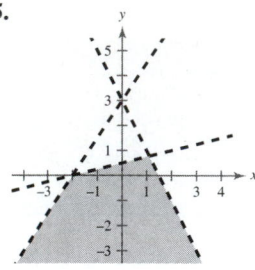

27. No solution

51. $x + \frac{3}{2}y \le 12$
  $\frac{4}{3}x + \frac{3}{2}y \le 15$
  $x \quad\ \ge 0$
  $y \ge 0$

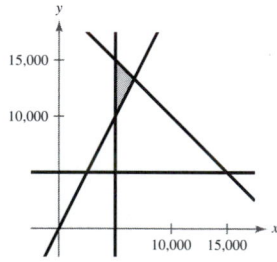

29.

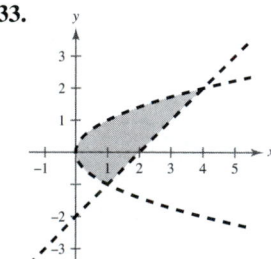

31.

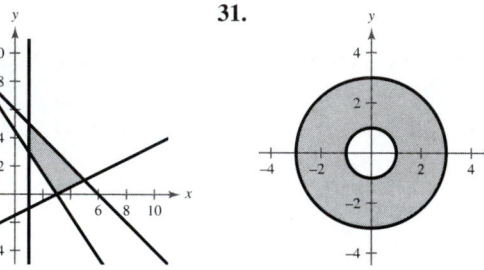

53. $x + y \le 20{,}000$
  $y \ge \quad 2x$
  $x \quad \ge 5{,}000$
  $y \ge 5{,}000$

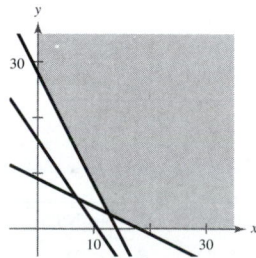

33.

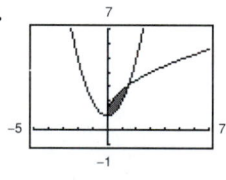

35.

37.

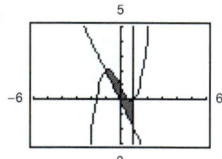

39.

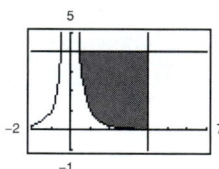

55. $20x + 10y \ge 280$
  $15x + 10y \ge 160$
  $10x + 20y \ge 180$
  $x \qquad \ge 0$
  $y \ge 0$

41. $\frac{1}{4}x + \frac{1}{4}y \le 1$ **43.** $y \ge 4 - x$
  $x \ge 0$ $y \ge 2 - \frac{1}{3}x$
  $y \ge 0$ $x \ge 0, \quad y \ge 0$
45. $x^2 + y^2 \le 16$ **47.** $2 \le x \le 5$ **49.** $y \le \frac{3}{2}x$
  $x \ge 0, \quad y \ge 0$ $1 \le y \le 7$ $y \le -x + 5$
                          $y \ge 0$

57. $xy \geq 500$
     $2x + \pi y \geq 125$
     $x \qquad \geq \quad 0$
     $\quad \quad y \geq \quad 0$

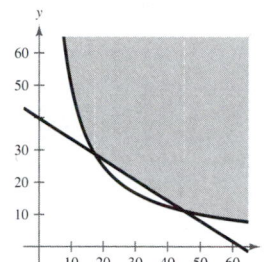

59. Consumer surplus: 1600
   Producer surplus: 400

61. Consumer surplus: 40,000,000
   Producer surplus: 20,000,000

63. Test a point on either side.

Section 9.5   *(page 717)*

Warm Up   *(page 717)*

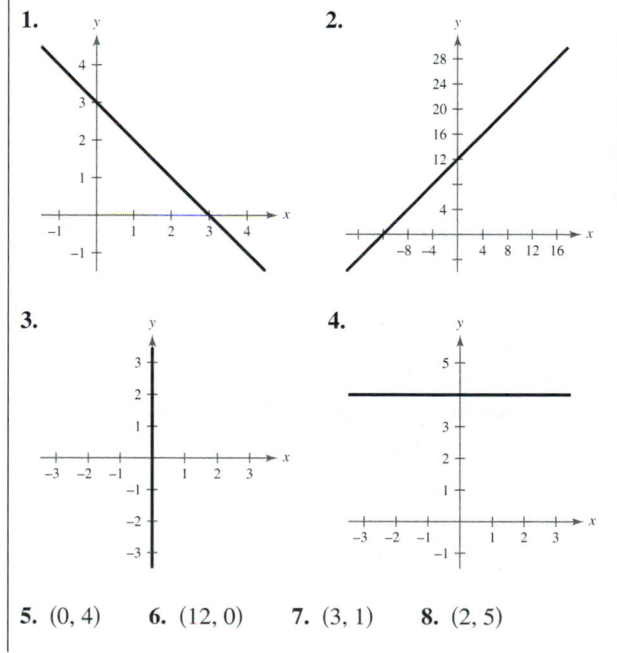

5. $(0, 4)$ **6.** $(12, 0)$ **7.** $(3, 1)$ **8.** $(2, 5)$

9.

10.

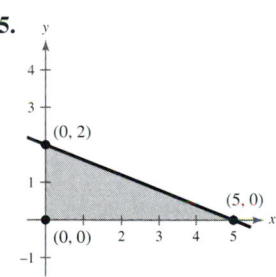

1. Minimum at $(0, 0)$: 0   **3.** Minimum at $(0, 0)$: 0
   Maximum at $(0, 6)$: 30   Maximum at $(6, 0)$: 60

5. Minimum at $(0, 0)$: 0   **7.** Minimum at $(0, 0)$: 0
   Maximum at $(3, 4)$: 17   Maximum at $(4, 0)$: 20

9. Minimum at $(0, 0)$: 0
   Maximum at $(60, 20)$: 740

11. Minimum at $(0, 0)$: 0
   Maximum at any point on the line segment connecting $(60, 20)$ and $(30, 45)$: 2100

13.

15.

Minimum at $(0, 0)$: 0   Minimum at $(0, 0)$: 0
Maximum at $(5, 0)$: 30   Maximum at $(0, 2)$: 48

17.

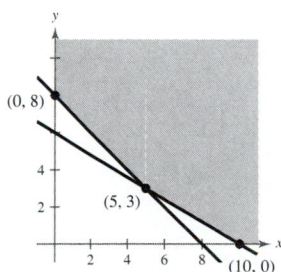

Minimum at $(5, 3)$: 35
No maximum

19.

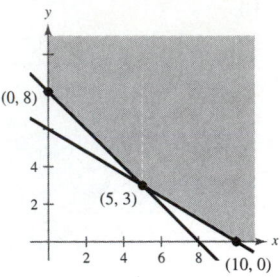

Minimum at $(10, 0)$: 20

No maximum

21.

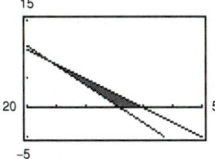

23.

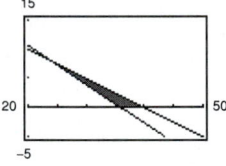

Minimum at $(24, 8)$: 104

Maximum at $(40, 0)$: 160

Minimum at $(36, 0)$: 36

Maximum at $(24, 8)$: 56

25. Maximum at $(3, 6)$: 12   **27.** Maximum at $(0, 10)$: 10

29. Maximum at $(0, 5)$: 25   **31.** Maximum at $\left(\frac{22}{3}, \frac{19}{6}\right)$: $\frac{271}{6}$

33. $z = x + 5y$ **35.** $z = 4x + y$

37. 750 units of model A

1000 units of model B

Maximum profit: $83,750

39. 200 units of the $250 model

50 units of the $400 model

Maximum profit: $11,500

41. Three bags of brand X

Six bags of brand Y

Minimum cost: $21.67 per bag

43. Eight audits

Eight tax returns

Maximum revenue: $18,400

45.

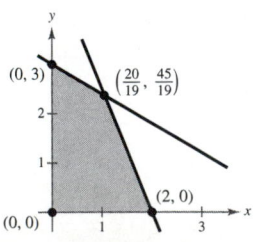

z is maximum at any point on the line segment connecting $(2, 0)$ and $\left(\frac{20}{19}, \frac{45}{19}\right)$.

47.

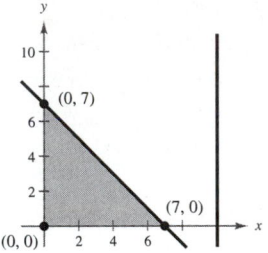

The constraint $x \le 10$ is extraneous. Maximum at $(0, 7)$: 14

49.

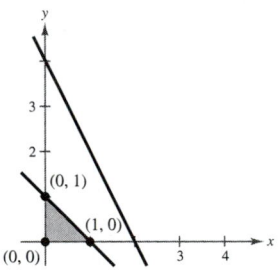

The constraint $2x + y \le 4$ is extraneous. Maximum at $(0, 1)$: 4

51. (a) $t > 9$ (b) $\frac{3}{4} < t < 9$ **53.** $\dfrac{9}{2(x + 3)}$

55. $\dfrac{x^2 + 2x - 13}{x(x - 2)}$

Focus on Concepts *(page 721)*

1. A solution of a system is an ordered pair that satisfies each equation in the system.

2. For a linear system the result will be a contradictory equation such as $0 = N$, where N is a nonzero real number. For a nonlinear system there may be an equation with imaginary roots.

3. There will be a contradictory equation of the form $0 = N$, where N is a nonzero real number.

4. The algebraic methods yield exact solutions.

5. (a) One   (b) Two   (c) Four

6. The system has no solution.

7. The lines are distinct and parallel.

$x + 2y = 3$

$2x + 4y = 9$

8. (a) Interchange any two equations.

(b) Multiply an equation by a nonzero constant.

(c) Add a multiple of one equation to any other equation in the system.

9. No. When -2 times Equation 1 is added to Equation 2, the constant is -11.

10. The first system is inconsistent, because -4 times Equation 1 added to Equation 2 yields $0 = -4$.

11. d   **12.** b   **13.** c   **14.** a

15. (a) The boundary would be included in the solution.

(b) The solution would be the half-plane on the opposite side of the boundary.

Review Exercises   *(page 722)*

1. $(1, 1)$ **3.** $(5, 4)$ **5.** $(0, 0), (2, 8), (-2, 8)$

7. $(0, 0), (-3, 3)$ **9.** $(4, 4)$ **11.** $\left(\frac{5}{2}, 3\right)$

13. $(-0.5, 0.8)$ **15.** $(0, 0)$ **17.** $\left(\frac{14}{5} + \frac{8}{5}a, a\right)$

19. $3x + y = 7$
$-6x + 3y = 1$ **21.** 4762 units

23. 75% solution: 40 liters
50% solution: 60 liters

25. 218.75 miles per hour, 193.75 miles per hour

27. $\left(\dfrac{500,000}{7}, \dfrac{159}{7}\right)$ **29.** $(4.8, 4.4, -1.6)$

31. $(3a + 4, 2a + 5, a)$ **33.** $(-3a + 2, 5a + 6, a)$

35. $2x + y - 2z = 1$
$x + y - z = 0$
$2x - 3y - 2z = 5$ **37.** $y = 2x^2 + x - 5$

39. $x^2 + y^2 - 4x + 4y - 1 = 0$

41. 10 gallons of spray X
5 gallons of spray Y
12 gallons of spray Z **43.** $y = 1.01x + 1.54$

45. (a) $y = 0.956x + 2.799$

(b)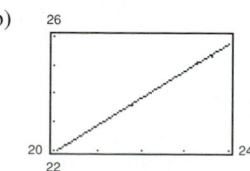

(c) Good model

(d) A 1-year change in x results in a 0.956-year change in y.

47.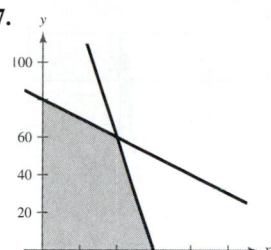

49.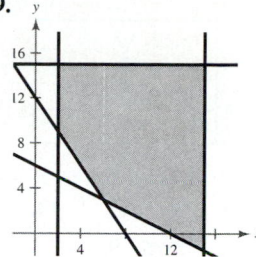

51.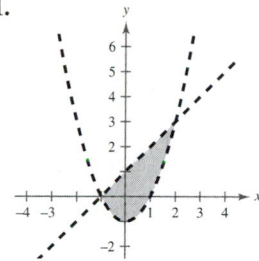

53.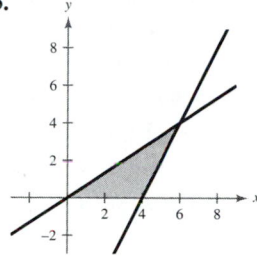

55. $-x + y \le 4$
$2x + y \le 22$
$-x + y \ge -2$
$2x + y \ge 7$

57. $x + y \le 1500$
$x \ge 400$
$y \ge 600$

59. Consumer surplus: 4,500,000
Producer surplus: 9,000,000

61. Maximum at $(5, 8)$: 47   **63.** Minimum at $(15, 0)$: 26.25

65. 20 perms   **67.** Three bags of brand X
Two bags of brand Y
Minimum cost per bag: $21

Chapter Test   *(page 728)*

1. $(2, -2)$ **2.** $(0, -1), (1, 0), (2, 1)$

3. $(8, 4), (2, -2)$ **4.** $(3, 2)$ **5.** $(-3, 0), (2, 5)$

6. $(3, 1)$ **7.** $(1, 5)$ **8.** $(2, -3, 1)$

9. $3x - y = 9$
$6x + y = 3$ **10.** $y = -\frac{1}{2}x^2 + x + 6$

11. 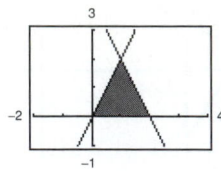   **12.**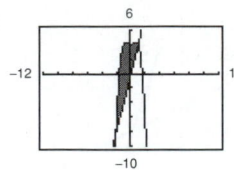

13. $x + 3y \le 45$ **14.** $(12, 0)$; $z = 240$

$7x + 3y \le 99$

$x \qquad \le 12$

$x \qquad \ge 0$

$\qquad y \ge 0$

15. \$275 model: 160 units

\$400 model: 140 units

CHAPTER 10

Section 10.1   *(page 740)*

Warm Up   *(page 740)*

1. -3 **2.** 30   **3.** 6   **4.** $-\dfrac{1}{9}$ **5.** Solution

6. Not a solution   **7.** $(5, 2)$ **8.** $\left(\frac{12}{5}, -3\right)$

9. $(40, 14, 2)$ **10.** $\left(\frac{15}{2}, 4, 1\right)$

1. 3×2 **3.** 3×1 **5.** 2×2

7. $\begin{bmatrix} 4 & -3 & : & -5 \\ -1 & 3 & : & 12 \end{bmatrix}$ **9.** $\begin{bmatrix} 1 & 10 & -2 & : & 2 \\ 5 & -3 & 4 & : & 0 \\ 2 & 1 & 0 & : & 6 \end{bmatrix}$

11. $x + 2y = 7$ **13.** $2x \qquad + 5z = -12$

$2x - 3y = 4$ $\qquad y - 2z = \quad 7$

$\qquad\qquad\qquad 6x + 3y \qquad = \quad 2$

15. Reduced row-echelon form

17. Not in row-echelon form

19. $\begin{bmatrix} 1 & 4 & 3 \\ 0 & 2 & -1 \end{bmatrix}$ **21.** $\begin{bmatrix} 1 & 1 & 4 & -1 \\ 0 & 5 & -2 & 6 \\ 0 & 3 & 20 & 4 \end{bmatrix}$

$\begin{bmatrix} 1 & 1 & 4 & -1 \\ 0 & 1 & -\frac{2}{5} & \frac{6}{5} \\ 0 & 3 & 20 & 4 \end{bmatrix}$

23. (a) $\begin{bmatrix} 1 & 2 & 3 \\ 0 & -5 & -10 \\ 3 & 1 & -1 \end{bmatrix}$ (b) $\begin{bmatrix} 1 & 2 & 3 \\ 0 & -5 & -10 \\ 0 & -5 & -10 \end{bmatrix}$

(c) $\begin{bmatrix} 1 & 2 & 3 \\ 0 & -5 & -10 \\ 0 & 0 & 0 \end{bmatrix}$ (d) $\begin{bmatrix} 1 & 2 & 3 \\ 0 & 1 & 2 \\ 0 & 0 & 0 \end{bmatrix}$

(e) $\begin{bmatrix} 1 & 0 & -1 \\ 0 & 1 & 2 \\ 0 & 0 & 0 \end{bmatrix}$

25. $\begin{bmatrix} 1 & 1 & 0 & 5 \\ 0 & 1 & 2 & 0 \\ 0 & 0 & 1 & -1 \end{bmatrix}$

27. $\begin{bmatrix} 1 & -1 & -1 & 1 \\ 0 & 1 & 6 & 3 \\ 0 & 0 & 0 & 0 \end{bmatrix}$ **29.** $\begin{bmatrix} 1 & 0 & 0 \\ 0 & 1 & 0 \\ 0 & 0 & 1 \end{bmatrix}$

31. $\begin{bmatrix} 1 & 2 & 0 & 0 \\ 0 & 0 & 1 & 0 \\ 0 & 0 & 0 & 1 \\ 0 & 0 & 0 & 0 \end{bmatrix}$

33. $x - 2y = \quad 4$ **35.** $x - y + 2z = \quad 4$

$\qquad\quad y = -3$ $\qquad\quad y - \ z = \quad 2$

$\quad (-2, -3)$ $\qquad\qquad\qquad z = -2$

$\qquad\qquad\qquad\qquad\quad (8, 0, -2)$

37. $(7, -5)$ **39.** $(-4, -8, 2)$ **41.** $(3, 2)$

43. $(4, -2)$ **45.** $\left(\frac{1}{2}, -\frac{3}{4}\right)$ **47.** Inconsistent

49. $(4, -3, 2)$ **51.** $(2a + 1, 3a + 2, a)$

53. $(5a + 4, -3a + 2, a)$ **55.** $(0, 0)$

57. $(0, 2 - 4a, a)$ **59.** $(1, 0, 4, -2)$ **61.** $(-2a, a, a)$

63. $x + \ y + \ 7z = -1$ **65.** \$800,000 at 8%

$\quad x + 2y + 11z = \quad 0$ \$500,000 at 9%

$2x + \ y + 10z = -3$ \$200,000 at 12%

67. $\dfrac{4x^2}{(x + 1)^2(x - 1)} = \dfrac{1}{x - 1} + \dfrac{3}{x + 1} - \dfrac{2}{(x + 1)^2}$

69. $y = x^2 + 2x + 5$

71. (a) $y = -0.004x^2 + 0.367x + 5$

(b)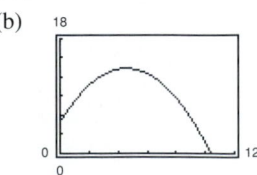

13 feet, 104 feet

(c) 13.418 feet, 103.793 feet

73. (a) $x_1 = s$, $x_2 = t$, $x_3 = 600 - s$, $x_4 = s - t$,

   $x_5 = 500 - t$, $x_6 = s$, $x_7 = t$

(b) $x_1 = 0$, $x_2 = 0$, $x_3 = 600$, $x_4 = 0$, $x_5 = 500$,

   $x_6 = 0$, $x_7 = 0$

(c) $x_1 = 0$, $x_2 = -500$, $x_3 = 600$, $x_4 = 500$,

   $x_5 = 1000$, $x_6 = 0$, $x_7 = -500$

75.

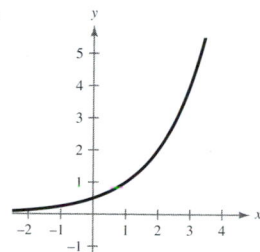

77.

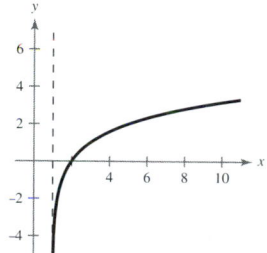

Section 10.2 *(page 755)*

Warm Up *(page 755)*

1. -5 **2.** -7 **3.** Not in reduced row-echelon form

4. Not in reduced row-echelon form

5. $\begin{bmatrix} -5 & 10 & : & 12 \\ 7 & -3 & : & 0 \\ -1 & 7 & : & 25 \end{bmatrix}$

6. $\begin{bmatrix} 10 & 15 & -9 & : & 42 \\ 6 & -5 & 0 & : & 0 \end{bmatrix}$ **7.** $(0, 2)$

8. $(2 + a, 3 - a, a)$ **9.** $(1 - 2a, a, -1)$

10. $(2, -1, -1)$

1. $x = -4$, $y = 22$ **3.** $x = 2$, $y = 3$

5. (a) $\begin{bmatrix} 3 & -2 \\ 1 & 7 \end{bmatrix}$ (b) $\begin{bmatrix} -1 & 0 \\ 3 & -9 \end{bmatrix}$ (c) $\begin{bmatrix} 3 & -3 \\ 6 & -3 \end{bmatrix}$

(d) $\begin{bmatrix} -1 & -1 \\ 8 & -19 \end{bmatrix}$

7. (a) $\begin{bmatrix} 7 & 3 \\ 1 & 9 \\ -2 & 15 \end{bmatrix}$ (b) $\begin{bmatrix} 5 & -5 \\ 3 & -1 \\ -4 & -5 \end{bmatrix}$ (c) $\begin{bmatrix} 18 & -3 \\ 6 & 12 \\ -9 & 15 \end{bmatrix}$

(d) $\begin{bmatrix} 16 & -11 \\ 8 & 2 \\ -11 & -5 \end{bmatrix}$

9. (a) $\begin{bmatrix} 3 & 3 & -2 & 1 & 1 \\ -2 & 5 & 7 & -6 & -8 \end{bmatrix}$

(b) $\begin{bmatrix} 1 & 1 & 0 & -1 & 1 \\ 4 & -3 & -11 & 6 & 6 \end{bmatrix}$

(c) $\begin{bmatrix} 6 & 6 & -3 & 0 & 3 \\ 3 & 3 & -6 & 0 & -3 \end{bmatrix}$

(d) $\begin{bmatrix} 4 & 4 & -1 & -2 & 3 \\ 9 & -5 & -24 & 12 & 11 \end{bmatrix}$

11. $\begin{bmatrix} -6 & -9 \\ -1 & 0 \\ 17 & -10 \end{bmatrix}$ **13.** $\begin{bmatrix} 3 & 3 \\ -\frac{1}{2} & 0 \\ -\frac{13}{2} & \frac{11}{2} \end{bmatrix}$

15. (a) $\begin{bmatrix} 0 & 15 \\ 6 & 12 \end{bmatrix}$ (b) $\begin{bmatrix} -2 & 2 \\ 31 & 14 \end{bmatrix}$ (c) $\begin{bmatrix} 9 & 6 \\ 12 & 12 \end{bmatrix}$

17. (a) $\begin{bmatrix} 0 & -10 \\ 10 & 0 \end{bmatrix}$ (b) $\begin{bmatrix} 0 & -10 \\ 10 & 0 \end{bmatrix}$ (c) $\begin{bmatrix} 8 & -6 \\ 6 & 8 \end{bmatrix}$

19. (a) $\begin{bmatrix} 6 & -21 & 15 \\ 8 & -23 & 19 \\ 4 & 7 & 5 \end{bmatrix}$ (b) $\begin{bmatrix} 9 & 0 & 13 \\ 7 & -2 & 21 \\ 1 & 4 & -19 \end{bmatrix}$

(c) $\begin{bmatrix} 20 & 7 & -8 \\ 24 & 7 & -2 \\ 2 & -5 & 30 \end{bmatrix}$

21. Not possible   **23.** $\begin{bmatrix} -1 & 19 \\ 4 & -27 \\ 0 & 14 \end{bmatrix}$ **25.** $\begin{bmatrix} 1 & 0 & 0 \\ 0 & 1 & 0 \\ 0 & 0 & \frac{7}{2} \end{bmatrix}$

27. $\begin{bmatrix} 60 & -20 & 10 & 60 \\ 72 & -24 & 12 & 72 \end{bmatrix}$ **29.** $\begin{bmatrix} 41 & 7 & 7 \\ 42 & 5 & 25 \\ -10 & -25 & 45 \end{bmatrix}$

31. $\begin{bmatrix} 151 & 25 & 48 \\ 516 & 279 & 387 \\ 47 & -20 & 87 \end{bmatrix}$ **33.** Not possible

35. $A = \begin{bmatrix} -1 & 1 \\ -2 & 1 \end{bmatrix}$ **37.** $A = \begin{bmatrix} 2 & 3 \\ 1 & 4 \end{bmatrix}$

$X = \begin{bmatrix} x \\ y \end{bmatrix}$ $X = \begin{bmatrix} x \\ y \end{bmatrix}$

$B = \begin{bmatrix} 4 \\ 0 \end{bmatrix}$ $B = \begin{bmatrix} 5 \\ 10 \end{bmatrix}$

$x = 4, y = 8$ $x = -2, y = 3$

39. $\begin{bmatrix} -4 & 0 \\ 8 & 2 \end{bmatrix}$ **41.** $\begin{bmatrix} 0 & 0 & 0 \\ 0 & 0 & 0 \\ 0 & 0 & 0 \end{bmatrix}$

43. $AC = BC = \begin{bmatrix} 2 & 3 \\ 2 & 3 \end{bmatrix}$ **45.** Not possible

47. Not possible   **49.** 2×2 **51.** Not possible

53. 2×3 **55.** $\begin{bmatrix} 72 & 48 & 24 \\ 36 & 108 & 72 \end{bmatrix}$

57. $AB = [\$1250 \quad \$1331.25 \quad \$981.25]$

The entries represent the profits from the two products at the three outlets.

59. $\begin{bmatrix} \$15,770 & \$18,300 \\ \$26,500 & \$29,250 \\ \$21,260 & \$24,150 \end{bmatrix}$

The entries are the wholesale and retail prices of the inventories at the three outlets.

61. $\begin{bmatrix} \$17.70 & \$15.00 \\ \$29.40 & \$25.00 \\ \$50.40 & \$43.00 \end{bmatrix}$

The entries are labor costs at each plant for each size of boat.

63. $A^2 = \begin{bmatrix} -1 & 0 \\ 0 & -1 \end{bmatrix}$

$A^3 = \begin{bmatrix} -i & 0 \\ 0 & -i \end{bmatrix}$

$A^4 = \begin{bmatrix} 1 & 0 \\ 0 & 1 \end{bmatrix}$

65. Diagonal matrix whose entries are the products of the corresponding entries of A and B.

Section 10.3 *(page 765)*

Warm Up *(page 765)*

1. $\begin{bmatrix} 4 & 24 \\ 0 & -16 \\ 48 & 8 \end{bmatrix}$ **2.** $\begin{bmatrix} \frac{11}{2} & 5 & 24 \\ \frac{1}{2} & 0 & 8 \\ 0 & 1 & 4 \end{bmatrix}$ **3.** $\begin{bmatrix} -5 & -2 \\ 4 & -13 \end{bmatrix}$

4. $\begin{bmatrix} -13 & 11 \\ -19 & 21 \end{bmatrix}$ **5.** $\begin{bmatrix} 1 & 0 \\ 0 & 1 \end{bmatrix}$ **6.** $\begin{bmatrix} 6 & 5 \\ 3 & -2 \end{bmatrix}$

7. $\begin{bmatrix} 5 & 6 & 2 \\ 4 & 7 & 2 \\ 24 & 36 & 11 \end{bmatrix}$ **8.** $\begin{bmatrix} 1 & 0 & 0 \\ 0 & 1 & 0 \\ 0 & 0 & 1 \end{bmatrix}$

9. $\begin{bmatrix} 1 & 0 & 3 & -2 \\ 0 & 1 & 4 & -3 \end{bmatrix}$

10. $\begin{bmatrix} 1 & 0 & 0 & -6 & -4 & 3 \\ 0 & 1 & 0 & 11 & 6 & -5 \\ 0 & 0 & 1 & -2 & -1 & 1 \end{bmatrix}$

1–7. $AB = I$ and $BA = I$

9. $\begin{bmatrix} \frac{1}{2} & 0 \\ 0 & \frac{1}{3} \end{bmatrix}$ **11.** $\begin{bmatrix} -3 & 2 \\ -2 & 1 \end{bmatrix}$ **13.** $\begin{bmatrix} 1 & -1 \\ 2 & -1 \end{bmatrix}$

15. Does not exist   **17.** Does not exist

19. $\begin{bmatrix} 1 & 1 & -1 \\ -3 & 2 & -1 \\ 3 & -3 & 2 \end{bmatrix}$ **21.** $\begin{bmatrix} 1 & 0 & 0 \\ -0.75 & 0.25 & 0 \\ 0.35 & -0.25 & 0.2 \end{bmatrix}$

23. $\begin{bmatrix} -\frac{1}{8} & 0 & 0 & 0 \\ 0 & 1 & 0 & 0 \\ 0 & 0 & \frac{1}{4} & 0 \\ 0 & 0 & 0 & -\frac{1}{5} \end{bmatrix}$ **25.** $\begin{bmatrix} -175 & 37 & -13 \\ 95 & -20 & 7 \\ 14 & -3 & 1 \end{bmatrix}$

27. $\frac{1}{2} \begin{bmatrix} -3 & 3 & 2 \\ 9 & -7 & -6 \\ -2 & 2 & 2 \end{bmatrix}$ **29.** $\frac{5}{11} \begin{bmatrix} 0 & -4 & 2 \\ -22 & 11 & 11 \\ 22 & -6 & -8 \end{bmatrix}$

31. Does not exist   **33.** $\begin{bmatrix} -24 & 7 & 1 & -2 \\ -10 & 3 & 0 & -1 \\ -29 & 7 & 3 & -2 \\ 12 & -3 & -1 & 1 \end{bmatrix}$

35. Answers will vary.   **37.** $(5, 0)$ **39.** $(-8, -6)$

41. $(3, 8, -11)$ **43.** $(2, 1, 0, 0)$ **45.** $(2, -2)$

47. No solution   **49.** $\left(3, -\frac{1}{2}\right)$ **51.** $(-1, 3, 2)$

53. No solution   **55.** $(5, 0, -2, 3)$

57. $10,000 in AAA-rated bonds

$5000 in A-rated bonds

$10,000 in B-rated bonds

59. $9000 in AAA-rated bonds

$1000 in A-rated bonds

$2000 in B-rated bonds

61. The inverse matrix remains the same for each system.

63. $I_1 = -3$ amperes

$I_2 = 8$ amperes

$I_3 = 5$ amperes

65. (a) Answers will vary.

(b) $A^{-1} = \begin{bmatrix} \frac{1}{a_{11}} & 0 & 0 & \vdots & 0 \\ 0 & \frac{1}{a_{22}} & 0 & \vdots & 0 \\ 0 & 0 & \frac{1}{a_{33}} & \vdots & 0 \\ 0 & 0 & 0 & \vdots & \frac{1}{a_{nn}} \end{bmatrix}$

67. Men: 1.171 minutes; Women: 1.239 minutes

69. $x = \dfrac{2 \ln 315}{\ln 3} \approx 10.47$ **71.** $x = 2^{6.5} \approx 90.51$

Section 10.4 *(page 775)*

Warm Up *(page 775)*

1. -22 **2.** 35 **3.** -15 **4.** $-\dfrac{1}{8}$ **5.** -45

6. -16 **7.** $\begin{bmatrix} 1 & -3 \\ 0 & 1 \end{bmatrix}$ **8.** $\begin{bmatrix} 1 & -3 \\ 0 & 1 \end{bmatrix}$

9. $\begin{bmatrix} 1 & 3 & 4 \\ 0 & 1 & 1 \\ 0 & 0 & 0 \end{bmatrix}$ **10.** $\begin{bmatrix} 1 & 2 & 4 \\ 0 & 1 & \frac{10}{7} \\ 0 & 0 & 0 \end{bmatrix}$

1. 5 **3.** 5 **5.** 27 **7.** -24 **9.** 6 **11.** 0

13. 0 **15.** -9 **17.** -0.002 **19.** 0

21. (a) $M_{11} = -5$, $M_{12} = 2$, $M_{21} = 4$, $M_{22} = 3$

(b) $C_{11} = -5$, $C_{12} = -2$, $C_{21} = -4$, $C_{22} = 3$

23. (a) $M_{11} = 30$, $M_{12} = 12$, $M_{13} = 11$, $M_{21} = -36$,

$M_{22} = 26$, $M_{23} = 7$, $M_{31} = -4$, $M_{32} = -42$, $M_{33} = 12$

(b) $C_{11} = 30$, $C_{12} = -12$, $C_{13} = 11$, $C_{21} = 36$, $C_{22} = 26$,

$C_{23} = -7$, $C_{31} = -4$, $C_{32} = 42$, $C_{33} = 12$

25. -75 **27.** 96 **29.** 170 **31.** -58 **33.** -30

35. -168 **37.** 0 **39.** 412 **41.** -126 **43.** 0

45. -336 **47.** 410 **49–53.** Answers will vary.

55. $-1, 4$ **57.** $8uv - 1$ **59.** e^{5x} **61.** $1 - \ln x$

63. (a) -3 (b) -2 (c) $\begin{bmatrix} -2 & 0 \\ 0 & -3 \end{bmatrix}$ (d) 6

65. (a) 2 (b) -6 (c) $\begin{bmatrix} 1 & 4 & 3 \\ -1 & 0 & 3 \\ 0 & 2 & 0 \end{bmatrix}$ (d) -12

67. $A = \begin{bmatrix} 1 & 3 \\ -2 & 4 \end{bmatrix}$, $B = \begin{bmatrix} -4 & 0 \\ 3 & 5 \end{bmatrix}$

$|A + B| = -30$, $|A| + |B| = -10$

69. A square matrix is a square array of numbers. A determinant of a square matrix is a real number.

71. $8x = (y - 3)^2$ **73.** $\dfrac{x^2}{64} + \dfrac{y^2}{28} = 1$

Section 10.5 *(page 787)*

Warm Up *(page 787)*

1. 1 **2.** 0 **3.** -8 **4.** x^2 **5.** 8 **6.** 60

7. $\begin{bmatrix} 7 & -3 \\ -2 & 1 \end{bmatrix}$ **8.** $\begin{bmatrix} 1 & 2 & -1 \\ -1 & -2 & 2 \\ 2 & 5 & 0 \end{bmatrix}$ **9.** $\begin{bmatrix} 0.16 \\ 0.36 \\ 0.48 \end{bmatrix}$

10. $[23 \quad 54 \quad 8]$

1. $(2, -2)$ **3.** $(-1, 3, 2)$ **5.** $\left(0, -\frac{1}{2}, \frac{1}{2}\right)$ **7.** 7

9. 14 **11.** $\frac{33}{8}$ **13.** $\frac{5}{2}$ **15.** 28 **17.** $\frac{16}{5}$ or 0

19. 250 square miles   **21.** Collinear   **23.** Not collinear

25. Collinear   **27.** $3x - 5y = 0$ **29.** $x + 3y - 5 = 0$

31. $2x + 3y - 8 = 0$ **33.** $x = -3$

35. Uncoded: $[20, 18, 15]$, $[21, 2, 12]$, $[5, 0, 9]$, $[14, 0, 18]$,

$[9, 22, 5]$, $[18, 0, 3]$, $[9, 20, 25]$

Encoded: $[-52, 10, 27]$, $[-49, 3, 34]$, $[-49, 13, 27]$,

$[-94, 22, 54]$, $[1, 1, -7]$, $[0, -12, 9]$,

$[-121, 41, 55]$

37. 1 −25 −65 17 15 −9 −12 −62 −119 27 51

48 43 67 48 57 111 117

39. −5 −41 −87 91 207 257 11 −5 −41 40 80 84

76 177 227

41. HAPPY NEW YEAR   **43.** SEND PLANES

45. MEET ME TONIGHT RON

Focus on Concepts *(page 790)*

1. Interchange two rows.

Multiply a row by a nonzero constant.

Add a multiple of a row to another row.

2. They are the same.

3. A matrix in row-echelon form is in reduced row-echelon form if every column that has a leading 1 has zeros in every position above and below its leading 1.

4. Consistent—an infinite number of solutions

5. Inconsistent

6. Consistent—a unique solution

7. Consistent—an infinite number of solutions

8. (a) The operation can be performed.

(b) The operation can be performed.

9. (a) The operation can be performed.

(b) The operation cannot be performed. The number of rows in B must be the same as the number of columns in A.

10. (a) The operation cannot be performed. The orders of A and B must be the same to perform the operations.

(b) The operation can be performed.

11. The matrix must be square and its determinant nonzero.

12. A square matrix is a square array of numbers, and a determinant is a real number associated with a square matrix.

13. No. The matrix must be square.

14. If A is a square matrix, the cofactor C_{ij} of the entry a_{ij} is $(-1)^{i+j} M_{ij}$, where M_{ij} is the determinant obtained by deleting the ith row and jth column of A. The determinant of A is the sum of the entries of any row or column of A multiplied by their respective cofactors.

15. No. Each matrix is in row-echelon form, but the third matrix cannot be achieved from the first or second matrix with elementary row operations.

Review Exercises *(page 791)*

1. $\begin{bmatrix} 3 & -10 & 15 \\ 5 & 4 & 22 \end{bmatrix}$

3. $\begin{aligned} 5x + y + 7z &= -9 \\ 4x + 2y &= 10 \\ 9x + 4y + 2z &= 3 \end{aligned}$

5. $\begin{bmatrix} 1 & 0 & 0 \\ 0 & 1 & 0 \\ 0 & 0 & 1 \end{bmatrix}$

7. $(10, -12)$ **9.** $(-0.2, 0.7)$ **11.** $(2, -3, 3)$

13. $\left(\frac{31}{42}, \frac{5}{14}, \frac{13}{84}\right)$ **15.** $\left(-2a + \frac{3}{2}, 2a + 1, a\right)$

17. Inconsistent

19. The part of the matrix corresponding to the coefficients of the system reduces to a matrix in which the number of rows with nonzero entries is the same as the number of variables.

21. $\begin{bmatrix} -13 & -8 & 18 \\ 0 & 11 & -19 \end{bmatrix}$ **23.** $\begin{bmatrix} 14 & -2 & 8 \\ 14 & -10 & 40 \\ 36 & -12 & 48 \end{bmatrix}$

25. $\begin{bmatrix} 44 & 4 \\ 20 & 8 \end{bmatrix}$ **27.** $\begin{bmatrix} 4 & 6 & 3 \\ 0 & 6 & -10 \\ 0 & 0 & 6 \end{bmatrix}$

29. $\begin{bmatrix} 48 & -18 & -3 \\ 15 & 51 & 33 \end{bmatrix}$ **31.** $\begin{bmatrix} 14 & -22 & 22 \\ 19 & -41 & 80 \\ 42 & -66 & 66 \end{bmatrix}$

33. $\begin{bmatrix} -14 & -4 \\ 7 & -17 \\ -17 & -2 \end{bmatrix}$ **35.** $\frac{1}{3}\begin{bmatrix} 9 & 2 \\ -4 & 11 \\ 10 & 0 \end{bmatrix}$

37. $\begin{aligned} 5x + 4y &= 2 \\ -x + y &= -22 \end{aligned}$

39. $\begin{bmatrix} \frac{1}{5} & \frac{1}{5} \\ \frac{1}{10} & -\frac{1}{15} \end{bmatrix}$ **41.** $\begin{bmatrix} \frac{1}{2} & -1 & -\frac{1}{2} \\ \frac{1}{2} & -\frac{2}{3} & -\frac{5}{6} \\ 0 & \frac{2}{3} & \frac{1}{3} \end{bmatrix}$ **43.** 550

45. 279   **47.** $(-3, 1)$ **49.** $(1, 1, -2)$

51. $(2, -4, 6)$ **53.** Inconsistent

55. Eight carnations, four roses   **57.** $y = x^2 + 2x + 3$

59. (a) $y = 59.9 + 3.6t$

(b)

(c) The median price was increasing by an average of $3600 per year.

(d) $113,900

61. 16 **63.** 7 **65.** $x - 2y + 4 = 0$

67. $2x + 6y - 13 = 0$ **69.** Answers will vary.

Chapter Test *(page 796)*

1. $\begin{bmatrix} 1 & 0 & 0 \\ 0 & 1 & 0 \\ 0 & 0 & 1 \end{bmatrix}$ **2.** $\begin{bmatrix} 1 & 0 & -1 & 2 \\ 0 & 1 & 0 & -1 \\ 0 & 0 & 0 & 0 \\ 0 & 0 & 0 & 0 \end{bmatrix}$

3. $\left(1, 3, -\frac{1}{2}\right)$ **4.** $y = -\frac{1}{2}x^2 + x + 2$

5. (a) $\begin{bmatrix} 1 & 5 & -2 \\ 0 & -4 & 3 \end{bmatrix}$ (b) $\begin{bmatrix} 15 & 12 & 12 \\ -12 & -12 & 0 \end{bmatrix}$

(c) $\begin{bmatrix} 7 & 14 & 0 \\ -4 & -12 & 6 \end{bmatrix}$

6. $\begin{bmatrix} 8 & -8 \\ 16 & -4 \\ 6 & 12 \end{bmatrix}$ **7.** $\begin{bmatrix} \frac{1}{2} & \frac{2}{5} \\ 1 & \frac{3}{5} \end{bmatrix}$ **8.** $(13, 22)$

9. -2 **10.** 7

CHAPTER 11

Section 11.1 *(page 806)*

Warm Up *(page 806)*

1. $\frac{4}{5}$ **2.** $\frac{1}{3}$ **3.** $(2n + 1)(2n - 1)$

4. $(2n - 1)(2n - 3)$ **5.** $(n - 1)(n - 2)$

6. $(n + 1)(n + 2)$ **7.** $\frac{1}{3}$ **8.** 24 **9.** $\frac{13}{24}$ **10.** $\frac{3}{4}$

1. 3, 5, 7, 9, 11 **3.** 2, 4, 8, 16, 32

5. $-2, 4, -8, 16, -32$ **7.** $2, \frac{3}{2}, \frac{4}{3}, \frac{5}{4}, \frac{6}{5}$

9. $3, \frac{12}{11}, \frac{9}{13}, \frac{24}{47}, \frac{15}{37}$ **11.** $0, 1, 0, \frac{1}{2}, 0$ **13.** $\frac{5}{2}, \frac{11}{4}, \frac{23}{8}, \frac{47}{16}, \frac{95}{32}$

15. $1, \frac{1}{2^{3/2}}, \frac{1}{3^{3/2}}, \frac{1}{4^{3/2}}, \frac{1}{5^{3/2}}$ **17.** $3, \frac{9}{2}, \frac{9}{2}, \frac{27}{8}, \frac{81}{40}$

19. $-1, \frac{1}{4}, -\frac{1}{9}, \frac{1}{16}, -\frac{1}{25}$ **21.** $\frac{2}{3}, \frac{2}{3}, \frac{2}{3}, \frac{2}{3}, \frac{2}{3}$ **23.** -73

25. 28, 24, 20, 16, 12 **27.** 3, 4, 6, 10, 18

29.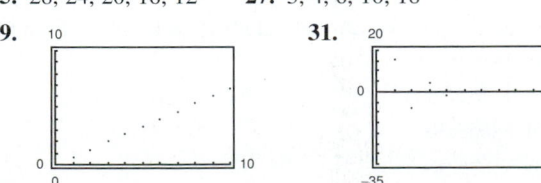

31.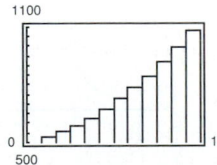

33.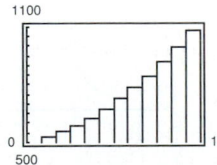

35. c **37.** d **39.** $\frac{1}{30}$ **41.** 90 **43.** $n + 1$

45. $\frac{1}{2n(2n + 1)}$ **47.** $a_n = 3n - 2$ **49.** $a_n = n^2 - 1$

51. $a_n = \frac{n + 1}{n + 2}$ **53.** $a_n = \frac{(-1)^{n+1}}{2^n}$ **55.** $a_n = 1 + \frac{1}{n}$

57. $a_n = \frac{1}{n!}$ **59.** $a_n = (-1)^{n+1}$

61. 6, 8, 10, 12, 14 **63.** 81, 27, 9, 3, 1

$\quad a_n = 2n + 4$ $\qquad a_n = \frac{243}{3^n}$

65. 35 **67.** 40 **69.** 30 **71.** $\frac{9}{5}$ **73.** 238

75. 30 **77.** 81 **79.** $\frac{47}{60}$ **81.** $\displaystyle\sum_{i=1}^{9} \frac{1}{3i}$

83. $\displaystyle\sum_{i=1}^{8}\left[2\left(\frac{i}{8}\right) + 3\right]$ **85.** $\displaystyle\sum_{i=1}^{6}(-1)^{i+1}3i$

87. $\displaystyle\sum_{i=1}^{20}\frac{(-1)^{i+1}}{i^2}$ **89.** $\displaystyle\sum_{i=1}^{5}\frac{2^i - 1}{2^{i+1}}$

91. (a) $A_1 = \$5100.00, A_2 = \$5202.00, A_3 = \$5306.04,$
$A_4 = \$5412.16, A_5 = \$5520.40, A_6 = \$5630.81,$
$A_7 = \$5743.43, A_8 = \$5858.30$

(b) $\$11,040.20$

93. $A_1 = 529.73, A_2 = 555.79, A_3 = 588.31, A_4 = 627.29,$
$A_5 = 672.73, A_6 = 724.63, A_7 = 782.99, A_8 = 847.81,$
$A_9 = 919.09, A_{10} = 996.83, A_{11} = 1081.03$

95. $\$11,131.5$ million

97. 1, 1, 2, 3, 5, 8, 13, 21, 34, 55, 89, 144
$2, \frac{3}{2}, \frac{5}{3}, \frac{8}{5}, \frac{13}{8}, \frac{21}{13}, \frac{34}{21}, \frac{55}{34}, \frac{89}{55}, \frac{144}{89}$

99. $\$500.95$ **101.** Answers will vary. **103.** True

Section 11.2 *(page 816)*

Warm Up *(page 816)*

1. 36	**2.** 240	**3.** $\frac{11}{2}$	**4.** $\frac{10}{3}$	**5.** 18
6. 4	**7.** 143	**8.** 160	**9.** 430	**10.** 256

1. Arithmetic sequence, $d = -2$

3. Not an arithmetic sequence

5. Arithmetic sequence, $d = -\frac{1}{4}$

7. Arithmetic sequence, $d = 4$

9. Arithmetic sequence, $d = 0.4$

11. 8, 11, 14, 17, 20

Arithmetic sequence, $d = 3$

13. $\frac{1}{2}, \frac{1}{3}, \frac{1}{4}, \frac{1}{5}, \frac{1}{6}$

Not an arithmetic sequence

15. 97, 94, 91, 88, 85

Arithmetic sequence, $d = -3$

17. $1, 4, \frac{7}{3}, \frac{7}{2}, \frac{17}{5}$

Not an arithmetic sequence

19. 15, 19, 23, 27, 31   **21.** 200, 190, 180, 170, 160

$a_n = 11 + 4n$ $a_n = 210 - 10n$

23. $\frac{3}{2}, \frac{5}{4}, 1, \frac{3}{4}, \frac{1}{2}$

$a_n = \frac{7}{4} - \frac{1}{4}n$

25. 5, 11, 17, 23 29   **27.** $-2.6, -3.0, -3.4, -3.8, -4.2$

29. 2, 6, 10, 14, 18   **31.** $-2, 2, 6, 10, 14$

33. $a_n = 1 + (n-1)3$ **35.** $a_n = 100 + (n-1)(-8)$

37. $a_n = x + (n-1)(2x)$ **39.** $a_n = 4 + (n-1)\left(-\frac{5}{2}\right)$

41. $a_n = 5 + (n-1)\left(\frac{10}{3}\right)$

43. $a_n = 100 + (n-1)(-3)$

45. b   **47.** c

49. 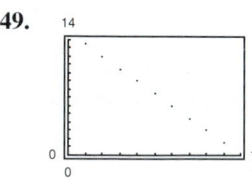   **51.**

53. Linear   **55.** 620   **57.** 4600   **59.** 265

61. 4000   **63.** 1275   **65.** 25,250   **67.** 355

69. 126,750   **71.** 520   **73.** 44,625   **75.** 10,120

77. 10,000   **79.** (a) $40,000   (b) $217,500

81. 2340   **83.** 405 bricks

85. (a) 4, 9, 16, 25, 36

(b) n^2

(c) $\frac{n}{2}[1 + (2n - 1)] = n^2$

87. 4

Section 11.3 *(page 825)*

Warm Up *(page 825)*

1. $\frac{64}{125}$	**2.** $\frac{9}{16}$	**3.** $\frac{1}{16}$	**4.** $\frac{5}{81}$	**5.** $6n^3$
6. $27n^4$	**7.** $4n^3$	**8.** n^2	**9.** $\frac{2^n}{81^n}$	**10.** $\frac{3}{16^n}$

1. Geometric sequence, $r = 3$

3. Not a geometric sequence

5. Geometric sequence, $r = -\frac{1}{2}$

7. Not a geometric sequence   **9.** Not a geometric sequence

11. 2, 6, 18, 54, 162   **13.** $1, \frac{1}{2}, \frac{1}{4}, \frac{1}{8}, \frac{1}{16}$

15. $5, -\frac{1}{2}, \frac{1}{20}, -\frac{1}{200}, \frac{1}{2000}$ **17.** $1, e, e^2, e^3, e^4$

19. $3, \frac{3x}{2}, \frac{3x^2}{4}, \frac{3x^3}{8}, \frac{3x^4}{16}$

21. 64, 32, 16, 8, 4   **23.** 4, 12, 36, 108, 324

$a_n = 128\left(\frac{1}{2}\right)^n$ $a_n = \frac{4}{3}(3)^n$

25. $6, -9, \frac{27}{2}, -\frac{81}{4}, \frac{243}{8}$

$a_n = 6\left(-\frac{3}{2}\right)^{n-1}$

27. $\left(\frac{1}{2}\right)^7$ **29.** $-\frac{2}{3^{10}}$ **31.** $100e^{8x}$ **33.** $500(1.02)^{39}$

35. 9   **37.** $-\frac{2}{9}$ **39.** a   **41.** b

43.   **45.**

47. Increased powers of real numbers between -1 and 1 approach zero.

49. (a) $2593.74

(b) $2653.30

(c) $2685.06

(d) $2707.04

(e) $2717.91

51. $22,689.45   **53.** 8, 4, 6, 5   **55.** 511   **57.** 43

59. 29,921.31   **61.** 6.4   **63.** 2092.60

65. $\displaystyle\sum_{n=1}^{7} 5(3)^{n-1}$ **67.** $7808.24   **69.** Answers will vary.

71. (a) $26,198.27   **73.** (a) $637,678.02

    (b) $26,263.88   (b) $645,861.43

75. Answers will vary.   **77.** 126 square inches

79. (a) $5,368,709.11

    (b) $10,737,418.23

    (c) $21,474,836.47

81. 2   **83.** $\frac{2}{3}$ **85.** $\frac{16}{3}$ **87.** 32   **89.** $\frac{4}{11}$ **91.** $\frac{7}{22}$

93.

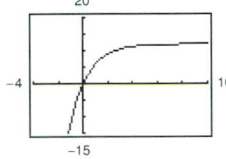

Horizontal asymptote: $y = 12$

Corresponds to the sum of the series.

95. (a) 152.42 feet

    (b) 19 seconds

Section 11.4 *(page 837)*

Warm Up *(page 837)*

1. 24	**2.** 40	**3.** $\frac{77}{60}$	**4.** $\frac{7}{2}$	**5.** $\frac{2k+5}{5}$
6. $\frac{3k+1}{6}$	**7.** $8 \cdot 2^{2k} = 2^{2k+3}$	**8.** $\frac{1}{9}$		
9. $\frac{1}{k}$	**10.** $\frac{4}{5}$			

1. $\dfrac{5}{(k+1)(k+2)}$ **3.** $\dfrac{(k+1)^2(k+2)^2}{4}$

5.–17. Answers will vary.

19. 210   **21.** 91   **23.** 979   **25.** 70   **27.** −3402

29. $\displaystyle\sum_{n=0}^{\infty}(1+4n)$ **31.** $\displaystyle\sum_{n=1}^{\infty}\left(\frac{9}{10}\right)^{n-1}$ **33.** $\displaystyle\sum_{n=2}^{\infty}\frac{1}{2n(n-1)}$

35.–47. Answers will vary.

49. See page 622.   **51.** 1, 3, 5, 7, 9

53. 4, 2, −2, −4, −2

55. 0, 3, 6, 9, 12

    First differences: 3, 3, 3, 3

    Second differences: 0, 0, 0

    Linear

57. 3, 1, −2, −6, −11

    First differences: −2, −3, −4, −5

    Second differences: −1, −1, −1

    Quadratic

59. 0, 1, 3, 6, 10

    First differences: 1, 2, 3, 4

    Second differences: 1, 1, 1

    Quadratic

61. 2, 4, 6, 8, 10

    First differences: 2, 2, 2, 2

    Second differences: 0, 0, 0

    Linear

63. 1, 2, 6, 15, 31

    First differences: 1, 4, 9, 16

    Second differences: 3, 5, 7

    Neither

65. $a_n = n^2 - n + 3$ **67.** $a_n = \frac{1}{2}n^2 + n - 3$

69. $(2, 4)$ and $\left(-\frac{1}{2}, \frac{1}{4}\right)$ **71.** $(1, 2, 1)$

Section 11.5 *(page 845)*

Warm Up *(page 845)*

1. $5x^5 + 15x^2$	**2.** $x^3 + 5x^2 - 3x - 15$	
3. $x^2 + 8x + 16$	**4.** $4x^2 - 12x + 9$	**5.** $\dfrac{3x^3}{y}$
6. $-32z^5$	**7.** 120 **8.** 336 **9.** 720 **10.** 20	

1. 10   **3.** 1   **5.** 15,504   **7.** 4950   **9.** 4950

11. The first and last numbers in each row are 1. Every other number in each row is formed by adding the two numbers immediately above the number.

13. 35   **15.** 56   **17.** $x^4 + 4x^3 + 6x^2 + 4x + 1$

19. $a^3 + 6a^2 + 12a + 8$

21. $y^4 - 8y^3 + 24y^2 - 32y + 16$

23. $x^5 + 5x^4y + 10x^3y^2 + 10x^2y^3 + 5xy^4 + y^5$

25. $r^6 + 18r^5s + 135r^4s^2 + 540r^3s^3 + 1215r^2s^4 + 1458rs^5 + 729s^6$

27. $x^5 - 5x^4y + 10x^3y^2 - 10x^2y^3 + 5xy^4 - y^5$

29. $1 - 6x + 12x^2 - 8x^3$

31. $x^8 + 20x^6 + 150x^4 + 500x^2 + 625$

33. $\dfrac{1}{x^5} + \dfrac{5y}{x^4} + \dfrac{10y^2}{x^3} + \dfrac{10y^3}{x^2} + \dfrac{5y^4}{x} + y^5$

35. $2x^4 - 24x^3 + 113x^2 - 246x + 207$

37. $32t^5 - 80t^4s + 80t^3s^2 - 40t^2s^3 + 10ts^4 - s^5$

39. $81 - 216z + 216z^2 - 96z^3 + 16z^4$

41. 1,732,104   **43.** 180   **45.** $-326{,}592$ **47.** 210

49. $n + 1$ terms   **51.** $x^2 + 12x^{3/2} + 54x + 108x^{1/2} + 81$

53. $x^2 - 3x^{4/3}y^{1/3} + 3x^{2/3}y^{2/3} - y$

55. $3x^2 + 3xh + h^2$ **57.** $\dfrac{1}{\sqrt{x+h} + \sqrt{x}}$ **59.** -4

61. $2035 + 828i$ **63.** 1   **65.** 0.273   **67.** 0.171

69. 1.172   **71.** 510,568.785

73. $g(x) = x^3 + 12x^2 + 44x + 48$

    Shifted four units to the left

75.–77. Answers will vary.

79. (a) $g(t) = 0.1506t^2 + 3.7481t + 43.5584$

    (b)

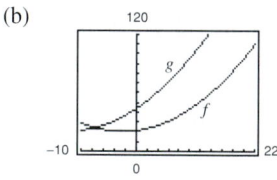

81.

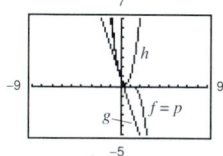

    $p(x)$ is the expansion of $f(x)$.

83. $g(x)$ is shifted three units to the right of $f(x)$.

85. $g(x)$ is the reflection of $f(x)$ in the x-axis.

Section 11.6 *(page 855)*

Warm Up *(page 855)*

1. 6656	**2.** 291,600	**3.** 7960	**4.** 13,800
5. 792	**6.** 2300	**7.** $n(n-1)(n-2)(n-3)$	
8. $n(n-1)(2n-1)$	**9.** $n!$	**10.** $n!$	

1. 6   **3.** 6   **5.** 3   **7.** 4   **9.** 48   **11.** 12

13. 64   **15.** 12   **17.** 6,760,000

19. (a) 900

    (b) 648

    (c) 180

    (d) 600

21. 64,000   **23.** (a) 720   (b) 48   **25.** 24

27. 336   **29.** 120   **31.** $n = 5$ or $n = 6$

33. 1,860,480   **35.** 970,200   **37.** 15,504

39. For some calculators the number is too great.

41. ABCD, ABDC, ACBD, ACDB, ADBC, ADCB, BACD, BADC, CABD, CADB, DABC, DACB, BCAD, BDAC, CBAD, CDAB, DBAC, DCAB, BCDA, BDCA, CBDA, CDBA, DBCA, DCBA

43. 120   **45.** 11,880   **47.** 420   **49.** 2520

51. AB, AC, AD, AE, AF, BC, BD, BE, BF, CD, CE, CF, DE, DF, EF

53. 4845   **55.** 3,838,380   **57.** 3,921,225

59. 21   **61.** (a) 70   (b) 30

63. (a) 70   (b) 54   (c) 16   **65.** 5   **67.** 20

69.–71. Answers will vary.

Section 11.7 *(page 867)*

Warm Up *(page 867)*

1. $\dfrac{9}{16}$	**2.** $\dfrac{8}{15}$	**3.** $\dfrac{1}{6}$	**4.** $\dfrac{1}{80{,}730}$	**5.** $\dfrac{1}{495}$
6. $\dfrac{1}{24}$	**7.** $\dfrac{1}{12}$	**8.** $\dfrac{135}{323}$	**9.** 0.366	**10.** 0.997

1. $\{(H, 1), (H, 2), (H, 3), (H, 4), (H, 5), (H, 6),$
    $(T, 1), (T, 2), (T, 3), (T, 4), (T, 5), (T, 6)\}$

3. $\{ABC, ACB, BAC, BCA, CAB, CBA\}$

5. $\{(A, B), (A, C), (A, D), (A, E), (B, C),$
    $(B, D), (B, E), (C, D), (C, E), (D, E)\}$

7. $\dfrac{3}{8}$ **9.** $\dfrac{7}{8}$ **11.** $\dfrac{3}{13}$ **13.** $\dfrac{3}{26}$ **15.** $\dfrac{1}{12}$ **17.** $\dfrac{11}{12}$

19. $\dfrac{1}{3}$ **21.** $\dfrac{1}{5}$ **23.** $\dfrac{2}{5}$ **25.** 0.3   **27.** 0.85

29. (a) 925,000   **31.** (a) 58%

    (b) 18%   (b) 95.6%

    (c) 17%   (c) 0.4%

33. (a) $\frac{672}{1254}$

(b) $\frac{582}{1254}$

(c) $\frac{548}{1254}$

35. $P(\{\text{Taylor wins}\}) = \frac{1}{2}$

$P(\{\text{Moore wins}\}) = P(\{\text{Jenkins wins}\}) = \frac{1}{4}$

37. (a) $\frac{21}{1292} \approx 0.016$ **39.** (a) $\frac{1}{3}$

(b) $\frac{225}{646} \approx 0.348$ (b) $\frac{5}{8}$

(c) $\frac{49}{323} \approx 0.152$

41. (a) $\frac{1}{120}$ **43.** (a) $\frac{1}{169}$

(b) $\frac{1}{24}$ (b) $\frac{1}{221}$

45. (a) $\frac{14}{55}$ **47.** (a) $\frac{1}{4}$

(b) $\frac{12}{55}$ (b) $\frac{1}{2}$

(c) $\frac{54}{55}$ (c) $\frac{9}{100}$

(d) $\frac{1}{30}$

49. (a) 0.9702   **51.** (a) $\frac{1}{1024}$

(b) 0.9998   (b) $\frac{243}{1024}$

(c) 0.0002   (c) $\frac{781}{1024}$

53. 0.1024   **55.** $\frac{7}{16}$

57. (a) As you consider successive people with distinct birthdays, the probabilities must decrease to take into account the birth dates already used. Because the birth dates of people are independent events, multiply the respective probabilities of distinct birthdays.

(b) $\dfrac{365}{365} \cdot \dfrac{364}{365} \cdot \dfrac{363}{365} \cdot \dfrac{362}{365}$

(c) Answers will vary.

(d) Q_n is the probability that the birthdays are *not* distinct, which is equivalent to at least two people having the same birthday.

(e)

n	10	15	20	23	30	40	50
P_n	0.88	0.75	0.59	0.49	0.29	0.11	0.03
Q_n	0.12	0.25	0.41	0.51	0.71	0.89	0.97

(f) 23

59. 0.454

Focus on Concepts *(page 872)*

1. Natural numbers

2. (a) Odd-numbered terms are negative.

(b) Even-numbered terms are negative.

3. True   **4.** True   **5.** True   **6.** True

7. (a) Each term is obtained by adding the same constant (common difference) to the previous term.

(b) Each term is obtained by multiplying the same constant (common ratio) by the previous term.

8. (a) Arithmetic. There is a constant difference between consecutive terms.

(b) Geometric. Each term is a constant multiple of the previous term. In this case the common ratio is greater than 1.

9. Each term of the sequence is defined in terms of the previous term.

10. Increased powers of real numbers between 0 and 1 approach zero.

11. d   **12.** a   **13.** b   **14.** c

15. The signs of the terms alternate in the expansion $(x - y)^n$.

16. Same

17. $_{10}P_6 > {}_{10}C_6$. Changing the order of any of the six elements selected results in a different permutation but the same combination.

18. $0 \le p \le 1$

19. $\frac{1}{3}$. The probability that an event does not occur is 1 minus the probability that it does occur.

20. Meteorological records indicate that over an extended period of time with similar weather conditions it will rain 60% of the time.

Review Exercises *(page 873)*

1. $8, 5, 4, \frac{7}{2}, \frac{16}{5}$ **3.** $72, 36, 12, 3, \frac{3}{5}$

5. 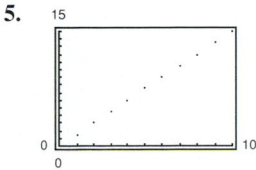   **7.**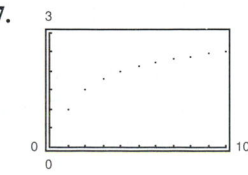

9. $\displaystyle\sum_{k=1}^{20} \frac{1}{2k}$ **11.** $\displaystyle\sum_{k=1}^{9} \frac{k}{k+1}$ **13.** 30   **15.** $\frac{205}{24}$

17. 6050   **19.** 418

21. 3, 7, 11, 15, 19   **23.** 1, 4, 7, 10, 13

25. 35, 32, 29, 26, 23   **27.** 9, 16, 23, 30, 37

     $a_n = 38 - 3n$ $a_n = 2 + 7n$

29. $a_n = 103 - 3n$; 1430

31. 80   **33.** 88   **35.** 25,250

37. (a) \$43,000   (b) \$192,500   **39.** $4, -1, \frac{1}{4}, -\frac{1}{16}, \frac{1}{64}$

41. $9, 6, 4, \frac{8}{3}, \frac{16}{9}$ or $9, -6, 4, -\frac{8}{3}, \frac{16}{9}$

43. $120, 40, \frac{40}{3}, \frac{40}{9}, \frac{40}{27}$ **45.** $25, -15, 9, -\frac{27}{5}, \frac{81}{25}$

$a_n = 120\left(\frac{1}{3}\right)^{n-1}$ $a_n = 25\left(-\frac{3}{5}\right)^{n-1}$

47. $a_n = 16\left(-\frac{1}{2}\right)^{n-1}$, 10.67 **49.** 127 **51.** 8

53. 12 **55.** 24.849

57. (a) $a_t = 120,000(0.7)^t$

(b) \$20,168.40

59. \$5111.82

61. Answers will vary.   **63.** Answers will vary.

65. 15 **67.** 6720

69. $\dfrac{x^4}{16} + \dfrac{x^3 y}{2} + \dfrac{3x^2 y^2}{2} + 2xy^3 + y^4$

71. $\dfrac{64}{x^6} - \dfrac{576}{x^4} + \dfrac{2160}{x^2} - 4320 + 4860x^2 - 2916x^4 + 729x^6$

73. $41 + 840i$ **75.** $118,813,760$ **77.** $\frac{1}{9}$

79. $P(\{3\}) = \frac{1}{6}$

$P(\{(1, 5), (5, 1), (2, 4), (4, 2), (3, 3)\}) = \frac{5}{36}$

81. $\frac{31}{32}$ **83.** 0.0475

Cumulative Test for Chapters 9–11
(page 878)

1. $(1, 2), \left(-\frac{3}{2}, \frac{3}{4}\right)$ **2.** $(2, -1)$

3. $(4, 2, -3)$ **4.** $(1, -2, 1)$

5.

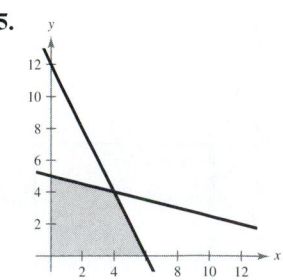

Maximum of $(4, 4)$: $z = 20$

6. $\begin{bmatrix} 9 & 1 \\ 7 & -7 \\ 15 & -10 \end{bmatrix}$ **7.** $\begin{bmatrix} -175 & 37 & -13 \\ 95 & -20 & 7 \\ 14 & -3 & 1 \end{bmatrix}$

8. 920 **9.** 6 **10.** Answers will vary.

11. $z^4 - 12z^3 + 54z^2 - 108z + 81$

12. 120 **13.** $\frac{1}{4}$

APPENDIX A

Section A.1   *(page A10)*

1.

Stems	Leaves
7	0 5 5 5 7 7 8 8 8
8	1 1 1 1 2 3 4 5 5 5 5 7 8 9 9 9
9	0 2 8
10	0 0

3.

Stems	Leaves
6	18 68 71 94
7	16 19 25 25 42 57 58 76 84 88
8	13 28 35 41 46 59 61 62 66 81 83 89 91 92 92
9	05 06 15 18 25 25 26 28 41 44 83 90 92
10	04 10 62 95
11	51 78 86
12	23
13	
14	
15	
16	26

5.

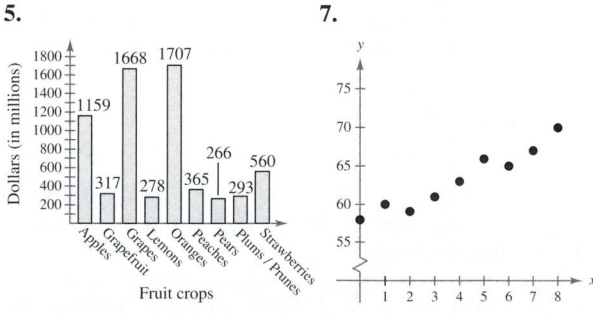

7.

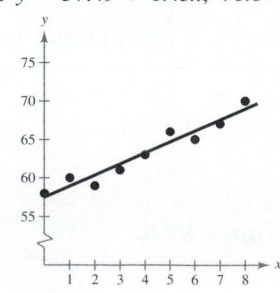

9. $y = 57.49 + 1.43x$; 71.8

11.

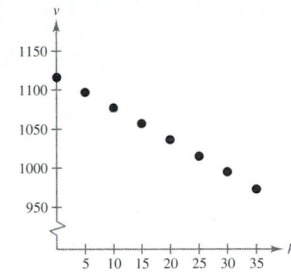

13. $v = 1117.3 - 4.1h$; 1006.6

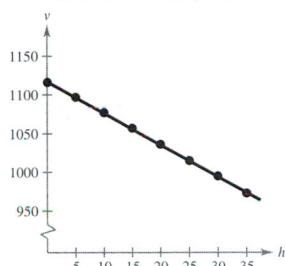

15. (a)

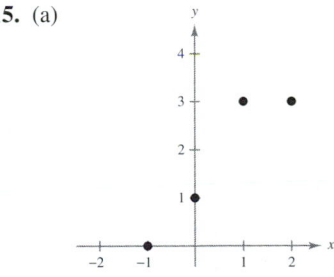

(b) $y = 1.1x + 1.2$; 0.7

(c) $y = \frac{11}{10}x + \frac{6}{5}$; 0.7

17. (a) and (c)

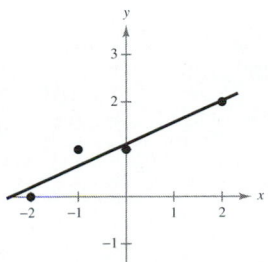

(b) $y = \frac{16}{35}x + \frac{39}{35}$

19. (a) and (c)

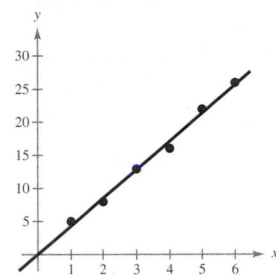

(b) $y = \frac{30}{7}x$

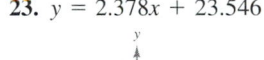

21. $y = -2.179x + 22.964$ **23.** $y = 2.378x + 23.546$

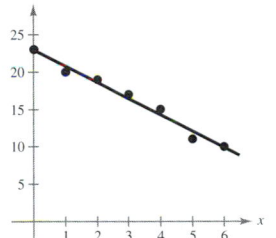   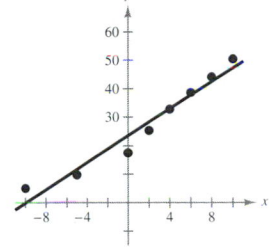

25. (a) $S = 384.1 + 21.2x$; $\$95{,}784.10$

(b)

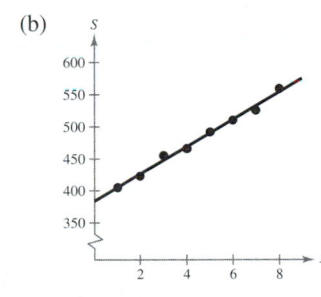

(c) $r = 0.996$

Section A.2 *(page A19)*

1. Mean: 8.86; median: 8; mode: 7

3. Mean: 10.29; median: 8; mode: 7

5. Mean: 9; median: 8; mode: 7

7. The mean is sensitive to extreme values.

9. Mean: $\$67.14$; median: $\$65.35$

11. Mean: 3.07; median: 3; mode: 3

13. One possibility: {4, 4, 10}

15. Mean: 76.6; median: 82; mode: 42
The median gives the most representative description.

17. $\bar{x} = 6$, $v = 10$, $\sigma = 3.16$ **19.** $\bar{x} = 2$, $v = \frac{4}{3}$, $\sigma = 1.15$

21. $\bar{x} = 4$, $v = 4$, $\sigma = 2$ **23.** $\bar{x} = 47$, $v = 226$, $\sigma = 15.03$

25. 3.42   **27.** 101.55   **29.** 1.65

31. $\bar{x} = 12$ and $|x_i - 12| = 8$ for all x_i

33. It will increase the mean by 5, but the standard deviation will not change.

35. With $\bar{x} = 235$ and $\sigma = 28$:

At least 75% of the scores in $[179, 291]$

At least 88.9% of the scores in $[151, 319]$

With $\bar{x} = 235$ and $\sigma = 16$:

At least 75% of the scores in $[203, 267]$

At least 88.9% of the scores in $[187, 283]$

INDEX OF APPLICATIONS

Chemistry and Physics

Construction

Time and Distance

U.S. Demographics

INDEX

A

Absolute value
  of a complex number, 645
  equations involving, 145
  function, 217
  inequalities involving, 157
  properties of, 5
  of a real number, 5
Acute angle, 457
Addition, 107
  of complex numbers, 134
  of a constant, 154
  of functions, 227
  of inequalities, 154
  of matrices, 746
  of polynomials, 28
  properties of vectors, 625
  of rational expressions, 48
  of real numbers, 6
  of a vector, 623
Additive identity
  for a complex number, 134
  for a matrix, 749
  for a real number, 7
Additive inverse
  for a complex number, 134
  for a real number, 6
Adjacent side
  of a right triangle, 467
Adjoining matrices, 761
Algebra of calculus, 58
Algebraic errors, 56
Algebraic expression, 6
  domain of, 45
Algebraic function, 394
Ambiguous case (SSA), 604
Amplitude
  of simple harmonic motion, 532
  of sine and cosine curves, 494
Amplitude modulation, 533
Angle, 456
  acute, 457

between two nonzero vectors, 637
central, 459
complementary, 458
conversions between degrees and radians, 460
coterminal, 456
degree, 457
initial side, 456
measure of, 457
negative, 456
obtuse, 457
of one degree, 457
positive, 456
reference, 481
standard position, 456
supplementary, 458
terminal side, 456
vertex, 456
Angular speed, of a particle, 461
Arccosine function, 519
Arc length, 461
Arcsine function, 517
Arctangent function, 519
Area
  common formulas for, 111
  of an oblique triangle, 606
  of a triangle, 781
Argument
  of a complex number, 646
Arithmetic sequence, 810
  common difference of, 810
  nth term of, 811
Associative property of addition
  for complex numbers, 135
  for matrices, 748
  for real numbers, 7
Associative property of multiplication
  for complex numbers, 135
  for matrices, 748
  for real numbers, 7

Asymptote
  horizontal, 335
  of a hyperbola, 367, 368
  of a rational function, 336
  slant, 288, 346
  vertical, 335
Augmented matrix, 731
Average of n numbers, A13
Axis (axes)
  coordinate, 64, 225
  of an ellipse, 364
  of a hyperbola, 366, 367
  imaginary, 645
  of a parabola, 257, 362
  real, 645
  of symmetry, 257

B

Back-substitution, 665, 687
Bar graph, 65, A4
Base, 13
  of an exponential function, 394
  of a logarithmic function, 406
Basic equation for partial fraction decomposition, 354
Basic rules of algebra, 7
Bearing, 530
Bell-shaped curve, 439
Bimodal, A13
Binomial, 27, 840
Binomial coefficient, 840
Binomial expansion, 843
Binomial Theorem, 840
Book value, 190
Bounded interval on the real number line, 3
Bounds for real zeros of a polynomial, 296
Boyle's Law, 320
Branch of a hyperbola, 366
Break-even point, 669

A199